ENCYCLOPEDIA OF ENVIRONMENTAL RESEARCH

(2 VOLUME SET)

ENVIRONMENTAL SCIENCE, ENGINEERING AND TECHNOLOGY

Additional books in this series can be found on Nova's website
under the Series tab.

Additional E-books in this series can be found on Nova's website
under the E-books tab.

ENVIRONMENTAL REMEDIATION TECHNOLOGIES, REGULATIONS AND SAFETY

Additional books in this series can be found on Nova's website
under the Series tab.

Additional E-books in this series can be found on Nova's website
under the E-books tab.

ENVIRONMENTAL SCIENCE, ENGINEERING AND TECHNOLOGY

ENCYCLOPEDIA OF ENVIRONMENTAL RESEARCH

ALISA N. SOUTER
EDITOR

Nova Science Publishers, Inc.
New York

LIBRARY OF CONGRESS CATALOGING-IN-PUBLICATION DATA

Encyclopedia of environmental research / [edited by] Alisa N. Souter.
p. cm.
Includes bibliographical references and index.
ISBN 978-1-61761-927-4 (hardcover : alk. paper)
1. Environmentalism--Research--Encyclopedias. I. Souter, Alisa N.
GE195.E53 2010
333.72--dc22
2010037348

Published by Nova Science Publishers, Inc. + *New York*

CONTENTS

PREFACE

This book presents important research advances in the study of environmental research, as well as grassland biodiversity, agricultural runoff, biodiversity hotspots, estuaries, freshwater ecosystems and conservation of natural resources. Topics discussed herein include sustainable waste management; carbon sequestration in soil; effects of habitat destruction on populations and communities; natural renewable water resources; economics of fisheries; biodiversity of tropical freshwater plankton; and ecosystem functioning of open/closed estuaries.

Short Communication A - Non point source pollutants such as nitrogen and phosphorus are transported via agricultural drainage ditches to receiving waters and result in eutrophication and downstream ecosystem degradation. Seasonal precipitation patterns, and land use contribute to variable concentrations of nitrogen and phosphorus loss from the landscape. This study investigated the differences in total inorganic-phosphate (TIP) and dissolved inorganic-nitrogen (DIN) concentrations in drainage ditches as a function of hydrology (stormflow vs. baseflow), land use (farmed vs. CRP vs. control) and season (growing vs. dormant). TIP and DIN ditch concentrations typically decreased from farms > CRP > control over both seasons. DIN concentrations were highest in baseflows during the growing season; however, mean DIN concentrations in baseflow were not significantly higher in the growing season than in the dormant season. Mean TIP concentrations and particulate phosphorus ratios increased in the dormant season with increased precipitation volumes and storm events. TIP concentrations were highest in growing season stormflows following fertilizer applications. A change in land use from agriculture to CRP decreases associated TIP and DIN in runoff and baseflow concentrations but residual phosphorus is still present in stormflows 10 yrs post the discontinuation of farming activity. This research highlights the episodic nature of nutrient contamination driven by land use, anthropogenic fertilization, and the temporal variability of storm events.

Short Communication B - Climatic change and local shifts in land-use have modified the Mediterranean landscape in the last decades, leading to an increase of the frequency and magnitude of geo-hydrological events, such as floods.

The more and more frequent trigger of floods during moderate rainfall intensity leads to wonder whether the main control factor is the climate dynamics or the land-use.

During the past twenty years, EU legislation has encouraged and subsidised major changes in land-use throughout the Mediterranean region (Faulkner et al., 2003), leading to an increase of arable land for cereal production in the hilly and low mountainous areas, traditionally mantled by shrub and low forest.

In Italy, the area affected by hydrogeological risk is 21,505 Km^2, which corresponds to the 7.1% of the whole territory. From 1991 to 2001 more than 1000 flood events occurred (Ministry of Environment, 2003). In only one year, 2003, major floods affected more than 300,000 people and caused damage up to 2,184 million euros.

For the Basilicata region, an area extending for 9,992 km^2 in southern Italy, both historical and recent information on floods is available. The area affected by flood risk is 260 Km^2 (2.6% of the whole territory) and mainly includes the hilly areas and the coastal belt (Metaponto flat), which are exploited in agriculture, the main regional economic source. On the hills vineyards and olive cultivation predominate, whereas in the Metaponto flat orchards and cereals are the main cultivation. The hilly area and the floodplain are connected by a dense network of deep gullies, running out into natural streams used by farmers for water supplying. Both gullies and streams have undergone great modifications in the last thirty years (Clarke and Rendell, 2000). Because of the advancements in earth-works machinery, gully walls with original slope angles > 45° never ploughed before, were levelled for the production of cereals and orchards, with a considerable increase of sediment production. At the same time natural streams connected with the gully outlets were canalized, with gradual decreasing of their beds and their flows. Thus, gullies became effective links for the transferring of runoff and sediment from uplands to valley bottom and pediment channels where they have exacerbated flooding.

Land-use map and digital elevation models of different years (1949, 1986, 2000) are available in a GIS-database, allowing topographic comparisons within the area under study. Simple buffering and map overlay indicate an increase of flooding in areas which have recently undergone land-use changes.

Owing to complex and stringent economical and social issues in European agriculture, the topic seems to be highly relevant.

Chapter 1 - The question of how climate change will affect carbon storage in grasslands is important, as it could lead to tipping points in the climate system. Most of the carbon in grasslands is stored in the soil, and this is also the most permanent store. The main question is how temperature affects soil organic matter development and decomposition. Some authors have found no effect, but this may be because the temperature under which the soils developed needs to be taken into account when assessing the temperature sensitivity of soil organic matter decomposition. Temperature decreases rather uniformly with altitude, and the authors used an altitudinal gradient as a temperature gradient. The authors sampled soils from an altitudinal gradient in four heath and grass covered Scottish mountains. Soil carbon content, carbon to nitrogen ratio and isotope ratios were correlated to altitude, plant cover and other soil parameters. The soils were then incubated for a year, and the change in these parameters over the incubation period where measured again and related to altitude and other factors. The results show that the most important environmental factors for the soil carbon and nitrogen content and isotope ratios were altitude and soil phosphorous. Plant species composition appeared unimportant. The changes over the course of the incubation were small, but they give weak support for the hypothesis that soils developed at low temperatures were least developed to start with, and therefore changed the most.

Chapter 2 - Riparian field margins (i.e. fenced areas adjacent to watercourses) are becoming more widespread in intensively managed grassland systems in the UK as a means of mitigating diffuse pollution associated with intensive livestock farming. Invertebrate assemblages were examined in a range of riparian margins to determine their potential to deliver additional biodiversity goals. It was found that ground beetle assemblages in wide (> 4 m) margins were more distinct from the adjacent field than those in narrow margins (≤ 2 m) or riparian zones open to grazing. Furthermore, both wide and narrow riparian margins were found to enhance the abundance of several groups of invertebrates that are important prey items of farmland birds (i.e. leatherjackets, sawfly larvae and harvestmen). While fenced margins had higher abundances of bird prey items, the denser vegetation typical of such ungrazed margins may result in the prey being less accessible to foraging birds. Consequently routine management of such margins may be required to open up the vegetation structure and thus optimize their potential benefit to wildlife. Appropriately managed riparian margins have therefore the potential to increase biodiversity at the field level, by increasing the abundance of key invertebrates, and also at the landscape level, by increasing habitat heterogeneity and thus supporting distinct assemblages of ground beetles.

Chapter 3 - Analysis of plant diversity aims to understand the organization and the variability of biological populations within ecosystems. In classical analysis, individual plants are firstly identified on the basis of morphological/cytological parameters, then biodiversity is evaluated tacking into account presence/absence, abundances and densities of plants. Although morphological parameters are easily accessible, they provide limited precision on the differentiation between individuals that share a high similarity. This is especially true in the plant world where a same species includes several varieties that could not be easily distinguished morphologically. Cytological parameters provide reliable qualitative information but are not suitable for a quantitative evaluation of biodiversity at ecosystemic scales. The high variability between and within plant taxa can be highlighted chemically through qualitative and quantitative analyses of secondary metabolite patterns in different plants. Quantitatively, the different patterns show variability based on occurrences of different metabolites which can have major or minor levels. In the case where two chemical patterns have the same metabolites, they can be distinguished from different relative levels (i.e. regulation ratios) of such metabolites. These different variability criteria (presence/absence, major/minor, relative levels) allow to extract reliable chemical fingerprints for chemotaxonomical classification of plants. Moreover, such chemical polymorphism can be correlated to different intrinsic and/or extrinsic conditions of the plants in order to evaluate its adaptative, selective and/or evolutive values. Finally, relationships between chemical patterns and environmental conditions provide efficient tools to survey qualitatively and quantitatively the fluctuations of plant diversity in time and space. This chapter provides a review on chemotaxonomical criteria helping to understand complex structures of plant diversity. It focuses particularly on the chemotaxonomic usefulness of phenolic compounds (phenylpropanoids, flavonoids, etc.) in analysis of chemical polymorphisms at different systematic levels (from family to variety and chemotype via genus and species). Five plant families are treated here consisting of Poaceae and Liliaceae (monocotyledons), and Asteraceae, Lamiaceae and Fabaceae (dicotyledons). These families are representative of a great biomass part in the grasslands, and their chemotaxonomic analyses can usefully help to manage the biodiversity in such ecosystems.

Chapter 4 - This chapter describes a methodological framework for integrating indigenous knowledge and ecological methods for promoting local communities' participation in the implementation of the Convention on Combating Desertification (CCD) and the Convention on Biological Diversity (CBD) at local community levels. The authors developed and tested the framework using herding communities in Northern Kenya. The methods for implementing the framework included semi-structured interviews and group discussions with key informants for generating information on livestock management, changes in vegetation indicators and historical changes in land use patterns. Other methods used included joint transect walks with knowledgeable herders to assess environmental change using ecological indicators (vegetation and soil) and herder anthropogenic indicators (i.e., landscape grazing potential and landscape grazing suitability), monitoring marked transects, using satellite images taken at different times, and herders' knowledge to evaluate long-term changes in vegetation cover around permanent settlements. Finally, a workshop was organized with herders and Environmental Management Committees (EMCs) in which they participated in informal discussions on issues addressed by the joint research project. The authors then synthesized the findings to evaluate the effectiveness of the framework for desertification assessment, monitoring and control. From the results they may conclude that local community participation in assessment and monitoring of environment change would contribute to the implementation of the CCD and the CBD. This chapter makes specific recommendations for applying the framework to achieve global goals for local actions at local community levels.

Chapter 5 - The Udzungwa Mountains are part of the Eastern Afromontane hotspot, one the 34 biodiversity hotspots in the world. Moreover, they are part of the ancient chain of crystalline blocks named "Eastern Arc" that represents one of the most important areas of Africa in terms of endemic animal and plant species. However, in the Eastern Arc, while forests have been widely investigated and are often safeguarded by the institution of protected areas, montane grasslands have received little attention yet, in terms both of biological research and of conservation, and are threatened by rapidly expanding of cultivations and timber plantations. The present study is aimed to improve knowledge about the composition of Amphibian and Reptile communities of two sites in the southern part of the Udzungwa plateau. The field research was carried out from December 2004 to January 2005 by means of VES and pitfall traps with drift fences, and also by opportunistically searches. The check-list of the first site (placed near Bomalang'ombe village) updates and integrates the preliminary data collected and published by Menegon et alii (2006). The check-list of the second site (placed near Mapanda village), on the authors' knowledge, represents the first herpetological check-list for the area. On the basis of the species accumulation curves, species richness estimates and effort simulations run for each site, the Bomalang'ombe check-list seems to characterize the community of the area satisfactorily in the considered season, while the Mapanda check-list has to be considered preliminary. However, both study sites exhibit a biological value worthy of attention by the conservation point of view, since both of them show the occurrence of taxa strictly endemic or near-endemic to the Eastern Arc (in some cases endemic to the Udzungwa block and also strictly endemic to the site that represents the type locality) and the occurrence of at least one species vulnerable to the extinction threat. Finally, the occurrence of highly specialized grass-dwelling species suggest that the montane grasslands of the Udzungwa have a long evolutionary history, therefore deserving more scientific and conservation efforts.

Chapter 6 - Steppe grassland spreading over eastern Europe to eastern Asia is one of the largest grassland ecosystems in the world, which is precious but fragile. Last four decades the grazing pressure has been increasing due to increase of animal and human population in Inner

Mongolia, China. It may affect on the grassland ecosystem, including grassland biomass, grass quality, and species diversity. But how can we maintain the valuable ecosystems and sustainable livestock farming at the same time? The authors examined the relationship between grazing pressure and grassland vegetation using drastically improved information technology such as remote sensing, Geographic Information Systems (GIS) and Global Positioning System (GPS). The experiment was carried out at Xilingol steppe, where the annual precipitation is around 300 mm, and annual mean air temperature is around 0°C. In the former part of the chapter, the authors discussed how to evaluate the grazing pressure under free ranging condition utilizing multiple livestock farmers in the extended grassland without fences and borders. Standing against this problem, the authors put GPS on the shoulder of sheep to trace the daily traveling distance. Sheep herd moved between 11 and 12 km in a day during summer season. GPS also showed us the place and time where the sheep herd spent. Newly developed bite-counter clarified the number of biting during grazing separated from rumination. The time series of biomass and grass quality (in term of crude protein content) changes were successfully monitored by satellite vegetation indices such as Terra MODIS-EVI and NOAA AVHRR-NDVI. In the latter part, the authors examined the effects of grazing pressure on steppe vegetation in the same district. In 1960's, several pipe lines were buried inside grassland from lake to supply drinking water for animal. Opening mouths were set at 5 or 6 km intervals. Livestock villages were formed around the opening, therefore, surrounding grassland came under the influence of gradation of grazing pressure concentrically. Started from certain opening, the authors put quadrates on 4 directions at 1 km intervals until 5 km of distance. In addition of botanical composition survey, biomass, soil conditions and spectral reflectance measurements were executed at each plot. Vegetation cover, biomass and plant height increased in accordance with distance from opening. Both NDVIs obtained from radiometer at ground and Landsat image showed similar trend as biomass change. Botanical composition also changed. Land was almost bare around the opening, then some low palatable herbaceous plants are remained near the opening. Three or four km apart from opening, the original vegetation and biodiversity recovered and biomass became constant. Finally, the authors discussed effects of grazing pressure on steppe vegetation from a comprehensive stand point.

Chapter 7 - The South Brazilian Campos grasslands (also known as only Campos) are unique ecosystems. Located in the southernmost part of Brazil, these ecosystems are rich in plant species, being more diverse than forest ecosystems in the same area. Due to their geographical position (humid subtropics), in a transitional region between tropical and temperate area, these grasslands show mixture of C3 and C4 grasses and a high diversity of other botanical families, with typical tropical and temperate species.

Since the climate is subtropical humid, with rainfall regularly distributed all over the year, forest physiognomies would be expected to cover the area. However, diverse grasslands occur in vast areas, in contact (or not) with forest, raising one of the most interesting questions: are these grasslands natural? Paleopalinological studies already showed that these grasslands were present in the region before forest. Until the Holocene, Campos grasslands dominated. Afterwards, climate changed (hotter and more humid), enabling forest expansion. Therefore, Campos grasslands are natural ecosystems. However, even with no edaphic restrictions and a climate more propitious to forest expansion, forests do not dominate nowadays, raising another important question: why are Campos grasslands still present?

Disturbance is probably the most important factor maintaining both grassland physiognomy and diversity. Both grazing and fire have important effects on grassland dynamics. South America does not have large native herbivores and cattle are the most important grazers in these grasslands. Jesuits introduced them in the XVII century and cattle raising is one of the most important economic activities in Southern Brazil. Before that, fire might have influenced vegetation dynamics and delayed forest expansion. Paleopalinological studies showed the presence of charcoal since the Holocene. Nowadays, fire still plays a great role on grassland dynamics, but is a very polemic issue.

In this chapter, I intend to give the reader an overview about this unique ecosystem, almost unknown by the majority of international scientific community and ignored for its ecological relevance even by most of Brazilian scientific community. I will also show the role of disturbance in maintaining Campos biodiversity and dynamics as well as the importance of its conservation.

Chapter 8 - In the past, semi-natural grasslands in Japan were maintained using traditional management practices such as mowing, burning, and grazing. However, grassland areas have been decreasing drastically recently due to land development or the abandonment of management. Thus, conservation measures are urgently required for organisms specific to semi-natural grassland habitats. To examine management schemes for conserving such organisms, the authors investigated plant species and seasonal fluctuation patterns in species and individuals of flowering plants and adult butterflies in a semi-natural grassland in central Japan. The authors' study sites were firebreaks where the grass was mowed and removed, plantation areas that were mowed, unpaved roads with mowed banks, abandoned grassland, and scattered scrub forest. Areas that had been abandoned for less than five years were dominated mostly by Arundinella hirta, while areas that had been abandoned for more than ten years were dominated by large tussocks of Miscanthus sinensis or scrub forests of Rhamnus davurica var. nipponica. The plant species composition differed between grassland sites and scrub sites, and that of firebreaks was different from other grassland sites. The sites under management sustained a larger number of flowers and butterflies than the sites without management. Furthermore, the firebreak sites sustained flowers in June and July, while the plantation and banks of unpaved road sites sustained flowers primarily in August and September. The number of butterflies increased in the firebreak in June and at the other sites in August and September, in relation to the occurrence of flowers at each site. These results suggest that management is essential to sustain this grassland and that different management regimes, such as mowing alone and mowing with grass removal, induce different plant species compositions and different numbers of flowers, and lead to different numbers of adult butterflies during a season. On the other hand, R. davurica var. nipponica, which composed the scrub forest, is a host plant for a threatened butterfly species. In conclusion, heterogeneous environments with different vegetation structures by season and seral stage under different management regimes support plant and butterfly diversity in semi-natural grassland habitats, and continuation of traditional management practices in the future may be important.

Chapter 9 - The processes able to regulate plant abundance and distribution have generally been studied by addressing aboveground interactions such as plant – plant competitive interactions, plant – herbivores and/or parasites relationships and disturbance creating new patches for plant colonization. Importantly, it is now well established that plant species dynamic is tightly interlinked with the development of soil community. The rhizosphere, i.e. the biologically active soil compartment where root-root and root-microbes communications occurred, is of particular importance in mediating plant fitness and community composition. This active soil zone promotes inter alia the proliferation of particular microbial communities strongly involved in plant nutrition capabilities. Nutrients are heterogeneously distributed in soil and also, nutrients uptake by plants results in depletion zone around roots. Therefore, plant species differential capacity (i) to perform strategies allowing them to colonize soil patches or, (ii) to select microbial communities that are specific in their genetic and functional diversities and/or (iii) to form efficient microbial symbiosis for acquisition of nutrients, could be considered as main biological factors to explain the spatial distribution and the maintenance of multi-species association in plant communities. This chapter aims at reviewing some of the recent advances made in understanding plant species composition mainly in grassland communities through the involvement of plant-soil-microbes feedbacks. The authors will focus on interactions between plants and soil heterogeneity and/or rhizosphere microbes in affecting plant species competitive dominance.

Chapter 10 - Arbuscular mycorrhizal fungi (AMF) habitat almost all terrestrial ecosystems, and play a critical role in many ecological processes. In grasslands of almost all types, AMF are demonstrated to increase the plant biodiversity in most experiments, although some experiments indicate a decreased biodiversity. This regulation of grassland biodiversity by AMF can be attributed to several mechanisms. The differential growth responses of host plants to AMF are the direct factor leading to the altered biodiversity, and the promoted uptake of mineral nutrients with different degree by AMF among host plants is regarded to underlie the differential growth responses. Furthermore, hyphal links established between host plants of the same or different species represent a transport passage for resources (such as water, nutrients, carbon), by which these resources flow freely and finally redistribute evenly within the community. In practice, this regulation can be employed to restore the degraded grasslands, to monitor the competition between plants in the managed grasslands, and even to construct grasslands with pre-designed community structure. Future research is proposed in relation to the emerging issues and the practical application.

Chapter 11 - Habitat fragmentation can be defined as a process where continuous areas of natural habitats are broken into small patches separated by other habitats different from the original ones. Today, habitat fragmentation is a common issue to almost every ecosystem in the world, since anthropogenic land uses transformed initially continuous habitats into mosaic landscapes represented by isolated native patches surrounded by man-altered environments .

Anthropogenic matrices usually act as selective filters to species movements among native patches in the landscape, and therefore, the persistence of animal and plant populations in fragmented habitats will depend to a great extent on the matrix permeability. Landscapes are commonly classified into continuous or fragmented however, the landscape is not a binary mosaic formed by natural habitat and matrix – or habitat and non-habitat – and the species certainly do not perceive it that way (presence/ absence of resources), as the authors will discuss later on in this chapter

Despite being a rather controversial issue, several authors have shown the importance of protecting small native patches resulted from habitat fragmentation, as in the landscape context they are able to keep a significant portion of local biodiversity. Andrèn (1994) states that landscape biodiversity may increase considerably when several small fragments are close to each other and permit animal and plant fluxes as in a continuous habitat. The ability of a species to move throughout the landscape is related to habitat connectivity (); it refers to the functional linkage among patches either due to patch proximity or to matrix permeability. Therefore, the degree of habitat connectivity, which depends on matrix quality, is essential to maintain native species in a fragmented landscape.

Chapter 12 - Genetic diversity and species diversity are both crucial for ecosystem stability. Moreover, genetic diversity within a population is fundamental for broadening gene pools in breeding programmes and for food security. For the efficient conservation of species richness as well as of genetic resources in permanent grassland (in-situ), relationships between genetic diversity within target species and species diversity at their original habitats should be considered. In order to identify valuable grassland types for targeted in-situ conservation, two important forage grasses, Festuca pratensis Huds. and Lolium multiflorum Lam. have been chosen as model species. Genetic diversity of 12 ecotype populations of F. pratensis and L. multiflorum was assessed by means of 22 and 24 SSR markers, respectively. Species composition and abundance were determined at their collection sites. Analysis of molecular variance revealed low and nonsignificant differences among the grassland types of both species. Genetic diversity within F. pratensis ecotype populations (expected heterozygosity HE) was negatively correlated with species diversity (mainly Shannon index and species evenness) at their collection sites. For L. multiflorum no association was found. Between the observed classes of grassland types, differences in genetic diversity within F. pratensis populations were not significant, but intermediately managed Heracleum-Dactylis grassland habitats held ecotype populations with significantly more rare alleles than extensively managed Mesobromion and Festuco Agrostion habitats. However, species diversity indicated by the Shannon index was significantly lower in Heracleum-Dactylis than in Mesobromion grassland habitats. These results indicate a possible conflict of the aims to maintain species-rich grassland types on the one hand, and to conserve ecotype populations of target species with a high genetic diversity and a high number of rare alleles on the other hand. This conflict should be resolved by including the aim of conserving genetic diversity of grassland species in national or international programmes of biodiversity management.

Chapter 13 - This chapter examines links between the management of waste and its impacts on climactic changes due in part to rising consumption levels. Differences between greenhouse gas emissions from developed and developing countries are examined, including the contributions of the 20 highest emitting countries between the period of 1950-2000, including that of China and the USA. The manner in which waste management and climatic changes are links is examined, as well as strategies for reducing the impacts including waste prevention and minimisation. The chapter concludes by looking ahead to future technological and market requirements.

Chapter 14 - As the world shrunk and expanded at the same time, as interdependence between (and within) the nature and human society became obvious, physics, information sciences and ecology have known (among others sciences) a radical evolution and have tried growingly to adapt to the complex, dynamic, interrelated and more and more uncertain socio-economic and natural systems. In this context, and through assessment and management of the Jervis Bay marine park, the authors question the concepts of quantum environmental information, of quantum environmental assessment and of quantum environmental knowledge.

Chapter 15 - Soil, together with water, is a basic resource for humanity, as it is stated in the first item of the European Soil Chart. Unlikely it has been thought for long time, soil is a limited resource, easily destroyable but not easily renewable at human life timescale, and therefore should be protected in order to preserve its fundamental functions, both biotic and abiotic. It is worth noting how the soil as a resource is poorly cared for under international conventions, in spite of its importance. The UN Convention on desertification (CCD), f. i., is based on regional development politics rather than sound scientific background, and it is disadvantaged by conceptual problems.The problem of soil conservation is one of the great emergencies in the XXI century. Recent estimates indicate losses of 20 ha/min/year soil, caused by different processes and mechanisms: wind and water erosion, deforestation and land degradation, salinisation, acidification, chemical contamination, etc. Rapid decline in quality and quantity of soil resource, and uncontrolled natural resources consumption, which affects about 33% of the Earth land surface, involving more than 1 billion people, stresses global agriculture with long term negative consequences. In the next twenty years, food security will emerge as a major global issue, forcing developed countries to provide more food, thus provoking imbalances in soil nutrient resources, soil pollution, land degradation, reduction of pedodiversity and biodiversity, increased occurrence of human health problems. New paradigms of soil science, therefore, are needed in order to encompass such dramatic global scenery. Soils are extremely important, f. i., in the global cycle of carbon. Nevertheless, consideration of soils was slow to emerge in the context of the UN Framework Convention of Climate Change (FCCC). Soil scientists could play an increasingly important role in understanding the global carbon cycle and to point out ways to reduce carbon dioxide levels in the atmosphere by storing carbon in ecosystems and producing bio-fuel. But the scientific community needs also to increase the visibility of soils in international environmental, social and political context. During the last decades, the focus of societies has changed from agricultural production and forestry towards environmental issues. Soil science is increasingly targeting environmental and cultural issues, such as sustainable land use, ecosystems restoration, remediation strategies for contaminated soils, protection of the food chain and of groundwater resources, protection of human health as well as protection of soil as a cultural and natural heritage. In the next years, these issues and other specific aspects, such as soil science for archaeological dating, forensic soil science, and other applications to increasing social and economic demands, will gain importance to achieve the goal of soil and land conservation in a changing world.

Chapter 16 -Anthropogenic emission of CO_2 and other greenhouse gases was rapidly increased with the Industrial Revolution and this event has caused a world interest in identfying strategies of reducing the rate of gaseous emission. The intergovernmental Panel on Climate Change shows that from 1850 and 1998 the emission from terrestrial ecosystem was about half of fossil fuel combustion. Agriculture can be a source or sink for atmospheric CO_2 because soil organic carbon pool (SOCP) in soil surface is sensitive to changes in land use and soil management practice. The carbon sink capacity of the world agricultural and degraded soils is 50-66% of the historic carbon loss that are of 42 to 78 Gt of carbon respectively. Carbon (C) sequestration implies transferring atmospheric CO_2 into long-lived pools and subsequent storage of fixed C as soil organic carbon (SOC). In this way the conservation of plant residues in agricultural soil play an important role in CO_2 sequestration. The mechanism by which crop residues contribute to SOC is through their chemical, phisical and biological stabilization. In this chapter the authors discussed the role of the plant residues in the carbon sequestration throughout plant tissue stabilization in soil, giving a new approach and understanding of the plant residue conservation in soil.

Chapter 17 - Use of collagen in food production is limited by its low nutritive value and deficient essential amino acids. Irrespective of gelatine and glue manufactured by boiling refined native collagen, industrial processing of collagen is connected with production of a considerable amount of difficultly utilizable protein waste. In the phase of refining collagen raw material, hydrolysates of keratin and accompanying proteins (albumins, globulins) that are isolated with ease go away; their following usage comes up against similar problems as with keratin. Use of collagen waste as of secondary industrial raw material is usually complicated by an excessive density of irreversible crosslinks introduced for stabilizing against chemical and microbial effects. The first stage in processing such waste is hydrolysis that is partially alkaline, acid, but above all enzymatic, and interesting for its low energy demands. Obtained collagen hydrolysates behave like gelatine, however, with a view to their molecular weight being 5-10 times lower, they exhibit higher hydrophility and formation of their gels requires higher dry substance concentration. Sole collagen hydrolysates proved themselves well as a component in urea-formaldehyde resins reducing formaldehyde emissions by their cured films.The same as polyamides, they form considerably stable networks when crosslinking epoxide oligomers; their potential disintegration through microbial procedures is advantageous when removing protective films prior to recycling plastic automobile or aircraft parts on termination of their life. Collagen hydrolysate, crosslinked with polymeric dialdehydes (e.g. dialdehyde starch) results in biodegradable or even edible packaging materials which, under suitable conditions, may be processed by technology of dipping, casting and drying as well as by procedures typical for processing synthetic polymers (extrusion and compression molding).

Chapter 18 - The Brazilian Atlantic forest is a hotspot of biodiversity, characterized by its high species richness, level of endemism and danger of extinction. The remnants of this biome stand in the most populated and developed region of the country and their conservation is in constant conflict with the strategies of development. The authors' study was developed in the State of Espírito Santo, in the municipalities of Linhares and Sooretama. Located at the north of Rio Doce, it was one of the last regions of the Atlantic forest to be deforested, and forest fragmentation took place mostly after 1950. The authors' purpose was: 1) characterize the evolution of the forest fragmentation in this region, the present situation of its remnants, and the main strategies, conflicts and potentials for conservation; 2) evaluate the conservation value of the forest fragments based on a study of the community of a group of insects – the community of Blattaria; 3) integrate these information to propose guidelines to resolve potential conflicts based on positive and non conflictual conservation recommendations. The landscape of Linhares and Sooretama is dominated by plantations, among which can be found two large reserves and fragments of diverse sizes, shapes and degrees of disturbance. Only in Sooretama there are 213 fragments, which, with the Sooretama Biological Reserve, cover 44% of the municipality area. Similarly to the reserves, the fragments were preserved by the imposition of the law and due to the interest of people of keeping the forest as a reserve of value. Fragments are used for several purposes, with some of them provoking strong disturbances that could lead to their massive disappearance in a near future. The analysis of the community of Blattaria showed that the number of species shared among sites is very low (between 22.9% and 39.6%), with a high number of species found only in single fragments and nowhere else; the richness per area in small fragments is not smaller than in large reserves; habitat diversity is stronger than the effect of area per se. These results show that even after more than 50 years of fragmentation and isolation, the situation of the biodiversity in this region is not lost: small and very disturbed fragments still shelter important biodiversity of insects, and can be of extreme interest for composing a set of remnants that maximizes the number of species to be protected. Based on it, the management and conservation of the biodiversity should first of all include the fragments, and for this reason should be conceived in the landscape or regional scale. At this scale, fragments should be envisaged in a management plan, in which restrictions and uses allowed are explicit. In such a context, the choice of priority fragments should be done taking in account the landscape diversity: it should include fragments from different hydrographic basin, soil types, vegetation facies, as well as the diversity of use.

Chapter 19 - The Araucaria forest, also called mixed ombrophilous forest is a particular ecosystem in the southern Brazilian highlands, characterized by the admixture of two distinct vegetations: the tropical afro-Brazilian and the temperate austro-Brazilian floras.

During at least 1000 years, after the post-glacial climate changes of the Quaternary, this ecosystem dominated the landscape of around 200,000 km2 of southern Brazilian highlands. However, the intensive exploitation process occurred mainly between the years 1940's to 1970's reduced the Brazilian Araucaria forest to about 3% of their original area.

The conservation of the forest remnants, the protection of the wild animals and the natural regeneration of many plant species may be promoted by the economical valorization of Araucaria forest, mainly through its sustainable management.

Similarly, the establishment of natural reserves and agroforestry systems can also promote forest conservation. In addition, these in situ conservation areas may serve as source of genetic material for ex situ conservation, plantation establishment and for the exploration of the adaptive capacity of different genetic resources through provenance trials.

However, a sustainable management of the forest remnants implies comprehension about the distribution of the genetic diversity of keystone species, allowing making decisions which circumvent the reduction of the forest adaptability to environmental variations, such as the climatic changes. Furthermore, it has been shown that biotechnological tools may be very useful in promoting the preservation of the remaining genetic diversity of the forest species.

Chapter 20 - The Mexican ichthyofauna is exceptionally rich, composed of 507 species represent 6% of the fish species known around the world, and 163 of them 32% are endemic of Mexico with endemism at a family, genera and species level. Result of complex climatological and geological history, producing high isolated basins and promoting speciation. Is common between Mexican ichthyofauna species with reduced distribution area, besides the environmental conditions become unsuitable for native fish by anthropogenic impact including; pollution, introduction of exotic species, severe fragmentation and destruction of habitat. A recent study conducted in the occident of the Mesa central in 2001 reveals that the principal causes of destruction of habitat includes principally: dryness 68%, piping 20%, and build of a spa 5%. The species in the list of danger include 185 species in a category of threat and 11 species reported as extinct. The factors that promote extinction in fish are: Channel incision of the habitat, limited physiographic distribution, biological specialization and fragmentation principally.

Chapter 21 - One of the main environmental impacts caused by fisheries is the emission of gases from the fuels used for the vessels. These gases contribute to global warming. The aim of this chapter is to use LCA (Life Cycle Assessment) to compare the global warming potential of artisanal and industrial fishing off the northern coast off Rio de Janeiro state (municipalities of Arraial do Cabo and Cabo Frio, respectively). For this analysis, one ton of common sea bream was adopted as the functional unit. The artisanal fishing data were obtained from interviews with local fishermen, while the industrial fishing data were collected from an industrial fishery. The study made use of the UmbertoTM software package. The result of this comparison showed that industrial fishing has a greater impact on global warming than artisanal fishing.

Chapter 22 - "What is true of natural systems is also true of human ones: much as we might like to turn back the clock and rearrange land use as if on a tabula rasa, there are complex historical reasons why people live and work in the places they do, and it is usually best to manage with those reasons in mind rather than wishing the world were otherwise."

Prior to human settlement the only species of wildlife (i.e. terrestrial vertebrates) found in the Hawaiian Islands were birds and one mammal (the Hawaiian hoary bat [Lasirus cinereus semotus]). Thus, due to the islands' extreme geographic isolation (ca. 3,500 km from any continental landmass), they were devoid of any mammalian herbivores, grazers, predators, amphibians, and reptiles . The absence of these particular groups of species meant that birds, as the dominant terrestrial vertebrate, were able to exploit and adapt to the numerous tropical niches available. In fact, among the 109 endemic species of birds known to exist only in the Hawaiian Islands, they assumed top roles as carnivores, seed dispersers, and pollinators. When the first Polynesians arrived in the Hawaiian Islands sometime between 300 and 600 A.D., they introduced a number of wild and domestic animal species, both intentionally and accidentally, including pigs (or pua'a; Sus scrofa vittatus), Pacific rats (Rattus exulans), and red junglefowl (or moa; Gallus gallus; Asquith 1995, Hawaii Audubon Society 1997). A millennium later (ca. 1778-1800) European explorers arrived and brought an additional suite of animals to the islands, including domestic livestock such as cattle, goats, sheep, and pigs, which were all large bodied mammalian grazers (i.e. ungulates). In the centuries since, people have continued to bring in new animal species for agriculture (e.g., expansion of goats and cattle), pets (e.g., cats and dogs), pest control (e.g., small Indian mongoose [Herpestes auropunctatus]), and hunting (e.g., mouflon sheep [Ovis musimon], wild turkey [Meleagris gallopavo]). In fact, the past two hundred years has seen an explosive increase in the number of all non-native species in Hawaii, such that today there are at least 200 (19 mammals, 53 birds, 28 amphibians and reptiles) introduced wildlife species in Hawaii.

The explosive increase in introduced wildlife, coupled with their widespread distribution across the ecosystems of Hawaii has led to a number of ecological repercussions, including the decline and/or extinction of many native plants and animals, habitat loss, and wholesale changes to the islands' ecosystems. These ecological repercussions can easily be understood in the context that native ecosystems of Hawaii were devoid of ground predators, competitors, large grazers, and many pathogens/diseases. In other words, many common types of ecological interactions present in other parts of the world (particularly continents) were either reduced or absent in Hawaii, and hence native species were not adapted to them. Consequently, birds and plants, along with the handful of insects that were able to colonize the Hawaiian Islands formed specialized relationships that were inherently susceptible to outside perturbations. Notably, though, no one knows exactly what fully intact ecosystems resembled or what all the species interactions might have been prior to human settlement, when very large bodied bird species existed and natural disturbance regimes were in place in Hawaii.

Chapter 23 - Animals and products derived from different organs of their bodies have constituted part of the inventory of medicinal substances used in various cultures since ancient times. Regrettably, wild populations of numerous species are overexploited around the globe, the demand created by the traditional medicine being one of the causes of the overexploitation. Mammals are among the animal species most frequently used in traditional folk medicine and many species of bovids are used as medicines in the world. The present work provides an overview of the global usage of bovids in traditional folk medicine around the world and their implications for conservation. The results demonstrate that at least 55 bovids are used in traditional folk medicine around the world. Most of species (n=49) recorded were harvested directly from the wild, and only six species of domestic animals. Of the bovids recorded, 50 are included on the IUCN Red List of Threatened Species and 54 are listed in the CITES. By highlighting the role played by animal-based remedies in the traditional medicines, the authors hope to increase awareness about zootherapeutic practices, particularly in the context of wildlife conservation.

Chapter 24 - The geologic record indicates that species have typically persisted for 1 to 10 million years. Yet as many as 5,000 to 100,000 species are thought to go extinct each year. As human population size and standards of living increase around the world, it is anticipated that the strain on natural resources will drive an extinction rate approximately 100 to 1,000 times greater than the 'natural' background extinction rate. Indeed, half of all living bird and mammal species are speculated to go extinct within 200 to 300 years. This trend in biodiversity has been dubiously termed 'the sixth extinction', suggesting that humankind's impact on the planet is comparable to the celestial cataclysm that eradicated the dinosaurs. It has been recently estimated that 10 to 40 million years of evolution will be needed to replenish Earth's biodiversity from the current extinction crisis.

Human-induced extinctions pose an ethical problem that is probably less compelling to most people than the extent to which our welfare depends on Earth's biodiversity. Approximately one quarter of all prescription drugs taken in the United States are derived from plants. Naturally occurring ecosystem services, such as water purification, pest and climate control, and crop pollination, were valued at $33 trillion US Dollars per year in 1997. The natural world is a source of future discoveries whose worth is unknowable. For example, as-yet unidentified plants, algae, or bacteria may prove to be model organisms for the industrial recovery of solar energy. Humanity's welfare is intimately linked to the intelligent use of available resources, and the urgency of implementing sustainable practices grows lock-step with the rise in global population and per-capita resource consumption.

Chapter 25 – Freshwater ecosystems, such as lakes, rivers, ponds, and wetlands, are a precious and critically important form of natural resource. They are important not only for humans by providing water for drinking and domestic use, but also for numerous aquatic animals that depend on water for their very existence. Moreover, freshwater ecosystems also provide valuable indirect services such as water purification and buffer against hurricanes. However, natural disasters and anthropogenic activities can unfavorably affect these ecosystems by pushing these ecosystems and the associated aquatic food webs into undesirable and potentially unstable regimes. Natural disasters such as floods and hurricanes often lead to sudden regime shifts, while changes caused by anthropogenic pollution effects are relatively slow but equally dangerous. Under such circumstances, external intervention by humans through management strategies is called for. Such strategies should aim not only to stabilize the ecosystems but also to make those inherently resilient so that destabilizing effects in future can be effectively managed. The broad theme of this work is the development of such management strategies. Since freshwater ecosystems are dynamic in nature, a time dependent management strategy is more desirable than a static one. Consequently, this work proposes to use optimal control theory for deriving time dependent management policies. The work considers an aquatic three species food chain model (Rosenzweig-MacArthur model) which has been frequently studied in theoretical ecology literature. The work identifies undesirable regimes for this model and applies optimal control theory (Pontryagin's maximum principle) to derive time dependent management strategies that achieve the desired regime change. Fisher information, a measure based on information theory, is proposed as the sustainability metric and used to formulate the objective function for the control problem. The work also compares the top-down and bottom-up control philosophies for the food chain model so that the most effective management option can be identified. Since natural ecosystems are frequently not well understood and hence associated with considerable uncertainties, the work utilizes efficient modeling techniques from finance literature for a robust analysis. An Ito process is used to model time dependent uncertainty and stochastic maximum principle is utilized to solve the optimal control problem. The results will highlight the role of systems theory in sustainable management of freshwater systems and ecosystems in general.

Chapter 26 - The authors highlight the need for a strong scientific approach to biomonitoring and conservation of fresh waters and identify key shortcomings in the present approach. For instance, mismatches between the identity of the stressors that the authors observe and the metrics that they use to measure their impacts are becoming increasingly evident. As water quality has improved in many areas new stressors have become important and the focus purely on responses to organic enrichment and eutrophication becomes increasingly untenable. While there are notable exceptions, many biomonitoring and management schemes are inappropriate, poorly designed, or based on weak science. The authors argue that there is an unfortunate tendency in this field to resort to anecdotal "evidence" and unsubstantiated inference, rather than to strong experimental evidence of cause-and-effect relations based on hypothesis testing. Further, the field has often developed in isolation from recent advances in ecological theory, which has stunted our ability to progress from understanding to knowledge to prediction of ecosystem responses to stressors. Pertinent examples of this come from the rapid advances made recently on the relationship between biodiversity and ecosystem processes and in food web research, neither of which have so far informed monitoring and management practices. This disconnection has led to some notable instances of 'pseudoscience' being used at the expense of sound scientific concepts, which the authors identify in a set of case studies. There are also other more subtle disadvantages of an overemphasis on repetitive monitoring and data collection for the purposes simply of compliance to environmental targets, rather than for science. Vast resources are being used for such purposes with almost no improvement in our mechanistic understanding of how freshwater ecosystems operate. Even if these issues are resolved, however, legislative problems still need to be addressed. Methodological inertia inhibits the implementation of new developments in ecological biomonitoring, because of the phobia of novelty and a willingness to stick with familiar systems, even if they are no longer completely fit-for-purpose and in need of overhauling.

Chapter 27 - The present problems that are related to water and sanitation in Sudan are many and varied, and the disparity between water supply and demand is growing with time due to the rapid population growth and aridity. The situation of the sewerage system in the cities is extremely critical, and there are no sewerage systems in the rural areas. There is an urgent need for substantial improvements and extensions to the sewerage systems treatment plants. The further development of water resources for agriculture and domestic use is one of the priorities to improve the agricultural yield of the country, and the domestic and industrial demands for water. This study discusses the overall problem and identifies possible solutions.

Chapter 28 - The Upper Delaware River represents a valuable ecological resource of New York and Pennsylvania. Recognized as one of the country's most scenic rivers, it is greatly appreciated for its high quality trout fisheries and the variety of outdoor recreation experiences it offers. The river system serves as a water supply for over 8 million people in New York City, via an out-of-basin transfer, and over 9 million people within the basin. The three New York City water supply reservoirs (Cannonsville, Pepacton and Neversink) are designed to withdraw more than 70% of the average annual volume from the top of the basin at three stems of the Delaware: the West Branch, the East Branch and the Neversink River. Severe reductions and fluctuations in river flows, channel alterations, and landscape modifications resulted in increased water temperature and changes in habitat suitability in the river. As a consequence of these alterations, the aquatic fauna in the upper Delaware River have been impaired.

This article summarizes (with minor modifications) a report published by Trout Unlimited in 2001 that helped galvanize efforts toward the improvement of management of the Delaware River. The complex ecological alterations caused by unnatural flow conditions, past environmental degradation, and numerous water supply demands on the basin necessitate a well-defined management scheme to optimize the use of the river as a resource and secure its long-term sustainability. Rehabilitation of the upper Delaware River requires both an immediate overhaul of the flow regime below the reservoirs, monitoring and evaluation, and research. In the initial phase, acute deficits should be addressed by increasing base flows, and reducing ramping rates and peak amplitudes. In the second phase, an intensive multidisciplinary study is proposed to quantify limiting ecosystem factors, to better understand groundwater and surface water interactions, and to determine management options. This information will provide a solid basis for evaluating specific options and an integrative, long-term management plan for the upper Delaware River.

Chapter 29 - Freshwater tidal wetlands are productive and often support high biodiversity. While there have been some quantitative studies of the effects of many invasive plants in North American freshwater wetlands, much is still assumed for a number of invasive species. The outcomes of invasive-native plant interactions, and the factors involved, are more subtle than assumed. The author reviewed the statistical and anecdotal results of published studies on the impacts of invasive plants with large potential ranges in freshwater tidal wetlands of North America. A number of studies reported little or no change in native species richness and diversity, and found outcomes differed depending upon several specific factors. The author recommends that more empirical research be conducted on the interactions of these species (both in field and greenhouse) with native plants in freshwater tidal wetlands because very little information is available and yet many management decisions have already been made.

Chapter 30 - Emergent macrophytes are very important contributors to sustenance of faunal communities and nutrient cycling. They also form prime feature of the landscape of almost every aquatic habitat. Material translocation dynamics between above- and below-ground components describe the life cycling and survival strategies of emergent plants. Trends of material translocation of emergent plants can be investigated by way of field observations and subsequently by the models for organ-specific growth. This paper reviews and describes the morphology and material translocation of emergent plants under a set of different environmental conditions, namely, seasonal, spatial, sediment, water depth and harvesting.

Chapter 31 - Community ecology searches for general rules to explain patterns in species' distribution, but to date, progress has been slow. This lack of progress has been attributed to the fact that ecological rules - and the mechanisms that underpin them - are contingent on the organisms involved, and their environment. Coupled with the vast complexity of biological systems, the result is that there are few rules that are universally true in community ecology. However, effective restoration management requires a thorough understanding of the ecological effects of degradation. This sets a challenge for ecological research to unveil the causal mechanisms underlying patterns in species' distribution. There is broad consensus that the way forward is to link pattern and process through species traits, as these provide the causal mechanisms explaining how abiotic and biotic factors set limits to species occurrences.

In a recently developed method, species traits were used to define groups of aquatic macroinvertebrate species with similar causal mechanisms underlying the species-environment relationships. By investigating interrelations between traits and interpreting their function, it was possible to define 'sets of co-adapted species traits designed by natural selection to solve particular ecological problems', which are termed life-history strategies.

In this chapter the authors discuss the position of life-history strategies in community ecology and its merits to conservation and nature management. By including the causal mechanisms for a species' survival under particular environmental conditions, life-history strategies can be used to explain species occurrences and generate testable predictions. As a species' identity is made subordinate to its biology, they may be used to compare water bodies found at a large geographical distance, which may comprise different regional species pools or span species distribution areas. By aggregating species in life-history strategies, biodiverse assemblages can be compressed into a few meaningful, easily interpretable relationships.

When the field of community ecology is envisaged as a continuum, two approaches can be considered the end points, focusing either on communities or individual species. While research on individual species is strongly rooted in causal mechanisms, the results are difficult to generalize to other species, complicating the extrapolation to whole communities. However, on the other end of the continuum, results are aggregated (e.g. indices of diversity, evenness and similarity) to such an extent that the causal mechanisms are obscured. The fresh approach of life-history strategies may provide the best of both worlds, aggregation information over many different species without sacrificing information on the causal mechanisms underlying a species' presence or absence.

Chapter 32 - The environmental impact of phosphorus (P) waste from the aquaculture industry is increasingly a matter of concern in Japan and elsewhere around the world. Over the past two decades, increasing concerns over excessive P loading have resulted in a large number of studies aimed at better understanding issues related to P output from aquaculture production. Most of these studies were fresh water fish species, however, particular attention of P loading in marine aquaculture species and adoption of nutritional approaches for overcoming that dilemma has not been extensively studied. Yellowtail aquaculture in Japan is the highest of any farmed fish species and 57% of total Japanese fish culture production consists of yellowtail. The reduction in P from yellowtail culture will lead to a great contribution towards the goal of reducing P from the aquaculture industry in Japan. There are some approaches have been adopted to reduce P effluent from yellowtail aquaculture based on series of long term research and this recent research demonstrates that technology can be effective to meet this end. Feed formulation improvements aimed at improving digestibility and retention of nutrients by fish are key to reducing dissolved and solid waste outputs. Careful selection of more highly digestible practical ingredients, highly available inorganic P sources and judicious use of additives, such as phytase can contribute to improve digestibility of certain feed nutrients and potentially reduce solid and P waste outputs. Excretion of non-fecal soluble P is greatly increased as dietary available P is increased above the requirement level and that excess P in diets cause environmental pollution is increasingly recognized as a serious ecological problem. Therefore, it is of importance not to feed more P than the fish require. In fish culture operations, the majority of feed is fed to larger fish. The dietary requirement of P for larger fish is less than to smaller fish. P requirement for large fish through non-fecal (balance study) approach may greatly help to formulate low P loading diet and ultimately contribute to reduce effluent P from aquaculture operation. Fish meal is the source of the most dietary P in fish commercial diets, wherein it exists as hydroxyapatite (bone P) or tricalcium phosphate which is poorly absorbed in fish intestine. Due to that low absorption of fish meal, currently feed manufacturer are interested to supplement readily available inorganic P in commercial feed along with fish meal source. However, for larger fish, P is needed only for the physiological maintenance rather than the growth. Study revealed that P supplementation is not necessary for the fish meal-based diet for large yellowtail and elimination of P supplementation to the fish meal-based diet resulted in a 34.5% reduction in non-fecal soluble P from yellowtail aquaculture effluent. The overall study suggested the key formulation of low P diet for yellowtail that will contribute to reduce the pollution in aquaculture effluents.

Chapter 33 - Aquaculture is responsible for a 20 million tonnes increase in fish production over the last decade, with an estimate of 48 million tones in 2005. The global increase in aquaculture production was made possible through the introduction of new areas, species, and practices and through increased production from existing systems. Nutrient pollution from aquaculture and intensification of fish production, in turn, may cause declines in aquaculture productivity by promoting outbreaks of disease among the fish. Antibiotics as feed additives, which played an important role for controlling diseases in the past, have been widely criticized for the negative impacts to the surround aquatic systems. The potential substitute for antibiotics are the so called nutraceuticals, multi-physiological, bioactive and pollution-free additives, acting as immuno-stimulants that improve resistance to diseases by enhancing non-specific defence mechanisms. In the present chapter the authors studied the effects of vitamin E supplementation in the process of induced wound healing in Nile tilapias Oreochromis niloticus. During a period of 60 days, fish (initial body weight=30g) were fed two experimental diets, supplemented (450 mg/kg diet) or not with vitamin E. Thereafter, all animals were anaesthetized and submitted to dermal wounds. The histomorphometric assessment was checked after 3, 7, 14, 21 and 28 days post-wounding. The cicatricial retraction and appearance of the wounds were monitored during the trial. Moreover, the histomorphometry of the mucous cells, chromatophores, inflammatory cells, fibroblasts, collagen fibers and scales were also used as indicators for the wound healing capacity. The rate of wound retraction was significantly higher in the vitamin E supplemented group. Such healing was a result of an increase of inflammatory cells, mucous cells, chromatophores and collagen fibers. The results indicate that fish fed vitamin E rich diet have enhanced dermal wound healing capacity.

Chapter 34 - Coastal zones have been focal points of social conflicts and environmental problems in Tabasco, Mexico. The cumulative conflicts arising from a heterogeneous regional development had led to an increased lack of coastal resources management, increased population, reduction of stocks of valuable species due to over-fishing, and environmental deterioration. The aim of this study was to assess fishermen access to resources in the coastal zone of Tabasco in terms of income and costs of fisheries production and determine the effect on aquaculture development. Results showed that fishing cooperatives have lost competitiveness due to a lack of integration into a market-oriented production, making fishery entrepreneurship a high-risk activity and incapable of producing efficiently. The conditions fishermen experience suggest that the hierarchical approach for resources management influenced by political decisions have trapped resource managers in programs that address symptoms rather than causes of basic fisheries management problems. Therefore, economic diversification in and outside the fishery sector needs to be implemented. Aquaculture could be developed as a means to improve the role of fishing cooperatives to support production. Although state intervention is not the only solution, the devolution of regulatory functions to local communities may help to restore the crucial qualities of collective action and sustain aquaculture.

Chapter 35 - The total ornamental fish industry worldwide (including dry goods) is valued at approximately US $15 billion dollars and it has been estimated that approximately one billion ornamental fish are commercialized every year with a value in the order of US $6 billion dollars. Freshwater species constitute the bulk of the trade; 90 percent of these are obtained from aquaculture and only 10 percent are wild captured. Around 800 to 1000 species and varieties are traded worldwide. Ornamental fish trade has been recognized as an important pathway for the introduction of non native species in several countries and present trends indicate that this pathway may turn into the main source of exotic invasive species in North America. Invasive species are characterized by posing different threats to the environment, the economy and human health. Among the main impacts provoked in the aquatic environment by invasive species are: competition with native species, hybridization, predation, introduction of diseases, habitat disruption and trophic webs modification. The introduction of exotic species has been related to the extinction of 54 percent of aquatic native species worldwide. Overall, 70 percent of the extinctions of North American fishes and 60 percent of those from Mexico are related to non native species, totally or partially. Aquarium trade has shown an accelerated increase during the last decade with a trade value of US $160 million dollars. This increase parallels the boost of exotic species in the country. In fact, in the 80's only 55 non native fish species were registered in Mexico and by 2004 the number raised to 118, of which 67 (58.26 percent) have turned invasive. Several facts contribute to explain this: i) The low amount of varieties cultured in Mexico (61 varieties pertaining to 19 species), compared to the huge number of varieties imported (more than 700 from 117 families), ii) The number of fish imported in Mexico; 40 million ornamental fish are traded annually, of which 45 percent are imported (nearly 18 million fish were imported in 2006) while 55 percent are captive bred iii) there is a lack of official regulations for the establishment and operation of farms producing ornamental fish and for the translocation of ornamental fish within the country. As a result, ornamental fish species have been established in 9 out of 10 continental aquatic regions of Mexico. Some of these species have already severely impacted the environment and the economy in most regions of the country.

Chapter 36 - Although biodiversity has often been indicated to increase towards the equator, taxonomic expertise and research efforts have focused mainly on temperate regions. Hence, biodiversity and limnology of tropical freshwaters are presently poorly understood. Also in Thailand, literature on limnology is scarce, and many of the studies carried out are published in the grey literature or in local (university) journals with extremely limited circulation.

Two outdoor microcosm studies were previously carried out in Thailand to evaluate the fate and effects of the insecticide chlorpyrifos (CPF) and the herbicide linuron (LIN) on tropical freshwater communities under semi-field conditions. The present chapter lists and discusses the periphyton, phytoplankton and zooplankton species communities encountered in the controls of these experiments. The periphyton community was studied in the LIN experiment using microscopic slides that were placed approximately 10 cm under the water surface of the microcosms. After an incubation period of four and six weeks, the periphyton community existed solely of Chamaesiphon sp. whereas in later stages (eight and ten weeks incubation), contributions of chlorophytes and diatoms increased. The LIN phytoplankton community was dominated by Chlorophyta with a radiation of Scenedesmus species, whereas a bloom of the cyanobacterium Microcystis aeruginosa was noted in the CPF experiment. The large cladoceran Diaphanosoma sp., one of the dominating cladocerans in the LIN experiment, was absent in the CPF experiment and replaced by smaller cladoceran species. Possible underlying influences of tropical seasonality and climatic conditions on the observed community structures, primary producers-zooplankton interactions and implications for conservation of tropical freshwaters are discussed.

Chapter 37 - The Industry of Aquaculture has been profiting from fish sperm cryopreservation for many years. Sperm from more than two hundred fish species, most of them with commercial value, has been cryopreserved, allowing year round availability of sperm and the creation of sperm banks from those individuals with special interest. However, to date, fish embryo cryopreservation has not been successfully achieved in any teleost species. The benefits of fish embryo cryopreservation are numerous, not only for aquaculture but also for conservation purposes. This technique will allow for the maintenance of a constant supply of animals, reduce the facilities required on the fish farm, facilitate animal transportation between different farms and enable the preservation of valuable lines. These practical advantages will also report a cost reduction and minimize the impact of epidemics on productivity. This chapter presents an overview of the recent advances in this field and their impact on Aquaculture.

Chapter 38 - Apart from representing the vast majority (71%) of South Africa's 258 functional estuaries, temporarily open/closed estuaries (TOCEs) are common in Australia, on the southeastern coasts of Brazil and Uruguay, the southwestern coasts of India and Sri Lanka, but are poorly represented in North America, Europe and much of Asia. The regular change between open and closed mouth phases makes their physico-chemical dynamics more variable and complicated than that of permanently open estuaries. Mouth states are driven mainly by interplay between wave or tide driven sediment transport and river inflow. Mouth closure cuts off tidal exchanges with the ocean, resulting in prolonged periods of lagoonal conditions during which salinity and temperature stratification may develop, along with oxygen and nutrient depletion. Mouth breaching occurs when water levels overtop the frontal berm, usually during high river flow, and may be accompanied by scouring of estuarine sediment and an increased silt load and turbidity during the outflow phase. Microalgae are key primary producers in TOCEs, and while phytoplankton biomass in these systems is usually lower than in permanently open estuaries, microphytobenthic biomass is often much higher in TOCEs than in permanently open systems. During the closed phase, the absence of tidal currents, clearer water and greater light penetration can result in the proliferation of submerged macrophytes. Loss of tidal action and high water levels, however, also result in the absence or disappearance of mangroves and have adverse effects on salt marsh vegetation. Zooplankton are primary consumers both in the water-column and within the upper sediment, due to diel migrations. A prolonged period of TOCE mouth closure leads to poor levels of zooplankton diversity, but also to the biomass build-up of a few dominant species. Benthic meiofaunal abundance is usually greater during closed phases and is generally dominated by nematodes. Macrobenthic densities, and occasionally even biomass, in TOCEs are higher than in permanently open systems. The dominance of estuarine and estuarine-dependent marine fish species in TOCEs is an indication of the important nursery function of these systems. Marine juvenile fish recruit into TOCEs not only when the mouth opens, but also during marine overwash events when waves from the sea wash over the sand bar at the mouth. The birds that occur in TOCEs are mostly piscivorous, able to catch a variety of fish species either from the surface or by diving underwater. Waders are absent or uncommon because of the infrequent availability of intertidal feeding areas when the mouth is closed. Addressing the challenges facing the sustainable management of TOCEs is critical, as in some cases their ecological integrity, biodiversity and nursery function have already been compromised.

Chapter 39 - Diatoms are an important and often dominant component of the microalgal assemblages in estuarine and shallow coastal environments. Given their ubiquity and strong relationship with the physical and chemical characteristics of their environment, they have been used to reconstruct paleoenvironmental changes in coastal settings worldwide. The quality of the inferences relies upon a deep knowledge on the relationship of modern diatom species and their ecological requirements, as well as on the taphonomic constrains that can be affecting their preservation in sediments. In Argentina, information on estuarine diatom ecology is scattered and fragmentary. Studies on estuarine diatoms from the 20th century have been mostly restricted to taxonomic descriptions of discrete assemblages. Given the lack of detailed studies on the distribution of modern diatoms in local estuarine environments and their relationship with the prevailing environmental conditions, most paleoenvironmental reconstructions were based on the ecological requirements of European diatoms. However, studies on diatom distribution along estuarine gradients from Argentina have increased in recent years, constituting a potential source of data for paleoecologists. In this chapter, the literature on modern estuarine diatoms from Argentina is revised in order to synthesize the available ecological information and to detect possible modern analogues for Quaternary diatom assemblages. The main objective is to build bridges between ecology and paleoecology, and to discuss the reaches and limitations of the different approaches to diatom-based paleoenvironmental reconstructions. Further studies exploring the relationship between estuarine diatom distribution and environmental characteristics are necessary in order to increase the precision of paleoenvironmental inferences in the region and to generate new hypothesis for further study.

Chapter 40 - This chapter aims to highlight possible pollution impacts in a protected estuarine ecosystem, Amvrakikos Gulf, considered one of the most important wetlands in

Greece. Although land-based discharges in the area are low, Amvrakikos Gulf receives inputs by riverine transport, mainly contaminants related to agricultural practices. In order to assess pollution impacts, biological effects (determined with biomarkers and bioassays) and contaminant levels were measured in sentinel species (mussels Mytilus galloprovincialis) over a two-year period. Biomarkers at the sub organism level (Scope for Growth [SFG]) revealed stress conditions and provided early warning signals of possible consequences at higher levels of biological organization. Biochemical markers (acetylcholinesterase, glutathione peroxidase and metallothionein) suggested the presence of organic contaminants but absence of elevated metal levels and indicated a risk for pesticide contamination. Likewise, the Microtox bioassay applied in fluid extracts from mussels demonstrated stress conditions and the presence of organic contaminants based on the production of distinctly different light level-time response curves of the luminescent bacteria (Vibrio fidceri). Chemical analysis in the mussel tissues confirmed contamination of agricultural origin showing moderately elevated ΣDTT concentrations, low ΣPCB and low heavy metal concentrations.

Chapter 41 - The Santos-São Vicente Estuarine System is located in the central coast of São Paulo State, Brazil, in a metropolitan region known as Baixada Santista. This region has a permanent population of over 1,200,000 and an estimated fluctuating population of over 780,000 (Hortellani et al., 2005), and comprehends the municipalities of Santos, São Vicente, Praia Grande, Cubatão and Guarujá. The area bears the largest commercial harbour in Latin America, the Port of Santos, as well as one of the most important petrochemical, chemical and metallurgical industrial complexes of Brazil, the region of Cubatão city. Baixada Santista region has become famous as one of the most relevant examples of degradation of coastal environments as a consequence of anthropogenic activities.

Climate in the region is hot and wet, with a yearly average temperature of 22°C and relative humidity rates that reach more than 80% throughout the year. Rainfall rates range between 2,000 to 2,500mm, with a more intense regime in summer months in the south hemisphere - January to March.

The Santos-São Vicente estuarine system is confined between the Serra do Mar cliff and the Atlantic Ocean. Rivers coming down from the Serra do Mar flow fast and intensely, losing energy when they reach the plains, which have little or no declivity. As a result, a complex pattern of streams and creeks are formed, transforming vast regions in wetlands covered by mangroves. The São Vicente Estuary is located at the Western outlet of the system and the Santos Estuary, where the Santos Port is situated, lies at the eastern outlet. Both estuaries are interconnected at the upper part of the system, where another port can be found, the Cubatão Maritime Terminal, a private port to the Companhia Siderurgica Paulista – Steelworks Company of São Paulo (COSIPA).

Chapter 42 - The authors describe water quality in two hypersaline negative estuaries in the Northern Gulf of California, along the coast of Sonora, Mexico, over a two-year period. In the Northern Gulf, non-mangrove salt marshes known as esteros (negative estuaries) cover 134, 623 ha. Esteros are characterized by an extreme tidal range, higher salinity at their head than at their mouth due to high evaporation, limited freshwater input and a mixed semi-diurnal tidal regime. Between 2005 and 2007, the authors sampled surface temperature, dissolved oxygen, salinity, pH, depth, chlorophyll, nutrients ($NH+4$, $NO3^-$, $NO2^-$, and $PO4$), and total suspended solids across one wetland, Estero Morúa. The authors also led a participatory monitoring effort, where oyster farmers took daily measurements of surface temperature, salinity, pH, and dissolved oxygen in both Estero Morúa (31°17′09" N; 113° 26′19" W) and Estero Almejas (31°10′15" N; 113°03′53" W). These sites allow comparisons between distinct habitats and levels of oceanic influence. Estero Morúa is a high energy lagoon with a narrow mouth and a prominent permanent channel restricted by spits, while Almejas is an open bay with a large intertidal mudflat. Among the main findings are 1) Surface temperatures follow a seasonal pattern, with highest temperatures in June to September and lowest from December to February. 2) Low rainfall and runoff together with high seawater input and evaporative loss results in high salinities. 3) High dissolved oxygen concentrations and low nutrient levels are indicative of the recharge rate between the wetlands and the sea, and are characteristic of oligotrophic systems. It is likely that residence time and tidal dynamics are the main factors dictating the physicochemical dynamics of these systems.

Chapter 43 - The southwestern coast of India was under the influence of local marine environments to a minor extent in the geological past. The sea level along this part of the coast stood around 60–100 m below the present MSL during the last glacial maxima (around 20,000 YBP) and the rivers flowing at that time incised their valley to this base level. Later a humid climate with maximum representation of mangrove vegetation around 10,000 YBP was reported in this area, which suggested strengthened Asiatic Monsoon till the first part of Atlantic period. Initial subsidence and consequent flooding due to transgression, which had occurred 8000-6000 YBP, destroyed the mangrove vegetation giving rise to peaty soil. The pattern of rivers and geomorphological set up suggested that coastline was much towards east during the geological past and the rivers were flowing to the west and debouching in to the sea. The occurrences of peat sequence in the sediments in the low-lying area around the present Vembanad Lake and their radio carbon dating studies indicated their formation was from submerged coastal forest, especially mangrove vegetation. An event of regression (5000- 3000 YBP) was occurred along Kerala coast during the late Holocene. Contemporaneous to this, one of the major backwater systems of the southwestern coast- the Vembanad Lake- was developed. The shell deposits of Kerala, which form a very rich source of carbonate, are formed by the accumulation of dead shell-bearing organisms and it is suggested that these organisms after being trapped in their ecological niche were destroyed due to marine regression. The present day backwaters and estuaries that occur behind the sandbars in Kerala thus possibly owe their origin to the regression during the past. Present study on the formation of palaeodeposit of sand in the Meenachil River basin lying along the southwest coast of India point towards the morphometric rearrangement and a multi-proxy analysis of the palaeodeposits could portray the Holocene geomorphological modifications of the southwestern coast. Sedimentological, geochemical, palynological and palaeobotanical analysis were done to evaluate the influence of palaeoclimatic factors on the modifications of the earth's surface configuration along this part of the Indian sub-continent and it is suggested that geomorphological modifications of the southwestern coast of India shall be classified into three categories (1) Pre-Vembanad Lake formation, (2) Contemporaneous to Lake formation and (3) Post-Lake formation.

Chapter 44 - The Changjiang (Yangtze) River is known to contribute significantly to the ecosystems of the Changjiang River estuary and adjacent waters. In this chapter the authors present some long-term data of freshwater discharge, sediment load, nutrient concentrations and compositions in the river and estuary waters, as well as the long-term response of the ecosystem in the estuary.

The freshwater discharge from the Changjiang River fluctuated but no trend of increase or reduction in the past six decades; however, sediment load decreased continuously from the 1950s to present, and a sharp decrease was observed after the closure of the Three Gorges Dam.

The concentrations of dissolved inorganic nitrogen and phosphate increased in the Changjiang River water by a factor of six from the 1960s to present, and a reduction in dissolved silicate by two thirds over the same period. Concomitantly, an increase in DIN concentration and a reduction in silicate concentration both by a factor of two were observed in the estuarine water.

As an ecological consequence to such nutrient changes, the chlorophyll a concentration increased by a factor of four since the 1980s, and both the number and affected area of harmful algal blooms increased rapidly since 1985 in the Changjiang River estuary and adjacent sea areas. Also, the area of hypoxic zone off the Changjiang River estuary increased in the recent decades.

The authors predict that, the freshwater discharge and sediment load from the Changjiang River would be reduced further due to continuous construction of dams and water division projects on the river; the symptoms of eutrophication associated with nutrients would worsen in the Changjiang River estuary in the near future due to continuously increasing of the riverine nutrient pressure.

Chapter 45 - The Tampa Bay region faces projected stress from climate change, contaminants, nutrients, and of human development on a natural ecosystem that is valued (economically, aesthetically and culturally) in its present state. With fast-paced population increases, conversion and development of open land, and other stresses, redressing past damage to Bay habitats and protecting them in the future will remain the greatest challenge for managers in this region. Maintaining water quality gains of recent decades and sustaining ecological services requires careful thought and planning to compensate for these stressors. Approaches piloted during this study will be applicable to many urbanized estuaries, particularly along the U.S. Gulf of Mexico coastline, because they face similar stresses.

Regional and local planners, managers, and decision-makers, like those in the Tampa region, require new or modified ways to address questions regarding the production, delivery, and consumption of ecosystem services under various projected future scenarios of climate change and urban development. Approaches that the authors are developing collaboratively must address human well-being endpoints (the focus of ecosystem services) as well as more traditional measures of ecosystem integrity, sustainability, and productivity. Predictive estimates of ecosystem services value require 1) Alternative future scenarios, especially population growth and effects from climate change 2) Models translating environmental conditions set by these scenarios into ecosystem services production and their sustainability at different scales from local neighborhoods to watershed and larger and 3) An approach to visualize and link model outputs to a common currency for prioritization and valuation. Collectively, these comprise tools that can be used to make well-informed, thoughtful, and publically transparent decisions.

Chapter 46 - The Santos Estuarine System (SES) is a complex of bays, islands, estuarine channels and rivers located on the Southeast coast of Brazil, in which multiple contaminant sources are situated in close proximity to mangroves and other protected areas. In the present study, the bottom sediment quality from the SES was assessed using the Sediment Quality Triad approach, which incorporates concurrent measures of sediment chemistry, toxicity and macrobenthic community structure. Elevated concentrations of metals were detected in the inner parts of the estuary, in the vicinity of outfalls, and in the eastern zone of Santos Bay. PAHs were found at high concentrations only in the Santos Channel. Anionic detergents were found throughout the system, with higher concentrations occurring close to the sewage outfall diffusers and in the São Vicente Channel. Sediments were considered toxic based on whole sediment tests with amphipods and porewater tests with sea urchin embryos. The observed toxicity appeared to coincide with proximity to contaminant sources. The macrobenthic community for the entire study area showed signs of stress, as indicated by low abundance, richness and diversity. The integrative approach suggested that both environmental factors and contaminants were responsible for the altered benthic community structure. The most critically disturbed area was the Santos Channel (upper portion), followed by the São Vicente and Bertioga Channels, and the immediate vicinity of the sewage diffusers.

Chapter 47 - The evaluation of feed ingredients is crucial to nutritional research and feed development for aquaculture species. In evaluating unconventional ingredients for use in aquaculture feeds, there are several important knowledge components that should be understood to enable the judicious use of a particular ingredient in feed formulation. This includes information on (1) ingredient digestibility, (2) ingredient palatability and (3) nutrient utilization and interference.

Diet design, feeding strategy, fecal collection method and method of calculation all have important implications on the determination of the digestible value of nutrients from any unconventional ingredient. There are several ways in which palatability of unconventional ingredients can be assessed, usually based on variable inclusion levels of the unconventional ingredient in question in a reference diet and feeding of those diets under apparent satiate or self-regulating feeding regimes. The ability of fish to use nutrients from the test ingredient, or defining factors that interfere with that process, is perhaps the most complex and variable part of the ingredient evaluation process. It is crucial to discriminate effects on feed intake from effects on utilization of nutrients from unconventional ingredients (for growth and other metabolic processes). To allow an increased focus on nutrient utilization by the animals, there are several experimental carried out, which are based on variations in diet design and feeding regime used to determine the optimum level of substitutions of unconventional feed stuffs ingredients. Other issues such as ingredient functionality influence on immune status and effects on organoleptic qualities are also important consideration in determining the value of unconventional ingredients in aquaculture feed formulations.

Chapter 48 - Poultry production in the United States has grown dramatically in recent years which has resulted in the generation of large quantities of poultry litter. The high nutrient content of poultry litter makes it an excellent soil nutrient amendment. However, concern has arisen regarding potential negative impacts on stream water quality. Numerous studies have evaluated the edge of field effects of litter applications through plot studies while others have evaluated instream effects. However, an integration of both instream (watershed scale) as well as edge of field (plot studies) effects of litter applications on water quality was needed. Beginning in 1994, four related studies were conducted in East Texas by Stephen F. Austin State University, the Texas Commission on Environmental Quality, and the Angelina Neches River Authority to determine these potential effects. This project included: 1) a study to determine the effects on runoff water quality from 12 experimental plots with four different rates of surface litter application, 2) an upstream/downstream stream gaging study to evaluate water quality of two tributary streams of the Attoyac Bayou in areas of intense poultry production, 3) an assessment of water quality conditions within the Attoyac Bayou, 4) the use of a computer simulation model (AGNPS) to determine possible water quality effects of litter applications in the study areas. Storm water samples collected from the runoff plots and watershed gaging stations were analyzed for nutrients (total phosphorus, orthophosphorus, nitrate-nitrogen, Total Kjeldahl nitrogen, and potassium), total suspended sediment, pH, and conductivity. At watershed gaging stations, weekly grab samples were also collected and analyzed for the above parameters along with dissolved oxygen, temperature, and bacteria. Surface plot data indicated that a vegetated filter strip of 4.5 m was effective in reducing nutrient losses at the edge of fields where litter is applied, though buffer effectiveness was seen to decline after multiple applications of litter at higher rates. For the stream sampling sites, significantly higher nitrogen concentrations were found from pastured sites receiving broiler litter than from upstream forested sites. However, these concentrations were still below levels that could result in adverse water quality impacts and likely result from multiple nonpoint pollution sources in the pastured watershed. Bacterial concentrations were higher in pastured watersheds, though significant contributions resulted from wildlife in the forested watersheds as well. Results from these studies indicate that with proper management, including the use of streamside buffers and appropriate application rates, poultry litter applications are not likely to result in water quality degradation in the Attoyac Bayou.

Chapter 49 - There is public concern worldwide about the impact of agriculture on the environment and the migration of agrochemicals from their target to nearby terrestrial and aquatic ecosystems and sometimes to the atmosphere and other times to the groundwater. To achieve the highest yields, farmers use many agrochemicals and practices, which frequently have repercussions for the nearby natural or/and semi-natural adjacent areas. Research is needed to identify the correct amount of fertilisers and the appropriate management to be applied in order to minimise non-target effects while allowing the farmers profitable yield from their agro-ecosystems. Some agrochemicals can leave the agro-ecosystem by the runoff water, by the percolation of soil water or by evaporating into the atmosphere as gases. The recent awareness about the relationship between agricultural activities and non-point source pollution is also growing in many parts of the world. The challenge for farmers, land managers, and land users is to maintain the quality of surface waters and the health of biological communities by reducing and managing the amount of sediment, nutrients, and other pollutants in agricultural runoff. Making changes to land use or correcting past abuses is often expensive. A considerable amount of research is underway by organisations with a commitment to, or responsibility for, managing the effects of land-use practices on the environment. Studies have been made concerning land-use practices to reduce environmental impact, such as more efficient fertiliser use, sustainable grazing practices, fencing of riverbank access, and re-planting of vegetation along streams. Finally, research has supported a range of monitoring activities to determine the quantity and fate of agro-materials in agricultural runoff; this information is used to assess the health and status of agro-ecosystems and form a basis for evaluating the effectiveness of land-management practices to reduce threats.

Chapter 50 - A full understanding of the N cycling in lotic ecosystems is crucial given the increasing influence of human activities on the eutrophication of streams. Knowledge of stream processes involved in nutrient dynamics and metabolism in human-altered ecosystems is currently limited. On the other hand, N uptake and metabolism are likely to be linked in pristine streams, fact that remains unclear in human-altered streams. The authors aimed to examine N dynamics and the degree of linkage to ecosystem metabolism in two streams draining catchments with contrasting land uses, dominated by forest (reference) and agricultural activity (affected by agricultural runoff). In the two streams, the authors selected two reaches located upstream and downstream of a WWTP effluent input (i.e., point source) to compare the effect of point versus diffuse sources on the chemical and functional attributes of the two study scenarios. To achieve these objectives, the authors estimated rates of different biogeochemical processes involving N (i.e., uptake, nitrification and denitrification) and examined relationships between measured and estimated (based on in situ metabolism measurements) N demand upstream and downstream of the point source in the two streams. All measurements were done on 8 and 9 samplings through 2001-2003. The authors' results showed that land uses modulate the effect of point sources on chemical and functional attributes of the receiving streams. In the agricultural stream, diffuse sources from adjacent agricultural fields were likely to overwhelm the local effect of the point source on nutrient concentrations. In contrast, nutrient concentrations significantly increased below the point source. These different effects on stream water chemistry were reflected on the effects of the point source on the functional attributes of the study streams. Rates of studied biogeochemical processes were higher downstream than upstream of the point source in the forested stream, whereas they were similar between the two reaches in the agricultural stream. On the other hand, ambient nutrient concentrations upstream of the point source were significantly higher in the agricultural than in the forested stream, except for NH_4^+-N. This trend was reflected on higher nitrate uptake and denitrification rates in the agricultural than in the forested stream. Finally, coupling between measured and estimated N demand becomes weaker with increasing nutrient inputs from human activities to streams. Overall, the authors' results allow us to introduce the concept of the agricultural stream syndrome, by providing insights of stream ecosystem function in scenarios affected by agricultural runoff.

Chapter 51 - In agricultural environments such as dairy farms, large volumes of dirty water are produced, which must be managed and disposed of. Dirty water is typically spread on to fields, but excess run off can lead to depleted oxygen levels, pollution and damage to aquatic life within surrounding water courses. In this chapter, the available techniques for treatment of agricultural run off will be reviewed and compared.

A case study of treatment of dairy water at a UK dairy farm is presented in detail. At the farm, water with biochemical oxygen demand of typically 2500 mg/l and total solids of typically 6000 mg/l was produced from yard run-off, requiring treatment before disposal by land spreading. The performance of aeration, reed beds, overland flow and soil percolation plots, are compared based on extensive performance data from the working dairy farm. Average removals of BOD5 were over 95 % for the intensive aeration plant, 55 – 65 % for the reed bed, 55 % for the percolating soil plot and 80 – 90 % for the overland flow plot. However, it was shown that the reed bed could be highly effective as a polishing step following prior treatment by aeration. For total solids removal, the overland flow plot achieved the highest removals (42 – 57 %), followed by the intensive aeration plant (33 – 40 %). High loadings of solids applied to the reed beds caused operational problems such as blockages. Removals of total nitrogen, total ammoniacal nitrogen, phosphorus and COD in the treatment systems are also reported and compared. The highest average removals of these pollution indicators of total nitrogen 87.3 %, total ammoniacal nitrogen 98.1 %, phosphorus 63.7 % and COD 82.6 % occurred in the intensive aeration plant, with the other systems displaying lower removals. The costs of building and operating the different treatment systems are estimated, in order to evaluate the economics of different treatment options in addressing the problem of agricultural discharge. Finally an outlook on future treatment technologies is presented, including considerations of costs of different treatment options.

Chapter 52 - Water stress in Northern China has intensified water use conflicts between upstream and downstream areas and also between agriculture and municipal/industrial sectors. North China Plain (NCP) is one of the most important grain cropping areas in China. It is a giant alluvial plain formed by deposition by the Yellow, Hai, and Luan Rivers and their tributaries. Furthermore, there are some megalopolis, such as Beijing and Tianjin. This region has changed from water-rich in the 1950's to water-poor area at present, which indicates various ecosystem degradations such as dry-out Yellow River, nearly closed Hai River, groundwater degradation and seawater intrusion in the NCP. The author has so far developed the process-based model, called NIES Integrated Catchment-based Eco-hydrology (NICE) model (Nakayama, 2008a, 2008b, 2008c, 2009; Nakayama and Watanabe, 2004, 2006a, 2006b, 2008a, 2008b, 2008c; Nakayama et al., 2006, 2007), which includes surface-unsaturated-saturated water processes and assimilates land-surface processes describing the variation in phenology with satellite data. In this research, the author simulated the effects of irrigation on groundwater flow dynamics in the North China Plain by coupling the NICE with DSSAT-wheat and DSSAT-maize, two agricultural models. This combined model (NICE-AGR) was applied to the Hai River catchment and the lower reach of the Yellow River (530 km wide by 840 km long) at a resolution of 5 km. It reproduced excellently the soil moisture, evapotranspiration and crop production of summer maize and winter wheat, correctly estimating crop water use. So, the spatial distribution of crop water use was reasonably estimated at daily steps in the simulation area. In particular, the NICE-AGR reproduced groundwater levels better than the use of statistical water use data. This indicates that the NICE-AGR does not need detailed statistical data on water use, making it very powerful for evaluating and estimating the water dynamics of catchments with little statistical data on seasonal water use. Furthermore, the simulation reproduced the spatial distribution of groundwater level in 1987 and 1988 in the Hebei Plain, showing a major reduction of groundwater level due mainly to over-pumping for irrigation. These results show that this model is very powerful to simulate the future crop productivity in relation to the water withdrawal in this area. This study is very important for evaluation of ecosystems and environments on the moving edge of freshwater and marine.

Chapter 53 - In the Mediterranean the best lands have gradually been dedicated to the growing of cereals, as this is the foundation of the public's diet, leaving the worst, most degraded and generally sloping areas for vineyards and olives. The traditional management of these two agricultural products also tends to keep the ground bare, thereby intensifying runoff, erosion and soil degradation.

The sector's economic importance is undeniable, given that in producing areas it is the foundation of the economy and rural employment. The use of alternative bare soil management methods for this type of crop grown on slopes is a priority. The vegetable coverings have already been tried out in more humid countries, but it´s not very common in semiarid regions. Thanks to the biomass generated by the coverings, it is an effective method for erosion control and to increase the organic matter content in soil. But one must keep in mind that the coverings necessarily establish a relationship with the crop. Among others, one must point out the competition for nutrients and, above all, for water at certain times of year. In semi-arid climates, this aspect is crucial.

This chapter aims to explore how these vegetable coverings affect water dynamics, measuring the soil moisture and the runoff generated over the course of one year, and if a clear competition effect exists for water, which could damage not irrigated vineyards.

The vineyard in this study is dryland, located in the center of Spain, near Madrid. Its surface (2 ha) was divided following three treatments: Traditional tillage, Grass sowing using Brachypodium distachyon, and using rye (Secale cereale).

The average soil moisture data for the different periods of the vine's growth cycle show that the live cover of Brachypodium treatment presented less moisture than the other two treatments at 35 cm depth in spring and during the autum buds due to water consumption. The grape production of this treatment is less than the other two.

The runoff annual totals showed that the Bare soil treatment is similar to that of the Brachypodium treatment, equivalent to 8% of the total rainfall; the runoff of the Secale treatment was 9%. Regarding the Bare soil, the consequences of a certain storm depend on the time elapsed since the soil was last tilled.

Chapter 54 - The nomenclature is not standardized, and various authors describe the same features using different names. This ambiguity is especially evident in the terminology used for the subzones of the shore and littoral areas. In the absence of a widely accepted standard nomenclature, coastal researchers would do well to accompany their reports and publications with diagrams and definitions to ensure that readers will fully understand their use of terms.

Beaches include both the shore and nearshore subzones. They are composed of the material – sediment – primarily provided by the erosion of beach cliffs and/or by rivers that drain lowland areas. Other sources of materials, like shell fragments and the remains of microscopic organisms that live in the coastal waters, are less common. Beaches on volcanic islands are frequently composed of dark fragments of the basaltic lava that make up islands, or of coarse debris from coral reefs that develop around islands in low latitudes. Thus, around the world, there are beaches composed of crushed coral, beaches made of quartz sand, beaches made of rock fragments, beaches made of black (or even green) volcanic material, beaches composed of shell fragments, and even artificial beaches composed of scrap metal dumped at the beach (Thurman and Trujillo, 1999).

In general sediments consisting of sand and gravel occur on the upper and middle shoreface, muds of fluvial origin occur on the shelf and mixed sands and muds are found in the lower shoreface zone (transition zone to shelf). Bed sediments generally fine in seaward direction, supplied by offshore-directed bottom currents (rip currents and storm-induced currents). Sometimes relatively coarse-grained relict sediments are found on the shoreface

Chapter 55 - The entrances and/or access channels of harbors and other facilities constructed on coasts and estuaries where sediment supply is abundant might suffer massive sedimentations during extreme conditions, such as storm surges. Such massive sedimentation events often occur in a short time period, causing serious navigation problems and considerable economic losses. China's coast is stricken each year by a number of typhoons and temperate cyclones, and in recent years several harbors did experience massive sedimentations. Nonetheless, due to the complex interactions between coastal forces including wind, tidal current, wave, sediment transport, human activities and topographic configuration, etc., the underlying causes for such rapid, massive sedimentations have not been well understood and universally-applicable mitigation engineering measures are still lacking. Sea waves that approach the coast are becoming increasingly asymmetric and finally breaking due to the topographic effects in surf zones. This imposes shear stress upon the sea bed and initiates sediment transport. When the shear stress becomes high enough, ripples and dunes could be smoothed out and a thin layer of high concentration sediments is intensely transported on the sea bed. Such a sheet-flow transport process has been considered one of the major processes responsible for the massive sedimentation events. Nevertheless, good formulas that are capable of predicting longshore sediment transport rate in sheet-flow regime by stronger waves/currents are still lacking, though a good many efforts have been made to understand the sheet-flow mechanics, and formulas of various complexity have been proposed in recent years. This chapter introduces a formula that has a sound theoretical ground. It is able to predict longshore sand transport rate in sheet-flow regimes with higher precision, as verified with a recently-conducted physical model test as well as comparisons with other existing formulas. To mitigate the adverse massive sedimentations it is necessary to construct breakwaters or other infrastructure to impound the longshore sediment transport. However, an improper breakwater layout might even worsen the scenario. This chapter explores the optimum breakwater layout based upon the physical model test and presents a general guidance on breakwater layout design for coastal facilities. A new research perspective in coming years is also suggested.

Chapter 56 - If no conservation work is carried out, subsiding coastal areas are prone to progressive submersion by the sea and to changes in their physiographical features.

In Italy, the Lagoon of Venice is a peculiar example on this situation, where the Relative Sea Level Rise (RSLR) is the result of the sea level rise and the land subsidence both natural and man-induced. In non-urbanised areas in the Lagoon, the mean rate of RSLR over the last 5000 years was about 1 mm/year, assuming the lowest figures in the last 1000 years. In built-up areas, where compaction of superficial marshy ground due to huge load of buildings must be added, the rate is between 1 and 2 mm/year. Over the last century, the RSLR has totalled about 23 cm, equal to a mean rate of nearly to 2 mm/year.

In the Lagoon of Venice, both natural and anthropogenic evidence of RSLR is revealed in sediments. Natural evidence goes back to the period prior to human settlement, and that due to man's activities dates back to historical times. The works in question were mainly carried out in order to avoid having to abandon settlements due to exceptional flooding.

In the archaeological areas of the Lagoon, human settlements act as proxies clarifying evolution. In some places, man has abandoned previous settlements: for example, the island of S. Lorenzo was urbanised in the first few centuries AD, remained inhabited until the XIV century, and was then abandoned, due to a new increase in sea level. At other times, man succeeded in preventing this process by infill works, or by raising paved levels above mean sea level, as on the island of S. Francesco del Deserto, which has been inhabited since the I century AD. Instead, other non-urbanised areas in the Lagoon, such as salt-marshes or beach cordons, have disappeared with the passing of time, and now lie well below mean sea level. This natural situation is clearly testified by foraminifera - useful proxies of past situations of environmental evolution and its local chronology.

In the urbanised areas, and chiefly in the historical centre of Venice, buildings have acted as proxies of the RSLR in historical times. In the last 100 years, technical improvements have allowed ground elevation and sea level to be regularly monitored. At present, inside the Lagoon, some areas lying at approximately sea level, such as high marshes, are undergoing a significant increase in flooding, mostly attributable to the RSLR, and are becoming perennially submerged areas due to the lack of intervention. In view of the considerable importance of the city of Venice, much conservation and restoration work is being undertaken, in a everlasting confrontation with the RSLR.

Chapter 57 - The action of waves and currents often generates bedforms in the coastal zone. These bedforms significantly affect the sediment transport. Bars on sandy beaches contribute to the natural protection of the shoreline. Feedback processes occur in the coastal area which forms a complex system. The coastal erosion is an important problem. Depending on the considered site, it may be more appropriate to let the erosion occur, to use "soft" methods such as beach nourishment to prevent the shoreline from erosion, or to protect the coastal zone with hard structures. The greatest care must be taken when hard structures, which may lead to significant erosion at their foot, are considered. A multidiscipline approach is essential to improve the forecast of the shoreline evolution, in particular within a context of climate change.

Chapter 58 - Modern technology of baked goods largely uses sourdoughs because of the many advantages offered over baker's yeast. Sourdough is characterized by a complex microbial ecosystem, mainly represented by lactic acid bacteria (LAB) and yeasts, whose fermentation confers to the resulting product its characteristic features such as palatability and high sensory quality. Raw materials used in baking are not heat-treated; thus, they bring their wild microorganisms to the production process. Investigation of the composition and evolution of microbial communities of sourdough and raw materials is relevant in order to determine the potential activities of sourdough microorganisms. LAB are the main factor responsible for flavor development, improvement of nutritional quality as well as stability over consecutive refreshments of sourdough. LAB also establish some durable microbial associations, and the cell-cell communication process is crucial in determining sourdough performance during fermentation. The central aim of this chapter is to report on the knowledge of LAB hosted in raw materials and sourdoughs, their biodiversity and evolution before and during fermentation, their useful properties, their quorum sensing mechanisms and the methods routinely applied to their detection and monitoring in sourdough ecosystems.

Chapter 59 - Microscopic green algae are among the most widespread microrganisms occurring in terrestrial environments. For more than two centuries, generalizations on the diversity and biogeography of these organisms have been based entirely on morphological species concepts. However, ultrastructural and molecular data produced in the last 30 years have revealed a scenario in substantial contrast with morphological classifications. It has become clear that these organisms have been affected by an extreme morphological convergence, which has restricted their morphology to a narrow range, not indicative of their great genetic diversity. Their habit is very simple and uniform, usually referable to a few types (unicellular, uniseriate filamentous, sarcinoid colony) and offers very few characters useful for taxonomic and systematic purposes. These factors make the identification of terrestrial green algae and a correct interpretation of their biogeography very difficult. "Flagship" taxa with easily recognizable habit are the ones for which the best generalizations are possible. Examples of such taxa include the order Trentepohliales (for the highest diversity occurs in humid tropical regions of central-south America and south-eastern Asia) and members of the order Prasiolales (which are typically associated with polar and cold-temperate regions). In consideration of the recent developments, it is clear that many basic concepts about the biogeography of terrestrial green algae will have to be reconsidered critically. A deep understanding of this topic will require considerable work on many aspects of the biology of these organisms (systematics, distribution, dispersal, physiology), in which species circumscriptions based on molecular data will be a mandatory requirement.

Chapter 60 - New Caledonia is a peculiar hotspot, a small-sized island (ca. 17,000 km^2), relatively isolated from any continent (ca. 1200 km from Australia), with moderately high mountains and complex orography. Its biota is very rich in endemic species and highly endangered. The authors' analysis of the number of references in systematics, ecology and conservation shows that its biota attracts attention of scientists since long ago. In a first and long period, references focused on the description of the biodiversity. More recently these descriptions were intensified and complemented by ecological and later by conservation studies. Recent researches on phylogenetics and biogeography indicated that the biota of New Caledonia is characterized by short range endemism and rarity, with three patterns of endemism: (I) species regionally endemic to New Caledonia distributed in the whole range of an ecosystem; (II) short range endemics with parapatric/allopatric distributions; (II) short range endemics with disjunct distributions. Researches on conservation showed that this biota is highly endangered due to three main threats: fire, mining and invasive species. In this chapter the authors detail these threats and elaborate a model based on their frequency and spatial distribution to understand how they could affect species with contrasting patterns of endemicity. The authors' analysis shows that the conservation of the biodiversity in a context where species are dominantly short range endemics and rare is a main problem to be faced by New Caledonian authorities as well as by scientific researchers that must provide the basis for political decisions. Regardless the biogeographical pattern of endemism the chances of survival of rare species with short ranges in the case of large scale habitat destruction are quite low. In the case of threats that are more restricted in area, the loss of a species disjunctly distributed is more problematic in terms of loss of phylogenetic diversity. In this case, speciation by niche conservatism can be hypothesized to be less frequent, thus each species can be implied to be more original and in stronger need of conservation by itself. In addition, due to the distances from one another, the number of closely related species in a small island can be much lower than in the case of species with parapatric/allopatric distribution. Fires, mining and introduced species need special control. The two firsts for the habitat destruction they promote over extensive areas and the later by the possibility of continuing to endanger even in areas officially protected.

Chapter 61 - The current study was carried out on a longitudinal reach of an intermediate zone of the River Allier (France) that incorporates a flooded gravel pit, representing a significant depositional zone. The survey compared the macroinvertebrate communities of erosional and depositional forms along the study section. Fourty-four taxa were recorded in the erosional forms and 46 in the depositional zones, 57 % being common. Globally, the biodiversity in the deepest areas including the former gravel pit, did not diverge significantly from the other parts of the studied section. Further analysis of 22 samples from each geoform type revealed that total and EPT richness per sample were higher in the erosional areas than in the others. In the two geoforms the feeding groups and locomotion/substratum relations differed also, both underlining ecological and functional differences. Body size of the four dominant taxa (Orthocladiinae, Chiromini, Hydropsyche and Psychomyia pusilla) present in both erosional and depositional zones was also analysed and only Orthocladiinae differed between geoforms and depth, the larger individuals being in depositional zones and deeper water. Overall these results suggest that in the context of a rapid but reliable water quality assessment for biomonitoring, surveys of shallow, and thus more accessible erosional areas of gravelbed rivers may provide adequate representation of whole system. The ability to omit sampling of deeper depositional zones without adversely affecting the reliability of the outcome may prove both practicable and cost effective. The fact that erosional forms were seen to contain higher biodiversity than depositional areas and to shelter many of the taxa generally considered the most pollution-sensitive, gives further confidence.

Chapter 62 - Current loss of biodiversity places a premium on the task of recognizing and formulating proposals on potential areas for biological conservation based on scientific criteria; among these tasks, identification of hotspots has a relevant role on conservation of biodiversity. Among the approaches used in their recognition, biogeographic methods have a relevant role. In this study, the authors discuss the application of different biogeographic methods to identify plant biodiversity hotspots, based on the congruence among areas of endemism, panbiogeographic nodes, and Pleistocene refugia. Land plants are one of the best known biological groups that can be used as a test model for studies in biological diversity. The authors' study is based on previous biogeographic analyses of mosses, gymnosperms and angiosperms carried out in different places around the world, where panbiogeographic nodes and areas of endemism have been proposed for these plants, also including refugia proposed worldwide based on different biological groups. A remarkable congruence among these areas recognized by the application of different biogeographic methods for these plant groups is noted in western North America, Appalachian, Mesoamerica, southern Chile, southeastern Brazil, central Africa, Japan, southeastern China, Tasmania, New Caledonia, northeastern Australia, New Guinea, and New Zealand; this congruence indicates that these areas deserve special status in plant biodiversity and conservation. The authors propose that these areas identified by different biogeographic methods are important biodiversity areas and can be recognized as hotspots; these areas have a relevant role in plant biodiversity and are important in conservation due to their climatic conditions (refugia), the historical factors that have been involved in their evolution (nodes), and the restricted distribution of some plant taxa that inhabit them (areas of endemism).

Chapter 63 - Rapid deforestation occurred in northern Thailand over the last few decades and is expected to continue. The consequences of deforestation on biodiversity are substantial and widely recognized. The objectives of this chapter were to 1) predict future land-use change, 2) generate ecological niches of large mammals, and 3) assess wildlife concentrations and their hotspots as priorities for biodiversity conservation. Three land use change scenarios between 2002-2050 were evaluated. Scenario 1 is a continuation of land use trends and predicts that forest cover will decrease from the present 57% to only 45% in 2050. Scenario 2 is an integrated-management scenario and scenario 3 is a conservation-oriented scenario directed by government policy to maintain respectively 50% and 55% forest cover (natural forest and plantation) and strictly prevent encroachment in protected areas. Geographic Information Systems (GIS) and a spatially-explicit model (CLUE-s) were employed to explore land use changes. In addition, a machine learning algorithm based on maximum entropy theory (MAXENT) was used to generate ecological niches of 18 large mammal species as a proxy of biodiversity. The likely occurrences of selected wildlife species were aggregated and classified as wildlife concentrations. In addition, the predicted deforestation areas were overlaid on high wildlife concentration to determine threats to wildlife or wildlife hotspots both inside and outside protected areas.

The results reveal that forest cover in 2050 will mainly persist in the west and upper north of the region, which is rugged and not easily accessible. In contrast, the highest deforestation is expected to occur in the lower north and remnant habitats will disappear in this area. Current suitable habitats for most wildlife species are located in the west, north and east of the region. The predicted areas of high wildlife concentration or richness (≥ 7 species likely present) are found in the western forest complex and in the northeast of the region, covering approximately 16,000 km2 or 9.3% of the region. In addition, wildlife hotspots derived from the trend scenario encompass approximately 3,100 km2 or 1.8% and 74% is predicted in protected area coverage. These areas are identified as high priority for biodiversity conservation. In contrast, the hotspots are predicted of less than 0.5% for the scenarios 2 and 3. Based on the model outcomes, the authors recommend conservation measures to minimize the impacts of future deforestation on wildlife hotspots.

Chapter 64 - Compared to other major regions of the North Temperate Zone, China harbours a large number of representatives of very ancient plant lineages. Relict taxa are generally survivors of formerly much more widespread lineages, having also suffered the extinction of their close relatives. Moreover, in some cases they have remained superficially unchanged for millions of years, thus being regarded as true 'living fossils'. Chinese relict lineages are mainly concentrated in the central and southern regions of the country (usually below latitudes of 35°N), where the occurrence of numerous and extensive glacial refugia has been hypothesized. Some well known examples of relict plants still surviving in China are Cathaya argyrophylla, Cyclocarya paliurus, Eucommia ulmoides, Ginkgo biloba, and Metasequoia glyptostroboides. Most of the abovementioned species are now threatened because of their small distribution areas and increasing habitat destruction and fragmentation. The extirpation of such relict taxa would imply the loss of unique evolutionary history.

Chapter 65 - The Semen Mountains, a U.N. World Heritage Site is located in the North Ethiopia, in the Gondar Administrative region. The park encompasses a wide altitudinal range: which extends from below 2000m to above 4000m asl. It includes extensive high plateau areas (roughly 3200 m to 4000m), steep escarpments and lower-lying parts (below 3000 to 2000 m) and flat areas. This results in a rich mosaic pattern of different habitats, which promotes species richness and high biodiversity. The Semen lowlands, primarily harbouring afro-montane vegetation, are richest in species, whereas the afro-alpine belt, although, the most spectacular part of the mountains, is the poorest. In general, species diversity decreases with increasing altitude. At present, approximately 550 taxa of angiosperms are recorded. The park is famous for their rare and endangered animals (including the endemic Walia ibex, Capra walie) also harbour endemic plant species, i.e. Rosularia semiensis and Maytenus cortii. The Semen Mountains is the historic plant collecting locality, and about 40% flowering plants originate from this region alone. But in the past few decades, the park is facing deterioration due to socio-political problems in the North of the country.

In: Encyclopedia of Environmental Research   ISBN: 978-1-61761-927-4
Editor: Alisa N. Souter   © 2011 Nova Science Publishers, Inc.

Short Communication A

SEASONAL PATTERNS OF NITROGEN AND PHOSPHORUS LOSSES IN AGRICULTURAL DRAINAGE DITCHES IN NORTHERN MISSISSIPPI

Robert Kröger[*,1], *Marjorie M. Holland*[2], *Matt T. Moore*[3]
and Charlie M. Cooper[3]

[1]Department of Wildlife and Fisheries, Mississippi State University, Starkville, Mississippi 39762, USA
[2]Department of Biology, University of Mississippi, University, Mississippi 38677, USA
[3]USDA-ARS, National Sedimentation Laboratory, Oxford, Mississippi 38655, USA

ABSTRACT

Non point source pollutants such as nitrogen and phosphorus are transported via agricultural drainage ditches to receiving waters and result in eutrophication and downstream ecosystem degradation. Seasonal precipitation patterns, and land use contribute to variable concentrations of nitrogen and phosphorus loss from the landscape. This study investigated the differences in total inorganic-phosphate (TIP) and dissolved inorganic-nitrogen (DIN) concentrations in drainage ditches as a function of hydrology (stormflow vs. baseflow), land use (farmed vs. CRP vs. control) and season (growing vs. dormant). TIP and DIN ditch concentrations typically decreased from farms > CRP > control over both seasons. DIN concentrations were highest in baseflows during the growing season; however, mean DIN concentrations in baseflow were not significantly higher in the growing season than in the dormant season. Mean TIP concentrations and particulate phosphorus ratios increased in the dormant season with increased precipitation volumes and storm events. TIP concentrations were highest in growing season stormflows following fertilizer applications. A change in land use from agriculture to CRP decreases associated TIP and DIN in runoff and baseflow concentrations but residual phosphorus is still present in stormflows 10 yrs post the discontinuation of farming activity. This research highlights the episodic nature of nutrient contamination

[*] Corresponding author: Department of Wildlife and Fisheries, Mississippi State University, Box 9690, Starkville, Mississippi 39762, USA; rkroger@cfr.msstate.edu; Tel +1 662-325-4731; Fax +1 662-325-8795

driven by land use, anthropogenic fertilization, and the temporal variability of storm events.

INTRODUCTION

Nutrients and sediments from agriculture are major contributors to non-point sources of pollution and the impairment of U.S. waters (Carpenter et al. 1998). Contaminated discharge in the Mississippi River Basin is transported via drainage ditches to receiving waters such as streams and rivers, and eventually impacts coastal ecosystems such as the Gulf of Mexico (Rabalais et al. 1996, Turner and Rabalais 2003). Surface drainage ditches act as a primary conduit for surface and subsurface flows from the agricultural landscape, and thus are a major source of agricultural non-point source (NPS) pollutants (Nguyen and Sukias 2002). Agricultural discharge into drainage ditches is episodic, and thus information is needed on the variability of nutrient input as a result of hydrology, fertilization and land use to evaluate the temporal contribution of nutrient pollutants receiving waters.

Nitrogen (N) and phosphorus (P) are major contributors of nutrient NPS pollution from farms. However, N and P are each variable in respect to hydrological transport pathways. Typically, soil cation exchange capacity tends to increase ammonium concentrations in silt, loam, and clay soils. Depending on the intensity of subsurface drainage, management practices and connectivity of macropore flow, negatively charged nitrate NO_3^- may be transported to tile drains and receiving waters through prominent pathways of subsurface-flow (Owens et al. 1992, Angle et al. 1993, Skaggs et al. 1994). Leaching of NO_3^- most likely occurs post fertilizer application, when it is in excess in the root zone of the crops. Furthermore, NO_3^- is subject to leaching through late fall (October), and winter (November-February) when crop uptake of N is minimal or the farms lie fallow (Angle et al. 1993).

The loss of P in surface runoff occurs in inorganic, organic, dissolved and particulate forms (Sharpley et al. 1994). For the majority, P adsorbs to the fine particulate fraction in soil profiles, and is transported via bedload or suspended-sediment in surface runoff and erosion to receiving waters. However, some studies have shown P to move with subsurface drainage (Skaggs et al. 1994, Sims et al. 1998), where P flows occurs from downward movement of P by leaching or preferential flow through macropores.

Change in land use out of agriculture has an effect on anthropogenic sources of nutrients. The Natural Resources Conservation Services' (NRCS) Conservation Reserve Program (CRP) puts farm lands adjacent to receiving water out of crop production to create a buffer between the agricultural landscape and aquatic systems. The lack of agriculture results in a lack of anthropogenic fertilizer application, which potentially decreases nutrient concentrations within drainage ditches, and subsequently downstream aquatic ecosystems.

Seasonal influences on rainfall amounts, intensity, and frequency will have an effect on nutrient concentrations especially in relation to timing of fertilizer application and the time lag between application and rainfall patterns. Farm management practices are also seasonally influenced. In the growing season planting and crop growth occurs, while over the dormant season crops are harvested and the farms lie fallow. This study investigated the differences in TIP and DIN concentrations in drainage ditches as a function of hydrology (stormflow vs. baseflow), land use (farm vs. CRP vs. control) and season (growing vs. dormant).

MATERIALS AND METHODS

Study Sites

Primary intercept drainage ditches were sampled for 2 yrs within three different land use categories (farm vs. CRP vs. control) to compare the seasonal losses of nitrogen and phosphorus in baseflow and stormflow (Figure 1).

The farm ditches sampled drained no-till cotton farmed in a summer row-crop, winter fallow sequence. The two farms had a single surface drainage ditch, each draining approximately 13 ha. The farms were underlined with Chenneby silt loam soils (Morris 1981), high water tables and had no subsurface tile drains. Farm drainage ditches were on average 440 m ($\pm$ 26) long, and consisted of eight sampling locations positioned equidistant ($\pm$ 50 m) along the drainage ditches.

The CRP ditch drained an area of farmland previously farmed in conventional tilled cotton and rotational soybean, but was converted 10 y ago into CRP. The CRP site had a similar surface geology and had no subsurface tile drains. The CRP ditch was situated in close proximity to the farm ditches so that weather and storm event patterns were consistent between ditches. Within the same county and watershed, the control ditch drained a watershed with negligible anthropogenic influences.

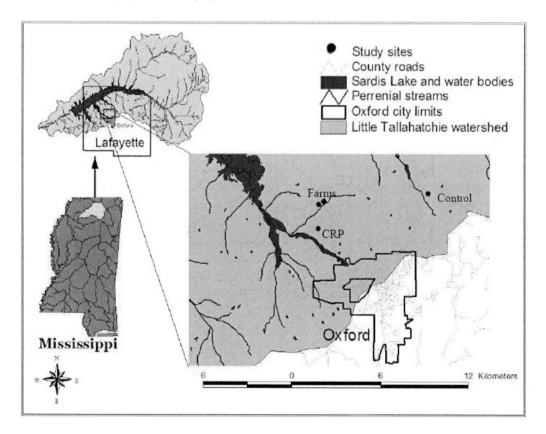

Figure 1. Study locations within the Little Tallahatchie watershed, northern Mississippi. GIS layers and data courtesy MARIS (www.maris.state).

The control ditch flowed continuously throughout the year out of a non-agricultural watershed (+/- 200 ha) and had a similar soil type to both farm and CRP ditches (Chenneby silt loam) (Morris 1981). CRP and control ditches were on average 600 m ($\pm$ 95) long, and consisted of seven sampling locations positioned equidistant ($\pm$ 90 m) along the drainage ditch.

Farms were annually fertilized with ammonium nitrate (NH_4NO_3) and phosphate (PO_4^{3-}) using a split fertilizer application. The first application occurred pre-planting in May of 112 kg ha^{-1} NH_4NO_3 and 56 kg ha^{-1} PO_4^{3-}. Subsequently, a second application of 56 kg ha^{-1} NH_4NO_3 was applied early July when the cotton was six nodes in height. Seasons were defined as growing (April – September) and dormant (October - March). April – September was used to adequately represent spring-bloomers such as *Juncus effusus* (March-May) and late summer blooms, e.g. *Sagittaria* sp.and goldenrods (*Solidago* and *Euthamia spp.*). The dormant season was classed as the remaining months from October – March.

Sampling of Baseflow Versus Stormflow Events

Surface flows typically as a result of a storm event are defined as stormflows. In contrast, baseflow is typically independent of a rainfall event, and is more than likely subsurface flows that may have been transformed by exposure through the soil profile (Haygarth et al. 2000). Sampling was divided into baseflow and stormflow samples. Grab baseflow samples were taken monthly at each sampling location to determine seasonal and annual differences in nutrient concentrations.

Stormflow samples were representative water samples generated by a specific storm event that elevated water levels within the ditch. Storm events were defined as rainfall events that generated sufficient surface runoff to elevate water volumes, and velocities within the drainage ditches. The storm water sample was obtained using a 400ml sampling container attached to a stake in the middle of the ditch, suspended 2 - 3cm above the baseflow water level. Storm samples were retrieved within 48 hours (usually 24 hrs) to negate the influence of evaporation on nutrient concentrations. The storm sample was a representative water and sediment sample of that specific storm event, which was more than likely a conservative measure as it was a mixed sample from the rising or first flush event, peak and falling limbs of the storm hydrograph. Refer to Kröger et al. (2007a, 2008) for hydrological and nutrient load data. All water samples were transported back to the lab in an ice chest, and kept at 4°C until chemical analysis took place. A total of 26 storm events over 2 y generated sufficient surface runoff to raise water levels in the respective ditches. Rainfall was recorded for the farms and CRP sites at the USDA-National Sedimentation Laboratory, Oxford MS, located 3 km from both CRP and farm sites. Control rainfall data was provided by the UMFS weather station on site of the control ditch.

Chemical Analyses for Nitrogen and Phosphorus

Unfiltered samples were analyzed for total inorganic phosphorus (TIP) using the ammonia persulfate digestion procedure with a colormetric molybdate reaction read at 880 nm on a spectrophotometer (Murphy and Riley 1962). The TIP analysis procedure converts organic dissolved and particulate forms to inorganic P during digestion. Samples filtered though a 0.45μm cellulose membrane were analyzed for dissolved inorganic phosphorus (DIP), ammonia (NH_3), nitrate (NO_3^-) and nitrite (NO_2^-). Subtracting DIP from TIP produced the particulate phosphorus (PP) fraction. Dissolved inorganic phosphorus was determined using the same colorimetric method as TIP, only after the sample was filtered. Nitrate and NO_2^- were analyzed using a Dionex Ion Chromatograph fitted with an anion conductivity detector (detection limit $>0.05mg\ l^{-1}$). Ammonia was determined by a standard phenate method (APHA 1998). Dissolved inorganic nitrogen (DIN) was the summation of NH_3, NO_3^- and NO_2^-.

Statistical Analyses

A one-way ANOVA was used to compare the means of nutrient concentrations between sites (land use) and season, and subsequently between base- and stormflow. All data were log transformed to meet the assumptions set forth by ANOVA. When sites (farm vs. CRP + control) were grouped together, or compared between base and storm flows they were compared using 2-sample, two tail, unequal variance student t-tests. All averages reported are +/- standard error at an alpha of 0.05. Sample location data were averaged for a ditch to provide a single estimate on nutrient concentrations, rather than violating the assumptions of independence by examining each sampling location independently within a single ditch.

RESULTS

Seasonal Rainfall Patterns

Variations in precipitation amounts and intensity of storm events occurred between growing and dormant seasons for 2004 and 2005 (Table 1). In both 2004 and 2005, approximately 50% of the rainfall occurred in short, intense rainfall events (stormflows) that generated significant amounts of overland surface runoff and elevated water levels within all ditches. In 2004, there was a greater overall stormflow within the dormant season, with more than double as much precipitation ($p < 0.0001$) falling over the 2004 dormant season than the growing season. In 2005 there were no significant differences ($p = 0.95$) in stormflow amounts, precipitation amounts ($p = 0.87$) and average storm size between the growing and dormant seasons (Table 1). Storm events were significantly larger ($p < 0.0001$) over the 2004 dormant season than the 2005 dormant season(81.9 ± 27 mm storm event^{-1}) (Table 1). There were no significant differences ($p = 0.082$) in total rainfall volumes between 2004 and 2005; however, the distribution of rainfall between growing and dormant seasons was quite marked.

Total Inorganic Phosphorus: Seasonal Vs. Flow Conditions Vs. Land Use

Figures 2 and 3 specify the differences in baseflow and stormflow TIP species respectively between farm, CRP and unfarmed drainage ditches in different seasons. Farm ditches consistently had significantly higher concentrations of all TIP species for base- and stormflows throughout both growing ($F_{Base\ 3,\ 254}=18.05$, $p < 0.0001$; $F_{Storm\ 3,\ 218} = 53.09$, $p < 0.0001$) and dormant seasons ($F_{Base\ 3,\ 239} = 42.48$, $p < 0.0001$; $F_{Storm\ 3,\ 357} = 63.09$, $p < 0.0001$) than control and CRP ditches.

Table 1. Differences in rainfall patterns (number of storm events) and precipitation amounts over the growing and dormant seasons between 2004 and 2005.

Year		Season	Σ (mm)	Mean (Σ) (mm/storm event)	n	S.E.
2004-05	Storm Events	Growing	221	36.84	6	8.17
		Dormant	573	81.9	7	27.49
	Precipitation	Growing	543.313	3.0184	180	0.68
		Dormant	1005.586	5.55	179	1.12
2005-06	Storm Events	Growing	369.67	52.81	7	4.33
		Dormant	322.07	53.67	6	4.75
	Precipitation	Growing	658.15	3.61	183	0.86
		Dormant	676.67	3.80	178	0.72

n represents the total number of storm or precipitation events for a particular season. Precipitation includes all rainfall events that did not produce significant runoff to elevate water levels within the ditch

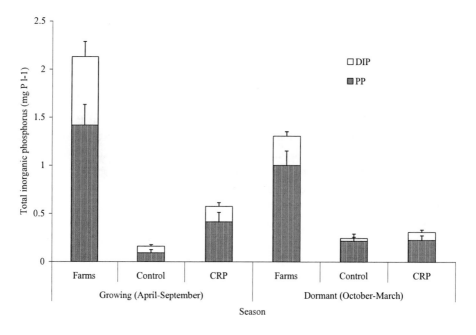

Figure 2. Mean ($\pm$ S.E.) seasonal TIP (DIP+PP) concentrations within surface drainage ditches for monthly baseflow conditions from June 2004 – May 2006.

Even though there were no significant differences between base- and stormflow DIP concentrations between growing and dormant seasons for either farm ($p = 0.1144$) or control ditches ($p = 0.3841$), the larger proportion of DIP over the growing season in farm ditches suggests an anthropogenic fertilizer application (Figure 2 and 3).

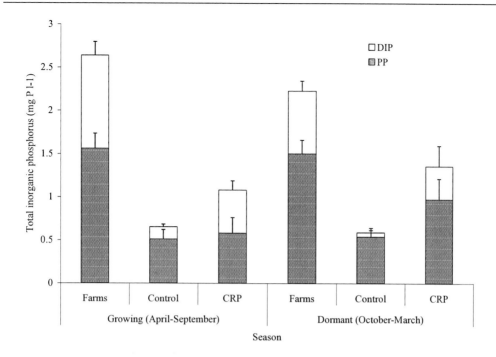

Figure 3. Mean ($\pm$ S.E.) seasonal TIP (DIP+PP) concentrations within surface drainage ditches for stormflow conditions from June 2004 – May 2006

The relative proportion of PP contributed to TIP in base- and stormflows increases from the growing season to the dormant season for both farms (67 – 77%) and controls (65 – 85%) most likely as a result of increased water volumes and sediment loads. The CRP ditch always had significantly higher baseflow TIP, PP and DIP concentrations ($p \leq 0.0001$) than the control ditch for base- and stormflows.

Stormflow (Figure 3) TIP concentrations for farm, CRP and control ditches were significantly higher for growing ($t_{farm}=5.629$, $t_{CRP}=3.826$, $t_{UMFS}=4.712$; $p < 0.001$) and dormant seasons ($t_{farm}=6.072$, $t_{CRP}=5.232$, $t_{UMFS}=4.571$; $p < 0.001$) than baseflow concentrations. Furthermore, the CRP ditch again, had significantly higher concentrations ($p < 0.001$) concentrations of DIP and PP when compared to the control ditch. The control ditch had no significant differences in TIP concentrations between growing and dormant seasons ($p = 0.167$). The growing season ratio of DIP: PP for farm baseflow was 0.48. The growing season ratio of DIP:PP for farm stormflow within the growing season was 0.69, suggesting a larger DIP contribution in stormflow than in baseflow.

Dissolved Inorganic Nitrogen: Seasonal Vs. Flow Conditions Vs. Land Use

Figures 4 and 5 specify the differences in baseflow and stormflow DIN species respectively between farm, CRP and control drainage ditches in different seasons. Farm ditches consistently have significantly higher concentrations of all DIN species when compared to the CRP and control ditches ($F_{Base3, 281}=18.05$; $p < 0.001$).

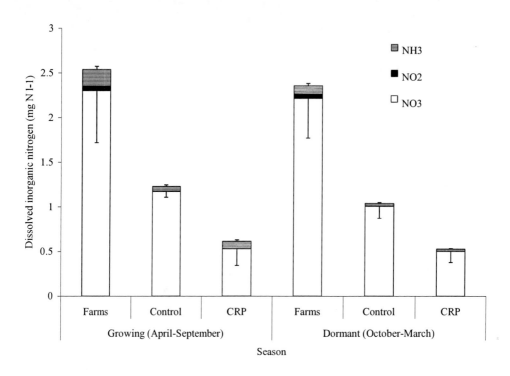

Figure 4. Mean ($\pm$ S.E) seasonal DIN ($NO_3+NO_2+NH_3$) concentrations within surface drainage ditches for monthly baseflow conditions from June 2004 to May 2006

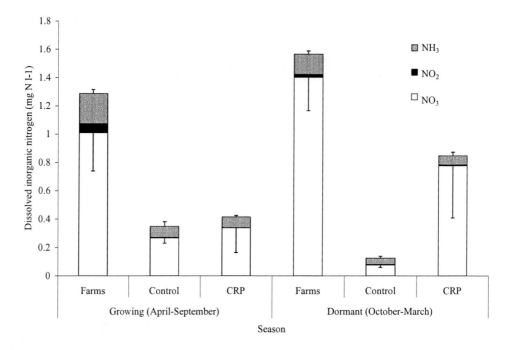

Figure 5. Mean ($\pm$ S.E) seasonal DIN ($NO_3+NO_2+NH_3$) concentrations within surface drainage ditches for stormflow conditions from June 2004 to May 2006

There were no significant differences between seasons for CRP and control ditches for NH_3 and NO_2^-. Nitrite contributed close to zero to the overall DIN concentrations for control and CRP ditches, and contributed less than 5% to DIN for farm ditches, where as NO_3^- was always at least 90% of DIN for both farm, CRP and control ditches. There were no significant differences ($p = 0.359$) in farm DIN baseflow concentrations between the growing and dormant seasons.

Stormflow (Figure 5) conditions had significantly lower concentrations of DIN, particularly NO_3^- for farms, UMFS and CRP ditches for both the growing ($F_{Storm\ 3,\ 218} = 2.38$, $p = 0.07$) and dormant seasons ($F_{Storm\ 3,\ 313} = 5.94$, $p = 0.006$) as compared to baseflow conditions. Dormant season farm stormflow DIN and NO_3^- concentrations were higher than the growing season, but were not significant ($p = 0.18$). Ammonia ($p = 0.001$) and NO_2^- ($p = 0.041$) farm concentrations were significantly more concentrated in the growing season than the dormant season (Figure 5).

DISCUSSION

Nutrient movement from the agricultural landscape into drainage ditches was a factor of variable precipitation events, surface runoff contributions and surrounding watershed land use. The contribution of applied fertilizers on the farms resulted in maximum concentrations of farm NO_3^-, DIN, DIP and TIP occurring in the growing season post fertilization in both baseflow and stormflow conditions in both years of study as a result of runoff and leaching. The exception occurred for DIN and NO_3^- for farm dormant season stormflows. A lack of NO_3^- leaching and growing season rainfall during 2004, coupled with large storm events over the dormant season would more than likely increase NO_3^- concentrations. Angle et al (1993) suggested that N fertilizer accumulation, and lack of leaching would result in higher concentrations of N being transported through late fall into winter. Furthermore it was hypothesized that the senescence of nutrient enriched above ground ditch vegetation could have resulted in an increase in DIN over the dormant season (Kröger et al. 2007b).

Lower concentrations of DIN in the dormant season baseflows suggest a decrease in N source with time after fertilizer application as a result of subsurface leaching and overland surface runoff (Owens et al. 1992). Significantly higher concentrations of DIN in baseflow over stormflow suggest a predominance of NO_3^- leaching through subsurface flows. Many studies support that baseflow will have higher concentrations of NO_3^- and other mobile constituents, than a stormflow that has generated overland surface runoff (Jackson et al. 1973, Timmons et al. 1977, Bauder and Schneider 1979, Owens et al. 1992). However, the conversion of land from agriculture into CRP, or the lack of agriculture has shown significantly lower, if not negligible DIN concentrations in surface drainage ditches. Thus removing sensitive land from agricultural production reduces DIN fertilizer applications which play a major role in the contribution toward NPS pollution of downstream ecosystems and ultimately coastal environments.

Higher concentrations of TIP (DIP and PP) in stormflows suggest that phosphorus, in the mid-South, is predominantly transported via overland surface pathways. A decrease in TIP concentrations with time highlights the effect of an anthropogenic fertilizer event on P concentrations in farm ditches and suggests temporal leaching post application. However, conversely, TIP concentrations increase with time from the growing to dormant season for the CRP and control ditches. The likelihood of increased precipitation in the dormant season, resulting in increased overland runoff suggests an increased sediment load, and higher stormflow TIP and PP concentrations than baseflow. CRP ditches had higher P concentrations in base- and storm flows when compared against the control or reference site within the watershed. Prior to CRP conversion the landscape had been actively farmed and fertilized in accordance with standard practices for cotton and soybean. Long term P fertilizer additions have the effect of concentration and accumulation in the soil, thus providing a potentially long term source of P leachate via stormflows (Haygarth and Jarvis 1999).

Understanding nutrient concentrations within the drainage ditch under variable land uses allows farm effluent loads to be determined (Kröger et al. 2007a, Kröger et al. 2008). It is important to remember that agricultural drainage ditches are wetland ecosystems that potentially play a mitigating role on influent nutrient loads further downstream. Determining the role of agricultural drainage ditches in their capacity to reduce nutrient loads is vital to understanding the effluent concentrations and loads reaching receiving waters.

CONCLUSION

Nutrient contamination from agricultural landscapes is characteristically episodic and driven by fertilization events, management practices, and hydrological variability of storm events. The order of TIP and DIN concentrations in base- and stormflows were farms > CRP > control, where the addition of annual fertilizers on farms elevated DIN and TIP concentrations with respect to CRP and control ditches. Nitrogen and P concentrations in base- and stormflows for all land uses were driven by the characteristics of storm events between the seasons influencing sub-surface and surface runoff.

REFERENCES

Angle, J. S.; Gross, C. M.; Hill, R. L.; McIntosh, M. S. *J. Environ. Qual.* 1993, *22*, 141-147.

APHA. *Standard methods for the examination of water and wastewater. 20th Edition.*; American Public Health Association, Washington D.C., 1998;

Bauder, J. W.; Schneider, R. P. *Journal of the Soil Science Society of America* 1979, *43*, 348-352.

Carpenter, S.; Caraco, N. F.; Correl, D. L.; Howarth, R. W.; Sharpley, A. N.; Smith, V. H. *Issues in Ecology* 1998, *3*, 1-12.

Haygarth, P. M.; Heathwaite, A. L.; Jarvis, S. C.; Harrod, T. R. *Advances in Agronomy* 2000, *69*, 153-178.

Haygarth, P. M.; Jarvis, S. C. *Advances in Agronomy* 1999, *66*, 195-249.

Jackson, W. A.; Asmussen, L. E.; Hauser, E. W.; White, A. W. *J. Environ. Qual.* 1973, *2*, 480-482.

Kröger, R.; Holland, M. M.; Moore, M. T.; Cooper, C. M. *J. Environ. Qual.* 2007a, *36*, 1646-1652.

Kröger, R.; Holland, M. M.; Moore, M. T.; Cooper, C. M. *Environmental Pollution* 2007b, *146*, 114-119.

Kröger, R.; Holland, M. M.; Moore, M. T.; Cooper, C. M. *J. Environ. Qual.* 2008, *37*, 107-113.

Morris, W. M. 1981. Soil Survey of Lafayette County, Mississippi. U.S. Department of Agriculture, Soil Conservation Service and Forest Service, Washington, D.C.

Murphy, R.; Riley, J. P. *Analytica Chimica Acta* 1962, *27*, 31-36.

Nguyen, L.; Sukias, J. *Agric. Water Manage.* 2002, *92*, 49-69.

Owens, L. B.; Edwards, W. M.; Van Keuren, R. W. *J. Environ. Qual.* 1992, *21*, 607-613.

Rabalais, N. N.; Turner, G. A.; Dortch, Q.; Wiseman, W. J., Jr.; Gupta, B. K. S. *Estuaries* 1996, *19*, 386-407.

Sharpley, A. N.; Chapra, S. C.; Wedepohl, R.; Sims, J. T.; Daniel, T. C.; Reddy, K. R. *J. Environ. Qual.* 1994, *23*, 437-451.

Sims, J. T.; Simard, R. R.; Joern, B. C. *Journal of Environmental Quality* 1998, *27*, 277-293.

Skaggs, R. W.; Breve, M. A.; Gilliam, J. W. *Crit. Rev. Env. Sci. Technol* 1994, *24*, 1-32.

Timmons, D. R.; Verry, E. S.; Burwell, R. E.; Holt, R. F. *Journal of Environmental Quality* 1977, *6*, 188-192.

Turner, R. E.; Rabalais, N. N. *Bioscience* 2003, *53*, 563-572.

In: Encyclopedia of Environmental Research
Editor: Alisa N. Souter

Short Communication B

GLOBAL CHANGE-INDUCED AGRICULTURAL RUNOFF AND FLOOD FREQUENCY INCREASE IN MEDITERRAENAN AREAS: AN ITALIAN PERSPECTIVE

Marco Piccarreta[*] and Domenico Capolongo

Dipartimento di Geologia e Geofisica, Università degli Studi di Bari,
via Orabona,4 Bari, Italy

ABSTRACT

Climatic change and local shifts in land-use have modified the Mediterranean landscape in the last decades, leading to an increase of the frequency and magnitude of geo-hydrological events, such as floods.

The more and more frequent trigger of floods during moderate rainfall intensity leads to wonder whether the main control factor is the climate dynamics or the land-use.

During the past twenty years, EU legislation has encouraged and subsidised major changes in land-use throughout the Mediterranean region (Faulkner et al., 2003), leading to an increase of arable land for cereal production in the hilly and low mountainous areas, traditionally mantled by shrub and low forest.

In Italy, the area affected by hydrogeological risk is 21,505 Km^2, which corresponds to the 7.1% of the whole territory. From 1991 to 2001 more than 1000 flood events occurred (Ministry of Environment, 2003). In only one year, 2003, major floods affected more than 300,000 people and caused damage up to 2,184 million euros.

For the Basilicata region, an area extending for 9,992 km^2 in southern Italy, both historical and recent information on floods is available. The area affected by flood risk is 260 Km^2 (2.6% of the whole territory) and mainly includes the hilly areas and the coastal belt (Metaponto flat), which are exploited in agriculture, the main regional economic source. On the hills vineyards and olive cultivation predominate, whereas in the Metaponto flat orchards and cereals are the main cultivation. The hilly area and the floodplain are connected by a dense network of deep gullies, running out into natural streams used by farmers for water supplying. Both gullies and streams have undergone great modifications in the last thirty years (Clarke and Rendell, 2000). Because of the advancements in earth-works machinery, gully walls with original slope angles > 45° never ploughed before, were levelled for the production of cereals and orchards, with a

---

[*] E-mail:marcopiccarreta@geo.uniba.it; capolongo@geo.uniba.it.

considerable increase of sediment production. At the same time natural streams connected with the gully outlets were canalized, with gradual decreasing of their beds and their flows. Thus, gullies became effective links for the transferring of runoff and sediment from uplands to valley bottom and pediment channels where they have exacerbated flooding.

Land-use map and digital elevation models of different years (1949, 1986, 2000) are available in a GIS-database, allowing topographic comparisons within the area under study. Simple buffering and map overlay indicate an increase of flooding in areas which have recently undergone land-use changes.

Owing to complex and stringent economical and social issues in European agriculture, the topic seems to be highly relevant.

2. STUDY AREA AND METHODS

The study area is located in the Pisticci territory within the hydrographical basin of the Basento river (Figure 1). The climate is typically Mediterranean, characterized by hot dry summers and mild wet winters. The yearly average rainfall is 574.17 mm and is mainly concentrated from November to January; the yearly average temperature is of 16° during summer and an average minimum of 8° during winter. Marine terraced deposits of sandy-conglomeratic nature widely outcrop in the area (Boenzi et al., 1976), which have been incised by deep gullies locally up to the clayey bedrock.

The gullies are perpendicular to the present shoreline in the hilly part, but subsequently they are captured Northwards (the Basento river), probably due to a tectonic disturb which is sub-paralell to the shoreline (Bentivenga et al., 2004). The gullies are characterised by vertical sidewalls and they are 10 – 30 m deep and 25 – 450 m wide, with a high degree of lateral expansion in relation to head retreat or linear advance due to the frequent failure of gully walls and the retreat of gullies towards drainage ways. Sediments mobilised from the walls are usually either removed by flowing water after high-intensity rainstorms or deposited on the gully bottom, which may lead to some degree of stabilisation.

The period ranging from the 1975s to the 1990s was significant in the region since the agricultural system experienced a substantial transformation after the advent of mechanisation. This change has led to the increase of land degradation and floods. Flood events occurring from 1951 to 2005 (Table 1) were available for the area (Caloiero and Mercuri, 1982; Progetto AVI; Clarke and Rendell, 2006). Data show that in the last ten years the area has undergone a strong increase of the frequency of flood events (Figure 2). For each recorded flood events, the daily rainfall (P1) and the previous cumulate rainfalls on two (P2), three (P3), five (P5), ten (P10), fifteen (P15), thirty (P30) and sixty (P60) days were calculated. From Table 1 clearly emerges that, for the last decade (except for November 2004), flood events were triggered by moderate intensity rainfall, even when they had been concentrated in five consecutive days, in the last decade. On the contrary, the events which had taken place in the first thirty years (1951 – 1980) were often triggered either by daily extreme events or heavy five days cumulates.

The research was based on a multitemporal analysis of pluviometrical data, remotely sensed data (aerial photos and orthophotos of 1954, 1986 and 2002) and digital terrain models (DEMs) (Figure 3), in order to:

- • assess the role of precipitation on the flooding processes of the study area;

- • assess the human impact;
- • estimate the erosion/deposition patterns dynamics

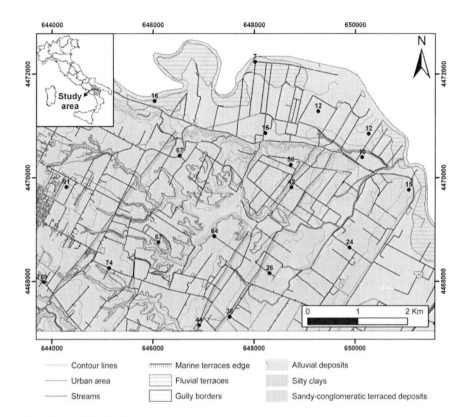

Figure 1. Location of the study area.

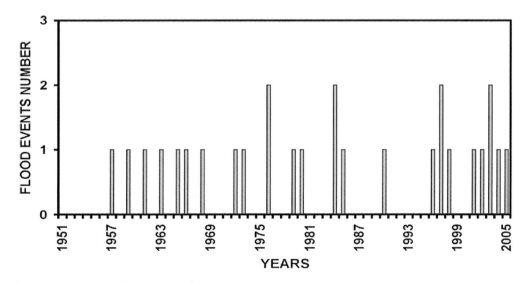

Figure 2. Frequency of flood events from 1951 to 2005.

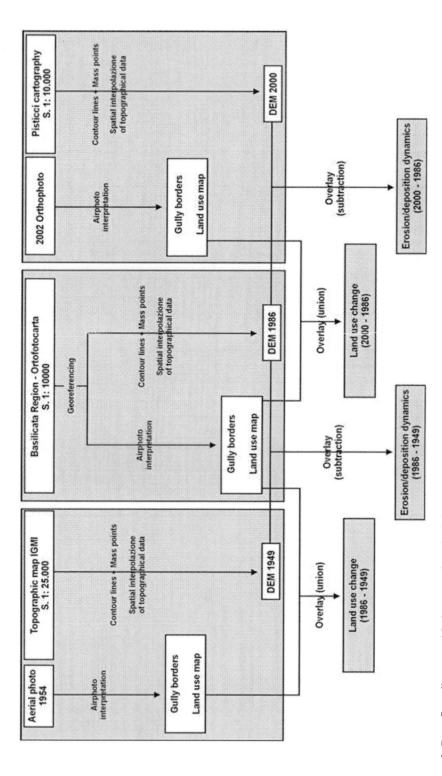

Figure 3. Data flow diagram with the general methodological process applied.

Table 1. Chronological listing of flood events in southeast Basilicata 1951–2005 and antecedent rainfall conditions over two days (P2), three days (P3), five days (P5), ten days (P10), fifteen days (P15), thirty days (P30) and sixty days (P60)

Date	24h	P2	P3	P5	P10	P15	P30	P60
21-24/11/1957	32.4	42.4	50.2	100.1	108.9	127.8	216.2	353.3
24-25/11/1959	314.7	375.7	391.5	391.5	407.7	514.8	571.0	619.4
15-17/1/1961	112.1	130.9	130.9	130.9	130.9	132.1	157.5	260.6
9-10/10/1963	54.8	97.4	97.4	97.4	142.5	149.7	161.8	193.1
21-25/9/1965	10.5	30.6	47.9	84.7	84.7	99.2	100.7	121.7
7-9/10/1966	35.3	83.5	93.5	93.5	99.9	100.1	130.7	131.9
12-15/12/1968	20.5	27.1	127.7	147.5	159.3	211.9	252.1	301.1
15-20/1/1972	34.8	119.0	151.4	155.1	209.2	209.7	260.6	274.6
30/12/1972-3/1/73	15.2	48.8	50.3	123.9	185.6	199.6	199.8	207.7
5-6/11/1976	89.2	96.0	97.2	100.6	109.7	153.8	222.8	227.5
18-20/1976	10.2	64.6	89.8	89.8	99.9	189.5	300.3	327.8
18-21/2/1979	15.3	35.0	35.0	89.9	90.4	117.5	147.4	159.9
11-13/1/1980	24.6	103.0	117.7	117.7	118.0	133.9	143.2	199.9
13-15/11/1984	32.5	131.4	172.8	172.8	172.8	172.8	196.1	267.8
29-30/12/1984	64.2	155.9	155.9	164.0	191.1	193.6	301.0	519.3
15-17/1/1985	18.8	31.7	31.7	31.7	50.1	55.9	250.6	401.6
25-27/12/1990	36.9	111.6	156.4	156.9	157.0	174.0	236.0	406.5
08/02/1996	30.4	31.6	31.7	31.7	99.3	135.6	137.3	224.6
28/10-1/11/1997	61.6	84.4	116.6	129.1	140.3	140.6	191.9	258.7
25/11/1997	33.8	42.2	42.2	65.2	80.0	106.1	237.7	300.5
04/02/1998	10.8	19.5	19.5	45.7	45.7	71.5	89.5	97.1
15/01/2001	8.7	71.3	71.3	71.3	74.9	134.2	158.0	222.6
01/12/2002	22.0	24.2	24.4	74.6	74.8	101.2	101.8	120.0
23-25/1/2003	13.6	54.6	63.0	63.2	71.8	92.6	121.8	394.0
26/11/2003	19.0	20.8	20.8	21.2	21.4	21.4	32.8	104.8
11-12/11/2004	163.4	208.2	208.4	225.4	236.0	236.0	241.0	272.4
07/10/2005	44.6	50.2	52.6	61.8	72.8	74.2	105.2	141.0

2.1 Precipitation Patterns

Daily pluviometric data from 1951 to 2005 relative to the Pisticci station were obtained from the National Hydrographic Service. In order to evaluate changes in the precipitation patterns and extremes during the whole period and in the last climatological normal (1976 – 2005), some indices recommended by the World Meteorological Organization were used:

- Total annual precipitation (AP);
- Number of wet days (P > 1mm) (NWD);
- Simple daily intensity index (SDII);
- Maximum 5d precipitation (P5dmax);
- Number of days with P > 90th percentile (very wet days), evaluated for the whole year (NDP90);
- Number of days with P > 95th percentile (extreme events), evaluated for the whole year (NDP95);
- Fraction of annual total precipitation due to events exceeding the 90th percentile (FP90);

- Fraction of annual total precipitation due to events exceeding the 95th percentile (FP95);
- Mean dry spell lengths (days) (DSM):
- Maximum number of consecutive dry days (CDD);
- Mean wet spell lengths (days) (WSM);
- Maximum number of consecutive wet days (CWD)

Trend significance was tested using the Mann-Kendall-Sneyers trend test (Sneyers, 1990).

2.2. Land Use Dynamics

Land use change is examined by using multitemporal analyses of aerial photos of 1954/55 (scale of 1:33, 000) and orthophotos of 1986 and 2002, following the same procedure as used by Piccarreta et al. (2006a). Nine typologies of land use/cover have been mapped (Figure 3): (a) built-up areas, (b) sown grounds, (c) streams, (d) vineyards, (e) olive, (f) woodland, (g) orchards, (h) greenhouse and (i) mediterranean scrubs.

2.3. Erosion/Deposition Processes Dynamics

The rates of channel incision and of sediment production were computed from the subtraction of the digital elevation models in grid format (1986–1949; 2000-1986), following the same procedure adopted by Martínez-Casasnovas (2003).

The 1949 DEM was constructed from 25-m spaced contour lines, mass points and break lines (streams incision) acquired from the Istituto Geografico Militare Italiano. The algorithm used for spatial interpolation was the Topogrid ArcInfo command, which uses an iterative finite difference interpolation technique that produces DEMs without losing the surface continuity of methods such as kriging or splines. The horizontal resolution given to the DEM was 20 m. The 1986 and 2000 DEM were constructed from 10-m contour lines, mass points and drainage lines acquired from the Technical Cartography of Basilicata Region. They were built with a horizontal resolution of 20 m, in order to make a better comparison with the 1949 DEM. The accuracy of the three DEMs was checked by means of a field survey using differential GPS.

The subtraction of the digital elevation models (1986-1949; 2000- 1986) produced a new grid with

the altitude difference for each cell of the grid. A negative value in cells of the difference grid was interpreted as erosion (surface lowering or gully deepening), a positive value as filling or aggradation and a very low or zero value as stable areas. The sediment production rate was calculated according to Eq. (1) (Martínez-Casasnovas, 2003).

$$SPR = (ED * GR^2) / AT$$

where SPR = sediment production rate (m year^{-1}), ED = sum of the elevation differences (m), GR = horizontal grid resolution (m) (20 m in the present case), A= surface (plane view) of the gully system area (m), T = time of the studied period (years).

3. RESULTS

3.1. Pluviometric Dynamics

All the pluviometric indices were analyzed for the pluviometric station of Pisticci; the results are reported in Table 2.

From 1951 to 2005 both total annual rainfalls and number of rainy days had strongly decreased, with a consequent reduction in rainfall intensity. The same statistically significant trend is confirmed for all the indices for extremes, but for the dry spell lentgh and the consecutive dry days. The obtained data depict a scenario characterized by a strong decrease in total annual rainfalls and in the frequency and magnitudo of extreme events. The dry spell length is more and more increasing, while the wet periods are decreasing.

In the last climatic normal (1976 – 2005) total annual rainfalls are increased, while a strong decrease in the number of rainy days occurred, leading to an amount in the daily precipitation intensity. The indices for extremes analyses show that as the frequency of extreme events descreases, their fraction and their concentrations in five days increase. Similarly to the macroscale analysis (1951 – 2005), the dry spell length increases (the consecutive dry days do not present any trend) while the wet periods decrease. The data indicate that in the last period a great change in pluviometric regime occurred, with an enlargement of the dry spells and a tendency of intense rainfalls to concentrate in macro-events of 3-4 consecutive wet days of great magnitudo.

3.2. Land Use Dynamics

During the 51-yr period examined in this study, soil management underwent remarkable changes in the study area (Figure 3). The results are given in Table 2. The aerial photos and orthophotos along with the field survey, have revealed that the main changes occurred from 1949 to 1986, whereas from 1986 to 2000 no remarkable shifts were recorded.

Table 2. Results of the Mann – Kendall test statistic to the pluviometric indices for the Pisticci rain gauge over the period 1951 – 2000

	AP	NWD	SDII	P5dmax	NDP90	FP90	NDP95	FP95	DSM	CDD	WSM	CWD
1951-2005	-3.53**	-4.0**	-1.84	-2.27*	-3.85**	-3.59**	-4.03**	-3.62**	1.69	0.24	-2.79**	-5.00**
1976-2005	0.55	-1.59	1.94	1.34	-1.20	0.55	-1.41	0.59	1.27	-0.16	-0.66	-2.12*

The asterisk (*) indicates the confidence interval (* 95% and ** 99%)

The main changes in the management from 1949 to 1986 involve the orchards (strawberry, apricot, peach, citrus fruit) which occupy a surface of 69.6, with a strong decrease of woodland and mediterranean scrubs areas (18.7 ha and 61.8 ha respectively). Moreover several areas previously devoted to cereal cultivation were transformed into orchards; anyway sown ground areas are weakly increased due to the reclamation of slopes characterized by mediterranean scrubs for cereal cultivation. The prevalent form of sown ground culture is the monoculture of durum wheat. This practice is mainly concentrated in the valley bottom and over the terraced surfaces where water supplies can be available more easily. It must be emphasized that ploughing and sowing are effected in October and November, the major plant growth occurs from March to May and the harvest takes place from June to August. As a consequence, the soil is deprived of vegetation cover from September to February and experiences degradation due to winter extreme storm events.

It is necessary to emphasize that the main land use changes took place after the '70s, when climate moved towards drier conditions.

3.3. Erosion/Deposition Processes Dynamics

The altitude difference maps as result of the subtraction of the 1949 and 1986 and of the 1986 and 2000 digital elevation models in the gullied area are shown in Figure 4. Negative cell values indicate erosion (surface lowering or gully deepening), positive cell values point to gully filling or aggradation, and a very low or zero value indicates stable areas. The erosion rate was calculated by Eq. 1. The DEM difference maps show that the pattern of erosion and deposition within the gully is very different in the two considered periods.

The sediment production from 1949 to 1986 was estimated in 9695.7 Mg (262.1 Mg / year); the deposition prevails on the erosion, and the deposition rate in the observed period is of 12.8 mm/year. The areas of sediment production are mainly located where the deepening processes of are more active, in correspondence of gully walls and gully heads as a consequence of downcutting and mass movements produced by the undercutting of the walls by stream flows. The deposition begins to take place near the bifurcation of the gully branches (cross section 1 in Figure 4) and become wider in the middle part (cross section 2 in Figure 4) and near the outlet (cross section 3 in Figure 4). As clearly shown by the three cross sections, land levelling could be considered in an embryonic phase; the modeling of slopes and surrounding areas through the use of heavy mechanization, led to decrease their slopes and to increase the production of loose sediments, which have been stored in the bottom of the gully.

It is important to remark that the obtained data have not to be related to the previous 1972 climatic characteristics, because the interpretation of aerial photos of 1972 has shown that the main changes in land use were subsequent to this date, in conjunction with a climate shift toward more arid condition with less effective rainfalls (Piccarreta et alii, 2004).

The same procedure was applied to compute the erosion rate in the gully during the period 1986 – 2000. The sediment production was estimated in 26867.4 Mg (1919.1 Mg / year). Erosion processes flatly domains on deposition, whit an erosion rate amounting to 94.1 mm / year, very close to those measured in other Mediterranean environments (Martínez-Casasnovas, 2003; Martínez-Casasnovas et al, 2003; 2004).

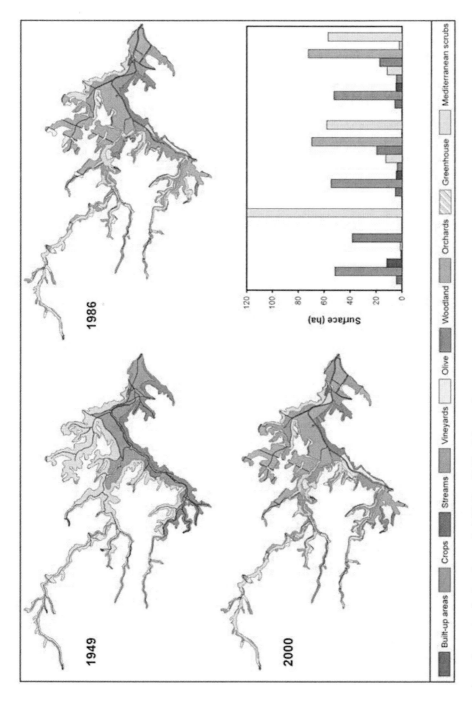

Figure 4. Land use dynamics during the period 1949 – 2000 of the study area.

During the studied period human action acquired a great importance. As shown in Figure 4 (cross section 1 and 3), the longitudinal profile of the gully in 2000 was almost completely flattened for cereals production.

4. DISCUSSION

The pluviometric regime underwent a radical change during the investigated period. Until 1965 rainfalls were abundant and well distributed in the year. Starting from 1970 an abrupt decrease of total annual rainfalls and a progressive increase of dry horizon have been recorded. In the last ten years of observation a new shift in pluviometric regime has been observed, whereas total rainfalls and dry periods are increasing with a strong decrease of rainy days and consecutive wet days. At the same time, the number of events exceeding the threshold of 90th and 95th percentile decreases while the fraction of total rainfalls exceeding the same thresholds increases. This dynamic shows a scenario in which extreme events of different magnitudo are concentrated in very short periods (3-4 consecutive days) separated by a long dry horizon. Human impact gets heavier after 1972, when the number of extreme rainstorms stongly decreases. From 1972 to 1986 eight flood events occurred. With the exception for the last event (1985), these flood events were mainly triggered by very extreme rainfalls falling in 24 or 48 hours. In this spell of time loose sediment were stored in the gully bottoms of the and near the outlet reaching a high degree of stabilization. At the same time natural streams connected with the gully outlets were canalized, with gradual decreasing of their beds and their flows.

In the last fifteen years the frequency of flood events have strongly increased, almost always due to rainfall events of moderate intensity. From 1986 to 2000 human impact is also important, but most of the sediment previously stored in the vally bottom were removed and trasported downstream to the pediment channels, were they accumulated again. The concentration of rainfalls in macro-events of 3-4 days, has exacerbated the phenomenon of "muddy floods " on the slopes and favoured the sediment transport downstream. The pediment canalized channels were filled by the sediment avoiding the normal water flow and favouring floodings in the surrounding areas.

5. CONCLUSION

Global change, expressed as a change in precipitation patterns and land use through time, influences geo-hydrological events in southern Italy. In the last years these events were triggered also by rainfalls of moderate intensity. The present paper contributes to the studies on floods in the Ionian belt of Basilicata, analyzing quali-quantitatively pluviometric, land use and erosion/deposition dynamics.

In Basilicata the flood risk appears intimately linked to the human impact, owing to the wrong application of some CAP measures, which have allowed the reclamation of woodland and bushy gully slopes for durum wheat and orchards cultivation favouring a great increase of sediment production in the gully bottom and of the frequency of muddy floods.

The floods frequency has increased in the last fifty years due to, also, the recent modification of the pluviometric regime, characterized by enlargement of the dry spells and by tendency of intense rainfalls to concentrate in macro-events of 3-4 consecutive wet days.

REFERENCES

Bentivenga, M., Coltorti, M., Prosser, G., Tavarnelli, E., 2004. A new Interpretation of Terraces in the Taranto Gulf: the Role of Extensional Faulting. Geomorphology 60: 383-402.

Boenzi, F., Palmentola, G., Valduga, A., 1976. Caratteri geomorfologici dell'area del Foglio ‹‹Matera››. Bollettino Società Geologica Italiana 95: 527 – 566

Borselli, L, Torri, D, Øygarden, L, De Alba, S, Martìnez-Casasnuevas, JA, Bazzoffi P, Jakab G. 2006. Land levelling. In Boardman J and Poesen J (eds). Soil Erosion in Europe. Wiley, Chichester; Chapter 2.12.

Brunetti, M, Colacino M, Maugeri, M, Nanni T. 2001. Trends in the daily intensity of precipitation in Italy from 1951 to 1996. *International Journal of Climatology* 21: 299–316.

Brunetti, M., Maugeri, M., Nanni, T., Navarra A. 2002. Droughts and extreme events in regional daily Italian precipitation series. *International Journal of Climatology* 22: 1455-1471.

Brunetti, M., Buffoni, L., Mangianti, F., Maugeri, M., Nanni, T. 2004. Temperature, precipitation and extreme events during the last century in Italy. Global and Planetary Change 40: 141 - 149.

Caloiero, D., Mercuri, T., 1982. Le alluvioni in Basilicata dal 1921 al 1980. C.N.R. – Istituto di Ricerca per la Protezione Idrogeologica, Cosenza, 28-53.

Capolongo, D., Pennetta, L., Piccarreta, M., Fallacara, G., Boenzi, F., 2008. Spatial and temporal variations in soil erosion and deposition due to land-levelling in a semi-arid area of Basilicata (Southern Italy). Earth Surface Processes Landforms 33: 364 – 379.

Capolongo, D., Diodato, N., Mannaerts, C.M., Piccarreta, M., Strobl, R.O. (in press). Analyzing temporal changes in climate erosivity using a simplified rainfall erosivity model in Basilicata, southern Italy. *Journal of Hydrology*. doi:10.1016 /j.jhydrol.2008.04.002

Clarke ML, Rendell HM. 2000. The impact of the farming practice of remodelling hillslope topography on badland morphology and soil erosion processes. Catena 40: 229–250.

Clarke ML, Rendell HM. 2006. Hindcasting extreme events: the occurrence and expression of damaging floods and landslides in Southern Italy. Land Degradation and Development 17: 365–380.

Faulkner H, Ruiz J, Zukowskyj P, Downward S. 2003. Erosion risk associated with rapid and extensive agricultural clearances on dispersive materials in southeast Spain. Environmental Science and Policy 6: 115–127.

Martínez-Casasnovas JA. 2003. A spatial information technology approach for the mapping and quantification of gully erosion. Catena 50: 293–308.

Piccarreta M, Capolongo D, Boenzi F. 2004. Trend analysis of precipitation and drought in Basilicata from 1923 to 2000 within a Southern Italy context. *International Journal of Climatology* 24: 907 – 922.

Piccarreta M, Capolongo D, Boenzi F, Bentivenga M. 2006. Implications of decadal changes in precipitation and land use policy to soil erosion in Basilicata, Italy. Catena 65: 138–151.

Sneyers R. 1990. On the statistical analysis of series of observation. WMO, Technical Note N. 143, Geneve, 192 pp.

RESEARCH AND REVIEW ARTICLES

In: Encyclopedia of Environmental Research   ISBN: 978-1-61761-927-4
Editor: Alisa N. Souter   © 2011 Nova Science Publishers, Inc.

Chapter 1

SOIL ORGANIC MATTER IN AN ALTITUDINAL GRADIENT

Bente Foereid[1] and Kim Harding[1]*
[1]The Macaulay Institute, Craigiebuckler, Aberdeen AB15 8QH, U.K.

ABSTRACT

The question of how climate change will affect carbon storage in grasslands is important, as it could lead to tipping points in the climate system. Most of the carbon in grasslands is stored in the soil, and this is also the most permanent store. The main question is how temperature affects soil organic matter development and decomposition. Some authors have found no effect, but this may be because the temperature under which the soils developed needs to be taken into account when assessing the temperature sensitivity of soil organic matter decomposition. Temperature decreases rather uniformly with altitude, and we used an altitudinal gradient as a temperature gradient. We sampled soils from an altitudinal gradient in four heath and grass covered Scottish mountains. Soil carbon content, carbon to nitrogen ratio and isotope ratios were correlated to altitude, plant cover and other soil parameters. The soils were then incubated for a year, and the change in these parameters over the incubation period where measured again and related to altitude and other factors. The results show that the most important environmental factors for the soil carbon and nitrogen content and isotope ratios were altitude and soil phosphorous. Plant species composition appeared unimportant. The changes over the course of the incubation were small, but they give weak support for the hypothesis that soils developed at low temperatures were least developed to start with, and therefore changed the most.

---

* Corresponding author: Department of Natural Resources, School of Applied Sciences, Building 53, Cranfield University, Cranfield, Bedford, MK43 0AL, UK, E-mail: b.foereid@cranfield.ac.uk, Tel: 01234 752776, Fax: 01234 752970

INTRODUCTION

Understanding the global carbon cycle is necessary to predict future biotic feedbacks to global change (Cox *et al.* 2000). In grasslands, the most important store of carbon is in the soil, as the plant carbon pool is small and turning over quickly. Carbon turnover in soils also has larger uncertainty than turnover in other biotic pools (Friedlingstein *et al.* 2006). Generally soil and litter decomposition are assumed to depend on moisture and temperature. This has been found to be broadly correct for a large number of environments (Couteaux *et al.* 1995; Cadish and Giller, 1997; Parton *et al.* 2007).

Model usually assume that decomposition rate depends on temperature (Parton *et al.* 1987; Parton *et al.* 1993; Jenkinson *et al.* 1987; Kirschbaum, 1995; Kättener *et al.* 1998; Kättener *et al.* 1998; Giardina and Ryan, 2000; Ågren and Bosatta, 2002). Most decomposition studies have focused on litter decomposition. However, for soil organic matter, one study found no temperature dependence (Giardina and Ryan, 2000). This can be explained as the quality of the organic matter has equilibrated to the prevailing condition, as the soils were all incubated at the mean annual temperature of the area they came from (Ågren and Bosatta, 2002). That could explain why nitrogen mineralisation measured under standard conditions, appeared to increase or stay unchanged with altitude (Morecroft *et al.* 1992a).

Temperature decreases with altitude in a fairly predictable manner, in Scotland this is approximately 1° C drop in temperature per 100 m increase in altitude (Harding, 1978). Although other abiotic factors varies with altitude as well, temperature is widely accepted to be the most important to shape the biological processes. An altitudinal gradient can therefore be regarded as temperature gradient.

Stable isotopes are increasingly being used to study element cycles (Ehlinger *et al.* 2000; Krull and Skjemstad, 2003; Staddon, 2004). δ^{13}C increases with depth or stays unchanged in forest soils, while δ^{15}N increases with depth (Natelhoffer and Fry, 1988). δ^{15}N and δ^{13}C appears to be an indicator of how decomposed or old the soil is, as microbes discriminate against it (Dijkstra *et al.* 2006; Dijkstra *et al.* 2008), although it may be different under waterlogged conditions (Wedin *et al.* 1995). δ^{13}C is known to change with growing conditions in C_3 plants (Farquar *et al.* 1989). This has also been observed as a gradient with increasing altitude (Morecroft *et al.* 1992b).

We aimed at understanding how the quality and degree of decomposition of organic matter varied with altitude. The hypothesis was that the higher the altitude the less humified or developed the soil would be, as it was developed at lower temperatures (Ågren and Bosatta, 2002). We sampled soils from altitudinal gradients and measured soil parameters thought to be indicators of soil age or degree of humification. We investigated how they varied with altitude and with other biotic and abiotic factors. We also incubated all the soils at the same temperature, to see if there was most change in those thought to be less developed or humified.

MATERIALS AND METHODS

Site Description

Four mountains Beinn a' Bhuird, Derry Cairngorm, Glas Maol and Glas Tulaichean, in the Scottish Grampian mountain range were sampled during autumn 2002. Samples were taken at 50 m altitudinal intervals along an altitudinal gradient from 700 m to 1000 m (at Glas Maol also at 1050 m). Site specific data is given in Table 1. Average yearly rainfall in the area was 910 mm year^{-1} measured at Breamar and average annual temperature was 0.8°C measured at Cairngorm (1245 m, 57°N, 3°W). Temperature data were averaged over 1989-1996 and rainfall data over 1992-2000. Plant species composition measurements at all sites were carried out during the summer of 2003. Plant species composition was recorded as percentage cover of each species. The species found are listed in Table 2.

Sampling and Experimental Procedures

The upper 5 cm of soil (excluding litter moss layer) was collected. Four samples were taken at each site, with 2 m between each sample. Soil samples were kept at a cool store (4 °C) until the start of the experiment in early 2003. Soil samples were incubated for one year at 16/10 °C (day/night). Samples were watered with demineralised water about twice a week, so that water content in the soil was kept close to field capacity. The soil trays were kept open to the air, and excess water was allowed to drain away. Samples were taken out for analysis before the experiment started and at the end of the experiment. Samples were ball-milled (0.5 mm mesh) and soil carbon and nitrogen as well as ^{15}N and ^{13}C were measured at a Delta plus Advantage isotope ratio mass spectrometer linked to a Flash EA Elemental Analyser (Thermo Electron, Bremen, Germany). Soil pH was measured in 0.1 M CaCl$_2$ on untreated soil samples. Total extractable P was determined after alkali fusion on milled samples.

Statistics and Data Analysis

Polynomial functions of total carbon and nitrogen, δ^{13}C and δ^{15}N C:N ratio phosphorous content and pH against altitude were fitted (SigmaPlot2001). The highest order polynomial that showed significant fit to the data was chosen. For total C and N, δ^{15}N, δ^{13}C and C:N ratio the differences between before and after the incubation were calculated. Also for the differences polynomial functions were adapted when possible. The four samples were autocorrelated and therefore an average was used in further statistical analysis.

A multivariate PCA (Canoco 4.0) analysis was carried out where the factors were plant cover data, altitude, pH, total C, N and P, δ^{15}N, δ^{13}C and change in total C and N, δ^{15}N and δ^{13}C during incubation. Negative values were transformed before analysis. δ^{13}C was transformed by multiplying by −1, as all values were negative. In the other parameters where negative values occurred (δ^{15}N and all the measures of change), the same number was added to all measurements. The analysis was also run without the plant cover data. As the inclusion of the plant cover data did not change the position of the loading for the soil variables, these data were left out of the further analysis. The sample from Glas Maol 700 m showed up as an outlier in the score plot (due to much higher pH than the other samples), so this sample was left out of the further analysis. A hybrid RDA (Canoco 4.0) was performed where all the variables measuring change during the incubation were taken to be response variables, while the soil variables measured at the start as well as altitude were taken to be explanatory variables. This analysis requires that all response variables are measured in the same units

(Leps and Smilauler, 2003). The response variables were therefore standardised so that the range was from 0 to 1 for all. As the PCA analysis showed that altitude and phosphorous content were virtually uncorrelated, and spanned the variation in the data set well, a multiple linear regression analysis (SigmaPlot2001) using these two as predictor variables was performed.

Table 1. Information about sampling sites.

Mountain	Altitude at summit (m)	Location	Transect aspect	Geology
Beinn a' Bhuird	1197	N57:05:15 W3:29:59	SSW	Biotite - Granite
Glas Tulaichean	1051	N56:51:56 W3:33:30	S	Caenlochan Schist
Derry Cairngorm	1155	N57:03:45 W3:37:21	S	Granite
Glas Maol	1068	N56:52:21 W3:22:06	ENE	Glas Moal Schist, An Soach-Cairnwell Transition, Creag Leacach Quartzite, Cairn of Claise Transition, Schiehallion Boulder Bed, Cairn Aig Mhala Limestone, Glen Callater banded group, Gleann Beag Schist, Baddoch Burn Dolomite

Table 2. Plant species (or ground cover) and numbers used in Figure 3.

No.	Plant species / ground cover	No.	Plant species / ground cover
1	*Alchemilla alpina*	30	*Juncus bufonius*
2	*Alchemilla molis*	31	*Juncus squarrosus*
3	*Agrostis canina*	32	*Juncus trifidus*
4	*Agrostis capillaris*	33	Lichen
5	Anemone nemorosa	34	*Leotodon sp.*
6	*Anthoxanthum oduratum*	35	*Listera cordata*
7	*Arctostaphylos uva-ursi*	36	*Loiseleuria procumbens*
8	Bare ground	37	*Luzula sp.*
9	*Calluna vulgaris*	38	*Lycopdium annotinum*
10	*Campula rotundifiolia*	39	*Melampyrum sp.*
11	*Carex bigelowii*	40	Moss
12	*Carex panicea*	41	*Nardus stricta*
13	*Carex pilulifera*	42	*Pinguicula vulgaris*
14	*Carex pulicaris*	43	*Polygonum viviparum*
15	*Carex sp.*	44	*Potentilla erecta*
16	*Cerastium fontanum*	45	Rock
17	*Deschampsia cespitosa*	46	*Rubus chamaeorus*

Table 2. (Continued)

No.	Plant species / ground cover	No.	Plant species / ground cover
18	*Deschampsia flexuosa*	47	*Rumex acetosa*
19	*Diphasiatrum alpinum*	48	*Salix herbacea*
20	*Empetrum nigrum*	49	*Scirpus cespitosa*
21	*Erica tetralix*	50	*Solidago viraurea*
22	*Eriophorum angustifolium*	51	*Succisa pratensis*
23	*Eriophorum spp.*	52	*Thalicatrum alpinum*
24	*Eriophorum vaginatum*	53	*Vaccinium myrtillus*
25	*Festuca ovina*	54	*Vaccinium uliginosum*
26	*Festuca vivipara*	55	*Vaccinium vitis-idaea*
27	*Galium saxatile*	56	*Viola riviniana*
28	*Gnaphalium supinum*	57	*Viola sp.*
29	*Huperzia selago*		

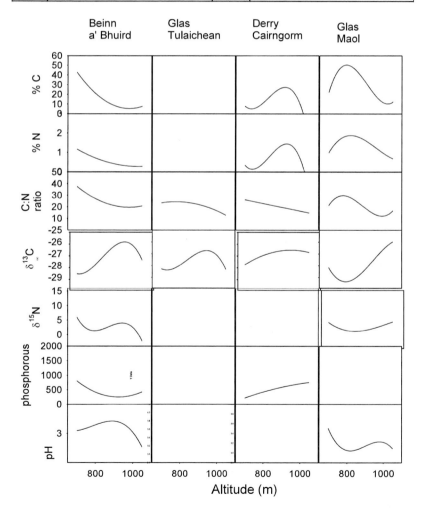

Figure 1. Scatter plots of measured parameter against altitude. The fitted graphs show the highest order polynomial to give significant fit (no graph indicates no significant fit to any polynomial).

RESULTS

There was a trend with altitude for most of the measured soil variables at most of the mountains (Figure 1). C:N ratio showed a stronger trend than total C and N separately. $\delta^{13}C$ showed a trend with altitude at all the mountains, while there were fewer trends for the other variables ($\delta^{15}N$, total P and pH) (Figure 1). There were also not so many significant trends when changes during the incubation were compared across altitudes (Figure 2). In general, changes over the incubation were small, and often not significant (data not shown).

The multivariate analysis showed that almost all the total variation in the data-set (98.5 %) was spanned by two axes. The first of these axes was almost identical to phosphorous content (85.2 %), the second almost identical to altitude (13.3 %) (Figure 3). The soil variables were more determined by the P-axis than the plant cover data (Figure 3). Plant cover depended both on altitude and soil (Figure 3). The plant cover data did not appear to have any effects on soil factors that were not described by the altitude and P-content. The hRDA analysis showed that most of the soil chemical parameters spanned the same axis, whilst soil pH appeared to be different (Figure 4). The increase in $\delta^{15}N$ and to $\delta^{13}C$ and decrease in total nitrogen during incubation was larger at higher altitudes (Figure 4), whilst changes in total carbon content were more related to soil chemical parameters.

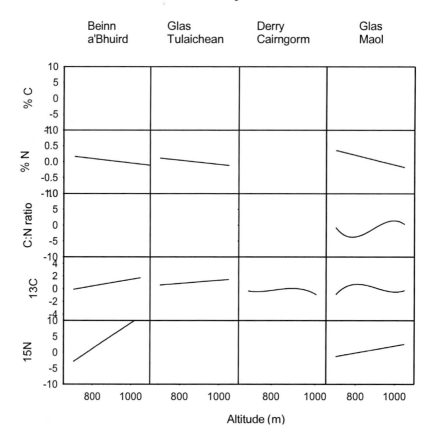

Figure 2. Scatter plots of change in measured parameter during incubation against altitude. The fitted graphs show the highest order polynomial to give significant fit (no graph indicates no significant fit to any polynomial).

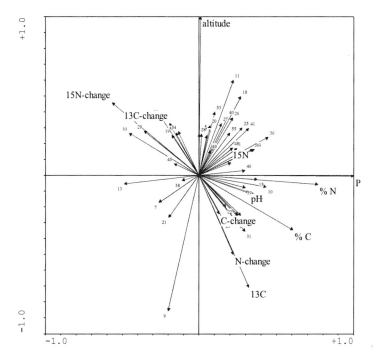

Figure 3. PCA of the measured soil parameters and plant cover data. Plant species are coded as numbers, the species can be found in table 4. Measured soil parameters are shown in larger fonts. PC 1 spans 85.2 % of the total variation, PC 2 spans 13.3 %.

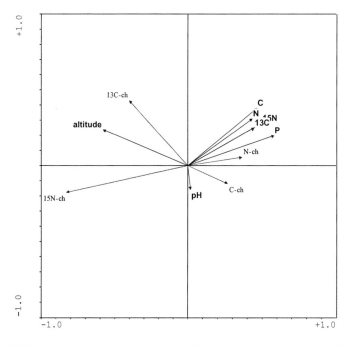

Figure 4. Hybrid RDA-plot of changes during incubation as response variables, soil parameters measured at the start and altitude as explanatory variables (Glas Maol 700 m is left out of the analysis). PC 1 spans 68.3 % and PC 2 14.3 % of the variance in the relation between explanatory and response variables.

Table 3. Linear regression of altitude and soil phosphorous against measured parameter (left column).Regression parameters are given as constant and coefficient for the two significant factors (altitude and soil phosphorous). Averages of the 4 samples at each site are used. Sample from Glas Maol at 700 m is considered an outlier and removed from the analysis.

	Regression parameters: Constant Altitude Phosphorous	Parameter p-value	Regression summary r2 and p
% N	0.357749 -0.000488383 0.00152542	0.5405 0.4629 <0.0001	0.61097212 <0.0001
% C	45.094 -0.0521118 0.0325175	0.0098 0.0087 <0.0001	0.53522354 <0.0001
N15			0.5741
C13	-32.8253 0.00720327 -0.00135419	<0.0001 <0.0001 0.0056	0.62926108 <0.0001
C/N	60.1154 -0.043693 8.55419e-005	<0.0001 <0.0001 0.9739	0.61396371 <0.0001
pH			0.3166
% N increase			0.9681
% C increase			0.8360
N15 increase	-6.23738 0.0130025 -0.00611225	0.0854 0.0030 0.0003	0.50887165 0.0001
C13 increase			0.1745
C/N increase			0.5318

The regression analysis showed a highly significant effect of both altitude and soil phosphorous on most of the measured soil variables (Table 3). Only for $\delta^{15}N$ and pH the regression model did not fit, for %N there was no significant effect of altitude and for C: N ratio there was no significant effect of soil phosphorous. There was generally little change over the course of the incubation, and only the change in $\delta^{15}N$ showed significant fit to the regression model (Table 3). Both the effects of altitude and soil phosphorous were significant.

DISCUSSION

The increase in C:N ratio with altitude could indicate less decomposed material, with more undecomposed litter. However, ^{15}N enrichment, normally a sign of decomposition, was not affected by altitude. It is possible that other factors have been more important for $\delta^{15}N$ distribution, particularly percentage of N coming from atmospheric deposition. The distribution of $\delta^{13}C$ shows the overriding influence of plant enrichment, which is depend on altitude (Morecroft et al. 1992b; Morecroft and Woodward, 1996).

The PCA analysis shows that all the variation, both in the measured soil parameters and in species composition at the sites, could be expressed as functions of only two factors: Altitude and total soil phosphorous. This seems to indicate that much of the scatter in the graphs of soil factors against altitude can be explained by soil phosphorous. Soil phosphorous explains more of the total variance than altitude, indicating that soil fertility may be more important than temperature. Soil phosphorous is also strongly correlated to site co-ordinates (data not shown); indicating that it is related to local geology. Generally soil P was more important for the measured soil parameters than for the plants, while at least some plants showed stronger dependence on altitude. Particularly soil nitrogen content correlates strongly with soil P, whilst soil carbon content was less correlated to soil P, and more dependent on altitude.

The small change in all samples and all variables over a year indicates that temperature was probably not the main limiting factor on mineralisation. Although soil P (or soil geochemistry) was an important factor, the regression analysis indicated that even when this factor was included, there were few significant effects. This raises the question what controls decomposition rates and keeps them low. The plant species that grows on the site (and produces the litter) appear to be unimportant, at least for the range of plant species on these sites. Soil pH is probably the most likely candidate. All sites but one in our sample had low pH. This sample was not included in the analysis because it was so different from all the others, but it did have higher carbon loss rate.

To the extent there was any significant trend with altitude, they confirm the hypothesis that changes during incubation would be largest in the samples from high altitudes (Table 2, Figure 2). If our hypothesis is correct, we would expect total C and N to decrease more in the samples from higher altitudes, C:N ratio to decrease more, $\delta^{13}C$ and $\delta^{15}N$ to increase more at higher altitudes. There are trends in that direction for % N, $\delta^{13}C$ and $\delta^{15}N$ for individual mountains (Figure 2), however, only for $\delta^{15}N$ the effect was found to be significant overall (Table 2). We can conclude that we have weak support for our hypothesis that soil organic material deposited at low temperatures (higher altitudes) was less decomposed than organic material deposited at higher temperatures (lower altitudes). However, other factors, possibly soil pH, seem to be more important to control decomposition rates in these systems overall.

CONCLUSION

We have found that both plant species composition and soil chemical parameters could to a large degree be predicted by altitude and soil phosphorous. Plant litter from high altitudes appears to be less decomposed than plant litter developed at lower latitudes, probably due to temperature. However, we have found that other ecological factors are more important. One other important factor is soil phosphorous. Plant composition appears not to be important for plant litter degradability.

ACKNOWLEDGMENT

The authors wish to thank Macaulay Institute analytical Services for soil analysis. This study was funded by the Scottish Executive Environment and Rural Affairs Department and the Macaulay Development Trust.

REFERENCES

Ågren, G. I. & Bosatta, E. (2002). *Soil Biol Biochem.*, 2002, *34*, 129-132.

Cadish, G. & Giller, K. E. (1997). *Driven by Nature: Plant litter Quality and Decomposition.* CAB International, Wallingford, UK.

Couteaux, M. M., Bottner, P. & Berg, B. (1995). *TREE*, *10*, 63-66.

Cox, P. M., Betts, R. A., Jones, C. D., Spall, S. A. & Totterdell, I. J. (2000). *Nature, 408*, 184-187.

Dijkstra, P., Ishizu, A., Doucett, R., Hart, S. C., Schwartz, E., Menyailo, O. V. & Hungate, B. A. (2006). 13C and 15N natural abundance of the soil microbial biomass. *Soil Biol Biochem.*, *38*, 3257-3266.

Dijkstra, P., LaViolette, C. M., Coyle, J. S., Doucett, R. R., Schwartz, E., Hart, S. C. & Hungate, B. A. (2008). *Ecol Lett.*, *11*, 389-397.

Ehlinger, J. R., Buchmann, N. & Flanagan, L. B. (2000). *Ecol Applic.*, *10*, 412-422.

Farquar, G. D., Ehleringer, J. R. & Hubiek, K. T. (1989). *Ann Rev Plant Physiol Plant Mol Biol.*, *40*, 503-37.

Friedlingstein, P., Cox, P., Betts, R., Bopp, L., von Bloh, W., Brovkin, V., Cadule, P., Doney, S., Eby, M., Fung, I., Bala, G., John, J., Jones, C., Joos, F., Kato, T., Kawamiya, M., Knorr, W., Lindsay, K., Matthews, H. D., Raddatz, T., Rayner, P., Reick, C., Roeckner, E., Schnitzler, K. G., Schnur, R., Strassmann, K., Weaver, A. J., Yoshikawa, C. & Zeng, N. (2006). *J Clim.*, *19*, 3337-3353.

Giardina, C. P. & Ryan, M. G. (2000). *Nature, 404*, 858-860.

Harding, R. J. (1978). *Geo Ann, Ser A.*, *60*, 43-49.

Jenkinson, D. S., Hart, P. B. S., Rayner, J. H. & Parry, L. C. (1987). *Modelling the turnover of organic matter in long-term experiments at Rothamsted.*, *INTECOL Bulletin*, *15*, 1-8.

Kättener, T., Reichstein, M., Andren, O. & Lomander, A. (1998). *Biol Fertil Soils*, *27*, 258-262.

Kirschbaum, M. U. F. (1995). *Soil Biol Biochem*, *27*, 753-760.

Krull, E. S. & Skjemstad, J. O. (2003). *Geoderma*, *112*, 1-29.

Leps, J. & Smilauler, P. (2003). *Multivariate Analysis of Ecological Data using CANOCO.* Cambridge University Press, Cambridge, 269.

Morecroft, M. D., Woodward, F. I. & Harris, R. H. (1992b). *Funct Ecol*, *6*, 730-740.

Morecroft, M. D. & Woodward, F. I. (1996). *New Phytol*, *134*, 471-479.

Morecroft, M. D., Marrs, R. H. & Woodward, F. I. (1992a). *J Ecol*, *80*, 49-56.

Natelhoffer, K. J. & Fry, B. (1988). *Soil Sci Soc Am J*, *52*, 1633-1640.

Parton, W. J., Schimel, D. S., Cole, C. V. & Ojima, D. S. (1987). *Soil Sci of Am J*, *51*, 1173-1179.

Parton, W. J., Scurlock, J. M. O., Ojima, D. S., Gilmanov, T. G., Scholes, R. J., Schimel, D. S., Kirchner, T., Menaut, J. C., Seastedt, T., Garcia Moya, E. Apinan, K. & Kinyamrio, J. I. (1993). *Global Biogeochem Cycl*, *7*, 785-809.

Parton, W., Silver, W. L., Burke, I. C., Grassens, L., Harmon, M. E., Currie, W. S., King, J. Y., Adair, E. C., Brandt, L. A., Hart, S. C. & Fasth, B. (2007). *Science*, *315*, 361-364.

Staddon, P. L. (2004). *TREE*, *19*, 148-154.

Wedin, D., Tieszen, L. L., Dewey, B. & Pastor, J. (1995). *Ecology*, *76*, 1383-1392.

In: Encyclopedia of Environmental Research
Editor: Alisa N. Souter

ISBN: 978-1-61761-927-4
© 2011 Nova Science Publishers, Inc.

Chapter 2

THE BIODIVERSITY POTENTIAL OF RIPARIAN FIELD MARGINS IN INTENSIVELY MANAGED GRASSLANDS

L. J. Cole[*], D. I. McCracken, D. Robertson and W. Harrison
SAC, Auchincruive, Ayr, KA6 5HW, UK

ABSTRACT

Riparian field margins (i.e. fenced areas adjacent to watercourses) are becoming more widespread in intensively managed grassland systems in the UK as a means of mitigating diffuse pollution associated with intensive livestock farming. Invertebrate assemblages were examined in a range of riparian margins to determine their potential to deliver additional biodiversity goals. It was found that ground beetle assemblages in wide (> 4 m) margins were more distinct from the adjacent field than those in narrow margins (≤ 2 m) or riparian zones open to grazing. Furthermore, both wide and narrow riparian margins were found to enhance the abundance of several groups of invertebrates that are important prey items of farmland birds (i.e. leatherjackets, sawfly larvae and harvestmen). While fenced margins had higher abundances of bird prey items, the denser vegetation typical of such ungrazed margins may result in the prey being less accessible to foraging birds. Consequently routine management of such margins may be required to open up the vegetation structure and thus optimize their potential benefit to wildlife. Appropriately managed riparian margins have therefore the potential to increase biodiversity at the field level, by increasing the abundance of key invertebrates, and also at the landscape level, by increasing habitat heterogeneity and thus supporting distinct assemblages of ground beetles.

INTRODUCTION

The countryside of South West Scotland is dominated by intensively managed livestock farming and consequently diffuse pollution from agriculture is a cause for concern from both a water quality and public health perspective (Vinten *et al.* 2004). The erection of fences

---

[*] Corresponding author: E.mail: Lorna.Cole@sac.ac.uk

along field margins adjacent to water courses (i.e. riparian margins) to mitigate diffuse pollution and protect riverbanks from erosion is encouraged by agri-environment schemes in the UK and is becoming widespread throughout South West Scotland (Hopkins *et al.* 2007). Many declining farmland species utilise agricultural field margins [e.g. bumblebees (Bäckman and Tiainen 2002), yellowhammers (Bradbury *et al.* 2000) and water voles (Rushton *et al.* 2000)] and riparian margins are known to be particularly rich in biodiversity as they support a diverse array of habitats, landforms and communities (Corbacho *et al.* 2003). Riparian margins also have the potential to facilitate faunal movement in an otherwise hospitable landscape through the creation of contiguous corridors of semi-natural vegetation. Such margins, therefore, not only provide a means of reducing diffuse pollution but also provide an opportunity to help offset the declines in farmland biodiversity associated with intensive livestock production. There is a however a lack of information on the factors influencing biodiversity within riparian field margins and a greater understanding is therefore required before effective management prescriptions can be developed.

METHODS

A total of 22 sampling locations were established on seven grassland dominated farms in South Ayrshire, Scotland (UK National Grid Reference: NS 53). Each sampling location was allocated to one of three categories: Narrow Sites (i.e. sites with narrow fenced off riparian strips, width ≤ 2 m), Wide Sites (i.e. sites with wide fenced off riparian strips, width >4 m) and Open Sites (i.e. sites with no fence between the field and watercourse). At each site sampling transects were established adjacent to the watercourse (Water) and 4-6 m into the field from the fenceline (Field). In the case of open sites the Field transect was established 4-6 m from the Water transect. For Wide sites a third transect was established, where feasible, at the midpoint between the fenceline and the Water transect (Middle). A total of seven different treatments were therefore investigated: Narrow Water, Narrow Field, Wide Water, Wide Middle, Wide Field, Open Water and Open Field (Figure 1).

Ground active invertebrates were monitored at each transect using a row of nine pitfall traps (75 mm diameter and 100 mm deep) installed at 2 m intervals. Sampling was conducted over a three year period (2004-2006) with two trapping periods, each of four weeks duration (June/July and July/August), per year. On collection, the nine pitfall samples in each line were pooled. Both the activity and abundance of invertebrates can influence pitfall catches and consequently the abundance of invertebrates caught by pitfall trapping is referred to as the activity abundance (Thiele, 1977). In addition to invertebrate sampling, data were collected at each transect on vegetation height, vegetation density, plant species richness, field and margin management, soil characteristics and physical attributes of the field margin.

RESULTS AND DISCUSSION

Multivariate analyses found that narrow and wide riparian margins supported distinct ground beetle assemblages from intensively managed grassland fields and this observation was most noticeable for wide margins (Cole *et al.* 2008a). The distinction between ground beetle assemblages of open riparian margins and intensively managed fields was not as clear indicating assemblages in open margins were similar to those found in the fields. Shade loving species such as *Leistus rufescens* and *Stomis pumicatus*, were typically found in fenced riparian margins (i.e. narrow and wide) but were rarely found in open riparian margins or in intensively managed grassland fields. It is likely that the tall dense vegetation typically found in these ungrazed fenced margins provided shadier, more humid condition, than open margins or the adjacent fields. The fields and open riparian margins had higher frequencies of species that are typical of lowland grassland such as *Clivina fossor* and *Trechus micros* further supporting the fact that open margins provided a similar habitat for ground beetles as grassland fields.

While the structure of ground beetle assemblages differed between wide riparian margins and intensively managed grassland fields, REML analyses found that the number of ground beetle species recorded, and ground beetle diversity did not differ significantly and the highest diversity of ground beetles was actually recorded in open margins (Cole *et al.* 2008a). The activity abundance of ground beetles was also higher in open margins and intensively managed grassland fields than fenced margins. Wide riparian margins therefore did not promote ground beetle diversity or abundance, however, by creating an alternative habitat that supports distinct ground beetle assemblages in an otherwise homogenous grassland landscape, fenced riparian field margins (particularly when >4m wide) promoted diversity at the level of the farm.

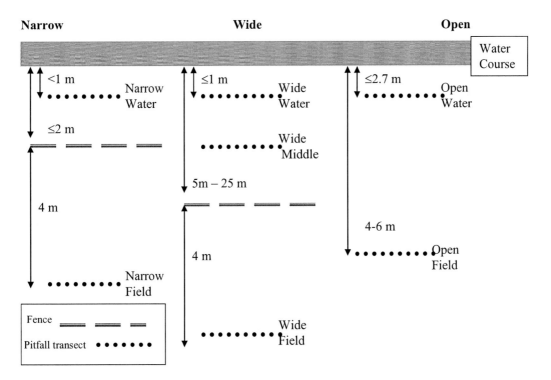

Figure 1. Experimental setup of the survey sites indicating the three types of riparian margins (Narrow, Wide and Open) and the sampling transects within each margin type thus providing the seven treatments. Adapted from Cole *et al.* (2008a).

While ground beetles were found in their lowest activity abundance in fenced riparian margins, the activity abundance of harvestmen (Opiliones), sawfly (Symphyta) larvae, leatherjakets (Tipulidae) and Cicadellidae plant bugs was greater in fenced margins (irrespective of width) than the adjacent fields (Cole *et al.* 2008b, Cole *et al.* 2008c). The activity abundance of these groups of invertebrates in open margins, on the other hand, did not differ from the adjacent fields indicating these invertebrate groups were favoured by the exclusion of livestock from the margins (Cole *et al.* 2008b, Cole *et al.* 2008c). Limacidae slugs showed a similar trend with the activity abundance being greater in wide margins (but not narrow margins) than the adjacent fields.

These groups of invertebrates have different ecological requirements and as such they are probably influenced by different aspects of the riparian margins. Phytophagous groups such as sawfly larvae and Cicadellidae plant bugs are likely to be favoured by the higher diversity of plants, and thus food, in the fenced riparian margins resulting in the margins supporting a greater diversity of phytophagous species. Slugs and harvestmen, on the other hand, are prone to desiccation and are thus likely to be influenced by differences in humidity. The taller denser vegetation, which was typical of the fenced riparian margins, would favour hydrophilic species by providing a more humid and stable habitat. If harvestmen were solely driven by humidity preferences, however, one may expect a positive relationship between activity abundance and vegetation density. This was not the case, and a negative relationship between activity abundance and vegetation density was actually observed (Cole *et al.* 2008b). Ground beetle activity abundance was also found to be negatively related to vegetation density (Cole *et al.* 2008a). As both ground beetles and harvestmen are predatory, it is possible that dense vegetation impeded hunting while sparser vegetation facilitates the detection and capture of prey (Telfer *et al.* 2000).

The taller, denser vegetation typically found in these fenced riparian margins is likely to impair not just predatory invertebrates, but also foraging birds. Consequently while a greater abundance of some invertebrate prey species occurred in the margins, the tall dense vegetation may decrease the visibility of prey items (Butler and Gillings, 2004) and impede the movement of the foraging bird (Devereux *et al.*, 2004). There is often a discrepancy between grassland management practices which favour the abundance of invertebrate prey and those which favour prey accessibility (McCracken and Tallowin 2004; Plantureux et al., 2005) and consequently it has been suggested that management prescriptions which promote a heterogenous mixture of cut and uncut areas within the margin may deliver the optimum benefit to foraging birds (Perkins *et al.* 2001).

CONCLUSION

The exclusion of livestock from riparian margins resulted in distinct ground beetle assemblages from intensively managed grassland fields and favoured several groups of invertebrates (e.g. sawfly larvae, Cicadellidae bugs and harvestmen) which are important dietary components for many farmland birds. While the abundance of invertebrate prey items was enhanced, the taller, denser vegetation of the fenced margins is likely to impede foraging birds. To ensure riparian margins maximise their biodiversity potential, management prescriptions should consider not simply enhancing the abundance of prey, but also the detectability and accessibility of prey to foraging birds.

REFERENCES

Bäckman, J. P. C. & Tiainen, J. (2002). Habitat quality of field margins in a Finnish farmland area for bumblebees (Hymenoptera: *Bombus* and *Psithyrus*). *Agriculture, Ecosystems & Environment, 89*, 53-68.

Bradbury, R. B., Kyrkos, A., Morris, A. J., Clark, S. C., Perkins, A. J. & Wilson, J. D. (2000). Habitat associations and breeding success of yellowhammers on lowland farmland *Journal of Applied Ecology, 37*, 789-805.

Butler, S. J. & Gillings, S. (2004). Quantifying the effects of habitat structure on prey detectability and accessibility to farmland birds. *Ibis, 146*, 123-130.

Cole, L. J., Robertson, D., Harrison, W. & McCracken, D. I. (2008c). The biodiversity value of riparian field margins in intensively managed grasslands. *SAC and SEPA Biennial Conference. Agriculture and the Environment VII. Land management in a changing Environment.* 26-27 March 2008, University of Edinburgh, Edinburgh

Cole, L. J., Robertson, D., Harrison, W., McCracken, D. I. & Tiley, G. E. (2008b). Promoting biodiversity on intensively managed grassland in Scotland. *European Grasslands Federation. Biodiversity and Animal Feed - future challenges for grassland production.* 9-12 June 2008, Uppsala, Swede. 90-92.

Cole, L. J., McCracken, D. I., Baker, L. & Parish, D. (2007). Grassland conservation headlands: Their impact on invertebrate assemblages in intensively managed grasslands. *Agriculture, Ecosystems & Environment, 122*, 252-258.

Cole, L. J., Morton, R., Harrison, W., McCracken, D. I. & Roberston, D. (2008a). The influence of riparian buffer strips on Carabid beetle (Coleoptera, Carabidae) assemblage structure and diversity in intensively managed grassland fields. *Biodiversity and Conservation, 17*, 2233-2245.

Corbacho, C., Sanchez, J. M. & Costillo, E. (2003). Patterns of structural complexity and human disturbance of riparian vegetation in agricultural landscapes of a Mediterranean area. *Agriculture, Ecosystems and Environment, 95*, 495-507.

Devereux, C. L., McKeever, C. U., Benton, T. G. & Whittingham, M. J. (2004). The effect of sward height and drainage on Starlings and Lapwings foraging in grassland habitats. *Ibis, 146*, 115-122.

Hopkins, J. J., Duncan, A. J., McCracken, D. I., Peel, S. & Tallowin, J. R. B. (2007). *High value grasslands: providing biodiversity, a clean environment and premium products.* The British Grassland Society, Cirencester, 352.

McCracken, D. I. & Tallowin, J. R. (2004). Sward and structure, the interactions between farming practices and bird food resources in lowland grassland. *Ibis, 146*, 108-114.

Perkins, A. J., Whittingham, M. J., Morris, A. J. & Bradbury, R. B. (2001). Use of field margins by foraging yellowhammers *Emberiza citronella. Agriculture, Ecosystems and Environment, 1884*, 1-8.

Plantureux, S., Peeters, A. & McCracken, D. (2005). Biodiversity in intensive grasslands, Effects of management, improvement and challenges. In: R. Lillak, R. Viiralt, A. Linke, & V. Geherman, (Eds.), *Integrating Efficient Grassland Farming and Biodiversity.* Estonian Grassland Society, Tartu, Estonia, 417-426.

Rushton, S. P., Barreto, G. W., Cormack, R. M. & MacDonald, D. W. (2000). Modelling the effects of mink and habitat fragmentation on the water vole. *Journal of Applied Ecology*, *37*, 475-490.

Telfer, M. G., Meek, W. R., Lambdon, P., Pywell, R. F. & Sparks, T. H. (2000). The carabids of conventional and widened field margins. *Aspects of Applied Biology*, *58*, 411-416.

Thiele, H. U. (1977). *Carabid beetles in their environment*. Springer Verlag, Berlin

Vinten, A. J. A., Crawford, C., Cole, L., McCracken, D. I., Sym, G., Duncan, A. & Aitken, M. N. (2004). Evaluating the impact of buffer strips and rural BMPs on water quality and terrestrial biodiversity. In: D. Lewis, & L. Gairns, (Eds.), *Agriculture and the Environment: Water Framework Directive and Agriculture*. Scottish Agriculutreal College, Edinburgh and Scottish Environmental Protection Agency, Edinburgh.

In: Encyclopedia of Environmental Research
Editor: Alisa N. Souter

ISBN: 978-1-61761-927-4
© 2011 Nova Science Publishers, Inc.

Chapter 3

CHEMOTAXONOMICAL ANALYSES OF HERBACEOUS PLANTS BASED ON PHENOLIC PATTERNS: A FLEXIBLE TOOL TO SURVEY BIODIVERSITY IN GRASSLANDS

Nabil Semmar[1, 2,], Yassine Mrabet[1] and Muhammad Farman[3]*

[1]ISSBAT, Institut Supérieur des Sciences Biologiques Appliquées de Tunis, Tunisia.
[2]Laboratoire de Pharmacocinétique et Toxicocinétique;
Pharmacy School of Marseilles, France.
[3]Department of Chemistry, Quaid-i-Azam University, Islamabad-45320, Pakistan.

ABSTRACT

Analysis of plant diversity aims to understand the organization and the variability of biological populations within ecosystems. In classical analysis, individual plants are firstly identified on the basis of morphological/cytological parameters, then biodiversity is evaluated tacking into account presence/absence, abundances and densities of plants. Although morphological parameters are easily accessible, they provide limited precision on the differentiation between individuals that share a high similarity. This is especially true in the plant world where a same species includes several varieties that could not be easily distinguished morphologically. Cytological parameters provide reliable qualitative information but are not suitable for a quantitative evaluation of biodiversity at ecosystemic scales. The high variability between and within plant taxa can be highlighted chemically through qualitative and quantitative analyses of secondary metabolite patterns in different plants. Quantitatively, the different patterns show variability based on occurrences of different metabolites which can have major or minor levels. In the case where two chemical patterns have the same metabolites, they can be distinguished from different relative levels (i.e. regulation ratios) of such metabolites. These different variability criteria (presence/absence, major/minor, relative levels) allow to extract reliable chemical fingerprints for chemotaxonomical classification of plants. Moreover, such chemical polymorphism can be correlated to different intrinsic and/or extrinsic

---
* Corresponding author: E-mail: nabilsemmar@yahoo.fr

conditions of the plants in order to evaluate its adaptative, selective and/or evolutive values. Finally, relationships between chemical patterns and environmental conditions provide efficient tools to survey qualitatively and quantitatively the fluctuations of plant diversity in time and space. This chapter provides a review on chemotaxonomical criteria helping to understand complex structures of plant diversity. It focuses particularly on the chemotaxonomic usefulness of phenolic compounds (phenylpropanoids, flavonoids, etc.) in analysis of chemical polymorphisms at different systematic levels (from family to variety and chemotype via genus and species). Five plant families are treated here consisting of Poaceae and Liliaceae (monocotyledons), and Asteraceae, Lamiaceae and Fabaceae (dicotyledons). These families are representative of a great biomass part in the grasslands, and their chemotaxonomic analyses can usefully help to manage the biodiversity in such ecosystems.

I. INTRODUCTION

Grasslands are complex herbal systems consisting of great mixture spaces containing many plant taxa which are very neighbors the ones with the others. Such highly dense mixtures are sources of great biodiversity which is not easy to evaluate.

More sensitive tools than the morphological (visual) parameters are needed to detect, distinct, classify, survey and manage the different plant taxa co-occurring in complex herbal mixtures of grasslands. Moreover, although genetic tools provide very precise information on the biodiversity, survey/management operations requires more accessible (rapid) tools in order to routinely evaluate the plant diversity within a reasonable time.

The phytochemistry brings a compromise solution between precision and rapidity through the analysis of secondary metabolites in plant tissues. Beyond single chemical descriptions of plant taxa, secondary metabolites can provide information on the living conditions of a plant because they are generally produced as responses or signals to different intrinsic and/or extrinsic conditions. In summary, analysis of secondary metabolites in plants represents a precious tool to analyze quantitatively and qualitatively the chemical polymorphism of a plant taxon and to link it to different plant (tissue, age) and environment conditions (seasons, climate, soil, altitude, biotic interactions, etc.). The identification of chemical patterns within and between plant taxa tacking into account intrinsic and/or extrinsic conditions, represents a central objective of chemotaxonomy.

In this chapter, we will particularly focus on the phenolic compounds (flavonoids and phenylpropanoid derivatives) as powerful and reliable tools to evaluate the chemical diversity of herbs in the grasslands. The interest of chemotaxonomy based on phenolic compounds will be illustrated by different herbal families belonging to monocotyledons and dicotyledons. Many phytochemical studies will be presented concerning two monocotyledon and three dicotyledon families in order to illustrate the phenolic patterns of these plant taxa. Monocotyledons will be illustrated here by two great families consisting of Poaceae (Gramineae) and Liliaceae. Dicotyledons will be illustrated by three important families represented by Asteraceae (Compositae), Lamiaceae (Labiatae) and Fabaceae (Papilionaceae or Leguminoseae). The set of all herbaceous families of the whole taxonomic system is over the content of this chapter.

Differences in phenolic patterns will be underlined in order to show how phytochemistry allows chemical distinction between plants at different taxonomic levels: families, tribes, genera, sections, species, varieties and chemotypes. Chemical patterns can be defined on the basis of several criteria leading to distinguish different plants by simple analysis of their chemical contents. Such chemotaxonomic criteria can be ranged either as qualitative or quantitative:

Qualitatively, plants can be separated by the presence/absence of some metabolites. High taxonomic levels, e.g. families, can be easily distinguished by some specific chemical skeletons (aglycons) which are synthesized in early metabolic steps. When two taxa share same aglycons, the substitutions of chemical groups on these skeletons can be frankly different leading to different final metabolites in the two taxa. In the case where aglycons and substitutions are similar in two plant taxa, a sharp chemotaxonomical criterion is brought by examining the position of the substitutions on the aglycons. Such criterion is often used at lower taxonomic levels, e.g. species of a same genus.

In addition to qualitative criteria, quantitative chemotaxonomic criteria can be efficiently used to separate more finely plant taxa containing same metabolites. For a given metabolite, major/minor/trace criterion can be used to classify the plants according to their biosynthesis ability for such a chemical compound. Thus, the plant will be classified as either highly or few productive of such compound. In the case where same metabolites are similarly major or minor in two different plant taxa, their ratio can be different in each taxon. The ratio between two metabolites is generally a very sensitive parameter allowing the distinction between taxa that have a low distinctness. This can be the case of varieties of a same plant species.

Beyond these elementary qualitative and quantitative criteria, the consideration of relative levels of all analyzed metabolites gives the metabolic profile of a plant taxon, which is richer in terms of chemotaxonomic information. The set of all profiles corresponding to several individual plants can be statistically treated by means of multivariate analyses, e.g. hierarchical cluster analysis and/or principle component analysis. Such statistical analyses help to highlight the structure of a complex chemical polymorphism between and within plant sub-populations. Chemical varieties of a same species can be highlighted from chemical profiles leading to define the concept of chemotype. The different chemotypes can be distinguished the ones the others by all the possible chemotaxonomical criteria (qualitative and quantitative). They can be correlated with intrinsic and/or extrinsic conditions of the plant in order to understand the biological/ecological significance of chemical polymorphism. The chemotaxonomic conclusions can usefully serve as a basis for better management of natural spaces as grasslands: although the high herbal density makes difficult the survey of biomass in grassland, phytochemistry and chemotaxonomy bring sensitive tools helping to distinguish between plants whatever their morphological similarities.

II. INTEREST OF PHYTOCHEMISTRY AND CHEMOTAXONOMY IN DIFFERENT EVALUATION ASPECTS OF BIODIVERSITY

The biodiversity represents a complex system where multiple interactions occur between many biological components, as well at macroscopic as at cellular and molecular levels. At macroscopic scales (ecology, botany), the biodiversity is studied from parameters that are easily accessible (e.g. morphometric measures, species number counting, etc.). This provides less precision compared to the molecular genetic parameters that provide precise systematic information especially at lower taxonomic levels (e.g. variety). However, genetic tools are less accessible or even not practical in routine analysis. In other words, there are a certain opposition between accessibility and precision of information. A compromise can be provided by chemical fileds as phytochemistry, metabolomics, chemotaxonomy, etc. which have the advantage to join easily these two information criteria (Figure 1). In plant analysis, phytochemistry is doubly advantageous because it provides:

- rapid method giving a reliable picture on the plant chemical diversity (Figure 2).
- possibility to analyse and to interpret the plant chemistry at different scales (genetic expression, metabolic regulation, ecological adaptation, etc.) (Figure 3).

By linking chemical contents of plants to different environment factors or biological states, phytochemical data can be reliably translated in terms of chemotaxonomical keys. Thus, the chemotaxonomy represents a rapid, informative and relatively precise tool to link plant chemical patterns to genetic, metabolic, physiological and/or ecological factors.

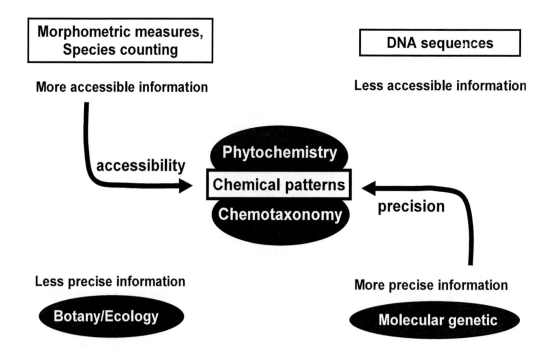

Figure 1. Requirement of phytochemistry and chemotaxonomy as rapid and reliable tools providing rich information to assessment and management of plant diversity.

II.1. Global Interest and Focuses of Chemotaxonomy

Chemotaxonomy aims to identify different plant on the basis of their chemical contents. Therefore phytochemical data representing the whole plant or a given tissue (e.g. leaf, stem, root, seed, flower, fruit, etc.) are qualitatively and quantitatively analysed in order to highlight original chemical traits between and within taxa (Figure 2).

In botany, the different plant taxa are identified by reference to a hierarchical classification extending from family to species or variety via tribe and genus. Chemotaxonomy tends to extract chemical fingerprints from which different taxonomical levels can be reliably identified. Such chemical fingerprints help both to distinguish between taxa (e.g. between genera or species) and to characterize the diversity within each taxon.

Beyond correspondence between chemistry and botany, the chemical fingerprint can be interpreted in relation to intrinsic factors (genetic control) and/or to extrinsic conditions (environment) (Figure 3):

When a chemical pattern is genetically controlled, it will be confirmed (i.e. remains unchanged) after a plant has been transplanted from environmental conditions to others. In this case, the different chemical patterns of individual plants will be interpreted as genetic fingerprints of the plant taxon independently of the environment conditions. Such characteristic makes the chemotaxonomical data to be good parameters in phylogenic analysis.

In the case where the chemical pattern is flexible under different environmental conditions, it will help to analyse the ecological adaptability of the plant taxon by reference to the local environmental conditions. Such flexibility of chemical patterns is useful to analyse the roles and ecological significances of secondary metabolites in plants. However, the chemical identity will not be affected by such flexibility, but it will be characterized by an environment-dependent chemical polymorphism.

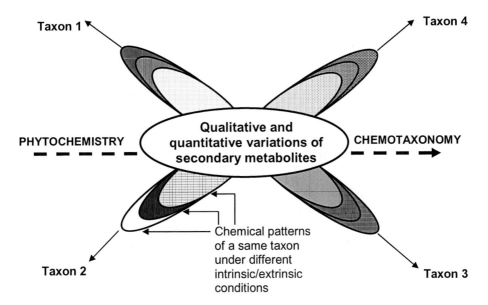

Figure 2. Intuitive representation showing how plant taxa can be classified from qualitative and quantitative analyses of phytochemical data leading to chemotaxonomical separations between and within taxa.

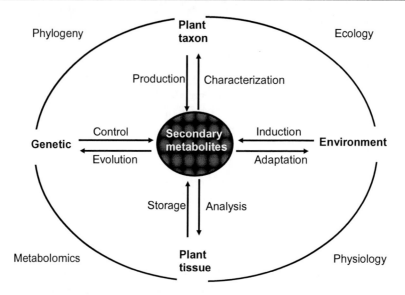

Figure 3. Schematic representation showing different biological poles under which phytochemical data can be interpreted leading to different chemotaxonomical values of plant chemical patterns

Moreover, the chemotaxonomical data provide several pictures on plant chemical polymorphism from which metabolic trends and biosynthesis pathways can be approached. Under such aspect, the chemotaxonomy becomes a useful tool for metabolomics.

Finally, the chemical patterns of secondary metabolites produced/stored by a given tissue can be analysed in relation to different environment conditions (e.g. seasons, soils, climates, etc.) in order to understand more the role of the tissue (of leaf, stem, root, flower, seed, etc.) in the plant life. This gives to the chemotxonomy a usefulness in physiology.

III. CHEMOTAXONOMIC PARAMETERS

The chemical features of different plant taxa can be distinguished from different criteria concerning the analysed metabolites:

- Occurrence, rarity, prevalence of a metabolite,
- Ratio between two metabolites,
- Profile of all the metabolites.

Combination of these different criteria provides a specific chemical patern that can be helpfully used as a fingerprint to identify each plant taxon.

Secondary metabolites are particularly used as chemotaxonomic markers because they are not constitutive molecules (i.e. not uniformly present), by opposition to the primary metabolites (sugars, fatty acids, amino acids, nucleic acids, etc.). Consequently, their occurrences, amounts and metabolic regulations vary significantly both between and within plant taxa leading to a concept of chemical polymorphism (Figure 4):

Occurrence analysis (qualitative analysis) of secondary metabolites requires the identification of their chemical structures. Then, such structural identification offers a first chemotaxonomical key helping to separate different plant taxa on the basis of presence/absence of compounds in their patterns.

The structural analysis of a secondary metabolite implies identification of its aglycone (skeleton) as well as the chemical substitutions attached on this aglycone, etc. (Figure 4):

- Two plant taxa can be easily separated when they produce different aglycone families (e.g. flavones for the one and flavonols for the other).
- However, when two plant taxa synthesise a same aglycon family, they can be distinguished by different substitutions on the aglycone. Such substitutions consist of different chemical groups (hydroxyl, methyl, methyl-ether, glycosyl, acyl, etc.) (Figure 4, 11).
- In the case where two plant taxa have the same substitutions on their similar aglycones, they can be chemically differentiated by the substitution positions on the aglycone (Figure 4, 6a, 9, 11).

Analysis of the content of secondary metabolites (quantitative analysis) leads to separate major from minor compounds in a same plant taxon. Major/minor compound criterion is generally a good chemotaxonomical key as major compounds in a plant taxon can be minor in another taxon, and vice versa (Figure 4).

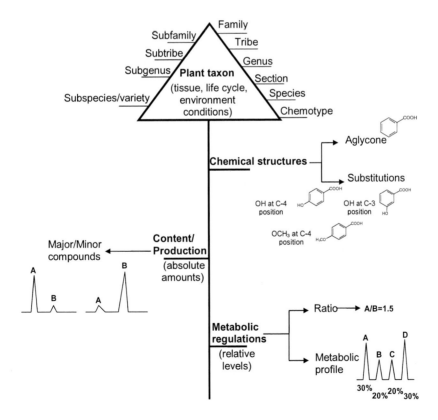

Figure 4. Different chemotaxonomical parameters on the basis of which plant taxa can be chemically characterized.

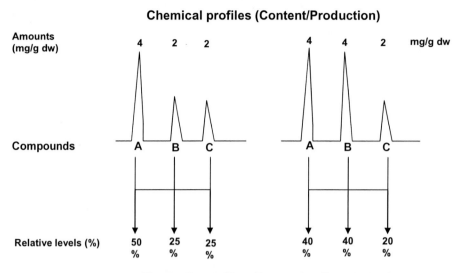

Figure 5. Conversion of a chemical profile (e.g. HPLC or GC) to a metabolic profile by standardizing the absolute amount (e.g. mg/g dw) of each compound in reference to the total amount (sum) of all the compounds. Note that the absolute amount (4mg/g dw) of compound A does not correspond to the same metabolic regulation in the two profiles, i.e. 50% in the first profile and 40% in the second.

Complementary to structure and content analyses, metabolic profile analysis considers each metabolite in relation to the others. Therefore, if two compounds are both major (minor) in two plant taxa, their ratio can be helpfully used to highlight a significant difference between the two taxa; the metabolic ratio is often taxon-dependent whereas the content (amount) is generally influenced by environment conditions. Beyond the metabolic ratio where only two compounds are considered, the set of all compounds defines the metabolic profile of the plant taxon (Figure 4). The shape of such profile can be characterized by the percentages of the different metabolites in reference to the sum of their amounts (Figure 5). Such metabolic profile provides a fine picture on the metabolic regulation within a plant, and can be reliably used to analyse the chemical polymorphism within a same plant species (variety level). When different metabolic profiles are highlighted for a same plant species, they define different chemotypes. In such case, the genetical or ecological significance of such chemotypes can be analysed *in situ*.

IV. POWERFUL CHARACTERISTICS OF SECONDARY METABOLITES FOR CHEMOTAXONOMY

Secondary metabolites have general characteristics that are opposite to primary metabolites (sugars, fatty acids, amino acids, nucleic acids, etc.). Such characteristics can be summarised by the fact that secondary metabolites are:

- Not constitutive (not universal), i.e. produced occasionally under some intrinsic/extrinsic conditions. Therefore, their presence/absence in a taxon has a chemotaxonomical value for the chemical identification of the studied taxon.
- Quantitatively minor compared to the primary metabolites, and consequently their variations result in significant differences within and between taxa.
- Structurally very diversified, resulting in a great number (several thousands) of compounds which are specific to different plant taxa.

The highly diversified world of secondary metabolites can be classified into three main classes: Phenolic derivatives, terpenoids and nitrogen-containing metabolites. Each of these three classes is highly diversified by different chemical subclasses. This chapter presents especially the class of phenolic derivatives including a wide range of phenolic acid and flavonoid structures.

V. CHEMICAL DIVERSITY OF PHENOLIC DERIVATIVES

Phenolic compounds can be essentially divided into two classes: (a) phenolic acids or phenylpropanoid derivatives and (b) flavonoids. From these two basic classes, many other phenolic structures can be directly derived (**Figure 6, 8, 10**).

V.1. Phenolic Acid Derivatives: Chemical Structures and Metabolic Pathways

A phenolic acid consists of a phenol (hydroxylated benzene) substituted by a lateral chain containing one to three carbons (C1-chain or C3-chain) with a terminal carboxylic group. They have the composition formula C6C1 and C6C3, respectively. The phenolic acids with C1-chain are derived from benzoic acid (Figure 6b); those with C3-chain are derived basically from cinnamic acid (Figure 6a).

Benzoic acid derivatives can be biogenetically synthesized from a β-oxidation of cinnamic derivatives (Figure 6a-b). Moreover, there are several other chemical structures which are derived from phenolic acids:

- Quinones can be linked to simple phenolic compounds from which they are derived by oxidative processes. They include benzoquinones, naphtoquinones and anthraquinones (highly colored) (Figure 6c).
- Internal cyclisation of the lateral chain of hydroxycinnamic acids leads to the synthesis of hydroxycoumarins (Figure 6d, 7). The introduction of a C_2 group into hydroxycoumarin yields furanocoumarin.

Some plants as peanuts, possess a stilbene synthase activity, by which *p*-coumaroyl CoA reacts with three molecules of malonyl CoA to give a stilbene (Figure 8). In the presence of chalcone synthase, such reaction yields chalcone, a flavonoid which represents the precursor of all the other flavonoids.

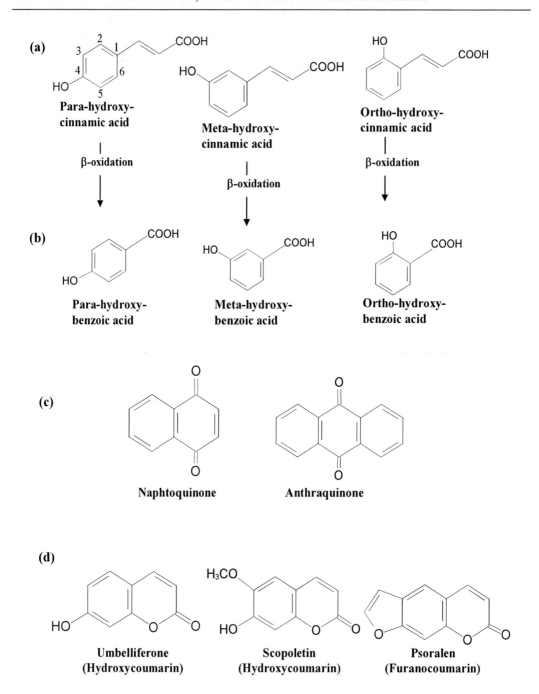

Figure 6. Chemical structures of phenylpropanoid derivatives. (a) cinnamic acids, (b) benzoic acids, (c) quinones, (d) coumarins.

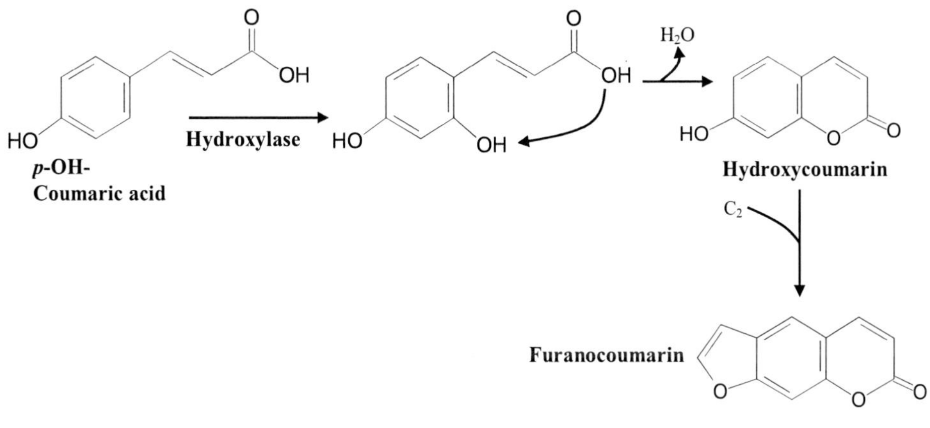

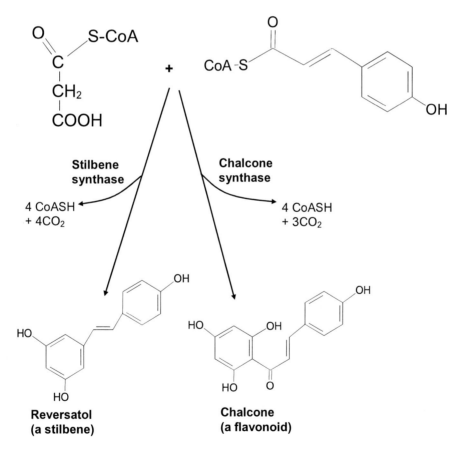

Figure 7. Biosynthesis of hydroxylcoumarins and furanocoumarins from *p*-coumaric acid.

Figure 8. Biosynthesis of chalcone (flavonoid) and reversatol (stilbene) from *p*-coumaric acid (phenylpropanoid) according to two enzymatic ways.

V.2. Flavonoids: Chemical Structures and Metabolic Pathways

The flavonoids consist of C6-C3-C6 basic skeleton in which two benzene rings (A- and B-rings) are separated by an oxygenated heterocycle or chain containing three carbons (Figure 9). According to the structure of the heterocycle, the flavonoids can be divided into several families, i.e. flavones, flavonols, flavanones, dihydroflavonols, chalcones, retrochalcones, aurones, leucoanthocyanidins, proanthocyanidins, anthocyanins, isoflavonoids, homoisoflavanones, pterocarpans, etc. (Figure 10).

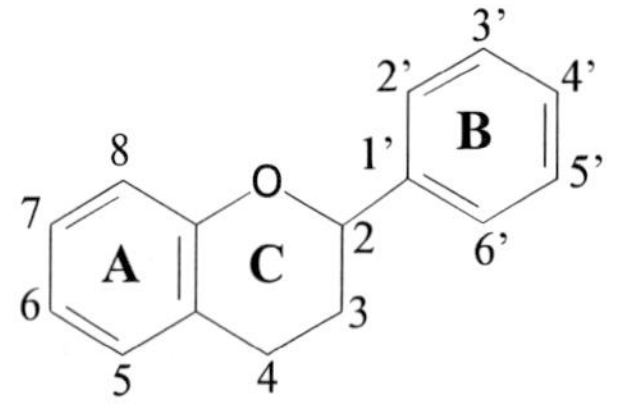

Basic flavonoid aglycon

Figure 9. Basic chemical skeleton of flavonoids.

Chalcone is a flavonoid characterized by an "open heterocycle" (i.e. absence of heterocycle) by opposition to all other flavonoid skeletons. It is synthesized by condensation of a phenylpropanoid with acetate molecules (Figure 10). Chalcone is the first flavonoid which appears within the flavonoid metabolic network.

Retrochalcones are unusual chalcones lacking oxygen functionality at heterocycle and clearly differ from normal chalcones in that they do not undergo isomerization into flavanones (Figure 10).

Aurones are based on the 2-benzylidene-coumaranone or 2-benzylidene-3(2H)-benzofuranone system, and characterized by the presence of a five-membered heterocyclic ring. The occurrence of aurones is limited to 16 plant families, especially Asteraceae and Fabaceae, in higher plants (Iwashina, 2000).

Flavanone results from isomerisation of chalcone by means of chalcone isomerase. The ring structure of flavanone is formed by the addition of a phenolic hydroxyl group to the double bond of the carbon chain connecting the two phenolic rings. Flavanone is a precursor of a variety of flavonoids (Figure 10).

A desaturation reaction forming a double bond between C-2 and C-3 of C-ring (Figure 9) is involved in the formation of both flavones and flavonols (Figure 10). Flavones have substitutions on A- and B-rings but lack oxygenation at the 3-position of C-ring, by opposition to flavonols. Flavones and flavonols are synthesized from flavanones and dihydroflavonols, respectively. Moreover, enzymatic hydroxylation of flavanone at the 3-position by means of flavanone 3-β-hydroxylase converts it into respective dihydroflavonol.

Leucoanthocyanidins are defined as monomeric flavonoids (flavan, 3,4-diols and flavan-4ols), which produce anthocyanidins by cleavage of a C-O bond on heating with mineral acid (Haslam, 1982; Porter, 1980).

The proanthocyanidins represent a major group of compounds that occur ubiquitously in woody and some herbaceous plants. They are flavan-3-ol oligomers that produce anthocyanidins by cleavage of a C-C interflavanyl bond under strongly acidic conditions (Figure 10) (Porter, 1980).

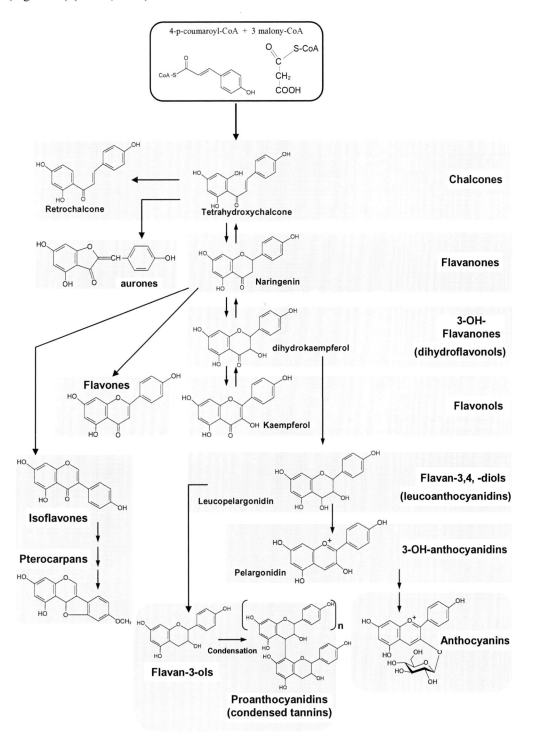

Figure 10. Flavonoid biosynthetic pathways.

The structures of anthocyanins are based on 2-phenylbenzopyrylium ion (flavylium cation); their glycosylation yields anthocyanins (Figure 10). From the taxonomic point of view, anthocyanins are of interest because of the variety of glycosidic and acylated combinations within these structures. Conjugations and different substitutions contribute to the stability of the pigment *in vivo* and may also determine subtle changes in colour properties (Harborne and Williams, 1994).

Principally found in legumes (Fabaceae), isoflavonoids are a group of compounds that originate from flavanones (Figure 10). The factor differenciating isoflavonoids from the other flavonoids is the linking of the B-ring to the C-3 rather than the C-2 position of the C-ring. Subsequent modifications can result in a wide range of structural variation, including the formation of additional heterocyclic rings such those of pterocarpans (Figure 10).

Homoisoflavanones are a special group of phenolic compounds which are related to isoflavonoids and contain a basic 3-benzyl-4-chromanone skeleton (Figure 42). They may be open chain (Figure 42a) or ring closed structures (Figure 42b-c).

(a)

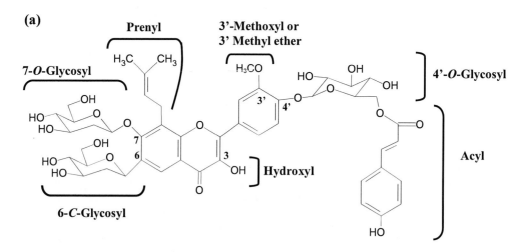

(b)

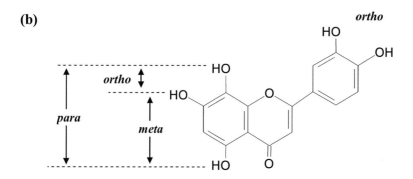

Figure 11. Different chemical substitutions responsible for the diversification of secondary metabolite structures

V.3. Metabolic Processes of Chemical Structure Diversification

The chemical diversity of the phenolic compounds is further increased by different chemical substitutions at different positions of the molecule. These substitutions generally consist of hydroxylation, methoxylation and glycosylation (Figure 11). Methoxyl (or methyl ether) group on the molecule results from the fixation of a methyl on a hydroxyl (Figure 12). In glycosylation, a sugar can be attached either to a hydroxyl group (*O*-glycosylation) or directly to a carbon of the aglycone (*C*-glycosylation) (Figure 11a). In some cases, aromatic and aliphatic acids, sulphate, prenyl or isoprenyl groups are also attached to the flavonoid nucleus and their glycosides (Figure 11a).

Taking into account the relative positions of the different substitutions on aglycone, three structural states can be defined: *orhto-* (*o*-), *meta-* (*m*-) and *para-* (*p*-) positions (Figure 11b): in benzene molecule, a position *ortho* refers to two substituents located on two successive carbons that are linked each one to the other. *Meta-* and *para-* positions refer to two substituents located on carbons that are separated by one and two atoms, respectively.

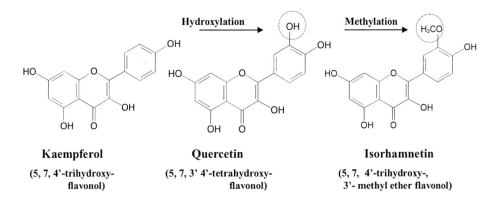

Kaempferol	**Quercetin**	**Isorhamnetin**
(5, 7, 4'-trihydroxy-flavonol)	(5, 7, 3' 4'-tetrahydroxy-flavonol)	(5, 7, 4'-trihydroxy-, 3'- methyl ether flavonol)

Figure 12. Diversification of chemical structures of flavonoids from hydroxylation and methylation processes.

VI. GENERAL LINES ON PRESENTED HERBACEOUS PLANT FAMILIES

The two next sections provide many illustrations concerning phytochemical and chemotaxonomical analyses of different herbaceous plant families. Five herbaceaous families will be presented in the next sections consisting of two monocotyledons and three dicotyledons (Figure 13, 14):

Monocotyledons will be illustrated by the families Poaceae and Liliaceae. The family Poaceae (or Gramineae) including wheat, rize, barley, etc. is agronomically important . This family essentially consists of grasses which are particularly abundant in grasslands. Chemically, the Poaceae provide many illustrative examples on chemotaxonomy based on *C*-glycosyl flavones as well as phenolic acid derivatives (in the leaves essentially). This family has been considered as a highly specialized herbaceous group on the basis of occurrence of particular secondary metabolites which are not widely distributed in herbaceous families (3-

deoxyanthocyanins, flavonoid sulphates, tricin, etc.) (Williams and Harborne, 1973; Harborne, 1977).

The family Liliaceae provides many chemotaxonomical study cases based on very diversified flavonoid structures: anthocyanins, flavonol glycosides, homoisoflavanones, alkaloid-flavonoids, phenylethyl glycosides, proanthocyanidins, etc. . Such chemical diversity is compatible with heterogeneity of the family which has been revised by the APG system (APG, 2003). Such heterogeneity will be illustrated by different phytochemical examples which will focus mainly on genera and species representative of *sensu stricto* Liliaceae by reference to the APG revision.

In dicotyledons, three herbaceous families have been selected to illustrate the interest of chemotaxonomy in biodiversity assessment: Asteraceae, Lamiaceae and Fabaceae (Figure 13).

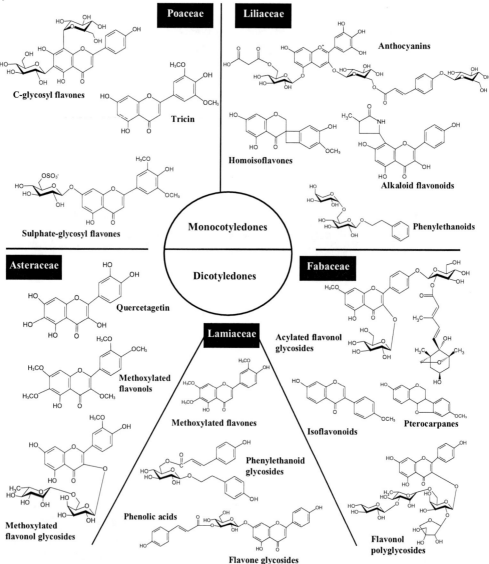

Figure 13. Chemical structures of phenolic comppunds characterizing different herbaceous families.

Lamiaceae are essentially characterized by a prevalene of flavone derivatives, whereas flavonol derivatives seem to occur more significantly in Asteraceae. The phytochemistry of these two families illustrate well how methoxylations of flavones or flavonols at different positions result in different chemical fingerprints of different plant taxa. Characteristic metabolites are also highlighted in these two families, e.g. quercetagenin (poly hydroxylated flavonol) in Asteraceae, or phenylethanoids in Lamiaceae.

Concerning the family of Fabaceae, it is particularly characterized by a high chemical variability based on isoflavonoids. Moreover, this family provides illustrations on chemotaxonomy based on original glycoside structures (e.g. flavonol polyglycosides, chalcone glycoside derivatives, flavone C-glycosides). Interestingly, glycosides containing apiose (a rarely observed sugar) seem to be a characteristic of different Fabaceae species (Iwashina, 2000). Moreover, original acylated structures are also highlighted within this family.

The different plant families presented here are representative of different phylogenic clusters (Figure 14).

Figure 14. Dendrogram representing phylogenic classification of different plant orders according to Angiosperm Phylogeny Group system (APG, 2003).

VII. MONOCOTYLEDONS

Monocotyledons occupy an important position among the angiosperms with up to 100 families, comprising roughly 56000 species sharing several biological features (Harborne. and Williams, 1994; Simpson, 2006). Among these, the traditionally considered one is their development from a seedling with a single cotyledon in contrast to a couple of cotyledons typical of dicotyledons. Many other concrete morphological characters make the distinction of these two groups such as parallel/reticulate leaves venation and presence/absence of sieve tube plastids, respectively. Commonly herbaceous, many monocotyledon families run the worldwide economy by producing the majority of agricultural biomass. A well-known example is the Poaceae (Gramineae) family, which includes grain crops such as rice, wheat, corn, etc.

Monocotyledon systematics has been the subject of several studies (Chase et al., 1993, 1995, 2004, 2006). The high genetical affinity between the members of monocotyledons has been highlighted by Angiosperm Phylogeny Group system (referred as APG II, 2003), which represents a consensus within a number of botanists (Figure 14). In the dendrogram resulting from phylogenic analysis, the monocotyledons showed a high distinctness separating them easily from the dicotyledons. Moreover, they showed both genetic homogeneity and diversity: homogeneity is represented by a single clade which separated early giving a frank identity to the monocotyledons; on the other hand, the interior of this clade showed different subclades revealing an important diversity within the monocotyledon class.

VII.1. Poaceae Family (or Gramineae)

The classification and identification of Poaceae is particularly difficult because of the general similarity of leaf shape, the limited variation in stem branching patterns and the reduced nature of the inflorescence and floral parts. Moreover, the Poaceae family has a sheer size consisting of 6000-9000 species leading to major survey problems (Harborne and Williams, 1976).

VII.1.1. Family level chemotaxonomy

In a chemotaxonomical study, Del Pero Martínez (1985) compared distribution of flavonoids in different herbaceous monocotyledons including Poaceae. Significant differences between plant families have been highlighted on the basis of the prevalence of flavones or flavonols (Figure 15). Moreover, different chemical substitutions (O- and C-glycosylations, methylation, sulphatation, etc.) have been proved to be chemotaxonomically usefull to characterize the Poaceae family:

Original occurrence profiles were observed in Poaceae (Williams and Harborne, 1971; Harborne and Williams, 1976), Juncaceae (Williams and Harborne, 1975), Cyperaceae (Harborne, 1971a; Williams and Harborne, 1977) where flavone O-glycosides prevail over those of flavonols. In addition, there are two other characters that could be considered to be specialization traits: these include methylation of the flavones and substitution at position 5 (Figure 15). In APG II system (APG, 2003), the Poaceae and Cyperaceae showed high phylogenic similarity as they aggregate within a same sub-clade (Poales) within the clade of monocotyledons (Figure 14). Such similarity has been confirmed by distribution profiles of flavonoids (Figure 15): these two plant families have high amounts of *C*-glycosylflavones (in 60-100% species), tricin (60-100%) and flavonol *O*-glycosides (10%). However, the chemical differentiation between these two families is brought by 5-*O*-glycosyl flavones which occurred significantly more in Poaceae than in Cyperaceae. In addition, the Poaceae possess sulphated flavonoids (Figure 22d) that are absent in the Cyperaceae and, conversely this latter family contains 6-hydroxyluteolin (Figure 16) which is absent in the Poaceae. The chemical diversity of Poaceae will be further illustrated in the next sections taking into account different taxonomic levels (beyond family level).

The Juncaceae showed a pattern fairly different from that of the previously described families. They have a large occurence of flavones (60-95% species) and 5-*O*-methyl (Me) flavones (60-90%) together with flavonols (20-30%) and lack *C*-glycosyl flavones.

In Commelinaceae, the flavonoid content appears to be relatively homogeneous because of a high frequency of flavone *C*-glycosides (Del Pero Martinez and Swain, 1985). These compounds were presented in 60-78% of the species examined.

Homogeneous flavonoid distribution is also observed in the family Zingiberaceae (Hutchinson, 1979; Cronquist, 1981). In this family, the plant species showed a prevalence of flavonol glycosides (60-100% species).

In Bromeliaceae, a greater diversity of chemical compounds was observed (Williams, 1978): A high level of common flavonols (40-52% species) was observed with an occurrence of common flavones and *C*-glycosyl-flavones in 13% and 10% of species, respectively. However, the most distinctive constituents are the 6- and 8-hydroxyflavonols (6-11%), the 6-hydroxyflavones (20-27%) and methylated structures (20-24%).

A chemical diversity was also highlighted from the flavonoid pattern of Restionaceae (Cronquist, 1981) (Figure 15). Although the common flavonol glycosides dominate the pattern (40-54% species), the family also contains several other compounds: *O*-glycosides, *C*-glycosides and 8-hydroxyflavonols (Figure 16) occurred in more than 20% of the examined species. However, the presence of rare methylated flavonols (larycitrin and syringetin) (Figure 16) can be attributed to a certain degree of biochemical specialization, which distinguishes the family from the rest.

Remarkable uniformity in leaf flavonoids has been found for Poaceae species (Harborne and Williams, 1976). Chemical screening of the grasses has indicated that the Poaceae family may be distinguished by a pattern based on the co-occurrence of tricin _ a rare methylated flavone tricetin (5, 7, 4'-trihydroxy-3', 5'-dimethoxy flavone _ (Figure 17a, b) with *C*-glycosides based on apigenin and luteolin (Figure 17c, d) (Harborne and Hall, 1964; Bate-Smith, 1968; Williams et al., 1974).

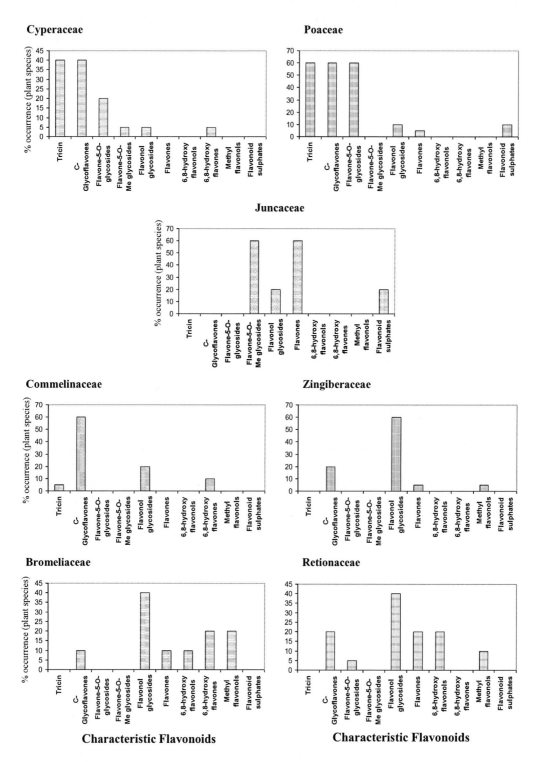

Figure 15. Profiles showing the relative distributions of characteristic flavonoids in different families of monocotyledons (Del Pero Martínez et al., 1985). Tops of rectangles correspond to inferior occurrence limits of flavonoids in plant families (e.g. 60% means the flavonoid is present in at least 60% species).

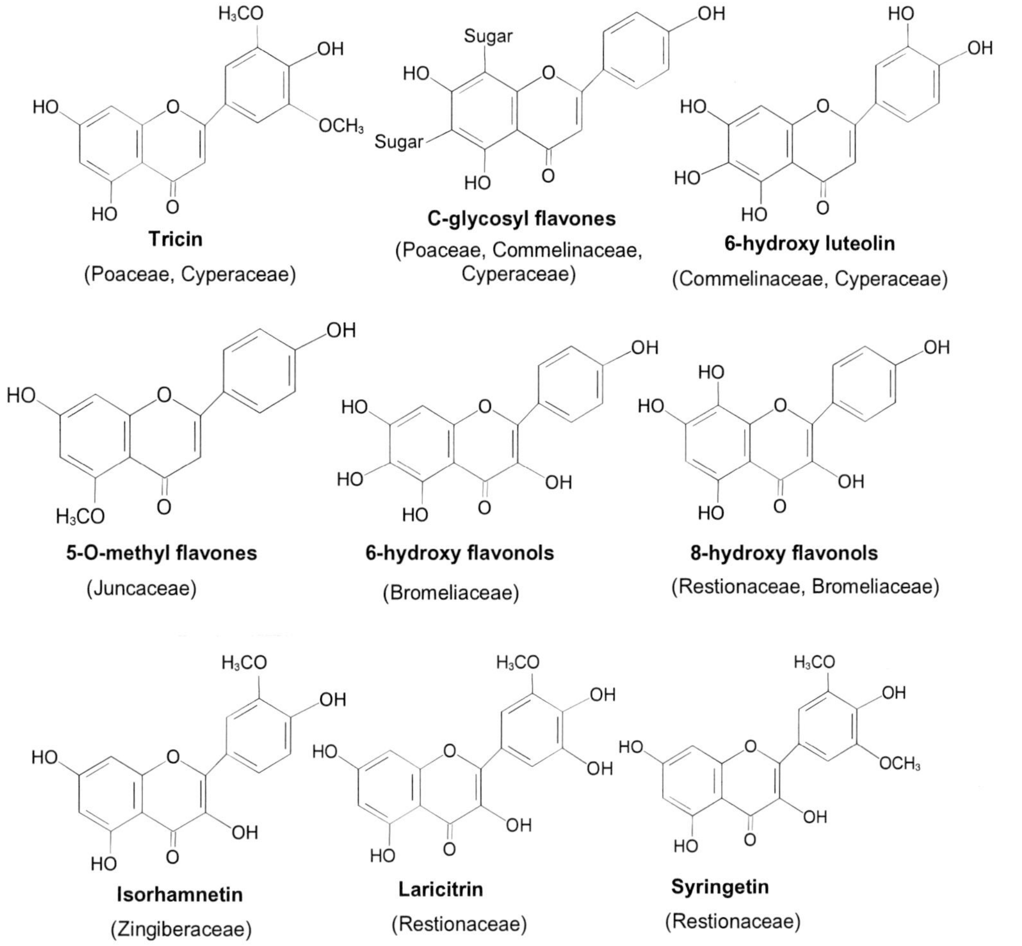

Figure 16. Chemical structures of some characteristic flavonoids distributed among different plant families of monocotyledons.

Harborne and Hall (1964) in a survey of 20 grass genera found flavone *C*-glycosides (Figure 18, 19) to be amongst the major flavonoid leaf constituents and their findings were supported by the work of Bate-Smith (1968) in a study of 69 grass species. In another work on the leaf flavonoids in Poaceae, Harborne and Williams (1976) concluded the abundance of flavone *C*-glycosides in 93% of 274 surveyed species. Several later works on flavonoid glycosides of Poaceae confirmed such chemical feature (Figure 18, 19) (Chopin et al., 1977; Waiss et al., 1979; Nawwar et al., 1980; Chari et al., 1980; Dellamonica et al., 1983; Besson et al., 1985; Gluchoff-Fiasson et al., 1989; Harborne et al., 1986a, 1986b; Rofi and Ponilio, 1987; Snook et al., 1994; Webby and Markham, 1994; Snook et al., 1995; Yoon et al., 2000; Norbaek et al., 2000; Bouaziz et al., 2001; Markham and Mitchell, 2003; Norbaek et al., 2003). Flavone *C*-glycosides have a sugar which is carbon-carbon linked to the A-ring in the 6- and/or 8-positions generally.

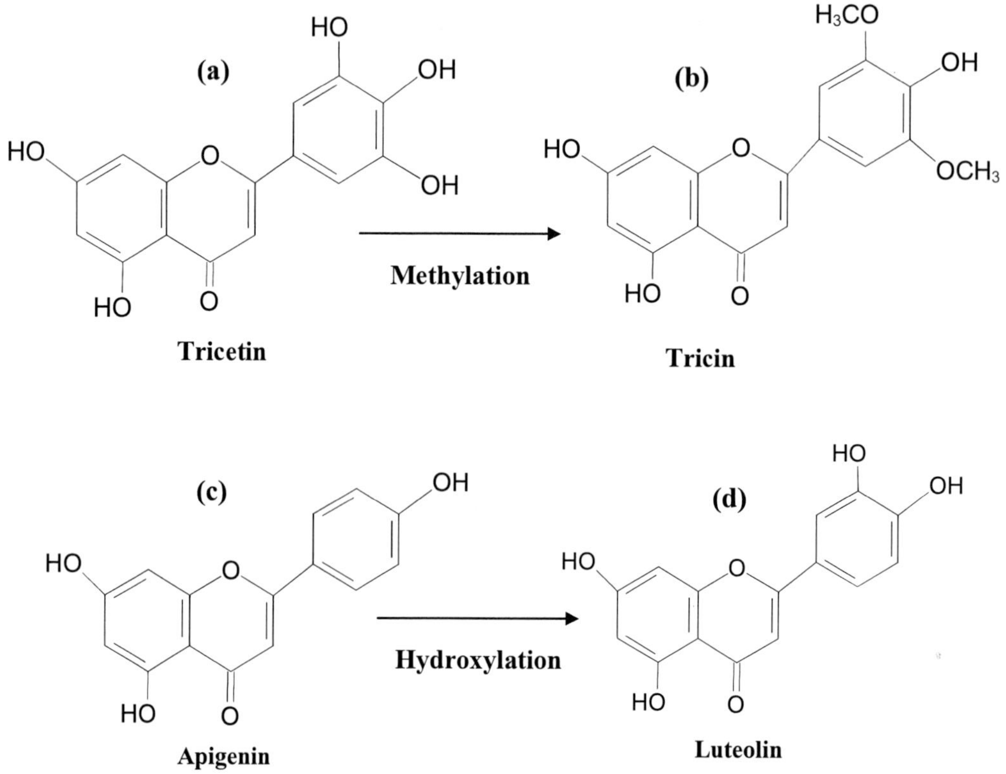

Figure 17. Chemical structures of some flavones commonly found in Poaceae; tricetin (a), its methylated derivative, tricin (b), apigenin (c) and luteolin (d). Although it is rare in the plant world, tricin is a common flavone in the family Poaceae.

Apigenin and luteolin *C*-glycosides (Figure 18, 19) were reported to be more abundant in Poaceae than tricin *C*-glycosides. Although they occurred in the leaves of 93% of the species examined, tricin *C*-glycosides were often detected only as trace constituents (Harborne and Williams, 1976). These compounds are chemotaxonomical keys of Poaceae by their simple presence and not by their amounts. Among the tricin *O*-glycosides, tricin 5-glucoside is the most commonly occurring glycoside (Figure 22a). The 7-glucoside (Figure 22b), 7-diglycoside, 7-neohesperidoside (Figure 22c) and 7-glucoside sulphate (Figure 22d) are also frequently present (Harborne and Williams, 1976).

VII.1.2. Subfamily/tribe level chemotaxonomy

The distribution of luteolin (Figure 17d) is of some interest in the subfamily Pooideae, where it was found to be in 15%, 43% and 43% of species of the tribes Poeae, Triticeae and Agrostideae, respectively. Further, it was found absent in the tribes Aveneae, Phalarideae and Stipeae (Figure 20) (Harborne and Williams, 1976).

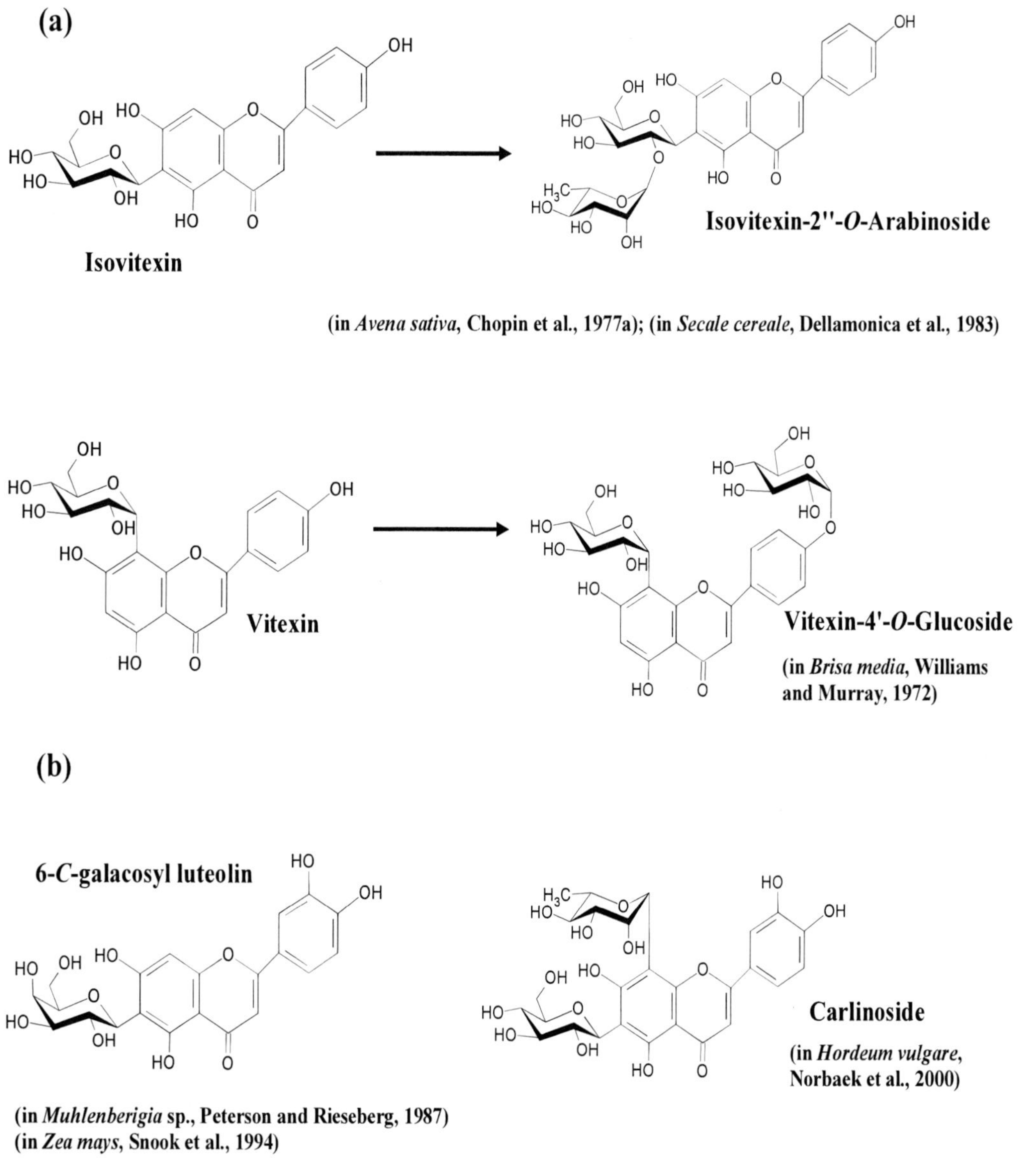

Figure 18. Chemical structures of some *C*-glycosyl flavones deriving from apigenin (a) and luteolin (b), commonly found in Poaceae family.

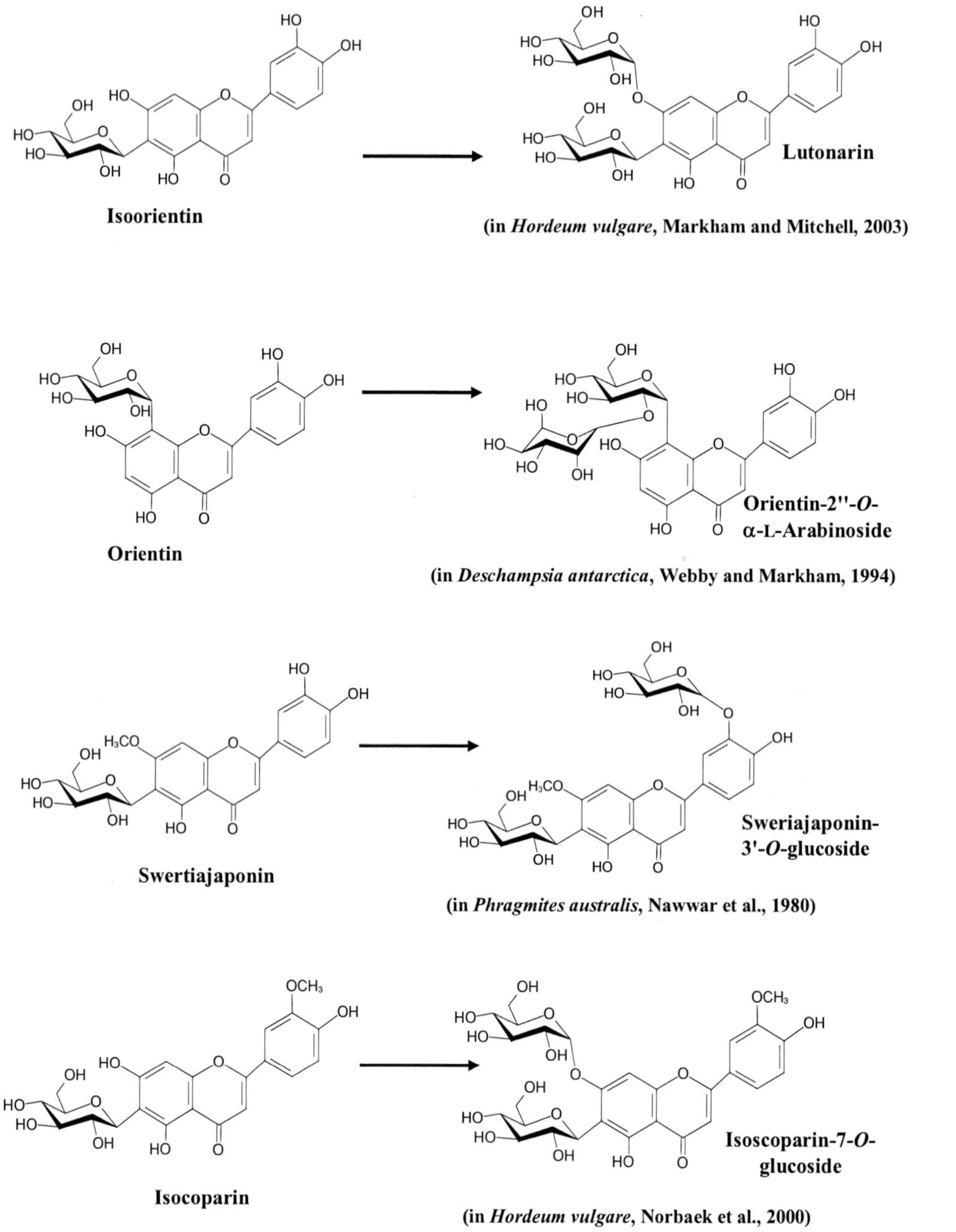

Figure 19. Chemical structures of some *C*-glycosyl flavones deriving from luteolin and commonly found in the family Poaceae.

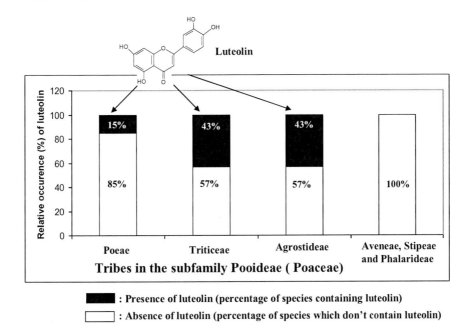

Figure 20. Relative occurrence of luteolin as a chemotaxonomical key at tribe level in Poaceae. The results are based on 274 surveyed species by Harborne and Williams (1976).

In general, there were no large differences between subfamilies based on major flavonoids in the leaves. However, minor constituents played an important role in chemotaxonomical differentiation between subfamilies (Figure 21):

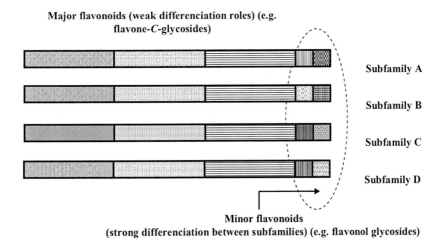

Figure 21. Schematic representation showing the role of minor flavonoids in chemical differentiation between different subfamilies of Poaceae.

Although it was fully confirmed fact that flavonols are unusual constituents in grass leaves, quercetin and kaempferol (Figure 12) (3% and 5% of the examined species, respectively) were confined entirely to the two subfamilies Festucoideae and Panicoideae

(Harborne and Williams, 1976). This gives a good example on the chemotaxonomic interest of minor compounds through their exclusive presence in some taxa.

The distribution of the tricin glycoside sulphates (Figure 22d) in the family Poaceae is of some systematic interest since they are present in genera of only three subfamilies (Harborne and Williams, 1976; Williams et al., 1974): the Panicoideae (18% of examined species), Chloridoideae (15%) and Arundinoideae (40%). The first two subfamilies originate exclusively from tropical or subtropical regions while the Arundinoideae, of more varied geographical distribution, are mainly from southern warm temperate regions.

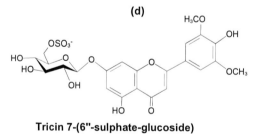

(a) Tricin-5-glucoside

(b) Tricin-7-glucoside

(c) Neohesperidose — Tricin-7-neohesperidoside

(d) Tricin 7-(6''-sulphate-glucoside)

Figure 22. Chemical structures of tricine *O*-glycosides in species of Poaceae.

The genus *Microlaena* placed by Bentham and Hooker (1883), Hackel (1887) and Hubbard (1959) in the Pooideae can be distinguished from all the other members of this subfamily by the presence of the negatively charged tricin glycoside. Such evidence therefore supports the removal of the genus from the Pooideae. Prat (1936) places *Microlaena* in a position of isolation within the Chloridoideae, in which negatively charged flavonoids have been found, but later (Prat, 1960) subjected it to a debateable classification. Then *Microlaena* has been included in the Arundinoideae, where flavonoid sulphates (Figure 22d) are also frequent.

Distribution of both proanthocyanidins and flavan-4-ols (Figure 10) in Poaceae family was exclusively confined to species of the tropical-subtropical subfamily Panicoideae with two occurrences in the subfamily Chloridoideae, viz. in the Brazilian species, *Gymnopogon foliosus* and the Nigerian species *Eragrostis atrovirens* (Harborne and Williams, 1976).

VII.1.3. Genus level chemotaxonomy

The genus *Sorghum* (subfamily Panicoideae, tribe Andropogoneae) seems to be characterised by the occurrence of luteolin (Figure 17d) (Harborne and Williams, 1976).

Although the major flavonoids in Poaceae (flavone *C*-glycosides) are not good taxonomic markers at subfamily and tribal levels, some individual compounds have proved useful at the generic level: for example, in the genus *Erianthus*, the presence of a luteolin di-*C*-glycoside distinguishes Old world from New world species (Williams et al., 1974).

Within the subfamilies, negatively charged flavonoids (flavonoid sulphates) (Figure 22d) are often characteristic constituents of some genera, e.g. *Panicum*, *Bothriochloa*, *Saccharum*, *Paspalum*, *Cortaderia* and *Microlaena* (Harborne and Williams, 1976; Barron et al., 1988).

Analysis of glycosylflavones in the genus *Briza* provided interesting information on the phylogenic state of this grass taxon (Williams and Murray, 1972): The remarkable feature was that the simple process of doubling up the chromosomes changes the glycosylflavone pattern in such that 4'-hydroxy apigenin-based compounds are largely replaced by their 3', 4'-dihydroxy luteolin-based counterpart, in the leaves. It should be stressed that this is essentially a quantitative change in flavonoid synthesis affecting the degree of B-ring hydroxylation. The change from diploid to autotetraploid, occurring both naturally and artificially by colchicin induction in *Briza media*, may have been the key to the successful spread of the genus from Europe to South America.

VII.1.4. Species level chemotaxonomy

The relative rarity of flavonol glycosides in the grasses leads some rare species to be chemically distinguished (Figure 21): Harborne (1967) found flavonols in only four among 50 surveyed species, viz. rutin (Figure 93j) in *Festuca pratensis*, quercetin 3-glucoside (Figure 93h) in *Poa pratensis* and kaempferol and quercetin glucosides in *Panicum bulbosum* and *Lolium perenne*. Similarly, Bate-Smith (1968) reported flavonols in only two among 69 examined Poaceae species. Saleh et al., (1971) isolated possible acylated derivatives of kaempferol and quercetin 3-glucosides (Figure 93g, 93h) from leaves of *Stipa lemmonii*.

An original flavonoid pattern in Poaceae concerns the species *Rottboellia exaltata* where flavonols replace flavones as the major leaf flavonoid components (Harborne and Williams, 1976). Three quercetin glycosides were isolated from this plant: the 3-glucoside (Figure 93h), the 3-rutinoside (Figure 93j) and a triglycoside in which the sugars were galactose, glucose and rhamnose.

The Poaceae species *Sorghum vulgare* Pers. can be strongly distinguished by the presence of 3-deoxyanthocyanidins, viz. apigeninidin (Figure 23a) and luteolinidin (Figure 23b) that are very rare anthocyanidins occurring in the higher plant world (Nip and Burns, 1969, 1971). Generally such compounds occur in lower plants: Musci (Bendz et al., 1962; Bendz and Mårtensson, 1963) and Pteridophytes (Harborne, 1966a; Crowden and Jaman,

1974). The occurrence of these compounds attributes to the productive species a high specialization statut within the Poaceae.

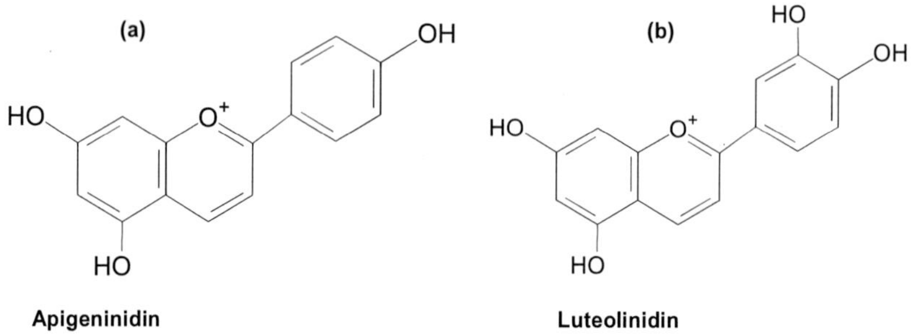

Figure 23. Chemical structures of some deoxyanthocyanidins found in *Sorghum vulgare* (Poaceae).

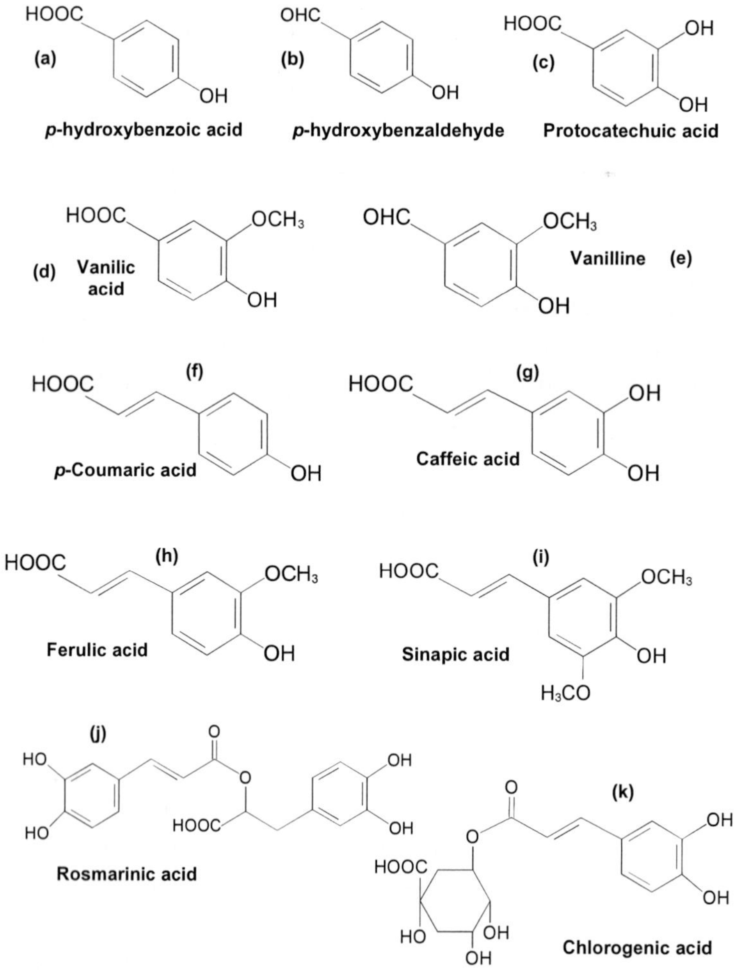

Figure 24. Chemical structures of different phenolic acid derivatives found in Poaceae.

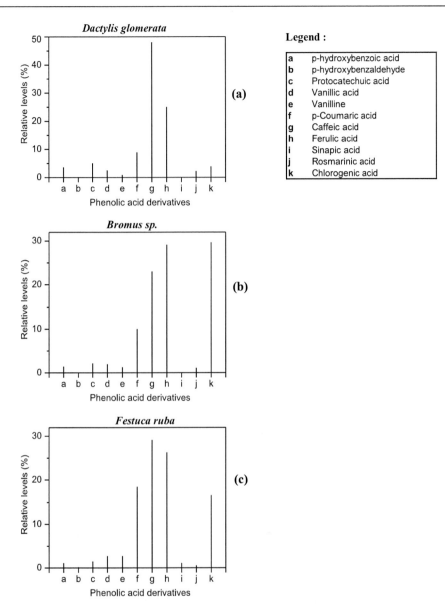

Figure 25. Phenolic acid derivative profiles of three species of Poaceae (Mika et al., 2005). The chemical structures are presented in figure 24.

Analysis of phenolic acids in different species of Poaceae highlighted the usefulness of such compounds in chemotaxonomic differentiation at species and variety levels. Míka et al. (2005) compared phenolic profiles between different species of Poaceae belonging to three different genera: *Dactylis glomerata*, *Festuca ruba* and *Bromus* sp. . Several individual plants of each species were examined and their phenolic profiles consisted of 11 phenolic acid derivatives (Figure 24); such profiles showed significant differences between species (genera) (Figure 25).

The phenolic profile of *Dactylis glomerata* was dominated by caffeic acid (50-60% of total metabolism) followed by ferulic acid (15-30%) (Figure 24g, h) (Figure 25a). Apart from these two major compounds, other compounds co-occur as minors (*p*-coumaric acid, Figure 24f) or as traces (chlorogenic acid, Figure 24k).

In the phenolic profile of *Bromus* sp. (Figure 25b), the chlorogenic acid is major (20-40%) by opposition to *Dactylis glomerata* (traces). It co-occurred with ferulic acid (20-40%) followed by caffeic acid (15-25%). Similarily to *D. glomerata*, *p*-coumaric acid was minor in *Bromus*. sp. (5-12%).

The phenolic profile of *Festuca ruba* was shared by four major compounds (Figure 25c): Chlorogenic, caffeic, ferulic and *p*-coumaric acids the relative levels of which varied between 15-30%. Although caffeic acid was the dominant compound, *p*-coumaric acid can be considered as the most characteristic of *Festuca ruba* because of its higher regulations (17-21%) compared to the two other species (≤12%).

VII.1.5. Infraspecific level chemotaxonomy

In addition to single analysis of 11 phenolic compounds in *Festuca ruba* and *Bromus* sp., Míka et al. (2005) surveyed the phenolic profiles at three sampling stages: heading (H), flowering (F) and post-flowering (PF) stages. The three relative profiles of a given species were compared in order to analyse the chemical variability of the 11 compounds in time (Figure 26a, b).

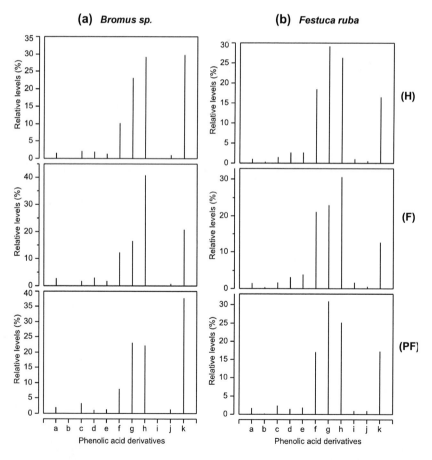

Figure 26. Phenolic profiles of *Bromus* sp. and *Festuca ruba* (Poaceae) at different sampling stages; heading (H), flowering (F) and post-flowering (PF) stages (Míka et al., 2005). The labels (a-k) of compounds and corresponding structures are presented in figure 24.

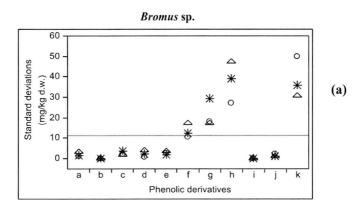

Bromus sp.

(a)

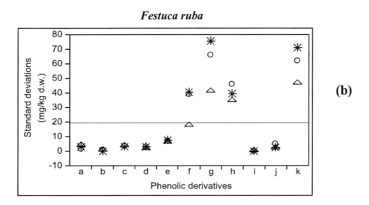

Festuca ruba

(b)

Figure 27. Dispersion diagrams of standard deviations of the amounts of the 11 phenolic compounds analysed in *Bromus* sp. (a) and *Festuca ruba* (b) at three sampling stages. *: heading stage; Δ: flowering stage; ○: post flowering stage.

The phenolic profiles of relative amounts (%) in *Bromus* sp. showed important chemical variability depending on the sampling stage (Figure 26a): for example, the ratio (R) between ferulic acid and chlorogenic acid is around 1 at heading stage; it doubles at flowering stage (R=2) then it decreases by moiety at post flowering stage (R=0.5). Such variability illustrates the usefulness of phytochemistry not only in matter of taxonomy, but also in matter of development state survey of a given taxon in time.

The phenolic profiles of *Festuca ruba* showed less variability in time, but the flowering stage seemed to be distinguished from the two former and later stages by a higher ratio between ferulic and caffeic acids (Figure 26b). The ratio ferulic acid/caffeic acid varied around 1.3 in (F) against 0.8-0.9 in (H) and (PF) stages.

Moreover, Míka et al. (2005) analysed the individual contents (mg/kg) of each plant taxon to calculate a standard deviation (SD) for each compound and at each sampling stage. Therefore, higher SD values can provide keys to identify what sampling stage correspond to the highest chemical diversity (Figure 27). By considering the 11 compounds, the dispersion diagrams showed that heading stage corresponded generally to the highest production variability (highest SD) in the two analysed taxa. On this basis, Míka et al. (2005) concluded that the heading stage was more appropriate to chemical polymorphism analysis of *Festuca ruba* and *Bromus* sp.

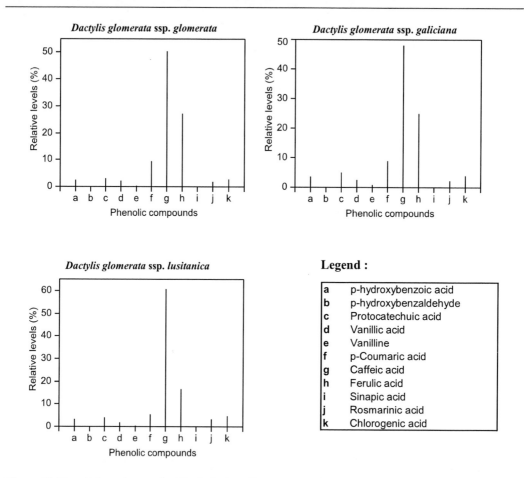

Figure 28. Regulation patterns (in %) of 11 phenolic compounds in three subspecies *Dactylis glomerata* (Poaceae) (Míka et al., 2005).

Infraspecific analysis has been also applied on the species *Dactylis glomerata* which was represented by three subspecies (Míka et al., 2005): ssp. *glomerata*, ssp. *galiciana* and ssp. *lusitanica*. Comparison of the relative profiles of phenolic compounds led to conclude that the three subspecies were similar with respect to all major and minor compounds (Figure 28). However, the subspecies *lusitanica* differed significantly from the two others by the ratio between caffeic and ferulic acids (the two major compounds): the ratio caffeic acid/ferulic acid was around 3.5-3.6 in ssp. *lusitanica*, whereas it was around 1.8-1.9 in the two other subspecies. Such differences in ratio could be linked to a ploidy factor as *Dactylis glomerata* is a complex polyploid with a diploid cytotype (Casler et al., 1996; Míka et al., 2005).

VII.2. Liliaceae Family

VII.2.1. Family level chemotaxonomy

The family Liliaceae has been reported to be a very diversified botanically, but a recent phylogenic revision highlighted a heterogeneity taxonomic composition (APG, 1998, 2003). Therefore, many species previously belonging to Liliaceae has been classified into

Asparagale order. The numbers of genera and species have been reduced from about 250 genera and nearly 3000 species to 10 and 420, respectively. This section presents the chemotaxonomy of Liliaceae under both the new and old classification aspects. It focuses on the species strictly classified in Liliaceae (after APG revision), but gives some examples on other species removed from the Liliaceae in order to show the usefulness of chemotaxonomy to highlight heterogeneities that can't be detected morphologically. Liliaceae species are widespread all over the word. Underground organs are bulbs, tubers or rhizomes. Flowers are usually conspicuously coloured and fragrant, organized singly or in cluster type of inflorescence.

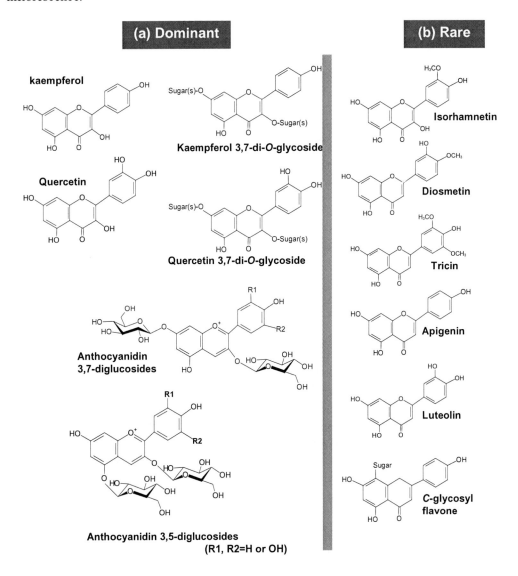

Figure 29. Frequent and rare flavonoid structures in Liliaceae. The frequent structures characterize the general feature of the family Liliaceae. Rare structures are interesting at lower taxonomic level (genus, species) where they can characterize some particular taxa.

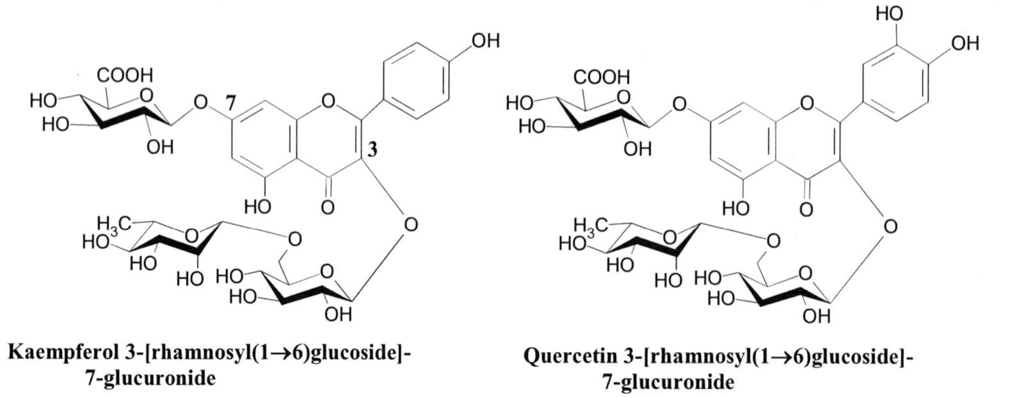

Kaempferol 3-[rhamnosyl(1→6)glucoside]-
7-glucuronide

Quercetin 3-[rhamnosyl(1→6)glucoside]-
7-glucuronide

Figure 30. Some common flavonol di- and triglycosides in Liliaceae.

Analysis of flavonoids in the flowers and leaves of 60 species of Liliaceae highlighted flavonol derivatives (Figure 29a) as dominant compounds (Skrzypczakowa, 1967). Williams (1975), investigating 168 species confirmed such chemical fingerprint, and found that quercetin and kaempferol (Figure 29a) were the most characteristic constituents of the Liliaceae (*sensu* Melchior) (Melchior, 1964) (40 and 42% of the sample, respectively). However, simple flavones (luteolin and apigenin) were found in only 24 and 20% of taxa. On the other hand, methylated flavonoids such diosmetin, tricin and isorhamnetin were of rare occurrence and flavone *C*-glycosides were found in very few species (Figure 29b). In previous works, the isorhamnetin has been reported to be a common pollen constituent in the Liliaceae (Kotake and Arakawa, 1956; Togasawa et al., 1966; Komissarenko et al., 1964).

The Liliaceae species were characterized by the presence of flavonoid di- and tri-*O*-glycosides and a rare occurrence of monoglycosides in leaf tissue (Williams, 1975). Flavonol 3, 7-diglycosides (Figure 29a; Figure 30) appear to be characteristic constituents of the Liliaceae (Budzianowski, 1991; Nagy et al., 1984, 1986; Bučkova et al., 1988; Williams, 1975).

Several anthocyanins have been reported to occur in various species of Liliaceae (Figure 31). Many of these have very original structures. Delphinidin, pelargonidin and cyanidin are the main anthocyanidin occurring in the Liliaceae (Figure 32) (Nørbæk and Kondo, 1999; Bloor, 2001; Toki et al., 1998; Hosokawa et al., 1995a, 1995b, 1995c; Torskangerpoll et al., 1999; Saito et al., 2003). In Liliaceae, anthocyanins with both aromatic and aliphatic acyl groups (Figure 33) have been characterized. The carbon 6 of glucose seems to be frequently acylated (Figure 31) such as acylations at other positions can have interesting chemotaxonomic aspect at low taxonomic levels. Malonic acid (Figure 33) is a common aliphatic acid associated with anthocyanins, and malonylated anthocyanins have been found regularly in members of the Liliaceae, Alliaceae and Poaceae.

VII.2.2. Subfamily/Tribe level chemotaxonomy

In the past, Liliaceae (lilies family) has been treated as a large assemblage (Liliaceae *sensu lato*), which has more recently been broken up into numerous segregate families. Dahlgren et al. (1985) underlined that the broad taxon of Liliaceae was grossly unnatural; they recognized as families only groups in which they had some confidence of monophyly;

Angiosperm Phylogeny Group (APG) also used this system (APG, 2003). The taxonomic heterogeneity within the broadest classification of Liliaceae (*sensu* Melchior) (Melchior, 1964) has been confirmed by an analysis of flavonoids in the leaves of 168 Liliaceae species (Williams, 1975):

The subfamilies Scilloideae, Asphodeloideae and Melanthioideae were shown to be the most chemically heterogeneous groups within the Liliaceae (old classification), because they contained both flavone and flavonol derivatives (Figure 29a, b). However, the subfamilies Wurmbaeoideae and Lilioideae appeared to be more chemically homogeneous because of the dominance of either flavones or flavonols, respectively.

More precisely, the chemical diversity of the subfamily Scilloideae was linked to the fact that it includes different genera producing flavone glycosides or flavonol glycosides. However, although flavone and flavonol co-occurred at the level of subfamily, Williams (1975) reported no co-presence at the level of genus.

The Melanthioideae was shown to be unusual because of the presence of flavone *C*-glycosides in addition to flavone and flavonol *O*-glycosides. However at the level of species, the flavonoid patterns were generally described as either flavone *O*-glycoside, flavonol *O*-glycosides or flavone *C*-glycosides. A recent taxonomic revision of the family Liliaceae by the APG system, led to classify the subfamily Melanthiodeae in the family Melanthiaceae (APG, 2003).

In the Asphodeloideae, the flavonoids seemed to be well distributed at a tribal level: the species of the tribe Dianelleae characteristically contained kaempferol (Figure 29a), and proanthocyanidins, viz. procyanidin (Figure 34a) and prodelphinidin (Figure 34b); in the tribe Asphodeleae, the flavones luteolin and apigenin (Figure 29b) occurred; in the tribes Aphyllantheae, Hemerocalleae and Aloeae, only flavonols were found. From a recent taxonomic revision of the family Liliaceae by the APG system, the subfamily Asphodeloideae has been classified in the family Asphodelaceae (APG, 2003).

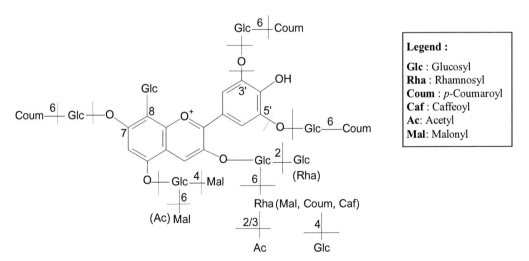

Figure 31. Backgroud chemical structure showing the different types of substitutions observed in anthocyanins of Liliaceae. Dotted lines indicate possible absence of corresponding substitutions according to the taxon. Substitutions of a same level are represented between brackets.

By considering the fact that flavones are less frequent than flavonols in the Liliaceae, the subfamily Wurmbaeoideae was shown to be original by production of only flavones (luteolin, apigenin and their glycosides) (Figure 29b).

The subfamily Lilioideae seemed to be chemically homogenous as flavonoid patterns consisted of quercetin and kaempferol glycosides (Figure 30) in 76 and 80% of the 168 species surveyed by Williams (1975). This subfamily has been retained as representative of Liliaceae after taxonomic revision by Dahlgren et al. (1985) and APG system (1998, 2003).

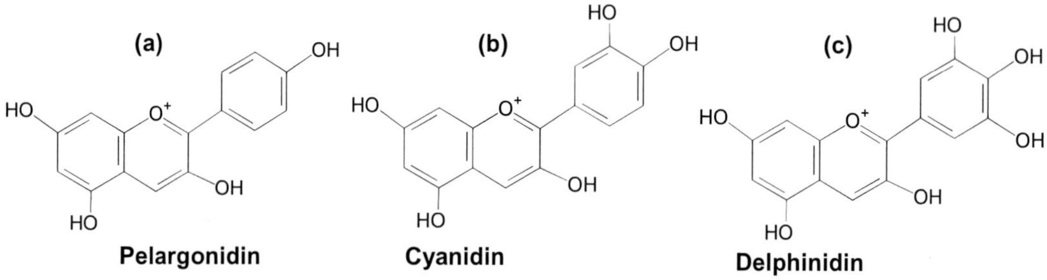

Figure 32. Chemical structures of common anthocyanidins in the family Liliaceae.

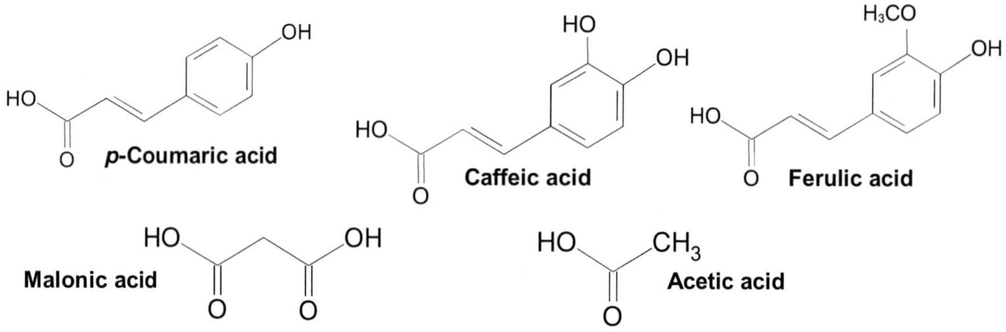

Figure 33. Chemical structures of frequent aromatic and aliphatic acyls in the anthocyanins of Liliaceae.

VII.2.3. Genus level chemotaxonomy

Lilies are popular ornamental plants with red, pink, orange, yellow or white flowers, often with dark red spots. Analysis of anthocyanins in ten cultivars and one species of the genus *Lilium* highlighted the production of a particular chemical structure by several distantly related species (Nørbæk and Kondo, 1999; Synge, 1980): anthocyanins from flowers of *Lilium* have been reported to contain 3-rutinoside-7-glucoside and 3-rutinoside of cyanidin (Figure 35).

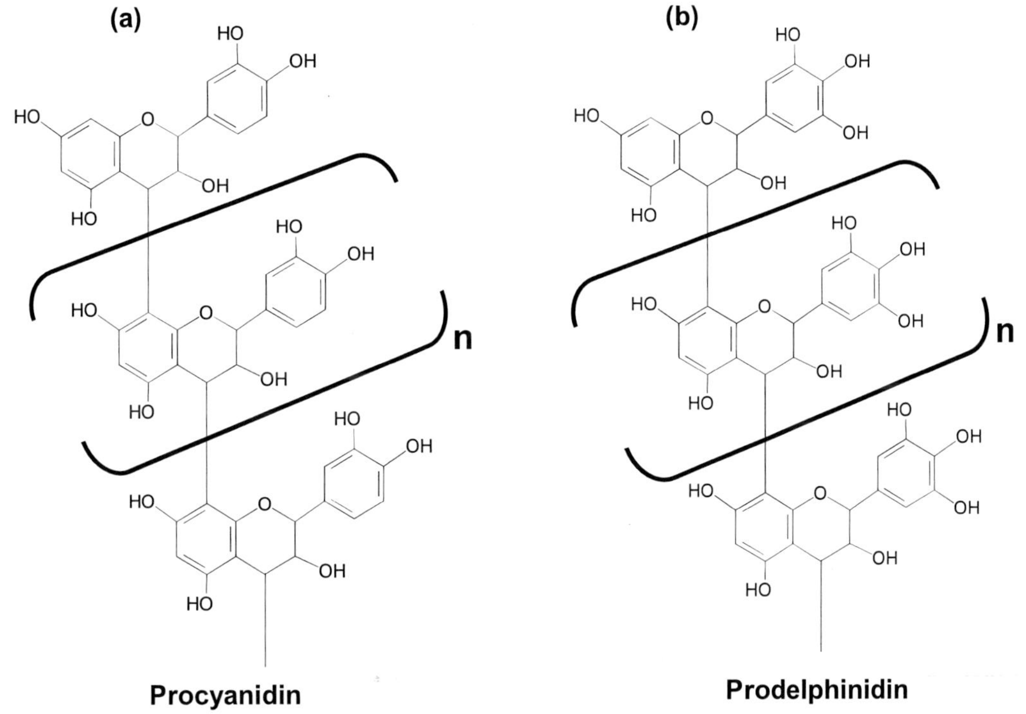

(a) Procyanidin

(b) Prodelphinidin

Figure 34. Chemical structures of proanthocyanidin units found in the tribe Dianelleae (Family Liliaceae) (Williams, 1975).

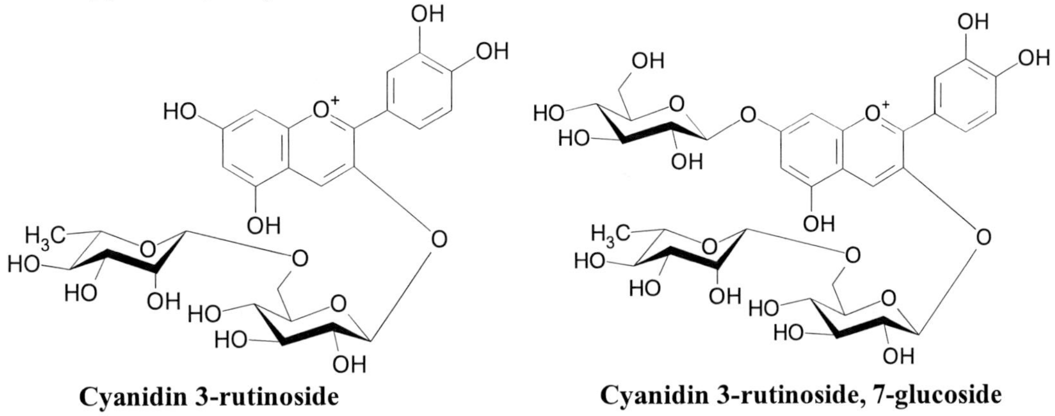

Cyanidin 3-rutinoside

Cyanidin 3-rutinoside, 7-glucoside

Figure 35. Chemical structures of anthocyanidin found in the genus *Lilium*.

Different studies on species *Tulipa* showed that some anthocyanin derivatives were uniformly present throughout the genus. The chemical feature of *Tulipa* was characterized by seven pigments: the 3-glucosides and 3-rhamnoglucosides of pelargonidin, cyanidin and delphinidin (Figure 36a-f) in addition to delphinidin 3,5-diglucoside (Figure 36g) (Halevy and Asen, 1959; Shibata, 1956; Shibata and Sakai, 1958; Shibata and Ishikura, 1960).

In a recent study, the flowers of 17 species and 25 cultivars of tulips were subjected to qualitative and relative quantitative examination for anthocyanins (Torskangerpoll et al., **2005**). Altogether five anthocyanins were identified as the 3-O-(6''-O-α-rhamnopyranosyl-β-

glucopyranoside) of pelargonidin (Figure 36d), cyanidin (Figure 36e) and delphinidin (Figure 36f), and the 3-O-[6"-O-(2'''-O-acetyl-α-rhamnopyranosyl)-β-glucopyranoside] of pelargonidin (Figure 37a) and cyanidin (Figure 37b). These five pigments represented 12%, 43%, 7%, 31% and 2%, respectively, of the total anthocyanin amount in the tepals of *Tulipa* species (Figure 38a), and 30%, 37%, 20%, 4% and 6%, respectively, in the cultivar tepals (Figure 38b). Nearly 50% of the samples contained acetylated anthocyanins (Figure 37a, b). Cyanidin-3-rutinoside (Figure 36e) had the widest distribution (71% of the species and 100% of the cultivars).

The anthocyanins containing acetyl groups at the position 2 or 3 of rhamnose (Figure 31) (Figure 37, 50) can be considered as very original structures because in anthocyanins of higher plant world, the acetyl groups are generally linked to glucosyl 6-positions (Torskangerpoll et al., 1999; Nakayama et al., 1999).

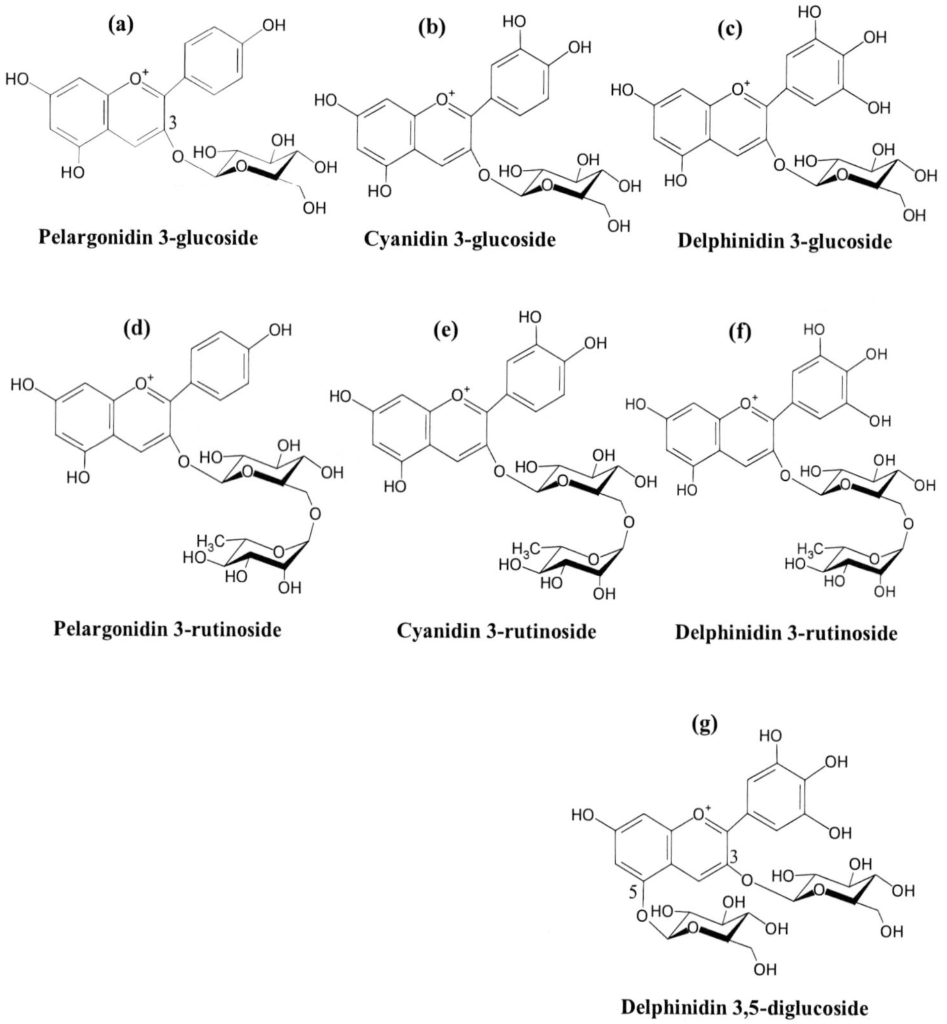

Figure 36: Chemical structures of glucosylated and rutinosylated anthocyanins characterizing the flowers of *Tulipa* (Liliaceae).

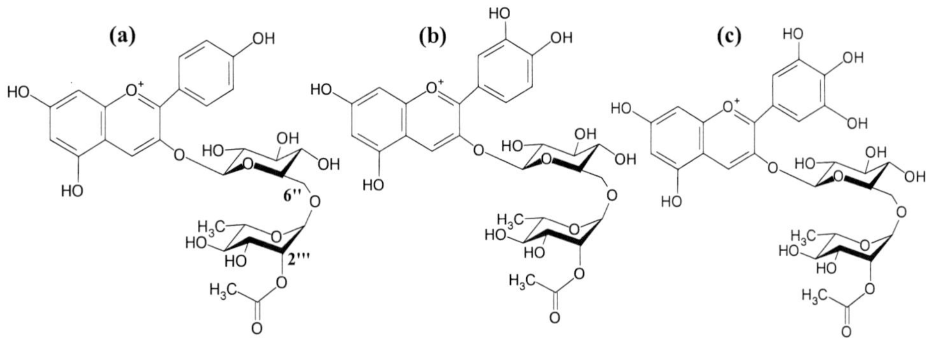

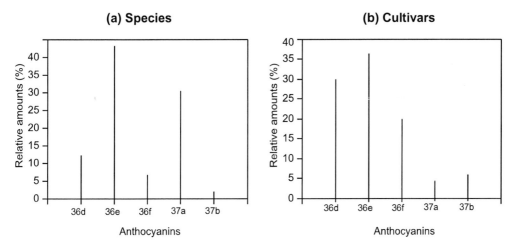

Figure 37. Chemical structures of acetylated 3-rutinosyl of pelargonidin (a), cyanidin (b) and delphinidin (c) found in the flowers of *Tulipa* species (Torskangerpoll et al., 2005; Nakayama et al. (1999).

(a) Species

(b) Cultivars

Figure 38. Average anthocyanin profiles of 17 species (a) and 25 cultivars (b) of *Tulipa* genus (Torskangerpoll et al., 2005). The compound labels (36d-f, 37a-b) on the abscissa axis correspond to chemical structures illustrated in figures 36 and 37.

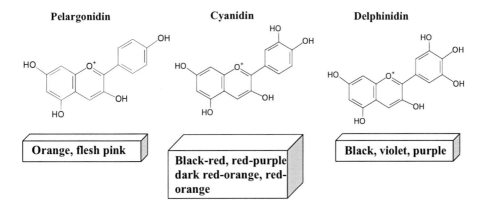

Figure 39. Variation of flower colours in *Tulipa* according to the abundance of the anthocyanidin type in the anthocyanin contents (Shibata and Ishikura, 1960).

VII.2.4. Subgenus/Section level chemotaxonomy

Robinson and Robinson (1931) found that garden tulips contained either a mixture of cyanidin and pelargonidin biosides or cyanidin bioside and delphinidin diglucoside. Then, Beal et al. (1941) found that a pelargonidin 3-pentose glycoside was restricted to species in the subgenus *Leiostemones* (26 species examined) and a delphinidin 3-pentose glycoside was restricted to species in the subgenus *Eriostemones* (6 species examined). More recently, Nieuwhof et al. (1990) showed that the two subgenera *Leiostemones* (17 species examined) and *Eriostemones* (9 species examined) can be separated almost completely on the basis of their carotenoid and anthocyanidin (pelargonidin, cyanidin, delphinidin) contents (Figure 39, 40).

After examination of the anthocyanidin content of 107 varieties of tulips by paper chromatography, Shibata and Ishikura (1960) concluded that garden varieties having black, black-purple, fade-sky, violet and purple flower were delphinidin type (i.e. the delphinidin content was more than 50% of the total anthocyanin content), those having red-purple, pink, black-red, deep-crimson, crimson, deep-red, dark red-orange and red-orange flower belonged to the cyanidin type, and those having orange and flesh pink flower belonged to the pelargonidin type (Figure 39). Later, van Eijk et al. (1987) listed 503 genotypes according to their type of pigmentation (carotenoid, anthocyanidin). They concluded that combinations of cyanidin and pelargonidin generally induced attractive red and orange colours, especially in combination with carotenoids, while the combination of delphinidin and carotenoids usually gave unattractive brownish colours (Figure 40).

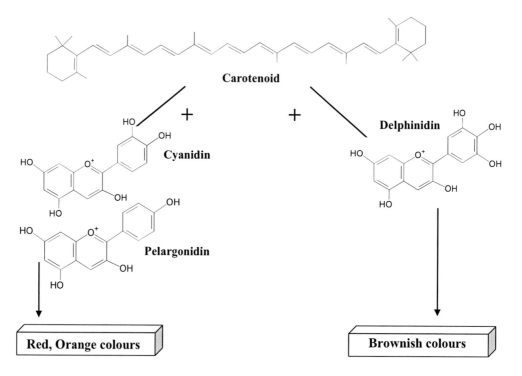

Figure 40. Combination between carotenoids and anthocyanins leading to different flower colours in *Tulipa* (van Eijk et al., 1987).

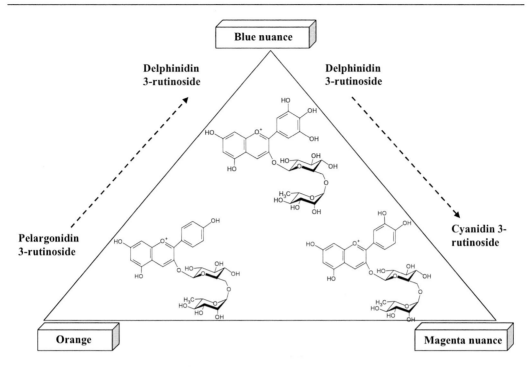

Figure 41. Variation of the petal colour of *Tulipa* in relation to the variation in the relative proportion of anthocyanidin 3-rutinoside the ones at the expense of others (Torskangerpoll et al., 2005).

In a recent study on anthocyanins, Torskangerpoll et al. (2005) analysed 17 species and 25 cultivars of *Tulipa*. By considering the different flower colours, the authors showed that all tepals classified with "blue nuances" concerned exclusively cultivars and contained delphinidin 3-rutinoside as major anthocyanin, but no or just trace of pelargonidin derivatives (Figure 41). The species and cultivars having "magenta nuances" showed anthocyanin content in which the relative amounts of cyanidin 3-rutinoside increased at the expense of delphinidin-3-rutinoside. Orange coloured tepals were to a large extent correlated with high relative amounts of the pelargonidin derivatives (non acetylated and acetylated pelargonidin 3-rutinoside) (Figure 36d, 37a). Acetylation of anthocyanins furnished a weak colour effect whereas aromatic acyl groups have been reported to be responsible for bluing effect (Torskangerpoll et al., 2005).

As a general chemical feature within the genus *Tulipa*, the two acetylated anthocyanins (Figure 37a, b) showed a co-occurrence trend, and if present, they were accompanied with non-acetylated analogous (Figure 36d, e) (Torskangerpoll et al., 2005).

The most striking difference between the total anthocyanin contents in the tepals of species and cultivars was revealed by the relative amounts of delphinidin 3-rutinoside (Figure 36f) and pelargonidin 3-(2'''-acetyl-rutinoside) (Figure 37a) (Torskangerpoll et al., 2005): in the species, these two pigments constituted 7% and 31% of the anthocyanins, respectively; in the cultivars, they constituted 20% and 4%, respectively (Figure 38).

Under traditional botanical aspect, the 17 examined tulip species (Torskangerpoll et al., 2005) may be divided into two subgenera: *Tulipa* and *Eriostemones*. Each of these two subgenera may further be divided into three sections: *Tulipa*, *Tulipanum* and *Eichleres* for the subgenus *Tulipa*, and *Saxatiles*, *Biflores* and *Australes* for *Eristemones*. A principle component analysis on the relative amounts of anthocyanins in the 17 species showed that all six species belonging to the section *Eichleres* grouped together into a small (i.e. homogeneous) space. Such homogeneity was due to the fact the six species were all characterized by relatively high amounts of the pelargonidin derivatives (Figure 36d, 37a), and showed similar tepal colouration (orange nuances) (Figure 41).

VII.2.5. Species level chemotaxonomy

There are several taxonomically reports concerning original flavonoid structures in different species of Liliaceae:

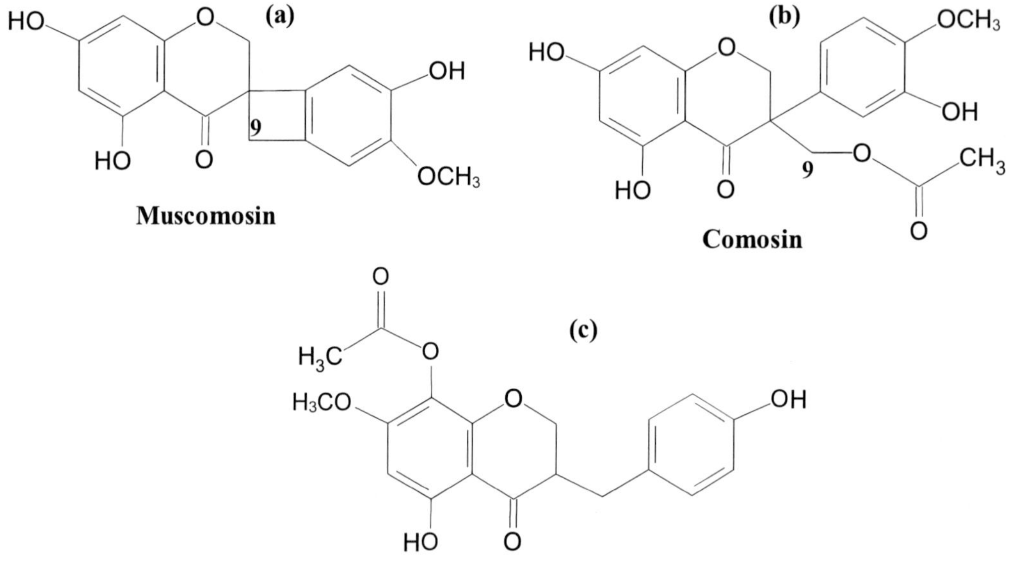

8-*O*-demethyl-8-*O*-acetyl-7-*O*-methyl-3,9-dihydropunctatin

Figure 42. Chemical structures of homoisoflavones isolated from the bulbs of *Muscari comosum* (Adinolfi et al., 1985a).

A number of homoisoflavanones characterised by the basic 3-benzyl-4-chromanone skeleton have been isolated from bulbs of *Muscari comosum* (Adinolfi et al., 1985a; Adinolfi et al., 1984; Adinolfi et al., 1985b; Finckh and Tamm, 1970; Kouno et al., 1973; Heller and Tamm, 1981). Noteworthy homoisoflavanone structures has been isolated consisting of muscomosin (Figure 42a), comosin (Figure 42b) and an acetylated dihydropunctatin derivative (Figure 42c): muscomosin is a homoisoflavanone of the rare scillascillin type (Kouno et al., 1973); comosin possesses an original homoisoflavanone skeleton since the 9-carbon is not benzylic; compound (c) was the first acetyl 3-benzyl-4-chromanone derivative hitherto found in nature (Adinolfi et al., 1985a).

In *Lilium candidum* growing in Czechoslovakia, unusual kaempferol structures substituted by methyl succinate (Figure 43a) and by an alkaloid (lilaline) (Figure 43b) at 8-position have been isolated from the white flowers (Bučkova et al. , 1988; Mašterová et al., 1987). The first compound appears to be biogenetically precursor of the second (lilaline).

Among the flavonoid glycosides, kaempferol 3-(glucosyl(1→2)galactoside (Nagy et al., 1984) and kaempferol 3-(xylosyl (1→2)glucoside (Nagy et al., 1986) were isolated from the flowers of *L. candidum* (Figure 44a, 44b).

Besides the flavonoid glycosides, a phenylethyl glycoside has been isolated from the flowers of *Lilium candidum* (Figure 45a) (Uhrín et al., 1989). In the same chemical family, the 2-phenylethyl palmitate was also isolated from this species (Figure 45b) (Uhrín et al., 1989).

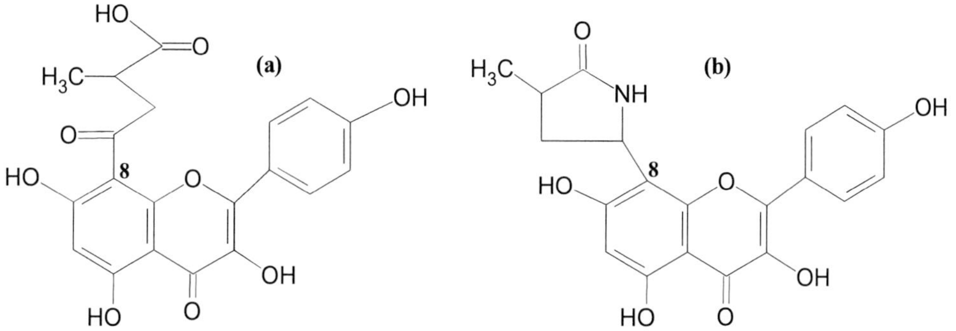

8-(3''-methyl-succinoyl)-kaempferol

Lilaline

Figure 43. Aliphatic acylated (a) and alkaloid (b) flavonoids isolated from the white flowers of *Lilium candidum*; (a) appear to be an obvious biogenetic precursor of (b) (Bučkova et al., 1988; Mašterová et al., 1987).

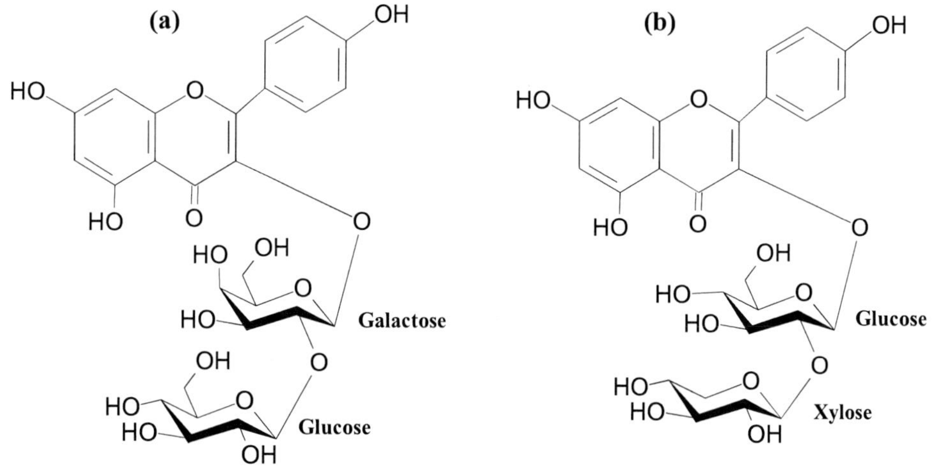

Figure 44. Kaempferol 3-diglycosides isolated from the white flowers of *Lilium candidum* (Nagy et al., 1984, 1986).

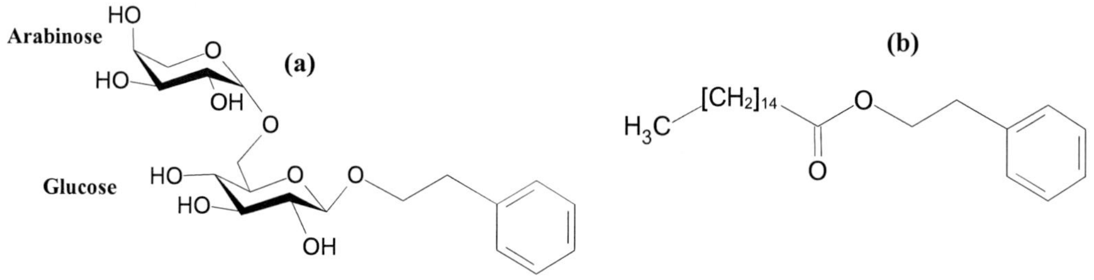

Figure 45. Phenylethyl glycoside (a) and phenylethyl palmitate (b) isolated from the flowers of *Lilium candidum* (Liliaceae) (Uhrín et al., 1989).

Saito et al. (2003) have isolated 8-*C*-glucosyl cyanidin 3-[6-(malonyl)]glucoside (Figure 46) from the purple flowers of *Tricyrtis formosana* cultivar Fujimusume, which is the first natural *C*-glycosyl anthocyanin to be reported.

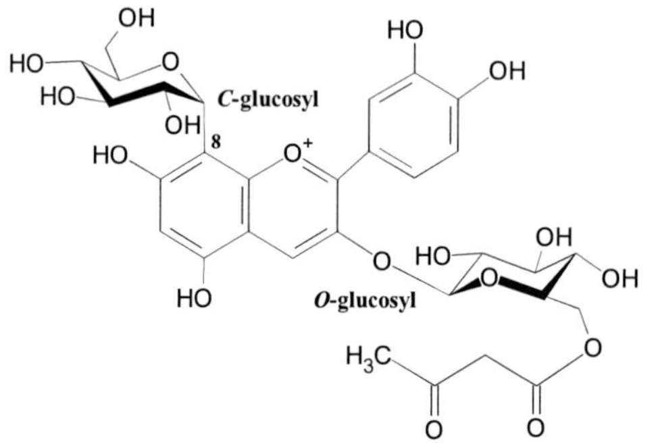

Figure 46. *C*-glucosylated cyaniding-based anthocyanin found in the purple flowers of *Tricyrtis formosana* cultivar Fujimusume (Saito et al., 2003).

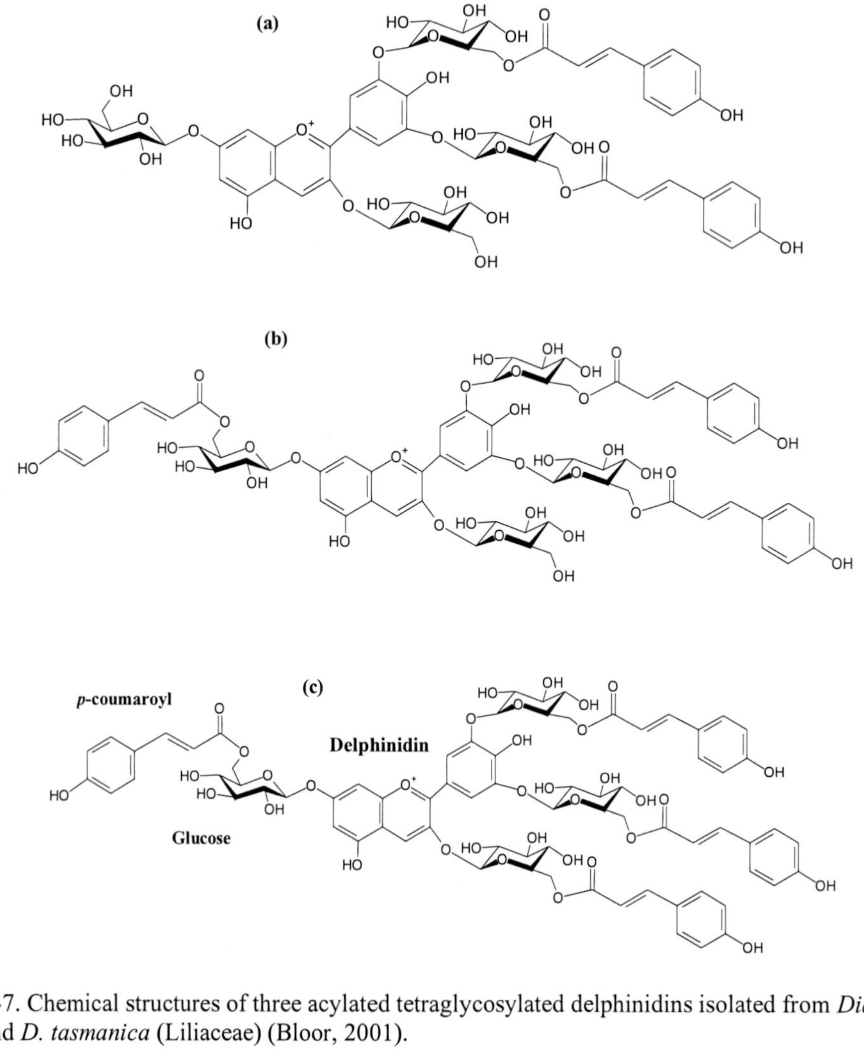

Figure 47. Chemical structures of three acylated tetraglycosylated delphinidins isolated from *Dianella nigra* and *D. tasmanica* (Liliaceae) (Bloor, 2001).

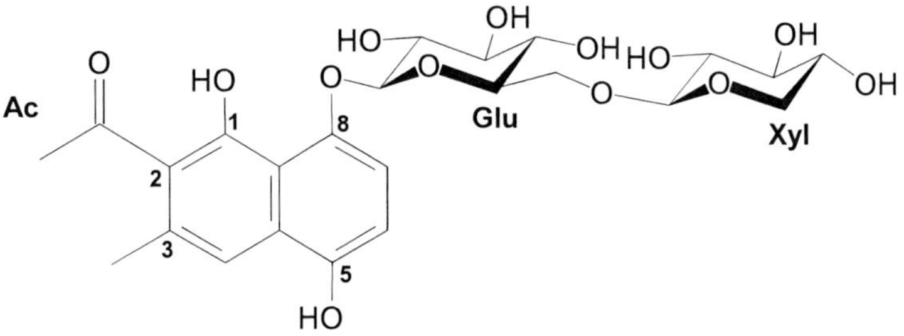

Figure 48. Chemical structure of acetylated diglycosylated naphthalene, a characteristic yellow pigment extracted from *Dianella* berries (Bloor, 2001). Glu: Glucosyl; Xyl: Xylosyl; Ac: Acetyl.

Tricin (Figure 17b), unusual in the *Liliaceae*, was found as the only flavonoid constituent in leaves of *Hyacinthus orientalis* (Williams, 1975); nine tricin glycosides were isolated. More recently, the genus *Hyacinthus* has been classified into the family of Hyacinthaceae (Asparagales) (APG, 2003).

Dianella species (Blue flax lily, blueberry lily) are rhizome herbs characterized by blue or blue-violet berries. In New Zealand, the major species is *Dianella nigra* Colenso featuring small, intensely blue, round berries. Of the more numerous Australian species, *Dianella tasmanica* Hook is remarkable for its larger and more elongated blue-vilolet fruits. Three acylated delphinidin 3,7,3',5'-tetraglucosides have been isolated from these two remarkably coloured species (Figure 47a-c) (Bloor, 2001). One of these pigments had four 6-coumarylglucoside units located at different aglycone positions (Figure 47c).

In addition to acylated anthocyanins, the extracts from the two blue *Dianella* berries contained a yellow non-anthocyanic pigment consisting of a characteristic acetylated diglycosylated naphthalene nucleus (Figure 48) (Bloor, 2001). Under acidic environment, such a chemical structure is known to be a quinone precursor making it potentially toxic (Colegate et al., 1986).

Other chemical studies have focused on the presence of naphtoquinones (Figure 6c) in the root of *Dianella nigra* and *Dianella revolute* R. Br. (Colegate et al., 1986; Briggs et al., 1975; Cooke and Sparrow, 1965). From APG revision, the genus *Dianella* has been classified into the family Hemerocallidaceae (Asparagales) (APG, 1998, 2003).

In the genus *Tulipa*, quercetin and kaempferol 3-gentiobioside, 7-glucuronide (Figure 49) have been isolated from perianths of *T. gesneriana* (Budzianowski, 1991).

Original acetylated anthocyanins in the genus *Tulipa* have been initially isolated from the orange-red *Tulipa* "Queen Wilhelmina" (Torskangerpoll et al., 1999). These compounds are 3-*O*-[6"-*O*-(2'''-*O*-acetyl-α-rhamnopyranosyl)-β-glucopyranoside] of pelargonidin and cyanidin (Figure 37a, 37b). These chemical structures were original because of the atypical presence of acetyl moiety on the 2-position of rhamnose (acetylated 6-position of glucose is generally observed). Moreover, the chemotaxonomic importance of these structures increased by the fact that they represent the first report of anthocyanins with an acyl group linked to axial sugar position (Torskangerpoll et al., 1999).

Apart from cyanidin and pelargonidin in which the rhamnose was acetyled at 2-position (Figure 37a, 37b), Nakayama et al. (1999) isolated from the purple anthers of *Tulipa gesneriana*, a delphinidin-based analogous compound (Figure 37c) as well as delphinidin with acetylated rhamnose at the position 3, viz. delphinidin-3-(3'''-acetylrutinoside) (Figure 50). However, Torskangerpoll et al. (2005) reported the absence of these compounds in the tepals of the *T. gesneriana* var. *macrospila*, highlighting a variability linked to the analysed plant part.

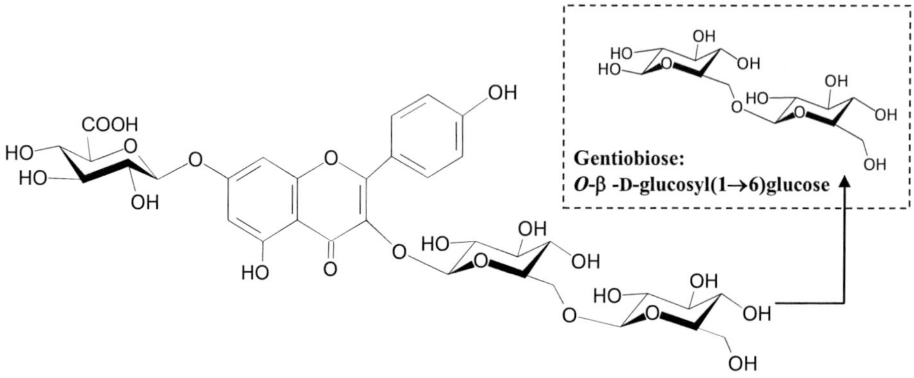

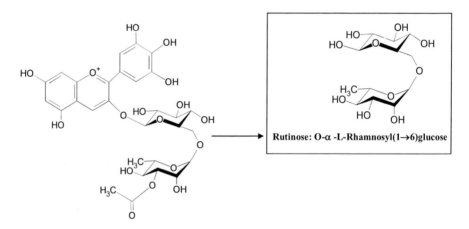

Figure 49. Chemical structure of kaempferol 3-gentiobioside, 7-glucuronide isolated from perianths of *Tulipa gesneriana* (Budzianowski, 1991).

Figure 50. Chemical structure 3-(3'''-acetyl-rutinoside) of delphinidin isolated from the purple anthers of *Tulipa gesneriana* (Nakayama et al., 1999).

Later, more investigations on 17 species and 25 cultivars of *Tulipa* growing in Norway and Denmark confirmed the importance of anthocyanidin 3-rutinosides (Figure 36d-f) and their acetylated homologuous (Figure 37a,b) as major compounds in the chemical feature of this genus (Torskangerpoll et al., 2005). Interestingly, *Tulipa didieri* and *T.* "Red Carpet" were the only species and cultivar, respectively, to contain all five pigments.

Contrary to *T. didieri* and *T.* "Red Carpet", other species and cultivars were chemically characterized by the exclusive presence of one anthocyanin: the species *T. orphanideae* and the cultivars *T.* "New Design" and *T.* "Fringed Apeldoorn" contained only cyanidin 3-rutinoside (Figure 36e) which was the most largely distributed anthocyanin in the analysed population (Torskangerpoll et al., 2005).

Triteleia bridgesii formerly considered as Liliaceae has been classified into Asparagales by APG system (APG, 1998, 2003). It is distributed in North and South America, and it is a popular ornamental with blue-purple flowers. From its flowers, delphinidin 3-[6-*p*-

coumaroylglucoside], 5-[6-malonylglucoside] has been isolated as a basic structure of major anthocyanins (Toki et al., 1998). One of these anthocyanins had additional glucosyl moiety connected to the 4-hydroxyl group of *p*-coumaric acid (Figure 51). Its structure was delphinidin 3-[glucosyl-(1→4)-coumaroyl-(3'→6)-glucoside], 5-[malonyl-(1→6)-glucoside].

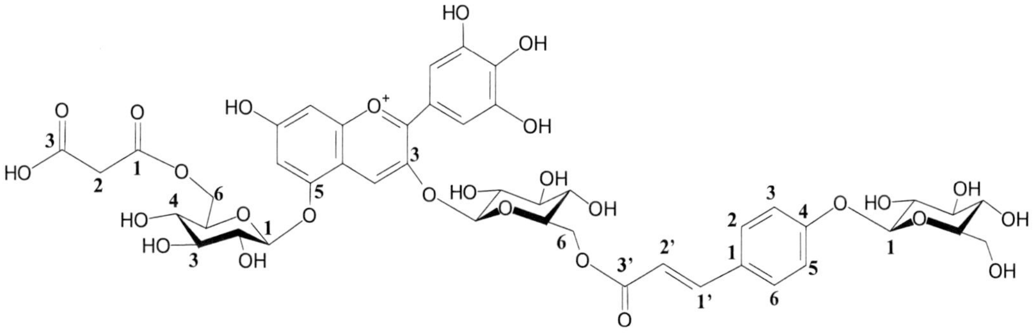

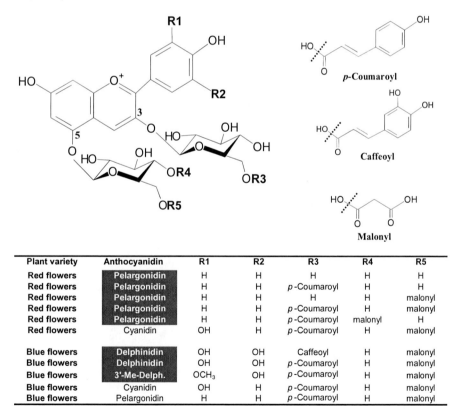

Figure 51. Original chemical structure of delphinidin-based anthocyanin isolated from the flowers of *Triteleia bridgesii* (Liliaceae: old classification; Asparagale: APG system), and acylated by both aliphatic and aromatic acyls and containing a terminal glucosyl moiety attached on the *p*-coumaroyl (Toki et al., 1998).

Plant variety	Anthocyanidin	R1	R2	R3	R4	R5
Red flowers	Pelargonidin	H	H	H	H	H
Red flowers	Pelargonidin	H	H	*p*-Coumaroyl	H	H
Red flowers	Pelargonidin	H	H	H	H	malonyl
Red flowers	Pelargonidin	H	H	*p*-Coumaroyl	H	malonyl
Red flowers	Pelargonidin	H	H	*p*-Coumaroyl	malonyl	H
Red flowers	Cyanidin	OH	H	*p*-Coumaroyl	H	malonyl
Blue flowers	Delphinidin	OH	OH	Caffeoyl	H	malonyl
Blue flowers	Delphinidin	OH	OH	*p*-Coumaroyl	H	malonyl
Blue flowers	3'-Me-Delph.	OCH$_3$	OH	*p*-Coumaroyl	H	malonyl
Blue flowers	Cyanidin	OH	H	*p*-Coumaroyl	H	malonyl
Blue flowers	Pelargonidin	H	H	*p*-Coumaroyl	H	malonyl

Figure 52. Chemical structures of anthocyanins isolated from the flowers of *Hyacinthus orientalis* (formerly Liliaceae; newly Hyacinthaceae, Asparagales), and essentially based on pelargonidin in red flowers and delphinidin in blue flowers (Hosokawa et al., 1995a, b, c).

VII.2.6. Infraspecific level chemotaxonomy

Hyacinthus orientalis L. is a popular ornamental plant with blue, red, pink, yellow or white flowers. Nearly 20 similar anthocyanidin 3,5-diglucosides with a cinnamic acid derivative located at the 6-position of the 3-*O*-glucosyl and possible malonyl or acetyl connected to the 5-*O*-glucosyl have been isolated from flowers of *H. orientalis* (Hosokawa et al., 1995a, b, c). Infraspecific differences have been found between varieties having different flower colours: red *Hyacinthus* petals contain pelargonidin derivatives (Figure 52), while the blue flowers have mainly delphinidin derivatives (Figure 52).

VIII. DICOTYLEDONS

VIII.1 Asteraceae Family

The family of Asteraceae (or Compositae) includes over 1000 genera with approximately 25000-30000 species widely spreaded in tropical and subtropical regions (Bremer, 1994).

VIII.1.1 Family level chemotaxonomy

The metabolism of Asteraceae seems to balance a major production of flavonoids against the production of other metabolites belonging to other chemical classes. The high production of flavonoids in Asteraceae has been shown to provide good chemotaxonomic markers for this plant family. This can be illustrated by chemical structures that are specific to Asteraceae, or rarely observed elsewhere:

The family Asteraceae is among the few families of higher plants that contain aurones (Figure 10) (Giannasi, 1975; Crawford and Stuessy, 1981; Harborne, 1966b; Iwashina, 2000). Another fingerprint molecule of Asteraceae is represented by quercetagetin (Figure 55) which is found almost entirely in this family (Harborne and Heywood, 1976). Some exceptional occurrences have been, however, reported in Fabaceae (Harborne, 1971b). Finally among the different flavonoids, the flavonols were relatively more reproted in the Asteraceae.

Apart from entire characteristic molecules, some chemical substitutions tend to be more observed in Asteraceae leading to other chemotaxonomical keys: in many cases, polymethoxylations have been reported to characterize the flavonoid exudates.

Some metabolic trends have been also reported about the Asteraceae: this plant family has been shown to present a negative correlation between methoxylation and *O*-glycosylation degrees of flavonoids (Figure 11a) (Emerenciano et al., 1987, 2001).

VIII.1.2. Subfamily/Tribe level chemotaxonomy

The subfamily Asteroideae (sensus Bremer, 1994, 1996) can be characterized by high flavonoid variability, compared to other subfamilies (Barnadesioideae, Carduoideae and Cichorioideae). In Asteroideae, the maximal diversification of flavonoid diversification can be seen in the tribes Anthemideae, Heliantheae, Helenieae and Eupatorieae. Moreover, the subfamily Asteroideae shows a high degree of *O*-methyl substitution for flavonoids, and particularly in the tribes Eupatorieae, Anthemideae, Inuleae, Plucheae and Astereae (Emerenciano et al., 2001). Also, the tribe Inuleae contains many genera (e.g. *Pulicaria*) producing 7-*O*-methyled flavonoids (Figure 58), whereas in other tribes, 6- and/or 6, 8- methylation is more usual (Wollenweber, 1994). It is also interesting to note that some tribes of Asteroideae are able to produce flavonoid sulphates which are relatively rare in nature; these tribes are Inuleae, Eupatorieae and Heliantheae (Williams and Harborne, 1994; Barron et al., 1988).

The tribes Heliantheae and Helenieae can be characterized by their ability to produce aurones and chalcones (Figure 10, 54c) (Emerenciano et al., 2001).

The tribe Lactuceae (subfamily: Cichorioideae = Lactucoideae) is an original case characterized by chemical patterns relatively rich in highly glycosylated and poorly methylated flavonoids (Emerenciano et al., 2001). It is interesting to note that this tribe was treated as a subfamily in the old classification system of Wagenitz (1976).

The flavonoid pattern of the subfamily Barnadesioideae is characterized by high levels of flavonols allied to the absence of flavones (Figure 10) (Emerenciano et al., 2001).

The subfamily Carduloideae, recognized as subfamily by Bremer (1996), showed advanced phytochemical status, viz. : a pattern with a high oxygenation degree at position 6 for flavones and flavonols, a low oxygenation degree at C-8 for flavones, and a flavone/flavonol ratio around of 146/59 (Emerenciano et al., 2001). The flavone/flavonol ratio was introduced by Harborne (1977) as an indication of evolutionary trends for a given taxon; from its high flavone/flavonol ratio, the subfamily Carduloideae can be classified within advanced group status, whereas the "exclusive" flavonol production in the subfamily Barnadesioideae places it at a basal position within the Asteraceae.

VIII.1.3. Subtribe/Genus level chemotaxonomy

Within different genera of Asteraceae, species show particular compounds giving chemotaxonomical significance to corresponding genus.

The occurrence of flavonol methyl ethers, especially 3- and 6-methyl ethers, appears to be a characteristic feature of the genus *Artemisia* (Delazar et al., 2007). Axillaroside (3-methyl ether flavonol) (Figure 53) and a number of other methylated flavonols have been reported from many species of the genus *Artemisia*, and also from other genera of the family *Asteraceae*, e.g. *Centaurea*, *Tagetes*, etc. (Delazar et al., 2007). Shilin et al., 1989 showed that the leaf exudates of *Artemisia* species are characterized by free flavonols, particularly simple and polymethoxylated flavonols (Figure 74).

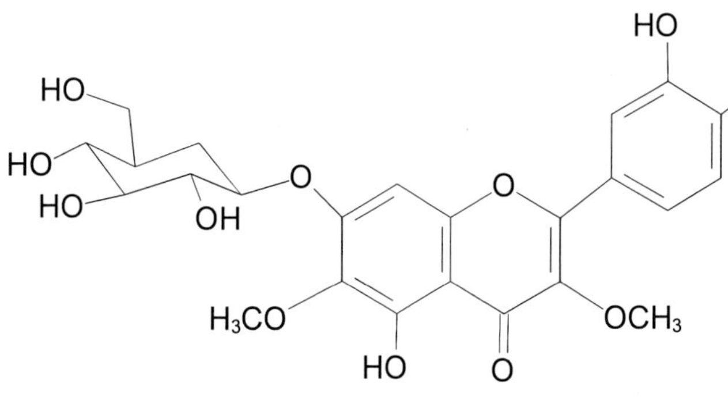

Figure 53. Chemical structure of 3-, 6-dimethyl ether flavonol, frequent in the genus *Artemisia*.

The flavanones pinocembrin and pinostrobin (Figure 54a) are commonly found in species of *Lychnophora* genus (Grael et al., 2005; Sakamoto et al., 2003; Costa et al., 1993; Bazon et al., 1997). *Lychnophora* is an endemic Brazilian genus and its species have a restricted distribution in the Bahia, Goiásand Minas Gerais states (Robinson, 1999; Moreira Alves et al., 2008).

(a)

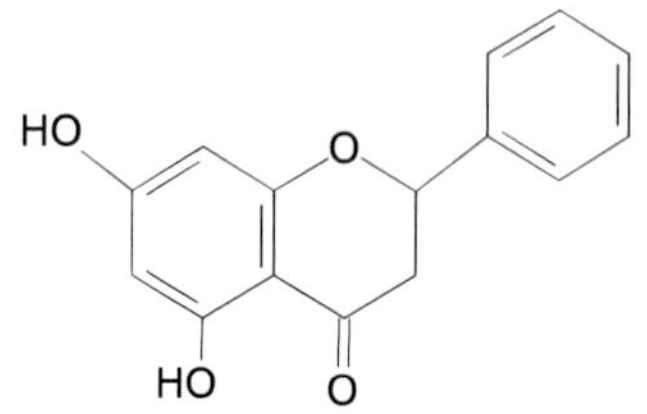

Pinocembrin (Flavanone) 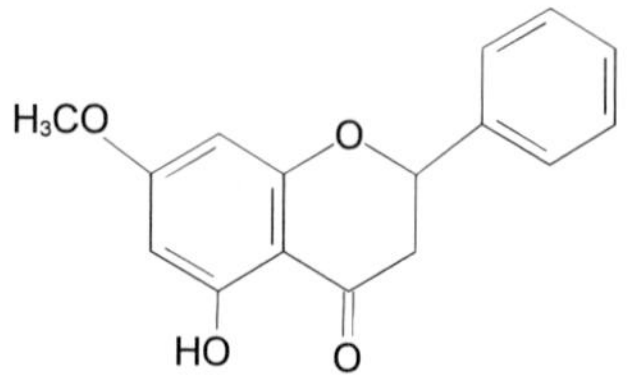   Pinostrobin (Flavanone)

(b)

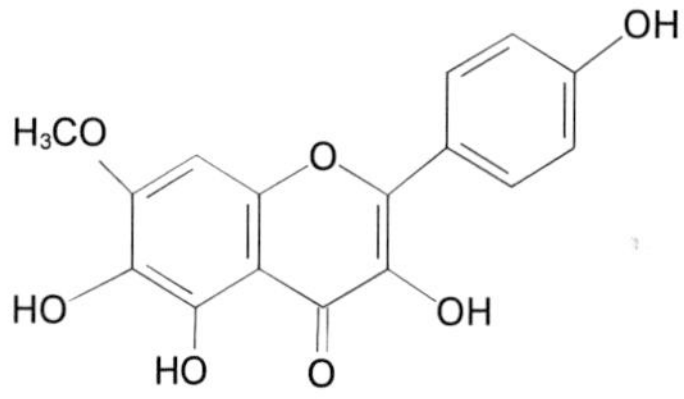

6-hydroxykaempferol 7-methyl ether 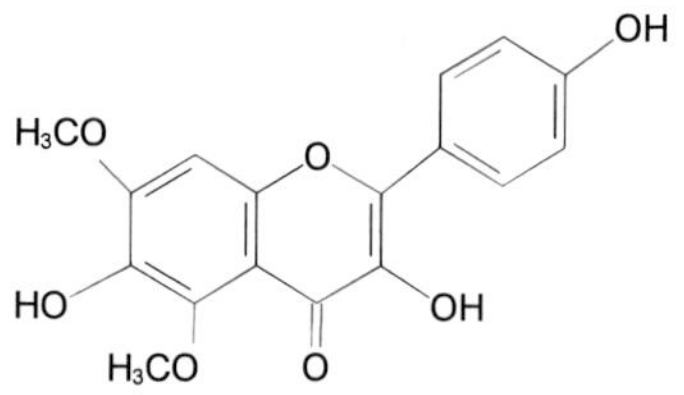   6-hydroxykaempferol 5,7-dimethyl ether

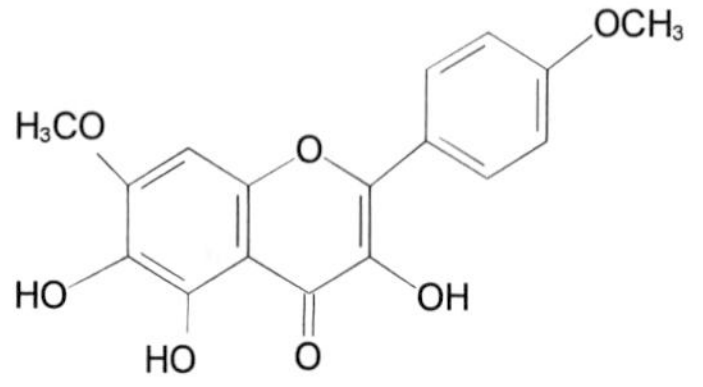

6-hydroxykaempferol 7,4'-dimethyl ether 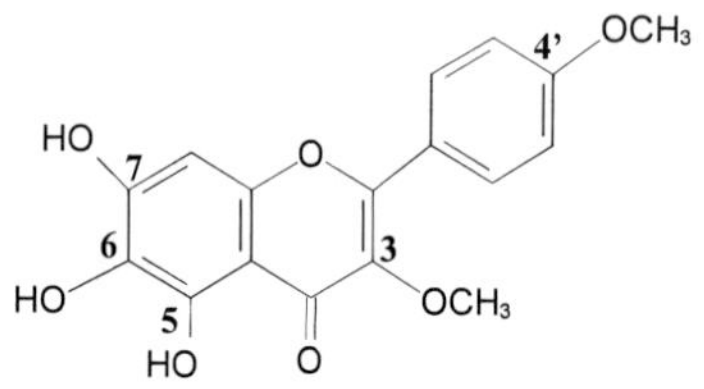   6-hydroxykaempferol 3,4'-dimethyl ether

(c)

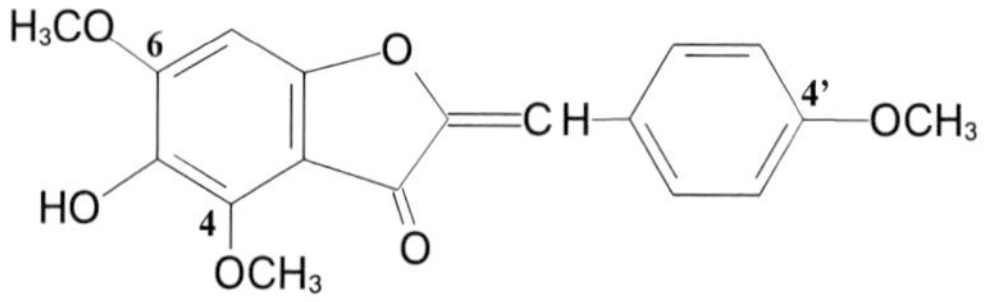

5-hydroxy-4,6,4'-trimethoxyaurone

Figure 54. Flavanones (a), methoxylated 6-hydroxykaempferol derivatives (flavonols) (b) and methoxylated aurone (c) characterizing *Lychnophora*, *Eupatorium* and *Helianthus* species (Asteraceae), respectively (Sakamoto et al., 2003; Herz et al., 1972; Alfatafta and Mullin, 1992).

The 6-hydroxykaempferol 7-methyl ether together with the corresponding 3,7-, 5,7-, 7,4'- and 3,4' dimethyl ethers (Figure 54b; 58c) have been reported from *Eupatorium* species (Herz et al., 1972). This genus comprises about 1200 species which are found in tropic and temperate zones (Ye et al., 2008).

The most of species of the genus *Achillea* are characterized by the predomination of 6-hydroxyflavones and 6-hydroxyflavonols and their methyl-ethers (Valant-Vetschera, 1987; Ivancheva and Tsvetkova, 2003).

Among the surface flavonoids, Williams et al. (2003) described quercetagetin 3,7,3'-trimethyl-ether (Figure 58d) as a major constituent in the leaves and inflorescences of four among the five known European species: *Pulicaria dysentica*, *P. paludosa*, *P. sicula*, *P. vulgaris*. The genus *Pulicaria* belongs to the tribe Inuleae, and contains ca. 80 species with a distribution from Europe into North Africa and Asia (Williams et al., 2003).

In the subtribe Coreopsidinae, the genera *Dahlia*, *Coreopsis*, *Cosmos*, and others, produce chalcones as yellow flower pigments (Giannasi, 1975; Crawford and Stuessy, 1981). Chalcones are flavonoids lacking a central heterocyclic ring (C-ring) (Figure 8, 10). Water-soluble yellow pigments are also produced by the flowers of some *Dahlia* and *Coreopsis* as aurone glycosides (Figure 10) (Giannasi, 1975).

In the tribe Heliantheae, aurones have been isolated from species belonging to the genus *Helianthus* (Figure 54c) (Alfatafta and Mullin, 1992).

VIII.1.4. Species level chemotaxonomy

The variations in the flavonoid profiles of *Chrysothamnus* species (*Asteraceae*) showed diagnostic values at the species level:

From comparison of leaf flavonoids between *Chrysothamnus nauseosus* and *C. viscidiflorus*, hydroxylation of the 6-position in flavonols appears to be a common feature in *C. viscidiflorus*: from *C. viscidiflorus* (whole plant), Urbatsch et al. (1975) isolated six *O*-methylated flavonols derived from 6-hydroxykaempferol, quercetagetin (6-hydroxyquercetin) and 8-methoxyquercetagetin (Figure 55a). In addition to 6-hydroxy-flavonols, Stevens et al. (1999) characterized *C. viscidiflorus* by a flavanone pattern containing methyl ethers of naringenin, eriodictyol and taxifolin-3-acetate (Figure 55b). For the non 6-OH substituted flavonols, quercetin and its methyl ethers were found present whereas kaempferol and its methylated derivatives were absent.

The leaves of *Chrysothamnus nauseosus* were characterized by flavone (essentially) and flavonol pattern combining methyl ethers of apigenin, isoscutellarein, luteolin (Figure 56) and kaempferol (Figure 12) (Stevens et al., 1999).

The leaves of *Chrosothamnus humilis* showed a simpler flavonoid pattern based on naringenin (Figure 56) and methyl ethers of kaempferol and quercetin (Figure 12) (Stevens et al., 1999).

In *Taraxacum mongolicum*, the flavanone glycoside hesperidin and 3',5,7-trihydroxy-4'-methoxyflavanone, as well as three isoetin derivatives (flavanones) were isolated for the first time from the genus *Taraxacum*, so they may be useful as chemotaxonomic markers for this species (Shi et al., 2008) (Figure 57).

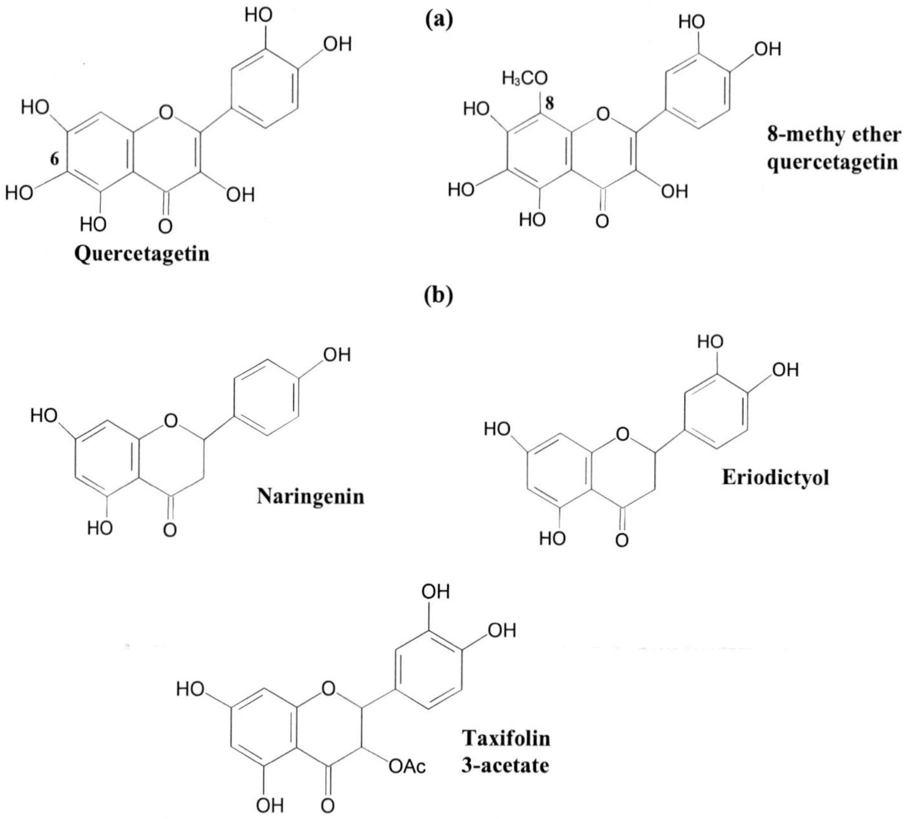

Figure 55. Chemical structures of 6-hydroxy flavonols (a) and flavanones (b) characterizing the flavonoid pattern of *Chrysothamnus viscidiflorus* (Asteraceae). The compounds are frequently found in methoxylated forms (Stevens et al., 1999). Occurrence of compounds in other plant taxa is mentioned in the text.

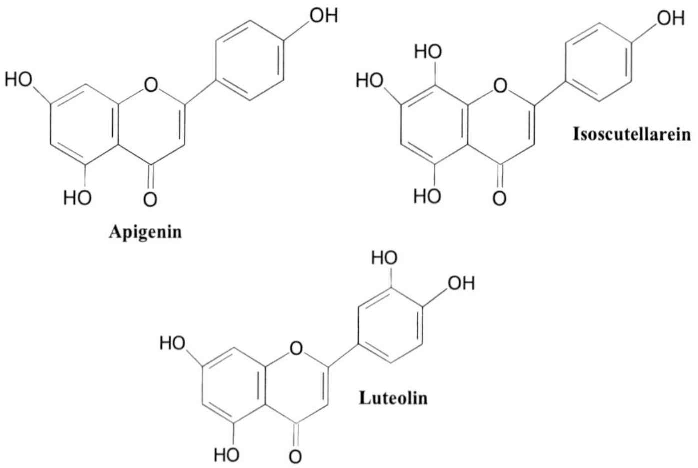

Figure 56. Chemical structures of some flavones characterizing the flavonoid pattern of *Chrysothamnus nauseosus*. Occurrence of chemical structures in other plant taxa is mentioned in the text.

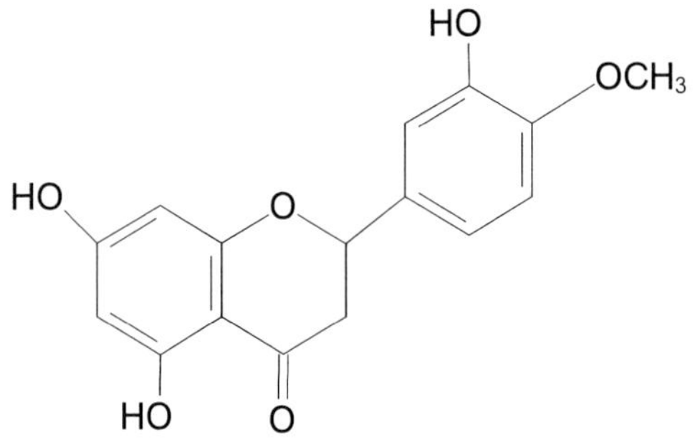

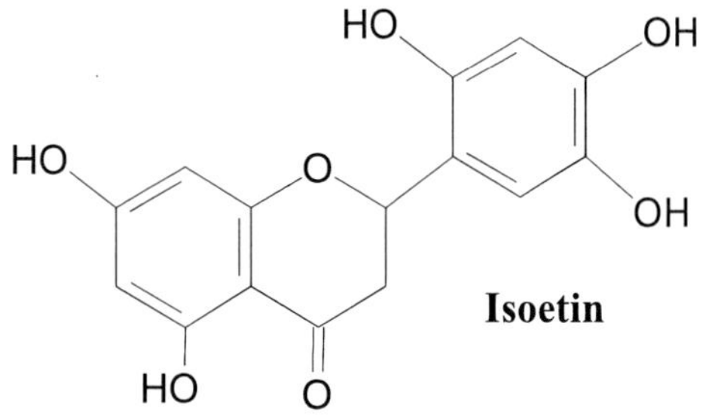

3',5,7-trihydroxy-4'-methoxyflavanone

Figure 57. Chemical structures of a 5,7,3'-trihydroxy,4'-methoxyflavanone and of isoetin (a pentahydroxyflavanone) characterizing the flavonoid pattern of *Taraxacum mongolicum* (Shi et al., 2008).

The surface (lipophilic) flavonoids in European *Pulicaria* species provide a good example on the importance of methylation position in differentiation between different botanical species. Methylated flavonoids (Figure 58) analysed from the leaves (Figure 59) and inflorescences (Figure 60) showed differences between the species:

Among the five known Europenan *Pulicaria* species, the main constituents found in *Pulicaria dysenterica* (England, Italy, France) were: 6-hydroxykaempferol 3-7-dimethyl ether (Figure 58c) and quercetagetin 3,7,3'-trimethyl-ether (Figure 58d) (Williams et al., 2000; Williams et al., 2003).

Pulicaria vulgaris has a simple surface flavonoid pattern: the major leaf components were scutellarein 6-methyl ether (Figure 58a) and quercetagetin 3,7,3'-trimethyl ether (Figure 58d); those of inflorescence were quercetagetin 3, 7, 3'-trimethyl ether (Figure 58d) and quercetagetin 3,7,3',4'-tetramethyl ether (Figure 58e) (Williams et al., 2003).

In *Pulicaria sicula*, a complex mixture containing a series of highly methylated quercetagetin derivatives characterized the leaf and inflorescence surfaces, including the fully methylated hexamethyl ether (Figure 58i) and 3,7,3',4'-tetramethyl ether (Figure 58e). Two further fluorescent blue (in UV light) quercetagetin derivatives were detected in small amounts as a penta- and a tetramethyl ether. The leaf additionally produced a fully methylated flavone, sinensetin (6-hydroxyluteolin 5,6,7,3',4'-pentamethyl ether) (Figure 58h), not detected in the other European species (Williams et al., 2003).

The main leaf surface flavonoid of *Pulicaria paludosa* was scutellarein 6-methyl ether (Figure 58a) with traces of other 6-hydroxyflavones, including 6-hydroxyluteolin 7-methyl ether (Figure 58b) and quecetagetin 3,5,7,3'-tetramethyl ether (Figure 58f) (Williams et al., 2003). By contrast, the major inflorescence flavonoid was a flavonol, quercetagetin 3,7,3'-trimethyl ether (Figure 58d).

Pulicaria odora differed from all the other European species in producing a series of highly methylated 6-hydroxykaempferol derivatives as the major surface flavonoids of both leaf and inflorescence, including the fully methylated 3, 5, 6, 7, 4'-pentamethyl ether (Figure 58g) (Williams et al., 2003). Only traces of quercetagetin derivatives (Figure 62) were detected in *P. odora* and there was also evidence of two minor flavanonols (or 3-OH-flavanones, Figure 10) in the exudates.

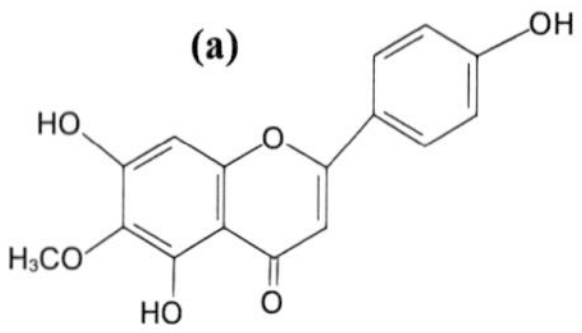

(a) Scutellarein 6-methyl ether

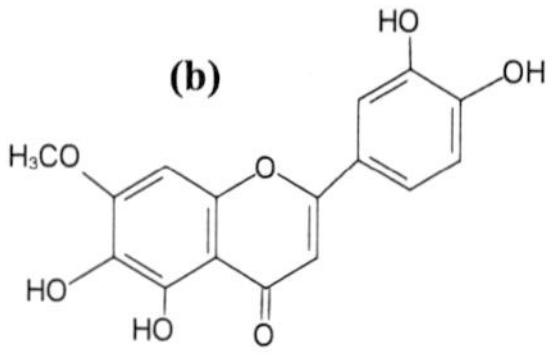

(b) 6-Hydroxyluteolin 7-methyl ether

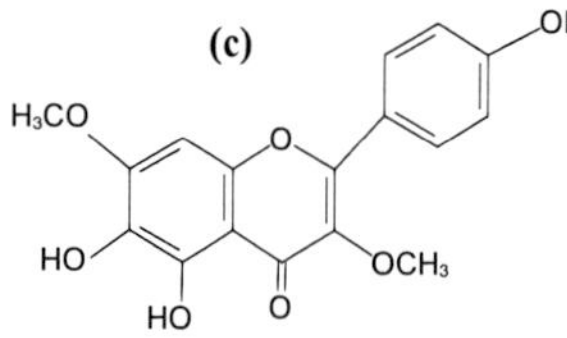

(c) 6-hydroxykaempferol 3,7-dimethyl ether

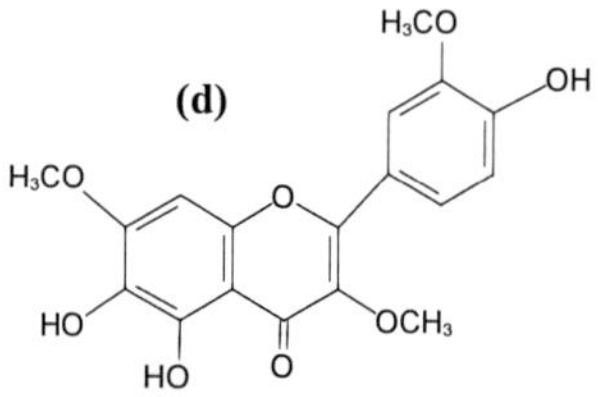

(d) Quercetagetin 3,7,3'-trimethyl ether

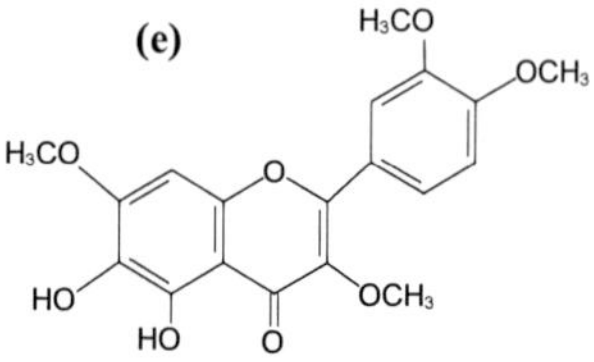

(e) Quercetagetin 3,7,3',4'-tetramethyl ether

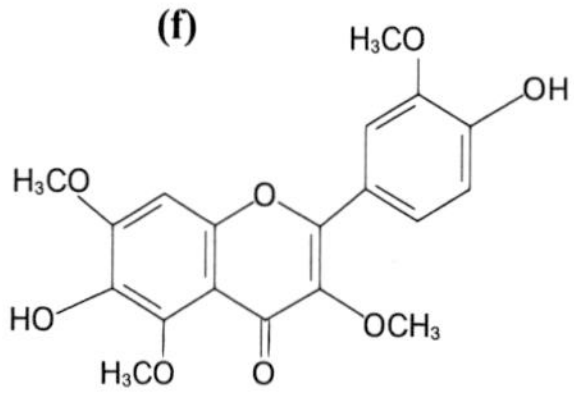

(f) Quercetagetin 3,5,7,3'-tetramethyl ether

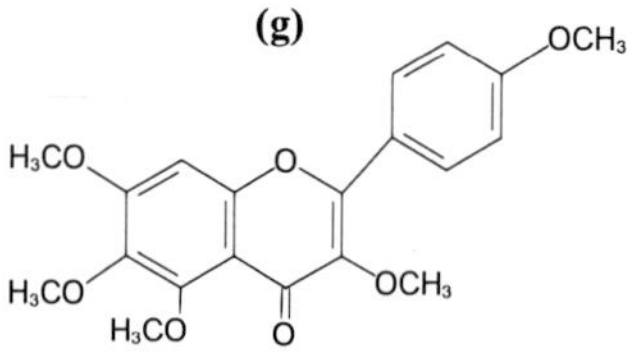

(g) 6-Hydroxykaempferol 3,5,6,7,4'-pentamethyl ether

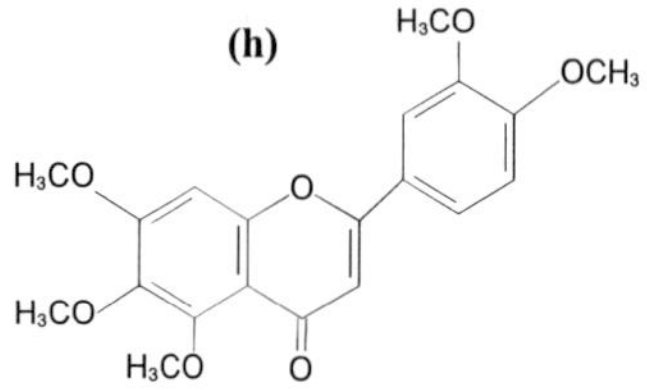

(h) 6-Hydroxyluteolin 5,6,7,3',4'-pentamethyl ether (i.e. Sinensetin)

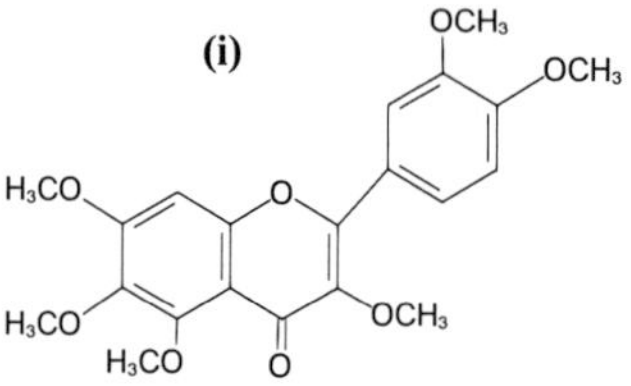

(i) Quercetagetin 3,5,6,7,3',4'-hexamethyl ether

Figure 58. Different methoxylated derivatives of flavones (a, b, h) and flavonols (c-g, i) produced by the leaves and inflorescences of different European *Pulicaria* species (Asteraceae, tribe Inuleae) (Williams et al., 2003). Occurences of compounds in other taxa are mentioned in the text.

These different surface flavonoid patterns show how European *Pulicaria* species can be finely distinguished chemically by considering specific and/or major/minor compounds, as well as by combining the leaf and inflorescence patterns of each species (Figure 59, 60). For example, *P. sicula* is characterised by more major compounds (Figure 60) than the other species. It was alone to produce compound (i) (Figure 58i) (Figure 59, 60). This was also the case for compound (g) (Figure 58g), specific to *P. odora*, but better represented on its inflorescence pattern (Figure 60) than its leaf pattern (Figure 59). Although the inflorescence flavonoid patterns (Figure 60) of *P. dysenterica* and *P. paludosa* were slightly separated by a minor compound (c) (Figure 58c) which is only present in *P. dysenterica*, their leaf patterns separate them better through different major compounds, viz. (d) and (a), respectively (Figure 58, 59). These two compounds were major together in the leaves of *P. vulgaris* (Figure 59), whereas each one excluded the other in *P. dysenterica* and *P. paludosa*.

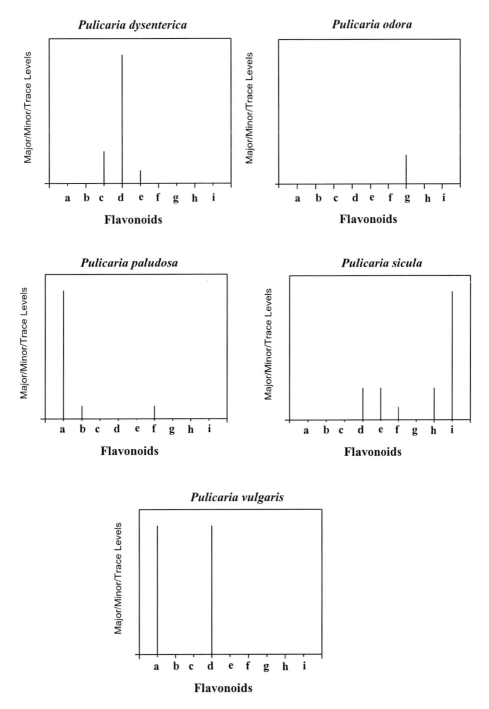

Figure 59. Profiles of methoxylated flavones and flavonols (a-i) produced by the leaves of different *Pulicaria* species (Asteraceae, tribe Inuleae). The chemical structures of flavonoids (a-i) are illustrated in figure 58 (Williams et al., 2003).

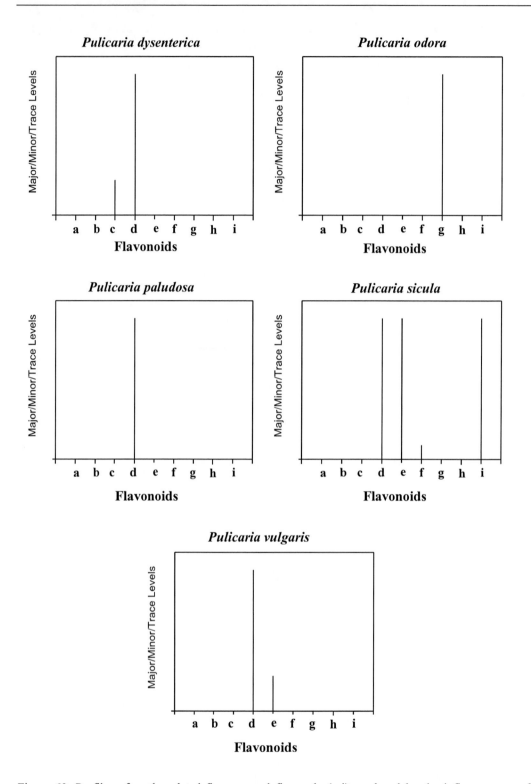

Figure 60. Profiles of methoxylated flavones and flavonols (a-i) produced by the inflorescences of different *Pulicaria* species (Asteraceae, tribe Inuleae). The chemical structures of flavonoids (a-i) are illustrated in figure 58 (Williams, et al., 2003).

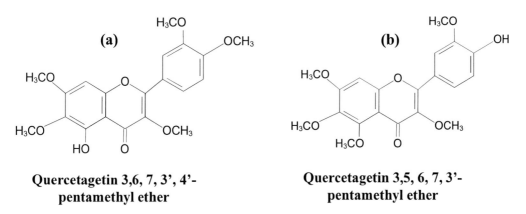

Dihydrokaempferol (a)

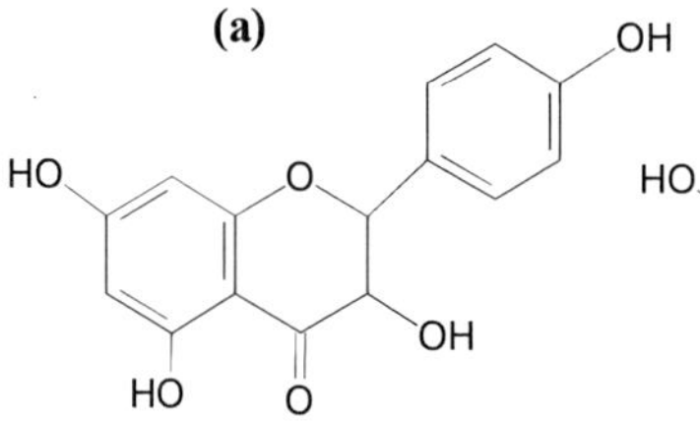

Dihydroquercetin (b)

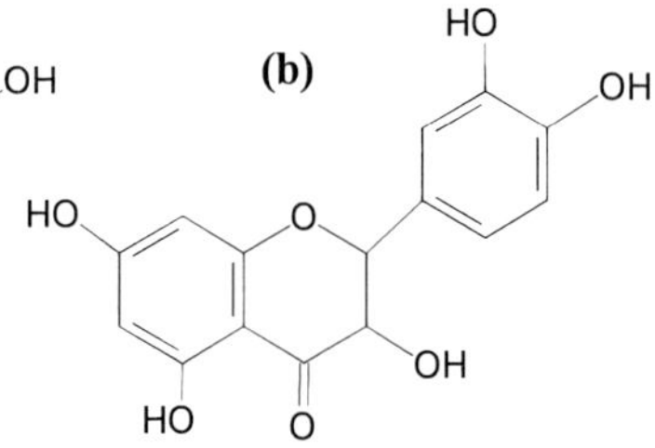

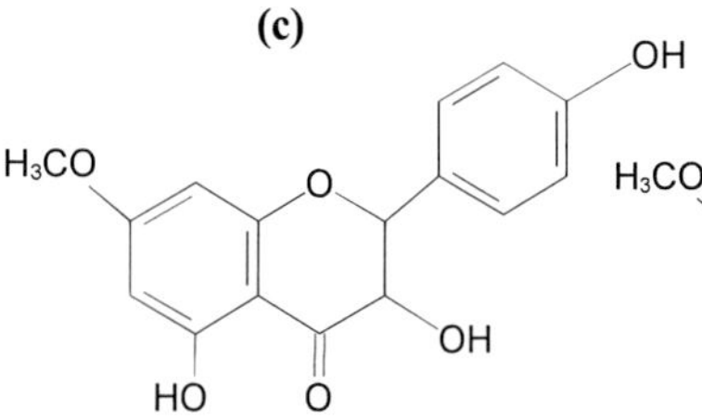

Dihydrokaempferol 7-methyl ether (c)

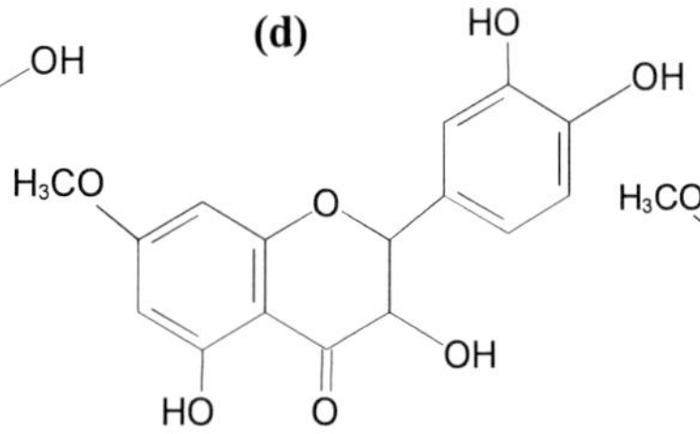

Dihydroquercetin 7-methyl ether (d)

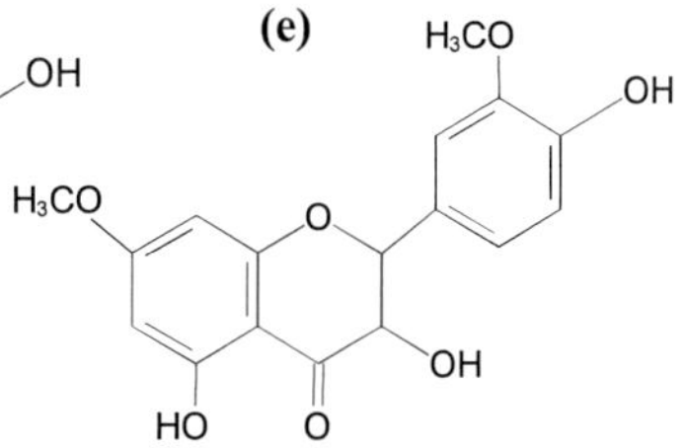

Dihydroquercetin 7,3'-dimethyl ether (e)

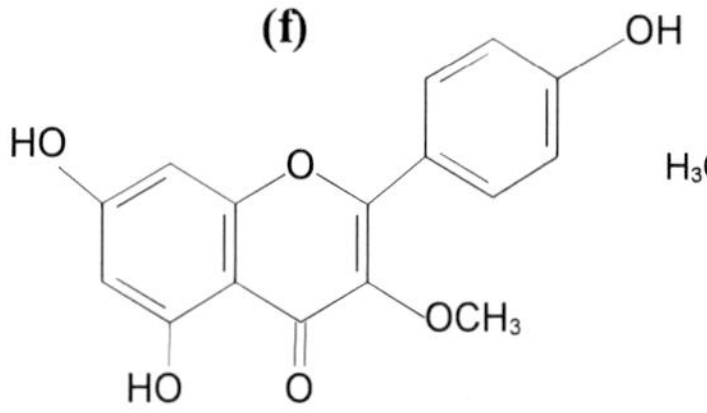

Kaempferol 3-methyl ether (f)

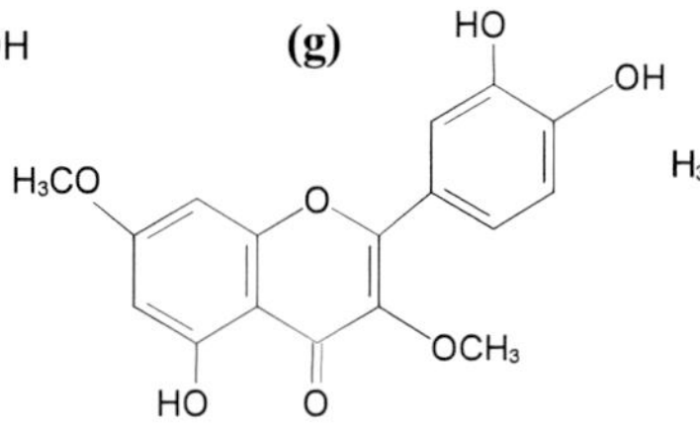

Quercetin 3,7-dimethyl ether (g)

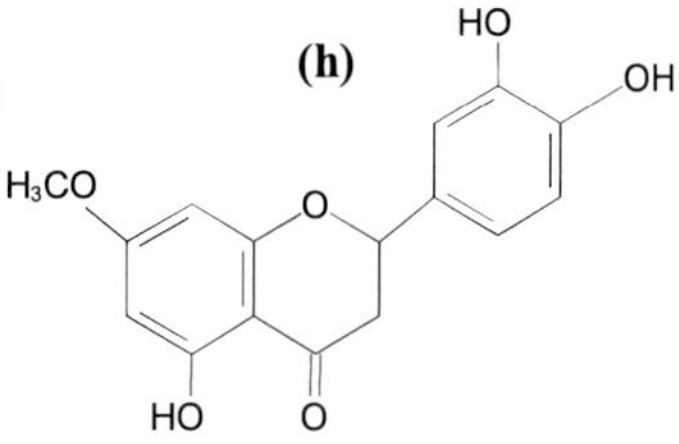

Eriodictyol 7-methyl ether (h)

Figure 61. Chemical structures of dihydroflavonols (a-b), methoxylated dihydroflavonols (c-e), flavonols (f-g) and a flavanone (h) characterizing the flavonoid pattern of *Pulicaria undulata* (Asteraceae) (Abdel-Moqib et al., 1989; Metwally et al., 1986). Occurence of compounds in other taxa is mentioned in the text.

Quercetagetin 3,6, 7, 3', 4'-pentamethyl ether (a)

Quercetagetin 3,5, 6, 7, 3'-pentamethyl ether (b)

Figure 62. Chemical structures of methoxylated derivatives of quercetagetin characterizing the flavonoid pattern of *Pulicaria arabica* (Asteraceae) (Melek et al., 1988).

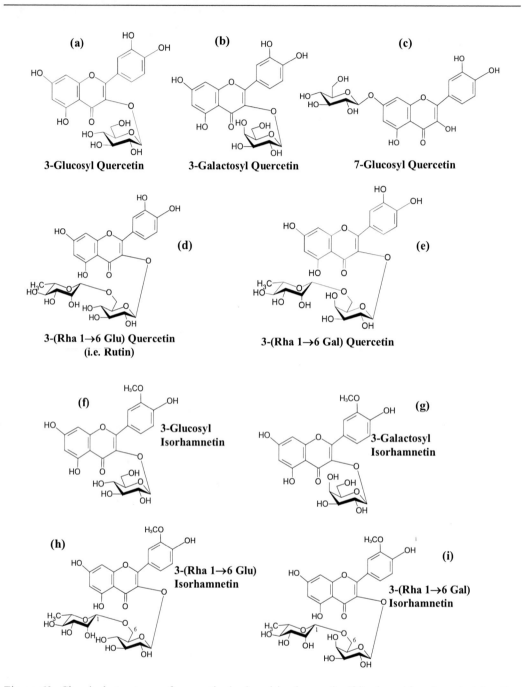

Figure 63. Chemical structures of quercetin (a-e) and isorhamnetin (f-i) glycosides characterizing the flavonol patterns of the leaves (a-e) and the inflorescence (a-i) of *Pulicaria padulosa* (Asteraceae) (Williams et al., 2003). Occurrence of compounds in other taxa is mentioned in the text.

Dihydrokaempferol, dihydroquercetin (Figure 61a-b) and their 7-methyl ethers (Figure 61c-e) have been reported from *Pulicaria undulata*, a species which has a distribution from North Africa to Saudi Arabia, The Yemen and Oman (Abdel-Moqib et al., 1989). Moreover, kaempferol 3-methyl ether and quercetin 3,7-dimethyl ether were identified (Figure 61f-g). In a previous study on *P. undulata* (Metwally et al., 1986), the flavanone: eriodictyol 7-methyl ether (Figure 61h) was also reported from the aerial parts. This report was original because the flavanones are not cited in *pulicaria* species.

From *Pulicaria arabica* (Melek et al., 1988), the highly methylated flavonols: quercetagetin 3,6,7,3',4'-penta- (Figure 62a), 3,5,6,7,3'-penta- (Figure 62b) and 3,5,6,7,3',4'-hexa- (Figure 58i) methyl ethers were isolated.

Analysis of the vacuolar (hydrophilic) flavonoids in the aerial parts of *Pulicaria* species showed both similarities and differences in the glycoside composition of the different species:

The most complex patterns were seen in the leaves and inflorescence of *Pulicaria paludosa* (Williams et al., 2003):

-   The leaves contained 3-glucoside, 3-galactoside and 7-glucoside of quercetin, as well as rutin and the corresponding 3-rhamnosylgalactoside (Figure 63a-e).
-   The inflorescence additionally contained quercetin 3-diglucuronide and isorhamnetin 3-glucoside, 3-galactoside, 3-rutinoside and 3-rhamnosylgalactoside (Figure 63f-i).

The vacuolar flavonoid pattern in the inflorescences of *Pulicaria vulgaris* differed from that of *P. dysenterica* in the fact that *P. vulgaris* produces quercetin-3-glucoside and 3-galactoside (Figure 63a, b) in addition to quercetin 3-glucuronide which was also observed in *P. dysenterica* (Figure 64) (Williams et al., 2003).

Pulicaria sicula had a similar quercetin monoglycoside profile to that of *P. paludosa*, but differed in the absence of rutin (Figure 63d), quercetin 3-rhamnosylgalactoside (Figure 63e) and isorhamnetin derivatives (Figure 63f-63i). Among the five European species, *P. sicula* was the only species to produce quercetin 7-glucuronide (Figure 65a) which co-occurred with the 7-glucoside (Figure 63c) in the inflorescence (Williams et al., 2003).

Pulicaria odora was unique by the production of the rare 7-glucoside of patuletin (quercetin 6-methyl ether) (Figure 65b) and 6-methyl ether of 6-hydroxykaempferol (Figure 65c) (Williams et al., 2003).

In *Pulicaria incisa*, the flavonoid pattern consists of kaempferol and quercetin 3-galactosides (Figure 66b, 63b), kaempferol and quercetin 3-methyl ethers (Figure 61f, 66a), quercetin 3,7-dimethyl ethers (Figure 61g) and dihydroquercetin 7-methyl ether (Figure 61d) were identified (Mansour et al., 1990).

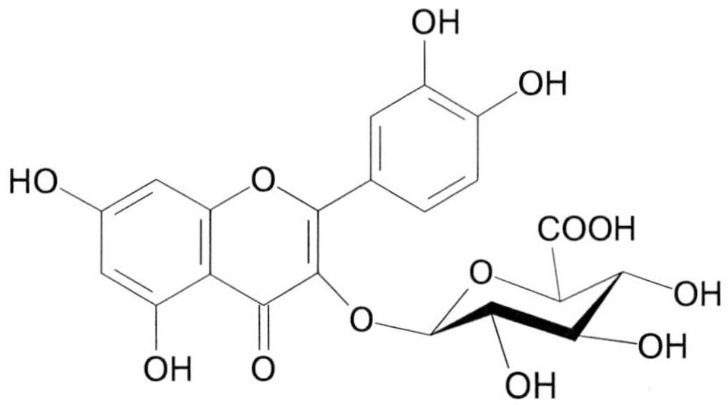

Figure 64. Chemical structure of quercetin-3-glucuronide among the characteristic glycosides of inflorescence vacuolar flavonoid patterns of *Pulicaria vulgaris* and *P. dysenterica* (Asteraceae) (Williams et al., 2003)

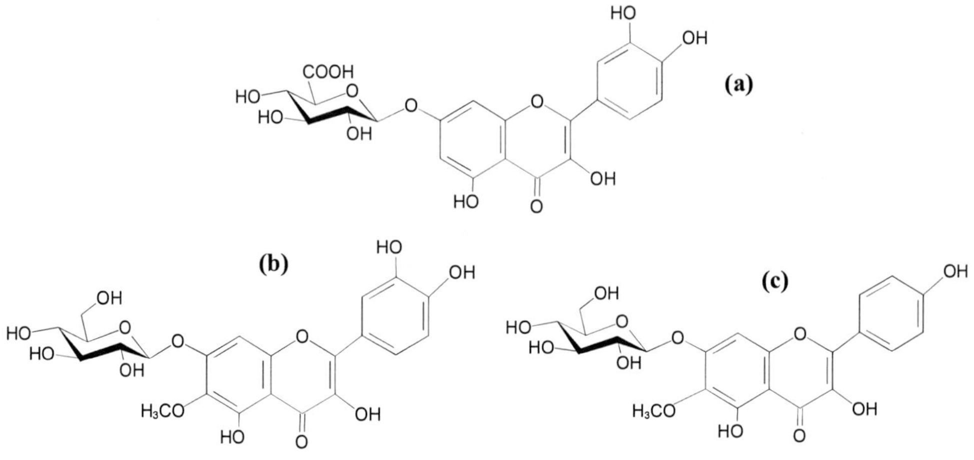

Figure 65. Chemical structures of quercetin-7-glucuronide (a), patuletin-7-glucoside (b) and 7-glucoside of the 6-methyl ether of 6-hydroxykaempferol (c) characterizing *Pulicaria sicula* (a) and *P. odora* (b, c), respectively, from other European *Pulicaria* species (Asteraceae) (Williams et al., 2003).

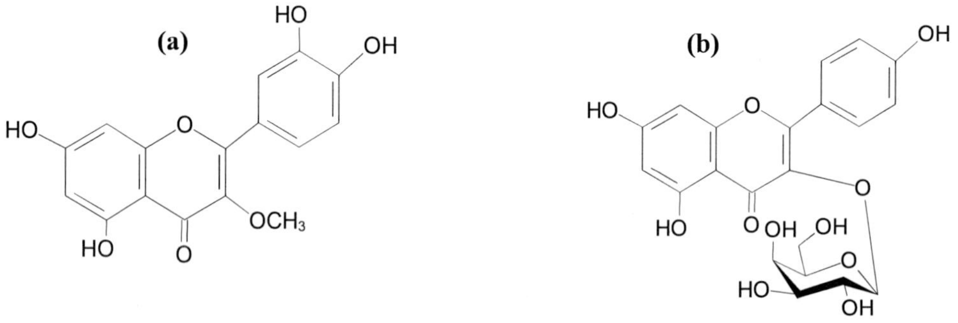

Figure 66. Quercetin 3-methyl ether (a) and kaempferol 3-galactoside (b) found in the characteristic flavonoid pattern of *Pulicaria incisa* (Asteraceae) in addition to other flavonoids (Mansour et al., 1990). Occurrence of compounds in other plant taxa is mentioned in the text.

Within the genus *Haplopappus*, the species *H. bustillosianus* has been quantitatively and qualitatively distinguished by an original surface flavonoid pattern. In contrast to the *Haplopappus* genus in which surface flavonoids are produced in only minute amounts (Morales et al., 2000; Urzúa et al., 1997, 2004; Urzúa, 2004), the flavonoids of the species *H. bustillosianus* represented more than 5% of the leaf surface compounds. Qualitatively, this species was also original by the fact that its main flavonoids were oxygenated at the position 6 of the aglycone (Figure 67), a substitution not frequently found in the genus *Haplopappus*, but frequent in the family Asteraceae. These flavonoids corresponded to 3,6,4'-trimethyl ether (Figure 67a) and 3,6-dimethyl ether (Figure 67b) of 6-hydroxykaempferol (i.e. 5,7-dihydroxy-3,6,4'-trimethoxyflavone and 5,7,4'-trihydroxy-3,6-dimethoxyflavone, respect-tively) (Urzúa et al., 2007).

Unusual 6-substituted flavone has been reported in the species *Bidens pilosa*: its aerial parts produce a *C*-6-ketopyrano-substituted flavone: the 5-*O*-methylhoslundin (Figure 68) (Sarker et al., 2000). Moreover, the occurrence of this unusual flavone in both *Bidens pilosa* (Asteraceae) and *Hoslundia opposita* (Lamiaceae) is worthy to note for chemotaxonomic and ecological studies of two very distantly related species (Sarker et al., 2000).

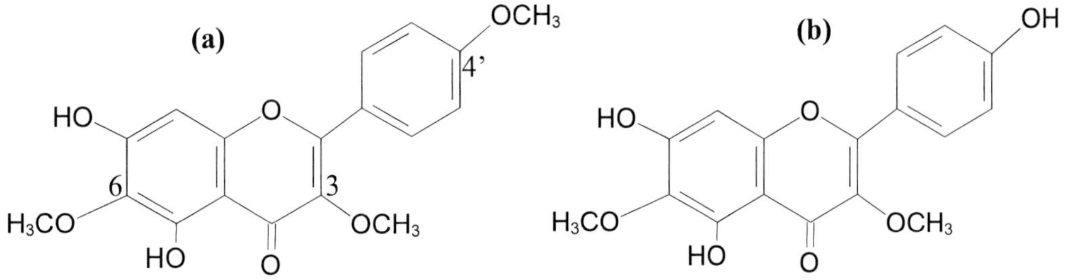

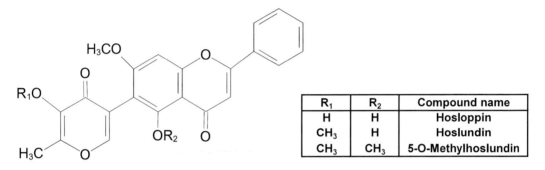

Figure 67. 6-Oxygenated and methoxylated flavonols characterizing the species *Haplopappus bustillosianus* (Asteraceae) within the genus *Haplopappus* (Urzúa et al., 2007).

R₁	R₂	Compound name
H	H	Hosloppin
CH₃	H	Hoslundin
CH₃	CH₃	5-O-Methylhoslundin

Figure 68. Chemical structures of flavones with ketopyrano-substitution. The 5-*O*-methyl hoslundin is a characteristic compound of the species *Bidens pilosa* in the Family Asteraceae.

Flamini et al. (2001a, b) reported the occurrence of flavonol sulphates in the aerial parts and roots of the species *Centaurea bracteata* Scop., an italian herbaceous plant (Figure 69). The presence of five flavonoid sulphates in the roots of this species, besides those isolated from its aerial parts, seems to be noteworthy, because such compounds are relatively rare in nature, and are generally found in aerial parts (Williams and Harborne, 1994).

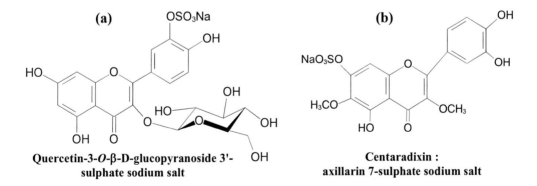

Quercetin-3-*O*-β-D-glucopyranoside 3'-sulphate sodium salt

Centaradixin :
axillarin 7-sulphate sodium salt

Figure 69. Chemical structures of some flavonoid sulphates isolated from the aerial parts (a) and roots (b) of *Centaurea bracteata* (Asteraceae) (Flamini et al., 2001a, b).

VIII.1.5. Infraspecific level chemotaxonomy

Apart from the flavonoids which are globally produced by a species leading to inter-specific chemical distinctions, other compounds can be present or major in only some individuals of a same species leading to an infraspecific chemical polymorphism. Some illustrations of chemical polymorphisms will be presented for different Asteraceae species in the next sections.

Four chemotypes I-IV were identified in *Pulicaria dysenterica*, sampled within Europe (Italy, France, England) on the basis of the surface flavonoids in aerial parts (Williams et al., 2000). The characteristic flavonoids of these chemotypes showed different degrees and positions of methylation on aglycones (kaempferol, quercetagetin) (Figure 70, 71):

chemotype I contained quercetagetin 3,7-dimethyl ether (Figure 70a); chemotype II contained 6-hydroxykaempferol-3,4'-dimethyl ether (Figure 70b); chemotype III produced 6-hydroxykaempferol-3,7-dimethyl ether (Figure 58c) with quercetagetin 3,7,3'-trimethyl ether (Figure 60) (Figure 70c); chemotype IV contained these two later compounds, as well as quercetagetin 3, 7, 3', 4'-tetramethyl ether (Figure 58e) and 6-hydroxykaempferol-3,7,4'-trimethyl ether (Figure 70d).

By contrast, the vacuolar (hydrophilic) flavonoid of all the analysed tissues (leaf, ray floret, disc floret) in all the individual plants of *Pulicaria dysenterica* was uniformly quercetin 3-glucuronide (Figure 64) (Williams et al., 2000).

Figure 70. Chemical structures of characteristic surface flavonoids in individual plants of *Pulicaria dysenterica* classified into four chemotypes I-IV (Williams et al., 2000).

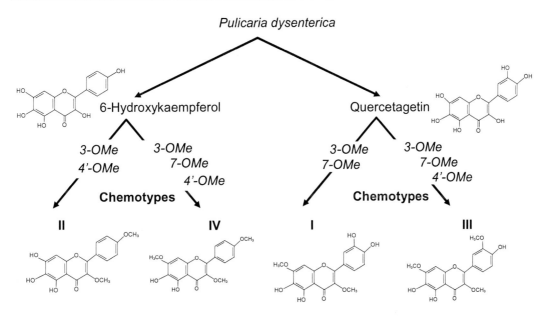

Figure 71. Differentiation between the four chemotypes of *Pulicaria dysenterica* (Asteraceae) from different methylation patterns of 6-hydroxykampferol and quercetagetin (Williams et al., 2000).

The first analyses of surface flavonoids in anthodium (inflorescence) of individual *Chamomilla recutita* plants led to the identification of two chemotypes characterized by a high or low contents of chrysosplenetin or jaceidin (methoxylated flavonols), respectively (Figure 72a, 72b) (Repčák et al., 1999).

In *Chamomilla recutita*, it has also been shown that the polyploidization results in increase of the percentage of both jaceidin (Figure 72b) and chrysosplenetin (Figure 72a) by approximately twice; this represents a significant increase in flavonoid content in comparison with an increase of other chamomile compounds (Figure 73a, b) (Repčák et al., 1999). However, the proportion of chrysosplenetin to jaceidin in the chrysosplenetin chemotype, or jaceidin to chrysosplenetin in the jaceidin chemotype was not influenced by polyploidy (Figure 73c) (Repčák et al., 1999). This provides a good example of the usefulness of metabolic ratio in chemotaxonomy.

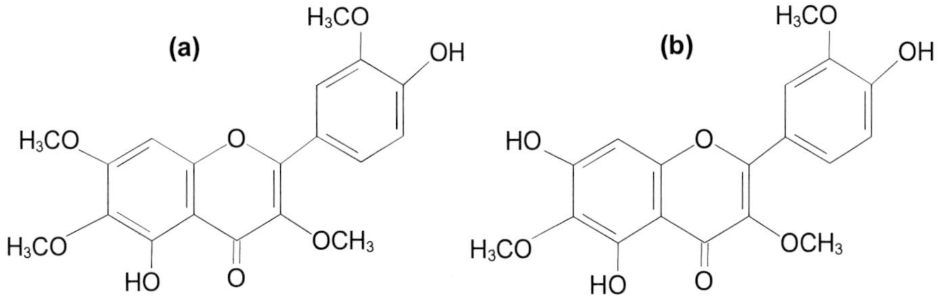

Chrysosplenetin   **Jaceidin**

Figure 72. Chemical structures of chrysosplenetin and jaceidin, two methylated quercetagetin-based flavonols characterising the two chemotypes of *Chamomilla recutita* (Asteraceae), respectively (Repčák et al., 1999).

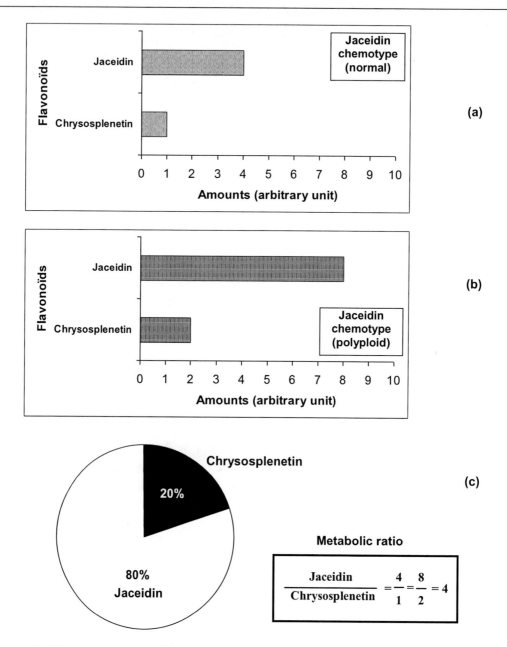

Figure 73. (Chrysosplenetin-Jaceidin) patterns of normal (a) and polyploid (b) "Jaceidin chemotype" of *Chamomilla recutita* (Asteraceae) showing the increasing effect of polyploidy on flavonol production. (c) Metabolic ratio between jaceidin and chrysosplenetin representing a stable characteristic of *C. recutita* whatever the ploidy level (Repčák et al., 1999).

The surface flavonoid aglycones of *Artemisia vulgaris* have been analysed into 40 plant populations distributed in Bulgaria (Nikolova, 2006). The leaf exudates of the studied populations accumulated methylated free aglycones based mainly on quercetin. Quercetin 3, 7, 3'-trimethyl ether appeared to be the main flavonoid aglycone (Figure 74a). Kampferol 3, 7-dimethyl ether was found in trace amounts. Quercetin, quercetin 3, 7-dimethyl ether (Figure 61g) and quercetin 3, 3'-dimethyl ether were also detected (Figure 74b). Beyond this global chemical pattern of the species *A. vulgaris*, three chemotypes were highlighted on the basis of presence/absence of highly methylated quercetagetin derivatives (Figure 74): quercetagetin 3, 6, 7, 3'-tetramethyl ether (Figure 74d) and quercetagetin 3, 6, 7, 3', 4'-pentamethyl ether (Figure 74e). Chemotype I yielded quecetagetin 3, 6, 7, 3'-tetramethyl ether (Chrysosplenetin chemotype). Chemotype II did not produce the latter compound in the leaf exudates, but quercetagetin 3, 6, 7, 3', 4'-pentamethyl ether prevailed (artemetin chemotype). In other populations, chrysosplenetin and artemetin (Figure 74d, e) were not detected yielding chemotype III (chemotype without quercetagetin derivatives). This latter chemotype appeared to be the most widespread (ca. 50% populations) in agreement with other studies on populations of *A. vulgaris* from Austria and Germany (Valant-Vetschera et al., 2003), where only chemotype III has been observed.

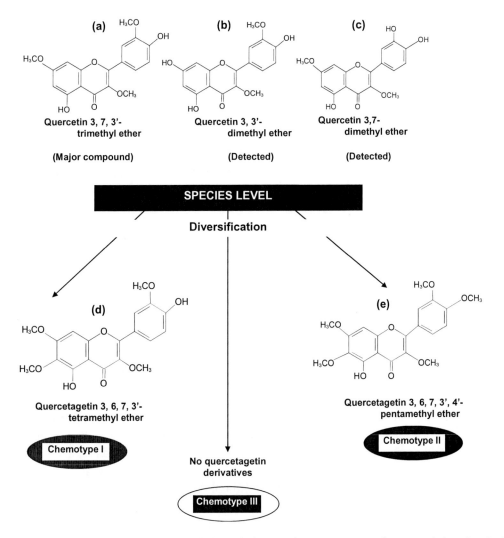

Figure 74. General chemical characteristics of the species *Artemisia vulgaris* and its chemical polymorphism based on the production or non production of methylated quercetagetin derivatives in the leaf exudates (Nikolova, 2006).

VIII.2. Lamiaceae Family

The family of Lamiaceae (Bentham, 1876) has a cosmopolitan distribution and highly varied morphological characters. According to a recent classification, the Lamiaceae consists of eight subfamilies (Cantino et al., 1992). This classification includes the former results of Wunderlich (1967), El-Gazzar and Watson (1970a) and El-Gazzar (1974) and was deduced by tacking into account 85 morphological characteristics in a cluster analysis. Leitner (1942) revealed a correlation between the number of nuclei in Lamiaceae pollen grains and number of colpi in their exines, viz. trocolpate grains are binucleate and hexacolpate are trinucleate. These palynological features led Erdtman (1945) to subdivide the family into two main subfamilies: Lamioideae with tricolpate 2-nucleate pollen grains, and Nepetoideae with hexacolpate 3-nucleate pollen grains. This two subfamily system seems to be well confirmed by significant differences in chemical patterns of the two taxa (Cantino and Sanders, 1986).

VIII.2.1. Family level chemotaxonomy

The distribution of the different external flavonoids was suitably used for chemotaxonomical analyses of Lamiaceae (Harborne et al., 1986c; Tomás-Barberán et al., 1988b, 1988c; Tomás-Lorente et al., 1988; Hermández et al., 1987; Adzet et al., 1988a, 1988b). As characteristic of the Lamiaceae, significantly more flavones than flavonols (Figure 10) have been reported (Tomás-Barberán and Wollenweber, 1990). The majority of these flavonoids were found only in a few instances resulting in chemotaxonomical values at low systematic levels (genera, species, subspecies).

The most interesting differences on the flavonoid nucelus have been found in the A-ring (Figure 9). Oxygenation on the A-ring at C-6 and/or C-6 with C-8 positions is nearly universal while single oxygenation at C-8 position is rather unusual within this family and has been described only in *Scutellaria* species (wogonin) (Figure 83e) and in *Salvia glutinosa* (herbacetin 3,8,4'-trimethyl ether) (Figure 102j) (Tomás-Barberán and Wollenweber, 1990; Tomás-Barberan et al., 1988a). A high degree of methylation of the hydroxyls (ability to produce active methyl transferases) is also observed, the hydroxyl at C-5 position generally remaining free (Tomás-Barberan and Gil, 1992; Tomás-Barberán and Wollenweber, 1990) (Figure 79, 80, 83-84, 87, 89, 95-96). For the B-ring a monosubstitution at C-4' or a disubstitution at C-3', C-4' are common in Lamiaceae (Tomás-Barberan and Wollenweber, 1990) (Figure 84, 86-91, 93, 95).

The different roles proposed for these external flavonoids in ecological biochemistry include antimicrobial, antifungal and insecticidal (or insect deterrent) activity by provision of a chemical barrier against pathogens as well as UV-screening and an adaptative mechanism to life in semi-arid or alpine habitats (Wollenweber and Dietz, 1981; Proksch and Prodriguez, 1985). The role of epicuticular layers as well as of dense mats of trichomes in reducing transpiration in desert plants from Baja California has been mentioned (Proksch and Prodriguez, 1985).

Many species of Lamiaceae contain glucosyl phenylpropanoids based generally on 3,4-disubstituted cinnamic acid derivatives (caffeic or ferulic acid) (Figure 75) (Pedersen, 2000). Rosmarinic acid (Figure 78) is also among the common constituents of the Lamiaceae (Janicsák et al., 1999).

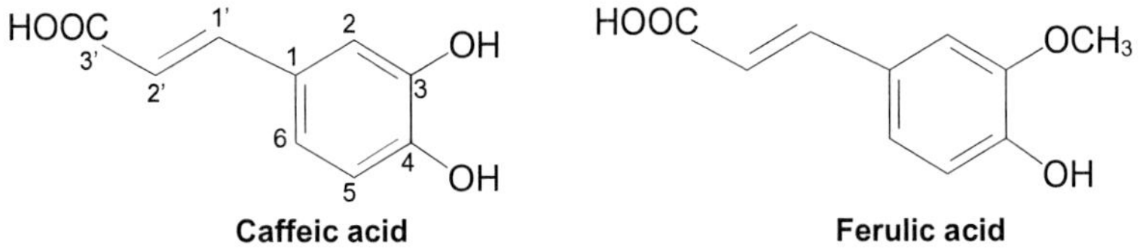

Caffeic acid   **Ferulic acid**

Figure 75. Chemical structures of caffeic acid and ferulic acid.

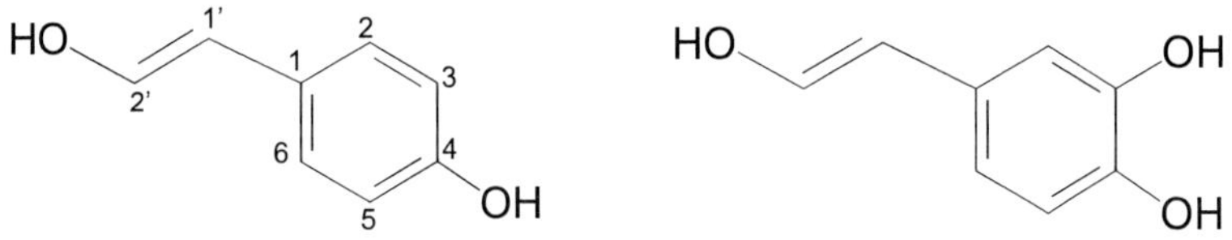

Figure 76. Basic chemical structures of phenylethanoids

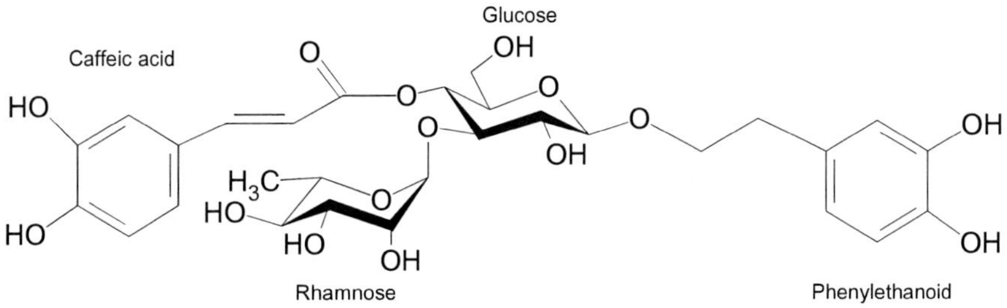

Figure 77. Chemical structure of acetoside, a common phenylethanoid glycoside in Lamiaceae.

Moreover, phenylethanoids (Figure 76) have been found to be widely distributed in many species belonging to different genera (Pedersen, 2000; Sudo et al., 1997; Hennebelle et al., 2008; Siciliano et al., 2005; Sahpaz et al., 2202).

Phenylethanoid glycosides, including acetoside (or verbascoside) (Figure 77, 82), have been isolated in relatively large quantities from several Lamiaceae species (Yalçin and Kaya, 2006; Janicsák et al., 1999; Calis et al., 1990, 1992a, b).

VIII.2.2. Subfamily/Tribe level chemotaxonomy

The family Lamiaceae is divided into two subfamilies Lamioideae and Nepetoideae, on the basis of pollen features (Erdtman, 1945); El-Gazzar and Watson, 1970b). Chemically, rosmarinic acid (Figure 78) has been shown to be an excellent chemotaxonomic marker because of its presence in the subfamily Nepetoideae and absence in the subfamily Lamioideae (Janicsák et al., 1999; Harborne, 1966c).

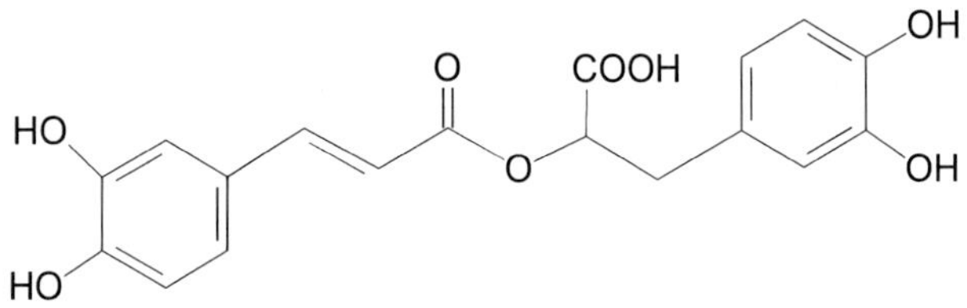

Figure 78. Chemical structure of rosmarinic acid.

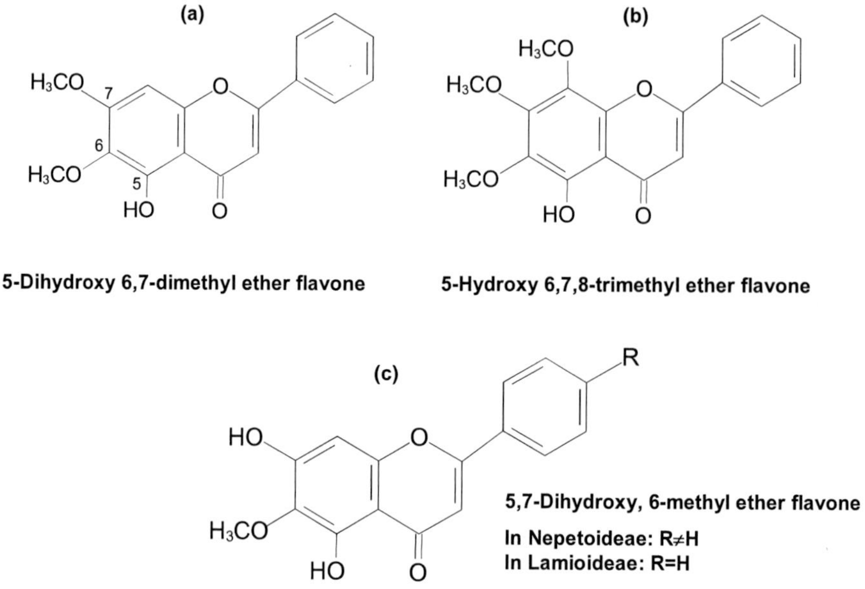

Figure 79. Common substitutions of the flavones in Lamioideae and Nepetoideae (Lamiaceae).

In addition to the presence/absence of rosmarinic acid, these two subfamilies are well distinguished by other chemical features: the Lamioideae is poor in essential oils and produces iridoid glycosides (monoterpene glycosides). By contrast, the Nepetoideae is rich in essential oils and lacks iridoids (with some exceptions) (Cantino and Sanders, 1986).

Members of these two subfamilies produce highly diversified structures of external (surface) flavonoids leading to efficient differentiation tools at different taxonomic levels (subfamily, genus, species). However, some genera are generally known by the absence of these compounds leading to another chemotaxonomic feature at the generic level (e.g. *Lavandula*, *Phlomis*, *Nepeta*, *Melissa*, etc) (Tomás-Barberán and Wollenweber, 1990). Both subfamilies produce 5-hydroxy-6,7-dimethoxy- (Figure 79a) and 5-hydroxy-6,7,8-trimethoxyflavones (Figure 79b) as a general rule, but the presence of 5,7-dihydroxy-6-methoxyflavones with a substituted B-ring (Figure 79c) is characteristic of the subfamily Nepetoideae, particularly of *Salvia*, *Rosmarinus* and *Ocimum* species (Figure 81) (Tomás-Barberán and Wollenweber, 1990); in Lamioideae, a 5,7-hydroxy 6-methyl ether flavone has been found in the genus *Scutellaria* but with unsubstituted B-ring.

In the subfamily Nepetoideae, the tribe Saturejeae can be chemotaxonomically characterized by the production of the 5,6-dihydroxy-,7,8-dimethoxyflavone (Figure 80) (Tomás-Barberán and Wollenberg, 1990). This tribe includes the genera *Thymus*, *Satureja*, *Micromeria*, *Acinos*, *Calamintha*, *Origanum* and *Mentha*.

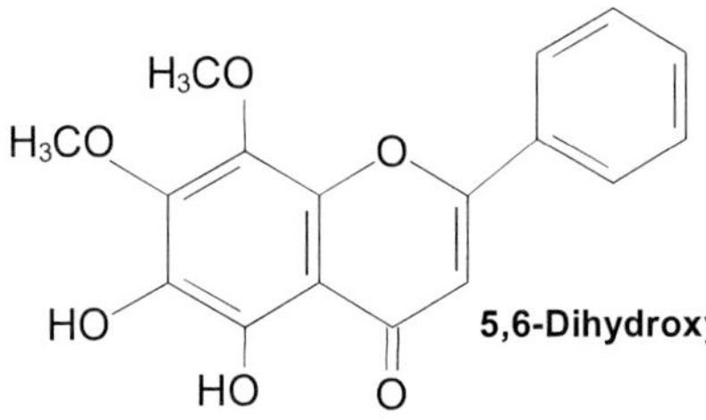

5,6-Dihydroxy 7,8-dimethyl ether flavone

Figure 80. A common substitution of the flavones in the tribe Saturejeae (Nepetoideae).

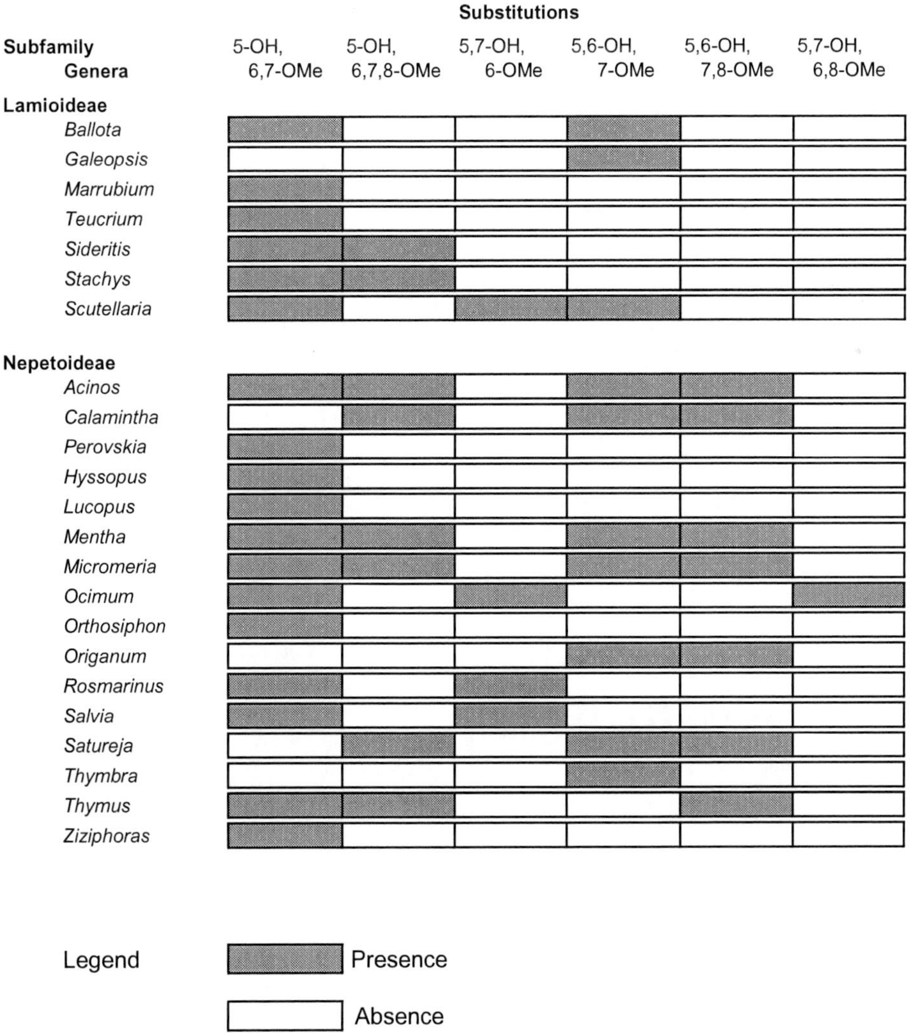

Subfamily / Genera	Substitutions 5-OH, 6,7-OMe	5-OH, 6,7,8-OMe	5,7-OH, 6-OMe	5,6-OH, 7-OMe	5,6-OH, 7,8-OMe	5,7-OH, 6,8-OMe
Lamioideae						
Ballota	Presence	Absence	Absence	Presence	Absence	Absence
Galeopsis	Absence	Absence	Absence	Presence	Absence	Absence
Marrubium	Presence	Absence	Absence	Absence	Absence	Absence
Teucrium	Presence	Absence	Absence	Absence	Absence	Absence
Sideritis	Presence	Presence	Absence	Absence	Absence	Absence
Stachys	Presence	Presence	Absence	Absence	Absence	Absence
Scutellaria	Presence	Absence	Presence	Presence	Absence	Absence
Nepetoideae						
Acinos	Presence	Presence	Absence	Presence	Presence	Absence
Calamintha	Absence	Presence	Absence	Presence	Presence	Absence
Perovskia	Presence	Absence	Absence	Absence	Absence	Absence
Hyssopus	Presence	Absence	Absence	Absence	Absence	Absence
Lucopus	Presence	Absence	Absence	Absence	Absence	Absence
Mentha	Presence	Absence	Absence	Presence	Presence	Absence
Micromeria	Presence	Absence	Absence	Presence	Presence	Absence
Ocimum	Presence	Absence	Presence	Absence	Absence	Presence
Orthosiphon	Presence	Absence	Absence	Absence	Absence	Absence
Origanum	Presence	Absence	Absence	Presence	Absence	Absence
Rosmarinus	Presence	Absence	Presence	Absence	Absence	Absence
Salvia	Presence	Absence	Presence	Absence	Absence	Absence
Satureja	Absence	Presence	Absence	Presence	Absence	Absence
Thymbra	Absence	Absence	Absence	Presence	Absence	Absence
Thymus	Presence	Presence	Absence	Absence	Presence	Absence
Ziziphoras	Presence	Absence	Absence	Absence	Absence	Absence

Legend   ▓ Presence   □ Absence

Figure 81. Presence-absence patterns of the hydroxylation and methoxylation at different positions (5, 6, 7, 8) of the A-ring of flavones in different genera of the subfamilies Lamioideae and Nepetoideae (Lamiaceae) (Tomás-Barberán and Wollenweber, 1990).

The subfamily Lamioideae seems to be more rich in the phenylethanoid glycosides (e.g. acetoside derivatives) (Figure 77; Figure 82) than Nepetoideae which is rather poor

112 Nabil Semmar, Yassine Mrabet and Muhammad Farman

(Pedersen, 2000; Tomás-Barberán and Wollenweber, 1990; Siciliano et al., 2005; Nishimura et al., 1991; Calis et al., 1990, 1992a; Ito et al., 2006).

In addition to acetoside-related phenylethanoids, coumaroylated flavone glycosides (Figure 85a-c) have been shown to be good chemotaxonomic markers for the subfamily Lamioideae (Tomas-Barberan et al., 1992).

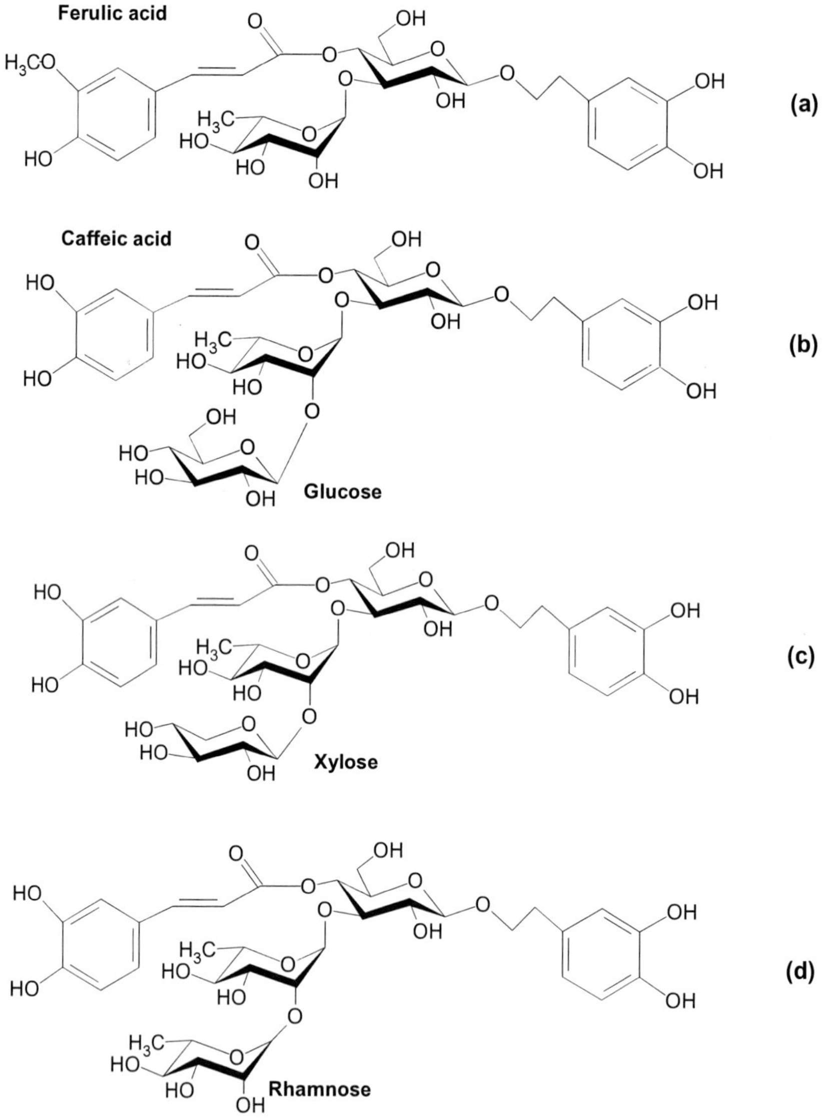

Figure 82. Chemical structures of some phenylethanoid glycosides occurring in Lamiaceae. (a) Leucosceptoside A from *Stachys macrantha*, *S. sieboldi* (Calis et al., 1992a; Nishimura et al., 1991) and *Lamium purpurenum* (Ito et al., 2006); (b), (c), (d) Phlinoside A, B, C, respectively, from *Phlomis linearis* (Calis et al., 1990).

VIII.2.3. Subtribe/Genus level chemotaxonomy

Analysis of the external flavonoids and phenylpropanoids in several Lamiaceae species showed a reliability of such compounds to differentiate between different genera belonging to the two subfamilies Lamioideae and Nepetoideae (Tomás-Barberán and Wollenweber, 1990). Moreover, it was shown that accumulation of external or surface compounds is concerned generally with species living in semi-arid or arid habitats; such chemical contents support the role of these compounds in the adaptation to such habitats. However, species growing in mesic or wet middles were not or less concerned by such accumulation.

In the subfamily Lamioideae:

- Members of the genus *Marrubium* accumulated only trace of external flavonoids.
- In the genus *Scutellaria*, the presence of flavonoid aglycones with unsubstituted B-ring (Figure 83a-f) has been reported as a characteristic feature, while in the other genera, monosubstituted and disubstituted B-rings are common (Barberán, 1986; Tomimori et al., 1985, 1986). The exudate flavonoids detected in this genus include chrysin (Figure 83a) , tectochrysin (Figure 83b), baicalein 6-methyl ether, and 7-methyl ether (Figure 83c, d), wogonin (Figure 83e) and wogonin 7- methyl ether (Figure 83f) (Yuldashev and Karimov, 2005; Miyaichi et al., 2006; Cho and Lee, 2004).
- In the the the genus *Sideritis*, di- to pentamethoxylated flavones are produced on the basis of a 5-hydroxy, 6,7-dimethyl ether flavone with a mono- or di-substituted B-ring (Figure 84a-l). According to increasing methylation levels, these flavonoids are cirsimaritin, cirsiliol, salvigenin, cirsilineol, eupatorin, xanthomicrol, sideritoflavone, gardenin-B, 5,3'-dihydroxy-6,7,8,4'-tetramethoxyflavone, 5,4'-dihydroxy-6,7,8,3'-tetramethoxyflavone, 5-hydroxy-6,7,3',4'-tetramethoxyflavone and 5-desmethyl-nobiletin (Barberán et al., 1985a; Tomás-Lorente et al., 1988; Tomás-Barberan et al., 1988c).
- In the genus *Teucrium*, the presence of cirsimaritin (Figure 84a), cirsiliol (Figure 84b), salvigenin (Figure 84c), cirsilineol (Figure 84d) and 5-hydroxy-6,7,3',4'-tetramethoxyflavone (Figure 84k) was widely reported (Harborne et al., 1986c). These compounds are externally deposited and are especially abundant in the species growing in xeric habitats of southern Spain (Tomás-Barberán and Wollenweber, 1990). From their flavonoid patterns, *Teucrium* spp. are clearly different from *Sideritis* as they produce in the A-ring only 5-hydroxy-6,7-dimethoxyflavones (Figure 84a-e), whereas *Sideritis* produce in addition flavonoids methoxylated at C-8 (5-hydroxy-6,7,8-trimethoxyflavones) (Figure 84f-j, 84l).
- In the genus *Ballota*, flavonoid *p*-coumaroyl glucosides (Figure 85a-c) and polymethoxylated flavonoids are generally considered as valuable chemotaxonomic markers, but interspecific variability has been reported (Hennebelle et al., 2008; Siciliano et al., 2005; Sahpaz et al., 2002; Agrawal; 1989; El-Ansari et al., 1995; Tomás-Barberán and Wollenberg, 1990). This genus consists of about 33 species growing mainly in the Mediterranean regions.

Different genera of Nepetoideae have been characterized by original flavonoid patterns:

- The presence of external (surface) flavonoids is nearly universal in the genus *Salvia* (Figure 102). The presence of salvigenin (Figure 83c) and cirsimaritin (Figure 83a) has been reported in several *Salvia* species (Wollenweber and Dietz, 1981). In addition, hispidulin (Figure 86a) nepetin (Figure 86b), jaceosidin (Figure 86c), genkwanin (Figure 86d) and luteolin methyl ethers were detected.

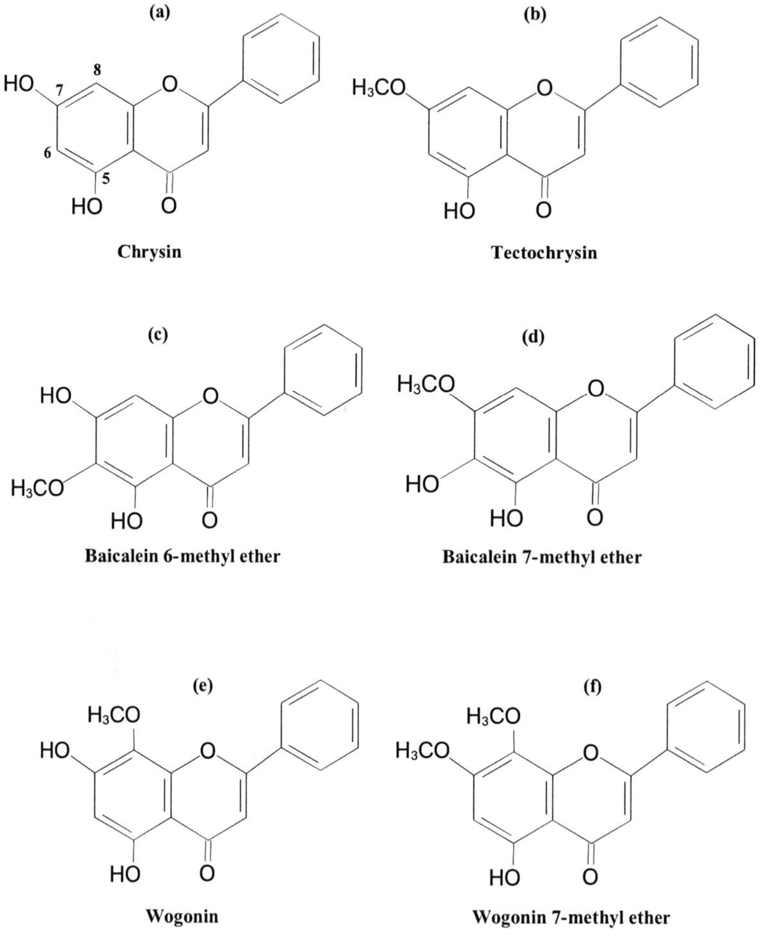

Figure 83. Chemical structures of unsubstituted B-ring flavones characterizing the genus *Scutellaria* (Lamiaceae) within the subtribe Lamioideae, and occurring in different genera of the subfamily Nepetoideae. Occurrence of compounds in other plant taxa is mentioned in the text.

- Tomás-Barberán and Wollenweber (1990) working on two species of *Rosmarinus* (*R. officinalis* and *R. eriocalix* Jordan and Fourr.) showed similar flavonoid patterns to that of the genus *Salvia*, characterized by the occurrence of salvigenin (Figure 84c), cirsimaritin (Figure 84a), cirsiliol (Figure 84b), hispidulin (Figure 86a), nepetin (Figure 86b) and jaceosidin (Figure 86c), and methyl ethers of luteolin and apigenin (genkwanin) (Figure 86d).
- Despite the chemical similarities between *Ocimum*, *Salvia* and *Rosmarinus*, the genus *Ocimum* seems to be the only one to produce external flavonoids with a 5,7-

dihydroxy-6,8-dimethoxy pattern (nevadensin) (Figure 87) (Figure 81) (Tomás-Barberán and Wollenberg, 1990).

- These authors reported from different *Perovskia* species, the occurrence of different exudate flavonoids including cirsimaritin (Figure 84a), salivigenin (Figure 84c), genkwanin (Figure 86d), apigenin-7,4'-dimethyl ether (Figure 88a) and traces of luteolin 7-methyl ether (Figure 88b).

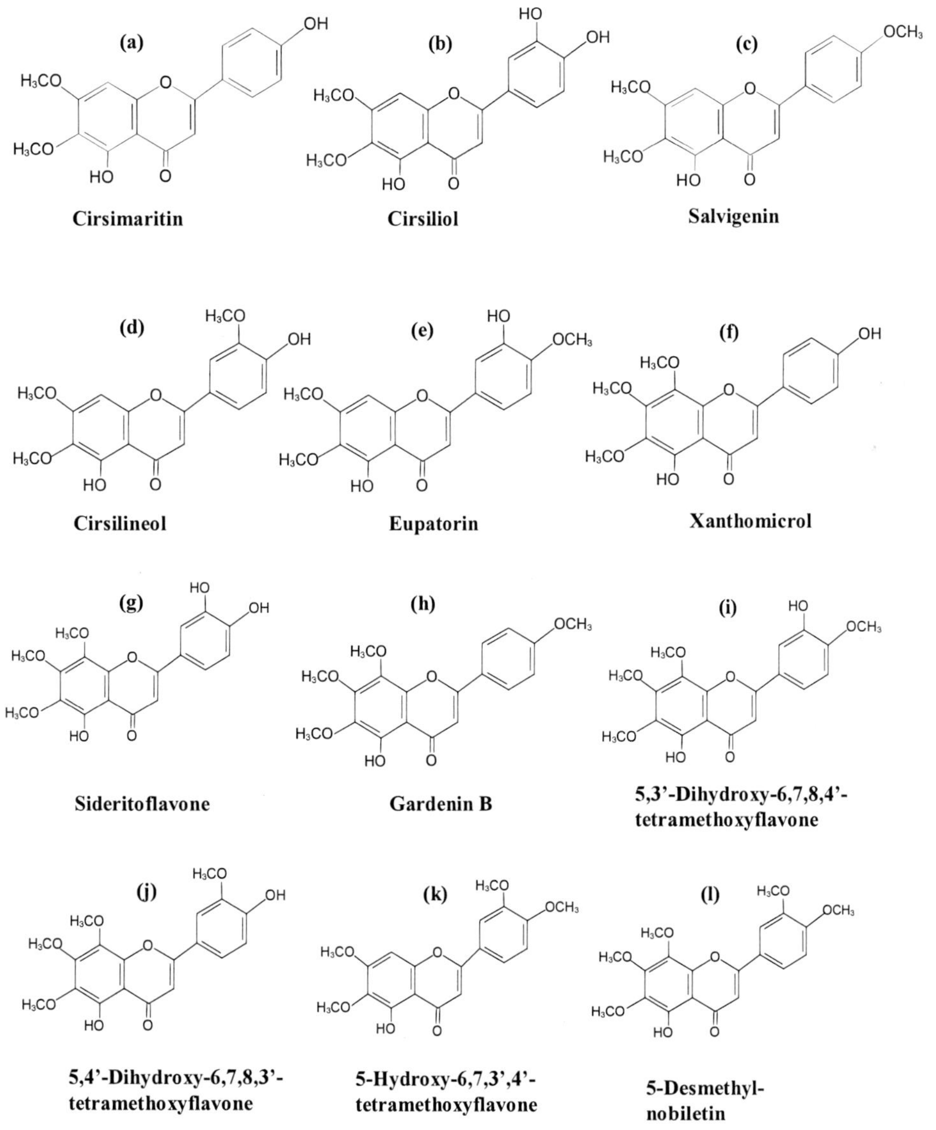

Figure 84. Methoxylated flavones based on a 5-hydroxy, 6,7-dimethyl ether structure with a mono- or disubstituted B-ring, occurring in different genera of Lamiaceae.

- Within the genus *Thymus*, several phytochemical works showed that the free flavone aglycones are the most usefull taxonomic markers among all the phenolic compounds (Adzet and Martinez, 1981; Van den Broucke, 1982; Tomas-Barberan et al., 1985; Ferreres et al., 1985; Voirin et al., 1985; Hernandez et al., 1987; Adzet et al., 1988). This genus *Thymus* is rich in external flavonoids, especially the species growing in semi-arid habitats (sections: *Pseudothymbra*, *Thumus*, *Piperella* and *Mastichina*) (Hermández et al., 1987). The external flavonoids of these sections are quite different in structures, the most remarkable feature being the presence of 5,6-dihydroxy-7,8-dimethoxyflavonoids (thymonin, pebrellin and thymusin) (Figure 80, 89a-c) and the absence of 5,6-dihydroxy-7-methoxyflavones (Figure 83d), which are normally co-occuring with the former in related genera (Figure 81).

(a)

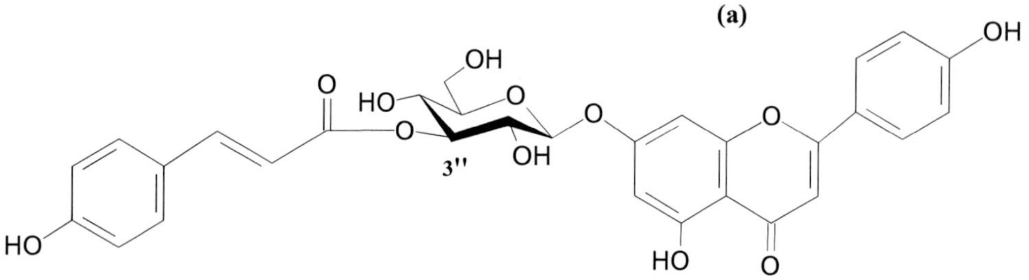

Apigenin-7-β-D-*O*-(3''-*O*-*p*-coumaroyl) glucopyranoside (in *Ballota larendana*)

(b)

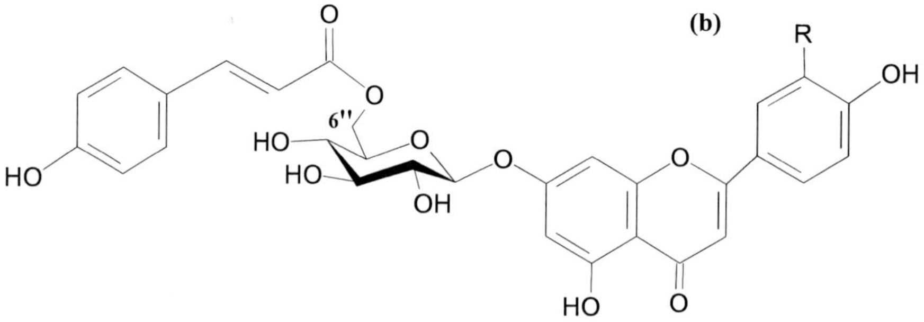

R=H: Apigenin-7-β-D-*O*-(6''-*O*-*p*-coumaroyl)glucopyranoside   (in *Ballota larendana*)

R=CH₃: Chrysoeriol-7-β-D-*O*-(6''-*O*-*p*-coumaroyl)glucopyranoside   (in *Ballota undulata*)

(c)

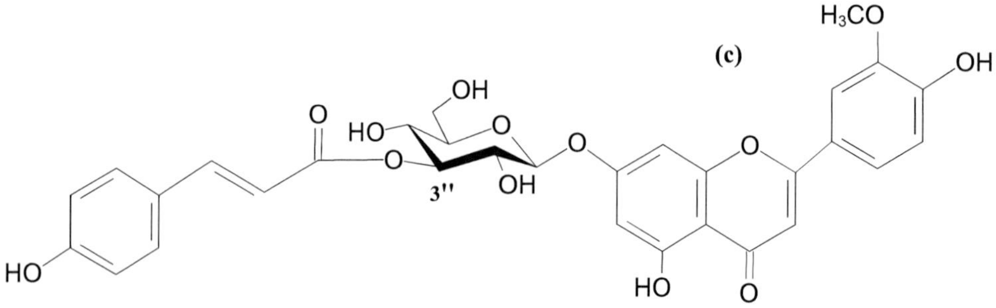

Chrysoeriol-7-β-D-*O*-(3''-*O*-*p*-coumaroyl)glucopyranoside

(in *Ballota pseudodictamnus*, *B. acetabulosa*)

Figure 85. Chemical structures of some flavone *p*-coumaroyl glucosides characterizing species of *Ballota* species (Hennebelle et al., 2008; Tiziana et al., 2005; Sahpaz et al., 2002).

In the subfamily Nepetoideae, several *Thymus*-related genera externally accumulate the 5,6-dihydroxy-7-methoxy (Figure 83d) and 5,6-dihydroxy-7,8-dimethoxyflavones (Figure 80, 89a-c). These genera are *Acinos, Micromenia, Satureja, Calamintha, Origanum* and *Mentha* (Tomás-Barberán et al., 1988b; Tomás-Barberán and Wollenweber , 1990), but in these cases both types of compounds are generally produced together (Figure 81):

-   In the genus *Acinos*, 8-methoxycirsilineol (sideritoflavone) (Figure 84g) and cirsilineol (Figure 84d) were present similarly.
-   In the genus *Calamintha*, the studied species also accumulated sideritoflavone (Figure 84g) and 5-desmethyl nobiletin (Figure 84l) externally.
-   In *Micromeria* exudates, sideritoflavone (Figure 84g) and 5-desmethyl nobiletin (Figure 84l) occur with 5-hydroxy-6,7,3',4'-tetramethoxyflavone (Figure 84k).

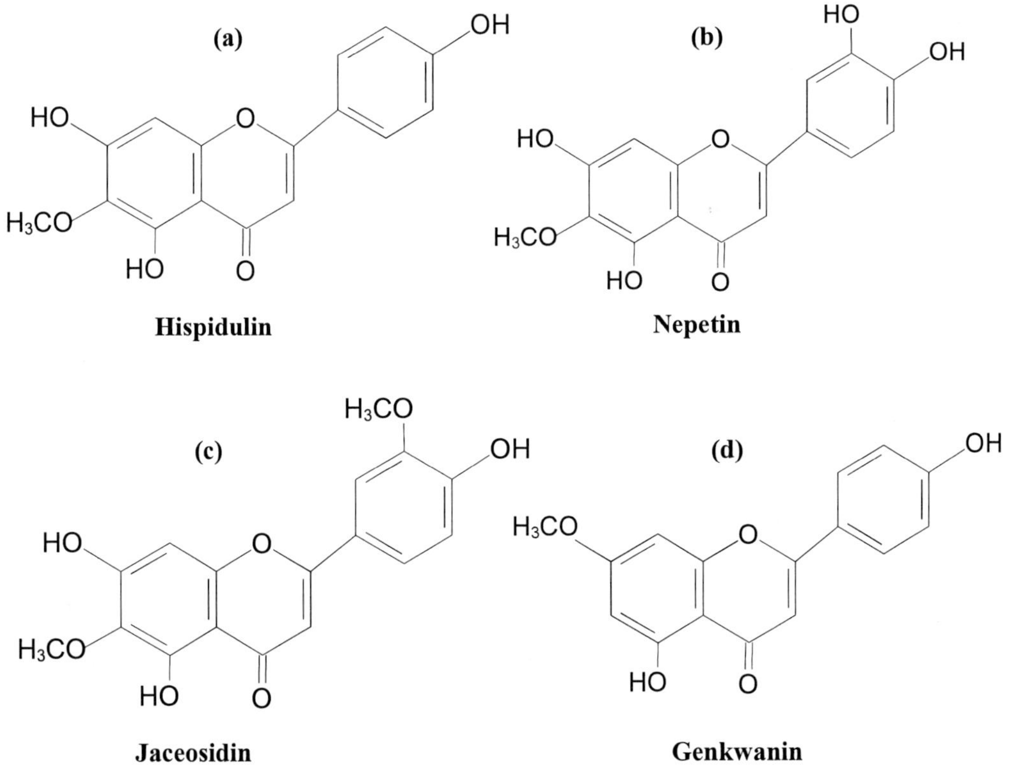

Figure 86. Chemical structures of some methoxylated flavones occurring in different genera (*Rosmarinus, Salvia, Ocimum*) of the tribe Nepetoideae (Lamiaceae). Occurrence of chemical structures in other plant taxa is mentioned in the text.

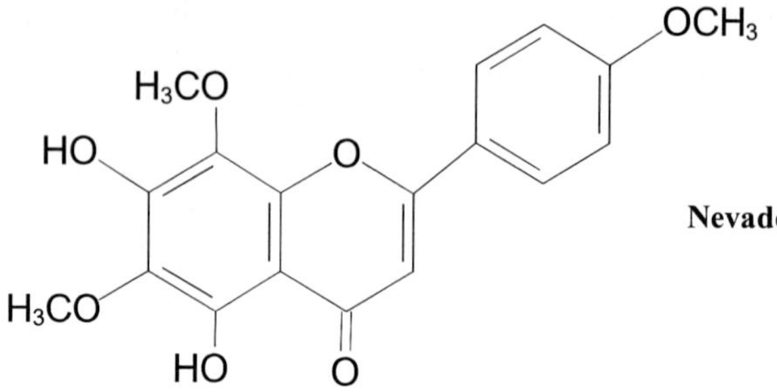

Figure 87. Chemical structure of nevadensin, a 5,7-dihydroxy-6,8-dimethoxy flavone characterizing the genus *Ocimum* from the genera *Salvia* and *Rosmarinus* (tribe Nepetoideae) in presence of several shared methoxylated flavones.

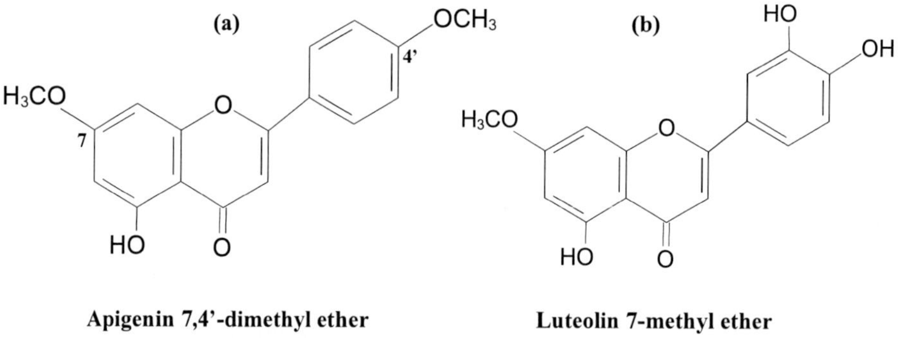

Figure 88. Chemical structures of some methoxylated apigenin and luteolin (flavones) reported as exudates flavonoids in species of the genus *Perovskia* (Lamiaceae) (Tomás-Barberán and Wollenweber, 1990). Occurrence of chemical structures in other plant taxa is mentioned in the text.

Contrary to the occurrence of surface flavonoids in several genera of the subfamily Nepetoideae, other genera were reported to be characterized by the absence of such compounds (Tomás-Barberán and Wollenweber, 1990): Species belonging to genera *Lavandula*, *Clinopodium*, *Nepeta*, *Melissa*, *Dracocephalum*, *Elsholzia*, *Glechoma*, *Horminum* and *Prunella* were generally devoid of such external compounds.

Origanum genus belongs to the subfamily Nepetoideae, tribe Mentheae. Analysis of the surface flavonoids in the leaves of nine *Origanum* species, collected from different Mediterranean countries (Greece, Egypt, Algeria) highlighted general characteristics that could be used as chemotaxonomical keys (Skoula et al., 2008) (Figure 90):

- Among the surface flavonoids of *Origanum* species, several of them showed 8-methoxy group.
- All the surface flavonoids bore a free 5-hydroxyl group.
- In the B-ring, most of compounds were 4'-methoxylated.

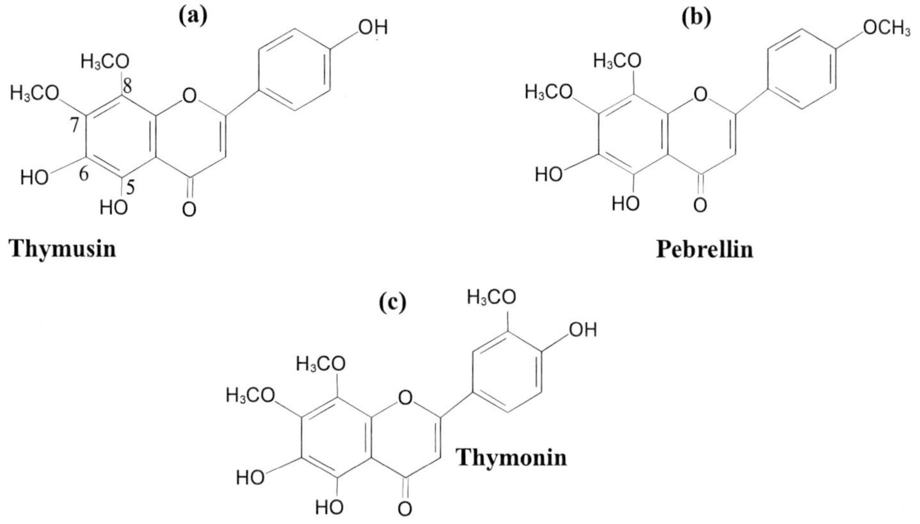

(a)

Thymusin

(b)

Pebrellin

(c)

Thymonin

Figure 89. Chemical structures of surface methoxylated flavones characterizing the genus *Thymus* and other related genera. The genus *Thymus* produces flavones particularly based on a 5,6-dihydroxy-7,8-dimethyl ether structure. Occurrence of compounds in other plant taxa is mentioned in the text.

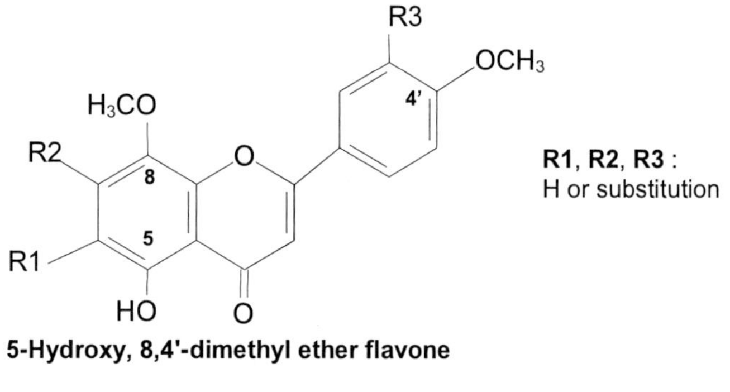

R1, R2, R3 :
H or substitution

5-Hydroxy, 8,4'-dimethyl ether flavone

Figure 90. Chemical structure of 5-hydroxy, 8,4'-dimethyl ether flavone, a surface flavonoid commonly produced by different *Origanum* species (Skoula et al., 2008).

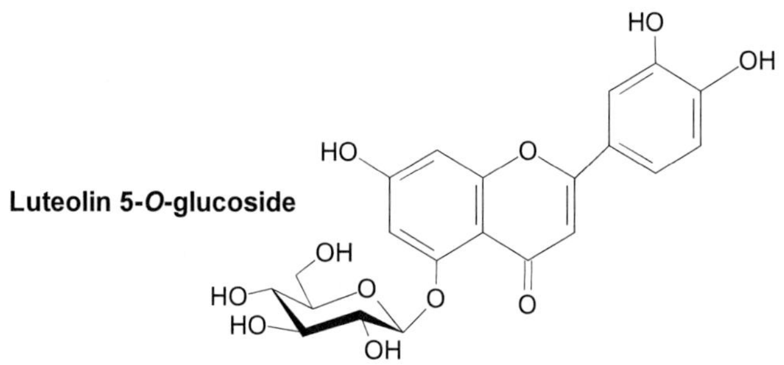

Luteolin 5-*O*-glucoside

Figure 91. Chemical structure of luteolin 5-*O*-glucoside, a compound rarely substituted at C-5 position, found in *Ocimum* genus but uncommon in Lamiaceae (Grayer et al., 2002).

Analysis of flavonoid glycosides in nine *Ocimum* species by Grayer et al. (2002) reported the occurrence of luteolin 5-*O*-glucoside (Figure 91) in all the investigated species. On this basis, the chemotaxonomical value of such flavone glucoside for the genus *Ocimum* was considered all the more since it is an uncommon constituent in the family Lamiaceae. Generally, flavone 5-*O*-glycosides are rare since 5-hydroxy group forms hydrogen bonding with the adjacent 4-carbonyl group (Iwashina et al., 1995a, 1995b).

VIII.2.4. Subgenus level chemotaxonomy

In the subfamily Lamioideae, the genus *Galeopsis* is divided into two subgenera: *Galeopsis* and *Ladanum*. Analysis of external flavonoids revealed that species belonging to subg. *Galeopsis* were devoid of such compounds (*G. tetrahit* L., *G. bifida* BOENM., *G. pubescens* BESS, *G. speciosa* MILLER), while species from subg. *Ladanum* accumulated them (*G. angustifolia* EHRH., *G. ladanum* L., *G. pyrenaica* BARTL, *G. segetum* NECKER) (Figure 92) (Tomás-Barberán and Wollenweber, 1990). Other species of the subgenus *Ladanum* showed single occurrence (minor or trace) of external flavonoids.

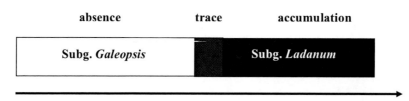

Occurrence and level of external flavonoids in the two subgenera of the genus *Galeopsis*

Figure 92. Schematic representation showing the usefulness of external flavonoids to distinguish between and within the two subgenera of the genus *Galeopsis* (Lamiaceae) on the basis of presence-absence and major-trace criteria.

The flavonoid glycosides in different species *Ocimum* were investigated taking into account their subgeneric classification. Grayer et al. (2002) analysed *Ocimum* sp. belonging to three subgenera: *Ocimum*, *Gymnocimum* and *Nautochilius*. A chemical variability was highlighted between these three subgenera on the basis of differences in hydroxylation and/or glycosylation levels and positions (Figure 94):

Among the *Ocimum* species, those belonging to subgenus *Gymnocimum* produced flavone glycosides containing glucuronic acid (Figure 93a, 93b) (Figure 94g), a sugar that was not detected in the subgenera *Ocimum* and *Nautichilus* (Figure 94a-f) (Grayer et al., 2002). Apart from the type of glycosidic part, both glycosylation position and aglycone type differed in this subgenus as compared to the two others: the species *O. tenuiflorum* representing this subgenus, accumulated flavone 7-*O*-glycosides (Figure 93a, b) rather than flavonol 3-*O*-glycosides (in the two other subgenera) (Figure 93h, j, k).

The major flavonoid glycosides in six *Ocimum* species belonging to subgenus *Ocimum* (among seven examined species) consisted of flavonol 3-*O*-glycosides (quercetin 3-*O*-glucoside and/or quercetin 3-*O*-rutinoside, i.e. rutin) (Figure 93h, 93j) (Grayer et al., 2002). In addition to these major compounds, the glycosylated flavonoid pattern of the subgenus *Ocimum* can be further identified by the occurrence of minor glycosides such as Vicenin-2

(*C*-flavone) (Figure 93l), luteolin and apigenin 5-*O* and 7-*O*-glucosides (Figure 93c-f) and kaempferol 3-*O*-glycosides (Figure 93g, 93i, 94).

The analysis of the flavonoids of *Ocimum lamiifolium* belonging to subgenus *Nautochilus*, showed that its major compound was also a flavonol 3-*O*-glycoside (i.e. quercetin 3-*O*-xylosyl(1'''→2'')galactoside) (Figure 93k). However, both the sugars and their interglycosidic linkage differed from those of the species belonging to the subgenus *Ocimum* (Grayer et al., 2002). Furthermore, *O. lamiifolium* lacked the flavone *C*-glycoside (Figure 93l), but produced larger amounts of flavone 5-*O*-glycosides (Figure 93e-f) than species belonging to subgenus *Ocimum* (Figure 94).

On the basis of morphological characters, the genus *Origanum* was divided into three subgeneric groups A, B, C (Ietswaart, 1980). By opposition to the monoterpenoids and sesquiterpenoids (Skoula et al., 1999), the distribution of surface flavonoids within the genus *Origanum* showed an agreement with the three subgeneric groups (Skoula et al., 2008):

Species in group C (*Origanum floribundum, O. vulgare* subsp. *glandulosum* and subsp. *Hirtum*) (collected from Bilda/Algeria and from Crete/Greece) lacked methoxylated flavanones which were generally substituted at positions 6 and 7 (Figure 96c-f) (Figure 97C); however, non-methylated flavanones (Figure 96a, b) were relatively more abundant in group C than the other groups (Figure 97C). Moreover, flavones with 5,6-dihydroxy,7,4'-dimethoxy substitution pattern were detected only in this group (Figure 95c, g, k) (Figure 97C) (Skoula et al., 2008).

Conversely, flavones with the 5,4'-dihydroxy,6,7-dimethoxy substitution pattern (e.g. (cirsimaritin; xanthomicrol; cirsilineol) (Figure 95d, h, j) were restricted to species in groups A and B (Figure 97A, B).

C-4'-methoxylation was absent only from group B species (*Origanum microphyllum, O. majorana, O. onites*, Greece; *O. syriacum*, Egypt) (Figure 95c, e, g, i, k) (Figure 97B). In contrast to flavone methyl ethers, flavanone methyl ethers were only detected in group B species: among the four studied species of group B, alls contained 7-methoxyflavanones while two (*O. microphyllum, O. onites*, collected from Crete, Greece) contained 6,7-dimethoxyflavanones (Figure 96c-f) (Figure 97B).

With the exception of *Origanum majorana*, species of groups B and C had generally higher ratios of flavanones (Figure 96) to flavones (Figure 95), compared to species of group A in which flavones were more abundant (Figure 97).

The lack of 6-methoxy substituted compounds in group C species suggests that flavonoid 6-*O*-methyltransferase activity is either absent or non-operative. Similarly, flavonoid 4'-*O*-methyltransferase appears to be absent or inactive in group B. In contrast, both 6- and 4'-methoxy substituted compounds are found in taxa of group A (Figure 95e, 95i), presumably due to the activity of the corresponding methyltransferases (Skoula et al., 2008).

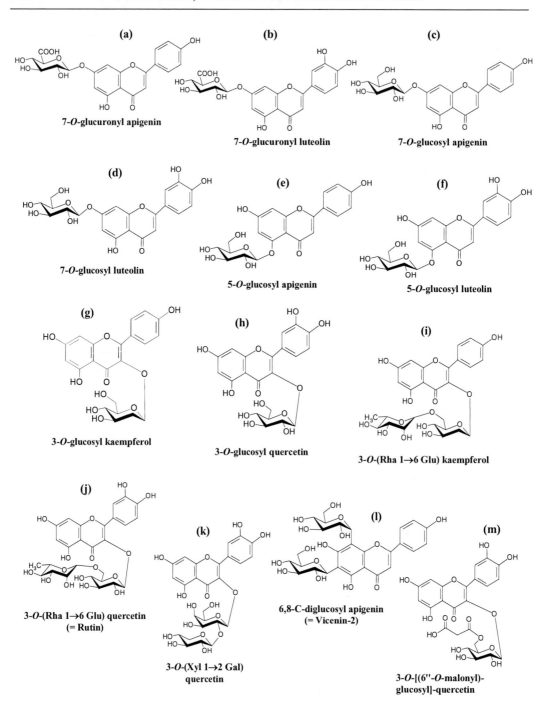

Figure 93. Chemical structures of different *O*-glycosides (a-k, m) and *C*-glycoside (l) occurring at different relative levels in species of genus *Ocimum* belonging to the three subgenera: *Ocimum*, *Gymnocimum* and *Nautochilus* (Grayer et al., 2002). Occurrence of compounds in other plant taxa is mentioned in the text.

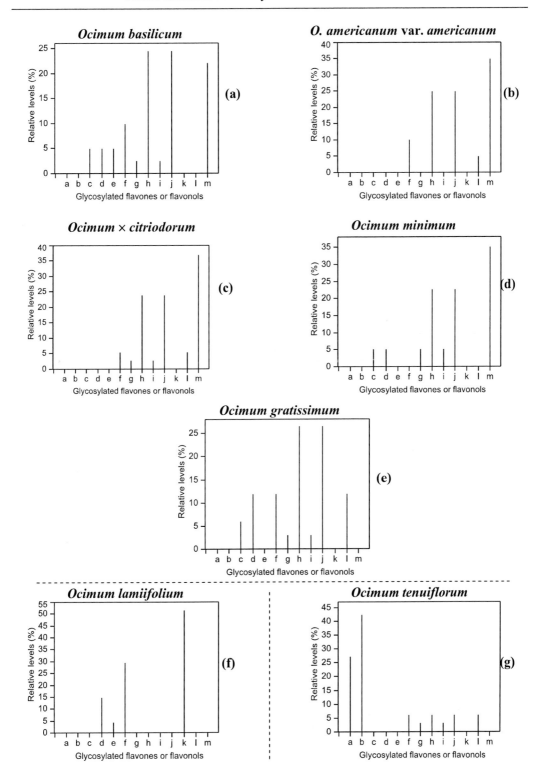

Figure 94. *O*-Glycosylated flavonoid patterns of species *Ocimum* belonging to different subgenera. (a-e) Subgenus *Ocimum*; (f) subgenus *Nautochilus*; (g) subgenus *Gymnocimum* (Grayer et al., 2002). The structures of compounds are presented in figure 93.

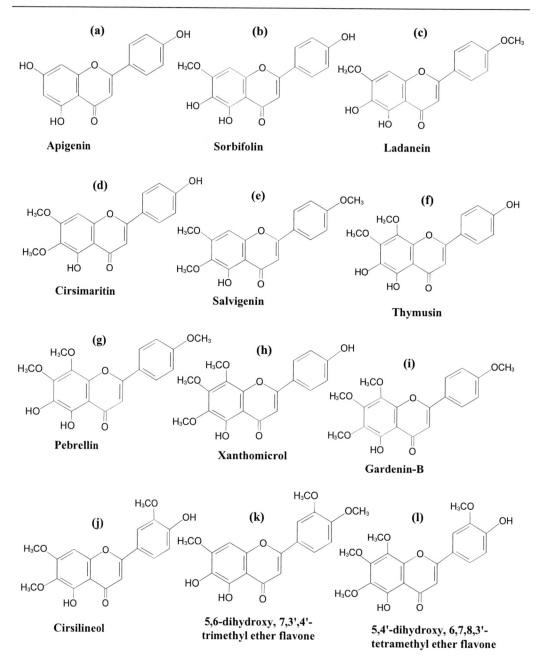

Figure 95. Chemical structures of surface flavones in different species *Origanum* (Skoula et al., 2008). Occurrence of chemical structures in other taxa is mentioned in the text.

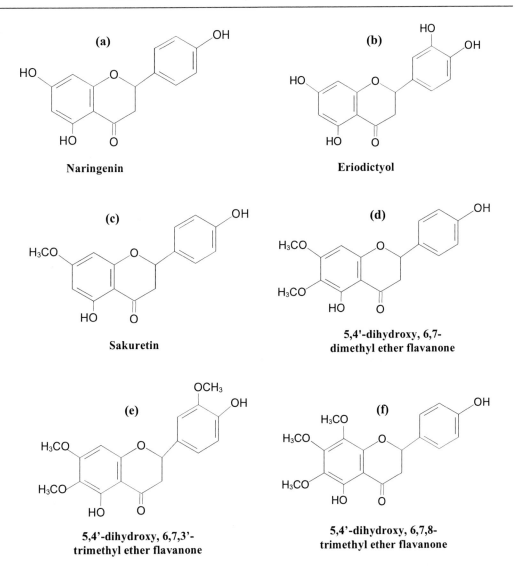

Figure 96. Chemical structures of surface flavanones in different species *Origanum* (Skoula et al., 2008).

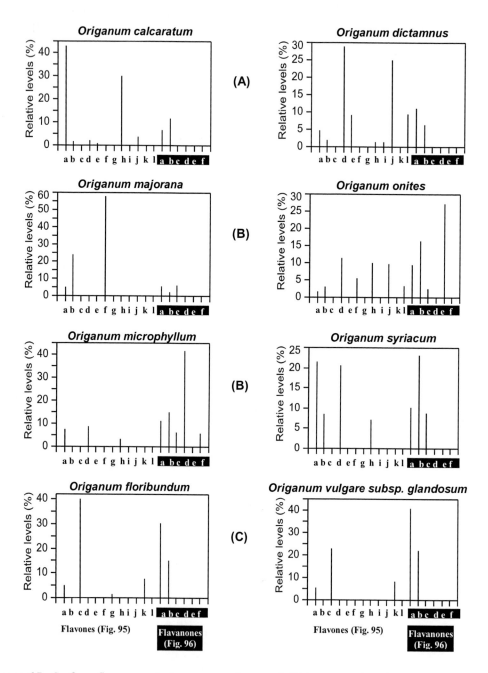

Figure 97. Surface flavone and flavanone patterns of different species *Origanum* (Lamiaceae) belonging to three different subgeneric groups A, B, C (Skoula et al. 2008). Chemical structures of compounds are presented in figures 95 and 96.

Species of the genus *Micromeria* showned a flavone glycoside pattern in agreement with the section to which they belong: acylated and non acylated glycosyl acacetin (Figure 98a, b) have been isolated from the leaves of different species of *Micromeria* (Marin et al., 2001). These two compounds however showed different relative levels which seemed to be significant at the section level (Figure 98): belonging to the botanical section *Pseudomelissa*

Benth., *Micromeria albanica*, *M. thymifolia* and *M. dalmatica* showed higher contents of both acacetin 7-rhamnosyl(1'''→6'')glucoside (Figure 98a) and 7-[6''''-O-acetylglucosyl (1''''→2'')]rhamnosyl(1'''→6'')glucoside (Figure 98b). However, *M. Juliana* and *M. cristata*, placed in section *Eumicromeria*, showed lower levels of these compounds (Figure 98). The different species were collected in Yugoslavia; many species *Micromeria* grow in open habitats in the Mediterranean area, and some are endemic to some regions of the Balkan Peninsula.

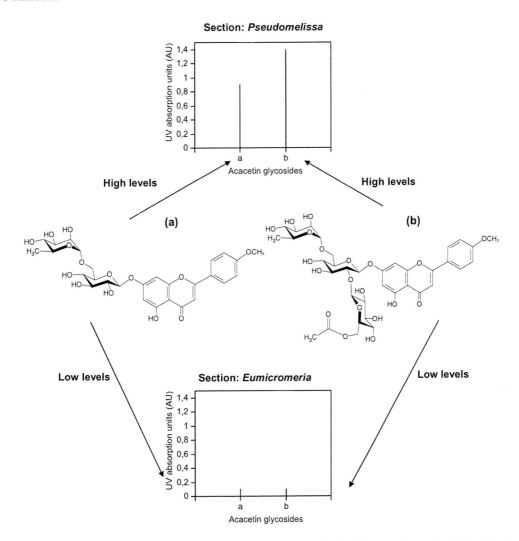

Figure 98. Chemical profiles of two acacetin glycosides occurring in *Micromeria* species belonging to two different sections, *Pseudomelissa* and *Eumicromeria* (Marin et al., 2001). The absorbances of compounds were measured on samples obtained from 5mg of dried leaves.

These flavonoid glycoside characters correlate well with results obtained from other studies, in which the two sections were distinguished by differences in the flavone aglycone profiles of their leaf surfaces (Tomás-Barberan et al., 1991) and micromorphological features (Husain et al., 1990).

VIII.2.5. Species level chemotaxonomy

Analysis of flavonoid and/or phenylpropanoid patterns (aglycones and glycosides) led to differenciate many species of a same genus in Lamiaceae. The different species were individually characterized by flavonoid features based on some specific compounds which co-occurred or excluded the ones the others. Moreover, ratios between co-occurred compounds have been shown to be good chemotaxonomical markers in some species.

In the genus *Marrubium*, the species *M. globasum ssp. Libanoticum* living in Lebanon was chemically distinguished by unusual flavanone derivative: in addition to flavonol and flavone glycosides, naringenin 7-*O*-β-D-glucoside (Figure 99) was found in this species, while it has never been documented earlier in *Marrubium* genus (Rigano et al., 2006). These authors reported a pecularity consisting of the occurrence of only glycosides and not free aglycones in *M. globasum ssp. Libanoticum*.

In the genus *Stachys*, the species *S. glutinosa* L. produces external flavonoid aglycones, which were identified as cirsiliol, sideritoflavone, cirsimaritin, cirsilineol, xanthomicrol and 5,4'-dihydroxy-6,7,8,3'-tetramethoxyflavone (Figure 84a-b, d, f-g, j) (Tomás-Barberán and Wollenweber, 1990). This flavonoid pattern is similar to that of several *Sideritis* spp., thus supporting a close taxonomical correlation between the two genera.

Among the European *Teucrium* species, *T. nuchense* L. has been distinguished by the occurrence of nuchensin (Figure 102b) (5,6,3'-trihydroxy-7,4'-dimethoxyflavone) (Barberán, 1986).

Within the genus *Ballota*, the presence, at least in significant amounts, of phenylethanoids and/or acylated flavone glycosides (Figure 77, 85, 100) provides different chemical features to distinguish between species:

For instance, phenylethanoids were not isolated from the methanolic extract of *Ballota larendana*, an endemic species of Turkey (Hennebelle et al., 2008) (Davis, 1982). However, *p*-coumaric acid substituted on 7-*O*-glucosyl-flavones was reported in this species (Figure 85a-b) (Hennebelle et al., 2008)

Naringenin-7-*O*-glucoside

Figure 99. Chemical structure of naringenin-7-*O*-glucoside, a characteristic flavanone glycoside of *Marrubium globasum ssp. Libanoticum* within the genus *Marrubium*.

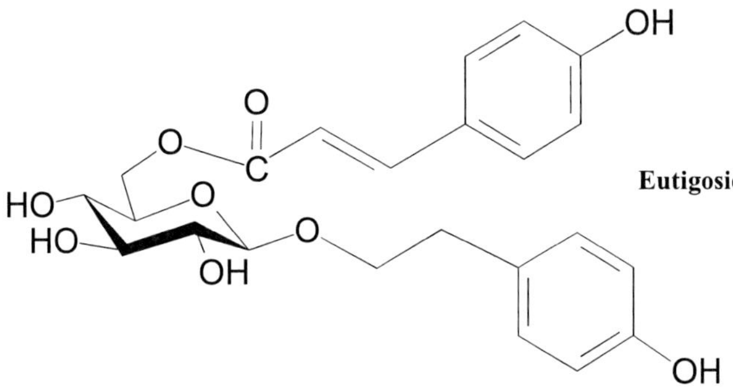

Figure 100. Chemical structure of eutigoside A, a characteristic phenylethanoid glycoside of *Ballota acetabulosa* (Sahpaz et al., 2002)

Only one but rare coumaroylated phenylethanoid glycoside, eutigoside A (Figure 100), was reported from the methanolic extract of *B. acetabulosa* (L.) Benth. (Sahpaz et al., 2002). Eutigoside A is a very rare compound leading it to be of a chemotaxonomic interest for this species.

Among the species producing coumaroylated and caffeoylated phenylethanoid glycosides, *Ballota pseudodictamnus* has been characterised by acetoside (Figure 77), forsythoside B (Figure 101a) and chrysoeriol-7-β-D-*O*-(3"-*O*-*p*-coumaroyl)glucopyranoside (Figure 85c) (Hennebelle et al., 2008).

In contrast, a species such as *Ballota nigra* L. contains acetoside-related phenylethanoids, where the glucose moiety is acylated by a caffeoyl residue (Figure 77) (Hennebelle et al., 2008; Seidel et al., 1999).

Ballota undulata (Sieb. Ex Fresen.) Benth., a perennial herb widespread in Jordan, particularly in waste places and mountains (Al Eisawi, 1998), showed more diversified acylation, i.e. both coumaroylated and caffeyolated glycosides (Siciliano et al., 2005; (Endo et al., 1982; Miyase et al., 1996; Andary et al., 1982; Liu et al., 1998): several acetoside derivatives (caffeoylated phenylethanoid glycosides) were idenitified (Figure 77, 101a-c) (Siciliano et al., 2005). Moreover, the co-occurrence of a coumaroylated flavonoid glycoside was reported in this species (Figure 85b).

Ballota hirsuta Benth., a typical representative of semi-arid habitats in southern Spain (Ferreres et al., 1986) showed a surface flavonoid pattern characterized by the occurrence of several aglycone structures: salvigenin (5-hydroxy-6,7,4'-trimethoxyflavone) (Figure 84c), kumatakenin (5,4'-dihydroxy-3,7-dimethoxyflavone) (Figure 102d), genkwanin (5,4'-dihydroxy-7-methoxyflavone) (Figure 86d), ladanein (5,6-dihydroxy-7,4'-dimethoxyflavone) (Figure 102a), nuchensin (5,6,3'-trihydroxy-7,4'-dimethoxyflavone) (Figure 102b) and isokaempferide (5,7,4'-trihydroxy-3-methoxyflavone) (Figure 102c).

Other *Ballota* spp. are known by only trace amounts of surface flavonoids as compared to *B. hirsuta*. The species list includes *Ballota pseudodictamnus* (L.) Benth. (Hennebelle et al., 2008), *B. nigra* subsp. *foetida* LAMK, and *B. acetabulosa* (L.) Benth. .

In the genus *Galeopsis*, the species *G. angustifolia* was described to be rich in surface flavonoids including ladanein (Figure 102a) (Tomás-Barberán and Wollenweber, 1990).

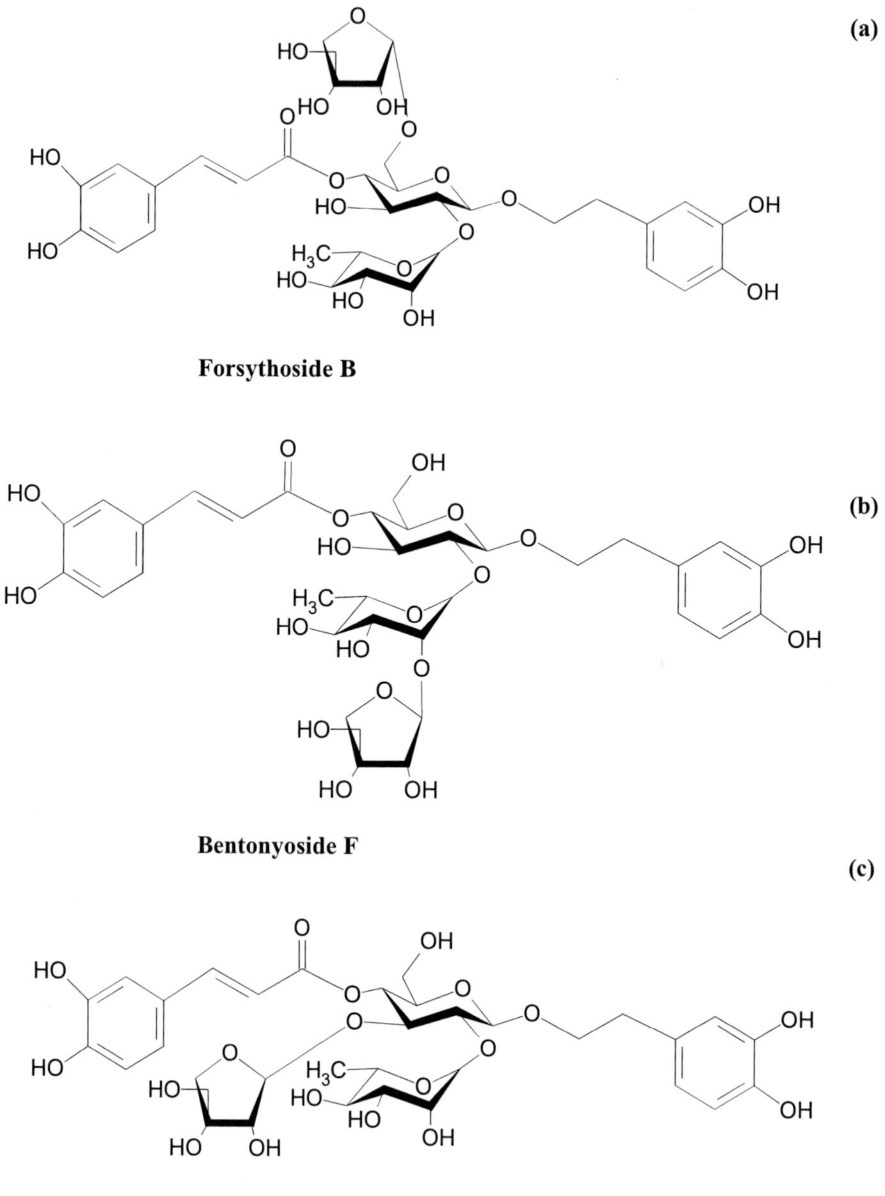

(a)

Forsythoside B

(b)

Bentonyoside F

(c)

Lysionotoside

Figure 101. Chemical structures of caffeoylated phenylethanoid glycosides found in different species of *Ballota* (Hennebelle et al., 2008; Siciliano et al., 2005).

In the subfamily Nepetoideae, several studies on the surface flavonoids of the species *Salvia glutinosa* reported a complex pattern based on several methoxylated flavones and flavonols (Tomás-Barberán and Wollenweber, 1990): apigenin (Figure 95a), genkwanin (Figure 86d), isokaempferide, kumatakenin, ayanin, retusin (Figure 102c-f), luteolin (Figure 104c), luteolin 7-methyl ether (Figure 88b), kaempferol (Figure 9), rhamnocitrin, ermanin, kaempferol 3,7,4'-trimethyl ether, herbacetin 3,8,4'-trimethyl ether (Figure 102g-j), quercetin 3-methyl ether (Figure 66a), quercetin 3-7-dimethyl ether, rhamnazin and pachypodol (Figure 102k-m).

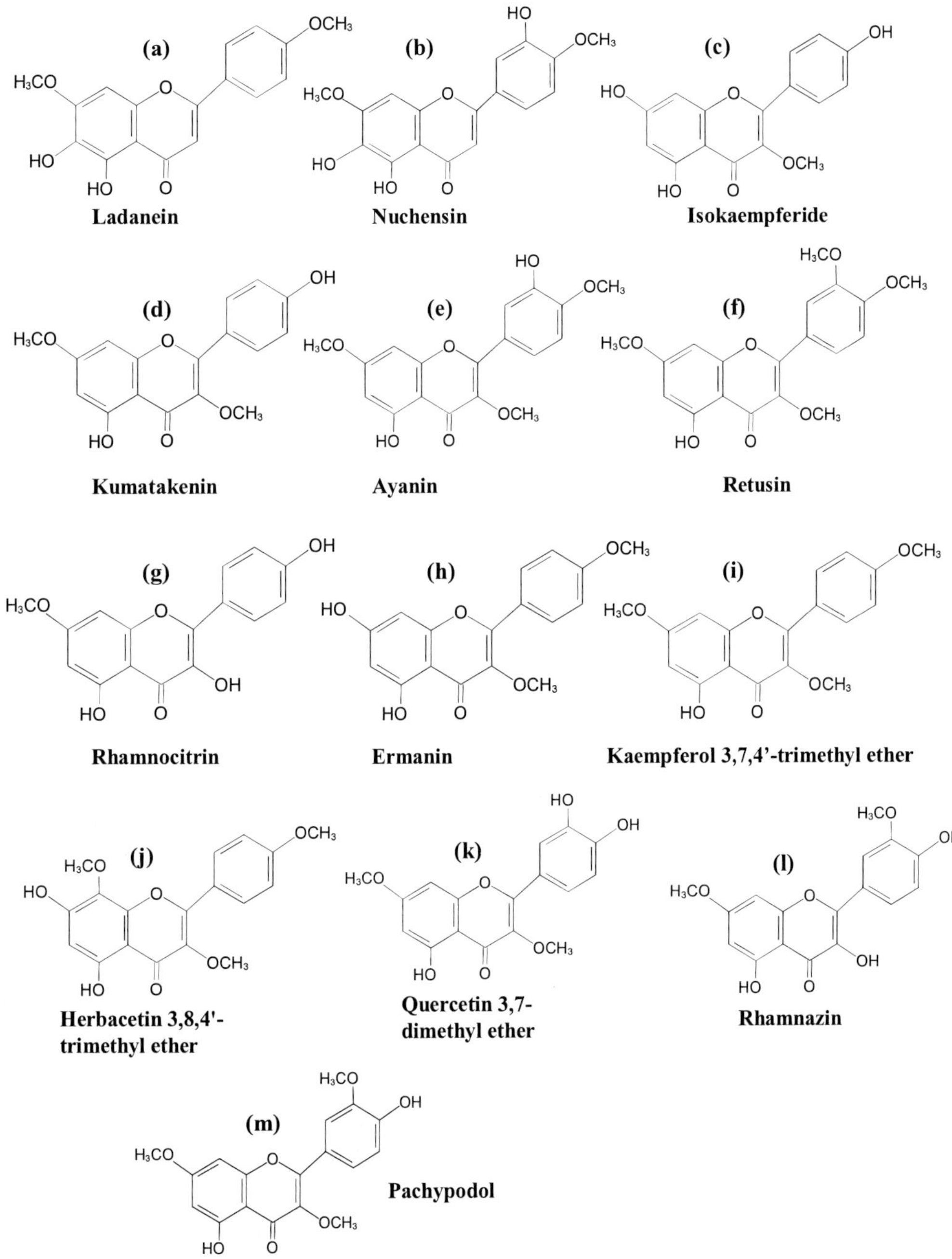

Figure 102. Chemical structures of methoxylated flavones and flavonols representing the complex surface flavonoid pattern of *Salvia glutinosa* (Tomás-Barberán and Wollenweber, 1990). Occurrence of compounds in other plant taxa is mentioned in the text.

The presence of nevadensin (Figure 87) and salvigenin (Figure 84c) was reported for *Ocimum canum* SIMS earlier (Xaasan et al., 1980). Their external accumulation and their co-occurrence with hispidulin (Figure 86a) were proved by Tomás-Barberán and Wollenweber (1990). Thereafter, the accumulation of flavones with a 5,7-dihydroxy-6,8-dimethoxy-A-ring (nevadensin) was reported in other members of the genus *Ocimum*.

5,6,4'-trihydroxy-7,3'-dimethoxyflavone   **5,6-dihydroxy-7,8,3',4'-tetramethoxyflavone**

Figure 103. Chemical structures of flavone methyl ethers found in *Mentha piperita* (Lamiaceae) (Jullien et al., 1984).

Although the genus *Thymus* was characterized by the presence of 5,6-dihydroxy-7,8-dimethoxyflavonoids (thymonin, thymusin and pebrellin) (Figure 89a-c) which exclude the 5,6-dihydroxy-7-methoxyflavones (Figure 83d), co-occurrence of these two types of compounds was observed in two *Thymus* species: *T. piperella* L. (Barberán et al., 1985b) and the North African *T. satureioides* COSS. (Voirin et al., 1985).

The species of the genus *Satureja* normally do not excrete flavonoids; however, xanthomicrol (Figure 84f) has been detected in some cases as well as a number of flavanones (e.g. naringenin) (Figure 55b) in *Satureja salzmanii* P.W.BALL exudates (Tomás-Barberán and Wollenweber , 1990).

The occurence of 5,6-dihydroxy-7,8,3',4'-tetramethoxyflavone (Figure 103b) has been reported for *Mentha piperita* L. (Jullien et al., 1984), where it was found along with sideritoflavone (Figure 84g), thymonin (Figure 89c), gardenin B (Figure 84h), 5-desmethylnobiletin (Figure 84l), xanthomicrol (Figure 84f), 5,4'-dihydroxy-6,7,8,3'-tetramethoxyflavone (Figure 84i) and 5,6,4'-trihydroxy-7,3'-dimethoxyflavone (Figure 103a).

Analysis of the surface flavonoids in different species *Ocimum* originary from different geographical area (USA, Canada, Brazil, UK) and cultivated in a same place (Purdue) showed significant differences between (a) their patterns, (b) their total amounts of exudate flavonoids (Figure 104a), and (c) the ratios between major and minor compounds (Grayer et al., 2004): from *Ocimum americanum* to *O. minimum* via *O. × citriodorum* and *O. basilicum*, the production of surface flavones and the ratio (nevadensin/salvigenin) (Figure 84c, 87) tend to decrease (Figure 104b):

Among individual plants *Ocimum americanum*, most examined specimens appeared to be characterized by the production of high quantities of surface flavones on average 7.4mg/g dried leaves, but varying from 0.4 to 50.6 mg/g (Figure 104a, whole variation range not represented in the figure) (Grayer et al., 2004), nevadensin (Figure 87) being the major (more than 50% of the total) and salvigenin (Figure 84c) the second major flavonoid (Figure 106). The third major surface flavonoid was often the unusual flavone, pilosin (Figure 105) (Viera et al., 2003). Other minor flavones (10-15 compounds) are produced in most specimens of *O. americanum* (Figure 106).

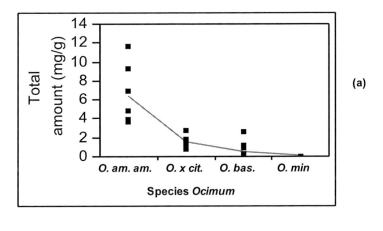

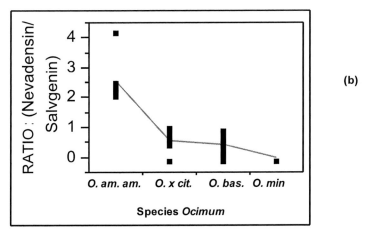

Legend:
O. am. am: Ocimum americanum var. americanum
O. × cit.: Ocimum × citriodorum
O. bas.: Ocimum basilicum
O. min.: Ocimum minimum

Figure 104. Variation of the total amount of surface flavonoids (a) and of the ratio between the levels of Nevadensin and Salvigenin (b) in relation to the species *Ocimum* (Grayer et al., 2004). The lines join the mean values of the four species.

Pilosin

Figure 105. Chemical structure of pilosin, an unusual flavone characterizing the species *Ocimum americanum* (Viera et al., 2003).

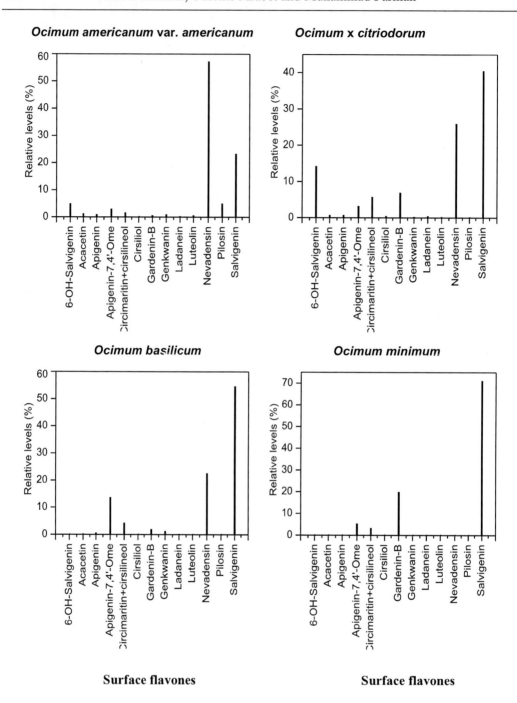

Figure 106. Average patterns of surface flavones in different species *Ocimum* (Grayer et al., 2004).

Although typical specimens of *O. × citriodorum* produced lower amounts of surface flavonoids than *O. americanum* (2.1 vs 7.4 mg/g dw) there was an overlap between individual plants (Grayer et al., 2004). However, these two species were better separated by different ratios between their two major flavonoids, nevadensin and salvigenin (Figure 106): in *Ocimum americanum* this ratio is typically between 2 and 5 whereas it is close to 1 in *O. × citriodorum* (Figure 104b).

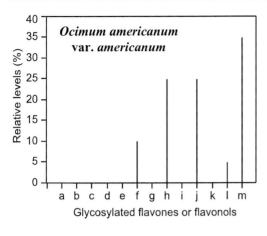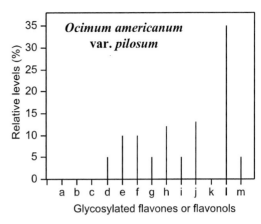

Figure 107. Distinction between two varieties of *Ocimum americanum* (var. *americanum* and *pilosum*) on the basis of their ratio between *C*- and *O*-glycosyl flavonoids (Grayer et al., 2002). The labels (a-m) correspond to glycosides presented in figure 93.

The amounts of surface flavones found in specimens of *O. basilicum* were on average lower than in *Ocimum* × *citriodorum*, being 0.6 mg/g (Figure 104a) (Grayer et al., 2004). Most of the specimens of *O. basilicum* produced much less nevadensin than salvigenin (Figure 106), but the ratios found varied from 0.14 to 1.89 (Figure 104b).

Specimens of *O. minimum* generally produced the lowest amount of surface flavones (typically 0.1-0.2 mg/g dried leaves) (Figure 104a). Salvigenin is usually the major flavone and nevadensin is either present in low amounts or absent (Figure 104b, 106) (Grayer et al., 2004).

VIII.2.6. Infraspecific level chemotaxonomy

Analysis of flavonoid glycosides in two varieties of *Ocimum americanum* showed reliable chemotaxonomical keys based on a ratio between flavone *C*-glycoside (Figure 93l) and flavonol 3-glycosides (Figure 93h, j, m) (Grayer et al., 2002) (Figure 107):

In *O. americanum* var. *pilosum*, the major flavonoid glycoside is vicenin-2 (a flavone *C*-glycoside) (Figure 93l). Its level was estimated to be twice as high as that of quercetin 3-*O*-rutinoside (Figure 93j) and 3-glucoside (Figure 93h). In contrast, *O. americanum* var *americanum* can be identified by a trace of vicenin-2, but an accumulation of quercetin 3-*O*-rutinoside and 3-*O*-glucoside (Figure 93h, j). Quercetin 3-*O*-malonyl glucoside (Figure 93m) was only present in the trace amounts in *O. americanum* var. *pilosum*, whereas its levels were comparable to those of quercetin 3-*O*-rutinoside and 3-*O*-glucoside in *O. americanum* var. *americanum*.

In other species of the genus *Ocimum*, infraspecific variation was highlighted on the basis of minor compounds: thus, *Ocimum minimum*, *O. gratissimum* and *O. tenuiflorum* showed infraspecific variations with respect to only minor flavonoid glycosides, while the profile of the major compounds was similar in all the accessions belonging to a same species (Grayer et al., 2002).

VIII.3. Fabaceae Family

VIII.3.1. Family level chemotaxonomy

The family Fabaceae is known to be rich in flavonoids showing an important variability of chemical structures. The occurrence of isoflavonoids (Figure 108-110, 113, 117-118, 124, 128) mostly concerns the Fabaceae. Species of this family are able to produce isoflavonoid phytoalexins as fungitoxic chemicals subsequently to the invasion or attempted invasion of their tissues by micro-organisms (Ingham, 1972; Van Etten and Pueppke, 1976). Such isoflavonoid phytoalexins include isoflavones, isoflavanones, isoflavans and pterocarpans (Figure 108-110).

Moreover, this family is known to produce many flavanone aglycones (Figure 124) as free state in bark, roots or root bark of plants (Iwashina, 2000). Another chemical characteristic of Fabaceae consists of the occurrence of aurones (not frequent pigments) (Figure 10, 123b).

VIII.3.2. Subfamily/Tribe level chemotaxonomy

Among the subfamilies of Fabaceae, the Papilionoideae is particularly rich in isoflavonoids (Dixon and Summer, 2003):

In the tribe Loteae, isovestitol (isoflavan) (Figure 108c) has been reported as a phytoalexin in response to pathogenic attacks (Ingham, 1977b). The species of Loteae may also accumulate vestitol and sativan (Figure 108b, d) (Ingham, 1977b; Bonde et al., 1973).

Compared to the tribe Trifolieae, Loteae differs in two important aspects: none of its species appear to produce medicarpin (Figure 109a) or any other pterocarpan phytoalexin and all accumulate demethylvestitol (Figure 108a), a compound not found in the Trifolieae.

A comparison between the tribes Vicieae and Trifolieae showed that species of the Vicieae characteristically produce phytoalexins, notably the pterocarpan pisatin (Figure 109c) that is apparently absent from the Trifolieae. However, the Trifolieae species accumulate pterocarpans (principally medicarpin and maackiain) (Figure 109a, 109b) which best are poorly represented in the Vicieae (Ingham, 1981). A comparable situation was shown for the constitutive isoflavones of these two tribes: Daidzein, pratensein, formononetin and biochanin A (Figure 110) were found in Trifolieae but not in Vicieae. However in the latter tribe, compounds as orobol and sayanedin have been recorded (Fig. 110e-f) (Charaux and Rabaté, 1939; Isogai et al., 1970).

VIII.3.3. Genus level chemotaxonomy

Generic surveys of the Fabaceae suggest that phytoalexin studies might provide information on plant taxonomy, chemistry, ecology and pathology. For instance, the fungus-inoculated leaves of *Melilotus* species accumulate only the pterocarpan: medicarpin (Figure 109a), whereas members of the allied genus *Medicago* often produce medicarpin together with one or more isoflavan derivatives (Figure 108) (Ingham, 1977a).

The intermediate genus *Trigonella* contains many species which, in terms of their phytoalexin production, resemble either *Meliotus* or *Medicago* (Ingham and Harborne, 1976).

In addition, other *Trigonella* species (8 of out examined 35, i.e. 23%) accumulate medicarpin (Figure 109a) with the related pterocarpan, maackiain (Figure 109b), a compound which is not associated to *Meliotus* or *Medicago* (Ingham, 1977a; Ingham and Harborne, 1976).

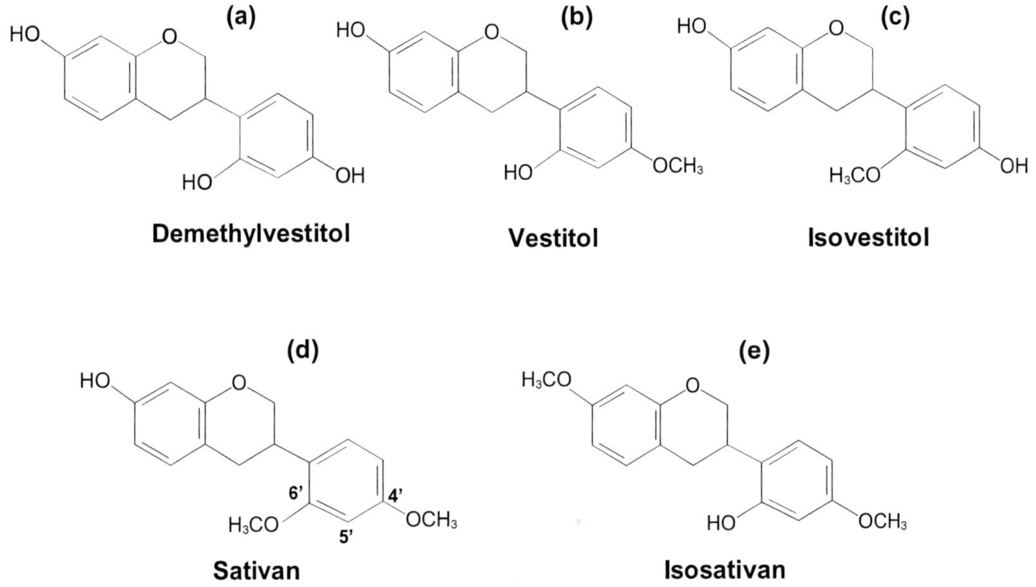

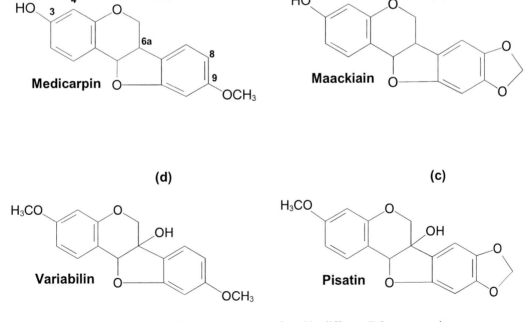

Figure 108. Chemical structures of some isoflavans found in different Fabaceae species.

Figure 109. Chemical structures of some pterocarpans found in different Fabaceae species.

The genus *Trifolium* is a large (ca 300 species) and agriculturally important taxon (Hutchinson, 1964; Zohary and Heler, 1984). It is distributed in temperate and subtropical regions of both hemispheres (Bisby et al., 1994) with a particular richness in the Mediterranean region (Zohary and Heler, 1984; Zohary, 1970). *Trifolium* has been reported to be chemically more variable than other genera belonging to the tribe Trifolieae (e.g. *Melilotus, Factorovskya, Melilotus, Parochetus* and *Ononis*) (Ingham, 1978). The author reported that 93% of examined *Trifolium* species (52 out of 55) contained stilbene (Figure 118c) or isoflavonoid (Figure 108-109) phytoalexins in the leaflet diffusates, after their inoculation by the fungus *Helminthosporium carbonum* Ullstrup.

Among the isoflavonoids, Medicarpin (Figure 109a) was reported to be the most widely distributed in *Trifolium* species. About 24% of examined species showed co-occurrence of medicarpin and maackianin (Figure 109a, b).

In *Trifolium*, the isoflavan (vestitol) (Figure 108b), was invariably present as a diffusate component (except section Paramesus) (Ingham, 1978). However, its 2'-O-methyl ether (sativan) (Figure 108d) was significantly less abundant in *Trifolium* members (37% of examined species) compared with *Medicago* (71%) and *Trigonella* (75%) (Ingham and Harborne, 1976). Ingham (1978) reported the replacement of sativan by its structural isomer, isosativan (Figure 108e), in the diffusates of 9 *Trifolium* species (out of 55).

An unusual feature of *Trifolium* is the co-occurrence of isoflavans (Figure 108) with maackiain (Figure 109b) in diffusates of some species. This is in contrast to *Trigonella* where maackiain has not been found together with isoflavan derivatives (Ingham and Harborne, 1976).

Apart from isoflavonoid derivatives, another study concerning the flavonoids of 57 species *Trifolium* reported the prevalence of quercetin (Figure 12) either alone or in mixture with a number of unidentified flavonoids (Oleszek and Stochmal, 2002). The concentration of quercetin was in the range 0.05-3mg/g. The authors suggested that the seeds of some *Trifolium* species may provide a potential source of beneficial phytochemicals in the human diet, due to their high concentration of quercetin.

The genus *Cicer* (Tribe Cicereae) consists of approximately 40 annual and perennial species (Ingham, 1981). After inoculation of stem with the non-pathogenic fungus, *Helminthosporium carbonum*, a substantial accumulation of medicarpin (150-720 µg/g fresh wt tissue) and maackiain (70-690 µg/g) (Figure 109a, b) has been reported in all the 15 *Cicer* species examined by Ingham (1981). However, the species were apparently unable to produce any of the simple isoflavan derivatives (e.g. vestitol and sativan) (Figure 108b, d) which frequently co-occur with medicarpin and maackiain in *Medicago, Trifolium* and *Trigonella* (Ingham, 1979; Ingham, 1978; Ingham and Harborne, 1976).

The importance of medicarpin and maackiain (Figure 109a, b) as chemical fingerprints in Fabaceae has also been underlined by the genus *Ononis*: a survey by Ingham (1978) highlighted a production of maackiain by the majority, and medicarpin by all the species examined.

In *Trifolium, Trigonella, Ononis* and *Cicer*, medicarpin was found to be present in quantities significantly greater than those of maackiain (Ingham, 1981, 1978; Ingham and Harborne, 1976).

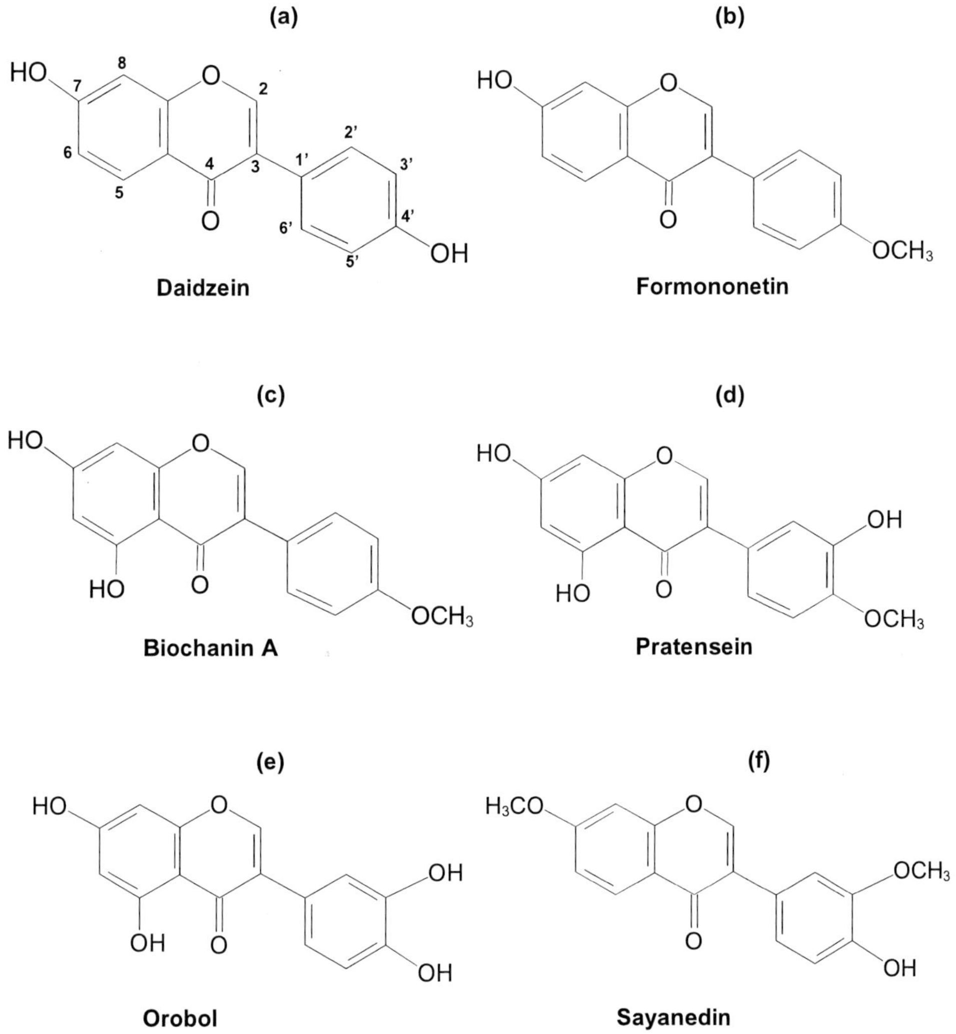

Figure 110. Chemical structures of some isoflavones found in Fabaceae species.

Comparison of the flavonoid contents between seven *Medicago* and five *Trigonella* species, showed that the major components in the genus *Medicago* were mainly flavone glycosides (Figure 111), whereas in *Trigonella*, flavonol glycosides (Figure 113) prevailed (Saleh et al. 1982). The 7-monoglucuronides of apigenin (Figure 111a), luteolin (Figure 111b), chrysoeriol (Figure 111c) and tricin (Figure 111d) appeared as the major flavonoids of *Medicago* species. The 3,7-diglucoside of kaempfeol being the most common glycoside present in *Trigonella* species (Figure 113). Moreover, *Medicago* species have been described by the absence or the rarity of isoflavones (Figure 10, 110) (Harborne, 1969; Saleh et al., 1982).

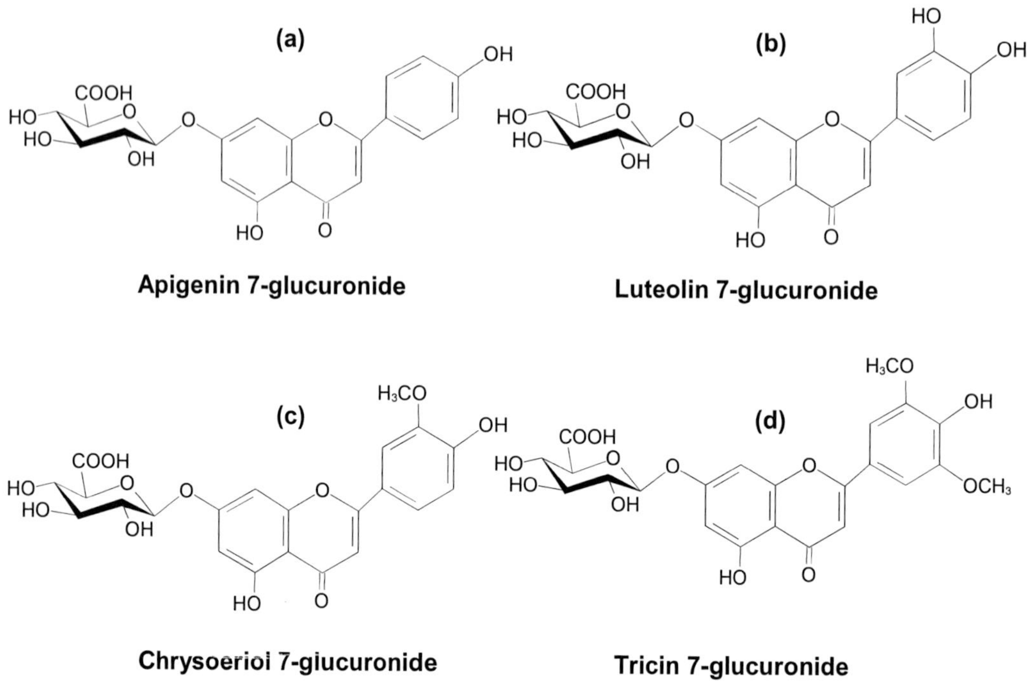

Figure 111. Chemical structures of flavones glucuronides found in the genus *Medicago* (Saleh et al., 1982).

Baptisia and *Thermopsis* are closely related genera containing the only North American members of the tribe Podalyrieae. *Thermopsis* is a widespread group occurring at high elevations from India across China, Russia, Alaska and western North of America, and with an isolated series of species in the Appalachian Mountains of the Eastern United States. Concerning *Baptisia*, it is restricted to Eastern North America where it occupies habitats which are considered as younger floristically than those of *Thermopsis* (Dement and Mabry, 1975). The two genera are quite similar morphologically being distinguished primarily by pod form. Turner (1967) has suggested that *Baptisia* arises from a *Thermopsis*-like ancestor by adaptation to the warm and moist habitats in which it now occurs.

Flavonoid data support the close relationship between the two genera since the flavonoid patterns of *Thermopsis* are similar to those of some *Baptisia* species (Dement and Mabry, 1975). By analysing the flavonoid distributions in *Thermopsis* and *Baptisia* species living in North America, Dement and Mabry (1975) extracted distinctive chemical features between the two genera: contrary to *Thermopsis*, *Baptisia* contained flavonols, flavonol glycosides, 6-hydroxyisoflavones (6-hydroxy-genistein, tectorigenin and afrormosin) (Figure 112a-c) and methylene dioxyisoflavones (pseudobaptigenin) (Figure 112d). However, *Thermopsis* was qualitatively characterized by the occurrence of chrysoeriol (Fig. 111c), pratensein (Figure 110d), 3'-methoxyorobol (Figure 112e) and their glucosides (absent in *Baptisia*).

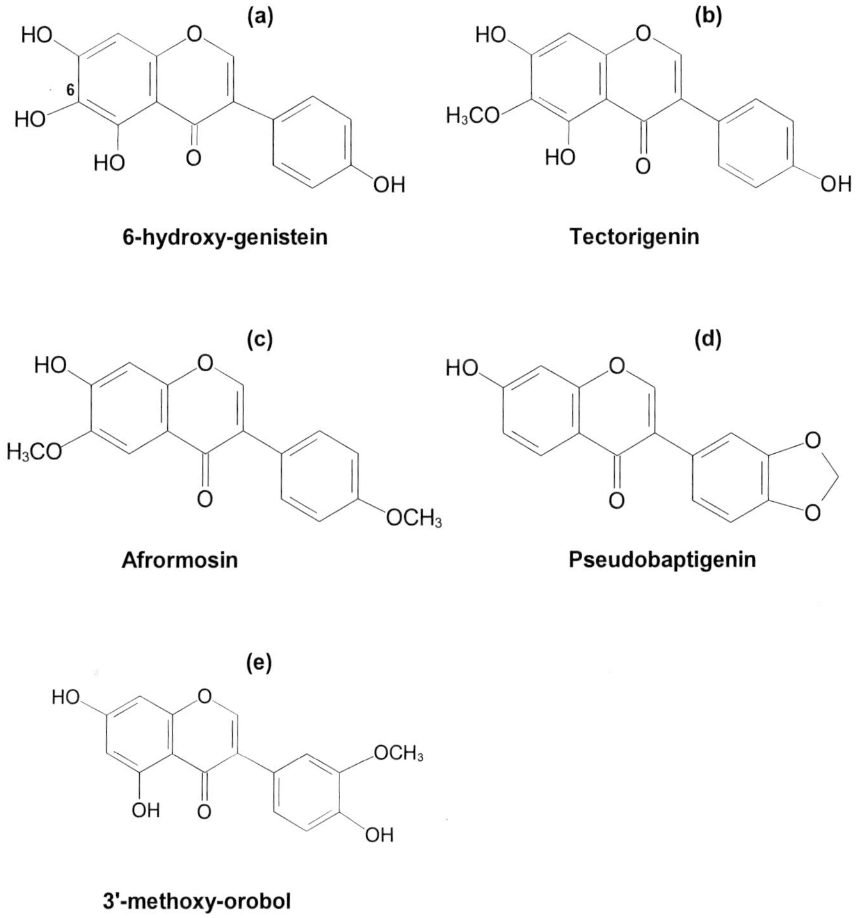

(a) 6-hydroxy-genistein

(b) Tectorigenin

(c) Afrormosin

(d) Pseudobaptigenin

(e) 3'-methoxy-orobol

Figure 112. Chemical structures of some isoflavones found in different Fabaceae species.

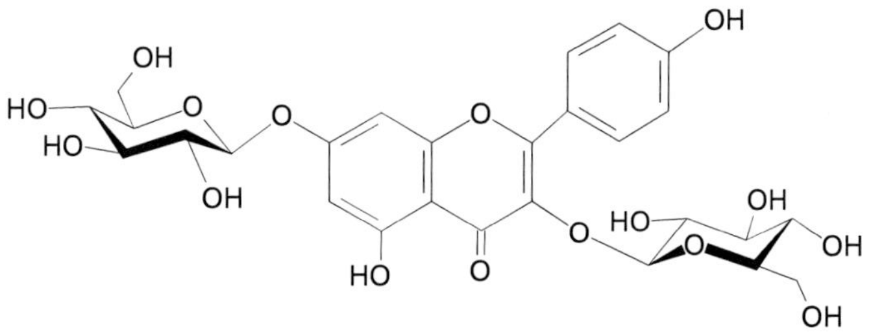

Kaempferol 3,7-diglucoside

Figure 113. Chemical structure of kaempferol 3,7-diglucoside, a frequent flavonol glycoside in the genus *Trigonella* (Saleh et al., 1982).

In addition, the genus *Baptisia* exhibits a higher variability of chemical patterns in comparison to *Thermopsis*. Flavonoid distribution data in the genus *Baptisia* have been helpfully used to highlight chemical differentiations within the genus (Markham et al., 1970), to document hybridization (Alston and Hemple, 1964; Alston et al., 1962) and to analyse hybrid populations (Alston and Tuner, 1962).

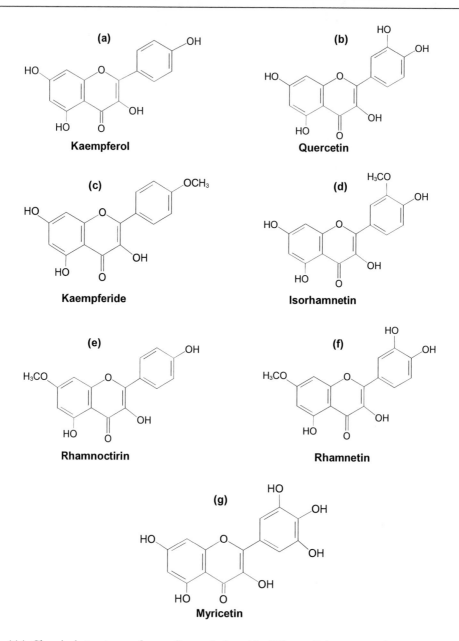

Figure 114. Chemical structures of some flavonols found in different Fabaceae species.

The flavonoid pattern of the genus *Astragalus* have been studied through several species (Semmar et al., 2005; Bedir et al., 2000; Bisby et al., 1994; Cui et al., 1991, 1993; Alaniya et al., 1975; Norris and Stermitz, 1970; Nakabayashi, 1952). The different works reported a predominance of flavonol glycosides. Thus, *O*-glycosides of kaempferol, quercetin, kaempferide, isorhamnetin, rhamnetin, rhamnocitrin, myricetin, etc. have been reported from different *Astragalus* species (Figure 114). The diversity of flavonol glycosides in the genus *Astragalus* can be illustrated by different degrees of glycosylation where mono-, di-, tri and tetraglycosides were identified in different species. Moreover, several *Astragalus* species showed the presence of apiose in their glycosides (Figure 125b, 126), while this sugar is not frequently observed in the glycosides of the different plant taxa (Semmar et al. 2001, 2002a; Bedir et al., 2000; Cui et al., 1993).

VIII.3.4. Subgenus/Section level chemotaxonomy

In the genus *Trifolium*, maackiain (Figure 109b) was apparently confined to the subgenera *Lotoideae* (section *Lotoidea*, 2 species), *Trifolium* (section *Trifolium*, 9 species) and *Falcatula* (1 species, *Trifolium ornithopodioides*) (Ingham, 1978).

The section *Chronosemium* (subgenus *Lotoideae*) was distinguished by the production of non-isoflavonoid phytoalexins; isoflavonoids (Figure 108, 110) seem to be uncommon in this section (Ingham, 1978).

VIII.3.5. Species level chemotaxonomy

Within the genus *Trifolium*, *T. repens* (white clover), *T. pratense* (red clover) and *T. hybridum* (alsike clover) have been reported to produce phytoalexins in response to pathogen attacks (Higgins and smith, 1972; Ingham, 1976a; Cruickshank et al., 1974).

In a later study (Ingham, 1978), several morphologically similar *Trifolium* species pairs were found similar in terms of isoflavonoid phytoalexin biosynthesis: *T. apertum*/*T. alexandrinum*, *T. occidentale*/*T. repens*, *T. resupinatum*/*T. tomentosum* (Figure 115).

In other cases, phytoalexins provided efficient chemical tools to distinguish between morphologically similar species. For instance, whilst vestitol and sativan (Figure 108b, d) were produced by *Trifolium medium* (zigzag clover), neither compound was isolated from the allied species *T. heldreichianum* (Figure 116). This is also the case for *T. medium* and *T. pratense* which are morphologically similar (and may occasionally be confused), but can be distinguished by comparison of their leaf phytoalexins. Similarly, the isoflavonoid pattern of *T. stellatum* (traces of medicarpin plus 3 isoflavans: vestitol, isosativan and arvensan) (Figure 109a, 108b, 108e, 118b) showed little resemblance with that of *T. dasyurum* (traces of medicarpin) (Figure 109a) (Ingham, 1978).

Pope et al. (1953) isolated biochanin A (Figure 110c) from red clover (*Trifolium pratense*). In 1965, Schultz showed that glycosides of biochanin A and formononetin (Figure 110b, c) are present in red clover. More recently, isoflavones, their glycosides, their malonated glycosides (Figure 117) and their acetyl glycosides were determined in the red clover extracts (de Rijge et al., 2001; Klejdus et al., 2001; Gu and Gu, 2001; Krenn et al., 2002). Moreover, several *Trifolium* species were original because of their unusual chemical features:

Trifolium hybridum was the only *Trifolium* species able to convert maackiain to its 4-methoxy derivative (Figure 118a) (Ingham, 1978).

Among the isoflavans, arvensan (Figure 118b) was isolated only from *Trifolium arvense* and *T. stellatum*. This compound can be considered as chemotaxonomically characteristic of these two species as much as extensive surveys proved its absence elsewhere in the Papilionoideae subfamily (Ingham, 1978).

Although the isoflavonoid phytoalexins are uniformly produced by *Trifolium* species, *T. campestre* and *T. dubium* (subgenus *Lotoideae*, section *Chronosemium*) were original as they accumulate only trans-resveratrol (Figure 118c), a mildly fungitoxic stilbene, rather than isoflavonoid-based phytoalexins (Ingham, 1976c; Langcake and Pryce, 1976).

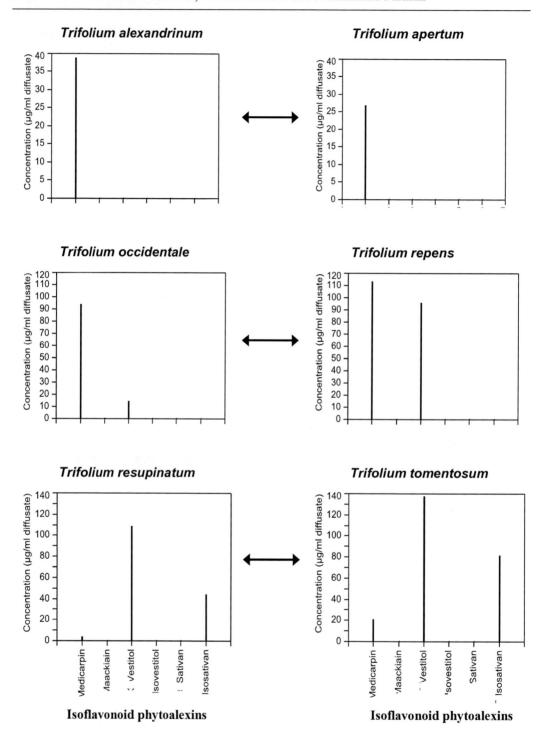

Figure 115. Concentration profiles of isoflavonoid phytoalexins produced in diffusate by different species pairs of *Trifolium* that are morphologically similar (Ingham, 1978).

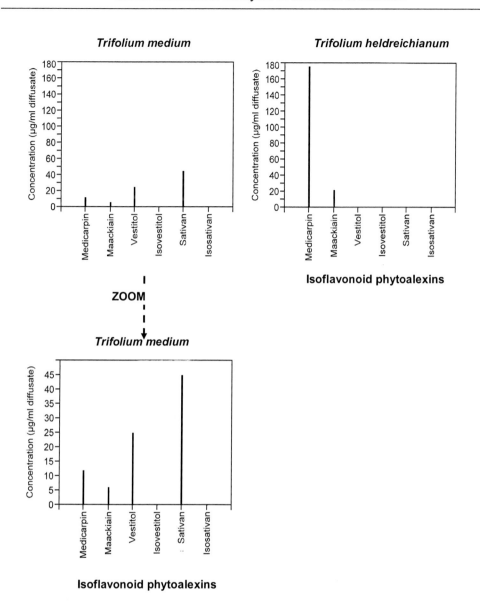

Figure 116. Concentration profiles of isoflavonoid phytoalexin produced by two morphologically comparable *Trifolium* species; the profiles highlight the usefulness of phytochemistry to distinguish between plant species despite their morphological similarity (Ingham, 1978).

In subgenus *Lotoideae*, the species *Trifolium aureum* was also characterized by a single compound phytoalexin pattern consisting of only vestitol (isoflavan) (Figure 108b) (Ingham, 1978).

Flavonoids, isoflavonoids and their glycosides have also been isolated from *Trifolium striatum, T. alexandrinum, T. subterraneum, T. echinatum, T. polyphyllum* C.A. Meyer, *T. resupinatum* L. var *microcephalum* Zoh., and several other *Trifolium* species (Tarr, 1993; Shehata et al., 1982; Wang et al., 1998; Shalashvili, 1993; Lukyanchikov and Kazakov, 1982; Isik et al., 2007; Toebes et al., 2005; Foo et al., 2000).

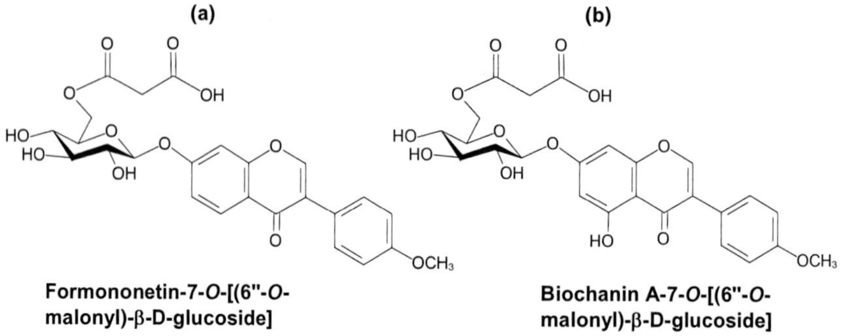

(a)

Formononetin-7-*O*-[(6"-*O*-malonyl)-β-D-glucoside]

(b)

Biochanin A-7-*O*-[(6"-*O*-malonyl)-β-D-glucoside]

Figure 117. Chemical structures of malonylated isoflavone glycosides found in *Trifolium pratense* (de Rijke et al., 2001).

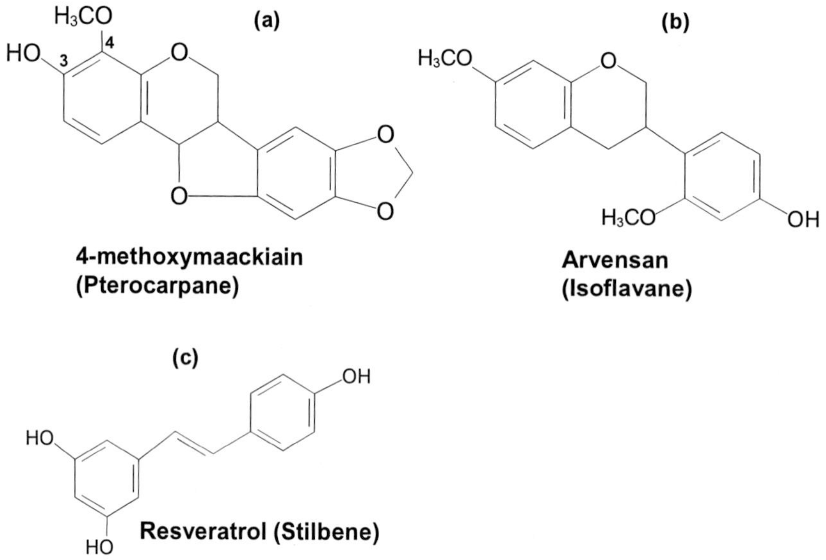

(a)

4-methoxymaackiain (Pterocarpane)

(b)

Arvensan (Isoflavane)

(c)

Resveratrol (Stilbene)

Figure 118. Chemical structures of different phenolic derivatives charcaterizing different species in Fabaceae family (for detail, see text).

Hofmann et al. (2000) identified flavonoids involved to ultraviolet-B-radiation (UV-B) in nine populations of white clover, *Trifolium repens*. The major flavonoids present in the leaves consisted of derivatives of quercetin and kaempferol (flavonols). The structures of the compounds involved in the response to UV-B were quercetin-3-*O*-β-D-xylopyranosyl-(1→2)-β-D-galactopyranoside (Figure 119b) and kaempferol-3-*O*-β-D-xylopyranosyl-(1→2)-β-D-galactopyranoside (Figure 119a).

Trifolium repens has been also chemically characterized by the occurrence of the polyhydroxylated chalcone (2',3',4',5',6'-pentahydroxy-chalcone) (Figure 119c) and chalcanol glucosides (trifochalcanoloside I, trifochalcanoloside II and trifochalcanoloside III) (Figure 119d-f) in the roots (Olesszek and Stochmal, 2002). Such compounds have been also isolated from the seeds of *T. alexandrinum* (Mohamed et al., 2000).

Two new biocoumarins named repensin A and repensin B, were isolated from *Trifolium repens*. They were identified as 7-methoxy-7',8'-dihydroxy-8,6'-biocoumarinyl and 7,5'-

dihydroxy-3,6'-biocoumarinyl, respectively (Zhan et al., 2003). In another study, tetra-and pentahydroxylated and methoxylated flavones have been isolated for the first time from *Trifolium repens* (Figure 120) (Ponce et al., 2004).

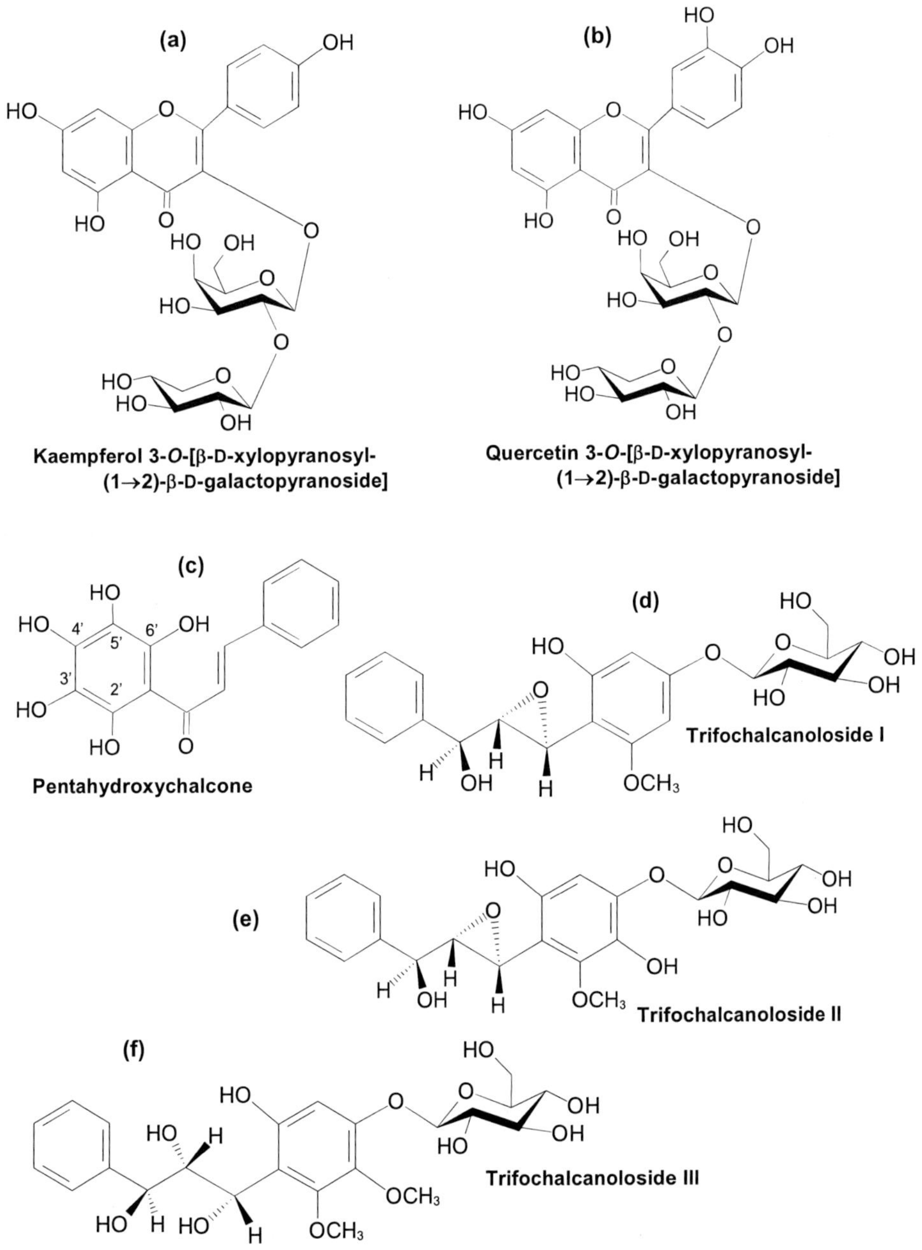

Figure 119. Flavonol 3-*O*-diglycosides isolated from the leaves of *Trifolium repens* and Chalcanol 4-*O*-glucosides isolated from the roots and the seeds of *T. repens* and *T. alexandrinum*, respectively (Hofmann et al., 2000; Mohamed et al., 2000; Olesszek and Stochmal, 2002). Note that the conventional carbon numbers in chalcones differ from the other flavonoids (see 119c).

5,6,7,8,4'-pentahydroxy, 3-methoxyflavone

Figure 120. Polyhydroxylated flavonoid found in *Trifolium repens* (Ponce et al., 2004).

Figure 121. Chemical structures of flavone 6-*C*, 8-*C* monoglucosides and 6,8-di-*C*-glucosides found in *Trigonella foenum-graecum* (a-f) and *T. corniculata* (a-g) (Seshadri et al., 1972, 1973).

Several studies on Trigonella were carried out on the basis of the flavonoids in the seeds. These flavonoids were mainly *C*-glycosides of flavones:

The seeds of *Trigonella foenum-graecum* proved to contain orientin, isoorientin, vitexin and vitexin-7-glucoside (Adamska and Lutomski, 1971), isovitexin (Seshadri et al., 1973), vicenin-1, vicenin-2 (Wagner et al., 1973) and vicenin-2"-*O*-*p*-coumarate (Sood et al., 1976) (Figure 121a-f).

The seeds of *Trigonella comiculata* also contained two *C*-glycosides, identified as acacetin-6,8-di-*C*-glucoside (Figure 121g) and its mono-acetate (Seshadri et al., 1972).

In the genus *Medicago*, *M. scutellata* is one of the few species that does not accumulate sativan (Figure 108d); alternatively, it produces isosativan (Figure 108e) as a phytoalexin (Ingham, 1978).

Analysis of flavonoids in *M. sativa* revealed the presence of 7-glucuronide and 7-diglucuronide of tricin and chrysoeriol 7-glucuronide (Figure 111c, d) (Harborne and Hall, 1964) as well as 4',7-dihydroxyflavone (Bickoff et al., 1965) and 3',4',7-trihydroxyflavone (Olah and Sherwood, 1971).

Medicago radiata was particularly original because it had both chemical features of *Medicago* and *Trigonella* genera (Saleh et al., 1982). In other words, it seems to be placed closer to *Medicago* with a strong link however, to *Trigonella*: the presence of apigenin and luteolin-7-glucuronide (Figure 111a, b) shows its link to most *Medicago* species. However, the fact that it contains kaempferol 3,7-diglucoside (Figure 113) as well as the isoflavones formononetin and daidzein (Figure 110a, b) provides an evidence on its affinity toward *Trigonella*. Saleh et al. (1982) suggested therefore, that *M. radiata* should be retained in the genus *Medicago*, with close affinities to *Trigonella*.

In the genus *Cicer*, study of pterocarpan isoflavonoids showed that different species can be chemically distinguished by decreasing medicarpin/maackiain ratios (Figure 109a, b): these ratios ranged from *ca.* 5:1 in *C. anatolicum* to approximately 1:1 in *C. judaicum* and *C. pinnatifidum*, via intermediate states (*ca.* 3:1) characterizing *C. arietinum*, *C. echinospermum*, *C. pungens* and *C. yamashitae* (Ingham, 1981) (Figure 122).

In the genus *Glycyrrhiza* (licorice), the species *G. echinata* (Furuya et al., 1971), *G. pallidiflora* (Fukai et al., 1990) and *G. inflata* (Kajiyama et al., 1992a, b; Demizu et al., 1992) showed original comparable flavonoid patterns due to the occurrence of dibenzoylmethanes and retrochalcones (rarely observed in nature) (Figure 124) (Ayabe et al., 1982). The first two species are taxonomically classified into the section *Echinatae*, whereas the third one is located in the section *Bucharicae* (Ammosov and Litvinenko, 2007). There are several species belonging to these two sections that need to be furthermore investigated chemically. Whether retrochalcones or dibenzoylmethanes occur in these taxonomically related species is worth investigation in order to fully understand the chemotaxonomy of the genus *Glycyrrhiza*.

The species *G. pallidiflora* was also characterized by an original isoflavonoid pattern: while most of isoflavonoids obtained from other *Glycyrrhiza* species are prenylated, those of *G. pallidiflora* were simple (Figure 124) (Kajiyama et al., 1993).

Moreover, a disctinctive chemical feature of *G. pallidiflora* can be obtained from its glycoside fraction which seems to be essentially derived from isoflavones (Figure 124). However, while a number of glycosides of liquiritigenin (flavanone) and isoliquiritigenin (chalcone) occur in other *Glycyrrhiza* species (e.g. *G. uralensis*, *G. glabra*, *G. inflata*), *G. pallidiflora* seems to be devoid of these glycosides (Figure 124) (Shibata et al., 1978).

The occurrence of 3-arylcoumarin skeletons is also noteworthy since these are quite rare members of the isoflavonoid family (Figure 123a). Such compounds were reported to be synthesised by the species *G. uralensis* and *G. pallidoflora* (Kinoshita et al., 1978; Zhu et al., 1984; Kajiyama et al., 1993).

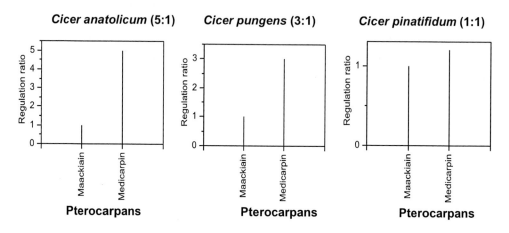

Figure 122: Variation of the metabolic regulations of medicarpin and maackiain, the one at the expense of other, in different species *Cicer* (Ingham, 1981).

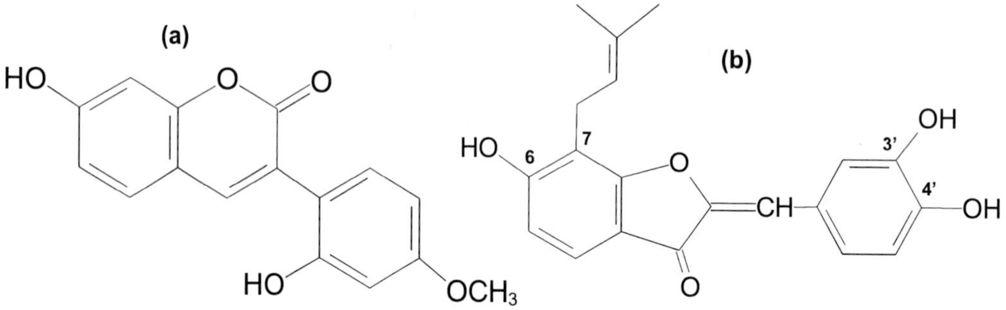

Figure 123. Chemical structure of 3-arylcoumarin (a) and isoprenylated aurone (b) found in *Glycyrrhiza uralensis* and *G. glabra*, respectively (Kinoshita et al., 1978; Zhu et al., 1984) (Asada and Yoshikawa, 2000).

In the genus *Glycyrrhiza*, the first aurone was isolated from the hairy roots of the species *G. glabra* (Figure 123b) (Asada and Yoshikawa, 2000). Such aurone, named licoagroaurone, is based on sulfuretin skeleton (6,3',4'-trihydroxyaurone) with a prenyl at the position 7.

Within the genus *Astragalus*, different species were chemically characterized by the occurrence and co-occurrence of different flavonol glycosides, viz. glycosides of kaempferol, quercetin, isorhamnetin, rhamnetin and rhamnocitrin in *A. caprinus* (Semmar et al., 2005), kaempferol, kaempferide, rhamnetin and myricetin in *A. complanatus* (Cui et al., 1991, 1993), kaempferol and isorhamnetin in *Astragalus sinicus* (Nakabayashi, 1952), quercetin and isorhamnetin in *A. miser* (Norris and Stermitz, 1970), kaempferol in *A. caucasicus* (Alaniya et al., 1975), isorhamnetin in *A. vulneraria* (Bedir et al., 2000), etc. (Figure 114d).

Beyond the aglycone pool, original structures have been identified in different species of *Astragalus* on the basis of different glycosylation and/or acylation types:

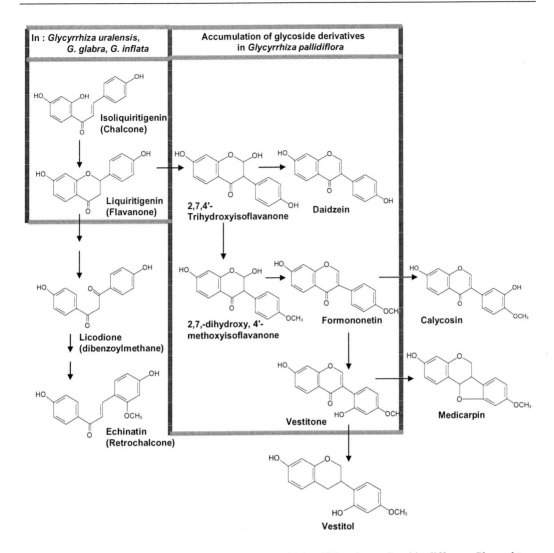

Figure 124. Biosynthesis pathways of isoflavonoids which will be glycosylated in different *Glycyrrhiza* species (Ayabe et al., 1982; Ammosov and Litvinenko, 2007).

In *Astragalus caprinus* Maire, tetraglycosides of kaempferol, quercetin and isorhamnetin have been isolated from the leaves (Figure 125b-c; 126a-b) (Semmar et al., 2001, 2002a, 2002b, 2005). Moreover, some tetraglycosides of kaempferol can be acylated by a *p*-coumaroyl or a feruloyl at different positions (3" or 4") of the galactose leading to four other new compounds (Figure 125d-g) (Semmar et al., 2002b). All kaempferol glycosides (Figure 125b-g) seem to be derived from a same precursor structure, consisting of a newly identified kaempferol triglycoside: kaempferol-3-*O*-(β-D-xylosyl-(1→3)-α-L-rhamnopyranosyl-(1→6)] -β-D-galactopyranoside (Figure 125a) (Semmar et al., 2002a). Other new original compounds have been isolated from the leaf material of *A. caprinus* (Semmar et al., 2002a, 2005), consisting of rhamnetin, rhamnocitrin and rhamnazin diglycosides which were acylated or not acylated by a methylated glutaroyl moiety (aliphatic acid) (Figure 126c-f).

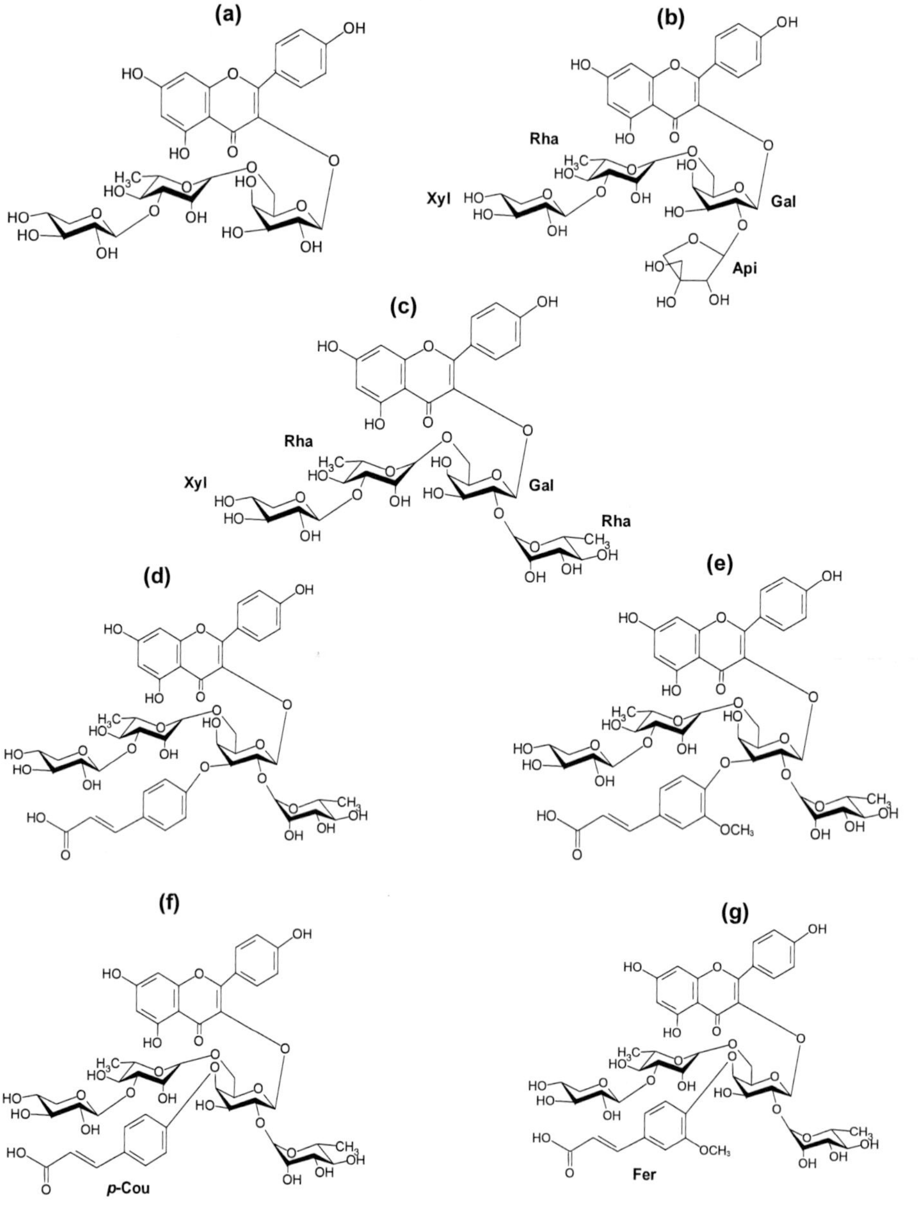

Figure 125. Chemical structures of non acyalated and acylated tri- and tetraglycosides of kaempferol isolated from the leaves of *Astragalus caprinus* (Semmar et al., 2001, 2002a, 2002b). Legend: Gal: galactosyl; Rha: rhamnosyl; Api: apiosyl; Xyl: xylosyl; p-Cou: p-coumaroyl; Fer: ferruloyl.

Astragalus vulneraria D.C. has been characterized by a triglycoside consisting of isorhamnetin-3-*O*-[β-D-apiofuranosyl-(1→2)]-[α-L-rhamnopyranosyl-(1→6)]-β-D-galactopyranoside (Figure 127a) (Bedir et al., 2000). Comparison between *A. caprinus* and *A. vulneraria* showed a strong similarity in the glycosidic sequence (3-*O*-[β-D-apiofuranosyl-(1→2)]-[α-L-rhamnopyranosyl-(1→6)]-β-D-galactopyranoside) (Figure 125b, 126a-b, 127a) highlighting an original chemical feature concerning the genus *Astragalus*.

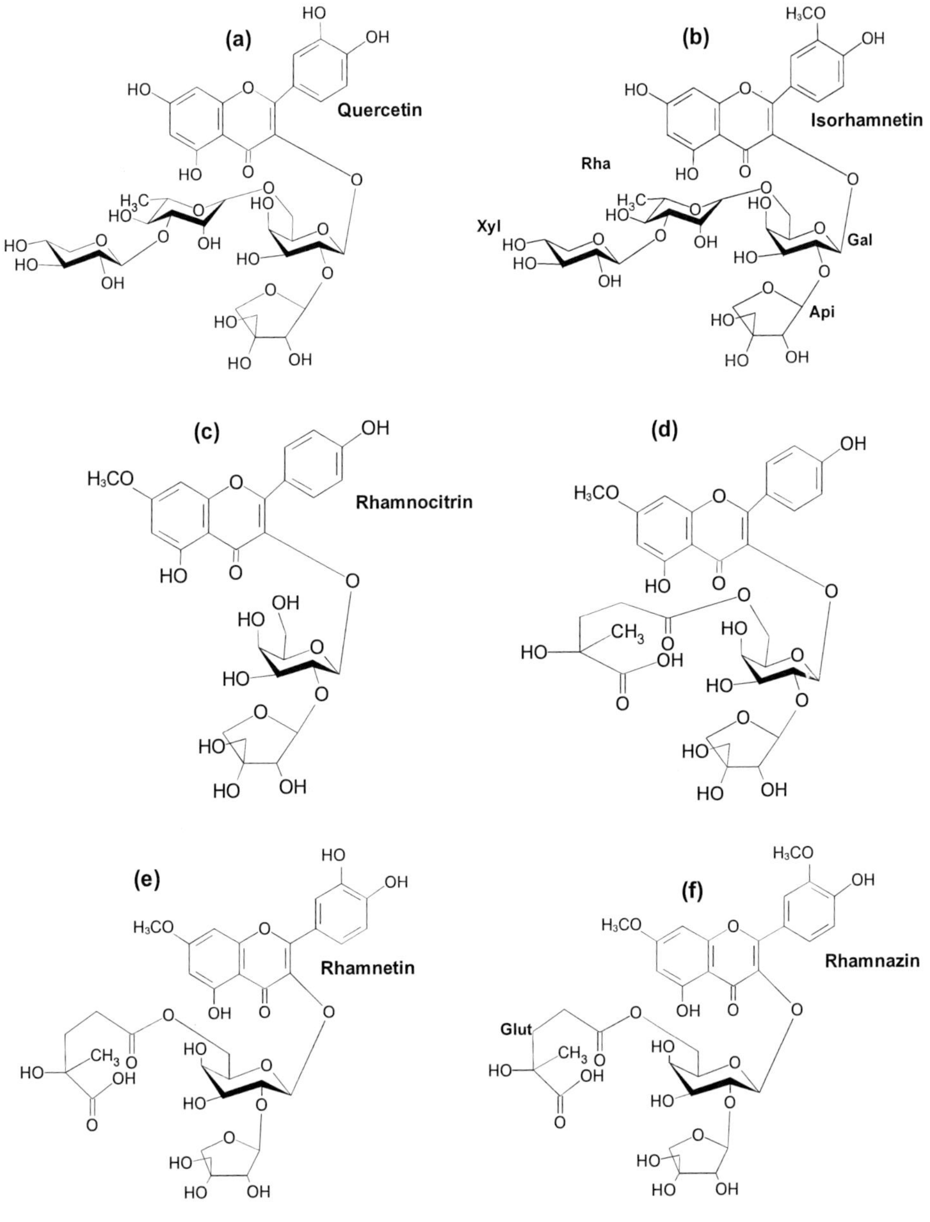

Figure 126. Chemical structures of querectin (a) and isorhamnetin (b) tetraglycosides, rhamnocitrin diglycoside (c) and acylated diglycosides of rhamnocitrin (d), rhamnetin (e) and rhamnazin (f) isolated from the leaves of *Astragalus caprinus*. (Semmar et al., 2001, 2002a, 2002b). Legend: Gal: galactosyl; Rha: rhamnosyl; Api: apiosyl; Xyl: xylosyl; Glut: glutaroyl.

The species *Astragalus shikokianus* has been characterized by a bidesmosidic (two glycosidic branches) tetraglycoside of kaempferol (Figure 127b) (Yahara et al., 2000).

In *Astragalus complanatus* R. Br., a very original kaempferol glycoside acylated by an abscissic acid-type sesquiterpene have been isolated from the seeds (Figure 127c) (Cui et al., 1991). This species was also characterized by the occurrence of myricetin (Figure 114g), which was not reported in other *Astragalus* species. Moreover, isoflavone glycosides have been isolated from the seeds (Figure 127e, 127f) (Cui et al., 1993).

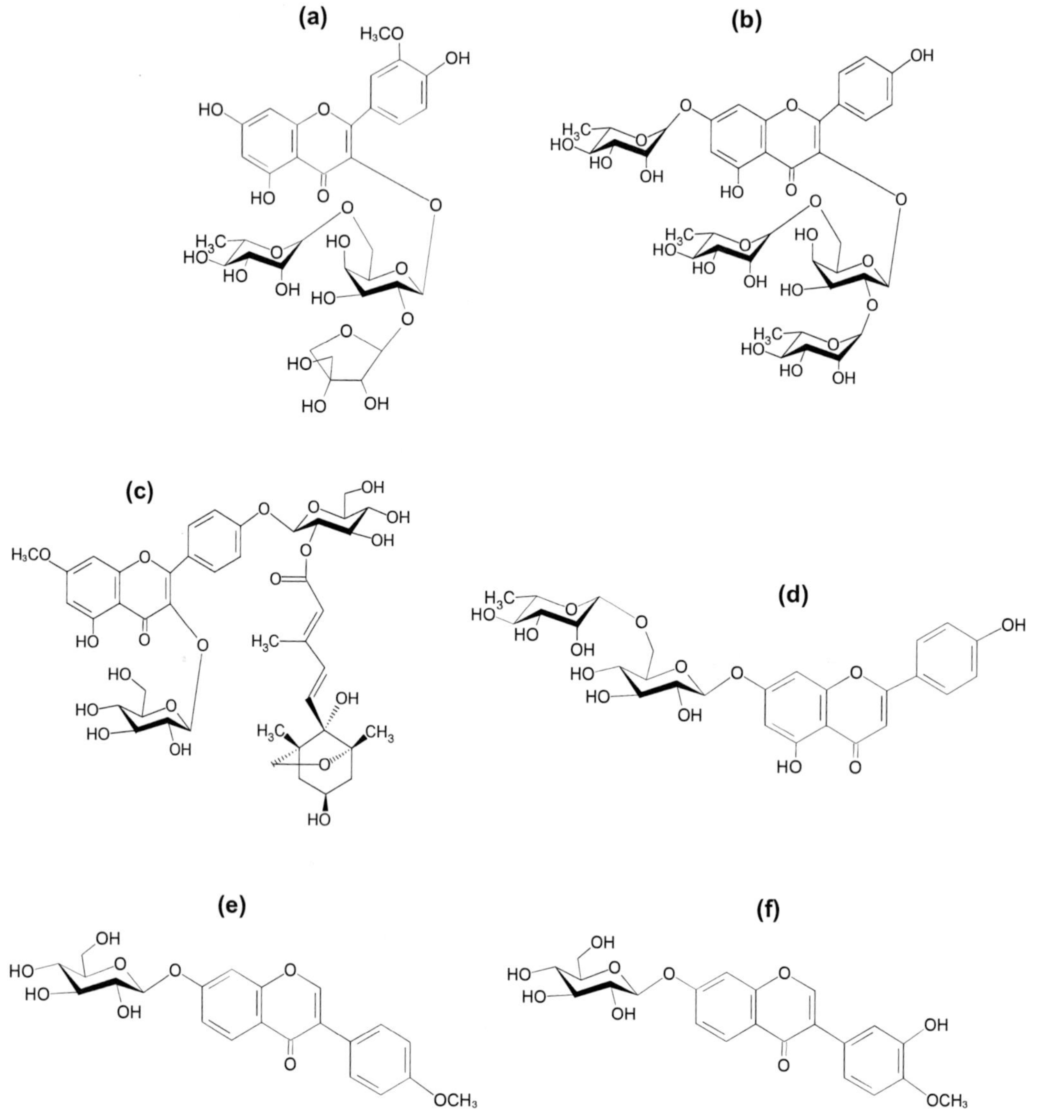

Figure 127. Chemical structures of mono-, di-, tri- and tetra-glycosides of flavonols, flavones and isoflavones in different Fabaceae species. Note the unusual acylated structure (c).

In addition to the flavonol glycosides, some *Astragalus* species have been described by their ability to synthesize flavone glycosides: Apigenin 7-*O*-rutinoside (Figure 127d) has isolated from *A. onobrychis* L. and represented a new compound for the whole genus *Astragalus* (Benbassat and Nikolov, 1995).

VIII.3.6. Infraspecific level chemotaxonomy

Bilton et al. (1976) have reported that leaflets of red clover (*Trifolium pratense*) inoculated with *Botrytis cinerea* (the grey-mould fungus) contained the isoflavonoids, pisatin and variabilin (Figure 109c, d) in addition to the previously described phytoalexins medicarpin and maackiain (Figure 109a, b) (Higgins and smith, 1972).

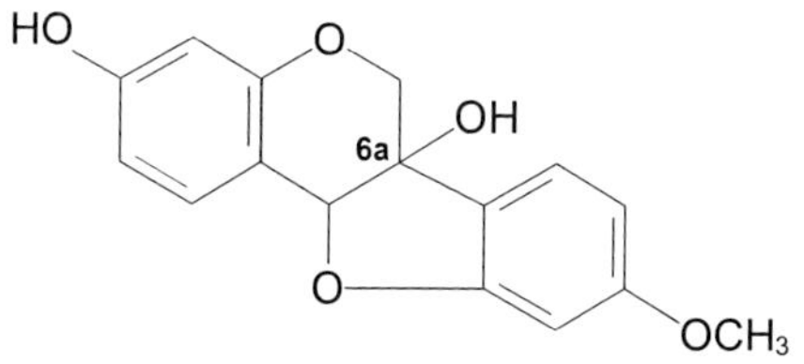

6a-Hydroxymedicarpin 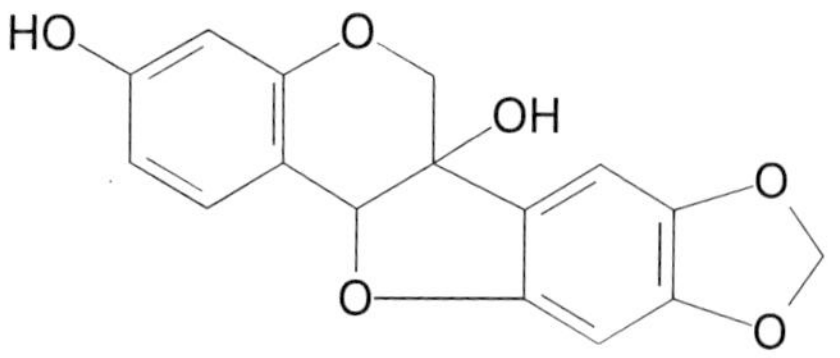   **6a-Hydroxymaackiain**

Figure 128. Chemical structures of pterocarpan phytoalexins deriving from medicarpin and maakiain, and found in the leaves of *Trifolium pratense* after inoculation by the fungus *Botrytis cinerea*.

However, neither pisatin nor variabilin were isolated from *Trifolium pratense* incoculated by *Helminthosporium carbonum* by opposition to the same species inoculated by *Botrytis cinerea* (Ingham, 1978). The parasite *B. cinerea* differs significantly from *H. carbonum* in that it has the proven ability to detoxify medicarpin and other pterocarpan phytoalexins via the introduction of a non-aromatic hydroxyl-group at C-6a (Ingham, 1976b; Van den Heuvel and Glazener, 1975; Van den Heuvel, 1976). Therefore, it is probable that pisatin and variabilin (Figure 109c, d) from *T. pratense* are fungal modification products of medicarpin and maackiain (Figure 109a, b) rather than new phytoalexins. This hypothesis was based on the fact that *B. cinerea* converts medicarpin and maakiain *in vitro* to give the corresponding 6a-hydroxy derivatives (Figure 128a, b) and that both compounds (which are presumably the immediate precursors of pisatin and variabilin respectively) occur in *B. cinerea*-infected red clover leaflets (Bilton et al., 1976) which actively synthesize medicarpin and maackiain.

This example gives an illustration on the usefulness of phytochemistry in identification of the parasitism mode that a plant species could confront, helping to bring appropriate care solutions.

A chemotaxonomical analysis has been carried out on the species *Astragalus caprinus* on the basis of the flavonol glycosides in the leaves (Semmar et al., 2005). The work concerned a set of 404 individual plants collected in North, Center and South of Tunisia. HPLC analysis of the flavonoid glycosides in the leaves highlighted the occurrence of 14 major compounds 13 of which have been identified by ^{1}H and ^{13}C NMR and MS (Figure 125, 126).

A multivariate analysis of the dataset (404 plants × 14 flavonoids) was performed by combining correspondence analysis (CA) with hierarchical cluster analysis (HCA): from the three first factors given by CA, a dendrogram has been obtained in two steps (Semmar et al., 2005):

- Computation of Euclidean distances between plants;
- Clustering of the plants within homogenous groups according to Ward aggregation algorithm.

The resulting dendrogram (Figure 129) highlighted four homogenous chemical groups corresponding to four chemotypes which separated at a high distinctness level. The metabolic patterns of these four chemotypes (Figure 130) had different shapes due to different major/minor and/or present/absent compounds:

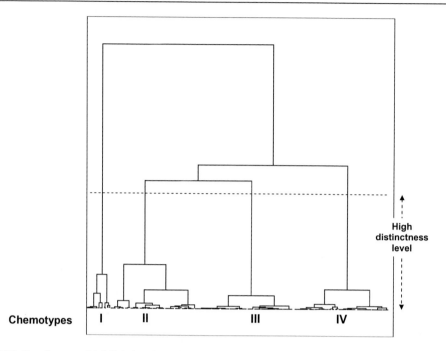

Figure 129. Dendrogram highlighting four chemotypes of *Astragalus caprinus* obtained from Euclidean distance and Ward aggregation algorithm applied on a dataset of 14 flavonol glycosides analysed in the leaves of 404 individual plants (Semmar et al., 2005).

- Chemotype I was exclusively able to produce the less polar glycosides, i.e. the non acylated and acylated diglycosides of rhamnocitrin (Figure 126c, d), and acylated diglycosides of rhamnetin and rhamnazin (Figure 126e, f). These four compounds were absent from the three other chemotypes (Figure 130). Moreover, quercetin and isorhamnetin tetraglycosides (Figure 126a, b) had high regulation ratios in chemotype I (Figure 130). Another characteristic of this chemotype concerned the occurrence of only the tetraglycoside of kaempferol having an apiose, among the kaempferol glycosides (Figure 125b, 130); the other kaempferol glycosides were absent or detected at trace levels. From this last characteristic, it is also interesting to note that all the compounds of chemotype I had apiose leading to deduce a biogenetic relationship between them: e.g. linked to high activity of apiosyl transferase.
- Chemotype II (Figure 130) was mainly characterized by high regulation levels of quercetin tetraglycoside (Figure 126a) favouring relatively its methylated derivative, isorhamnetin tetraglycoside (Figure 126b). Moreover, the apiosylated tetraglycoside of kaempferol (Figure 125b) occurred with significant relative levels leading to sustain the hypothesis about high apiosyl-transferase activity. However, the diglycosides of rhamnocitrin, rhamnetin and rhamnazin (Figure 126c-f) were absent, probably because of a non-expression of methyl-transferase (in aglycone biosynthesis) and/or acyl-transferase.
- By opposition to chemotypes I and II, chemotype III was characterized by a prevalence of kaempferol glycosides, and particularly kaempferol tetraglycosides which had not apiose (Figure 125c-g) (except the one already cited, Figure 125b). The highest relative levels concerned the non acylated kaempferol tetraglycoside (Figure 125c), followed by its acylated derivatives (by *p*-coumaric or ferulic acid)

(Figure 125d-g). All these compounds had a rhamnose at the 2-position of galactose instead of apiose.

- Chemotype IV consisted of a second chemotype based on kaempferol glycoside. Its characteristic compound was kaempferol triglycoside (Figure 125a) in presence of a high regulation of its apiosylated derivative (Figure 125b) (Figure 130). This second compound was particularly the most occurred in all the chemotypes and could be an index of the importance of apiosyl transferase in the species *A. caprinus*. However, the pattern of chemotype IV was as much pronounced as the relative levels of the characteristic compound, kaempferol triglycoside (Figure 125a), were higher.

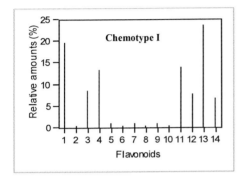
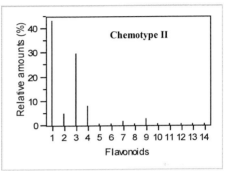

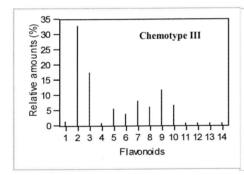
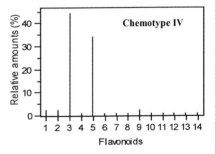

Legend:

Chemical structures of compounds 1, 4, 12, 13, 11, 14 are presented in figure 126 (a-f), respectively.

Chemical structures of compounds 5, 3, 2, 7, 9, 8, 10 are presented in figure 125 (a-g), respectively

Compound 6 (acylated kaempferol glycoside) was not entirely identified because of insufficient amounts

Figure 130. Four relative profiles of flavonol glycosides corresponding to four chemotypes of *Astragalus caprinus* (Semmar et al., 2005). Along the abscissa axis, the flavonoids were ranked according to their elution order in C18 HPLC, i.e. from the most polar (1) to the less (14) polar compound.

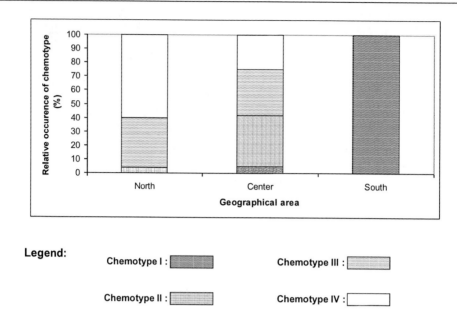

Figure 131. Relative distribution of the four chemotypes of *Astragalus caprinus* in three geographical area of Tunisia (North, Center, south) (Semmar et al., 2005).

Analysis of the distribution of the four chemotypes showed significantly differences between the three sampling sites of Tunisia (North, Center, South) (Figure 131):

- The South contained exclusively chemotype I which is characterized by the occurrence of the less polar glycosides (Figure 126c-f) absent elsewhere. This geographical area corresponds to arid climate which could exert a selective pressure leading to favour only chemotype I.
- The northern plants corresponded essentially to chemotypes III and IV. The North of Tunisia is characterised by higher humidity and rainfalls compared with the South and the Center. On this basis, the development of chemotypes III and IV in the North could translate good adaptative conditions for the expression of these two chemical patterns.
- Finally the Center of Tunisia was original in the sense that it showed the highest chemotypic diversity. The four chemotypes co-occurred with a particular well development of chemotype II. From such co-occurrence, the region of Center could be considered as an evolutive space for all the chemotypes of *Astragalus caprinus*. This is in agreement with the intermediate location and climatic conditions of Center region.

Correlation analysis between the fourteen compounds showed negative trends between glycosides based on different aglycones (Semmar et al., 2001; 2007). Positive correlations concerned as well glycoside having a same aglycone (e.g. kaempferol glycosides) as glycosides belonging to a same aglycone family (e.g. quercetin and isorhamnetin, or methylated aglycones at 7-position between them). Further analysis of these correlations has been performed through a simulation metabolomic approach (Semmar et al., 2007). This new approach combined *in silico* the four chemotypes for several generations in order to extract a metabolic backbone from which the chemical polymorphism of *A. caprinus* has been further analysed. The simulation results confirmed the preliminary correlation analysis and led to conclude that the four chemotypes of *A. caprinus* emerged from competition between and within three metabolic pathways (Figure 132):

- Pathway I gathered the acylated diglycosides of rhamnocitrin, rhamnetin and rhamnazin, the high regulations of which led to the development of chemotype I.
- Pathway II concerned tetraglycosides of quercetin and isorhamnetin, and appeared to be competitive toward pathway III representing kaempferol glycosides. Its high expression led to the development of chemotype II.
- Pathway III included all the kaempferol glycosides (Figure 125) which shared strongly positive correlations except the triglycoside of kaempferol (Figure 125a) which showed fairly positive correlations. Such fairly positive correlations could be explained by the branched location of the kaempferol triglycoside (Fig. 132): subsequently, this compound can be glycosylated either by a rhamnose or an apiose to give (through two subpathways) two different tetraglycosides (Figure 125b, c). A high expression of rhamnosyl-transferase is favourable for the development of chemotype III characterized by non-acylated and acylated tetraglycosides containing rhamnose instead of apiose. By default, chemotype IV will be favoured leading to co-occurrence of kaempferol triglycoside and its apiosylated tetraglycoside derivative (Figure 130; 125a, b).

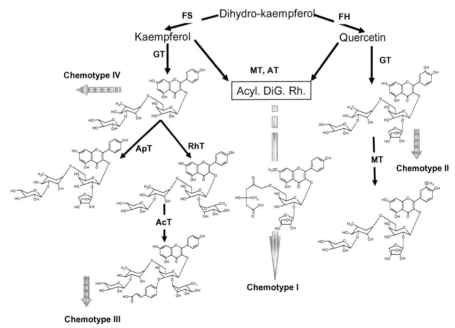

Legend:
AcT: Acyl Transferase; ApT: Apiosyl Transferase; FH: Flavonol Hydroxylase; FS: Flavonol Synthase; GT: Glycosyl Transferase; MT: Methyl Transferase; RhT: Rhamnosyl Transferase; Acyl. DiG. Rh.: Acylated Diglycosides of Rhamnocitrin, Rhamnetin or Rhamnazin.

Figure 132. Links between metabolic pathways of flavonol glycosides and the four chemotypes of *Astragalus caprinus*, resulting from a simulation-based metabolomic approach (Semmar et al., 2007).

REFERENCES

Abdel-Moqib, M., Davidar, A. M., Metwally, M. A. & Abou-Elzhab, M. (1989). Flavonols of *Pulicaria undulata*. *Pharmazie*, *44*, 801.

Adamska M. & Lutomski J. (1971). C–Flavonoidglykoside In Den Samen Von *Trigonella Foenum–Graecum.Planta Med.* 20, 224-229.

Adinolfi, M., Barone, G., Belardini, M., Lanzetta, R., Laonigro, G. & Parrilli, M. (1984). *Phytochemistry*, *23*, 2091-2093.

Adinolfi, M., Barone, G., Belardini, M., Lanzetta, R., Laonigro, G. & Parrilli, M. (1985a). Homoisoflavanones from *Muscari comosum* bulbs. *Phytochemistry*, *24*, 2423-2426.

Adinolfi, M., Barone, G., Lanzetta, R., Laonigro, G., Mangoni, L. & Parrilli, M. (1985b). Three 3-benzyl-4-chromanones from *Muscari comosum*. *Phytochemistry*, *24*, 624-626.

Adzet, T. & Martinez, F. (1981). Flavonoids in the leaves of *Thymus*: a chemotaxonomic survey. *Biochem. Syst. Ecol.*, *9*, 293-295.

Adzet, T., Cañigueral, S. & Iglesias, J. (1988a). Chromatographic survey of polyphenols from *Salvia* species. *Biochem Syst. Ecol.*, *16*, 29-32.

Adzet, T., Vila, R. & Cañnigueral, S. (1988b). Chromatographic analysis of polyphenols of some Iberian *Thymus*. *J. Ethnopharmacol.*, *24*, 147-154.

Agawal, P. K. (1989). *Carbon-13 NMR of Flavonoids*. Elsevier, New York.

Alaniya, M. D., Komissarenko, N. F. & Kermertelidze, E. P. (1975). Ascaside, a new flavonol glycoside of *Astragalus caucasicus*. *Khim Prir. Soedin.*, *11*, 351-354.

Al-Eisawi, D. M. (1998). *Field Guide to Wild Flowers of Jordan and Neighbouring Countries*. The National Library, Amman.

Alfatafta, A. A. & Mullin, C. A. (1992). Epicuticular terpenoid and an aurone from flowers of *Helianthus annuus*. *Phytochemistry*, *31*, 4109-4113.

Alston, R. E. & Tuner, B. L. (1962). New Techniques In Analysis of Complex Natural Hybridization. *Proc. Natn. Acad. Sci.*, USA, *48*, 130-137.

Alston, R. E., Turner, B. L., Lester, R. N. & Horne, D. (1962). Chromatographic Validation of Two Morphologically Similar Hybrids of Different Origins. *Science*, *137*, 1048-1050.

Alston R.E. and Hemple K. (1964). Chemical Documentation of Interspecific Hybridization. *J. Heredity*, *60*, 267-269.

Ammosov, A. S. & Litvinenko, V. I. (2007). Phenolic compounds of the genera *Glycyrrhiza* L. and *Meristotropis* Fish. Et §ey. (review). *Pharmaceutical Chemistry Journal*, *41*, 372-395.

Andary, C., Wylde, R., Laffite, C., Privat, G. & Winternitz, F. (1982). Structure of verbascoside and orobanchoside, caffeic acid sugar esters from *Orobanche rapum-geniste*. *Phytochemistry*, *21*, 1123-1127.

Angiosperm Phylogeny Group (APG) (1998). An ordinal classification of the families of flowering plants. *Ann. Missouri Bot. Gard.*, *85*, 531-553.

Angiosperm Phylogeny Group (APG) (2003). An update of the Angiosperm Phylogeny Group classification for the orders and families of flowering plants: APG II. *Botanical Journal of the Linnean Society*, *141*, 399-436.

Asada, Li. W. & Yoshikawa, T. (2000). Flavonoid constituents from *glycyrrhiza glabra* hairy root cultures. *Phytochemistry*, *55*, 447-456.

Ayabe, S., Yoshikawa, T., Kobayashi, M. & Furuya, T. (1982). Biosynthesis of a retrochalcone, echinatin: involvement of O-methyltransferase to licodione. *Phytochemistry*, *19*, 2331-2336.

Barberán, F. A. T. (1986). The flavonoids from the Labiatae. *Fitoterapia*, *57*, 67-95.

Barberán, F. A. T., Hernandez, L, Ferreres, F. & Tomás, F. (1985b). A highly methylated 6-hydroxyflavones and other flavonoids from *Thymus piperella*. *Planta Med.*, 1985, 452-454.

Barberán, F. A. T., Núñez, J. M. & Tomás, F. (1985a). An HPLC study of flavones from some Spanish *Sideritis* species. *Phytochemistry*, *24*, 1285-1288.

Barron, D., Varin, L., Ibrahim, R. K., Harborne, J. B. & Williams, C. A. (1988). Sulphated flavonoids: an update. *Phytochemistry*, *27*, 2375-2395.

Bate-Smith, E. C. (1968). The phenolic constituents of plants and their taxonomic significance. *J. Linn. Soc.*, *60*, 325-383.

Bazon, J. N., Lopes, J. L. C., Vichnewski, W. & Lopes, J. N. C. (1997). Constituents of *Lychnophora brunioides*. *Fitoterapia*, *68*, 92.

Beal, G. H., Price, J. R. & Sturgess, V. C. (1941). A survey of the anthrocyanins. VII. The natural selection of flower colour. *Proc. Roy. Soc.*, (London), *B130*, 113-126.

Bedir, E., Çalis, I., Piacente, S., Pizza, C. & Khan, I. A. (2000). A new flavonol glycoside from the aerial parts of *Astragalus vulneraria*. *Chem. Pharm. Bull*, *48*, 1994-1995.

Benbassat, N. & Nikolov, S. (1995). Flavonoids from *Astragalus onobrychis*. *Planta Med.*, *61*, 100.

Bendz, G. & Mårtensson, O. (1963). Moss pigments II. The anthocyanins of *Bryum rutilans* Brid. and *Bryum weigelii* Spreng. *Acta Chem. Scand.*, *17*, 266.

Bendz, G., Mårtensson, O. & Terenius, L. (1962). Moss pigments I. The anthocyanins of *Bryum cryophilum* O. Mårt. *Acta Chem. Scand.*, *16*, 1183-1190.

Bentham, G. & Hooker, J. D. (1883). *Genera Plantarum*, Vol. III, Part II, Reeve et al., Williams and Norgate, London.

Bentham, G. (1876). Labiatae. In: G. Bentham, & K. D. Hooker (Eds.), *Genera Planetarum, vol 2*. Reeve & Co., London (1876), 1160-1233.

Besson, E., et al. (1985). *C*-glycosylflavones from *Oryza sativa*. *Phytochemistry*, *24*, 1061-1064.

Bickoff, E. M., Livingston, A. L. & Witt, S. C. (1965). *Phytochemistry*, *4*, 523.

Bilton, J. N., Debnam, J. R. & Smith, I. M. (1976). 6a-hydroxypterocarpans from red clover.*Phytochemistry*, *15*, 1411-1412.

Bisby, F. A., Buckingham, J. & Harborne, J. B. (1994). *Phytochemical Dictionary of the Leguminoseae*. Chapman and Hall, London.

Bloor, S. J. (2001). Deep blue anthocyanins from blue *Dianella* berries. *Phytochemistry*, *58*, 923-927.

Bonde, M. R., Millar, R. L. & Ingham, J. L. (1973). Induction and identification of sativan and vestitol as two phytoalexins from *Lotus corniculatus*. *Phytochemistry*, *12*, 2957-2959.

Bouaziz M., Simmonds M.S.J., Grayer R.J., Kite G.C. & Damak M. (2001). Flavonoids from *Hypparhenia hirta* Stapf (Poaceae) growing in Tunisia, *Biochem. Syst. Ecol.*, *29*, 849-851.

Bremer, K. (1994). *Asteraceae: Cladistics and Classification*. Timber Press, Portland, OR.

Bremer, K. (1996). Major clades and grades of the Asteraceae. In D. J. N. Hind, & H. J. Beentje (Eds.), *Compositae: Systemetics Proceedings of the International Compositae Conference, Vol. 1*. Royal Botanic Gardens, Kew, 1-7.

Briggs, L. H., Briggs, L. R. & King, A. W. (1975). New Zeland phytochemical survey. 14. Constituents of *Dianella nigra* Col. *N. Z. J. Sci.*, *18*, 559-563.

Bučkova, A., Eisenreichová, E., Haldová, M., Uhrin, D. & Tomko, J. (1988). A new acylated kaempferol derivative from *Lilium candidum* L. *Phytochemistry*, *27*, 1914-1915.

Budzianowski, J. (1991). Six flavonol glucuronides from *Tulipa gesneriana.Phytochemistry*, *30*, 1679-1682.

Calis, I., Basaran, A. A., Saracoglu, I., Sticher, O. & Rüedi, P. (1990). Phlinosides A, B and C, three phenylpropanoid glycosides from *Phlomis linearis*. *Phytochemistry*, *29*, 1253-1257.

Calis, I., Basaran, A. A., Saracoglu, I. & Sticher, O. (1992a). Iridoid and phenylpropanoid glycosides from *Stachys macrantha*. *Phytochemistry*, *31*, 167-169.

Calis, I., Ersöz, T., Tasdemir, D. & Rüedi, P. (1992b). Two phenylpropanoid glycosides from *Leonurus glaucescens*. *Phytochemistry*, *31*, 357-359.

Cantino, P. & Sanders, R. W. (1986). Substantial classification of Labiatae. *Syst. Bot.*, *11*, 163-185.

Cantino, P. D., Harley, R. M. & Wagstaff, S. J. (1992). Genera of Labiatae; status and classification. In: R. M. Harley, & T. Reynolds (Eds.), *Advances in Labatae Science*. Royal Botanic Gardens, Kew, 511-522.

Charaux, C. & Rabaté, J. (1939). Constitution chimique de l'orobol. *Bull. Soc. Chim. Biol.*, *21*, 1330-1333.

Chari, V. M., Harborne, J. M. & Williams, C. A. (1980). Identification of 6″-*O*-acetyl-8-*C*-galactosylapigenin in *Briza media*. *Phytochemitsry*, *19*, 983-984.

Chase, M. W. (2004). Monocot relationships: an overview. *American Journal of Botany*, *91*, 1645-1655.

Chase, M. W., Soltis, D. E., Olmstead, R. G., Morgan, D., Les, D. H., Mishler, B. D., Duvall, M. R., et al. (1993). Phylogenetics of seed plants: an analysis of nucleotide sequences from the plastid gene rbcL. *Annals of the Missouri Botanical Garden*, *80*, 528-580.

Chase, M. W., Stevenson, W. D. W., Wilkin, P. & Rudall, P. J. (1995). Monocot systematics: a combined analysis. In: P. J. Rudall, P. J. Cribb., D. F. Cutler, & C. J. Humphries (Eds.), *Monocotyledons: Systematics and evolution*. Royal Botanic Gardens. Kew, 685-730.

Chase, M. W., Fay, M. F., Devey, D., Maurin, O., Rønsted, N., Davies, J., Pillon, Y., Petersen, G., et al. (2006). Multigene analyses of monocot relationships: a summary. In J. T. Columbus, E. A. Friar, J. M. Porter, L. M. Prince, & M. G. Simpson, (Eds.), *Monocots: comparative biology and evolution (excluding Poales)*. Rancho Santa Ana Botanical Garden, Claremont, Ca, 63-75.

Cho, J. & Lee, H. K. (2004). Wogonin Inhibits Ischemic Brain Injury in a Rat Model of Permanent Middle Cerebral Artery Occlusion. *Biol. Pharm. Bull.*, *27*, 1561.

Chopin J., Dellamonica G., Bouillant M. L., Besset A., Popovici G. & Weissenböck G. (1977). *C*-Glycosylflavones from *Avena sativa*. *Phytochemistry*, *16*, 2041-2043.

Colgate, S. M., Dorling, P. R. & Huxtable, C. R. (1986). Dianellidin, stypandrol and dianellinone: an oxidation-related series from *Dianella revoluta*. *Phytochemistry*, *25*, 1245-1247.

Cooke, R. G. & Sparrow, L. G. (1965). Colouring matters of Australian plants. XII. Quinones from *Dianella nigra* and *Stypandra grandis*. *Aust. J. Chem.*, *18*, 218-225.

Corticchiato, M., Bernardini, A., Costa, J., Bayet, C., Saunois, A. & Voirin, B. (1995). Free flavonoid aglycones from *Thymus herba-barona* and its monoterpenoid chemotypes. *Phytochemistry*, *40*, 115-120.

Costa, F. B., Dias, D. A., Lopes, J. L. C. & Vichnewsky, W. (1993). Flavonoids and heliangolides from *Lychnophora diamantinana*. *Phytochemistry*, *34*, 261-263.

Crawford, D. J. & Stuessy, T. F. (1981). The taxonomic significance of anthochlors in the subtribe Coreopsidinae (Compositae, Heliantheae). *Amer J. Bot.*, *68*, 107-117.

Cronquist, A. (1981). *An Investigated System of Classification of Flowering Plants*. Columbia University Press, New York.

Crowden, R. K. & Jarman, S. J. (1974). 3-Deoxyanthocyanins from the fern *Blechnum procerum*. *Phytochemistry*, *13*, 1947-1948.

Cruickshank, I. A. M, Veeraraghavan, J. & Perrin, D. R. (1974). Some physical factors affecting the formation and/or net accumulation of medicarpin in infection droplets on white clover leaflets. *Aust. J. Pl. Physiol.*, *1*, 149-156.

Cui, B., Inoue, J., Nakamura, M. & Nohara, T. (1991). A novel acylated flavonoid glycoside from *Astragalus complanatus*. *Tetrahedron Lett.*, *32*, 6135-6138.

Cui, B., Nakamura, M., Kinjo, J. & Nohara, T. (1993). Chemical constituent of Astragali semen. *Chem. Pharm. Bull.*, *41*, 178-182.

Dahlgren, R. M. T., Clifford, H. T. & Yeo, P. F. (1985). *The Families of Monocotyledons*, Springer-Verlag, Berlin.

Damtoft, S., Franzyk, H. & Jensen, S. R. (1995). Biosynthesis of secoiridoids in *Fontanesia*. *Phytochemistry*, *38*, 615-621.

Davis, P. H. (1982). *Flora of Turkey and East Aegean Islands, vol. 7*. Edinburgh University Press, Edinburgh.

de Rijge, E., Zarfa Gomez, A., Arise Udo, F. Th., Brinkman, A. & Gooljer, C. (2001). Determination of isoflavone glucoside malonates in *Trifolium pratense* L. (red clover) extracts. *J. Chromatogr. A.*, *932*, 55-64.

Del Pero Martinez, M. A. & Swain, T. (1985). Flavonoids and chemotaxonomy of the Commelinaceae. *Biochem. Syst. Ecol.*, *13*, 381-402.

Delazar, A., Naseri, M., Nazemiyeh, H., Talebpour, A. H., Imani, Y., Nahar, L. & Sarker, S. D. (2007). Flavonol 3-methyl ether glucosides and a tryptophylglycine dipeptide from *Artemisia fragrans* (Asteraceae). *Biochem. Syst. Ecol.*, *35*, 52-56.

Dellamonica, D., et al. (1983). Two isovitexin 2″-O-glycosides from primary leaves of *Secale cereale*. *Phytochemistry*, *22*, 2627-2628.

Dement, W. A. & Mabry, T. J. (1975). Biological implications of flavonoid chemistry in *Baptisia* and *Thermopsis*. *Biochem. Syst. Ecol.*, *3*, 91-94.

Demizu, S., Kajiyama, K., Hiraga, Y., Kinoshita, K., Koyama, K., Takahashi, K., Tamura, Y., Okada, K. & Kinoshita, T. (1992). Prenylated dibenzoylmethane derivatives from the root of *Glycyrrhiza inflata* (Xinjiang licorice). *Chem. Pharm. Bull.*, *40*, 392-395.

Dixon, R. A. & Summer, L. W. (2003). Legume natural products: Understanding and manipulating complex pathways for human and animal health. *Plant Physiol.*, *131*, 878-885.

El-Ansari, M., Nawwar, M. A. & Salej, N. A. M. (1995). Stachysetin, a diapigenin-7-glucoside-*p,p'*-dihydroxy-truxinate from *Stachys aegyptiaca*. *Phytochemistry*, *40*, 1543-1548.

El-Gazzar, A. & Watson, L. (1970a). Some economic implications of the taxonomy of Labiatae: Essential oils and rusts. *New Phytol.*, *69*, 487-492.

El-Gazzar, A. & Watson, L. (1970b). A taxonomic study of Labiatae and related genera. *New Phytol.*, *69*, 451-486.

El-Gazzar, A. (1974). Numerical taxonomy of the Verbenaceae: a reassessment. *Egypt J. Bot.*, *17*, 69-83.

El-Naggar, L. J. & Beal, J. L. (1980). Iridoids : A Review. *J. Nat. Prod.*, *46*, 649-707.

Emeranciano, V. P., Ferreira, Z. S., Kaplan, M. A. C. & Gottlieb, O. R. (1987). A chemosystematic analysis of tribes of Asteraceae involving sesquiterpene lactones and flavonoids. *Phytochemistry*, *26*, 3103-3115.

Emerenciano, V. P., Militão, J. S. L. T., Campos, C. C., Romoff, P., Kaplan, M. A. C, Zambon, M. & Brant, A. J. C. (2001). Flavonoids as chemotaxonomic markers for Asteraceae. *Biochem. Syst. Ecol.*, *29*, 947-957.

Endo, K., Takahashi, K., Abe, T. & Hikino, H. (1982). Structure of forsitoside B, an antibacterial principle of *Forythia koreana* stems. *Heterocycles*, *19*, 261-271.

Erdtman, G. (1945). Pollen morphology and Plant taxonomy . IV. Labiatae, Verbenaceae, and Avicenniaceae. *Svensk Bot. Tidskr*, *29*, 279-285.

Ferreres, F., Barberan, F. A. T. & Tomas, F. (1985). 5,6,4'-trihydroxy-7,8-dimethoxyflavone from Thymus membranaceus. *Phytochemistry*, *24*, 1869-1871.

Ferreres, F., Tomas, F., Barberan, F. A. T. & Hernandez, L. (1985). *Plantes Méd. Phytothérapie XIX 2*, 89.

Ferreres, F., Tomás-Barberán, F. A. & Tomás-Lorente, F. (1986). Flavonoid compounds from *Ballota hirsuta*. *J. Nat. Prod.*, *49*, 654-655.

Finckh, R. E. & Tamm, C. (1970). The Homo-Isoflavanones III – Isolation and Structure of Punctatin, 3,9-Dihydropunctatin, 4'-*O*-methyl-3,9-dihydropunctatin, 4'-Demethyleucomin and 4'-Demethyl-5-*O*-methyl-3,9-dihydro-eucomin. *Experientia*, *26*, 472.

Flamini, G., Pardini, M. & Morelli, I. (2001a). A flavonoid sulphate and other compounds from the roots of *Centaurea bracteata*. *Phytochemistry*, *58*, 1229-1233.

Flamini, G., Antognoli, E. & Morelli, I. (2001b). Two new flavonoids and other compounds from the aerial parts of *Centaurea bracteata* from Italy. *Phytochemistry*, *57*, 559-564.

Foo, L. Y., Molan, A. L., woodfield, D. R. & McNabb, W. C. (2000). The phenols and prodelphinidins of white clover flowers. *Phytochemistry*, *54*, 539-548.

Fujai, T., Wang, Q. H., Inami, R. & Nomura, T. (1990). Structures of prenylated dihydrochalcon, gancaonin J and homoisoflavanone, gancaonin K from *Glycyrrhiza pallidiflora*. *Heterocycles*, *31*, 643-650.

Furuya, T., Matsumoto, K. & Hikichi, M. (1971). Echinatin, a new chalcone from tissue culture of *Glycyrrhiza echinata*. *Tetrahedron Lett.*, *12*, 2567-2569.

Giannasi, D. E. (1975). The flavonoid systematics of the genus *Dahlia* (Compositae). *Mem. New York Bot. Gard.*, *26*, 1-125.

Gluchoff-Fiasson, K, Jay, M. & Viricel, M. R. (1989). Flavone *O*- and *C*-glycosides from *Setaria italica*. *Phytochemistry*, *28*, 2471.

Grael, C. F. F., Albuquerque, S. & Lopes, J. L. C. (2005). Chemical constituents of *Lychnophora pohlii* and trypanocidal activity of crude plant extracts and of isolated compounds. *Fitoterapia, 76*, 73-52.

Grayer, R. J., Chase, M. W. & Simmonds, M. S. J. (1999). A comparison between chemical and molecular characters for the determination of phylogenetic relationships among plant families: An appreciation of Hegnauer's "Chemotaxonomie der Pflanzen". *Biochem. Syst. Ecol., 27*, 369-393.

Grayer, R. J., Kite, G. C., Beitch, N. C., Eckert, M. R., Marin, P. D., Senanayake, P. & Paton, A. (2002). Leaf flavonoid glycosides as chemosystematic characters in *Ocimum*. *Biochem. Syst. Ecol., 30*, 327-342.

Grayer, R. J., Viera, R. F., Proce, A. M., Kite, G. C., Simon, J. E. & Paton, A. J. (2004). Characterization of cultivars within species of *Ocimum* by exudates flavonoid profiles. *Biochem. Syst. Ecol., 32*, 901-913.

Gu, L. V. & Gu, W. Y. (2001). Chgaracterisation of soy isoflavones and screening for novel malonyl glycosides using high-performance liquid chromatography-electrospray ionisation-mass spectrometry. *Phytochem. Anal., 12*, 377-382.

Hackel, (1887). In *Die Naturlichen Pflanzenfamilien*, A. Engler, & K. Prantl (Eds.), *2(ii)*, 1-97.

Halevy, A. H. & Asen, S. (1959). Identification of the anthocyanins in petals of tulip varieties Smiling Queen and Pride of Haarlem. *Plant Physiol., 34*, 494-499.

Harborne, J. B. & Hall, E. (1964). Plant polyphenols—XII. The occurrence of tricin and of glycoflavones in grasses. *Phytochemistry, 3*, 421-428.

Harborne, J. B. (1966a). Comparative biochemistry of the flavonoids-II. 3-Deoxyanthocyanins and their systematic distribution in ferns and gesnerads. *Phytochemistry, 5*, 589-600.

Harborne, J. B. (1966b). Comparative biochemistry of the flavonoids-I. distribution of chalcone and aurone pigments in plants. *Phytochemistry, 5*, 111-115.

Harborne, J. B. (1966c). Caffeic acid Ester distribution in higher plants. *Z. Naturforsch., 21b*, 604-605.

Harborne, J. B. (1967). *Comparative Biochemistry of the Flavonoids*. Academic Press, London.

Harborne, J. B. (1969). Chemosystematics of the Leguminosae. Flavonoid and isoflavonoid patterns in the tribe Genisteae. *Phytochemistry, 8*, 1449-1456.

Harborne, J. B. (1971a). Distribution and taxonomic significance of flavonoids in the leaves of the cyperaceae. *Phytochemistry, 10*, 1569-1574.

Harborne, J. B. (1971b). In *Chemotaxonomy of the Leguminoseae*. J. B. Harborne, et al. (Eds.), Academic Press, London, 31-72.

Harborne, J. B. & Williams, C. A. (1976). Flavonoid Patterns in Leaves of the Gramineae. *Biochem. Syst. Ecol., 4*, 267-280.

Harborne, J. B., Heywood, V. H. & King, L. (1976). Evolution of yellow flavonols in flowers of anthemideae. *Biochem. Syst. Ecol., 4*, 1-4.

Harborne, J. B. (1977). Flavonoids and evolution of the Angiosperms. *Biochem. Syst. Ecol., 5*, 5-22.

Harborne, J. B., Boardley, M., Fröst, S. & Holm, G. (1986a). The flavonoids in leaves of diploid *Triticum* species (Gramineae). *Plant Syst. Evol., 154*, 251-257.

Harborne, J. B., (1986b). The natural distribution in angiosperms of anthocyanins acylated with aliphatic dicarboxylic acids. *Phytochemistry, 25*, 1887-1894.

Harborne, J. B., Tomás-Barberán, F. A., Williams, C. A. & Gil, M. I. (1986c). A chemotaxonomic study of flavonoids from European *Teucrium* species. *Phytochemistry, 25*, 2811-2816.

Harborne, J. B. & Williams, C. A. (1994). Recent advances in the chemosystematics of the monocotyledons. *Phytochemistry, 37*, 3-18.

Haslam, E. (1982). Proanthocyanidins. In: *The Flavonoids. Advances in Research.* J. B. Harborne, & T. J. Mabry (Eds.), Chapman & Hall, London.

Heller, W. & Tamm, C. (1981). Homoisoflavanones and biogenetically related compounds. *Fortschr. Chem. Org. Naturst., 40*, 105-152.

Hennebelle, T., Sahpaz, S., Ezer, N. & Bailleul, F. (2008). Polyphenols from *Ballota larendana* and *Ballota pseudodictamnus*. *Biochem. Syst Ecol., 36*, 441-443.

Hernandez, L., Tomas-Barberan, F.A. & Tomas-Lorente, F. (1987). A chemotaxonomic study of free flavone aglycones from some Iberian *Thymus* species. *Biochem. Syst. Ecol., 15*, 61-67.

Herz, W., Gibaja, S., Bhat, S. V. & Srinivasan, A. (1972). Dihydroflavonol and other flavonoids of *Eupatorium* species. *Phytochemistry, 30*, 1269-1271.

Higgins, V. J. & Smith, D. G. (1972). Separation and identification of two pterocarpanoid phytoalexins produced by red clover leaves. *Phytopathology, 62*, 235.

Hofmann R.W., Swinny E.E., Bloor S.J. et al., 2000. Responses of nine *Trifolium repens* L. populations to ultraviolet-B radiation: differential flavonol glycoside accumulation and biomass production. *Annal Bot. 86*, 527-537.

Hosokawa K., Fukunaga Y., Fukushi E. & Kawabata J. (1995b). Five acylated pelargonidin glucosides in the red flowers of *Hyacinthus orientalis*. *Phytochemistry, 40*, 567.

Hosokawa, K., et al. (1995c). Seven acylated anthocyanins in the blue flowers of *Hyacinthus orientalis*. *Phytochemistry, 38*, 1293-1298.

Hosokawa, K. Fukunaga, Y., Fukushi, E. & Kawabata, J. (1995a). Acylated anthocyanins from red *Hyacinthus orientalis*. *Phytochemistry, 39*, 1437-1441.

Hubbard, C. E. (1959). In *The Families of Flowering Plants*, Vol. II Monocotyledons, J. Hutchinson (Ed.), Clarendon, Oxford, 710-741.

Husain, S. Z., Marin, P. D., Silic, C., Quaiser, M. & Petcovic, B. (1990). A micromorphological study of some representative genera in the tribe Saturjeae (Lamiaceae). *Botanical Journal of the Linnaean Society, 103*, 59-80.

Hutchinson, J. (1979). *The Families of Flowering Plants*. Oxyford University Press. Oxford.

Hutchinson, (1964). *The Genera of Flowering Plants*, Vol. 1. Clarendon Press, Oxford.

Ietswaart, J. H. (1980). A taxonomic revision of the genus *Origanum* (Labiatae). In: *Leiden Botanical Series, vol. 4.*, Leiden University Press, The Hague.

Ingham, J. L. (1972). Phytoalexins and other natural products are factors in plant disease resistance. *Botan. Rev., 38*, 343-424.

Ingham, J. L. (1976a). Isosativan: an isoflavan phytoalexin from *Trifolium hybridum* and other *Trifolium* species. *Z. Naturforsch, 31c(5-6)*, 331.

Ingham, J. L. (1976b). Fungal modification of pterocarpan phytoalexins from *Melilotus alba* and *Trifolium pratense*. *Phytochemistry, 15*, 1489-1495.

Ingham, J. L. (1976c). 3,5,4'-trihydroxystilbene as a phytoalexin from groundnuts (*Arachis hypogaea*). *Phytochemistry, 15*, 1791-1793.

Ingham, J. L. (1977a). *Z. Naturforsch., 32c*, 449.

Ingham, J. L. & Harborne, J. B. (1976). Phytoalexin induction as a new dynamic approach to the study of systematic relationships among higher plants. *Nature, Lond., 260*, 241-243.

Ingham, J. L. (1977b). Isoflavan phytoalexins from *Anthyllis, Lotus* and *Tetragonolobus*. *Phytochemistry, 16*, 1279-1282.

Ingham, J. L. (1978). Isoflavonoid and stilbene phytoalexins of the genus *Trifolium*. *Biochem. Syst. Ecol., 6*, 217-223.

Ingham, J. L. (1979). Isoflavonoid phytoalexins of the genus *Medicago. Biochem. Syst. Ecol., 7*, 29-34.

Ingham, J. L. (1981). Isolation and identification of *Cicer* isoflavonoids. *Biochem. Syst. Ecol., 9*, 125-128.

Isik, E., Sabudak, T. & Oksuz, S. (2007). Flavonoids from *Trifolium resupinatum* var. *microcephalum. Chem. Nat. Comp., 43*, 614-615.

Isogai, Y., Komoda, Y. & Okamoto, T. (1970). Plant Growth Regulators in the Pea Plant (*Pisum sativum* L.). *Chem. Pharm. Bull., 18*, 1872.

Ito N., Nihei T., Kaduka R., Yaoita Y. & Kikuchi M., 2006. Five new phenylethanoid glycosides from the whole plants of *Lamium purpureum* L.. *Chem. Pharm. Bull. 54*, 1705-1708.

Ivencheva, S. & Tsvetkova, R. (2003). In: Imperato Filippo (Ed.), *Advances in Phytochemistry*. Research Signpost. Trivandrum, India, 85-96.

Iwashina, T. (2000). The structure and distribution of the dlavonoids in plants. *J. Plant Res., 113*, 287-299.

Iwashina, T., Matsumoto, S., Nishida, M. & Nakaike, T. (1995a). New and rare flavonol glycosides from *Asplenium* trichomanes-ramosum as stable chemotaxonomic markers. *Biochem. Syst. Ecol., 23*, 283-290.

Iwashina, T., Kadota, Y., Ueno, T. & Ootani, S. (1995b). Foliar flavonoid composition in Japanese *Cirsium species* (Compositeae), and their chemotaxonomic significance. *J. Jap. Bot., 70*, 280-290.

Janicsák, G., Máthé, I., Miklóssy-Vári, V. & Blunden, G. (1999). Comparative studies of the rosmarinic and caffeic acid contents of Lamiaceae species. *Biochem. Syst. Ecol., 27*, 733-738.

Jullien, F., Voirin, B., Bernillon, J. & Favre-Bonvin, J. (1984). Highly oxygenated flavones from *Mentha piperita. Phytochemistry, 23*, 2972-2973.

Junior, P. (1990). Recent Developments in the Isolation and Structure Elucidation of Naturally Occurring Iridoid Compounds. *Planta Med., 56*, 1.

Kajiyama, K. Demizu, S., Higara, Y., Kinishita, K., Koyama, K., Takayashi, K., Tamura, Y., Okada, K. & Kinoshita, T. (1992a). Two prenylated retrochalcones from *Glycyrrhiza*. *Phytochemistry, 31*, 3229-3232.

Kajiyama, K., Demizu, S., Higara, Y., Kinishita, K., Koyama, K., Takayashi, K., Tamura, Y., Okada, K. & Kinoshita, T. (1992b). New prenylflavones and dibenzoylmethane from *Glycyrrhiza inflata. J. Nat. Prod., 55*, 1197-1203.

Kajiyama, K., Higara, Y., Takayashi, K., Hirata, S., Kobayashi, S., Sankawa, U. & Kinoshita, T. (1993). Flavonoids and isoflavonois of chemosytematic significance from *Glycyrrhiza pallidiflora* (Leguminoseae). *Biochem. Syst. Ecol., 21*, 785-793.

Kinoshita, T., Saitoh, T. & Shibata, S. (1978). A new 3-arylcoumarin from licorice root. *Chem. Pharm. Bull., 26*, 135-140.

Klejdus, B., Vitamvasova-Sterbova, D. & Kuban, V. (2001). Identification of isoflavone conjugates in red clover (*Trifolium pratense*) by liquid chromatography-mass spectrometry after two-dimensional solid-phase extraction. *Anal. Chim. Acta*, *450*, 81-97.

Komissarenko, N. G., Chernobai, V. T. & Kolesnikov, D. G. (1964). *Dokl. Akad. Nauk. SSSR.*, *158*, 904.

Kotake, M. & Arakawa, H. (1956). Narcissin (Isorhamnetin-3-rutinosid) aus den Pollen von *Lilium auratum* Lindle. *Naturwissenshaften*, *43*, 327-328.

Kouno, I., Komori, T. & Kawashi, T. (1973). Zur struktur der neuen typen homo-isoflavanone aus bulben von *Scilla scilloides druce*. *Tetrahedron Letters*, *14*, 4569-4572.

Krenn, L., Unterrieder, I. & Ruprechter, R. (2002). Quantification of isoflavones in red clover by high-performance liquid chromatography. *J. Chromatogr. B*, *777*, 123-128.

Langcake, P. & Pryce, R. J. (1976). The production of resveratrol by *Vittis vinifera* and other members of the Vitaceae as a response to infection or injury. *Physiol. Pl. Pathol.*, *9*, 77-86.

Leitner, J. (1942). Ein Beitrag zur Kenntniss der Pollenkörner der Labiatae. *Österr. Bot. Z.*, *91*, 29-40.

Liu, Y., Wagner, H. & Bauer, R. (1998). Phenylpropanoids and flavonoid glycosides from *Lysionotus pauciflorus*. *Phytochemistry*, *48*, 339-343.

Lukyanchikov, S. M. & Kazakov, A. L. (1982). Flavonoids from *Trifolium polyphyllum*. *Khimiya Prirod Soed*, *2*, 251-252.

Mansour, R. M. A., ahmed, A. A., Melek, F. R. & Saleh, N. A. M. (1990). The flavonoids of *Pulicaria incisa*. *Fitoterapia*, *61*, 186-187.

Marin, P. D., Grayer, R. J., Veitch, N. C., Kite, G. C. & Harborne, J. H. (2001). Acacetin glycosides as taxonomic markers in *Calamintha* and *Micromeria*. *Phytochemistry*, *58*, 943-947.

Markham, K. R. & Mitchell, K. A. (2003). The mis-identification of the major antioxidant flavonoids in youn barley (*Hordeum vulgare*) leaves. *Z. Naturforsch.*, *58c*, 53-56.

Markham, K. R., Mabry, T. J. & Swift, W. T. (1970). Distribution of flavonoids in the genus baptisia (leguminosae). *Phytochemistry*, *9*, 2359-2364.

Mašterová, I., Urhín, D. & Tomko, J, (1987). Lilaline − A flavonoid alkaloid from *Lilium candidum*. *Phytochemistry*, *26*, 1844-1845.

Melchior, H. (1964). In *Syllabus der Pflanzen-familien*. Gebruder Borntraeger, Berlin, 515.

Melek, F. R., El-Ansari, M., Hassan, A., Regaila, A., Ahmed, A. A. & Mabry, T. J. (1988). Methylated flavonoid aglycones from *Pulicaria arabica*. *Rev. Latinoam. Quim.*, *19*, 119-120.

Metwally, R. M. A., Dawidar, A. A. & Matwally, S. (1986). A new thymol derivative from *Pulicaria undulata*. *Chem. Pharm. Bull.*, *34*, 378-379.

Michalsk, K., Szneler, E. & Kisiel, W. (2007). Sesquiterpenoids of *Picris koreana* and their chemotaxonomic significance. *Biochem. Syst. Ecol.*, *35*, 459-461.

Míka, V., Kubáň, B., Klejdus, B. & Odstrčilová and Nerušil, P. (2005). Phenolic compounds as chemical markers of low taxonomic levels in the family *Poaceae*. *Plant Soil Environ.*, *51(11)*, 506-512.

Miyaichi, Y., Morimoto, T., Yaguchi, Y., Kizu, H., Tomimori, T. & Vetschera, K. (1991). Studies on the Constitutents of *Scutellaria* Species. XIV. on the Constituents of the Roots and the Leaves of *Scutellaria alpina* L. . *Chem. Pharm. Bull.*, *39*, 199-201.

Mohamed, K. M., Hassanean, H. A., Ohtani, K., Kasai, R. & Yamasaki, K. (2000). Chalcanol glucosides from seeds of *Trifolium alexandrium*. *Phytochemistry*, *53*, 401-404.

Morales, G., Sierra, P., Borquez, J. & Loyola, L. A. (2000). *J. Chil. Chem. Soc.*, *49*, 137.

Moreira Alves, K. C., Gobbo-Neto, L. & Lopes, N. P. (2008). Sesquiterpene lactones and flavonoids from *Lychnophora reticulata* Gardn. (Asteraceae). *Biochem. Syst. Ecol.*, *36*, 434-436.

Murakami, T. & Tanaka, N. (1988). Occurences, structure and taxonomic implications of fern constituents. In W. Herz, H. Grisebach, G. W. Kirby, & Ch. Tamm (Eds.), *Progress in the Chemistry of Organic Natural Products*, *54*, Springer, Wien, 231-269.

Myase, T., Yamamota, R. & Veno, A. (1996). Phenylpropanoid glycosides from *Stachys officinalis*. *Phytochemistry*, *43*, 475-479.

Nagy, E., Neszmelyi, A. & Verzar-Petri, G. (1986). Characterization flavonoids of *Lilium candidum* L. and their distribution in different plant parts. *Stud. Org. Chem.*, (Amsterdam) *23*, 265-271.

Nagy, E., Seres, I., Verzar-Petri, G. & Neszmelyi, A. (1984). Kaempferol-3-O-[β-D-glucopyranosyl-(1→2)-β-D-galactopyranoside] a new flavonoid from *Lilium candidum* L. *Anorg. Chem Org. chem.*, *36*, 1813-1815.

Nakabayashi, T. (1952). A flavonoid pigment astragalin (kaempferol-3-glucoside) in *Astragalus sinicus* flowers. *J. Agric. Chem. Soc.*, *26*, 539-541.

Nakayama, M., Yamagushi, M., Urashima, O., Kan, Y., Fukui, Y., Yamagushi, Y. & Koshioka, M. (1999). Anthocyanins in the dark purple anthers of *Tulipa gesneriana*: identification of two novel delphinidin 3-O-(6-O-(acetyl-α-rhamnopyranosyl)-β-glucopyranosides). *Biosci. Biotechnol. Biochem.*, *63*, 1509-1511.

Nawwar, M. A. M, El Sissi, H. I. & Barakat, H. H. (1980). The flavonoids of *Phragmites australis* flowers. *Phytochemistry*, *19*, 1854-1856.

Nieuwhof, M., Van Raamsdonk, L. W. D. & Van Eijk, J. P. (1990). Pigment composition of flowers of *Tulipa* (Liliaceae) as a parameter for biosystematic research. *Biochem. Syst. Ecol.*, *18*, 399-404.

Nikolov, S. D., Elenga, P. A. & Panova, D. (1984). Flavonoides de l'espèce *Astragalus glycyphyllos* L. Groupe Polyphenols (1984). Institut Supérieur de l'Industrie Alimentaire, Plovdiv, Bularia, 114.

Nikolova, M. (2006). Infraspecific variability in the flavonoi composition of *Artemisia vulgaris* L. *Acta Bot. Croat.*, *65*, 13-18.

Nip, W. K. & Burns, E. E. (1969). Pigment characterization in grain *Sorghum*. I. Red varieties. *Cereal Chem.*, *46*, 490-495.

Nip, W. K. & Burns, E. E. (1971). Pigment characterization in grain *Sorghum*. II. White varieties. *Cereal Chem.*, *48*, 74-80.

Nishimura, H., Sasaki, H., Inagaki, N., Chin, M. & Mitsuhashi, H. (1991). Nine phenethyl alcohol glycosides from *Stachys sieboldii*. *Phytochemistry*, *30*, 965-969.

Norbaæk, R. & Kondo, T. (1999). Anthocyanins from flowers of *Lilium* (Liliaceae). *Phytochemistry*, *50*, 1181-1184.

Norbæk, R., Brandt, K. & Kondo, T. (2000). Identification of flavone *C*-glycosides including a new flavonoid chromophore from barley leaves (*Hordeum vulgare* L.) by NMR techniques, *J. Agric. Food Chem.*, *48*, 1703-1707.

Norbæk, R., Aaboer D. B. F., Bleeg I. S., Christensen B. T., Kondo T. & Brandt K. (2003). Flavone *C*-glycoside, phenolic acid and nitrogen contents in leaves of barley subject to organic fertilization treatments. *J. Agric. Food Chem.*, *51*, 809-813.

Norris, F. A. & Stermitz, F. R. (1970). 4'-*O*-Methylquercetin 3-glucoside from *Astragalus miser* var. *Oblogifolius*. *Phytochemistry*, *9*, 229-230.

Olah, A. F. & Sherwood, R. T. (1971). Flavans, isoflavones, and coumestans in alfalfa infected by *Ascochyta imperfecta*. *Phytopathology*, *61*, 65-69.

Oleszek, W. & Stochmal, A. (2002). Triterpene saponins and flavonoids in the seeds of *Trifolium* species. *Phytochemistry*, *61*, 165-170.

Pedersen, J. A. (2000). Distribution and taxonomic implications of some phenolics in the family Lamiaceae determined by ESR spectroscopy. *Biochem. Syst. Ecol.*, *28*, 229-253.

Ponce, M. A., Scervino, J. M., Erra-Balsells, R., Ocampo, J. A. & Godeos, A. M. (2004). Flavonoids from shoots and roots of *Trifolium repens* (white clover) grown in presence or absence of the arbuscular mycorrhizal fungus *Glomus intraradices*. *Phytochemistry*, *65*, 1925-1930.

Pope, G. S., Elcoate, P. V., Simpson, S. A. & Andrews, D. G. (1953). Isolation of an oestrogenic isoflavone (biochanin A) from red clover. *Chem. Indus.*, *10*, 1092.

Porter, L. J. (1988). Flavans and proanthocyanidins. In: *The Flavonoids. Advances in Researches since 1980*. J. B. Harborne (Ed.), Chapman & Hall, chap. 2.

Prat, H. (1936). La systématique des Graminées. *Ann. Sci. Nat. Bot.*, *18*, 165-258.

Prat, H. (1960). Vers une classification naturelle des Graminées. *Bull. Soc. Bot. France*, *107*, 32-79.

Proksch, P. & Rodriguez, E. (1985). Ein Modell für chemoökologische Adaptationen von Wüstenpflanzen. *Biologie in unserer Zeit*, 75-80.

Repčák, M., Švehlíková, V. & Imrich, J. (1999). Jaceidin and chrysosplenetin chemotypes of *Chamomilla recutita* (L.) Rauschert. *Biochem. Syst. Ecol.*, *27*, 727-732.

Rigano, D., Arnold, N. A., Bruno, M., Formisano, C., Grassia, A., Piacente, S., Piozzi, F. & Senatore, F. (2006). *Biochem. Syst. Ecol.*, *34*, 256-258.

Robinson, G. M. & Robinson, R. (1931). A survey of anthocyanins. I. *Biochem. J.*, *25*, 1687-1705.

Robinson, H. (1999). Generic and subtribal classification of American Vernonieae. *Smithson. Contrib. Bot.*, *89*, 1-116.

Rofi, R. D. & Pomilio, A. B. (1987). Luteolin 6-C-beta-ristobioside from *Poa annua*. *Phytochemistry*, *26*, 859-860.

Sahpaz, S., Skaltsounis, A. L. & Bailleul, F. (2002). Polyphenols from *Ballota acetabulosa*. *Biochem. Syst. Ecol.*, *30*, 601-604.

Saito, N., et al. (2003). The first isolation of *C*-glycosylanthocyanin from the flowers of *Tricyrtis formosana*. *Tetrahedron Lett.*, *44*, 6821-6823.

Sakamoto, H. T., Flausino, D., Castellano, E. E., Stark, C. B. W., Gates, P. J. & Lopes, N. P. (2003). *J. Nat. Prod.*, *66*, 693-695.

Saleh, N. A., Bohm, B. A. & Maze, J. R. (1971). Flavonoids of *Stipa lemmonii*. *Phytochemistry*, *10*, 490-491.

Saleh, N. A. M, Boulos, L., El-Negoumy, S. I. & Abdall, M. F. (1982). A comparative study of the flavonoids of *Medicago radiata* with other *Medicago* and related *Trigonella* species. *Biochem. Syst. Ecol.*, *10*, 33-36.

Sampaio-Santos, M. & Kaplan, M. A. C. (2001). Biosynthesis Significance of Iridoids in Chemosystematics. *J. Braz. Chem. Soc.*, *12(2)*, 144-153.

Sarker, S. D., Bartholomew, B., Nash, R. J. & Robinson, N. (2000). 5-O-methylhoslundin: An unusual flavonoi from *Bidens pilosa* (Asteraceae). *Biochem. Syst. Ecol.*, *28*, 591-593.

Schultz, G. (1965). Isoflavone glucoside formononetin-7-glucoside and biochanin A-7-glucoside in *Trifolium pratense* L., *Naturwissenschaften*, *52*, 517.

Seidel, V., Bailleul, F. & Tillequin, F. (1999). Terpenoids and phenolics in the genus *Ballota* L. (Lamiaceae). *Recent Res. Dev. Phytochem.*, *3*, 27-39.

Semmar, N., Fenet, B., Glucoff-Fiasson, K., Comte, G. & Jay, M. (2002b). New flavonol tetraglycosides from *Astragalus caprinus*. *Chem.. Pharm. Bull.*, *50*, 981-984.

Semmar, N., Fenet, B., Glucoff-Fiasson, K., Hasan, A. & Jay, M. (2002a). Four new flavonol glycosides from the leaves of *Astragalus caprinus*. *Journal of Natural Products*, *65*, 576-579.

Semmar, N., Fenet, B., Lacaille-Dubois, M. A., Glucoff-Fiasson, K., Chemli, R. & Jay, M. (2001). Two new glycosides from *Astragalus caprinus*. *Journal of Natural Products*, *64*, 656-658.

Semmar, N., Jay, M. & Chemli, R. (2001). Chemical diversification trends in *Astragalus caprinus* (Leguminosae) based on the flavonoid pathway. *Biochemical Systematics and Ecology*, *vol. 29(7)*, 727-738.

Semmar, N., Jay, M., Farman, M. & Chemli, R. (2005). Chemotaxonomic analysis of *Astragalus caprinus* (Fabaceae) based on the flavonic patterns. *Biochem. Syst. Ecol.*, *33*, 187-200.

Semmar, N., Jay, M. & Nouira, S. (2007). A new approach to graphical and numerical analysis of links between plant chemotaxonomy and secondary metabolism from HPLC data smoothed by a simplex mixture design. *Chemoecology*, *vol. 17*, 139-156.

Seshadri, T. R., Sood, A. R. & Varshney, I. P. (1972). *Indian J. Chem.*, *10*, 26.

Seshadri, T. R., Varshney, I. P. & Sood, A. R. (1973). *Curr. Sci. (India)*, *42*, 421.

Shalashvili, K. G. (1993). Flavonoids of *Trifolium echinatum* and *T. diffusum*. *Khimiya Prirod Soed.*, *33*, 468-469.

Shehata, M. N., Hassan, A. & elshazly, K. (1982). Identification of the estrogenic isoflavones in fresh and fermented berseem clover (*T. alexandrinum*). *Aust. J. Agric. Res.*, *33*, 951-956.

Shi, S. Y., Zhang, Y. P., Huang, K. L., Zhao, Y. & Liu, S. Q. (2008). Flavonoids from *Taraxacum mongolicum*. *Biochem. Syst. Ecol.*, *36*, 437-440.

Shibata, M. (1956). Uber das Anthocyanin in den dunkelpurpurnen Bluten von *Tulipa gesneriana* L. Studien uber die Phisiologie von Liliaceen I. *Bot. Mag. Tokyo*, *69*, 462-468.

Shibata, M. & Sakai, E. (1958). On the anthocyanin in the blood-red flower of *Tulipa gesneriana* L. Studies on the physiology of Liliaceae II. *Bot. Mag. Tokyo*, *71*, 6-11.

Shibata, M. & Ishikura, N. (1960). Paper chromatographic survey of anthocyanin in tulip-flowers. *I. Jap. J. Bot.*, *17*, 230-238.

Shibata, S. & Saitoh, T. (1978). Flavonoid compounds in licorice root. *J. Indian Chem. Soc.*, *55*, 1184-1191.

Shilin, Y., Roberts, M. F. & Phillipson, J. D. (1989). Methoxylated flavones and coumarins from *Artemisia annua*. *Phytochemistry*, *28*, 1509-1511.

Siciliano, T., Bader, A., Vassallo, A., Braca, A., Morelli, I., Pizza, C. & De Tommasi, N. (2005). Secondary metabolites from *Ballota undulata* (Lamiaceae). *Biochem. Syst. Ecol.*, *33*, 341-351.

Simpson, M. G. (2006). *Plant systematics*. Elsevier Academic Press, London, 138-221.

Sinskaya, E. N. (1950). *Flora of cultivated Plants of the U.S.S.R.*, *vol. 13*, Oldbourne Press, London, 7.

Skoula, M., Gotsiou, P., Naxakis, G. & Johnson, C. B. (1999). A chemosystematic investigation on the mono- and sesquiterpenoids in the genus *Origanum* (Labiatae). *Phytochemistry*, *52*, 649-657.

Skoula, M., Grayer, R. J., Kite, G. C. & Veitch, N. C. (2008). Exudate flavones and flavanones in *Origanum* species and their interspecific variation. *Biochem. Syst. Ecol.*, *36*, 646-654.

Skrzypczakova, L. (1967). Flavonoids in the family Liliaceae. I. Analysis of glucoside and aglycons. *Diss. Pharm. Pharmacol.*, *19*, 537-543.

Snook, M. E., et al. (1994). New flavone C-glycosides from corn (*Zea mays* L.) for the control of the corn earworm (*Helicoverpa zea*), *ACS Symp. Ser.*, *557*, 122.

Snook, M. E., et al. (1995). New C-4''-hydroxy derivatives of maysin and 3'-methoxymaysin isolated from corn silks (*Zea mays*). *J. Agric. Food Chem.*, *43*, 2740-2745.

Sood, A. R., Boutard, B., Chdenson, M., Chopin, J. & Lebreton, P. (1976). A new flavone C-glycoside from *Trigonella foenum graecum*. *Phytochemistry*, *15*, 351-352.

Stevens, J. F., Wollenweber, E., Ivancic, M., Hsu, V. L., Sundberg, S. & Deinzer, M. L. (1999). Leaf surface flavonoids of *Chrysothamnus*. *Phytochemistry*, *51*, 771-780.

Sudo, H., Takushi, A., Ide, T., Otsuka, H., Hirata, E. & Takeda, Y. (1997). Premnethanosides A and B: Phenylethanoids from leaves of *Premna subscandens*. *Phyrochemistry*, *46*, 1147-1150.

Synge, P. M. (1980). *Lilies: A revision of Elwes' Monograph of the Genus Lilium and its Supplements*. B.T. Batsford, London.

Tarr, F. (1993). Isoflavone glycosides and flavanol glycosides in *Trifolium repens* extract. *Pharm. Biol.*, *42*, 656-658.

Toebes, A. H., De Boer, V., Verkleij, J. A., Lingeman, H. & Ernst, W. H. (2005). Extraction of isoflavone malonyl glucosides in *Trifolium striatum* L. *Magyar Kem Foly*, *99*, 89-92.

Toki, K., Saito, N. & Honda, T. (1998). Acylated anthocyanins from the blue-purple flowers of *Triteleia bridgesii*. *Phytochemistry*, *48*, 729-732.

Tomas-Barberan, F. A. T., Hernandez, L., Ferreres, F. & Tomas, F. (1985). Highly Methylated 6-Hydroxyflavones and other Flavonoids from *Thymus piperella*. *Planta Med.*, 452-454.

Tomás-Barberán, F. A., Grayer-Barkmeijer, R. J, Gil, M. I. & Harborne, J. B. (1988a). Distribution of 6-hydroxy, 6-methoxy- and 8-hydroxyflavone glycosides in the Labiatae, the Scrophulariaceae and related families. *Phytochemistry*, *27*, 2631-2645.

Tomás-Barberán, F. A., Husain, S. Z. & Gil, M. I. (1988b). The distribution of methylated flavones in the Lamiaceae. *Biochem. Syst. Ecol.*, *16*, 43-46.

Tomás-Barberán, F. A., Husain, S. Z., Rejdali, M., Harborne, J. B. & Heywood, V. H. (1988c). External and vacuolar flavonoids from Ibero-NorthAfrican *Sideritis* species. A chemosystematic approach. *Phytochemistry*, *27*, 165-170.

Tomás-Barberán, F. A. T. & Wollenweber, E. (1990). Flavonoid aglycones from the leaf surfaces of some *Labiatae* species. *Pl. Syst Evol.*, *173*, 109-118.

Tomás-Barberán, F. A., Gil, M. I., Marin, P. D. & Tomás-Lorente, F. (1991). Flavonoids from some Yugoslavian *Micromeria* species:chmotaxonomical aspects. *Biochem. Syst. Ecol.*, *19*, 697-698.

Tomás-Barberán, F. A. & Gil, M. I. (1992). Chemistry and natural distribution of flavonoids in the Labiatae. In: R. M. Harley, & T. Reynolds (Eds.), *Advances in Labiate Science*. Royal Botanic Gardens, Kew, 299-305.

Tomás-Barberán, F. A., Gil, M. I., Ferreres, F. & Tomas-Lorente, F. (1992). Flavonoid p-Coumaroyl and 8-Hydroxyflavone Allosylglucosides in Some Labiatae. *Phytochemistry*, *31*, 3097-3102.

Tomás-Lorente, F. A., Harborne, J. B. & a,d Self, R. (1987). Twelve 6-oxygenated flavone sulfates from *Lippia nodiflora* and *L. canescens*. *Phytochemistry*, *26*, 2281-2284.

Tomás-Lorente, F., Ferreres, F., Tomás-Bárberán, F. A., Rivera, D. & Obón, C. (1988). Some flavonoids and the diterpene borjatriol from some Spanish *Sideritis* species. *Biochem. Syst. Ecol.*, *16*, 33-42.

Tomimori, T., Miyaichi, Y., Imoto, Y., Kizu, H. & Namba, T. (1985). Studies on Nepalese crude drugs. 5. On the flavonoid constituents of the root of *Scutellaria discolor* COLEBR. (1). *Chem. Pharm. Bull.*, *33*, 4457-4463.

Tomimori, T., Yukinori, M., Imoto, Y., Kizu, H. & Namba, T. (1986). Studies on Nepalese crude drugs. 6. On the flavonoid constituents of the root of *Scutellaria discolor* COLEBR. (2). *Chem. Pharm. Bull.*, *34*, 406-408.

Torskangerpoll, K., Fossen, T. & Andersen, Ø. M. (1999). Anthocyanin pigments of tulips. *Phytochemistry*, *52*, 1687-1692.

Torskangerpoll, K., Nørbæk, R., Nodland, E., Øvstedal, D. O. & Andersen, Ø. M. (2005). Anthocyanin content of *Tulipa* species and cultivars and its impact on tepal colours. *Biochem. Syst. Ecol.*, *33*, 499-510.

Turner, B. L. (1967). Plant Chemosystematics and Phylogeny. *Pure Appl. Chem.*, *14*, 189.

Urbatsch, L. E., Bacon, J. D. & Mabry, T. J. (1975). Flavonol methyl ethers from *Chrysothamnus viscidiflorus*. *Phytochemistry*, *14*, 2279-2282.

Urhín, D., Bučkova, A., Eisenreichová, E., Haladová, M. & Tomko, J. (1989). Constituents of *Lilium candidum* L. *Chem Papers*, *43*, 793-796.

Urzúa, A., Mendoza, L., Andrade, L. & Miranda, B. (1997). Diterpenoids in the Trichome Resinous Exudate from *Haplopappus shumannfi*. *Biochem. Syst. Ecol.*, *25*, 683-684.

Urzúa, A. (2004). Secondary metabolites in the epicuticle of *Haplopappus foliosus* dc. (Asteraceae). *J. Chil. Chem. Soc.*, *49*, 137-141.

Urzúa, A., Contreras, R., Jara, P., Avila, F. & Suazo, M. (2004). Comparative chemical composition of the trichome secreted exudates and of the waxy coating from *Haplopappus velutinus*, *H. illinitus*, *H. shumanni* and *H. uncinatus*. *Biochem. Syst. Ecol.*, *32*, 215-218.

Urzúa, A., Iturra, B., Sebastián, B. & Muñoz, M. (2007). Chemical components from the surface of *Haplopappus bustillosianus*. *Biochem. Syst. Ecol.*, *35*, 794-796.

Valant-Vetschera, K. M. (1987). Flavonoid glycoside accumulation trends of *Achillea nobilis* L. and related species. *Biochem. Syst. Ecol.*, *15*, 45-52.

Valant-Vetschera, K., Fischer, R. & Wollenweber, E. (2003). Exudate flavonoids in species *Artemisia* (Asteraceae-anthemideae): new results and chemosystematic interpretation. *Biochem. Syst. Ecol.*, *31*, 487-498.

Van den Broucke, C. O., Domisse, R. A., Esmans, E. L. & Lemli, J. A. (1982). Three methylated flavones from *Thymus vulgaris*. *Phytochemistry*, *21*, 2581-2583.

Van den Heuvel, J. & Glazener, J. A. (1975). Comparative abilities of fungi pathogenic and nonpathogenic to bean (*Phaseolus vulgaris*) to metabolize phaseollin. *Neth. J. Pl. Pathol.*, *81*, 125-137.

Van den Heuvel, J. (1976). Sensitivity to, and metabolism of, phaseollin in relation to the pathogenicity of different isolates of *Botrytis cinerea* to bean (*Phaseolus vulgaris*). *Neth. J. Pl. Pathol.*, *82*, 153-160.

van Eijk, J. P., Nieuwhof, M., van Keulen, H. A. & Keijzer, P. (1987). Flower colour analyses in tulip (*Tulipa* L.). The occurrence of carotenoids and flavonoids in tulip tepals. *Euphytica*, *36*, 855-862.

Van Etten, H. D. & Pueppke, S. C. (1976). In *Biochemical Aspects of Plant Parasite Relationships*. Friend J. & Threfall D.R. (Eds.), Academic Press, London, 239-289.

Viera, R. F., Grayer, R. J. & Paton, A. J. (2003). Chemical profiling of *Ocimum americanum* using external flavonoids. *Phytochemistry*, *63*, 555-567.

Voirin, B., Viricel, M. R., Favre-Bonvin, J., Van Den Broucke, C.O. & Lemli, J. (1985). 5,6,4'-Trihydroxy-7,3'-dimethoxyflavone and other methoxylated flavonoids isolated from *Thymus satureioides*. *Planta Med.*, 523-525.

Wagenitz, G. (1976). Systematics and phylogeny of the Compositae. *Plant Syst. Evol.*, *25*, 673-674.

Wagner, H., Iyengar, M. A. & Hörhammer, L. (1973). Vicenin-1 and -2 in the seeds of *Trigonella foenumgraecum*. *Phytochemistry*, *12*, 2548.

Waiss, A. C., Chan, Jr., B. G., Elliger, C. A., Wiseman, B. R., McMillian, W. W., Widstrom, N. W., Zuber, M. S. & Keaster A. J. (1979). Maysin, a flavone glycoside from corn silks with antibiotic activity toward corn earworm. *J. Econ. Entomol.*, *72*, 256-258.

Wang, S. F., Ridsdill-Smith, T. J. & Ghisalberti, E. L. (1998). Role of isoflavonoids in resistance of subterranean clover trifoliates to the redlegged earth mites. *Phytochemistry*, *52*, 601-605.

Webby, R. & Markham, K. R. (1994). Isoswertiajaponin 2''-O-β-arabinopyranoside and other flavone-C-glycosides from the Antarctic grass *Deschampsia antarctica*. *Phytochemistry*, *36*, 1323-1326.

Williams, C. A. (1975). Biosystematics of the Monocotyledoneae – Flavonoid Patterns in Leaves of the Liliaceae. *Biochem. Syst. Ecol.*, *3*, 229-244.

Williams, C. A. (1978). The systematic implications of the complexity of leaf flavonoids in the Bromeliaceae. *Phytochemistry*, *17*, 729-734.

Williams, C. A. & Harborne, J. B. (1971). Flavonoid patterns in the monocotyledons. Flavonols and flavones in some families associated with the Poaceae. *Phytochemistry*, *10*, 1059-1063.

Williams, C. A. & Harborne, J. B. (1973). Negatively charged flavones and tricin as chemosystematic markers in the palmae. *Phytochemistry*, *12*, 2417-2430.

Williams, C. A. & Harborne, J. B. (1975). Luteolin and daphnetin derivatives in the juncaceae and their systematic significance. *Biochem. Syst. Ecol.*, *3*, 181-190.

Williams, C. A. & Harborne, J. B. (1977). Flavonoid chemistry and plant geography in the Cyperaceae. *Biochem. Syst. Ecol.*, *5*, 45-51.

Williams, C. A. & Harborne, J. B. (1994). Flavone and flavonol glycosides. In: *The flavonoids; Advances in Research since 1986*. J. B. Harborne (Ed.), Chapman & Hall, Cambridge, 344.

Williams, C. A., Harborne, J. B. & Smith, P. (1974). The taxonomic significance of leaf flavonoids in *Saccharum* and related genera. *Phytochemistry*, *13*, 1141-1149.

Williams, C. A., Harborne, J. B. & Greenham, J. (2000). Geographical variation in the surface flavonoids of *Pulicaria dysenterica*. *Biochem. Syst. Ecol.*, *28*, 679-687.

Williams, Ch. A., Harborne, J. B., Greenham, J. R., Grayer, R. J., Kite, G. C. & Eagles, J. (2003). Variations in lipophilic and vacuolar flavonoids among European *Pulicaria* species. *Phytochemistry*, *64*, 275-383.

Williams, C. A. & Murray, B. G. (1972). Flavonoid variation in the genus *Briza*. *Phytochemistry*, *11*, 2507-2512.

Wollenweber, E. (1994). Flavones and Flavonols. In *The flavonoids: Advances in Research Since 1984*, J. B. Harborne (Ed.), Chapman and Hall, London, 259-335.

Wollenweber, E. & Dietz, V. H. (1981). Occurence and distribution of free flavonoid aglycones in plants . *Phytochemistry*, *20*, 869-932.

Wollenweber, E. (1978). The distribution and chemical constituents of the farinose exudates in gymnogrammoid ferns. *Amer. Fern J.*, *68*, 13-28.

Wunderlich, R. (1967). Ein Vorschlag zu einer natürlichen Gliederung der Labiaten and Grund der Pollenkörner, der Samenentwicklung und des reifen Samens. *Oesterr. Bot. Z.*, *114*, 383-483.

Xaasan, C. C., Cilmi, C. X., Faarax, M. X., Passananti, S., Piozzi, F. & Paternostro, M. (1980). Unusual flavones from *Ocimum canum*. *Phytochemistry*, *19*, 2229-2230.

Yahara, S., Kohjyouma, M. & Kohoda, H. (2000). Flavonoid glycosides and saponins from *Astragalus shikokianus*. *Phytochemistry*, *53*, 469-471.

Yalçin, F. N. & Kaya, D. (2006). Ethnobotany, Pharmacology and Phytochemistry of the Genus *Lamium* (Lamiaceae). *FABAD J. Pharm. Sci.*, *31*, 43-52.

Ye, G., Huang, X. Y., Li, Z. X., Fan, M. S. & Huang, C. G. (2008). A new cadinane type sesquiterpene from *Eupatorium lindleyanum* (Compositae). *Biochem. Syst. Ecol.*, *36*, 741-744

Yoon, K. D., Kim, C. Y. & Huh, H. (2000). The flavone glycosides of *Sasa borealis*, *Saengyak Hakhoechi*, *31*, 224-227.

Yuldashev, M. P. & Karinov, A. S. (2005). Flavonoids of *Scutellaria ocellata* and *S. nepetoides*. *Chem. Nat. Compd.*, *37*, 431-433.

Zhan, Q. F., Xia, Z. H., Wang, J. L. & Lao, A. N. (2003). Two new biocoumarins from *Trifolium repens* L.. *J. Asian Nat. Prod. Res.*, *5*, 303-306.

Zhu, D. Y., Song, G. O., Jiang, F. X., Chang, X. R. & Guo, W. B. (1984). Studies on chemical constituents of *Glycyrrhiza uralensis* Fisch.: the structure of isolicoflavonol and glycycoumarin. *Acta Chim. Sin.*, *42*, 1080-1084.

Zohary, M. & Heler, D. (1984). *The genus of Trifolium*. Ahva Printing Press, Jerusalem, 67.

Zohary, M. (1970). Trifolium L. In: *Flora of Turkey and the East Aegean Islands, vol. 3*, P. H. Davis (Ed.). University Press, Edinburgh, 384-448.

In: Encyclopedia of Environmental Research ISBN: 978-1-61761-927-4
Editor: Alisa N. Souter © 2011 Nova Science Publishers, Inc.

Chapter 4

FRAMEWORK FOR INTEGRATING INDIGENOUS KNOWLEDGE AND ECOLOGICAL METHODS FOR IMPLEMENTATION OF DESERTIFICATION CONVENTION

*Hassan G. Roba[1] and Gufu Oba[2]**

[1] The National Museums of Kenya, PO Box 40658 00100 Nairobi, Kenya.
[2] Noragric, Department of International Environment and Development Studies,
Norwegian University of Life Sciences, PO Box 5003, N-1432 Ås, Norway.

SUMMARY

This chapter describes a methodological framework for integrating indigenous knowledge and ecological methods for promoting local communities' participation in the implementation of the Convention on Combating Desertification (CCD) and the Convention on Biological Diversity (CBD) at local community levels. We developed and tested the framework using herding communities in Northern Kenya. The methods for implementing the framework included semi-structured interviews and group discussions with key informants for generating information on livestock management, changes in vegetation indicators and historical changes in land use patterns. Other methods used included joint transect walks with knowledgeable herders to assess environmental change using ecological indicators (vegetation and soil) and herder anthropogenic indicators (i.e., landscape grazing potential and landscape grazing suitability), monitoring marked transects, using satellite images taken at different times, and herders' knowledge to evaluate long-term changes in vegetation cover around permanent settlements. Finally, a workshop was organized with herders and Environmental Management Committees (EMCs) in which they participated in informal discussions on issues addressed by the joint research project. We then synthesized the findings to evaluate the effectiveness of the framework for desertification assessment, monitoring and control. From the results we may conclude that local community participation in assessment and monitoring of environment change would contribute to the implementation of the CCD and the CBD.

---

* Corresponding author: E-mail: gufu.oba@umb.no

This chapter makes specific recommendations for applying the framework to achieve global goals for local actions at local community levels.

Keywords: Biodiversity; global environmental conventions; herder indicators; ecological indicators; land degradation; local participation.

1. INTRODUCTION

A major challenge faced in the implementation of global environmental conventions such as the Convention on Combating Desertification (CCD) and Convention for Biological Diversity (CBD) is lack of a methodological framework for achieving local community participation. We have in this chapter synthesized our own experiences and those of others that in the past made attempts to develop participatory environmental assessment and monitoring that have relevance for the implementation of the Global Environmental Conventions (GECs). We will suggest that assessment and monitoring of environmental degradation in the past relied exclusively on conventional ecological methods and indicators selected by scientists. This happens in spite of the global environmental conventions (GECs) such as CCD placing strong emphasis on the need for participation by local communities (UNCED, 1992). Among the factors that have impeded local participation in the implementation of the GECs is a lack of integration of indigenous knowledge and ecological methods (Seely and Moser, 2004; Oba et al., 2008a,b). An integrated methodological approach is therefore needed to address environmental degradation, particularly that which is associated with changes in land use intensification and desertification (Lusigi, 1981). One of the principal causes of desertification in the parts of sub-Saharan Africa has been sedentarization of formerly mobile populations of pastoralists and greater concentrations of land use pressure.

Sedentarization of pastoral populations is either voluntary, as an adaptation to changes in economic and environmental conditions, or through forceful settlements by states for purposes of development or economic rehabilitation (Salzman, 1980). Whatever the cause, pastoral sedentarization has environmental consequences. In Sahelian Africa, over-exploitation of vegetation resources around settlements is reported to be responsible for inducing desertification (Swift 1975; Mabbutt 1984; Mabbutt 1985; Thomas et al., 2000). Pastures are said to be depleted, showing dramatic declines along gradients of land use pressure by livestock grazing. Depletion of woody species around settlements is reported to be associated with over-exploitation of woody plants for the construction of livestock night enclosures (Lamprey and Yussuf 1981), the collection of wood for fuel (Benjaminsen, 1993), cultivation (Lamprey, 1976), and over-browsing by livestock (Oba, 1998; Oba et al., 2000a). The changes are pertinent to the principal goals of the GECs, which are concerned with reversing land degradation through improved land management.

The chapter proposes a framework for the integration of indigenous knowledge and ecological methods for assessment and monitoring of land degradation at the local level. In the first section, a brief description of the participatory research is presented and the frequently used terms in the essay are defined in the context of their use. In section two, the theoretical perspectives for integrating indigenous knowledge and scientific methods are discussed. In section three, the methodological perspectives for achieving the integration of indigenous knowledge and ecological methods are described. In section four, the mechanisms for developing integration are given, using a schematic framework that links global and local goals. The fifth section provides a synthesis of ongoing research that would show how the framework was implemented with the implications for the Global goals for combating desertification and reversing the loss of biodiversity.

2. INTEGRATION OF GLOBAL GOALS AND LOCAL ACTIONS

Participatory research is rooted in the shift in theories from modernization theory associated with top-down technological transfer, to neo-populist theory that advocates for local people participation—which uses bottom-up approaches (Sillitoe, 1998). The proponents of bottom-up or local participation approaches, present convincing arguments that local people have accumulated a wealth of knowledge over time, based on long-term experiences, that can complement scientific knowledge in environmental conservation (Richards, 1980; Knight, 1980). Major emphasis is placed on the roles indigenous knowledge and local management play in conservation (Warren, 1992; Berkes et al., 2000), and protection of the land from degradation (see below for definitions). Additionally, there is growing interest in how indigenous knowledge and management practices can be used in collaboration with standard scientific methods for improving understanding of environmental change (Dahlberg, 2000; Reed et al., 2007). Global environmental problems and the need for local participation were discussed during an international conferences related to UNCOD and UNCED, where participating nations agreed on joint action plans. Since environmental changes are attributed to a multitude of factors (Geist, 2005), appropriate methods and sensitive indicators are needed for assessment and monitoring.

For the above reasons, Agenda 21 of the Rio Conference recognizes the role local communities play in environmental assessment and management (UNCED, 1992). According to Agenda 21, partnerships with local communities aim to achieve the global goals for the sustainable use of natural resources. For example, the role of indigenous knowledge in combating desertification and droughts is contained in Agenda 21, Chapter 12 part 18 (d), which states that the United Nations seeks to "promote participatory management of natural resources, including rangeland, to meet both the needs of rural populations and conservation purposes, based on innovative or adapted indigenous technologies". Further, Agenda 21, Chapter 12, subsection 23 (a), states that government should "integrate indigenous knowledge related to forests, forest lands, rangeland and natural vegetation into research activities on desertification and drought". In relation to the conservation of biological diversity, Agenda 21, Chapter 15, part 4 (g) states that global partners should "recognize and foster the traditional methods and the knowledge of indigenous people...relevant to the conservation of biological diversity and the sustainable use of biological resources" (for recent discussions also see, Johnson et al. 2006).

Various nations are signatories to the global conventions through the ratification of the different articles. The national goals are the tools for implementing the conventions by means of National Action Programs (NAPs), which in turn, comprise strategies and methods for implementing the global goals. Considering the broad geographical, and ecological variability, and the socio-economic factors linked to the process of environmental change, there is no single indicator that may be used for the assessment and monitoring thereof. The goals at the national level therefore include linking scientific assessment and monitoring activities with the environmental management practices of local communities. For example, according to Article 10 (f) of UNCCD (1994), parties to the convention should: "provide for effective participation of local, national and regional levels of non-governmental organization and local population…particularly resource users including farmers and pastoralists and their respective organization in policy planning, decision making and review of national programs."

Additionally, according to the Convention on Biological Diversity (CBD), each member state party to the convention is obliged to promote local participation in the management of biological diversity. Article 8j of the CDB states:

> "Subject to its national legislation, respect, preserve and maintain knowledge, innovations and practices of indigenous and local communities embodying traditional lifestyles relevant for the conservation and sustainable use of biological diversity and promote their wider application with the approval and involvement of the holders of such knowledge, innovations and practices and encourage the equitable sharing of the benefits arising from the utilization of such knowledge, innovations and practices" (UNCCD 1994).

The NAPs can benefit from global mechanisms (such as the Global Environmental Facility) that make technology and funds available for initiating implementation activities by means of local participation (Johnson et al. 2006). By tapping into local knowledge systems, environmental assessment and monitoring in response to anthropogenic and natural ecosystem drivers at local levels can be achieved (Krugman, 1996). Previously, the use of indigenous knowledge for promoting local participatory assessment and monitoring of environmental change was constrained by the lack of integration of local knowledge systems with scientific methods. Conducting evaluations and monitoring human impacts on the environment are pre-requisites for accomplishing the implementation of the GECs.

The purpose of this essay is therefore to shift the approach to enable local communities, such as herders, and ecologists to be partners in the assessment and monitoring of the implementation of the GECs concerned with CCD at community levels, using traditional systems of land use. Rather than discussing the actual implementation of the GECs, our main purpose is to discuss the potential application of integrated methods for implementing the goals of NAPs at local community levels. The success of the integration of local knowledge in environmental assessment and monitoring is influenced by shifts in theoretical viewpoints on environmental dynamics. Such discourses in turn determine the type of research methods that can be used to link management objectives of local communities and conservation goals. Participatory methods face challenges in terms of questions relating to "how" local knowledge can be used – how to collaborate with conventional scientific methods and understand the perspectives of indigenous knowledge, particularly how it functions in relation to environmental change. Whereas the need for integrating participatory knowledge with conventional scientific methods is often demanded as a prerequisite for achieving global participation, any frameworks for achieving such integration are poorly documented (for a recent attempt, see Dougill and Reed 2006; Oba et al., 2008a,b).

A schematic framework for understanding how the objectives of the GECs could be used to guide national and local actions to achieve global and local goals is proposed in this essay. Integrated methods (see Figure 1) based on local environmental knowledge and ecological methods could be tested to show how the broader global objectives for the implementations of GECs such as the CCD and CBD could be tackled at local community levels. Before discussing the approaches that might be used for achieving these goals at community levels, it is necessary for the reader to be familiar with how some of the terminologies and concepts were applied in this essay.

2.1. Definition of Terms and Concepts

In this essay different terms are used in relation to the assessment and monitoring of land degradation and biodiversity loss. The terms are explained within the context of their use, and therefore universal agreement is not presumed. For example, the knowledge held by the local communities has been described using different terms including: "indigenous knowledge", "local knowledge", "folk knowledge", "indigenous technical knowledge" (ITK), "traditional ecological knowledge" (TEK) and "indigenous ecological knowledge" (IEK), among others. Each of the terms has a different connotation in terms of importance and the application of knowledge by different communities. We have used some of the terms interchangeably to describe the knowledge held by local communities. These knowledge systems are used by individuals but represent the sum total of knowledge used by particular local communities (Roba and Oba, 2008). However, reference is made to the most popular terms, such as "indigenous knowledge/local knowledge" and "traditional ecological knowledge" to describe the roles local communities play in environmental management. According to Warren (1991), the term "indigenous knowledge" describes the knowledge developed by a given community, which is different from scientific knowledge systems generated through universities or at government research stations. Indigenous ecological knowledge refers to experiences acquired over a lifetime through observations and in relation to social norms and institutions that shape human interaction with the environment (Berkes et al., 2000; Fernandez-Gimenez, 2000). Such knowledge is useful, for example, in describing concepts such as land degradation.

"Land degradation" is a composite term and the definition depends on the context used by both scientists (Stocking and Murnaghan, 2001; Warren 2002) and local communities (Oba and Kotile, 2001). In general, land degradation is defined as the loss of utility or potential in relation to biological organisms as well as changes in the physical environment that would alter the functions of natural systems (Abel and Blaikie, 1989; Barrow, 1991). According to Dodd (1994), degradation may refer to a decrease in plant productivity or unfavorable changes in species composition, but does not imply that changes are permanent. In the rangelands, Abel and Blaikie (1989), defined land degradation as a permanent decline in land for the yielding of livestock products under a given system of production. This means that in terms of pastoral production, where milk and meat are major products, land degradation leads to a downward spiral in livestock productivity. In pastoral systems where multiple livestock species are managed, better insight about the processes of environmental change can be gained by considering herders' perceptions. Herders define land degradation in relation to livestock productivity. A degraded environment, according to herders, does not support livestock productivity at optimal levels due to the loss of important fodder species. Accordingly, a landscape that is degraded for one species of livestock e.g. grazers, may not be so for browsers (Oba and Kaitira, 2006). Throughout the article, a broader approach to understanding land degradation is adopted, rather than narrow ecological definitions alone.

Ecologists also use other terms such as "desertification" when referring to extreme levels of land degradation. The concept of desertification implies both temporal and spatial perspectives. Internationally, desertification is defined as "land degradation in arid, semi-arid and dry sub-humid areas resulting from various factors, including climatic variations and human activities" (UNCCD, 1994). The nature, extent and reversibility of the changes associated with desertification have been at the center of scholarly debates (e.g., Helldén, 1988; Stiles, 1995; Thomas and Middleton, 1994). We use the terms desertification and land degradation concurrently, but often "desertification" implies permanent and irreversible changes in vegetation and soil conditions, whereas "land degradation" is used to describe changes which are reversible with management or when the anthropogenic pressures are removed. Given that the classification of both degradation and desertification might rely on similar indicators (see below), the issue is at what levels changes would be described as simple degradation or as processes that cause desertification. One of the important indicators for evaluating land degradation and desertification is biodiversity.

Biological diversity is a broad concept that describes the variety of all living life forms encompassing species, genetics and ecosystems. Our focus on biodiversity is mainly on the diversity of plant species of the grazing lands. In the assessment and monitoring of changes in plant species diversity, conservation interests are based on the assumption that loss of species diversity (number of different species) and species richness (number of species per unit area) are important criteria for assessing ecosystem degradation. In the grazing lands, herders are concerned with satisfactory livestock production. Therefore, the concept of the total plant species pool, which is important from the conservation viewpoint, does not adequately capture herders' requirements. According to herders, changes in biodiversity in the grazing lands refer to changes in plant species composition in relation to livestock fodder requirements (Mapinduzi et al., 2003; Oba and Kaitira, 2006, Roba and Oba, 2008). By focusing on key forage species, the herders use utilitarian definitions of biodiversity. The concept of invasive species is well known to ecologists, while herders refer to "bad" or "good" biodiversity in terms of the extent of unpalatable plant species, such as those that might be associated with bush encroachment. In this essay, the use of the term "biodiversity" refers to both the conservation and utilitarian values in terms of how it is assessed and monitored.

"Assessment" and "monitoring" of vegetation are relative terms used for understanding environmental dynamics. "Assessment" refers to observation of the status of various indicators that influence environmental health. It involves evaluation at one point in time to generate baseline data on vegetation and soil physical characteristics. For herders, assessment is done more frequently across the grazing landscape to ensure an acceptable quality and quantity of fodder for multiple livestock species. Ecological assessment of grazing ecosystems is less frequent and on limited spatial scales. Observations are usually made at a series of sampling transects and plots in order to generate data that are used to generalize the status of grazing resources. By comparison, "monitoring" is an evaluation process conducted several times over long periods to determine responses to management and other environmental factors such as rainfall (Holecheck et al., 1995). Monitoring of important environmental variables, such as change in species composition, is conducted using similar methods, for the same objectives and over a long period of time. The observed trends in the variables monitored can then be related to different drivers. Herders' assessment and monitoring of environmental change is not limited to environmental indicators, but includes livestock production performances such as volume of milk and animal body condition. We have used assessment and monitoring to describe activities conducted by herders and ecologists to determine the suitability of the grazing environment on short-term and long-term basis. Assessment and monitoring must be conducted by range managers to understand how management has influenced the condition of the rangelands (short term) and trends (long term changes) in environmental indicators.

Vegetation "condition" and "trends" are terms commonly used to describe "the state of range health" (NRC, 1994). Range condition is scaled in terms of the "climax" vegetation. The use of the concept is limited due to difficulties in establishing a climax vegetation composition for determining change in arid zones (Ellis and Swift 1988). The idea of using climax vegetation for determining "healthy" range condition shows that the term is closely related to the equilibrium ecological theory. The concept is used in relation to the "range condition status" which the herders consider as being optimal for livestock production, while ecologists use the term to mean a departure from the assumed "climax" vegetation status. Thus the term "condition" is used in relation to utilization. "Trend" refers to the direction of change of range condition, which can be rated as upward, downward or stable (Holecheck et al., 1995). To describe trend, an observation of change in species composition is required. In pastoral systems, herders have accumulated knowledge of the direction of vegetation change more so than ecologists.

Another term that appears in the chapter is "integration" which, as used here, implies the combined use of local knowledge and scientific methods to understand environmental change. Integration of assessment and monitoring of environmental change is achieved by the simultaneous use of multiple indicators used by local people and ecologists. The local communities use indigenous knowledge and composite indicators (hereafter also referred to as "anthropogenic" indicators), while ecologists use ecological methods and ecological indicators. The integration of local knowledge and ecological methods improves understanding of environmental change, as the two systems complement each other. Thus, could integration be achieved by asking the local informants questions about the environment and using ecological methods to measure required variables? Or does integration involve a whole range of processes of environmental assessment and monitoring and decision-making? In the way the term is used here, integration is taken as a process that finally leads to rational decisions. Creating a situation for sharing information and understanding different viewpoints by local communities as well as by scientifically trained technicians, result in a common forum for discussing the problems of the environment. By jointly using the different methods it would be possible to understand the perceptions about the reversibility of land degradation, which is related to the concept of "resilience".

The concept of ecological resilience described by Holling (1973) refers to changes in ecosystems when subjected to perturbation. Resilience is defined as the capacity of a system to buffer or resist change in response to a given magnitude of disturbance, before losing the capacity to respond (Gunderson, 2000; Perrings and Walker 1995). The level of resilience can be defined in terms of systems potential. In grazing lands, for example, different landscapes disclose varied "potential" in relation to soils and vegetation. The potential of the resource system describes the capacity of the system to resist degradation. The concept of ecological resilience is used to understand change in species composition and vegetation structure. Long-term fluctuations in vegetation variables can be examined to see if they present a characteristic of resilience under unpredictable rainfall regimes, or if they portray a more linear change as presented by the equilibrium ecological model. In addition, variables such as species inventory, change species frequency, and species cover over long periods of time are important for understanding the resilience of an ecosystem after years of droughts and sustained grazing pressure. The capacity of the system to spring back is what allows land degradation in arid ecosystems to be reversible; the lack of resilience would cause desertification (Binns, 1990; Oba et al., 2008). Herders are aware of the resilient property of arid ecosystems, based on several years of observation. In arid ecosystems, the concept of resilience has helped herders to develop adaptive management strategies that enable them to modify their management strategy according to prevailing environmental conditions. Using their knowledge of individual landscapes and their capacity to cope with grazing pressures (similar to what they call landscape grazing potential-LGP) as well grazing preferences of specific livestock species (i.e. landscape grazing suitability-LGS) , herders regulate grazing movements allowing the land to regenerate, even after heavy use. This shows that management strategies have an influence on the resilience of an ecosystem (e.g. Perrings and Walker, 1997).

Each of the concepts or terms discussed here elicits different discourses in terms of environmental change. It is important to mention that no single theoretical viewpoint can adequately address the different discourses, necessitating the application of an interdisciplinary approach for understanding the processes that result in environmental changes associated with desertification. The terms are for discussing different theoretical and methodological perspectives related to local community participation in the implementations of the GECs related to the CCD and the CBD. The concepts are important for understanding continuous shifts in scientific paradigms of environmental change from deterministic cause-effect views to multi-directional approaches that acknowledge the influence of environmental variability and the importance of indigenous knowledge. We next describe the theoretical perspectives for better understanding of the framework.

3. THEORETICAL PERSPECTIVES

Major environmental discourses have played a central role in shaping how environmental problems are perceived. The dominant environmental discourses such as desertification and biodiversity loss have attempted to portray a crisis scenario, especially in arid environments (Lamprey, 1983). Increases in human and livestock populations have been associated with adverse effects on the environment. The crisis narratives are rooted in different environmental theories, which predict the relationships between biotic and abiotic, social-ecological and economic components of an ecosystem. For example, different viewpoints of land degradation in arid ecosystems have been influenced by different scientific theories. The ecological theories reflect *a priori* environmental functions in terms of processes, explained in terms of deductive relationships between causes and effects of land degradation. Management is often not part of the theory, although the impact of management on the environment uses the theory to analyze the effects. This implies that local knowledge in environmental assessment and monitoring is usually not part of the theory description and verification. This was until recently, when ecologists re-evaluated existing ecological theories for guiding management decisions, particularly in arid lands (Behnke and Scoones, 1993), for the following reasons. Firstly, ecological theories do not explain all the outcomes of environmental change, particularly where management or human decisions in land use are involved. Secondly, for ecosystems such as arid lands, earlier ecological theories had assumed stability, while the system behavior is better described by variability. Empirical evidence in support of spatial and temporal variability questioned the value of using ecological theories that prescribed stability and predictability (Ellis and Swift 1988). The stable and predictable systems failed to acknowledge the management systems of local resource users, while the variable systems considered the rationale of local resource use. These developments therefore became the impetus for the integration of indigenous knowledge and ecological methods.

Various viewpoints from the perspective of the dynamics of vegetation in arid ecosystems in response to land use by local pastoralists are reviewed here. Since the early days of range management as a science (e.g., Clements 1916), ecologists considered proper management in terms of the equilibrium between grazer populations and allowable forage utilization (Heady 1975). This theory holds that range managers should maintain the numbers of grazers on a given range commensurate with its potential carrying capacity—which is defined as the maximum stock carried per unit of land in a given time (Pratt and Gwynne, 1977; Bartels et al., 1993). The theory presumes that it is the grazers that drive the changes in plant composition and therefore by regulating stocking levels, the dynamics of vegetation can be maintained at desirable levels of plant composition to promote environmental sustainability.

The equilibrium theory, which is described above in a simplified version, is deterministic and uni-directional, since it does not take into consideration environmental drivers, such as rainfall variability, as the principal control agents for driving range production. Indeed, the theory, which was designed under humid conditions, might work adequately in ecosystems with predictable rainfall, but problems arose in arid environments (Ellis and Swift, 1988; Oba et al. 2000b). Using the responses of vegetation parameters to herbivore populations, the theory would predict negative changes when the population exceeds the carrying capacity or the stocking potential (Lamprey and Yussuf, 1981; Lamprey, 1983; Sinclair and Fryxell, 1985). Therefore, based on the equilibrium theory, land degradation that occurs in arid lands, would be blamed on management systems that ignore the "equilibrial" relationships between nature's functions and use.

The equilibrium theory influenced early pastoral development throughout sub-Saharan Africa in a significant manner, with adverse consequences for the environment and production systems (Sandford, 1983). One such adverse consequence was sedentarization of former nomads that resulted in the breakdown of traditional systems of land use, causing precisely those environmental problems which the theory was meant to guard against (Sinclair and Fryxel, 1985). The proponents of the theory excluded local indigenous knowledge and purposely focused on the use of ecological indicators (see below) for environmental assessment and monitoring. There are, however, critical similarities between the equilibrium theory and its explanation of vegetation changes in relation to grazing pressures, and local knowledge of herders on how vegetation indicators respond to sustained grazing. The equilibrium theory postulates that sustained grazing pressure induces shifts in plant species composition, where some species that are more sensitive decrease, while others, that are unpalatable or more resistant to grazing pressure, increase or remain stable. Careful analysis of indigenous knowledge shows that herders have comparable understanding about the species that would decrease, and others that would increase or remain stable. The qualifying difference is that according to the local herders, the shifts in plant species vary with the type of livestock. The implication is that land degradation cannot be a universal problem in grazing lands. The land degraded for grazing livestock would still be sustainable for browsing stock and vice versa. Ecologists rarely took such views into account.

Shifts in ecological thinking (e.g., Ellis and Swift 1988) propose that vegetation changes in highly variable environments, such as in arid lands, are more sensitive to environmental drivers including rainfall variability, than they are to grazer populations. The non-equilibrium theory proposes that in environments with high coefficients of variation, the spatial-temporal dynamics of vegetation resources cannot be accounted for by grazing alone (Fernandez-Gimenez and Allen-Diaz 1999; Oba et al., 2000b; Sullivan and Rhode, 2002). Rather, the rhythms of range production are closely related to spatial and temporal rainfall variability. This means that the same environment might experience a "boom" at one time, and "burst" production at another, in response to varied rainfall regimes. Rainfall varies from season to season, resulting in substantial differences in range production according to seasons (temporally) and spatially by sites or geographical distribution of plant production. The non-equilibrium model of rangeland dynamics therefore puts emphasis on the unpredictable nature of ecosystems and the inability on the part of management to develop practical plans based on prior knowledge for manipulating stocking rates. Pastoral production, which involves livestock mobility, is adapted to variability (Fernandez-Gimenez and Allen-Diaz 1999; Oba et al., 2000b) by tracking the variable resources opportunistically (Behnke and Scoones, 1993).

Herders are aware that plant species composition and cover change with variation in rainfall and across heterogeneous landscapes. Herders use this knowledge of seasonal and spatial variability of grazing resources to promote mobility. Local knowledge of resource management could therefore potentially contribute to the assessment and monitoring of the vegetation dynamics aimed at understanding the mechanisms described by different ecological models. The role of local knowledge is deemed to be even more relevant when the dynamics of arid ecosystems are described using social-ecological resilience viewpoints.

According to the social-ecological resilience viewpoint, ecological processes are closely linked to social activities, including decisions on land use and livestock management (Berkes et al., 1998; Oba et al., 2008). In the social-ecological resilience model, adaptive management is central to pursuing livestock production goals. Pastoral adaptive management involves making adjustments to production variability by means of mobility. Thus, the explanation offered by the non-equilibrium theory about the dynamics of arid lands has closer parallels with the socio-ecological resilience theory. By means of the latter, the functions of indigenous knowledge in the management of variable environments can be justified. Based on their detailed knowledge of seasonal variations in species composition, herders might suggest that a particular plant species is not present at the time of assessment (e.g. in dry season), but will be seen again in the wet season. Thus, herders believe that arid ecosystems are highly resilient and in their view degradation occurs only when livestock mobility is curtailed and heavy grazing is sustained over a long period that would result in loss of key forage species.

The importance of local knowledge in addressing complex environmental problems has been supported further by interdisciplinary studies that used the principles of political ecology. Political ecology underscores the importance of environmental narratives and discourses for addressing desertification and biodiversity loss (Leach and Mearns, 1996; Batterbury et al., 1997; Laris, 2004). Studies in human-environment relationships have shown that many of the environmental crises debated at global level are exaggerated (Bassett and Zuéli, 2000). The narratives advanced at global levels on the status of environment should be related to the counter narratives of local people (e.g. Bassett and Crummey, 2003). Political ecology and environmental history therefore provide important links for analyzing human-environmental interactions (Benjaminsen and Lund, 2001). Local communities' narratives on causes and trajectories of environmental change are important for the implementation of the GECs, as well as in mitigating processes such as desertification and biodiversity loss.

Based on the above discussion, it is evident that processes of environmental change involve more than ecological changes and are influenced by social factors, including decision making by local land users (Blaikie and Brookfield, 1987; Oba et al., 2008a, b). In the light of the recognition of the close interaction between social and ecological systems, there is a need to adopt a more holistic approach that goes beyond the disciplinary divides, and in addition co-opts local knowledge and practices using assessment and monitoring of environmental changes at local levels. To achieve this objective, indigenous knowledge and conventional scientific methods should not be portrayed as competing binary opposites, but rather, as complementary sets of research and management tools for addressing environmental problems (Agrawal, 1995; Nygren, 1999).

Advances in interdisciplinary methods have transformed environmental research from purely ecological perspectives to incorporate wider social, political and economic aspects (Blaikie and Brookfield, 1987). This knowledge interplay underscores the role of local communities in studies on environmental change. The approach is similar to the conceptual socio-economic and ecological model (SEEM) that has been described elsewhere (Oba et al., 2008b). The SEEM model proposes that analysis of environmental change should consider local socio-economic and ecological drivers that may inform decision makers on the relationship between land use and environmental change. In the SEEM model, appropriate anthropogenic and ecological indicators are used to describe environmental changes. This paper utilizes the SEEM approach for integrating indigenous and ecological methods by conducting environmental assessment and monitoring jointly with herders in the arid zones of Africa. The integration process demands the development of efficient methods to accommodate indigenous knowledge variables and ecological methods for environmental assessment and monitoring.

4. METHODOLOGICAL PERSPECTIVES

The integration of indigenous knowledge in research faces some methodological challenges (Scoones and Thompson, 1993). An important methodological issue is the selection of knowledgeable members of the community. It has been argued that indigenous knowledge is not homogenous among the local people due to stratification into gender, age, social classes and other disparities in power relations (Sillitoe, 2002). Therefore, researchers are obliged to find a balance between participation by highly heterogeneous groups, while at the same time maintaining rigor demanded by standard scientific methods, such as random selection of informants. Another problem of integration arises because of epistemological differences between indigenous and standard scientific methods (Purcell and Onjoro, 2002). In scientific research, the focus is on cause and effect relationships and developing predictive models about future outcomes in relation to propositions made by different scientific theories (Freeman, 1992). The scientific method is deductive as well as reductive and may be deterministic in predicting processes of environmental change. By comparison, by its very nature, indigenous knowledge is rooted in local contexts, practices and beliefs (Woodley, 2005). It is better understood using cognitive anthropology, focusing on the relationships between human culture and human thoughts and beliefs. Unlike the traditional ethnographic approach, where an outsider uses his/her worldviews to study certain cultures, cognitive anthropology stresses how people make sense of reality according to their own indigenous cognitive categories. The cognitive anthropology approach aims at understanding the local peoples' worldviews using the *emic* perspective, as opposed to traditional ethnographic outsiders' *etic* views.

The cognitive anthropological approach of eliciting local knowledge uses formal questionnaires. However, the questionnaires are developed with literate informants in mind and are often inappropriate in rural settings where research is conducted with local people (Antweiler, 2004). Most researchers using questionnaires are not comfortable with the responses elicited, because they are void of capturing real experiences, and there is the added disadvantage of the time it takes (Kumar, 2002). Although questionnaires are prepared based on presumed problems and sets of objectives that may not address the goals of local people, they may be useful for soliciting generalized information, to be followed up by more participatory approaches. The time drawback has led to the adoption of more pragmatic, rapid participatory methods. Quick and popular methods such as Participatory Rural Appraisal (PRA) and Rapid Rural Appraisal (RRA) have been used to bring together local people and researchers/development agents to address local issues (Chambers, 1999). These participatory approaches use visual learning such as drawings and maps so that non-literate and less articulate members of the local community can participate in exercise-oriented discussions (Kumar, 2002). RRA focuses on the use of observation and verbal interactions including semi-structured interviews, transect walks, and group discussions (Chambers, 1999). By comparison, PRA stresses the use of shared visual representation by local people. The RRA and PRA methods allow local people to share knowledge and practice with external development and research agents to implement locally relevant development objectives (Chambers, 1994). Rapid participatory methods are, however, disadvantageous because of the lack of empirical and theoretical grounding (Antweiler, 2004).

The data collected through rapid participatory methods represent mostly the local contexts in which they are generated, with limited wider application. The most appropriate forum for using participatory methods is organized workshops involving both local and technical groups, for one-to-one discussions concerning pertinent environmental or social issues. For example, in herding communities, information on the local environment is held in both the public domain and also by some key knowledgeable members of the society, including the elders and herder scouts. Due to their long-term experience, the elders have more detailed information on historical environmental changes including land use patterns, levels of livestock productivity and the extent and composition of vegetation in the grazing rangelands. The elders' knowledge is continuously updated during daily assessment and monitoring of the status of grazing resources. Individuals rely on accumulated knowledge within the community, suggesting that there is uniformity of information, regardless of whether individuals or groups were involved in environmental assessments. The indigenous knowledge has two sides – the knowledge related to the production systems, such as livestock, that is often used as a barometer of environmental change, and the knowledge of the environment itself in terms of grazing landscapes, soils, vegetation types and the history of land use. Thus, the herders have accumulated *knowledge* of the interaction of livestock production systems with biophysical resources. This knowledge is put into everyday *practical* use to regulate livestock-environment interactions.

When selecting research methods, it is therefore important that collaborating researchers adequately understand the knowledge and practice aspects of indigenous knowledge systems. The basic resource management units are landscapes. Assessment and monitoring of environmental change is based on the knowledge of a landscape that has been accumulated over the long term and enriched by continuous observations using different environmental indicators. For the herders, the process has utilitarian value, which is aimed at maximizing livestock productivity through appropriate and timely management decisions (Oba et al., 2008a, b). The knowledge of the local environment is therefore part of daily practice embedded in the local cultures and influenced by the requirements of the production systems.

The scale of assessment and monitoring and the indicators used all have relevance to local production systems. Previously, different approaches have been used to elicit participation, including questionnaires, semi-structured interviews, workshops and collaborative fieldwork (Dougill et al., 2006; Huntington, 2000; Stringer et al., 2006). These methods have different levels of flexibility in accessing indigenous knowledge in different local contexts. Structured questionnaires are useful when quantification is desired and the interviewer has a clear understanding of the information required. For example, in pastoral environments, data on the number of livestock managed by different households in different management systems (e.g. mobile versus home herds) can be collected using a structured questionnaire. Semi-structured interviews are less restrictive and may use sketch guidelines, while the participants and the interviewer engage in open-ended discussions. Semi-structured discussions with individuals or groups help the interviewer to strike a rapport with principal participants in more informal settings, thereby allowing a free exchange of information. Similarly, workshops on specific environmental issues involving researchers and participants can result in fruitful discussions. In workshops, different issues raised earlier during interviews could be deliberated further to generate new insights on the topics under discussion. Interview methods are important for eliciting knowledge engrained in local land use practices. The approach requires developing personal relationships to promote dialogue and the exchange of ideas as part of participatory research.

Scientific methods (hereafter referred to as "ecological methods") are standardized and strongly empirical. In vegetation assessments, ecological methods include the use of transects to measure the density and composition of vegetation occurring along a line or belt transect, or a road may be used as transect. Additionally quadrants or plots are used to sample vegetation in circumscribed areas (Sorrells and Glenn, 1991). There are, however, some important requirements in ecological sampling, including having representative samples by randomizing sampling units, including transects and quadrants/plots. The sampling units must also be of the right size in order to capture the diversity of plant life forms such as herbs, shrubs and trees. The selection of sampling strategies needs to consider variability along ecological or anthropogenic gradients. Another factor central to ecological monitoring is the use of the right scales for measuring the vegetation variables from landscapes to larger regional levels. Ecological methods linked to different scales of vegetation change are useful for testing hypotheses. Sampling designs must incorporate spatial and temporal factors that may influence vegetation changes. Spatial limitations of quadrant and transect methods in the assessment and monitoring of vegetation have been solved using remote sensing technology which has a wider spatial coverage. Satellite imagery can be used to compare vegetation covers at different times in order to analyze the temporal dynamics (Roba 2008). The integration of indigenous knowledge and ecological methods for the implementation of GECs should harmonize the various methodological orientations. Before discussing the framework for integration of the two methods for addressing local environmental problems, it is imperative to situate participatory actions at the local level within the context of the broader plans for the implementations of GECs.

5. FRAMEWORK FOR INTEGRATING LOCAL KNOWLEDGE AND ECOLOGICAL METHODS

The implementation of GECs is linked to the activities at global, national and local levels. At the global level, conventions are negotiated as part of collective global responsibilities as mentioned earlier. At the national level, each country implements the conventions according to the articles, by setting priorities according to national goals and implementation strategies. Implementation at the local level necessitates consideration of the diversity of ecological, production and social-cultural systems, and the use of local knowledge for resource assessment and monitoring. Local participation as recommended at the global level, and implementation plans through national and local levels are described using a conceptual framework (Figure 1).

5.1. Indigenous Knowledge

Local communities have accumulated important knowledge for biodiversity monitoring and assessment (Vermeulen and Koziell, 2002; Hellier et al., 2004) which is associated with local strategies to sustain livelihood systems. For example, herders describe the status of biodiversity in relation to livestock grazing resources. Local vegetation is monitored for changes in plant species composition that may affect key fodder species for livestock grazing (Bollig and Schulte 1999; Mapinduzi et al., 2003; Gemedo et al., 2006; Oba et al. 20008a). Monitoring focuses on land use history where past experiences serve as a baseline for comparing current changes. Equally valuable is knowledge of land degradation. Since livestock is the main source of livelihood, herders' perceptions of land degradation are influenced by livestock production requirements (Bollig and Schulte, 1999; Oba and Kotile, 2001; Okoti et al., 2006; Gemedo et al., 2006; Reed and Dougill 2002).

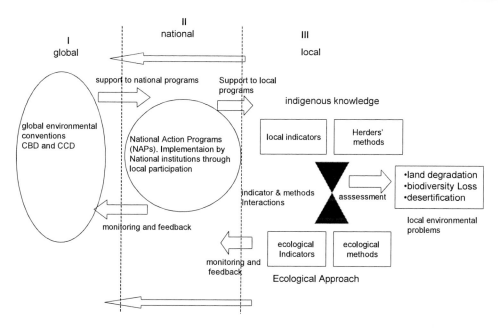

Figure 1. Framework for the implementation of Global Environmental Conventions, showing integration of indigenous knowledge and ecological methods for the assessment and monitoring of environmental change.

The framework has three interrelated components (hereafter referred to as "stages I-III"). In Stage I, global environmental problems such as desertification, land degradation and biodiversity loss are described. Stage II of the conceptual framework is concerned with the implementations of NAPs (Figure 1). In this essay it is assumed that national action plans (NAPs) may be concerned with reversing land degradation and desertification, while at the same time conserving biodiversity using both conventional methods and local knowledge (Stage III). This may be done by working through technical departments that are represented at the local level. Thus the approach used here is the very first step in evaluating the implementation process at local levels. It is at Stage III that the integration of herder indigenous knowledge and ecological methods is most critical for conducting assessment and monitoring of environmental change.

Herders not only monitor the trends of vegetation change over the long term, but they also make inferences from livestock production performance. In terms of vegetation, herders monitor both the quantity and the quality of fodder. The status of vegetation guides herder decisions for livestock management (Oba et al., 2008a, b). Therefore, herder perceptions of land degradation in terms of livestock performance involve more than just describing the biophysical attributes, such as the physical characteristics of the soil and the quantity and quality of the vegetation. It involves understanding complex sets of perceptions about inherent factors in the environment that cannot be measured directly. Herders in eastern Africa use concepts such as '*finn*', used by the Oromo speaking group (Oba, 1985a) and '*mirr*' used by the Rendille herders (Roba, 2008), to describe range health that is not directly related to the status of biophysical resources in the grazing lands. For ecological experts, such attributes that cannot be quantified, would not warrant further analysis because they are "unscientific", but in reality, these are the silent features of local knowledge that have enabled sustained livestock production in arid ecosystems. Such knowledge has often been used for regulating livestock populations and the landscape potential for grazing. Herders' assessment and monitoring of land degradation based on qualitative vegetation attributes related to livestock production has the potential for understanding environmental change in conjunction with ecological methods.

Additionally, local communities classify landscapes according to utility of the vegetation and suitability of the soils (Oba and Kaitera, 2006; Verlinden and Dayot, 2005). Herders classify pasture and landscapes for purposes of daily and seasonal livestock grazing movements. They determine landscape grazing suitability (LGS) based on the forage available and grazing preferences of different livestock species, while for determining long-term livestock use, they consider landscape grazing potential (LGP). The concepts are widely applied among global pastoralists. For example, Mongolian herders classify grazing landscapes in terms of ecological zones, which are utilized by different livestock types during different seasons of the year – such as winter versus summer pastures (Fernandez-Gimenez, 2000), while the Maasai classify pastures according to different livestock classes and agro-ecological zones (Mapinduzi et al., 2003; Oba and Kaitira 2006; Hodgson and Schroeder, 2002). The Borana herders of southern Ethiopia and northern Kenya classify landscapes according to both seasonal use and daily herding practices (Oba and Kotile, 2001; Dabasso, 2006). Knowledge of landscapes is based on soils and vegetation types (Oba et al., 2000c). The knowledge of different landscapes in relation to vegetation characteristics and their potential provides important guidelines for managing each type of grazing landscape and each type of livestock species.

Herders base their knowledge of landscapes on the history of land use. Environmental history is about past events that describe the interaction between the environment and society (Hughes, 2001). Environmental change at the local level is, however, not a one-time experience, but a process with historical dimensions on the causes and magnitude of change. Local communalities have accumulated information on the events that altered land use patterns, including extreme climatic factors and settlement patterns that might have changed the local vegetation. Local environmental history therefore provides an important source of data for understanding past environmental status and for better appreciation of management and environmental drivers that might have adversely affected the environment.

5.2. Ecological Methods

Ecological methods are important for evaluating progress towards management objectives (Havstad and Herrick, 2003; West 2003). One purpose of ecological methods is to collect baseline data for ecosystems against which the impact of management strategies on species loss can be compared (Spellerberg, 2005). In assessing and monitoring vegetation change, the most common sampling methods are plots and transects. The plot method is useful for measuring vegetation in a fixed area. The plots may be "nested" by sub-dividing them into smaller units to capture plants of different sizes (e.g. herbaceous, shrubs and trees). In the plots, vegetation cover may be estimated using visual observations or by using the point-step method. In the point-step method, the observer paces across the range and a metallic loop is placed at the tip of the boot (Hill et al., 2005). Individual species, litter, bare ground or rocks are then recorded. Plots may be randomly or systematically placed along marked transects according to land use gradients or vegetation types. The transect method involves the establishment of a systematic sampling procedure designed for monitoring vegetation change. Transects marked along gradients of land use may be used (e.g Fernandez-Gimenez and Allen-Diaz, 2001; Landsberg et al., 2003; Todd, 2006).

The importance of ecological assessment and monitoring in range management is constrained by the disparities between the scales at which monitoring is conducted and the scale of management (Oba et al. 2003; Oba et al. 2008b). Considering the existence of larger spatial scales and complex ecological interactions, it is not always possible to monitor systems from the perspective of land use alone. Monitoring therefore has two inter-related functions: understanding ecological change and responses by livestock production, and the protection of key biodiversity. Monitoring on a broad scale may be less sensitive for analyzing processes occurring at fine scales, but it offers opportunities for wider spatial coverage, particularly when aerial photography and satellite imagery are used. For example, the initial reports of the advancing desert margins in the Sahel region were based on aerial photographs (Lamprey, 1988). Additionally, recent observations on the greening of the Sahel were based on observations from satellite imagery (Anyamba and Tucker, 2005; Herrmann et al., 2005; Olsson et al., 2005). Satellite imagery has often been used to elucidate change in land cover over long time periods (e.g. Ringrose et al., 1996; Shoshany et al., 1996; Palmer and van Rooyen 1998; Virtanen et al., 2002; Bouma and Kobryn 2004; Kiunsi and Meadows 2006; Shalaby and Tateishi 2007). Although satellite imagery solves the problem of monitoring spatial coverage, it fails to reveal finer details with regard to changes in species composition (Trodd and Dougill, 1998). Satellite observations therefore need to be supported further by ground surveys to discover details of the processes of ecological change in terms of the effects on individual plant species.

Furthermore, details on the causes of environmental changes observed from satellite data need to be supported by information from local land use history. Both fine-scale and coarse-scale environmental assessment and monitoring methods need to specify the indicators used. Assessment and monitoring methods used by herders and ecologists therefore share important attributes, including the use of vegetation indicators. The two differ, however, in how the observations are interpreted and the scales at which the field assessments are done. The differences in interpretation represent a bottleneck for herder participation in joint monitoring and assessment exercises. We therefore address the methodological challenges in integrating indigenous knowledge with ecological methods for the assessment and monitoring of environmental change in grazing lands (Stage III, Figure 1).

5.3. The Selection of Indicators

The main constraint of the application of local knowledge in environmental assessment and monitoring is the lack of acknowledgement of suitable indicators (Krugman, 1996; Hambly, 1996), which are pre-requisites for achieving participatory research. The indicators must be those that are recognized locally (Fraser et al., 2006; Stringer and Reed, 2007). Herder indicators are divided into anthropogenic and biophysical indicators. The anthropogenic indicators include changes in livestock health, body conditions, calving and mating rates, among other factors related to livestock production (Gemedo et al., 2006; Wasonga et al., 2003). The biophysical indicators are related to livestock production objectives, which is satisfactory herd growth and production performance (Reed and Dougill 2002).

Moreover, herders make distinctions between different grazing resources in terms of their spatial variation, which is usually in terms of heterogeneous landscapes. Herders have two viewpoints in terms of heterogeneous landscapes. Firstly, they consider that different landscapes have varied potential for grazing. Secondly, they infer that each landscape discloses grazing suitability for a given species of livestock, as compared to other species. Landscape grazing potential (LGP) is an inherent property used by herders to describe the ability of a landscape to withstand continuous grazing, while landscape grazing suitability (LGS) describes the relative preference by different livestock species (Oba and Kotile, 2001; Oba et al., 2008b). Both LGP and LGS are local management terms used by the herders to determine the importance of different landscapes for livestock grazing management. Anthropogenic indicators may be used as herder indicators—referring to ecological indicators, livestock production indicators and herder perceptions. The selection of herder indicators is based on long-term interactions with the herding peoples in different regions of eastern Africa (Mapinduzi et al., 2003; Oba and Kotile 2001; Oba et al., 2000c; Oba and Kaitira 2006; Oba et al., 2008a, b; Angassa and Oba, 2008) and southern Africa (Sheuyange et al., 2005).

For vegetation monitoring, herders use both quantitative and qualitative indicators. Vegetation cover is used to estimate the amount of biomass available for livestock grazing. Low vegetation cover is however not always a sign of land degradation. Thus, according to herders, an area with less vegetation cover might support livestock production even more than an environment that appears to be better endowed. For the herders, it is important to consider not only the vegetation cover present, but some inherent property of the grazing land that the livestock seem to desire, as reflected by their body condition. Indeed, herders make deductions more from livestock grazing behavior than from pre-determined knowledge. Taking a cue from livestock grazing behavior for making management decisions appears not to conflict with the wealth of knowledge accumulated by herders. Rather, it suggests that the sources of such knowledge can be diverse.

Additionally, herders monitor changes in the quality of grazing resources using the presence or absence of key fodder species. The relative availability of key fodder species is an important indicator of land degradation (Bollig and Schulte 1999; Mapinduzi et al., 2003; Roba and Oba, 2008). According to herders, different landscapes have characteristic key fodder species that are monitored on continuous basis (Oba and Kaitira 2006). By monitoring the proportions of useful plant species and less useful species in the grazing landscapes, herders are able to make appropriate herding decisions. When the availability of key species declines, herders make management adjustments to minimize the problem of land degradation. This knowledge is complemented by observed changes in species composition, from herbaceous and grass species appropriate for grazers, to shrubs and trees, that might indicate a corresponding shift in the type of livestock managed – from grazers to browsers. This type of knowledge has allowed herders to apply vegetation changes as indicators for making management decisions.

Another indicator that herders commonly use is soil physical characteristics (Oba 2001). Loose and dusty soils are considered to be vulnerable to grazing-induced degradation (Oba and Kaitira, 2006). Local people also use qualitative attributes including color and texture to classify soil and to determine its appropriateness for livestock kraaling (Gray and Morant,

2003; Oudwater and Martin, 2003). In herding communities, soils are used together with vegetation indicators to determine landscape suitability for grazing by different livestock species. The herders also use concepts such as "warm" or "cold" soils to determine the suitability for grazing and establishing pastoral camps. Herder indicators discussed in this section; share common functions with ecological indicators.

Ecological assessment and monitoring are also concerned mostly with changes in vegetation and soil conditions. Soil nutrients might be used as indicators of desertification (Oba at al. 2008c). Monitoring is a selective process, but for purposes of comparability, the ecological indicators that are most sensitive to anthropogenic pressures should be selected. Thus, monitoring should be able to account for environmental dynamics that differentiate between the effects of episodic events associated with rainfall variability, and impacts that might be associated with land use pressure. Moreover, what to monitor and the scale at which monitoring is conducted, depend on the overall objective of monitoring. At the landscape scale, the ecological indicators monitored are mainly vegetation, soil physical characteristics (Havstad and Herrick, 2003) and nutrients (Oba et al. 2008c). Attributes of vegetation change, including abundance, density, frequency, cover, richness and biomass, are monitored by observing changes over time, as well as the direction of change relative to the baseline data (West, 2003).

Among the ecological indicators, plant density is defined as the number of individual plants per unit area. It is determined by counting species in each plot and taking the average when more than one sample is considered. For conservation purposes, an increased number of individuals per unit area is often considered to be indicative of a healthy ecosystem (but this might not be the preference of herders). Species frequency is influenced by plot size. Plant cover represents the area occupied by vegetation relative to bare ground. In ecological assessments, vegetation cover is associated with biomass production. Reduction in cover is closely linked to the process of land degradation. Species richness and diversity are important measures of ecosystem health. Reduction in species richness and diversity are generally viewed as indicating a loss of ecological potential. When livestock grazing patterns are considered, the indicators of vegetation change include a shift in species balance between fodder and non-fodder categories (Oba et al. 2008a; Roba and Oba 2009). In remote sensing, changes in vegetation cover at different times are compared. An increase in vegetation cover in areas observed from earlier imagery as being bare is indicative of vegetation recovery, while the opposite is true where there is loss of cover (Roba 2008).

6. APPLYING THE METHODOLOGICAL FRAMEWORK

6.1. Land Degradation in Northern Kenya

In northern Kenya, pastoral sedentarization has been associated with land degradation and desertification (Lamprey and Yussuf, 1981; Lusigi 1984). The problem has been captured within the global debate of environmental degradation and desertification. Many of the scholarly research conducted in Northern Kenya during the 1970s emphasized environmental degradation, both in the sub-humid zones and the arid lowlands, where land use associated with pastoralists' sedentarization was blamed. However, previous attempts to assess and monitor land degradation induced by pastoralists' sedentarization in northern Kenya used scientific methods alone (Lusigi et al., 1986, Hary et al., 1996, Keya, 1997; Oba et al., 2003).

In the earlier attempts, local community knowledge on environmental change was ignored and, no effort was made to integrate local people's knowledge with ecological assessments. However, the local communities were usually blamed for contributing to the process of environmental degradation around sedentary settlements (Field, 1981). Due to the exclusion of local people, the findings from environmental assessments, including those of northern Kenya, have remained contradictory and the implementation of the GECs has been poorly addressed (see section 7.2) (Figure 2).

During the previous four decades, the Ariaal and Rendille pastoralists' systems of livestock management in northern Kenya have been transformed from mobile to sedentary systems (Oba, 1994). The process of settlement by nomads was accelerated as a result of drought disasters that impoverished many herders in the 1970s, forcing them into famine relief camps. In the sub-humid zone, the settlements in Karare, in the Marsabit District, were initially developed to rehabilitate destitute nomads by means of crop cultivation, while in the arid lowlands, the Rendille settled around the main towns of Kargi and Korr to receive food relief (Figure 2). Later, the government and missionaries initiated development programs that encouraged pastoralists to settle around these towns, which had grown into administrative, educational and security centers that provided watering facilities for humans and livestock. Despite the changed patterns of land use, the Ariaal and Rendille pastoralists maintained mobility of their herds, while the greater proportion of the human population remained in the sedentary settlements. The main concerns for the scholarly discussions were the impact of sedentary land use on the scarce vegetation resources. The scientists of the UNESCO-Integrated Project on Arid Lands (IPAL), who conducted much of the earlier work, were concerned about the accelerating land degradation and loss of vegetation cover, due to the over-exploitation of woody vegetation for the construction of night enclosures (*boma*), as well as overgrazing associated with high livestock stocking densities around the settlements. One estimate suggests that within the settlement rangelands, livestock stocking densities exceeded 25 Tropical Livestock Units km^{-1} whereas in the remote rangelands the stocking densities were much lower. The heavy stocking was linked to the lack of woody plant regeneration (Lamprey, 1976).

The research goals of the UNESCO-IPAL project were to understand the state of land degradation and livestock production within 18,000 km^2 of the home range of the Rendille and the Ariaal pastoralists. Using vegetation maps, the grazing lands were delineated according to vegetation units (hereafter referred to as *range units*) that corresponded with grazing landscapes used for seasonal grazing by the multi-species livestock comprising sheep, goats, cattle and camels. During the initial surveys conducted throughout the grazing home range of the Ariaal and the Rendille pastoralists, the conclusion was that the conditions of the rangelands varied from "fair" to "poor", except for a few vegetation types in the sub-humid zone of Marsabit mountain that were rated as "good" (Lusigi, 1984). Around the settlements, increased extraction of woody vegetation for building human shelters and livestock enclosures resulted in reduced woody cover, while livestock grazing pressure was reported to have had negative impacts on the herbaceous vegetation and woody species regeneration (Walther and Herlocker, 1980; Lamprey and Yussuf, 1981). In the lowlands, vegetation around settlements was over-utilized and the vegetation communities were mapped as "man-made deserts", which in the arid lowlands, extended up to a distance of 8 km from the settlements (Lusigi, 1984). The general scientific perception at the time was that the man-made deserts around the settlements were expanding and threatening the surrounding grazing lands. The perception was influenced by discourses at the time on the Sahelian desertification (Lamprey, 1983).

Although an interdisciplinary team of researchers conducted the investigations aimed at placing the people central to the problem, there was little evidence that the ecological research included the local herders in conducting environmental assessments.[1] Despite this lack, the anthropological research uncovered explicitly the wealth of knowledge that the pastoralists had in managing their arid environments (O'Leary, 1985). However, mostly the scientific empirical ecological research findings were used to influence the recommendations proposed by the management plan for the Western Marsabit District (Lusigi, 1984).

The management plan considered policy recommendations for the development of the Ariaal and Rendille rangelands, emphasizing the need for considering development according to local socio-economic needs and priorities set by the local people. However, it gave little weight to the role of indigenous knowledge in resource management, by recommending that "range areas should be developed, conserved, and managed in accordance with the ecological principles of proper land use" (Lusigi, 1984:484). At the time, the thinking of the IPAL research team was influenced by the equilibrium ecological theory that readily blamed the pastoralists for causing desertification due to high livestock stocking densities around settlements resulting in over-exploitation of vegetation resources (Field, 1981). Interestingly enough, whereas the ecological component of the project concluded that the environmental changes were permanent, leading to "desertification", the social scientists in the team placed the problem within the wider socio-economic and political problems (e.g., O'Leary, 1984).

In the remaining part of the chapter we present a synthesis of the findings on how we fully integrated indigenous knowledge of the Ariaal and the Rendille herders and ecological methods to assess and monitor rangelands that were reported to be degraded or desertified in the earlier IPAL reports. Our research replicated the methodological framework using two neighbouring pastoral communities-the Ariaal and the Rendille, who inhabited the sub-humid area on Marsabit Mountain and the arid lowlands, respectively. The two systems, due to differences in the levels of rainfall, altitude and topography, display differences in vegetation characteristics in terms of species composition and cover, regardless of levels of use. Since the two groups use different ecosystems, it was assumed that their knowledge of environmental change would differ; thus preference was given to using the knowledge of the two pastoral groups to understand environmental changes in their local land use contexts. Nonetheless, the selection of the sub-humid (equilibrium) and the arid zones (non-equilibrium) was useful in understanding the characteristics of vegetation changes that might be explained in terms of existing ecological theories. Furthermore, the use of different ecosystems was useful in investigating whether concepts such as land degradation and desertification might be relative to (a) use consistent with indigenous knowledge and (b) who is assessing and monitoring – the herders or the ecologists by applying the framework shown by Figure 3. The framework elaborates how Stage III in Figure 1 operates through participation by local herders. This involves the selection of appropriate indicators. The key findings are synthesized following the steps described. The steps were conducted simultaneously, but each is highlighted here to show the implementation process.

---

[1] Much of the environmental awareness was, however, conducted by means of training local herders (see Oba, 1985a, 1985b; Lusigi, 1984).

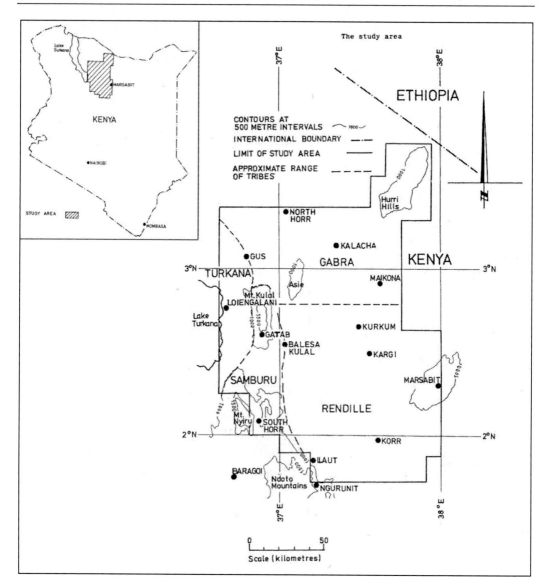

Figure 2. Location of Kargi and Korr in inset map of Kenya showing the location of the former UNESCO-IPAL study area, modified from Map 2 in Lusigi (1984).

6.1.1. Step I: Interviews

To implement Step I, semi-structured interviews were conducted with key informants to gather general information on herders' landscape knowledge, changes in vegetation composition and cover, and household livestock management strategies. Group discussions with community elders were used to elicit information on communal land use patterns and changes in vegetation characteristics around settlements. Herder narratives of environmental change and livestock management were described in detail.

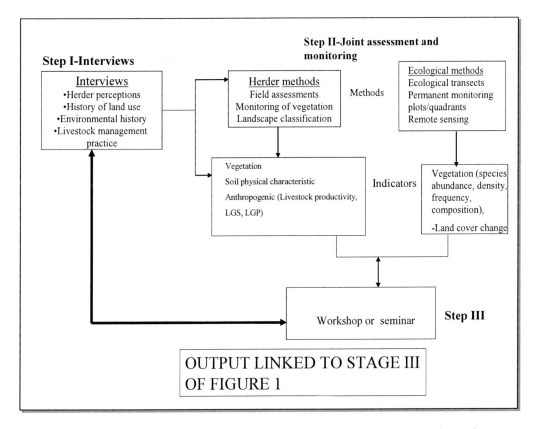

Figure 3. Methods for integrating herder knowledge and ecological methods for assessing and monitoring land degradation for the implementation of Global Environmental Conventions at local community levels. Note the processes describe the implementation of the interactions used in Stage III of Figure 1.

6.1.2. Step II: Joint field assessments

In Step II, the integration of local knowledge and scientific methods involved joint assessments with selected key herders and the ecological team. The integration was achieved by considering concurrently the methods and indicators used by the herders and ecologists. Herder assessment and monitoring of environmental change at the landscape scale were integrated with ecological methods using transects. Herders described the macro-landscape in which transects were located, using the anthropogenic indicators (LGP and LGS). The transect walks were conducted with herders across heterogeneous landscapes to describe vegetation and soil indicators. At each stop along transects, "plots" are placed to represent landscape patches, which are traditionally used as units for conducting environmental assessment and monitoring. Herders and ecologists then describe the range condition across the landscape patches as "good" or "poor" based on the assessment of vegetation and soil indicators. Herders described the vegetation trends as "stable" or "declining", using their knowledge of past land use history. Herders used soil physical conditions in each of the landscape patches. It is usual for the herders to classify soils as "warm" or "cold", or as "degraded" or "stable" (Oba and Kaitira, 2006). We selected the settlements of Kargi (N 02° 31 275` E 037° 34`) and Korr (N 02 00° 200` E 037° 30`) that were associated with *in situ* desertification (Lusigi, 1981). We established 4 km transects set in four compass directions from the centres of the two settlements. The settlements showed different patterns of pastoral camps. In Kargi, the pastoral camps formed a central cluster within 0-1 km radius of the settlement, while for the Korr site, the pastoral camps were located > 4 km from the settlement. In Kargi settlement there was evidence of sand dune movements in the areas of pastoral camps, while in Korr soil movements were evident 4 km from the settlement. Along the transects, soils were sampled at 200 m intervals from 0-20 cm depth (n = 80 samples for each site). Soils were mixed and about 250 g analyzed for total nitrogen (% N), total organic carbon (% C), extractable phosphorus (% P) and electro-conductivity (Ec) using the standard laboratory methods. Woody cover was also estimated. Soil nutrient gradients and woody cover were analysed using linear constrained ordination in CANOCO. Redundancy Analysis (RDA) was used with soil nutrients and woody cover as response variables and distance as explanatory variable.

In terms of vegetation attributes, herders described the species in each landscape as either "resident" or "foreign". In addition, trends of key fodder species are described as "increasing", "decreasing" or "stable", with a preference index of each species for different livestock categories. Herders and ecologists then jointly identified plant species and conduct a species census, while ecologists estimate cover in different patches. The same procedure is repeated if there are permanently marked transects where baseline data exists. Plant cover, species richness and frequency were analyzed by seasons and the data from ecological monitoring used to corroborate with the data from herders' monitoring.

6.1.3. Remote sensing

Long-term land cover changes were evaluated using two pairs of Landsat images, one set taken on 5 January 1986 (ID: ETP168R58_5T19860105) and the other taken on 21 February 2000 (ID: ELP168R058 7T20000221). The images covered the study areas (Kargi and Korr), which had been georeferenced and projected (Reference Datum and Ellipsoid was WGS 84; Projection UTM; Zone 37). The 1986 images are appropriate for understanding vegetation status at the peak of the "desertification" described by IPAL (Lusigi, 1984), while the 2000 images served as the most recent available evidence of vegetation change. The two pairs of images were both taken during the dry season, thus reducing possible effects of seasonal variations on land cover. The Landsat images (one pair of 1986 and the other of 2000) were created in ERDAS IMAGINE (Leica, 2006). The 4, 3, 2 band combination was used for Red, Green and Blue (RGB), respectively. This was in order to enhance the features of interest, i.e. the green vegetation. Some filtering was done to the images to enhance the vegetation features. Making the 2000 image pairs the "master", and the 1986 pair the "slave", image-to-image registration was done. This was important so that any shift between images of different dates could be eliminated. Taking the two towns (Kargi and Korr) as focal points, a square Area of Interest (AOI) was created with a 5 km radius from each center. This was used to clip the area of interest in all four images. The image clips were exported, in Tagged Image File (tif) format. Using Ilwis software (ILWIS, 2005), the image clips were imported into Ilwis image format. Using pre-determined classes, supervised image classification was done for the different images. The ArcView Geographical Information System (GIS) was used to overlay the different layers (images and shape files) together in a GIS environment (ArcView, 2006).

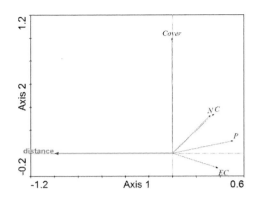 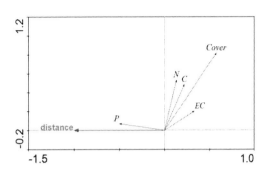

Figure 4. Soil nutrients and woody cover ordination from the settlements: (A) Kargi and (B) Korr in Northern Kenya.

The same sites were ground surveyed in 2005 and field data was collected (Geographical Positioning Systems points and their corresponding photos on the images). Based on our knowledge of the land cover of the study area from field visits, the visual out-look of the images and their spectral signatures, we assigned classes to the training pixel samples for each image. Four classes (bare ground, dwarf shrubs, woody vegetation and rock outcrops) were used for training pixel samples. These were used later to classify the images and to obtain the four maps (two for each site) (Figure 4), showing the land cover types in 1986 and the changes over time up to the year 2000. The maximum likelihood classifier (algorithm) was applied in the classification in order to avoid unknowns and the bias experienced with other classifiers. The 1986 and 2000 maps for each area were crossed to obtain the statistics to detect the changes. This was important in order to obtain a clear picture of how the land cover had changed. The shifts in cover indicators for all the cover types were then reported. In this study, we describe trees as woody, differentiating them from shrubs. The distinction was for convenience in presenting different cover classes (see keys to Figure 4).

6.1.4. Step III: The workshop

A workshop was organized for key participants to discuss the roles played by the environmental management practices of local communities in terms of the recovery of vegetation cover around settlements. The information gathered in Steps I and II was then synthesized and clarified with the herders in the workshop. The workshop discussion was informal and provided feedback from the joint field survey methods. The workshop also provides the herders with a forum to express their views on the drivers of environmental change related to the problems of land degradation, desertification and land rehabilitation. Factors affecting herders' participation in environmental management were discussed with the elders who represent pastoral herders from the camps and the representatives of the Environmental Management Committee (EMC). The workshop then discussed what role herders could play in the implementation of environmental conservation in and around the settlements specifically, but also on the grazing home range in general. The factors that constrain the implementation and how these could be overcome in the view of the communities were also discussed.

7. SYNTHESIS

7.1. Impact of Pastoral Sedentarization on Vegetation

In northern Kenya, pastoral sedentarization has been associated with land degradation and desertification (Lamprey and Yussuf, 1981). The assumption is that the processes result from high human and livestock populations around sedentary centers that have negative impacts on the local environment. The evidence was that although sedentarization has reduced human mobility and the grazing home range has shrunk during previous decades, livestock management in both the arid lowlands and the sub-humid zone has remained mobile, which might have reduced the risks of land degradation (Roba and Oba, 2008; Roba and Oba, 2009). The Ariaal and the Rendille pastoralists have continued to practice herd splitting into mobile *fora* stock and home herds (Roba and Oba, 2008). In the sub-humid (Ariaal) and the semi arid lowlands (Rendille), more livestock are managed in the *fora* mobile camps than in the home herds. In the sub-humid lowlands (Ariaal), 63% of cattle and 9% of small stock are managed in the *fora* herd. In the arid zone (Rendille), 98% of cattle, 91% of goats, 92% of sheep and 74% of camels are managed at *fora* (Roba, 2008). However, in the sub-humid zone, due to the reduced extent of grazing on the home range, the stocking density was twice that in the *fora* area. Nonetheless our analysis shows that the risks of environmental degradation have been reduced through the continuous exchange of livestock between the two systems. This observation underscores the roles played by herders in livestock management decisions. Livestock mobility helps to reduce over-exploitation of the vegetation and makes it possible to maintain periodic high livestock populations near settlements. The re-adjustment in land use patterns within the reduced grazing areas while maintaining traditional herd mobility shows that the Ariaal and Rendille pastoralists have adapted to the changing grazing resources, making the systems more resilient than earlier reported (Roba and Oba 2008).

In the findings, there is no evidence of grazing-induced permanent losses of vegetation around the settlements, both in the sub-humid and in the arid lowlands (Roba and Oba, 2008; Roba, 2008). In the sub-humid zone, vegetation conditions did not show spatial variation across radial distances from settlements and were comparable with the benchmark-the forest reserve (Roba and Oba 2008). Woody vegetation cover did not show variation across radial distances from settlements, but the herbaceous cover was greater in the benchmark compared to the settlements rangeland (Table 1)

Range "condition" between the settlement rangelands and the benchmark, did not differ according to the herders ($F_{1,\ 168} = 0.45$, $P = 0.502$) and ecologist ($F_{1,\ 168} = 0.68$, $P = 0.41$) ratings and was positively correlated with "condition" ratings by ecologists. Vegetation "trends" based on long-term observations by herders were rated as "stable" in the benchmark, while in the settlements rangeland, they were rated as "declining". Declining vegetation trends imply that over the years, vegetation composition and cover are lower than they were in earlier times (Roba and Oba 2008). The lack of spatial variation in vegetation cover may be an indication that within sampling distances, the vegetation has experienced similar levels of impact. However, increased herbaceous cover in the benchmark implies that the settlement rangelands have experienced greater grazing pressures. Herbaceous and woody species richness was higher in the settlement rangeland compared to the benchmark. This observation implies that some fodder species are either resistant to grazing or have been stimulated by grazing impacts (e.g., Oba et al., 2000d).

Table 1. Variations of environmental indictors between the settlement rangelands and the benchmark site in Karare location, Marsabit District, northern Kenya.

Sites	Herbaceous cover (%)	Bush cover (%)	Herbaceous species richness	Woody species richness
Settlements	32.19 ± 2.28^b	20.89 ± 1.50^a	6.92 ± 0.18^a	3.16 ± 0.14^a
Benchmark	62.12 ± 3.49^a	20.44 ± 2.19^a	5.72 ± 0.22^b	1.8 ± 0.19^b

Means $\pm$ SE in the same column followed by different letters are significantly different at P = 0.05

The findings have important management implications from the perspective of biodiversity conservation. The assumption that livestock grazing contributes to loss of biodiversity was not supported. The settlements in the sub-humid zone showed evidence of increased bush cover, which the herders associated with continuous use, episodic rainfall and the ban on the use of fire for managing the vegetation (Roba and Oba, 2008). The changes in species composition were considered as degradation by the herders, since they reduced livestock productivity, especially affecting the grazers. Herders responded to increased tree and shrub diversities by shifting to browsing livestock species (Thomas and Twyman, 2004). For this reason, in the sub-humid zone, the small stocks are becoming increasingly important in response to the problems of bush encroachment. Other researchers have also associated changes in species composition from herbaceous to woody/shrub with heavy grazing pressure (Moleele et al., 2002)

In the arid lowlands of the Rendille, in contrast to the earlier desertification narratives, trends of vegetation recovery were observed around the settlements of Kargi and Korr (Roba, 2008). The areas close to the settlements that were rested from livestock grazing and tree exploitation for more than a decade showed recovery. The same sites were earlier described as desertified (Walter and Helocker, 1981; Lamprey and Yussuf, 1981). Along the road transects from Korr, important fodder species and the herbaceous and shrub cover were significantly greater in the proximity of the settlement than at distances further away. Greater tree density was observed within 1–2 km from Kargi, while greater tree cover was evident within 1–2 km from the Korr settlement. The evidence indicates that management of livestock grazing and anthropogenic uses have contributed to increased woody vegetation cover and composition around the settlements. By contrast, the areas around the pastoral camps (5–8 km outside the settlements) that sustained heavy livestock grazing pressure disclosed sustained degradation (Roba, 2008).

The evidence from the studies has three important implications. Firstly, the sedentarization-desertification hypothesis that assumes that livestock remain permanently stationed in the settlement rangelands is linked to land degradation. Secondly, in areas where herders have been involved in environmental conservation, recovery has occurred. Thirdly, land degradation will continue to occur in areas where heavy livestock grazing and human over-harvesting of vegetation has been sustained, such as those around the pastoral camps. Applying ecological theories, we might deduce that the arid lowlands have shown non-equilibrium behavior, where both management and environmental stochasticity of rainfall influence vegetation dynamics. The evidence of reversibility of land degradation confirms that the arid lowlands are resilient. The vegetation dynamics in the sub-humid zone show that heavy grazing pressure shifted vegetation thresholds towards greater woody vegetation cover and a decline in herbaceous cover, ecological characteristics which are described by the equilibrium ecological theory (Roba and Oba 2008).

7.2. Herder Landscape Classification, Vegetation Assessment and Monitoring

Herder assessment and monitoring of vegetation change was conducted at landscape levels. Herders classified landscapes using vegetation and landforms (Roba, 2008; Roba and Oba, 2009). In the sub-humid zone, the Ariaal herders classified the grazing landscapes around the settlements into six micro-landscapes, while in the arid lowlands the Rendille herders identified three grazing landscapes (Roba and Oba, 2009). Across the landscape systems, herders monitored changes in vegetation on a continuous basis. Assessment of vegetation "condition" was based on the level of vegetation cover, the degree of grazing impact, and the proportion of fodder to non-fodder species.

Herders monitored qualitative changes in species composition across the grazing landscapes. Landscapes were associated with key fodder species that influenced use by different livestock species. In each landscape, fodder species were described as "increasing", "decreasing" or "stable", depending on changes in abundance in relation to livestock fodder requirements. In the sub-humid rangelands (Ariaal system), 66% of species were described as stable, 12% as increasing and 14% as decreasing. In the arid lowlands (Rendille system), 75% of the recorded species were described as stable, 5% as increasing and 20% as decreasing. In both zones, the decreasing species were those that are most important for livestock fodder, while the increasing species were those that are least useful (Roba and Oba, 2009; Roba, 2008). Monitoring species abundance in relation to livestock fodder requirements makes herder assessment and monitoring more relevant for livestock production, compared to ecological assessment which focuses more on the total species pool that is important from the conservation perspective. The "presence" or "absence" of key fodder species across landscapes was used as an indication of environmental change. The percentage loss of key fodder species varied across the different landscape types and the landscapes showed different levels of vulnerability to grazing pressure. For the herders, the landscape approach to biodiversity assessment and monitoring is linked to livestock management decisions. Herders continuously monitor landscape condition to avoid over-exploitation that may result in total degradation (Roba and Oba, 2009).

The herders' practice of biodiversity monitoring at the landscape scale provides an important basis for participatory biodiversity management for two reasons. Firstly, landscape serves as an important scale for conventional biodiversity conservation (Pearson et al., 1996). The logic is that different landscapes have a unique assemblage of the flora. Disappearance of the key fodder species may result in dysfunctional landscapes for the purposes of livestock grazing. Monitoring the key fodder species by the herders is comparable with the ecological concept of keystone species for monitoring the functioning of ecosystems (e.g. Walker et al., 1999). In terms of participatory management of biodiversity, herders provide an early warning in terms of the loss of key forage species, since they conduct continuous assessment and monitoring for purposes of livestock production. Secondly, herders' assessment and monitoring of vegetation at the landscape scale is important for livestock management across local landscapes (Roba, 2008). Their detailed assessment and monitoring of biodiversity at the landscape scale makes an important contribution to the implementation of the CBD at the local level. This is also the level at which to incorporate local communities into the implementation of the national action plans (NAPs).

7.3. Herders' Perceptions of Land Degradation

Herders used multiple indicators to assess and monitor land degradation (Roba and Oba, 2008; Roba and Oba, 2009; Roba, 2008). Herder objectives were to sustain livestock productivity, which is influenced by vegetation and is linked to anthropogenic indicators including landscape grazing potential (LGP) and landscape grazing suitability (LGS). According to the herders, a "good" rangeland supports optimal livestock productivity as opposed to a "bad" one. In a good environment, livestock productivity indicators that were considered optimal included the volume of milk produced, mating frequency and fur condition. Livestock performance is used as an important barometer for monitoring change in the environment. Changes in the environment from "good" to "bad" result from changes in vegetation cover, composition and soil physical (Roba and Oba 2008; Roba and Oba, 2009) and chemical properties.

Monitoring processes of desertification using herder perceptions of soil degradation and soil nutrient indicators have not been widely reported. Furthermore, research on desertification is usually concerned with the impacts of agricultural and pastoral production but less on the ecosystem processes (Tongway & Whitford, 2002). A common assumption is that desertification results in loss of soil fertility in the grazing lands. Assumed linkages between desertification processes and soil nutrient loss around settlements in the arid zones of Africa (Lusigi, 1981) has however not been confirmed by research. We investigated if there is a gradient of nutrient loss from two pastoral settlements in Northern Kenya. Herders classified soils as "warm" or "cold". "Warm" soils were preferred for grazing livestock compared to "cold" soils (Table 2). Preliminary analysis indicated that the soils classified as "warm" had high organic carbon compared to "cold" soil (Table 2).

Additionally, soils were classified into categories of "degraded" and "stable". Degraded soils were avoided when selecting sites for livestock grazing and for establishing settlements (Roba 2008). Thus, comparable to farmer knowledge in farming communities (e.g., Gray and Morant, 2003; Oudwater and Martin, 2003) the relationships between herder perceptions of soil physical and nutritional qualities could make an important area of future research. In understanding the potential use of soil nutrients as indicators of desertification, we had assumed that processes that caused desertification contribute to the losses of soils. These however appear not to have been confirmed by our preliminary results from around settlements.

**Table 2. Herder described soil types (as cold and warm) and
their corresponding nutrients.**

Local soil type	Nutrient Composition			
	% C	% N	% P	Ec
Cold Soils	2.21 ± 0.23	0.27 ± 0.04	0.087 ± 0.01	0.09 ± 0.01
Warm Soils	2.78 ± 0.11	0.33 ± 0.01	0.083 ± 0.008	0.089 ± 0.004
Warm-Cold Soils	2.23 ± 0.1	0.26 ± 0.02	$0.085 \pm 0.0.01$	0.07 ± 0.009
F-statistics	$F_{2, 112} = 4.84$, $P = 0.01$	$F_{2, 112} = 3.91$, $P = 0.02$	$F_{2, 112} = 0.01$, $P = 0.98$	$F_{2, 112} = 1.73$, $P = 0.18$

Total nitrogen, total organic carbon and Ec were negatively correlated with distances from the settlements, but the correlations were not significant ($p > 0.05$) (Figure 4A-B). Extractable phosphorous was negatively correlated for the Kargi but positively for the Korr settlement ($p < 0.05$). In Kargi woody cover showed no spatial patterns, while for Korr, woody cover decreased with increasing distance from the settlement. Woody cover showed no correlation with soil nutrients except for Ec for the Korr site ($r = 0. 20$, $p < 0.05$). In both settlements, the patterns of extractable Phosphorus varied according to the locations of the pastoral camps suggesting that the changes were related more to livestock activities than to losses attributable to degradation. The distributions of total nitrogen and total organic matter were also greater around the settlements (albeit being insignificant). The results showed that soil nutrients in the settlements did not directly reflect the losses often linked to the processes of desertification. The responses of different nutrients along degradation gradients appeared to reflect the positive roles played by livestock in nutrient transport into the pastoral camps. The evidence showed that settlements in the arid zones of Northern Kenya accumulated nutrients contrary to the common expectations. The observation provides an important area of future research in addressing land degradation using proximate indicators such as soil nutrients. The evidence showed that the herders indirectly inferred conditions of soils that reflected degradation and suitability of grazing landscapes.

Besides quantitative and qualitative analysis of vegetation indicators, herders used qualitative changes for which they used concepts such as '*mirr qabdo*' and '*mirr maqabdo*' to determine the use of landscapes for livestock grazing. In the *mirr qabdo* environment, even with reduced vegetation cover, livestock performance was considered to be satisfactory, while in the case of *mirr maqabdo*, good vegetation cover did not imply satisfactory livestock performance. Herders reported occurrence of pests that affect livestock performance in the *mirr maqabdo* environment (Roba 2008). This evidence shows that herders' understanding of changes in the rangelands in terms of livestock production encompasses more than the description of vegetation quantity and quality alone. Considering the complex interactions in ecological systems, which are rarely observed and monitored using ecological methods, herders' knowledge and experiences accumulated over many years of using and managing the grazing environment are therefore important.

The Ariaal and the Rendille herders consider reductions in vegetation cover and change in species composition (e.g. from most valuable to less valuable fodder species) as degradation. They insisted that changes in vegetation cover were due to livestock grazing pressure and the impact of droughts. Once the grazing pressure was reduced, herders expected the vegetation cover to improve, especially with an increase in rainfall. Reversibility in the vegetation indicators is less well understood from an ecological point of view, due to the lack of long-term monitoring. The ecological management of vegetation usually assumes linear changes in vegetation attributes in relation to grazing pressure, as described by the equilibrium theory. Herder perceptions of the reversibility of change in vegetation structure are comparable with the non-equilibrium theory that acknowledges the role of external factors such as rainfall variability (Ellis and Swift, 1988), and ecological resilience, which emphasizes multiple thresholds (Folke, 2006).

In terms of grazing, the anthropogenic indicators for assessing and monitoring environmental change, in particular the landscape grazing potential (LGP), are linked to vegetation and soil characteristics. Landscape grazing suitability (LGS) is determined in terms of preferred plant species for different livestock classes (Roba and Oba, 2009). The landscape grazing potential and grazing suitability are therefore important indicators for making management decisions (Oba et al., 2008b). Herders' reliance on multiple indicators for the assessment and monitoring of environmental change for livestock production provides useful information for understanding land degradation at the local level. It is evident that environmental change at the local level is better addressed by considering not only ecological indicators, but also anthropogenic indicators used mostly by land users. The local communities' perception and use of indicators are crucial for the successful implementations of the GECs at the local level (Roba, 2008).

7.4. Long Term Environmental Dynamics

The lack of long-term data has constrained the understanding of environmental dynamics in arid ecosystems. Hypotheses on environmental change at the global level (e.g. land degradation, desertification and biodiversity loss) are based on short-term data or extrapolation of data from different scales. There is a need therefore for long-term monitoring of environmental change to establish the nature, extent and reversibility of the processes of degradation (Dodd, 1994; Dean et al., 1995). The research emphasizes the importance of long-term ecological monitoring for understanding environmental dynamics. In the arid lowlands system, vegetation indicators over a period of two decades showed no permanent losses of plant species (Roba, 2008). Changes in species frequency and cover did not vary over a 24 year period. The variations were characterized by fluctuations between wet (1982 and 2006) and dry season sampling (1983 and 2005). Dissimilarities in species composition were more evident in the wet season samples (1982/2006) compared to the dry season samples, an indication of the temporary influence of rainfall, as opposed to permanent degradation from grazing impacts. Ecological monitoring was corroborated by herder monitoring that provided details on the history of land use, changes in the seasons of use in the different range units. According to the herders, the patterns of land use in the majority of the range units have changed from seasonal to year-round use. Herders described the vegetation trends as "stable", despite changes in land use regimes, and they associated the stability with the potential of the rangelands (Roba, 2008).

Long-term monitoring of vegetation changes around permanent settlements also showed that changes described earlier as desertification were actually reversible, according to satellite data (Figure 5). Between 1986 and 2000, the tree and shrub cover increased from 14.7% to 54.9% around Kargi and from 27.6% to 36.6% around the Korr settlement (Roba, 2008). The observed vegetation cover changes support the recent evidence of the "greening" of the Sahel (e.g., Rasmussen et al., 2001; Anyamba and Tucker 2005; Herrmann et al., 2005). Although the causes of vegetation recovery in the Sahel and northern Kenya were probably different, both regions were earlier described as desertified (Lusigi, 1984; Hermann et al., 2005; Olsson et al., 2005).

In Northern Kenya, improved vegetation cover around settlements was associated with management interventions and episodic rainfall. Around the Kargi and Korr settlements, the high density of woody vegetation, especially of *Acacia tortilis,* resulted from seed dispersal by livestock and protection of mature trees by the community (Roba, 2008). The evidence shows that herders have played important roles in vegetation recovery, although for long time they were viewed as environmental villains (O'Leary, 1984). In the Kargi and Korr settlements, with the help of international NGOs, the local communities formed Environmental Management Committees (EMCs) that were responsible for protecting trees around the settlements. The sustainability of recovery depends, however, on the continuation of such vegetation protection. The practice of vegetation protection was built on the experiences from the IPAL trials in which tree painting was used as a deterrent to tree cutting (see also Lusigi, 1984).

From the above observations, it may be deduced that the arid environment in northern Kenya is resilient in relation to long term changes in vegetation structure. The earlier perceived denudations were evidence of only a temporary impact of vegetation over-exploitation. The monitoring of vegetation change has been informed by the participation of the local herders, who provided details of past and current land use practices. Albeit on a limited scale, local communities have achieved the implementation of the GECs related to the CCD and the CBD through their participation and management practices. A critical part of the role they played was that they became responsible for the well-being of their own environments. Because of the protection of the trees, they are experiencing benefits in terms of harvesting *Acacia* pods, which are being fed to their goats. In degraded environments, such resources were not accessible to them. Accordingly, local communities must have a reason, which is often linked to their livelihoods, in order to become involved in environmental protection (Roba, 2008). Successful local participation for vegetation management around settlements is an important contribution to NAPs, and therefore more effort is needed from the national officials to achieve full participation of the herders to enhance further vegetation recovery.

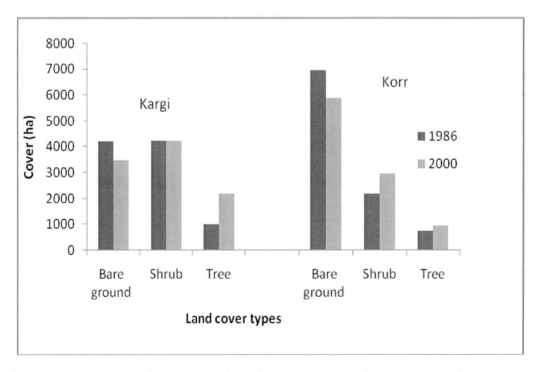

Figure 5. Change in land cover types around Kargi and Korr settlement in northern Kenya between 1986 and 2000.

8. IMPLICATIONS FOR GLOBAL ENVIRONMENTAL GOALS

In this section we shall generalize the outcome in implementing the different steps of research methods for integrating herder knowledge and ecological methods. The implementation of the framework proposed in this chapter can be expected to become a process for addressing the global goals. At the global level, the Articles of the conventions would guide the process. At the national level, the methods and indicators for assessing and monitoring environmental changes are a priority for the NAPs, whose function is to translate the global goals into local actions. Information about the NAPs and their mechanisms for implementing the global goals require additional research. In contextualizing our methods within the global goals and local actions, we have conceptualized environmental problems within broader objectives. Three areas where actions are needed at local community levels are in addressing problems of land degradation, desertification and biodiversity loss. This proposed approach suggests selected pastoral herders/agro-pastoralists as the focus group because the problems of environmental degradation are usually associated with their forms of land use.

At local land use levels, the methods and environmental indicators are specific to types of land use, which in this chapter was grazing by livestock. If the focus is on farming communities, different methods and indicators might be chosen. Our framework is however flexible and could be used to fit new priorities. For the successful implementation of the GECs, activities at local levels are linked to the global goals (Stage I) and national actions (Stage II). Technical departments working with local communities implement NAPs. The participation of local communities could be increased by working through local institutions such as Community Based Organizations (CBOs) that facilitate the implementation of the GECs by organizing awareness workshops and linking local activities to local production systems. The strategy of using existing institutions to reach local communities could be successful as far as creating awareness of environmental problems is concerned, including desertification and biodiversity loss. However, sustaining the participation of local communities is often not been possible, partly because the problems addressed were not linked to local systems of livestock and natural resource management. Another problem that the NAPs have not solved is the matter of who is responsible for the environment. The proposed framework (Figure 1) will make it easier for the various stakeholders to assume the responsibility for the implementation of the GECs. We have suggested this by using CBOs, including Environmental Management Committees (EMCs), as the focal point.

In the joint venture of resource management, local communities would gain new ideas, including awareness of problems such as land degradation. The NAPs officials could utilize information generated with the support of local people to develop integrated management plans, which would ensure the sustainable use of resources, according to the provisions of the GECs. Integrating local people in environmental management would also involve addressing socio-economics issues that may currently constrain the sustainable use of resources. Although willing to participate in environmental assessment, monitoring and management of resources according to the requirements of the GECs, local communities are confronted by challenges, such as the lack of alternative materials for the construction of livestock enclosures. The NAPs need to consider programs that address socio-economic problems that might aggravate land degradation. Although the proposed conceptual framework provides guidelines on how local participation for the implementation of the GECs may be achieved, a major challenge is garnering support at global and national levels to facilitate local actions. In the future, research should investigate how Stage II involving NAPs and global actions could be used to promote local participation (see Figure 1). This is a policy issue, which should take cognizance of the rights and obligations of all the stakeholders, while maintaining global obligations.

9. RECOMMENDATIONS

In this chapter we have developed a framework for translating global environmental convention goals into local actions, through the integration of indigenous knowledge and ecological methods. The framework demonstrates how local participation in the implementation of the GECs could be achieved by focusing on different action areas of the global goals. Translating the goals into local actions requires the development of appropriate methods and the selection of sensitive environmental indicators. Locally used indicators were sensitive to environmental changes, while at the same time being amenable to proxy indicators of livestock production performance. In terms of the implementation of the CCD at local levels, it is recommended that conventional indicators and methods for assessment and monitoring should be reconciled with those used by local communities. In order to implement the CBD at local levels, it is recommended that biodiversity conservation agenda should be considered in line with local communities' utilitarian objectives of vegetation assessment and monitoring, for the purposes of livestock management. Joint assessment and monitoring for the implementation of the CBD may involve establishing an inventory of plant species that occur in local landscapes. Local communities should be trained in terms of collecting sample specimens to be kept in a community herbarium. With such an inventory and specimens in place, it would be possible for taxonomists to conduct proper classifications, thereby providing important baseline information for monitoring the status of plant species in the local landscapes at some future dates. The plant samples could be used to create local herbaria to support the training of local communities, EMC members as well as schools, in order to promote the spirit of the implementations of GECs.

ACKNOWLEDGMENTS

The research was funded by the Norwegian Science Council through the project "Local Community Participation in Implementations of Global Environmental Conventions and Poverty Alleviations" Project no. 16139/S30.

REFERENCES

Abel, N. O. J. & Blaikie, M. (1989). Land degradation, stocking rates and conservation policies in the communal rangelands of Botswana and Zimbabwe. *Land Degradation and Rehabilitation*, *1*, 101-123.

Agrawal, A. (1995). Indigenous and Scientific Knowledge: Some Critical Comments. *IK Monitor, 3(3)*, 33-41

Angassa, A. & Oba, G. (2008). Herder perceptions on impacts of range enclosures, crop farming, fire Ban and Bush Encroachment on the rangelands of Borana, Southern Ethiopia. *Human Ecology, 36*, 201-215.

Antweiler, C. (2004). Local Knowledge Theory and Methods: An Urban Model from Indonesia. In: A. Bicker, P. Sillitoe, & J. Pottier (Eds.), *Investigating local Knowledge, New directions, New Approaches*. Ashgate, England, 1-34.

Anyamba, A. & Tucker, C. J. (2005). Analysis of Sahelian vegetation dynamics using NOAA-AVHRR NDVI data from 1981-2003. *Journal of Arid Environments, 63*, 596-614.

Awere-Gyekye, K. (1984). Major Range Types in the South-Western Marsabit District of northern Kenya. *In: Basic Map*, UNESCO-IPAL Nairobi.

Bake, G. (1983). An analysis of climatological data from the Marsabit District of nnorthern Kenya. *IPAL Technical Report No. B-3.*

Barrow, C. J. (1991). *Land Degradation: Development and Breakdown of Terrestrial Environments*. Cambridge University Press, Cambridge.

Bartels, G. B., Norton, B. E. & Perrier, G. K. (1993). An examination of the carrying capacity concept. In: R. H. Behnke, I. Scoones, & C. Kerven, (Eds.), *Range ecology at disequilibrium, new models of natural variability and pastoral adaptation in African savannas* ODI, London, 89-103.

Bassett, T. J. & Crummey, D. (2003). Contested Images, Contested Realities. Environment and Society in Africa Savannas. In: T. J. Bassett, & D. Crummey (Eds.), *African Savannas. Global Narratives and Local Knowledge of Environmental Change*. James Currey, Oxford, 1-31.

Bassett, T. J. & Zuéli, K. B. (2000). Environmental Discourses and the Ivorian Savanna. *Annals of the Association of American Geographers, 90*, 67-95.

Batterbury, S., Forsyth, T. & Thomson, K. (1997). Environmental transformations in the developing countries: hybrid research and democratic policy. The *Geographical Journal, 163*, 126-132.

Behnke, R. H. & Scoones, I. (1993). Rethinking range ecology: Implications for rangeland management in Africa. In: R. H. Behnke, I. Scoones, & C. Kerven (Eds.), *Range ecology at disequilibrium, new models of natural variability and pastoral adaptation in African savannas* ODI, London, 1-30.

Benjaminsen, T. A. & Lund, C. (Eds.). (2001). *Politics, Property and Production in the West African Sahel: Understanding Natural Resources Management*. Nordiska Afrikainstitutet, Uppsala.

Benjaminsen, T. A. (1993). Fuel wood and desertification: orthodoxies discussed on the basis of field data from Gourma region in Mali. *Geoforum, 24*, 397-409.

Berkes, F., Colding, J. & Folke, C. (2000). Rediscovery of traditional ecological knowledge as adaptive management. *Ecological Applications, 10*, 1251-1262.

Berkes, F., Kislalioglu, M., Folke C. & Gadgil, M. (1998). Exploring the basic ecological unit: Ecosystem-like concepts in traditional societies. *Ecosystems, 1*, 409-415.

Binns, T. (1990). Is Desertification a Myth? *Geography, 75(2)*, 106-113.

Blaikie, P. & Brookfield, H. (1987). *Land Degradation and Society*. London: Methuen.

Bollig, M., A. & Schulte, A. (1999). Environmental change and pastoral perceptions: Degradation and indigenous knowledge in two African pastoral communities. *Human Ecology*, *27*, 493-514.

Bouma, G. A. & Kobryn, H. T. (2004) Change in vegetation cover in East Timor, 1989-1999. *Natural Resources Forum*, *28*, 1-12.

Chambers, R. (1994). The origins and practice of participatory rural appraisal. *World Development*, *22*, 953-969.

Chambers, R. (1999). *Whose reality Counts? Putting the first last*. Intermediate Technology Publication, London.

Clements, F. E. (1916). *Plant succession: an analysis of the development of vegetation*. Washington: Carnegie Institute Pub., *242*, 1-512.

Dabasso, B. H. (2006). Herder knowledge of landscape scale biodiversity monitoring in Marsabit central, northern Kenya. MSc.thesis - Universitetet for miljø- og biovitenskap, Ås.

Dahlberg, A. C. (2000). Interpretations of environmental changes and diversity: a critical approach to indications of degradation – the case of Kalamate, north West Botswana. *Land Degradation and Development*, *11*, 549-562.

Dean, W. R. J., Hoffman, M. T., Meadows, M. E. & Milton, S. J. (1995). Desertification in the semi-arid Karoo, South Africa: Review and reassessment *Journal of Arid Environment*, *30*, 247-264.

Dodd, J. L. (1994). Desertification and degradation in sub-Saharan Africa. The role of livestock. *Bioscience*, *44(1)*, 28-34.

Dougill, A. J. & Reed, M. S. (2006) Framework for community –based rangeland sustainability assessment in the Kalahari, Botswana, In: J. Hill, A. Terry, & W. Woodland (Eds.), *Sustainable Development: National Aspirations, Local Implementation*, Ashgate Publishing, 31-50.

Dougill, A. J., Fraser, E. D. G, Holden, J. Hubacek, K., Prell, C., Reed, M, S., Stagl, S. & Stringer, L. C. (2006). Learning from Doing Participatory Rural Research: Lessons from the Peak District National Park. *Journal of Agricultural Economic*, *57*, 259-275

Ellis, J. E. & Swift, D. M. (1988). Stability of African pastoral ecosystems: alternate paradigms and implications for development. *Journal of Range Management*, *41*, 450-459.

Fernandez-Gimenez, M. E. & Allen-Diaz, B. (1999). Testing a non-equilibrium model of rangeland vegetation dynamics in Mongolia. *Journal of Applied Ecology*, *36*, 871-885.

Fernandez-Gimenez, M. E. & Allen-Diaz, B. (2001). Vegetation change along gradients from water sources in three grazed Mongolian ecosystems. *Plant Ecology*, *157*, 101-118.

Fernedez-Gimenez, M. A. (2000). The role of Mongolian nomadic pastoralists' ecological knowledge in rangeland management . *Ecological Application*, *10*, 1318-1326.

Field, A. C. (1981). An introduction to the IPAL livestock programme. IPAL Technical Report A-5, Unesco, Nairobi.

Folke, C. (2006). Resilience: The emergence of a perspective for social-ecological systems analyses. *Global Environmental Change*, *16*, 253-267.

Fraser, E. D., Dougill, A. J., Mabee, W. E., Reed, M. & McAlpine, P. (2006). Bottom up and top down: Analysis of participatory processes for sustainability indicator identification as a pathway to community empowerment and sustainable environmental management. *Journal of Environmental Management*, *78*, 114-127.

Fratkin, E. (1986). Stability and resilience in East African pastoralism: The Rendille and the Ariaal of northern Kenya. *Human Ecology*, *10*, 269-287.

Fratkin, E. (1987). The Organization of labor and production among the Ariaal Rendille, nomadic pastoralists of northern Kenya. Phd thesis, Catholic University of America.

Fratkin, E. & Smith, K. (1994). Labor, Livestock, and Land: The Organization of Pastoral Production. In: E. Fratkin, K. A. Galvin, & E. A. Roth (Eds.), *African Pastoralist Systems; an integrated approach*, Lynne Rienner, Boulder, 69-89.

Freeman, M. M. R. (1992). The nature and utility of traditional ecological knowledge *Northern Perspectives*, *20(1)*, 9-12.

Geist, H. (2005). *The Causes and Progression of Desertification*. Ashgate, England.

Gemedo, D., Isselstein, J. & Maass, B. L. (2006). Indigenous ecological knowledge of Borana pastoralists in southern Ethiopia and current challenges. *International Journal of Sustainable Development & World Ecology*, *13*, 113-130.

Gray, L. & Morant, P. (2003). Reconciling Local Perception with Scientific Assessment of Soil Quality Changes in Southern Burkina Faso, *Geoderma*, *111*, 425-437.

Gunderson, L. H. (2000). Ecological Resilience - in Theory and Application. *Annual Review of Ecology and Systematics*, *31*, 425-439.

Hambly, H. (1996). Introduction. In H. Hambly, & T. O. Agura (Eds.), *Grassroots Indicators for Desertification . Experience and Perspectives from Eastern and Southern Africa*. IDRC, Ottawa. 1-7.

Hary, I., Schwartz, H. J., Pielert, V. H. C. & Mosler, C. (1996). Land degradation in African pastoral systems and the destocking controversy. *Ecological Modelling*, *86*, 227-233.

Havstad, K. M. & Herrick, J. E. (2003). Long-Term Ecological Monitoring. *Arid Land Research and Managements*, *17*, 389-400.

Heady, H. F. (1975). *Rangeland Management*. McGraw-Hill. New York.

Helldén, U. (1988). Desertification Monitoring: Is the Desert Echroaching? *Desertification Control Bulletin*, *17*, 8-12.

Hellier, A., Newton, A. C. & Gaona, S. C. (2004). Use of indigenous knowledge for rapidly assessing trends in biodiversity: a case study from Chiapas, Mexico. *Biodiversity and Conservation*, *8*, 869-889.

Herlocker, D. & Walther, D. (1991). Condition of Marsabit Range Lands. In : *Marsabit District: Range Management Handbook of Kenya*, *Vol. II*, No. 1. H. Schwartz, S. Shaabani, & D. Walther (Eds.), Nairobi: Ministry of Livestock Development, 51-52.

Herrmann, S. M., Anyamba, A. & Tucker, C. J. (2005). Recent trends in vegetation dynamics in the African Sahel and their relationship to climate. *Global Environmental Change*, *15*, 394-404.

Hill, D., Fasham, M., Tucker, G., Shewry, M. & Shaw P. (2005). *Handbook of Biodiversity methods. Survey, Evaluation and Monitoring*. Cambridge University Press, Cambridge.

Hodgson, D. L. & Schroeder, R. A. (2002). Dilemmas of counter-mapping community resources in Tanzania, *Development & Change*, *33(2002)*, 79-100.

Holecheck, J. L., Pieper, R. D. & Herbel, C. H. (1995). *Range Management. Principles and Practices*. Prentice Hall, New Jersey.

Holling, C. S. (1973). Resilience and stability of ecological systems. *Annual Review of Ecology and Systematics*, *4*, 1-23.

Hughes, J. D. (2001). *An environmental history of the world: Humankinds' changing role in the community of life*. Routedge, London.

Huntington, H. P. (2000). Using Traditional Ecological Knowledge in Science: Methods and Applications. *Ecological Applications, 10*, 1270-1274.

Keya, G. A. (1997). Effects of herbivory on the production ecology of the perennial grass *Leptothrium senegalense* (Kunth.) in the arid lands of northern Kenya. *Agriculture, Ecosystems & Environment, 66*, 101-111.

Kiunsi, R. B. & Meadows, M. E. (2006). Assessing land degradation in the Monduli District, northern Tanzania. *Land Degradation and Development, 17*, 509-525.

Knight, C. G. (1980). Ethnoscience and the African Farmer: Rationale and Strategy (Tanzania). In: D. Brokensha, D. M. Warren, & O. Werner (Eds.), *Indigenous Knowledge Systems and Development*. University Press of America. New York, 203-229.

Krugman, H. (1996). Towards improved indicators to measure desertification and monitor the implementation of the Desertification Convention. In: H. Hambly, & T. O. Agura (Eds.), *Grassroots indicators for desertification. Experiences and perspectives from eastern and southern Africa*. IDRC, Ottawa, 20-37.

Kumar, S. (2002). *Methods for community participation: a complete guide for practitioners*. ITDG, London.

Lamprey H. (1976). *The UNEP-MAB Integrated Project in Arid Lands: Phase III*. Reg. Proj. Doc. Pp. 1101-77.UNEP, Nairobi.

Lamprey, H. (1983). Pastoralism yesterday and today: The overgrazing problem. In F. Bourlier, (Ed.), *Tropical savannas*. Elsevier Press, Amsterdam.

Lamprey, H. & Yussuf, H. (1981). Pastoral and desert encroachment in northern Kenya. *Ambio, 10*, 131-134.

Lamprey, H. F. (1988). Report on the desert encroachment reconnaissance in northern Sudan. *Desertification Control Bulletin, 17*, 1-7.

Landsberg, J., James, C. D., Morton, S. R., Muller, W. J. & Stol, J. (2003). Abundance and composition of plant species along grazing gradients in Australian rangelands. *Journal of Applied Ecology, 40*, 1008-1024.

Laris, P. (2004). Grounding Environmental Narratives: The Impact of a Century of Fighting Against Fire in Mali. In: W. G. Moseley, & B. I. Logan (Eds.), *African Environment and Development. Rhetoric, Programs and Realities*. Ashgate, England., 63-85.

Leach, M. & Mearns, R. (Eds.). (1996). *The Lie of the Land: Challenging Received Wisdom on the African Environment*. Cambridge University Press, Cambridge.

Lusigi, W. J. (1981). Combating desertification and rehabilitating degraded production systems in northern Kenya. IPAL Technical Report A-4, UNESCO, Nairobi.

Lusigi, W. J. (Ed.). (1984.) *Integrated Resource Assessment and Management Plan for Western Marsabit District, northern Kenya*. Part I and II, Integrated Project in Arid Lands (IPAL) Technical Report No. A-6.

Lusigi, W. J., Nkurunziza, E. R., Awere-Gyekye, K. & Masheti, S. (1986). Range resource assessment and management strategies for the south-western Marsabit, northern Kenya. IPAL Technical Report D-5. UNESCO, Nairobi.

Mabbutt, J. A. (1984). A new Global Assessment of the Status and Trends of Desertification. *Environmental Conservation, 11*, 103-113.

Mabbutt, J. A. (1985). Desertification of the World's Rangelands. *Desertification Control Bulletin, 12*, 1-11.

Mapinduzi, A. L., Oba, G ., Weladji, R. B. & Colman, J. E. (2003). Use of indigenous ecological knowledge of the Maasai pastoralists for assessing rangeland biodiversity in Tanzania. *African Journal of Ecology, 41*, 329-336.

Moleele, N. M., Ringrose, S., Matheson, W. & Vanderpost, C. (2002). More woody plants? The status of bush encroachment in Botswana's grazing areas. *Journal of Environmental Management, 64*, 3-11.

NRC (1994). *Rangeland Health. New Methods to Classify, Inventory, and Monitor Rangelands* National Academy Press, Washington D.C.

Nygren, A. (1999). Local Knowledge in the Environment –Development: From dichotomies to situated knowledge. *Critique of Anthropology, 19*, 267-288.

O'Leary, M. F. (1984). Ecological villains or economic victims: the case of the Rendille of northern Kenya. *Desertification Control Bulletin, 11*, 17-21.

O'Leary, M. F. (1985). Economics of pastoralism in northern Kenya: The Rendille and Gabra. Integrated Projects in Arid Lands (IPAL) Technical Report F-3, UNESCO, Nairobi.

Oba, G. (1985a). Perception of environment among Kenyan pastoralists: Implications for development. *Nomadic Peoples, 19*, 33-57.

Oba, G. (1985b). Local participation in guiding extension programs: a practical proposal. *Nomadic Peoples, 18*, 27-45.

Oba, G. (1994). *The role of indigenous range management knowledge for desertification control in northern Kenya.* Research Report No.4 .EPOS, Uppsala University.

Oba, G. (1998). Effects of excluding goat herbivory on *Acacia tortilis* woodland around pastoralist settlement in northwest Kenya. *Acta Oecologica, 19*, 395-404.

Oba, G. (2001). Indigenous Ecological Knowledge of Landscape Change in East Africa. *IALE Bulletin, 19(3)*, 1-3.

Oba, G. & Kaitira, L. M. (2006). Herder knowledge of landscape assessments in arid rangelands in northern Tanzania. *Journal of Arid Environments, 66*, 168-186.

Oba, G. & Kotile, D. G. (2001). Assessments of landscape level degradation in southern Ethiopia: Pastoralists versus ecologists. *Land Degradation and Development, 12*, 461-475.

Oba, G., Byakagaba, P. & Angassa, A. (2008a). Participatory monitoring of biodiversity in East African grazing lands. *Land degradation & Development, 19*, 636-648.

Oba, G., Sjaastad, E. & Roba, H. G. (2008b). Framework for participatory assessments and implementation of global environmental conventions at the local level. *Land Degradation & Development, 19*, 65-76.

Oba, G., Mengistu, Z. & Stenseth, N. C. (2000d). Compensatory Growth of the African Dwarf Shrub *Indigofera spinosa* Following Simulated Herbivory. *Ecological Applications, 10*, 1133-1146.

Oba, G., Post, E., Stenseth, N. C. & Lusigi, W. J. (2000a). Role of small ruminants in arid zone Environments: A Review on research perspectives. *Annals of Arid Zone, 39*, 305-332.

Oba, G., Post, E., Syvertsen, P. O. & Stenseth, N. C. (2000c). Bush cover and range condition assessments in relation to landscape and grazing in southern Ethiopia. *Landscape Ecology, 15*, 535-546.

Oba, G., Stenseth, N. C. & Lusigi, W. (2000b). New perspectives on sustainable grazing management in arid zones of sub-saharan Africa. *BioScience, 50*, 35-51.

Oba, G., Weladji, R. B., Lusigi, W. J. & Stenseth, N. C. (2003). Scale-dependent effects of grazing on rangeland degradation in northern Kenya: A test of equilibrium and non-equilibrium hypotheses. *Land Degradation and Development*, *14*, 83-94.

Oba, G., Weladji, R. B., Msangameno, D. J., Kaitira, L. M. & Stave, J. (2008c). Scaling effects of proximate desertification drivers on soil nutrients in northeastern Tanzania. *Journal of Arid Environments, Vol. 72*, 1820-1829.

Okoti, M., Keya, G. A., Esilaba, A. O. & Cheruiyot, H. (2006). Indigenous technical knowledge for resource monitoring in northern Kenya. *Journal of Human Ecology, 20*, 183-189.

Olsson, L., Eklundh, L. & Ardö, J. (2005). A recent greening of the Sahel-trends, patterns and potential causes. *Journal of Arid Environments*, *63*, 556-566.

Ottichilo, W. K. (1990). Report of the Kenya pilot study FP/6201-87-04 using the FAO/UNEP methodology for assessment and mapping of desertification. In: *Desertification revisited. Proceedings of an Ad Hoc Consultative Meeting on the Assessment of Desertification*, Odingo RS (Ed.). UNEP-DC/PAC: Nairobi; 123-178.

Oudwater, N. & Martin, A. (2003). Methods and issues in exploring local knowledge of soils. *Geoderma*, *111*, 387-401.

Palmer, A. R. & van Rooyen, A. F. (1998). Detecting vegetation change in the southern Kalahari using Landsat TM data. *Journal of Arid Environments*, *39*, 143-153.

Pearson, S. M., Turner, M. G., Gardner, R. H. & O'Neill, R. O. (1996). An Organism-based perspective of habitat fragmentation. In: R. C. Szaro, & D. W. Johnston (Eds.), *Biodiversity in managed landscapes. Theory and Practice*. Oxford University Press, New York, 77-93.

Perrings, C. & Walker, B. H. (1995). Biodiversity loss and the economics of discontinuous change in semi-arid rangelands. In: C. Perrings, K. G. Mäler, C. Folke, C. S. Holling, & B. O. Jansson (Eds.), *Biological Diversity: Economic and Ecological Issues*, Cambridge, Cambridge University Press, 190-210.

Perrings, C. & Walker, B. (1997). Biodiversity, resilience and the control of ecological-economic systems: the case of fire-driven rangelands. *Ecological Economics, 22*, 73-83.

Pratt, D. J. & Gwynne, M. D. (1977). *Rangeland management and ecology in East Africa*. Hodder and Stoughton, London.

Purcell, T. & Onjoro, E. A. (2002). Indigenous knowledge, power and parity. In: I. Sillitoe, A. Bicker, & J. Pottier (Eds.), *Participating in Development. Approaches to Indigenous Knowledge*. Taylor and Francis Group, London and New York, 162-188.

Rasmussen, K., Fog, B. & Madsen, J. E. (2001). Desertification in reverse? Observations from northern Burkina Faso. *Global Environmental Change, 11*, 271-282.

Reed, M. S. & Dougill, A. J. (2002). Participatory selection process for indicators of rangeland condition in the Kalahari. *The Geographical Journal, 168*, 224-234.

Reed, M. S., Dougill, A. J. & Taylor, M. J. (2007). Integrating local and scientific knowledge for adaptation to land degradation: Kalahari rangeland management options. *Land Degradation and Development, 18*, 249-268.

Richards, P. (1980). Community Environmental Knowledge in African Rural Development. In: D. Brokensha, D. M. Warren, & O. Werner (Eds.), *Indigenous Knowledge Systems and Development*. University Press of America. New York, 181-194.

Ringrose, S., Vanderpost, C. & Matheson, W. (1996). The use of integrated remotely sensed and GIS data to determine causes of vegetation cover change in southern Botswana. *Applied Geography, 16*, 225-242.

Roba, H. G. & Oba, G. (2008). Community participatory landscape classification and biodiversity assessment and monitoring grazing land in northern Kenya. *Journal of environmental management*, doi:10.1016/j.jenvman.2007.12.017.

Salzman, P. C. (1980). Process of sedentarization among the nomads of Baluchistan. In: P. C. Salzman (Ed.), *When nomads settle: Process of sedentarization as adaptation and response*. Praeger, New York, 95-110.

Sandford, S. (1983). *Management of Pastoral Development in the Third World*. John Wiley and Sons, New York.

SAS. (2003). *The SAS systems for Windows*, Release version 9.1. SAS, USA.

Schlee, G. (1991). Traditional pastoralists: land use strategies. In: *Range management handbook of Kenya. Volume II. Marsabit District*. Nairobi, Ministry of Agriculture and Livestock Development.

Scoones, I. & Thompson, J. (1993). Rural Peoples Knowledge, Agricultural Research and Extension practice. *IIED Research series, 1(1)*, 1-20.

Seely, M. & Moser, P. (2004). Connecting Community Action and Science to Combat Desertification: Evaluation of a Process. *Environmental Monitoring and Assessments, 99*, 33-55.

Shalaby, A. & Tateishi, R. (2007). Remote sensing and GIS for mapping and monitoring land cover and land-use changes in the northwestern coastal zone of Egypt. *Applied Geography, 27*, 28-41.

Sheuyange A., Oba, G. & Weladji, R. B. (2005). Effects of anthropogenic fire history on savanna vegetation in northeastern Namibia. *Journal of Environmental Management, 75*, 189-198.

Shoshany, M., Kutiel, P. & Lavee, H. (1996). Monitoring temporal vegetation cover changes in Mediterranean and arid ecosystems using a remote sensing technique: case study of the Judean Mountain and the Judean Desert. *Journal of Arid Environments, 33*, 9-21.

Sillitoe, P. (1998). What Know Natives? Local Knowledge in Development. *Social Anthropology, 6(2)*, 203-220.

Sillitoe, P. (2002). Participatory observation to participatory development. Making anthropology work. In: I. Sillitoe, A. Bicker, & J. Pottier (Eds.), *Participating in Development. Approaches to Indigenous Knowledge*. Taylor and Francis Group, London, 1-23.

Sinclair, A. R. E. & Fryxell, J. M. (1985). The Sahel of Africa: Ecology of a Disaster. *Canadian Journal of Zoology, 63*, 987-94.

Sorrells, L. & Glenn, S. (1991). Review of Sampling Techniques used in Studies of Grassland Plant Communities. *Proceedings of Oklahoma Academy of Science, 71*, 43-45.

Spellerberg, I. F. (2005). *Monitoring ecological change*. Cambridge University Press, Cambridge.

SPSS. *Statistical Packages for Social Science*. Version 12. SPSS Inc.USA.

Stiles, Daniel. (1995). *Social Aspects of Sustainable Dryland Management*. John Wiley & Sons, New York.

Stocking, M. A. & Murnaghan, N. (2001). *Handbook for the field assessment of land degradation*. Earthscan Publications, London.

Stringer, L. C. & Reed, M. S. (2007). Land degradation assessment in southern Africa: integrating local and scientific knowledge bases. *Land Degradation and Development*, *18*, 99-116.

Stringer, L. C., Dougill, A. J., Fraser, E., Hubacek, K., Prell, C. & Reed, M. S. (2006). Unpacking "participation" in the adaptive management of social-ecological systems: a critical review. *Ecology and Society*, *11*, 39.

Sullivan, S. & Rhode, R. (2002). On non-equilibrium in arid and semi-arid grazing systems. *Journal of Biogeography*, *29*, 1595-1618.

Swift, J. (1975). Pastoral Nomadism as a form of land-use: The Twareg of the Adrar n Iforas. In: T. Monod (Ed.), *Pastoralism in Tropical Africa*. Oxford University Press, London, 443-454.

Ter Braak, C. J. F. & Šmilauer, P. (1998). CANOCO Reference Manual and User's Guide to Canoco for Windows. *Microcomputer Power*, Ithaca, USA.

Thomas, D. S. G., Sporton, D. & Perkins, J. (2000). The environmental impact of livestock ranches in the Kalahari, Botswana: Natural resource use, ecological change and human response in a dynamic dryland system. *Land Degradation and Development*, *11*, 327-341.

Thomas, D. S. G. & Twyman, C. (2004). Good or bad rangeland ? Hybrid knowledge, science and local understanding of vegetation dynamics in the Kalahari. *Land Degradation and Development*, *15*, 215-231.

Thomas, D. S. G. & Middleton, N. J. (1994). *Desertification: Exploding the Myth*. Chichester: Wiley.

Todd, S. W. (2006). Gradients in vegetation cover, structure and species richness of Nama-Karoo shrublands in relation to distance from watering points. *Journal of Applied Ecology*, *43*, 293-304.

Trodd, N. M. & Dougill, A. J. (1998). Monitoring vegetation dynamics in semi-arid African rangelands. Use and limitation of Earths observation data to characterize vegetation structure. *Applied Geography*, *18*, 315-330.

UNCCD. (1994). *United Nations Convention to Combat Desertification*, Geneva: United Nations.

UNCED. (1992). *United Nations Conference on Environment and Development Agenda 21*. Rio de Janeiro, June.

Verlinden, A. & Dayot, B. (2005). A comparison between indigenous environmental knowledge and a conventional vegetation analysis in north central Namibia. *Journal of Arid Environments*, *62*, 143-175.

Vermeulen, S. & Koziell, I. (2002). *Integrating global and local values. A review of biodiversity assessment*. IIED, Natural Resource Issue, paper 3.

Virtanen, T., Mikkola, K., Patova, E. & Nikula, A. (2002). Satellite image analysis of human caused changes in the tundra vegetation around the city of Vorkuta, north-European Russia. *Environmental Pollution*, *120*, 647-658.

Walker, B., Kinzig, A. & Langridge, J. (1999). Plant attribute diversity, resilience and ecosystem function: The nature and significance of dominant and minor species. *Ecosystems*, *2*, 95-113.

Walther, D. & Herlocker, D. J. (1980). A preliminary study of the relationship between vegetation, soils and land use with South-Western Marsabit District. IPAL Technical Report A-3, UNESCO, Nairobi.

Warren, A. (2002). Land degradation is contextual. *Land Degradation and Development, 13*, 449-459.

Warren, D. (1992). *Indigenous knowledge, biodiversity conservation and development.* Keynote address at the International Conference on Conservation of Biodiversity in Africa: Local initiatives and institutional roles, 30 August-3 September 1992, Nairobi, Kenya.

Warren, D. M. (1991). *Using Indigenous Knowledge in Agricultural Development.* World Bank Discussion Paper No.127. Washington, D.C.: The World Bank.

Wasonga, V. O., Ngugi, R. K. & Kitalyi, A. (2003). Traditional Range Condition and Trend Assessment: Lessons from Pokot and Il Chamus Pastoralists of Kenya. *Anthropologist, 5*, 79-88.

West, N. (2003). History of Rangeland Monitoring in the U.S.A. *Arid Land Research and Management, 17*, 495-545.

Woodley, H. (2005). Indigenous Knowledge: A Conceptual Framework and a Case from Solomon Islands. In: J. Gonsalves, T. Becker, A. Braun, D. Campilan, H. de Chavez, E. Fajber, M. Kapiriri, J. Rivaca-Caminade, & R. Vernooy (Eds.), *Participatory Research and Development for Sustainable Agriculture and Natural Resource Management: A Sourcebook. Volume 1: Understanding Participatory Research and Development.* IDRC, Ottawa. 65-75.

In: Encyclopedia of Environmental Research
Editor: Alisa N. Souter

ISBN: 978-1-61761-927-4
© 2011 Nova Science Publishers, Inc.

Chapter 5

MONTANE GRASSLANDS OF THE UDZUNGWA PLATEAU, TANZANIA: A STUDY CASE ABOUT ITS HERPETOLOGICAL IMPORTANCE WITHIN THE EASTERN AFROMONTANE HOTSPOT

Roberta Rossi[1], Raffaele Barocco[2], Sebastiano Salvidio[3] and Michele Menegon[4]

[1]Dipartimento di Biologia Cellulare e Ambientale – Università degli Studi di Perugia, via Elce di Sotto, 06123 Perugia, Italy.
[2]Dipartimento di Biologia Applicata – Università degli Studi di Perugia, borgo XX Giugno, 06121 Perugia, Italy
[3]DIPTERIS – Università di Genova, Corso Europa 26, I-16132 Genova, Italy.
[4]Museo Tridentino di Scienze Naturali, via Calepina 14, C.P. 393, I-38100 Trento, Italy & Udzungwa Mountains Monitoring Centre, c/o Udzungwa Mountains National Park, P.O. Box 99 Mang'ula, Tanzania.

ABSTRACT

The Udzungwa Mountains are part of the Eastern Afromontane hotspot, one the 34 biodiversity hotspots in the world. Moreover, they are part of the ancient chain of crystalline blocks named "Eastern Arc" that represents one of the most important areas of Africa in terms of endemic animal and plant species. However, in the Eastern Arc, while forests have been widely investigated and are often safeguarded by the institution of protected areas, montane grasslands have received little attention yet, in terms both of biological research and of conservation, and are threatened by rapidly expanding of cultivations and timber plantations. The present study is aimed to improve knowledge about the composition of Amphibian and Reptile communities of two sites in the southern part of the Udzungwa plateau. The field research was carried out from December 2004 to January 2005 by means of VES and pitfall traps with drift fences, and also by opportunistically searches. The check-list of the first site (placed near Bomalang'ombe village) updates and integrates the preliminary data collected and published by Menegon et alii (2006). The check-list of the second site (placed near

Mapanda village), on our knowledge, represents the first herpetological check-list for the area. On the basis of the species accumulation curves, species richness estimates and effort simulations run for each site, the Bomalang'ombe check-list seems to characterize the community of the area satisfactorily in the considered season, while the Mapanda check-list has to be considered preliminary. However, both study sites exhibit a biological value worthy of attention by the conservation point of view, since both of them show the occurrence of taxa strictly endemic or near-endemic to the Eastern Arc (in some cases endemic to the Udzungwa block and also strictly endemic to the site that represents the type locality) and the occurrence of at least one species vulnerable to the extinction threat. Finally, the occurrence of highly specialized grass-dwelling species suggest that the montane grasslands of the Udzungwa have a long evolutionary history, therefore deserving more scientific and conservation efforts.

Keywords: Amphibians, Reptiles, Tanzania, Eastern Arc, biodiversity hotspot, Udzungwa plateau, montane grasslands, check-lists.

INTRODUCTION

In order to detect the most important areas for conserving biodiversity, British ecologist Norman Myers introduced the biodiversity "hotspot" concept in 1988 (Myers, 1988). To be qualified as a hotspot, a region must meet two strict criteria: it must hold at least 1500 endemic species of vascular plants, and it has to have lost at least 70% of its original habitat. To date 34 biodiversity hotspots have been identified. The total number of terrestrial vertebrates endemic to the hotspots is 12066, representing 42% of all terrestrial vertebrate species. Reptiles and Amphibians are more prone to hotspot endemism than are the more wide-ranging Mammals and Birds (Mittermeier et al., 2004). For Reptiles and Amphibians, many fewer hotspot endemics are considered threatened or extinct than it would be expected on base of habitat loss; however, this mismatch is small in temperate hotspots, suggesting that many threatened endemic species in the poorly known tropical hotspots have yet to be included on the IUCN Red Lists (Brooks et al., 2002). Therefore it would be necessary the investment of far greater effort in the tropics in both the compilation of existing data and in the collection of new data in the field (da Fonseca et al., 2000; Brooks et al., 2002).

The Udzungwa Mountains are part of the Eastern Afromontane hotspot (Mittermeier at al., 2004); in particular they are one of the 13 blocks of the "Eastern Arc" (Lovett, 1985), a chain of ancient mountains considered one of the most important regions of Africa for endemic animals and plants. However, in the Eastern Arc considerable areas of forest and grassland habitats have been converted to agriculture and plantations (Burgess et al., 2007). The majority of the forest habitat of the Eastern Arc, that is estimated to be less than 30% of its original extent (Newmark, 1998), is found within various different categories of Forest Reserve in Tanzania (Burgess et al., 2007), most of which were originally created for their water catchment functions and to prevent landslides and flooding (Burgess et al., 2002). Among the 13 blocks of the Eastern Arc, the Udzungwa Mountains, together with the Uluguru Mountains and the East Usambara Mountains, have the largest number of single-block endemics (found in no other blocks), the largest number of Eastern Arc endemics, the largest number of the combination of near-endemics and strict endemics. The importance of each Eastern Arc block is highly correlated with the area of forest and the amount of survey effort expended, and these two variables are also themselves intercorrelated (Burgess et al., 2007). Conversely, the afromontane grasslands that are interspersed within the forests in the southern highlands of Tanzania, in areas such as the Kitulo Plateau and the Udzungwas, and that have important water catchment functions for the Great Ruaha and Kilombero rivers with the exception of those areas included in the Udzungwa Mountains National Park, have no land-use action plan and no legal conservation status (Salvidio et al., 2004) falling within the "Gaps in Tanzania's Protected Area Network" (Rodgers, 1998). These habitats have been neglected from a biological point of view and are increasingly threatened by fragmentation due to a rapid increase in extent of cultivation and timber plantations (Salvidio et al., 2004). Indeed, very few faunal surveys have been carried out in these habitats and, to our knowledge, the only printed check-list of Amphibian and Reptile species for this habitat type in the Udzungwa Mountains, and in the whole Eastern Arc, is the one provided by Menegon et al. (2006) for a little area placed north-west of Bomalang'ombe village and investigated during various surveys between 1998 and 2002.

The present study is aimed to contribute in increasing the knowledge about Amphibian and Reptile occurrence in the Udzungwa montane grasslands; in particular, it is aimed to investigate the composition of the herpetological community of two sites placed in the southern part of the block, providing:

- an herpetological check-list of species for the site placed north-west of the Bomalang'ombe village, in order to update and integrate the preliminary data provided by Menegon et al. (2006);
- the first herpetological check-list of species for a grassland area placed between Mapanda and Igeleke villages.

STUDY AREA

The Udzungwa Mountains are situated in central-southern Tanzania and cover an area of about 10000 km^2. They arise northwards from the Great Ruaha valley at 300 m a.s.l. as hills resolving in an undulating highland over 1200 m a.s.l. and reaching peaks of 2800 m a.s.l.; they form southwards a steep scarp, created in consequence of the tectonic movements that generated the Kilombero Valley system (Rodgers & Homewood, 1982).

In the Udzungwa Mountains the climate is influenced by the November-May Southeast monsoon winds and the rainfall is unimodal in pattern, starting in November with a peak in April and ending in June (Wasser & Lovett, 1993). High altitudes above 1600 m receive more rain than the lower slopes (Shangali et al., 1998). The rainfall in that area ranges from 1800 to 2000 mm per year (Rodgers & Homewood, 1982) and sometimes reaches over 3000 mm per year in the wetter areas (Shangali et al., 1998).

The vegetation of the Udzungwa Mountains is constituted by sub-montane forest, montane forest, upper montane forest and montane grasslands with tree clumps (Lovett, 1993). At lower elevations, the submontane forest grades into the "transitional" forest (Burgess et al., 2007; "lowland forest" and "dry lowland forest" according to Lovett, 1993), that are often grouped with the lowland Coastal Forests found along the eastern seabord of Africa from Somalia to Mozambique in the south (Burgess et al., 2007). The grassland forest edge is often maintained by fire (Lovett, 1993).

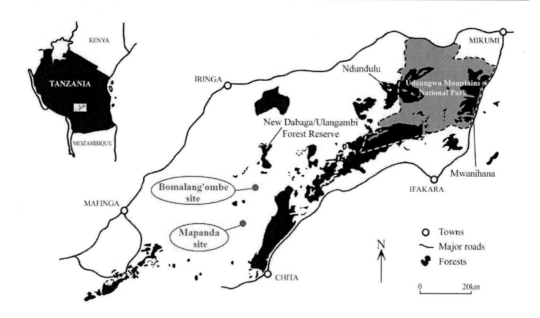

Figure 1. The Udzungwa Mountains with the Bomalang'ombe and the Mapanda sites in evidence. Towns, major roads and forests are represented (adapted from Marshall et al., 2005).

In the present study two grassland sites were investigated: the site of Bomalang'ombe and the site of Mapanda (from the names of the adjacent villages respectively). They are placed in the southern part of Udzungwa *plateau*, and are about 16 km far from each other. Both of them are about 10 km far from the forests of the Udzungwa scarp and less than 10 km far from some small forested patches included into Forest Reserves (Figure 1; Sheet 233/2, "Dabaga"; Sheet 233/3, "Mgunga"; Sheet 233/4, "Idewa").

The site of Bomalang'ombe (Figure 2) covers an area of 10 km^2 and is placed at north-west of Bomalang'ombe village (Coordinates 36L 0815473 9090814; altitude 1975 m a.s.l.) at altitudes included between 1870 m and 1951 m. The site comprehends the 1 km^2 area of Gendavaki Valley intensively investigated by Menegon et al. (2006) between 1998 and 2002 (the valley named "Gendawaki" in that paper is called "Gendavaki" in the present paper according to the Sheet 233/2, "Dabaga"). During the field activities the bottom of Munzu valley and Gendavaki valley was swampy (as indicated in the Sheet 233/2 "Dabaga", where in correspondence of these valleys is represented the symbol of seasonal swamps). The site lacks superficial rocks. Some cultivations were present in the area (both in the bottom of the valleys and in the slopes of the hills characterizing the plateau), that was interested by cow pasturage too (about ten cows during the sampling period were observed). The commonest plant taxa in the site, preliminarly identified by Roy Gerreau (Africa and Madagascar Section of the Botanical Garden of Missouri, USA, pers. comm.) on the basis of plant specimens collected during the field research, include the following Genus: *Aloe, Andropogon, Brachiaria, Cyanotis, Cyperus, Drosera, Eragrostis, Impatiens, Lycopodium, Pentas, Protea* (Barocco et al., 2008). Other recorded Genus are: *Buchnera, Buddleja, Bulbine, Chamaecrista, Chlorophytum, Cirsium, Clematis, Clerodendrum, Commelina, Crassula, Ctenitis, Cycnium, Cyphostemma, Dicoma, Digitaria, Dissotis, Dolichos, Emilia, Eriocaulon, Eriopogon, Eriosema, Eugenia, Fadogia, Fimbristylis, Gerbera, Helichrysum, Heteropogon, Holostoma, Hyparrhenia, Hypoxis, Isoete, Justicia, Kyllinga, Lannea, Loudetia,*

Macrotyloma, Melinus, Multidentia, Osyris, Otiophora, Plectranthus, Polygala, Polygonum, Pycnostachys, Rhynchospora, Sacciolepis, Satureja, Scleria,, Senecio, Sopubia, Spermacoce, Syzygium, Tephrosia, cf. *Vangueria, Vernonia, Vitex, Xyris* (Barocco et al., 2008). Specimens belonging to Genus *Gloriosa* and *Striga* were also recorded and photographed (by B.R. and R.R.) but not collected. Some other specimens were identified just at family level and belong to the following taxa: Acanthaceae, Apiaceae, Asteraceae, Iridaceae, Lamiaceae, Lythraceae, Orchidaceae, Poaceae, Scrophulariaceae. Specimens belonging to Pteridophyta were also recorded and collected (Barocco et al., 2008).

The site of Mapanda (Figure 3) covers an area of 23 km^2 and is situated from north-west to north-east respect to Mapanda village (Coordinates 36L 0805187 9069852, altitude 1895 m a.s.l.) at altitudes included between 1708 m and 1832 m. The site shows the presence of superficial rocks and lacks seasonal swamps (Sheet 233/3, "Mgunga"; Sheet 233/4, "Idewa"). Some cultivations were present in the area and no cow pasturing was observed during the sampling period.

Figure 2. The Bomalang'ombe site.

Figure 3. The Mapanda site.

MATERIALS AND METHODS

The field activities were carried out from December 2004 to January 2005, during the rainy season. Coordinates and altitudes were recorded by means of a GPS GARMIN E-TREX, used together with the following topographic maps (1:50 000): Sheet 233/2 "Dabaga", Sheet 233/3 "Mgunga" and Sheet 233/4 "Idewa". Amphibians and Reptiles were sampled by means of Visual Encounter Surveys (VES, Heyer et al., 1994); the effort applied totalled 208 hour x operator (divided into 52 sampling units) in the site of Bomalang'ombe, and 50 hour x operator (divided into 9 sampling units) in the site of Mapanda. The searches were conducted during both daily and nightly to sample the highest number of species. In the site of Bomalang'ombe also pitfall traps (PFT, 11 20-litres baskets per drift fence) with drift fences (four drift fences, each one 50 meters long) were used (Heyer et al., 1994). Three lines of pitfall traps with drift fence had been located in the Munzu valley (one along the edge of the swampy bottom of the valley, one in the low part and one in the high part of a slope bordering the valley) in a parallel direction respect to the Munzu river; another line was located south-east of Munzu valley in the high part of the slopes of the undulating highland. The traps were checked twice a day; the effort applied to this method amounted to 88 days x drift fence. Moreover some specimens were opportunistically collected into Bomalang'ombe village. Collected specimens were determined by means of keys and descriptions available in literature (Schiøtz, 1971, 1975; Poynton & Broadley, 1985a, b, 1987, 1988; Lambiris, 1989; Branch, 1998; Schiøtz, 1999; Channing, 2001; Broadley & Branch, 2002; Salvidio et al, 2004; Spawls et al., 2004; Poynton et al., 2005;); in particular, for *H. viridiflavus* group, containing numerous forms considered by some authors full species, by others subspecies and treated by others as vernacular names (Schiøtz, 1971, 1975; Richards, 1981; Laurent, 1983; Wieczorek et al., 1998; Schiøtz, 1999; Wieczorek et al., 2001), we refer to Schiøtz (1971) considering *H. goetzei* Ahl, 1931 as a full species. Some voucher specimens were preserved in 70% alcohol. The samples were compared with the East-African samples included in the Collection of the *Museo Tridentino di Scienze Naturali* in Trento, where they have been deposited. Each voucher specimen has been labelled by an individual alphanumeric code that was inserted into the informatic database of the Museum.

Species accumulation curves, species richness estimates and effort simulations were run for each site by means of the software EFFORT PREDICTOR V 1.0, in order to evaluate the significance of the studied communities characterization. Since the order in which species are included in a species accumulation curve influences its overall shape (Magurran, 2005) and therefore also the evaluation of the species richness, 1000 randomizations were run for species accumulation curves and 100 randomizations were run for species richness estimate of each investigated site. These evaluations were run on the basis of the specimen number recorded per each species per VES sampling unit (Appendix I and II).

The conservation *status* of the recorded species was updated to the 2009 IUCN Red List of Threatened Species (IUCN, 2009).

RESULTS

Check-lists. In the site of Bomalang'ombe the presence of 28 species (15 Amphibians and 13 Reptiles) belonging to four Amphibian and six Reptile families was recorded. Eight Amphibian and four Reptiles species were revealed only by means of VES, one Amphibian and two Reptiles only by means of pitfall traps, five Amphibians and five Reptiles by means both of VES and of PFT, one of Amphibian and two Reptiles only opportunistically into or near the village of Bomalang'ombe (Table I). This stresses again that using several sampling methods is the best approach to inventory herpetological species in montane grassland habitats. The check-list might be modified as consequence of the identification at the species level of *Phrynobatrachus* sp., *Hyperolius* sp.1 (represented by the voucher specimens MTSN 8805 and MTSN 8798) and *H.* sp.2 (MTSN 8786), that were not possible to determine only by means of keys and descriptions available in literature. In fact, both the genera *Phrynobatrachus* and *Hyperolius* are taxonomically problematic groups (Poynton & Broadley, 1985, for genus *Phrynobatrachus*; Wieczorek et al., 2001, for genus *Hyperolius*). Also *Ptychadena* sp. (MTSN 8739) is waiting for a species-level identification.

The preliminary check-list of Amphibian and Reptile species provided for a small area of the Gendavaki valley, that was part of the study area of this research, included 27 species (13 Amphibians and 14 Reptiles) (Menegon et al., 2006). In the present study, further nine species (five Amphibians and four Reptiles) were recorded (Table I). Conversely, the presence of one Amphibian species (*Ptychadena grandisonae* Laurent, 1954) and five Reptile species (*Philothamnus hoplogaster* (Günther, 1863); *Lycophidion uzungwense* (Loveridge, 1932); *Pseudaspis cana* (Linnaeus, 1758); *Duberria lutrix shirana* (Boulenger, 1894); *Dasypeltis scabra* (Linnaeus, 1758) recorded by Menegon et al. (2006) was not confirmed by this study.

In the site of Mapanda the presence of 12 species (eight Amphibians and four Reptiles) belonging to four Amphibian and four Reptile families, respectively, was recorded (Table II). The only species already recorded from the area before this study was *Cordylus ukingensis* (Broadley & Branch, 2002).

Species accumulation curves, species richness estimates and effort simulations. The species accumulation curve of the site of Bomalang'ombe is represented in Figure 4a. Interpolating experimental data with an exponential function, it results that the species richness amounts approximately to 22 species ($R^2 = 0.989$; Figure 5a). The corresponding effort simulation is shown in Figure 6a. Indeed, interpolating experimental data with a hyperbolic function, it results that the species richness amounts approximately to 27 species ($R^2 = 0.998$; Figure 5b). The corresponding effort simulation is shown in Figure 6b.

The species richness estimated for the site of Bomalang'ombe by means of the exponential function corresponds to the number of species recorded by means of VES and, since the evaluation is based on field data obtained through this sampling method, it suggests that all the species estimated to be recordable by means of VES were revealed.

Table I. Amphibian and Reptile taxa recorded in the Bomalang'ombe site during the present research. The asterisk (*) indicates the species/subspecies already recorded in the site by Menegon et al. (2006). VES = species/subspecies recorded by means of Visual Encounter Survey. PTF = species/subspecies recorded by means of pitfall traps with drift fences. O = species opportunistically recorded into or near Bomalang'ombe village.

Taxa	Sampling method
AMPHIBIA	
ANURA	
PIPIDAE	
Xenopus petersii Bocage, 1895 *	VES
HYPEROLIIDAE	
Hyperolius goetzi Ahl, 1931	VES
Hyperolius pictus Ahl, 1931 *	VES; PTF
Hyperolius puncticulatus (Pfeffer, 1893)	O
Hyperolius sp.1	VES
Hyperolius sp.2	VES
Leptopelis bocagii (Günther, 1864) *	VES; PTF
BUFONIDAE	
Amietophrynus gutturalis (Power, 1927) *	VES
Mertensophryne uzunguensis (Loveridge, 1932) *	VES
RANIDAE	
Amietia angolensis (Bocage, 1866) *	VES
Strongylopus fuelleborni (Nieden, 1910) *	PTF
Ptychadena perplicata Laurent, 1964	VES
Ptychadena porosissima (Steindachner, 1867) *	VES; PTF
Ptychadena uzungwensis (Loveridge, 1932)	VES; PTF
Phrynobatrachus sp.	VES; PTF
REPTILIA	
SQUAMATA	
SCINCIDAE	
Melanoseps loveridgei Brygoo & Roux-Estève, 1982	PTF
Trachylepis megalura (Peters, 1878) *	PTF
Trachylepis varia (Peters, 1867) *	VES; PTF
CORDYLIDAE	
Chamaesaura miopropus (Boulenger, 1894) *	VES
GERRHOSAURIDAE	
Tetradactylus ellenbergeri (Angel, 1922) *	VES; PTF
Tetradactylus udzungwensis Salvidio, Menegon, Sindaco & Moyer, 2004 *	VES; PTF
CHAMAELEONIDAE	
Chamaeleo goetzi goetzi Tornier, 1899	VES; PTF
Chamaeleo tempeli Tornier, 1899	O
Chamaeleo werneri Tornier, 1899	O
COLUBRIDAE	
Psammophylax variabilis Günther, 1893 *	VES; PTF
Crotaphopeltis hotamboeia Laurenti, 1768 *	VES
VIPERIDAE	
Causus rhombeatus (Lichtenstein, 1823)	VES
Bitis arietans (Merrem, 1820) *	VES

Table II. Amphibian and Reptile taxa recorded in the Mapanda site during the present research. The asterisk (*) indicates the species already recorded in the site (Broadley & Branch, 2002).

Taxa
AMPHIBIA
ANURA
PIPIDAE
Xenopus petersii Bocage 1895
Xenopus laevis (Daudin, 1802)
HYPEROLIIDAE
Afrixalus brachycnemis (Boulenger, 1896)
Afrixalus morerei Dubois, 1985
Hyperolius puncticulatus (Pfeffer, 1893)
Hyperolius pictus Ahl, 1931
BUFONIDAE
Amietophrynus gutturalis (Power, 1927)
RANIDAE
Amietia angolensis (Bocage, 1866)
REPTILIA
SQUAMATA
GEKKONIDAE
Lygodactylus capensis (Smith, 1849)
SCINCIDAE
Trachylepis varia (Peters, 1867)
CORDYLIDAE
Cordylus ukingensis (Loveridge, 1932) *
AGAMIDAE
Agama agama (Linnaeus, 1758)

Indeed, the species richness estimated for the same site by means of the hyperbolic function suggests that the species recorded in this research represent the 80.64% of the estimated species richness (approximately 27 species). The effort simulation based on the hyperbolic function indicates that further 68 sampling units (in addition to the 52 sampling units already performed) should have been necessary in order to reveal the occurrence of only further 10% of species.

The species richness estimated by means of the hyperbolic function is more reliable than the other one on statistical basis (R^2 value); moreover, only the data obtained by means of VES were taken in consideration to run species richness estimates, but in the research 6 further species were recorded by means of sampling method different from VES. On the basis of the previous observations and considering that PFTs permit to detect also rare or fossorial species, such as *Melanoseps loveridgei*, the most realistic conclusion is that the Amphibian and Reptile community of the site of Bomalang'ombe has been satisfactorily investigated.

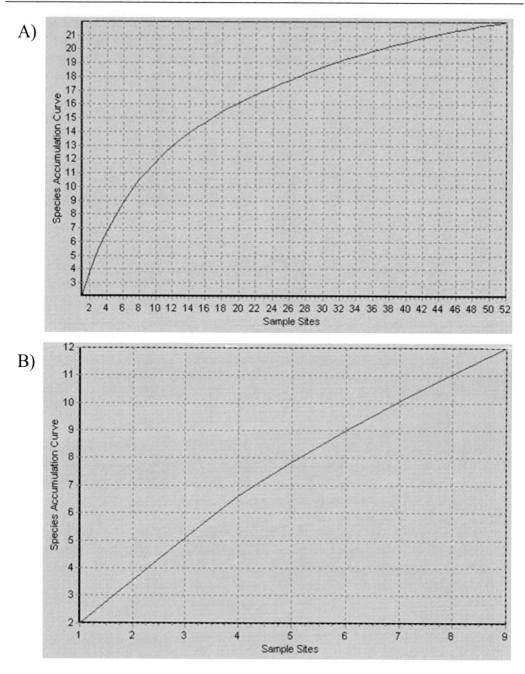

Figure 4. Species accumulation curve of the Bomalang'ombe (A) and of the Mapanda sites (B). The abscissa indicates the number of sampling effort units; it is named "Sample Sites" by software default, but in this case it represents the number of VES sampling units.

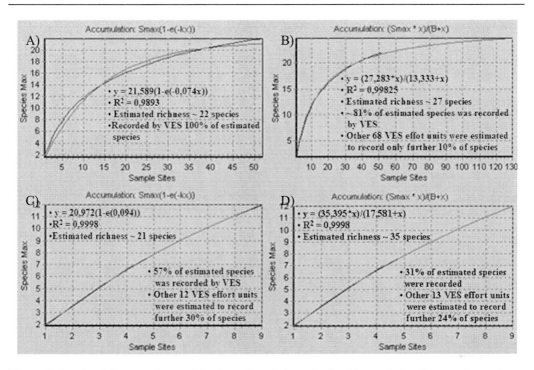

Figure 5. Species richness estimate of the investigated sites, obtained interpolating the experimental data with an exponential function (Bomalang'ombe site = A; Mapanda site = C) and with a hyperbolic function (Bomalang'ombe site = B; Mapanda site = D). The abscissa indicates the number of sampling effort units; it is named "Sample Sites" by software default, but in this case it represents the number of VES sampling units.

The species accumulation curve of the site of Mapanda is represented in Figure 4b. Interpolating experimental data with an exponential function, it results that the species richness amounts approximately to 21 species ($R^2 = 0.999$; Figure 5c). The corresponding effort simulation is shown in Figure 6c. Indeed, interpolating experimental data with a hyperbolic function, it results the species richness amounts approximately to 35 species ($R^2 = 0.999$; Figure 5a). The corresponding effort simulation is shown in Figure 6d. Thus, the species richness estimated suggests that the 12 species recorded in this research represents only 57.22% of those estimated by means of the exponential function (approximately 21 species) or 30.90% of those estimated by means of the hyperbolic function (approximately 35 species). The effort simulation based on the exponential function indicates that further 12 sampling units (in addition to the nine already run) should have been necessary in order to detect the presence of only about further 30% of species. Instead, the effort simulation based on the hyperbolic function indicates that further 13 sampling units should have been necessary to detect the occurrence of only about further 25% of species. The species richness estimated by means of the exponential function and by means of the hyperbolic function are identically reliable on statistical basis (R^2 value). However, the previous observations and the great difference between the two species richness estimates (21 and 35 maximum species) suggests that the check-list of the site of Mapanda should be considered as preliminary and that a greater sampling effort should be invested in the area in order to obtain a satisfactory characterization of the composition of Amphibian and Reptile community.

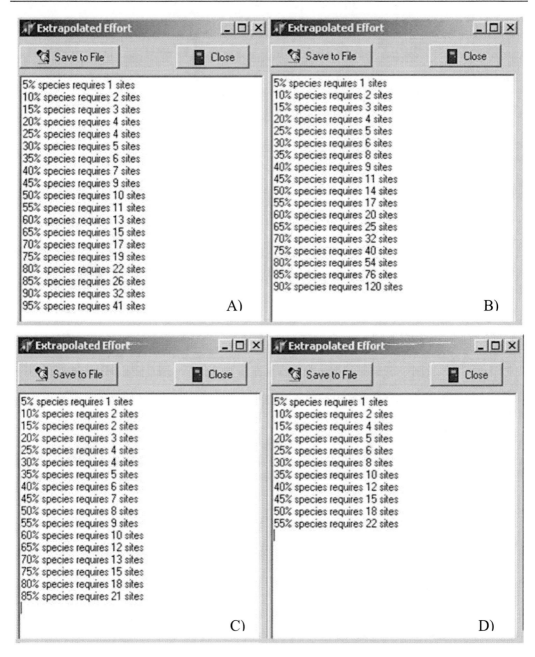

Figure 6. Effort simulations corresponding to the species richness estimates obtained interpolating experimental data with an exponential function (Bomalang'ombe site = A; Mapanda site = C) and with a hyperbolic function (Bomalang'ombe site = B; Mapanda site = D). The number of "sites" represents the number of effort units required to detect the occurrence of various percentages of the estimated species.

Endemism and Conservation Status

Among the taxa recorded in the site of Bomalang'ombe, six (two Amphibian species, three Reptile species and one Reptile subspecies) are strictly endemic or near-endemic to the Eastern Arc; in particular (referring to Burgess et al., 2007, if not differently specified):

- *Mertensophryne uzunguensis*, *Chamaeleo goetzi goetzi* and *C. tempeli* occur into the Eastern Arc only in the Udzungwa block; outside the Eastern Arc, they occur only in the Southern Rift Mountains;
- *Chamaeleo werneri* is strictly endemic to the Eastern Arc, occurring only in the Udzungwa, Nguu, Nguru, Uluguru and Ukanguru blocks;
- *Hyperolius puncticulatus* occurs into the Eastern Arc in the West Usambara, East Usambara, Nguu, Nguru, Uluguru, Ukanguru, Rubeho, Mahenge and Udzungwa; outside the Eastern Arc, it occurs in the area named "Coastal Forests from Kenya to Mozambique" and in the Southern Rift;
- *Tetradactylus udzungwensis* is strictly endemic to the Udzungwa Mountains; its presence had been recorded only from the type locality, a little area in the Gendavaki valley at north-west of Bomalang'ombe village (Salvidio et al., 2004).

Among the Amphibian species, one (*Hyperolius puncticulatus*) is listed as Endangered because its extent of occurrence is less than 5,000 km^2, all individuals are in fewer than five locations, and there is continuing decline in the extent and quality of its habitat (IUCN, 2009); another one (*Mertensophryne uzunguensis*) is listed as Vulnerable because its extent of occurrence is probably less than 20,000 km^2, its distribution is severely fragmented, and the extent of its habitat in southern Tanzania is declining (IUCN, 2009).

No data are available on the conservation *status* of the Reptiles of the Eastern Arc (IUCN, 2009; Burgess et al., 2007).

The check-list of the site of Mapanda comprehends two Amphibians endemic or near-endemic to the Eastern Arc; in particular:

- *Afrixalus morerei* is strictly endemic to the Udzungwa Mountains (Burgess et al., 2007), since it occurs only in open grasslands in the southern highlands of the block (Channing & Howell, 2006);
- *Hyperolius puncticulatus* occurs into the Eastern Arc in the West Usambara, East Usambara, Nguu, Nguru, Uluguru, Ukanguru, Rubeho, Mahenge and Udzungwa; outside the Eastern Arc, it occurs in the area named "Coastal Forests from Kenya to Mozambique" and in the Southern Rift (Burgess et al., 2007).

We think that also *Cordylus ukingensis* should be considered near-endemic to the Eastern Arc on base of Burgess et al. (2007) criteria since its occurrence have been recorded just in two localities in the Udzungwa Mountains and in the type locality, placed in the Ukinga Mountains (Spawls et al., 2004). Among the Amphibians, one (*Afrixalus morerei*) is listed as Vulnerable because its extent of occurrence is less than 20,000 km^2, it is known from fewer than ten locations, and there is continuing decline in the extent and quality of its habitat in the Udzungwa Mountains (IUCN, 2009). As already reminded, no data are available on the conservation *status* of the Reptiles of the Eastern Arc (IUCN, 2009; Burgess et al., 2007).

Biological Notes

The research has permitted to enlarge the knowledge about the habitat and/or the distribution of the following species: *Cordylus ukingensis, Causus rhombeatus, Chamaeleo werneri, Melanoseps loveridgei* and *Tetradactylus udzungwensis*.

Cordylus ukingensis is a poor-known Reptile, of which only three records are known in literature: one record from Tandala (Ukinga Mountains, Tanzania), the type locality; one from Kibenga (Udzungwa Mountains, Tanzania); one from an area about 10 Km south-east of Igeleke village (Udzungwa Mountains, Tanzania) (Spawls et al., 2004). The latter place corresponds to the north-eastern part of the site of Mapanda investigated during this study. On the basis of the previous observations, the specimen of *C. ukingensis* collected during our field research assumes a great importance. This specimen was found on the ground into a hole of a rock and it supports the opinion of Broadley and Branch (2002) according to who *C. ukingensis*, found by Moyer in rocky habitat in montane grassland, is not an arboreal species, as suggested by Spawls et al. (2004). Moreover, this lizard (MTSN 1700) represents the third collected specimen of *Cordylus ukingensis*; the other two are the olotype (MCZ 30761), conserved in the Museum of Comparative Zoology of the Harvard University (Cambridge) and the one (NMZB 11578) conserved in the Natural Hystory Museum of Zimbabwe (Bulawayo) (Broadley & Branch, 2002).

Causus rhombeatus was known to be distributed in Tanzania in the south-western part of the Country; it had not been recorded in south-central Tanzania. Isolated records for the species include the Chyulu Hills in Tsavo and around Kibwezi, and the lower southern slopes of Meru Mountain and Kilimanjaro Mountain (Spawls et al., 2004). The presence of the species in the Bomalang'ombe site slightly enlarges its known distribution range, that probably still remains underestimated.

Chamaeleo werneri is considered a closed forest species vulnerable to habitat change and loss (Spawls et al., 2004). During this study several *C. werneri* were collected on trees and shrubs in the Bomalang'ombe village; it suggests that the species is more tolerant than previously thought.

Melanoseps loveridgei is a fossorial Reptile living in moist savanna at low altitude (Spawls et al., 2004); its presence was recorded also in closed forest (Udzungwa Scarp Forest Reserve) at altitudes included between 1400 and 1600 m, and between 1800 and 1900 m (Menegon & Salvidio, 2005). In the present research one sample of *M. loveridgei* was caught in a pitfall trap in the site of Bomalang'ombe: it represents the first record of the species in montane grassland habitat at an altitude of about 1920 m.

Tetradactylus udzungwensis is a recently-described species (Salvidio et al., 2004) that have been recorded only for the type locality (Gendavaki valley, Udzungwa Mountains, Tanzania) up to now. During this research, the species was recorded in another, contiguous valley, the Munzu valley, near the type locality (about 1.5 km far from).

CONCLUSION

Extensive agricultural practices, timber plantation and heavy livestock grazing are expanding at the expenses of the destruction of the grassland vegetation in the Udzungwa

Mountains (Salvidio et al., 2004) and in other afromontane grassland habitats (e.g., Ndang'ang'a et al. 2002). However, the patterns of species richness of fauna and flora in these habitats remain difficult to interpret and provide conflicting interpretation of age of these grasslands (Meadows & Linder, 1993). In several sites from the southern Afromontane region, palynological evidence has revealed that many grasslands are of considerably greater age than commonly argued: they were present and widespread as long as 12000 BP and, therefore, they are not derived from forest clearence through recent human economic and cultural activities. These findings cannot automatically be extrapolated to the whole Afromontane region (Meadows & Linder, 1993). However, some of the Reptiles detected in the present study, such as *Chamaesaura miopropus*, *Tetradactylus ellenbergeri, T. udzungwensis, Trachylepis megalura,* are highly specialized grass-dwelling species and suggest that, at least, some of the montane grasslands of the Udzungwa plateau have a long evolutionary history. All these arguments support the necessity to increase scientific and in particular conservation effort in these areas. In both the studied sites cultivations and timber plantation are likely to have the most significant negative impact on Amphibians and Reptiles and interested both the bottom of valleys and the slopes of the hills characterizing the plateau. The main crops observed in the area are represented by mais. Near the Bomalang'ombe site timber plantations were observed too; they often comprehend alloctonous species, such as *Pinus patula* (authors' observations; Salvidio et al., 2004). Although Amphibian populations are declining worldwide at cause of different natural and anthropogenic causes (e.g., Houlahan et al., 2000; Beebee and Griffiths, 2005; Di Rosa et al., 2007), habitat destruction remains the principal exctinction threat in Africa.

Therefore, considering that the study area is part of a biodiversity hotspot that has already lost more than 70% of its original habitat extent, the problem of habitat loss and deterioration is the most serious threat. This appears even more truthful considering that Amphibians and Reptiles are more prone to endemism than other Tetrapod Classes in the biodiversity hotspots (Mittermeier et al., 2004) and that their threatened status is suspected to be underestimated (Brooks et al., 2002).

Gaps of knowledge about Amphibian and Reptile biodiversity in the Eastern Arc montane grasslands, where few faunal surveys have been carried out (Salvidio et al., 2004), weakens the possibility of effective conservation actions. In conclusion, the investment of a greater scientific and conservation effort on the montane grasslands of the Udzungwa plateau is not only a necessity, but an urgence.

AKNOWLEDGMENTS

We would like to thank the staff of CEFA N.G.O. for helping us in various ways during our journeys in Tanzania; the Government of Tanzania; the Tanzania Wildlife Research Institute (TAWIRI) and the Tanzania Commission for Science and Technology (COSTECH) for providing permission to study Reptiles and Amphibians of the Udzungwa Mountains (COSTECH − permits NO 2004-335-ER-98-13 and NO 2004-275-NA-2004-106); the CITES management authority in Tanzania and the Wildlife Division, that issued the necessary permits for the export of the specimens; Roy Gerreau (Africa and Madagascar

Section of the Botanical Garden of Missouri, USA) for the preliminary identification of plant specimens collected during field surveys. We thank also Mario Mearelli and Gianandrea La Porta for some useful suggestions about data elaboration.

REFERENCES

Barocco, R., Rossi, R., Gigante, D. & Venanzoni, R. (2008). Aspetti del paesaggio vegetale di un'area studio della comunità di rettili e anfibi nelle praterie montane dei Monti Udzungwa (Tanzania, Africa Orientale). Abstract of 44th Meeting of Società Italiana di Scienza della Vegetazione. 27th-29th February 2008, Ravenna.

Beebee, T. C. J. & Griffiths, R. A. (2005). The amphibian decline crisis: a watershed for conservation biology? *Biological Conservation*, *125*, 271-285.

Branch, B. (1998). Field Guide to Snakes and other Reptiles of southern Africa. Cape Town: Struik Publishers.

Broadley, D. G. & Branch, W. R. (2002). A review of the small east African *Cordylus* (Sauria: Cordylidae), with the description of a new species. *African Journal of Herpetology*, *51(1)*, 9-34.

Brooks, T. M., Mittermeier, R. A., Mittermeier, C. G., da Fonseca, G. A. B., Rylands, A. B., Konstant, W. R., Flick, P., Pilgrim, J., Oldfield, S., Magin, G. & Hilton-Tailor, C. (2002). Habitat Loss and Extinction in the Hotspots of Biodiversity. *Conservation Biology*, *16(4)*, 909-923.

Burgess, N. D., Cordeiro, N., Doggart, N., Fieldså, Hansen, L. A., Howell, K., Kilahama, F., Nashanda, E., Menegon, M., Moyer, D., Perkin, A., Stanley, W. & Stuart, S. (2007). The biological importance of the Eastern Arc Mountains of Tanzania and Kenya. *Biological Conservation*, *134*, 209-231.

Burgess, N. D., Doggart, N. H. & Lovett, J. C. (2002). The Uluguru Mountains of eastern Tanzania: the effect of forest loss on biodiversity. *Orix*, *36*, 140-152.

Channing, A. (2001). Amphibians of Central and southern Africa. USA: Cornell University Press.

Channing, A., & Howell, K. M. (2006). Amphibians of East Africa. Ithaca, London and Frankfurt: Cornell University Press and Edition Chimaira.

Da Fonseca, G. A. B., Balmford, A., Bibby, C., Boitani, L., Corsi, F., Brooks, T., Gascon, C., Olivieri, S., Mittermeier, R. A., Burgess, N., Dinerstein, E., Olson, D., Hannan, L., Lovett, J., Moyer, D., Rahbek, C., Stuart, S. & Williams, P. (2000). Following Africa's lead in setting priorities. *Nature*, *405*, 393-394.

Di Rosa, I., Simoncelli, F., Fagotti, A. & Pascolini, R. (2007). The proximate cause of frog declines? *Nature*, *447*, E4-E5.

Heyer, W. R., Mc Diarmid, R. W., Donnelly, W. R., Foster, M. S. & Hayek, L. C. (1994). Measuring and Monitoring Biological Diversity: Standard Methods for Amphibians. Washington and London: Smithsonian Institution Press.

Houlahan, J. E., Findlay, C. S., Schmidt, B. R., Meyer, A. H. & Kuzmin, S. L. (2000). Quantitative evidence for global amphibian population declines. *Nature*, *404*, 752-755.

IUCN (2009). IUCN Red List of Threatened Species. Version 2009.1. <www.iucnredlist.org>. Downloaded on 28 May 2009.

Lambiris, A. J. L. (1989). The frogs of Zimbabwe. Torino: Museo Regionale di Scienze Naturali.

Laurent, R. F. (1983). La superspèce *Hyperolius viridiflavus* (Duméril & Bibron, 1841) (Anura Hyperoliidae) en Afrique Centrale. *Italian Journal of Zoology, 1(18)*, 1-93.

Lovett, J. C. (1985). Moist forests of Tanzania. *Swara, 8(5)*, 8-9.

Lovett, J. C. (1993). Eastern Arc moist forest flora. In J. C. Lovett, & S. K. Wasser (Eds.), *Biogeography and Ecology of the Rain Forests of Eastern Africa* (33-57). Cambridge: Cambridge University Press.

Magurran, A. E. (2004). *Measuring Biological Diversity.* Oxford: Blackwell Publishing.

Marshall, A. R., Brink, H., Fanning, E. & Topp-Jørgensen, J. E. (2005). Monkey abundance and social structure in two high elevation forest reserves in the Udzungwa Mountains of Tanzania. *International Journal of Primatology, 26(1)*, 127-145.

Meadow, M. E. & Linder, H. P. (1993). A paleoecological perspective on the origin of Afromontane grasslands. *Journal of Biogeography, 20*, 345-355.

Menegon, M. & Salvidio, S. (2005). Amphibian end reptile diversity in the Southern Udzungwa Scarp Forest Reserve, South-Eastern Tanzania. In B. A. Huber, B. J. Sinclair, & K. H. Lampe (Eds.), African Biodiversity: Molecules, Organisms, Ecosystems. *Proceedings of the 5th International Symposium of Tropical Biology.* Museum Koenig. Bonn

Menegon, M., Moyer, D. & Salvidio, S. (2006). Reptiles and Amphibians from a montane grassland: Gendawaki Valley, Udzungwa Mountains, Tanzania. African Herp News, *40*, 8-14.

Mittermeier, A. R., Robles Gil, Hoffman, Pilgrim, J. M., P., Brooks, T., Mittermeier, C. G., Lamoreux, J. & da Fonseca, G. A. B. (2004). *Hotspots Revisited: Earth's Biologically Richest and Most Endangered Terrestrial Ecoregions.* Mexico City: Conservation International.

Myers, N. (1988). Threatened biotas: "hotspots" in tropical forests. *The Environmentalist, 8*, 187-208.

Ndang'ang'aa,b, P. K., du Plessis M. A., Ryana, P. G. & Bennun, L. A. (2002). Grassland decline in Kinangop Plateau, Kenya: implications for conservation of Sharpe's longclaw (*Macronyx sharpei*). *Biological Conservation, 107*, 341-350.

Newmark, W. D. (1998). Forest area, fragmentation, and loss in the Eastern Arc Mountains: implications for the conservation of biological diversity. *Journal of the East African Natural History, 87*, 29-36.

Poynton, J. C. & Broadley, D. G. (1985a). Amphibia Zambesiaca 1: Scolecomorphidae, Pipidae, Microhylidae, Hemisidae, Arthroleptidae. *Annals of the Natal Museum, 26(2)*, 503-553.

Poynton, J. C. & Broadley, D. G. (1985b). Amphibia Zambesiaca 2: Ranidae. *Annals of the Natal Museum, 27(1)*, 115-181.

Poynton, J. C. & Broadley, D. G. (1987). Amphibia Zambesiaca 3: Rhacophoridae and Hyperoliidae. *Annals of the Natal Museum, 28(1)*, 161-229.

Poynton, J. C. & Broadley, D. G. (1988). Amphibia Zambesiaca 4: Bufonidae. *Annals of the Natal Museum, 29(2)*, 447-490.

Poynton, J. C., Menegon, M. & Salvidio, S. (2005). *Bufo uzungwensis* of Southern Tanzania (Amphibia: Anura): a history of confusion. *African Journal of Herpetology, 54(2)*, 159-170.

Richards, C. M. (1981). A new color pattern variant and its inheritance in some members of the superspecies *Hyperolius viridiflavus* (Duméril & Bibron) (Amphibia Anura). *Italian Journal of Zoology, 16(15)*, 337-351.

Rodgers, W. A. & Homewood, K. M. (1982). Biological values and conservation prospects for the forests and primate populations of the Udzungwa Mountains, Tanzania. *Biological Conservation, 24*, 285-304.

Rodgers, W. A.(1998). Planning a Protected Area Network for Tanzania. Kakakuona, Tanzania Wildlife Magazine, 8.

Salvidio, S., Menegon, Moyer, D., M. & Sindaco, R., (2004). A new species of elongate seps from Udzungwa grasslands, southern Tanzania (Reptilia, Gerrhosauridae, *Tetradactylus* Merrem, 1820). *Amphibia-Reptilia, 25*, 19-27.

Schiøtz, A. (1971). The superspecies *Hyperolius viridiflavus* (Anura). Videnskabelige Meddelelser Dansk Naturhistorisk Forening, *134*, 21-76.

Schiøtz, A. (1975). *The treefrogs of Eastern Africa*. Copenhagen: Steenstrupia.

Schiøtz, A. (1999). *Treefrogs of Africa*. Frankfurt am Main: Edition Chimaira.

Shangali, C. F., Mabula, C. K. & Mmari, C. (1998). Biodiversity and human activities in the Udzungwa Mountain forests, Tanzania: Ethnobotanical survey in the Udzungwa Scarp Forest Reserve. *Journal of East African Natural History, 87*, 291-318.

Spawls, S., Howell, K., Drewes, R. & Ashe, J. (2004). A field giude to the Reptiles of East Africa. San Diego, California: Academic Press.

Wasser, S. K. & Lovett, J. C. (1993). Introduction to the biogeography and ecology of the rain forests of eastern Africa. In J. C. Lovett, & S. K. Wasser (Eds.), *Biogeography and Ecology of the Rain Forests of Eastern Africa* (3-7). Cambridge: Cambridge University Press.

Wieczorek, A. M., Channing, A. & Drewes, R. C. (2001). Phylogenetic relationships within the *Hyperolius viridiflavus* complex (Anura: Hyperoliidae), and comments on taxonomic status. *Amphibia-Reptilia, 22*, 155-166.

Wieczorek, A. M., Channing, A. & Drewes, R. C. (1998). A review of the taxonomy of the *Hyperolius viridiflavus* complex. *Herpetological Journal, 8*, 29-34.

CARTOGRAPHY

Sheet 233/2, "Dabaga" (1982). Printed by Survey and Mapping Division, Ministry of Lands, Housing and Urban Development, Tanzania.

Sheet 233/3, "Mgunga" (1982). Printed by Survey and Mapping Division, Ministry of Lands, Housing and Urban Development, Tanzania.

Sheet 233/4, "Idewa" (1982). Printed by Survey and Mapping Division, Ministry of Lands, Housing and Urban Development, Tanzania.

Appendix I. Number of samples counted per species/subspecies per effort unit by VES in the Bomalang'ombe site.

Species/subspecies	Effort units																	
	1	2	3	4	5	6	7	8	9	10	11	12	13	14	15	16	17	18
Ptychadena uzungwensis	1	0	0	0	0	6	0	0	0	0	0	0	0	0	1	0	0	0
Hyperolius pictus	2	0	2	0	0	0	0	0	0	0	1	0	0	0	0	1	0	0
Chamaeleo goetzi subsp. goetzi	1	0	0	0	2	0	0	0	0	0	0	0	0	0	0	0	0	0
Psammophilax variabilis	0	1	0	0	0	0	0	0	0	0	0	0	0	0	0	0	0	0
Phrynobatrachus sp.	0	1	0	0	0	8	0	4	0	0	0	0	2	0	0	1	0	0
Leptopelis bocagii	0	0	0	2	0	0	0	0	0	0	0	0	0	2	0	0	0	0
Amietophrynus gutturalis	0	0	0	5	0	0	2	0	0	0	0	0	0	5	0	0	0	0
Tetradactylus ellenbergeri	0	0	0	0	2	0	0	0	0	0	0	0	0	0	0	1	1	0
Tetradactylus udzungwensis	0	0	0	0	1	0	0	0	0	0	0	0	0	0	0	0	0	0
Ptychadena sp.	0	0	0	0	0	1	0	0	0	0	0	0	0	0	0	0	0	0
Trachylepis varia	0	0	0	0	0	0	0	1	0	0	4	2	0	0	1	0	0	2
Mertensophryne uzunguensis	0	0	0	0	0	0	0	0	1	0	0	0	0	2	0	0	2	0
Amietia angolensis	0	0	0	0	0	0	0	0	0	1	0	0	0	0	0	0	0	0
Ptychadena porosissima	0	0	0	0	0	0	0	0	0	0	0	0	0	1	0	0	0	0
Hyperolius sp. 1	0	0	0	0	0	0	0	0	0	0	0	0	0	0	1	0	0	0
Causus rhombeatus	0	0	0	0	0	0	0	0	0	0	0	0	0	0	0	0	0	0
Crotaphopeltis hotamboeia	0	0	0	0	0	0	0	0	0	0	0	0	0	0	0	0	0	0
Hyperolius goetzi	0	0	0	0	0	0	0	0	0	0	0	0	0	0	0	0	0	0
Hyperolius sp. 2	0	0	0	0	0	0	0	0	0	0	0	0	0	0	0	0	0	0
Bitis arietans	0	0	0	0	0	0	0	0	0	0	0	0	0	0	0	0	0	0
Chamaesaura miopropus	0	0	0	0	0	0	0	0	0	0	0	0	0	0	0	0	0	0
Xenopus petersii	0	0	0	0	0	0	0	0	0	0	0	0	0	0	0	0	0	0

Appendix I. (Continued)

Species/subspecies	Effort units																
	19	20	21	22	23	24	25	26	27	28	29	30	31	32	33	34	35
Ptychadena uzungwensis	0	1	0	1	0	0	0	1	0	0	0	1	0	0	0	0	0
Hyperolius pictus	0	0	0	0	0	2	0	0	0	0	0	0	0	6	1	1	0
Chamaeleo goetzi subsp. *goetzi*	0	0	0	0	1	0	0	0	0	0	0	0	0	1	0	0	1
Psammophilax variabilis	0	0	0	0	1	0	0	0	0	0	0	0	0	0	0	0	0
Phrynobatrachus sp.	0	1	0	0	0	0	0	2	0	0	0	0	0	0	1	0	0
Leptopelis bocagii	0	0	4	0	0	0	0	0	0	1	2	0	0	0	0	0	0
Amietophrynus gutturalis	0	0	3	0	0	11	0	0	0	1	4	0	0	1	0	1	0
Tetradactylus ellenbergeri	0	0	0	0	0	0	0	0	0	0	0	0	0	0	0	0	0
Tetradactylus udzungwensis	0	0	0	0	0	0	0	0	0	0	0	0	0	0	0	0	0
Ptychadena sp.	0	0	0	0	0	0	0	0	0	0	0	0	0	0	0	0	0
Trachylepis varia	0	0	0	0	2	0	1	0	2	0	0	1	0	0	0	0	0
Mertensophryne uzunguensis	0	0	3	0	0	0	0	0	0	0	0	0	0	0	0	1	0
Amietia angolensis	0	0	1	1	0	0	0	0	0	0	0	0	0	0	0	0	0
Ptychadena porosissima	0	0	1	1	0	0	0	0	0	0	0	0	0	0	0	1	0
Hyperolius sp. 1	0	0	0	0	0	1	0	0	0	0	0	0	0	1	0	0	0
Causus rhombeatus	0	0	0	0	0	0	0	0	1	0	0	0	0	0	0	0	0
Crotaphopeltis hotamboeia	0	0	0	0	0	0	0	0	0	0	1	0	0	0	0	0	0
Hyperolius goetzi	0	0	0	0	0	0	0	0	0	0	0	0	0	2	0	0	0
Hyperolius sp. 2	0	0	0	0	0	0	0	0	0	0	0	0	0	0	0	0	0
Bitis arietans	0	0	0	0	0	0	0	0	0	0	0	0	0	0	0	0	0
Chamaesaura miopropus	0	0	0	0	0	0	0	0	0	0	0	0	0	0	0	0	0
Xenopus petersii	0	0	0	0	0	0	0	0	0	0	0	0	0	0	0	0	0

Appendix I. (Continued)

Species/subspecies	Effort units																
	36	37	38	39	40	41	42	43	44	45	46	47	48	49	50	51	52
Ptychadena uzungwensis	0	1	1	0	0	0	0	0	0	0	0	0	0	0	0	1	0
Hyperolius pictus	0	0	1	1	0	0	0	0	0	0	1	0	5	0	0	0	0
Chamaeleo goetzi subsp. goetzi	2	0	0	0	0	0	0	0	0	0	0	0	4	0	0	0	0
Psammophilax variabilis	0	0	0	0	0	0	0	0	0	0	0	0	0	0	0	0	0
Phrynobatrachus sp.	0	0	1	9	0	0	0	0	0	0	0	0	0	0	0	0	0
Leptopelis bocagii	0	0	0	0	0	0	0	0	0	0	0	0	0	2	1	0	0
Amietophrynus gutturalis	1	0	0	0	0	0	0	0	0	0	0	2	0	2	3	0	0
Tetradactylus ellenbergeri	0	1	0	0	0	0	0	0	0	0	0	0	0	0	0	0	0
Tetradactylus udzungwensis	0	0	0	0	0	0	0	0	0	0	0	0	0	0	0	0	0
Ptychadena sp.	0	0	0	0	0	0	0	0	0	0	0	0	0	0	0	0	0
Trachylepis varia	0	0	0	1	0	2	0	0	2	0	2	0	0	0	0	0	0
Mertensophryne uzunguensis	0	0	0	0	0	0	1	1	0	1	0	0	0	0	0	0	0
Amietia angolensis	0	0	0	0	0	0	0	0	0	0	0	0	0	0	0	0	0
Ptychadena porosissima	0	1	1	0	0	0	0	0	0	0	0	0	0	0	1	0	0
Hyperolius sp. 1	0	0	0	0	0	0	0	0	0	0	0	0	0	0	0	0	0
Causus rhombeatus	0	0	0	0	0	1	0	0	0	0	0	0	0	0	0	0	0
Crotaphopeltis hotamboeia	0	0	0	0	0	0	0	0	0	0	0	1	0	0	0	0	0
Hyperolius goetzi	0	0	0	0	0	0	0	0	0	0	0	0	3	0	0	1	0
Hyperolius sp. 2	0	0	1	0	0	0	0	0	0	0	0	0	0	0	0	0	0
Bitis arietans	0	0	0	0	1	0	0	0	0	0	0	0	0	0	0	0	0
Chamaesaura miopropus	0	0	0	0	0	0	0	0	2	0	0	0	0	0	0	0	0
Xenopus petersii	0	0	0	0	0	0	0	0	0	0	0	0	1	0	0	6	0

Appendix II. Number of samples counted per species per effort unit by VES in the Mapanda site.

Species	Effort units								
	1	2	3	4	5	6	7	8	9
Amietia angolensis	1	0	0	0	0	2	0	0	0
Xenopus laevis	0	2	0	0	0	0	0	0	0
Amietophrynus gutturalis	0	0	2	0	0	0	0	0	0
Xenopus petersii	0	0	0	11	3	0	0	0	0
Hyperolius puncticulatus	0	0	0	1	0	10	0	0	0
Afrixalus brachycnemis	0	0	0	1	0	0	0	0	0
Trachylepis varia	0	0	0	1	1	0	0	0	1
Afrixalus morerei	0	0	0	0	3	0	0	0	0
Hyperolius pictus	0	0	0	0	0	0	0	0	1
Agama agama	0	0	0	0	0	0	0	0	2
Lygodactylus capensis	0	0	0	0	0	0	0	0	4
Cordylus ukingensis	0	0	0	0	0	0	0	0	1

In: Encyclopedia of Environmental Research
Editor: Alisa N. Souter

ISBN: 978-1-61761-927-4
© 2011 Nova Science Publishers, Inc.

Chapter 6

EVALUATION OF GRAZING PRESSURE ON STEPPE VEGETATION BY SPECTRAL MEASUREMENT

Tsuyoshi Akiyama[1], Kensuke Kawamura[2], Ayumi Fukuo[3], Toru Sakai[4], Zuozhong Chen[5] and Genya Saito[6]*

[1]River Basin Research Center, Gifu University, 1-1 Yanagido, Gifu 501-1193, Japan.
[2]Graduate School for International Development and Cooperation, Hiroshima University, 1-5-1 Kagamiyama, Higashi Hiroshima 739-8529, Japan.
[3]Japan International Research Center for Agricultural Science, 1-1 Ohwashi, Tsukuba, Ibaraki 305-8686, Japan.
[4]Research Institute for Human and Nature, 457-4 Motoyama, Kamigamo, Kita-ku, Kyoto 603-8047, Japan.
[5]Institute of Botany, Chinese Academy of Sciences, Beijing 100093, China.
[6]Graduate School of Agricultural Science, Tohoku University, Miyagi 981-8555, Japan.

ABSTRACT

Steppe grassland spreading over eastern Europe to eastern Asia is one of the largest grassland ecosystems in the world, which is precious but fragile. Last four decades the grazing pressure has been increasing due to increase of animal and human population in Inner Mongolia, China. It may affect on the grassland ecosystem, including grassland biomass, grass quality, and species diversity. But how can we achieve for maintaining the valuable ecosystems and sustainable livestock farming at the same time? We examined the relationship between grazing pressure and grassland vegetation using drastically improved information technology such as remote sensing, Geographic Information Systems (GIS) and Global Positioning System (GPS). The experiment was carried out at Xilingol steppe, where the annual precipitation is around 300 mm, and annual mean air temperature is around 0°C. In the former part of the chapter, we discussed how to evaluate the grazing pressure under free ranging condition utilizing multiple livestock farmers in the extended grassland without fences and borders. Standing against this problem, we put GPS on the shoulder of sheep to trace the daily traveling distance. Sheep

* Corresponding author: Tel: +81-75-707-2457, Fax: +81-75-707-2531, E-mail: torus@chikyu.ac.jp.

herd moved between 11 and 12 km in a day during summer season. GPS also showed us the place and time where the sheep herd spent. Newly developed bite-counter clarified the number of biting during grazing separated from rumination. The time series of biomass and grass quality (in term of crude protein content) changes were successfully monitored by satellite vegetation indices such as Terra MODIS-EVI and NOAA AVHRR-NDVI. In the latter part, we examined the effects of grazing pressure on steppe vegetation in the same district. In 1960's, several pipe lines were buried inside grassland from lake to supply drinking water for animal. Opening mouths were set at 5 or 6 km intervals. Livestock villages were formed around the opening, therefore, surrounding grassland came under the influence of gradation of grazing pressure concentrically. Started from certain opening, we put quadrates on 4 directions at 1 km intervals until 5 km of distance. In addition of botanical composition survey, biomass, soil conditions and spectral reflectance measurements were executed at each plot. Vegetation cover, biomass and plant height increased in accordance with distance from opening. Both NDVIs obtained from radiometer at ground and Landsat image showed similar trend as biomass change. Botanical composition also changed. Land was almost bare around the opening, then some low palatable herbaceous plants are remained near the opening. Three or four km apart from opening, the original vegetation and biodiversity recovered and biomass became constant. Finally, we discussed effects of grazing pressure on steppe vegetation from a comprehensive stand point.

Keywords: Grazing, Biomass, Inner Mongolia, Land degradation, Spectral reflectance, Satellite, Steppe vegetation.

1. INTRODUCTION

Steppe grassland spreading over east Europe to east Asia is one of the largest grassland in the world, which is precious but fragile ecosystem. Large area of the steppe is suffering from on-going desertification due to human activities such as overgrazing (Li et al., 2000). Mr. Xu, the Director-general for the Science and Technology Agency of Inner Mongolia, told that the desertification area in China is increasing 2460 km^2 per year. The Xilingol Steppe is a major husbandry region in Inner Mongolia. Many studies have been conducted concerning the effects of grazing on vegetation in this region. Akiyama and Kawamura (2007) got published review paper putting focus on grassland degradation in China.

There are three types of steppes in Xilingol: meadow steppe, typical steppe and desert steppe. Typical steppe occupies most of the area. Meadow steppe is mainly dominated by *Stipa baicalensis*, *Filifolium sibiricum* and various forbs, whereas typical steppe can be grouped into *Leymus chinensis* steppe and *Stipa grandis* steppe (Li et al., 1988). In order to avoid desertification and maintain steppe grassland in good condition for sustainable animal husbandry, it is important to evaluate grazing intensity accurately and timely. However, detecting the grazing intensity in a steppe for free grazing is the most difficult.

As shown in Figure 1, grassland conditions can be determined by the balance of grass production (GP) and herbage intake (HI) by animals. When HI is larger than GP, grasslands will be degraded. However, if GP is superior to HI, grasslands will be conserved and the land will recover (Akiyama and Kawamura, 2007). GP may regulated by soil fertility and topography of the area, and pasture management and climate condition of each year. On the other hand, HI is affected by grazing intensity, animal behavior, palatability and so on.

There are several methods to evaluate GP and HI relationship. In recent years, sensors onboard satellite had been developed memorably. Since around 2000, sensors like Quickbird (3.2 m) and IKONOS (4 m) attained high resolution compared with conventional SPOT/HRV (10 m) sensor. Spectral resolution also improved as be seen in EO-1/Hyperion (220 bands) or Terra/MODIS (36 bands) compared with conventional Landsat/ETM+ (8 bands). In addition, MODIS revisits same area in one day intervals. It means resolution improved in spatial, spectral and temporal view points. It may helpful for monitoring environmental changes like steppe degradation.

Satellite observation is effective to take general view on wide area. But it has difficulties to clarify causal link happening inside ecosystem, such as interaction of grazing by herbivores and reaction of plant communities. So, another approach is encouraged to examine by field experiments. For example, the distance from the nearest village is generally used as an indicator of grazing pressure (Nakamura et al., 2000) under the free grazing condition. The line transect method and long-term point research are useful for investigating the impacts of grazing on vegetation, and some results have been achieved (Li, 1996; Li et al., 1994; Nakamura et al., 1998, 2000). *Leymus chinensis* and *Stipa grandis* steppes, which are the major original vegetation of typical steppes in Inner Mongolia, shift to an *Artemisia frigida* type steppe under long-term heavy grazing, and to a *Potentilla acaulis* type steppe under sustained overgrazing (Li, 1996; Li and Wang, 1999). However, steppes under moderate grazing achieved the highest species diversity and ANPP (aboveground net primary production), suggesting that a sustainable utilization of steppe is possible if it is kept under careful management (Chen and Wang, 2000; Li, 1996; Wang et al. 1999). As another approach of field experiment, Yiruhan et al. (2005) evaluated the adaptability of herbage species to environmental variation through a long-term grazing experiment in Japan.

Trials for the measurement of bio-information on ground using spectral radiometers had been attempted on many crops and pasture plants. Such spectral data provided us information of plant quantity and quality (e.g. Evri et al., 2008; Sakai et al., 2002). Thus, spectral reflectance measurement is a promising method for monitoring the condition of the grassland in a non-destructive way and one that is connected with satellite image analysis.

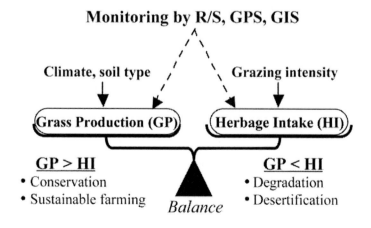

Figure 1. Concept of real-time steppe monitoring system using satellite remote sensing, GPS and GIS.

As satellites onboarding super-high resolution and hyper-spectral sensors like IKONOS or EO-1/Hyperion have become ready at hand in recent years (e.g. Thenkabail, 2003), the importance is increasing to collect the site specific information as of the ground truth data. This is the reason that we choose spectro-radiometric method corresponding to field data.

This chapter consists of two parts for evaluating effects of animal grazing on wide steppe vegetation under free grazing. In the first half part, we briefly introduce recent achievements to estimate grassland biomass and quality as grassland factors (GP), and Global Positioning System (GPS) and bite counter (BC) putting on sheep for animal factors (HI). So we intended real-time monitoring of GP and HI using satellite data, and Geographic Information Systems (GIS).

The second one is reflectance measurement using handy type spectrometer according to the distance from water tank. We tried to measure the gradation of grazing pressure from changes of reflectance and botanical composition.

2. Evaluation of Grazing Intensity from Satellite Data

The Xilingol steppe is situated in the northeastern part of Inner Mongolia, which is located about 400 – 600 km north of Beijing. All the experiments in this section were carried out in the Baiyinxile Livestock Farm. It covers a total area 3,750km^2, which is one of the farms of the Xilingol League (Figure 2) during 1998 to 2004. The average altitude is ranging 1,300 to 1500 meters above sea level. The mean annual air temperature is around 0 degree C and the total annual precipitation ranged from 250 – 350 mm (1982 – 1998). The dominant soil types are Chestnut and Chernozem (XLSFWS, 1991; Chen and Huang, 1988).

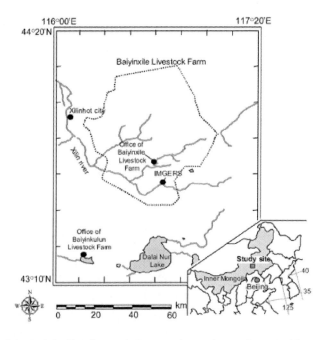

Figure 2. Location of the Baiyinxile Livestock Farm and experimental area, with an outset map showing the location of the study site in northern China. IMGERS: Inner Mongolia Grassland Ecosystem Research Station, Chinese Academy of Sciences.

2.1 Changes in Grassland Type in Xilingol Steppe

We analyzed five Landsat scenes acquired on summer between 1979 and 1997 using ERDAS/Imagine software, five landcover types could be discriminated in the study area. They are three steppe vegetation classes such as high productive meadow steppe, medium productive typical steppe, and overgrazed desert steppe, in addition, cropland class and sand dune class. During these 20 years, the areas of productive meadow steppe decreased. Instead, areas of croplands and low productive desert steppe were increased (Figure 3, Akiyama and Kawamura, 2003; Fukuo et al., 2008). The changes were notable for the period between 1979 and 1989. But it was difficult to identify more precise vegetation types from time series Landsat images.

2.2 Detection of Biomass and Quality Change

Kawamura et al. (2003a; 2003b) found high correlations between Enhanced Vegetation Index (EVI, Huete et al., 1997) derived from Terra/MODIS and grassland biomass or grass quality including live biomass (R^2=0.744, P < 0.001), total biomass (R^2=0.707, P < 0.001), and standing crude protein (R^2=0.728, P < 0.001). Applying these relationships on four sheep farms with different degree of grazing, seasonal biomass and quality changes are shown in Figure 4. Non-grazed plot, NG in Figure 4 (top) is a mowing site, and increase grazing intensity in order of light grazing (LG), intermediate grazing (MG) and heavy grazing (HG), equivalent to 2, 4 and 6 sheep per hectare, respectively. The highest biomass in summer was attained in NG, but rashly dropped crude protein content at the same time. On the other hand, HG plot with 6 sheep kept low biomass but high crude protein content throughout the seasons. The maximum values of live biomass were 266.7, 166.4, 99.7 and 63.8g DM m^{-2} in NG, LG, MG and HG, respectively. The minimum value of crude protein content in NG, LG, MG and HG were 9.8%, 10.5%, 11.4% and 12.0%, respectively. Thus, EVI can produce biomass map and crude protein distribution map of whole study area.

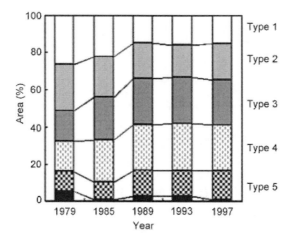

Figure 3. Changes in steppe vegetation in Xilingole steppe observed by five time series of Landsat images between 1979 and 1997. Type 1: Meadow steppe, 2: Typical steppe, 3: Grazing land, 4: Cropland, 5: Sand dune or Village.

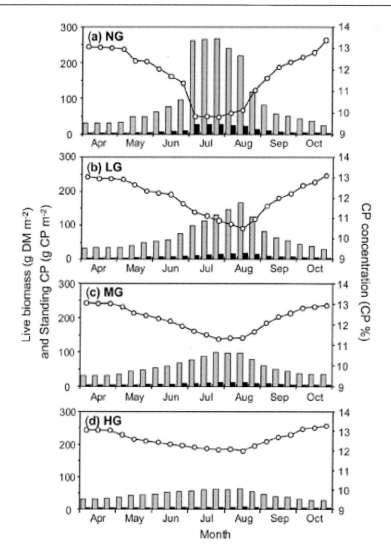

Figure 4. Seasonal changes in estimated live biomass (gray bars), standing crude protein (filled bars) and crude protein concentration (sequential line) in NG (non-grazed grassland), LG (lightly grazed grassland), MG (intermediately grazed grassland and HG (heavily grazed grassland) in Xilingol steppe.

2.3 Monitoring of Animal Behavior

By mean of satellite based Global Positioning System (GPS), accurate positioning data of herbivores can be provided (Kawamura et al., 2003b; 2005). In addition, a neckband bite counter (BC) attached to a collar on herbivores, to count biting jaw movements was evaluated (Kawamura et al., 2005; Umemura et al., 2003). These are excellent opportunity to expand knowledge on the foraging behavior of free ranging herbivores. Figure 5 shows image putting GPS and BC on a sheep. Daily traveling course of sheep herd, the spending time, and bite count could be recorded (Kawamura et al. 2005). Figure 6 represents daily tracking course of sheep herd in the experiment farm. It was clarified that sheep traveled 11 to 12 km in a day in this district during summer season.

Figure 5. Setting GPS receiver and Bite counter (BC) onto the sheep; (a) Handy type GPS and BC, (b) GPS collar and BC, (c) sheep of handy type GPS and BC was fixed, and (d) view of start grazing in the early morning.

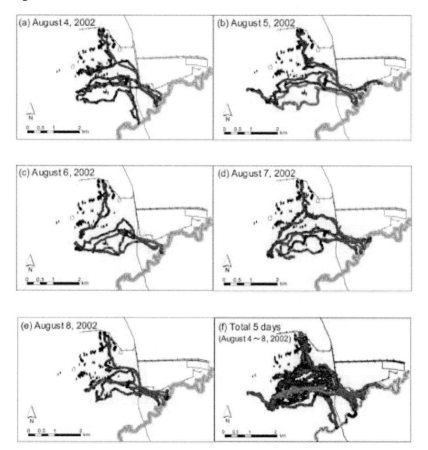

Figure 6. Daily tracking of sheep group at 1-min intervals with GPS over 5 days (August 4~8, 2002). Average tracking distance was 11 to 12 km in this farm on this season.

2.4 Real-time Steppe Monitoring System

Combining these kinds of satellite technology, we might build real-time steppe management system. It permits to estimate grassland production (GP), and herbage intake (HI) by herbivores with the help of GIS. GIS database consists of 3-D topography derived from Terra/ASTER image, maps of biomass and forage quality obtained by daily MODIS images, grazing information by GPS and BC on the sheep,

Figure 7 shows relation between a kind of grazing intensity (a) and steppe biomass (b) in early August (Kawamura et al., 2005). Three sheep farms were using this area. Each grid size is 250 x 250 meter to adjust MODIS pixel size. To quantify the distribution of grazing pressure (a) during test period, grazing distribution map was created using a grid cell method from the tracking data recorded by GPS. Biomass distribution (b) can be estimated from MODIS vegetation index. Here, the MODIS/NDVI reduced with dense grazing except for Xilin river side areas (lower right). The equation for NDVI is as follows.

$$NDVI = (NIR - Red) / (NIR + Red)$$

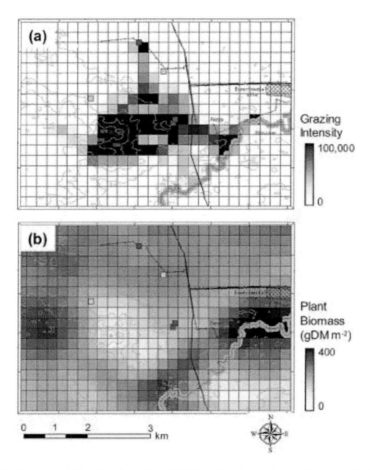

Figure 7. Distribution maps of grazing intensity (GI) (a) and estimated plant biomass (b) during 5 days (August 4 – 8, 2002). Distribution map of (a) GI was calculated from GPS data, and (b) plant biomass was calculated from MODIS/NDVI.

Here, NIR is the intensity of near infrared reflectance, and Red means red reflectance. The NDVI (Normalized Difference Vegetation Index) is a popular spectral vegetation index proposed by Rouse et al. (1974). It comes from the principle that chlorophyll in green leaf absorbs red light, while mesophyll cells reflect the near infrared wavelengths. Therefore, it relates to the biomass and vigor of vegetation.

3. CHANGE OF STEPPE VEGETATION IN ACCORDANCE WITH THE DISTANCE FROM THE WATER TANK

Satellite observation is effective to take general view on wide area. But it has difficulties to clarify causal link happening inside ecosystem, such as interaction of grazing by herbivores and reaction of plant communities. Here in this section, we introduce more precisely a trial to detect spectral changes under the gradation of grazing pressure (Fukuo et al., 2008).

We chose an area specified by the Landsat/TM image in the Xilingol steppe, where the distance from the village can be used as an effective indicator for grazing intensity since the water tank in the village is the main drinking place for animals. We therefore analyzed the relationships between the extent of grazing degradation and spectral reflectance factors as a function of the distance from the water tank and combined with a vegetation survey. The purpose of this study was to test whether the spectral method is effective for detecting grassland degradation in a non-destructive way in the Inner Mongolia steppe.

3.1 Methods

3.1.1 Experimental site, baiyinkulun livestock farm, pipelines and branch No.4

The Baiyinkulun Livestock Farm is one of the farms of the Xilingol League and is located in the southern Xilingol Steppe (Figure 8). The area of the Baiyinkulun Livestock Farm extends approximately 50 km east to west and 30 km north to south.

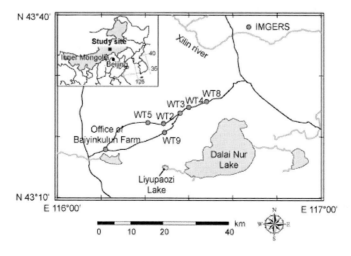

Figure 8. Location of Baiyinkulun Livestock Farm and water tanks (WT) along the pipeline, with an inset map showing the location of the study site in northern China.

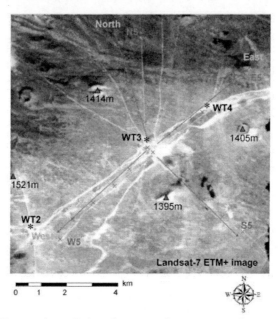

Δ: Mountain, *: WT; Water tank, ×: Points for vegetation survey and spectral measurement; WT2: Water tank No.2 at power substation; WT3: Water tank No.3 at the center of Silian village (Branch No.4); WT4: Water tank No.4 at Silian village (about 2.5 km northeastern from center); Landsat-7 ETM+ satellite data acquired on August 14, 2001. (R:G:B = ETM5:ETM8:ETM2).

Figure 9. Setting of test sites for vegetation, soil and spectral measurements at Branch No.4 (WT-3) of the Baiyinkulun Livestock Farm in the Xiligol steppe, overlaid on Landsat/ETM+ image acquired on August 14, 2001. The red arrows show two courses from southwest to northeast (W-E course), and north to southeast (N-S course) for the field surveys.

The steppes are subjected to grazing or mowing spread around the surrounding small villages with water tanks to supply water for living and animal husbandry. The Baiyinkulun Livestock Farm occupies 141,650 ha of grassland in total of which 138,000 ha is used for livestock (grazing and mowing) and 3,650 ha for cropland.

Securing water for animals is the most serious problem which this district faces because the hard base rock prevents the digging of water wells. From 1965 to 1969, pipelines were constructed from the Liyupaozi Lake for transporting fresh water at a distance of 73.2 km. Here, in the Baiyinkulun Livestock Farm, 10 water tanks were built at 5 to 7 km intervals for supplying water for daily living as well as for animals (Figure 8). Therefore, nomadic people built their houses near the tanks, forming small villages. As thousands of animals went to the tank and back every day to drink water, the land around the drinking place became bare. The gradient of grazing pressure might mitigate in concentric circles inversely proportional to the distance from the village center.

Among 10 branches of Baiyinkulun Livestock Farm, we are watching with interest Branch No.4 (villagers call it Silian) at water tank No.3 (WT-3). Most of the residents are raising sheep, cattle or goats. Figure 9 is the study site displayed on a Landsat/ETM+ image acquired on August 14, 2001. The red cross symbols (×) represent the sampling points for 1999. The total area of Branch No.4 is about 20,000 ha, of which 19,000 ha were available grasslands, consisting of 13,000 ha for grazing and 6,000 ha for mowing according to the statistics of 1997. No croplands nor protective fences for separating grazing from mowing grasslands in this Branch were observed during the experiment period (1998 to 1999). In 1973, Branch No. 4 consisted of only 35 families (172 people). However, this has since increased to 199 families (430 people) in 1999. Accordingly, the total livestock number has increased to 35,600 HT (sheep equivalent animal unit) in 1999, where 1 horse is 6 HT, 1 cattle is 5 HT, and 1 goat is 0.8 HT, according to Zhang and Liu (1992).

3.1.2 Setting of the test site

Branch No.4 (Silian village) is one of the villages located on the pipeline running east to west. WT-3 (water tank No. 3) at Branch No.4 is situated at 116°28'10"E, 43°23'56"N. Along the pipeline, a local beaten road passing from southwest to northeast can be found; we call it as W-E course (Figure 9). To the east from WT-3, WT-4 (water tank No. 4) and several farm houses are located between the 2.5-km (E2.5) and 3-km points (E3). WT-2 (water tank No. 2) and a village called power substation is located at 6-km west along the pipeline. More than half of the village families raise animals. On this road, large-sized motor-trucks come and go during the summer season for shipping out produced domestic animals and for carrying in wintering feed. A lot of nomadic people and animals go back and forth along the road. Meanwhile, in the north-south direction (N-S course), small local paths are running but it is difficult to pass full-sized trucks, so there is no farm house nearby.

In the four directions shown in Figure 9, we used GPS equipment to decide on the test sites, as described below: site-00 was a drinking place near the WT-3. Twenty-one test sites were chosen every 1 km until the 5-km point from site-00 in 4 directions. For instance, at the 5-km point to the east, we called this site E5. Such a configuration in space is illustrated in Figure 9. Addition to the 21 test sites, site E2.5 was chosen at a distance of 2.5 km east from site-00, where there is another tank (WT-2). Besides, S0.1 was chosen at the edge of central circle of bare land, about 100-m south of site-00, to ascertain tolerant species under the highest grazing intensity. No vegetation was found at site-00, while reflectance was not measured at E2.5 and N5 because of low solar radiation during the surveyed hours.

Soil and vegetation surveys as well as spectral reflectance measurements were carried out during the summers of 1998 and 1999 around Branch No.4.

3.1.3 Vegetation and soil survey

The vegetation survey was conducted according to the method proposed by Braun-Blanquet (1964). Five 1-m^2 quadrates were examined at each site. Firstly, a central quadrate (Q_c) was chosen, then another 4 quadrates (Q_e, Q_w, Q_s, Q_n) chosen that were stationed 10 m away from Q_c in 4 directions (east, west, south and north, respectively) determined using a compass. The canopy height (cm) and total coverage (%) of each quadrat were then recorded, and the mean values of the five quadrates were used to represent the plant height and plant coverage of the site. In addition, within the central quadrat (Q_c), the coverage (C, %) and height (H, cm) of each species was recorded. The species frequency (F, %) was measured using the point method of multiple contacts (100 points along two 20-meter lines across the plot, i.e. from Q_e to Q_w and from Q_s to Q_n), where F (%) is the number of species touched among the 100 points (Goldall, 1952). We used SDR (the summed dominance ratio) to indicate the dominance of the species in the community (Numata and Yoda, 1957).

The SDR was then calculated using the equation: SDR = (H'+C'+F')/3, where H' is the relative plant height (cm), C' is the relative coverage (%) and F' is the relative frequency (%); H' (C' or F') = H (C or F) of the species/maximum value of H (C or F) in the quadrate, respectively. Due to the frequency of animal foraging and trampling, there appeared a radius 100-m circle of bare soil around the site-00. For the vegetation survey, S0.1 was therefore used instead of site-00, 100-m south of site-00, which was able to grow some plants in an extremely sparse coverage and low canopy height. All of the plants were named according to Liu and Liu (1988).

After the vegetation survey and spectral measurements, all aboveground parts were removed at the soil surface, and the total fresh weight (FW, g) in situ was measured using a portable electric balance (Model-BL2200H, Shimadzu Co. Ltd., Kyoto, Japan). Parts of the samples were brought back to IMGERS to be oven dried at 80°C for the estimation of their dry weight (DW, g) of the biomass. A survey of species composition was carried out in 1998, and other quantitative information on the plant community was measured during the summer seasons of 1998 and 1999.

The soil hardness (SH, mm) was measured at nine points for each site using a Yamanaka standard type soil hardness tester (Fujiwara Scientific CO.LTD., Tokyo) after the removal of aboveground plant materials. The surface color and moisture of the soil were also recorded, but it was too difficult to relate this to the land characteristics because they were strongly affected by the previous rainfall event.

3.1.4 Spectral reflectance measurement

In order to detect the grazing degradation of grasslands by a non-destructive method, spectral reflectance measurements were carried out. A field-type spectral radiometer (MD-01 type) was used in this experiment which was manufactured by PREDE CO.LTD., Tokyo. This device is able to measure reflectance at 4 visible (450, 545, 650, 699 nm) and 2 near-infrared wavelengths (750 and 850 nm). The measurer stands facing the sun, stretching the optical fiber over the plant canopy keeping a distance of about 1 m between the tip of the fiber and the canopy. As the instant view angle of this equipment is $10°$, if keeping a 1-m distance, the reflectance value inside 35.3 cm in diameter can be measured. Measurements were carried out during clear daytime between 9:00 and 15:00 local time to get strong and constant solar radiation. The spectral reflectance was corrected to a relative value using a standard white board (Spectralon, Labsphere CO.LTD. USA). The spectral measurements were repeated 7 times for each quadrate. After omitting the maximum and minimum values, the remaining 5 data points were averaged. The measurements were repeated for all 5 of the quadrates for each site. In order to compare the quantitative biological information of grasslands, we used NDVI, a spectral vegetation index. NDVI = (NIR - Red) / (NIR + Red).

3.2 Vegetation Change with Distance from Water Tank

3.2.1 Changes in species composition

Figure 10 shows the changes in dominance (SDR, %) of major species with increasing distance from site-00 in the N-S and W-E directions (see Figure 8). As described above, site-00 is the drinking place for animals near the water tank No.3 (WT-3) of the Silian village (Branch No. 4). Due to frequent animal foraging and trampling, site-00 displayed a bare cover of plants. We illustrated the result of S0.1 in Figure 10 to indicate vegetation under an extremely high grazing pressure 100 m from site-00 in a region where some plants are able to grow in an extremely low coverage of 20% and a plant height of 3.4 cm (also see Figure 11a). With increasing distance from site-00 in all 4 directions (N, S, W, E), there was a similar tendency in terms of species dominance changes with distance from site-00. That is, negative indicators of grazing intensity suggested by Li et al. (1999) (see discussion for details), such as *Stipa grandis* and *Filifolium sibiricum*, increased in dominance (SDR, %) with distance from site-00. Whereas positive indicators of grazing intensity, such as *Cleistogenens squarrosa* and *Agropyron cristatum*, decreased in dominance with distance from site-00. Such a tendency in the W-E direction seemed not as clear as in the N-S direction as a result of disturbances due to the existence of other water tanks (WT-2 and WT-4) and farm houses along the pipeline. The areas inside a 1-km diameter from site-00 were thought to be subject to heavy grazing, where positive indicators for grazing intensity, i.e. *A. cristatum* and *C. squarrosa*, dominated the communities. Whereas negative indicators for grazing intensity, i.e. *S. grandis* and *F. sibiricum*, were absent or with lower SDR values at sites a distance of 1 km from site-00. In the meantime, a widespread weedy species, *Plantago asiatica*, appeared at these sites, also indicating the effects of human daily life. Because of another water tank (WT-4) located near E3, sites E2.5 and E3 exhibited a reduced SDR of *S. grandis* and an increased SDR of *A. cristatum*, as well as the presence of the weedy species *P. asiatica*.

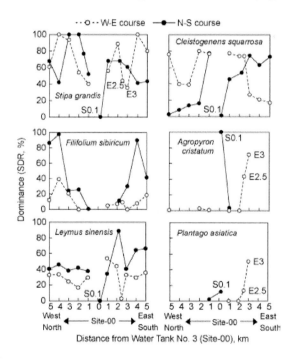

Figure 10. Changes in dominance (SDR, %) of the major species with distance from the village of Branch No.4 (WT-3: site-00) in the Xilingol steppe (July, 1999). The W-E course is from site-00 to the southwestern and northeastern 5-km points, and the N-S course is from site-00 to the northern and southeastern 5-km points (detailed see Figure 9). Inserted E2.5 and E3 indicates that these two sites are kind of special due to the fact that they are near WT-4 (water tank No. 4). Site-00 is bare land, therefore the results of S0.1, 100-m south of site-00, is illustrated to represent vegetation under extreme grazing pressures.

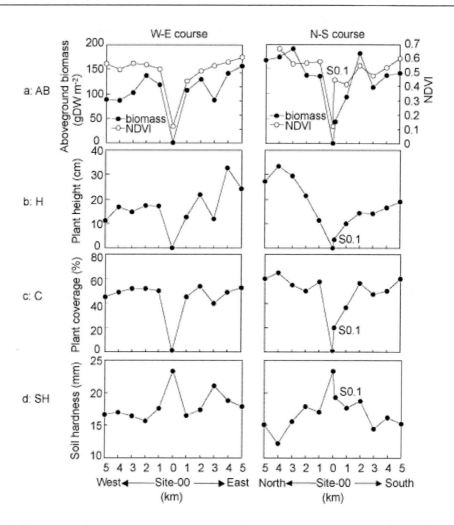

Figure 11. Changes in plant quantity information and soil hardness according to the distance from the water tank at Branch No.4 of the Baiyinkulun Livestock Farm. a: aboveground biomass (AB) and NDVI, b: plant height (H), c: plant coverage (C) and d: soil hardness (SH). NDVI at E2.5 and N5 were not measured because of the low solar radiation during surveyed hours.

Similarly, the grazing pressure seemed to decline with distances of greater than 2 km from site-00, since the biomass increased from 42.9 g DW m^{-2} (S0.1 at 100 m away from site-00) to more than 180 g DW m^{-2} (see Figure 11a). For sites at distances of between 2 and 3 km away from site-00, *Stipa grandis* and *Leymus chinensis* were found, and dominated the community with increasing SDR with distance away from site-00. Meanwhile, *Cleistogenes squarrosa* decreased in SDR with increasing distance away from site-00 for distances of more than 2 km, and this also was indicative of a decreased grazing intensity, especially for distances greater than 2 km away from the village.

For distances of 4 or 5 km from the village in the N-S direction, meadow steppe appeared instead of typical steppe. This area was mainly used for mowing or for light grazing by horses. *F. sibiricum* and *S. grandis* (data mixed with *Stipa baicalensis*) dominated the communities and with more richness in species, including many forbs.

3.2.2 Changes in quantitative biological parameters with distance from water tank

Figure 11 shows quantitative biological information on the grasslands such as the aboveground biomass (a: AB), the plant height (b: H), the plant coverage (c: C) and the soil hardness (d: SH) as a function of distance from WT-3 (site-00) at Branch No.4 to the 5-km points for the four directions. Because there was a similar trend in biomass changed with a distance from site-00 towards to the four directions between 1998 and 1999, we only illustrated the results of 1999 in Figure 11 for explanation.

In the N-S direction of Figure 11a, AB increased linearly with increasing distance from site-00, and remained at over 160 g DW m^{-2} for distances of between 3 and 5 km (N3, N4 and N5). A similar trend is observed in the W-E direction, however, for the same reason as the changes in the species composition described above, it is not as clear as in the case of the N-S direction because of disturbances by the existence of other water tanks and farm houses along the pipeline. On the same graph (Figure 11a), the changes in the NDVI values are drawn. The NDVI curves appeared to be almost proportional to AB, C and H.

Slow but constant increments were observed from site-00 to N4 in H of Figure 11b and reached a maximum value (34 cm) at N4 in the N-S direction. In the W-E direction, the maximum value reached was 33 cm at E4, but the curve had roughness and hollows.

The coverage (C) in Figure 11c showed almost the same patterns for the N-S and E-W directions. In both directions, steep V shapes formed between site-00 at the 1-km points, securing 50% coverage, then it became almost constant out to the 5-km points. The maximum coverage was 66% at N4. Hollows at E3 were commonly found in all three sets of data. This result might have been affected by farm houses near WT-4.

3.2.3 Soil hardness (SH)

The changes in SH are shown in Figure 11d in relation to the distance from site-00. The values are over 20 mm (very tough) around site-00, but soften with distance to 12 mm at N4. This trend was clear in the northerly direction, but was obscure in the W-E direction. The SH curve had another peak at the E3 site.

3.3 Discussion

3.3.1. Floristic composition changes after long-term heavy grazing

Li et al. (1999) established a grazing experimental plot (21 ha) with 21 subplots in IMGERS and compared it with different stocking rates in an *Artemisia frigida* steppe about 30-km north of Branch No. 4 in the Xilingol steppe. This study began in 1990 with the aim of investigating short and long-term effects of grazing on vegetation and soils. The results of the first 3 – 5 years have been reported involving changes in species composition, species diversity, as well as the above and belowground biomass with grazing management (Li and Wang, 1999). Li (1996) suggested six ecological groups in relation to grazing intensity in the steppe, *i.e.* decreaser, disappearer, increaser, invader, species adapting to moderate grazing and fluctuating species. Decreaser and disappearer are negative indicators for grazing pressure such as *Achnatherum sibiricum*, *Stipa grandis* and *Filifolium sibiricum*, while increaser and invader are positive indicators such as *Carex duriuscula*, *Artemisia frigida*, *Cleistogenes squarrosa*, *Agropyron cristatum* and *Potentilla acaulis* (Nakamura et al., 1998).

In this study, such a change in species composition with distance from the village was clear in the N-S direction, but was disturbed by the existence of farm houses along the W-E direction. For example, due to the WT-4 located at 2.7-km east of the Silian village, these sites (E3 and E2.5) exhibited a reduced dominance of *Stipa grandis* and an increased dominance of *Agropyron cristatum*, as well as the appearance of the weedy species *Plantago asiatica*. A similar tendency was observed at the W4 and W5 sites close to WT-2 (Figure 10). However, along the N-S direction, there were no farm houses found in the areas at distances of between 1 km and 5 km from the site-00. It was obvious that the changes in species composition along the N-S direction could be explained by long-term grazing effects with decreasing grazing intensity with increasing distance away from site-00. Moreover, it indicated that areas at a distance of 3 km away from the village (i.e. at S3 or N3) may be under a relatively moderate grazing pressure, reaching higher biomass and dominated with *Stipa grandis,* with the highest SDR in the N-S direction (Figs.10 and 11a).

3.3.2 Relations between NDVI and biological parameters

Figure 12 shows the relationships between NDVI and several biological parameters, such as the aboveground biomass (AB), the coverage (C), the plant height (H) and one soil parameter (soil hardness, SH) measured during the 1999 experiment using the same data as in Figure 11.

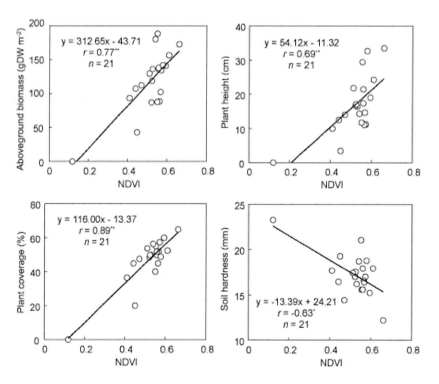

Figure 12. Relationships between NDVI and several plant quantity parameters observed in 1999 at Branch No.4. Each point coordinate data used 21 test sites in Figure 11. Upper left; aboveground biomass (AB) vs. NDVI; Upper right: plant height (H) vs. NDVI; Bottom left; plant coverage (C) vs. NDVI; and Bottom right; soil hardness (SH) vs. NDVI. * P < 0.01; ** P < 0.001.

This includes all sites on the W-E and N-S courses. A high correlation was observed, particularly between NDVI and C ($r = 0.89$, $P < 0.001$, $n = 21$). The correlations of NDVI vs. AB ($r = 0.77$, $P < 0.001$, $n = 21$), and NDVI vs. H ($r = 0.69$, $P < 0.001$, $n = 21$) were also significant, although the correlation coefficient was not as high as that of NDVI vs. C. The same tendency was reported using satellite data analysis (Kawamura et al., 2003b), that the correlation was much higher in the coverage (%) than in plant height because remote sensors sharply sense the percentage plant coverage in the nadir angle, so it is more sensitive to horizontal information than to vertical.

Xiao et al. (1997) analyzed Landsat/TM data for detecting steppe vegetation types in Xilingol using 1987 image. In the multispectral classification, 14 vegetation cover types were recognized. Among them, the total area of degraded and desertificated grasslands, summed from the *Stipa krylovii* steppe, *Artemisia frigida* steppe and desertificated grasslands, accounted for 26.4% of the total 29,440 km^2 of land area.

3.3.3 Relation of spectral reflectance and grazing degradation

Land degradation triggered in arid/semiarid grasslands in the world is mainly caused by overgrazing (UNEP, 1984). However, by raising freely several thousands of flock on several thousands of hectares of grassland, it is difficult to quantify grazing pressure in each area. Without such a grassland management record, how can we judge objectively the extent of the overgrazing? A simple ratio of the total animal numbers 35,600 HT by the grassland area yields 2.74 head ha^{-1} in average over 13,000 ha at Branch No.4. Although it is similar to the optimal stocking rate (2.67 head ha^{-1}) obtained by the grazing experiments done by Wang et al. (1999) at IMGERS, the area around the village is suffering from overgrazing of hundreds of times this level.

In semi-arid districts, a deterioration of grassland is easily brought about by overgrazing, meaning a frequent removal of herbage. In addition, the treading by animal causes a destruction of aboveground organs and root systems as a result of soil compaction (Kubo and Akiyama, 1977).

The possible mechanisms of grazing degradation are summarized in Table 1. Herbage biomass decreases and plants become dwarfed, and plant litters are removed by grazing animals and strong wind. Herbage production reduces by the serious and continuous removal of assimilation organs under ceaseless heavy grazing, sometimes resulting in bare land. This is why it is easy to dry up the soil surface. At the same time, the floristic composition become poor and simple, but some tolerant species prosper forming rugged patches consisting of unpalatable species. Meanwhile, the soil becomes compact by frequent trampling, and defecation brings about an enrichment of the soil chemical components by supplying large amounts of feces and urine exhausted by animals. At the ecosystem level, the soil surface becomes dry, resulting in a decline in herbage productivity and species diversity. Most of these changes in plant and soil conditions are related to an enhancement in the brightness captured by spectral-radiometer or satellite sensors which are reflected in NDVI values in this experiment. These are the characteristics of the direct deterioration of grassland caused by overgrazing.

Table 1. Effects and mechanisms of land degradation caused by overgrazing and its detection using spectral reflectance.

Animal actions	Effects on plant & soil	Changes of plant & soil	Ecosystem level
Foraging	Plant height & biomass	Dwarf and bare*	Decline in productivity
Foraging	Litter accumulation	Dissappear*	Soil drying
Foraging	Composition	Deforage tolerant species	Floristic deterioration
Foraging	Unpalatable community	Rugged surface	Floristic deterioration
Trampling	Lodging and organ dying	Loss of leaf area*	Decline in productivity
Trampling	Soil (physical effect)	Soil harden*	Soil degradation
Defecation	Soil (chemical effect)	Enrichment	Weedy species

*: Increase reflectance or decrease NDVI.

The extent of land degradation occurred so rapidly around the village, that the biomass and the floristic composition had changed drastically step by step. Landsat/TM data of middle-resolution can not follow such sudden changes. In that meaning, getting the field spectral data together with the biological information are useful for using super-high resolution satellite data in the near future. Under similar point of view, Harris and Asner (in print) detected grazing gradient with airborne imaging spectroscopy in Utah, USA, and showed that detection of persistent grazing gradients with imaging spectroscopy is feasible. But they did not refer about changes in the floristic composition.

The spectral reflectance data could not detect the changes in floristic composition directly in this experiment. We measured some unpalatable species communities such as *Iris*, *Potentilla* and *Caragana microphylla* using the same spectro-radiometer of this study, and found that there existed weak relationships between vegetation indices and plant coverage ratio of the species. However, it was difficult to generalize for such unpalatable communities. In the meanwhile, inedible species for animals are apt to form distinctive and pure communities near the village, which may be detectable from the difference of texture in the highly resolution satellite images. By this method, Yamamoto (personal communication, 2000) detected weed communities in grasslands using roughness index of satellite image. If we can find a spectral index able to specify plant indicators for grazing degradation, spectral detection of grazing degradation may improve to detect the changes in floristic composition due to overgrazing. Nevertheless, our results indicated that NDVI achieved by spectral reference could well reflect the growth parameters such as aboveground biomass, plant coverage and height along grazing intensity gradients. This strongly suggested that spectral detection of grazing degradation with a spectral vegetation index (e.g. NDVI) is effective in the Xilingol steppe in a non-destruction way.

4. CONCLUSION

In order to detect the grazing pressure on steppe vegetation, spectral reflectance was measured using earth observation satellite and portable spectral radiometer in the Xilingol steppe, Inner Mongolia.

Time series of Landsat images depicted vegetation change since 1970's. It delineated the increase of the low productive grasslands in Xilingol steppe. Vegetation indices derived from

Terra/MODIS data, such as EVI and NDVI, were highly correlated (R^2>0.7) with steppe biomass and grass quality in terms of crude protein amount. In order to estimate herbage intake by herbivores, GPS and Bite Counter were put on sheep. GPS clarified daily travelling course and distance of sheep. Monitoring with satellite image will be a mighty method for management of grassland and animal behavior timely and widely.

Supply of drinking water is an important for nomadic peoples in Xilingol. So the distance from the village can be used as an effective indicator for grazing intensity since the water tank in the village is the main drinking place for animals. We therefore analyzed the relationships between the extent of grazing degradation and spectral reflectance factors as a function of the distance from the water tank and combined with a vegetation survey.

There is only one main drinking place (site-00) for animals at the center of Branch No.4 of Baiyinkulun Livestock Farm. In addition to the measurement of surface reflectance, vegetation surveys were carried out involving floristic composition and plant growth parameters as a function of the distance from site-00 at every 1-km interval to each four directions until the 5-km point. It becomes clear that the effects of grazing on vegetation and soil mitigated according to the distance from site-00. Negative plant indicators of grazing intensity such as *Stipa grandis* and *Filifolium sibiricum* increased in dominance with distance from site-00, whereas positive indicators of grazing intensity, such as *Cleistogenens squarrosa* and *Agropyron cristatum,* decreased in dominance. Changes in growth parameters along the environmental gradients were reflected on NDVI (Normalized Difference Vegetation Index), in which NDVI was positively correlated with the aboveground biomass (r = 0.77), plant height (r = 0.69) and coverage (r = 0.89), respectively. These results suggested that a spectral vegetation index is effective in the detection of the degradation of grazing grassland in a non-destruction way.

REFERENCES

Akiyama, T. & Kawamura, K. (2003). Vegetation changes and land degradation of steppe in Inner Mongolia. *Journal of Japan Agricultural Systems Society, 19(3),*12-22.**

Akiyama, T. & Kawamura, K. (2007). Grassland degradation in China: Methods of monitoring, management and restoration. *Grassland Science, 53,* 1-17.

Braun-Blanquet, J. (1964). Pflanzensoziologie 3, Aufl. Wien. Springer-Verlag. 865, Suzuki, T. (Translated) *Plant Sociology, 1,* Asakura Shoten, Tokyo*

Chen, Z. Z. & Wang, S. P. (2000). *Typical Steppe Ecosystems in China.* Beijing, P.R: Science Press, 125-172.***

Chen, Z. Z. & Huang, D. H. (1988). The characteristics of meadow grassland and its effects on Chernozem formation process in the Xilin River Valley, Inner Mongolia. *Scientia Geographica Sinica, 8,* 38-46. ***

Evri, M., Akiyama, T. & Kawamura, K. (2008). Optimal visible and near-infrared waveband used in hyperspectral indices to predict crop variables of rice. *Journal of Japan Agricultural Systems Society, 24,* 19-29.

Fukuo, A., Akiyama, T., Mo, W., Kawamura, K., Chen, Z. & Saito, G. (2008). Spectral detection of grazing degradation in the Xilingol Steppe, Inner Mongolia. *Journal of Integrated Field Science, 5,* 29-40.

Goldall, D. W. (1952). Some considerations in the use of point quadrate methods for the analysis of vegetation. *Australian Journal of Biological Science*, 5, 1-41.

Huete, A. R., Liu, H. Q. & Batchily K. and van Leeuwen, W. (1997). A comparison of vegetation indices global set of TM images for EOS-MODIS. *Remote Sensing of Environment*, 59, 440-451.

Kawamura, K., Akiyama, T., Watanabe, O., Hasegawa, H., Zhang, F. P., Yokota, H. & Wang, S. P. (2003a). Estimation of aboveground biomass in Xilingol steppe, Inner Mongolia using NOAA/NDVI. *Grassl. Sci*, 46, 1-9.

Kawamura, K., Akiyama, T., Yokota, H., Tsutsumi, M., Watanabe, O. & Wang, S. (2003b). Quantification of grazing intensities on plant biomass in Xilingol steppe, China using Terra MODIS image. International Archives of Photogrammetry, *Remote Sensing and Spatial Information Sciences*, Vol.44, Part 7/W14, C-5, 1-8.

Kawamura, K., Akiyama, T., Yokota, H., Yasuda, T. & Umemura, K. (2005). Use of the bite counter on sheep for detecting the jaw movements. *Jpn J. Grassl Sci*, 52, 144-148.

Kubo, S. & Akiyama, T. (1977). Treading on the grassland. photosynthesis and dry matter production of pressed pasture plants. *Bulletin of National Grassland Research Institute*, 10, 15-22.**

Li, B., Yong, S. P. & Li, Z. H. (1988). The vegetation of the Xilin River Basin and its utilization. *Research on Grassland Ecosystem No. 3* (eds. IMGERS). 181-225. Beijing Science Press, Beijing.***

Li, S. G., Harazono, Y., Oikawa, T., Zhao, H. L., He, Z.Y. & Chang, X. L. (2000). Grassland desertification by grazing and the resulting micrometeorological changes in Inner Mongolia. *Agricultural and Forest Meteorology*, 102, 125-137.

Li, Y. H. (1996). Ecological variance of steppe species and communities on climate gradient in Inner Mongolia and its indication to steppe dynamics under the global change. *Acta Phytoecologica Sinica*, 20, 193-206.

Li, Y. H. & Wang, S. P. (1999). Response of plant and plant community to different stocking rates. *Grassland of China*, 3, 11-19.***

Li, Y. H., Chen, Z. Z. Wang, S. P. & Huang, D. H. (1999). Grazing experiment for sustainable management of grassland ecosystem of Inner Mongolia steppe: Experimental design and the effects of stocking rates on grassland production and animal liveweight. *Acta Agrestia Sinica*, 7, 173-182.

Li, Y. H., Mo, W. H. Ye, B. & Yang, C. (1994). Aerial biomass and carrying capacity of steppe vegetation in Inner Mongolia and their relations with climate. *Journal of Arid Land Resources and Environment*, 8, 43-50.***

Liu, S. R. & Liu, Z. L. (1988). Outline of flora of the Xilin River Basin, Inner Mongolia. *Research on Grassland Ecosystem No. 3* (eds. IMGERS). 268-274. Beijing Science Press, Beijing.***

Nakamura, T., Go, T., Li, Y. H. & Hayashi, I. (1998). Experimental study on the effects of grazing pressure on the floristic composition of a grassland of Baiyinxile, Xilingole, Inner Mongolia. *Vegetation Sciences*, 15, 139-145.

Nakamura, T., Go, T. & Wuyunna and Hayashi, I. (2000). Effects of grazing on the floristic composition of grasslands in Baiyinxile, Xilingole, Inner Mongolia. *Grassland Science*, 45, 324-350.

Numata, M. & Yoda, K. (1957). Structure of artificial grassland communities and succession. *Journal of Japanese Grassland Science*, 3, 4-11.*

Rouse, J. W., Haas, R. W., Schell, J. A., Deeering, D. A. & Harlan, J. C. (1974). Monitoring the vernal advancement and retrogradation (Greenwave effect) of natural vegetation. Greenbelt, MD: NASA/GSFCT Type III Final Report.

Sakai, T., Jia, S., Kawamura, K. & Akiyama, T. (2002). Estimation of aboveground biomass and LAI of understory plant (*Sasa senanensis*) using a hand-held spectro-radiometer. *Journal of the Japan Society of Photogrammetry and Remote Sensing, 41*, 27-35.**

Thenkabail, S. P., Hall, J., Lin, T., Ashton, S. M., Harris, D. & Enclona, A. E. (2003). Detecting floristic structure and pattern across topographic and moisture gradients in a mixed species Central African forest using IKONOS and Landsat-7 ETM+ images. *International Journal of Applied Observation and Geoinformation, 4*, 255-270.

Umemura, K., Sudo, K., Ogawa, Y. & Watanabe, N. (2003). Estimation of herbage intake using collar-type bite counter. *Japanese Journal of Grassland Science, 49(Ex)*, 216-217.*

UNEP (1984) Desertification Control Bulletin No.10.

Wang, S. P., Li, Y. H. & Chen, Z. Z. (1999). The optimal stocking rate on grazing system in Inner Mongolia steppe. II. Based on relationship between stocking rate and aboveground net primary productivity. *Acta Agrestia Sinica, 7*, 192-197.***

Xiao, X. M., Ojima, D. S., Ennis, C. A., Schimel, D. S. & Chen, Z. Z. (1997). Land cover classification of the Xilin River Basin, Inner Mongolia, using Landsat TM imagery. *Research on Grassland Ecosystems,* 240-252. Beijing Science Press, Beijing.

XLSFWS (Xilingol League Soil Fertilizer Working Station) (1991). Soils of Xilingol League. People's Publishing of Inner Mongolia, Huhhot, China, 61-87.***

Yiruhan, Shiyomi, M., Takahashi, S., Okubo, T., Akiyama, T., Koyama, N. & Tsuiki, M. (2005). Evaluating the adaptability of herbage species to environmental variation through a long-term grazing experiment. *Grassland Science, 51*, 287-295.

Zhang, Z. T. & Liu, Q. (1992). Rangeland resources of the major livestock regions in China and their development and utilization. Beijing: Science and Technology Press of China, Beijing, 115-116.***

*In Japanese only.
**In Japanese with English summary.
***In Chinese with English summary.

In: Encyclopedia of Environmental Research
Editor: Alisa N. Souter

ISBN: 978-1-61761-927-4
© 2011 Nova Science Publishers, Inc.

Chapter 7

SOUTH BRAZILIAN *CAMPOS* GRASSLANDS: BIODIVERSITY, CONSERVATION AND THE ROLE OF DISTURBANCE

Alessandra Fidelis

Laboratory of Landscape Ecology and Conservation, Department of Ecology,
Universidade de São Paulo, Brazil

ABSTRACT

The South Brazilian *Campos* grasslands (also known as only *Campos*) are unique ecosystems. Located in the southernmost part of Brazil, these ecosystems are rich in plant species, being more diverse than forest ecosystems in the same area. Due to their geographical position (humid subtropics), in a transitional region between tropical and temperate area, these grasslands show mixture of C_3 and C_4 grasses and a high diversity of other botanical families, with typical tropical and temperate species.

Since the climate is subtropical humid, with rainfall regularly distributed all over the year, forest physiognomies would be expected to cover the area. However, diverse grasslands occur in vast areas, in contact (or not) with forest, raising one of the most interesting questions: are these grasslands natural? Paleopalinological studies already showed that these grasslands were present in the region before forest. Until the Holocene, *Campos* grasslands dominated. Afterwards, climate changed (hotter and more humid), enabling forest expansion. Therefore, *Campos* grasslands are natural ecosystems. However, even with no edaphic restrictions and a climate more propitious to forest expansion, forests do not dominate nowadays, raising another important question: why are *Campos* grasslands still present?

Disturbance is probably the most important factor maintaining both grassland physiognomy and diversity. Both grazing and fire have important effects on grassland dynamics. South America does not have large native herbivores and cattle are the most important grazers in these grasslands. Jesuits introduced them in the XVII century and cattle raising is one of the most important economic activities in Southern Brazil. Before that, fire might have influenced vegetation dynamics and delayed forest expansion. Paleopalinological studies showed the presence of charcoal since the Holocene. Nowadays, fire still plays a great role on grassland dynamics, but is a very polemic issue.

In this chapter, I intend to give the reader an overview about this unique ecosystem, almost unknown by the majority of international scientific community and ignored for its ecological relevance even by most of Brazilian scientific community. I will also show the role of disturbance in maintaining *Campos* biodiversity and dynamics as well as the importance of its conservation.

THE SOUTH BRAZILIAN *CAMPOS* GRASSLANDS: VEGETATION CHARACTERIZATION AND HISTORY

Campos Grasslands: An Overview

Grasslands comprise one-fourth of total vegetation of Earth's plant cover (Kucera 1981), covering African and South American savannas, steppes and prairies from North America, South America and Eurasia, and other partly anthropogenic meadows and pastures (Jacobs et al. 1999; Kucera 1981). Floristic diversity of grasslands varies broadly (Sala et al. 2001). Some grassland physiognomies are within ecosystems considered to be hotspots of biodiversity, such as the Cerrado vegetation (Myers et al. 2000).

A large spectrum of grassland physiognomies can be found in South America: from Pampas, situated mostly in Argentina and with high plant diversity (between 32° and 38°S, Bredenkamp et al. 2002) until the Andean alpine formation like the Llanos, with a poorer flora (Barthlott et al. 1996). The tropical savannas, as described by Sarmiento ("tropical vegetation where certain forms of grasses dominate and where seasonal droughts and frequent fires are normal ecological factors", 1984), cover large areas of South America, from Venezuela until the northern part of South Brazil.

Grasslands physiognomies are represented in all Brazilian biomes in more or less extent (IBGE 2004). The Brazilian *Campos* grasslands cover 13.7 million ha (ca. 23% of total area in southern Brazil, Overbeck et al. 2007). *Campos* grasslands are found throughout three Brazilian states: Paraná (1.4 million ha), Santa Catarina (1.8 million ha) and Rio Grande do Sul (10.5 million) (Overbeck et al. 2007). Since the greatest portion of *Campos* grasslands are located in this last state, as well as most of the studies, this chapter will focus on researches conducted in Rio Grande do Sul.

In the northern part of Rio Grande do Sul, grasslands are found until up to 1000 m in association with Araucaria forest (Boldrini 1997; Pillar and Quadros 1997), belonging thus to the Mata Atlântica biome (Figure 1A). Vast areas in central and southwestern part of the State are dominated by grasslands rich in species, where cattle grazing are the principal economic activity and management (Boldrini 1997; Nabinger et al. 2000). These areas belong to the Pampa biome (Figure B) and its physiognomy and plant diversity are similar to the grasslands found in northern Argentina and Uruguay.

Figure 1. A – *Campos* at the northern part of Rio Grande do Sul (São José dos Ausentes, Bioma Mata Atlântica), showing the typical forest-grassland mosaic in the region. B – *Campos* at the southern part of Rio Grande do Sul (Biome Pampa) with the presence of *Aristida jubata* (Picture: Ilsi Boldrini)

Biodiversity of Campos Grassland

Campos grasslands are very rich in plant species. Klein (1975, 1984) estimated ca. 4000 species, whilst Boldrini (1997) estimated 3000 species. In comparison to the Cerrado biome, *Campos* grasslands are poorer in species, but the area of both ecosystems should be considered (Cerrado comprises 2 million km2, with a flora of ca. 6000 plant species, Mendonça et al. 1998). However, exact numbers for the entire *Campos* grasslands can not be given, since most of studies concentrate in Rio Grande do Sul state. Only for this state, Boldrini et al. (2009) estimated 2200 plant species. Recent studies for the northern part of these grasslands (belonging to the Mata Atlântica biome) identified more than 1000 plant species, being 8.5% endemic species and 5% already endangered species (Boldrini et al. 2009).

Using only published studies from *Campos* grasslands, Cerrado (only grassland physiognomy, Mendonça et al. 1998), Argentinean Pampa (Perelman et al. 2001; Ragonese 1967) and Uruguayan grasslands (Altesor et al. 1998; Altesor et al. 2005), I compiled available plant species lists from these ecosystems, in order to compare taxa, genera and botanical families. Unfortunately, few are the published studies with species lists and thus, comparisons are difficult to be performed. Therefore, these are only preliminary results and further compilations should be carried out, in order to complete them. In addition, these results should be analyzed carefully, since Cerrado species lists are the most complete in comparison to Pampa and Uruguayan grasslands. In a total of 1750 species recorded, *Campos* grasslands have 19% of species common to the Cerrado, and 14% and 7% common to Pampa and Uruguayan grasslands respectively. More botanical families are common to *Campos* grasslands and Cerrado (ca. 76%), as well as genera (ca. 47%). However, 31% of all families are also found in the Argentinean Pampa and 22% in the Uruguayan grasslands.

Though preliminary, there results are very interesting, because they show that *Campos* grasslands are a transitional area between tropical (Cerrado) and subtropical (Argentinean Pampa and Uruguay) ecosystems and therefore, they have both tropical and temperate species. Since most of the studies from *Campos* grasslands were carried out near Porto Alegre (in the border between Mata Atlântica and Pampa biome), fewer species are common to Argentinean Pampa and Uruguayan grasslands, since these ecosystems are an extension of the Brazilian Pampa and are entirely located in the subtropics. If the only flora list used in the southern part of Rio Grande do Sul is considered (Girardi-Deiro et al. 1992), more species are common to the Argentinean Pampa (29%) and Uruguayan grasslands (23.75%), than to the Cerrado (20.25%). The characteristic species of grasslands in the Pampa biome according to Overbeck et al. 2007 and their presence in the Argentinean Pampa, Uruguayan grasslands and Cerrado are shown on Table 1.

This important characteristic as a transitional zone is also reflected in the grasses. There is a mixture of C_3 and C_4 grasses found in these grasslands: ca. 70% of grass species are C_4 and 30% are C_3. The principal genera are *Paspalum, Panicum, Aristida, Axonopus* and *Andropogon* (all C_4) and *Briza, Piptochaetium* and *Stipa* (all C_3). The coexistence of C_3 and C_4 grasses in the same ecosystem turns these grasslands unique and worthy to be conserved.

Although a complete analysis of *Campos* flora is not yet possible, due to the lack of studies with species lists for the entire biome, some estimations can be made about the most species–rich botanical families, such as Asteraceae (ca. 600 species), Poaceae (400 – 500), Leguminosae (ca. 250 species), and Cyperaceae (ca. 200 species) (Araújo 2003; Boldrini 1997, 2002; Longhi-Wagner 2003; Matzenbacher 2003; Miotto and Waechter 2003).

Campos grasslands can also show a high fine-scale diversity at plot level, as demonstrated by Overbeck et al. (2005) for grasslands near Porto Alegre (Rio Grande do Sul). They found an average of 34 species on a plot (0.75 m^2) and a total of 450 species in a 220ha area of *Campos* grassland (Overbeck et al. 2006). In a grazed area in Eldorado do Sul (Rio Grande do Sul), I found in average 10 species in very small plots (0.04 m^2), whilst in larger plots (0.25 m^2), 42 plant species could be found. Such results show that *Campos* grasslands are very rich in plant species, even in fine-scale, at plot level.

Past Overview of Campos Grasslands: Vegetation History

Lindman (1906) was one of the first naturalists to describe the *Campos* grasslands vegetation. At that time, he already noticed the existence of grasslands in a climate propitious to forest. Several authors discussed the fact that *Campos* are relict from a drier period (Klein 1960, 1984; Rambo 1942, 1953, 1956). Rambo (1953, 1956) concluded that *Campos* grasslands were older than forests and, furthermore, woody species and the Araucaria forest were still invading grassland areas, changing the grass-dominated landscape to a shrubland. However, some authors, like Pawels (1954) proposed that climate was a determinant factor on formation of plant communities and the expansion of forests over *Campos* grasslands was not happening, because the climate was still not humid enough.

Table 1. Characteristic species and families in Brazilian *Campos* grasslands (region belonging to the Pampa biome) and their presence in the Argentinean Pampa, Uruguayan grassland and Cerrado. Asterisk means that species is endemic to *Campos* grasslands.

Taxa	Families	Pampa	Uruguay	Cerrado
Eryngium horridum Malme	Apiaceae		x	x
Eryngium sanguisorba Cham. & Schltdl.	Apiaceae			x
Aspilia montevidensis (Spreng.) Kuntze	Asteraceae		x	
Aster squamatus (Spreng.) Hieron	Asteraceae	x	x	x
Baccharis coridifolia Spreng.	Asteraceae	x	x	x
Baccharis dracunculifolia DC.	Asteraceae		x	x
Baccharis trimera (Less.)DC.	Asteraceae		x	x
Eupatorium buniifolium Hook. et Arn.	Asteraceae		x	
Gamochaeta spicata (Lam.) Cabrera	Asteraceae	x	x	
Senecio brasiliensis (Spreng.) Less.	Asteraceae	x		x
Vernonia flexuosa Sims.	Asteraceae	x	x	
Cyperus luzulae (L.) Retz	Cyperaceae			x
Hypoxis decumbens L.	Hypoxidaceae		x	x
Juncus capillaceus Lam.	Juncaceae	x	x	
Adesmia bicolor (Poir.) DC. *	Leguminosae	x	x	
Desmodium incanum DC.	Leguminosae			x
Macroptilium prostratum (Benth.) Urb.	Leguminosae			x
Rhynchosia diversifolia M. Micheli	Leguminosae	x		
Trifolium polymorphum Poir *	Leguminosae	x	x	
Andropogon lateralis Nees	Poaceae	x	x	x
Andropogon ternatus (Spreng.) Nees	Poaceae		x	
Aristida jubata Arechav.	Poaceae			x
Aristida spegazzinii Arech.	Poaceae	x		
Axonopus affinis Chase	Poaceae	x	x	
Bothriochloa laguroides (DC.)Herter	Poaceae	x	x	
Bouteloua megapotamica (Spreng.) O. Kuntze *	Poaceae	x		
Briza subaristata Lam.	Poaceae	x	x	
Coelorachis selloana (Hack.) Camus	Poaceae		x	
Melica rigida Cav.*	Poaceae		x	
Dichanthelium sabulorum (Lam.) Gould & C.A. Clark	Poaceae	x	x	
Paspalum dilatatum Poir.	Poaceae	x	x	x
Paspalum nicorae Parodi	Poaceae	x		
Paspalum notatum Fl.	Poaceae	x	x	x
Paspalum pumilum Nees	Poaceae	x		
Piptochaetium lasianthum Griseb.	Poaceae		x	
Piptochaetium stipoides (Trin. & Rupr.)Hackel	Poaceae	x	x	
Stipa megapotamia Sprengel ex Trin.	Poaceae		x	
Stipa philippii Steud.*	Poaceae	x	x	
Stipa setigera C.Presl.	Poaceae		x	
Borreria verticillata (L.) G.F.W.Mey	Rubiaceae			x
Richardia humistrata (Cham. et Schlecht.) Steud.	Rubiaceae		x	

Several palynological studies confirmed the hypothesis from Rambo about the grasslands. During the Late Glacial, grasslands were dominant in the southern part of South America, which is an evidence for colder and drier climates (Behling et al. 2002). Grasslands prevailed throughout the Late Glacial and Holocene (Behling 1997; Behling and Pillar 2007). The northern part of the State of Rio Grande do Sul, which is in contact with the Araucaria forest, was probably treeless (Behling and Pillar 2007), due to the cold climate with strong frosts (Behling 2002) and fire was rare at that time (Behling 2002; Behling et al. 2004). The same tendency of climate could be found in palynological studies in the central Argentinean Pampa, where dry grasslands dominated at 10,500 BP (Prieto 1996). Until the early and mid-Holocene, there were vast areas with grasslands, although floristic composition had changed due to warm and dry climate (Behling 1997; Ledru et al. 1998). Araucaria forest expansion began after 1000 BP. (Behling 2002; Behling et al. 2004), reflecting a more humid climate. Charcoal records could also be found in these studies, confirming thus, the occurrence of fire at that time. Fire began to be more frequent at ca. 7400 cal yr BP, changing floristic composition (Behling and Pillar 2007; Behling et al. 2004).

More recently, studies in the southern Brazilian highlands (Mata Atlântica biome) also confirmed the existence of grasslands before the expansion of Araucaria forest. Dümig et al (2008) used stable carbon isotope ratios ($\partial^{13}C$) and ^{14}C activity to trace the presence of forest or grassland in grassland areas, Araucaria forest and forest patches (capões in Portuguese) within the grasslands. Their results showed clearly that all grassland areas were always grasslands and not the result of deforestation. Araucaria forest and forest patches had C_4-derived soil organic carbon (SOC) in the subsoil, a good indicator for former grasslands in the area. They conclude, that in the area, Araucaria forest expansion began after 1500-1300 yr BP, which more or less agrees with the palynological data.

The scientific evidence of the presence of grasslands before forest is more important than it seems, since most of people, even some environmental authorities, think that these grasslands are the result of deforestation. Brazil is widely known for its forests. Although grasslands are also species-rich ecosystems, they are usually more neglected than forests.

The Role of Disturbance on the Maintenance of *Campos* Grasslands

Disturbance is any event in time that partially or totally destroys plant biomass (Grime 1979) and thus, opens up free spaces (gaps) for further colonization by individuals of the same or different species (Lavorel et al. 1997).

Fire and grazing are examples of common types of disturbance affecting grasslands all over the world. They can be both natural (e.g. fire caused by lightning; rabbit grazing) and anthropogenic. In both cases, they will affect not only plant populations, but also vegetation community as a whole, and their responses will depend on disturbance intensity and timing.

Although both fire and grazing remove biomass and use complex organic molecules and convert them to other products (organic and mineral, e.g. ashes and dung, Bond and Keeley 2005), fire differs from grazing, since it did not "choose" the biomass that will be removed by being more palatable or having more proteins. Fire needs fuel to spread and usually, grasslands have great quantities of fuel (e.g. dead biomass, grasses, etc). Therefore, several grassland ecosystems are also considered to be dependent on fire ("flammable ecosystems"), such as the North American Prairies and Australian savannas. On the other hand, grasslands can be excellent pastures, being the source of protein for different animals, e.g. cattle, rabbits.

Campos grasslands are under the effect of both fire and grazing. Cattle raising is a very important economical activity for the region, whilst fire is set usually by local farmers to stimulate new resprouts for cattle during winter and early spring, and by local people in order to "clean" the area.

Brazilian Campos Grasslands and Grazing: A Sustainable System?

Cattle grazing is the major management activity in temperate grasslands, with economic and conservation objectives (Bullock et al. 2001). In Southern Brazil, pastures occupy more than 18 million ha (IBGE 2006). Cattle was introduced in Southern Brazil in the XVII Century by Jesuits and nowadays, it is one of the most important economical activities in the region, with a beef cattle population of 26 million heads (Nabinger et al. 2000).

In *Campos* grasslands, the already mentioned coexistence of C_3 and C_4 grasses is an advantage for cattle grazing, since there is a balance of forage production through the different seasons. However, this balance is affected by farmer's management, since they burn their pastures, usually in winter and early spring due to the high amount of dead biomass (Heringer and Jacques 2002). This practice is known to be detrimental for C_3 grasses (Llorens and Frank 2004) and therefore, the use of fire is very criticized.

Dense areas of perennial grasses and forbs constitute grazed grasslands in Uruguay (Teixeira and Altesor 2009), which are similar to grazed areas in *Campos* grasslands. Grazed grasslands in southern Brazil are composed by a mosaic of grazed and ungrazed patches, in areas where grazing intensity is not very high. Grazed patches are dominated by stoloniferous and rhizomatous grasses, such as *Paspalum notatum*, whilst ungrazed patches have several unpalatable species. However, such patches can serve as grazing refuges for other palatable species, as showed by Fidelis et al. (2009b). *Eryngium horridum*, a spiny rosette species, which is usually not browsed by cattle, grows within these ungrazed patches. Within its leaves, several small forbs and other palatable grasses grow and even produce flowers. Some species cannot be found outside these ungrazed patches, where this rosette species is present. Therefore, they concluded that, although *E. horridum* is considered an obnoxious species by several farmers, they have an important ecological role in maintaining biodiversity in *Campos* grasslands, serving as diaspore pools of different grassland species, which otherwise are browsed by cattle.

Grazing intensity is an important factor influencing grassland dynamics in southern Brazil. Blanco et al. (2007) found that in intensively grazed areas in southern Brazil, there was an absence of woody species and a preference by grazers correlated to low proportions of senescent leaves, that means, an optimization for better forage quality. Sosinski & Pillar (2004) observed that in areas with low grazing intensity, plants with high leaf resistance were dominant, whilst in heavily grazed areas, the opposite situation was found: the dominance of species with low leaf resistance.

Intensively grazed areas can lead to soil erosion in areas with vulnerable soil conditions (Overbeck et al. 2007). Overgrazing in the southwestern part of Rio Grande do Sul led to processes of erosion and even desertification (Trindade et al. 2008), bringing severe economical problems to the region. In addition, vegetation is replaced by less productive species, with low forage quality (Nabinger et al. 2000). On the other hand, low intensities of grazing can lead to a shrub encroachment, decreasing the quality of pastures. Furthermore, there is an increase of tall grasses, with low nutritional values, and obnoxious species, such as *Baccharis* (Asteraceae, shrub species, Nabinger et al. 2000).

Though their high quality and sustainable capacity for cattle grazing, natural pastures have been replaced in the last decades by cultivated pastures (see below for exact numbers). The main species (usually from Africa and Europe) used are *Panicum maximum, Hiparrhenia rufa,* several species of *Brachiaria (B. brizantha, B. decumbens, B. humidicola, B.*

dictioneura), *Cynodon* sp., *Lolium multiflorum, Avena strigosa,* among other exotic grasses (Nabinger et al. 2000). Some legumes are also used, such as *Lotus corniculatus, Trifolium repens* and *Medicago sativa* (Nabinger et al. 2000). More recent, the introduction of *Eragrostis plana*, known as "Capim-Annoni" has become a huge problem in southern Brazilian grasslands. This African grass is a plague, which spread all over Rio Grande do Sul, decreasing pastures quality. It is estimated that *E. plana* has already invaded 10% of the total area of *Campos* grasslands (Medeiros et al. 2004), being thus one of the greatest problems for the conservation and sustainable management of these grasslands.

Therefore, the correct management of *Campos* grasslands is of crucial importance in order to prevent the invasion of exotic species, maintain plant diversity and the best conditions for cattle grazing, with a great quantity of forage species. A sustainable use of these grasslands is viable, since natural grasslands in southern Brazil can maintain more than 500 g of daily weight gain per animal and furthermore, during growing season, pastures can attain 150 to 180 kg LW/ha (Nabinger et al. 2000).

Fire in Brazilian Campos Grasslands: The Good or the Bad Guy?

Fire is an important event in many ecosystems. It shapes vegetation, maintains physiognomy and structure, as well as diversity (Bond and Keeley 2005). However, its effects are mostly dependent on which temperature fire reaches, the quantity of available fuel and the nature of biomass (Bond and van Wilgen 1996; Whelan 1995).

Fire is an important factor in tropical savannas since thousands of years (Ramos-Neto and Pivello 2000). Brazilian Cerrados can be defined as pyrophitic or fire-adapted and even fire-depending ecosystems (Furley 1999), because fire stimulates nutrient recycling, resprouting, fruiting, and seedling of several plant species (Coutinho 1982; Ramos-Neto and Pivello 2000). The absence of fire leads to an increase in woody species (Moreira 2000), as also observed in other "flammable ecosystems" in the world (Bond et al. 2005). Though the importance of fire in this ecosystem, management practices do not include fire as a relevant tool for vegetation management. In the opposite, fire is considered to be detrimental by most of people and researches involving its use are difficult to be approved by environmental agencies.

Brazilian *Campos* grasslands are influenced by fire. Most fires are anthropogenic and set carelessly in most areas. Therefore, it is also a polemic issue in these grasslands, being only its negative effects on vegetation pointed out. Charcoal records confirmed the presence of fire in *Campos* grasslands: they became more frequent at ca. 7400 cal yr BP, changing floristic composition (Behling and Pillar 2007; Behling et al. 2004). Until nowadays, some areas are regularly burned by local farmers, in order to stimulate resprouting when there is a high accumulation of dead biomass, mostly in winter periods (Heringer and Jacques 2002).

Unfortunately, few are the studies in *Campos* grasslands about the effects of fire on vegetation dynamics. Eggers and Porto (1994) studied the effects of fire on native grasslands structure and diversity. A species with a high capacity of regeneration after fires is *Eryngium horridum* (Apiaceae). Population biology studies showed the positive effect of fire on increasing and maintaining populations of this rosette species (Fidelis et al. 2008). After aboveground biomass removal, new rosettes could be observed after few weeks, and sometimes, up to five new rosettes could be counted (Fidelis et al. 2008).

Exclusion of disturbance leads to shrub encroachment, as already mentioned for other ecosystems under fire influence. Oliveira & Pillar (2004) observed the expansion of forest over grassland in areas excluded from both fire and grazing since more than 20 years. The same pattern was found by Müller (2005). After disturbance exclusion, there is an increase in tall grasses, which shade small forbs. There is thus, a decrease in plant diversity, as observed by Overbeck et al. (2005). Shrub cover increases and after sometime, shrubs shade grasses. There is a great accumulation of dead biomass, which is highly flammable. If fire does not occur and forest areas are close, the establishment of forest species (mostly trees) can be found, beginning from the borders in direction to the grassland. Another way of forest expansion over excluded grassland matrix is by the facilitation of isolated established trees in the grassland. *Araucaria angustifolia* is a good example of nurse tree that facilitates the establishment of other forest trees below its canopy, initiating thus, the process of nucleation of forest patches within the grassland (Duarte et al. 2006). As already mentioned, soil analyses from samples inside these patches confirmed that they were formerly grasslands (see Dümig et al. 2008).

Table 2 shows the differences in biomass from one month after fire until 15 years of exclusion. As one can observe, dead biomass increases a lot after 15 years of exclusion. Graminoid biomass is still high, due to the dominance of tall tussock grasses, such as *Andropogon lateralis*. On the other hand, forb biomass decreases drastically already after five years of fire exclusion. The same results can be found when one analyzes plant diversity and vegetation structure: there is a decrease in forb diversity and cover (Fidelis 2008; Overbeck et al. 2005).

Therefore, forbs seem to be the most affected functional group by the lack of disturbance in these grasslands. In grasslands excluded since 12 years, a rosette plant, *Eryngium horridum* served as a facilitator for forb species, since forbs could establish and grow within its leaves, which opened up space within the dense grass matrix, providing thus, more light to the lower strata (Fidelis et al. 2009b).

Table 2. Dead biomass, graminoid and forb biomass (kg/ha) measured in natural
***Campos* grasslands in different times of exclusion of fire: one month,**
three months, one year, five years (data from Fidelis et al. 2006)
and 15 years (data from Skiba 2009).

	dead biomass	graminoids	forbs
1 month	819.29	376.96	346.56
3 months	374	603.00	598.50
1 year	1121.35	1426.11	181.67
5 years	-	1490.85	51.29
15 years	8056.77	1673.16	40.29

One month after fire, several plant species flowered and produced seeds (Fidelis & Blanco, in preparation). Even C_3 grasses, which have already flowered before fire (grasslands burned during summer) showed inflorescences after biomass removal, showing their rapid capacity of biomass recover and allocation of nutrients to flower production (Fidelis & Blanco, in preparation). Many species from *Campos* show developed belowground systems from different types: rhizomes, rhizophores and tuberous roots (typical storage organs), as well as xylopodia (no storage organ, Fidelis et al. 2009a). Xylopodia is many times confounded with lignotubers. Differently from lignotubers, which are typical storage organs originated from cotyledons and with suppressed buds or epicormic strands on the stem (Burrows 2002; James 1984), xylopodium has no storage parenchyma tissue, only normal xylem parenchyma (Appezzato-da-Glória and Estelita 2000). Nevertheless, these structures are usually combined with tuberous roots, which possess storage parenchyma tissues (Appezzato-da-Glória and Estelita 2000; Milanez and Moraes-Dallaqua 2003), providing thus, the reserve nutrients necessary for plant to resprout after fire or drought events. However, a very important function of xylopodium is its great gemmiferous potential (Appezzato-da-Glória et al. 2008), since several buds that can form new stems after organ injuries caused by fire, for example, cover this structure. Xylopodia are common structures found in plants from Cerrado (Rizzini 1965) and South American tropical savannas (Sarmiento 1983). These ecosystems are characterized by the presence of a dry period and thus, plants with developed belowground systems would be favoured in such events. As already mentioned, *Campos* do not have a markedly dry season and precipitation is well distributed all over the year. Would the presence of such structures be a relict from drier periods? If so, why do so many species still have developed belowground systems? I hypothesize that, the presence of belowground systems with the function to storage nutrients and/or great gemmiferous potential would favour plant species in disturbed environments, since there would be a rapid allocation of nutrients for aerial biomass recover after disturbance caused by both fire and grazing.

In conclusion, even considered by most people and even researches as the "bad guy", fire should be taken into account when one discusses about grassland dynamics and management. The fact that fire is seen as detrimental to vegetation is due to the lack of long-term studies about its effect on vegetation dynamics and diversity and also because its erroneous use by local farmers. There are no sufficient studies to support the unconditional use of fire as a tool on vegetation management; therefore, studies about the effects of fire on vegetation should be supported.

Conservation of Campos Grasslands

Besides its importance as relict grassland and also due to its biodiversity, *Campos* grasslands conservation has been neglected. Due to anthropogenic pressure, areas of natural grasslands have decreased in the last decades. In the Cerrado biome for example, grasslands and shrublands still have ca. 24% of the total remaining area (MMA 2007b). Nowadays, the Pampa biome still has 23% of its original grassland cover. Unfortunately, less than 1% of *Campos* area is under legal protection (Overbeck et al. 2007), although other areas have already been signed as priority for conservation. However, 48% of its total area has already been changed (MMA 2007a), mostly into soybeans and maize plantation. More recently, exotic trees plantations (mostly *Pinus* and *Eucalyptus)* are increasing in area, threatening thus, native grasslands.

Table 3. Total area (1000 ha) used for crops and pastures in southern Brazil from 1970 to 2006 (data from IBGE 2006).

Land use	1970	1975	1980	1985	1995	2006
crops	11028.453	12991.459	14571.446	14523.479	12306.292	18313.631
pastures	21612.679	21159.758	21313.458	21432.343	20696.549	18145.573

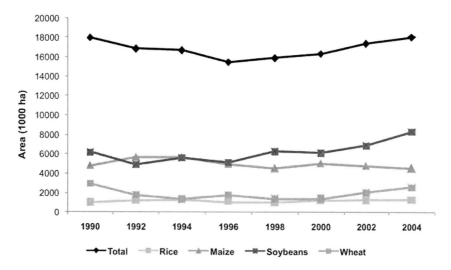

Figure 2. Area used for the most important crops in southern Brazil: rice, maize, soybeans and wheat, as well as the total crop area from 1990 to 2004 (data from IBGE 2004).

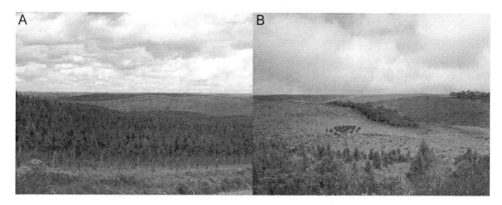

Figure 3.A – Pinus plantation in the region of Cambará do Sul, Rio Grande do Sul. B – Young pinus planted in vast areas over natural grasslands in Rio Grande do Sul. Notice that forest areas are maintained and plantation is exclusively over grassland areas.

Since 1970, the total crop area has increased in 66%, whilst the pasture areas decreased 16% (Table 4). Natural pastures decreased 25% from 1970 to 1996 and cultivated pastures increased in more than 100% in the same period (IBGE 2004, 2006). Soybeans and rice were the crops that increased the most in area in southern Brazil, mostly over native areas of grasslands (Figure 2): 35% and 27% respectively from 1990 to 2004 (IBGE 2004). Other typical cultivations for the region decreased in the same period, as maize (-3.5%) and wheat (-12%).

Besides the great increase in cultivated area of soybeans, which is a general tendency in Brazil, another type of cultivation has raised some concern in grassland ecologists and conservation biologists: the expansion of cultivated forests of *Pinus* and *Eucalyptus*. From 1990 to 2003, the production of wood in southern Brazil and Rio Grande do Sul increased 84% and 150% respectively (IBGE 2004). In 1996, the cultivated area in southern Brazil already covered 2 million ha and there are projects, mostly in Rio Grande do Sul, to increase this area, mostly over native grasslands areas. Local government and some farmers support the introduction of *Eucalyptus* trees in the southermost part of Rio Grande do Sul, and *Pinus* in the highlands (Figure 3A and B), in the northern part, since a great return on investment for the area is expected. Besides the economical point of view, part of the population and small farmers concern about this drastic change in the traditional land use. "Gaúchos", as people from this region are known, are deeply attached to their culture, which is based on cattle grazing. Change of landscape (from grassland to forest), as well as change in land use (cattle to silviculture) can account to not only ecological, but also to cultural consequences.

Protected areas in Brazil usually do not manage the vegetation. Inside these areas, people are forbidden to live, cattle is excluded and ecological processes run free without intervention. There are two main problems about this approach: 1) if one intends to conserve *Campos* grasslands, management of vegetation should be implemented. The total exclusion of disturbance will lead to loss of grassland species and changes in vegetation physiognomy. 2) accumulation of dead biomass is high and after some years of exclusion, catastrophical wild fires reaches the area, killing several animals and threatening vegetation regeneration, since fire intensities will be much higher than in areas with less accumulated dead biomass. If Protected Areas manage vegetation correctly, such events could be avoided. Mowing might also be an efficient tool instead of fire, although its costs are much higher and implementation is more complicated. Probably the best solution would be the creation of a mosaic of different management practices, as for example with grazed, mowed and burned patches, at different times. But as already mentioned above, it is too soon to give any advice for conservation biologists and environmental agencies about management practices in *Campos* grasslands, since more long-term studies are needed.

Campos are not a priority for both local government and environmental authorities. Several people consider grassland a result of deforestation and thus, grassland areas should be reforested. Brazil is a "dendromaniac" country, where forests are more valuable ecosystems than others such as Caatinga and the grasslands. As an example, the Brazilian law to protect native vegetation is called "Código Florestal" (Forest Code). Therefore, efforts should be directed to environmental education of local people in order to valorise the ecosystem where they are living in. Government should support initiatives aiming the conservation of *Campos*, as for example, cattle grazing in southern Brazil instead of encouraging changes in land use, planting vast areas of native grasslands with exotic tress.

ACKNOWLEDGMENTS

I would like to thank Jörg Pfadenhauer for reviewing this manuscript and supporting the initial idea of writing this chapter. I am grateful thankful to Eduardo Vélez, Gerhard Overbeck and Carolina Blanco for their valuable suggestions.

REFERENCES

Altesor, A., DiLandro, E., May, H. & Ezcurra, E. (1998). Long-term species change in a Uruguayan grassland. *Journal of Vegetation Science*, *9*, 173-180.

Altesor, A., Oesterheld, M., Leoni, E., Lezama, F. & Rodríguez, C. (2005). Effect of grazing on community structure and productivity of a Uruguayan grassland. *Plant Ecology, 179*, 83-91.

Appezzato-da-Glória, B. & Estelita, M. F. M. (2000). The development anatomy of the subterranean system in *Mandevilla ilustris* (Vell.) Woodson and *M. velutina* (Mart. ex Stadelm.) Woodson (Apocynaceae). *Revista Brasileira de Botânica, 23*, 27-35.

Appezzato-da-Glória, B., Hayashi, A. H., Cury, G., Soares, M. K. M. & Rocha, R. (2008). Underground systems of Asteraceae species from the Cerrado. *The Journal of the Torrey Botanical* Society, *135*, 103-113.

Araújo, A. C. (2003). Cyperaceae nos campos sul-brasileiros. 54o. Congresso Nacional de Botânica; Belém, Pará. 127-130.

Barthlott, W., Lauer, W. & Placke, A. (1996). Global distribution of species diversity in vascular plants: towards a world map of phytodiversity. *Erdkunde, 50*, 317-328.

Behling, H. (1997). Late Quartenary vegetation, climate and fire history of the Araucaria forest and campos region from Serra Campos Gerais, Paraná State (South Brazil). *Review of Palaeobotany and Palynology, 97*, 109-121.

Behling, H. (2002). South and southeast Brazilian grasslands during Late Quaternary times: a synthesis. *Palaeogegraphy, Palaeoclimatology, Palaeoecology, 177*, 19-27.

Behling, H., Arz, H. W., Pätzold, J. & Wefer, G. (2002). Late Quaternary vegetational and climate dynamics in southeastern Brazil, inferences from marine cores GeoB 3229-2 and GeoB 3202-1. *Palaeogegraphy, Palaeoclimatology, Palaeoecology, 179*, 227-243.

Behling, H. & Pillar, V. D. (2007). Late quaternary vegetation, biodiversity and fire dynamics on the southern Brazilian highland and their implication for conservation and management of modern Araucaria forest and grassland ecosystems. *Philosophical Transactions of the Royal Society B, 362*, 243-251.

Behling, H., Pillar, V. D., Orlóci, L. & Bauermann, S. G. (2004). Late Quaternary *Araucaria* forest, grassland (campos), fire and climate dynamics, studied by high-resolution pollen, charcoal and multivariate analysis of the Cambará do Sul core in southern Brazil. *Palaeogegraphy, Palaeoclimatology, Palaeoecology, 203*, 277-297.

Blanco, C. C., Sosinski, E. E. Jr., dos Santos, B. R. C., da Silva, M. A. & Pillar, V. D. (2007). On the overlap between effect and response plant functional types linked to grazing. *Community Ecology, 8*, 57-65.

Boldrini, I. I. (1997). Campos do Rio Grande do Sul: caracterização fisionômica e problemática ocupacional. *Boletim do Instituto de Biociências da Universidade Federal do Rio Grande do Sul, 56*, 1-39.

Boldrini, I. I. (2002). Campos sulinos: caracterização e biodiversidade. In: E. L. Araújo, A. N. Noura, E. V. S. B. Sampaio, & J. M. T. Carneiro (Eds.), Biodiversidade, conservação e uso sustentável da Flora do Brasil. Recife: Sociedade Botânica do Brasil, Universidade Federal Rural do Pernambuco, 95-97.

Boldrini, I. I., Egger, L., Mentz, L., Miotto, S. T. S., Matzenbacher, N., Longhi-Wagner, H. M., Trevisan, R., Schneider, A. A. & Setubal, R. B. (2009). *Flora*. In: I. I. Boldrini (ED.). Biodiversidade dos campos do planalto de araucárias. Brasília: MMA. 38-94.

Bond, W. J. & Keeley, J. E. (2005). Fire as a global "herbivore": the ecology and evolution of flammable ecosystems. *Trends in Ecology and Evolution, 20*, 387-394.

Bond, W. J. & van Wilgen, B. W. (1996). *Fire and plants*. London: Chapman ˜Hall. 263.

Bond, W. J., Woodward, F. I. & Midgley, G. F. (2005). The global distribution of ecosystems in a world without fire. *New Phytologist*, *165*, 525-538.

Bredenkamp, G. J., Spada, F. & Kazmierczak, E. (2002). On the origin of northern and southern hemisphere grasslands. *Plant Ecology*, *163*, 209-229.

Bullock, J. M., Franklin, J., Stevenson, M. J., Silvertown, J., Coulson, S. J., Gregory, S. J. & Tofts, R. (2001). A plant trait analysis of responses to grazing in a long-term experiment. *Journal of Applied Ecology*, *38*, 253-267.

Burrows, G. E. (2002). Epicormic strand structure in *Angophora, Eucalyptus* and *Lophostemon* (Myrtaceae) - implication for fire resistance and recovery. *New Phytologist*, *153*, 111-131.

Coutinho, L. M. (1982). Ecological effects of fire in Brazilian Cerrado. In: B. J. Huntley, & B. H. Walker (Eds.), *Ecology of tropical savannas*. Berlin: Springer Verlag, 273-291.

Duarte, L. d. S., Dos-Santos, M. M. G., Hartz, S. M. & Pillar, V. D. (2006). Role of nurse plants in Araucaria Forest expansion over grassland in south Brazil. *Austral Ecology*, *31*, 520-528.

Dümig, A., Schad, P., Rumpel, C., Dignac, M. F. & Kögel-Knabner, I. (2008). *Araucaria* forest expansion on grassland in the southern Brazilian highlands as revealed by ^{14}C and $\partial^{13}C$ studies. *Geoderma*, *145*, 143-157.

Eggers, L. & Porto, M. L. (1994). Ação do fogo em uma comunidade campestre secundária, analisada em bases fitossociológicas. *Boletim do Instituto de Biociências*, *53*, 1-88.

Fidelis, A. (2008). Fire in subtropical grasslands in Southern Brazil: effects on plant strategies and vegetation dynamics. Freising: Technische Universität München. 151.

Fidelis, A., Appezzato-da-Glória, B. & Pfadenhauer, J. (2009a). A importância da biomassa e das estruturas subterrâneas nos Campos Sulinos. In: V. D. Pillar, S. C. Müller, Z. M. S. Castilhos, & A. V. A. Jacques (Eds.), Campos Sulinos - conservação e uso sustentável da biodiversidade. Brasília: Ministério do Meio Ambiente, 85-97.

Fidelis, A., Müller, S., Pillar, V. D. & Pfadenhauer, J. (2006). Efeito do fogo na biomassa aérea e subterrânea dos Campos Sulinos. In: X. R. G. Campos (Ed.), Desafios e oportunidades do Bioma Campos frente à expansão e intensificação agrícola; Pelotas: EMBRAPA - Clima Temperado.

Fidelis, A., Overbeck, G., Pillar, V. D. & Pfadenhauer, J. (2008). Effects of disturbance on population biology of a rosette species *Eryngium horridum* Malme in grasslands in southern Brazil. *Plant Ecology*, *195*, 55-67.

Fidelis, A., Overbeck, G. E., Pillar, V. D. & Pfadenhauer, J. (2009). The ecological value of *Eryngium horridum* in maintaining biodiversity in subtropical grasslands. *Austral Ecology, 34*, 558-566.

Furley, P. A. (1999). The nature and diversity of neotropical savanna vegetation with particular reference to the Brazilian cerrados. *Global Ecology & Biogeography*, *8*, 223-241.

Girardi-Deiro, A. M., Gonçalves, J. O. N. & Gonzaga, S. S. (1992). Campos naturais ocorrentes nos diferentes tipos de solo no Município de Bagé, RS. 2: fisionomia e composição florística. *Iheringia*, *42*, 55-79.

Grime, J. P. (1979). *Plant Strategies and vegetation processes*. Chichester, UK.: John Wiley and Sons.

Heringer, I. & Jacques, A. V. A. (2002). Acumulação de forragem e material morto em pastagem nativa sob distintas alternativas de manejo em relação às queimadas. *Revista brasileira de Zootecnia, 31*, 599-604.

IBGE, Instituto Brasileiro de Geografia e Estatística. 2004. *Mapa de Biomas.* Available at: http://mapas.ibge.gov.br/website/biomas2/viewer.htm2008. Accessed April 2009.

IBGE, Instituto Brasileiro de Geografia e Estatística. 2006. *Censo Agropecuário.* Available at: http://www.ibge.gov.br/home/estatistica/economia/agropecuaria/censoagro/2006/default. shtm. Accessed June 2009.

Jacobs, B. F., Kingston, J. D. & Jacobs, L. L. (1999). The origin of grass-dominated ecosystems. *Annals of the Missouri Botanical Garden, 86*, 590-643.

James, S. (1984). Lignotubers and burls - their structure, function and ecological significance in Mediterranean ecosystems. *The Botanical Review, 50*, 225-266.

Klein, R. M. (1960). O aspecto dinâmico do pinheiro brasileiro. *Sellowia, 12*, 17-44.

Klein, R. M. (1975). Southern Brazilian phytogeographic features and the probable influence of upper Quaternary climatic changes in the floristic distribution. *Boletim Paranaense de Geociências, 33*, 67-88.

Klein, R. M. (1984). Aspectos dinâmicos da vegetação do sul do Brasil. *Sellowia, 36*, 5-54.

Kucera, C. L. (1981). *Grassland and fire.* General Technical Report WO-26. United States Forest Service, 90-111.

Lavorel, S., McIntyre, S., Landsberg, J. & Forbes, T. D. A. (1997). Plant functional classifications: from general groups to specific groups based on response to disturbance. *Trends in Ecology and Evolution, 12*, 474-478.

Ledru, M. P., Salgado-Labouriau, M. L. & Lorscheitter, M. L. (1998). Vegetation dynamics in southern and central Brazil during the last 10,000 yr. B.P. *Review of Palaeobotany and Palynology, 99*, 131-142.

Lindman, C. A. M. (1906). A vegetação do Rio Grande do Sul. Porto Alegre: Universal, 356.

Llorens, E. M. & Frank, E. O. (2004). El fuego en la provincia de La Pampa. In: C. Kunst, S. Bravo, & J. L. Panigatti (Eds.), Fuego en los ecosistemas argentinos. Santiago del Estero: Instituto Nacional de Tecnología Agropecuaria.

Longhi-Wagner, H. M. (2003). Diversidade florística dos campos sul-brasileiros: poaceae. 54o. Congresso Nacional de Botânica; Belém, Pará, 117-120.

Matzenbacher, N. I. (2003). Diversidade florística dos campos sul-brasileiros: asteraceae. 54o. Congresso Nacional de Botânica; Belém, Pará, 124-127.

Medeiros, R. B., Pillar, V. P. & Reis, J. C. L. (2004). Expansão de *Eragrostis plana* Ness (capim-annoni-2) no Rio Grande do Sul e indicativos de controle. Reunión del grupo técnico regional del Cono Sur en mejoramiento y utilización de los recursos forrajeros del área tropical y subtropical, Grupo Campos; Salto: Memorias, 208-211.

Mendonça, R. C., Felfili, J. M., Walter, B. M. T., Silva Júnior, M. C., Rezende, A. V., Filgueiras, T. S. & Nogueira, P. E. (1998). Flora Vascular do Cerrado. In: S. M. Sano, & S. P. Almeida (Eds.), Cerrado: ambiente e flora. Planaltina: EMBRAPA, 289-556.

Milanez, C. R. D. & Moraes-Dallaqua, M. A. (2003). Ontogênese do sistema subterrâneo de *Pachyrhizus ahipa* (Weed.) Parodi (Fabaceae). *Revista brasileira de Botânica, 26*, 415-427.

Miotto, S. T. S. & Waechter, J. L. (2003). Diversidade Florística dos campos sul-brasileiros: fabaceae. 54o. Congresso Nacional de Botânica; Belém, *Pará*, 121-124.

MMA, Ministério do Meio Ambiente. 2007a. Mapas de Cobertura Vegetal. Brasília, 16.

MMA, Ministério do Meio Ambiente. 2007b. Mapeamento de cobertura vegetal do bioma cerrado. Brasília, 93.

Moreira, A. G. (2000). Effects of fire protection on savanna structure in Central Brazil. *Journal of Biogeography*, *27*, 1021-1029.

Müller, S. C. (2005). Padrões de espécies e tipos funcionais de plantas lenhosas em bordas de floresta e campo sob influência do fogo. *Porto Alegre: Universidade Federal do Rio Grande do Sul.*, 135.

Myers, N., Mittermeier, R. A., Mittermeier, C. G., Fonseca, G. A. B. & Kent, J. (2000). Biodiverity hotspots for conservation priorities. *Nature*, *403*, 853-858.

Nabinger, C., Moraes, A. d. & Maraschin, G. E. (2000). Campos in Southern Brazil. In: G. Lemaire, J. Hodgson, A. d. Moraes, C. Nabinger, & P. C. F. Carvalho (Eds.), *Grassland Ecophysiology and Grazing Ecology: CAB International*, 355-376.

Oliveira, J. M. & Pillar, V. D. (2004). Vegetation dynamics on mosaics of Campos and Araucaria forest between 1974 and 1999 in Southern Brazil. *Community Ecology*, *5*, 197-202.

Overbeck, G. E., Müller, S. C., Fidelis, A., Pfadenhauer, J., Pillar, V. D., Blanco, C., Boldrini, I. I., Both, R. & Forneck, E. D. (2007). Brazil´s neglected biome: the Southern Campos. *Perspectives in Plant Ecology and Systematics*, *9*, 101-116.

Overbeck, G. E., Müller, S. C., Pillar, V. D. & Pfadenhauer, J. (2005). Fine-scale post-fire dynamics in southern Brazilian subtropical grassland. *Journal of Vegetation Science*, *16*, 655-664.

Overbeck, G. E., Müller, S. C., Pillar, V. D. & Pfadenhauer, J. (2006). Floristic composition, environmental variation and species distribution patterns in a burned grassland in southern Brazil. *Brazilian Journal of Biology*.

Pawels, G. (1954). Algumas notas sobre a distribuição do campo e mata no sul do país e a fixidez do limite que os separa. In: C. N. d. Geografia (ED.). Aspectos da geografia riograndense. Rio de Janeiro: Serviço Gráfico do Instituto Brasileiro de Geografia e Estatística, 32-38.

Perelman, S. B., Leon, R. J. C. & Oesterheld, M. (2001). Cross-scale vegetation patterns of Flooding Pampa grasslands. *J Ecology*, *89*, 562-577.

Pillar, V. D. & Quadros, F. L. F. (1997). Grassland-forest boundaries in Southern Brazil. *Coenoses*, *12*, 119-126.

Prieto, A. R. (1996). Late Quaternary Vegetational and Climatic Changes in the Pampa Grassland of Argentina. *Quaternary Research*, *45*, 73-88.

Ragonese, A. E. (1967). Vegetación y ganadería en la República Argentina. Buenos Aires: Coleccion Cientifica del I.N.T.A. 218.

Rambo, B. (1942). A fisionomia do Rio Grande do Sul. São Leopoldo: Editora Unisinos, 473.

Rambo, B. (1953). História da flora do Planalto Riograndense. *Anais Botânicos do Herbário Barbosa Rodrigues*, *5*, 185-232.

Rambo, B. (1956). A flora fanerogâmica dos Aparados riograndenses. *Sellowia*, *7*, 235-298.

Ramos-Neto, M. B. & Pivello, V. R. (2000). Lightning fires in a Brazilian savanna national park: rethinking management strategies. *Environmental Management*, *26*, 675-684.

Rizzini, C. T. (1965). Estudos experimentais sobre o xilopódio e outros órgãos tuberosos de plantas do cerrado. *Anais da Academia brasileira de Ciências*, *37*, 87-113.

Sala, O. E., Austin, A. T. & Vivanco, L. (2001). Temperate grassland and shrubland ecosystems. *Encyclopedia of Biodiversity*, *5*, 627-635.

Sarmiento, G. (1983). The savannas of tropical America. In: F. Bourliére (ED.). Tropical Savannas. Amsterdam: Elsevier, 245-288.

Skiba, A. (2009). Impact of an invasive species (*Pinus* spp.) on subtropical grasslands in Southern Brazil. Freising: Technische Universität München, 70.

Sosinski, E. E. & Pillar, V. D. (2004). Respostas de tipos funcionais de plantas à intensidade de pastejo em vegetação campestre. *Pesquisa agropecuária brasileira*, *39*, 1-9.

Teixeira, M. & Altesor, A. (2009). Small-scale spatial dynamics of vegetation in a grazed Uruguayan grassland. *Austral Ecology*, *34*, 386-394.

Trindade, J. P. P., Quadros, F. L. F. & Pillar, V. D. (2008). Vegetação campestre de areais do Sudoeste do Rio Grande do Sul sob pastejo e com exclusão do pastejo. *Pesquisa agropecuária brasileira*, *43*, 771-779.

Whelan, R. J. (1995). *The ecology of fire*. Cambridge: Cambridge University Press, 346.

In: Encyclopedia of Environmental Research
Editor: Alisa N. Souter

ISBN: 978-1-61761-927-4
© 2011 Nova Science Publishers, Inc.

Chapter 8

RELATIONSHIP OF MANAGEMENT PRACTISES TO THE SPECIES DIVERSITY OF PLANTS AND BUTTERFLIES IN A SEMI-NATURAL GRASSLAND, CENTRAL JAPAN

Masako Kubo[1],*, Takato Kobayashi[2], Masahiko Kitahara[3] and Atsuko Hayashi[4]

[1]National Institute for Land and Infrastructure Management, Ministry of Land, Infrastructure and Transport and Tourism, 1 Asahi, Tsukuba, Ibaraki 305-0804, Japan.
[2]Utsunomiya University, 350 Minemachi, Utsunomiya, Tochigi 321-8505, Japan.
[3]Yamanashi Institute of Environmental Sciences, 5597-1 Kenmarubi, Fujiyoshida, Yamanashi, 403-0005, Japan
[4]Yamanashi Forest Research Institute, Saishoji 2290-1, Masuho, Yamanashi 400-0502, Japan

ABSTRACT

In the past, semi-natural grasslands in Japan were maintained using traditional management practices such as mowing, burning, and grazing. However, grassland areas have been decreasing drastically recently due to land development or the abandonment of management. Thus, conservation measures are urgently required for organisms specific to semi-natural grassland habitats. To examine management schemes for conserving such organisms, we investigated plant species and seasonal fluctuation patterns in species and individuals of flowering plants and adult butterflies in a semi-natural grassland in central Japan. Our study sites were firebreaks where the grass was mowed and removed, plantation areas that were mowed, unpaved roads with mowed banks, abandoned grassland, and scattered scrub forest. Areas that had been abandoned for less than five years were dominated mostly by *Arundinella hirta*, while areas that had been abandoned for more than ten years were dominated by large tussocks of *Miscanthus sinensis* or scrub

---

* Corresponding author: E-mail: k.masako@poppy.ocn.ne.jp

forests of *Rhamnus davurica* var. *nipponica*. The plant species composition differed between grassland sites and scrub sites, and that of firebreaks was different from other grassland sites. The sites under management sustained a larger number of flowers and butterflies than the sites without management. Furthermore, the firebreak sites sustained flowers in June and July, while the plantation and banks of unpaved road sites sustained flowers primarily in August and September. The number of butterflies increased in the firebreak in June and at the other sites in August and September, in relation to the occurrence of flowers at each site. These results suggest that management is essential to sustain this grassland and that different management regimes, such as mowing alone and mowing with grass removal, induce different plant species compositions and different numbers of flowers, and lead to different numbers of adult butterflies during a season. On the other hand, *R. davurica* var. *nipponica*, which composed the scrub forest, is a host plant for a threatened butterfly species. In conclusion, heterogeneous environments with different vegetation structures by season and seral stage under different management regimes support plant and butterfly diversity in semi-natural grassland habitats, and continuation of traditional management practices in the future may be important.

INTRODUCTION

In Japan, semi-natural grasslands have traditionally been anthropogenically sustained and managed as grazing and meadow habitats (Ohkubo and Tsutida 1998). The grasses in these grasslands have been mowed and used as fertilizer for crops, forage for domestic livestock, and material for residential building (Washitani 2003). There have been drastic changes in Japanese agriculture, forestry, and livestock farming since ca. 1950, including a decrease in the number of farmhouses and a decline of the sericulture industry (Tsunekawa 2003), and recently semi-natural grasslands have been abandoned in many regions. As a consequence, various plant species that depend on such grasslands have also become threatened (Environment Ministry of Japan 2000). The importance of sustaining the biodiversity of rural areas and satoyama including semi-natural grasslands has been raised by Japan's national biodiversity strategy (Nature Conservation Bureau, Ministry of the Environment, Government of Japan 2008).

The survival of many butterfly species has been closely related to anthropogenic activities in semi-natural grasslands (Hiura 1973). The numbers of butterfly species and individuals depending on such grasslands has been decreasing (Ae et al. 1996). Butterfly species composition can differ between habitats with different vegetation structure or management regimes (Holl 1995; Spitzer et al. 1997; Inoue 2003; Kitahara and Watanabe 2003). Since butterflies are phytophagous insects through the larval and adult stages, butterfly distribution sensitively reflects changes in vegetation (Ehrlich et al. 1972; Weiss et al. 1987; Hill et al. 1995; Blair and Launer 1997; Wood and Gillman 1998). Some studies have suggested that the distribution of adult butterflies reflects the availability of nectar sources more than the presence of suitable host plants (Grossmueller and Lederhouse 1987; Loertscher et al. 1995). The abundance of adults of most butterfly species is closely associated with the abundance of flowers of key nectar source species (Feber et al. 1996; Steffan-Dewenter and Tscharntke 1997; Bergman et al. 2008; Kitahara et al. 2008). When anthropogenic management activities affect the survival and flowering of grassland plant species, the species diversity of butterflies in semi-natural grasslands is also likely affected.

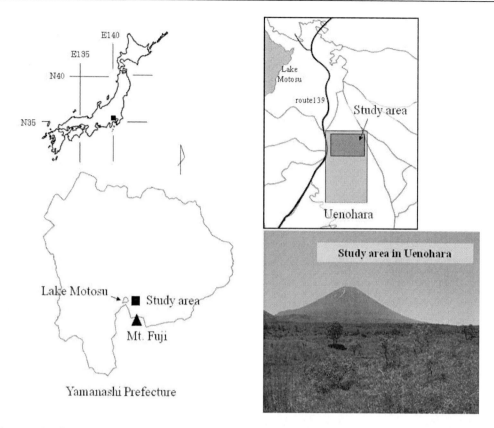

Figure 1. Study area.

In this report, we describe the species diversity of plants and butterflies through the growing season in a semi-natural grassland in central Japan. We investigated plant species composition and seasonal fluctuation patterns in the numbers and species of flowers of nectar sources and adult butterflies under different vegetation structures and management regimes. The vegetation structure varied from semi-natural grassland to scattered scrub forest, and management included mowing, mowing with removal of grass, and abandonment. We discuss the relationship between management and vegetation, and between the distribution of flowers and butterflies, and also suggest ways to conserve the species diversity of both plants and butterflies in semi-natural grassland.

STUDY SITES

The semi-natural grassland studied was located in the Uenohara area at the north-western foot of Mt. Fuji, central Japan (980 m above sea level, Figure 1). This grassland, which includes scattered scrub forest, covers about 43 ha, and the study area was about 6 ha. The topography of the area is almost flat, with irregular undulations of ca. 8 m in elevation. The soil in the area is composed mainly of scoriaceous lava and volcanic ash that were deposited during past eruptions of Mt. Fuji.

This grassland is dominated mostly by poaceous grasses such as *Miscanthus sinensis*, *Arundinella hirta*, and *Spodiopogon sibiricus*. Various herbaceous plant species are also present. The vegetation is representative of semi-natural grasslands and includes plant species that are threatened in Japan. Numerous individuals and various butterfly species (Photo 1), including species that are threatened in Japan, inhabit the area. The scattered scrub forests within the grassland are dominated by *Rhamnus davurica* var. *nipponica*. The area surrounding the grassland is secondary forest with deciduous broad-leaved trees or planted *Larix kaempferi* forests.

Gonepteryx maxima with *Cirsium tanakae*   *Plebejus argus* with *Ranunculus japonicus*

Strymonidia mera with *Zanthoxylum piperitum*   *Aeromachus inachus*

Leptalina unicolor with *Ixeris dentata*   *Hesperia florinda*

Photo 1. Various species of butterflies and flowers.

This area appears to have been used as a source of grass for fuel, forage, and grazing until ca. 50 years ago (Mayumi Takahashi, personal communication). Although the area was afforested 50 years ago, it experiences severe climate and soil conditions such as strong winds, low temperatures, and frozen soils in winter, which hinder tree survival (Kubo et al. 2005). As a consequence, there has been continuous anthropogenic management involving mowing of the surrounding plantations. Large areas of the grassland have been abandoned for several years or decades.

METHODS

Investigation of Management and Vegetation

To investigate the relationship between management and vegetation, we used a version of the Minna de GIS software (http://www13.ocn.ne.jp/%7Eminnagis/). The GIS inputs included a dominant species-based vegetation map in the form of vector data and vectorized information on the number of years since abandonment provided from a ledger of afforestation records. We covered these layers with a 5-m interval grid and collected the traits of all layers for the grid nodes in the study area. The study area was about 6 ha, and the vegetation map of the area included five plant communities, defined by dominant species such as the herbaceous species *Arundinella hirta* and *Miscanthus sinensis*, the scrub tree species *Rhamnus davurica* var. *nipponica*, bamboo, and bare ground.

Establishment of Study Sites

We selected six study sites in the semi-natural grassland that represented five different conditions, i.e., different vegetation structures and management strategies, as follows:

(1) Firebreaks at two sites: Firebreak 1 and Firebreak 2 (Photo 2-a and b)
    A firebreak about 6 m wide was established on the border between the grassland and the surrounding forest in 1959, where the grass has been mowed and removed annually in autumn since 1961. Therefore, the vegetation structure of these sites is grassland, and the treatment is mowing with grass removal.
(2) Plantation area with mowing in the grassland: Mowing area (Photo 2-c)
    Parts of the grassland in the plantations have been mowed annually in autumn since 1998, and have thus been mowed for more than 6 years. Therefore, the vegetation structure is grassland, and the treatment is mowing alone.
(3) Unpaved road with adjacent mowing: Road (Photo 2-d)
    Unpaved roads with banks traverse the grassland, including an unpaved, 2-m-wide road with little vegetation, exposed soil, and, in some cases, stones of scoriaceous lava and some rainwater pools. The banks on the sides of this road are covered with grass that is mowed annually in autumn. This road was established in 1987 and mowing of the banks began before 1998; thus mowing has continued for more than 6 years. Therefore, the vegetation structure is grassland, and the treatment is mowing alone.

(4) Abandoned grassland dominated by poaceous grass: Abandoned grassland (Photo 2-e)
    These portions of grassland have been abandoned for more than three years; thus the
    vegetation structure is grassland and the treatment is 'without management.'

(5) Scattered scrub forest: Scrub (Photo 2-f)
    Scattered scrub forests within the grassland are dominated by *Rhamnus davurica* var.
    nipponica. Many parts of the scrub forests have been abandoned for several years or
    decades. However, due to a lack of records, we could not ascertain the number of
    years since abandonment. Therefore, the vegetation structure is scrub and the
    treatment is 'without management.'

a. Firebreak 1 in May

b. Firebreak 1 in June

c. Mowing area in June

d. Road in September

e. Abandoned grassland in September

f. Scrub in June

Photo 2. Study sites.

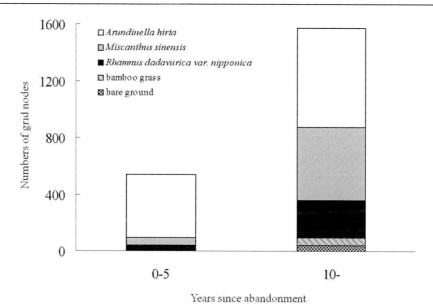

Figure 2. Vegetation of the study area, and number of years since abandonment.

Investigation of Plants and Flowers

We established 10 square plots (1×1 m) at about 20-m intervals at each site. At the Scrub site, the plots were established on the forest floor. In all plots, we investigated the height of vegetation, including dried grass, in May 2004, and plant species and vegetation height, excluding dried grass, in September 2004. We also investigated the numbers of flowers or inflorescences of each species serving as a nectar source for adult butterflies in each plot twice a month from May to September in 2004 and 2005 (but only in 2005 in Firebreak 2 and Road). Flower species of nectar sources were determined by recording butterfly visits during our investigation and from other studies (e.g., Kitahara 2000) conducted in the neighborhood of our study site.

Investigation of Butterflies

Observations were made using the line transect method (Pollard 1977, 1984; Thomas 1983; Gall 1985) at each site. Transect counts were performed twice a month during the adult flight season from May to September in 2004 and 2005, within the period from 10:00 to 12:00 under good weather conditions (but only in 2005 in Firebreak 2). While walking at a steady pace along the transect line, we recorded the number of adult individuals of each butterfly species sighted within a belt approximately 6 m wide. Those individuals that could not be identified immediately by sight were captured by net and released after identification. The transect lines for the butterfly investigation were 200 m long.

RESULTS

Vegetation and Years Since Abandonment

There were 2116 grid nodes in the study area. Of these, 541 were at sites that had been abandoned for less than 5 years, including 445 nodes in the *Arundinella hirta* plant community, 55 nodes in the *Miscanthus sinensis* community, 41 nodes in the *Rhamnus davurica* var. *nipponica* community, and 0 nodes in the bamboo and bare ground communities. There were 1575 nodes at sites that had been abandoned for more than 10 years; these included 702 *A. hirta* nodes, 515 *M. sinensis* nodes, 261 *R. davurica* var. *nipponica* nodes, 54 bamboo nodes, and 43 bare ground nodes. The *A. hirta* plant community was larger in nodes that had been abandoned for less than five years, and those of *M. sinensis* and *R. dadavurica* var. *nipponica* were larger in nodes that had been abandoned for more than ten years (chi-square test, $\alpha < 0.05$, Figure 2). No nodes had been abandoned for 6–9 years.

Table 1. The numbers of species and individuals of plant flowers and butterflies at each study site.

	Firebreak 1	Firebreak 2	Mowing area	Road	Abandoned grassland	Scrub
Plant						
Mean number of plant species (/m²)	27.7 ± 3.8ᵃ	27.3 ± 5.3ᵃ	30.3 ± 3.8ᵃᶜ	25.7 ± 3.6ᵃᵈ	24.1 ± 4.0ᵃᵈ	20.4 ± 3.4ᵇ
Mean number of annual species (/m²)	3.0 ± 1.2ᵃ	2.5 ± 1.1ᵃ	0.8 ± 0.6ᵇ	1.2 ± 0.8ᵇ	0.7 ± 0.8ᵇ	0.9 ± 0.6ᵇ
Mean number of perennial species (/m²)	22.6 ± 2.5ᵃᶜ	20.8 ± 4.4ᵃᵉ	25.1 ± 2.7ᵃᶠ	19.6 ± 2.5ᵃᵈᵉ	20.0 ± 3.9ᵃᵉ	15.9 ± 3.0ᵇ
Mean number of woody species (/m²)	1.1 ± 1.1ᵃᶜ	2.0 ± 1.7ᵃ	3.3 ± 1.4ᵇ	3.1 ± 1.0ᵈ	2.7 ± 1.3ᵈ	2.2 ± 0.6
Vegetation height						
Mean vegetation hight in May (cm)	6.0 ± 2.1ᵃᶜ	6.0 ± 2.1ᵃᶜ	9.0 ± 2.1ᵃᵈ	6.0 ± 2.1ᵃᶜ	16.5 ± 3.4ᵇ	17.5 ± 2.6ᵇ
Mean vegetation hight in September (cm)	80.0 ± 23.6	56.0 ± 12.6ᵃ	57.0 ± 14.9ᵃ	60.0 ± 13.3ᵃ	113.0 ± 41.4ᵇ	77.0 ± 10.6ᵃ
Flower						
Mean number of flowering species (/m²)	18.7 ± 4.4ᵃ	11.0 ± 2.7ᵇ	14.3 ± 5.3ᶜ	11.7 ± 2.4ᵇ	8.3 ± 5.9ᵇ	6.0 ± 3.0ᵇᵈ
Mean number of flower (/m²)	51.6 ± 21.9	57.1 ± 24.0	61.0 ± 44.1	80.4 ± 51.4ᵃ	17.6 ± 12.7ᵇ	29.3 ± 32.0
Butterfly						
Number of individual (/200m)	230.0	228.0	133.5	142.0	95.0	99.5
Number of species (/200m)	31.0	26.0	24.0	34.0	19.0	24.0

All flower and butterfly data are averages per year. Mean values ± standard deviation per plot are given for numbers of plant species, including annual, perennial, and woody species, flowering species, flowers, and vegetation height. Different letters (a and b, c and d, e and f) indicate significant differences between sites ($P < 0.05$). Significant differences for plant species including annual, perennial, and woody species, and vegetation height in May were obtained from an analysis of variance (ANOVA) and Fisher's protected least significant difference test (Fisher's PLSD). Significant differences for flowering species, flowers, and vegetation height in September were obtained from Kruskal-Wallis and Scheffé tests.

Plant Species and Vegetation Height

The mean number of plant species per plot was higher in every grassland site than in Scrub, and higher in the Mowing area than in the Road and Abandoned grassland sites (ANOVA, Fisher's PLSD, $P < 0.05$, Table 1). The mean numbers of annual plant species in Firebreaks 1 and 2 were significantly higher than at all other sites (ANOVA, Fisher's PLSD, $P < 0.05$). The mean number of perennial plant species was significantly lower in Scrub than at all other sites, that in Firebreak 1 was significantly higher than that in Road, and that in the Mowing area was significantly higher than that in Firebreak 2, Road, and Abandoned grassland (ANOVA, Fisher's PLSD, $P < 0.05$). The mean number of woody species in the Mowing area site was significantly higher than that in Firebreak 1 and Firebreak 2, while that in the Road and Abandoned grassland sites was significantly higher than that in Firebreak 1 (ANOVA, Fisher's PLSD, $P < 0.05$).

The mean vegetation height including dried grass in May was greater in Abandoned grassland and Scrub than in all other sites, and lower in Firebreaks 1 and 2 and Road than in Mowing area (ANOVA, Fisher's PLSD, $P < 0.05$). The mean vegetation height, excluding dried grass, in September was significantly greater in Abandoned grassland than in Firebreak 2, Mowing area, Road, and Scrub (Kruskal-Wallis test, Scheffé test, $P < 0.05$).

Flowering of Nectar Sources

The mean number of flowering species per plot was significantly higher in Firebreak 1 than in Firebreak 2, Road, Abandoned grassland, and Scrub, and that of Scrub was significantly lower than that of Mowing area (Kruskal-Wallis, Scheffé-test, $P < 0.05$, Table 1). The mean number of flowers per plot was largest in Road, followed by Mowing area, Firebreak 2, Firebreak 1, Scrub, and Abandoned grassland (Table 1). The mean number of flowers per plot in Road was larger than that in Abandoned grassland (Kruskal-Wallis, Scheffé-test, $P < 0.05$).

Plant Species Composition

We used detrended correspondence analysis (DCA) to analyze the plant species composition of each plot (Figure 3). The axis 1 and axis 2 eigenvalues were 0.280 and 0.211, respectively. The axis 1 score for every grassland site was significantly different from the Scrub score, and that of Firebreak 1 was significantly different from that of Mowing area (Kruskal-Wallis test, Scheffé test, $P < 0.05$, Table 2). The axis 2 scores of Firebreaks 1 and 2 differed significantly from those of Mowing area, Road, and Scrub, and that of Road was significantly different from that of Mowing area, Abandoned grassland, and Scrub (ANOVA, Fisher's PLSD, $P < 0.05$, Table 2).

The axis 1 score was positively correlated with vegetation height in May, and negatively correlated with the number of annual and perennial species (Table 3). The axis 2 score was positively correlated with the number of annual species and negatively correlated with the number of woody species (Table 3).

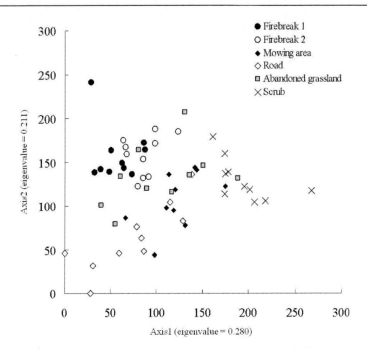

Figure 3. Ordination of the sites by detrended correspondence analysis (DCA).

Table 2. Mean ± standard deviation of ordination axis scores for plant species composition of different treatment types by detrended correspondence analysis (DCA).

Study site	axis1	axis2
Firebreak 1	57.8 ± 21.2^{ac}	158.7 ± 31.5^{a}
Firebreak 2	86.5 ± 18.3^{a}	158.5 ± 23.1^{a}
Mowing area	122.3 ± 29.0^{ad}	106.3 ± 32.0^{bc}
Road	75.2 ± 45.8^{a}	63.4 ± 38.5^{bd}
Abandoned grassland	105.0 ± 47.4^{a}	133.4 ± 35.0^{c}
Scrub	194.9 ± 30.9^{b}	129.7 ± 24.2^{bc}

Different letters (a and b, c and d) indicate significant differences between sites ($P < 0.05$). Significant differences for axis 1 were obtained from Kruskal-Wallis and Scheffé tests, and those for axis 2 were obtained from an analysis of variance (ANOVA) and Fisher's protected least significant difference test (Fisher's PLSD).

Number and Species of Adult Butterflies

The number of species of butterflies was largest at Road sites, followed by Firebreak 1, Firebreak 2, Mowing area/Scrub, and Abandoned grassland (Table 1). The number of individual butterflies was largest at Firebreak 1, followed by Firebreak 2, Road, Mowing area, Scrub, and Abandoned grassland.

Table 3. The correlation between axis 1 and 2 scores and environmental conditions by detrended correspondence analysis (DCA) for plant species composition.

	axis 1		axis 2	
Vegetation hight in May	r = 0.604	P<0.001	r = 0.037	ns
Vegetation hight in September	r = -0.126	ns	r = -0.039	ns
Number of annual species	r = -0.318	P<0.05	r = 0.326	P<0.05
Number of perennial species	r = -0.299	P<0.05	r = -0.065	ns
Number of woody species	r = 0.037	ns	r = -0.536	P<0.001

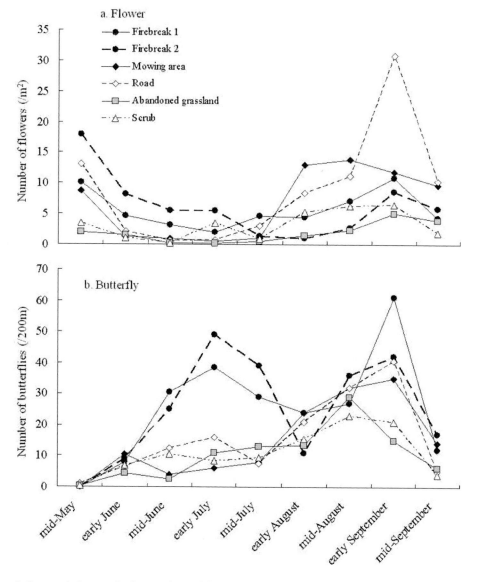

Figure 4. Seasonal changes in the numbers of flowers and butterflies at each study site (revised from Kubo et al. 2009). (a) Flower abundance, where the number of flowers is the mean number per plot per year at each study site. (b) Butterfly abundance, where the number of butterflies is the mean number per year at each study site.

Table 4. List of the species of flowering plants observed in the present study, and the numbers of flowers at each study site (revised from Kubo et al. 2009).

Flowering species	Firebreak 1	Firebreak 2	Mowing area	Road	Abandoned grassland	Scrub	Total
Compositae							
Aster ageratoides ssp. *ovatus*	0.20 ± 0.63	0.60 ± 1.90	2.40 ± 5.34	0.20 ± 0.63	0.30 ± 0.63	0.00 ± 0.00	37.0
Aster scaber	6.30 ± 7.17	5.00 ± 5.83	2.40 ± 5.42	11.1 ± 25.5	2.80 ± 4.30	0.00 ± 0.00	276.0
Erigeron annuus	5.75 ± 6.13	1.90 ± 3.38	4.10 ± 4.88	13.2 ± 26.2	0.35 ± 1.10	0.00 ± 0.00	253.0
Solidago virga-aurea ssp. *asiatica*	0.80 ± 2.06	0.50 ± 1.59	0.00 ± 0.00	0.00 ± 0.00	2.00 ± 6.32	1.40 ± 4.43	47.0
Inula salicina var. *asiatica*	0.00 ± 0.00	0.00 ± 0.00	1.10 ± 1.91	0.60 ± 1.07	0.65 ± 0.82	0.00 ± 0.00	23.5
Eupatorium lindleyanum	2.15 ± 2.71	0.60 ± 1.26	0.75 ± 2.20	0.50 ± 0.85	0.20 ± 0.35	0.00 ± 0.00	42.0
Cirsium japonicum	0.05 ± 0.16	0.40 ± 1.26	0.00 ± 0.00	0.00 ± 0.00	0.00 ± 0.00	0.00 ± 0.00	4.5
Cirsium tanakae	0.40 ± 0.52	2.50 ± 2.37	1.65 ± 2.32	0.60 ± 1.58	0.80±1.06	0.50 ± 0.53	64.5
Cirsium nipponicum var. *incomptum*	0.05 ± 0.16	0.60 ± 1.35	0.00 ± 0.00	0.00 ± 0.00	0.00 ± 0.00	0.00 ± 0.00	6.5
Picris hieracioides	0.10 ± 0.32	0.00 ± 0.00	0.00 ± 0.00	0.20 ± 0.63	0.10 ± 0.32	0.00 ± 0.00	4.0
Ixeris dentata	6.15 ± 7.48[a]	6.80 ± 5.01[c]	1.60 ± 3.11	0.30 ± 0.95[d]	0.00 ± 0.00[bd]	0.60 ± 0.88[d]	154.5
Valerianaceae							
Patrinia scabiosaefolia	0.15 ± 0.47	0.00 ± 0.00	0.00 ± 0.00	0.00 ± 0.00	1.15 ± 2.65	0.05 ± 0.16	13.5
Dipsacaceae							
Scabiosa japonica	0.00 ± 0.00	0.30 ± 0.67	0.00 ± 0.00	0.00 ± 0.00	0.00 ± 0.00	0.00 ± 0.00	3.0
Scrophulariaceae							
Veronica rotunda var. *petiolata*	0.45 ± 0.69	0.00 ± 0.00	0.00 ± 0.00	0.00 ± 0.00	0.30 ± 0.67	0.00 ± 0.00	7.5
Labiatae							
Prunella vulgaris var. *lilacina*	0.00 ± 0.00	0.00 ± 0.00	0.00 ± 0.00	0.00 ± 0.00	0.40 ± 1.26	0.00 ± 0.00	4.0
Stachys japonica var. *intermedia*	0.00 ± 0.00	1.10 ± 3.48	0.00 ± 0.00	5.20 ± 9.33	0.00 ± 0.00	7.90 ± 8.23	142.0
Clinopodium chinense var. *parviflorum*	0.00 ± 0.00[a]	0.00 ± 0.00[a]	1.55 ± 4.90	0.30 ± 0.95[a]	0.30 ± 0.95[a]	6.20 ± 5.80[b]	83.5
Mosla dianthera	0.00 ± 0.00	1.10 ± 3.48	0.00 ± 0.00	0.00 ± 0.00	0.00 ± 0.00	0.00 ± 0.00	11.0
Salvia japonica	7.95 ± 14.6	1.20 ± 3.79	22.1 ± 44.9	8.50 ± 17.9	1.60 ± 5.06	7.30 ± 23.1	486.5
Primulaceae							
Lysimachia clethroides	0.00 ± 0.00	0.00 ± 0.00	0.10 ± 0.21	0.00 ± 0.00	0.00 ± 0.00	0.00 ± 0.00	1.0
Lysimachia vulgaris var. *davurica*	0.15 ± 0.47	0.00 ± 0.00	0.00 ± 0.00	0.00 ± 0.00	0.00 ± 0.00	0.00 ± 0.00	1.5
Onagraceae							
Oenothera biennis	0.00 ± 0.00	0.00 ± 0.00	0.10 ± 0.32	0.00 ± 0.00	0.00 ± 0.00	0.00 ± 0.00	1.0
Leguminosae							
Lespedeza bicolor	1.95 ± 2.65	1.40 ± 2.46	1.40 ± 3.43	5.70 ± 7.87	0.15 ± 0.47	0.00 ± 0.00	106.0
Geraniaceae							
Geranium yesoense var. *nipponicum*	0.55 ± 0.14[a]	0.40 ± 0.70[a]	1.60 ± 1.47	3.50 ± 3.34[b]	0.35 ± 0.53[a]	0.15 ± 0.34[a]	65.5

Table 4. (Continued)

Flowering species	Firebreak 1	Firebreak 2	Mowing area	Road	Abandoned grassland	Scrub	Total
Rosaceae							
Potentilla fragarioides var. *major*	0.35 ± 0.67	1.00 ± 1.63	0.10 ± 0.32	0.10 ± 0.32	0.70 ± 1.89	0.00 ± 0.00	22.5
Potentilla freyniana	9.5 ± 13.1	17.0 ± 10.6	9.55 ± 12.4	14.6 ± 11.9	2.05 ± 3.95	4.55 ± 5.94	572.5
Sanguisorba officinalis	0.10 ± 0.32^a	0.00 ± 0.00^a	2.30 ± 2.64^a	8.20 ± 5.94^b	1.35 ± 2.12^a	0.55 ± 1.30^a	125.0
Agrimonia nipponica	3.75 ± 11.9	1.30 ± 3.77	7.75 ± 18.4	6.80 ± 11.1	0.45 ± 0.96	0.00 ± 0.00	200.5
Saxifragaceae							
Astilbe microphylla	0.40 ± 0.61	0.70 ± 1.56	0.10 ± 0.32	0.00 ± 0.00	0.35 ± 1.11	0.00 ± 0.00	15.5
Ranunculaceae							
Ranunculus japonicus	4.35 ± 2.69	12.7 ± 15.9^a	0.35 ± 0.75^b	0.80 ± 1.75^b	1.10 ± 2.04^b	0.00 ± 0.00^b	193.0
Iridaceae							
Iris sanguinea	0.00 ± 0.00	0.00 ± 0.00	0.00 ± 0.00	0.00 ± 0.00	0.05 ± 0.16	0.00 ± 0.00	0.5
Iris ensata var. *spontanea*	0.00 ± 0.00	0.00 ± 0.00	0.00 ± 0.00	0.00 ± 0.00	0.10 ± 0.21	0.05 ± 0.16	1.5

Numbers indicate means ± standard deviation per plot per year. Different letters (a and b, c and d) indicate significant differences between sites ($P < 0.05$). Significant differences were obtained from Kruskal-Wallis and Scheffé tests.

Seasonal Fluctuation Patterns in the Number of Flowers and Butterflies

The seasonal fluctuation pattern of the mean number of flowers per year differed among sites (Figure 4a). Firebreaks 1 and 2 had flowers from May to September. Although there were flowers in May in Mowing area and Road, they decreased in June and July and increased in August and September more than at other sites. The number of flowers in Scrub increased from July, and that in Abandoned grassland remained low.

The seasonal fluctuation pattern of the mean number of butterflies per year also differed among sites (Figure 4b). The number of butterflies in Firebreaks 1 and 2 was high, especially in June and July, whereas other sites had few butterflies during this period. The number of butterflies increased in August and September at all sites, but the number of butterflies in Abandoned grassland and Scrub was smaller than that in grassland sites with management.

Seasonal Fluctuation Patterns of Species of Flowers and Butterflies

We found a mean of 2970 flowers of 32 species per year (Table 4). Figure 5(a) shows the seasonal fluctuation pattern in the number of flowers of the nine most abundant species and all other species across all sites. The nine most abundant species produced a mean of 2403 flowers per year, which accounted for 81% of all flowers per year. Many flowers of *Potentilla freyniana* and *Ranunculus japonicus* were found in May and June, while those of *Ixeris dentata* were found in June and July. Many flowers of herbaceous species of the Compositae and Labiatae including *Erigeron annuus*, *Stachys japonica* var. *intermedia*, and *Salvia japonica* were found in the longer flowering period from July to September. Many species included among the "other species" flowered in August and September.

Adult butterflies appeared in early June, and we found a mean of 928 individuals of 46 species per year (Table 5). Figure 5(b) shows the seasonal fluctuation pattern in the number of butterfly individuals for the nine most abundant and all other species across all sites. The nine most abundant species accounted for a mean of 725 butterflies per year, or 78% of all individual butterflies per year. Only a few species were recorded in June, although there were numerous individuals of *Plebejus argus*, *Ypthima argus*, and *Leptalina unicolor*. The numbers of species and individuals increased in August and September.

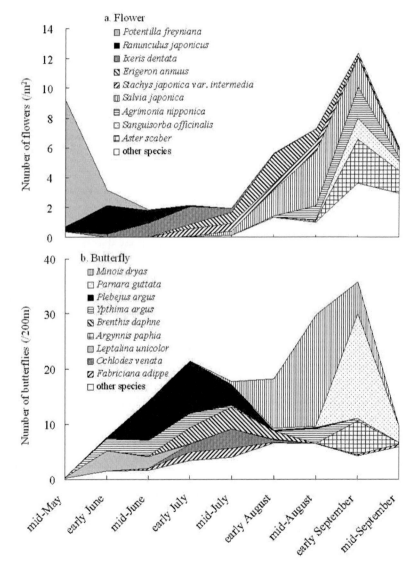

Figure 5. Seasonal changes in the numbers of flowers and butterflies of each species across all sites (revised from Kubo et al. 2009). (a) Flower abundance, where the number of flowers is the mean number per plot per year. (b) Butterfly abundance, where the number of butterflies is the mean number per year. The ten different patterns in the graph legend indicate the nine most abundant species per year and all others in both (a) and (b).

The Abundance of Each Flower and Butterfly Species at Each Site

The number of flowers of each species differed among sites (Table 4, Figure 6a). *R. japonicus* comprised 88% of the mean number of flowers in Firebreaks 1 and 2. The mean number of *Ixeris dentata* was largest in Firebreaks 1 and 2, and accounted for 84% of the flowers there. The mean number of *Sanguisorba officinalis* was largest in Road, where it accounted for 66% of the flowers. The number of *Erigeron annuus* was largest in Road (with 52% of the total), that of *Stachys japonica* var. *intermedia* was largest in Scrub (56%), that of *Salvia japonica* was largest in Mowing area (45%), that of *Agrimonia nipponica* was largest in Mowing area and Road (73%), and that of *Aster scaber* was largest in Road (40%).

Table 5. List of the butterfly species observed in the present study, and the number of adult individuals at each study site (revised from Kubo et al. 2009).

Butterfly species	Firebreak 1	Firebreak 2	Mowing area	Road	Abandoned grassland	Scrub	Total
Papilionidae							
Parnassius citrinarius	0.5	0.0	0.0	0.0	0.0	0.0	0.5
Papilio machaon	0.5	1.0	0.5	1.5	0.5	0.5	4.5
Papilio xuthus	0.5	0.0	1.5	0.0	0.0	0.0	2.0
Papilio macilentus	0.0	0.0	0.5	0.5	0.0	0.0	1.0
Papilio bianor	0.0	1.0	0.0	0.0	0.0	0.0	1.0
Pieridae							
Colias erate	3.0	6.0	1.0	4.5	2.5	0.0	17.0
Eurema mandarina	6.0	7.0	1.5	2.0	2.5	6.5	25.5
*Gonepteryx maxima	3.0	2.0	0.5	1.5	0.0	2.0	9.0
Pieris rapae	1.0	0.0	0.0	0.5	0.0	0.0	1.5
Pieris spp.	2.0	7.0	2.5	2.0	1.0	5.5	20.0
Lycaenidae							
Strymonidia mera	3.0	0.0	0.0	1.0	0.5	10.5	15.0
Rapara arata	0.0	0.0	0.0	0.0	0.0	0.5	0.5
Lycaena phlaeas	1.5	0.0	0.0	0.0	0.0	0.0	1.5
Lampides boeticus	3.0	1.0	0.0	1.0	1.0	0.0	6.0
Pseudozizeeria maha	0.0	0.0	0.0	0.5	0.0	0.0	0.5
Celastrina argiolus	0.5	2.0	0.0	0.5	0.0	0.0	3.0
Everes argiades	2.0	0.0	0.0	1.0	0.0	0.0	3.0
*Plebejus argus	56.0	48.0	7.0	4.5	8.0	2.0	125.5
Libytheidae							
Libythea celtis	0.0	0.0	0.0	0.5	0.0	0.0	0.5
Nymphalidae							
*Brenthis daphne	11.0	9.0	9.5	5.5	10.5	6.0	51.5
Argyronome laodice	6.5	7.5	5.9	2.4	2.0	0.3	24.5
Argyronome ruslana	1.2	1.0	0.7	0.0	0.0	1.0	3.9

Table 5. (Continued)

Butterfly species	Firebreak 1	Firebreak 2	Mowing area	Road	Abandoned grassland	Scrub	Total
Nephargynnis anadyomene	0.0	0.0	0.0	0.0	0.0	0.0	0.0
Argynnis paphia	13.5	11.0	4.1	7.5	1.0	3.0	40.1
Fabriciana adippe	7.6	11.5	2.6	3.5	0.0	0.8	26.0
Argyreus hyperbius	0.2	1.0	0.2	0.5	0.0	0.0	1.9
Ladoga glorifica	1.5	2.0	0.0	0.0	0.5	1.0	5.0
Ladoga camilla	0.0	1.0	0.0	0.0	0.0	2.0	3.0
Neptis sappho	0.0	0.0	0.0	0.5	0.0	0.0	0.5
Polygonia c-aureum	4.0	5.0	8.5	3.0	3.0	0.5	24.0
Nymphalis xanthomelas	0.5	0.0	0.0	0.5	0.0	0.0	1.0
Inachis io	1.5	4.0	0.5	1.5	0.0	1.0	8.5
Cynthia cardui	0.0	2.0	1.0	0.5	0.0	0.0	3.5
Satyridae							
Ypthima argus	12.0	14.0	11.0	16.5	5.0	8.5	67.0
Minois dryas	28.5	33.0	42.0	47.0	37.0	25.0	212.5
Hesperiidae							
Daimio tethys	0.5	0.0	0.0	1.0	0.0	0.0	1.5
**Leptalina unicolor*	6.0	9.0	6.5	4.5	4.0	4.5	34.5
**Aeromachus inachus*	5.0	0.0	0.0	0.0	1.5	1.0	7.5
Thoressa varia	0.0	0.0	0.0	0.5	0.0	0.0	0.5
Thymelicus sylvaticus	0.0	0.0	0.0	0.5	0.0	0.0	0.5
**Hesperia florinda*	0.0	0.0	1.5	0.0	0.0	0.0	1.5
Ochlodes venata	11.5	12.0	1.0	1.0	4.0	3.0	32.5
Ochlodes ochracea	0.0	0.0	0.5	0.5	0.0	0.0	1.0
Potanthus flavum	0.0	2.0	0.0	0.0	0.0	1.0	3.0
Parnara guttata	36.5	28.0	23.0	23.5	10.5	13.5	135.0

All data show the total numbers of individuals per year at each site. * Endangered species noted by the Environment Ministry of Japan (2006).

The number of butterflies of each species differed among sites (Table 5, Figure 6b). The number of *Plebejus argus* in Firebreaks 1 and 2 (83%) was larger than in the other sites. Numbers of *Ochlodes venata* and *Fabriciana adippe* were also larger in Firebreaks 1 and 2. Although most of the top nine species were also found in Abandoned grassland and Scrub, the number of individuals was smaller than in the managed sites.

DISCUSSION

Relationship between Management and Vegetation

Vegetation height in May was greater in Abandoned grassland and Scrub than in any other site (Table 1), because dried grasses from the previous year had not been removed. Among sites with management, the ground surface in Firebreaks 1 and 2, where grass was mowed and removed, was exposed from autumn to spring. In contrast, grass was not removed from Mowing area and Road, and thus dried grass covered the soil surface. Abandonment and management regimes such as mowing with or without grass removal led to different types of dried grass at our study site, which, in turn, led to differences in plant species abundance and plant species composition.

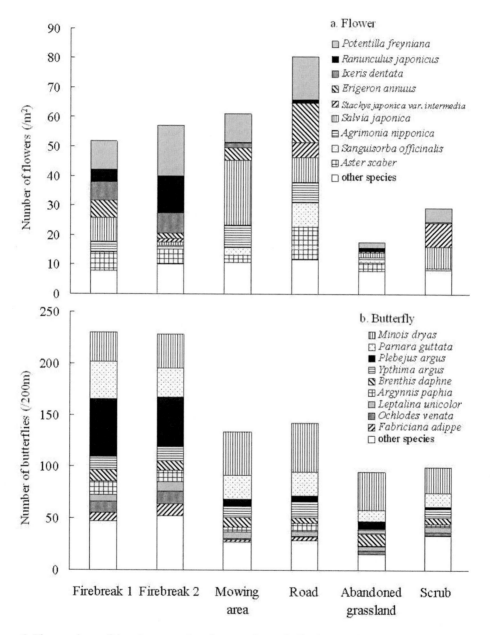

Figure 6. The numbers of the nine most abundant species and all other species of flowers and butterflies at each study site. (a) Flower abundance, where the number of flowers is the total number per plot per year at each study site. (b) Butterfly abundance, where the number of butterflies is the total number per year at each study site. The ten different patterns in the graph legend indicate the nine most abundant species per year and all others in both (a) and (b).

Plant species composition differed between grassland sites (Firebreak 1, Firebreak 2, Mowing area, Road, and Abandoned grassland) and scrub sites (Scrub), and between sites with mowing and grass removal (Firebreaks 1 and 2) and sites with mowing alone (Mowing area and Road, Table 2, Figure 3). The number of annual plant species was larger in Firebreaks 1 and 2 (Table 1), because the soil surface was exposed, allowing annual plants to become established. Conversely, the number of woody plant species was larger in the Mowing area and Road (Table 1), indicating that mowing alone did not eliminate woody plants. The number of total plant species and perennial plant species in Abandoned grassland was equal to that in other grassland sites with management (Table 1), suggesting that plant species composition did not change soon after abandonment. Moreover, there was a reduced number of perennial plant species in Scrub (Table 1).

These results suggest that when mowing with grass removal is stopped, the number of annual plant species decreases. Moreover, although plant communities dominated by perennial species may remain for several years after abandonment, woody species will eventually invade and establish. The total number of plant species and perennial plant species will decrease when scrub dominates. Annual plant species in particular may decline during succession (Stadler et al. 2007). Herbaceous plants that reproduce vegetatively may persist for several years after abandonment (Ohno 2003). Our results suggest that different management regimes and/or lengths of time since abandonment affect site conditions that in turn lead to different plant species compositions.

The dominant species in abandoned areas were mostly poaceous grasses such as *Miscanthus sinensis* and *Arundinella hirta*. In particular, *M. sinensis* is a deep-rooted species that spreads quickly (Yano 1965); in abandoned grasslands, it tends to form large tussocks, which produce unsuitable conditions, intensely suppressing light and physically excluding other herbaceous species. Soil conditions also change with changes in plant species. Scattered scrub forests in the abandoned grassland are categorized by various sized colonies of *Rhamnus davurica* var. *nipponica* that spread concurrently with the dominance of *M. sinensis*. In this grassland, *Rhamnus* colonies appear to spread about 10 years after abandonment (Figure 2).

Relationship between Management and Flowering

The number of flowers that were the only nectar source species for adult butterflies was clearly larger in the managed grassland than in the abandoned grassland (Table 1), although there were similar flowering species at both sites (Table 4, Figure 6a). Moreover, the fluctuation pattern of the total number of flowers and the number of each species in the managed sites differed between treatments such as mowing alone and mowing with grass removal (Figure 4a).

Firebreaks 1 and 2 had flowers throughout the period from May to September. These sites had large numbers of flowers in June and July, when other sites had few (Figure 4a); plants of *Ranunculus japonicus* and *Ixeris dentata* flowering in June and July (Figure 5a) were found mostly in Firebreaks 1 and 2 (Figure 6a). Flowers of the Labiatae and Compositae, such as *Erigeron annuus*, *Salvia japonica*, *Agrimonia nipponica*, *Sanguisorba officinalis*, and *Aster scaber*, were found in August and September (Figure 5a), and were concentrated in Mowing area and Road (Figure 6a).

We suggest that grass removal after mowing provides suitable conditions for some plant species to flower in spring, and other species emerging from the litter can come into flower even with mowing alone. Litter accumulation causes dark conditions (Ohno 1996) and physical suppression of germination (Sydes and Grime 1981). *Ranunculus japonicus* and *I. dentata* come into flower in June and July. There was relatively little vegetation in the firebreaks in June and July because dried grasses were removed in autumn and new vegetation had not yet fully recovered. Such conditions would benefit flowering species such as *R. japonicus* and *I. dentata*. The flowerscapes of some species of Compositae and Labiatae, which were abundant in Mowing and Road sites, where the grass had only been mowed, were able to emerge through the litter and current grass.

In abandoned grassland, the number of flowers was considerably, but not significantly, smaller than at other managed sites (Table 1). Several plant species can survive even under unsuitable conditions after management is abandoned, but typically fail to propagate and therefore disappear over time (Ohno 2003). At our study site, although plant species composition and flowering species in abandoned grassland were similar to those in mowed grassland sites (Table 1, Figure 3), the abandonment of management led to a decrease in the number of flowers. Plant species composition and flowering species will change in the near future if the area continues to develop without management or other disturbance. Moreover, the number of flowering species in Scrub was smaller (Table 1) and the major flowering species differed from those in the other grassland sites (Table 4).

Our results show that management sustains numerous flowers (Table 1), and different treatments sustain different numbers of each flowering species (Table 4, Figure 6a), inducing seasonal changes (Figs. 4a and 5a). Fruiting is important for plants that can reproduce sexually by means of seeds. The decrease in flowers can be an indicator of site conditions (Valverde and Silvertown 1998; Endels et al. 2005). The future dynamics of grasslands may depend on whether there are suitable conditions for plant species to flower, which is affected by anthropogenic management such as mowing alone or mowing with grass removal.

Relationship between the Numbers and Species of Flowers and Butterflies

We found more adult butterflies associated with a larger number of flowers providing nectar in the managed grassland sites than in the abandoned grassland, and the number of butterflies in abandoned grassland was similar to that in scrub forest (Table 1). The seasonal fluctuation pattern in the number of butterflies was correlated with that of flowers in the managed sites (Figure 4), which implies that the micro-distribution of adult butterflies depends on the seasonal fluctuation pattern of the flowers controlled by management.

Firebreaks 1 and 2 had large numbers of individual butterflies through the growing season, and butterflies were numerous even in June and July, when other study sites had few individuals (Figure 4b). *Plebejus argus*, which emerged mostly in June and July (Figure 5b), comprised a large proportion of the butterfly populations at Firebreaks 1 and 2 (Figure 6b). Moreover, *Ochlodes venata*, which emerged mostly in July, was most abundant at the Firebreak sites (Figure 6b) when the flowering species were *Ixeris dentata*, *Erigeron annuus*, and *Stachys japonica* var. *intermedia*, with *I. dentata* being the most abundant in the Firebreaks (Figure 6a).

The flowering species that appeared during the period of *P. argus* abundance were *R. japonicus* and *I. dentata* (Figure 5a), which were found mostly in Firebreaks 1 and 2 (Figure 6a). *Plebejus argus*, moreover, lays its eggs along the margins between bare ground and vegetation (Thomas 1985), and therefore, would likely benefit from management such as mowing and grass removal. *P. argus* reportedly depends on suitable habitat being continuously available within relatively large areas of the biotope, and if an entire metapopulation becomes extinct, natural recolonization is unlikely to occur if the nearest *P. argus* source is more than a few kilometers away (Thomas and Harrison 1992; Thomas et al. 1992). Sustained treatment of the firebreak would thus be essential for *P. argus* in our study sites, since the area within a few kilometers is mostly woodland.

The numbers of butterflies increased in August and September across sites. Numbers increased in Mowing area and Road as much as in Firebreaks 1 and 2, and the numbers of flowers also increased in Mowing area and Road (Figure 4). The flowers of Labiatae and Compositae are small and were counted to determine which traits had contributed to the increases in the total number of flowers at the Mowing area and Road sites compared to Firebreaks 1 and 2. The number of flowers in Firebreaks 1 and 2 increased slightly in August and September, and the number of butterflies increased noticeably (Figure 4). Various flowering species described as "other species" would be nectar sources in August and September in Firebreaks 1 and 2.

Numbers of butterflies were lower in abandoned grassland than in managed grassland and smaller numbers of both species and individual butterflies reflected the decrease in flowers (Figure 4). The numbers of species of butterflies and plants were found to have decreased 3 years after abandonment in a field in Germany (Steffan-Dewenter and Tscharntke 1997). The abandoned grassland in our study had been abandoned for more than 3 years. In contrast, Erhardt (1985) showed that the number of butterfly species did not increase in a mowed area of grassland with disturbance in a sub-alpine area of Switzerland, suggesting that mowing may destroy flowers and lead to a decline of butterflies in earlier stages. If undisturbed grassland is also important for butterflies in earlier stages, abandoned grassland would be important for the conservation of butterflies.

The number of individual butterflies was as high in Scrub as in Abandoned grassland and the number of flowers was also similar, but the number of flowering species in Scrub was smaller than that in Abandoned grassland (Table 1). The dominant tree species in the scattered scrub forest in the study area was *Rhamnus davurica* var. *nipponica*, which is an important host plant for larvae of *Gonepteryx maxima*, a threatened species, and *Strymonidia mera* (Table 5). For this reason, although it provides a smaller number of flowers and flowering species, the scrub forest in the grassland provides habitat for species of butterflies depending on seral stages that differ from the grassland.

Our results imply that heterogeneous conditions with different management regimes and different vegetation structures support habitat for various species of butterfly depending on different seasons and different seral stages. Previous reports have suggested that landscape heterogeneity is important for butterfly diversity and species composition (Natuhara et al. 1999; Weibull et al. 2000; Hamer et al. 2003). Our study site encompassed managed grasslands including different treatments and abandoned grasslands of various ages within a semi-natural grassland, creating a mosaic of different vegetation in various seral stages, which allowed successive flowering through the growing season. Our results suggest that landscape heterogeneity at smaller scales is important for the conservation of flowers and adult butterflies in semi-natural grasslands.

Management for Conservation of Plant and Butterfly Diversity

We propose four management measures to sustain plant and butterfly diversity in the semi-natural grassland studied here: 1) continue to maintain firebreaks via mowing with grass removal; 2) establish several mowing areas, where mowing is conducted annually for more than 6 years, and rotate the areas; 3) maintain a partial scrub forest but do not allow it to dominate the grassland; and 4) retain the abandoned grassland in the area.

The longer successive periods of management in the firebreaks may have led to their different plant species composition and greater diversity and abundance of flowers and butterflies compared to other sites. Lindborg and Eriksson (2004) reported that plant species diversity was related to the degree of habitat connectivity 100 years ago, but not to present-day connectivity. Stadler et al. (2007) emphasized the importance of ancient grassland communities through investigation of dry grasslands in central Europe. Cousins and Lindborg (2008) suggested that remnant habitats function as source communities for grassland specialist species. If the duration of succession has indeed affected our study site, it would imply that management in the firebreak is essential to restoring the vegetation.

The response of insect communities to the effects of habitat loss may be more sensitive or quicker than that of plant communities (Taki and Kevan 2007); thus, butterflies that depend on nectar plants and/or special ground conditions for oviposition (such as *Plebejus argus*) may be the first organisms to disappear following a major change in the habitat or vegetative structure, because many plants can survive for several years after abandonment without flowering. The relationship between plant communities and/or ground conditions and butterflies in younger stages and/or during oviposition remains to be clarified, and specific management strategies should be devised accordingly.

Management is essential for the conservation of plant and butterfly diversity in semi-natural grasslands, and we suggest that different treatments are preferable: our study supports mowing alone and mowing with grass removal. The different treatments would lead to different seasonal changes and species diversity, and abandoned areas at different seral stages could also enhance this diversity. Moreover, these treatments should be conducted continuously.

CONCLUSION

Our study site was located in public land in Yamanashi Prefecture, Japan. Members of the Yamanashi Prefectural Government have sought to manage this grassland for the conservation of biodiversity. The area is about 43 ha, including the core area and our study site, which covered about 6 ha. The management strategy suggested by this research is to maintain firebreaks via mowing with grass removal; to establish several mowing areas where mowing will be conducted for at least 6 years and to rotate these areas; and to retain the abandoned grassland in the area. These treatments should be conducted in the core area; in addition, a partial scrub forest should be maintained around the core area. Moreover, successive monitoring and examination of the management effects are preferable, and more adaptive management should be approved.

The demands for and purposes of management are changing. Historically, Japan has been associated with semi-natural grasslands, and these areas have influenced not only the livelihoods of its residents but also the national culture. Grassland scenes are depicted in old Japanese poems (waka) with various flowers, old pictures, and Japanese folk tales from ancient periods. Although it would be difficult to conserve every area of semi-natural grassland, it is necessary to protect some core areas. In the past, every aspect of life and culture in semi-natural grassland ecosystems was connected through seasons and periods, and now we must connect them to the future.

ACKNOWLEDGMENTS

We thank M. Kawanishi and A. Ichisawa for assistance with the field investigations. We thank M. Takahashi for providing historic information on our study site. We are also grateful to members of the Yamanashi Forest Research Institute and the Yamanashi Institute of Environmental Sciences. We are grateful to R. Tanaka, who made the effort to develop the management plan for the Uenohara area, and members of the Yamanashi Prefectural Government, who make the effort to carry out the management plan. We are grateful to H. Koike at Yokohama National University for support of GIS analysis. We are also grateful to S. Yamamoto at the National Institute for Agro-Environmental Sciences for reviewing this paper.

REFERENCES

Ae, S. A., Hirowatari, T., Ishii, M. & Brower, L. P. (Eds.) (1996). *Decline and Conservation of Butterflies in Japan III*. Osaka, The Lepidopterological Society of Japan.

Bergman, K. O., Ask, L., Askling, J., Ignell, H., Wahlman, H. & Milberg, P. (2008). Importance of boreal grasslands in Sweden for butterfly diversity and effects of local and landscape habitat factors. *Biodiversity and Conservation, 17*, 139-153.

Blair, R. B. & Launer, A. E. (1997). Butterfly diversity and human land use: species assemblages along an urban gradient. *Biological Conservation, 80*, 113-125.

Cousins, S. A. O. & Lindborg, R. (2008). Remnant grassland habitats as source communities for plant diversification in agricultural landscapes. *Biological Conservation, 141*, 233-240.

Ehrlich, P. R., Breedlove, D. E., Brussard, P. F. & Sharp, M. (1972). Weather and the regulation of subalpine populations. *Ecology, 53*, 243-247.

Endels, P., Jacquemyn, H., Brys, R. & Hermy, M. (2005). Rapid response to habitat restoration by the perennial *Primula veris* as revealed by demographic monitoring. *Plant Ecology, 176*, 143-156.

Environment Ministry of Japan (2000). *Threatened Wildlife of Japan, Volume 8*. Tokyo, Japan Wildlife Research Center.

Environment Ministry of Japan (2006). *Threatened Wildlife of Japan, Volume 5*. Tokyo, Japan Wildlife Research Center.

Erhardt, A. (1985). Diurnal Lepidoptera: sensitive indicators of cultivated and abandoned grassland. *Journal of Applied Ecology, 22*, 849-861.

Feber, R. E., Smith, H. & Macdonald, D. W. (1996). The effects on butterfly abundance of the management of uncropped edges of arable fields. *Journal of Applied Ecology, 33*, 1191-1205.

Gall, L. F. (1985). Measuring the size of Lepidopteran populations. *Journal of Research on Lepidoptera, 24*, 97-116.

Grossmueller, D. W. & Lederhouse, R. C. (1987). The role of nectar source distribution in habitat use and oviposition by the tiger swallowtail butterfly. *Journal of the Lepidopterists' Society, 41(3)*, 159-165.

Hamer, K. C., Hill, J. K., Benedick, S., Mustaffa, N., Sherratt, T. N., Maryati, M. & Chey, V. K. (2003). Ecology of butterflies in natural and selectively logged forests of northern Borneo: the importance of habitat heterogeneity. *Journal of Applied Ecology, 40*, 150-162.

Hill, J. K., Hamer, K. C., Lace, L. A. & Banham, W. M. T. (1995). Effects of selective logging on tropical forest butterflies on Buru, Indonesia. *Journal of Applied Ecology, 32*, 754-760.

Hiura, I. (1973). *Butterfly crossing the sea*. Tokyo, Soujyu (in Japanese).

Holl, K. D. (1995). Nectar resources and their influence on butterfly communities on reclaimed coal surface mines. *Restoration Ecology, 3(2)*, 76-85.

Inoue, T. (2003). Chronosequential change in a butterfly community after clear-cutting of deciduous forests in a cool temprate region of central Japan. *Entomological Science, 6*, 151-163.

Kitahara, M. (2000). Food resource usage patterns of adult butterfly communities in woodland habitats at the northern foot of Mt. Fuji, central Japan. *Japanese Society of Environmental Entmology and Zoology, 11*, 61-81 (in Japanese with English summary).

Kitahara, M. & Watanabe, M. (2003). Diversity and rarity hotspots and conservation of butterfly communities in and around the Aokigahara woodland of Mount Fuji, central Japan. *Ecological Research, 18*, 503-522.

Kitahara, M., Yumoto, M. & Kobayashi, T. (2008). Relationship of butterfly diversity with nectar plant species richness in and around the Aokigahara primary woodland of Mount Fuji, central Japan. *Biodiversity and Conservation, 17*, 2713-2734.

Kubo, M., Kobayashi, T., Kitahara, M. & Hayashi, A. (2009). Seasonal fluctuations in butterflies and nectar resources in a semi-natural grassland near Mt. Fuji, central Japan. *Biodiversity and Conservation, 18(1)*, 229-246.

Kubo, M., Matsutani, J. & Hayashi, A. (2005). Factors of the failure of planting in Uenohara-area at the foot of Mt. Fuji. *Bulletin of Yamanashi Forest Research Institute, 24*, 61-67 (in Japanese with English summary).

Lindborg, R. & Eriksson, O. (2004). Historical landscape connectivity affects present plant species diversity. *Ecology, 85(7)*, 1840-1845.

Loertscher, M., Erhardt, A. & Zettel, J. (1995). Microdistribution of butterflies in a mosaic-like habitat: the role of nectar sources. *Ecography, 18*, 15-26.

Natuhara, Y., Imai, C. & Takahashi, M. (1999). Pattern of land mosaics affecting butterfly assemblage at Mt Ikoma, Osaka, Japan. *Ecological Research, 14*, 105-118.

Nature Conservation Bureau, Ministry of the Environment, Government of Japan (ed) (2008). *Our lives in the web of life*. Tokyo, Nature Conservation Bureau, Ministry of the Environment, Government of Japan.

Ohkubo, K. & Tsutida, K. (1998). Conservation of the semi-natural grassland. In: M. Numata, (Ed.), *Handbook of the Natural Conservationv* (432-476). Tokyo, Asakura (in Japanese).

Ohno, K. (1996). Herbaceous life history in the deciduous forest. In: M. Hara, (Ed.), *Natural history in beech forest* (113-156). Heibon-sya, Tokyo (in Japanese).

Ohno, K. (2003). Process of species decline and loss based on the current living condition of plants. In: Natural history museum and institute, Chiba (ed), *The transision of wild flowers in Boso, Japan* (108-119). Kisarazu, Urabesyobou (in Japanese).

Pollard, E. (1977). A methods for assessing changes in the abundance of butterflies. *Biological Conservation*, *12*, 116-134.

Pollard, E. (1984). Synopic studies on butterfly abundance. In: R. I. Vane-Wright, & P. R. Ackery (Eds.), *The Biology of Butterflies* (59-61). London, Academic Press.

Spitzer, K., Jaroš, J., Havelka, J. & Lepš, J. (1997). Effect of small-scale disturbance on butterfly communities of an Indochinese montane rainforest. *Biological Conservation*, *80*, 9-15.

Stadler, J., Trefflich, A., Brandl, R. & Klotz, S. (2007). Spontaneous regeneration of dry grasslands on set-aside fields. *Biodiversity and Conservation*, *16*, 621-630.

Steffan-Dewenter, I. & Tscharntke, T. (1997). Early succession of butterfly and plant communities on set-aside fields. *Oecologia*, *109*, 294-302.

Sydes, C. & Grime, J. P. (1981). Effects of tree leaf litter on herbaceous vegetation in deciduous woodland. *Journal of Ecology*, *69*, 237-248.

Taki, H. & Kevan, P. G. (2007). Does habitat loss affect the communities of plants and insects equally in plant-pollinator interactions? Preliminary findings. *Biodiversity and Conservation*, *16*, 3147-3161.

Thomas, C. D. (1985). Specializations and polyphagy of *Plebejus argus* (Lepidoptera: Lycaenidae) in North Wales. *Ecological Entomology*, *10*, 325-340.

Thomas, C. D. & Harrison, S. (1992). Spatial dynamics of a patchily distributed butterfly species. *Journal of Animal Ecology*, *61*, 437-446.

Thomas, C. D., Thomas, J. A. & Warren, M. S. (1992). Distributions of occupied and vacant butterfly habitats in fragmented landscapes. *Oecologia*, *92*, 563-567.

Thomas, J. A. (1983). A quick method for estimating butterfly numbers during surveys. *Biological Conservation*, *27*, 195-211.

Tsunekawa, A. (2003). Transition of satoyama landscapes in Japan. In: K. Takeuchi, R. D. Brown, I. Washitani, A. Tsunekawa, & M. Yokohari (Eds.), *Satoyama, the traditional rural landscapes of Japan* (41-51). Tokyo, Springer.

Valverde, T. & Silvertown, J. (1998). Variation in the demography of a woodland understorey herb (*Primula vulgaris*) along the forest regeneration cycle: projection matrix analysis. *Journal of Ecology*, *86*, 545-562.

Washitani, I. (2003). Satoyama landscapes and conservation ecology. In: K. Takeuchi, R. D. Brown, I. Washitani, A. Tsunekawa, & M. Yokohari (Eds.), *Satoyama, the traditional rural landscapes of Japan* (16-23). Tokyo, Springer.

Weibull, A., Bengtsson, J. & Nohlgren, E. (2000). Diversity of butterflies in the agricultural landscape: the role of farming system and landscape heterogeneity. *Ecography*, *23*, 743-750.

Weiss, S. B., White, R. R., Murphy, D. D. & Ehrlich, P. R. (1987) Growth and Dispersal of larvae of the checkerspot butterfly *Euphydryas editha*. *Oikos*, *50*, 161-166.

Wood, B. & Gillman, M. P. (1998). The effects of disturbance on forest butterflies using two methods of sampling in Trinidad. *Biodiversity and Conservation*, 7, 597-616.

Yano, N. (1965). On the subterranean organ of wild plants of grassland, I. *Miscanthus sinensis* ANDERSON. *Grassland Science*, *11(1)*, 48-54 (in Japanese with English summary).

In: Encyclopedia of Environmental Research
Editor: Alisa N. Souter

ISBN: 978-1-61761-927-4
© 2011 Nova Science Publishers, Inc.

Chapter 9

TOWARDS THE INFLUENCE OF PLANT-SOIL-MICROBES FEEDBACKS ON PLANT BIODIVERSITY, GRASSLAND VARIABILITY AND PRODUCTIVITY

*A. Sanon[1, 2, 3], T. Beguiristain[3], A. Cébron[3], J. Berthelin[3],
I. Ndoye[1,2], C. Leyval[3], Y. Prin[4], A. Galiana[4],
E. Baudoin[5] and R. Duponnois*[*1, 5,]

[1] Laboratoire Commun de Microbiologie IRD/ISRA/UCAD,
Centre de Recherche de Bel Air. BP 1386. Dakar. Sénégal
[2] Université Cheikh Anta Diop (UCAD). Faculté des Sciences et Techniques.
Département de Biologie Végétale. BP 5005. Dakar. Sénégal
[3] LIMOS, Laboratoire des Interactions Microorganismes - Minéraux - Matière Organique
dans les Sols – UMR 7137 CNRS-UHP. Faculté des Sciences et Techniques.
BP 70239, 54506 Vandoeuvre-les-Nancy Cedex. France
[4] CIRAD. UMR 113 CIRAD/INRA/IRD/SUP-AGRO/UM2.
Laboratoire des Symbioses Tropicales et Méditerranéennes (LSTM). TA10/J,
Campus International de Baillarguet. Montpellier. France
[5] IRD. UMR 113 CIRAD/INRA/IRD/SUP-AGRO/UM2.
Laboratoire des Symbioses Tropicales et Méditerranéennes (LSTM). TA10/J,
Campus International de Baillarguet. Montpellier. France

ABSTRACT

The processes able to regulate plant abundance and distribution have generally been studied by addressing aboveground interactions such as plant – plant competitive interactions, plant – herbivores and/or parasites relationships and disturbance creating new patches for plant colonization. Importantly, it is now well established that plant species dynamic is tightly interlinked with the development of soil community. The rhizosphere, i.e. the biologically active soil compartment where root-root and root-

---

* Corresponding author: Tel: + 221 33 849 33 22 ; Fax : + 221 33 849 33 02. Email: Robin.Duponnois@ird.fr

microbes communications occurred, is of particular importance in mediating plant fitness and community composition. This active soil zone promotes *inter alia* the proliferation of particular microbial communities strongly involved in plant nutrition capabilities. Nutrients are heterogeneously distributed in soil and also, nutrients uptake by plants results in depletion zone around roots. Therefore, plant species differential capacity (*i*) to perform strategies allowing them to colonize soil patches or, (*ii*) to select microbial communities that are specific in their genetic and functional diversities and/or *(iii)* to form efficient microbial symbiosis for acquisition of nutrients, could be considered as main biological factors to explain the spatial distribution and the maintenance of multi-species association in plant communities. This chapter aims at reviewing some of the recent advances made in understanding plant species composition mainly in grassland communities through the involvement of plant-soil-microbes feedbacks. We will focus on interactions between plants and soil heterogeneity and/or rhizosphere microbes in affecting plant species competitive dominance.

Key words: grasslands; plant diversity; belowground competition; soil heterogeneity; soil microflora; mycorrhizae.

INTRODUCTION

Since plant biodiversity and species composition affect the functioning and stability of terrestrial ecosystems (Tilman et al., 1996), one of the major goal in plant ecology is to explain how plant coexist in natural ecosystems (Grime, 2001; Wardle, 2002). According to Begon et al. (1996), plant coexistence is defined as *"the living together of two species within the same habitat such that neither tends to be eliminated by the other"*. Some factors as nutrient availability, herbivory, environmental characteristics, etc are known to determine plant community structure and to change ecosystem processes (Tilman, 1988; Kliromonos, 2002). However the biological processes that regulate plant diversity and species composition are not yet fully understood. The understanding of these mechanisms is of major importance in order to enhance the performances of rehabilitation programmes for the conservation and restoration of natural ecosystems. Numerous studies have been focused on habitat and plant characteristics to explain plant species coexistence. It is often admitted that plant species coexistence is based on resource heterogeneity of the abiotic environment (Tilman, 1988) as well as plant characteristics such as rooting depth, phenology and germination requirements (Grubb, 1977; Berendse, 1981). However other factor could be involved in plant coexistence processes. For example, it has been demonstrated that herbivores, pathogens and mutualist could determine how plant species coexist in natural ecosystems (Clay and Holah, 1999; van der Heijden, 2002). It is now well admitted that soil microflora plays a key role in the functioning and stability of ecosystems (Tillman et al., 1996) and more particularly the mutualistic interactions between plants and microbes (Bruno et al., 2003). For instance, il has been clearly demonstrated that belowground diversity of arbuscular mycorrhizal fungi (AMF) was involved in the maintenance of plant biodiversity and in ecosystem functioning (van der Heijden et al., 1998a).

The aim of this chapter is to review some of the recent advances in the understanding of the ecological processes involved in the maintenance of plant species composition, mainly in grassland communities, with special reference to the plant-soil-microbiota feedbacks.

Interactions between plants and soil heterogeneity and/or rhizosphere microbes that affect plant species competitive dominance will be particularly explored (Fig. 1).

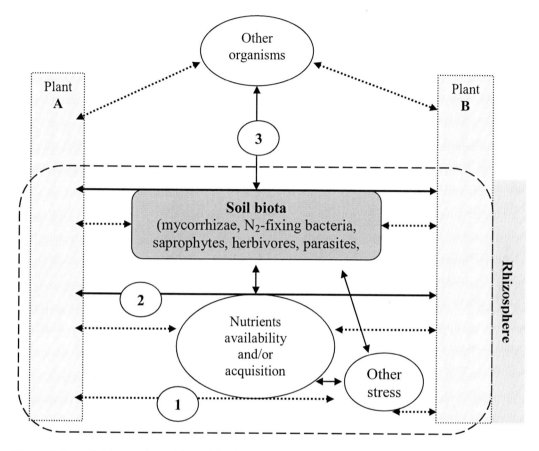

Figure 1. Potential interactions able to drive individual plant performance thereby regulating interference between plant species. Dash arrows (1) represent differential effects on distinct plant species. Solid bold arrows (2) represent influence on competitive abilities that may result from differential effects on plant species. Solid arrows (3) describe interactions that may not necessary involve plants but may affect them.

1. INTERACTIONS IN PLANT-SOIL-MICROBES CONTINUUM

Although there is established evidence that aboveground flora (as primary provider of resources that integrate the soil food web) can have important effects on the belowground subsystem, for a feedback to occur it is necessary that belowground organisms can influence aboveground community structure and functioning. As a result, soil microbes display a large genetic and functional diversity and their abundances may vary greatly in space and time and, specificity in associations with plant may also occur.

1.1. Diversity, Abundance and Function

Soil-dwelling microorganisms encompass a diversity of phylogenetic groups and all three major functional groups (producer, consumer, and decomposer) (Prosser, 2002; Reynolds et al., 2003; Thies, 2008). Soil and rhizosphere microorganisms vary both in time and space and are responsible for a wide range of ecosystem services, including decomposing organic matter, cycling and immobilizing nutrients, aggregating soil, filtering and bioremediating pollutants, suppressing and causing plant disease, and producing and releasing greenhouse gasses. A long-standing challenge for studies in soil and rhizosphere ecology has been developing effective methods that can be used to describe the diversity, abundance and function of soil and plant-associated microbial populations (Prosser, 2002; Thies, 2008; Neumann et al., 2009). Torsvik et al. (1990 a,b) estimated that in 1 g of soil there are 4000 different bacterial "genomic units" based on DNA–DNA reassociation. Yet, our knowledge of soil microbial diversity is limited in part by our inability to study soil microorganims (Tate, 1997; Kirk et al., 2004). Interactions between soil microbes and plants span the range from mutualistic to pathogenic (Nyvall, 1999; Reynolds et al., 2003; Smith & Read, 2008). As decomposers, soil microbes are directly involved in soil biogeochemistry cycles and therefore mediate the bulk of terrestrial vegetation's nutrient demand (Schlesinger, 1991; Marschner, 1997; Carpenter-Boggs, 2000). In turn, plant matter is the major source of photosynthetically fixed carbon for decomposers. Because plant species differ in both the quantity and quality of resources that they return to soil, individual plant species may have important effects on components of the soil biota and the processes that they regulate. In view of this, one of the most fascinating hot spots of activity and diversity in soil is the rhizosphere (Jones & Hinsinger, 2008). Yet microbes and plants also compete for soil nutrients (Hodges et al., 2000), making their relationship simultaneously mutualistic and competitive (Harte & Kinzing, 1993). Contrastingly, nitrogen-fixing bacteria and mycorrhizal fungi enhance host plant fitness by improving provision and uptake of nutrients or by providing protection against others pests (Duponnois & Cadet, 1994; Founoune et al., 2001; Johansson et al., 2004; Duponnois et al., 2005; Alikhani et al., 2006; Gentilli & Jumpponen, 2006; Smith & Read, 2008). Additionaly, non-mycorrhizal fungi, various rhizosphere bacteria, protozoa, and nematodes have also been reported to provide protection to plants from soilborne enemies like fungi, bacteria, actinobacteria, protozoa, nematodes, and viruses (Nyvall, 1999; Lavelle & Spain, 2001).

1.2. Specificity of Association and Specificity of Response

Plants are capable of promoting particular rhizospheric microbial populations to facilitate uptake of a limiting soil resource but in the other side, they may also deal with microbes lessining their fitness (Hamilton et al., 2001). Plants affect indigenous microbial populations in soil and, each plant species is thought to select specific microbial populations. Root exudates are a driving force in this process, but researchers are only beginning to understand the role of single compounds in mediating belowground interactions (Bais et al., 2006; Haichar et al., 2008; Berg & Smalla, 2009). For instance, grassland plant species differ in the composition of microbial communities around their roots (Bardgett et al., 1999), which helps to explain why soils planted with different grassland species support different abundances of soil microbes and microbe-feeding fauna (Griffiths et al., 1992). Plants not only provide nutrients for microorganisms, but some plants species also contain unique antimicrobial metabolites in their exudates. Importantly, many invasive plant species can have major effects on microbial communities in soils by releasing secondary compounds in their rhizosphere, which will ultimately alters belowground diversity (Callaway & Ridenour, 2004; Wolfe & Klironomos, 2005; Van der Putten et al., 2007a). In addition, different soil types are assumed to harbour specific microbial communities, as reported from a continental-scale study of soil bacterial communities (Fierer & Jackson, 2006) and, the same authors documented that the microbial biogeography may be strongly controlled by edaphic variables, especially by pH. There is no doubt that factors such as soil properties and plant species, influence the structure and function of microbial communities. However, the extent to which both factors contribute to control microbial communities is not fully understood (Berg & Smalla, 2009).

Mutualistic and pathogenic associations between plants and soil microbes are known to range in their specificity from highly specific associations (e.g., association between some mycorrhizal fungi or pathogens and their host plant) to cosmopolitan associations of plant roots with rhizosphere bacteria. One could distinguish the specificity of association (i.e., the ability to form specific associations) and the specificity of the plant and microbe responses to their association (i.e., the dependence of relative fitness on specific associations, as estimated by specificity of growth responses) (Reynolds et al., 2003). Interactions that have relatively high specificity of association, such as the association between ectomycorrhizal plants and ectomycorrhizal fungi, might also be expected to show relatively high specificity of response. However, plant-microbe interactions that show low specificity of association would not necessarily show low specificity of response. The interaction between plants and AM fungi shows relatively low specificity of association, but the plant response to individual species of AM fungi can vary greatly depending on the plant-AM fungal combination (van der Heijden et al., 1998a). Similarly, the relative growth rates of AM fungi also depend greatly on the identity of the plant with which they are associated (Eom et al., 2000; Bever, 2002).

We know that microbial populations are abundant in the rhizospheres surrounding plant roots (Paul & Clark, 1996), where they may be influenced by the quality and quantity of plant-derived soil inputs (Rovira, 1969; Hamilton & Frank, 2001; Wolfe & Klironomos, 2005). Thus, the possibility arises that specific plant-microbe and/or plant-microbial-enzyme associations could evolve (Grayston et al., 1998; Westover & Bever, 2001; Berg & Smalla, 2009).

2. THE SOIL, A TEMPORALLY- AND SPATIALLY-VARYING ENVIRONMENT

We discuss here the evidence for nutrient heterogeneity in soil and whether resource heterogeneous partitioning affects the composition and distribution of plant communities. We address the later question by considering some of differential adaptative strategies evolved by plant species to colonize soil patches, such as differential root system architecture or further distinct proliferation of roots in soil nutrient-rich zones (as species differ both in their ability to select such patches and in the speed with which they can take up the available nutrients), without explicitly taking into account the substantial involvement of rhizospheric microbes in nutrients' acquisition by plants, which will be discussed later.

2.1. Evidence for Soil Nutrients Heterogeneous Distribution

A major feature of soils is their temporal and spatial heterogeneities from the nm to the km scales (Pierret et al., 2007). Soils are complex assemblages of extremely diverse habitats, which certainly explain why they harbour such a diversity of organims and, in which water and nutrient concentrations vary both temporally and spatially. Nutrient heterogeneity, in particular, is ubiquitous within natural habitats (Gross et al., 1995; Ryel et al., 1996; Cain et al., 1999) and occurs at a wide range of scales, including those relevant to the root systems of individual plants (Lechowicz & Bell, 1991; Robertson et al., 1993; Farley & Fitter, 1999a). Heterogeneity in nutrient partitioning may result from environmental factors such as abiotic (e.g., geology, topography, climate) and/or biological (e.g., biological transformations, uptake or replenishment in the rooting zone) processes (Tilman et al., 1997a; Barot & Gigoux, 2004; Tongway & Ludwig, 2005). As a consequence of nutrient heterogeneity in soil, plant species differ greatly in many ways, including nutrient uptake rates, productivity, decomposition rates, tissue allocation, and root deployment patterns, and thus the dynamic of nutrients in the rhizosphere is likely to vary among plant growth forms, genera, and species (Schenk & Jackson, 2002; Meinders & van Breemen, 2005). In addition, when considering microbial activities in soil, a heterogeneous distribution of nutrients may result from the variability in microbe density and function. As soil nitrogen (N) and phosphorus (P) are the most common limiting nutrients in terrestrial ecosystems (Chapin et al., 1986), we mainly focus on these resources in our discussion. These nutrients exist in a variety of inorganic (e.g., ammonium, phosphate) and organic (e.g., amino acids, nucleic acids) pools made available to plants through the action of soil enzymes (e.g., proteases, ribonucleases), the bulk of which are thought to come from bacteria and fungi (Tabatabai & Dick, 2002). To a large-scale spatial level, it could also be expected that microbially mediated partitioning of soil nutrients to occur most strongly in low productivity habitats, where soil resources are relatively more limiting than light (Tilman, 1988).

Such spatial variation creates a mosaic of patches differing in size and nutrient availability (Farley & Fitter, 1999b). Therefore, a plant root is likely to experience a range of nutrient concentrations as it grows through the soil but also, soil patches displaying distinct nutrients availability (i.e., different in the type, the chemical nature and in the quantity) may be expected for plant species colonization. The timing and degree of response to soil nutrient heterogeneity differs among species (Crick & Grime, 1987; Eissenstat & Caldwell, 1988; Gross et al., 1993; Fransen et al., 1998) and also depends on the nature of nutrient in presence (Drew, 1975; Jackson & Caldwell, 1989) causing some to predict that soil nutrient heterogeneity may affect the outcome of interspecific interactions or species coexistence (Grime et al., 1991; Gross et al., 1993; Black et al., 1994; Mou et al., 1997).

2.2. Competitive Coexistence in Heterogeneous Environments

Having discussed the evidence of soil heterogeneity in space and time, we now explore whether this heterogeneity could influence plant species distribution and dynamic. Plants coexistence has been used by ecologists to describe a mixture of species in a biotic community and such coexistence constitutes a biological riddle because the tendency towards competitive exclusion should favor a monoculture (Hart et al., 2003). Theories have tried to explain plant coexistence through interactions among species (Grime, 1973; Callaway, 1997; Rees et al, 2001) whereas the role of spatial segregation and disturbance in regulating plant coexistence has been addressed through non-interaction theories (Drew, 1975; Pacala & Crawley, 1992; Fransen et al., 2001; Amarasekare, 2003). Nutrient heterogeneity could directly alter belowground interactions between neighboring plants (Casper & Jackson, 1997) and differentially affecting whole plant growth rates and aboveground interactions (Cahill & Casper, 1999). Plant species may differ in the types of patches to which they respond by root proliferation (Farley & Fitter, 1999b). Proliferation can occur through either increased rates of fine root formation or decreased root death rates (Gross et al., 1993). Therefore, differences between plant species in their ability to exploit the heterogeneity in terms of nutrients availability in soil may affect their distribution, and could be a mechanism that reduces interspecific root competition. Given that the roots of co-occuring plants can intermingle (Pechnáčková et al., 1999), if species with coarse root systems forage widely but insensitively they may leave small nutrient-rich areas that precise foragers can exploit, and such differentiation might enable plants to avoid competition for nutrients (Campbell et al., 1991). However, Campbell et al. (1991) used in their study plant species that were unlikely to co-occur, and the hypothesis has yet to be tested on species exploiting a common pattern of soil heterogeneity (Farley & Fitter, 1999b). The potential advantage of being able to exploit nutrient-rich or nutrient-poor patches by localized proliferation of roots may depend on the nature of the root system. Plant species present different roots system architecture, i.e. how the roots branch and how they are arranged in the soil. Root architecture can affect the efficiency of nutrient acquisition, thereby influencing plant community composition. Small patches and low concentration patches offer less return in terms of nutrients to the plant, and so the species that failed to respond or showed a reduced response may ensure a greater return for their investment of root growth into patches. From their studies analyzing the differing responses of a group of co-occurring plant species to heterogeneity in nutrients availability in soil, Farley & Fitter (1999b) suggested that those with copious roots are both more likely to encounter and better able to exploit nutrient-rich areas and that, the response may involve proliferation of roots or change in root architecture or a combination of both. The same authors concluded that plant species that co-occur in a habitat, and which therefore are likely to encounter similar patterns of heterogeneity, differ both in the way in which they respond to patches and to the type of patch to which they respond. These differences may affect species distribution within the habitat and may be an element of the niche differentiation that permits co-existence (Farley & Fitter, 1999b; Amarasekare, 2003). Considering this assumption, coexistence therefore requires some form of niche difference or partitioning between species that increases the strength of intra-specific competition relative to that of inter-specific competition (Amarasekare, 2003).

Again, when considering belowground competition between plants of different species exhibiting different foraging abilities, two scenarios may occur. If the larger species is better able to capture nutrients from enriched patches than the smaller species, the foraging behavior of the larger species will exacerbate the size difference with the larger smaller species and increase its competitive ability. Alternatively, if the smaller species is the better forager, it may have access to nutrient patches that are unavailable for the larger species by which the size differences between the species will be reduced and competitive abilities become more equal (Weiner et al., 1997). This latter scenario is reminiscent of the suggestion that small, subordinate plant species are capable of "high-precision" foraging by which they are better able to exploit small patches of high nutrient availability than the larger dominant species (Campbell et al., 1991). Therein, coexistence requires that competing species exhibit a strict dominance hierarchy, i.e., superior competitors can displace inferior competitors from occupied patches, but inferior competitors cannot displace superior competitors (Nee & May, 1992; Tilman et al., 1994). It is this feature that allows spatial niche differences between species. As competition involves displacement, inferior competitors do not have access to patches that superior competitors occupy. Because of their superior colonization ability however, inferior competitors can establish in patches that superior competitors do not colonize. Thus, a trade-off between displacement ability and colonization ability allows competing species to partition space or another limiting factor (e.g., nutrients) that varies in space (Amarasekare, 2003).

Hence, in patchy environments, both plant size and foraging behavior in interaction are expected to determine the competitive ability belowground (Fransen et al., 2001). Importantly, experimental studies indicated that the response of a plant species to soil nutrient heterogeneity may vary greatly when plants are grown individually or when grown with neighbors (Caldwell et al., 1991, 1996; Cahill & Casper, 1999). Caldwell et al. (1991, 1996) found that the root exploitation of nutrient patches by a plant of one species depends on the species identity of neighboring roots and on the size of competing plants, then more complicated the matter.

3. CONSEQUENCES OF PLANT SPECIES DIFFERENTIAL ABILITIES TO ASSOCIATE WITH SOIL MICROBIOTA ON PLANT COMMUNITY COMPOSITION AND DISTRIBUTION

As discussed below, nutrient patchiness in soil constitutes solely a driving force to mediate plant community structure. But this question has been documented whithout the explicit involvement of soil microbes. Importantly, nutrients availability and dynamic in soil are closely related to microbial community functions and these two parameters (nutrients cycling and microbial community functions) will difficultly be separated in natural ecosystems. Additionnaly, soil microbes are also strongly involved in nutrients acquisition by plants. Therein, we report here the complex interactions occuring in plant-soil-microbes interface (i.e., the rhizosphere) and which may be more helpful for understanding processes able to regulate plant community composition and dynamic.

Since microbes and their enzymatic activities are differentially associated with plant species, a kind of resource partitioning may arise in which different plant species access different pools, thus avoiding competition for the same pools as postulated by Reynolds et al. (2003). These same researchers also argued that plants might then experience a single nutrient axis (e.g., N or P), as a diversity of axes, thereby vastly increasing the number of niches available for competing plant species. On the surface, this might simply be the standard niche-partitioning model, but the niche partitioning may be enabled in this case by soil microbes (Reynolds et al., 2003).

In this way, a *"super microbe hypothesis"* has been suggested under the simplest form of microbial mediation in niche partitioning and, this hypothesis stated that a single microorganism was able to produce the entire range of enzymes and different plant species cause this "super microbe" to express different enzymes (Reynolds et al., 2003). However, inevitable physiological trade-offs and the diversity of microorganisms with distinct enzymatic capabilities (Thies, 2008; Berg & Smalla, 2009) make this super microbe hypothesis untenable. A more realistic hypothesis has been suggested in the literature: plant species associate with different groups of microbes, so gaining differential access to nutrient pools (Griffiths et al., 1992; Grayston et al., 1998; Johansson et al., 2004; Duponnois et al., 2005; Alikhani et al., 2006; Gentilli & Jumpponen, 2006; Smith & Read, 2008; Berg & Smalla, 2009). Evidence supporting this last hypothesis would confirm the important implication that soil microbial diversity and microbial dynamics regulates plant species coexistence.

3.1. Implication of Soil Microbiota in Mediating Plants' Nutrition and Fitness

Microbial interactions with roots may involve either endophytic or free living microorganisms and can be symbiotic, associative or casual in nature. Beneficial symbionts include N_2-fixing bacteria (e.g., rhizobia) in association with legume and interaction of roots with mycorrhizal fungi. Microbes contribute to the nutrition of plants through a variety of mechanisms including direct effects on nutrients availability (e.g., N_2-fixation by diazotrophs and P-mobilization by many organisms), enhancement of root growth (plant growth promoting rhizobacteria or PGPR), as antagonists of root pathogens (Raaijmakers et al., 2009) or as saprophytes that decompose soil detritus and subsequently increase nutrient availability through mineralization and microbial turnover. Additionnaly, indicative reports suggested that the most nutrient-efficient microorganisms may be efficienty selected by plants in the rhizosphere to cope with the scarcity of many nutrients in sols (Marschner et al., 2006; Calvaruso et al., 2007; Robin et al., 2008). Such processes are likely to be of greater significance for nutrient availability in the rhizosphere where there is increased supply of readily metabolizable carbon and where mobilized nutrients can be more easily captured by roots (Johansson et al., 2004; Gentili & Jumpponen, 2006; Alikhani et al., 2006; Richardson et al., 2009).

N_2-fixing bacteria are classified as being either symbiotic (rhizobia and *Frankia* species) or as free-living (associative) and/or root endophytic microorganisms (Cocking, 2003; Richardson et al., 2009). Rhizobia develop symbiotic relationships with host legumes and through atmospheric N_2 fixation within nodules can provide up to 90% of the N requirements of the plant (Franche et al., 2009). This group of microorganisms has also been reported to exhibit phosphate-solubilizing capabilities (Chabot et al., 1996; Alikhani et al., 2006). Phosphate solubilization activity mediated by rhizobia may be of crucial importance because P deficiency can severely limit plant growth and productivity, in particular in legumes, where both the plants and their symbiotic bacteria are affected and, P scarcity may have deleterious effect on nodule formation, development and function (Robson et al., 1981). Free-living N_2-fixers also have the potential for providing N to host plants but so far, the direct contribution of N-fixation by diazotrophs to the N nutrition of plants and subsequent growth promotion still remains in discussion (Richardson et al., 2009).

Mycorrhizal associations are potentially key players in mediating plant nutrition. Although it is well accepted that mycorrhiza assist in the acquisition of mineral nutrients already in the soil solution, there has recently been increasing interest that mycorrhizal mycelia, either by themselves, or altogether with bacteria or other fungi, may actively release nutrients from mineral particles and rock surfaces through weathering (Landeweert et al., 2001; Wallander, 2006; Finlay, 2004, 2008). Low molecular weight (LMW) organic acids have been proposed to play a role in weathering of minerals and all types of mycorrhizal fungi have been suggested to exhibit phosphatase activity (Marschner, 1995). In sterile soil, AM grasses showed increased growth and P uptake compared to noninfected plants when supplied with organic P sources as ribonucleic acid (RNA) (Jayachandran et al., 1992) and phytate (Tarafdar & Marschner, 1994). Moreover, Duponnois et al. (2005) observed that inoculation of *Acacia holosericea* with the AM fungus, *Glomus intraradices*, significantly improve phosphorus nutrition and growth of plants subsequent to rock phosphate amendment. Additionally in their experiment, Remigi et al. (2008) reported an increase in nutrient (N and P) content in the soil collected from areas where plants inoculated with the ectomycorrhizal fungus, *Pisolithus albus* IR100, have previously been cultured in comparison to un-inoculated plants. Among explicative reasons of nutrient content stimulation in ectomycorrhized plant substrate of culture, the authors suggested greater biomass of litter returning to soil, promotion of the development of bacterial communities potentially beneficial to the symbiosis and to the plant (e.g., organic phosphate-solubilizing bacteria; Frey-Klett et al., 2005). Ecto-, endo- and ericoid mycorrhizae are also capable of breaking down complex organic N and taking up dissolved organic N such as amino acids (Ahmad & Hellebust, 1991; Hawkins et al., 2000; Hodge et al., 2001; Smith & Read, 2008), facilitating host plants access to direct organic N (Turnbull et al., 1996; Reynolds et al., 2003). Recent studies have focussed on the production of extracellular proteases by ectomycorrhizal fungi under controlled conditions (Nygren et al., 2007). The production of these hydrolytic enzymes and their impact on nutrient mobilization are very variable at the intra- and inter-specific levels (Buée et al., 2009). Nevertheless, evidence about the contribution of AM fungi to the N nutrition of plants remains conflicting. In some experiments mycorrhizal plants had higher N concentrations than nonmycorrhizal controls (Johansen et al., 1994; Tobar et al., 1994); whereas in other studies no differences could be detected in the N status of inoculated and uninoculated plants (Johansen, 1999; Hawkins et al., 2000). This is particularly true for N acquisition from organic sources by AM fungi because the degradation, of organic compounds, the uptake and transfer of organic N to the plant, have for a long time been considered a prerogative of ectomycorrhizal fungi (Tibbett et al., 2000; Tibbett & Sanders, 2002; Müller et al., 2007). Importantly, Cappellazzo et al. (2008) described and functionally characterized in a yeast mutant, an amino acid permease (*GmosAAP1*) from the AM fungus *Glomus mosseae*. They reported that *GmosAAP1* mRNA was detected in the extraradical mycelium, the fungal structure that explores soil resources and that, organic N supplied as the amino acid pool induced an increase in the *GmosAAP1* transcript levels. These findings definitively underline that *GmosAAP1* plays a role in the first steps of amino acid acquisition, allowing direct amino acid uptake from the soil and extending the understanding of molecular machinery by which AM fungi exploit soil N resources.

Lindahl et al. (2005) reviewed enzymatic activities of mycorrhizal mycelia and concluded that wider recognition of the ability of many mycorrhizal fungi to mobilize nutrients from complex organic sources is a necessary step in the further development of nutrient cycling models, particularly in ecosystems with low nutrient availability. In natural communities, studies also well documented that mycorrhizae could strongly mediate plant nutrient acquisition (Simard et al., 1997, 2002; Nara & Hogetsu, 2004; Van der Putten et al., 2007b; Lambers et al., 2008; Remigi et al., 2008; Smith & Read, 2008), thereby influencing plant competitive dominance. Sequestration of N and P by mycorrhizal fungi from a range of biologically relevant substrates such as pollen (Perez-Moreno & Read, 2001a), dead nematodes (Perez-Moreno & Read, 2001b), collembola (Klironomos & Hart, 2001), and saprophytic mycelia (Lindahl et al., 1999) has been reported and the evidence has been reviewed by Read & Perez-Moreno (2003). The implication of different groups of mycorrhizal fungi in microbial mobilization-immobilization cycles results in mobilization of N and P from microbial, micro-faunal, meso-faunal, and plant litter, enabling the development of distinctive plant communities along altitudinal or latitudinal gradients (Finlay, 2008).

Community complexity increases during succession as plant, microbial, and animal species diversity increase, and this leads to increases in the quantity and kinds of nutrient pools. This is expected to be particularly true for primary succession, but could also operate to a lesser extent in certain kinds of secondary succession, such as those initiating after severe fire, where much of the above- and belowground resources may be volatilized and homogenized (Reynolds et al., 2003). In addition, as succession succeds, microbe-mediated niche diversification is also expected to be strongly affected by the specificity of the associations between plants and microbes.

3.2. EVIDENCE FOR SOIL MICROBIAL MECHANISMS IN REGULATION OF PLANT DIVERSITY AND COMMUNITIES' VARIABILITY

Soil microbiota and their feedback effect on plant growth and survival can strongly influence the relative abundance of plant species within a community and this may occur both in terms of promotion of multispecies assemblage and in terms of species replacement as succession proceeds.

3.2.1. Maintenance of Multispecies Assemblage

Established results document very clearly the role of soil microorganisms in regulating availability and acquisition of nutrients by plants. A crucial interrogation that may arise concerns the ability of this microbial community to facilitate differential access of plant species to different pools of nutrients, mainly N and P (Reynolds et al., 2003) in such a way that they could reduce competition between plant and thereby promote their coexistence.

The development of microbes (e. g., free-living bacteria) and their enzymes could locally regulate nutrients availability and the form of nutrients available in the soil. Different plant species could then access different forms of nutrients (Reynolds et al., 2003) and such mechanisms could favor the establishment and the development of different plant species according to their nutrition requirements.

Interactions between plant species and soil microbes can contribute to the maintenance of diversity in plant communities through feedbacks on plant growth resulting from changes in the composition of the soil community (Bever et al., 1997; Bever, 1999; Reynolds et al., 2003). Feedbacks result from the community dynamics generated by the specificity of response in plant–microbe interactions. Because growth rates of microbes are known to be host dependent, the composition of the soil community will likely be influenced by the identity of the local host plant. As a result, unique soil communities may form under different plant species (Bever et al., 1996; Westover et al., 1997; Wolfe & Klironomos, 2005) and under plant communities that differ in composition and abundance (Zak et al., 2003; Johnson et al., 2004). Moreover, as the growth promotion effect of soil microbes is also host-specific, the change in the soil microflora composition will likely alter the relative performance and local abundance of different plant species (Bever et al., 1997). The feedback represents the dynamics of the net direct effects of the microbial community on individual plant types (Bever et al., 1997; Reynolds et al., 2003). Contrasting dynamic feedback interactions (either positive, negative, or neutral) could occur between plants and the microbial communities that develop around their roots (Bever et al., 1997) but positive and negative feedbacks remain more documented when addressing plant community dynamics.

Positive feedbacks occur when plant species promote microbes near their roots that have beneficial effects on the plant that cultivate them, such as mycorrhizal fungi and nitrogen fixers. Positive feedbacks are thought to lead in main cases to a loss of local community diversity (Bever et al., 1997; Bever, 2003) by enhancing competitive performance of the host plant species. An illustrative example concerns the attempts at growing pine that failed until trees were inoculated with compatible ectomycorrhizal fungi, and now these pines are increasingly perceived as a potentially invasive species (Rejmanek & Richardson, 1996). A similar pattern has been documented with the invasive olive (*Elaeagnus angustifolia*) and its N-fixing *Frankia* mycosymbiont (Richardson et al., 2000). The success of plants that are dependent upon their symbiotic partners (mycorrhizal fungi or N-fixing bacteria) are dependant upon the initial abundance of their symbionts (Medve, 1984; Larson & Siemann, 1998) and nevertheless, plant coexistence could be promoted by local-scale positive feedback (Reynolds et al., 2003). Indeed, Molofsky et al. (2001) described multispecies assemblage mediated by plant species ability to associate with arbuscular vs ectomycorrhizal fungi.

Negative feedbacks generally occur when plant species promote pathogenic microbes in their rhizospheres, creating conditions that are increasingly hostile to the plants that cultivate the pathogens (Bever, 1994; Klironomos, 2002). However, negative feedback can also result from host-specific changes in the composition of the soil community that improve the growth of a second plant species (i.e., an indirect facilitation mediated by alterations in the microbial community) (Reynolds, 2003). In this concern, one might note that the composition of mycorrhizal fungal community might change in a manner that improves the growth of a competing plant species (Bever, 1999, 2002). There is accumulating evidence of negative feedback important role in structuring plant communities. Negative plant-soil biota feedbacks appear to be predominant in natural systems, and they provide frequency-dependent regulation of populations and help to maintain plant species diversity by increasing species turnover rates (Florence, 1965; van der Putten et al., 1993; Bever et al., 1997; Bever, 2003; Klironomos, 2002). For instance in agricultural settings, the accumulation of species-specific soil pathogens is though to drive crop rotation (Reynolds et al., 2003). Much of the evidence for negative feedback in unmanaged communities has come from efforts to test the Janzen-Connell hypothesis that high tropical forest tree diversity results from negative density-dependent mortality resulting from species-specific seed predation or herbivory (Janzen, 1970; Connell, 1971). Experimental studies documented that seedling mortality in tropical forests has been repeatedly found to increase with the density of conspecifics (e.g., Wills et al., 1997; Harms et al., 2000) and with proximity to mature conspecifics (Augspurger, 1992; Condit et al., 1994). Evidence suggests that these effects could derive from the accumulation of soilborne pathogens (Augspurger, 1988; Klironomos, 2002). The high mortality of black cherry (*Prunus serotina*) seedlings near adult conspecifics and at high seedling densities has been shown to result from accumulation of soil pathogens in the genus *Pythium* (Packer & Clay, 2000).

In grassland communities, examples of negative soil community feedbacks involvement in regulating plant community are also conclusive. In greenhouse assays of soil community feedback, several complementary microbial mechanisms of negative feedback have been identified, including the accumulation of host-specific soil pathogens in the genus *Pythium* (Mills & Bever, 1998; Westover & Bever, 2001), host-specific shifts in the composition of rhizosphere bacteria (Westover & Bever, 2001) and, host-specific changes in the composition of the AM fungal community (Bever, 2002). In addition, Van der Putten et al. (1990) and Van der Putten & Troelstra, (1990) have also identified complementary microbial mechanisms, with accumulation of root feeding nematodes working synergistically with pathogenic fungi, to decrease plant growth.

Exotic invasive plant species are generally though to experience positive feedbacks with the soil community. In their study, when Wolfe & Klironomos (2005) grown five of North America most notorious exotic invaders in soil that had been cultured by each of the five species, a positive growth effect was observed compared with growth in soil that had been cultured by a different species. They argued that changes in the soil community as a result of the presence of these plants would not induce negative growth effects on the same plant. But, when five rare native species were treated in the same way, a negative growth effect was observed when growing in their own soil compared with the growth of these plants in the soil of others species. It suggested that the plants accumulated pathogens in their local soil community. Hence, they concluded that exotic plants, and in some cases widespread native plants, could be abundant within native communities because they do not experience the same negative feedback with soil biota as do rare native species (Klironomos, 2002). The feedbacks between plants and the soil community can therefore explain the relative abundance of plant species, with the most abundant species having positive or neutral feedbacks with the soil (Klironomos, 2002; Wolfe & Klironomos, 2005).

Plants and soil microorganisms have a limited ability to move towards nutrient-enrichided zones, compared with animals. To cope with those conditions, they have therefore evolved a whole range of strategies for accessing nutrient resources in the soil (Marschner, 1995; Lambers et al., 2006; Lambers et al., 2008; Smith & Read, 2008). Differences in the efficiency by which different plant species take up and use nutrients could also determined the make up of plant communities. Plants have several contrasting and complementary strategies for increasing the acquisition of both N and P, and for broadening the options of uptake from resources of different chemical composition (Read & Perez-Moreno, 2003; Shane & Lambers, 2005; Lambers et al., 2008). Plants can improve the availability and acquisition of nutrients through mechanisms such as development of various types of mycorrhizae, N_2-fixing symbioses, cluster roots, root preferential proliferation in soil nutrient-rich patches, etc (Francis & Read, 1994; Lambers et al., 2008). Certain plant species combine more than one strategy to successfully establish (e.g. Casuarinaceae and Fabaceae are both mycorrhizal and cluster-bearing; Lambers et al., 2006). It has been documented that cluster roots may be more efficient in acquiring P in very P-impoverished soil because these structures could produce large amounts of carboxylates, which release P from strongly sorbed forms by either replacing P bound to Al or Fe in acid soils or to Ca in alkaline soils or by local reduction of pH in highly alkaline soils. Cluster roots could also produce phosphatases and these compounds could strongly mine organic sources and release P (Lambers et al., 2008). Lambers et al. (2006) documented in South Western Australia that non-mycorrhizal *Proteaceae* with cluster roots occupy the most P-impoverished soils, whereas the mycorrhizal *Myrtaceae* without root clusters predominate on soils that have somewhat higher P levels. Considering the soil as a spatially heterogeneous environment, plant species coexistence could be maintained through plant differential abilities to colonize distinct soil patches. This distribution pattern may seem surprising, since mycorrhizae are generally thought beneficial for P acquisition from nutrient-poor soils (Smith & Read, 2008). However, it is now firmly established that the strategy of species with cluster roots allows mining of P from soils where the availability of P is very low, and most P is sorbed, whereas mycorrhizal plants predominantly scavenge P that is in solution (Lambers et al., 2006, 2008). Nevertheless, further works are needed to fully appreciate the more intricate relationships between soils and roots and to address the potential for species with highly efficient P-acquisition strategies to benefit their less efficient neighbors.

Also regarding plant species differential abilities to mobilize scarce nutrients in sol, Raynaud et al. (2008) reported that large diffusion rates of root exudates my result in increase bioavailability of nutrients such as phosphate for neighboring plant roots which do not exhibit same exudation potential, thereby contributing to facilitation rather than competition between plant species.

Agent-mediated coexistence is a non-interaction theory that has been proposed as a mechanism for maintaining multi-species assemblages in plant communities (Pacala & Crawley, 1992). Among the agents promoting plant coexistence, mycorrhizal fungi have been proposed as efficient drivers for the maintenance of biodiversity within plant communities (Janos, 1980; Allen & Allen, 1990; Zobel & Moora, 1995). It is clear from various studies that mycorrhizae have large influences on plant community structure and are an important factor in the stability of plant species composition, as evidenced by the large magnitude of changes in plant communities reported in response to the presence *vs* absence of mycorrhizal symbionts or to the particular species of mycorrhizal fungi present (Grime et al., 1987; van der Heijden et al., 1998 a,b; Hart et al., 2003). Spatial heterogeneity of mycorrhizal fungal infectivity may increase plant species diversity, allowing nonmycotrophic and mycotrophic species to coexist in patches of low and high mycorrhizal soil potential, respectively (Allen, 1991). Hartnett & Wilson (2002) further extend Allen's hypothesis and suggest that, in habitats with significant spatial heterogeneity in mycorrhizal soil infectivity, there will be a positive relationship between interspecific variability in host plant mycotrophy and plant species diversity. A related hypothesis is that spatial heterogeneity in mycorrhizal soil infectivity, coupled with variation in host species mycorrhizal dependency for colonization and establishment (e.g., regeneration niche differences) may also enhance species diversity in plant communities (Hartnett et al., 1994). Importantly, the mycorrhizal enhancement of germination and establishment of many tallgrass prairie plant species may not only promote species diversity, but may also increase the rate of succession by allowing highly mycotrophic species to more rapidly displace ruderal species (Smith et al., 1998).

Previous studies have been carried out with AM fungi and focused almost exclusively on a few AM fungal species in a single genus (*Glomus*). Afterwards, substantial host specificity of AM fungi with plant hosts has been demonstrated (Bever et al., 1996; van der Heijden et al., 1998a). This host specificity, in combination with AM fungal diversity, have been involved in the maintenance of plant community diversity (van der Heijden et al., 1998a; Bever et al., 2001) although the specific mechanisms remain not well understood (Reynolds et al., 2003). In the experiments carried out by van der Heijden et al. (1998a), increasing fungal diversity resulted in greater species diversity and higher productivity, suggesting that changes in belowground diversity of mycorrhizal symbionts can drive changes in aboveground diversity and productivity. The mechanism behind these effects is likely to be differential effects of specific plant-fungus combinaitions on the growth of different plant species. Regarding this, some authors suggested that higher AM fungi diversity could lead to higher plant coexistence simply by increasing the probability of individual plant species associating with a compatible and effective AM fungi partner (Hart et al., 2003). This is consistent with the idea emerging from many molecular studies, that the degree of mycorrhizal specificity may be higher than Hitherto supposed. If the addition of new fungal species leads to increases in the survival and vigour of more plant species that are responsive to mycorrhizal colonization, there may be a positive feedback on the mycorrhizal fungi, leading to more efficient resource utilization and increases in overall productivity (Finlay, 2004, 2008). Nevertheless, these effects may be context dependent and Vogelsang et al. (2006) suggested that plant diversity and productivity are more responsive to AM fungi identity than to AM fungi diversity *per se*, and that AM fungal identity and P environment can interact in complex ways to alter community-level properties. Whatever, increased AM fungi species richness is argued to be beneficial both in terms of host compatibility (Sanders & Fitter, 1992; Bever et al., 1996; van der Heijden et al., 1998b; Eom et al., 2000; Klironomos, 2000; Klironomos et al., 2000) and in terms of multifunctionality of AM fungi (Newsham et al., 1995; Klironomos, 2000). Ultimately, as previously underlined by van der Heijden & Scheublin (2007), defining functionally distinct AM fungal groups is essential to fully understand the interactions between plant and AM fungal communities in agricultural and natural ecosystems.

Interestingly, if AM fungi are capable of direct access to pools of soil nutrients (e.g., phosphorus) not necessarily or really inaccessible to nonmycorrhizal plants, then it may be possible to predict that different AM fungal species might facilitate differential access of plant species to these pools, thus promoting plant species coexistence through nutritional niche partitioning (Reynolds et al., 2003). Although AM fungi have traditionally been thought as non-specific in their association with plants, increasing evidence has demonstrated substantial specificity in plant host response to different AM fungal species (Klironomos, 2003; van der Heijden et al., 1998b). Later, Reynolds et al. (2006) reported that nutrient partitioning may be a less advantageous strategy than an ability to take advantage of multiple forms of nitrogen or phosphorus, especially for sessile organisms in a temporally and spatially heterogeneous world. Thereby, partitioning along any number of other niche dimensions (Tilman & Pacala, 1993) or competitive equivalence (Hubbell, 2001; Silvertown,

2004; Tilman, 2004) must account for the coexistence of largely co-dominant plant species, reflecting the "fundamental" but not the "realized" niches of these species (Reynolds et al., 2006). Additionally, plant mycorrhizal dependence and their position in the local dominance hierarchy (Urcelay & Diaz, 2003) in combination with AM fungal presence and abundance in soil may strongly influence plant community dynamics. In this context, it has been suggested that if an otherwise less competitive plant species is infected by more AM fungi than is a highly competitive plant species, then AM fungi should promote coexistence by increasing the ability of less competitive species to access nutrients (Zobel & Moora, 1995; Moora & Zobel, 1996). Consistent with this, Facelli & Facelli (2002) documented that mycorrhizal symbiosis has the potential to strongly influence plant community structure by favoring coexistence with mycorrhizal plants when soil nutrient distribution is heterogeneous because it promotes pre-emption of limiting resources. However, one may note that if a highly competitive plant species is also more infected by AM fungi, then AM fungi could simply reinforce competitive dominance by that species (West, 1996).

Different plant species can be compatible with the same species of mycorrhizal fungi and be connected to one another by a common mycelium and to a large extent, plants within communities can be interconnected and form a common mycorrhizal network based on their shared mycorrhizal associates (Francis & Read, 1984; Simard et al., 1997; Selosse et al., 2006; Richard et al., 2009). Common mycorrhizal networks either could originate from fungal genets colonizing neighboring roots during their growth, but also from hyphal fusions uniting previously separated mycelia (Selosse et al., 2006). Transfer of carbon, nitrogen, and phosphorus (Finlay & Read, 1986; Newman & Eason, 1993; Simard & Durall, 2004; Selosse et al., 2006) are well documented between interconnected plants and, exchanges may occur either with conspecific and interspecific plants. One important consequence of nutrient transfer through common mycorrhizal networks is that adult plants could favor early establishment of conspecific or interspecific seedlings (Newman, 1988; Simard & Durall, 2004) but also, subordinate species could be maintained in plant community owing nutrients uptake from mycorrhizal mycelia guilds (Hoeksema & Kummel, 2003; Smith & Read, 2008). Mycoheterotrophy reflects an example of mycorrhizal networks-mediated plant species coexistence. Mycoheterotrophic plants neither photosynthesize nor have chlorophyll, but are mycorrhized with fungi that are also connected to other green plants, which constitute the ultimate source of carbon for the mycoheterotroph (Bidartondo, et al., 2002; Selosse et al., 2002; Leake, 2004). Also, Grime et al. (1987) hypothesized that their detected increase in plant species diversity in turfgrass microcosms in response to the presence of mycorrhizae was due to extensive mycelial networks facilitating interplant resource transfer *via* hyphal connections. They postulated that, allowed small, suppressed plants to obtain carbohydrates from the larger, dominant species *via* shared mycorrhizal hyphae, ultimately increased the equitability of species abundances. These same authors then suggested that the "export of assimilate from 'source' (canopy dominants) to 'sink' (understory components) through a common mycelial network may be an important mechanism for the maintenance of multi-species assemblages in infertile soils". Nevertheless, Pfeffer et al. (2004) found no evidence to support the movement of carbon between interconnected roots of AM plants.

Contrastingly, a key point to note is that mycorrhizal networks can, sometimes, lead to lower diversity if one species in the network is the dominant sink for nutrients (Connell & Lowman, 1989; Allen & Allen, 1990). This case is well illustrated by the ability of some nonnative invasive plant species to divert to their advantage carbon and nutrients (P) transfers through the common mycelial network, thus severely reducing the growth of native plants (Zabinski et al., 2002; Carey et al., 2004).

The results of these various experiments indicate the potential that the movement of plant resources through hyphal interconnections may be an important mechanism influencing plant species interactions and community structure in grasslands and other plant communities. However, results are equivocal and much further studies are needed to determine whether this mechanism plays a significant role in patterns of species abundances and diversity in natural communities. In addition, the methodological limitations and difficulties in measuring the patterns and consequences of this phenomenon in the field will be the greatest challenge (Hartnett & Wilson, 2002).

Mycorrhizal symbioses have recently been reported to strongly counterbalance the allelopathic effects of exotic fast growing trees against native herbaceous plant species communitiy adjacent to stands of exotic plants. In experiments carried out in greenhouse in sahelian conditions, Sanon et al. (2006) and Kisa et al. (2007) observed that previous inoculation of exotic trees, *Gmelina arborea* or *Eucalyptus camaldulensis*, with the AM fungus *Glomus intraradices* significantly increased the growth of native herb species in the mesocosms where the trees were cultivated; thus favoring coexistence of exotic trees and native herbaceous plant species community. In the same line of ideas, from experiments carried out in greenhouse conditions, we observed that the growth of native *Acacia* species was severely reduced in the soil invaded by *Amaranthus viridis*, an annual weed native from Central America (Sanon et al., unpublished data). Interestingly, the inoculation of *G. intraradices* was highly beneficial to the growth and nodulation of *Acacia* species. From both studies (inoculation of exotic fast growing trees to alleviate their allelopathic effect on endogenous communities and inoculation of native *Acacia* species to favour their reestablishment in soils displaying severe alteration of chemical and microbial characteristics due to invasion of *A. viridis*), the authors postulated that the beneficial effect of AM fungus inoculation may result from either the well-developed mycelial network owing equalization of distribution of soil resources among competitively dominant and subdominant plant species (Grime et al., 1987; Wirsel, 2004) or allelochemical-mediating effects from AM fungi which altogether with their mycorrhizosphere microbial communities are known to inactivate or catabolize toxic compounds (Pellissier & Souto, 1999; Blum et al., 2000). Similar results regarding restoration of native plant diversity by utilizing AM fungi have also been reported in southern California by Vogelsang et al. (2004). In these ecosystems, most native weed plant species were dependent on AM symbioses for optimal growth. Conversely, growth of many of the pernicious exotic weedy species was not improved, and could be reduced by these fungi. Then, inoculation of AM fungi enabled speeded establishement of vigorous locally adapted vegetation in these areas, which will ultimately reduced exposure of natural communities to the forces of erosion.

Globally, these results highlight the role of the AM symbiosis in the processes involved in plant species coexistence and in ecosystem management programmes that target restoration of native plant diversity.

Additionally, the potential for microbially mediated niche partitioning may be of great importance in plant coexistence mediation by microbes; and this partitioning of the niche by soil microbiota may even be greater in ecto- and ericoid mycorrhizae which are thought to be more diverse than AM fungi and to exhibit greater host specificity (Reynolds et al., 2003). In this concern, Turnbull et al. (1995) documented that isolates of four ectomycorrhizal species from northern Australian forests had different capabilities to use a range of amino acids, protein and inorganic N sources and this was affected by host plant identity.

In combination to their direct effects on their host plants, mycorrhizal fungi may influence plant communities indirectly through their effects on interactions between plants and their herbivores, pathogens, pollinators, and other microbial mutualists (Finlay & Soderstrom, 1989; Fitter & Garbaye, 1994 ; Wurst et al., 2004 ; Cahill et al., 2008 ; Gehring & Bennett, 2009). In turn, these interactions can indirectly and, occasionally directly influence mycorrhizal fungal communities and their functions (Eom et al., 2001; Gehring & Bennett, 2009). Plants often experience simultaneous and potentially conflicting effects from mycorrhizal fungi and herbivores, pathogens, and other symbionts, and all of these organisms jointly influence important plant resources and costs and benefits. Therefore, strong interactions and indirect effects of mycorrhizal fungi in plant communities should be expected (Gehring & Bennet, 2009). Although the direct effects of mycorrhizae through symbiotic uptake of mineral nutrients have been well explored, accumulating evidence suggests that these strong indirect effects of mycorrhizae may be of great relevance in plant communities composition and dynamics and should not be ignored (Hartnett & Wilson, 2002). Such complex, multitrophic interactions still remain not enough documented and may strongly vary in time and space. In addition to plant P and N content, other plant compounds, such as secondary metabolites (Gange & West, 1994; Koide, 2000), are affected by AM fungi root colonisation. Changes in foliar chemistry may influence plant herbivore interactions and, herbivore performance has been reported to be positively affected (Gange & West, 1994; Gange et al., 1999) or negatively (Gange & West, 1994; Gange, 2001) by AM fungus colonisation, depending on plant, herbivore and fungal species present. Ecosystem multitrophic interactions may result ultimately in subsequent effects on plant fitness, community composition and distribution. These informations underscore the importance of above-and belowground linkages and indicate that alterations in mycorrhizal and rhizosphere processes can have large indirect effects on plant communities through their effects on plant responses to above- and belowground consumers.

Overall, evidence suggests that the presence or absence of mycorrhizal fungi, the growth responses (nutrient uptake improvement but certainly mycorrhizal mediation of other biotic interactions such as plant-herbivore or plant-pathogen interactions also) or mycorrhizal dependency of host plant species, and/or differential plant species responses to particular species of mycorrhizal fungi may be potentially important factors shaping the performance and relative abundance of plant species within local communities.

Complemantary experimental works must be undertaken for a better understanding of whether AM community density impacts on plant community composition and dynamic. At a high concentration of AM fungi inoculum, infection by AM fungi is thought to become detrimental rather than beneficial because heavily infected plants might experience a large carbon removal that outweighs any benefit (Gange & Ayres, 1999; Hart et al., 2003). Moreover, it is important to note that soil phosphorus availability in soil could determine whether AM fungi will affect the outcome of competition among neighboring plants. In condition of low soil P availability resulting in strong competition between plant species for this resource, the effect of AM fungi might be more efficient by increasing the ability of mycorrhizal plants to acquire nutrients. In the other side, Hart et al. (2003) documented that the benefit of increased P uptake due to mycorrhizal infection of mycotrophic plant species may probably be negligible in soil with high P availability. As a result, the cost of the AM association might exceed the benefit they provide (Johnson et al., 1997; Purin & Rillig, 2008), reducing the competitive capabilities of the host plant and therefore its ability to coexist with non-mycorrhizal neighbors.

Nitrogen-fixing bacteria are other important drivers of ecosystem processes through the "Agent-mediated coexistence" hypothesis, because nitrogen is one of the main elements that limits plant productivity in natural ecosystems. Recent studies have shown that legumes play a key role in species-rich grassland by increasing plant productivity and nitrogen capture (Tilman et al., 1997b; Mulder et al., 2002; Hooper & Dukes, 2004; van der Heijden et al., 2006). There are several ways in which rhizobia can influence plant communities. Rhizobia may facilitate the growth of other plant species by increasing nitrogen availability in the soil. This is possible when nitrogen transfer from legumes to nonlegumes occurs (through root exudates or mycorrhizal hyphal networks) or when legume roots or root nodules decompose (Heichel, 1987; Mårtensson et al., 1998; Simard et al., 2002). Intercropping of legumes and nonlegumes, a common practice in agriculture, is based on this principle (Trenbath, 1974; Vandermeer, 1989; Akinnifesi et al., 2006) and finally, some studies have stated that rhizobia directly promote the growth of various nonhosts (Yanni et al., 1997; Antoun, et al., 1998; van der Heijden et al., 2006). In view of this, van der Heijden et al. (2006) observed that the legume species in their studies required rhizobia to successfully coexist with nonlegumes but also that, symbiotic interactions between legumes and rhizobia contribute to plant productivity, plant community structure and acquisition of limiting resources in dune grassland. They particularly observed that, in the presence of rhizobia, competitive interactions between legumes and nonlegumes did not occur. Instead, niche separation and resource partitioning were observed. Legumes utilized symbiotically fixed atmospheric nitrogen, whereas nonlegumes obtained nitrogen from the soil. Also in the part of their study implemented in the field, the relative abundance of the nitrogen isotope ^{15}N, the δ^{15}N value, in legumes collected from the field was close to zero, the signature of atmospheric nitrogen; strongly indicating that legumes and nonlegumes in the field also seemed to use different nitrogen sources (van der Heijden et al., 2006).

3.2.2. Species Replacements

Species replacement could be viewed as a mechanism to ensure plant diversity; mainly in the situation where an abundant species is progressively replaced by a single one or by several other species. We explicitly avoid extreme situations such as those in which for instance, a species A is completely displaced by a species B as in this case we assist to species replacement during succession rather than long-term coexistence (Van der Putten et al., 1993).

Plant ability to associate with particular symbionts could be crucial in mediating plant species replacement during succession. The importance of positive feedback dynamics between plants and N-fixing microbes in primary successions is well documented. For instance, N-fixers such as *Dryas* are among the first vascular plants to establish during primary succession at Glacier Bay (Crocker & Major, 1955). Moreover, death and decomposition of N-fixing species results in soil enrichment in nutrients. Thus, the benefits of N-fixation are not restricted to the plant host (Chapin et al., 1994), and a positive feedback dynamic facilitates transitions towards additional plant species (Reynolds et al., 2003). Such mechanisms could also been observed with N-fixing associations in secondary successions on infertile soil (Reynolds et al., 2003).

In contrast to N-fixing associations, mycorrhizal associations are relatively ubiquitous and act to increase access to rather than add nutrients to soil. The availability of P and other nutrients often decrease in late succession (Walker & Syers, 1976; Lambers et al., 2008). In this concern, Reeves et al., (1979) and Janos (1980) envisioned a shift from nonmycorrhizal to obligately mycorrhizal plants from early to late secondary succession, with low levels of facultatively mycorrhizal associations throughout. Contrarily to these previous observations, more recent studies rather suggest, that very mycotrophic plants species would be pioneers in early successions in certain environmental conditions, acting as "nurse plants" as termed by Carrillo-García et al. (1999) to promote "fertility islands" (Garner & Steinberger, 1989) or "resource islands" (Reynolds et al., 1990; Schlesinger et al., 1996) where facilitation and replacement among plants species may be highly fostered (Callaway, 1995, 1997). Again, Smith & Read (2008) also reported in sand dune and many other successional communities a shift later in succession from herbaceous plant species involved in obligate symbioses with AM fungi to woody species involved in obligate symbioses with ecto- or ericoid mycorrhizae, coincident with a shift in predominance of inorganic versus organic N (Read, 1993). Nevertheless, positive feedback, operating alone, would be expected to lead to monocultures, or even, in the absence of soil building or temporal shifts in forms of nutrients (e.g., inorganic to organic N), arrested successions (Reynolds et al., 2003). Some authors additionally postulated that higher host specificity of ecto- compared to endomycorrhizae leads to dominance by ectomycorrhizal species (Connell & Lowman, 1989).

Negative feedback dynamics are expected to be less important during the initial stage of succession, which is characterized by harsher conditions and lower host densities that are favorable to most disease organisms. As plant host densities build and modify the abiotic environment, conditions should become more favorable to soil pathogens, leading to an increasing role for negative feedback in driving species replacements in succession (Reynolds et al., 2003). Van der Putten et al. (1993) pointed out that soilborne diseases can drive successional change in a foredune community: individuals showed reduced biomass in the soil of their successors but not in the soil of their predecessors. The traits of early successional species themselves may make them particularly vulnerable to negative feedback by soil pathogens. Rapid growth is a well-known trait of ruderal species, and it is also established that rapid growth trades off with allocation to herbivore defense (Coley et al., 1985; Poorter, 1990). Considering that belowground patterns of growth and defense reflect aboveground patterns, ruderal species, by their own success, increase the likelihood that they will be replaced by slower growing species, better defended against pathogens (Reynolds et al., 2003). Differences in quality as well as in quantity of antiherbivore defense have been documented for early and late successional species, defenses or ruderal tend to be against generalist herbivores while defenses of climax species tend to be against specialists (Coley et al., 1985). Specialization of plant defenses against soil pathogens would then increase over successional time and as a result increased specialization of pathogens to overcome sophisticated plant defenses should occur (Reynolds et al., 2003), leading to reciprocal negative feedback that promotes plant species coexistence or replacement. Furthermore, opportunities for pathogen dispersal to a site increase over succession, leading to greater pathogen diversity and increasing the chances for reciprocal negative feedback.

Resource partitioning mediated by soil microbes might play a role in species replacements if the forms of available nutrients (N or P) change through succession. Gorham et al. (1979) implicated fungal symbionts in such sequential partitioning. Indeed, the example of a shift from arbuscular to ecto- or ericoid mycorrhizal plant species with change in inorganic to organic forms of N, previously reported in our discussion, well illustrated this process. Others shifts in forms of nutrients over succession (e.g., nitrate to ammonium, or protein to chitin) would also provide opportunity for additional species replacements on the basis of soil resource specialization (Reynolds et al., 2003). Again, Reynolds et al. (2003) also document that as the diversity of microbes and nutrient inputs increase over succession, the opportunity for microbially mediated differentiation in resource use would increase, promoting increase plant community diversity over succession. Supporting this idea, ectomycorrhizal infectiveness and diversity has been found to increase over a successional gradient (Boerner et al., 1996). For instance, in the case of soil pathogen and mycorrhizal fungi, associations typically exhibit higher levels of specificity as successions proceeds. Additionally, certain researchers envisioned the same pattern for all microorganisms globally involved in mediating resources differentiation, and predicted that greater opportunities for species coexistence (i.e., greater specialization of resource niches) may arise as succession proceeds.

However, more research studies on such interactions and their relevance for the maintenance of plant diversity remain needed for mycorrhizae and others microbes in general.

CONCLUSION

All these results showed that soil microbiota is of particular importance in the functioning and stability of terrestraial ecosystems and support the idea that plant-microbe interactions are an integral part of ecosystems. Hence it emphasizes that soil microbiota (and more particularly mutualic microorganisms such as rhizobia and AM fungi) must be conserved, protected and managed to ensure the success of rehabilitation programmes of terrestrial ecosystems. Moreover ecological mechanisms that govern plant-soil-microbiota interactions have to be fully investigated in order to understand how plant coexist in natural ecosystems.

REFERENCES

Ahmad, I. & Hellebust, J. A. (1991). Enzymology of nitrogen assimilation in mycorrhiza. *Methods in Microbiology, 23*, 181-202.

Akinnifesi, F. K., Makumba, W. & Kwesiga, F. R. (2006). Sustainable maize production using *Gliricidia/Maize* intercropping in southern Malawi. *Experimental Agriculture, 42*, 441-457.

Alikhani, H. A., Saleh-Rastin, N. & Antoun, H. (2006). Phosphate solubilization activity of rhizobia native to Iranian soils. *Plant & Soil, 287*, 35-41.

Allen, E. B. & Allen, M. F. (1990). The mediation of competition by mycorrhizae in successional and patchy environments. In : J. B. Grace, & G. D. Tilman (Eds), *Perspectives on plant competition* (pp. 367–389). New York, USA : Academic Press.

Allen, M. F. (1991). *The ecology of mycorrhizae.* Cambridge, UK : Cambridge University Press.

Amarasekare, P. (2003). Competitive coexistence in spatially structured environments: a synthesis. *Ecology Letters, 6*, 1109-1122.

Antoun, H., Beauchamp, C. J., Goussard, N., Chabot, R. & Lalande, R. (1998). Potential of *Rhizobium* and *Bradyrhizobium* species as plant growth promoting rhizobacteria on non-legumes : effects on radishes (*Raphanus sativus* L.). *Plant & Soil, 204*, 57-67.

Augspurger, C. K. (1992). Experimental studies of seedling recruitment from contrasting seed distributions. *Ecology, 73,* 1270-1284.

Augspurger, C. K. (1988). Impact of plant pathogens on natural plant populations. In A. J. Davy, M. J. Hutchings, & A. R. Watkinson (Eds), *Plant population ecology* (pp. 413-433). Oxford, UK: Blackwell Scientific,

Bais, H. P., Weir, T. L., Perry, L. G., Gilroy, S. & Vivanco, J. M. (2006). The role of root exudates in rhizosphere interactions with plants and other organisms. *Annual Review of Plant Biology, 57,* 234266.

Bardgett, R. D., Mawdsley, J. L., Edwards, S., Hobbs, P. J., Rodwell, J. S. & Davies, W. J. (1999). Plant species and nitrogen effects on soil biological properties of temperate upland grasslands. *Functional Ecology, 13,* 650-660.

Barot, S. & Gignoux, J. (2004). Mechanisms promoting plant coexistence: can all the proposed processes be reconciled? *Oikos, 106,* 185-192.

Begon, M., Harper, J.L. & Townsend, C.R. (1996). *Ecology: individuals, populations and communities*, 3rd edn. Oxford, UK: Blackwell.

Berendse F. 1981. Competition between plant populations with different rooting depth. II. Pot experiments. *Oecologia, 48*, 334-341.

Berg, G. & Smalla, K. (2009). Plant species and soil type cooperatively shape the structure and function of microbial communities in the rhizosphere. *FEMS Microbiology Ecology, 68,* 1-13.

Bever, J. D. (2002). Negative feedback within a mutualism: Host-specific growth of mycorrhizal fungi reduces plant benefit. *Proceedings of the Royal Society of London, 269,* 2595-2601.

Bever, J. D. (1994). Feedback between plants and their soil communities in an old field community. *Ecology, 75,* 1965-1977.

Bever, J. D., Morton, J. B., Antonovics, J. & Schultz, P. A. (1996). Host-dependent sporulation and species diversity of arbuscular mycorrhizal fungi in a mown grassland. *Journal of Ecology, 84,* 71-82.

Bever, J. D., Westover, K. M. & Antonovics, J. (1997). Incorporating the soil community into plant population dynamics: The utility of the feedback approach. *Journal of Ecology, 85,* 561-573.

Bever, J. D. (1999). Dynamics within mutualism and the maintenance of diversity: inference from a model of interguild frequency dependence. *Ecology Letters, 2,* 52-61.

Bever, J. D., Schultz, P., Pringle, A. & Morton, J. (2001). Arbuscular mycorrhizal fungi: more diverse than meets the eye and the ecological tale of why. *BioScience, 51,* 923-931.

Bever, J.D. (2002). Negative feedback within a mutualism: host-specific growth of mycorrhizal fungi reduces plant benefit. *Proceedings of the Royal Society of London, 269*, 2595-2601.

Bever, J. D. (2003). Soil community feedback and the coexistence of competitors: conceptual frameworks and empirical tests. *New Phytologist, 157*, 465-473.

Bidartondo, M. I., Redecker, D., Hijri, I., Wiemken, A., Bruns, T. D., Dominguez, L., Sérsic, A., Leake, J. R. & Read, D. J. (2002). Epiparasitic plants specialized on arbuscular mycorrhizal fungi. *Nature, 419,* 339-392.

Black, R. A., Richards, J. H. & Manwaring, J. H. (1994). Nutrient uptake from enriched soil microsites by three great basin perennials. *Ecology, 75,* 110-122.

Blum, U., Statman, K. L., Flint, L. J. & Shaefer, S. R. (2000). Induction and/or selection of phenolic acid-utilizing bulk-soil and rhizospheric bacteria and their influence on phenolic acid phytotoxicity. *Jounal of Chemical Ecology, 26,* 2059-2078.

Boerner, R. E. J., DeMars B. G. & Leicht, P. N. (1996). Spatial patterns of mycorrhizal infectiveness of soils along a successional chronosequence. *Mycorrhiza, 6,* 79-90.

Buée, M., De Boer, W., Martin, F., van Overbeek, L. & Jurkevitch, E. (2009). The rhizosphere zoo: An overview of plant-associated communities of microorganisms, including phages, bacteria, archaea, and fungi, and of some of their structuring factors. *Plant & Soil,* doi 10.1007/s11104-009-9991-3.

Bruno, J.F., Stachowicz, J.J. & Bertness, M.D. (2003). Inclusion of facilitation into ecological theory. *Trends in Ecology and Evolution, 18,* 119-125.

Cahill, JR, J. M. & Caper, B. B. (1999). Growth consequences of soil nutrient heterogeneity for two old-field herbs, *Ambrosia artemisiifolia* and *Phytolacca americana*, grown individually and in combination. *Annals of Botany, 83,* 471-478.

Cahill, J. F., Elle, E., Smith, G. R. & Shore, B. H. (2008). Disruption of a belowground mutualism alters interactions between plants and their floral visitors. *Ecology, 89,* 179-1801.

Cain, M. L., Subler, S., Evans, J. P. & Fortin, M. -J. (1999). Sampling spatial and temporal variation in soil nitrogen availability. *Oecologia, 118,* 397-404.

Caldwell, M. M., Manwaring J. H. & Durham S. L. (1991): The microscale distribution of neighbouring plant roots in fertile soil microsites. *Functional Ecology, 5,* 765-772.

Caldwell, M. M., Manwaring J. H. & Durham S. L. (1996). Species interaction at the level of fine roots in the field: Influence of soil nutrient heterogeneity and plant size. *Oecologia, 106,* 440-447.

Callaway, R. M. (1995). Positive interactions among plants. *Botanical Review, 61,* 306–349.

Callaway, R. M. (1997). Positive interactions in plant communities and the individualistic-continuum concept. *Oecologia, 112,* 143-149.

Callaway, R. M. & Ridenour, W. M. (2004). Novel weapons: invasive success and the evolution of increased competitive ability. *Frontiers in Ecology and the Environment, 2,* 436-443.

Calvaruso, C., Turpault, M. P., Leclerc, E. & Frey-Klett, P. (2007). Impact of ectomycorrhizosphere on the functional diversity of soil bacterial and fungal communities from a forest stand in relation to nutrient mobilization processes. *Microbial Ecology, 54,* 567-577.

Campbell, B. D., Grime, J. P. & Mackey, J. M. L. (1991). A trade-off between scale and precision in resource foraging. *Oecologia, 87,* 532-538.

Cappellazzo, G., Lanfranco, L., Fitz, M., Wipf, D. & Bonfante, P. (2008). Characterization of an amino acid permease from the endomycorrhizal fungus *Glomus mosseae*. *Plant Physiology, 147,* 429-437.

Carey, E. V., Marler, M. J. & Callaway, R. M. (2004). Mycorrhizae transfer carbon from a native grass to an invasive weed: evidence from stable isotopes and physiology. *Plant Ecology, 172,* 133-141.

Carpenter-Boggs, L., Kennedy, A. C. & Reganold, J. P. (2000). Organic and biodynamic management: effects on soil biology. *Soil Science Society of America Journal, 64,* 1651-1659.

Carrillo-García, A., León de la Luz, J. L., Bashan, Y. & Bethlenfalvay, G. J. (1999). Nurse plants, mycorrhizae, and plant establishment in a disturbed area of the Sonoran desert. *Restoration Ecology, 7,* 321-335.

Casper, B. B. & Jackson, R. B. (1997). Plant competition underground. *Annual Review of Ecology & Systematics, 28,* 545-570.

Chabot, R., Antoun, H. & Cescas, M. P. (1996). Growth promotion of maize and lettuce by phosphate solubilizing *Rhizobium leguminosarum* biovar phaseoli. *Plant & Soil, 184,* 311-321.

Chapin III, F. S., Walker, L. R., Fastie, C. L & Sherman, L. C. (1994). Mechanisms of primary succession following deglaciation at Glacier Bay, Alaska. *Ecological Monographs, 64,* 149-175.

Clay, K. & Holah, J. (1999). Fungal endophyte symbiosis and plant diversity in successional fields. *Science, 285,* 1742-1744.

Cocking, E. C. (2003). Endophytic colonization of plant roots by nitrogen-fixing bacteria. *Plant & Soil, 252,* 169-175.

Coley, P. D., Bryant, J. P & Chapin III., F. S. (1985). Resource availability and plant antiherbivore defense. *Science, 230,* 895–899.

Condit, R., Hubbell, S. P. & Foster, R. B. (1994). Density dependence in two understory tree species in neotropical forest. *Ecology, 75,* 671– 680.

Connell, J. H. & Lowman, M. D. (1989). Low-diversity tropical rainforests: some possible mechanisms for their existence. *American Naturalist, 134,* 88–119.

Connell, J. H. (1971). On the role of natural enemies in preventing competitive exclusion in some marine animals and in rainforest trees. In P. J. den Boer, & G. R. Gradwell (Eds), *Dynamics of populations* (pp. 294 –310). Wageningen, The Netherlands: Center for Agricultural Publishing and Documentation.

Crick, J. C. & Grime, J. P. (1987). Morphological plasticity and mineral nutrient capture in two herbaceous species of contrasted ecology. *New Phytologist, 107,* 403-414.

Crocker, R. L. & Major, J. (1955). Soil development in relation to vegetation and surface age at Glacier Bay, Alaska. *Journal of Ecology, 43,* 427-448.

Drew, M. C. 1975. Comparisons of the effects of a localized supply of phosphate, nitrate, ammonium and potassium on the growth of the seminal root system, and the shoot in barley. *New Phytologist, 75,* 479-490.

Duponnois, R. & Cadet, P. (1994). Interactions of *Meloidogyne javanica* and *Glomus* sp. on growth and N_2 fixation of *Acacia seyal. Afro Asian Journal of Namatology, 4,* 228-233.

Duponnois, R., Colombet, A., Hien, V. & Thioulouse, J. (2005). The mycorrhizal fungus *Glomus intraradices* and rook phosphate amendment influence plant growth and microbial activity in the rhizosphere of *Acacia holosericea. Soil Biology & Biochemistry, 37,* 1460-1468.

Duponnois, R. (2006a). Mycorrhiza Helper Bacteria: Their ecological impact in mycorrhizal symbiosis. In M. K. Rai (Ed.), *Handbook of Microbial Biofertilizers* (pp. 117-135). Binghamton, NY : Food Products Press.

Duponnois, R. (2006b). Bacteria Helping Mycorrhiza Development. In K. G. Mukerji, C. Manoharachary, & J. Singh (Eds.), *Microbial Activity in the Rhizosphere* (pp, 297-310). Berlin: Springer-Verlag.

Duponnois, R., Galiana, A. & Prin, Y. (2008). The mycorrhizosphere effect: a multitrophic interaction complex improves mycorrhizal symbiosis and plant growth. In Z. A. Siddiqui, M. S. Akhtar, & K. Futai (Eds.), *Mycorrhizae: Sustainable Agriculture and Forestry* (pp. 227-238). Dordrecht, The Netherlands: Springer.

Eissenstat, D. M. & Caldwell, M. M. (1988). Seasonal timing of root growth in favorable microsites. *Ecology, 69*, 870-873.

Eom, A. H., Hartnett, D. C. & Wilson, G. W. T. (2000). Host plant species effects on arbuscular mycorrhizal fungal communities in tall grass prairie. *Oecologia, 122*, 435-444.

Eom, A. H., Wilson, G. W. T. & Hartnett, D. C. (2001) Effects of ungulate grazers on arbuscular mycorrhizal symbiosis and fungal community composition in tallgrass prairie. *Mycologia, 93*, 233-242.

Facelli, E. & Facelli, J. M. (2002). Soil phosphorus heterogeneity and mycorrhizal symbiosis regulate plant intra-specific competition and size distribution. *Oecologia, 133*, 54-61.

Farley, R. A. & Fitter, A. H. (1999a) Temporal and spatial variation in soil resources in a deciduous woodland. *Journal of Ecology, 87*, 688-696.

Farley, R. A. & Fitter, A. H. (1999b). The response of seven co-occuring woodland herbaceous perennials to localized nutrient-rich patches. *Journal of Ecology 87*, 849-859.

Fierer, N. & Jackson, R. B. (2006). The diversity and biogeography of soil bacterial communities. *Proceedings of the National Academy of Sciences of USA, 103*, 626-631.

Finlay, R. & Read, D. J. (1986). The structure and function of the vegetative mycelium of ectomycorrhizal plants. I. Translocation of ^{14}C-labeled carbon between plants interconnected by a common mycelium. *New Phytologist, 103*, 143-156.

Finlay, R. D., & Soderstrom, B. (1989) Mycorrhizal mycelia and their role in soil and plant communities. In M. Clarholm, & L. Bergstrom (Eds), *Ecology of Arable Land* (pp. 139-148). Amsterdam : Kluwer.

Finlay, R. D. (2004). Mycorrhizal fungi and their multifunctional roles. *Mycologist, 18,* 91-96.

Finlay, R. D. (2008). Ecological aspects of mycorrhizal symbiosis: with special emphasis on the functional diversity of interactions involving the extraradical mycelium. *Journal of Experimental Botany, 59,* 1115-1126.

Fitter, A. H. & Garbaye, J. (1994). Interactions between mycorrhizal fungi and other soil organisms. *Plant & Soil, 159*, 123-132.

Florence, R. G. (1965). Decline of old-growth redwood forests in relation to some soil microbiological processes. *Ecology, 46*, 52-64.

Founoune, H., Duponnois, R., Bâ, A. M. & El Bouami, F. (2001). Influence of the dual arbuscular endomycorrhizal / ectomycorrhizal symbiosis on the growth of *Acacia holosericea* (A. Cunn. ex G. Don) in glasshouse conditions. *Annals of Forest Sciences, 59*, 93-98.

Franche, C., Lindström, K. & Elmerich, C. (2009). Nitrogen-fixing bacteria associated with leguminous and non-leguminous plants. Plant Soil doi:10.1007/S11104-008-9833-8

Francis, R. & Read, D. J. (1984). Direct transfer of carbon between plants connected by vesicular–arbuscular mycorrhizal mycelium. *Nature 307*, 53-56.

Francis, R. & Read, D. (1994). The contributions of mycorrhizal fungi to the determination of plant community structure. *Plant & Soil, 159*, 11-25.

Fransen, B., de Kroon, H. & Berendse, F. (1998). Root morphological plasticity and nutrient acquisition of perennial grass species from habitats of different nutrient availability. *Oecologia, 115*, 351-358.

Fransen, B., De Kroon, H. & Berendse, F. (2001). Soil nutrient heterogeneity alters competition between two perennial grass species. *Ecology, 82*, 2534-2546.

Frey-Klett, P., Chavatte, M., Clausse, M. L., Courrier, S., Le Roux, C., Raaijmakers, J., Martinotti, M. G., Pierrat, J. C. & Garbaye, J. (2005). Ectomycorrhizal symbiosis affects functional diversity of rhizosphere fluorescent pseudomonads. *New Phytologist, 165*, 317-328.

Gange, A. & West, H. M. (1994). Interactions between arbuscular mycorrhizal fungi and foliar-feeding insects in *Plantago lanceolata* L. *New Phytologist, 128*, 79-87.

Gange, A. C. & Ayres, R. L. (1999). On the relation between arbuscular mycorrhizal colonization and plant "benefit". *Oikos, 87*, 615-621.

Gange, A. C., Bower, E. & Brown, V. K. (1999). Positive effects of an arbuscular mycorrhizal fungus on aphid life history traits. *Oecologia, 120*, 123-131.

Gange, A. C. (2001). Specie-specific responses of a root- and shoot-feeding insect to arbuscular mycorrhizal colonization of its host plant. *New Phytologist, 150*, 611-618.

Garner, W. & Steinberger, Y. (1989). A proposed mechanism for the formation of fertile islands in the desert ecosystem. *Journal of Arid Environments, 16*, 257-262.

Gehring, C. & Bennett, A. (2009). Mycorrhizal fungal-plant-insect interactions: the importance of a community approach. *Environmental Entomology, 38*, 93-102.

Gentili, F., & Jumpponen, A. (2006). Potential and possible uses of bacterial and fungal biofertilizes. In: M. K. Rai (Ed) *Handbook of Microbial Biofertilizers* (pp 1-28). Binghamton, NY, USA: Food Products Press.

Gorham, E., Vitousek P. M. & Reiners, W. A. (1979). The regulation of chemical budgets over the course of terrestrial ecosystem succession. *Annual Review of Ecological Systems, 10*, 53-84.

Grayston, S. J., Wang, S., Campbell, D. C. & Edwards A. C. (1998). Selective influence of plant species on microbial diversity in the rhizosphere. *Soil Biology & Biochemistry, 30*, 369-378.

Griffiths, B. S., Welschen, R., Arendonk, J. J. C. M. & Lambers, H. (1992). The effect of nitrate-nitrogen on bacteria and the bacterial-feeding fauna of different grass species. *Oecologia, 91*, 253-259.

Grime, J. P. (1973). Competitive exclusion in herbaceous vegetation. *Nature, 242*, 344-347.

Grime, J. P., Mackey, J. M. L., Hillier, S. H. & Read, D. J. (1987). Floristic diversity in a model system using experimental microcosms. *Nature, 328*, 420-422.

Grime, J. P., Campbell, B. D., Mackey, J. M. L. & Crick, J. C. (1991). Root plasticity, nitrogen capture and competitive ability. In: D. Atkinson (Ed), *Plant root growth : an ecological perspective* (pp. 381-397). Oxford: Blackwell Scientific Publications.

Grime, J.P. (2001). *Plant strategies, vegetation processes and ecosystem properties.* Chichester, UK: John Wiley and Sons.

Gross, K. L., Peters, A. & Pregitzer, K. S. (1993). Fine root growth and demographic responses to nutrient patches in four old-field plant species. *Oecologia, 95,* 61-64.

Gross, K. L., Pregitzer, K. S. & Burton, A. J. (1995). Spatial variation in nitrogen availability in three successional plant communities. *Journal of Ecology, 83,* 357-367.

Grubb, P. (1977). The maintenance of species richness in plant communities: the importance of the regeneration niche. *Biological Reviews, 52,* 107-145.

Haichar, F. E., Marol, C., Berge, O., Rangel-Castro, J. I., Prosser, J. I., Balesdent, J., Heulin, T. & Achouak, W. (2008). Plant host habitat and root exudates shape soil bacterial community structure. *The ISME Journal, 2,* 1221-1230.

Hamilton, E., W., III & Frank, D. A. (2001). Can plants stimulate soil microbes and their own nutrient supply? Evidence from a grazing tolerant grass. *Ecology, 82,* 2397-2402.

Hart, M. M., Reader, R. J. & Klironomos, J. N. (2003). Plant coexistence mediated by arbuscular mycorrhizal fungi. *TRENDS in Ecology & Evolution, 18,* 418-423.

Harte, J. & Kinzig, A. P. (1993). Mutualism and competition between plants and decomposers: implications for nutrient allocation in ecosystems. *American Naturalist, 141,* 829-846.

Hartnett, D. C., Samenus R. J., Fischer, L. E & Hetrick B. A. D. (1994). Plant demographic responses to mycorrhizal symbiosis in tallgrass prairie. *Oecologia, 99,* 21-26.

Hartnett, D. C. & Wilson, G. W. T. (2002). The role of mycorrhizas in plant community structure and dynamics : lessons from grasslands. *Plant & Soil, 244,* 319-331.

Hawkins, H. J., Johansen, A. & George, E. (2000). Uptake and transport of organic and inorganic nitrogen by arbuscular mycorrhizal fungi. *Plant & Soil, 226,* 275-285.

Heichel, G. H. (1987). Legume nitrogen: symbiotic fixation and recovery by subsequent crops. In Z. R. Helsel (Ed), *Energy in Plant Nutrition and Pest Control* (pp. 227-261). Amsterdam: Elsevier Science Publishers.

Hodge, A., Robinson, D. & Fitter, A. (2000). Are microorganisms more effective than plants at competing for nitrogen? *Trends in Plant Science, 5,* 304-308.

Hodge, A., Campbell, C. D. & Fitter, A.H. (2001). An arbuscular mycorrhizal fungus accelerates decomposition and acquires nitrogen directly from organic material. *Nature, 413,* 297-299.

Hoeksema, J. D. & Kummel, M. (2003). Ecological persistence of the plant-mycorrhizal mutualism: a hypothesis from species coexistence theory. *American Naturalist, 162,* 40-50.

Hooper, D. U. & Dukes, J. S. (2004). Overyielding among plant functional groups in a long-term experiment. *Ecology Letters, 7,* 95-105.

Hubbell, S. P. (2001). *The unified neutral theory of biodiversity and biogeography.* Monographs in population biology. Princeton: Princeton University Press.

Jackson, R. B. & Caldwell, M. M. (1989). The timing and degree of root proliferation in fertile-soil microsites for three cold-desert perennials. *Oecologia, 81,* 149-153.

Janos, D. P. (1980). Mycorrhizae influence tropical succession. *Biotropica, 12,* 56-54.

Janzen, D. H. (1970). Herbivores and the number of tree species in tropical forests. *American Naturalist, 102,* 592-595.

Jayachandran, K., Schwab, A. P. & Hetrick, B. A. D. (1992). Mineralization of organic phosphorus by vesicular-arbuscular mycorrhizal fungi. *Soil Biology & Biochemistry, 24,* 897-903.

Johansen, A. (1999). Depletion of soil mineral N by roots of *Cucumis sativus* L. colonized or not by arbuscular mycorrhizal fungi. *Plant & Soil, 209*, 119-127.

Johansen, A., Jakobsen, I. & Jensen, E. S. (1994). Hyphal N transport by a vesicular–arbuscular mycorrhizal fungus associated with cucumber grown at three nitrogen levels. *Plant & Soil, 160*, 1-9.

Johansson, J. F., Paul, L. R. & Finlay, R. D. (2004). Microbial interactions in the mycorrhizosphere and their significance for sustainable agriculture. *FEMS Microbiology Ecology, 48,* 1-13.

Johnson, D., Vandenkoornhuyse, P. J., Leake, J. R., Gilbert, L., Booth, R. E., Grime, J. P. & Young, J. P. W. (2004). Plant communities affect arbuscular mycorrhizal fungal diversity and community composition in grassland microcosms. *New Phytologist, 161*, 503-515.

Johnson, N. C., Graham, J. H. & Smith, F. A. (1997). Functioning of mycorrhizal associations along the mutualism-parasitism continuum. *New Phytologist, 135*, 575-585.

Jones, D. L. & Hinsinger, P. (2008). The rhizosphere: complex by design. *Plant & Soil, 312*, 1-6.

Kirk, J. L., Beaudette, L. A., Hart, M., Moutoglis, P., Klironomos, J. N., Lee, H. & Trevors, T. T. (2004). Methods of studying soil microbial diversity. *Journal of Microbiological Methods, 58*, 169-188.

Kisa, M., Sanon, A., Thioulouse, J., Assigbetse, K., Sylla, S., Spichiger, R., Dieng, L., Berthelin, J., Prin, Y., Galiana, A., Lepage, M. & Duponnois, R. (2007). Arbuscular mycorrhizal symbiosis can counterbalance the negative influence of the exotic tree species *Eucalyptus camaldulensis* on the structure and functioning of soil microbial communities in a sahelian soil. *FEMS Microbiology Ecology, 62*, 32-44.

Klironomos, J. N. (2000). Host-specificity and functional diversity among arbuscular mycorrhizal fungi. In C. R. Bell, Brylinsky, M., & Johnson-Green, P. (Eds), *Microbial Biosystems : New Frontiers*, In Proceedings of the 8th International Symposium on Microbial Ecology (pp. 845-851). Halifax: Atlantic Canada Society for Microbial Ecology.

Klironomos, J. N. (2002). Feedback with soil biota contributes to plant rarity and invasiveness in communities. *Nature, 417*, 67-70.

Klironomos, J. N. (2003). Variation in plant response to native and exotic arbuscular mycorrhizal fungi. *Ecology, 84*, 2292-2301.

Klironomos, J. K., McCune, J., Hart, M. & Neville, J. (2000). The influence of arbuscular mycorrhizae on the relationship between plant diversity and productivity. *Ecology Letters, 3*, 137-147.

Klironomos, J. N. & Hart, M. M. (2001). Animal nitrogen swap for plant carbon. *Nature, 41*, 651-652.

Koide, R. T. (2000). Mycorrhizal symbiosis and plant reproduction. In Y. Kapulnik, & D. D. Douds (Eds), *Arbuscular Mycorrhizas: Physiology and Function* (pp. 19–46). Dordrecht: Kluwer Academic Publishers.

Lambers, H., Shane, M. W., Cramer, M. D., Pearse, S. J. & Veneklaas E. J. (2006). Root structure and functioning for efficient acquisition of phosphorus: matching morphological and physiological traits. *Annals of Botany, 98*, 693-713.

Lambers, H., Raven, J. A., Shaver, G. R. & Smith S. E. (2008). Plant nutrient-acquisition strategies change with soil age. *TRENDS in Ecology & Evolution, 23*, 95-103.

Landeweert, R., Hofflund, E., Finlay, R. D. & van Breemen, N. (2001). Linking plants to rocks: Ectomycorrhizal fungi mobilize nutrients from minerals. *TRENDS in Ecology & Evolution, 16,* 248-254.

Larson, J. L. & Siemann E. (1998). Legumes may be symbiont-limited during old-field succession. *American Midland Naturalist, 140,* 90-95.

Lavelle, P., & Spain, A. V. (2001). *Soil ecology.* Boston, Massachusetts, USA: Kluwer Academic.

Leake, J. R. (2004). Myco-heterotroph/epiparasitic plant interactions with ectomycorrhizal and arbuscular mycorrhizal fungi. *Current Opinion in Plant Biology, 7,* 422-428.

Lechowicz, M. J. & Bell, G. (1991). The ecology and genetics of fitness in forest plants. II. Microspatial heterogeneity of the edaphic environment. *Journal of Ecology, 79,* 687-696.

Lindahl, B., Stenlid, J., Olsson, S. & Finlay, R. D. (1999). Translocation of ^{32}P between interacting mycelia of a wood decomposing fungus and ectomycorrhizal fungi in microcosm systems. *New Phytologist, 144,* 183-193.

Lindahl, B. D., Finlay, R. D., & Cairney, J. W. G. (2005). Enzymatic activities of mycelia in mycorrhizal fungal communities. In: J. Dighton, P. Oudemans, & J. White (Eds) *The fungal community, its organization and role in the ecosystem* (pp. 331-348). New York: Marcel Dekker.

Marschner, H. (1995). *Mineral nutrition of higher plants.* California, USA: Academic Press.

Marscher, H. (1997). *Mineral nutrition of higher plant* (2^d edition). London, England: Academic Press.

Marschner, P., Solaiman, Z. & Rengel, Z. (2006). Rhizosphere properties of *Poaceae* genotypes under P-limiting conditions. *Plant & Soil, 283,* 11-24.

Mårtensson, A. M., Rydberg, I. & Vestberg, M. (1998). Potential to improve transfer of N in intercropped systems by optimising host-endophyte combinations. *Plant & Soil, 205,* 57-66.

Medve, R. J. (1984). The mycorrhizae of pioneer species in disturbed ecosystems in western Pennsylvania. *American Journal of Botany, 71,* 787-794.

Meinders, M., & van Breemen, N. (2005). Formation of soil–vegetation patterns. In: G.M. Lovett, C. G. Jones, M. G. Turner, & K. C. Weathers (Eds), *Ecosystem Function in Heterogeneous Landscapes* (pp. 207-227). New York: Springer.

Mills, K. E. & Bever, J. D. (1998). Maintenance of diversity within plant communities: soil pathogens as agents of negative feedback. *Ecology, 79,* 1595-1601.

Molofsky, J. , Bever, J. D., & Antonovics, J. (2001). Coexistence under positive frequency dependence. *Proceedings of the Royal Society of London, 268,* 273-277.

Moora, M. & Zobel, M. (1996). Effect of arbuscular mycorrhiza and inter- and intraspecific competition of two grassland species. *Oecologia, 108,* 79-84.

Mou, P. U., Mitchell, R. J. & Jones, R. H. (1997). Root distribution of two tree species under a heteregeneous nutrient environment. *Journal of Applied Ecology, 34,* 645-656.

Mulder, C. P. H., Jumpponen, A., Högberg, P. & Huss-Danell, K. (2002). How plant diversity and legumes affect N dynamics in experimental grassland communities. *Oecologia, 133,* 412-421.

Müller, T., Avolio, M., Olivi, M., Benjdia, M., Rikirsch, E., Kasaras, A., Fitz, M., Chalot, M. & Wipf, D. (2007). Nitrogen transport in the ectomycorrhiza association: the *Hebeloma cylindrosporum-Pinus pinaster* model. *Phytochemistry, 68,* 41-51.

Nara, K. & Hogetsu T. (2004). Ectomycorrhizal fungi on established shrubs facilitate subsequent seedling establishment of successional plant species. *Ecology, 85*, 1700-1707.

Nee, S. & May, R. M. (1992). Dynamics of metapopulations: habitat destruction and competitive coexistence. *Journal of Animal Ecology, 61*, 37-40.

Neumann, G., George, T. S. & Plassard, C. (2009). Strategies and methods for studying the rhizosphere - the plant science toolbox. *Plant & Soil*, doi 10.1007/s11104-009-9953-9.

Newman, E. I. (1988). Mycorrhizal links between plants: their functioning and ecological significance. *Advances in Ecological Research, 18*, 243-270.

Newman, E. I. & Eason, W. R. (1993). Rates of phosphorus transfer within and between ryegrass (*Lolium perenne*) plants. *Functional Ecology, 7*, 242-248.

Newsham, K., Fitter, A. H. & Watkinson, A. R. (1995). Arbuscular mycorrhiza protect an annual grass from root pathogenic fungi in the field. *TRENDS in Ecology & Evolution, 83*, 991-1000.

Nygren, C. M., Edqvist, J., Elfstrand, M., Heller, G. & Taylor, A. F. S. (2007). Detection of extracellular protease activity in different species and genera of ectomycorrhizal fungi. *Mycorrhiza, 17*, 241-248.

Nyvall, R. F. (1999). *Field Crop Diseases* (3^d edition). Iowa, USA: Iowa State University Press.

Pacala, S. & Crawley, M. J. (1992). Herbivores and plant diversity. *American Naturalist, 155*, 435-453.

Paul, E. A., & Clarck, F. E. (1996). *Soil microbiology and biogeochemistry*. San Diego, California, USA: Academic Press.

Pecháčková, S., During, H. J., Rydlová, V. & Herben, T. (1999). Species-specific spatial pattern of below-ground plant parts in a montane grassland community. *Journal of Ecology, 87*, 569-582.

Pellissier, F. & Souto, X. (1999). Allelopathy in Northern temperate and boreal semi-natural woodland. *Critical Reviews in Plant Sciences 18*, 637-652.

Perez-Moreno, J. & Read, D. J. (2001a). Exploitation of pollen by mycorrhizal mycelial systems with special reference to nutrient cycling in boreal forests. *Proceedings of the Royal Society of London, 268*, 1329-1335.

Perez-Moreno, J. & Read, D. J. (2001b). Nutrient transfer from soil nematodes to plants: a direct pathway provided by the mycorrhizal mycelial network. *Plant, Cell & Environment, 24*, 1219-1226.

Pfeffer, P. E., Douds, D. D., Bücking, H., Schwartz, D. P. & Shachar-Hill, Y. (2004). The fungus does not transfer carbon to or between roots in an arbuscular mycorrhizal symbiosis. *New Phytologist, 163*, 617-627.

Pierret, A., Doussan, C., Capowiez, Y., Bastardie, F. & Pagès, L. (2007). Root functional architecture: a framework for modeling the interplay between roots and soil. *Vadose Zone Journal, 6*, 269-281

Poorter, H. (1990). Interspecific variation in relative growth rate: on ecological causes and physiological consequences. In: H. Lambers, M. L. Cambridge, H. Konings, & T. L. Pons (Eds), *Causes and consequences of variation in growth rate and productivity of higher plants* (pp. 45– 68). The Hague, The Netherlands: SPB Academic Publishing.

Prosser, J. I. (2002). Molecular and functional diversity in soil micro-organisms. *Plant & Soil, 244*, 9-17.

Purin, S. & Rillig, M. (2008). Parasitism of arbuscular mycorrhizal fungi: reviewing the evidence. *FEMS Microbiology Letters 279*, 8-14.

Raaijmakers, J. M., Paulitz, T. C., Steinberg, C., Alabouvette, C. & Moënne-Loccoz, Y. (2009). The rhizosphere: a playground and battlefield for soilborne pathogens and beneficial microorganisms. *Plant Soil* doi: 10.1007/s11104-008-9568-6

Read, D. J. & Perez-Moreno, J. (2003). Mycorrhizas and nutrient cycling in ecosystems – a journey towards relevance? *New Phytologist, 157*, 475-492.

Read, D. J. (1993). Mycorrhiza in plant communities. *Advances in Plant Pathology, 9*, 1-31.

Rees, M., Condit, R., Crawley, M. J., Pacala, S. & Tilman, D. (2001). Long-term studies of vegetation dynamics. *Science, 293*, 650-655.

Reeves, F. B., Wagner, D., Moorman, T. & Kiel, J. (1979). The role of endomycorrhizae in revegetation practices in the semi-arid Est. I. A comparison of incidence of mycorrhizae in severely disturbed vs. natural environments. *American Journal of Botany, 66*, 6-13.

Rejmanek, M. & Richardson D. M. (1996). What attributes make some plant species more invasive? *Ecology, 77*, 1655-1661.

Remigi, P., Faye, A., Kane, A., Deruaz, M., Thioulouse, J., Cissoko, M., Prin, Y., Galiana, A., Dreyfus, B. & Duponnois, R. (2008). The exotic legume tree species *Acacia holosericea* alters microbial soil functionalities and the structure of the arbuscular mycorrhizal community. *Applied & Environmental Microbiology, 74*, 1485-1493.

Raynaud, X., Jaillard, B. & Leadley, P. W. (2008). Plants may alter competition by modifying nutrient bioavailability in rhizosphere: a modeling approach. *American Naturalist, 171*, 44-58.

Reynolds, J. F., Virginia, R. A. & Cornelius, J. M. (1990). Resource island formation associated with the desert shrubs, creosote bush (*Larrea tridentata*) and mesquite (*Prosopis glandulosa*) and its role in the stability of desert ecosystems: a simulation model. *Supplemental Bulletin of the Ecological Society of America, 70*, 299-300.

Reynolds, H.L., Packer, A., Bever, J.D. & Clay, K. (2003). Grassroots ecology: plant-microbe-soil interactions as drivers of plant community structure and dynamics. *Ecology, 84*, 2281-2291.

Reynolds, H. L., Vogelsang, K. M., Hartley, A. E., Bever, J. D. & Schultz, P. A. (2006). Variable response of old-field perennials to arbuscular mycorrhizal fungi and phosphorus source. *Oecologia, 147*, 48-358.

Richard, F., Selosse, M. A. & Gardes, M. (2009). Facilitated establishment of *Quercus ilex* in shrub-dominated communities within a mediterranean ecosystem: do mycorrhizal partners matter ? *FEMS Microbiology Ecology, 68*, 14-24.

Richardson, D. M., Allsopp, N, D'Antonio, C. M., Milton, S. J. & Rejmanek, M. (2000). Plant invasions- the role of mutualisms. *Biological Reviews of the Cambridge Philosophical Society, 75*, 65-93.

Richardson, A. E., Barea, J. M., McNeill, A. M. & Prigent-Combaret, C. (2009). Acquisition of phosphorus and nitrogen in the rhizosphere and growth promotion by microorganisms. *Plant & Soil* doi : 10.1007/s11104-009-9895-2

Robertson, G. P., Crum, J. R. & Ellis, B. G. (1993). The spatial variability of soil resources following long-term disturbance. *Oecologia, 96*, 451-456.

Robin, A., Vansuyt, G., Hinsinger, P., Meyer, J. M., Briat, J. F. & Lemanceau, P. (2008). Iron dynamics in the rhizosphere: consequences for plant health and nutrition. *Advances in Agronomy, 99*, 183-225.

Robson, A. D., O'Hara, G. W. & Abbott, L. K. (1981) Involvement of phosphorus in nitrogen fixation by subterranean clover (*Trifolium subterraneum* L.). *Australian Journal of Plant Physiology, 8*, 427-436.

Rovira, A. D. (1969) Plant root exudates. *Botanical Review, 35*, 35-56.

Ryel, R. J. , Caldwell, M. M. & Manwaring, J. H. (1996). Temporal dynamics of soil spatial heterogeneity in sagebrush–wheatgrass steppe during a growing season. *Plant & Soil, 184*, 299-309.

Sanders, I. R. & Fitter, A. H. (1992). The ecology and functioning of vesicular-arbuscular mycorrhizas in co-existing grassland species. I. Seasonal patterns of mycorrhizal occurrence and morphology. *New Phytologist, 120*, 517-524.

Sanon, A., Andrianjaka, Z. N., Prin, Y., Bally, R., Thioulouse, J., Comte, G. & Duponnois R. (2009). Rhizosphere microbiota interfers with plant-plant interactions. *Plant & Soil*, doi 10.1007/s11104-009-0010-5.

Sanon, A., Martin, P., Thioulouse, J., Plenchette, C., Spichiger, R., Lepage, M. & Duponnois, R. (2006). Displacement of an herbaceous plant species community by mycorrhizal and non-mycorrhizal *Gmelina arborea*, an exotic tree, grown in a microcosm experiment. *Mycorrhiza, 16*, 125-132.

Schenk, H. J. & Jackson, R. B. (2002). Rooting depths, lateral root spreads and below-ground/above-ground allometries of plants in water-limited ecosystems. *Journal of Ecology, 90*, 480-494.

Schlesinger, W. H. (1991). Biogeochemistry: an analysis of global change. San Diego, California, USA: Academic Press.

Schlesinger, W. H., Raikes, J. A., Hartley, A. E. & Cross, A. F. (1996). On the spatial pattern of soil nutrients in desert ecosystems. *Ecology, 77*, 364-374.

Selosse, M. A., WEIß, M., Jany, J. L. & Tillier, A. (2002). Communities and populations of sebacinoid basidiomycetes associated with the achlorophyllous orchid *Neottia nidus-avis* (L.) L.C.M. Rich. and neighbouring tree ectomycorrhizae. *Molecular Ecology, 9*, 1831-1844.

Selosse, M. A., Richard, F., He, X. & Simard, S.W. (2006). Mycorrhizal networks : des liaisons dangeureuses ? *TRENDS in Ecology & Evolution, 21*, 621-628.

Shane, M. & Lambers, H. (2005). Cluster roots: a curiosity in context. *Plant & Soil, 274*, 101-125.

Silvertown, J. (2004). Plant coexistence and the niche. *TRENDS in Ecology & Evolution, 19*, 604-611.

Simard, S. W., Perry, D. A., Jones, M. D., Myrold, D. D., Daniel M. Durall, D. M. & Molina, R. (1997). Net transfer of carbon between ectomycorrhizal tree species in the field. *Nature, 388*, 579-582.

Simard, S. W., Jones, M. D. & Durall, D. M. (2002). Carbon and nutrient fluxes within and between mycorrhizal plants. In M. G. A. van der Heijden & I. R. Sanders (Eds) *Mycorrhizal Ecology. Ecological Studies 157* (pp. 33–74). Heidelberg: Springer Verlag.

Simard, S. W. & Durall, D. M. (2004). Mycorrhizal networks: a review of their extent, function, and importance. *Canadian Journal of Botany, 82*, 1140-1165.

Smith, M. R., Charvat, I. & Jacobson, R. L. (1998). Arbuscular mycorrhizae promote establishment of prairie species in tallgrass prairie restoration. *Canadian Journal of Botany, 76*, 1947-1954.

Smith, S. E., & Read, D. J. (2008). *Mycorrhizal symbioses* (3rd edition). London: Academic Press.

Tabatabai, T., & Dick, W. A. (2002). Enzymes in soil. In: R. G. Burns, & R. P. Dick (Eds), *Enzymes in the environment* (pp. 567-596). New York, USA: Marcel Dekker.

Tarafdar, J. C. & Marschner, H. (1994). Phosphatase activity in the rhizosphere and hyphosphere of VA mycorrhizal wheat supplied with inorganic and organic phosphorus. *Soil Biology & Biochemistry, 26,* 387-395.

Tate, R. L. III. (1997). Soil microbial diversity: whither to now? *Soil Science, 162,* 605-606.

Thies, J. E. (2008). Molecular methods for studying microbial ecology in the soil and rhizosphere. In C. S. Nautiyal, & P. Dion (Eds), *Molecular Mechanisms of plant and microbe coexistence* (pp, 411-436). Berlin Heidelberg: Springer-Verlag.

Tibbett, M., Hartley, M. & Hartley, S. (2000). Comparative growth of ectomycorrhizal basidiomycetes (*Hebeloma* spp.) on organic and inorganic nitrogen. *Journal of Basic Microbiology, 40,* 393-395.

Tibbett, M. & Sanders, F. E. (2002). Ectomycorrhizal symbiosis can enhance plant nutrition through improved access to discrete organic nutrient patches of high resource quality. *Annals of Botany, 89,* 783-789.

Tilman, D. (1988). *Plant strategies and the dynamics and structure of plant communities.* New Jersey, USA: Princeton University Press.

Tilman, D., & Pacala, S. (1993). The maintenance of species richness in plant communities. In R. E. Ricklefs, & D. Schluter (Eds.), *Species diversity in ecological communities : historical and geographical perspectives* (pp. 13-25). Chicago: University of Chicago Press.

Tilman, D., May, R. M., Lehman, C. L. & Nowak, M. A. (1994). Habitat destruction and the extinction debt. *Nature, 371,* 65-66.

Tilman, D., Wedin, D. & Knops, J. (1996). Productivity and sustainability influenced by biodiversity in grassland ecosystems. *Nature, 379,* 718-720.

Tilman, D., Knops, J., Wedin, D., Reich, P., Ritchie, M. & Siemann, E. (1997a). The influence of functional diversity and composition on ecosystem processes. *Science, 277,* 1300-1302.

Tilman, D., Lehman, C. L. & Thomson, K. T. (1997b). Plant diversity and ecosystem productivity. Theoretical considerations. *Proceedings of the National Academy of Sciences of the USA, 94,* 1857-1861.

Timan, D. (2004). Niche tradeoffs, neutrality, and community structure : a stochastic theory of resource competition, invasion, and community assembly. *Proceedings of National Academy of Sciences USA, 101,* 10854-10861.

Tobar, R., Azcón, R. & Barea, J. M. (1994). Improved nitrogen uptake and transport from ^{15}N-labelled nitrate by external hyphae of arbuscular mycorrhiza under water-stressed conditions. *New Phytologist, 126,* 119-122.

Tongway, D. J., & Ludwig, J. A. (2005) Heterogeneity in arid and semiarid lands. In G. M. Lovett, C. G. Jones, M. G. Turner, & K. C. Weathers (Eds), *Ecosystem Function in Heterogeneous Landscapes* (pp. 9-30). New York: Springer.

Torsvik, V., Goksoyr, J. & Daae, F.L. (1990a). High diversity in DNA of soil bacteria. *Applied & Environmental Microbiology, 56,* 782 -787.

Torsvik, V., Salte, K., Soerheim, R. & Goksoeyr, J. (1990b). Comparison of phenotypic diversity and DNA heterogeneity in a population of soil bacteria. *Applied & Environmental Microbiology, 56,* 776-781.

Trenbath, B. R. (1974). Biomass productivity of mixtures. *Advances in Agronomy, 26,* 177-210.

Turnbull, M. H., Schmidt, S., Erskine, P. D., Richards, S. & Stewart, G. R. (1996). Root adaptation and nitrogen source acquisition in natural ecosystems. *Tree Physiology, 16,* 941-948.

Urcelay, C. & Diaz, S. (2003). The mycorrhizal dependence of surbordinates determines the effect of arbuscular mycorrhizal fungi on plant diversity. *Ecology Letters, 6,* 388-391.

van der Heijden, M. G. A. , Klironomos, J. N., Ursic, M., Moutoglis, P., Streitwolf-Engel, R., Boller, T., Wiemken, A. & Sanders. I. R. (1998a). Mycorrhizal fungal diversity determines plant biodiversity, ecosystem variability and productivity. *Nature, 396,* 69-72.

van der Heijden, M. G. A., Boller, T., Wiemken, A. & Sanders I. R. (1998b). Different arbuscular mycorrhizal fungal species are potential determinants of plant community structure. *Ecology, 79,* 2082-2091.

van der Heijden, M.G.A. (2002). Arbuscular mycorrhizal fungi as a determinant of plant diversity: in search for underlying mechanisms and general principles. In: van der Heijden MGA, Sanders IR, eds. *Mycorrhizal ecology.Ecological studies 157.* Berlin, Germany: Springer Verlag, 243–265.

van der Heijden, M. G. A., Bakker, R., Verwaal, J., Scheublin, T. R., Rutten, M., van Logtestijn, R. & Taehelin, C. (2006). Symbiotic bacteria as a determinant of plant community structure and plant productivity in dune grassland. *FEMS Microbiology Ecology, 56,* 178-187.

van der Heijden, M. G. A. & Scheublin, T. R. (2007). Functional traits in mycorrhizal ecology: their use for predicting the impact of arbuscular mycorrhizal fungal communities on plant growth and ecosystem functioning. *New Phytologist, 174,* 244-250.

Van der Putten, W. H. & Troelstra, S. R. (1990). Harmful soil organisms in coastal foredunes involved in degeneration of *Ammophila arenaria* and *Calammophila baltica. Canadian Journal of Botany, 68,* 1560-1568.

Van der Putten, W. H., Maas, P. H. T., Van Gulik, W. J. M. & Brinkman, H. (1990). Characterization of soil organisms involved in the degeneration of *Ammophila arenaria. Soil Biology & Biochemistry, 22,* 845-852.

Van der Putten, W. H., Van Dijk, C. & Peters, B. A. M. (1993). Plant-specific soil-borne diseases contribute to succession in foredune vegetation. *Nature, 362,* 53-56.

Van der Putten, W. H., Klironomos, J. N. & Wardle, D. A. (2007a). Microbial ecology of biological invasions. *The ISME Journal, 1,* 28-37.

Van der Putten, W .H., Kowalchuk, G. A., Brinkman, E. P., Doodeman, G. T. A., van der Kaaij, R. M., Kamp, A. F. D., Menting, F. B. J. & Veenendaal, E. M. (2007b). Soil feedback of exotic savanna grass relates to pathogen absence and mycorrhizal selectivity. *Ecology, 88,* 978-988.

Vandermeer, J. H. (1989). *The Ecology of Intercropping.* New York: Cambridge University Press.

Vogelsang, K. M., Bever, J. D., Griswold, M. & Schulz, P. A. (2004). The use of mycorrhizal fungi in erosion control applications. Final report for Caltrans. California Department of Transportation Contract no. 65A0070, Sacramento (California).

Vogelsang, K. M., Reynolds, H. L. & Bever, J. D. (2006). Mycorrhizal fungal identity and richness determine the diversity and productivity of a tallgrass prairie system. *New Phytologist, 172*, 554-562.

Walker, T. W. & Syers, J. K. (1976). The fate of P during pedogenesis. *Geoderma, 15*, 1-19.

Wallander, H., Nilsson, L. O., Hagerberg, D. & Bååth, E. (2001). Estimation of the biomass and seasonal growth of external mycelium of ectomycorrhizal fungi in the field. *New Phytologist, 151*, 752-760.

Wardle, D.A. (2002). *Communities and ecosystems: linking the aboveground and belowground components*. Princeton, NJ, USA: Princeton University Press.

Weiner, J., Wright, D. B. & Castro, S. (1997). Symmetry of below-ground competition between *Kochia scoparia* individuals. *Oikos, 79*, 85-91.

West, H. M. (1996). Influence of arbuscular mycorrhizal infection on competition between *Holcus lanatus* and *Dactylis glomerata*. *Journal of Ecology, 84*, 429-438.

Westover, K. M. & Bever J. D. (2001). Mechanisms of plant species coexistence: complementary roles of rhizosphere bacteria and root fungal pathogens. *Ecology, 82*, 3285-3294.

Westover, K., Kennedy, A. & Kelley, S. (1997). Patterns of rhizosphere microbial community structure associated with co-occurring plant species. *Journal of Ecology, 85*, 863-873.

Wolfe, B. E. & Klironomos, J. N. (2005). Breaking new ground: soil communities and exotic plant invasion. *BioScience, 55*, 477-487.

Wurst, S., Dugassa-Gobena, D., Langel, R., Bonkowski, M. & Scheu, S. (2004). Combined effects of earthworms and vesicular-arbuscular mycorrhizas on plant and aphid performance. *New Phytologist, 163*, 169-176.

Yanni, Y. G., Rizk, R. Y., Corich, V., Squartini, A., Ninke, K., Philip-Hollingsworth, S., Orgambide, G., de Bruinj, F., Stoltzfus, J., Buckley, D., Schmidt, T. M., Mateos, P. F., Ladha, J. K. & Dazzo, F. B. (1997). Natural endophytic association between *Rhizobium leguminosarum* bv *trifolii* and rice roots and assessment of its potential to promote rice growth. *Plant & Soil, 194*, 99-114.

Zabinski, C. A., Quinn, L. & Callaway, R. M. (2002). Phosphorus uptake, not carbon transfer, explains arbuscular mycorrhizal enhancement of *Centaurea maculosa* in the presence of native grassland species. *Functional Ecology, 16*, 758-765.

Zak, D. R., Holmes, W. E., White, D. C., Peacock, A. D. & Tilman, D. (2003). Plant diversity, soil microbial communities, and ecosystem function: Are there any links? *Ecology, 84*, 2042-2050.

Zobel, M. & Moora, M. (1995). Interspecific competition and arbuscular mycorrhiza : importance for the coexistence of two calcareous grassland species. *Folia Geobotanica Phytotaxon, 30*, 223-230.

In: Encyclopedia of Environmental Research   ISBN: 978-1-61761-927-4
Editor: Alisa N. Souter   © 2011 Nova Science Publishers, Inc.

Chapter 10

ARBUSCULAR MYCORRHIZAL FUNGI: A BELOWGROUND REGULATOR OF PLANT DIVERSITY IN GRASSLANDS AND THE HIDDEN MECHANISMS

Qing Yao[1] and Hong-Hui Zhu[2]*

[1]College of Horticulture, South China Agricultural University, Guangzhou, China, 510642.
[2]Guangdong Institute of Microbiology, Guangzhou, China, 510070.

ABSTRACT

Arbuscular mycorrhizal fungi (AMF) habitat almost all terrestrial ecosystems, and play a critical role in many ecological processes. In grasslands of almost all types, AMF are demonstrated to increase the plant biodiversity in most experiments, although some experiments indicate a decreased biodiversity. This regulation of grassland biodiversity by AMF can be attributed to several mechanisms. The differential growth responses of host plants to AMF are the direct factor leading to the altered biodiversity, and the promoted uptake of mineral nutrients with different degree by AMF among host plants is regarded to underlie the differential growth responses. Furthermore, hyphal links established between host plants of the same or different species represent a transport passage for resources (such as water, nutrients, carbon), by which these resources flow freely and finally redistribute evenly within the community. In practice, this regulation can be employed to restore the degraded grasslands, to monitor the competition between plants in the managed grasslands, and even to construct grasslands with pre-designed community structure. Future research is proposed in relation to the emerging issues and the practical application.

---

* Corresponding author: E-mail: yaoqscau@scau.edu.cn, Tel: 86-20-85286902, Fax: 86-20-85280228.

1. INTRODUCTION

Grasslands are the major and important components of terrestrial ecosystems which comprise nearly 50% of land area on the globe (Sims and Bradford, 2001). Although the climate-driven redistribution of grasslands has been continuing and the world area of grasslands is expected to increase in the future (Parton et al., 1995), the decease in grassland area has been occurring simultaneously in some areas. In China, grasslands have been declining at approximately 15000 km^2 per year since the early 1980s, mainly due to various forms of degradation (Liu and Diamond, 2005). Biodiversity of the ecosystems is of great importance to the stability of these ecosystems, including the grasslands. It is well acknowledged that high plant diversity is necessary for the grasslands for the protection from the degradation (Frank and McNaughton, 1991; Tilman and Downing, 1994). So far, many aboveground factors have been demonstrated to affect the plant diversity of grasslands, such as grazing (herbivore), topography, mankind activity (Buschmann et al., 2005; Bakker et al., 2006; Marini et al., 2009; Guo et al., 2007). These factors have been intensively studied and the responsible mechanisms have been elucidated. Recently, research also shed light on the belowground factors, such as the soil quality, soil microbes (van der Heijden et al., 1998; Stohlgren et al., 1999), suggesting that the ecology goes belowground (Copley, 2000).

Among the belowground factors, arbuscular mycorrhizal fungi (AMF) are of special significance to the plant diversity of the grasslands, which has been repeatedly confirmed recently (van der Heijden et al., 1998; Urcelay and Díaz, 2003; Vogelsang et al., 2006; Karanika et al., 2008). AMF are obligate symbiotic soil fungi, classified as the eumycotan fungal phylum Glomeromycota (Schüßler et al. 2001; formerly as the Zygomycota). Upon the establishment of symbiosis relationship – arbuscular mycorrhizae, AMF benefit from plants by utilizing the plant-derived carbohydrates, and in turn as exchange, provide mineral nutrients to plants (Smith and Read, 1997). Arbuscular mycorrhizae are structurally characterized by the formation of arbuscles in root cortex, which are the sites for nutrition exchange. Although the carbohydrate sink of AMF represents 3%-20% of host plant photosynthate (Jakobsen and Rosendahl, 1990), plants normally perform better in vigor in most cases. This growth promotion effect is mainly derived from the improved nutrients (P in particular) uptake by AMF (Ferrol et al., 2002). Besides the nutritional effect, other physiological effects have also been demonstrated to contribute to the enhanced growth in the inoculated plants, e.g. the altered phytohormone level (Yao et al., 2005), the modified root system architecture (Yao et al., 2009). Based on these nutrition-dependent or -independent effects, AMF affect many physiological aspects of individual plant, and further exert influence on the plant ecosystem processes (van der Heijden, 2002; Hart et al., 2003; and literature therein).

2. EFFECT OF AMF ON THE PLANT DIVERSITY IN GRASSLANDS

2.1. Increasing vs Decreasing the Diversity

It is widely accepted that AMF can affect plant diversity (Read 1998; Johnson et al., 2004). Pioneering research have shown a positive relationship between the presence of AMF and plant diversity (Grime et al. 1987; Gange et al. 1993; van der Heijden et al. 1998). However, some studies reporting the opposite effect also existed (Newsham et al. 1995; Hartnett & Wilson 1999). In general, it appears that the evidence supporting the positive effect dominate in the literature.

Increasing the plant diversity

Grime et al. (1987) explored the influence of AMF on floristic plant diversity in a model microcosm. They found that the diversity in the microcosm inoculated with AMF was significantly higher than that in the non-inoculated control. The increase was due to the promotion of the biomass of the subodinate species relative to that of the dominant (Grime et al., 1987). Similarly, in two independent experiments using microcosm and macrocosm systems respectively, plant community composition and structure in microcosms fluctuated greatly depending on the AMF taxa at low AMF diversity, and plant diversity as well as productivity in macrocosms increased significantly with increasing AMF species richness (van der Henijden et al., 1998). These results clearly demonstrate the increased plant diversity in simulated grasslands by AMF. Recently, some experiments in field conditions have further confirmed the positive effects of AMF on the plant diversity. In a mountainous herbaceous grassland of northern Greece, AMF community was suppressed with the application of fungicide benomyl to evaluate their influence on the plant diversity (Karanika et al., 2008). In the fungicide-treated plots, there was a decline in the plant diversity although the plant productivity was not affected, suggesting that AMF increase the plant diversity in this grassland independent of the primary productivity. Contrastingly, however, a greenhouse experiment showed that inoculation of AMF decreased the productivity by 15%-24% at the community level, but concomitantly increased the plant diversity (Stein et al., 2009).

In conclusion, the evidence, whatever in greenhouse or in field conditions, strongly demonstrate that AMF can increase the plant diversity in spite of the plant productivity, and that AMF community with high diversity may result in plant community with high diversity.

Decreasing the biodiversity or no effect

Interestingly, even in few number but some reports indicate a negative relationship between plant diversity and AMF in grasslands. For example, in a 5-year field experiment in tallgrass prairie, benomyl was applied to suppress the mycorrhizal activity for evaluating the influence of AMF on the plant diversity. Suppression of mycorrhizal symbiosis resulted in a large increase in plant diversity, indicating the negative relationship between AMF and plant diversity (Hartnett and Wilson, 1999). In a field experimental mesocosm, where the dominant species are non-mycorrhizal and the subordinate species are mycorrhizal, plant diversity was unaffected by the presence of AMF in the low water table treatment, but was significantly decreased in the presence of AMF in the high water table treatment (Wolfe et al., 2006). They attributed the neutral or negative effect to the shorter time-scale compared with others. It seems that the time-scale of experiments may affect the relationship between AMF and plant diversity. Some greenhouse studies coincide well with the field research. In a greenhouse experiment, Landis et al. (2005) found that there was no relation between plant diversity and the number of AMF species, and no significant correlation between plant diversity and AMF was observed, indicating that AMF do not appear to play a role in plant community composition.

2.2. Influence of AMF Identity on the Plant Diversity

Evidence from both greenhouse and field experiments clearly demonstrate that AMF can regulate the plant diversity in grasslands. However, further study reveal that not only the presence of AMF but also the AMF identity can affect the plant diversity, which deepens our understanding of the relationship between AMF and plant diversity.

In a rhizobox experimental unit, Oliveira et al. (2006) used *Conyza bilbaoana* (a dominant plant species with high mycorrhizal dependency) and *Salix atrocinerea* (a subordinate plant species with low mycorrhizal dependency) to evaluate the different effects of 4 native AMF on the coexistence of these two plant species. According to the prediction of the conceptual model proposed previously (Urcelay and Díaz, 2003), they predicted a decreased plant diversity in the presence of AMF. However, the results indicated different effects in plant coexistence depending on the AMF species present, indicating the dependence of the plant diversity on the AMF identity. In microcosms of thirty liter, the non-mycotrophic plant species *Atriplex sagittata* contributed nearly 70% to the total plant biomass without AMF inoculation, contrasting to only about 10% in the presence of three AMF mixture. Furthermore, three individual AMF species also exerted different influences on the growth of *A. sagittata* and other two co-existing plant species. The authors suggested that not only the presence but also the AMF identity regulated the community structure (Püschel et al., 2007) and thus the plant diversity. Another microcosm experiment comparing the native and exogenous AMF reached a similar conclusion (Yao et al., 2008).

Stampe and Daehler (2003) conducted a field experiment to assess the influence of AMF identity on the plant community structure. They found that, although *Melinis repens* remained as the dominant plant species in all treatments, the Shannon index of diversity varied with AMF species identity, and the increased diversity appeared to be related to decreased dominance by *M. repens* in the presence of certain AMF species. They indicated that the composition of the AMF community belowground can influence the structure of the plant community aboveground.

It is well understood that the growth of an individual plant species can be promoted at different extents by different AMF species. Therefore, it is not surprising that AMF identity can exert an important influence on the plant diversity. Research have clarified that increasing AMF richness promoted plant diversity (van der Heijden et al., 1998; Vogelsang et al., 2006), but it seems that this AMF richness effect was small relative to the effects of individual AMF species, and that plant diversity is more responsive to AMF identity than to AMF diversity per se (Vogelsang et al., 2006). Moreover, the extent to which AMF influence plant diversity or coexistence depends on the specific AMF isolate (Scheublin et al., 2007). Therefore, when determining the effect of AMF on plant diversity or coexistence, not only the presence of AMF but also the composition of AMF communities must be taken into account (van der Heijden et al., 1998; 2003).

2.3. Other Factors Involved in the AMF-regulated Plant Diversity

As one of the important factors regulating the plant diversity in grasslands, AMF always interact with other factors of both aboveground and belowground. The most documented factors interacting with AMF to regulate the plant diversity include grazing, light, atmosphere

CO_2 level, waterlogging, soil nutrient availability and etc (Eriksson, 2001; Maurer et al., 2006; Landis et al., 2005; Johnson et al., 2003; Wolfe et al., 2006; Daleo et al., 2008; van der Heijden et al., 2008).

Using greenhouse chamber systems, Johnson et al. (2003) demonstrated that AMF reduced the plant richness at ambient CO_2 but increased it at elevated CO_2. This was resulted from the low mortality rates of several C_3 forbs at ambient CO_2 but the high one at elevated CO_2. They suggest that CO_2 enrichment ameliorates the carbon cost of some AM symbioses, which further regulates the AMF-driven plant community structure. Other research revealing the interaction between AMF and CO_2 level in regulating the plant diversity (Tang et al., 2009) shows similar results. However, recent research comparing the abrupt rise and the gradual rise in CO_2 level argued that those experiments with abrupt rise model may overestimate the some community responses to increasing CO_2 level, because the response of AMF to the higher CO_2 level increased gradually through 6 years was similar to that to the ambient CO_2 level (Klironomos et al., 2005). Among belowground factors, nutrient availability is of special importance in that AMF activity and functioning are strongly regulated by it (Bradley et al., 2006; Gryndleret al., 2006). A greenhouse microcosm experiment consisting of grasses and legumes was established to evaluate the interaction between AMF and N enrichment (van der Heijden et al., 2008). Results showed the proportion of legume biomass out of total shoot biomass at high N supply was 19% with AMF and only 3% without AMF. This occurred because AMF and N enrichment had a big impact on plant community composition, but with opposite effects (van der Heijden et al., 2008). Field experiment in a salt-marsh plant community also demonstrated an active interaction between AMF and nutrient availability (Daleo et al., 2008). AMF increased the plant growth of *Spartina densiflora* at low nutrient levels but reduced it at high nutrient levels, while *Spartina alterniflora* was not colonized by AMF. *S. alterniflora* displaced *S. densiflora* at the nutrient or fungicide applications. These results suggest that nutrient supply can regulate the direction of the AMF influence on the plant community structure (Daleo et al., 2008).

3. MECHANISMS INTERPRETING THE EFFECT

Since AMF (in the terms of both the presence of AMF and the AMF identity) can regulate the plant diversity in grasslands, the mechanisms underlying it have been of great significance to plant ecologists. During the past decades, several mechanisms have been proposed to directly or indirectly interpret the evidence gathered from the greenhouses or the fields. The main mechanisms include the mycorrhizal dependency, the nutrient exploration, and the hyphal links, which stand in different aspects and also complement well each other.

3.1. Mycorrhizal Dependency

Gerdermann (1975) first raised the term "mycorrhizal dependency", and defined it as "the degree to which the plant is dependent on the mycorrhizal condition to produce maximum growth or yield at a given level of soil fertility". Two formulas for calculating this index were subsequently put forward by Menge et al. (1978) and Plenchette et al. (1983), both of which are identical in nature and reflect the growth response of plants to AMF colonization. If the mycorrhizal dependency is positive, it means that the plant benefits from AMF, and if the mycorrhizal dependency is negative, it means that AMF reduces the growth of the plant under the test conditions (van der Heijden, 2002).

In a tallgrass prairie, AMF were suppressed by benomyl application (Hartnett and Wilson, 1999). This suppression led to a decrease in the abundance of the dominant species (obligately mycotrophic C_4 tall grasses) and a compensatory increase in the abundance of many subordinate facultatively mycotrophic C_3 grasses and forbs. More early investigation indicated that the mycorrhizal dependency of the dominant (>99%) was higher than that of the subordinates (22%-63%) (Hetrick et al. 1990). Although the mycorrhizal dependency of plants under the competition condition might be modified, Hartnett and Wilson (1999) regarded it as an important mechanism contributing to the altered plant diversity. As described before, both increase and decrease in plant diversity induced by AMF have been reported (Grime et al. 1987; van der Heijden et al. 1998; Newsham et al. 1995; Hartnett & Wilson 1999). Based on the research, Urcelay and Díaz (2003) presented a conceptual model to explain the relationship between plant diversity and AMF, and suggested that it is the mycorrhizal dependency of the subordinate species that determines the effect of AMF on plant diversity. However, this model is simple and neglects the necessity of comparing the mycorrhizal dependencies of the dominant and the subordinate species.

In a grassland ecosystem, even though both dominant and subordinate are mycorrhizal, three outcomes concerning the plant diversity can occur theoretically: A) If the mycorrhizal dependency of the dominant is higher than that of the subordinate, AMF will enhance the growth of the dominant more than that of the subordinate, resulting in a decrease in the plant diversity. B) If the mycorrhizal dependency of the dominant is lower than that of the subordinate, AMF will enhance the growth of the dominant less than that of the subordinate, resulting in an increase in the plant diversity. C) Sometimes, if the mycorrhizal dependency of the dominant is equal to that of the subordinate, AMF will enhance equally the growth of the dominant and the subordinate, leaving the plant diversity stable. In this context, the influence of AMF on plant diversity can be summarized in Table 1.

This prediction applies not only to the experimental system but also to the field condition. In the former case, AMF or AMF community is normally simplified. In the latter case, however, the native AMF community is always complicated. The native AMF community should be considered as a whole and the mycorrhizal dependency should be referred to the growth response of the plant to the AMF community rather than a particular AMF species. The native AMF community composition might fluctuate with time proceeding (Helgason et al., 1999), and correspondingly, the effect of AMF will also change. This complicates the study in field conditions, especially in long term.

3.2. Nutrient Exploration

AMF promote the nutrient uptake of plants in that their extensive external hyphae can explore more soil nutrient and translocate to the roots (Smith and Read, 1997). In this context, sufficient hyphae density in soil appears the prerequisite for the AMF functions. Other AMF characteristics relative to it may also influence the ecological functions of AMF.

Table 1. Theoretical prediction of the influence of AMF on the plant diversity with full consideration of the mycorrhizal dependency of the dominant and the subordinate species.

Prerequisites	Mycorrhizal dependency	Direction of AMF effect on the diversity of plant community
Both dominant species and subordinate species are mycotrophic.	$D > S$ [*]	- [**]
	$D = S$	0
	$D < S$	+
Dominant species are non-mycotrophic and subordinate species are mycotrophic.	$S > 0$	+
	$S = 0$	0
	$S < 0$	-
Dominant species are mycotrophic and subordinate species are non-mycotrophic.	$D > 0$	-
	$D = 0$	0
	$D < 0$	+

[*] D and S indicate the mycorrhizal dependency of dominant species and subordinate species, respectively.

[**] -, 0 and + indicate the negative, nil and positive effect of AMF colonization on the diversity of pant community, respectively.

In the macrocosm system, the increase in AMF diversity elevated not only the plant diversity but also the hyphal density and the nutrient depletion in soil (van der henjiden et al., 1998). They contributed the increased plant diversity partly to the enhanced nutrient exploration by AMF. When investigating the coexistence of two plant species, the AMF species with greater external hyphal lengths were confirmed to support higher coexistence ratio and even more seed production (Oliveira et al., 2006). The external hyphal length closely correlated with the coexistence ration (y=0.083x +23, R^2=0.514, P<0.001), indicating that more nutrient exploration (normally resulted from higher external hyphal length) is beneficial to higher plant diversity. On the other hand, plants can increase their competitive advantage through the AMF partnership. Walling and Zabinski (2004) compared the external hyphae length in the rhizosphere soil of an invasive plant species and a native plant species and found that the hyphae was more in rhizosphere soil of the invasive species than that of the native species. This increase in external hyphae length represents increased soil volume and nutrients explored for the host plant.

However, more nutrient exploration and higher external hyphae density are not necessarily associated with high plant diversity. Bingham and Biondini (2009) indicated that external hyphal length increased when plant communities were dominated by species with high root density, high root to shoot ratios, and high nitrogen use efficiency. Therefore, it seems more reasonable that the conception of nutrient exploration can be employed to explain why AMF can affect the plant diversity, but not be used to tell the direction in which AMF affect the plant diversity. For example, in a given grassland ecosystem where dominant species is more mycotrophic than the subordinate, AMF will decrease the plant diversity. In this case, if the nutrient exploration by AMF is promoted by applying sesquiterpene lactones

(Akiyama et al., 2005) or whatever other agents, the dominant will be more vigorous and plant diversity will be further decreased.

3.3. Hyphal Links

AMF are symbiotic soil fungi in association with most terrestrial plants with low host specificity (Sanders, 2003). It means that the external hyphae spreading from a colonized root can colonize the adjacent roots of different species. Therefore, the extensive belowground hyphal networks are able to link together different host plants by means of hyphae growing into the soil and colonizing the roots of the neighboring plant of diverse species (Graves et al., 1997; Read, 1998; van der Heijden et al., 1998). The hyphae connect the plants of the same or different species are referred as hyphal links. Hyphal links are regarded as one mechanism explaining the regulated plant diversity by AMF, because soil resource, e.g. C, P, N, and water, can be redistributed among the neighbors via hyphal links (Hartnett and Wilson, 1999, Meding and Zasoski, 2008).

Using ^{14}C labeling technique, Carey et al. (2004) observed the transfer of C from *Festuca idahoensis* to an invasive weed, *Centaurea maculosa*, via hyphal links. The carbon transferred contributed up to 15% of the aboveground carbon in *C. maculosa*. In fact, the movement of plant photosynthates to AMF is rapid and accounts for 3.4% of the carbon initially fixed by the plants within 70 hours (Johnson et al., 2002). Wilson et al. (2006) reported that interplant transfer of ^{32}P via hyphal links accounted for >50% of the total ^{32}P acquisition by *Sorghastrum nutans*, but accounted for only 20% of ^{32}P uptake into *Artemisia ludoviciana*, indicating that the degree of the contribution of hyphal links depends on the plant species. In fact, the contribution can be very low or zero in some cases. For example, in the ryegrass-clover combination, the interplant transfer of P or N was not appreciable (Yao et al., 2003) or even nil (Rogers et al., 2001). It is also suggested that the allocating pattern of nutrients via hyphal links is strongly controlled by the AMF identity (van der Henijden et al., 2003). More recently, AMF hyphal links has been demonstrated to transfer the water (Egerton-Warburton et al., 2007), which implies the possibility of AMF in regulating the plant diversity in the semi-arid grasslands.

Based on the evidence gathered, it is hypothesized that AMF can smooth out the differences of plants in the acquisition of resource, and competition for resource is even further reduced (Eriksson, 2001). Undoubtly, the redistribution of resource via hyphal links can substantially reshape the plant community structure and alter the plant diversity. For example, due to the transfer of carbon from the dominant species to the subodinate species through hyphal links, the plant diversity was increased by AMF in a microcosm system (Grime et al., 1987).

4. PRACTICAL SIGNIFICANCE

4.1. Alteration of Plant Community Structure Using AMF

Since AMF can regulate the diversity and community structure of plant ecosystem, the alteration of plant community structure in an expected direction may be achieved by inoculating appropriate AMF species into the ecosystem of interest. In a subtropical orchard, the growth responses of native cover crops to indigenous and exogenous AMF species were estimated (Yao et al., 2008). They found that the indigenous AMF greatly enhanced the growth of the dominant plant species *Ageratum conyzoides*, but not the subordinate species *Cyperus difformis*, implying the critical role of the indigenous AMF in the construction of the community structure of cover crops. *A. conyzoides* is regarded as beneficial plants in orchards in that they can attract predatory mite *Amblyseius newsami* and thus protect the orchards from severe damage from pests (Kong et al., 2005). In this situation, any practice to stimulate the indigenous AMF will favor the vigorous of *A. conyzoides* instead of manual seeding, which has been practiced so far but with much labor. This strategy can also be employed to other ecosystem, which is intended to change its plant community structure with an orientation.

4.2. Competition in Agroecosystem Regulated by AMF

In plant ecosystems, competition between individuals is ubiquitous and inevitable. In cover-cropping orchards, however, competition between cover crops and fruit trees is not plausible, because it can greatly reduce the growth vigor of fruit trees (Tworkoski and Glenn, 2001), especially in the young orchards. During the early stage, roots of fruit trees are shallow and thus compete severely with the roots of cover crops for water and mineral nutrients. In a greenhouse experiment, Yao et al. (2005) tried to alleviate the competition between citrus seedlings and a leguminous cover crop, *Stylosanthes gracilis*, using AMF inoculation. However, AMF increased the P uptake and plant growth of *S. gracilis* but showed negligible effect on citrus seedlings. Inoculation also promoted more roots of *S. gracilis* entering the soil volume occupied by the roots of citrus seedlings, implying that more severe competition between roots of two species occurs. They attributed this difference to the high mycorrhizal dependency of *S. gracilis* than citrus seedlings. Similarly, in another experiment where intercropping system with tomato and bahia grass (*Paspalum notatum* Flügge) were monitored as mycorrhizal or non-mycorrhizal, Sylvia et al. (2001) found that AMF increased the P uptake and plant growth of tomato by 30%. The different results of two experiments may be attributed to the mycorrhizal dependency.

In agro-ecosystems, it is necessary to alleviate the competition between different species in some cases and to strengthen it in others. The two experiments described above indicate that the regulation of the competition by AMF can be successful (Sylvia et al., 2001) or unsuccessful (Yao et al., 2005), mainly depending on the mycorrhizal dependency of the plants.

4.3. Restoration of Degraded Grasslands

Degradation of ecosystems is increasingly severe around the world, mainly due to the mankind activity. Among these degraded ecosystems, the degradation of grasslands is of special importance, because they are key components in the agro-ecosystems and landscape. In Japan, active volcanos sometimes seriously damage large areas. The newly deposited volcanic materials are very low in nutrients available for plants and very susceptible to erosion. To restore the vegetation on the slope of Mt. Fugendake (32°45' N, 130°19' E), about

3000 bags of unwoven polyester fabric containing seeds of various wild grass and shrub species, AMF inocula, slow-release chemical fertilizer, and some carriers, were placed on the degraded area. The grass plants that germinated from the bag were highly colonized with AM fungi. Six years after application, investigation showed that the inoculated AMF species were still proliferating, and the site where the bags were located became a base from which the plants revegetated the site and prevented serious erosion (Saito and Marumoto, 2002).

5. FUTURE PROSPECT

AMF, as a type of widely spread symbiotic soil fungi, can drive effectively the plant ecological processes and regulate the plant diversity, which has been supported by accumulating experimental evidence. However, more knowledge is required to deepen our understanding of this regulation.

Firstly, we now realize that the direction and the degree of the regulation of plant diversity by AMF strongly depend on both the plant identity and the AMF identity. From the ecological point of view, we can answer why benomyl application decreases the plant diversity in this field but increases it in another field, or why this AMF species increases the plant diversity in the microcosm but another AMF species decreases it. However, the explanatory information at the molecular physiological level is still very limited. Interpretation of the difference in the eco-physiological properties of plants, AMF and their interactions is still to be explored.

Secondly, hyphal links emerge as increasingly important factor in plant ecological processes (Selosse et al., 2007). The redistribution of C, P, N and water among plant community via hyphal links has attracted much attention for decades, because these resources readily become constraints in some plant ecosystems. Along with the emergence of novel environmental constraints, AMF may also function well via hyphal links in these ecosystems. For example, can heavy metal ions be transferred via hyphal links? How if a hyper accumulator coexists in a plant community?

Finally, since AMF can play great role in regulating the plant ecosystem processes, the utilization of AMF in practice need to be urged. According to the performance of AMF in fields, more problems may occur and provide new directions for the research.

ACKNOWLEDGMENTS

This work was supported by Natural Science Foundation of China (30200006) to QY, and by Natural Science Foundation of Guangdong Province (E05202480) to HHZ.

REFERENCES

Akiyama, K; Matsuzaki, K; Hayashi, H. Plant sesquiterpenes induce hyphal branching in arbuscular mycorrhizal fungi. *Nature*, 2005, 435, 824-827.

Bakker, ES; Ritchie, ME; Olff, H; Milchunas, DG, Knops, JMH. Herbivore impact on grassland plant diversity depends on habitat productivity and herbivore size. *Ecology Letters*, 2006, 9, 780-788.

Bingham, MA; Biondini, M. Mycorrhizal hyphal length as a function of plant community richness and composition in restored northern tallgrass prairies (USA). *Rangeland Ecology & Management*, 2009, 62, 60-67.

Bradley, K; Drijber, RA; Knops, J. Increased N availability in grassland soils modifies their microbial communities and decreases the abundance of arbuscular mycorrhizal fungi. *Soil Biology and Biochemistry*, 2006, 38, 1583-1595.

Buschmann, H; Keller, M; Porret, N; Dietz, H; Edwards, PJ. The effect of slug grazing on vegetation development and plant species diversity in an experimental grassland. *Functional Ecology*, 2005, 19, 291-298.

Carey, EV; Marler, MJ; Callaway, RM. Mycorrhizae transfer carbon from a native grass to an invasive weed: Evidence from stable isotopes and physiology. *Plant Ecology*, 2004, 172, 133-141.

Copley, J. Ecology goes underground. *Nature*, 2000, 406, 452-454.

Daleo, P; Alberti, J; Canepuccia, A; Escapa, M; Fanjul, E; Silliman, BR; Bertness, MD; Iribarne, O. Mycorrhizal fungi determine salt-marsh plant zonation depending on nutrient supply. *Journal of Ecology*, 2008, 96, 431-437.

Egerton-Warburton, LM; Querejeta, JI; Allen, MF. Common mycorrhizal networks provide a potential pathway for the transfer of hydraulically lifted water between plants. *Journal of Experimental Botany*, 2007, 58, 1473-1483.

Eriksson, Å. Arbuscular mycorrhiza in relation to management history, soil nutrients and plant species diversity. *Plant Ecology*, 2001, 155, 129-137.

Ferrol, N; Barea, JM; Azcón-Aguilar, C. Mechanisms of nutrient transport across interfaces in arbuscular mycorrhizas. *Plant & Soil*, 2002, 244, 231-237.

Frank, DA; McNaughton, SJ. Stability increases with diversity in plant communities: Empirical evidence from the 1988 Yellowstone drought. *Oikos*, 1991, 62, 360-362.

Gange, AC; Brown, VK; Sinclair, GS. Vesicular-arbuscular mycorrhizal fungi: a determinant of plant community structure in early succession. *Functional Ecology*, 1993, 7, 616-622.

Gerdemann, JW. Vesicular-arbuscular mycorrhizae. In: Torrey JG, Clarkson DT editors. *The Development and Function of Roots*. London: Academic Press; 1975; 575-591.

Graves, JD; Watkins, NK; Fitter, AH; Robinson, D; Scrimgeour, C. Interspecific transfer of carbon between plants linked by a common mycorrhizal network. *Plant & Soil*, 1997, 192, 153-159.

Grime, JP; Mackey, JML; Hillier, SH; Read, DJ. Floristic diversity in a model system using experimental microcosms. *Nature*, 1987, 328, 420-422.

Gryndler, M; Larsen, J; Hršelová, H; Řezáčová, V; Gryndlerová, H; Kubát, J. Organic and mineral fertilization, respectively, increase and decrease the development of external mycelium of arbuscular mycorrhizal fungi in a long-term field experiment. *Mycorrhiza*, 2006, 16, 159-166.

Guo, ZG; Long, RJ; Niu, FJ; Wu, QB; Hu, YK. Effect of highway construction on plant diversity of grassland communities in the permafrost regions of the Qinghai-Tibet plateau. *The Rangeland Journal*, 2007, 29, 161-167.

Hart, MM; Reader, RJ; Klironomos, JN. Plant coexistence mediated by arbuscular mycorrhizal fungi. *Trends in Ecology & Evolution*, 2003, 18, 418-423.

Hartnett, DC; Wilson, GWT. Mycorrhizae influence plant community structure and diversity in tallgrass praire. *Ecology*, 1999, 80, 1187-1195.

Helgason, T; Fitter, AH; Young, JPW. Molecular diversity of arbuscular mycorrhizal fungi colonizing *Hyacinthoides non-scripta* (bluebell) in a semi-natural woodland. *Molecular Ecology*, 1999, 8, 659-666.

Hetrick, BAD; Wilson, GWT; Todd, TC. Differential responses of C_3 and C_4 grasses to mycorrhizal symbiosis, phosphorus fertilization, and soil microorganisms. *Canadian Journal of Botany*, 1990, 68, 461-467.

Jakobsen, I; Rosendahl, L. Carbon flow into soil and external hyphae from roots of mycorrhizal cucumber plants. *New Phytologist*, 1990, 115, 77-83.

Johnson, D; Leake, JR; Read, DJ. Transfer of recent photosynthate into mycorrhizal mycelium of an upland grassland: short-term respiratory losses and accumulation of ^{14}C. *Soil Biology & Biochemistry*, 2002, 34, 1521-1524.

Johnson, D; Vandenkoornhuyse, PJ; Leake, JR; Gilbert, L; Booth, RE; Grime, JP; Young, JPW; Read, DJ. Plant communities affect arbuscular mycorrhizal fungal diversity and community composition in grassland microcosms. *New Phytologist*, 2004, 161, 503-515.

Johnson, NC; Wolf, J; Koch, GW. Interactions among mycorrhizae, atmospheric CO_2 and soil N impact plant community composition. *Ecology Letters*, 2003, 6, 532-540.

Karanika, ED; Mamolos, AP; Alifragis, DA; Kalburtji, KL; Veresoglou, DS. Arbuscular mycorrhizas contribution to nutrition, productivity, structure and diversity of plant community in mountainous herbaceous grassland of northern Greece. *Plant Ecology*, 2008, 199, 225-234.

Klironomos, JN; Allen, MF; Rillig, MC; Piotrowski, J; Makvandi-Nejad, S; Wolfe, BE; Powell, JR. Abrupt rise in atmospheric CO_2 overestimates community response in a model plant-soil system. *Nature*, 2005, 433, 621-624.

Kong, C; Hu, F; Xu, X; Zhang, M; Liang, W. Volatile allelochemicals in the *Ageratum conyzoides* intercropped citrus orchard and their effects on mites *Amblyseius newsami* and *Panonychus citri*. *Journal of Chemical Ecology*, 2005, 31, 2193-2203.

Landis, FC; Gargas, A; Givnish, TJ. The influence of arbuscular mycorrhizae and light on Wisconsin (USA) sand savanna understories. I. Plant community composition. *Mycorrhiza*, 2005, 15, 547-553.

Liu, J; Diamond, J. China's environment in a globalizing world. *Nature*, 2005, 435, 1179-1193.

Marini, L; Fontana, P; Klimek, S; Battisti, A; Gaston, KJ. Impact of farm size and topography on plant and insect diversity of managed grasslands in the Alps. *Biological Conservation*, 2009, 142, 394-403.

Maurer, K; Weyand, A; Fischer, M; Stöcklin, J. Old cultural traditions, in addition to land use and topography, are shaping plant diversity of grasslands in the Alps. *Bological Conservation*, 2006, 130, 438-446.

Meding, SM; Zasoski, RJ. Hyphal-mediated transfer of nitrate, arsenic, cesium, rubidium, and strontium between arbuscular mycorrhizal forbs and grasses from a California oak woodland. *Soil Biology & Biochemistry*, 2008, 40, 126-134.

Menge, JA; Johnson, ELV; Platt, RG. Mycorrhizal dependency of several citrus cultivars under three nutrition regimes. *New Phytologist*, 1978, 81, 553-559.

Newsham, KK; Watkinson, AR; West, HM; Fitter, AH. Symbiotic fungi determine plant community structure: changes in a lichen-rich community induced by fungicide application. *Functional Ecology*, 1995, 9, 442-447.

Oliveira, RS; Castro, PML; Dodd, JC; Vosátka, M. Different native arbuscular mycorrhizal fungi influence the coexistence of two plant species in a highly alkaline anthropogenic sediment. *Plant & Soil*, 2006, 287, 209-221.

Parton, WJ; Scurlock, JMO; Ojima, DS; Schimel, DS; Hall, DO. Impact of climate change on grassland production and soil carbon worldwide. *Global Change Biology*, 1995, 1, 13-22.

Plenchette, C; Fortin, JA; Furlan, V. Growth responses of several plant species to mycorrhizae in a soil of moderate P-fertility. I. Mycorrhizal dependency under field conditions. *Plant & Soil*, 1983, 70, 199-209.

Püschel, D; Rydlová, J; Vosátka, M. Mycorrhiza influences plant community structure in succession on spoil banks. *Basic & Applied Ecology*, 2007, 8, 510-520.

Read, DJ. The ties that bind. *Nature*, 1998, 396, 22-23.

Rogers, JB; Laidlaw, AS; Christie, P. The role of arbuscular mycorrhizal fungi in the transfer of nutrients between white clover and perennial ryegrass. *Chemosphere*, 2001, 42, 153-159.

Saito, M; Marumoto, T. Inoculation with arbuscular mycorrhizal fungi: the status quo in Japan and the future prospects. *Plant & Soil*, 2002, 244, 273-279.

Sanders, IR. Preference, specificity and cheating in the arbuscular mycorrhizal symbiosis. *Trends in Plant Science*, 2003, 8,143-145.

Scheublin, TR; van Logtestije, RSP; van der Heijden, MGA. Presence and identity of arbuscular mycorrhizal fungi influence competitive interactions between plant species. *Journal of Ecology*, 2007, 95, 631-638.

Schüssler, A; Schwarzott, D; Walker, C. A new fungal phylum, the Glomeromycota: phylogeny and evolution. *Mycological Research*, 2001, 105, 1413-1421.

Selosse, MA; Richard, F; He, X; Simard, SW. Mycorrhizal networks: des liaisons dangereuses? *Trends in Ecology & Evolution*, 2006, 21, 621-628.

Sims, P; Bradford, JA. 2001. Carbon dioxide fluxes in a southern plains prairie. *Agricultural & forest meteorology*, 109, 117-134.

Smith, SE; Read, DJ. Mycorrhizal Symbiosis. 1st Edition. Cambridge: Academic Press; 1997.

Stampe, ED; Daehler, CC. Mycorrhizal species identity affects plant community structure and invasion: A microcosm study. *Oikos*, 2003, 100, 362-372.

Stein, C; Rißmann, C; Hempel, S; Renker, C; Buscot, F; Prati, D; Auge, H. Interactive effects of mycorrhizae and a root hemiparasite on plant community productivity and diversity. *Oecologia*, 2009, 159, 191-205.

Stohlgren, TJ; Schell, LD; Heuvel, BV. How grazing and soil quality affect native and exotic plant diversity in Rocky mountain grasslands. *Ecological Applications*, 1999, 9, 45-64.

Sylvia, D; Alagely, A; Chellemi, D; Demchenko, L. Arbuscular mycorrhizal fungi influence tomato competition with bahiagrass. *Biology & Fertility of Soils*, 2001, 34, 448-452.

Tang, J; Xu, L; Chen, X; Hu, S. Interaction between C_4 barnyard grass and C_3 upland rice under elevated CO_2: Impact of mycorrhizae. *Acta Oecologica*, 2009, 35, 227-235.

Tilman, D; Downing, JA. Biodiversity and stability in grasslands. *Nature*, 1994, 367, 363-365.

Tworkoski, TJ; Glenn, DM. Yield, shoot and root growth, and physiological responses of mature peach trees to grass competition. *HortScience*, 2001, 36, 1214-1218.

Urcelay, C; Díaz, S. The mycorrhizal dependence of subordinates determines the effect of arbuscular mycorrhizal fungi on plant diversity. *Ecology Letters*, 2003, 6, 388-391.

van der Heijden MGA. Arbuscular mycorrhizal fungi as a determinant of plant diversity: In search of underlying mechanisms and general principles. In: van der Heijden MGA, Sanders IR, editors. *Mycorrhizal ecology*. Berlin: Springer; 2002; 243-265.

van der Heijden, MGA; Klironomos, JN; Ursic, M; Moutoglis, P; Streitwolf-Engel, R; Boller, T; Wiemken, A; Sanders, IR. Mycorrhizal fungal diversity determines plant biodiversity, ecosystem variability and productivity. *Nature*, 1998, 396, 69-72.

van der Heijden, MGA; Verkade, S; de Bruin, SJ. Mycorrhizal fungi reduce the negative effects of nitrogen enrichment on plant community structure in dune grassland. *Global Change Biology*, 2008, 14, 2626-2635.

van der Heijden, MGA; Wiemken, A; Sanders, IR. Different arbuscular mycorrhizal fungi alter coexistence and resource distribution between co-occurring plant. *New Phytologist*, 2003, 157, 569-578.

Vogelsang, KM; Reynolds, HL; Bever, JD. Mycorrhizal fungal identity and richness determine the diversity and productivity of a tallgrass prairie system. *New Phytologist*, 2006, 172, 554-562.

Walling, SZ; Zabinski, CA. Host plant differences in arbuscular mycorrhizae: Extraradical hyphae differences between an invasive forb and a native bunchgrass. *Plant & Soil*, 2004, 265, 335-344.

Wilson, GWT; Hartnett, DC; Rice, CW. Mycorrhizal-mediated phosphorus transfer between tallgrass prairie plants *Sorghastrum nutans* and *Artemisia ludoviciana*. *Functional Ecology*, 2006, 20, 427-435.

Wolfe, BE; Weishampel, PA; Klironomos, JN. Arbuscular mycorrhizal fungi and water table affect wetland plant community composition. *Journal of Ecology*, 2006, 94, 905-914.

Yao, Q; Li, XL; Ai, WD; Christie, P. Bi-directional transfer of phosphorus between red clover and perennial ryegrass via arbuscular mycorrhizal hyphal links. *European Journal of Soil Biology*, 2003, 39, 47-54.

Yao, Q; Wang, LR; Zhu, HH; Chen, JZ. Effect of arbuscular mycorrhizal fungal inoculation on root system architecture of trifoliate orange (*Poncirus trifoliata* L. Raf.) seedlings. *Scientia Horticulturae*, 2009, 121, 458-461.

Yao, Q; Zhu, HH; Chen, JZ. Growth responses and endogenous IAA and iPAs changes of litchi (*Litchi chinensis* Sonn.) seedlings induced by arbuscular mycorrhizal fungal inoculation. *Scientia Horticulturae*, 2005, 105, 145-151.

Yao, Q; Zhu, HH; Hu, YL; Li, LQ. Differential influence of native and introduced arbuscular mycorrhizal fungi on growth of dominant and subordinate plants. *Plant Ecology*, 2008, 96, 261-268.

In: Encyclopedia of Environmental Research
Editor: Alisa N. Souter

ISBN: 978-1-61761-927-4
© 2011 Nova Science Publishers, Inc.

Chapter 11

CARNIVOROUS MAMMALS IN A MOSAIC LANDSCAPE IN SOUTHEASTERN BRAZIL: IS IT POSSIBLE TO KEEP THEM IN AN AGRO-SILVICULTURAL LANDSCAPE?

Maria Carolina Lyra-Jorge, Giordano Ciocheti, Leandro Tambosi, Milton César Ribeiro and Vânia Regina Pivello
Departamento de Ecologia, Instituto de Biociências, Universidade de São Paulo, Brazil

INTRODUCTION: HABITAT FRAGMENTATION AND ITS EFFECTS ON BIODIVERSITY

Habitat fragmentation can be defined as a process where continuous areas of natural habitats are broken into small patches separated by other habitats different from the original ones (Wilcove *et al.* 1986; Andrén 1994). Today, habitat fragmentation is a common issue to almost every ecosystem in the world, since anthropogenic land uses transformed initially continuous habitats into mosaic landscapes represented by isolated native patches surrounded by man-altered environments (Nagendra *et al.* 2003).

Anthropogenic matrices usually act as selective filters to species movements among native patches in the landscape (Gascón *et al.* 1999), and therefore, the persistence of animal and plant populations in fragmented habitats will depend to a great extent on the matrix permeability (Ricketts 2001). Landscapes are commonly classified into continuous or fragmented (Fahrig 2003) however, the landscape is not a binary mosaic formed by natural habitat and matrix – or habitat and non-habitat – and the species certainly do not perceive it that way (presence/ absence of resources), as we will discuss later on in this chapter (Fahrig, 2003)

Despite being a rather controversial issue, several authors have shown the importance of protecting small native patches resulted from habitat fragmentation, as in the landscape context they are able to keep a significant portion of local biodiversity (Saunders *et al.* 1991; Lindenmayer & Nix 1993; Bodin et al. 2006). Andrèn (1994) states that landscape biodiversity may increase considerably when several small fragments are close to each other and permit animal and plant fluxes as in a continuous habitat. The ability of a species to move throughout the landscape is related to habitat connectivity (); it refers to the functional linkage among patches either due to patch proximity or to matrix permeability (With 1997; **Uezu** *et al.* 2005). Therefore, the degree of habitat connectivity, which depends on matrix quality, is essential to maintain native species in a fragmented landscape (Forman & Gordon 1986; With 1997; Tischendorf & Fahrig 2000; Ewers & Didham 2005).

Besides affecting species movements throughout the landscape, matrix quality also controls the permanence time of individuals in it, according to the resources it offers (Aberg *et al.* 1995), thus matrices of good quality may characterize a type of habitat effectively used by the species both in search for resources or traveling among preferential habitats (Smallwood & Fitzhugh 1995; Wagner & Fortin 2005). The quality of a matrix, however, is differently perceived by different species; some species may benefit from agricultural lands while others may be excluded (Gehring *et al.* 2003; Laurance 1994). As examples of the former case, some studies show the regular use of coffee plantations in Mexico (Moguel *et al.* 1999), banana and cocoa plantations in Costa Rica (Harvey *et al.* 2006), cocoa plantations in Brazil (Faria *et al.* 2006) and subsistence agriculture in Nepal (Acharya 2006) by the native fauna. On the other hand, species that required large territories and have small populations are particularly vulnerable to habitat fragmentation and can be locally extinct (Crooks 2002).

The increasing fragmentation and loss of natural areas, associated to changes in ecological processes and species extinctions demand urgent integration of human needs and the preservation of essential ecosystem processes. Approaches focused on the interactions between nature and man should be the basis for a transition to a more sustainable agriculture (Bignal 1998). The challenge of achieving development in a sustainable way was first globally discussed in the Brundlandt Commission (World Commission on Environment and Development – WCED), in 1983, and it has become a central question ever since. However, practical sustainable actions are being implemented very slowly although numerous studies have shown that the preservation of many species could be ensured if agricultural systems incorporated ecological concepts. On the other hand, conservationists have tried to find ways to integrate human land uses and native fauna needs (Vandermeer *et al.* 1997; Bignal 1998).

In this chapter, we intend to show the use of both natural and agricultural (silvicultural) habitats by the native carnivore fauna, and to demonstrate the possibility to maintain these populations in a fragmented landscape, provided that some large native patches are left and the matrix is permeable to the native fauna.

STUDY REGION

São Paulo, in southeastern Brazil, is the most developed and urbanized state in the country. Despite representing less than 3% of the Brazilian territory, São Paulo State accounts for more than one third of the national gross domestic product and more than 21% of the country's population (IBGE 2007a; IBGE 2007b). The state's rapid development started about two centuries ago, when a strong agricultural expansion was implemented, based especially on coffee cultivation, which by 1950-60 was replaced by industrial development and agribusiness. Today, the state's economy is based on industrial products and export commodities such as biofuel from sugarcane, paper and beef cattle (Igari et al. 2009).

As an expected consequence, São Paulo has lost more than 90% of its natural habitats, originally composed of the Atlantic dense rainforest on the east (alongside the Atlantic Ocean), a complex of savanna formations (regionally named *Cerrado*) in the central part of the state, and seasonal forests to the west. This strong process of habitat fragmentation in the state resulted in a few remaining patches of native vegetation, usually small and isolated (Metzger & Rodriges 2008).

In the northeast of São Paulo State, where our study was carried out (Santa Rita do Passa-Quatro and Luiz Antônio municipalities: 21°31'15''S - 47°34'42''W; 21°44'24''S - 47°52'01''W), a similar pattern of land occupation described for São Paulo State was observed: from 1962 to 1992, the region lost 60% of its original vegetation cover due to agriculture expansion (Kronka et al. 1993), and since 1992 agriculture and forestry are still expanding in the region but at a much slower pace. However, what distinguishes this region from the rest of São Paulo state is some large remnants of *cerrado* and seasonal forest. Our study region comprises the largest cerrado patches of the state, which are protected as nature preserves: Cerrado Pé-de-Gigante (1,212.9 ha), located in the Vassununga State Park (Korman 2003) and the Jataí Ecological Station (9,010.7 ha) (Decree 47.096/SP, from 18/September/2002). In addition, there are also in the study region four patches of seasonal forest in the study region, also part of the Vassununga State Park (sizes ranging from 12.1 ha to 327.8 ha) (Korman 2003) (Figure 1).

Therefore, the present land cover in the study region comprises a mosaic of natural formations and extensive monocultures, especially sugar cane and *Eucalyptus* species (Figure 2).

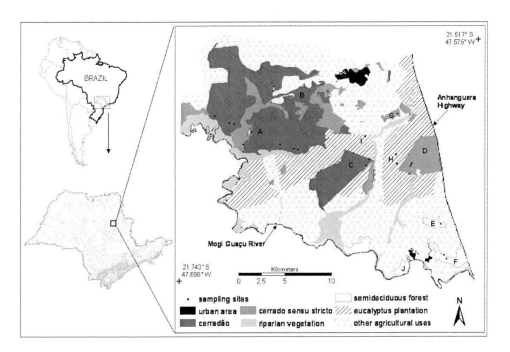

Figure 1. The study region (São Paulo State, Brazil): location and land use/ land cover classes. Dots represent the sampling sites; A, B and C = Jataí Ecological Station (EEJ) patches; D, E, F and J = Vassununga State Park (PEV) patches; G = private area with cerrado vegetation; H and I = eucalyptus plantation.

Figure 2. The mosaic of natural formations and extensive monocultures, especially of sugar cane and *Eucalyptus* species. (Photographed by Dr. Luciano Verdade)

cerrado "sensu lato"

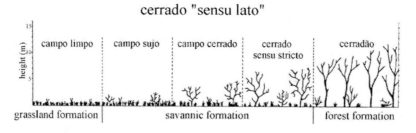

Figure 3. The *Cerrado* physiognomic gradient (modified from Coutinho 1978).

The native vegetation includes patches of seasonal forest and of different savanna physiognomies in an increasing density of trees, from *campo-sujo* (grassy field with scattered trees) to *cerradão* (sclerophyllous woodland), being *cerrado-sensu-stricto* an intermediate form (typical *cerrado* with grasses, shrubs and many trees) (Coutinho 1978; Oliveira & Marquis 2002; Shida 2005) (Figure 3).

The climate in the region is Cwa (according to Köppen 1948) or type II (following Walter 1986), which is the typical tropical savanna climate with wet summers (October to March) and dry winters (May to August); the annual rainfall is approximately 1,300 mm. The relief is gently rolling, formed by extensive and flat-topped hills.

METHODS

To analyze the use of different habitats in the study region by the native fauna we sampled eight patches of native vegetation (three of *cerradão*, three of *cerrado-sensu-stricto*, and two of seasonal forest), as well as two homogeneous plantations of *Eucalyptus* species (Figure 1). We concentrated our analyses on mammal carnivores, as they are top predators and respond for several ecological processes in the community, therefore they may be used as indicator species of community resources and equilibrium (Crooks 2002; Miller et al. 2001).

Data were collected in three-day monthly field trips, throughout 18 months (August/2004 to January 2006). In the field, two systematic methods were used to obtain the data: camera trapping and track plot recording.

Track plot recording (Lyra-Jorge 1999; Pardini et al. 2003) is based on the identification of the animal species through footprints in a plot, and allows estimating animal occurrence and richness. We randomly selected 21 sampling points from a larger group of points that met the following requisites: located along pre-existing trails or dirt roads, and had previously recorded footprints in the soil, indicating that the animals effectively used that area. Twenty-one track plots of 10 m X 2 m were installed in the *Eucalyptus* plantation and native vegetation patches; the number of plots in the patches was proportional to their sizes, resulting in nine plots in *cerradão*, six in *cerrado-sensu-stricto*, two in seasonal forest and four in the *Eucalyptus* plantations (Figure1). The sandy ground was used to create each track plot. The track plots were visited every day, during the field trips, in order to identify the footprints in the soil and to clear the ground for new records. Ambiguous footprints were ignored and footprints of the same species, in the same plot, and in the same day were considered as if they were from a single individual.

The camera trapping method (Wemmer *et al.* 1996; Tomas & Miranda 2003) is based on the identification of the animal species through photographs taken by an automatic camera triggered by the animal body heat and/or movement. It also permits to estimate animal occurrence and richness. To distribute the camera traps in the area, we used the same criteria as those used to place track plots. In addition, cameras were protected from direct sunshine (as they would set off if exposed to intense heat). The sampling points were visited every field trip to change the films and batteries of the cameras, which remained activated in the field during the entire sampling period.

We assumed that all sampling points containing camera traps and track plots were homogeneous in detecting carnivores, and that all carnivore species were equally detectable by both methods.

We used data from both methods to calculate a species accumulation curve in order to express carnivore richness. The curves were randomized 5,000 times through a rarefaction process (Santos 2003). Species richness was also estimated through the Bootstrap technique (Smith & Van Belle 1984; Santos 2003) with 5,000 randomizations.

The relative frequency (FR) of the carnivores recorded by camera traps and track plots was calculated according to the model (Crooks 2002):

i/N , where: i = number of occurrences of species I;
N= total occurrences in the physiognomy.

We compared the distribution of species records in the *cerradão*, *cerrado-sensu-stricto* and *Eucalyptus* plantation through the Kruskal-Wallis test (Zar 1999); we removed the seasonal forest from this analysis due to the very small sample size (N=2). We also calculated and compared species diversity in each vegetation physiognomy (except for seasonal forest, due to small sample size) with the Shannon-Wiener diversity index (Magurran 1988) and Kruskal-Wallis test (Zar 1999).

The similarity in the carnivore assemblage in each vegetation form was tested through a Multi-Response Permutation Procedure (MRPP) analysis (McCune 2002) using data randomized 1,000 times and the Bray-Curtis index (Beals 1984). We used the numbers of species records in each sampling point.

To verify a possible influence of the landscape structure on the intensity of habitat use by the carnivores we created a carnivore habitat use map based on a model generated by stepwise regression analyses using the data on species richness and species occurrence obtained only through camera trapping, as well as information on habitat type (land use/ land cover class) and landscape structure. We assumed a direct correlation between the number of animal occurrences and the intensity of habitat use. We generated a land use/land cover map from a Landsat5-TM satellite image (February/2005, spatial resolution of 30 m) to obtain the landscape indices, and we then located the sampling points on that map and calculated the following landscape metrics: percentage of remaining habitat in the landscape (PLAND), edge density (ED) and habitat patch shape (SHP), in a multi-scale approach (Wagner & Fortin 2005; Fortin & Dale 2005; McAlpine et al. 2006), as for each point (and for each pixel) we used search radii of 250, 500, 1,000 and 2,000 m (following Umetsu et al. 2008). Landscape metrics were chosen according to our perceptions on the animals ecological needs based on field experience. They were calculated using the moving windows option of FRAGSTATS software (McGarigal & Marks 1995).

The response variable (habitat use intensity) was analyzed through stepwise regression. The animal occurrences were set as a dependent variable, while the metrics (PLAND, ED and SHP) of the four land use/land cover classes (cerrado sensu stricto, cerradão, seasonal forest and Eucalyptus plantation) in the four scales (radii of 250, 500, 1,000, and 2,000 m) were the independent variables. The stepwise regression analysis followed the general model below:

$$HUI = \beta1*M(d,c1) + \beta2*M(d,c2) + \beta3*M(d,c3) + \beta4*M(d,c4) + \varepsilon$$

where: HUI = habitat use intensity; $\beta1...\beta4$ = regression parameters; $M(d,ci)$ = landscape metrics (PLAND, ED and SHP) calculated for the distance d (specific radius scale) and the land use/land cover class ci (cerrado-sensu-stricto, cerradão, seasonal forest and Eucalyptus plantation); ε = error

The inclusion of each independent variable in the model was determined based on its statistically significant additional contribution.

The best model was the one with the highest coefficient of determination (R^2), and it was used to generate the map of habitat use intensity. To produce the map, we used the algebra map option of SPRING GIS (Cordeiro et al. 2008), where degrees of habitat use intensity were associated to color intensity.

RESULTS AND DISCUSSION: HABITAT USE BY CARNIVORES IN THE FRAGMENTED LANDSCAPE

During the 18 months of sampling with both track plots (1,864 hours of exposure) and camera traps (12,960 hours of exposure), we were able to record ten carnivore species in the study region, which belonged to four different families (Table 1). Nine of these ten species were recorded by camera traps and seven species were recorded in the track plots. Two species of small felines (Leopardus tigrinus and Puma yagouaroundi) could not be distinguished by track plot recording, as their footprints were very much alike, but they could be recognized in the camera trap photographs. For this reason, these two species were grouped as "small felines" in some analyses. *Procyon cancrivorus* was not recorded by camera traps and was only found in the seasonal forest, while *Nasua nasua* was not recorded in the track plots (Table 1).

We believe we obtained a good representation of the local carnivore species richness, as the species accumulation curve calculated based on data from both methods approached the asymptote after nine months of sampling, and the obtained species richness (N= 9.0) was similar to the value estimated by Bootstrap (N= 9.16 ±0.16).

The species composition of the carnivore assemblage found in our study is in accordance with the species geographic distributions cited in the literature (Emmons 1997; Einsenberg & Redford 1999) and also agrees with other surveys carried out in the same region (Gargaglioni *et al.* 1998; Lyra-Jorge 1999; Talamoni *et al.* 2000). However, some carnivore species expected to be found in the region (according to Emmons 1997; Gargaglioni *et al.* 1998; Einsenberg & Redford 1999; Lyra-Jorge 1999; Talamoni *et al.* 2000) could not be detected. This could be either because some of those species are presently rare or even extinct in the region, or the methods used to sample local richness were not directed to some types of habitats and species niches. For example, *Lycalopex vetulus* and *Leopardus wiedii* – not detected – are naturally rare species, and their population densities are usually very low (Azevedo 1996; Jácomo *et al.* 2004); *Panthera onca* has not been seen in the region for more than 50 years and is probably locally extinct, as it was a favorite game animal; *Lontra longicaudis* and *Galictis cuja* are species associated to aquatic habitats, therefore their habitats or territories were not possible to be sampled with the methods we used (Emmons 1997; Pardini 1998).

Puma concolor and *Chrysocyon brachyurus* had the highest relative frequencies regardless of the sampling method (table 1), and there are several possible explanations for that. First, they are the largest carnivores in the region, have large home ranges and move constantly in search for food (Dietz 1984; Dickson & Beier 2002), and because of their vagility the same individuals might have been repeatedly recorded by the camera traps. Also, it has been shown that camera traps perform better in detecting large bodied animals (Carbone *et al.* 2002; Silveira *et al.* 2003) and consequently, these large animals may have been oversampled by the methodology here adopted (Lyra-Jorge *et al.* 2008).

An initial comparison of the habitat use by the species sampled through camera traps and track plots in *cerradão*, *cerrado-sensu-stricto*, and *Eucalyptus* plantation shows that although most species seem to prefer some habitats – *L. pardalis*, *N. nasua* and *C. semistriatus* were more frequent in *cerrado* physiognomies, whereas *C. thous* and *E. barbara* were more frequently found in the *Eucalyptus* plantation (table 1) – the distribution of all species records in these three vegetation forms was not statistically different (Kruskal-Wallis; p=0,943). This result was confirmed by the MRPP, which showed similarity in species composition among the different habitats (p=0.65; A=-0.013; expected Δ= 0.51; observed Δ= 0.52). Species diversity assessed by Shannon-Wiener index also showed no significant differences among those three habitat types (Kruskal-Wallis; p=0.31).

Table 1. Carnivore species recorded by camera traps and in track plots. (NR= number of species records in all vegetation forms; CD= *cerradão*, SS= *cerrado-sensu-stricto*, SF= seasonal forest, EP= *Eucalyptus* plantation; FA= relative frequency obtained through camera trap data; FC= relative frequency obtained based on track plot data; FT= relative frequency obtained with both sampling methods.)

Species	Family	NR	Percentage of NR in the vegetation form				Relative frequency (%)		
			CD	SS	SF	EP	FT	FA	FC
Puma concolor (Linnaeus, 1771)	Felidae	74	49	22	9	20	29.4	30.1	29.0
Leopardus pardalis (Linnaeus, 1758)	Felidae	39	56	33	3	8	15.5	23.1	11.1
Small cat	Felidae	11	0	55	27	18	4.4	4.0	8.0
Chrysocyon brachyurus (Illiger, 1811)	Canidae	78	47	18	8	27	31.0	24.1	32.3
Cerdocyon thous (Linnaeus, 1716)	Canidae	27	37	15	7	41	10.7	3.2	12.0
Nasua nasua (Linnaeus, 1766)	Procyonidae	02	50	50	0	0	0.8	5.9	0.0
Procyon cancrivorus (Cuvier, 1798)	Procyonidae	03	0	0	100	0	1.2	0.0	4.1
Conepatus semistriatus (Boddaert, 1784)	Mephitidae	11	64	27	0	9	4.4	6.1	1.8
Eira barbara (Linnaeus, 1758)	Mustelidae	07	29	0	14	57	2.8	3.0	1.8
Total		252					100.2	99.5	100.1

Similarly to the reported by other authors (Chinchila 1997; Oliveira 1998; Nuñez *et al.* 2000; Jácomo *et al.* 2004), these results indicated that even though most species maintained peculiar habitat preferences, the carnivore community as a whole was similar in the study region. This result supports the idea that carnivores in fragmented landscapes are more generalists than populations living in continuous and preserved areas and explore the entire region, not being restricted to the native vegetation patches (Azevedo 1996; Franklin *et al.* 1999; Donadio *et al.* 2001).

However, when we added information on the size and spatial arrangement of the different vegetation patches in the landscape the results were rather different. Although this analysis was performed using only camera trap data (Table 2), the species ratio recorded in the physiognomies were comparable in both data collecting methods. The model which included landscape structure data, yields high predictive power, with a significant coefficient of determination (R^2 = 0.88; p < 0.01). The best model selected took into account the occurrences of all the nine species recorded by camera traps (table 2), the four sampled habitats (*cerrado-sensu-stricto*: SS, *cerradão*: CD, seasonal forest: SF, *Eucalyptus* plantation: EP), the three landscape metrics (percentage of the habitat in the landscape: PLAND, edge density: ED, and patch shape: SHP), and only the radius of 250 m, as the independent variables generated on other scales (i.e., radii of 500, 1,000 and 2,000 m) did not make a statistically significant contribution to the model:

$$HUI = 0.04219*PLAND_{CD} + 0.01174*PLAND_{SS.} + 5.04857* SHP_{EP} - 0.26911*ED_{EP}$$

The independent variable that most contributed to the model was the percentage of *cerradão* in the landscape, *PLAND_{CD}* (partial $R^2 = 0.50$; $p < 0.01$).

The map of habitat use generated from the model above shows the *Eucalyptus* plantation as the most intensely used habitat, followed by *cerradão*. Seasonal forest, on the other hand, was the least used habitat (Figure 4). Therefore, although we had noticed habitat preferences by some species when using exclusively information about land use/ land cover classes, these preferences only became detectable when considering landscape parameters were included, which refined the analysis. This indicates the importance of considering spatial parameters to build more accurate predictive models especially aimed at fauna conservation, as also noticed by other authors (Andrén 1994; Fahrig 1998 McAlpine *et al.* 2006).

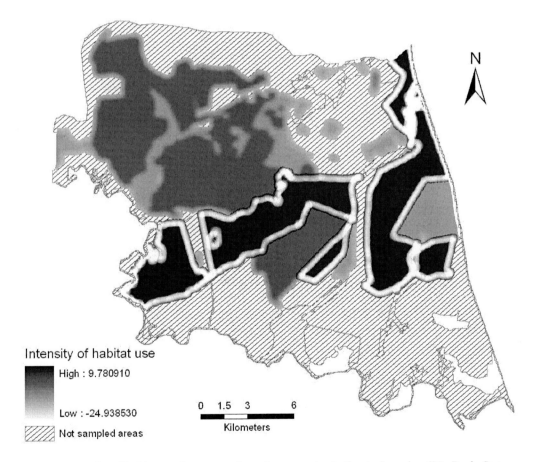

Figure 4. Intensity of habitat use by mammal carnivore species in the study region (São Paulo State, Brazil). (Darker to lighter = higher to lower intensity of use; see legend of vegetation physiognomies in Figure 1).

Table 2. Carnivorous species recorded by camera traps in the study region (CD= *cerradão*, SS= *cerrado-sensu-stricto*, SF= seasonal forest, EP= *Eucalyptus* plantation).

Species	Family	Land use/ land cover class				Total records
		CD	SS	SF	EP	
Puma concolor	*Felidae*	7	2	1	8	18
Leopardus pardalis	*Felidae*	9	2	0	2	13
Puma yagouaroundi	*Felidae*	0	0	0	1	1
Leopardus tigrinus	*Felidae*	0	1	0	0	1
Chrysocyon brachyurus	*Canidae*	7	1	0	6	14
Cerdocyon thous	*Canidae*	2	0	0	0	2
Eira barbara	*Mustelidae*	1	1	0	2	4
Conepatus semistriatus	*Mephitidae*	3	1	0	0	4
Nasua nasua	*Procyonidae*	1	1	0	0	2
Total		30	9	1	19	59

Figure 5. Carnivores mammals in the study region. 1: Leopardus pardalis; 2: Eira Barbara; 3: Chrysocyon brachyurus; 4: Conepatus semistriatus; 5: Puma concolor.

The *Eucalyptus* stands in the study region are surrounded by large protected patches of *cerradão* (Jataí Ecological Station-EEJ), and both the silviculture and the *cerradão* hold similarities in terms of vegetation structure (arboreal habitats with not very open canopy, sparse herbaceous layer, etc). Moreover, *Eucalyptus* cultures are based on 7-year cycles and are not as much intensely managed as the other crops in the region (ex. sugar-cane). We can then infer that carnivore species use *Eucalyptus* plantations at least to move around, although we do not know what other resources this vegetation may offer to the carnivores. Some generalist species are able to adapt to man-modified environments when their original habitats are severely reduced, and to find vital resources in such new habitats (Sánchez-Hernandéz et al. 2001; Reznick *et al.* 2004; Tabeni *et al.* 2005; Morán-López *et al.* 2006; McDougall *et al.* 2006). In this study region, *Puma concolor, Chrysocyon brachyurus, Cerdocyon thous* and *Eira barbara* seem to be relatively adapted to silvicultural *Eucalyptus* plantations, as they frequently use them, confirming the reported by other authors (Bisbal 1986; Jácomo *et al.* 2004).

Cerradão, the native physiognomy most intensely used by carnivores, comprises the largest native vegetation patches in the study region, all protected as natural preserves. These patches in the study region, all protected as natural preserves. These patches are mostly contiguous to the Mogi-Guaçu River riparian forests and swamps. These features may represent good quality habitats for the local fauna. On the other hand, seasonal forest patches are small and surrounded by sugar-cane plantations, which are highly managed throughout the year and receive large quantities of pesticides. Similarly, *cerrado sensu stricto* patches are small (except from Vassununga State Park, PEV, Figure 1 D) and surrounded by sugar-cane fields.

Therefore, our data show that top predator carnivores of medium to large size can be maintained in a highly fragmented agricultural landscape provided some large patches of native vegetation in good condition remain (performing as source patches, according to Donavan *et al.* 1995), and they are surrounded by a permeable matrix. Other authors came to similar conclusions in relation to other animal groups elsewhere (McAlpine *et al.* 2006; Baldissera *et al.* 2008; Marsden & Symes 2008). A highly permeable matrix is essential to connect habitats, permitting animal movements throughout the landscape and the maintenance of processes that are essential to the populations' persistence, such as dispersion and gene flow, in addition to animal foraging and housing (Elmhagen & Angerbjörn 2001; Hensen *et al.* 2005). In this study region, *Eucalyptus* plantations act as permeable matrix connecting patches of native vegetation, and this may be the reason for the permanence of a still rich carnivore assemblage in such an agricultural landscape (Lyra-Jorge & Pivello 2005).

CONCLUSION

A landscape is comprised of different types of habitats, used by fauna with different objectives – from foraging to reproduction – according to resource availability and quality. Habitats more intensely used are generally those with enough quality to maintain species residence and/or movement, and they usually either offer a great deal of vital resources or connect areas that have such resources, in that case acting as biodiversity corridors. In this sense, the manner and intensity of use of remnant habitats by native fauna is a relevant issue for biodiversity conservation.

In a scenario where it is no longer possible to exclude the intensive use of land by man and to keep large preserved natural areas, it is fundamental to contemplate means of turning agricultural matrices into good quality habitats for fauna. We showed in this chapter that it is possible to balance human land uses and fauna requirements provided that some ecological principles are taken into account, especially related to landscape structure. In the case of large and medium carnivores – as they are vagile animals that explore large territories – the maintenance of a net system of protected areas of different sizes that keep connectivity through a permeable matrix is essential.

REFERENCES

Acharya, K. P. Linking tree on farms with biodiversity conservation in subsistence farming in Nepal. *Biodiversity and Conservation, 15*, 631-646.

Andrén, H. Effects of habitat fragmentation on birds and mammals in landscapes with different proportions of suitable habitat: A Review. *Oikos, 71(3)*, 355-366.

Arberg, J., Jansson, G., Swenson, J. E. & Angelstam, P. (1995). The effect of matrix on the occurrence of hazel grouse in isolated habitat fragments. *Oecologia, 103*, 265-269.

Azevedo, F. C. C. (1996). Notes on behavior of the margay *Felis wiedii* in the Brazilian Atlantic forest. *Mammalia, 60*, 325-328.

Baldissera, R., Ganade, G., Brescovit, A. D. & Hartz, S. M. (2008). Landscape mosaic of Araucaria forest and forest monocultures influencing understorey spider assemblages in southern Brazil. *Austral Ecology* 33: 45-54. BEALS, E.W. 1984. Bray-Curtis ordination: A strategy for analysis of multivariate ecological data. *Advances in Ecological Research, 14*, 1-56.

Bignal, E. M. (1998). Using an ecological understanding of farmland to reconcile nature conservation requirements, EU agriculture policy and world trade agreements. *Journal of Applied Ecology, 35*, 949-954.

Bisbal, F. J. (1986). Food habits of some neotropical carnivores in Venezuela. *Mammalia, 50(3)*, 329-339.

Bodin, O; Tengö, M., Norman, A., Lundberg, J. & Elmquist, T. (2006). The value of small size: Loss of forest patches and ecological thresholds in southern Madagascar. *Ecological Applications, 16(2)*, 440-451.

Carbone, C., Conforti, C., Coulson, T., Franklin, N., Ginsberg, J. N., Grivths, M., Holden, J., Kinnaird, M., Laidlaw, R., Lynam, A., MacDonald, D. W., Martyr, D., McDougal, C., Nath, L., O'Brien, T. O., Seidnsticker, J., Smith, D. J., Tilson, R. & Shahruddin, W. N. (2002). The use of photograph rates to estimate densities of tigers and other cryptic mammals: response to Jannelle *et al. Animal Conservation, 5*, 121-132.

Chinchila, F. A. (1997). La dieta del jaguar (*Panthera onca*), el puma (*Felis concolor*) y el manigordo (*Felis pardalis*) em el Parque Nacional Corcovado, Costa Rica. *Ver Biologica Tropical, 459*, 1223-1229.

Cordeiro, J. P. C., Câmara, G., Freitas, U. M. & Almeida, F. (2008). Yet Another Map. *GeoInformatica*, DOI 10.1007/s10707-008-0045-4.

Coutinho, L. M. (1978). O conceito de cerrado. *Revista Brasileira de Botânica*, 1, 115-117.

Crooks, K. (2002). Relative sensitivities of mammalian carnivores to habitat fragmentation. *Conservation Biology, 16*, 488-502.

Dickson, B. G. & Beier, P. Home range and habitat selection by adult cougars in southern California. *Journal of Wildlife Management, 66(4)*, 1235-1245.

Dietz, J. (1984). Ecology and social organization of the maned wolf (*Chrysocyon brachyurus*). *Smithsonian Contribution Zoology, 392*, 1-51.

Donadio, E., Di Martino, S., Aubone, M. & Novaro, A. J. (2001). Activity patterns, home range, and habitat selection of the common hog-nosed skunk, *Conepatus chinga* in northwestern Patagonia. *Mammalia, 65*, 49-54.

Donavan, T., Lamberson, R., Kimber, A., Thompson, R. & Faaborg, J. (1995). Modeling the effects of habitat fragmentation on source and sink demography neotropical migrant birds. *Conservation Biology, 9(6)*, 1397-1407.

Eisenberg, J. F. & Redford, K. H. (1999). *Mammals of the Neotropics.* University of Chicago Press.USA. 609.

Elmhagen, B. & Angerbjörn, A. (2001). The applicality of metapopulation theory to large mammals. *Oikos, 94*, 89-100.

Emmons, L. (1997). *Neotropical rainforest mammals: a field guide.* University of Chicago Press. USA. 307.

Ewers, R. & Didham, R. (2006). Confounding factors in the detection of species responses to habitat fragmentation. *Biological Reviews, 81*, 117-142.

FAHRIG, L. 1998. When does fragmentation of breeding habitat affect population survival? *Ecological Modelling, 105*, 273-92.

Faria, D., Laps, R. R., Baumgarten, J. & Cetra, M. (2006). Bat and bird assemblages from forests and shade cacao plantations in two contrasting landscapes in the Atlantic forest of southern Bahia, Brazil. *Biodiversity and Conservation, 15*, 587-612.

Farigh, L. (2003). Efffects of habitat fragmentation on biodiversity. *Annual Review of Ecology, Evolution, and Systematics, 34*, 487-515.

Forman & Gordon (1986). *Landscape Ecology.* John Willey and Sons.

Fortin, M. J. & Dale, M. R. T. (2005). *Spatial Analysis: A Guide for Ecologists.* Cambridge University Press, Cambridge.

Franklin, W., Johnson, W., Sarno, R. & Iriarte, J. A. (1999). Ecology of the Patagonia puma in southern Chile. *Biological Conservation, 19*, 33-40.

Gargaglioni, L. H., Batalhão, M. E., Lapenta, M. J., Carvalho, M. F., Rossi, R. V. & Veruli, V. P. (1998). Mamíferos da Estação Ecológica de Jataí, Luiz Antônio, SP. *Papéis Avulsos de Zoologia, 40*, 267-287.

Gascon, C., Lovejoy, T. E., Bieregaard, R. O., Malcom, J. R., Stouffer, P. C., Vasconcelos, H., Laurance, W., Zimmerman, B., Tocher, M. & Borges, S. (1999). Matrix habitat and species richness in tropical Forest remnants. *Biological Conservation, 91*, 223-229.

Gehring, T. M. & Swihart, R. K. (2003). Body size, niche breath, and ecologically scaled responses to habitat fragmentation: mammalian predators in an agricultural landscape. *Biological Conservation, 109*, 283-295.

Harvey, C., Gonzalez, J. & Somarriba, E. Dung beetle and terrestrial mammal diversity in forests, indigenous agroforestry systems and plantain monocultures in Talamanca, Costa Rica. *Biodiversity and Conservation, 15*, 555-585.

Hensen, A., Knight, R. L., Marzluff, J. M., Powell, S., Brown, K., Gude, P. &, Jones, K. (2005). Effects of exurban development on biodiversity: patterns mechanisms and research needs. *Ecological Application, 15*, 1893-1905.

I.B.G.E. (2007a). Contas nacionais – Produto interno bruto dos municípios 2002-2005. Instituto Brasileiro de Geografia e Estatística - IBGE, Rio Janeiro, Brasil, 230.

I.B.G.E. (2007b). Contagem da população 2007. Instituto Brasileiro de Geografia e Estatística - IBGE, Rio Janeiro, Brasil, 311.

Igari, A. T., Tambosi, L. R. & Pivello, V. R. (2009). In press. Agribusiness opportunity costs and environmental legal protection: Investigating this trade-off on a hotspot preservation in the State of São Paulo – Brazil. Environental Management.

Jácomo, A. T., Silveira, L. & Diniz-Filho, A. F. (2004). Niche separation between the maned wolf (*Chrysocyon brachyurus*), the crab-eating fox (*Dusicyon thous*) and the hoary fox (*Dusicyon vetulus*) in Central Brazil. *Journal Zoology of London, 262*, 99-106.

Korman, V. (2003). *Proposta de interligação das glebas do Parque Estadual de Vassununga (Santa Rita do Passa Quatro, SP)*. MSc Thesis. Escola Superior de Agricultura Luiz de Queiroz, Universidade de São Paulo, Piracicaba, SP., Brazil.

Kronka, F. J. N., Matsukuma, C. K., Nalon, M. A., Del Cali, I. H., Rossi, M., Matos, I. F. A., Shin-Ike, M. S. & Pontinhas, A. S. (1993). *Inventário Florestal do estado de São Paulo*. Instituto Florestal. São Paulo. 199.

Laurance, W. F. (1994). Rainforest fragmentation and the structure of small mammal communities in tropical Queensland. *Biological Conservation, 69*, 23-32.

Lindenmayer, D. B. & Nix, H. (1993). Ecological principles for the design of wildlife corridors. *Conservation Biology, 55*, 77-92.

Lyra-Jorge, M. C., Ciocheti, G. & Pivello, V. R. (2008). Carnivore mammals in a fragmented landscape in northeast of São Paulo State, Brazil. *Biodiversity and Conservation, 17*, 1573-1580.

Lyra-Jorge, M. C. & Pivello, V. R. (2005). Mamíferos. 135-148. In V. R. Pivello, & E. M. Varanda (Eds.), *O Cerrado Pé-de-Gigante (Parque Estadual de Vassununga, São Paulo) - Ecologia e Conservação*. São Paulo, Secretaria de Estado do Meio Ambiente.

Lyra-Jorge, M. C. (1999). *Avaliação do potencial faunístico da A.R.I.E. Cerrado Pé-de-Gigante (Parque Estadual do Vassununga, Santa Rita do Passa Quatro – SP) com base na análise de habitats*. MsC thesis. Instituto de Biociências, Universidade de São Paulo.São Paulo, SP, Brazil.

Magurran, A. E. (1998). *Ecological diversity and its measurement*. University Press, Cambridge. UK. 178.

Marsden, S. J. & Symes, C. T. (2008). Bird richness and composition along an agricultural gradient in New Guinea: The influence of land use, habitat heterogeneity and proximity to intact forest. *Austral Ecology, 33*, 784-793.

Mcalpine, C. A., Bowen, M. E., Callaghan, J. G., Lunney, D., Rhodes, J. R., Mitchell, D. L., Pullar, D. V. & Poszingham, H. P. (2006). Testing alternative models for the conservation of koalas in fragmented rural–urban landscapes. *Austral Ecology, 31*, 529-544.

McCune, B. & Grace, J. B. (2002). *Analyses of ecological communities*. Software design, Gleneden Deach. USA.

McDougal, P. T., Réale, D., Sol, D. & Reader, S. M. (2006). Wildlife conservation and animal temperament: causes and consequences of evolutionary change for captive reintroduced and wild populations. *Animal Conservation, 9*, 39-48.

McGarigal, K. & Marks, B. (1995). *FRAGSTATS: spatial pattern analysis. Program to quantifying landscape structure*. Department of Agriculture, Forest Service. USA. 122.

Miller, B., Dugelby, B., Foreman, D., Rio, C. M., Noss, R., Phillips, M., Reading, R., Soulé, M. E., Terborgh, J. & Willcox, L. (2001). The importance of large carnivores to healthy ecosystems. *Endangered Species UPDATE, 18(5)*, 202-210.

Moguel, P. & Toledo, V. M. (1999). Biodiversity conservation in traditional coffee systems of Mexico. *Conservation Biology, 13(1)*, 11-21.

Morán-López, R., Guzmán, J. M., Borrego, E. C. & Sánchez, A. V. (2006). Nest-site selection of endangered cinereous vulture populations affect by anthropogenic disturbance: present and future conservation implications. *Animal Conservation, 9*, 29-37.

Nagendra, H., Munroe, D. & Southworth, J. (2004). From pattern to process: landscape fragmentation and the analysis of land use/land cover change. *Agriculture, Ecosystems & Environment, 101(2-3)*, 111-115.

Nuñez, R., Miller, B. & Lindzey, F. (2000). Food habits of jaguars and pumas in Jalisco, Mexico. *Journal of Zoology of London, 252*, 373-379.

Oliveira, P.S. & Marquis, R. J. (2002). *The cerrados of Brazil. Ecology and natural history of a neotropical savanna*. Columbia University Press, New York, 398.

Oliveira, T. G. (1998). *Herpailurus yagouaroundi. Mammalian Species, 578*, 1-6.

Pardini, R. (1998). Feeding ecology of the neotropical river otter *Lontra longicaudis* in an Atlantic forest stream, south-eastern Brazil. *Journal of Zoology of London, 245*, 385-391.

Pardini, R., Ditt, E. H., Cullen, L. JR., Bassi, C. & Rudran, R. (2003). Levantamento rápido de mamíferos terrestres de médio e grande porte. 181-202. In L. Cullen Jr., R. Rudran, & C. Valladares-Pádua, (Eds.), *Métodos de estudo em biologia da conservação e manejo de vida silvestre*. Editora da UFPR; Fundação O Boticário de Proteção à Natureza. Curitiba.

Reznick, D., Rodd, H. & Nunney, L. (2004). Empirical evidence for rapid evolution. In: R. Ferrière, U. Diekman, & D. Couvert, (Eds.), *Evolutionary Conservation Biology*. Cambridge University Press. Cambridge, U.K. 428.

Ricketts, T. (2001). The Matrix Matters: Effective Isolation in Fragmented Landscapes. *American Naturalist, 158*, 87-99. de alteración em el sureste de México. *Acta Zoológica del Mexico, 84*, 35048.

Sánchez-Hernandez, C., Romero-Almaraz, M. L., Colín-Martinez, H. & García-Estrada, C. (2001). Mamíferos de cuatro áreas com diferente grado de alteración em el sureste de México. *Acta Zoológica del Mexico, 84*, 35048.

Santos, J. A. (2003). Estimativa de riqueza em espécies. 19-42. In L. Cullen Jr., R. Rudran, & C. Valladares-Pádua (Eds.), *Métodos de estudo em biologia da conservação e manejo da vida silvestre*. Editora da UFPR. Curitiba.

Saunders, D. A., Hobbs, R. J. & Margulis, C. R. (1991). Biological consequences of ecossystem fragmentation: a review. *Conservation Biology, 5*, 18-32.

Shida, C. N. (2005). Caracterização física do cerrado Pé-de-Gigante e uso das terras na região. Evolução do uso das terras na região. 25-47 In V. R. Pivello, & E. aranda, (Eds.), *O Cerrado Pé-de-Gigante. Parque Estadual de Vassununga. Ecologia e Conservação*. SEMA, São Paulo.

Silveira, L., Jácomo, A. T. & Diniz-Filho, J. A. (2003). Camera trap, line transect census and track surveys: a comparative evaluation. *Biological Conservation, 114*, 351-355.

Smallwood, K. S. & Fitzhugh, E. L. (1995). A track count for estimating Mountain lion *Felis concolor californica* population trend. *Biological Conservation, 71*, 251-259.

Smith, E. P. & Van Belle, G. (1984). Non parametric estimation of species richness. *Biometrics, 40*, 119-129.

Tabeni, S. & Ojeda, R. A. (2005). Ecology of the desert small mammals in disturbed and undisturbed habitats. *Journal of Mammalogy, 70(2)*, 416-420.

Talamoni, S. A., Motta-Júnior, J. C. & Dias, M. M. (2000). Fauna de mamíferos da Estação Ecológica de Jataí e Estação Experimental de Luiz Antônio. 317-319. In J. E. Santos, &

J. S. R. Pires, (Eds.), *Estudos Integrados em Ecossistemas.Estação Ecológica de Jataí.* RiMa Editora. São Carlos. 346.

Tischendorf, L. & Fahrig, L. (2000). How should we measure landscape connectivity? *Landscape Ecology, 15*, 633-641.

Tomas, W. M. & Miranda, G. H. B. (2003). Uso de armadilhas fotográficas em levantamentos populacionais. 243-268. In L. Cullen Jr., R. Rudran, & C. Valladares-Pádua (Eds.), Métodos de estudo em biologia da conservação e manejo da vida silvestre. Editora UFPR.

Uezu, A., Metzeger, J. P. & Vielliard, J. M. E. (2005). Effects of structural and functional connectivity and patch size on the abundance of seven Atlantic forest bird species. *Biological Conservation, 123(4)*, 507-519.

Umetsu, F., Metzeger, J. P. & Pardini, R. (2008). Importance of estimating matrix quality for modeling species distribution in complex tropical landscapes: a test with Atlantic forest small mammals. *Ecography, 31*, 359-370.

Vandermeer, J. & Perfecto, I. (1997). The agroecosystem: a need for the conservation biologist's lens. *Conservation Biol*ogy, *11(3)*, 591-592.

Wagner, H. & Fortin, M. J. (2005). Spatial analysis of landscapes: Concepts and statistics. *Ecology, 86(8)*, 1975-1987.

Wemmer, C., Kunz, T., Lundie-Jekins, G. & McShea, W. (1996). Mammalian Sign. 157-176. In D. E. Wilson, F. R. Cole, J. D. Nichols, R. Rudran, & M. S. Foster, (Eds.), *Measuring and monitoring biological diversity. Standard methods for mammals.* Smithsonian Intitution Press.

Wilcove, D. S., McLellan, C. H. & Dobson, A. P. (1986). Habitat fragmentation in the temperate zone. In: Soulé, M. E. *Conservation Biology.* M. A. Sunderland, K. With, R. Gardner, & M. Turner (Eds.), Landscape connectivity and population distributions in heterogeneous environments. *Oikos, 78*, 151-169.

Zar, J. H. (1999). *Biostatistical analysis.* Prencinton Hall, New Jersey. 409.

In: Encyclopedia of Environmental Research
Editor: Alisa N. Souter

ISBN: 978-1-61761-927-4
© 2011 Nova Science Publishers, Inc.

Chapter 12

GENETIC DIVERSITY OF FESTUCA PRATENSIS HUDS. AND LOLIUM MULTIFLORUM LAM. ECOTYPE POPULATIONS IN RELATION TO SPECIES DIVERSITY AND GRASSLAND TYPE

Madlaina Peter-Schmid, Roland Kölliker and Beat Boller[*]

Agroscope Reckenholz-Tänikon, Research Station ART, Reckenholzstrasse 191,
8046 Zürich, Switzerland.

ABSTRACT

Genetic diversity and species diversity are both crucial for ecosystem stability. Moreover, genetic diversity within a population is fundamental for broadening gene pools in breeding programmes and for food security. For the efficient conservation of species richness as well as of genetic resources in permanent grassland (in-situ), relationships between genetic diversity within target species and species diversity at their original habitats should be considered. In order to identify valuable grassland types for targeted in-situ conservation, two important forage grasses, *Festuca pratensis* Huds. and *Lolium multiflorum* Lam. have been chosen as model species. Genetic diversity of 12 ecotype populations of *F. pratensis* and *L. multiflorum* was assessed by means of 22 and 24 SSR markers, respectively. Species composition and abundance were determined at their collection sites. Analysis of molecular variance revealed low and nonsignificant differences among the grassland types of both species. Genetic diversity within *F. pratensis* ecotype populations (expected heterozygosity H_E) was negatively correlated with species diversity (mainly Shannon index and species evenness) at their collection sites. For *L. multiflorum* no association was found. Between the observed classes of grassland types, differences in genetic diversity within *F. pratensis* populations were not significant, but intermediately managed *Heracleum-Dactylis* grassland habitats held ecotype populations with significantly more rare alleles than extensively managed *Mesobromion* and *Festuco Agrostion* habitats. However, species diversity indicated by the Shannon index was significantly lower in *Heracleum-Dactylis* than in *Mesobromion* grassland habitats. These results indicate a possible conflict of the aims to maintain

---

[*] Corresponding author: E-mail: beat.boller@art.admin.ch

species-rich grassland types on the one hand, and to conserve ecotype populations of target species with a high genetic diversity and a high number of rare alleles on the other hand. This conflict should be resolved by including the aim of conserving genetic diversity of grassland species in national or international programmes of biodiversity management.

INTRODUCTION

Permanent grasslands are of great value since they contain a high number of both common and threatened plant and animal species (Pärtel et al. 2005) as well as a rich genetic variability within plant species (Fjellheim and Rognli 2005; Van Treuren et al. 2005; Peter-Schmid et al. 2008). Changes in agricultural production methods during the last century were accompanied by a marked reduction and fragmentation of the area of European semi-natural grassland. Besides the loss of natural habitats (e.g. permanent grassland), many species disappeared from areas where they had previously been abundant. This was probably a consequence of restricted population sizes and reduced gene flow because of the increased isolation among the populations (Barrett and Kohn 1991; Ellstrand and Elam 1993).

In recent years, several biodiversity studies underlined the important role of species diversity for ecosystem stability and productivity (Tilman and Downing 1994; Hector et al. 1999; Tilman et al. 2006; Kirwan et al. 2007). Therefore, conservation of biodiversity has long been focused on protecting natural areas of high species richness. Moreover, the maintenance of species-rich, extensively managed permanent grassland types (e.g. *Mesobromion, Festuco-Agrostion*) has often been encouraged through financial support in agricultural systems (Anonymous 1998, 2001). Such permanent grassland types are known to provide essential refuges for a number of common as well as endangered plants and animals (Hohl 2006; Peter 2006).

Besides the importance of species diversity in permanent grassland, within-population genetic diversity is an equivalent component of biodiversity and has been shown to be essential for plant fitness, population viability and for the ability of species to respond to changing environmental selection pressures (Barrett and Kohn 1991; Young et al. 1996; Reed 2005). Highly variable and well adapted forage grass ecotype populations collected in permanent grassland have therefore extensively been used as a source of genetic variation to broaden breeding pools and to integrate specific traits (e.g. disease resistances) in breeding programmes (Breese and Hayward 1972; Boller et al. 2005). Until recently, the maintenance of genetic diversity for forage grass species was promoted exclusively by *ex-situ* conservation of populations in gene banks. However, in the Convention on Biological Diversity and in the resulting global action plan on plant genetic resources (PGR), the need to conserve and to sustainably use genetic diversity of plant species in their natural habitats (*in-situ* conservation) has been emphasised (Anonymous 2002). For a comprehensive conservation of biodiversity in grassland habitats, the interactions between species diversity in typical environments and their genetic diversity should also be taken into account. Until recently, these two aspects of species diversity have been studied in most cases independently by community ecologists and population genetics (Vellend 2005). The study presented here integrated both aspects, focusing on the identification of valuable permanent grassland sites which allow for a targeted *in-situ* conservation of *Lolium multiflorum* Lam. and *Festuca pratensis* Huds..

Italian ryegrass (*Lolium multiflorum* ssp. *italicum* Volkart ex Schinz et Keller) is an important species for forage production in temporary grassland. Although Italian ryegrass is considered a short-lived perennial, it thrives well in permanent grassland in mild and moist regions of Central Europe. Stebler and Schröter (1892) concluded from literature studies that the species had been introduced as a forage plant into Swiss agriculture north of the Alps between 1811 and 1840, and had since "wildered out everywhere". Stebler and Volkart (1913) state that Italian ryegrass was introduced to Swiss agriculture in 1820 and to Great Britain in 1831. This corresponds with the date for introduction given for Great Britain by Borrill (1976). Stebler and Schröter (1892) stated that "nowadays, Italian ryegrass is a steady companion of intensive meadow culture of flatter Switzerland up to about 700 to 800 m". We can thus assume that naturalized Italian ryegrass populations have adapted to this region for at least 100 and up to nearly 200 years. Stebler and Volkart (1913) were the first to report on a more frequent occurrence of the species, and they described a particular permanent meadow near Zurich where it contributed 36 % of total yield. Dietl and Lehmann (1975) observed that the proportion of Italian ryegrass in intensively managed meadows had increased markedly and proposed the name *Lolietum multiflori* for typical plant associations dominated by Italian ryegrass with yield proportions of 40 to 80 %.

Unlike *L. multiflorum* which apparently was introduced rather recently from the South, meadow fescue (*Festuca pratensis* Huds.) is regarded truly indigenous to Europe north and east of the Alps. It is distributed throughout the climatic regions of oceanic North-West Europe and the transitional oceanic/continental zone of central Europe (Borrill et al., 1976). Meadow fescue constitutes a significant component of species-rich permanent pastures and hay fields in alpine regions and in Eastern Europe. It was probably introduced to Scandinavia from Europe and West Asia, and has since become naturalized, and it was also introduced into North America, Japan, Australia and New Zealand. In a recent study based on AFLP diversity, 30 local Norwegian populations and 13 Nordic cultivars showed little variation between both local populations and cultivars (Fjellheim and Rognli 2005). Contrary to this, Swiss ecotype populations of *F. pratensis* studied by Peter-Schmid et al. (2008) were clearly separated from cultivars, indicating that the natural populations of meadow fescue in the Alps are much older than populations in Scandinavia.

Nowadays, the two species differ in their prevalence in Switzerland in that the abundance of *L. multiflorum* has strongly increased in certain areas which are suitable for intensive agriculture, whereas the abundance of the *a priori* more widespread species *Festuca pratensis* has often decreased as a consequence of intensified fertilization and more frequent utilization of permanent grassland. In order to elucidate the question if conservation efforts in more extensively managed grassland habitats are also beneficial for a high genetic diversity within species, the aim of the present study was to evaluate the relationship between genetic diversity and species diversity of the two forage grasses. Furthermore, parameters determining genetic and species diversity of different grassland types were investigated to evaluate their value for targeted *in-situ* conservation. Conclusions are drawn as to how aims of nature conservation can be combined with efforts to maintain valuable plant genetic resources for food and agriculture in grassland.

MATERIAL AND METHODS

Twelve ecotype populations of both *F. pratensis* and *L. multiflorum* have been sampled in permanent grasslands at different sites across Switzerland (Table 1), in the Swiss plateau, Jura and the northern foothills of the Alps (400 m a.s.l. to 1400 m a.s.l.) covering an area of roughly 15'000 km^2 and representing a range of differently managed types of grassland. Care was taken to select sites which had not been resown with commercial seed for at least 10 years and were therefore considered to be natural or seminatural. Initial ecotype populations consisted of 60 single tillers, each sampled at the habitat of origin from 60 separate plants close to anthesis. Tillers were sampled from plants which were separated by a distance of at least 3 m. Sampling areas covered 300 to 5'000 m^2, depending on homogeneity of vegetation and landscape. Tillers were isolated in growth chambers or greenhouse cabinets and were allowed to cross-pollinate only within individual populations. Seed harvested separately from each single tiller served to raise 60 individual plants per population.

One *F. pratensis* ecotype population was excluded (BÄR), due to its high similarity to the cultivars, as revealed by morphological as well as by genetic analysis. Twenty-three plants of each population were genotyped by means of 22 (*F. pratensis*) and 24 SSR markers (*L. multiflorum*). SSR analysis as well as the statistical analyses were described in detail in Peter-Schmid et al. (2008). Of the 34 polymorphic SSR markers used in total for analysis of 368 individual plants per species, 12 SSR markers revealed amplification products in both species, 10 only in *F. pratensis* and 12 only in *L. multiflorum*. At least one SSR marker was located on each linkage group (LG) and in *L. multiflorum* mostly two SSR markers were found on each LG. The 22 SSR loci used for F. pratensis detected a total of 151 alleles with an average of 6.9 alleles per locus. For *L. multiflorum*, 335 alleles were detected at 24 SSR loci, resulting in an average of 14 alleles per locus. To assess the genetic variation (expected heterozygosity, H_E) within and among ecotype populations as well as within and among cultivars, analysis of molecular variance AMOVA (Excoffier et al., 1992) was performed with the software package ARLEQUIN, version 3 (Schneider et al., 2000).

Besides a detailed habitat characterisation including geographic location (longitude, latitude), relief (altitude, aspect and slope), soil characteristics (pH, nutrient content, soil composition) and management intensity, the presence and relative abundance of each species at all collection sites was determined. Species abundance was assessed by the yield fraction estimation method (Dietl 1995) within an area of 5x5 m in the years 2004 and 2006. Species diversity was characterised by species richness, Shannon-Weaver diversity index (Shannon index) and species evenness (Magurran 2004). Based on the abundance data of both years, the collection sites were assigned to grassland types (Dietl and Jorquera 2003). *F. pratensis* habitats were grouped into four grassland types (*Mesobromion, Festuco-Agrostion, Lolio-Cynosuretum, Heracleum-Dactylis*) and similarly, *L. multiflorum* habitats into three (*Heracleum-Dactylis, Trifolio-Alopecuretum, Lolietum multiflori*).

To visualise the variation in species composition of the different sites, a correspondence analysis (CA) was performed on abundance data using CANOCO for Windows 4.5 (Ter Braak and Smilauer 2002). Pairwise differences in species composition (based on the first three scores obtained from the correspondence analysis of both years) between the habitats were calculated using Euclidean distances. To test whether similarities in species composition of the original habitats were correlated to genetic diversity of *F. pratensis* and *L. multiflorum* ecotype populations, distance matrices based on correspondence analysis and distance matrices based on molecular marker data were compared using the MXCOMP procedure of NTSYS-pc software version 2.02 (Rohlf 2000). The significances of the correlations between these matrices were tested using the normalized Mantel Z-statistics. Spearman's rank correlation coefficients were calculated between genetic diversity values (expected heterozygosity H_E, number of alleles and number of rare alleles) and species diversity values (abundance of target species, species richness, Shannon index and species evenness).

Table 1. Characteristics of the 12 *F. pratensis* and 12 *L. multiflorum* habitats investigated.

Origin	Acronym	Management intensity	Longitude (°E)	Latitude (°N)	Elevation (m a.s.l.)
F. pratensis					
Bärau	BÄR[a]	extensive	7°48'54"	46°55'53"	750
Birmensdorf	BIR	extensive	8°27'15"	47°21'09"	540
Blaachli	BLA	intermediate	7°29'46"	46°36'49"	1470
Boppelsen	BOP	extensive	8°25'05"	47°28'09"	580
Brandösch	BRA	intermediate	7°53'36"	46°58'17"	920
Gibswil	GIB	extensive	8°54'39"	47°18'53"	770
Grandval	GRA	intermediate	7°25'32"	47°17'18"	660
Hasliberg G.	HAG	intermediate	8°12'01"	46°44'43"	1130
Hasliberg K.	HAK	extensive	8°11'37"	46°44'46"	1170
Moron	MOR	intermediate	7°17'12"	47°15'57"	1190
Mösli	MÖS	extensive	8°32'57"	47°44'29"	810
Oberehrendingen	OBE	extensive	8°21'22"	47°29'13"	610
L. multiflorum					
Bazenheid	BAZ	intermediate	9°03'40"	47°25'56"	600
Doppleschwand	DOP	intensive	8°03'00"	47°00'33"	880
Gommiswald	GOM	intensive	9°01'40"	47°14'48"	760
Hüttlingen	HÜT	intermediate	8°59'11"	47°34'39"	430
Lenzen	LEN	intermediate	8°55'39"	47°21'27"	690
Littau	LIT	intermediate	8°14'48"	47°03'04"	460
Niederurnen	NIE	intensive	9°04'06"	47°07'24"	430
Root	ROO	intensive	8°23'55"	47°07'11"	470
Türlen	TÜR	intermediate	8°30'35"	47°16'08"	670
Weiningen	WEI	intermediate	8°26'26"	47°25'33"	530
Wernetshausen	WER	intensive	8°52'13"	47°18'14"	760
Wolhusen	WOL	intensive	8°04'22"	47°02'05"	650

[a] *F. pratensis* population BÄR was excluded from the calculations since it was considered to be a cultivar rather than an ecotype population.

RESULTS

• Variation in Grassland Types

The correspondence analysis based on the species abundances determined for *F. pratensis* and *L. multiflorum* habitats, showed a separation of habitats reflecting well the grassland types (Figure 1). The first two components of the correspondence analysis accounted for 65.4 % (*F. pratensis*) and for 28.5 % (*L. multiflorum*) of the total variation. For *F. pratensis*, very extensively managed and dry *Mesobromion* habitats were distinctly different from the extensively grazed but more humid *Festuco-Agrostion* and from the intermediately managed *Heracleum-Dactylis* habitats (Figure 1A). For *L. multiflorum*, the first axis separated the habitats into the two grassland types, humid *Trifolio-Alopecuretum* and *Lolietum multiflori* (Figure 1B).

• Genetic Differentiation of Ecotype Populations in Relation to Species Composition

Analysis of molecular variance (AMOVA) revealed most of the observed variability to be due to variation within the populations (Table 2). Only small, nonsignificant amounts of the total variation were explained by the different grassland types, namely 0.9 % for *F. pratensis* and 0.1 % for *L. multiflorum*. Highly significant components of variance were attributed to the variation among populations within grassland types. However, by far the largest part of variation (95.8 % and 98.3 % for *F. pratensis* and *L. multiflorum*, respectively) was due to variation within populations. For both species, a weak and nonsignificant Mantel correlation was found between the matrices of Euclidean distances based on species composition at the habitats and the genetic data (*F. pratensis*: r = 0.20, P = 0.08 and *L. multiflorum*: r = 0.16, P = 0.11).

Table 2. Analysis of molecular variance (AMOVA) based on 22 or 24 SSR markers and 23 individuals per population for 11 and 12 ecotype populations of four *F. pratensis* (*F.p.*) and three *L. multiflorum* (*L.m.*) grassland types respectively.

Source of variation	d.f.		Variance component[a]				Variance (%)	
	F.p.	*L.m.*	*F.p.*		*L.m.*		*F.p.*	*L.m.*
Among grassland types	3	2	0.05	n.s.	0.01	n.s.	0.9	0.1
Among populations/ within grassland types[b]	7	9	0.16	***	0.13	***	3.3	1.6
Within populations	495	540	4.82	***	7.92	***	95.8	98.3

[a] ** P < 0.01, *** P < 0.001; the probability of obtaining a more extreme random value computed from nonparametric procedures (1'000 data permutations).

[b] For *F. pratensis* four grassland types (*Mesobromion, Festuco-Agrostion, Lolio-Cynosuretum, Heracleum-Dactylis*) and for *L. multiflorum* three types (*Heracleum-Dactylis, Trifolio-Alopecuretum, Lolietum multiflori*)

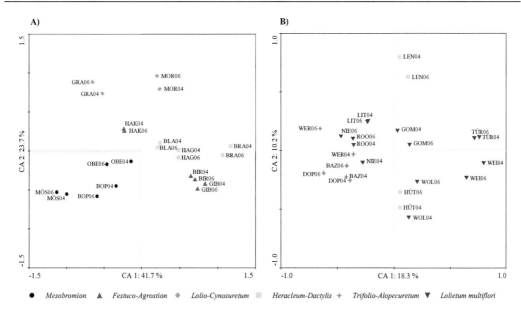

Figure 1. Correspondence analysis (CA) for the variation in species composition among the *F. pratensis* (A) and *L. multiflorum* (B) habitats evaluated in 2004 and 2006. Habitats were assigned to six grassland types: *Mesobromion, Festuco-Agrostion, Lolio-Cynosuretum, Heracleum-Dactylis, Trifolio-Alopecuretum, Lolietum multiflori* based on abundance data (Dietl and Jorquera, 2003).

Table 3. Spearman's rank correlation coefficients between genetic diversity (H_E), number of alleles, number of rare alleles, abundance of the target species, species richness, Shannon index and species evenness in 18 *F. pratensis* (upper triangle) and in 19 *L. multiflorum* (lower triangle) ecotype populations or habitats. Significant correlations: * $P \le 0.05$, ** $P \le 0.01$, *** $P \le 0.001$.

	H_E		No. of alleles		No. of rare alleles		Abundance of target species		Species richness		Shannon index		Species evenness	
H_E			0.39		0.19		0.10		-0.19		-0.44	*	-0.47	*
No. of alleles	0.53	**			0.50	*	0.03		-0.42		-0.33		-0.09	
No. of rare alleles	0.37		0.88	***			0.13		-0.35		-0.46	*	-0.42	
Ab. of target species	-0.03		0.18		0.23				-0.17		-0.28		-0.24	
Species richness	-0.16		-0.35		-0.15		-0.49	*			0.82	***	0.17	
Shannon index	-0.12		-0.29		-0.08		-0.57	**	0.90	***			0.67	***
Species evenness	-0.03		-0.15		0.01		-0.53	**	0.68	***	0.91	***		

- **Genetic Diversity within Ecotype Populations as Compared to Species Diversity**

For *F. pratensis* a significant negative correlation was observed between genetic diversity (H_E) and species diversity (Shannon index and species evenness; Table 3). The number of rare alleles within *F. pratensis* populations was negatively correlated with the Shannon index.

No relationship was found between genetic diversity of *L. multiflorum* ecotype populations and the species diversity at their habitats of origin. The abundance of *L. multiflorum* at the collection sites was found to be negatively correlated with all parameters of species diversity.

- **Genetic Diversity within Ecotype Populations in Relation to the Grassland Type**

For both species, within-population genetic diversity (H_E) was very similar for all grassland types, whereas the absolute values were clearly lower for *F. pratensis* compared to *L. multiflorum* (Table 4). Significantly more rare alleles were detected within *F. pratensis* ecotype populations collected in *Lolio-Cynosuretum* and *Heracleum-Dactylis* grassland than in *Mesobromion* and *Festuco-Agrostion*. Concerning species diversity, the Shannon index of *Heracleum-Dactylis* grassland was significantly lower compared to the remaining grassland types. Although the lowest numbers of species were found in *Heracleum-Dactylis* grassland, *Lolio-Cynosuretum* as well as *Festuco-Agrostion* grasslands did not differ significantly in species richness from *Heracleum-Dactylis* grassland. For *L. multiflorum*, differences in genetic and species diversity between the grassland types were not significant.

Table 4. Mean values for genetic diversity (H_E), number of rare alleles, abundance, species richness, Shannon index and species evenness calculated for the four *F. pratensis* and three *L. multiflorum* grassland types. Mean values followed by the same letter are not significantly different (Pairwise t-tests; P < 0.05).

	H_E		No. of rare alleles		Abundance (%)		Species richness		Shannon index		Species evenness	
F. pratensis												
Mesobromion	0.43	a	4.7	b	5.1	a	45	a	3.28	a	0.87	a
Festuco-Agrostion	0.44	a	4.3	b	9.0	a	33	b	3.02	b	0.86	a
Lolio-Cynosuretum	0.43	a	7.0	a	6.6	a	41	ab	3.23	a b	0.87	a
Heracleum-Dactylis	0.44	a	8.3	a	9.6	a	29	b	2.78	c	0.84	a
L. multiflorum												
Heracleum-Dactylis	0.66	a	17.0	a	13.8	a	22	a	2.49	a	0.81	a
Trifolio-Alopecuretum	0.66	a	17.7	a	20.3	a	18	a	2.30	a	0.80	a
Lolietum multiflori	0.67	a	19.6	a	22.5	a	22	a	2.54	a	0.82	a

DISCUSSION

The genetic characterisation of the two model species *F. pratensis* and *L. multiflorum* provided evidence that sites in various geographic regions, managed at different intensities should be considered for targeted *in-situ* conservation (Peter-Schmid et al. 2008). Taking the grassland type as a further indicator of potential conservation sites of high value, only a low and nonsignificant genetic differentiation among such grassland types resulted for both species by analysis of molecular variance. Nevertheless, in the correspondence analysis, the grouping of *F. pratensis* ecotype populations regarding the species composition at their collection sites (Figure 1A) was visually congruent to the cluster-analysis of genetic data (Figure 1A: Peter-Schmid et al. 2008). Similarly, a Mantel test also revealed a weak (but not significant) relationship between genetic differentiation of these populations and the differentiation of their collection sites regarding species composition. This indicates that environmental factors prevailing on sites (e.g. geographic region, management) may similarly affect genetic diversity within ecotype populations and species composition on site; this was also suggested by Linhart and Grant (1996).

As expected for outcrossing, wind-pollinated forage grasses, most of the genetic variation was found within the populations. Genetic diversity within each individual population (H_E) was similar among the grassland types of both species. Nevertheless, the loose but significant negative correlation between genetic diversity of *F. pratensis* ecotype populations and species diversity at their collection sites (Shannon index and species evenness) indicated that in this study the genetic diversity was lower in populations of habitats with higher species diversity. Until recently, species diversity and genetic diversity have been treated independently and their interactions have therefore rarely been investigated (Vellend and Geber 2005; Wehenkel et al. 2006). Studies of *Ranunculus acris* L. and of *Anthoxanthum odoratum* L. populations reported no relationships between species diversity in grassland habitats and genetic diversity within the populations, while a significant and slightly positive relationship was found for *Plantago lanceolata* L. (Odat 2004; Odat et al. 2004). Contrastingly, Vellend and Gerber (2005) found results similar to ours, showing a significant negative relationship between species and genetic diversity in their data of *Fagus silvatica* L. populations in Germany. Such associations of species diversity and within-population genetic diversity are obviously not unambiguous and may depend on the species of interest, due to their different responses to environmental conditions (Vellend 2005). In this study, the negative association found for *F. pratensis* could be a consequence of the higher competition between the species in extensively managed, species-rich grassland, compared to more intensively managed grassland types resulting in a reduction of population size in extensive types. This is supported by the low abundance of *F. pratensis* observed in *Mesobromion* (Table 4) and the known weak competition abilities of *F. pratensis* in mixtures due to its lower shoot vigour, its sensitivity to drought and to shading (Gügler 1993; Carlen et al. 2002). Moreover, for its best performance, *F. pratensis* requires a moderate fertilisation, which is often lacking in extensively managed grassland (Gügler 1993).

For a successful utilisation of plant genetic resources in breeding programmes, not only the genetic diversity within populations is of interest but also the occurrence of rare alleles, since the presence of rare alleles may be related to potential adaptive mutations (Neel and Cummings 2003). *F. pratensis* ecotype populations collected in intensively managed

Heracleum-Dactylis meadows contained the highest number of rare alleles, but these sites had also the lowest species diversity (Shannon index). Species diversity at *Lolio-Cynosuretum* habitats was almost as high as in extensively managed *Mesobromion* habitats, but *F. pratensis* populations from *Lolio-Cynosuretum* habitats harboured significantly more rare alleles although they did not reach their abundance in *Heracleum-Dactylis* meadows. Therefore, *Lolio-Cynosuretum* habitats seem to be interesting for the simultaneous maintenance of high species diversity as well as for the conservation of a high number of rare alleles, while *Heracleum-Dactylis* habitats are mainly interesting for the maintenance of plant genetic resources for breeding because of a lower species diversity coupled with highest numbers of rare alleles. Considering the results of *L. multiflorum*, no differences among the grassland types were obtained for genetic diversity values nor for species diversity indices. This is not surprising because the structure between *L. multiflorum* ecotype populations based on genetic diversity was much less pronounced than for *F. pratensis* ecotype populations (Peter-Schmid et al. 2008).

These observations suggest that a characterisation of the floristic composition of potential collecting and conservation sites of grassland genetic ressources would be helpful in selecting promising sites and in maximizing genetic diversity among ecotype populations. A similar conclusion was drawn by Boller et al. (2009) when assessing agronomic performance of the 12 *L. multiflorum* ecotype populations of the study presented here along with 8 additional ecotype populations and 4 reference cultivars. They found that *L. multiflorum* populations from *Arrhenatherion* habitats performed markedly poorer than populations from non-*Arrhenatherion* habitats in terms of yield, snow mould and *Xanthomonas* resistance, whereas populations from *Arrhenatherion* were less susceptible to crown rust (*Puccinia coronata* Corda). Thus, characterisation and classification of the floristic composition might be an easy means of assessing an assembly of environmental and management factors which are expected to affect natural selection within ecotype populations of grassland species, and finally their genetic diversity among grassland sites. This could be an alternative approach to the statistical evaluation of the effects of environmental factors on the structuring of genetic diversity, such as the one presented by Sampoux and Huyghe (2009) with a large collection of fine fescues (aggregates of *F. rubra* and *F. ovina*).

The present data of *F. pratensis* and those of Boller et al. (2009) with *L. multiflorum* suggest a possible trade-off between the goal of conserving species-rich grassland types and the goal of maintaining valuable plant genetic resources. Conservation efforts focusing exclusively on habitats with high species diversity (e.g. protected natural areas or extensively managed grassland) may not be appropriate for an adequate conservation of genetic diversity within *F. pratensis* ecotype populations. Similarly, protecting only *Arrhenatherion* sites as habitats for conserving high grassland and micro-fauna species richness will fail to help conserve the most valuable genetic resources of *L. multiflorum* for food and agriculture, which can be found on sites which are of no interest for nature conservation. A conservation concept taking these considerations into account was presented for Switzerland by Weyermann (2007), suggesting to describe and eventually protect five to nine agriculturally used surfaces for each of 16 vegetation units (associations) within each of 7 bio-geographic regions. While the results of the study presented here provided the basis for the development of this concept, it's more general validity has yet to be confirmed by a targeted collection of ecotype populations of selected grassland species and their subsequent characterisation and evaluation. Such a study is currently under way under the auspices of the Swiss National Plan of Action (NAP) for the conservation and sustainable utilisation of plant genetic resources.

REFERENCES

Anonymus (1998). Verordnung über die Direktzahlungen an die Landwirtschaft (SR-910.13). Available at http://www.bzo.ch/Downloads/Direktzahlungen_VO_910.13.pdf.

Anonymus (2001). Verordnung über die regionale Förderung der Qualität und der Vernetzung von ökologischen Ausgleichsflächen in der Landwirtschaft (SR-910.14). Available at http://www.admin.ch/ch/d/as/2007/6117.pdf.

Anonymus (2002). *Handbook of the Convention on Biological Diversity (CBD)*. Available at http://www.cbd.int/handbook/

Barrett, S. C. H. & Kohn, J. R. (1991). Genetic and evolutionary consequences of small population size in plants: implications for conservation. Pp. 3-30 in Genetics and conservation of rare plants, (D. A. Falk, & K. E. Holsinger, Eds.), Oxford University Press, New York, NY.

Boller, B., Schubiger, F. X., Tanner, P., Streckeisen, P., Herrmann, D. & Kölliker, R. (2005). La diversité génétique dans les prairies naturelles suisses et son utilisation en sélection. *Fourrages*, *182*, 263-274.

Boller, B., Peter-Schmid, M., Tresch, E., Tanner, P. & Schubiger, F. X. (2009). Ecotypes of Italian ryegrass from Swiss permanent grassland outperform current recommended cultivars. *Euphytica*, DOI 10.1007/s10681-009-9963-y

Borrill, M., Tyler, B. F. & Morgan, W. G. (1976). Studies in *Festuca* VII. Chromosome atlas (Part 2). An appraisal of chromosome race distribution and ecology, including *F. pratensis* var. *apennina* (De Not.) Hack, - tetraploid. *Cytologia*, *41*, 219-236.

Borrill, M. (1976). Temperate grasses: *Lolium, Festuca, Dactylis, Phleum, Bromus (Gramineae)*. Pp. 137-141 in Evolution of Crop Plants (N. W. Simmonds, Ed.),, Longman, London.

Breese, E. L. & Hayward, M. D. (1972). The genetic basis of present breeding methods in forage crops. *Euphytica*, *21*, 324-336.

Carlen, C., Kölliker, R., Reidy, B., Lüscher, A. & Nösberger, J. (2002). Effect of season and cutting frequency on root and shoot competition between *Festuca pratensis* and *Dactylis glomerata*. *Grass Forage Sci*, *57*, 247-257.

Dietl, W. (1995). Wandel der Wiesenvegetation im Schweizer Mittelland. *Zeitschrift für Ökologie und Naturschutz*, *4*, 239-249.

Dietl, W. & Jorquera, M. (2003). Wiesen- und Alpenpflanzen: erkennen und den Blättern, freuen an den Blüten. Österreichischer Agrarverlag Leopoldsdorf, Austria.

Dietl, W. & Lehmann, J. (1975). Standort und Bewirtschaftung der Italienisch-Raigras-Matten. *Mitteilungen für die Schweizerische Landwirtschaft*, *10*, 185-194.

Ellstrand, N. C. & Elam, D. R. (1993). Population genetic consequences of small population size: implications for plant conservation. *Annu Rev Ecol Syst*, *24*, 217-242.

Excoffier, L., Smouse, P. E. & Quattro, J. M. (1992). Analysis of molecular variance inferred from metric distances among DNA haplotypes: application to human mitochondrial DNA restriction data. *Genetics*, *131*, 479-491.

Fjellheim, S. & Rognli, O. A. (2005). Molecular diversity of local Norwegian meadow fescue (*Festuca pratensis* Huds.) populations and nordic cultivars - consequences for management and utilisation. *Theor Appl Genet, 111,* 640-650.

Gügler, B. (1993). Die Konkurrenz zwischen Wiesenschwingel (*Festuca pratensis* Huds.) und Knaulgras (*Dactylis glomerata* L.) bei verschiedener Bewirtschaftung. PhD thesis, Swiss Federal Institute of Technology, Zurich, Switzerland.

Hector, A., Schmid, B., Beierkuhnlein, C., Caldeira, M. C., Diemer, M., Dimitrakopoulos, P. G., Finn, J. A., Freitas, H., Giller, P. S., Good, J., Harris, R., Högberg, P., Huss-Danell, K., Joshi, J., Jumpponen, A., Körner, C., Leadley, P. W., Loreau, M., Minns, A., Mulder, C. P. H., O'Donovan, G., Otway, S. J., Pereira, J. S., Prinz, A., Read, D. J., Scherer-Lorenzen, M., Schulze, E. D., Siamantziouras, A. S. D., Spehn, E. M., Terry, A. C., Troumbis, A. Y., Woodward, F. I., Yachi, S. & Lawton, J. H. (1999). Plant diversity and productivity experiments in European grasslands. *Science, 286,* 1123-1127.

Hohl, M. (2006). Spatial and temporal variation of grasshopper and butterfly communities in differently managed semi-natural grasslands of the Swiss Alps. PhD thesis, Swiss Federal Institute of Technology, Zurich, Switzerland.

Kirwan, L., Lüscher, A., Sebastia, M., Finn, J., Collins, R., Porqueddu, C., Helgadottir, A., Baadshaug, O., Brophy, C., Coran, C., Dalmannsdottir, S. & Delgado, I. (2007). Evenness drives consistent diversity effects in intensive grassland systems across 28 European sites. *J Ecol, 95,* 530-539.

Linhart, Y. B. & Grant, M. C. (1996). Evolutionary significance of local genetic differentiation in plants. *Annu Rev Ecol Syst, 27,* 237-277.

Magurran, A. E. (2004). Measuring Biological Diversity. Wiley, Weinheim, Germany.

Neel, M. C. & Cummings, M. P. (2003). Effectiveness of conservation targets in capturing genetic diversity. *Conserv Biol, 17,* 219-229.

Odat, N. (2004). On the relationship between biotic and abiotic habitat diversity and genetic diversity of *Ranunculus acris* L. (*Ranunculaceae*), *Plantago lanceolata* L. (*Plantaginaceae*), and *Anthoxanthum odoratum* L. (*Poaceae*) within and between grassland sites. PhD thesis, Friedrich-Schiller- Universität, Jena, Germany.

Odat, N., Jetschke, G. & Hellwig, F. H. (2004). Genetic diversity of *Ranunculus acris* L.(*Ranunculaceae*) populations in relation to species diversity and habitat type in grassland communities. *Mol Ecol, 13,* 1251-1257.

Pärtel, M., Bruun, H. & Sammul, M. (2005). Biodiversity in temperate European grasslands: origin and conservation. Pp. 1-14 in Proceedings of the 13th International Occasional Symposium of the European Grassland Federation: Integrating efficient grassland farming and biodiversity (R. Lillak, R. Viiralt, A. Linke, & V. Geherman, Eds.). Tartu, Estonia, 29-31 August 2005.

Peter, M. (2006). Changes in the floristic composition of semi-natural grasslands in the Swiss alps over the last 30 years. PhD thesis, Swiss Federal Institute of Technology, Zurich, Switzerland.

Peter-Schmid, M. K. I., Boller, B. & Kölliker, R. (2008). Habitat and management affect genetic structure of *Festuca pratensis* but not *Lolium multiflorum* ecotype populations. *Plant Breed, 127,* 510-517.

Reed, D. H. (2005). Relationship between population size and fitness. *Conserv Biol, 19,* 563-568.

Rohlf, F. J. (2000). NTSYS-pc: numerical taxonomy and multivariate analysis system, version 2.2. Exeter Publishers, Setauket, NY.

Sampoux, J. P. & Huyghe, C. (2009). Contribution of ploidy-level variation and adaptive trait diversity to the environmental distribution of taxa in the 'fine-leaved fescue' lineage (genus *Festuca* subg. *Festuca*). *Journal of Biogeography*. DOI: 10.1111/j.1365-2699.2009.02133.x

Schneider, S., Roessli, D. & Excoffier, L. (2000). ARLEQUIN: a software for population genetics data analysis. *Genetics and Biometry Laboratory*, University of Geneva, Switzerland.

Stebler, F. G. & Schröter, C. (1892). Die besten Futterpflanzen, Teil I. 2. Auflage. K.J. Wyss, Bern.

Stebler, F. G. & Volkart, A. (1913). Die besten Futterpflanzen. Erster Band, 4. Auflage. K. J. Wyss, Bern.

Ter Braak, C. J. F. & Smilauer, P. (2002). CANOCO reference manual and CanoDraw for Windows user's guide: software for canonical community ordination (version 4.5). Microcomputer Power, Ithaca, NY.

Tilman, D. & Downing, J .A. (1994). Biodiversity and stability in grasslands. *Nature, 367*, 363-365.

Tilman, D., Reich, P. B. & Knops, J. M. H. (2006). Biodiversity and ecosystem stability in a decade-long grassland experiment. *Nature, 441*, 629-632.

Van Treuren, R., Bas, N., Goossens, P. J., Jansen, J. & Van Soest, L. J. M. (2005). Genetic diversity in perennial ryegrass and white clover among old Dutch grasslands as compared to cultivars and nature reserves. *Mol Ecol, 14*, 39-52.

Vellend, M. (2005). Species diversity and genetic diversity: parallel processes and correlated patterns. *American Naturalist, 166*, 199-215.

Vellend, M. & Geber, M. A. (2005). Connections between species diversity and genetic diversity. *Ecol Lett, 8*, 767-781.

Wehenkel, C., Bergmann, F. & Gregorius, H. R. (2006). Is there a trade-off between species diversity and genetic diversity in forest tree communities? *Plant Ecol, 185*, 151-161.

Weyermann, I. (2007). Konzept zur in situ Erhaltung von Futterpflanzen. Bundesamt für Landwirtschaft, Bern, http://www.cpc-skek.ch/pdf/ConceptFutterpflanzen_Vprovisoire.pdf

Young, A., Boyle, T. & Brown, T. (1996). The population genetic consequences of habitat fragmentation for plants. *Trends Ecol Evol, 11*, 413-418.

In: Encyclopedia of Environmental Research
Editor: Alisa N. Souter

ISBN: 978-1-61761-927-4
© 2011 Nova Science Publishers, Inc.

Chapter 13

SUSTAINABLE WASTE MANAGEMENT AND CLIMATIC CHANGE

*Terry Tudor**

School of Applied Sciences, University of Northampton, Northampton, UK

ABSTRACT

This chapter examines links between the management of waste and its impacts on climactic changes due in part to rising consumption levels. Differences between greenhouse gas emissions from developed and developing countries are examined, including the contributions of the 20 highest emitting countries between the period of 1950-2000, including that of China and the USA. The manner in which waste management and climatic changes are links is examined, as well as strategies for reducing the impacts including waste prevention and minimisation. The chapter concludes by looking ahead to future technological and market requirements.

INTRODUCTION

Defining Sustainable Waste Management

The central feature of the shift towards greater resource conservation over the past 10 to 15 years has been the 'rebirth' of the concept of 'sustainable development' as a means of better management and utilisation of the earth's resources. While according to O'Riordan (1993), the concept of 'sustainability' has existed for thousands of years, the notion of sustainable development perhaps gained most prominence after it was mooted in the very influential report 'Our Common Future' (WCED, 1987). This report resulted from the World Commission on Environment and Development Conference in 1987 (which was chaired by the then Minister of the Environment and later Prime Minister of Norway, Mrs. Gro Brundtland). The theme of 'sustainable development' was first adopted at the United Nations

* Tel: +44 01604 892398, Fax: +44 01604 720636, Email: terry.tudor@northampton.ac.uk.

Conference on Environment and Development (The Earth Summit) in Rio, Brazil in 1992. The concept was promoted in a holistic framework embracing a range of social, economic, environmental and more recently institutional factors. A central plank of the concept was the preservation and protection of current resources for future use, as embodied in what has now become the often quoted definition of sustainable development as (WCED, 1987:43):

DEVELOPMENT THAT MEETS THE NEEDS OF THE PRESENT, WITHOUT COMPROMISING THE ABILITY OF FUTURE GENERATIONS TO MEET THEIR OWN NEEDS

Since the Brundtland definition, various authors have also sought to argue for the incorporation of various additional factors including the 'institution', as well as 'socio-political' and 'cultural' constructs (UNCSD, 2001; Lehtonen, 2005; Tudor, 2006). Hence, sustainability can be considered to have four dimensions: social, economic, environmental and institutional. Based on this therefore, the sustainable management of waste could be defined as the management of resources for current and future generations within a context that meets social, and environment needs, but is decoupled from economic growth.

Rising Consumption Levels

Attention to the growing trend of consumerism in society begun to receive serious worldwide attention in the 1950s and 1960s. According to Welford (1998) during this period it was felt that growth, development and environmental protection could not co-exist. As a result, most of the development and economic theories during this time were 'anti-growth'. In his seminal book 'The Waste Makers' (Packard, 1960), the American journalist, social critic and author Vance Packard describes how American productivity had impacted on the national character. He criticises for example, the concept of 'Planned Obsolescence' made popular in the 1950s by American industrial designers whereby consumer products were deliberately designed and built to fail, thereby encouraging shorter product lives. Indeed, global per capita consumption has steadily increased over the past two decades and is predicted to continue to increase in future in line with increasing gross domestic product (GDP) (OECD, 2002). An excellent example of this, is rising global energy demand. This increasing demand leads to higher consumption and disposal levels along the lifecycle of products from extraction of raw materials to treatment and disposal of the waste generated.

Climatic Changes

Climate change refers to (IPCC, 2007:22):

Any change in climate over time, whether due to natural variability or as a result of human activity

Despite some initial skepticism, the issue of whether or not the environment is warming is indisputable. Sceptics initially put forward three main arguments, namely: (1) the earth was too vast to change (2) any warming that was taking place would be slow and easy to cope with, and (3) there was limited evidence to support a link. In addition, it has been argued that the climate changes naturally and has done so for 1000s of years. The environment and indeed mankind have simply adapted to these changes as they have occurred. However, beginning in the 1970s and 80s with dramatic rises in global temperatures, global greenhouse gas (GHG) emissions have risen steadily. Crucially, temperatures are predicted to continue to rise. Indeed, climate change models predict global temperature rises of approximately 1.4°C - 5.8°C during the period 1990 to 2100 (Houghton *et al.*, 2001). This will give rise to problems globally, but particularly in developing countries (Monni *et al.*, 2006). GHGs in the atmosphere have resulted in a 0.6°C increase in the earth's temperature over the past 100 years, with the 10 warmest years in the 20th Century having all occurred in the past 15 years (Defra, 2008). Many developing countries, particularly those in South East Asia and the Indian Subcontinent are increasingly at risk from flood potential. There is greater likelihood of extreme events and also of water stress. Significantly, these climatic changes are expected to affect human health, ecosystems, agriculture and economic activity among other issues (Stern, 2006; IPCC, 2007). In addition, according to the IPCC (2007), an analysis of the data since the 1970s suggests that anthropogenic warming has had a discernible influence on many biological and physical systems. The most appropriate manner in which to tackle these issues is to stabilise and/or reduce the concentration of GHGs in the atmosphere.

Global Climatic Impacts

Only a relatively few countries account for the majority of the worldwide's GHG emissions (Tables 1 and 2). It can be seen that generally, in 2000, developed countries (e.g. the USA, Canada and Australia) ranked highest, with developing countries (e.g. China, India and Indonesia) ranking lowest (CRI, 2007). In 2000, the USA produced 20% of the overall quantity, followed by China (15%). The 25-nation European Union (EU) which can report and meet its obligations collectively was third highest (14%). Apart from the USA and China, no single country contributed above 6%. Only nine countries emitted 2% or more. The top six countries contributed 60% of the overall emissions of the 185 countries worldwide, with the next 11 contributing another 15% (CRI, 2007). Developed countries (Annex 1 on the UNFCCC list) accounted for 72% of the total CO_2 emissions between 1950 and 2000 (WRI, 2007). It is important to note though the growing economic influence of the soc-called 'Bric' countries, namely Brazil, Russia, India and China, who are expected to be the leading global economies by 2050. Rising consumption levels in these countries will therefore result in rising levels of GHG emissions. This is based on rising consumption levels of energy (including fossil fuel based energy).

Table 1. GHG emissions for the top 20 emitting countries worldwide up to 2000.

Country	Annex 1	2000 (MMTCE)	% of world	1990 (MMTCE)	Difference 1990-2000 (%)	2000 Per capita GHG emissions (tonnes carbon/person)	1950-2000 cumulative energy CO_2 emissions (MMTCE)
United States	Y	1,876	20.4	1,631	13.1	6.6	58,107
China	N	1,355	14.7	1,029	24.1	1.1	19,587
EU	Y	1,294	14.1	1,369	-5.8	2.9	48,018
Russia	Y	523	5.7	780	-49.1	3.6	21,048
India	N	516	5.6	366	29.1	0.5	5,123
Japan	Y	369	4.0	328	11.1	2.9	10,193
Germany	Y	277	3.0	332	-19.9	3.4	12,918
Brazil	N	232	2.5	187	19.4	1.3	2,031
Canada	Y	187	2.0	154	17.6	6.1	4,757
UK	Y	180	2.0	198	-10.0	3.0	8,122
Italy	Y	145	1.6	132	9.0	2.5	3,926
Mexico	N	144	1.6	118	18.1	1.5	2,564
South Korea	N	142	1.5	79	44.4	3.0	1,892
France	Y	142	1.5	150	-5.6	2.4	5,100
Indonesia	N	138	1.5	97	29.7	0.7	1,253
Australia	Y	134	1.5	111	17.2	7.0	2,508
Ukraine	Y	132	1.4	253	-91.7	2.7	5,668
Iran	N	130	1.4	79	39.2	2.0	1,627
South Africa	N	114	1.2	97	14.9	2.6	2,784
Spain	Y	104	1.1	79	24.0	2.6	2,099
Poland	Y	102	1.1	119	-16.7	2.6	4,336
TOTAL		6,942	75.2	6,319	9.0		175,643
World		9,201		8,335			217,820

Source: WRI (2008)

As demonstrated in Table 2 despite the fact that most of the GHG emissions arose from developed countries, the overall percentage from developing countries is increasing. In addition, while the USA was the leading emitter within the developed countries (and overall), China was the leading emitter amongst developing countries. Both China and the USA produced around one third of the overall emissions total in 2000. Increasing industrialisation amongst developing countries, particularly the Bric countries will inevitably lead to increased levels of GHGs.

Table 2. A comparison of GHG emissions from developed and developing countries

INDICATOR	DEVELOPED COUNTRIES (%)	DEVELOPING COUNTRIES (%)	TOP 20 NATIONS (%)
1990 GHG emissions (excl land use)	58.0	40.1	75.8
2000 GHG emissions (excl land use)	51.5	47.2	75.4
200 GHG emissions (with land use)	41.3	57.6	70.4
Cumulative energy-related CO_2 emissions 1950-2000 (excl land use)	71.0	26.4	80.6
Cumulative energy-related CO_2 emissions (with land use)	51.5	46.7	73.9

Source: WRI (2008)

Waste Management and Climate Change

Though small, the management of waste is an important contributor to climate change, primarily through GHG emissions, and it is expected that with increasing worldwide populations and rising GDP, these emissions will rise (Monni *et al.,* 2006). However, it is important to note that impacting on climate change is only one environmental impact that the management of waste can have. Other key influences include, health effects due to air pollutants such as NOx and Sox, emission of ozone depleting substances, disamenity and noise (AEA Technology, 2001). Fig. 1 illustrates the manner in which waste management could influence climatic changes.

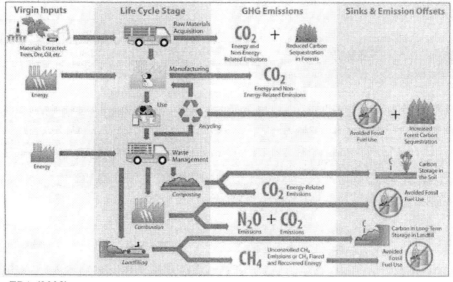

Source: EPA (2008)

As can be seen GHGs can be produced at any stage from extraction of raw materials through to the manufacture, transport, consumption and final disposal of the products. A significant point to note is that the factors work together rather than singly. For example, marine pollution and chemical runoff from agriculture often combine with increases in water temperatures and acidification of the oceans (IPCC, 2007).

Some forms of waste management (e.g. landfilling of biodegradable waste) can lead to the production of CO_2 (in the presence of air) and methane (in the absence of air) (Table 3). Methane accounts for around 80% of all GHG emissions from the waste sector and is 23 times more powerful than CO_2. This comparison is based on a Global Warming Potential (GWP) over a 100 year as calculated by the Intergovernmental Panel on Climate Change (IPCC). One tonne of biodegradable waste leads to the production of 200-400m^3 of landfill gas (Defra, 2008). The treatment of fossil fuel derived materials such as plastics, (e.g. through incineration) results in emissions to air. If a landfill is engineered, methane can be captured and converted into electricity in the form of syngas. However, methane capture does not always occur, for various reasons including logistics and costs.

Table 3. Key waste management processes that lead to net GHG fluxes.

KEY WASTE MANAGEMENT PROCESSES THAT PRODUCE GHGS
Emissions from landfilling of biodegradable waste
Emission of fossil-derived CO_2 from the combustion of plastics and textiles
NO_x from incineration
Emissions from the collection, transport and processing of wastes
Emissions from high halogenated compounds used in WEEE (as refridgerants and insulation in freezers and fridges)

Addressing the Impacts: Waste Management Strategies

Prevention and Minimisation

Prevention or minimisation of waste is the most appropriate strategy to reduce the production of GHGs, as it reduces the demand for the raw materials, thus for example, the carbon stocks in plants are preserved. Emissions from transportation of resources are also reduced. Thus there is a cumulative saving of GHG emissions throughout the entire lifecycle of resource production and consumption.

Source Segregation

An analysis of the emission potential of the main waste management treatments (AEA Technology, 2001) concluded that source segregation of MSW followed by recycling of dry recyclables and composting and/or anaerobic digestion of biodegradable wastes gives rise to the lowest net flux of GHGs.

Energy Recovery

Combined heat and power (CHP) can provide net savings in GHGs from bulk MSW, but the 'robustness' of the option is dependent largely on the energy source being replaced, while mechanical biological treatment (MBT) offers significant advantages over landfilling of bulk MSW or contaminated biodegradable waste with respect to net GHG flux (AEA Technology, 2001).

Recycling, Composting and Landfilling

Composting and anaerobic digestion (AD) of biodegradable waste and recycling of paper can produce the highest overall reduction in net flux of GHGs, largely due to the avoidance of emissions of landfill from the recycling of the materials. This avoidance can result in a decrease in the net flux of GHGs of around 260-470kg CO_2eq/tonne of municipal solid waste (MSW) (AEA Technology, 2001). However, these advantages are dependent on the level of engineering within the landfill. If there is only limited gas control then there is a net benefit of around 50 – 280kg CO_2eq/tonne. This is reduced to around 50kg (biodegradable materials) – 200kg CO_2eq/tonne (paper) and actually incurs a deficit of around 20-30kg CO_2eq/tonne if the landfill has a restoration layer in place and carbon sequestration is also taken into account. It is important to note, however, that these results should be set within their rightful context, as landfilling is an 'end of pipe' solution only. Indeed, the recycling of various dry recyclables, including glass (30kg CO_2eq/tonne) and aluminum (95kg CO_2eq/tonne) offer net benefits irrespective of the engineering of the landfill. Use of high-temperature treatment technologies, if also used for energy recovery, can save on the need for fossil fuels and thus reduce CO_2 emissions. Recycling and recovery/reuse can therefore serve to prolong the lifecycle of the resources and reduce the demand for new raw materials.

Composting allows the materials to be managed in aerobic conditions, thus reducing methane. Composting also displaces the need for the use of peat or fertilisers (AEA Technology, 2001). The CO_2 from AD can be converted into a useful energy source. Advanced thermal treatment systems such as gasification and pyrolysis can be used to produce alternative fuel sources apart from fossil fuels. In addition, the diversion of biodegradable waste from landfill, stabilisation of the waste before landfilling using mechanical biological treatment (MBT), or use of composting can all result in sequesteration (storage) of non-fossil carbon in the earth's surface for longer than the 100-year time horizon that is employed in measuring global warming potential of various processes.

Addressing the Impacts: Wider Strategies

Some countries such as those within the European Union (EU), have implemented targets for diversion of biodegradable waste away from landfill. This has been done under the Landfill Directive and also enables the countries to meet their Kyoto Protocol targets, through various means including taxes, incentives, direct banning and pre-treatment of waste.

Implementation of the Landfill Directive was therefore aimed at ensuring that waste management fully contributed to the goal of sustainable development by reducing the growth of waste, increasing the diversion of waste away from landfill, engaging the public to reuse and recycle more, promoting greater awareness of the environmental impacts of products at the end of their life and developing markets for recycled materials.

The Kyoto Protocol and the EU Action Plan on Climate Change are amongst two of the most widely known global approaches to managing climatic change. In 1992, the United Nations Framework Convention on Climate Change (UNFCCC) was developed at the United Nations Conference on Environment and Development in Rio de Janeiro, Brazil. A total of 181 governments and the European Commission (EC) signed the Framework Convention.

The Kyoto Protocol was adopted by consensus at the third Conference of Parties to the Framework Convention on Climate Change in Japan in 1997 and entered into force in 2005, having been ratified by around 150 countries (UNFCCC, 2007). An important point to note is that the Kyoto Protocol has not been signed by the USA, China and Australia. It is the only existing international agreement of commitment towards the reduction of GHG and provides legally binding targets for developed countries to reduce their overall emission of GHG by 5.2% below 1990 levels during the First Commitment Period (2008-2012). It requires developed countries to reduce their GHG emissions by at least 5% of 1990 levels by 2012, but does not set targets for 'developing' countries. Under the Kyoto Protocol developed countries have pledged to take the lead on addressing the impacts of climate change. Within this, however, there has been much debate surrounding what actions should be taken to address the issue and which of these should be 'common' (employed by all) and which should be 'differentiated' (employed by developed countries) (CRI, 2007). Common actions include the responsibility to monitor and report emissions. Alternatively, differentiated actions include a commitment to reduce emissions in line with the 1990 baseline levels, under the UNFCCC, and under the Kyoto Protocol for there to be individually specified targets for Annex 1 countries to be met between 2008-2012. Non-Annex 1 countries were exempt from these restrictions, however, they could voluntarily opt into the schemes. Discussions on the merits and demerits of the schemes are beyond the realms of this chapter, however, see WRI (2008) for a fuller discussion.

Some authors such as WRI (2008) argue that simply evaluating the emissions issue on the basis of whether it is a developing or developed country, as well as on a 1990 baseline fails to focus on the major emitters, and has differential impacts on various countries. For example, it discriminates against developing countries wishing to develop, places the economic burden for meeting targets solely on the developed countries and gives an unfair advantage to economies such as the Soviet Bloc and Eastern Europe that have recently experienced economic contractions. WRI (2008) argues that rather than using absolute values, a metric of percapita values would provide a more realistic representation of contribution. Percapita measures would also enable some leeway for economic growth amongst countries from the developing world. Another metric suggested it is that of carbon/greenhouse gas intensity (the efficiency with which nations utilise their energy in the production of goods and services, and the use of alternatives such as nuclear power).

CONCLUSION

Rising GHG emissions can be strongly linked in part to consumption patterns and by extension the quantity and types of waste that are produced. Greater efforts to reduce consumption levels are therefore crucial if current and future impacts are to be effectively avoided, reduced or delayed. While GHG emission levels are currently generally higher among developing countries, changing consumption patterns (most notably amongst the Bric countries), will result in an increasing shift towards these countries due to rising levels of industrialisation.

Understanding the issues surrounding the management of waste and consumption patterns and their impacts on climatic change requires a holistic approach encompassing a range of fields including psychology, chemistry, physics, biology, geography, economics, epidemiology and oceanography. It is only through an inter-disciplinary and multi-disciplinary approach that it will be possible to fully address the impacts.

A recent report (IPCC, 2007), suggested the need for a mix of strategies that includes mitigation, adaptation, technological development and research, combined with policies (e.g. incentive-based schemes), and action from all stakeholders in society (i.e. from individuals to governments). Addressing the issues of resource consumption and sustainable management of waste has therefore to incorporate a range of socio-economics (e.g. per capita population growth), resource use (i.e. energy, water), spatial (e.g. physical infrastructures and land use patterns), and technological factors.

Strategies need to encourage the development of advanced treatment technologies, be undertaken on a phased basis, be flexible to accommodate differences in economic standing and involve a mixture of initiatives. A number of options for adaptation are available including:

- Technological (e.g. treatment types)
- Behavioral (e.g. changes in consumption and production habits)
- Policy (diversion from landfill)
- Managerial (organisational sizes and structures)
- Legislative (e.g. taxes)

Table 4 illustrates the mix of strategies employed by the Government in England and Wales in their Waste Strategy 2000 (DETR, 2000). The measures included a mixture of regulatory, market based, planning, promotion and information approaches to achieve greater sustainability in the management of waste.

Table 4. The five main strategies of the Government White Paper on sustainable waste management in the England and Wales, in 2000.

Strategy	Examples
Regulatory	Increased costs of disposal, IPPC (Integrated Pollution and Prevention Control), Producer Responsibility
Market based	Tradable permits, public procurement, Landfill Tax
Planning	Integrated approach and consultation between stakeholders, using PPG (Planning Policy Guidance) notes
Promotion	Campaigns such as 'Are you doing your bit', focusing on producers, development of the Environmental Technology Best practice Programme (ETBPP)
Information	Targets, indicators, baselines, encouraging use of environmental management systems such as EMAS and ISO 14000

With respect to technologies, increasingly the low-emission technology of carbon dioxide capture and storage (CCS) is being considered as a potential option for mitigation. CCS employs the capture of CO_2 emissions from large sources (e.g. power stations) and storing it in, rather than allowing it to escape into the atmosphere. Current estimates suggest that non-commercial CCS should be available by 2014, with commercial plants by around 2016 to 2020. However, meeting these targets will be heavily dependent on the necessary regulatory, technical, legal issues, as well as the governmental policies being put in place. According to WRI (2008) reducing GHG emissions can involve use of market-based mechanisms that focus either on the level of emissions or the costs of compliance, with the market place determining the balance. For example, tradable permits can set the levels of the emissions, with the market place determining the worth of the permits. Similarly, a carbon tax that sets the maximum costs for reducing emissions could be employed, with the market place determining how much emission levels are actually reduced by.

There are a number of environmental, economic, social and institutional challenges that have to be overcome in order to effectively implement adaptation (IPCC, 2007). Adaptation will be important to address both the impacts from past emissions, as well as future emissions. Different options also have varying levels of influence, particularly for higher levels of warming, as well as to assess the influence of various groups. Hence it is important for there to be long-term strategic planning. However, adaptation alone is not expected to mitigate against long-term projected rises in the magnitude of warming.

Whilst mitigation is important, the focus for measures to combat climate change should be on resource efficiency/conservation and minimisation. This will require shifts in policies, research and development, technological change and provision of resources (economic and personnel) in order to adequately meet the challenges ahead. There should also be the development of a regulated technical standards and emissions programme with the accompanying quotas across a range of business sectors. Programmes should adopt a long-term approach and be grounded in regional and national perspectives. While this chapter has focused on the climatic change impacts, it is important to end by highlighting and recognising the number of wider potential benefits to be accrued from the implementation of more sustainable approaches. These include new manufacturing opportunities, partnerships, novel designs, and closer engagement with both internal and external stakeholders. However, as noted by Tudor (2009) these can only be successfully achieved if the approach is multi-sectoral, actively involves all relevant stakeholders, is economically viable and employs a 'holistic' approach.

REFERENCES

AEA Technology (2001). Waste management options and climate change. Available at www.ec.europa.eu/environment.

CRI (Congressional Research Institute) (2007). Greenhouse gas emissions: perspectives on the top 20 emitters and developed versus developing nations. *Encyclopedia of Earth*.

Defra (2008). Climate change and waste management: the link. Available at www.defra.gov.uk.

DETR (Department of the Environment, Transport and the Regions) (2000). Waste Strategy 2000. HMSO. London: UK.

EPA (2008). Lifecycle of greenhouse gas emissions. Available at www.yosemite.e pa.gov/oar/globalwarming.bsf/content/ActionsWasteBasicInfoGeneral.html

Houghton JT., Ding Y., Griggs DJ., Noguer M., van der Linden P.J., Maskell K., Johnson C.A. (eds.) (2001). Climate change 2001 – the scientific basis. Contribution of Working Group 1 to the Third Assessment Report of the Intergovernmental Panel on Climate Change. Cambridge University Press.

IPCC (Intergovernmental Panel on Climate Change) (2007). Working Group II contribution to the Intergovernmental Panel on Climate Change. Fourth assessment report, climate change 2007. Climate change impacts, adaptation and vulnerability. Available at www.

Lehtonen M. (2005). The environmental – social interface of sustainable development: capabilities, social capital, institutions. *Ecological Economics*. 49 (2): 1999-214.

Monni S., Pipatti R., Lehtila A. (2006). Global climate change mitigation scenarios for solid waste management. VTT Publications. Finland.

OECD (2002). Towards sustainable household consumption? Trends and policies in OECD counties. OECD policy brief. July, 2002: 3-4. Paris: France.

Packard VO. (1960). The waste makers. David McKay Co.

Stern N. (2006). The economics of climate change. HM Treasury. Available at www.hm-treasury.gov.uk/independent_reviews/stern_review_economics_climate_change/sternrevi ew_summary.

Tudor TL. (2006). an examination of the influencing factors and policies for sustainable waste management: a case study of the Cornwall NHS. Unpublished PhD thesis. University of Exeter, UK.

Tudor TL. (2009). The role of organisations in enhanced global environmental management: perspectives on climate change and waste management strategies. The

International Journal of Environment and Waste Management. Forthcoming.

UNCSD (United Nations Commission on Sustainable Development). (2001). Indicators Of Sustainable Development: Guidelines and methodologies. United Nations. Available at www.un.org/esa/sustdev/natlinfo/indicators/indisd-mg2001.pdf.

WCED (World Commission on Environment and Development). (1987). Our common future. Oxford University Press. Oxford.

Welford R. (1998). Corporate environmental management. Systems and Strategies. (2nd ed.) (ed.). Earthscan Publications Ltd.

WRI (World Resources Institute) (2007). Climate Analysis Indicators Tool. Washington DC.

WRI (World Resources Institute) (2008). Greenhouse gas emissions: perspectives on the top 20 emitters and developed versus developing nations. Washington DC.

In: Encyclopedia of Environmental Research   ISBN: 978-1-61761-927-4
Editor: Alisa N. Souter   © 2011 Nova Science Publishers, Inc.

Chapter 14

ENVIRONMENTAL UNCERTAINITY AND QUANTUM KNOWLEDGE?

*Christine Henon**

Charbonnières les Bains - FRANCE

ABSTRACT

As the world shrunk and expanded at the same time, as interdependence between (and within) the nature and human society became obvious, physics, information sciences and ecology have known (among others sciences) a radical evolution and have tried growingly to adapt to the complex, dynamic, interrelated and more and more uncertain socio-economic and natural systems. In this context, and through assessment and management of the Jervis Bay marine park, we question the concepts of quantum environmental information, of quantum environmental assessment and of quantum environmental knowledge.

Keywords: assessment, biodiversity, complexity, dynamic and interrelated systems, environmental integrated co-operative management, ever-changing, explicit, information, knowledge, natural environment, quantum, tacit, uncertainty.

A. INTRODUCTION. PRESENTATING THE JERVIS BAY PROBLEMATIC VERSUS SCIENTIFIC EVOLUTION

As the world became at the same time smaller and bigger, as interdependence between (and within) the nature and human society became increasingly obvious, many sciences have known a radical evolution. Physics has lead to quantum physics. Information and knowledge sciences have emerged. And, from the analyses and descriptions of simple typology to the theories and strategies for environmental integrated management, ecology tried growingly to

---

∗ christinehenon@hotmail.com

adapt to the complex, dynamic, interrelated and uncertain socio-economic and natural systems.

In this context, we present the issue of the assessments underlying the environmental integrated and co-operative management of the Jervis Bay marine park. These assessments are based on numerous data, either quantitative either qualitative, but coming from very various fields and sources. These data are emerging as well from explicit but also from tacit knowledge, the later often being affected by the phenomenon of the (Aboriginal) "stolen generations". And in addition to uncertainties arising from the lack of information and from human and societal perceptions and reactions, many others are the reflection of a complex ever-changing natural environment and therefore modify during the measurement. In any event, while having to be taken into account, they are not always measurable and can not be synthesized and used in models. In this context, even the reference used for what is called "pure water" is of a doubtful validity.

Given this situation, we try to link the previously highlighted aspects of quantum physics, of information and knowledge theories, of biodiversity's complexity to the questioning of the existence of a potential quantum environmental information and, maybe, of its implications on environmental assessment and knowledge.

B. QUANTUM THEORY, ECOLOGY, INFORMATION AND ADAPTATION TO UNCERTAINTY. WHEN PARADIGMATIC SHIFT MEETS ENVIRONMENTAL CHANGE.

In 1962, Kuhn used the word "paradigm" for referring to the practices that define a scientific discipline during a specific period of time (Kuhn, 1962). In 2001, the United Nations scientific panel studying climate change reported that humanity had only "likely" played a role in the potential warming trend (IPCC, 2001). In 2007, the same group admitted not only the "unequivocal" warming trend but as well the "very likely" responsibility of human activity for that trend over the last 50 years (IPCC, 2007). The world changes, science evolves, and vision of the world and of sciences shifts, and sometime meets. Through the following chronology, we would try to show the theoretical inter-evolution which we will use to look at the case developed afterwards.

In 1804, Dalton develops the Atomic Theory, showing that all elements are composed of indestructible and indivisible particles: the *atoms* (Smith, 1856). In 1900, among other scientists, Planck introduces the idea that energy is quantized and that the total energy of atoms is made up of indistinguishable energy elements - quanta of energy (Planck, 1901). Quantum *mechanics* takes off. *Quantum theory* describes the emission and absorption of these *energy quanta*.

In 1925, the *exclusion principle* states that two identical particles (elementary like the electron or composite like the proton) may not occupy the same quantum (energy) state simultaneously (Pauli, 1947). The *wave function* (De Broglie, 1925) calculated by the Schrödinger equation (Schrödinger, 1926) determines then the probability of the presence of this energy element and represents it by fuzzy clouds. And it shows that this element can sometime behave like a particle of matter and showing certain *wavelike properties* when in motion. This is the principle of wave-particle duality. This element is thus no longer localized in a given place or region but more or less spread out in infinite probable positions. Measured, it acts like a particle. Free, it acts like a wave. And it is how, in the natural environment, the electrons stay in unidentified orbital paths around the nucleus. Arising from this equation, the *superposition principle* affirms then that the characteristics of an atom, of a particle or of a quantum system, constitute a state. However, when a system has several possible states, the sum of all these states is also a possible state: the system is in a *superposition of state*. This is the *many-worlds* (or *multiverse*) *principle*. It holds that as soon as any element of energy or particle might be in any state, the universe of that element transmutes into a series of states / parallel universes equal to the number of possible states in which the element can exist, with each universe containing a unique single possible state of that element. This element can thus be in two places at the same time, an atom can be in a state of superposition of energy. The simple fact of measuring provokes the disappearance of the superposition of state, giving way to a single state. In 1927, according to the *principle of incertitude* (Heisenberg, 1927), it seems now impossible to know precisely both the position and the speed of this element. This one does not have an ordinary trajectory, therefore neither its position neither its speed can be simultaneously specified. Knowledge of the position and of the speed is complementary.

In 1935, Einstein, Podolski and Rosen proposed the famous paradox known as EPR entangled pairs. This paradox implies that interfering (by measure) with the properties of one part α of quantum system (this part being the first of entangled particles) can cause an instantaneous change in the state (or properties) of the other distant part β (this other part being the second of the entangled particle) in another place, whatever this place is (Einstein and al, 1935).

In 1936, Alan Turing gives a definition of processing and publishes the first ideas on computing (Turing, 1936).

From 1945, big changes revolutionize *ecology science*. The field moves indeed first from *typology* to *function*, then from *qualitative description* to *quantitative description*, and when ideas about nature begin to be expressed within theories and mathematical form (Taylor, 1988 ; Gragson, 2005). Through the *concept of ecosystem* and the *cybernetic view* of their functionally related and self-regulated parts, Eugene and Howard Odum (Odum, 1953, 1971) help indeed ecology to become a *discipline* (Golley, 1993). But soon, this discipline splits between *two schools*. The first one, the *school of ecosystem*, is based on a deterministic but holistic point of view. The second one, with the *population ecology's perspective*, is oriented on mathematical analyze of relationships community stability / species diversity and communities stabilities / complexity.

In 1958, Foskett begins to work at its definitions of data, information and knowledge (Foskett, 1958, 1979). *Data* are the first basic elements of thought and consciousness acquired by people through sensations or perceptions' recognition, basic existing elements that, at the beginning, are not interrelated and have no specific meaning. Through conscious mental activity, these data are then organized in concepts, interrelated within systems of ideas and become *information*. Therefore, if data are much more numerous than information, their tenor has far less meaning. And further than information, *knowledge* builds itself only through time and space, by plunging into many systems of ideas, many networks of facts, many subjectivities and / or by estimating, trying, judging through all those. Knowledge is presented as a unique information system that has been absorbed by a unique human mind and is thus an inherent part of a personality, impossible to communicate (when it is not the case of information). And in 1990, Bonitz argues as well that knowledge belongs to a specific person in a specific place at a specific time (Bonitz, 1990).

In 1966, Polanyi makes the distinction between *explicit knowledge* and *tacit knowledge*. He argues indeed that acts of exploration and creativity combine critical reasoned interrogations with motivations that ground on personal commitment and feelings, on something that the explorator or the creator knows but can not necessarily explain. It is tacit knowledge, very personal and hard to formalize, difficult to communicate or share with others, by opposition to explicit knowledge, which is codified and can be transmitted in a much more formal systematic language (Polanyi, 1966).

In 1973, Holling defines *resilience* as related to the amount of disturbance that an ecosystem can sustain before its control or structure changes. He helps as well to understand *time and space variability* within uncertainty and complexity contexts (Holling, 1973), so that the two schools of ecology meet, building new points of view on environmental management. And it is then the economical and ecological approaches that meet, officially in 1987, over the concept of *sustainable development,* meeting "the needs for the present generations without compromising the needs for the future generations" (WCE & D, 1987), recognizing the complex environmental interactions.

In 1982, Feynman suggested that *individual quantum systems* could be used for *computation* (Feynman, 1982).

In 1985, Deutsch presents the first description of a *quantum computer*. He describes how quantum computers could bring speed and power (Deutsch, 1985). And soon, many institutions work on a wide range of quantum physics application within information processing, measurement and computing, including, with rapid success, cryptography (Bennett et al, 1982) (Ekert, 1991). But quantum co-processors, teleportation (Bouwmeester et al, 1997), etc, are supposed to be close at hand.

In parallel, appears the *environmental evaluation*. It is a planning instrument calling upon the public participation while allowing an integration of the ecological considerations. It offers as well a method of identification of the possible environmental effects of development projects and enables project promoters to modify their plans, in order to mitigate or eliminate these effects (Unep-Unido, 1991). And in turn, the *integrated environmental evaluation* tries to allow several scientific disciplines to interpret, to combine and to exchange, in order to facilitate the understanding of complex phenomena.

At the same time, new notions help society to understand the effects that perturbations born from changes and uncertainty might have onto the ecosystems and their health. Consequently, they help society to manage these ecosystems with the utmost prudence. *Persistence*, for example, is related to the lasting of a variable before its change into a new value (Pimm, 1991). And hierarchical but unpredictable *self-organizing systems* build the ecosystems throughout food chains, communities and landscapes (Perry, 1995). That means that, even if ecosystems may tend to a kind of equilibrium, this equilibrium is very different from the traditional representation of a so called "balance of nature." The evolutionary and creatively spontaneous world, within its *macroscopic uncertainty,* has some *infinite attributes* (Pahl-Wostl 1995). And within their scale of time and space, within the complexity of the interrelated systems, these attributes have effects which are still *impossible to apprehend*. This situation leads directly to the problems of the *heterogeneity of the knowledge* and of *inherent residual uncertainties*. And it implies the need for adaptation measures and, therefore, for a type of *knowledge* that would takes these uncertainties into account, that would includes them as some indispensable elements of the future (Pahl-Wostl 1995).

In 1996, Nicolescu presents his vision of transdisciplinarity, with a *transdisciplinary vision* that proposes to consider a multidimensional reality, structured on many levels that would be coherent between them (Nicolescu, 1996). This reality replaces the traditional one dimension / one level reality.

In 1998, Holling summarizes the differences between *analytical ecology* and *integrative[1] ecology*, clearly showing prevalence of the mathematical modeling methodology in the last one. Throughout biotechnology, analytical ecology promises indeed economic benefit and health but as well shifting social values. It is thus seen as *too disciplinary* and experimental, too reductionist. A contrario, integrative ecology is seen as *interdisciplinary* throughout its systems approach and evolutionary biology, leading to a *new environmental management* that will face the *ever changing systems* (Holling, 1998).

Ever changing systems, ever changing information... *Quantum information* has as well a physical reality which is different from the one of traditional information: it is impossible to copy it, contrary to traditional information i.e. of a non-physical nature. In the quantum world, according to the *principle of non cloning,* information whose state is not known in its totality cannot be duplicated. If one tries to do it, one will destroy the quantum state of the information. The second point linking information to the quantum theory is that if the *superposition principle* implies that a piece of information can take several values at the same time, the *principle of non localizability* implies that the observation of information at a specific place can influence instantaneously the state of information at another place. Both pieces of information are correlated perfectly. It is the *property of entanglement* (Bennett & DiVincenzo, 2000).

In 2002, Gunderson and Hollings try to rationalize the interplay change / persistence and predictable / unpredictable by presenting an integrative theory: the *Panarchy[2]*. They argue that all *adaptive systems* have a set of *nested cycles* going on at different scales, a set that makes them resilient and able to absorb shocks and change (Gunderson & Holling, 2002). For the understanding of these systems, for the communication around and within, and for the determination of systemic sustainability, they propose a *dynamic and prescriptive process*. This process comes with a non-static monitoring, connected to actions and policies but as well to different scenarios for future, scenarios that include the uncertain and the unpredictable.

In 2004, at Oxford, Briggs launches the setting up of the *Quantum Information Processing Interdisciplinary Research Collaboration* (QIP IRC, Oxford) to carry out basic collaborative research and bring academics and industry together.

In 2007, Pahl-Wolst, Kabat and Möltgen note that awareness of the impacts of global and climate change push water management to be *more flexible, participative and adaptative* in front of increasing uncertainties (Pahl-Wolst et al, 2007). With the help of hard and soft systems analysis (the last been focused on socially constructed reality and subjective perceptions), this new management has first to improve the conditions for *inter and trans-disciplinary, system-oriented research* (Pahl-Wolst, 2007). It has then to systematically implement the *learning* from all outcomes of existing management strategies (Pahl-Wolst et al, 2007).

1 i.e. comprising economic, ecological, and social systems
2 Named so after the unpredictable Greek god Pan.

C. Biodiversity, Aboriginal Knowledge and Western Civilization. When Environmental Degradation Meets Informational Complexity.

The Jervis Bay example as it is used in here is derived from a 3 years research done by the author, in particular, between April 1999 and April 2000, in Australia and in France (Henon, 2001). Most of the theoretical documentation was collected in connection with the Centre for Retrospective Research of Marseille (University of Marseille III) and with the Centre for Oceanography of Marseille (University of Marseille II). Australian Data (in Jervis Bay, Canberra and Sydney) was collected in connection with the Cooperative Research Centre for Sustainable Tourism (University of Canberra) through documentary and field material and analysis and through informal, unstructured or semi-structured interviews of communities spokesmen and of environmental federal and national managers and scientists. These last persons included the managers of the Jervis Bay Marine Protected Area, of the Jervis Bay National Park, of the Bouderee Federal Park, of the Shoalhaven district, geo-administrative entity of Jervis Bay, and the Principal Scientist for the New South Wales Fisheries (his management incorporating the complementary practices of the responsible for the Marine protected areas of the New South Wales State).

For thousand years, the site of the bay has been inhabited by three aborigine tribes, the Wandandians, the Wodiwodi and the Jerringa. But at the end of the eighteen-century, Europeans arrive in Australia and soon consider the aboriginal populations as primitive, illiterate and fragile, on the point of extinction. However, contacts between Europeans and Aboriginals being more and more frequent, the number of half-bred children increases. And the European communities feel now endangered by this new cultural and biological proximity. They begin to believe necessary to protect progress and civilization by eliminating aboriginality in its hybrid form. A first attempt is done by trying to regulate the movements and relationships of the Aboriginals. A second one aims to "assimilation", by "extracting the color from the spirits and the bodies" so the half-bred children should forget their origins. These children are thus removed from their family, sent into institutions or adopted by white families. Contacts with their families, people and culture are prohibited. It is not before the end of the Sixties that half-bred children removal, known as the « stolen generation(s) », is officially stopped (HREOC, 1995, 1996, 1997). At this time, Australia new inhabitants also begin to become aware of the environmental brutal degradation. The natural environment remained indeed relatively preserved and divers during the centuries of aboriginal presence. But two hundred years only of colonial presence pushed this environment toward destruction. It becomes therefore suddenly interesting to use the aboriginal environmental knowledge. And the aboriginal life style regarded until then as primitive is now seen as ecological.

In 1997, in Jervis Bay, a Marine Protected Area is created. Its management is based on the transverse and co-operative uses of knowledge, of decision and of education, between scientists in charge, Aborigines and Euro-Australian communities and interest groups. In collaboration with other conservation organizations (the Jervis Bay National Park, the Bouderee Federal Park and the Shoalhaven district), *Programs of Monitoring and Research* are put in place. They aim to assess the quality of the environment, the impacts of human activities, the respect the public might have for the law, the perception this same public has of the protected areas, the effect of the conservation. In brief, these programs are supposed to improve the environmental knowledge. Co-operative environmental management has indeed to be fed with control information (human activities), organization information (administrative structure), planning information (of the uses, of the remedies), implantation information (of the resources). Monitoring is still the ultimate source of knowledge, feeding the reports of co-operative environmental management. It is as well a feedback of this co-operative environmental management.

In 1998, Martin Berg, from the Shoalhaven district, wants to be able to predict the effects of human activities on water quality and resources. He tries thus to set up a monitoring integrated program: the Cumulative Impacts Monitoring Program. This CIMP aims to identify the environmental changes, especially changes concerning water quality and quantity, and the causes of these changes, natural or entropic. It is as well supposed to ground the environmental co-operative integrated management of all the conservation organizations of the bay, for action, reaction and prevision, in order to increase the sustainable use, conservation, and management of all Jervis Bay's renewable natural resources, especially water. First of all, Marin Berg considers that a basic knowledge of the physical environment and of the natural and entropic influences might help to give a precise definition of "pure water", in order to make it the goal. But with all the different activities, land and water use types, land management practices, on both on-site and off-site, many influences are modifying this "pure water", as well as the causes of the regarding changes. Therefore, Martin Berg revises quickly the CMIP in order to determine the quality of this water by measuring and monitoring all the components of all the level of the biodiversity's complexity.

The first problem is that it includes an amazingly wide and transverse range of data and information. First, those are all the simple quantitative data, mostly of a chemical, physical, biological, geo-morphological, etc, kind (they include rainfall, runoff, and water quality -e.g., suspended sediments, nutrients and pesticides-). Second, they are the qualitative monitored data and information coming from the catchment, sub-catchment and communities level, i.e. some bio-physical data (Soil erosion occurrence, rainfall intensity, biological indicators – seaweed, etc.- occurrence, etc.) but mostly very transverse economic and socio-cultural data concerning the type and sector of human activities in Jervis Bay, the intensity of the human activities, the economic importance of the activities, the communities' members levels of income, the socio-cultural importance of the activities, the individual communities' members' perception of water quality, of what was considered as "pure water", an evaluation of the importance of the "pure water", of the practiced activity, of the conservation of the bay, the respect of the legislations by the communities and the public, the perception they had of the reserves and parks, the effect of the protection, the perception of water conservation, some ways to avoid soil erosion, some ways to avoid water pollution regarding specific activities, etc. It means a measurement and a monitoring of all the components of the biodiversity's complexity and of its sociological, cultural, economical interactive environment, including explicit and tacit local knowledge of Aboriginal and Euro-Australian communities. But the Aborigines community has lost a lot of its socio-cultural roots and, therefore, a lot of its traditional ecological knowledge through the phenomenon of the "stolen generation". And because of its immigration past, the Euro-Australian community has very different socio-cultural histories and memories. In this context, even the reference *"pure water"* has to face either a drastic lack of information either very various understandings of the concept, revealing thus of doubtful validity.

The second problem is that, although environmental data has been collected from many CIMP catchments for the last year, it is more and more obvious that the first monitoring and analysis lack of effectiveness. For example the data participating to the description of the impact of human activities on the water resources (degradations such as eutrophication of rivers and lakes) changes during the process of monitoring, reflecting the natural resources change during this process. Since analysis and interpretation lack, data collection continues without any critical evaluation of adequacy, accuracy, and reliability. But these collected data and information ground the program. And this one has then to establish relationships between all the obtained parameters and values in order to connect them to the co-operative environmental management of water and to determine the changes, their causes, the ecological answer to come and the human response to give. In this context, it becomes impossible to know if the monitoring is still valid, if all the biodiversity's complexity's impacting factors have effectively been taken into account.

Moreover, a part of the CIMP staff has difficulties to deal with data i.e. beyond its daily activities and experience. Adds to that the fact that the CMIP scientific contributors come mostly from unidisciplinary fields of research and intervene only monthly on the data collected. Each of them has then to give a report summarizing the environmental issues relevant to her / his scientific specialty. But they do limited effort to inter-collaborate, i.e. to cooperate with each other especially by sharing data and data analysis. Lack of inter-science cooperation into environmental analysis becomes then a major limitation to the quality of the reports that would need multi-disciplinary and teamwork approach. And following these difficulties, other questions are then raised. Who would be able to collect these data, information, and knowledge, to use them "well" and to control then the actions? Is it possible to teach all the knowledge related to the environmental protection and how to do it? Is it possible to make this person credible and how to do it? The experts surrounding Martin Berg still do not recognize transdisciplinarity, neither do they believe in the efficiency of a "general environmental practitioner". Until such a practitioner is trained, until attitudes change, Martin Berg tries to find all the *adequate competence*. But to join together varied and dispersed explicit or tacit knowledge is not simple: the most enthusiastic people are not always the most qualified, validated skills are already mobilized elsewhere, or knowledge remains too tacit even to be recognized.

D. CONCLUSION. QUESTIONING THE CONCEPT OF QUANTUM ENVIRONMENTAL INFORMATION

In the kaleidoscopic vision presented above, first, of a chronology of theories and concepts surrounding environment and / or information management, second, of the example of Jervis Bay's related management problematic, we question now the relationship which could be established between both these conceptual and experimental presentations and the existence and the particularities of a potential quantum environmental information.

The Jervis Bay environmental management asks for the recognition, collection and treatment of a whole host of information and knowledge, known and unknown, measurable and non measurable, changing and unchanging. In this case as in the many cases worked on by some of the authors presented in the first paragraphs, it shows how complex are measurement and testing of human activities' impact on the environment (and especially on water resources). In Jervis Bay, the causes are multiple: measurement and testing require a significant number of very varied external participants, the environmental data analysis have a very transverse quantitative and qualitative nature, many collected data are not suitable for any meaningful analysis, others modify perpetually.

Is the solution in the way to deal with uncertainty i.e. to test, to validate, to test again and to revise any models by the way of a trial-and-error learning ? Or is it in the agent-based approach, where the output patterns are generated by rules which are not often really related to nature's processes? But in the presented situation, the problem is not only in the model, which could maybe be permanently revised, but as well in the lack of adequate, accurate and reliable data and information. Or is it in the reminding that, perhaps, nature is fundamentally irreducible (Wolfram, 2001)?

Is the solution in a new way to see environmental information, environmental knowledge? Is it in the questioning of the limit between the integrative environmental vision of the world and its plain-disciplinary vision, between explicit but sometime obsolete fragments of data, "inference-laden signifiers of natural phenomena to which we have incomplete access" (Oreskes et al. 1994), and tacit environmental information and knowledge?

Is the solution in some quantum characteristics of environmental information? Environmental information is an abstract but represented entity, result of the integration of different data that might come from several thematic fields. In a given time, this information can thus be represented within one or several themes, even if, in its reality, it belongs to the single family of environmental information. Is it thus possible to say that its *representation of belonging* competes with its *reality of belonging?* Can this environmental information become a *quantum abstract entity,* diffuse within the interdisciplinary environmental field and within time? Can the quantum characteristic of the *indeterminism* be applicable to environmental information?

Is the solution in the resolution of the *problem of the environmental information measurement?* What is the process of the disturbance of an environment by the acquisition of some information? In this context, does information have systematically a *reality* which is connected to the reality of the data used (measured) for its building, but *disconnected* from the *new reality* of the data perturbed (and modified) by the measurement?

Is the solution in the determination of the different characteristics and states of quantum environmental information? Does a piece of environmental information have several states, depending of the time of measurement but as well of the various thematic fields to which the initial data (that make up this information) belongs. Is this piece of information able to be in a state of superposition of values, able to take *several values* at the same time? And, consequently, in some situations, can the observation of a piece of environmental information at a specific place influence the state of another piece of environmental information at another place? Is there something circulating between these two pieces of information? Does this "something" concern the material world, the energetic world, the conceptual world? And do these two pieces of information remain also correlated, *tangled up?*

Is the solution in these properties of superposition of state and of *entanglement,* properties which would then give integrated environmental evaluation a much greater *power of representation* of the environmental realities? Can therefore environmental information become quantum? Can a unit of quantum environmental information constitute a piece of potential knowledge or an *environmental photo* of one of the reality / state (and not only a scenario), a *coherent superposition* of various states of the assessed environment? Can the obtained results provide then not one but several holistic environmental assessments, leading to an embryonic *quantum environmental knowledge*?

REFERENCES

[1]   Bennett, C.H., Brassard, G., Breidbart, S. and S. Wiesner, 1982, Quantum cryptography, or unforgeable subway tokens, *Advances in Cryptology: Proceedings of Crypto 82,* Plenum Press, 267–275

[2]   Bennett, C.H. and D.P. DiVincenzo, 2000, Quantum information and computation, *Nature*, 404: 247-255

[3]   Bonitz, M., 1990, Information - knowledge – informatics, *IFID,* (15) 2: 3-7

[4]   Bouwmeester, D., Pan, J.W., Mattle, K., Eibl, M., Weinfurter, H. and A. Zeilinger, 1997, Experimental quantum teleportation, *Nature,* 390: 575–579

[5]   De Broglie, L., 1925, Thèse de doctorat, Recherches sur la théorie des quanta, *Annales de Physique*, 3, Paris: Masson & C (ed.)

[6]   Deutsch, D., 1985, Quantum Theory, the Church-Turing Principle and the Universal Computer, London: Proceedings of the royal society of London (400) 1818: 97-117

[7]   Einstein, A., B. Podolsky, and N. Rosen, 1935, Can quantum-mechanical description of physical reality be considered complete? *Physical Review,* 47: 777

[8]   Ekert, A., 1991, Doctoral thesis, Oxford, *Physical Review Letters,* 67: 661-663

[9]   Feynman, R., 1982, *Simulating Physics with Computers*, International Journal of Theoretical Physics, 21: 467-488

[10] Foskett, D.J., 1958, *Information service in libraries*, London: Crosby Lockwood.

[11] Foskett, D.J., 1979, Librarianship and the "One World", In: *Peebles '79: Libraries at large: a survey of some national and international in librarianship,* Glasgow: Alan G.D. White (ed), 45-52.

[12] Golley, F.B., 1993, *A History of the Ecosystem Concept in Ecology: More than the sum of the parts,* New Haven: Yale University Press.

[13] Gragson, T.L., 2005, Time in Service to Historical Ecology, *Ecological and Environmental Anthropology,* 1(1): 2-9.

[14] Gunderson, L.H. and C.S. Holling, 2002, *Panarchy. Understanding transformations in human and natural systems,* Washington DC: Island Press.

[15] Heisenberg, W., 1927, Über den anschaulichen Inhalt der quantentheoretischen Kinematik und Mechanik, *Zeitschrift für Physik,* 43: 172-198.

[16] Henon, C., *Le paradigme Gombessa : l'écologie cognitive pour l'environnement,* University of Canberra / CRRM of Marseille, Doctoral thesis.

[17] Holling, C.S., 1973, Resilience and stability of ecological systems, *Annual Review of Ecology and Systematics*, 4: 1-23.

[18] Holling, C.S., 1998, Two cultures of ecology, *Conservation Ecology,* 2(2): 4.

[19] HREOC, Human Rights and Equal Opportunity Commission, 1995, *Battles small and great: The first twenty years of the racial discrimination act,* Canberra: AGPS.

[20] HREOC, Human Rights and Equal Opportunity Commission, 1996, *Understanding racism in Australia*, Canberra: AGPS.

[21] HREOC, Human Rights and Equal Opportunity Commission, 1997, *Bringing them home: The report of the national inquiry into the separation of aboriginal and Torres Strait islander children from their families,* Sydney: Sterling Press.

[22] IPCC, The intergovernmental panel on climate change, WMO/UNEP, 2007, *Climate Change 2007: Synthesis Report,* Geneva: Pachauri, R.K. and the Core Writing Team.

[23] IPCC, The intergovernmental panel on climate change, WMO/UNEP, 2001, *Climate Change 2001: Synthesis Report, Watson,* Geneva: R.T. and the Core Writing Team.

[24] Kuhn, T.S., *The Structure of Scientific Revolutions,* 1962, Chicago: Univ. of Chicago Press.

[25] Nicolescu, B., 1996, *La transdisciplinarité, Manifeste,* Paris : Le Rocher.

[26] Odum, E.P., 1953, *Fundamentals of Ecology,* Philadelphia: Saunders.

[27] Odum, H.T., 1971, *Environment, Power and Society,* New York: Wiley.

[28] Oreskes, N., Shrader-Frechette, K., Belitz, K., 1994, Verification, validation, and confirmation of numerical models in the earth sciences, *Science,* 263: 641-646.

[29] Pahl-Wostl, C., 1995, *The dynamic nature of ecosystems: Chaos and order entwined,* Chichester: John Wiley & Sons.

[30] Pahl-Wostl, C., 2007, The implications of complexity for integrated resources management, *Environmental Modelling & Software archive.* Elsevier, 22 (5): 561-569.

[31] Pahl-Wostl, C., Kabat, P., and J. Möltgen, 2007, *Adaptive and Integrated Water Management: Coping with Complexity and Uncertainty,* New York: Springer-Verlag.

[32] Pauli, W., 1947, *Exclusion principle and quantum mechanics,* Neuchatel: Editions du Griffon.

[33] Perry, D.A., 1995, Self organizing systems across scales. *Trends Ecol. Evol.,* Elsevier. 10(6): 241-244.

[34] Pimm, S., 1991, *The Balance of Nature? Ecological issues in the conservation of species and communities,* Chicago: University of Chicago Press.

[35] Planck, M., 1901, Über das Gesetz der Energieverteilung in Normalspectrum. *Annalen der Physik,* 4: 553-563.

[36] Polanyi, M. 1966. *The Tacit Dimension.* New-York: Doubleday.

[37] Schrödinger, E., 1926, Quantizierung als Eigenwertproblem (Erste Mitteilung), *Annalen der Physik,* 79: 361-376.

[38] Schrodinger, E., 1926, Quantizierung als Eigenwertproblem (Zweite Mitteilung), Annalen der Physik, 79: 489-527.

[39] Smith, R.A., 1856, *Memoir of John Dalton and History of the Atomic Theory,* London: H. Bailliere.

[40] Taylor, P.J., 1988, Technocratic optimism. H.T. Odum, and the partial transformation of ecological metaphor after World War II, *Journal of the History of Biology,* 21: 213-244.

[41] Turing, A., 1936, On computable numbers, with an application to the Entscheidungsproblem, *Proceedings of the London Mathematical Society,* 2 (42): 230–265.

[42] Unep-Unido, 1991, *Audit and reduction manual for industrial emissions and wastes,* Technical Report, Paris.

[43] WCE & D, World Commission on Environment & Development, 1987, *Our common future,* Oxford: Oxford University Press.

[44] Wolfram, S., 2001, *A new kind of science.* Champaign, ILL: Wolfram Media Inc.

In: Encyclopedia of Environmental Research   ISBN: 978-1-61761-927-4
Editor: Alisa N. Souter   © 2011 Nova Science Publishers, Inc.

Chapter 15

SOIL: A PRECIOUS NATURAL RESOURCE

*C. Bini**

Department of Environmental Sciences,
University of Venice - Dorsoduro, 2137
30123 - Venezia, Italy

ABSTRACT

Soil, together with water, is a basic resource for humanity, as it is stated in the first item of the European Soil Chart. Unlikely it has been thought for long time, soil is a limited resource, easily destroyable but not easily renewable at human life timescale, and therefore should be protected in order to preserve its fundamental functions, both biotic and abiotic. It is worth noting how the soil as a resource is poorly cared for under international conventions, in spite of its importance. The UN Convention on desertification (CCD), f. i., is based on regional development politics rather than sound scientific background, and it is disadvantaged by conceptual problems.The problem of soil conservation is one of the great emergencies in the XXI century. Recent estimates indicate losses of 20 ha/min/year soil, caused by different processes and mechanisms: wind and water erosion, deforestation and land degradation, salinisation, acidification, chemical contamination, etc. Rapid decline in quality and quantity of soil resource, and uncontrolled natural resources consumption, which affects about 33% of the Earth land surface, involving more than 1 billion people, stresses global agriculture with long term negative consequences. In the next twenty years, food security will emerge as a major global issue, forcing developed countries to provide more food, thus provoking imbalances in soil nutrient resources, soil pollution, land degradation, reduction of pedodiversity and biodiversity, increased occurrence of human health problems. New paradigms of soil science, therefore, are needed in order to encompass such dramatic global scenery. Soils are extremely important, f. i., in the global cycle of carbon. Nevertheless, consideration of soils was slow to emerge in the context of the UN Framework Convention of Climate Change (FCCC). Soil scientists could play an increasingly important role in understanding the global carbon cycle and to point out ways to reduce carbon dioxide levels in the atmosphere by storing carbon in ecosystems and producing bio-fuel. But the scientific community needs also to increase the visibility

---

* E-mail: bini@unive.it.

of soils in international environmental, social and political context. During the last decades, the focus of societies has changed from agricultural production and forestry towards environmental issues. Soil science is increasingly targeting environmental and cultural issues, such as sustainable land use, ecosystems restoration, remediation strategies for contaminated soils, protection of the food chain and of groundwater resources, protection of human health as well as protection of soil as a cultural and natural heritage. In next years, these issues and other specific aspects, such as soil science for archaeological dating, forensic soil science, and other applications to increasing social and economic demands, will gain importance to achieve the goal of soil and land conservation in a changing world.

Keywords: natural resources, soil, sustainable land use, global change, new paradigms.

Man knows better what he has above its head (the sky)
than what he has under its feet (the soil).
Leonardo da Vinci

1. INTRODUCTION

Talking of soil, people often refer to different concepts:

- a physical substrate for plants, in agriculture and forestry;
- a fastidious weathering cover of rocks and sediments, in geology;
- an areal for buildings and urban infrastructures, in architecture;
- a claim for catastrophic events, in engineering geology.

Instead, soil is a living natural body, not only for the billions of microorganisms that live within the soil, but also for the reactions (dissolution, precipitation, oxidation, reduction, weathering, hydrolisis, chelation) that occur at the interface rock-soil-biosphere, and are responsible for horizon differentiation. (fig. 1)

As declared more than thirty years ago in the European Chart of Soil (European Council, 1972), soil is a fundamental resource for humanity, and plays major functions, both biological (biomass producer, ecological filter, genic reserve, habitat for plants and animals, including humans) and abiological (physical substrate for infrastuctures, source of inert materials, cultural and historical sink).Therefore, its conservation and sustainable use is a major concern for decision makers and Public Authorities, as it was depicted by Ambrogio Lorenzetti (half of XIII century) in the paintings of *good* and *bad land government* in the Municipal Palace of Siena (Fig. 2), and it is suggested by the European Union (2002).

Yet, as stressed by Jenny (1941, 1989), soil development is the result of a series of natural and anthropic processes which take place under the control of the factors of soil formation. With few exceptions, natural soils have horizons. In most cases, the typical "soil" has its distinct O, A, and B horizons, which provide evidence that this "soil profile" formed on a relatively stable location, and evolved on, within and concurrently with its landscape. By examining and analyzing a soil profile and its position in the landscape, it is possible to acquire information about the succession of natural events that took place at a given site.

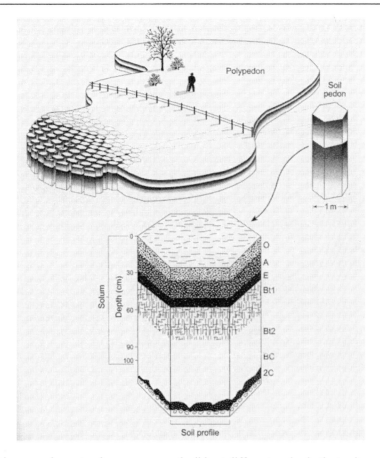

Figure 1. Soil is a complex natural resource reproducible at different scales in the territory: from polypedon (soil geography) to pedon (soil taxonomy), and from soil profile (soil morphology) to soil horizons (soil characters). (modified after Blum, 2006).

Figure 2. One of the first representation of the land is the picture by Ambrogio Lorenzetti in the Municipal Palace of Siena (Italy), devoted to *The effects of bad and good land government.*

It is something like a "phylogenesis" that drives soil evolution from soil profile to soil sequences and soil landscape (Fig. 3).

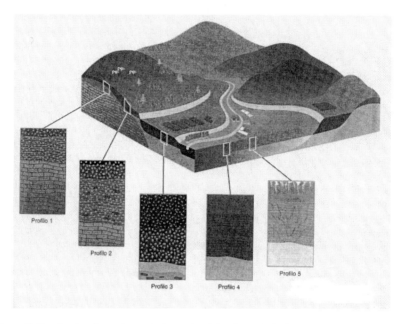

Figure 3. The soil distribution in a given territory (*soil landscape*) is the result of the soil evolution into phylogenetic sequences following the ClORPT equation (Jenny, 1947).

In the same sense, soils exemplifying pedogenetic processes responsible for current soil formation (e.g. humification, weathering, leaching, podsolization), or regulating delicate environmental equilibria (e.g. coastal dunes, wetland and badland areas, Fig. 4),

Figure 4. The typical landscape of badlands from clayey sediments (*calanchi*) in central Italy. Mass and water erosion determine intensive land degradation with significant soil losses. (Photo Bini).

or displaying a perfect harmony of different agricultural fields with the anthropogeographic landscape (Fig. 5) are of relevant interest, being "sites showing at best the natural processes" (Bini and Costantini, 2007). In this perspective, soils and soilscapes may be considered as worthy of conservation.

Figure 5. Human intervention determines significant modification of the natural landscape. New crops and new agricultural techniques contribute to the *anthropogeographyc* landscape, like this one in the vineyardsof the Chianti region (Tuscany, Italy). (Photo Bini).

Soil is one of the fundamental resources for agricultural food production, life and the environment, and therefore its functions and quality must be maintained in a sustainable condition. Once its quality or functions are compromised, remediation can be extremely difficult and expensive (Barberis et al., 2000). However, since the regeneration of soil through weathering of underlying rocks requires a long time, soil must be considered as a finite and not easily renewable resource.

Intensive competition exists between the three ecological soil and land functions, and the use of land for infrastructure and raw materials. Based on the shared experience, the ECC defined 8 main threats to land and soil (Montanarella, 2006), namely:

- Soil sealing through urbanisation and industrialisation;
- Soil contamination (local and diffuse);
- Erosion by water and wind;
- Compaction and other forms of physical degradation;
- Decline in soil organic matter;
- Loss of productivity and biodiversity;
- Salinisation and alcalinisation;
- Landslides and floods.

The impacts of these 8 threats on soil can be classified into irreversible damages and reversible ones, through a classification in order of urgency. Defining irreversibility, based on the time span of 100 years (about 4 human generations), sealing through urbanisation and industrialisation, intensive local and diffuse contamination, erosion by water and wind, compaction and landslides can be classified as irreversible, whereas decline in soil organic matter, loss of biodiversity, salinisation and alcalinisation can probably be handled as reversible damages.

The definition of sustainable use of soil resources as a spatial or temporal harmonisation of the different land uses in a given area, minimizing irreversible impacts, shows clearly that this is not a scientific but a political issue, which can be handled by top-down and bottom-up decisions (Blum, 2002).

The concept of multiple soil function and competition is crucial in understanding current soil-protection problems and their multiple impacts on the environment. Accordingly, a conceptual assessment framework has been developed applying the DPSIR approach adopted by the EEA to soil issues (Fig. 6) in the Directive on soil protection strategy (EUC, 2006). This approach requires the development of policy relevant indicators on soil issues which describe the interconnections between economic activities and society's behaviour affecting environmental quality.

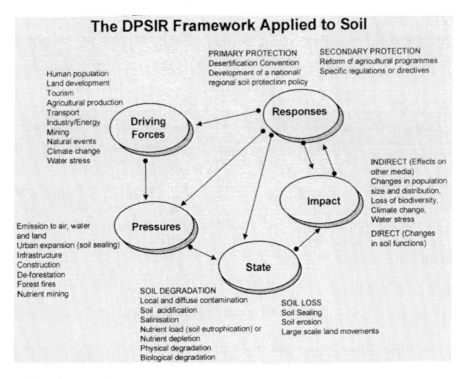

Figure 6. Flow-diagram of the DPSIR framework proposed by EU. (modified after Blum, 2006).

The identification of suitable indicators, representing specific environmental matrix, is based on the utilisation of patterns able to relate the pressures to the status of the matrix and the impacts to the possible responses.

The DPSIR model describes the issues Driving forces, Pressures, States, Impacts and Responses. It permits to represent all the elements and relations that characterise environmental processes, putting them in relation with environmental policies.

The components of the DPSIR framework can be defined as follows:

- Driving Forces. A driving force" is a necessity. Example of primary driving forces are the needs for house, food and water, while examples of secondary driving forces are the need for mobility, entertainment and culture. For an industrial sector, a driving force could be profit and to produce at low cost, while for agriculture a driving force could be food production.
- Pressures. Driving forces lead to human activities such as infrastructure construction, or transportation, and result in meeting a necessity. These human activities exert pressures on the environment, as a result of production or consumption processes, which can be divided into three main types: excessive use of resources, change in land use and emissions. In agriculture, a pressure could be organic matter decline, which indicates loss of fertility.
- States. As a result of pressures, the state of the environment, i.e the quality of the various environmental compartments in relation to their functions, is affected. The state of the soil is the combination of the physical, chemical and biological conditions, i.e. the soil quality.
- Impacts. The changes in the physical, chemical and biological state of the soil may have environmental and economic impacts. Changes in the state of the environment compartments may be referred to the functioning of the ecosystem, to the human health and to economic and social threats (e.g. decline in food production).
- Responses. A response by society or decision makers is the reaction of an undesired impact, and can affect any part of the chain between Driving Forces and Impacts. A response related to environmental pressures could be the development of national/local soil protection policy, or specific regulation concerning permissible heavy metal levels in soils.

Concerning soil, the problem of soil degradation is mainly driven by activities such as intensive agriculture or human population increase, which lead to pressures on the environment (e.g. emissions to air/water/land, urban development, or deforestation). As a consequence, these pressures directly affect the state of the environment, for example in terms of a degradation of the soil quality due to emissions of hazardous substances or topsoil loss due to erosion. Hence, information about these pressures on the environment are of great importance. Changes in the state may lead to impacts (changes in the population size and distribution, changes in crop yields), finally resulting in society's responses, such as the reform of the Common Agriculture Policy or the UN-CCD Convention. In turn, these responses will again affect each part of the DPSIR assessment framework

Applying the DPSIR model to soil, the state of soil (chemical) degradation could be for example a nutrient depleted soil; a driving force could be the insufficient market conditions for farmers (low prices); the pressure could be to supply nutrients replacing those lost with harvesting; the direct impact could be a considerable loss of fertility, leading to continuously lower harvests, and an indirect effect could be changes in population size and distribution, because people have to move to other, more friendly areas. The responses to this problem could be economic (e.g. changing market conditions, so that farmers receive enough income and replace nutrients), or technique (e.g. distribution of fertilisers to farmers) or legal (e.g. public incentives forcing farmers to select new crops), cross-linking cultural, social and economic drivers to technical and ecological drivers (Blum, 2006).

To face the approach to such a complex problem, four different issues are selected, which represent particular soil aspects, broadly correlated:

1) soil quality - it concerns the summation of soil intrinsic properties, that best characterise it as a natural matrix able to perform numerous and well-known functions. Soil quality definitions currently follow two concepts: the first is the "capacity of the soil to function", and the second is "fitness for use". Capacity to function refers to inherent soil properties derived from soil forming factors as defined by the CLORPT equation (Jenny, 1941); fitness for use is a dynamic concept and relates to soils as influenced by human use and management.

2) Physical degradation – It considers the degradation aspects of the soil matrix that risk to determine both a soil loss and a deterioration of part of its functions (loss of structure, compaction, decline in organic matter, etc.), as a consequence of processes that could be considered irreversible, at least in the human temporal scale. Human activities such as agriculture, industry, urban development and tourism give rise to soil degradation, the extent of which is determined, among other things, by the physical, chemical, and biological properties of the soil. The most severe causes of soil degradation in terms of irreversibility are erosion, desertification and soil pollution.

3) Diffuse contamination – it considers those qualitative soil aspects that could be progressively compromised by inappropriate soil use, in ways that do not respect the natural recovery times. The diffuse contamination affects the soil functions most in its buffering, filtering and transforming capacity. Currently the most important problems are soil acidification, mainly due to emissions from vehicles, power stations and other industrial processes. High concentration of heavy metals may occur due to high natural contents or to anthropic influences, causing threats to the food chain. Nutrient surplus is mainly due to overapplication of fertilizers, with high phosphorous and nitrogen contents leading to eutrophication of groundwater and waterways through soil erosion or surface run-off.

4) Local contamination – it considers one of the most serious concerns of last decades, the increase of strong soil contamination by human activities on well-defined areas. Local contamination is characteristic of regions where intensive industrial activities, inadequate waste disposal, mining, military activities or accidents pose a special stress on soil (Bini and Reffo, 2004; Bini et al, 2008). If the natural soil functions of buffering, filtering and transformation are overexploited, a variety of negative environmental impacts arise. Therefore, restoration interventions are needed, in order to restore soil functionality. Modern civilization is dependent on the managed exploitation of natural ecosystems: the challenge for future is to reconcile the demand of human development with the tolerance of nature.

2. THE GREAT EMERGENCIES OF XXI CENTURY

2.1. The Main Aspects of Soil Degradation

Several natural and anthropic causes, in different economic sectors, play an important part in contributing to soil degradation. Mineral natural enrichment, mine cultivation, forest fires, soil acidification/alcalinisation, fertiliser and pesticide application to agricultural soils, urban waste disposal, industrial activities, are the most common causes of soil degradation. Soil sealing and erosion are considered a major concern of irreversible soil losses, in relation to the time needed for soil to regenerate itself. Estimates related to the last decades indicate soil losses due to erosion of 20 ha/min/year, i.e. 200 square kilometres each year (up to 22% of the emerged land affected), with an economic cost of 15 Mld $/year.

Soil degradation due to local and diffuse contamination is another important threat to the agriculture over the long term even though the share of affected areas seems relatively small. Moreover, it can be reversed, if adequate measures are taken, such as clean-up and remediation plans.

Nevertheless, the problem of soil degradation could be aggravated by a combination of unfavourable natural conditions including the high proportion of steep land, heavy rainfall in autumn and winter when land cover is reduced to a thin topsoil layer. Loss of soil productivity in the eroded areas is a major problem, as is sedimentary deposition downstream, with erosion triggering sometimes irreversible degradation and desertification. For example, in Italy, the cost to society of sediment yield from agricultural land to off-farming areas is perceived high, particularly in terms of stream degradation and disturbance to wildlife habitat (Barberis et al., 2000).

Other threats of soil degradation have been recorded recently in EU (Montanarella, 2006):

- An estimated 115 million hectares or 12% of EU's total land area are subject to water erosion, and 42 million are affected by wind erosion.
- An estimated 45% of European soils have low organic matter content, principally in southern Europe.
- The number of potentially contaminated sites in EU is estimated at approximately 3.5 million hectares.
- Compaction: around 36% of European subsoils have high or very high susceptibility to compaction.
- Salinisation affects around 3.8 million ha in Europe.
- Landslides often occur more frequently in areas with highly erodible soils, clayey sub-soil, steep slopes, intense and abundant precipitation and land abandonment, such as the Alpine and the Mediterranean regions.
- Sealing: the area of the soil surface covered with an impermeable material, is around 9% of the total area in EU. During 1990-2000 the sealed area in EU increased by 6%.
- Biodiversity decline: soil biodiversity means not only the diversity of genes, species, ecosystems and functions, but also the metabolic capacity of the ecosystem and the

loss of pedodiversity. Soil biodiversity is affected by all the degradation processes listed above.

- Soil degradation during the last 40 years caused a decrease of about 30% in their water holding capacity, and a proportional shortening of the return time of catastrophic hydrological events.
- Soil degradation has also caused an impairment of several other eco-services: geomorphological fragility, land use, agriculture production, plant stress, food quality decline.

Various factors contribute to the "soil resource" decline and loss: urban development, erosion, pollution and agricultural production. The latter, which is intrinsically connected to the use of soil as a resource, has contributed in the last decades to its degradation and has reduced its productive capacity. On the other hand, agriculture could contribute to contrast natural degradation phenomena when conducted within the contest of a sustainable development. Important research dealing with the problem of soil degradation due to natural and anthropic causes has been carried out recently throughout Europe (Blum, 2002, 2006; Montanarella, 2006). For example, the European Soil Chart, emitted by the European Council in 1972, has recently found new attention with the establishment of a framework "Towards a soil protection strategy" in the Directive headed by the EU (22/09/2006). Another important document regarding soil resource management in terms of sustainability and sensitivity is the second edition of the report on environmental conditions, published in 1997 by the Italian Environmental Minister (Buscaroli et al., 2000). This document pointed out that soil, which is considered a natural resource, as are water, air, flora and fauna, can be subjected to negative impacts capable of causing its relatively rapid decline (Table 1).

Table 1. Summary of soil degradation causes.

Chemical degradation	Metal contamination
	Organic matter loss
	Salinisation
	Acidification
	Waste and sewage sludge spreading
Physical degradation	Compaction
	Hardsetting
	Superficial crust formation
Resource loss	Sealing by urbanisation
	Superficial erosion
	Landslide

2.2. Soil Vulnerability

The *vulnerability* concept has been widely applied to groundwater, and it is related to the protective effect operated by soil in the unsaturated zone (Civita and De Maio, 2000).

From the point of view of land vulnerability, the problem is more complex. There is an extensive literature on lands variously described as fragile, marginal, vulnerable, problem, and more- or less favoured. Early papers (e.g. Dent, 1990) focused on purely biophysical aspects of vulnerability including steep slopes, arid and semi-arid lands with highly variable rainfall patterns, areas that are poorly drained, too cold, or are of low inherent fertility. More recent contributions (Dumanski et al., 2002; Jones, 2002; Dumanski, 2006) have recognized that a major development preoccupation is not with the biophysical-chemical characteristics of land as such, but the susceptibility of different types of land to biophysical-chemical degradation on a temporary or long term basis, as a consequence of human activity (Nachtergaele, 2000).

Three main causes of land vulnerability to natural disasters are recorded. The first and most impressive is the close link between demographic growth rate and poverty, resulting in an accelerated urbanisation process particularly in developing countries: in 2015 the world population will be urbanized for 54% against 45% in 1995, according to the UN's estimates (UN, 1996). The more worrying aspect is that the majority of this urban expansion is previewed to go on completely without being regulated or planned. According to the Int. Red Cross (1995), 40 out of the 50 cities with greatest expansion rates are located in zones with a high seismic risk (Nachtergaele, 2000). Similarly, very seldom inundation, erosion or landslide risks are taken into account by planners when urban expansion takes place.

The second cause of enhanced vulnerability to natural disasters is directly linked to the unadapted use of the land resources with consequent disastrous off-site effects. The increasing expansion of infrastructures and constructions covering soils with cement (i.e. soil sealing), has an exponential effect on the inundation risk, since it results in less soil water storage, leading to enhanced run-off. This underlines the importance of investigating on soil knowledge to find any possible vulnerability factor that will permit to evaluate the soil deterioration risk concerning a particular impact and/or a particular soil function.

The third cause is the poor or inexistent prevention policy undertaken by most governments. Indeed the World Bank estimates that it is possible to reduce the cost of natural disasters by 280 thousand million dollars a year simply by investing one seventh of that amount in preventive measures, and enforcing already existing regulation for urban expansion.

There is little EU and National legislation that directly addresses the problems of soil degradation and loss, unlike air and water (EEA, 1999). Several directives have been established (e.g. nitrates, sewage sludge, heavy metals) which protect soil to some degree, but they have been set primarily to protect other environmental compartments (water, air and the food chain primarily) than soil. Moreover, information already existing on soil in EU is patchy, dissimilar (ranging from mining and waste disposal to land use planning, agriculture and bio-diversity), and not always easily available, since it is often held by a variety of organizations/authorities, and the distribution of information may be organised quite differently. This makes soil data collection and evaluation of soil concerns especially difficult at EU level.

2.3 Soil Erosion

Erosion is a complex phenomenon resulting from numerous interacting factors: soil, topography, land cover and climate (Wischmeier and Smith, 1978), and is one of the most significant forms of land degradation (soil truncation, loss of fertility, slope instability), greatly influenced by land use and management.

In Europe, the natural land cover has been extensively modified by agriculture activities. The effects of changed land use on soil loss rates and sediment yields are important for watershed and forest management as well as for the control of diffused sources of water pollution (Erskine et al., 2002). Soil losses and sediment yields from cropped areas have been estimated several times greater than native forest (Neil and Fogarty, 1991); order of magnitude increases in soil loss rates between retired land, pasture, wheat and bare fallow have been recorded (Erskine et al., 2002; Rey, 2003).

The Mediterranean region is particularly prone to erosion, since it is subject to long dry periods followed by heavy and erosive rainfall, falling on steep slopes with fragile soils, resulting in considerable soil losses (Van der Knijff et al., 1999).

In parts of the Mediterranean region, and particularly in Italy, where the climate characteristics, and the geological features, directly influence land properties (Federici and Rodolfi, 1994; Sorriso Valvo et al., 1995), erosion has reached a stage of irreversibility. With a very slow rate of soil formation, any soil loss of more than 1 t/ha/yr can be considered as irreversible within a time span of 50 – 100 years.

Figure 7. Sheet erosion in a vineyard in Tuscany (Italy). This is the first step of hillslope water erosion: water flows at surface with laminar flow, removing soil fine particles and driving them downwards in suspension. (Photo Bazzoffi).

Once the erosion process has started with laminar or sheet erosion (Fig. 7), it rapidly goes on to rill erosion (Fig. 8), and, in absence of any form of land protection or, at least, mitigation, progresses to a catastrophic gully erosion (Fig. 9). Losses of 20 to 40 t/ha in individual storms, that may happen once every two or three years, are measured regularly in Europe, with losses of more than 100 t/ha in extreme events (Morgan, 1992). In the European context with cool temperate climate and ondulating landscape, soil type, land use and climate pattern in relation to topography should really be considered as key controlling factors of hydrological and soil erosion processes (Bazzoffi, 2002). Further developments in field measurements under controlled conditions will yield valuable information for the understanding of such processes, and will help policy-makers develop appropriate programs and reliable methods of soil protection

Figure 8. Rill erosion in a vineyard in Tuscany (Italy). Rain water, after initial splash, is conveyed downwards in small rills. Climate, parent material and agrotechnical practices contribute greatly to soil erosion. (Photo Bazzoffi).

Figure 9. Catastrophic erosion (gully erosion) in a vineyard in Tuscany (Italy). Heavy precipitation digs profound gullies (up to 2 meters) on silty-clayey materials, determining important soil losses. (Photo Bazzoffi).

2.4. Land and Soil Contamination

Soil and environmental contamination is an issue whose importance, unlike air and water, has bee perceived only since recent years, and constitutes one of the great emergencies of current century, particularly considering that society deserves more and more attention to the effects of contamination on human health (Abrahams, 2002), and increasing consciousness of disease hazards as a consequence of exposition to chemicals like asbestos, uranium, dioxin, benzene. In most industrialised countries, the problem of characterising contaminated sites, and of their restoration is a major concern in the sector of soil and environmental protection.

A site is defined *contaminated* when it presents "chemical, physical or biological alterations of soil and subsoil, or of surface or subterranean waters, in such away that a danger to public health or to the natural or even anthropic environment could be envisaged". Areas previously occupied by highly contaminant industries present high contamination levels with both organic and inorganic substances. Many of the organic substances that are responsible for ecosystem contamination, are often very noxious to living organisms health, including humans. Similarly, many metals, if present at determined concentration levels, are toxic and may be absorbed by various organisms through the food chain. Risk assessment for human health, therefore, is assuming increasing importance in solution of problems connected with soil remediation. In areas subjected to high contamination, direct and indirect health risks make urgent the intervention, and acceptable the cost to sustain for remediation. In areas with moderate to low contamination levels, low health risk, or where cost is estimated to be excessive with respect to future benefits, remediation techniques may mitigate environmental risk and favour the recovery of previously degraded area to productive land uses.

Metal accumulation in the environment may occur at some locations, owing to different sources. Possible "natural' accumulation may be related to heavy metal-bearing rocks (e.g. Ni and Cr in serpentine) or to mineralized areas (e.g. Pb and Zn from mixed sulfide mines), while anthropogenic accumulation is related to industrial activities (e.g. Cd in metallurgy, Cr in varnish and leather factories), agriculture and urban sewage sludge (e.g. Zn and Cu from fertilizers; Cd, Pb, Cr from sludge).

Especially the last item is paying great attention at present, since increasing quantities of urban sludge are produced and extensively introduced in the environment (Fig. 10). Moreover, atmospheric input from industrial emissions, heat power plants, heavy traffic and acid rains may account for increasing heavy metal concentration in soils (Norra et al., 2006). Therefore, identification of the sources responsible for soil contamination is an important issue, since high loads of heavy metals applied to soils, or stored in soils, may determine soil quality degradation, surface and groundwater pollution, accumulation in plants, phytotoxicity and successive transfer to the food chain.

All trace elements are toxic if their intake through ingestion or inhalation is excessive. In particular Ag, As, Be, Cd, Ce, Ge, Hg, Pb, Tl are good examples of potentially harmful elements (PHEs) that have no proven essential functions, and are known to have adverse physiological effects at relatively low concentrations (Abrahams, 2002).

Figure 10. Solid urban waste is frequently buried at depth in soils, in such a way that soil characters may change strongly. Contamination is likely to occur at these sites. (Photo Dazzi).

While naturally enriched metals contribute to environmental contamination diffused on large areas, with generally moderate pollution level, contamination of anthropic origin is generally important and preoccupying, since it presents high pollution levels, although limited at small areas. The productive activities most involved in environmental contamination are:

- Industry (chemicals, electronics, metallurgy, varnish, tannery etc.);
- Emissions and spread off (traffic, heating, power plants etc.);
- Composting from urban waste;
- Landfilling;
- Agriculture (fertilisers, pesticides, sewage sludge).

At global level, 12% of the total surface occupied by degraded soils is affected by chemical soil degradation. The results of a research conducted recently in five European countries (Table 2) recorded approx 9000 contaminated industrial sites (0.2% of the land), that require immediate intervention to safeguard public health, or that in any case have severe land use limitations imposed by contamination. Other 14.000 sites require further investigations to ascertain the actual health risk. In Switzerland, over 100.000ha agricultural land have Zn concentration encompassing the National guidelines (Frossard, 2006), and more than 300.000ha present elevated levels of Cd and Pb, and 50.000ha of Cu.

Table 2. Number of contaminated sites in selected European countries (Adriano et al., 1995).

Country	Contaminated sites (total)	Contaminated sites in critical conditions
Germany	32500	10000
Belgium	8300	2000
Italy	5600	2600
Netherland	5000	4000
Denmark	3600	3600

In the last ten years, chemical soil degradation has progressively increased. The more recent data (EEA, 2006) indicate that:

- In the world, soil pollution involves totally 2 billions ha;
- In Europe over 500.000 contaminated sites have been identified; more of 20.000 of them with critical conditions;
- In Italy, contaminated sites are more than 10.000, and 2600 are in critical conditions. The Environmental Ministry has published a provisional priority list of sites where remediation is urgent; the first site to clean up is the petrol-chemical industry at Porto Marghera (Venice, Fig. 11), where a Master Plan aimed at protecting and restoring the unique scenery of the lagoon of Venice has been recently approved (Zonta et al., 2007).

Soil chemical contamination constitutes not only a social and sanitary problem, but also an economic one, since it implies relevant costs in terms of productivity decrease, food product quality and monetary value decline.

Moreover, the cost for remediating contaminated land, in particular with heavy metals, is very relevant, and only few industrialised countries (USA, Great Britain, the Netherland, Germany, Australia) have started remediation acts, while developing countries have serious difficulties.

Estimates related to the last decades indicate soil losses of more than 3 ha/min/year due to contamination, with a net economic damage of more than 3 billions $/year (Pierzynski, 2003).

Concerning particularly heavy metals, it is reported that approx 1 billion persons are affected by diseases due to lead contamination, approx 500.000 to Cd, and more than 100.000 to As, not considering the Asian regions (Bangladesh, Pakistan) where water pollution by As is dramatic, and risk to population is very high.

Figure 11. General view of the industrial area at Porto Marghera (Venice, Italy), a heavily contaminated site, where a Master Plan aimed at protecting and restoring the unique scenery of the lagoon of Venice has been recently approved. (from Bini and Reffo, 2004).

2.5. Decline in Organic Matter, Loss of Productivity

Soil organic matter content is an aspect of soil degradation, strictly correlated with acidification. Both natural factors (climate, parent material, vegetation) and human-induced factors (land use, atmospheric deposition, acid rains) contribute to soil organic matter content. For instance, large quantities of nitrogen are introduced into the atmosphere by industrial activities (over 30kg /ha/y in central Europe). High nitrogen content in the organic layer of forest soils is observed at sites receiving high atmospheric deposition loads, in comparison to areas receiving low deposition loads (Van Mechelen et al., 1997; Bini and Bresolin, 1998; Bini et al., 2006). The major function of SOM is carbon sequestration; however, other soil functions are strictly related to SOM (e.g. filtering, water retention, structure stability, etc). Therefore, decline in organic matter content has become a major concern.

The annual rate of loss of organic matter can vary greatly, depending on the above factors. A general trend in Europe is a decrease from north-west to south-east, i.e. from areas with lower mean annual temperature and higher effective moisture, to areas with higher mean annual temperature and less moisture. Results of a survey carried out in Europe indicate that 74% of the land in southern Europe is covered by soils containing less than 3.5% organic matter in the surface horizon , as a result of intensive cultivation, whereas in other European countries (e.g. England) the percentage of land affected by organic matter decline under 3.5% has increased to 42% (EUC, 2002).

2.6. Soil Compaction

Compaction is a frequent threat to soil (almost one third of soils in Europe are highly susceptible to compaction), and is determined by traditional agricultural practices (continuous cultivation, deforestation, mechanisation) combined with climatic conditions and negative soil physical properties (texture, structure, clay mineralogy, bulk density, organic matter content, water retention capacity). Once compaction has occurred, an overall deterioration of soil structure may result, and this may have serious consequences on the soil behaviour and functions (Fig. 12):

- accelerated effective runoff;
- increased risk of soil erosion;
- increased soil phosphorus losses with excess water;
- accelerated pollution of surface water by agrochemicals and organic wastes;
- increased green house gas production and nitrogen losses;
- decreased capacity of buffering and filtering;
- decreased biophysical fertility;
- decreased agricultural yield;
- increased management costs.

Current research into the causes and effects of compaction in soils of Europe (Jones, 2002) has demonstrated that it may be difficult and expensive to alleviate damages, and that the recorded effects go far beyond agricultural concerns.

Figure 12. Waterlogging in an olive grove as a consequence of a silty soil compaction by agricultural machinery (Tuscany, Italy). Decline in soil porosity, increased bulk density, decreased permeability, soil structure losses, are the main effects of compaction. (Photo Bazzoffi).

2.7. Soil Acidification

Soil acidification by atmospheric deposition of anthropogenic origin has been recorded particularly in the central part of Europe since the '80s (Ulrich, 1995). A major concern with acid deposition on soils is that their pH will ultimately decrease to levels at which toxic concentration of metals are reached. A soil survey carried out in European countries (Van Mechelen et al., 1997; EEC, 2002) showed that approximately half of the observed soil types in central Europe are affected by primary acidity (i.e. developed on acidic parent materials), and therefore with low buffering capacity against atmospheric (acid) deposition (Fig. 13). Moreover, a common characteristic of these soils is a low (less than 20%) base saturation. Instead, pH and average base saturation values of carbonate-free soils and calcareous soils (*eutric cambisols*, *gleysols*, *luvisols* and *calcisols*) increase from north to south, and from west to east in Europe, following the climatic gradient of a precipitation surplus in north-western Europe, to an evapotranspiration deficit in eastern Europe and in Mediterranean area. In areas with moderate rainfall, a concentration of basic cations may occur at the surface of acid soils, as a result of nutrient cycling, primarily through root absorption of basic cations from the parent material, and return to the soil surface through litter fall. A pH decrease of 0.5 – 1.0 units was observed at some sites in eastern Europe.

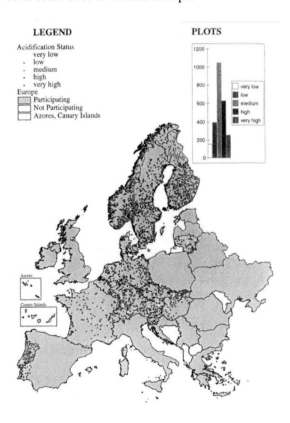

Figure 13. Soil acidification status of selected European countries. Updating and new data from other countries are currently available at FSCC, University of Gent, Belgium (modified after Van Mechelen et al., 1997).

Removal of basic cations from the exchangeable complex is a common process associated with acidification. A progressive acid input results in a lowering of the base saturation. Soils of southern Europe may be affected by acidity of a "secondary origin", due to leaching (Bini and Bresolin, 1998); however, low pH and low base saturation are quite uncommon in such soils. In areas with evapotranspiration deficit, it is unlike that basic cations, capable of neutralising atmospheric acid input, could be leached from the surface layers. Therefore, in these soils pH may range between 6.5 and 8.5, and base saturation ranges between 50% and 100%.

2.8. Salinisation/Alkalinisation

In recent years, there arose the consciousness, among soil scientists, that soil salinisation/alkalinisation is increasing more and more, in consequence of several processes, both of natural (salts in the parent material) and anthropic origin (marine water intrusion in coastal aquifers, strong groundwater uptake, irrigation with salt-enriched waters, heavy fertilisation, impeded drainage, salt spreading on roads to prevent ice formation during winter), frequently connected to global change.

At planetary level, salinisation affects 460M ha (Dazzi, 2002), of which 4M ha in EU, particularly in the east and south Europe, where evapotranspiration prevails over precipitation. Soils with *salic*, *gypsic* or *natric* horizons are present in 6, 5 and 7, respectively, of the 12 orders of the Soil Taxonomy, in different climatic regions. Soils with *saline* phases account for 144M ha, and soils with *sodic* phases for 134M ha, according to the FAO world soil map. Many biophysical and chemical soil qualities are affected by salinization: root penetration, structure stability, drainage impeded or slowed, root absorption, loss of fertility, crop yield decline, plant toxicity. The decline of ancient civilisation in Mesopotamia (4-5000 b.C.), as well as, more recently, that in Peru, India, Pakistan and even the indigenous Hohokam in the Salt River valley (Arizona), is attributed to secondary land salinisation in consequence of irrigation practices with not suitable water. Current FAO estimates indicate that every year 10M ha of irrigated land in the world are abandoned due to negative effects caused by salinisation/alkalinisation processes.

2.9. Soil Sealing by Urbanisation and Industrialisation

A new threat has emerged in the last decades to soils: land consumption due to uncontrolled urbanisation. The world population is expected to grow from current 6 billion to at least 8 billion in the year 2025, and the irrational development of urban space, for both civil buildings and industrial or commercial activities, and for the construction of infrastructure and new transport highways is rapidly consuming the soil resource, blocking the soil functions important to the ecology of the landscape, and altering, therefore, environmental equilibria. Of the traditional "anthropogeographic" landscape, as was designed by Sestini in the early 1947, today every sign is disappearing. The hinterland of the major cities has seen a consumption of agricultural land higher than 50% in the last 25 years, with high capability soils involved (Madrau, 2002). Rapid development of tourist residences and villages has occurred at sites with high aesthetic value, or with high capability for agriculture, provoking a loss of 37% soils of first capability class, and even 52% of III class. On the other hand, coastal towns have developed linearly, frequently in areas with high inundation risk. A sort of arable land erosion, therefore, occurs as an actual "hidden desertification", irrespective of the climatic conditions. A recent investigation on soil quality indicators (Nappi, 2000) estimated in over 10% the amount of "artificial areas" in north-eastern Italy, one of the most urbanised area in the country: land use change and consequent soil sealing contribute substantially to global change, with increased surface run off, water holding decline, evapotranspiration, emissions, etc.

3. APPLICATIONS

In order to serve as effective tools, methodological approaches to soil degradation research must have, first of all, an homogeneous information database in order to define basic environmental and territorial units, capable of supporting research and aimed at solving problems at the local level. Incomplete or fractional data can often render planning decisions of local government administration inadequate. Now and in the future this will become increasingly apparent as these same governments attempt to balance the often conflicting issues of quality of life with economic/productive instances and the need for bio-diversity with resource maintenance. A comprehensive program of land use management and urban planning at all levels of government, therefore, is needed, particularly at the municipal government level, where comprehensive land use data and reports detailing the causes of soil vulnerability are required.

The following list represents common understanding and experience in the study of soil degradation items:

1) analysis of land use changes in order to evaluate soil losses related to urbanisation and erosion;
2) natural fertility loss due to the impact of agricultural practices that alter physico-chemical properties;
3) aquifer vulnerability due to the infiltration of pollutants (e.g. agrochemicals, herbicides, livestock slurry, sewage sludge, etc.);
4) soil contamination as a result of the progressive concentration of harmful substances, such as heavy metals and other industrial by-products.
5) Soil acidification by atmospheric deposition, which affects especially surface layers of soils having low buffering capacity.

3.1. Land Use Changes

The current research methodology has permitted in several land surveys the evaluation of changes in land use that are primarily the result of human activities and interventions. The methodology compares land use conditions at two distinct periods of time, in order to register changes that have occurred. Suitable information is obtained from the analysis of aerial photographs taken during aerial surveys made at different times, which provide analogous and real images of the area at the time of the study. In addition, land use information is obtained from consulting existing cartography, such as maps at the scale 1:25,000 which illustrate land use using widely accepted symbols.

Information collected from different sources leads to a fixed number of land use classes, in order to permit a comparison between the two different situations. Once the data from each time period are homogenized, it is possible to use a cross referencing and reclassifying matrix to identify land use changes or continuities during the period under consideration. Each possible cross combination is evaluated and reclassified according to the following criteria: natural change dynamics, abandonment, extensivation, intensification, unchanged, anomalies, urban intensification and urban continuity.

Case Study 1 – Bisenzio River

The Bisenzio river is a noteworthy watershed located in Tuscany (Italy), flowing from the Apennines fringe, and has suffered important land use changes from main forest land to industrialised land in the '70-90s. The first step has been examination of aerophotographs taken in different periods, to observe the land use differences, and to evaluate their dynamics (Table 3).

Table 3. Land use dynamics in the upper Bisenzio river valley in the period 1954-2000.

Year 1954	•	Forest cover (mixed woods) prevails (70%) over agricultural land (30%) on slopes at elevation >500 m. Urbanisation is very little
	•	Grazing is diffused in grassland at the summit (Calvana Mount, 600m)
	•	Terraced agricultural land prevails on slopes at elevation <500m; olive groves are the main crops
'70-90	•	Frequent abandonment of terrace agriculture; ¼ of the land is abandoned and forest expands
	•	Numerous delineations of previous forest, olive groves or agricultural land turn to urban space
	•	Arable land devoted to herbaceous crops, in many delineations, is substituted by arboreal crops by 5%
present	•	Increased abandonment of agriculture practices, turn to urban and industrial land use; some delineations come back to forest
(2000)		
	•	Increased urban land use (by four times with respect to the '50s), often in areas not suitable for urbanisation (e.g. river banks)
	•	Forest plays still a prevailing role (up to 50%), following afforestation programs

The second step has been to draft the soil landscape map. Each mapping unit was described according to 5 characters:

Physiography: describes synthetically the site morphology, distinguishing various portions of the slope, and the slopes from the flats.

Substratum: describes the lithological types prevailing in the study area: sandstone, arenaceous flysch, calcareous flysch, clays, alluvial sediments.

Soil type and characters: describes and classifies the soil type for every class of physiography and substratum; *Inceptisols* (*lithic, dystric* and *typic eutrudepts* and *dystrudepts*) are the prevailing soil types, with subordinate *Entisols* and *Alfisols*.

Vegetation: describes the vegetation cover that characterizes each mapping unit.

Unit: is the synthetic code to identify on the map each soilscape delineation.

The third step has been to derive, from the soil landscape map, different thematic maps (e.g. land capability for agriculture, land suitability for forestry and grazing, soil erosion risk assessment) useful in land planning.

Case Study 2 – Ticino River Park (Buscaroli et al., 2000)

The Ticino river park is located in the provinces of Milan and Novara, in northern Italy. In order to identify land use changes that have occurred since the beginning of XX century, existent cartography at the scale 1:25,000 (edited by IGM, 1895) was used. From this analysis it is evident that land use changes are related to different geological and soil conditions: the Ticino river hydrological system developed in quaternary moraine cover to the north, and in alluvial sediments to the south, and has preserved its original form, characterised by wide river beds with steep banks, and a band of karst springs which create interesting micro-environments. The territory has undergone progressive and often intense urban expansion, particularly in the region of Milan. Urban development has been the primary and determining factor in the destruction of the karst springs. Furthermore, it has had a widespread and often negative impact on the regional hydrology due to the numerous wells that have been drilled for domestic water consumption. Overall 25% of the land is dedicated to urban areas. As a consequence, the number of karst springs has been reduced from 280 at the beginning of the century, to no more than 20 at present, with most of these occurring within illicit landfills.

3.2. Changes in the Physico-Chemical Characteristics

Substantial changes in demography and the nature of rural areas since the 1950's, due in part to advances in farming technology and agricultural management, have caused significant structural modifications to soil resource. The abandonment of marginal agricultural land on the one hand, and the concentration and intensification of agricultural activities in more restricted areas on the other hand, has been a determining factor in the alteration of soil characteristics. This is more evident in those situations in which agricultural practice entail to soils a great contribution of nutrients and fertility elements. Research into this problem has shown that, in addition to significant modifications in land use trends, physico-chemical changes, such as increases in alkalinization, lower levels of organic matter and reduction of permeability, have also occurred. As a result of preliminary experience, further investigations were carried out in selected farms of the Euganean hills (Padua, northern Italy), in order to create a predictive model capable of analysing the impact of the continued agricultural practices and other current forms of human intervention on the physico-chemical nature of soils. Furthermore, within the study area comparisons were made between the different land use dynamics in order to quantify the influence of territorial changes on variations in the average physico-chemical data. Soil samples were further analysed in order to compare physico-chemical parameters with land characteristics. The correlation matrix among the variables examined (land use, sand, silt, clay, pH, SOM, N-NO$_3$) shows (Table 4) that all data higher than an absolute value 0.2626 are significant at 95% (data by Pannetto, 2007).

Table 4. Correlation coefficients among original variables.

	Land use	sand%	silt%	clay%	pH	S.O.	N-NO3
sand%	0,44	1					
silt%	-0,29	-0,89	1				
clay%	-0,48	-0,79	0,42	1			
pH	-0,04	0,14	-0,21	0,01	1		
S.O.	0,31	0,54	-0,50	-0,39	-0,12	1	
N-NO3	0,29	-0,10	0,14	0,00	-0,13	0,26	1

Moreover:

• all the parameters with the exception of pH are related to land use;
• the three granulometric classes are correlated to land use with a very high coefficient;
• the amount of SOM is intimately linked to the three granulometric classes;
• pH and N-NO3 do not show particular correlation with the other parameters.

Furthermore, the average pH is included in a narrow range (6.7-6.9), although the absolute minimum is 5.4 (in forest soils), and the absolute maximum is 7.6 (in olive groves soils). However, the pH variability with respect to land use is very low. Within the recorded range, pH change risk is relatively remote, and should occur only as a consequence of strong environmental impacts (e.g. heavy fertilisation). The soil buffer ability (i.e. its aptitude to counteract possible pH variations) is related to the colloidal fraction of the soil itself.

Forest soils present the highest levels of SOM (mean 85,5 g/kg), while arable land the least (30.7 g/kg), as expected.

3.3. Environmental Vulnerability Due to Infiltration

The recent UE reports concerning environmental conditions (EEA, 1999, 2006) illustrate how effective planning and management of underground water resources is closely linked to soil conservation and the monitoring of human activities that impact these resources. These studies concluded that the protection of soil and water as natural resources is becoming increasingly important. On the basis of intrinsic land vulnerability it is possible to determine levels of vulnerability with particular reference to different contamination sources. Different approaches are followed in tackling the theme of the assessment of the land, namely:

-   safeguarding of the underground water against pollution (cfr EEC directive 91/676 known as Nitrates Directive);
-   regulation of the spreading of livestock effluents, configuring their use as agronomic practice and not as waste disposal.

The reasons for the existing different approaches may be attributed to a series of causes, among which the following may be mentioned:

1. The extreme variety of the geologic and pedologic situations;
2. The ratio between quantity and quality of data necessary for a correct definition of vulnerability maps, and the cost of both hydrologic and soil data;
3. The limited amount of existing experience of crossed processing of hydrologic and soil data which make an overall assessment of the intrinsic vulnerability of the aquifer and of the soil vulnerability,

Of particular significance are those methodologies which investigate the possible transfer of pollutants from land surface to the uppermost levels of the water-table and, as a consequence, the possible migration of these contaminants to deeper aquifers from where drinking water is taken, as it occurs in alluvial plains.

During the last ten years, urgent problems linked to water pollution and land resource degradation have led to the creation of many survey models for studying environmental impacts. These could be classified into two main typologies: numeric models and parametric models. The most used are parametric models, in which the land is divided in classes of vulnerability. The development of computer technology and particularly that of GIS has greatly simplified all phases of parametric studies while at the same time introducing the possibility to evaluate the impact of human interventions on the environment by means of predictive models.

The model mostly applied to these studies, whose acronym is SINTACS (Civita and DeMaio, 2000), derives from information and regulations used in EU and is based on a simple parametric rating system, in which:

* a group of environmental parameters that influence infiltration is defined;
* every category for each parameter is evaluated with an appropriately weighted vulnerability value; these values range from 1 to 5, corresponding, respectively, to minimum and maximum levels of infiltration; for every parameter the region is mapped and reclassified into analogous areas having equal infiltration properties and therefore equally weighted values of vulnerability;
* by cross referencing the different reclassified areas, a sum of weighted values is made relative to each area;
* resulting values are reclassified to distinguish each environmental vulnerability with respect to infiltration classes.

The data collected with specific studies of a pedologic, hydrologic and agronomic nature, are processed to provide a land classification with reference to soil vulnerability on one hand, and aquifer vulnerability on the other. By cross referencing these data we obtain the assessment of the soil suitability for the spreading of livestock effluents, or for groundwater protection.

The model, therefore, evaluates the possibility of water (or effluent) passing through the soil to the water-table below. The soil parameters that are analysed are:

- depth to groundwater
- effective infiltration
- unsaturated zone depuration characteristics
- surface texture
- aquifer hydrogeology
- hydraulic conductivity
- topography

The vulnerability index is obtained by summation of points attributed to each parameter, multiplied by the weighted values:

$I_{SINTACS} = \sum P_j * W_j$
where P_j = points attributed to parameter j and W_j = weighted value of parameter j.

The value assumed by ($I_{SINTACS}$) may range between 26 and 260 points, according to the environmental characters. For determined intervals of ($I_{SINTACS}$), a determined class of vulnerability is assigned, as a function of limiting factors (e.g. soil texture, depth to groundwater, etc.), as shown in Table 5.

Table 5. Vulnerability Index and classes of land vulnerability for groundwater and livestock effluent spreading (adapted from Buscaroli et al., 2000 and Baracco et al., 2000).

(ISINTACS)	Vulnerability class	Soil Suitability for groundwater protection	Soil Suitability for spreading effluents	Quantity of effluents kg/ha/y Nitrogen
<100	Low	Very suitable	Very suitable	340
100 - 140	Medium	Suitable	Suitable	340
141 - 186	High	Moderat. suitable	Moderat.suitable	250
187 – 210	Very High	Slightly suitable	Slightly suitable	170
>210	Extremely high	Not suitable	Not suitable	0

Among the advantages of the use of parametric methods such as SINTACS there is a greater reproducibility of the results and the possibility of comparing hydrologic situations that may be quite different from one another. Parametric models are, therefore, particularly suitable for the redaction of vulnerability maps. However, they have the disadvantage of requiring a considerable hydrologic data base, so they are applicable only after profound hydrologic investigation or in areas where complete data sets are already available.

The maps of the intrinsic vulnerability to contamination are an operative tool which allows assessment of the suitability of the subsoil to absorb a polluting substance. This form of mapping has been widely used in many parts of the world. From 1980 onwards, numerous examples of these maps (Fig. 14) have been developed.

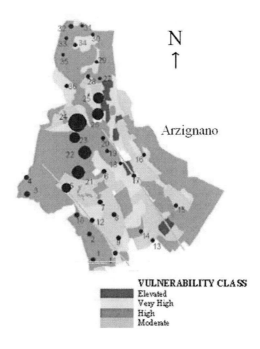

Figure 14. Soil vulnerability to groundwater in the tannery district of Arzignano (Vicenza, Italy), with indication of chromium concentration (dots are proportional to Cr concentration) in surface soil horizons. (from Bini et al., 2008).

Case Study - Irrigation Water (Buscaroli et al., 2000)

This case study explores the application of the vulnerability for the use of irrigation water due to infiltration in small towns in northern Italy, where the problem of the quality of irrigation water raises particular concern because of its use primarily in the agricultural sector. Throughout the study area several canals from which irrigation water is derived, have been sampled. Ten cases of water table contamination caused by high levels of ammonia have been registered. Concentrations of nitrate in the aquifer are greater than 15-20 mg/l. Contamination often derives from the spread of sewage sludge, use of soluble fertilizers in irrigation water, and irrigation practices using water containing high levels of organic matter. Sampling was undertaken on June. The differences recorded reflect the distinct geographical position, particularly at sites close to the canal which collects most of the waste originating from the high plain, where a high concentration of nutrients was detected. Pollution levels decrease downstream, since there are no factories or livestock farms in the area.

Case Study – Breeding Sewage Discharge (Buscaroli et al., 2000)

This case study explores the application of vulnerability to infiltration due to breeding sewage discharge in a small municipality in the region of Milan (northern Italy). Animal breeding load at that site is quite high (up to 100,000 pigs) and is linked to sewage discharge problems. With respect to local regulation (Regional Law 37/93), the municipality falls within the low soil vulnerability and high zootechnical load area. This regulation stipulated that livestock farms in this area must present an agricultural utilization plan, when the amount of live animal weight exceeds 3 tons/hectare. The soil vulnerability to infiltration was evaluated subdividing the territory into four classes of suitability (suitable, moderately suitable, slightly suitable, not suitable). Most of the study area is only slightly suitable, with the area situated close to riverside being not suitable. This situation poses a concern to the municipality for discharging sewage.

A sewage disposal plant could represent a possible solution. The treatment plant could deserve adjacent municipalities with a total load of around 100,000 pigs, which corresponds to 125,000 q sewage sludge. By-products such as the processed solid waste could be used as fertilizer to the agricultural sector, and biogases could be reused by the plant to generate its own electricity. Livestock activities and sewage discharge would become completely separated from territorial issues.

Case Study: *Spreading of Livestock Effluents* (Baracco et al., 2000)

The zone chosen for experimental spreading of livestock effluents is located in the province of Venice, where a large part of the territory belongs to the basin that drains into the lagoon of Venice, an area with considerable naturalistic and environmental value, but with a precarious ecological equilibrium. The overall territory is a prevalently farming area, with the following types of soils: Alfisols, Inceptisols, Mollisols, Entisols, Histosols.

Six limiting factors for the spreading of livestock effluents in relation to the characteristics of the soils were detected (namely: surface cracking, peat depth, skeleton, texture, depth to permeable layer, internal drainage). With this system, a greater grading of the classes of soil suitability for the spreading of livestock effluents was found, with a decrease of unsuitable soil units and an increase of the units belonging to the intermediate suitability classes. As regards the hydrologic aspect, the application of the SINTACS method to this area envisages the determination and assessment, on the whole territory, of the following seven hydrologic parameters: depth to groundwater, effective infiltration action, unsaturated attenuation capacity, hydraulic conductivity, soil overburden attenuation capacity, hydrologic characteristics of the aquifer, topographic slope. For each parameter a rate was assigned (variable from 1 to 10) which increased with the vulnerability. The rates for the seven selected parameters are then multiplied by a weight related to the environmental conditions of the area. All the processing was based on precise data banks, and in particular: archive of geognostic tests, subterranean waters monitoring network, archive of piezometric measurements.

In the area examined, the vulnerability is between medium and exceedingly high. The most vulnerable areas correspond to the areas with greater hydraulic conductivity, coinciding with old river beds and with sandy aquifers with a sub-surfacing table drained artificially by water scooping machines.

3.4. Metal Distribution and Concentration in Soils

The presence of heavy metals in soils, as a consequence of human activity, occurs when air loaded with atmospheric dust resettles on the ground, or when they are diffused by water (the superficial flowing of polluted water). In other cases it may result from industrial/chemical products discharge on soil, or by their presence in products such as fertilizers compost or soil conditioners, or in livestock sewage and depuration muds.

Soils that are contaminated by heavy metals present generally a low possibility of reducing the concentration of contaminants, primarily because of their negligible migratory capacity through the soil profile, and secondarily due to their low absorption rate by plants. Experiments carried out till now (see f.i. Bini et al., 2003) have targeted the diffusion and the concentration of some heavy metals and micronutrients (As, Cd, Co, Cr, Cu, Hg, Mn, Ni, Pb, Ti and Zn) present in soils as a function of the different kinds of added biomass (e.g. zootechnical sewage, depuration muds, compost) in order to create a database that correlates different geological and pedological typology with microntrients and heavy metals concentrations.

Special attention has been given to those elements with levels higher than the limits imposed by the national and local regulations and to standardize the analytical procedures for their determination, including total and extractable fractions.

Case Study: Venice Lagoon Watershed (Bini, 2008).

Legislation on maximum admissible levels of heavy metals in the environment in the EU is rather confusing. Indeed, a general regulatory guideline on the maximum trace element concentration in soils has not yet been established, the current references being related to the total metal content in waste and sewage sludge to be spread on soil (Adriano et al., 1995). Moreover, there is little agreement among the members in their implementation of the EC Directive of 1986. Several attempts have been made to adopt background values in order to obtain more viable reference values for regulatory decision. Although there are some similarities, standard criteria on the background level of metals in soils, however, are not yet established in many EU countries.

A profound insight of the above mentioned items, therefore, is needed, in order to achieve the following finalisations:

a) knowledge on the distribution and circulation of trace elements in the different geochemical domains (lithosphere, pedosphere, hydrosphere, biosphere, atmosphere) contributes to better understand the natural processes responsible for the soil genesis and evolution, the relations with landscape and vegetation, and the ecosystems equilibria;

b) human activities, technological processes and modern industrial products pose major environmental pollution concerns. Water, soil, vegetation may be appreciably affected by toxic or critical substances emitted in the atmosphere, or introduced in surface and groundwater, by these activities. Successively, after a variable time interval, they may be deposited at the earth surface, posing environmental hazard. Potentially harmful elements (As, Cd, Cr, Cu, Hg, Ni, Pb, Zn, etc.) may have toxic effects on living organisms, including humans;

c) Research on living organisms occurring at sites naturally and/or anthropically
    contaminated may allow identification of individuals (plants and animals), that are
    indicators of degraded environmental systems, or that may be utilized for restoration
    of contaminated sites, e.g. with phytoremediation techniques.

Based on the above mentioned assumptions, as a part of the intervention programs
against pollution of the lagoon of Venice, a survey of agricultural, industrial and undisturbed
(forest, grassland and wetland) areas, including the lagoon, was carried out in the Venice
drainage basin and the conterminous territory.

The Lagoon of Venice is a shallow transitional environment located in a densely
populated industrial/agricultural area (population 1,500,000), in the northern part of the
Adriatic sea, from which it is separated by some flat and narrow islands (Fig. 15).

Figure 15. Aerial photograph of Venice and its lagoon. (from Bini and Reffo, 2004).

Three inlets allow water exchange between the lagoon and the Adriatic sea. The whole area of the lagoon and the conterminous land has historically undergone severe anthropogenic pressure, with direct inputs of pollutants from both industrial and urban discharges. Other sources of contamination are river inputs, which collect industrial, domestic and agricultural polluting substances coming from the drainage basin of the lagoon.

The objectives of this work were:

- To evaluate the background level of heavy metals in soils of the Venice drainage basin and conterminous areas;
- To estimate the environmental impact of agricultural and industrial activity;
- To ascertain metal mobility and possible contamination of some sites, and the related environmental hazard, with special reference to the pollution of the Venice lagoon, which is a unique and delicate ecosystem.

Pollution sources could not be detected with certainty at present. Metal concentration decreases with increasing distance from the lagoon, and there are differences between undisturbed, agricultural and urban soils. Surface horizons are enriched with Pb, Cu, Cd, Hg, and this enrichment coincides with an enrichment in soil organic matter, suggesting Pb and Cu to enter, at least in part, the biogeochemical cycle; conversely, Cd and Hg should be introduced in the environment by human activities. This is in agreement with a general scheme proposed for heavy metals in undisturbed soils: the anthropogenic elements at surface are more likely to be bound to organic matter, while the geogenic elements would mainly be associated with silicates and Fe oxy-hydroxides, with a sensitivity to redox conditions.

Comparison of the correlations enables inferences concerning the interactions of trace elements with soil properties. This allows a first explanation of the distribution of heavy metal contents in soils. Metals are essentially associated with the soil organo-mineral fraction in the surface horizon, and have positive correlation with soil properties like clay and CEC, and negative correlation with sand, suggesting a lithogenic origin (i. e. they are pedogeo-chemical fingerprints).

Principal component analysis supports the differentiation between lithogenic and anthropogenic origin of trace elements, and the association between metals and fine-size grains, and that of Pb with organic matter, is evident in the principal components of PCA of both agricultural and forest soils. The positive loadings for most metals on PC1 indicate that this component is associated with a lithogenic origin, and the negative loadings of bioavailable metals at agricultural sites may be considered as an index of overall contamination of anthropic origin.

The lagoon acts as a sink where heavy metals contributed by agricultural land or originated by waste disposal or by industrial emissions and atmospheric deposition, are conveyed. The ultimate fate of heavy metals, therefore, is to contribute substantially to the lagoon and sediment pollution, and pose environmental hazard for living organisms, including humans.

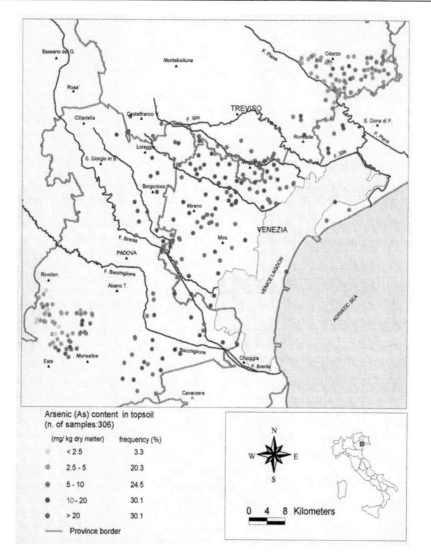

Figure 16. Map of arsenic concentration at different sites in the Venice lagoon watershed. (source: Environmental Protection Agency for Veneto Region, Italy).

The ecological risk as expressed by the hazard quotient (HQ = metal in organisms/metal in soils or sediments) posed by single elements is little for As (HQ <2 at Porto Marghera, <1.5 in central lagoon; <1 on the mainland, on the basis of 10 points), although at some places its concentration is higher than the regulatory threshold (Fig. 16), and very high for Hg and Cd (HQ <2.5 in central and outer border, <2 in the inner margin).

The cumulative toxic risk for As+Cd+Hg (+Zn) ranges between 6 and 9 toxic units at some hot spots in the central and southern lagoon (between Porto Marghera and Chioggia harbors), indicating a relevant bioaccumulation of trace elements in the lagoon ecosystem, and ranks <3 at other sites on the mainland.

Contamination is a major concern to the lagoon, and is probably a residual effect of severe historical inputs since the '50s. Instead, on the mainland slight diffuse contamination occur, and is likely due to anthropic as well as lithogenic origin.

Case Study: Chromium From Tannery District (Bini et al., 2008)

A research program aimed at assessing the actual chromium accumulation in soils and plants of a tannery industrial district in NE Italy was carried out in recent years, with the objective of highlighting possible contamination of soils and plants, and to ascertain the potential risk to human health.

Large differences in Cr concentration were observed in the area investigated, with a very scattered distribution. Mean Cr concentration in soils is 210 mg/kg (range 50-10.000 mg/kg). Most of the investigated sites present surface Cr concentrations higher than subsurface, suggesting local sources of Cr to be responsible for soil contamination (Fig. 17). Mean concentration of available (EDTA-extracted) Cr in soils is low (12.77 mg/kg in the topsoil, and 7.94 mg/kg in the subsoil), suggesting that most Cr is in a nearly immobile form. An explanation could be a multi-source Cr pollution at different places in relation to the historical tannery activity. Significant accumulation of the metal was recorded at sites located downstream, both in topsoil and in subsoil, suggesting heavy contamination from Cr waste disposal to have occurred in the last 50 years. Conversely, Cr concentrations at the sites located upstream correspond to the geochemical background.

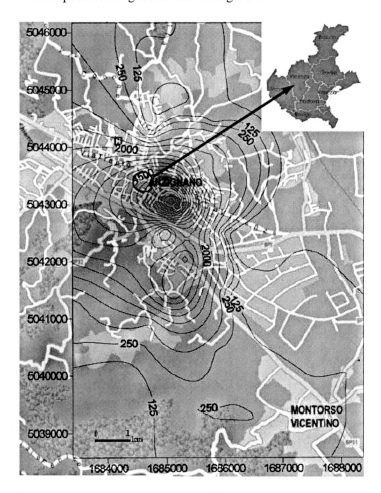

Figure 17. – Isolines of Cr concentration in the tannery district of Arzignano (VIcenza). (from Bini et al., 2008).

These observations are consistent with a Cr distribution pattern in the waste land, finding its origin in:

- a major pollution source of Cr directly derived from the recent industrial activity, or trapped in the fine fraction after leaching related to the former site activity, and
- a secondary pollution source linked to the environment evolution, much more diffuse and associated to Cr-bearing phases of the soil matrix.

A comparison between the soil Cr concentration and the soil vulnerability map shows that the most contaminated soils correspond to sites with high vulnerability to groundwater (see Fig. 14).

At those sites where Cr concentration is highest, chromium availability to plants and its ability to migrate into groundwater is the least, as demonstrated by the minor uptake by plants growing on the most contaminated soils. Indeed, chromium concentrations in selected plants (*Taraxacum officinale, Plantago lanceolata*) present wide ranges in both the species examined (0-222 mg/kg, mean 26 mg/kg), with little translocation from roots to leaves. Actually, the mean biological absorption coefficient (BAC) for the two species is 7.8, suggesting a root barrier effect to occur. Unlike what happens for essential elements, little amounts of chromium, as well as other heavy metals, are transferred from soil to root cells and then to aerial parts, thus preserving plants from toxic effects. Therefore, in the investigated area there is little toxicity hazard deriving from vegetables and other agricultural products, and reduced toxic risk to humans who utilize wild *Taraxacum* as a foodstuff or a medicinal plant.

Although potential toxic risk to humans is limited, current land contamination suggests precaution and control acts, to avoid major environmental injuries. Potential risk for human health by direct contact with soil, ingestion, and inhalation, however, should be taken in consideration by local Authorities. Moreover, considering the high soil vulnerability for groundwater, tannery by-product utilization in agriculture (e.g. hydrolyzed leather), and Cr-bearing sludge discharge in the territory should be strictly controlled, in order to avoid more land and water contamination.

Case Study Incinerator System (Buscaroli et al., 2000)

In collaboration with the Regional Environmental Protection Agency, heavy metal concentrations in the soil-plant system were evaluated as a result of the resettling of atmospheric emissions from an incinerator plant. The study began by generating a theoretical resettling map of pollutants scatter obtained using a mathematical model ISCLT undertaken by the Physics Dept of the University of Bologna (Italy). The RPA provided a passive monitoring, locating *Lolium multiflorum* cultivated plots, regularly mowed on a monthly basis from January to May, 1998. In addition, adequate soil samples at fixed depths were made next to the *Lolium multiflorum* plots and at different fixed distances from the incinerator system.

Concerning heavy metals, analytical results showed the existence of a good correlation with the provisional relapsing model; in fact, site 2, which is inside the area considered at highest risk, revealed the highest concentrations for all the elements considered and for all soil depth conditions. The reliability of the results was confirmed by the trend correspondence of calculated averages between metal concentrations into the different test locations grouped for the different positions with respect to the incinerator.

There is in fact (mainly for Cd, Pb, Cu and Ni) a negative gradient starting from the south-east towards the north-west, passing crosswise from the southwest where the average concentrations are intermediate levels relative to the other direction's average calculated values.

Even if at present the actual metal relapsing levels are not dangerous in this area, the situation at plots 1 and 2, in particular with respect to Ni, Cd requires a permanent phyto-pedological and meteorological monitoring.

3.5. Soil Acidification in Europe.

The main object of this European project is the assessment of the soil (chemical) conditions in relation to atmospheric deposition. The survey area has been divided in four regions based on the spatial distribution of the measured acid deposition load in Europe in the '90s, and was carried out on approx 5000 plots from 23 countries. Approximately half of the observed soil types (WRB, 1998) are Podzols and Cambisols, Leptosols and Regosols being also common. Soils with mor humus are dominant in northern Europe, while those with mull humus are frequent in southern Europe. Results show a correlation between soil chemical properties and atmospheric deposition. Atmospheric deposition has resulted in acid topsoil conditions (i.e. base saturation 20% or less, or pH values <3.5 in 42% of the plots), a higher nitrogen content in the organic layer of areas receiving a high deposition load, and high levels of heavy metals in strongly industrialised areas, particularly in cool and humid regions of central Europe (Van Mechelen et al., 1997).

In soils having low buffering capacity (i.e. soils with *dystric* properties) the pH will ultimately decrease, thus enhancing Al release to toxic concentrations, and also heavy metal availability occur. The coincidence of low pH, high acid load and low buffering capacity suggests that acid deposition is the cause for the high soil acidity at these sites, irrespective of the historical land utilisation types. Moreover, soils highly sensitive to acidification are usually derived from coarse textured and/or acidic parent materials (sands, sandstones, acidic crystalline rocks, while soils derived from clays, limestone, volcanic or metamorphic basic rocks are able to buffer a high input of acids.

The spatial distribution of soil acidity shows a NW-SE gradient, with countries having a continental or Mediterranean climate (i.e. Portugal, Italy, Slovenia, Hungary, Slovak Republic) that have more than 50% of the sites (mostly Fluvisols, Lixisols and calcareous soils) classified in the "low" or "very low" acidification status, and countries with more humid climate (Sweden, the Netherlands, Belgium) having 50% of sites (mostly Histosols, Podzols and Arenosols) in the classes "high" or "very high" (Van Mechelen et al., 1997).

4. NEW PARADIGMS OF SOIL SCIENCE

The above considerations and related applications not only do not open the future to essays of optimism, but, instead, illustrate the various threats concerning future land planning and soil management. In the last fifty years, the territory has undergone strong degradation, and soil fertility is continuously declining. Problems will aggravate with the rapidly increasing world population, unless adequate measures of control are taken. Soil conservation issues are already demanding more attention in most parts of the world. Therefore, it is time to find out new strategies for the protection and valorisation of this fundamental natural resource that is soil, as expressed in the recent EU framework Directive for Soil Protection (22/09/2006).

The challenge for XXI century is to harmonize the demand of human development with the need of soil protection.

Challenges directly related to soil protection are given below, together with some priority research areas:

Soil erosion:

-   Analysis of the forces which drive erosion and ecological and socio-economic effects;
-   Influence of land use and climate change on desertification and management;
-   Application of soil information systems for risk assessment on different scales;
-   Development of new conservation and remediation methods.

Soil compaction

-   Definition of soil conditions that are sensitive to compaction;
-   Analysis of compaction effects on soil quality;
-   Assessment of trends in agricultural machinery causing deep compaction;
-   Implementation of methods for predicting stress transmission and soil deformation;
-   Development of management tools to reduce soil compaction.

Soil contamination

-   Identification and quantification of contamination sources (geogenic and anthropogenic);
-   Improving methods for measurement of air-born contaminants;
-   Investigation of the fate and the long-term contaminants behaviour;
-   Identification of potentially dangerous substances in the soil-plant-sediment-water system;
-   Bioavailability of contaminants for humans, animals and plants;
-   Risk assessment for outputs from soil;
-   Improvement of techniques for remediation of contaminated soil.

Soil organic matter and biodiversity:

-   Definition of SOM in relation to soil functions and the potential carbon sequestration;
-   Development of standardized methods characterizing soil biodiversity;
-   Relationship between biodiversity and soil functioning;
-   Identification of combined management practices to optimize SOM and soil biodiversity.

Nutrient availability and fertilization:

-   Optimization of methods for adapting the N and P fertilization to the crop nutrient demand;
-   Combination of mineral and organic fertilizers in view of optimum SOM conditions;
-   Identification of driving forces on excess fertilization with N and P and quantification of their ecological and economic effects;
-   Definition of environmentally friendly levels of livestock densities;

Soil alkalinisation

-   Investigation of the factors which make a soil sensitive to salinization/sodification;
-   Influence of different water flow conditions on alkalinization;
-   Investigation of the (ir)reversibility of soil degradation processes caused by alkalinization;
-   Assessment of soil management and irrigation water quality under different climate;
-   Interrelationships between alkalinization and desertification and strategies for salt reclamation.

Soil sealing

-   Effect on the water flow in urban, suburban and rural areas;
-   Impacts on local, landscape and global level;
-   Establishment of methods to survey sealing with respect to area quality and quantity.

The goals of the EU soil thematic strategy (EEA, 2006), in particular, are:

-   Soil erosion prevention;
-   Soil compaction prevention;
-   Soil contamination prevention;
-   Increase SOM levels in soils;
-   Protect soil structure with adequate measures;
-   Reduce glass-house gas emissions.

This is the way a modern agriculture and land use planning are moving towards.

4.1. Towards a Sustainable Land Use

The increased environmental sensitivity of citizens in the last decades, joined to the changing economic perspectives, has addressed new tendencies in the agriculture policy in many countries, especially in EU. The Common Agricultural Policy, started in the '90s with actions aimed at ameliorating market conditions for farmers, has found new development points in the soil-agriculture compartment of the Directive *Agenda 2000*, with the use of appropriate technologies (e.g. minimum tillage), the introduction of environmental policies that enable the practice of sustainable agriculture (e.g. set-aside, directive nitrates, etc.), the recognition of farmers as safeguards of environmental and cultural heritage, the creation of green ways, agritourism, wine roads.

The foundation of sustainable agriculture, in particular, is considered a healthy land resource base and may serve as vehicle to achieve the goal of soil protection.

According to the American Society of Agronomy, *sustainable agriculture* (also called eco-compatible or integrated) is the agriculture which:

- Provides food and fibres for human needs;
- Is economically valid;
- Improves the natural resources of the agricultural farm and the whole environmental quality;
- Improves the life quality and wellbeing of farmers and the whole society.

This kind of agricultural management points to the objective of satisfying economic needs (food for consumers and income for farmers) without damaging the environment, which is civil society's heritage and a resource for future generations. In crops and livestock it utilizes as far as possible natural processes and renewable energy resources available in the farm, thus reducing the environmental impact coming from synthetic agrochemicals (pesticides, fertilizers, hormones, antibiotics). Intensive ploughing, monoculture seeding, sewage sludge disposal, residues from crops, oil mill and livestock effluents indiscriminate discharge on soil should be avoided while applying sustainable agriculture.

4.2. Land and Soil Protection Strategies

The global terrestrial environment consists of a mosaic of ecologically linked, natural and human land use ecosystems. The health of these ecosystems, defined as ecosystem integrity, depends on the ecosystem components and the synergy of processes that pass between them (Dumanski, 2006). For instance, Mutual interactions between soils and plants are already known, but the precise mechanisms are only partly understood. Knowledge about the precise soil requirements of many plants is lacking. Such information is important to ensure ecosystem stability especially where the land use is changed. A landscape approach, therefore, is required to study the role of soil as an integral component of natural and managed ecosystems (Dumanski, et al. 2002). Landscape studies promote understanding of soil ecological functions, and their interactions with socio-economic aspects, linking local benefits to global environmental resources.

Focus should be put, for instance, on afforestation (EU Directive 2080/98), reestablishment of wetlands (Ramsar Convention), transformation of conventional to ecological agriculture (EU Directive on nitrates) and other kinds of land use, in order to improve knowledge about such land modifications and to avoid adverse effects on the ecosystems, such as air and water pollution. Many environmental problems are so complex that they require cooperation with other scientists such as biologists, chemists and specialists in computer modelling (Borggaard, 2006). This development makes it necessary for soil scientists to change the focus of their research from themes concerned with increasing agricultural and forestry production towards environmental impact assessments and how to solve environmental problems such as soil contamination, erosion, carbon sequestration, nutrient leaching.

The tools to achieve significant results in land and soil protection are *knowledge* and *prevention*.

Knowledge

There is still lack of information on soil at global level. Better knowledge of soils, aimed at their use according to the proper suitability, may be achieved by chemical, physical and biological soil characterisation, soil data collection in digital banks, soil and product quality ratio. Moreover, information on crop yield and agricultural management enhances evaluation of soil suitability for different land uses.

Prevention

Land protection is operated starting from a correct soil management and land use planning that should take into consideration primarily land suitability for agricultural or forest land uses, but also those suitable for civil engineering (landfills, waste disposal, etc.), recreation, tourism and entertainment. Soil mapping based on georeferenced soil datasets, contributing to the knowledge of soils, therefore, is an important tool for correct land use and consequent human wellbeing (Mc Bratney et al., 2000; 2003). Soils vary gradually in geographic space and through time and form complex patterns in dependence of multiple interrelated environmental factors and anthropogenic and natural forcing functions. Soil research has been focused on the genesis of soils, their composition, factors that influence them, and their geographical distribution. It will be important that soil scientist play an active role in generating datasets and information but also in transfer and share knowledge with stakeholders, decision makers, land use planners, politicians and others. Soil becomes a partner in the earth, ecological and environmental sciences.

Soil erosion prevention

Soil erosion prevention may be achieved by conservative practices, namely:

- Avoiding terrain movements and geomorphic surface levelling;
- Controlling hillslope hydrology;
- Selecting friendly agricultural practices, compatible with soil protection:
    - Ploughing
    - Splitting
    - Hoeing
- Minimum tillage works:
    - Harrowing
    - Drigging
- No tillage
- Sod-seeding.

Soil Compaction Prevention

This threat affects soils with adverse chemical-physical characteristics (decrease of soil porosity, increase of bulk density, low structure stability, low SOM levels, coarse to medium texture), and is related to the use of agricultural heavy machinery. Prevention is achieved by the following interventions:

- Herb seeding;
- Utilize less powered tractors and operating machinery;
- Avoid overgrazing by defining livestock units in relation to soil type and humidity conditions;
- Replace intensive agriculture by re-introduction of traditional rotation practices and precision agronomy.
- Avoid deep ploughing by adopting simple agricultural works (minimum tillage, no tillage, sod-seeding).

Soil Fertility Amelioration

Nutrient availability and fertility levels may be achieved by:

- restoration of organic matter level with recycled biomass (urban residues, compost etc.);
- herb seeding;
- control of relationships soil-product quality;
- chemical, physical and biological soil characterisation;
- soil data banks.

4.3. Soil Reclamation and Remediation

Improved methods are needed to reclaim soils that are degraded because of erosion, salinization or because of contamination with organic and inorganic pollutants. Better on-site and off-site methods and strategies should be developed and optimized for cleaning soils polluted by heavy metals and organic xenobiotica. Remediation strategies for soils saturated with N and P because of excess fertilization for many years also need further attention. At the same time, increased efforts should be put in disseminating existing and new knowledge about these issues to prevent spreading of soil degradation and to reclaim degraded soils. Reclamation of salt-affected soils is a major concern too, considering the decreased general water availability, and the fact that good quality groundwater reserves are continuously decreasing at worldwide level, because of marine water intrusion, metal contamination, atmospheric dust input, etc.

Among the numerous soil reclamation (or remediation) techniques, phytoremediation is an emerging technology that holds great potential in cleaning up contaminants that: 1) are near the surface, 2) are relatively non-leachable, 3) pose little imminent risk to human health or the environment, and 4) cover large surface areas. Moreover, it is cost-effective in comparison to current technologies, and environmental friendly.

Phytoremediation is particularly suitable at sites where contamination is rather low and diffused over large areas, its depth is limited to the rhizosphere or the root zone, and when there are no temporal limits to the intervention. It has been calculated, indeed, that with present accumulator plants, at least 3-5 years are needed to have appreciable results in clean up a moderately contaminated soil (Bini, 2008).

One of the peculiar properties of phytoremediation is the economic aspect, since it presents costs much lower relative to the other technologies. Current estimates give costs of only 80 $/m^3 of soil cleaned with phytoextraction, and 250 $/m^3 with soil washing. According to Cunningham and Berti (1997), costs for phytodepuration would be 100000 $/ha, and excavation and landfilling would amount 500000 $/ha, while costs for traditional technologies of in-situ treatment would range between 500000 and 1000000 $/ha. However, actual costs for phytoremediation are still highly variable, and will be better estimated when this innovative technology will be really effective. Presently, indeed, the paucity of full scale application makes difficult to obtain reliable indication on remediation time and costs. A comparison of costs for different remediation technologies is reported in Table 6. It is noteworthy to point out that many ex-situ treatments require the combined use of different items reported in the table, and therefore the whole cost would be higher than the single item.

Table 6. Comparative costs of different remediation technologies per soil unit.

In-situ treatment	Costs (U.S. $/m^3)
Soil flushing	50-80
Bioremediation	50-100
Phytoremediation	10-35
Ex-situ treatment	
Exavation and transport to landfill	30-50
Disposal in sanitary landfill	100-500
Incineration or pyrolysis	200-1500
Soil washing	150-200
Bioremediation	150-500
Solidification	100-150
Vetrification	Up to 250

However, phytoremediation is not yet ready for full scale application, despite favourable initial cost projections, which indicate expansion of clean-up market to be likely in next years (Table 7).

Table 7. Phytoremediation market projection for past and future years in U.S. (millions U.S. $) (adapted from Bini, 2008).

Site categories	2000	2005	2010
Groundwater polluted with organic compounds	2-3	10-15	20-45
Heavy metal contaminated soils	1-2	15-22	40-80
Radionuclide contaminated soils	0.1-0.5	2- 5	25-30
Waste water polluted with heavy metals	0.1-1.5	1- 3	3- 5
Other sources	0-1	2- 5	12- 20
Whole U.S. phytoremediation market	3-8	30-50	100-180

4.4. Global Change

During the last century mankind has appropriated a large proportion of the environmental resources. These changes are summarized under the term "global change" (Frossard, 2006). The various aspects of global change will be further exacerbated as world population will increase from 6 to 9 billion. Food production must be doubled by 2050 to meet the needs of the growing population. All these changes will directly or indirectly affect soil properties and functions.

Whereas the consequences of global change on climate, air quality, on water quality/quantity and on biodiversity are largely discussed in the scientific community and in the public, and receive attention from funding bodies, the importance of the soil is less recognized. In industrialized countries this lack of awareness is related to several factors. Yet, soil is frequently considered a inert substratum for plants, or a fastidious cover of rocks, or a natural body subject to landslides, flooding, landfilling and other environmental disasters to remediate, and therefore it is estimated to have very low economic value. In contrast, soils can be traded as substratum for buildings and streets, in some instances at very high prices. Finally, the other functions of soil in addition to food production are hardly known to the broad public unless they are lost (Frossard, 2006). Yet, soils are natural resources that are not renewable at the human life timescale. The recent development of the European Strategy for Soil Protection is in this respect a good example.

Soil is a living medium in equilibrium with the other compartments of environment and in perpetual renewal at various scales of time. Possible impact of climatic change and human activities changes increase the question of the protection of soil resources from short to long term scale.

Observation of changes in soil components and soil properties under the impact of new environmental constraints (short term) and modelisation of soil evolution through pedogenetic processes allow to assess long term soil modifications. It is essential to evaluate how soils are resistant against changes, or resilient, or evolving out of control, to identify possible feed-back between soil components (King, 2006). In particular, to understand the possible effects of global change on the soil properties and functions (e.g. on weathering, on water and element fluxes, on soil biodiversity etc.) is a major concern of soil science, in connection with other disciplines as geochemistry, hydrology, physiology, agronomy, so as to adapt land use and management to new situations. The results of these interdisciplinary studies will need to be further integrated, not only conceptually, but also in numerical models (Grunwald, 2006).

Global connectivity, knowledge and information sharing have motivated holistic studies that focus on understanding functional relationships among ecosystem components. In this context, soil science plays a major role providing knowledge on soil patterns, processes and landscape dynamics. Ecosystem services characterize the functions that are useful to humans and contribute to ecosystem stability, resilience, sustainability and integrity. These services are diverse ranging from physical (e.g. best management practices that reduce nutrient leaching) to socioeconomics (e.g. crop production, cultural values) and aesthetic aspects.

Ecosystem services provided by multi-functional and multi-use landscapes are affected by the type, intensity, and spatial arrangement of land use and human activities as well as soil-landscape properties.

There are four major areas that have contributed to a gradual shift from qualitative to more quantitative soil-landscape characterization (Grunwald, 2006):

1. Novel mapping tools and techniques (infrared spectroscopy, remote sensing etc.);
2. Data management (GIS and database management systems);
3. Computing power to process multidimensional datasets;
4. Advanced multivariate geostatistical methods.

A more quantitative approach to soil science will enable to overcome knowledge gaps and improve our understanding of pedogenic processes at micro-, meso- and macro-scales, non-linear behaviour of ecosystem processes, biogeochemical cycling at multiple spatial and temporal scales, and assess effects of human activities and natural forcing functions on soil quality.

CONCLUSIONS

The previous analysis has allowed to characterise issues of soil systems which have shown to be vulnerable to natural and anthropic factors. This allows to elaborate meaningful and quantitative data: in the middle and long term new problems might arise concerning the soil degradation not only because of the use, instead of manure, of not traditional organic substances as solid urban waste, but also of sewage sludge, and the use of waste water and low quality water. Particularly this last item appears to be an aspect of great importance, due to the decreased general water availability. One important aspect is that the human pressure on soil and water resources will increase with a growing population. The threats are numerous and well documented and include loss of organic matter and fertility, erosion, pollution, losses connected to urban development, losses of soil functions and services such as water storage and nutrient cycling. The future health of soils calls for more involvement of soil scientists towards sustainable development.

Global Food Security
The global average cereal grain yield of 2.64 Mg/ha in 2000 must be increased to at least 3.60 Mg/ha by 2025 and 4.30 Mg/ha by 2050. With possible changes in dietary habits in emerging economies such as China and India, the average cereal grain yield will have to be increased to 4.40 Mg/ha by 2025 and 60 Mg/ha by 2050.

Soil and the Future of Human Civilization
Feeding world population of 6.5 billion in 2006, 7 billion in 2010, 8 billion by 2025 and 10 billion by 2050 and beyond mandates that soil quality be restored and enhanced. Food insecure population of 850 million in 2006 and increasing, along with several billions suffering from hidden hunger, leave no cause for complacency. The projected food grain deficit of 23 million Mg by 2010 must be met through improved systems of soil management.

Soil-Water Issues
Agriculture has more need in non-conventional water resources (e.g. waste water, recycled water, saline water). Following the necessary treatments, their use must be evaluated with attention not only considering the benefits for fertility and with the hope that the soil will

work as a dynamic and vital filter, but, above all, in the long period the risk of degradation might reduce the quality of soil. This should be done with a global and integrated approach improving in particular the criteria of soil evaluation with reference to the irrigation goals.

Fires

With reference to the burning practice, more research should be conducted on the vegetation-soil relationship, in order to study the formation of hydrorepellent substances, not only for the effects that burning determines in forest or pasture areas, but also in agriculture, which still use the practice of controlled fire as means of agronomic improvement against infesting herbs.

Sustainable Agriculture

It has to be underlined how almost all the land surfaces interested by processes of soil degradation act on the agroforestry systems, directly increasing or possibly solve many problems.

In fact, the new guidelines of community agricultural politics, as expressed by Agenda 2000, transfer the productivity and social agriculture of the Treatise of Rome to a new multifunctional role: to produce more healthy food with increased quality with respect to environment and consumer health, and towards the maintenance of the interacting environmental and territorial resources. Therefore, the concept of a sustainable agriculture is being accepted.

In a similar scenery it is difficult to establish whether agriculture in Europe, and particularly in Mediterranean area countries, above all the non irrigating areas, will evolve towards middle input and middle-high output systems or towards lower input and middle-low output: this will depend to a large extent from the economic environmental and social factors that will have to be solved at regional and local level.

Agriculture seems to return to more conservative management systems and this will be more feasible in the regions of Southern Europe. Practices that have already been used for many centuries in the long agricultural history in the countries of the Mediterranean basin (around 8,000 years), as for instance terrace-cultivation, crop rotations, green manure, organic enrichments, together with tillage of recent conception (minimum tillage, no tillage, ripper-tillage, etc.) will be used again.

Crop production is by far the biggest industry in the world and one of the foundations of culture and society. Agriculture, however, too often regards the soil as an entity in itself, rather than as a part of ecosystems that provide services to humankind, such as the water cycle and nutrient cycle, not to mention the vegetation. This is especially true when considering natural or semi-natural systems used for grazing or areas covered with forests.

It is worth noting how the soil as a resource is poorly cared for under international conventions, in spite of its importance. The UN Convention on desertification (CCD) is based on regional development politics rather than sound scientific background, and it is disadvantaged by conceptual problems. Soils are extremely important in the global cycle of carbon. Still, consideration of soils has been slow to emerge in the context of the UN Framework Convention of Climate Change (FCCC).

Human health issues call for increased attention to soil as a vector of diseases (by inhalation, ingestion, respiration, contact, or by transferring toxic elements to the food chain), while soil and water conservation issues are already demanding more attention in most parts of the world (Abrahams, 2002). With severely degraded areas emerging each day, ecological restoration, one of the most exploited subjects of science and technology today, where soil science plays a major role, will become more important. Soil science will continue to be important for dealing with global change and maintaining biodiversity.

Nevertheless, with reference to the above mentioned items, it is evident that, at least for some environments of Europe, the application *tout court* of the EU regulations foreseeing the introduction of extensive crop systems, can be a degrading factor in areas with particularly sensitive characteristics such as soil structure, SOM, water holding capacity, etc. Such introduction should be carefully evaluated in order to assess the proper land suitability and possible environmental impacts arising in the future. Experimentally this has been shown by different authors (see f.i. Barberis et al., 2000; Bazzoffi, 2002; Jones, 2002; Pagliai et al., 2004). The realisation of soil maps (1:250.000) extended to the whole EU territory can be useful for the recognition of these vulnerable soilscapes, and consequently be used for the introduction of more proper agricultural policy, taking into consideration the environmental and social context, and not only the more or less favourable market conditions.

Problems will aggravate with the rapidly increasing world population unless adequate measures of control are taken. Therefore, multidisciplinary cooperation of soil scientists with geological, biological, physical, toxicological, hydrological, geographical, geo-information, engineering, social, economic and political sciences is essential. Policy makers are finally requested to develop rational land use and management policies, including pro-soil measures.

In conclusion, an effective environmental and natural resources protection is possible only through a correct soil management. This objective can be achieved by:

- Improving soil knowledge (soil data base, soil maps, pedometrics…);
- convincing farmers to adopt sustainable agricultural practices to prevent soil and land degradation;
- Educating public opinion to soil conservation aspects;
- Increasing the soil visibility of soils in international context.

In next years, these issues and other specific aspects such as soil science application to archaeological sites, forensic soil science and other applications to increasing social and economic demands, such as soils of fragile ecosystems, soils in delicate environmental equilibrium, soils with cultural or aesthetic value, will gain importance to achieve the goal of soil and land conservation in a changing world.

REFERENCES

Abrahams P. W. (2002) - Soils: their implications to human health. *Sci Tot. Envir.*, 291, 1-32.
Adriano, D. C., Chlopecka, A., Kapland, D. I., Clijsters, H., Vangrosvelt, J. (1995) – Soil contamination and remediation philosophy, science and technology. In: R. Prost, editor,*Contaminated Soils*, 466-504, Paris, INRA.
Baracco L., Bassan V., Basso B., Rosetti P., Vitturi A., Zangheri P. (2000) – Aquifer vulnerability and soil vulnerability. An application in the provinces of Padua and Venice

of the Veneto regional regulation concerning mapping of soil suitability applied to spreading of livestock effluents. Boll. Soc. It. Sci. Suolo, 49, (1-2), 193-218.

Barberis R., Nappi P., Boschetti P. (2000) – Knowing the soil to protect the vulnerable and sensitive areas. The role of the national thematic centre for soil and contaminated sites. Boll. Soc. It. Sci. Suolo, 49, (1-2),235-246.

Bazzoffi P. (2002) – Impact of human activities on soil loss. Direct and indirect evaluation. In: Sustainable Land Management – Environmental Protection (M. Pagliai and R. Jones eds). Advance in Geoecology, 35, 429-442. Catena Verlag.

Bini C., (2008) - Soil restoration: remediation and valorisation of contaminated soils. *Francis & Taylor* (in press).

Bini C., (2008) - Fate of heavy metals in the Venice lagoon watershed and conterminous areas (Italy). Novapublisher (in press).

Bini C., Bresolin F. (1998) - Soil acidification by acid rain in forest ecosystems. A case study in northern Italy. Sci. Total Envir., 222: 1-15.

Bini C., Costantini E.A. (2006) - Soils as part of our cultural heritage and past environment records. Proc. XI Forum Unesco. Florence, 11-15 sept., 2006. (CD-Rom).

Bini C., Reffo S. (2004) - Restoration of an heavily contaminated site in the southeastern fringe of the Venice lagoon (Italy). *Proc. 1th Eur. Geosci. Un. Congr., Nice,* 6, A-01269.

Bini C., Zilioli D. (2008) – Is soil a cultural heritage? Proc. XI IPSAPA, Aquileia (UD). CD-rom.

Bini C., Gemignani S., Zilocchi L. (2006) – Effect Of Different Land Use On Soil Erosion In The Pre-Alpine Fringe (North-East Italy): Ion Budget And Sediment Yield. *Sci. Tot. Envir., 369, 433-440.*

Bini, C., Maleci, L., Romanin, A. (2008) – The chromium issue in soils of the leather tannery district in Italy. *Jour. Geoche. Expl.*, 96, 2-3: 194-202.

Bini C. , Gemignani S., Sartori G., Zilocchi L. (2003) - Fate of trace elements in the pedosphere of forest soils in alpine environment, Italy. *7th Int. Conf. Biogeoc. of Trace Elements,Uppsala.* Proc., 1 (SP01), 26-28.

Blum W. (2002) – Environmental protection through sustainable soil management, a holistic approach. In: Sustainable Land Management – Environmental Protection (M. Pagliai and R. Jones eds). Advance in Geoecology, 35, 1-8. Catena Verlag.

Blum W. (2006) – The future of soil science. In: The future of soil science (A. Hartemink ed.), 16-18. IUSS, Wageningen.

Borggaard O.K. (2006) – Future of soil science. In: The future of soil science (A. Hartemink ed.), 19-21. IUSS, Wageningen.

Buscaroli A., Gherardi M., Vianello G. (2000) – Investigations about soils and environmental vulnerability applied to the realisation of municipal plan instruments. Boll Soc. It. Sci. Suolo, 49, (1-2), 139-160.

Civita M., De Maio M. (2000) – Sintacs R5, a new parametric system for the assessment and automatic mapping of ground water vulnerability to contamination. Pitagora Editrice, Bologna, pp. 226.

Cunningham S.D., & Berti W.R. (1997) – Phytoestraction or in-place inactivation: technical, economic and regulatory consideration on the soil-lead issue. Proc IV ICOBTE Berkeley Cal. 627 – 628.

Dazzi C. (2002) – Salinità e qualità del suolo. Boll Soc. It. Sci. Suolo, 51 (1-2), 81-104.

Dent F.J. (1990) – Problems soils of Asia and the Pacific. Report on the expert consultation of the Asian network on problem soils. Bangkok, August, 1989. RAPA, Bangkok.

Dumanski, J. (2006) – Soil science, global environments and human wellbeing. In: The future of soil science (A. Hartemink ed.), 37-39. IUSS, Wageningen.

Dumanski, J., P.A. Bindraban, W.W. Pettapiece, P. Bullock, R.J.A. Jones, A. Thomasson (2002) - Land classification, sustainable land management, and ecosystem health. In: Encyclopedia of food and agricultural sciences. Encyclopedia of life support systems. EOLSS Publishers, Oxford, UK.

EEA (1999) – Environment in the EU at the turn of the century. Copenhagen (DK).

EEA (2006) - Soil protection for natural soil protection. EC workshop, Monte Verità, Ascona, October, 1-5, 2006.

Erskine W.D., Mahmoudzadeh H.A., Myers C. (2000) - Land use effects on sediment yields and soil loss rates in small basins of Triassic sandstone near Sidney, NSW, Australia. Catena, 49, 4; 271-287.

Federici R., Rodolfi G. (1994) - I processi naturali ed antropici responsabili della degradazione e mutazione del paesaggio italiano. Atti Acc. Georgofili, ser.VII, 41: 13-46.

Frossard E. (2006) – The role of soils for the society and the environment. In: The future of soil science (A. Hartemink ed.), 46-48. IUSS, Wageningen.

Grunwald, S. (ed.), 2006. Environmental soil-landscape modeling – geographic information technologies and pedometrics. p. 488. CRC Press, New York.

Grunwald S. (2006) – Future of soil science. In: The future of soil science (A. Hartemink ed.), 51-53. IUSS, Wageningen.

Jenny H. (1941) – Factors of soil formation. Dover Pub. Inc., New York, pp. 281.

Jenny H. (1989) – The soil resource. Springer Verlag, New York, pp. 425.

Jones R. (2002) – Assessing the vulnerability of soils to degradation. In: Sustainable Land Management – Environmental Protection (M. Pagliai and R. Jones eds). Advance in Geoecology, 35, 33-44. Catena Verlag.

King D. (2006) - Research for sustainable soil management. In: The future of soil science (A. Hartemink ed.), 68-70. IUSS, Wageningen.

Madrau S. (2002) – Il consumo di suolo per urbanizzazione in Sardegna negli anni 1954-1997. Boll. Soc. It. Sci. Suolo, 51, (1-2), 571-587.

McBratney, A.B, I.O.A. Odeh, T.F.A. Bishop, M.S. Dunbar, and T.M. Shatar, 2000. An overview of pedometric techniques for use in soil survey. Geoderma, 97: 293-327.

McBratney, A.B., M.L. Mendonça Santos, and B. Minasny, 2003. On digital soil mapping. Geoderma, 117: 25-49

Montanarella L. (2006) - The new European thematic strategy for soil protection. Hopes and risks. In: Soil protection for natural soil protection. EC workshop, Monte Verità, Ascona, October, 1-5.

Morgan R. P. C. Soil Erosion in the Northern Countries of the European Community. EIW Workshop: Elaboration of a Framework of a Code of Good Agricultural Practices, Brussels, 21-22 May 1992.

Nachtergaele F. O. (2000) – Soil vulnerability evaluation and location fragility assessment. Boll Soc. It. Sci. Suolo, 49, (1-2), 31- 50.

Nappi P. (2000) – Rappresentare la qualità del suolo mediante indicatori e indici: l'esperienza del Centro Tematico Nazionale suolo e siti contaminati. Rend. Acc. Naz. Sci., Mem. Sci. Fis. Nat., 118, 249-274.

Neil D.T., Fogarty P. (1991) - Land use and sediment yield on he southern Tablelands of NSW, Australia. Austr. J. Soil Water Conserv., 42: 33-39.

Norra, S., Lanka-Panditha, M., Kramar, U., Stuben, D. (2006) – Mineralogical and geochemical patterns of urban surface soils, the example of Pforzheim, Germany. *Appl. Geochem.*, 21: 2064-2080.

Pagliai M., Vignozzi N., Pellegrini S. (2004) - Soil structure and the effect of management practices. Soil and Tillage Research, 79: 131-143.

Pannetto P. (2007) - Parco regionale dei Colli Euganei: qualita' del suolo in un'ottica pedologico-ambientale. Bach. thesis, University of Venice (unpublished).

Pierzinsky G. (2003) – Trace element chemistry, contamination and ecotoxicity. 7th *Int. Conf. Biogeoc. of Trace Elements,Uppsala.* Proc., 1 (1), 14-15.

Rey F. (2003) - Influence of vegetation distribution on sediment yield in forested marly gullies. Catena, 50, 2-4: 549-562.

Sorriso-Valvo M., Bryan R.B., Yair A., Iovino F., Antronico L. (1995) - Impact of afforestation on hydrological response and sediment production in a small Calabrian catchment. Catena, 25: 89-104.

Tesi P.C. (2000) – Land planning for sustainable development: rural lands in the general land planning of the Italian Communes. Experience in the Sienese Chianti. Boll Soc. It. Sci. Suolo, 49, (1-2), 161-182.

Ulrich B. (1995) – The history and possible causes of forest decline in central Europe, with particular attention to the German situation. Environ. Reviews, Canada. Invited paper.

Van Der Knijff J. M., Jones R. J. A., Montanarella L. Soil Erosion Risk Assessment in Italy. Office for Official Publications of the European Communities; EUR 19022 EN, pp. 45. Luxembourg, 1999.

Van Mechelen L., Groenemans R., Van Rast E. (1997) – Forest soil conditions in Europe. EC-UN/ECE Report, Brussels, Geneva, pp. 261.

Wischmeier W. H., Smith D. D. Predicting rainfall erosion losses. USDA Agricultural Handbook No. 537. Washington D. C., 1978.

Zonta, R., Botter, M., Cassin, D., Pini, R., Scattolin, M., Zaggia, L. (2007) – Sediment chemical
Contamination of a shallow water area close to the industrial zone of Porto Marghera (Venice Lagoon, Italy). *Marine Pollution Bulletin*, 55, 529-542.

Reviewed by M. Pagliai, Director of the Research Centre for Agrobiology and Pedology of the Agricultural Research Council, Florence, Italy. Officer of the International Union of Soil Sciences (IUSS). President of the Italian Soil Science Society.

In: Encyclopedia of Environmental Research
Editor: Alisa N. Souter

ISBN: 978-1-61761-927-4
© 2011 Nova Science Publishers, Inc.

Chapter 16

CARBON SEQUESTRATION IN SOIL: THE ROLE OF THE CROP PLANT RESIDUES

Silvia Salati, Manuela Spagnol and Fabrizio Adani

Dipartimento di Produzione Vegetale – DiProVe
Università degli Studi di Milano, Milan, Italy

ABSTRACT

Anthropogenic emission of CO_2 and other greenhouse gases was rapidly increased with the Industrial Revolution and this event has caused a world interest in identfying strategies of reducing the rate of gaseous emission.The intergovernmental Panel on Climate Change shows that from 1850 and 1998 the emission from terrestrial ecosystem was about half of fossil fuel combustion. Agriculture can be a source or sink for atmospheric CO_2 because soil organic carbon pool (SOCP) in soil surface is sensitive to changes in land use and soil management practice. The carbon sink capacity of the world agricultural and degraded soils is 50-66% of the historic carbon loss that are of 42 to 78 Gt of carbon respectively. Carbon (C) sequestration implies transferring atmospheric CO_2 into long-lived pools and subsequent storage of fixed C as soil organic carbon (SOC). In this way the conservation of plant residues in agricultural soil play an important role in CO_2 sequestration. The mechanism by which crop residues contribute to SOC is through their chemical, phisical and biological stabilization. In this chapter we discussed the role of the plant residues in the carbon sequestration throughout plant tissue stabilization in soil, giving a new approach and understanding of the plant residue conservation in soil.

1. INTRODUCTION

1.1. Carbon Cycle

Carbon plays an important role in supporting life and all living organisms are based on the carbon atom. Among the common elements of the earth's surface, the carbon atom is one of few that can form long organic chains and rings: this ability is the foundation of organic chemistry. Carbon compounds can be gas, liquid, or solid under conditions generally found on the earth's surface, and the characteristics of the carbon atom make possible the existence of all organic compounds essential to life on earth.

The carbon cycle describes the exchange of carbon atoms between various reservoirs within the earth system. The carbon cycle is one of a number of geochemical cycles and since it involves the biosphere it is sometimes referred to as a bio-geochemical cycle. Carbon cycle is the mean by which carbon is exchanged between the terrestrial biosphere including soils, hydrosphere, lithosphere and atmosphere of the earth. The global carbon budget is the balance of the exchanges (gains and losses) of carbon between the carbon reservoirs or between one specific pool (e.g., atmosphere $\leftrightarrow$ biosphere) of the carbon cycle. For example the atmospheric CO_2 is incorporated, by photosyntesis, into the living tissue of green plants and turned in carbohydrates with the following chemical reaction:

$$energy + 6CO_2 + H_2O \implies C_6H_{12}O_6 + 6O_2$$

Successively plants, animal and soil microbes consume the carbon contained in the biomass returning carbon dioxide in the atmosphere through the following reaction:

$$C_6H_{12}O_6 \ (organic \ matter) + 6O2 \implies 6CO_2 + H_2O + energy$$

This kind of C cycle represent a simple short cycle. Anyway C cycle in more complex as other C exchanges occurred between different existing pools.

A source of carbon towards atmosphere, coming mainly from anthropogenic activity, is represented by the burning of fossil fuels, oils, coal, fuel wood and by the conversion of forest to agricoltural soils. These human activities releases carbon into the atmosphere more rapidly than its removing through photosintesis, so that this imbalance determines atmospheric CO_2 increase.

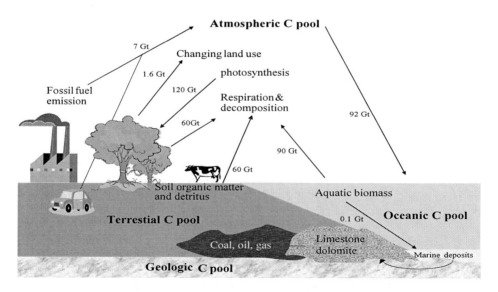

Figure 1. Carbon cycle and carbon fluxes throughout different C pools.

Lal (2008a) suggested there are five global carbon pools (Figure 2): i. atmospheric C-pool; ii. oceanic-C pool; iii. geological-C pool; iv. biotic C-pool; v. pedologic C-pool. The atmospheric C-pool contains about 760 Gt of C as CO_2 and it has been estimated, this pool increases of about 3.5 Gt C yr^{-1}.

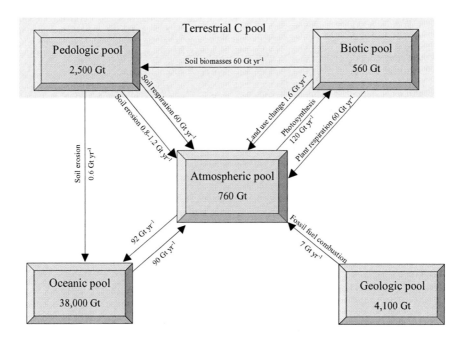

Figure 2. Principal global C pools and fluxes between them (from Lal, 2008a, modified).

Oceanic-C pool, that is estimated at 38,000 Gt, loose about 2 Gt C per year; as consequence of that on the scales of millennia, the oceans determine the atmospheric CO_2 concentration, not vice versa (Falkowski et al. 2000). Geological-C pool, that comprise gas, oil and coal (fossil fuels), is estimated at 4,100 Gt. Coal and oil represent approximately 40% of global CO_2 emissions (Schrag 2007). Every year 7 Gt of C are emitted in the atmosphere by fossil fuel combustion.

Both biotic and pedologic pools are called the terrestrial C pool. The smallest among the global C pools is the biotic pool (vegetative carbon) that can be further categorized into carbon in aboveground biomass (shoot), belowground biomass (root), and necromass (Hamburg 2000). The pedologic-C pool with 2,500 Gt at 1 m depth, can be divided into two fractions: soil organic carbon (SOC) pool estimated at 1,500 Gt and soil inorganic carbon (SIC) pool estimated at 950 Gt (Batjes 1996). The SOC pool is constituted of plants, animals and microorganisms at different stages of decomposition, root and microbial exudates, substances synthesized by living soil population (Schnitzer 1991). The SIC pool includes elemental C, calcium carbonate minerals (limestone) and dolomite, and comprises both primary or lithogenic inorganic carbonates, and secondary or pedogenic inorganic carbonates. Primary minerals are originated from the weathering of parent rock, secondary carbonates are formed through dissolution of primary carbonates and re-precipitation of weathering products. The reaction of atmospheric carbon dioxide (CO_2) with water (H_2O) and calcium (Ca^{2+}) and magnesium (Mg^{2+}) in the upper horizons of the soil, leaching into the subsoil and subsequent re-precipitation, results in formation of secondary carbonates and in the sequestration of atmospheric CO_2.

Considering the entire lithosphere, comprising the underground rock, the amount of C reservoir is about 75,000,000 Gt, that is much more than ocean pool (ca 38,000 Gt C) and the terrestrial biosphere (ca. 2500 Gt C) (Falkowski et al., 2000; Lal, 2004).

The terrestrial and atmospheric C pools are strongly linked (Figure 2); 120 Gt C yr^{-1} were absorbed by plants via photosynthesis but most of which returned back to the atmosphere through fossil fuel combustion, soil erosion, plant and soil respiration and through deforestation and land use change wich determines a soil depletion.

If carbon gained is more than C lost the pool is considered a carbon sink, on the contrary it is considered a carbon source.

Long-lived trees, limestone (carbon-containing shells of small sea creatures that build up into thick deposits), plastic, fossil fuels and carbon coat are considered carbon sinks. The burning of organic matter such as wool and fuels, the respiration of living organisms and the weathering of limestone rocks are carbon sources. Release agents include volcanic activity, forest fires, and many human activities.

2. PLANTS RESIDUES AND CARBON SEQUESTRATION

2.1 Carbon Sequestration in Terrestrial Ecosystems

Because of its magnitude, changes in the soil carbon pool are critical to the global C budget and could even affect global climate via altered atmospheric CO_2 concentrations (Jenkinson, 1991; Schlesinger, 1997). The CO_2 concentration in atmosphere has increased by 30 percent since the industrial era (IPCC, 2001), this increase is attributed principally to fossil fuels combustion, land use conversion and deforestation. Many researcher are in agreement to consider this increase as a cause of possible consequences on global climate change, so stabilizing atmosperic concentration of CO_2 is of great interest for scientific community and, at the same time, there is growing attention in studying the contribution of sequestered carbon in ecosystems (Wang et al., 2004).

There are three strategies to reduce the CO_2 emission in the atmosphere (Schrag, 2007): i. the use of low or no-carbon fuels; ii. the reducing the global energy use and iii. the sequester of CO_2 by engineering principles and natural processes.

Engineering processes concern to stabilize and minimize CO_2 with its injection at great depths in the oceans, because of their great sink capacity (5000-10000 Gt C), but sequestration of CO_2 would change the pH of the entire ocean and small perturbations in CO_2 or pH may thus have adverse effects for the ecology of the deep-see biota and for the global biogeochemical cycles dependent on deep-sea ecosystems (Seibel and Walsh, 2001).

Another technique is the injection of industrial CO_2 into deep geological strata. The main risks of this technique are caused by the density of CO_2 that is lighter than surrounding fluids, resulting in buoyancy driven flow, by viscosity that is much lower than surrounding fluids and by chemical interactions that could alter permeability near the well or at locations with large pressure changes (Tsang et al., 2002)

It's clear that cost and leakage of oceans and geological sequestration are principal issue witch need to be resolved (Lal, 2008a).

Natural processes concern terrestrial carbon sequestration, i.e. atmospheric CO_2 transfer into biotic and pedologic pools. These processes consisted in photosyntesis, OM stabilization in soil, biomass pyrolysis with the formation of charcoal and the formation of secondary carbonates. Terrestrial ecosystems represent the major carbon sink attributable above all to plant-C and C storage in soil organic matter (Lal, 2008a).

There are three main terrestrial C sequestration components: i. C stored in forests biomass; ii. C stored in wetlands; and iii. C stored in soil (Figure 3). Forests sequester C in biomacromolecules by phototosinthesis; forestation is indicated as one of the options to sequester C (Lamb et al., 2005). Wetlands represents a minimal area of the land earth but contain about 20% of the total soil C pool (Lal, 2007), and they can be considered a source of CO_2 owing to the drainage of peatlands and their cultivation but with the restoration wetlands could be an important C sink (Lal, 2008a).

Carbon sequestration from the soil represents our focal point and regards mainly the accumulation of organic matter in soil by chemical association, physical sequestration and biochemical recalcitrance, charcoal formation derived from partial combustion of wood, and C sequestered in the primary and secondary carbonates.

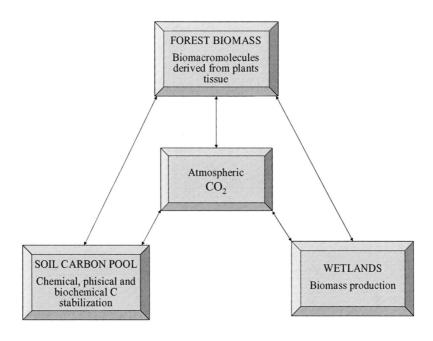

Figure 3. Carbon exchanges between the terrestrial ecosystems.

2.2. Carbon Sequestration in Soils

The main form by wich carbon enters in the terrestrial biosphere is by CO_2 assimilation by autotrophic photosynthesising organism. The assimilated C represents the gross primary production (GPP). About half of GPP is used to support autotrophic respiration and so it return to the atmosphere as CO_2. On the other hand the remainder GPP-CO_2 is fixed as net primary production (NPP) in vegetative biomass (Janzen, 2004) and successively enters into the soil by plant litter at 0.5-1 m of depth (Jones and Donnelly, 2004; Shrestha and Lal, 2006).

Lovett (2006) defined net ecosystem production (NEP) the difference between GPP and respiration from ecosystem. Therefore, NEP "represent the total amount of organic carbon in an ecosystem available for storage, export as organic carbon, or non-biological oxidation processes".

A soil contains about 75% of the terrestrial carbon pool playing an important role in the global C cycle. Carbon sequestration in soils occurs through inorganic and organic fixation of atmosferic CO_2. Inorganic soil carbon sequestration occurs with chemical reactions that convert CO_2 into soil inorganic compounds. Direct plants organic carbon sequestration and subsequent soil C sequestration are processes by which plants remove, through photosynthesys, atmospheric CO_2 that is successively stored into plants biomass. With plant-residues returning to the soil and subsequent stabilization processes, C biomass is indirectly incorporate into soil as organic carbon (SOC) with other nutrients such as N, P and S (Figure 4) (Lemus and Lal, 2005).

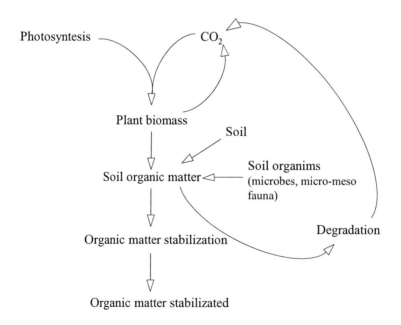

Figure 4. Carbon cycle in soil.

2.3. The Role of Crop Residues in Carbon Sequestration *Vs* Renewable Energies Production

Before 1900s, agriculture and forests were the major source of energy. Since the beginning of 1900 petroleum increasingly replaced vegetable oil, starch and cellulose as a feedstock for energy and industrial products (Morris and Ahmed, 1992). Fossil fuels are the dominant source of primary energy in the world (Figure 6) and with oil, coal and gas supplied more than 80 % of the total world request. Renewable energy sources represent only 13 % of the total primary energy supply. In particular biomass is the only renewable energy source that can supply the liquid transport fuel (Lal, 2008b).

From 1970s to nowadays the demand for energy has grown incredibly (Figure 5) due to growing economy, increasing population, and rising standard of living; at the same time no-renewable sources such as fossil fuels began to run low and increase their price so much to encourage the development of new technologies producing energy.

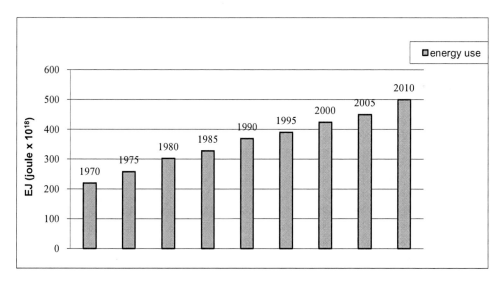

Figure 5. Energy use in the last 40 years (modified from Lal, 2008b).

The recent increase of the demand for liquid fuels obtained from agricultural crops (biofules) as transport fuels has reasserted the linkages between energy and agricultural output markets (FAO, 2008)

Many crop species (maize, rye, wheat etc.) can be used to produce ethanol from grain (first generation fuels) and from plant residue or the entire plant (second generation fuel). Also starch and glucose coming from sugar beet and sugarcane produce ethanol by fermentation, that is used normally in blends with gasoline (Sims et al., 2006)

World agricultural crop production is estimated at about 4 bilion Mg for all crops (Lal, 2008b); Weisz, (2004) evaluated the potential energy value of 1 Mg biomass in 3 Mkcal that, considering all crop residues used for energy production, represent about 15% of world energy needs (Lal, 2008b). So, biofuels production from biomass is among the potential strategies to reduce the use of non-renewable fuel sources.

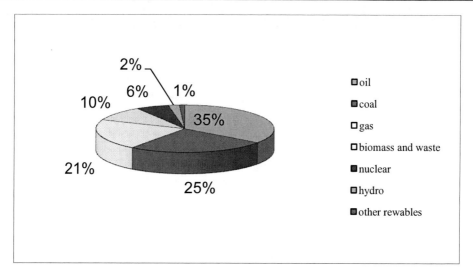

Figure 6. World primary energy demand by source (IEA, 2007).

The use of crop and crop residues to produce biofules become of great interest, also, because of the potential long-term influence on global climate change (WMO, 2006). Some authors are in agreement to consider the plant biomass for biofules production as a "carbon neutral" because there is a perfect CO_2 balance between the release in processing and the capture by the plant by photosynthesis (Sims et al., 2006).

On the other hand a great part of scientists reported negative impacts of biofuels from crop. Plant residue removing from soil, determine C depletion with soil fertility decreasing a low crop production. Energy crops compete on fertile soil with food and feed production (Pimentel, 1991; Giampietro et al., 1997). The conversion of natural land into arable energy crops determine a decrease in biodiversity, increase in pollution from fertilizers and pesticides uses and an increase of soil erosion (Pimentel, 2003). It should be considered, also, that C input to soil limited to unharvestable biomass, such as roots, may cause a decrease in SOC and subsequent soil depletion (Wilhelm et al., 2004). So removing crop residues to produce bioethanol cannot be a sustainable way, and biofuels must be produced through estabilishing biofuels plantations in marginal soils (Lal, 2008b).

2.4. Crop Residues Quality

The mechanism by which crop residues contributes to C sequestration in soil is through their return to the soil as litter, root detritus and exudates and their subsequent stabilization in soil

Cell wall, that represent the 70% of plant tissues, is the main resource for organic matter in soil (Pauly et al., 2008). It has been estimated that the net CO_2 fixation by land plants per year is approximately 56 Gt and that the worldwide crop biomass production is about 4.7 Gt, 80% of which is plant residue. The rate of decomposition of plant material depend on chemical composition of residues, environmental conditions and microbial communities. knowledge of the chemical composition of plant can be useful to explain its contribution to organic carbon in soil.

Major components of above and below-ground plant residues are polysaccharides, lignin, lipids, and tannins and their relative abundance varies widely between crop residues (Table 1).

Cellulose is a polymer of β 1,4- linked glucose units and occurs in a crystalline state as well as in an amorphous form (Carpita and McCann, 2000). The crystalline cellulose is highly resistant to chemical and biological hydrolysis because of its structure (Himmel *et al.*, 2007): tendency to form intra and inter-molecular hydrogen bonds by the hydroxyl groups on these linear cellulose chains leads to a fibrillar structure with crystalline properties. Cellulose is hydrolyzed by synergistic action of different enzymes: endoglucanases, exoglucanases and β-glucosidases.

Hemicelluloses are a group of polysaccharides of different composition and are branched polymers of hexoses, pentoses and uronic acids. The chains of hemicellulose have never formed exclusively from one type of monosaccharide, but they contain different types, in widely different ratio and in many combinations. The hemicellulose are decomposed more quickly that the cellulose from microorganisms above all actinomycetes (Steptomyces), fungi (Penicillum) and bacteria (Clostridium, Pseudomonas, Achromobacter). In the cell-wall the hemicelluloses are linked to the surface of crystalline cellulose forming the microfibril network..

**Table 1. Macromolecular composition of different crop residues
(data from Cestaat, 1990).**

Crop residues	Protein	Lipid	Fiber†	Ash	Extractive no-N‡
	(g kg dm⁻¹)				
Maize stem	40-50	5-10	350-400	50-70	450-500
Maize-cob	30-40	6-10	400-500	20-30	700-800
Grain Leguminous	80-120	10-30	350-450	50-80	350-450
Potato	80-100	30-40	250-300	100-130	400-500
Sugar-Beets	100-150	20-30	150-200	200-250	400-450
Sunflower	20-30	10-20	500-600	200-250	100-200
Soybean	70-80	10-15	470-480	80-90	340-360
Rice	400-500	10-20	350-400	100-150	400-450
Wheat	20-40	20-30	400-450	70-100	400-450
Tobacco	60-70	5-10	700-750	20-30	150-250

†Fiber: cellulose+emicellulose+lignin.

‡Extractive no-N: sugar, pectine, organic acid.

Lignin is a complex polymer of phenylpropane units and is composed of three different monolignols: coniferyl alcohol, sinapyl alcohol and paracoumaryl alcohol. Depending on the taxonomical affinity of the plant, three main type of lignin can be distinguished. Gymnosperm lignin consists mainly of coniferyl alcohol moieties, while lignin of angiosperms contains considerable amount of sinapyl alcohol units in addition to coniferyl alcohol. Within the angiosperms, the lignin of monocotyledones also contain paracoumaryl alcohol. Lignin contain no hydrolytic bonds but only aliphatic-, alcylaryl- and biaryl-bonds. Because of its molecular composition and complexity, lignin is recalcitrant to enzymatic attack and acts also protective fore the more labile fraction. Therefore lignin is a major determinant of the decomposition and humification of plant material (Adani et al., 2006). Syringil/Guayacil (S/G) ratio of lignin is commonly used in order to represent the lignin content and from this it seems to depend the chemical-physical nature of the same one (Matsui et al, 2000).

Cutin and suberin are the two major lipid-based polymers of vascular plants. Cutin is a biopolyester composed of mono- and polihydroxy and epoxy fatty acids whit a C_{16} and C_{18} chain length. Suberin, on the contrary, is composed of polyphenolic and polyaliphatic domains and contains monomers whit a higher chain length of C_{20} and C_{30} (Kögel-Knabner, 2002; Rasse et al., 2005)

Tannins are polyphenols that occur in higher plants and are differentiated in two categories: condensed tannins and hydrolysable tannins. Non hydrolysable tannins are formed by different flavan units connected by carbon linkages whereas the hydrolysable tannins are composed of a molecule of carbohydrate, generally glucose, to which gallic acid or similar acids are attached by ester linkages.

The above-ground part of the plant has received more attention in scientific studies than the below-ground plant, indeed there are lack of information about the role of root residues on accumulation of SOC and on the conservation of carbon. Many studies show that the contribution of below-ground plant residues to the pool carbon is very important (Balesdent–Balabane, 1996; Allmaras et al., 2004; Wilts et al., 2004; Rasse et al., 2005; Hooker et al., 2005; Johnson, 2006); in particular, in cropped system characterized by a little shoot-C returning to the soil (i.e. tillage), it was found that the carbon derived from roots contributes 1.5-3 times more than the above-ground plant and that residence time in soil of root-derived C is 2.4 folder that of shoot-derived C.

According to Silver and Miya (2001) the chemical composition of root tissues is the primary factor that affects their decomposition rates. Gale and Cambardella (2000), Puget and Drinkwater (2001) and Abiven et al. (2005), showed that higher content of lignin, cellulose and suberine (chemical composition) (Table 2) of root residues cause their slower decomposition than shoot residues. Degradation rate is also affected by interactions between soil and residues that determined the formation of aggregates that physically protected the biodegradation of organic C.

Table 2. Macromolecular composition for some of the major crop residues (data from Abiven et al., 2005).

Crop residues	Tissue	Hemicellulose[†]	Cellulose[†]	Lignilike[†][¶]	Soluble cell material[†]
			(g kg dm[-1])		
	steam+leaf[‡]	210	293	58	439
Maize					
	root[§]	247	324	139	290
	steam	300	420	90	190
Sorgum					
	root	230	270	190	310
	steam	360	340	30	270
Rice					

Table 2. (Continued).

	root	350	320	160	170
	steam	140	410	120	330
Soybean					
	root	210	410	260	120
	steam	270	520	60	90
Wheat					
	root	420	330	80	170

† van Soest method, 1991
¶Ligninlike: lignin, suberin and cutin
‡Salati, 2007
§Papa, 2008

2.5. Crop Residues Decomposition in Soil

Residue decomposition is dependent on extrinsic factors (i.e. climate, tillage, particle size) and on intrinsic factors (i.e. biochemical composition). The quality of crop residues become important when other factor are held constant (Johnson et al., 2007a; Heal et al., 1997). Generally the degradation rate is linked to C/N ratio: many studies show that residues with high C/N ratio, such as cereals, decompose slowly than those with low C/N ratio (Table 3) (Trinsoutrot, 2000). Additional the high lignin content (or lignin/N ratios) retard decomposition (Heal et al., 1997), although no universal relationship has been established (Wang et el., 2004).

Abiven et al., (2005) and Trinsoutrot et al., (2000) showed that short term C mineralization of crop residues was mainly related to the C presents in water soluble fraction, under conditions of non-limiting N, so the concentrations of the different polymers in the plant tissue are the most important factor that influence the C decomposition in soils.

Table 3. C/N ratio and lignin/N ratio for some crop residues.

Plant sample	C/N	Lignin/N	Reference
Maize	52	5	Adani et al., 2006
Sorghum	14	6	Kumar and Goh, 2000
Wheat	42	5	Kumar and Goh, 2000
Sugarbeet	15	-	Cestaat, 1990
Potato	24	-	Cestaat, 1990
Soybean stems	107	39.7	Johnson, 2007a

At the same time the plant maturity of crop residues affected the decomposition rate. Adani et al. (2006) observed that young maize plant (4-5 leaves), have a high content of cellular material such as membrane lipids (high content of alkyl C) and protoplast molecules (e.g. proteins) (high content of N-alkyl-C) (Fig. 8). On the contrary at the plant senescence maize residue was mainly formed of polymers such as lignin (high aromatic-C content) and carbohydrates (hemicelluloses and cellulose) (high O-alkyl-C content), and less of lipids (cutin) (alkyl-C) and protein (e.g. structural proteins) (N-alkyl-C) (Table 4) (Figure 7). Consequently, increased maturity leads to a drop in soluble cell components and a greater contribution of the cell wall to the total dry matter. Many studies have shown that the soluble fraction is quickly decomposed by soil microorganisms, while the fraction decomposed more slowly is represented by constituents of the cell wall (Bertrand et al., 2009).

Table 4. Macromolecular composition for maize plant at different stages of maturity (Adani et al., 2006).

Plant sample	Hemicellulose[†]	Cellulose[†]	Lignin[†]	soluble cell material[†]
	(g kg dm^{-1}) (ash free)			
Plant[‡]	254.6±6.3a	154.9±2.7a	17.7±0.6a	572.9±6.3b
Plant[§]	318.9±13.2b	306.7±9.7b	20.5±4.3a	354.0±13.9a
Plant[¶]	313.5±29.6b	306.3±14.6b	41.0±0.9b	339.3±29.6a

[†] van Soest method

[‡] plant post emergence (4-5 leaves); [§] plant maturity; [¶] plant senescence.

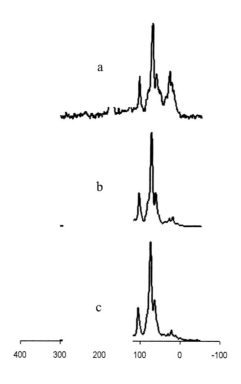

Figure 7. ^{13}C-CP-MAS-NMR of maize plant residues at post emergency (**a**), waxy maturity (**b**) and senescence (**c**) (Adani et al., 2006).

Studies conducted in our laboratories have demonstrated that not only the chemical composition but, also, the physical structure affects the plant residues degradation. Most of the hydrolytic enzymes have a molecular weight of 20-25 kD, corresponding to a size of about 4 nm, that hardly penetrate into micropores and small mesopores of plant structures, so that their activities are confined only to a portion of the total surface (Zimmerman et al., 2004; Chesson, 1997) (Table 5).

Recent our work showed a higher degradability of maize plant residue characterized by higher available surface (pores > 4-5- nm) than the alkali-soluble fraction of the cell wall characterized by low available surface (Table 5).

Table 5. Meso surface area, available surface area to enzimatic attack and % of degradation of different plant.

Pant Sample	Meso surface area ($m^2 g^{-1}$)	Available surface area ($m^2 g^{-1}$)	% degradation[§]
Maize steam+leaf[†]	1.86[†]	0.89[†]	48[‡]
Soluble cell wall maiz e[†]	0.32[†]	0.24[†]	15[‡]
Lucerne[¶]	2.1	0.8	n.d.
Maize steam[¶]	2.4	0.9	n.d.
Grass (Timothy)[¶]	4.9	2.6	n.d.
Wheat straw[¶]	3.2	1.7	n.d.

[†] Spagnol, 2005
[‡] Adani et al., 2006
[§] degradation after one year of incubation
[¶] Chesson, 1997

2.6. Crop Residues Stabilization and Sequestration in Soil

The amount of carbon in a soil is directly related to C inputs from primary biomass production, and to the C outputs by mineralization. Depending on environmental conditions and land use, soils may act as sources of sinks for carbon (Marschner et al., 2008). Different mechanisms contribute at the same time to SOM protection against decomposition: biochemical recalcitrance, chemical association and physical sequestration (Figure 8).

Von Lützov et al, (2006) differentiate between primary recalcitrance of plant litter and rhizodeposits as a function of their chemical characteristics, and secondary recalcitrance of microbial products, humic polymers and charcoal.

Primary recalcitrance can be defined also as the natural resistance of plant tissues to microbial and enzymatic deconstruction (Himmel et al., 2007); secondary recalcitrance could be defined as a long-term conservation in soil. The biochemical recalcitrance is linked to the both chemical and molecular conformation of plant residues: aromatic and alkyl compounds such as lignin, cutin, suberin and tannins, are resistant to biodegradation. Generally to the aromatic compound (lignin) is attributed the major chemical recalcitrance, but recent studies have suggested that, also, condensed tannins and polyphenols, cutin and suberin, contributed to SOC pool (Rasse et al., 2005).

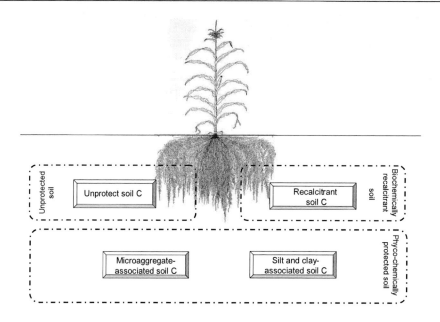

Figure 8. Mechanism of SOM stabilization

Some authors indicate that two pools of lignin coexist in the soil: a fast decomposing pool that represents about 93% of total lignin, derived from fresh-litter lignin, and a slow decomposing pool of lignin stabilized in SOM (Glaser, 2005; Rasse et al., 2006; Nierop and Filley, 2007). Recent studies, on the contrary, show that lignin is altered relatively quickly and does not appear stabilized in the long term in any soil fraction (Baldock and Nelson, 2000; Kögel-Knabner, 2000; Kiem and Kögel-Knabner, 2003).

Alkyl-C compounds (i.e. suberin and cutin) accumulates over the time scale of soil formation (Baldock et al., 2004) but selective preservation is due, above all, to surface interactions with minerals or hydrophobicity (Kögel-Knabner et al., 2008; Bachmann et al., 2008).

The consequence of persistence of unaltered plants components in soil is the formation of stable OM in soil. Stabilized OM in soil has been, always, indicated as soil humus. Humus represent a "dark brown or black amorphous material" characterized by a large surface area, high charge density, high affinity for water (Lal, 2007). While previously thought of as being high molecular weight polymers, today HS are thought to be compounds of relatively low molecular weight that form supramolecular associations clusters by hydrophobic interactions and by hydrogen bonding (Piccolo, 2002; Sutton and Sposito, 2005). The new vision of supramolecular structure of humic substances formulated from Piccolo (2002) is commonly accepted (Wershaw, 2004; Sutton and Sposito, 2005).

The process by which plant residues are transformed into humic substances (HS) is the humification process The main routes for humic substances formation include the lignin theory, the sugar–amine theory and the lignin–protein theory (Stevenson 1994). These theories hypothesize that biomacromolecules first break up into small constituents, and then subsequently recombine to give more complex organic matter (Hedges and Oades 1997). On the other hand the many studies conducted in both soil and marine ecosystems (Kögel-

Knabner et al. 1992; Hedges and Keil 1995; Adani et al., 2006, 2007) indicated that it is the preservation and modification of plant tissue that provides the humification pathway. On reviewing the origin of soil organic matter, Kögel-Knabner (2002) indicated humification to be the prolonged stabilization of organic substances, countering biodegradation, and to involve plant biopolymers such as lignin, cutin and suberin. Preservation of biomacro-molecules is due to biochemical recalcitrance caused by the inherent property of the molecular structure (Kögel-Knabner 2002). Therefore it can be concluded that humification consists in the preservation of bio-macromolecules in soil (Kelleher, 2006).

According to this concept the soil HS are a complex of distinct of faunal, microbial and plant biopolymers and proteins, lignin, polysaccharides and aliphatic polymers (Lorenz et al., 2007; Kelleher, 2006). Therefore, humification or better, the preservation of molecules due to their biological recalcitrance represents the first step of the OM stabilization in soil, although a such mechanism has not been completely explained.

This convincement is supported by data that showed how plant residue macromolecules directly isolated from maize plant residue, are preserved after incubation in soil, contributing directly to the soil humus fraction (Adani and Ricca, 2004; Adani et al., 2006; Adani et al., 2007) and that distinct molecules such as plant and microbial biopolymer has been identified in soil (Lehmann et al., 2008). Based on these hypotheses humification process, and subsequent OM stabilization in soil, begin with the plant photosynthesis in which carbon dioxide is fixed in plant organic compounds that are successively - when returned to the soil as crop residues - selectively preserved in soil because of their chemical and physical characteristics.

If biological recalcitrance allows plant molecules to be preserved in soil, long term stabilization of organic C, implies more complex mechanism not yet well understood such as chemical association and physical sequestration with mineral components of soil (Figure 9).

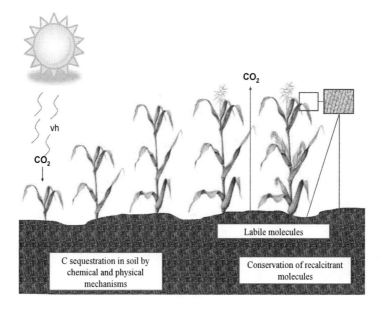

Figure 9. Hypotesis of crop residue C preservation and successive sequestration in soil.

Marschner et al. (2008) showed that stabilization of SOM against microbial degradation is due not only by preservation of recalcitrant compound but also by physical or chemical association whit organic or soil minerals. Black and fossil C are the only fraction of SOM that were persistent in soil without mineral association, on the contrary the soluble fraction of SOM (i.e. Na-pyrophosphate-extractable OM) had the highest turnover rates.

Depending on soil mineralogical composition, intermolecular interactions between mineral and organic matter contribute to chemical stabilization of SOM trough different mechanisms: ligand exchange, polyvalent cation bridges and weak interactions (von Lutzow, 2006).

Silt and clay fraction are quantitatively more effective in sequestering SOC than the coarser fractions (the rate of mineralization is slower). Six et al. (2002) found that the types of clay (i.e. 1:1 and 2:1 clays) have different capacity to adsorb organic materials, related to differences in CEC and specific surface area. Indeed, 2:1 clay minerals whith a high CEC and surface area (i.e. montmorillonite and vermiculite), have a higher binding potential than clay minerals whit a lower CEC and smaller specific surface (i.e. illite). Furthermore Fe- and Al-oxides, predominant in 1:1 minerals soils, are strong flocculants and they can reduce the available surface for adsorption of SOM and, at the same time, they might co-flocculate SOM stabilizing it.

It can be stated that non-crystalline amorphous mineral (e.g. allophone, recent formed Fe-hydroxide etc.) characterized by high surface area and CEC, have more affinity for SOM than well organized and crystalline mineral (e.g. kaolinite, gibbsite etc.) (Torn et al., 1997).

The relationship between silt, clay and SOM differs among soil types and land use, and soil cultivation reduces the silt and clay aggregates of SOM pool (Lorenz et al., 2008) causing a release of C (Six et al., 2002). Gale and Gambardella (2000) observed that soil managements promoting the return of residues to the soil, such as cropping and no-till, determine an increase of SOC and aggregation.

Soil organic matter is encapsulated whit micro and macro-aggregates and is spatially protected against decomposition, forming a physical separation of the microorganism from the substrate.

Macroaggregate (>250 μm) structures exerts a minimal amount of physical protection because of their suscebtiblity to management practices, whereas small micro aggregates (<250 μm), which are hardly break up by land management, are rich in small mesopore (2-10 nm) that protected OM due to size exclusion of enzymes (Mayer, 1994). Thanks to these characteristic and to the fact that the hydrolytic exoenzymes are as small as 3-4 nm, microaggregates are able to host and shelter organic matter (Mayer, 1994)

3. CONCLUSION

Article 3.4 of Kyoto protocol indicated cropland management as a human-induced strategy of CO_2 sequestration in agricultural soils (Smith, 2005).

Sequestration of C from plant biomass into soil organic matter is a key sequestration pathway in agriculture (Johnson et al. 2007b) because agricultural residues are the precursors for soil organic matter, the main store of carbon in the soil (Smith et al., 2008).

With the advent of industrial era, the combustion of the biomasses, the reclamation of the swamps, the deforestation and the agricultural mismanagement have determined a great reduction of the organic substance in the soil and of the plants biomass, with consequent increase of carbon dioxide in atmosphere.

Severe depletion of soil organic carbon pool degrades soil quality and adversely impacts water quality. Stover and other crop biomass are frequently referred to agricultural waste but when returned to the soil, crop residues replenish C that has been reduced of 30% to 50% of pre-cultivation levels (Schlesinger, 1985). Worlds soils could re-sequestered about 75 to 80 % of the C lost, but the rate of SOC sequestration is limited by ecological factors and management practices (Wojick, 1999).

Scientists are in agreement to propose recommended management practice (RMPs) (Figure 10) as a main provision to sustain soil organic carbon because of the major part of agricultural soils contain less carbon respect their ecological potential.

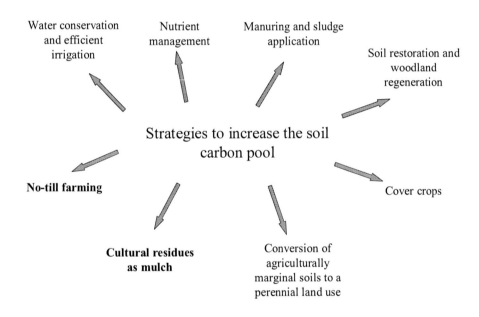

Figure 10. Recommend management practices scheme.

In particular the strategy of soil carbon management concern to raise the rates of crop residues to the soil. The conservation of the agricultural residues (no-till management) in the soil determine a multiple benefits besides theirs ability to increase the SOC: enhance in biodiversity, decrease in water and nutrient losses, minimize soil disturbance, moderate water retention and the thermal regimes, and, therefore, concur actively to the formation of a stabile soil structure and reduce risks of soil erosion.

The agricultural biomass residues are moreover principal source of nutrient and energy for the microorganisms of the soil that actively contribute to the formation and stabilization of soil aggregates, and for the plants since they take-up the micro nutrients necessary to enhance biomasses productivity and for humification of the same ones (Burgess et al., 2002; NREL, 2003; Mubarak et al., 2002).

Scientific data indicated the important role that land use and soil management plays in the global carbon budget (Soussana et al., 2004). So as showed in Table 6, agricultural land-management practices could considerable mitigate atmospheric carbon dioxide emission.

Table 6. Proposed correct management practices to mitigate CO_2 emission from agricultural ecosystems (modified from Smith et al., 2008).

Measure	Examples	Net evidence mitigation
Cropland management	Agronomy	**
	Nutrient management	**
	No-Tillage/residues management	**
	Water management	*
	Agroforestry	*
	Set aside-land use change	***
Grazing land management/pasture improvement	Grazing intensity	*
	Increased productivity	*
	Nutrient management	**
	Fire management	*
	Species introduction	**
Management of organic soils restoration of degraded lands	Avoid drainage of wetlands	**
	Erosion control, organic and nutrient amendments	**

Table 7. Global coltivated areas, world production of major crop and relative agricultural residues (data from FAO, 2008; Lal, 2005).

Crop	Global area	Global production	Residue production
	Mha	Gt	Gt
Wheat	215	602	903
Rice	150	630	945
Maize	145	711	711
Sorghum	45	59	88
Sugarcane	20	1300	325
Sugarbeet	5.4	248	62

Under no-tillage condition Halvorson et al (2002) observed a reduction in SOC of 9% respect to the conventional tillage practices. Campbell et al. (1998) showed a decrease in soil C content with straw removal. Lugato and Berti (2008) demonstated that human activities and agricultural soil management could affect C balance more strongly than climate change, so that SOC change was more marked among the RMPs than amon different climate change.

Actually no-till farming is adopted only on 6% of the total world cropland (Derpsch, 2005); extending this management practice to all agricultural lands determined a additional C sequester in soil of 05.-1 Gt C yr^{-1}(Pacala and Socolow, 2004).

Increase soil C pool by 1 Gt is equivalent to reduce CO_2 concentration in atmosphere of 0.47 ml L^{-1} and vice versa, in this way, plant residues in agricultural and management practices soil play an important role in CO_2 sequestration (Lal, 2007). In table 2 are reported global coltivated areas, world production of major crop and relative agricultural residues.

Actually the fate of crop residues could greatly contribute to increasing or reducing atmospheric CO_2 depending on its use. In developed countries, maize residues return to the soil, and the potential for ethanol production from cellulose has been suggested recently using crop residues as the main source of raw material. On the other hand, in developing countries, crop residues do not return to the soil and the C balance is negative (Rasmussen et al. 1998).

REFERENCES

Abiven, S.; Recous, S.; Reyes, V.; Oliver R. *Biol Fertil Soils*. 2005, 42, 119-128.

Adani, F.; Ricca, G. *Chemosphere* 2004, 56, 13-22.

Adani, F.; Spagnol, M.; Genevini, P. *Biogeochemistry* 2006, 78, 85-96.

Adani, F.; Spagnol, M.; Nierop, K.G.J. *Biogeochemistry* 2007, 82, 55-65.

Allmaras, R.R.; Linden, D.R.; Clapp, C.E. *Soil Sci Soc Am J*. 2004, 68, 1366-1375.

Bachmann, J.; Guggenberger, G.; Baumgarti, T.; ellerbrock, R.H.; Urbanek, E.; Goebel, M.O.; Kaiser, K.; Horn, R.; Fischer, W.R. *J Plant Nutr Soil Sci*. 2008, 171, 14-26.

Baldock, J.A; Nelson, P.N. In: *Handbook of Soil Science*. CRC Press, Boca Raton, FL, 2000, pp 25-84.

Baldock, J.A.; Masiello, C.A.; Gélinas, Y.; Hedges, J.I. *Mar Chem*. 2004, 92, 39-64.

Balesdent, J. and Balabane, M. *Soil Biol Bioch*. 1996, 9, 1261-1263.

Batjes, N. H. *Eur.J. Soil Sci*. 1996,. 47, 151–163.

Bertrand, I.; Prevot, M.; Chabbert, B. *Biores Techn*. 2009, 100 (1), 155-163.

Burgess, M.S.; Mehuys, G.R.; Madramootoo, C.A. *Can J Soil Sci*. 2002, 82, 127-38.

Campbell,C.A., McConkey, B.G., Biederbeck, V.O.,. Zentner, R.P, Curtin, D., and Peru, M.R. *Soil Till. Res*. 1998, 46,135–144.

Carpita, N.C. and McCann, M. In *Biochemistry and Molecular Biology of Plants*. Buchanan, B.B., Gruissem, W. and Jones, R.L. Eds.; Rockville, MD, 2000, pp. 52-108

Cestaat Impieghi dei Sottoprodotti Agricoli e Industriali 1990, vol.1, pp.311.

Chesson, A. In *Driven by nature Plant Litter Quality and Decomposition*. Cadisch, G.; Giller, K.E. Eds.; CAB International: Wallinford, UK 1997, pp 47-66.

Derpsch, R., 2005. The extent of conservation agriculture adoption worldwide: implications and impact. In Third World Congress on Conservation Agriculture, 3–7 October 2005, Nairobi, Kenya.

Falkowski, P.; Scholes, R.J.; Boyle, E.; Canadell, J.; Canþeld, D.; Elser, J.; Gruber, N. *Science* 2000, 290, 291–296.

FAO the state of food and agriculture 2008.

Gale, W.J.; Cambardella, C.A. *Soil Sci Soc Am J*. 2000, 64, 190-195.

Giampietro, M.; Ulgiati, S.; Pimentel, D. *Bio Sci*. 1997, 47, 587–600.

Glaser, G. *J Plant Nutr Soil Sci*. 2005, 168, 633-648.

Halvorson, A.D., Wienhold, B.J.; Black, A.L *Soil Sci. Soc. Am. J*. 2002, 66, 906–912.

Hamburg S.P. *Miti Adapt Strat Global Change* 2000, 5(1), 25–37

Heal, O.W.; Anderson, J.M.; Swift, M.J. In *Driven by Nature: Plant Litter Quality and Decomposition* Caddisch, G.; Giller, K.E Eds.:CAB International: Wallinford, UK. 1997, pp 3-30

Hedges, J.J.; Keil, R.G. *Mar Chem*. 1995, 49, 81-115.

Hedges, J.J.; Oades, J.M. *Org Geochem*. 1997, 27, 319-361.

Himmel, M.E.; Ding S.Y.; Johnson, D.K.; Adney, W.S.; Nimlos, M.R.; Brady, J.W.; Foust, T.D. *Sci*. 2007, 315, 804-807.

Hooker, B.A.; Morris, T.F.; Peters, R.; Cardon, Z.G. *Soil Sci Soc Am J*. 2005, 69,188-196.

IEA (International Energy Agency) 2007. *World Energy Outlook 2007*. Paris.

IPCC *Climate Change: Impacts, Adaptation and Vulnerability* 2001. IPCC Third Assessment Report.

Janzen, H.H. *Agric Ecosyst Environ*. 2004, 104, 399–417.

Jenkinson, D.S.*Nature* 1991, 351, 304–306.

Johnson, J.M.F.; Allmaras, R.R.; Reicosky, D.C. *Agron J*. 2006, 98, 622-636.

Johnson, J.M.F.; Barbour, N.W.; Weyers, S.L. *Soil Sci Soc Am J*. 2007a, 71, 155-162.

Johnson, J.M.F.; Franzluebbers, A.J. , Lachnicht ,W.S. ; Reicosky, D.C. *Environ Pollut*. 2007b, 150, 107-124.

Jones, M.B.; Donnelly, A. *New Phytol* 2004, 164, 423–39.

Kelleher, B. P.; Simpson, A. J. *Environ. Sci. Technol*. 2006, 40, 4605-4611.

Kiem, R.; Kögel-Knabner, I. *Soil Biol Biochem*. 2003, 35, 101-118.

Kögel-Knabner, I., Leeuw, J.W.; Hatcher, P.G.; Leeuw, J.W. *Sci Total Environ*. 1992, 117 (118), 175-185

Kögel-Knabner, I. *Org Geochem*. 2000, 31, 609-625.

Kögel-Knabner, I. *Soil Biol Biochem*. 2002, 34: 139-162.

Kögel-Knabner, I.; Guggenberger, G.; Kleber, M.; Kandeler, E.; Kalbitz, K.; Scheu, S.; Eusterhues, K.; Leinweber, P. *J Plant Nutr Soil Sci*. 2008, 171, 61-82.

Kumar, K.; Goh, K.M. *Adv Agron*. 2000, 68, 197-319.

Lal, R. *Science* 2004, 304, 1623-1627.

Lal, R. *Environ Int* 2005, 31, 575– 584

Lal, R. *Miti Adapt Strat Global Change* 2007, 12, 03-322.

Lal, R. *Phil Trans R Soc B*. 2008a, 363, 815–830.

Lal, R. *Waste Manage* 2008b, 28, 747–758.

Lamb, D.; Erskine, P.D.; Parrotta, J.A. *Science* 2005, 310, 1628–1632.

Lehmann J., Solomon D., Kinyangy J., Dathe L., Wirick S., Jacobsen C. *Nature Geosci*. 2008 1, 238-242.

Lemus, R.; Lal R. *Crit Rev Plant Sci*. 2005, 24, 1–21

Lorenz, K.; Lal, R.; Preston, C.M.; Nierop, K.G.J. *Geod*. 2007,142, 1-10.

Lorenz, K.; Lal, R.; Shipitalo, M.J. *Biol Fertil Soils*. 2008, 44, 1043-1051.

Lovett, G.M.; Cole, J.J.; Pace ML. *Ecosystems* 2006, 9,152–5.

Lugato, E.; Berti A. *Agr Ecosyst and Environ* 2008, 128, 97–103

Marschner1, B.; Brodowski, S.; Dreves, A.; Gleixner, G.; Gude, A.; Grootes, P.M.; Hamer1, U.; Heim, A.; Jandl, G.; Ji, R.; Kaiser, K.; Kalbitz, K.; Kramer, C.; Leinweber, P.; Rethemeyer, J.; Schäffer, A.; Schmidt, M.W.I.; Schwark, L.; Wiesenberg, G.L.B. *J Plant Nutr Soil Sci*. 2008, 171, 91-110.

Matsui, N.; Chen, F.; Yasuda, S.; Fukushima, K. *Planta* 2000, 210 (5): 831-835.

Mayer, L.M. *Geochim Cosmocchim Acta*. 1994, 58, 1271-1284.

Morris, D.; Ahmed, I. The Carbohydrates Economy: Making Chemicals and Industrial Materials from Plant Matter. Institute for Local Self-Reliance, 1992, Washington, DC, USA.

Mubarak, A.R.; Rosenani, A.B.; Anuar, A.R.; Zauya, S. Communications in *Soil Sci Plant Anal* 2002, 33, 609-22.

Nierop, K.G.J.; Filley, T.R. *Org Geochem.*2007, 38, 551-565.

NREL. Biomass feedstock composition and properties database [online] available at http://www.ott.doe.gov./biofuels/properties_database.html.

Pacala, S.; Socolow, R. *Science* 2004, 365, 968–972.

Papa, G. Studio del ruolo di biomolecole recalcitranti di radici di mais nei processi di umificazione nel suolo. Ms.D *Thesis,* 2008 – University of Milan, Italy.

Pauly, M. and Keegstra, K. *Plant J.* 2008, 54, 559-568.

Piccolo, A. *Adv Agron.* 2002, 75, 58-134.

Pimentel, D. *J Agr Environ Ethics.* 1991, 4, 1–13.

Pimentel, D. *Nat Res.* 2003, 12, 127–134.

Puget, P.; Drinkwater, L.E. *Soil Sci Soc Am J.* 2001, 65, 771-779.

Rasmussen, P.E.; Goulding, K.W.T.; Brown, J.R.; Grace, P.R.; Janzen, H.H.; Körschen, M. *Science* 1998, 282, 893-896

Rasse, D.; Rumpel, C.; Dignac, M.F. *Plant Soil.* 2005, 269, 341-356.

Rasse, D.; Dignac, M.F.; Bahari, H.; Rumpel, C. ; Mariotti, A. ; Chenu, C. *Eur J Soil Sci.* 2006, 57, 530-538.

Salati, S. Influenza delle modificazioni nelle vie biosintetiche della lignina in piante di mais *wild type* e *brown midrib (bm3)* sulla formazione e conservazione nel suolo di molecole umo-simili. *Ph.D. Thesis,* 2007, University of Milan, Italy.

Schlesinger,W.H. In Changes in soil carbon storage and associated properties with disturbance and recovery; Trabalha, J.R.; Reichle, D.E.; Eds; The changing carbon cycle: A global analysis. Springer-Verlag: New York, USA, 1985, pp. 194–220

Schlesinger, W.H. *Biogeochemistry: an analysis of global change*, 2nd edn. Academic Press.: San Diego, CA, USA, 1997.

Schnitzer, M. 1991, *Soil Sci.* 151, 41–58.

Schrag, D. P. *Science* 2007, 315,812–813.

Seibel, B.A.; Walsh, P.J. *Science* 2001, 294, 319–320.

Shrestha, R. K.; Lal, R. *Environ In.t* 2006, 32, 781–796

Silver, W.L.; Miya, R.K. *Oecol.* 2001, 129, 407-419.

Sims, R.E.H; Hastings A.; Schalamandinger B.; Taylor G.; Smith P. *Global Change Biol.* 2006, 12, 2054–2076.

Six, J.; Conant, R.T.; Paul, E.A.; Paustian, K. *Plant Soil.* 2002, 241, 155-176.

Smith, P. *Eur J Soil Sci.* 2005, 56, 673–680

Smith, P.; Martino, D.; Cai, Z.; Gwary, D.; Janzen, H.. *Phil Trans R Soc B.* 2008, 363, 789–813

Soussana, J.F.; Loiseau, P.; Vuichard, N.; Ceschia, E.; Balesdent, J.; Chevallier, T.*Soil Use Manage* 2004, 20 (Suppl.),19–23.

Spagnol, M. Contributo delle parti epigee della pianta di *Zea mays* L. Alla formazione di acidi umici del suolo *PhD Thesis,* 2005, University of Milan, Italy.

Stevenson, F.J. Humic chemistry genesis, composition, reactions. 2nd ed. John Wiley and Sons, New York, 1994.

Sutton, R.; Sposito, G. *Environ Sci Technol.* 2005, 39 (23), 9009-9015.

Torn, M.S., Trumbore, S.E., Chadwick, O.A., Vltousek, P.M., Hendricks, D.M. *Nature* 1997, 389, 170-173.

Trinsoutrot, I.; Recous, S.; Bentz, B.; Linères, M.; Chènebey, D.; Nicolardot, B. *Soil Sci Soc Am J.* 2000, 64, 918-926.

Tsang, C.F.; Benson, S.M.; Kogelski, B.; Smith, R.E. *Environ Geol.* 2002, 42, 275–281.

van Soest, P.J.; Robertson, J.B.; Lewis, B.A. *J Dairy Sci.* 1991, 74, 3583-3597.

von Lützow, M.,; Kögel-Knabner, I.; Ekschmitt, K.; Matzner, E.; Guggenberger, G.; Marschner, B.; Flessa, H. *Eur J Soil Sci.* 2006, 57, 426-445.

Wang, W.J.: Baldock, J.A.; Dalal, R.C.; Moody, P.W. *Sol Biol Biochem.* 2004, 36, 2045-2058.

Weisz, P.B. *Physics Today* 2004, (July), 47–52.

Wershaw, R.L. *Evaluation of conceptual models of natural organic matter (humus) from a consideration of the chemical and biochemical processes of humification.* U.S. Geological Survay, Reston, Virginia, USA. 2004

Wilhelm, W.W.; Johnson, J.M.F.; Hatfield, J.L.; Voorhees, W.B.; Linden, D.R. *Agron J.* 2004, 96, 1–17.

Wilts, A.R.; Reicosky, D.C; Allmaras, R.R.; Clapp C.E. *Soil Sci Soc Am J.* 2004, 68, 1342-1351.

WMO, *Greenhouse gas bulletin: the state of greenhouse gases in atmosphere using global observation up to December 2004.* World Meteorological Organization: Geneva, CH, 2006.

Wojick, D.E. *Carbon Storage in Soil: The Ultimate No-Regrets Policy?* A report to Greening Earth Society 1999.

Zimmerman, A.R.; Goyne, K.W.; Chorover, J, Komarneni, S.; Brantley, S.L. *Org Geochem.* 2004, 35, 355-375.

In: Encyclopedia of Environmental Research
Editor: Alisa N. Souter

ISBN: 978-1-61761-927-4
© 2011 Nova Science Publishers, Inc.

Chapter 17

COLLAGEN WASTE AS SECONDARY INDUSTRIAL RAW MATERIAL

*F. Langmaier** and P. Mokrejs[+]*

Institute of Polymer Engineering, Faculty of Technology,
T. Bata University, The Czech Republic

ABSTRACT

Use of collagen in food production is limited by its low nutritive value and deficient essential amino acids. Irrespective of gelatine and glue manufactured by boiling refined native collagen, industrial processing of collagen is connected with production of a considerable amount of difficultly utilizable protein waste. In the phase of refining collagen raw material, hydrolysates of keratin and accompanying proteins (albumins, globulins) that are isolated with ease go away; their following usage comes up against similar problems as with keratin. Use of collagen waste as of secondary industrial raw material is usually complicated by an excessive density of irreversible crosslinks introduced for stabilizing against chemical and microbial effects. The first stage in processing such waste is hydrolysis that is partially alkaline, acid, but above all enzymatic, and interesting for its low energy demands. Obtained collagen hydrolysates behave like gelatine, however, with a view to their molecular weight being 5-10 times lower, they exhibit higher hydrophility and formation of their gels requires higher dry substance concentration. Sole collagen hydrolysates proved themselves well as a component in urea-formaldehyde resins reducing formaldehyde emissions by their cured films. The same as polyamides, they form considerably stable networks when crosslinking epoxide oligomers; their potential disintegration through microbial procedures is advantageous when removing protective films prior to recycling plastic automobile or aircraft parts on termination of their life. Collagen hydrolysate, crosslinked with polymeric dialdehydes (e.g. dialdehyde starch) results in biodegradable or even edible packaging materials which, under suitable conditions, may be processed by technology of dipping, casting and drying as well as by procedures typical for processing synthetic polymers (extrusion and compression molding).

---

* langmaier@ft.utb.cz.

+ mokrejs@ft.utb.cz.

1. COLLAGEN

Collagen, a protein of connective tissues of live organisms with rather small differences in its sequence of amino acids (primary structure) but more in its supramolecular conformation, easily adapts to different biomechanical and biochemical functions in animal organisms. Beside its own supportive and ligamentous function, collagen controls interaction of live cells with connective tissue, and cell differentiation during embryonic development of animal organisms.

At present, eighteen or nineteen types of collagen or known, and these mainly differ in their supramolecular structure, less in basic protein chains. Supramolecular structure of fibrillar types of collagen (collagen types I, II, III, V and IX) is characterized by a triple super-spiral (super-helix). Collagen type IV is the collagen in walls of endothelial and epithelial cells (basal membranes) and its structure comprises both proteins and proteoglycans. With other types, structure of the collagen super-spiral is interrupted by non-helical segments that impart greater flexibility; still other types have globular domains attached to collagen chains.

Table 1. Particular collagen types and their occurrence.

Type	Most significant regions of occurrence
I	Skin, tendons, bones, muscles, intervertebral plates, cartilage, aorta, placenta
II	Skin, tendons, dentin, scars
III	Skin, muscles, aorta, placenta, lungs, liver
IV	Basal cellular membranes
V	Placenta, liver
VI	Placenta, uterus, aorta, long tendons of abattoircattle
VII	Anchorage of dermis and epidermis fibers (long chain –LC– collagen)
VIII	Endothelial collagen, basal membranes
IX	Fibril-associated collagen with interrupted triple helix, cartilage
X	Transition between bone and cartilage
XI	Cartilage
XII	Collagen fibrils, supers-spiral interrupted by non-helical structure
XIII	Epidermis, muscles, cartilage
XIV	Skin, cartilage: collagen super-spiral interrupted by non-helical structure
XV	Cellular walls: skin, uterus, lungs, smooth muscles
XVI	Cartilages
XVII	Skin
XVIII	Skin, heart, brain, lungs
XIX	Data so far problematic

Collagen types of minor importance (minor collagens: type VI – intima collagen, VII – LC– long-chain collagen, type VIII – endothelial collagen, type IX and/or X – cartilage collagen) are interesting more with a view to studying connective tissues of live organisms and their diseases or to medical research, and their industrial significance is quite small. An orientational survey of particular collagen types and their occurrence is apparent from data in Tab. 1.

Properties, especially those of fibrillar collagens (types I, II, III, V and XI), are strongly affected by intermolecular (inter-chain) crosslinking of polymer chains. Crosslinks arise through the reaction of aldehyde groups generated through enzymatic oxidation (lysil-oxydase) of amino groups of lysine side-chains to allysine. Produced aldehyde groups subsequently react with neighboring amino group of side chains of another lysine or hydroxy-lysine to aldimine (also known as azomethine or Schiff base). Such bonds – particularly in an acid environment – are more resistant to hydrolysis than the peptide bond characteristic of actual protein chains. Bonds produced from lysine-aldehyde may be split in an environment of low pH level (e.g., 0.5 M acetic acid), splitting of bonds based on hydroxylysine-aldehyde (produced from hydroxyl-lysine) is more difficult.

With aging of organism, crosslinks of aldimine type convert to more stable bonds (hystidyl-hydroxylysino-norleucine – HHL – current in collagen and tendons) through a reaction with neighboring hystidine residue. Conversely, in collagen of cartilage (group of collagen FACIT = Fibril-Associated-Collagen with Interrupted Triple-Helix) there are crosslinks of hydroxy-lysil-pyridinoline type. In collagen of basal membranes (type IV), disulphide (inter-chain) crosslinks arise through an aldehyde mechanism initiated by lysil-oxydase. Variability of crosslinks seems to depend more on type of connective tissue than on specific type of collagen. The basic pathway is primarily determined by hydroxylation of end telopeptides of collagen chains and by domain of lysine residues in the triple spiral. Problems of particular collagen types, of possibilities, formation and of character of inter-chain crosslinks in them are discussed in detail, e.g., by Eyre and Jian-Jiu Wu [1] or Heidemann [2].

Crosslink density in native collagen (particularly of fibrillar types) grows with the age of connective tissues of organisms and is cause of very strong changes in its properties. Its increase reduces the fraction of soluble collagen, swelling ability of collagen fibers (particularly in an acid environment), and sensitivity of collagen to proteolytic enzymes goes down; on the other hand, mechanical strength of collagen fibers and their stiffness grows [3].

Crosslinks introduced into fibrillar collagens artificially during industrial processing have a similar effect. Optimal crosslink density provides final collagen materials with resistance to chemicals and enzymes as well as convenient physical and chemical properties. Surpassing their optimal density shows through excessive stiffness and fragility (particularly in grain layer of leather containing finest collagen fibers), which leads to worse shop processing in following leather goods manufacture. Similar effects are also produced by oxidation of collagen fibers, which also essentially leads to undesired cross-link density [2, 3].

2. INDUSTRIAL USE OF COLLAGEN

As a protein of connective tissues, collagen (particularly fibrillar) is essentially a side-product of meat production. From point of view of industrial processing, the chief source of collagen is the skin of abattoir animals, making up 10-12 % of their live weight. Collagen content in bones of abattoir animals (together with parts of cartilages and joints) is estimated at 29-35 % of their live weight and represents a certain not overly important culinary application (food additives). A greater part of it is processed into bone oil; a degreased residue is then used for making lower-class types of glues, bone flour or spodium. A potential source of fibrillar collagen are tendons of abattoir cattle, making up 0.2-0.5 % according to estimates by various authors, or intestines (under 0.2 %) of live weight of abattoir cattle. While intestines are used as natural edible casings of meat products, short and long tendons of abattoir cattle are not industrially used at all. The leather industry, chief consumer of abattoir cattle leathers, requires their at least approximate shaping during which trimmings of small size, irregular shape and difficult processing are scrapped. These might be regarded as a certain source of collagen for other processing besides tanning.

Owing to deficient essential amino acids (above all, tryptophane, tyrosine and cysteine) and thus to extremely low nutritive value, collagen has only very limited significance for actual manufacture of foodstuffs. Utilizing the low nutritive value of collagen in diets for reducing body weight was quite extensively discussed in the seventies and eighties of the last century and it was found the unbalanced amino acid composition (excess glycine and arginine together with insufficient tryptophane, tyrosine and cysteine) may develop clinical problems for patients, for example, cardiac arrhythmia, dehydration and hypocalcaemia or amyotonia, which in extreme cases can even lead to their death. Efforts appeared aiming to correct amino acid unbalance of collagen by adding other proteins (or their hydrolysates) with a suitable content of deficit amino acids [1, 4], but practical success of such solutions was not too great.

The unbalance of collagen amino acid composition is a major problem even for application in feed mixtures for farm animals or poultry. Collagen proteins for such applications need not be rid of proteins of other types accompanying them in abattoir waste. As a rule, it is recommended to process protein waste from meat production as a whole, through partial hydrolysis, thickening and subsequent granulation of the product. The remaining problem consists in costs of evaporating excess water before granulation of the product. Necessity of combining such hydrolysates with hydrolysates of other proteins, if adequate nutritive quality is to be ensured, continues. A certain complication appears in tightened requirements for hydrolysis (in particular, high temperature or pressure) if the hazard of prion transfer (mad cow disease = Kreuzfeld-Jacob syndrome) to feed mixtures is to be eliminated. Some existing fragmentation in meat production, thus in production of protein waste, may be regarded as a complication of economic character.

Collagen, as a biopolymer, differs from synthetic polymers above all by its reaction with tissues of live organisms [5]. It is biodegradable, displays weak antigenecity [6] and possesses outstanding biocompatibility. A number of overall as well as specially focused works [7] were published on collagen medicinal applications. Fibrous structure of collagen conditions its high strength characteristics for which it is extensively used in live tissue engineering – including matrices for cultivating cell systems, replacing blood vessels [8-10]. Collagen films, collagen fibers processed to resemble cotton wool (collagen sponges) are extensively employed as styptic material [11] and material to cover deep wounds and burns. Collagen fibers are also the basis of surgical catgut.

Collagen soluble in an aqueous environment (not bonded by natural crosslinks) is finding application in cosmetic preparations for skin treatment. However, its incorporation in cosmetic formulations of this type is associated with their possibility of lower stability or life. Moisturizing effect for the skin is connected with formation of a collagen film on the treated skin, slowing diffusion of water through epidermis. Such an occlusion effect of film-forming hydratants is essentially analogous to the effect of collagen films applied to cover extensive open wounds or deep burns. With its quite high molecular weight, soluble collagen has limited capacity to penetrate deeper layers of epidermis and a moisturizing effect remains more bound to surface. Although soluble collagen is appreciated quite highly in such formulations, collagen hydrolysates, of ten- to twenty fold lower molecular weight, obviously have significant advantages when regarded from both viewpoints mentioned above.

The partial hydrolysate of collagen – gelatine – holds a special position. It lacks insolubility of fibrous collagen in an aqueous environment and permits in it formation of thermo-reversible gels. Gelatine, as a biodegradable and edible material, is employed more often than collagen itself as a biodegradable (edible) packaging material [12, 13]. Hard (HGC) and soft gelatine capsules (SGC) with a different content of hydrophilic plasticizers are currently used in pharmaceutics as a vehicle of various medicaments including antibiotics, and their increased crosslink density, mostly by means of low-molecular aldehydes, makes possible time-controlled concentration of medicine in the blood circulation. Increased crosslink density is above all achieved by means of low-molecular aldehydes, which, nevertheless, have a certain disadvantage in their toxicity to live organisms.

In addition, gelatine gels are used in the food industry for adapting the consistence of various syrups, creams, puddings and ice creams. As a food additive (thickening agent, stabilizer), gelatine is also employed in canned meat and fish, and there is a report describing use of gelatine to prolong taste of some food products (chewing gum).

However, the chief part of collagen is processed industrially into flat materials for shoe, clothing and fancy goods industry. Such types of manufacture are mostly economically quite lucrative, their principal negative remains high water consumption, chiefly for purifying collagen raw material, and production of a considerable amount of protein waste, comprising both proteins of non-collagen type (albumins, globulins, keratin and the like) which collagen must be essentially disposed of prior to every processing if quality of the final product is to achieve adequate grade, as well as waste of collagen type arising in final surface shaping. An imperfect estimate of suitable crosslink density in the final collagen product affects properties of the obtained material in fundamental manner, and in extreme cases becomes the source of further (tanned) collagen waste. Summarized, these circumstances pose quite a serious problem, because according to some authors the yield of producing leather fluctuates around 40-50 % related to starting protein raw material [14, 15].

3. WASTES OF COLLAGEN INDUSTRIAL PROCESSING

Collagen for industrial processing is obtained in the first place from hides of abattoir cattle. Properties of hide are to considerable extent individual, and the condition for industrial processing is the classification of hides in accordance with visually discernible damage (healed or unhealed wounds, cuts or cracks caused during flaying, damage to hide by skin disease, etc) or other criteria (type of abattoir cattle, weight of hide and the like). Degree of hide natural crosslinking, which is rather individually dependent mainly on age of the cattle, cannot be usually quite well considered. Variations in hide quality then show in variations in the quality and yield of products of their industrial processing (leathers, gelatine or glues, and also products for application in medicine, pharmaceutical and/or cosmetic manufacture).

Besides, obtaining hides at abattoirs is a discontinuous process, and securing a smooth supply of collagen raw material for following industrial processing presents a certain organizational problem. Consequently, adequate stability of hides against biodeterioration and even chemical hydrolysis has to be secured because the immunity system of abattoir cattle ceases to function after flaying.

Systematic experiments proved that hide biodeterioration at 29-32 °C shows after 156 hours at the latest, at 15 °C hide can be stored without harm for 2-3 days [16]. Hence, slowing biodeterioration processes may be achieved by cooling hides, after temperature after flaying was lowered (e.g. by cold water) to at least 20 °C. Preservation of hides with bactericide solutions (also in combination with a lower concentration of salt), by radiation or X-rays are efficient enough under certain conditions, but utilization in practice is barred for economic reasons. Salting either with solid NaCl or its concentrated solutions is the most widespread mode of hide conservation since about the half of the 19th century. Sufficiently conserved hide should not contain more than 12-20 % salt (per dry substance). Under favorable climatic conditions, conservation combining salting and drying can be applied but the procedure is not too widespread. The time needed to reduce water content in hide below a limit excluding biodeterioration may get prolonged due to formation of a proteinic surface film preventing diffusion of moisture from inside hide.

Industrial processing of hide, in general, comprises two basic stages: collagen final purification (preparation of pelt) and, if possible, controlled increase of natural crosslink density so that the created product is technically stable enough and thus technically applicable. The second phase of these mentioned is well-known as collagen "tanning".

Conservation of hides after flaying consequently requires that their original water content is renewed before processing, and that washing disposes them of coarse impurities. This starting operation is designated soaking; it is accelerated by mechanically moving hides in water bath or by adding surfactants or other substances that facilitate wetting the surface of conserved hides with water.

3.1. Degreasing Hides – Waste Fat

Before liming (possibly also before soaking), hides of a higher fat content (sheepskins, pigskins) have to be disposed of lipoid components that complicate collagen processing or of waste lime liquors through increased chemicals, water and energy consumption. In practice, degreasing of hides is performed by extracting lipoid components of raw hides with organic solvents, dispersing them in water baths of anionic surfactants or of their mixture with nonionic surfactants. Efficiency of degreasing collagen raw material may be to some extent influenced by pH of aqueous environment or ionic strength of the bath. Surfactants have the chief role in such baths by dispersing lipoid substances in an aqueous environment, they actually affect hide degreasing in negative manner. That is caused by their oriented sorption onto phase interface leading to formation of a diffusion barrier (surface film) on cellular membranes of fat cells [17].

Launching industrial production of lipolytic enzymes from raw materials of microbial origin led to a remarkable decrease in production costs of lipolytic enzymatic preparations. That removed the main obstacle to expanding biotechnological procedures for degreasing collagen raw material, designed in the first place for application in human medicine, pharmacy or cosmetics. In actual production of leather (flat materials for footwear, clothing and fancy goods manufacture) such procedures continue to be applied, above all for economic reasons, rather exceptionally.

Lipoids in collagen raw material are deposited in cells of fat whose walls are made of proteins. The combination of proteases and lipases enhances efficiency of leather enzymatic degreasing, and most commercially available enzymatic degreasing preparations contain both types of enzymes in a certain ratio. Hydrolysis of neutral triglycerides with lipolytic enzymes leads to formation of partial glycerides and free fatty acids (their salts in an alkaline environment) which then act as dispersing agents (surfactants) of neutral triglycerides. Typical examples of commercially successful products featuring a majority of lipolytic enzymes of microbial origin (Aspergillus) are Greasex^{TM}, NovoCor AD, or Lipolase^{TM} (Novozymes, Bagsvaerd, Denmark) initially developed for facilitating removal of fat spots by washing preparations. Maximal efficiency may be achieved by enzymatic preparations in acid, neutral or alkaline region of pH [18].

Fat released from hides is isolated, according to character of degreasing baths, by distilling off solvent; with dispersions stabilized with anionic surfactants (including enzymatic degreasing baths where dispersing agents are partial glycerides or free fatty acids and their salts) by changing level of pH. When nonionic surfactants are in majority, salting out may be employed.

Fats obtained may be utilized as a secondary raw material for manufacturing surfactants. Their rather irregular composition is a certain disadvantage, affected among others also by type of hides being processed. Very tolerable results were obtained by processing such fats into surfactants of mono- or diglycerol sulfates through a quite simple procedure undemanding on investment and manufacturing plant [19].

A number of preparations used for fat-liquoring leather in its production are based on this. Such preparations have an advantage in their low ability to migrate into the grain of such leathers. Migration of fat-liquoring preparations is regarded as the main cause of the origin of fatty spots mainly in the grain of greased (clothing) leathers.

3.2. Preparation of Pelt and Waste Proteins of Liming Liquors

In addition, industrial processing of collagen requires separation of collagen from other accompanying proteins, mainly keratin from epidermis, and albumins and globulins (from remaining body liquids) and elastin. This is mostly performed by a procedure based in principle on splitting disulfide crosslinks of keratin with strongly alkaline suspensions of reducing agent (unhairing hides with sulfide-calcium pulp). For economic reasons, use of sodium sulfide (Na_2S) or sodium hydrogen sulfide (NaHS) with calcium hydroxide prevails. Agents formerly used, sulfides of calcium (CaS) or arsenic (As_2S_3) have been practically abandoned today, the same as thioglycolic acid ($HS-CH_2-COOH$) - for unpleasant odor rather applied in the form of its sodium salt – and employed amino compounds (e.g., dimethylamine sulfate). Even the oxidation method for splitting disulfide crosslinks of keratin (for example, by means of chlorites or chloramines) is not in use today, also because of possibly released toxic gases (ClO_2), neither procedures applying various proteases because of the complicated control of such a process.

After erosion of keratin of the hair follicles, hair and skin are mechanically released, and separation of collagen from keratin of the epidermis (or of other accompanying proteins - albumins, globulins) is finished in liming liquors (3-4 % calcium hydrate and 1-3 % technical sodium sulfide) with mechanical movement of liquor. Liming is usually finished in 14-24 hours.

Relatively high content of calcium in collagen (consequence of Ca concentration in liming liquor) has to be reduced by washing with water and deliming in solutions containing substances changing poorly soluble Ca^{2+} compounds in pelt into water-soluble salts (lactic acid or its salts; hydrochloric acid, sulfuric acid, sulfonaphthalene acids and others). The last washing baths are heated to approx. 36 °C. Thus, apart from increasing the deliming effect, heating collagen reaches the temperature necessary for enzymatic removal of elastin or residues of other non-collagen proteins. Baths of pancreatic enzymatic enzymes (trypsin, papain and others) are utilized to this purpose, all exhibiting maximal efficiency at the required temperature. Simultaneously, a certain opening of the collagen structure is effected, which facilitates diffusion of tanning substances.

Such a method of purifying collagen (pelt preparation) is most usual; regardless of the way finally purified collagen is further used. The procedure is demanding on water consumption, and its by-products are difficultly recyclable lime liquors containing sludges, essentially insoluble calcium compounds (particularly $Ca(OH)_2$) and dissolved hydrolysates of keratin, albumins and possibly globulins, which are together called proteins of lime liquors. These increase ammonia pollution of tanning wastewaters and contribute to restricting development of pelt production especially there where pollution of the environment is a problem strictly observed.

Discussions on "clean" tanning technologies often deal with possibilities of hide liming in a closed cycle (with respect to Ca^{2+} compounds – including calcium salts of fatty acids – and protein components) Proposed solutions are mostly based on separating excess solid calcium hydroxide by centrifuging or by energetically less demanding sedimentation. Filtration techniques are complicated by the presence of protein hydrolysates and calcium soaps. Separated $Ca(OH)_2$, however, is difficult to use directly for preparing new liming liquors, mainly for the hazard of its microbial contamination that could negatively affect the result of liming. For this reason, annealing $Ca(OH)_2$ to CaO with the unpopular increase in energy demands of recycling is usually preferred. Economic barriers then keep the whole problem rather on the level of theoretical discussion.

Protein hydrolysates may be coagulated from liming liquors disposed of excess suspended $Ca(OH)_2$, by acidifying to pH 4.0-3.5 [20]. Coagulate may be separated to advantage by sedimenting or centrifugation. Its needed microbial stability is achieved by drying to a water content > 10 % (w/w). Energy required for drying coagulate, together with consumption of acid for coagulation, shifts costs of producing proteins from liming liquors to at least 0.9-1.5 Euro/kg air-dry product (in prices of 2005).

Proteins from liming liquors are products of no particular quality. Their ash content (particularly Ca^{2+} compounds) depends on applied method of separating excess $Ca(OH)_2$.

While the product isolated by centrifugation contains 0.5-2 % ash substance (depending on centrifugation conditions), with separation by sedimentation (technique less demanding on energy) it may reach 20-22 %. Content of S^{2+} found in their dry substance usually does not exceed 0.01 % (w/w).

A significant characteristic of isolated proteins is the presence of virtually all essential amino acids. Content of essential amino acids (from cattle hide liming) compared with their content in collagen [21-23] and in a convenient protein mix for feeding poultry [4] is shown in Tab. 2. Proteins from liquors reasonably correspond to requirements for a well-balanced content of essential amino acids in feed mixtures and may ensure good growth for chicken (broilers).

Table 2.Percentual presence of amino acids in protein isolated from tanning liquors (liming cattle hides) compared to that in a feed mixture ensuring (according to Chvapil) optimal poultry growth.

Amino acid	Lime-liquor protein	Optimal feed mixture /after Chvapil/	Keratin	Collagen
Glycine	5.97		9.5	21.7
Alanine	16.3		6.0	9.5
Valine *	10.74	4.37	6.0	7.4
Leucine *	0.7	3.93	7.7	3.4
i-Leucine *	5.4		4.3	1.5
Proline	6.8		7.6	16.0
Fenylalanine *	5.7	6.52 (Fenylal.+Tyr.)	2.5	2.3
Tyrosin	1.2		3.0	0.86
Tryptofan *	2.0	1.00	-	-
Serine	3.1		9.4	11.79
Threonine	3.98	3.48	5.8	2.17
Cystine/2	7.5	3.74 (Meth.+Cys./2)	11.9	0.25
Methionine *	4.5		0.4	0.68
Arginine *	6.2	6.08	6.9	9.38
Histidine *	0.65	2.00	0.9	0.82
Lysine *	2.7	5.43	2.8	4.16
Asparagic acid	13.7		5.8	6.13
Glutamic acid	11.5		9.6	1.27
Hydroxyproline	-		-	12.8
Hydroxylysine	-		-	1.3
Amide N_2	15.31		16.47	17.76
S	2.8		3.86	0.19

* essential amino acids

Contents of particular amino acids compared with total content of ammonia nitrogen and sulfur indicate that polypeptides isolated from liming liquors are made up of approx. 55-70 % (w/w) hydrolysates of keratin and of approx. 24-26 % (w/w) hydrolysates of originally present albumins. Hydrolysates of globulin may be represented by approx. 5-15 % (w/w) and least represented are hydrolysates of collagen (approx. 1.5-2.5 % w/w).

Considerable attention has lately been given to proteins of keratin type and their hydrolysates. Most works, however, are focused on keratin from feathers of poultry processed by the food industry, or from sheep wool. Disulfide bonds of keratin (which is also the majority component in proteins of liming liquors) complicate its processing into biodegradable films and foils. Keratin (mostly obtained from poultry feathers) is usually little degraded. Solutions of reduced keratin for casting foils or films, which are obtained by extracting with 2-mercaptoethanol, are not sufficiently stable, although Tsai et al. [24] report the possibility of preparing more stable solutions. Films obtained by casting such solutions, without added hydrophilic plasticizers, are too fragile. A certain improvement in mechanical properties may be achieved by adding chitosan. Keratin-chitosan composite films then also display antibacterial activity [25]. Oxidative erosion of keratin disulfide bonds – e.g. by action of $Na_2S_2O_5$ (sodium disulfide) which transforms disulfide bonds to the (S) sulfo- form [26] leads to more stable film-forming solutions. Films obtained from them by casting alter their characteristics with a change in moisture content, and without added hydrophilic plasticizers remain relatively fragile. Films obtained from keratin by extruding display somewhat better characteristics [27]. Fibers obtained by extruding keratin fibers combined with high-density polyethylene (HDPE) are sufficiently thermally stable up to temperatures of 200 °C [28].

The effort to technically utilize proteins from liming liquors has so far been less successful. Isolated proteins may be admittedly dissolved in a strongly alkaline medium (pH=11-12), but usually up to 21 % undissolved residue is found in them. The same as with keratin extracted from poultry feathers with reducing agents (2-mercaptoethanol), content of insoluble fraction increases in time with the aging of their solutions. Even though biodegradable films may be prepared even from dispersions containing a certain insoluble fraction, without added hydrophilic plasticizers they do not possess too good mechanical properties (fragility) and the uncertainty concerning potential microbial contamination, the same as relatively high production costs, limit their potential applications.

Grafting proteins from waste liming liquors with monomers of the acrylate or methacrylate family was initially realized rather to the purpose of utilizing such products in surface finishing of leathers and similar flat materials [29]. Films prepared through casting of their dispersions are yellow to light yellow, which poses a certain disadvantage for such applications. Nevertheless, their fragility has been reduced by grafted acrylate chains to such an extent that their application as packing materials for farming chemicals (packing for fertilizers, herbicides, insecticides and the like) can be taken into account. Their slow kinetics of swelling or dissolving in an aqueous environment, which such films and foils retain, can play a role in time-controlled release of potent farming chemicals, regardless of the fertilizing effect of the films themselves.

4. WASTE OF REFINED COLLAGEN TYPE

Purified collagen (pelt) is industrially used above all for producing flat materials (leathers), processed in footwear, clothing and fancy goods manufacture. A difficultly processable fraction in such manufactures (irregular scraps from leather shaping, splits and the like) are subsequently used to produce edible casings of meat products, or gelatine which is successful not only in manufacture of edible packages for food additives but also as blood plasma expander, carrier of some drugs, biodegradable packages of cosmetic preparations etc, or (in deeper degraded form) as a water-soluble adhesive for porous adherends (glue).

Casings of meat products are made from purified collagen through repeated swelling in an alkaline and acid medium and mechanical processing into a collagen dough (gel containing less than 12 % dry substance), which receives added hydrophilic plasticizers and is extruded into pre-tanning liquors of lower-molecular aldehydes (glutaraldehyde, formaldehyde) into formations of cylindrical shape. A modification of this "dry" process is disintegrating collagen in an acid medium into a viscous suspension of dry substance concentration approx. in the 4-5 % range, adding hydrophilic plasticizer and extruding into a coagulation bath of sodium chloride. Tanning is again performed, usually with glutaraldehyde. The wet process has an advantage of relatively lower energy consumption. In time, a number of modifications of both basic procedures [30] were proposed. Stability of obtained material (casings) is determined by the crosslinking degree (pre-tanning) with diluted solutions of lower-molecular aldehydes (formaldehyde, glyoxal, glutardialdehyde), all of which, however, show certain toxicity. Density of introduced crosslinks also determines to decisive extent rheological properties, digestibility or other quality characteristics. Excessive density leads to a product with unsatisfactory mechanical (high fragility) and rheological properties, while on the opposite, density too low negatively affects hydrolytic stability of the product. Crosslink density is more regulated empirically, which rather easily results in waste of tanned collagen type, whose utilization or further processing is quite problematic. For this reason, attention has lately focused on biotechnological possibility of crosslinking collagen with enzymes of amino-transferase type, in particular, trans-glutaminases of microbial origin (e.g. Streptoverticullium sp., Streptomyces sp.) [31, 32], whose efficiency does not depend on concentration of Ca^{2+} ions in the system. Trans-glutaminases usually catalyze transfer of acyl between the γ-carboxy-amide peptide-bound group and amino group of side chains (e.g., of ε-amino lysine). Crosslinking efficiency of proteins with higher glutamine content (in flexible region of protein chains) is then, as a rule, higher. Mechanism of such collagen crosslinking is near to crosslinking of native collagen during its aging, but efficiency of the process generally also depends on the type of amino-transferase as well as protein [33].

Extracting delimed pelt with water at temperatures up to 95 °C yields gelatine. Alkaline environment in pelt preparation leads to partial disamidation of asparagine and glutamine side chains, and such gelatine (type B - basic), therefore, has its isoelectric point shifted to the pH region of 4.8-5.0. Collagen obtained, for example, from bones of abattoir cattle disposed of inorganic fraction with hydrochloric acid by extraction with water, leads to type A gelatine (acid gelatine) which, on the contrary, displays an isoelectric point in the pH range of 7.0-9.0. The acid method is applied also when obtaining gelatine from pigskins [34].

According to Bailey et al. [35], collagen from hide of immature animals contains crosslinks of dehydroxynorleucine type, collagen of their bones then crosslinks of hydroxyisonorleucine type. Such crosslinks transform through aging to more stable bonds of histidinhydroxy-isonorleucine and pyridoline type. With aging of abattoir cattle, density of collagen crosslinks grows and their stability increases. Crosslinks make denaturation and hydrolysis of collagen as well as its transformation to gelatine difficult [36].

Peptide bonds of collagen chains are less resistant to hydrolysis than inter-chain crosslinks. Higher density of collagen crosslinks, which leads to necessarily higher water extraction temperatures (e.g. 85-95 °C), is thus associated with a decrease in mean molecular weight of gelatine and stiffness of its gel, hence with a drop in quality of obtained gelatine. High quality gelatine is obtained through extraction at temperatures up to 70-75 °C, quality of gelatine extracted at higher temperatures falls so much that it is rather employed as glue – natural adhesive for porous adherends, or as protective colloid in some systems. Quality of gelatine and its yield are thus somewhat contradictory requirements, associated (apart from others) with age (crosslinking degree) of starting raw material (pelt).

While interest in high quality gelatine (biodegradable and edible packaging material) at present increases, glues (cheap natural adhesives) are gradually superseded by adhesives based on synthetic polymers. Increasing production of cosmetic formulations for skin protection and care is stimulating interest in collagen hydrolysates used in such formulations as a humectant. Their mean molecular weight, much lower than with water-soluble collagen (originally recommended to this purpose) accelerates diffusion of humectant into treated epidermis. The majority of such cosmetic formulations have the character of dispersions whose tendency to instability in the presence of high-molecular substances like soluble collagen is quite strong and has to be respected during preparation (e.g. sequence of adding components, changes in pH of dispersions through their aging, etc). Replacing soluble collagen with collagen hydrolysate reduces these dangers.

Such development led to proposing a temperature limit to gelatine extraction at 70 °C (obtaining high quality gelatine of gel rigidity > 300 Bloom) and instead of increasing extraction temperature (connected with lower quality of obtained gelatine), to hydrolyze unextractable residue under these conditions with commercially available proteases of microbial origin into collagen hydrolysates of mean molecular weight around 20-30 kDa, which find very good application in cosmetic formulations for skin protection and care [37].

Processing Collagen Waste from Abattoir

An untraditional and relatively little utilized source of collagen are short and/or long tendons of abattoir cattle. To a certain limited extent they are used for preparing some collagen preparations (soluble collagen, collagen for food additives); separation of both kinds of tendons is performed rather exceptionally and the prevailing effort is to process them together with other abattoir waste into feed mixes or as fertilizer. However, short tendons are a source of quite pure collagen and such procedure is not economically (and ecologically) most appropriate.

While industrial processing of pelts applies the traditional method of purifying collagen, and biotechnological purifying procedures are utilized (for economic reasons in the first place) rather sporadically, development in the field of proteolytic and lipolytic enzymatic preparations of microbial origin created conditions for biotechnological procedures of purifying collagen from untraditional sources (short and long tendons of abattoir cattle, trimmings of hides during their raw shaping). Commercially available enzymatic preparations, initially designed for hide degreasing (e.g. NovoCor®, Greasex® from Novozymes, Bagsvaerd, Denmark, or some others) found ready use in this field.

Short tendons of abattoir cattle (musculus extensor communis, musculus flexor digitorum, musculus flexor digitorum profundis) consisting of relatively pure collagen, contain about 4.5-5% (per weight of processed material) lipoid components which easily pass into an acid or alkaline degreasing bath containing such enzymatic preparations. Work employs 1-% solution of enzymatic preparation NovoCor AD- in a medium of 1-% acetic acid, or 1-% solution of preparation Greasex™ in a medium of 1-% cyclohexylamine. The process is conducted at 38 °C without mechanical stirring of reaction mixture required; in 48 hours, sufficiently pure collagen for any other application is obtained.

When both preparations are applied, released lipoid substances are dispersed in degreasing baths without other added dispersing agent. For its function is taken over by fatty acids or partial glycerides released from starting material. Approximately 98 % of originally present (4.5-5 %) lipoids in starting material passes into purifying bath so their content in purified material decreases to detection limit Dry substance of acid purifying bath contains 56 % lipoids and 45.5 % proteins, with lipoids comprising approx. 2.8-3 % polar lipids (e.g. fosfolipids), 13 % free fatty acids, 47.5 % partial glycerides and 35-37 % triglycerides. The quite high content of protein components is obviously associated with higher solubility of weakly crosslinked collagen of short tendons in an acid environment [38].

Purifying bath containing 1 % (per weight of starting material) lipolytic preparation Greasex™ (from Novozymes, Bagsvaerd, Denmark) in an alkaline environment (pH level adjusted to 9.5 with cyclohexylamine) leads to a practically same degree of purifying collagen from short tendons. Substances unambiguously dominant in dry substance of degreasing bath are lipoids (82.5 %) containing polar lipids (approx. 6.2 %) and roughly similar contents of free fatty acid (27.5 %), partial glycerides (35.9 %) and triglycerides (30.7 %). Nevertheless, content of present proteins does not exceed 17.5% (per starting material), which is interpreted through lower solubility of weakly crosslinked collagen in an alkaline environment. Yield of collagen hydrolysate in this procedure attains 89-92 % (w/w, per weight of starting material). Efficiency of purifying collagen of short tendons in both ways, including composition of dry substance of both purifying baths, is shown for illustration in Tab. 3.

Table 3. Purifying collagen from short tendons by commercial enzymatic preparations.

Material	Starting	Degreased	
		1-% Greasex™	1-% Novocor AD
pH of bath	-	11.8	4.5
Proteins in dry substance of material	94.1	98.2	97.7
Lipoids in dry substance of material	4.90	< 0.01	< 0.05
Degreasing bath:			
Dry matter of purifying bath %(w/w)		1.60	2.12
In that:			
% lipoid substances		1.32	1.20
% proteins		0.27	0.97

Diluted acetic acid (0.05-0.15 mol/L) enables to extract soluble collagen from refined (degreased) short tendons, with following centrifugation then isolating 5.3-5.8 % (w/w, per starting material) of collagen, exhibiting mean molecular weight 269-329 kDa (through osmometry). Remaining solid fraction, following short denaturation (95 °C, 15 min), may be hydrolyzed with 1 % (w/w, per starting material) commercial protease (Neutrase 0.5L®, Novozymes, Bagsvaerd, Denmark) of microbial origin (Bacillus subtilis), which displays maximal proteolytic efficiency at pH 5.5-7.5 and a temperature of 45 °C. After 3-hour action by Neutrase at a temperature of 45 °C, and inactivation through short heating (95 °C, 15 min), a solution of collagen hydrolysate is obtained, whose thickening to a concentration of 15-20 % dry substance and drying on spray drier produces powdered collagen hydrolysate with a yield of 85-87 % (w/w, per starting material); the product is well soluble in water. Closer characteristics of this collagen hydrolysate from short tendons (refined through both discussed procedures) are shown in data of Tab. 4.

Table 4. Characteristics of collagen hydrolysate from short tendons of abattoir cattle for cosmetic preparations of skin care and protection.

Lipoid components, % in dry substance	*0.013*
Ammonia N$_2$, % in dry substance	*17.61*
Glucosaminoglycans	*not detected*
Aminosaccharides	*not detected*
Mean molecular weight, M$_w$, kDa	*0.7–2.0*

Mean molecular weight of collagen hydrolysate, ranging in the 1-2 kDa region, enables its use as a very efficient hydratant in preparations for skin care as well as starting material for surfactants of acyl-amide (Lamepon) type, known for their particularly favorable dermatological properties. Mean molecular weight of collagen hydrolysate around 1.0 kDa is the basic assumption for achieving good solubility of such surfactants in an aqueous environment and for their good yield. In the case of collagen hydrolysates obtained, for example, from tanned collagen waste whose mean molecular weight is in the 20-45 kDa region, obtained hydrolysate has to be further degraded, best in an acid environment which is the most effective for such a reaction [39].

In long tendons of abattoir cattle (musculus atlantis et capitis, musculus longus capitis), collagen is conversely accompanied with a considerable fraction of elastin, differing from collagen in a greater content of amino acids with nonpolar side chain (93 % amino acids consist of glycine, alanine, valine, leucine) and in low content of hydroxyproline and dicarbonic amino acids (aspartic acid and glutamic acid). In addition, elastin (and also its hydrolysates) contains approx. 0.4 mol% desmosine and same amount of isodesmosine, which are quite absent in collagen. Amino acid composition of elastin conditions its special tertiary structure (double helix structure), its much lower capacity to swell in an aqueous environment. It is cause of its greater resistance to chemicals and proteolytic enzymes, for which it is sometimes designated "inert" protein. Presence of elastin can be easily detected by histological coloring techniques. Unlike collagen, elastin intensively binds to histological dyes of fenolic type (e.g. orceine). Purifying and isolating of elastin seldom occurs in industrial practice because the practically single important industrial application of elastin is its utilization as hydratant in cosmetic skin and hair care preparations. Some previous pharmaceutical studies of κ-elastin (alkaline hydrolysate after degreasing elastin by organic solvents) indicate its dermo-structural activity and positive influence in stimulating biosynthesis in the epidermis of human skin [40-42]. Elastin hydrolysates are characterized, as a rule, by a very wide molecular weight distribution (0.5-150 kDa), and their contamination with lipoid components and collagen hydrolysates is a frequent problem.

High fraction of elastin in long tendons somewhat complicates their industrial application. In view of the different tendency of collagen and elastin to hydrolysis, it is more advantageous to conduct degreasing of long tendons in an alkaline environment (1.5-% aqueous solution of enzymatic preparation of microbial origin Greasex™ from Novozymes, Bagsvaerd, Denmark with maximal efficiency in the alkaline region. An alkaline environment (pH 11.8) can be advantageously made by adding 1-% (v/v) cyclohexylamine. This reduces undesirable collagen losses (dissolution in acid environment). Material disintegrated into particles of approximate size 3x3 cm can be satisfactorily purified in such bath at 40 °C in 48 hours. Content of lipoid components in long tendons ranges in somewhat higher limits than with short tendons (7-8 %, w/w, per starting material). Under such conditions, fraction of protein components in dry substance of degreasing baths does not exceed 5.5 %, highest contents there being fatty acids (57-58 % from dry substance) and 28-31 % triglycerides, while partial glycerides do not exceed 2.5-3.0 %. Dispersion of lipoids is apparently associated with Na salts of fatty acids (or their ammonium salts).

It is again advantageous for enzymatic hydrolysis of long tendons to denature purified protein material by shortly heating up to 90 °C (approx. 5 min); actual enzymatic hydrolysis, due to different tendency of collagen and elastin to hydrolysis, has to be conducted as two-stage fractional hydrolysis. Easier hydrolyzable collagen is hydrolyzed under milder conditions (1-% solution of commercial protease Neutrase 0.5L from Novozymes, Bagsvaerd, Denmark, based on Bacillus subtilis, in neutral environment, at 40 °C), and with a yield around 38 % (w/w, per starting material) a water-soluble collagen hydrolysate is obtained of mean molecular weight 0.8-1.02 kDa, which can be easily separated from water-insoluble residue. Its purity can be controlled by hydroxyproline content (% HyPro in dry substance~10.6 % w/w) and by absence of aminosaccharides or glucosaminoglycanes.

The water-insoluble residue, mainly including elastin, may be easily separated by filtration and hydrolyzed under much more drastic conditions, in a bath containing 1 % (v/v) cyclohexylamine and 5% protease of bacterial origin (Bacillus lichenformis) (Alcalase 2.4L, from Novozymes, Bagsvaerd, Denmark) at 70 °C. Under such conditions, hydrolysis is over in 5 hours. Inactivation of Alcalase may be effected by heating up reaction mixture to 95 °C for 5 min. Solution of elastin hydrolysate is separated from insoluble residue by filtration, then thickened to an approx. 20-% concentration of dry substance, and processed on spray drier. Yield of elastin hydrolysate of mean molecular weight 0.48-0.83 kDa ranges around 48 % (w/w, per starting weight of long tendons). Amount of HyPro found in elastin hydrolysate was 1.6 % (w/w), which gives evidence of its satisfactory quality. After hydrolysis, the water-insoluble residue (around 3.5% w/w per starting material), separated from elastin hydrolysate solution best by filtration, produces difficultly hydrolyzable segments of elastin and/or collagen, probably containing difficultly hydrolyzable crosslinks [43].

Irregular trimmings of hide, obtained by rough shaping for tanning purposes, may be processed in similar manner following their disintegration to particles measuring approx. 2.5-3 cm. Their processing is somewhat complicated by the irregular shape of trimmings as well as by present subcutaneous connective tissue, which is source of fluctuating content of proteins and also lipids. With removal of subcutaneous tissue primarily in view, it is advantageous to combine mechanical work (cutting to shape like in hide processing) with the effect of final purification bath. Final purification baths may be based on same enzymatic preparations as in the case of long and short tendons, but it is of advantage to work with at least 5-% concentrations of enzymatic preparations (5 % w/w per purified material). Time of purifying action (48 hours at 40 °C) then remains essentially retained. Dry substance of these final purifying baths is somewhat higher than in tendon processing (9.9-19.5 % w/w, per dry substance of starting material) and its fluctuation depends to some extent on previous mechanical processing of purified material. However, ratio of lipoid and protein components in dry substance of final purifying baths remains in these cases essentially retained.

Collagen hydrolysate, after denaturing by a short (5 min) heating to 95 °C, is obtained from finally purified solid phase separated by filtration, through partial hydrolysis with commercial proteolytic preparation (1 % Neutrase DH by Novozymes, Bagsvaerd, Denmark, of maximal efficiency in neutral environment) of a yield around 71 % (w/w per starting material); after filtering off water-insoluble residue, thickening aqueous phase to dry substance around 20 % and spray drying, its characteristics correspond to collagen hydrolysate from long tendons.

Following hydrolysis of elastin contained in filtered-off solid phase can be successfully performed under conditions that may again be more drastic: 5 % (w/w per starting material) commercial proteolytic preparation Esperase (Novozymes, Bagsvaerd, Denmark) at pH level 11.0 (adjusted by cyclohexylamine) and 60 °C with a reaction time of 4 hours. Yield of elastin hydrolysate (after filtration, thickening to a dry substance level of around 20 % and spray drying) is in a range of 6-10 % (w/w per weight of dry substance).

Solid phase (water-insoluble), non-hydrolyzed under these conditions, consists of keratin (on the average, 5.8-8.4 %, per weight of starting material), whose hydrolysis with current proteases is quite difficult.

Proteases easier attack peptide bonds than disulfide bonds and produced hydrolysates have the character of strongly branched polymers with a considerable fraction of disulfide bonds. Specific enzymes attacking disulfide bonds of keratin are quite difficultly accessible and relatively costly. Methods of chemically eroding disulfide bonds of keratin lead to products of somewhat unclear biological activity, some are even suspected of carcinogenic properties; reduction products of disulfide bonds are inclined to regressive oxidation of sulfide groups by atmospheric oxygen into disulfide bridges, whose formation then negatively influences solubility of keratin hydrolysates in water as well as their storability and thus their applicability in industrial practice. The issue of industrially employing keratin proteins is of growing significance especially with growing production of poultry, and it seems it will be just problems of utilizing keratin proteins that should receive increased attention in the future.

5. WASTE OF TANNED COLLAGEN TYPE

Except for gelatin and glue manufacture, industrial use of collagen is always associated with a purposeful increase in crosslink density, which especially increases resistance of collagen to microbial and hydrolytic deterioration. Estimating the required crosslink density to obtain a product with satisfactory stability and also rheological and physico-mechanical properties is on a rather empirical basis. Excessive crosslink density , for example with leathers, leads to unduly stiff materials, with casings it leads to materials that can be chewed and also digested only with difficulty. On the contrary, materials of low crosslink density show insufficient stability against microbial or hydrolytic deterioration. Estimating adequate crosslink density is in most cases subject to empiricism and often causes considerable occurrence of difficultly processable and usable tanned collagen waste. The problem of tanned collagen waste stands out most strongly in leather production, being in the range of 24.5-25.5 % (per weight of starting material). Although similar data from casings manufacture and other processing fields are scarcely available and, as a rule, incomplete, the frequently discussed problems of no-waste (clean) technologies for processing collagen, the same as studies of potentially applying tanned collagen waste as a secondary industrial raw material, suggest these problems have considerable importance.

Efforts of earlier date pursuing visualized application of tanned collagen waste, particularly from leather production (shavings, irregular pieces of split leather or leather, etc) as a heat- or sound-insulating construction material, did not prove too successful. Tanned collagen waste easily sorbs 30-35 % (w/w) moisture from its surroundings, and consequently changes its volume to considerable extent, as well as its thermal insulation capacity. Dimensional and thermal-insulating instability were the principal reason why such materials did not succeed in practice. With relatively less crosslinked casings (aldehyde crosslinking), efforts focused more on use in the way of farming fertilizers or additions of their hydrolysates to feed mixes. In both cases, increased crosslink density rather reduces utility of such waste for reasons that are quite understandable.

5.1. Collagen Hydrolysates

Higher crosslinking degree of tanned waste collagen stimulates interest in its processing through partial (controlled, if possible) hydrolysis, which is tied with the problem of separating tanning substances from actual protein hydrolysate. Crosslinks in collagen (both native and tanned) are usually more resistant to hydrolysis than peptide bonds of collagen chains. Partial hydrolysates of collagen, therefore, in many respects behave (see, e.g. viscosity characteristics) like solutions of strongly branched polymers [44]. With a view to the visualized industrial processing of collagen hydrolysates, the hydrolysis should not be too deep. The greatest fraction of tanned collagen comes from leather production, mostly from leathers tanned with basic salts of chromium, and that is why attention first focused on alkaline hydrolysis employing CaO or MgO. Under these conditions, soluble Cr^{3+} salts transform into $Cr(OH)_3$, soluble in water with difficulty and enabling quite easy separation of protein components from tanning substances. Hydrolysate already obtained through such procedure under quite mild conditions was recommended as a secondary raw material for preparing surfactants. Of course, there was the problem of a certain amount of chromium compounds in isolated hydrolysate, which are somewhat controversial regarding biological impacts. Processing sludge of reaction mixtures containing most released chromic compounds was complicated by present collagen polypeptides (around 10 %, w/w per dry substance), which increased difficulty of filtering off sludge from reaction mixture. Difficult separation of these phases contributes to worsening the economy of utilizing tanned collagen waste in this way.

In acid hydrolysis, chromic compounds pass into solution and chromic salts have to be separated from collagen hydrolysate by precipitating. That, together with necessary following filtration, poses a complication of no minor significance.

Processing chrome-tanned collagen waste through the action of oxidizing agents (e.g. hydrogen peroxide) in an alkaline environment, recommended by some authors [45, 46], transforms Cr^{3+} compounds into potentially cancerogenic compounds of Cr^{6+}, easily passing into solution together with collagen hydrolysate. Difficultly controllable hydrolysis of collagen takes place at the same time. Separating collagen hydrolysate of satisfactory purity for further industrial processing is thus made more complicated from viewpoint of both ecology and economy. Authors propose using obtained hydrolysate for filling (grain) or re-tanning leathers, and employing separated Cr^{6+} compounds for preparing new tanning baths. Even though such experiments hardly exceed laboratory dimensions, an altogether reserved attitude of leather manufacturers to such application may be registered.

Introduction of proteases of microbial origin into industrial practice stimulated interest in enzymatic hydrolysis of tanned collagen waste which has, compared to discussed procedures, the advantage of a lower time and energy demand. In virtually selective manner, proteases hydrolyze peptide bonds of collagen chains, and bonds of tanning substances to protein chains remain essentially intact, which enables to separate quite easily tanning substances from the reaction mixture from water-soluble collagen hydrolysate as insoluble sludge (solid phase). Sludge comprises tanning substances bound to short protein chains – hence a certain fraction [47]. The advantage of enzymatic hydrolysis is in a certain possibility of controlling mean molecular weight of obtained collagen hydrolysate through hydrolysis reaction time (under otherwise constant reaction conditions).

Such procedures were verified on pilot-plant scale; they usually employ commercially available proteases of microbial origin (e.g. Alcalase DNL from Novozymes, Bagsvaerd, Denmark). An example of working procedure applicable in practical operation is the preparation of collagen hydrolysate from chrome-tanned leather waste, which is specified, for example, in [48] as follows: Fifty kg water, 0.5 kg magnesium oxide (MgO) and 10 kg tanned collagen waste disintegrated into fragments of approx. size 2x2 cm are added into reactor. Gradually, under good stirring, 0.3 kg commercial protease Alcalase DNL is added and the reaction mixture is heated to 70 °C. Under these conditions, hydrolysis of tanned collagen waste proceeds for 4-5 hours. Following vacuum filtration produces a clear solution of collagen hydrolysate and filter cake containing around 5 % (w/w) of water-insoluble protein. The clear solution of collagen hydrolysate is vacuum-thickened to a concentration of approx. 30 % dry substance and spray-dried.

Prolonged reaction time brings about a decreased residual content of Cr^{3+} compounds and increased content of primary amino groups in collagen hydrolysate. Content of ash, or of amide nitrogen in hydrolysate, does not practically change with changed reaction time. A decrease in mean molecular weight of collagen hydrolysate occurs with a prolonged reaction time, which is a fact corresponding to the concept that bonds of Cr^{3+} tanning compounds are more resistant to enzymatic hydrolysis than peptide bonds of collagen chains. Compounds of Cr^{3+} bound to segments of protein chains thus accumulate in the sludge of reaction mixture, which facilitates separating residues of tanning substances by filtration or some other procedure from the solution of collagen hydrolysate [49]. Characteristics of collagen hydro-lysate obtained though such enzymatic hydrolysis after reaction times of 2, 4 or 6 hours are summarized for illustration in Tab. 5.

Table 5. Characteristics of powdered collagen hydrolysate of chrome-tanned leather waste, after 2-6 hours enzymatic hydrolysis with commercial proteolytic preparation Alcalase DNL.

Reaction time of hydrolysis, hours	2	4	6
Amide nitrogen, % (w/w, in dry substance)	16.3	16.0	15.7
Ash content, % (w/w, in dry substance)	3.3	2.5	3.0
Cr^{3+} content, ppm (in dry substance)	54.5	21.3	13.6
Mean molecular weight – numeric mean (M_N, kDa)	36.1	19.2	18.9
– weighed mean (M_w, kDa)	55.7	49.1	43.0
Polydispersity degree, $P = M_N/M_W$	1.5	2.2	2.0

The dominant portion of collagen waste comes from leather production and is almost exclusively tanned with compounds of Cr^{3+}. These possess a somewhat controversial character. While Cr^{3+} compounds are generally regarded from the environmental (or health) point of view as unobjectionable (a number of them being contained in plant growth stimulators or medicaments for diabetics), Cr^{6+} compounds to which most of them easily transform through oxidation are attributed the character of cancerogenic substances. When treating drinking water, oxidizing procedures are used practically exclusively and, therefore, presence of chromic compounds in secondary industrial raw materials, the same as in wastewaters, is considered to be undesirable. The possibility to isolate collagen hydrolysate from chromic compounds with relative ease is, therefore, a very significant factor in the enzymatic hydrolysis of tanned collagen waste.

Enzymatic hydrolysis of collagen waste, tanned in practically any other way including, for example, pre-tanning with lower-molecular aldehydes used in casings manufacture, has the same effect so that potential problems with controversial residues of tanning substances become negligible with this procedure.

Collagen hydrolysates resemble gelatine or glues in the ability of their aqueous solutions to transform (at certain concentrations of dry substance) into gels. Due to lower mean molecular weight, their necessary concentrations of dry matter are of course higher than with glues and rigidities of created glues (regarded as criterion of glue quality) are usually lower. Comparison with hide glues of current production and relevant information are given in data of Tab. 6.

According to [50], gel rigidity of hydrolysates (the same as melting temperature and some other characteristics) is a multifunctional dependency. It depends on mean molecular weight and polydispersity degree of collagen hydrolysates as well as on concentration of dry substance in gels and possibly on character of accompanying substances they contain.

Table 6. Characteristics of gels of enzymatic collagen hydrolysates.

Hydrolysate designation	Molecular weight, M_N (kDa)	Polydispersity degree, $P = M_N/M_W$	Concentration of gel formation, % (w/w)	Gel rigidity (at conc. of gel formation), Bloom
Glue K4	72.4	1.13	11.3	577
Glue K2	61.3	1.11	12.5	235
H 1	19.8	2.38	29.6	292.5
H 2	22.7	2.17	27.8	278.8
H 3	20.4	2.41	29.7	236.6
H 4	15.3	2.42	51.4	13.7

Quality of adhesives based on gels of enzymatic collagen hydrolysates is evidently lower than that of current gels and limits application merely to cases where low adhesive bond strength is sufficient and emphasis is laid on easy erosion by water, for example, to labels on glass (or other) bottles, etc.

Easily proceeding biodegradation of hydrolysate solutions may in some cases complicate applying their solutions as water-soluble adhesive, as coagulating agent of polymer latexes or as agricultural fertilizer. Problems associated with long-term storage of such solutions may be resolved, of course, by adding usual conservation agents, in the case of particularly thickened solutions by reducing water content (spray-drying thickened solutions), but admittedly at a price of increased production costs.

5.2. Collagen Hydrolysates as Secondary Industrial Raw Material

Utilizing collagen hydrolysates of tanned collagen waste as a farming fertilizer, ingredient in feed mixes, water- soluble adhesive or latex coagulant remains, for the obviously higher production costs as compared to hydrolysates of untanned collagen waste, somewhat problematic. Their wider industrial application is more associated with specific reactivity of such hydrolysates or with their transformation into biodegradable (possibly edible) packaging materials.

5.2.1. Curing Urea-Formaldehyde Adhesive Films

An interesting industrial application of collagen hydrolysates is potentially based on high reactivity of their primary amino groups with formaldehyde, which is gradually released by cured urea-formaldehyde adhesive films into the environment. Urea-formaldehyde (UF) adhesives are chiefly favored in the wood-working industry and textile industries for their solubility in an aqueous (cheap and safe) environment, economically undemanding manufacture and good parameters of produced adhesive bonds. The problematic property of such adhesive films merely remains in gradual release of cancerogenic formaldehyde (after curing – e.g. particle boards, furniture, carpets, etc) which is very negatively regarded especially in closed spaces. Solutions proposed for reducing such formaldehyde emissions usually involve a reduced formaldehyde-urea ratio for preparing such adhesives (UF adhesives with reduced formaldehyde content) or subsequently added urea to the adhesive, shortly before its application [51]. Such solutions, however, are linked with potential insufficient cure of such adhesive films, with consequently reduced strength of adhesive bonds thereby produced. Other authors recommend reducing formaldehyde emissions with added melamine-formaldehyde condensates [52] or with amines, ammonia or some other substances [53]. Despite considerable attention devoted to these problems it is not clear yet to what extent such subsequently added substances act merely as "traps" of free formaldehyde or to what extent they influence reaction mechanism of UF resin curing. Collagen hydrolysates with their relatively high content of primary amino groups may obviously alternate most of such chemicals, and their use will probably also be of better economic advantage.

Studies of the curing process of UF resin films executed through technique of thermal analysis (TGA, DSC) by means of dimethylol urea as model substance, proved their curing is accompanied with formation of less stable oxymethylene ($-CH_2-O-CH_2-$) as well as more stable methylene ($-CH_2-$) crosslinks in approximate molar ratio 1:1. Less stable oxymethylene bonds then transform to more stable methylene bonds with release of formaldehyde. It is highly probable this mechanism is the cause of formaldehyde emissions from incompletely cured UF adhesive films. Addition of collagen hydrolysate shifts the molar ratio of arising oxymethylene and methylene crosslinks to the 1:2 region benefiting more stable methylene bonds and thus reducing potential formaldehyde emissions from cured films by approx. 25 % [54, 55].

In an acid environment (necessary to cure UF adhesive films) the transition of unstable oxymethylene crosslinks is noticeably accelerated by heating films to 150-170 °C. Studies of the breakdown kinetics of oxymethylene bonds through TGA technique unambiguously proved importance of maintaining this reaction range, the same as keeping an acid environment in the cure of UF adhesive films, and a connection between these parameters and formaldehyde emitting by cured films was also proved. It was proved that observing the mentioned curing temperature interval, and producing a required acid environment with acid that does not decompose or volatilize under these temperature conditions may achieve a reduced formaldehyde emission level by cured UF films of up to another 50% of original emission level [56]. Mainly from the viewpoint of nowadays intensely discussed problems of purity of the atmosphere, such industrial utilization of collagen hydrolysates offers interesting possibilities.

5.2.2. Curing Epoxide Resins (Type of Diglycidyl-Ethers Of Bisfenol A)

Easily processing epoxide resins are often applied as binding polymers for protective coatings in the automobile and aeronautical industries. Resistance of their films to chemicals is advantageous in surface protection of materials but complicates recycling of plastic parts after their life. Modifying some of their inconvenient properties usually consists in incorporating soft polymer segments into epoxide chains or, more frequently, in selecting suitable crosslinking agents or their relative mutual concentrations. Agents that proved successful as convenient crosslinking agents were polyamides of nylon or caprolactam type, which increase modulus of elasticity of the final polymer. Because of their limited solubility, they are often combined with oligomers of diglycidyl –ether type in no-solvent systems ensuring, among others, minimal contamination of the atmosphere [57]. Gorton [58] was probably the first to point out that polyamides react with the oxirane ring of epoxides through their amide nitrogen. Collagen hydrolysates contain a number of functional groups capable of reacting with the oxirane ring of epoxide oligomers, and most significant of those are doubtlessly primary amino groups, even though, according to Zhong and Guo [59], amido groups may also participate in such crosslinking reactions and a certain participation by hydroxyl groups of collagen hydrolysate side chains cannot be ruled out either.

The same as polyamides, even systems of epoxide oligomer (diglycyl ethers of bisfenol A of mean molecular weight 348-480 Da) with collagen hydrolysate incline to utilizing no-solvent systems, all the more so that a clear reaction of both components was detected through DSC technique in temperature range 197-225 °C [60]. Reaction heat of such a crosslinking reaction displays clear dependency on the weight fraction of epoxide oligomer in reaction mixture, and its limit value of 519.2 J g^{-1} ($\approx \Delta H$=101.2 kJ mol^{-1} related to oxirane ring) is in good agreement with data for polyamides by other authors. Kinetics of this reaction, characterized by pre-exponential factor A (mean value 65-70 min^{-1}) and activation energy of the reaction E ($\approx$ 612-643 kJ mol^{-1}) display a reasonable, approx. 10-% scatter. Curing of epoxide oligomers with collagen hydrolysate requires, similarly to curing of polyamides, a reaction temperature of 200-220 °C. Breakdown temperature of collagen hydrolysate is at least 30 °C higher, and employing it as a crosslinking agent for epoxide oligomers of bis- diglycyl ethers of bisfenol A is quite realistic.

Linkage of epoxide segments through a protein chain offers potential degradation of the crosslinked polymer with proteolytic enzymes under relatively mild conditions. This enables to break down epoxide polymers and subsequently easily separate epoxide surface finishes of plastic automobile and/or aircraft parts after their life, which is advantageous for recycling plastic parts of products from these industries.

5.2.3. Biodegradable (Edible) Materials Based on Collagen Hydrolysate

Collagen hydrolysates, although to somewhat lesser extent, possess capacity to form hydrogels, which is otherwise typical of gelatine. Proteinic hydrogels particularly exhibiting good mechanical properties – strength in the first place, permeation capacity for gases or vapors and a certain resistance to moisture – are widely used as biodegradable (possibly even edible) packaging materials for products of the food, cosmetic and pharmaceutical industry, and also in human medicine. Environmental aspects of biodegradable packaging materials are accentuated in the first place by growing production of packaging materials based on synthetic plastics that are degradable with such difficulty in landfills, and whose recycling gets complicated by gradually deteriorating mechanical parameters and quite high costs.

Collagen hydrolysates of chrome-tanned leather wastes are a starting material not too suitable for such applications owing to its certain (though small) content of controversial Cr^{3+} compounds (see Tab. 5). By contrast, waste from casings manufacture (collagen pre-tanned with aldehydes or partly crosslinked by means of aminotransferases) which can be processed into collagen hydrolysate through the same procedure, lacks this difficulty, and its use in biodegradable and edible packaging materials is prevented (compared to gelatine) only by the lower mean molecular weight of hydrolysate or problems ensuing therefrom: higher concentration of dry substance required to form hydrogels and their lower, particularly mechanical, parameters. Higher content of Ca^{+2} or Mg^{+2} compounds for such applications does not matter. Basic model characteristics of collagen hydrolysate from such a source are presented in Tab. 7.

Table 7. Basic characteristics of enzymatic collagen hydrolysate from casings.

Dry substance, %	*92.99*
Amide nitrogen in dry substance, %	*14.85*
Ash in dry substance, %	*4.94*
Ca content in dry substance, ppm	*27,456.6*
Mg content in dry substance, ppm	*4,798.0*
Cr content in dry substance, ppm	*not detected*
Primary amino groups in dry substance, mmol –NH$_2$ g^{-1}	*0.216*
Average molecular mass (numerical mean, M$_N$), kDa	*17.75*

When processing such hydrolysates into usable biocompatible and biodegradable materials, their approx. tenfold lower molecular weight compared to that of traditional material for these purposes (gelatine) has to be taken into account. The lower molecular weight increases hydrophilic character of hydrolysate, its higher solubility in an aqueous environment is apparent as well as higher sensitivity of its films and foils or fibers to aerial moisture, accompanied by particularly worsened physical and mechanical characteristics. Such unfavorable properties of collagen hydrolysates may be eliminated by increasing density of their crosslinks.

The reaction currently used for increasing crosslink density involving (primary) amino groups of hydrolysate with lower-molecular aldehydes (formaldehyde, glyoxal, glutaraldehyde) has the disadvantage of substantial toxicity for biological subjects. For this reason, attention was more focused on polymeric aldehydes derived from various polysaccharides [61-63], but in the first place on commercially most easily available dial-dehyde starch (DAS). DAS demonstrates very low toxicity [64] and is even recommended as an absorbent for urea in chronic diseases of kidneys, uremia and others [65]. Moreover, some authors suppose that polymeric aldehydes may strongly plasticize protein films, foils and fibers [66]. However, with protein materials it is generally quite difficult to separate the effect of hydrophilic plasticizers from plasticizing effect of water, which is their natural plasticizer. A DSC study of the DAS plasticizing effect on film of collagen hydrolysate crosslinked with DAS showed that the influence of DAS alone (if film does not contain moisture) on reducing glass transition temperature (Tg, °C) is but very slight. Glass transition temperature of such films is within interval 177±3.9 – 199.1±2.1 °C which, compared to film without DAS (Tg = 189.5± 2.52 °C) or to the figure for non-crosslinked gelatine (containing structurally bound water) Tg = 175±10 °C and waterless gelatine, 193±3 °C [67] makes evident the minimal plasticizing effect of DAS itself and supports the opinion that with proteins and hydrophilic plasticizers, the active substance proper is more likely sorbed water [68].

Draye et al. [61] were probably the first to notice the increased aging tendency of gelatine hydrogels crosslinked with polymeric dialdehydes. They based their opinion on decelerated dissolution of such gels in an aqueous environment with time, which they particularly noticed during the first week following gel preparation. With gelatine gels, aging was essentially detected already earlier through mechanical and dynamic measurements by Nijenthuis [69]. Kozlov and Burdygina [70] demonstrated that gelatine containing less than 2 % water (w/w) behaves like a fragile, highly crosslinked polymer of poor water solubility, and they attribute this property to high density of interchain (hydrogen) crosslinks.

The crosslinking reaction of collagen hydrolysate with dialdehyde starch, performed for acceptable solubility in an alkaline environment, depends both on concentration of hydrolysate as well as on concentration of dialdehyde starch in the reaction mixture [71]. At concentrations not exceeding 27.5 % (w/w) collagen hydrolysate and 15 % (w/w per collagen hydrolysate) dialdehyde starch, thermo-reversible gels arise in the reaction mixture, gradually increase their strength through aging at room temperature (25 °C) and lose thermo-reversibility.

Gels prepared at higher concentrations of hydrolysate and starch dialdehyde in the reaction mixture lose thermo-reversibility already on cooling of reaction mixture, or also at the temperature (60 °C) at which the crosslinking reaction is conducted. Increasing rigidity of gels is associated with higher concentration of collagen hydrolysate in the reaction mixture, but chiefly with increased concentration of dialdehyde starch. Solubility in an aqueous environment, however, remains preserved with thermo-irreversible gels, merely its kinetics is considerably reduced. A similar effect can be found with films obtained from thermo-irreversible gels by casting and drying. This fact is obviously related to increased crosslink density in such gels, and preserved solubility in an aqueous environment supports the opinion that such crosslinks have the character of hydrogen bonds found with gelatine films containing less than 2 % water [71].

Aging of gels of collagen hydrolysate crosslinked with dialdehyde starch, usually accompanied with loss of thermo-reversibility, limits the time interval for their processing into biodegradable packaging materials through the technique of dipping, widespread during processing of gelatine into soft (SGC) or hard (HGC) gelatine capsules. This limitation is not valid for films, foils or other packaging materials prepared by the technology of casting (spraying) and drying. Time-limited drying conducted at a temperature of 105 °C (in an interval of 1-4 hours) permits to regulate in quite wide limits the disintegration time of these films in an aqueous environment (regardless of its pH level) and thus to control to considerable extent the rate of releasing active substances from such packages (bioavailability of medicines, vitamins and the like) [72].

On the contrary, thermo-irreversible gels of collagen hydrolysate crosslinked with dialdehyde starch (prepared at a higher concentration of both reactants than corresponds to thermo-reversible gels) may be processed into biodegradable gels through extrusion technology usual in processing of synthetic thermoplastics. Temperature of gel-sol transition, detected by DSC technique, does not overly differ with both gel types. In such determination, however, evaluation of the gel-sol transition is interfered with by the wide endothermal peak of evaporating water. Thermal coordinates of the minimum of such a wide endothermal peak allow to estimate that extruding thermo-irreversible gels of collagen hydrolysate crosslinked with dialdehyde starch may be carried out in temperature range 65-80 °C, but applying certain pressure to process such gels is necessary. Works in this field have not been finished yet, nevertheless, the path indicated already appears to be very promising.

6. Summary

Collagen, protein of connective tissues, produced by the food industry (meat production) is not too well utilizable as a raw material for foodstuffs or for preparing feed mixes for farm animals. Its occurrence is generally estimated at 20-30 % weight of abattoir cattle.

Industrially, most collagen is used as a raw material for flat materials (leathers) for clothing, footwear and fancy goods manufacture, a significant part is used as raw material for biodegradable (edible) packaging materials for foodstuffs, cosmetic and pharmaceutical products. A minor volume of collagen is utilized for medical purposes.

In view of industrial processing, an important characteristic of collagen is its crosslink density. With starting (untanned, native) collagen, it defines the yield of soluble collagen, quality and yield of gelatin or other products of its hydrolysis.

Beside products of collagen hydrolysis (production of soluble collagen and gelatine of various quality), industrial processing of collagen is always connected with increasing crosslink density, which gives final products adequate resistance to microbial and chemical deterioration, or mechanical and physical characteristics. Disproportionately high crosslink density of collagen-based industrial products deteriorates their utile properties and leads to the rise of "tanned" collagen waste whose further processing is practically always associated with its partial hydrolysis.

The most advantageous procedure in collagen partial hydrolysis is controlled enzymatic hydrolysis, which leads to collagen hydrolysates of mean molecular weight 15-40 kDa. During viscometric measurements, their molecules behave like branched polymers.

Such partial hydrolysates find application as a secondary raw material in production of surfactants of acyl amide (Lamepon) type, as material restricting formaldehyde emissions of cured urea-formaldehyde adhesive films, in manufacture of crosslinked epoxide resins degradable by proteolytic enzymes and by polymeric aldehydes after partial crosslinking (mostly by dialdehyde starch), of biodegradable (depending on character of initial crosslinked collagen product) and also edible packaging materials whose importance is growing proportionately to needs of polymeric packaging materials with long biodegradation time.

Biodegradable materials of this kind are marked by a strong tendency to aging produced by increased density of hydrogen (also interchain) crosslinks, which by their properties form a transition between (physical) hydrogels and chemogels. While physical hydrogels and chemogels received substantial attention in the past and differences between them were quite exactly specified, gels of transitional character received relatively smaller attention despite some of their properties being interesting from the application point of view.

Transitional gels have in the first place a marked tendency to aging, which shows in time dependency of rigidity, accompanied by gradual loss in thermo-reversibility. That complicates to some extent processing of thermo-reversible gels through dipping technology, typical of physical hydrogels based on gelatine. Films and foils of thermo-reversible gels may be processed to greater advantage into biodegradable (edible) packaging materials by casting technology and drying. Their disintegration in an aqueous environment is associated with moisture they contain and that can be successfully controlled in quite a wide range by drying conditions. At the same time, we may rightly assume controlled "bioavailability" of active substances covered under such packings.

Thermo-irreversible gels with a higher hydrogen crosslink density admittedly lack thermo-reversibility which is so characteristic for physical hydrogels, but their spontaneous solubility in an aqueous environment remains also preserved, even though considerably decelerated from the kinetic point of view. The advantage here is their potential processing through techniques usual with synthetic thermoplastics (extrusion). Even though the gel-sol phase transition observed by thermal analysis technique may be detected only with difficulty, due to it being covered by the endothermal peak of evaporating water, we may assume that temperatures of 65-80 °C under usual pressures are sufficient for such processing. Concrete parameters, of course, depend on type of extruded material and on ratio of crosslinking agent (dialdehyde starch) to collagen hydrolysate in starting reaction mixture.

Development in this field has not been quite finished. However, results hitherto obtained indicate that gels of collagen hydrolysate crosslinked (in general) with polymeric dialdehydes display very interesting behavior and may substantially expand potential industrial application of (crosslinked) collagen waste. In the nearest future, therefore, works in this field should intensify.

REFERENCES

[1]   Eyre D.R., Jian-Jiu Wu: Collagen cross-links. *Top Curr. Chem.* 2005 (247) 207-229, Springer, Berlin, Heidelberg >DOI 10.1007/b103828<.

[2]   Heidemann E.: Fundamentals of leather manufacturing. E. Roether Verlag, Darmstadt, 1993, pp 75-142.

[3]   Bienkiewicz K.: Physical chemistry of leather making. R.E. Krieger, Malabar Fl/U.S.A. 1983, pp 24-66.

[4]   Chvapil M.: Industrial uses of collagen, In: Parr D.A.D., Creamer L.K. (Eds.): Fibrous Proteins. Scientific, industrial and medical aspects, Vol. 1. Acad. Press, London 1979, pp 247-269.

[5]   McPherson J.M., Sawamura S., Armstrong R.: An examination of the biologicresponse to injectable, glutaraldehyde cross-linked collagens implants. *J. Biol. Mater.* 20, 93-107 (1986).

[6]   Maeda M.,Tani S., Sano A., Fujioka K.: Microsturcture and release characteristics of the minipellet a collagen base drugs delivery system for controlled release of protein drugs. *J. Controlled Rel.* 62, 313-324 (1999).

[7]   Lee Chi H., Singla A., Lee Y.: Biomedical applications of collagen. Int. *J. Pharmaceutics* 221, 1-22 (2001).

[8]   Chvapil M., Speer D.P., Holubec H., Chvapil T.A., King D.H.: Collagen fibers as a temporary scaffold for replacement of ACL in goats. *J. Biol. Mat. Res.* 27 (3), 313-325 (1993).

[9]   Auger F.A., Roubhia M., Goulet F., Berthold F., Moulin V., Germain L.: Tissue engineered human skin substitutes developed from collagen populated hydrated gels: clinical and fundamental applications. *Med. Biol. Eng. Comput.* 36, 801-812 (1998).

[10] Huynh T., Abraham G., Murray J., Brockbank K., Hagen P.O., Sullivan S.:Remodeling of an acellular collagen graft into a physiologically responsive nervovessel. *Nat. Biotechnol.* 17, 1083-1086 (1999).

[11] Rao K.P.: Recent development of collagen based materials for medical applications and drug delivery systems. *J. Biol. Sci* 7 (7), 623-645 (1995).

[12] Fonseca M.J., Alsina M.A., Reig F.: Coating liposomes with collagen Mw 50.000 increases uptake into liq. *Biochim. Biophys. Acta* 1279 (2), 259-265 (1996).

[13] Tcharodi D., Rao K.P.: Rate-controlling biopolymer mechanism as fundamental delivery systems for nifedipine: Development and in vitro evaluations. *Biomaterials* 17, 1307-1311 (1996).

[14] Buljan J., Reich G., Ludvik J.: Mass balance in leather processing. Proceedings of theCentenary Congress of the International Union of Leather Technologists and Chemists Societies, London 1997, pp. 138-156.

[15] Alexander K.: Developments in clean technology and eco-labeling. Proceedings of the conference "Clean technology 95 ", BLC Northampton 1995, pp.1-7.

[16] Heidemann E.: Fundamentals of leather manufacturing. E. Roether, Darmstadt 1993, pp.168-218.

[17] Addy V., Covington A.D., Watts A.: Enzyme degreasing of animal skins: A biochemical and misroscopical study. Proceedings of the Centenary Congress of the International Union of Leather Technologist and Chemists Societies, London 1997, pp. 478-489.

[18] Sternbjerg-Olsen B.: Enzymes, environmentally friendly compounds in the manufacture of leather. Proceedings of conference "Topical questions of environment protection in leather manufacturing", Partizanske (Sk), 14.-15. Nov. 1996, pp.12/1-3.

[19] Smjechowski K., Grabkowski M., Langmaier F., Kolomaznik K., Mladek M.: Possibleuse of waste grease from leather making (in Rus.). Proceedings of the international conference Moskow State University of Design and Technology, Moskow, 19-21. 4. 2000, pp. 137-145.

[20] Langmaier F., Kolomaznik K., Mladek M.: Lime liquor proteins. Leather Science and Engineering (PRC), 12 (4), 3-10 (2002).

[21] Schrooyen P.: Feather keratins. Modification and film formation. PhD thesis, University of Twente, The Netherlands, 1999, p. 9.

[22] Blazej A., Galatik A., Mladek M.: Technology of leather and fur (in Czech), SNTLPrague 1984, p. 108.

[23] Bienkiewicz K.: Physical chemistry of leather making, R. E. Krieger, Malabar Fl/U.S.A. 1983, pp. 34-40.

[24] Tsai G.J., Wu Y.Y., Su W.H.: Antibacterial activity of chitooligosaccharide mixture prepared by cellulose digestion of shrimp chitosan and its application to milk preservation. *J. Food Prot.* 63, 747-752 (2000).

[25] Tanabe T., Okitsu N., Tachibana A., Yamauchi K.: Preparation and characterization of keratin-chitosan composite film. *Biomaterials* 23, 817-825 (2002).

[26] Katoh K., Tanabe T., Yamauchi K.: Novel approach to fabricate keratin sponge scaffolds with controlled pore size and porosity. *Biomaterials* 25, 4255-4262 (2004).

[27] Katoh K., Shiayama M., Tanabe T., Yamauchi K.: Preparation and physicochemical properties of compression molded keratin films. *Biomaterials* 25, 2265-2272 (2004).

[28] Barone J.R., Schmidt W.E., Liebner Ch.F.E.: Compounding and molding of polyethylene composites reinforced with keratin feather fiber. *Comp. Sci. Technol.* 65, 683-692 (2005).

[29] Langmaier F., Blazej A., Galatik A.: Foils from tannery waste proteins. CZ Pat. 39 392 (1969).

[30] Osburn W.N.: Collagen casings. In Gennadios A. (Ed.): Protein based films and coatings. CRC Press, Boca Raton Fl/U.S.A. 2002, pp 445-465.

[31] Tang Ch.H., Jiang Y., Wen Q.B., Yang X.Q.: Effect of transglutaminase treatment on cast films of soy protein isolates. *J. Biotechnology* 120, 296-307 (2005).

[32] Carvalho R.A., Grosso C.F.R.: Characterisation of gelatin based films modified with transglutaminase, glyoxal and formaldehyde. *Food Hydrocolloids* 18, 717-726 (2004).

[33] Gerrard J.A.: Protein-protein crosslinking in food. Methods, consequences, applications. *Trends Food Sci. Technol.* 13, 391-399 (2002).

[34] Veis A.: The macromolecular chemistry of gelatin. Acad. Press, N.Y., London 1964, p. 125.

[35] Bailey A.J., Paul R.G., Knott L.: Mechanism of maturation and aging of collagen. Mechanism Ageing Develop. 106, 1-56 (1998).

[36] Reich G., Walther S., Stather F.: The influence of the age of cattle and pigskin on the yield and quality of gelatins obtained after acid conditioning process. In: Investigation of collagen and gelatine IV. Vol 18, Deutsches Lederinstitut, Freiberg/Sa, FRG, pp. 24-30.

[37] Mokrejs P., Langmaier F., Mladek M., Janacova D., Kolomaznik K., Vasek V.: Extraction of collagen and gelatin from meat industry by-products for food and non-food uses. Waste Management Research 2007,>DOI:10.1177/0734242X7081483<.

[38] Langmaier F., Mladek M., Kolomaznik K., Sukop S.: Collagenous hydrolsates fromuntraditional sources of proteins. *Int. J. Cosmetic Sci.* 23, 193-199 (2001).

[39] Langmaier F., Mladek M., Kolomaznik K., Maly A.: Degradation of chromed leather waste hydrolysates for the production of surfactants. Tenside, Surfactants, Detergents, 39, 31-34 (2002).

[40] Menasche M.: Pharmaceutical studies on elastine peptides (κ-elastine): blood clearance, percutaneous penetration and tissues distribution. *Path. Biol.* 29, 548-554 (1981).

[41] Forlot P.: Elastin in cosmetic formulations: hypotheses of action and activity. *J. Appl. Cosm.* 3, 114-119 (1985).

[42] Martini M.C., Seiller M. (Coord.): Actifs et additifs en cosmetologie. Lavoisier- Tech. Doc., Paris 1992, pp. 107-108.

[43] Langmaier F., Mladek M., Kolomaznik K., Sukop S.: Isolation of elastin and collagenpolypeptides from long cattle tendons as raw material for cosmetic industry. Int. *J. Cosm. Sci.* 24, 273-279 (2002).

[44] Heidemann E., Hein A., Moldehn R.: Untesuchung über Möglichkeiten Chromfalzpäne aufzuarbeiten. Leder 42, 133-143 (1991).

[45] Cot J., Marsal A., Manich A.M., Celma P., FRernandes-Hervas F., Cot-Gores J.: Transformation plant converting chromium waste into chemical product for leather Industry. *JALCA* 103 (3), 103-113 (2008).

[46] Cot J., Manich A.M., Marsal A. Fort M., Celma P., Carrio R., Chaque R., Cabeza L.F.: Processing collagenic residues isolation of gelatin by action of perchromates. *JALCA* 94, 115-123 (1999).

[47] Cabeza L.F., Taylor M.M., Brown E.L., Marmer W.N.: Treatment of sheepskin chrome shavings. Isolation of high valuie protein products and reuse of chromium in the tanning process. *JALCA* 94, 268-287 (1999).

[48] Kolomaznik K., Mladek M., Langmaier F., Janacova D.: Experiences in industrial practice of enzymatic dechromation of shavings. *JALCA* 95, 55-63 (2000).

[49] Langmaier F., Kolomaznik K., Mladek M.: Products of decomposition of chrome-tanned leather waste. *JSLTC* 83, 187-195 (2001).

[50] Langmaier F., Stibora M., Mladek M., Kolomaznik K.: Gel-sol transformation of chrome-tanned leather hydrolysates. JSLTC 85, 100-105 (2001).

[51] Pizzi A., Lipschitz L.,Valenzuelan J.: Theory and practice of preparation of flow formaldehyde emission and content. Holzforschung 589, 254-261 (1994).

[52] Dunky M.: Urea-formaldehyde adhesive resins for wood. *Int. J. Adh. Adhesives* 18, 95-107 (1998).

[53] Gessler H.D.: The reduction of indoor formaldehyde gas emanating from urea-formaldehyde foam insulation. *Evirn. Int.* 10, 305-307 (1984).

[54] Langmaier F., Sivarova J., Kolomaznik K., Mladek M.: Curing adhesives of urea-formaldehyde type with collagen hydrolysates of chrome-tanned leather waste. *J. Thermal Anal.* Calor. 75, 205-219 (2004).

[55] Langmaier F., Sivarova J., Kolomaznik K., Mladek M.: Curing of urea-formaldehydetype adhesives with collagen hydrolysates under acid condition. *J. Thermal Anal.* Calor. 76, 1015-1023 (2004).

[56] Langmaier F., Kolomaznik K., Mladek M., Sivarova J.: Curing urea-formaldehydeadhesives with hydrolysates of chrome-tanned leather waste from leather production. Int. *J. Adhesion Adhesives* 25, 101-108 (2005).

[57] Parra D.F., Mercury J.R., Matos H., Brito H.F., Romano R.R.: Thermal behavior ofepoxy powder coatings using thermogravimetry/differetial thermal analysis coupled gas chromatogr. spectrometry (TGA/DTA-GCMS) technique: identification of degradation products. *Thermochimica Acta* 386 (2), 143-151 (2002).

[58] Gorton B.S.: Interaction of nylon polymers with epoxy resins in adhesive blends. *J. Appl. Polymer Sci.* 8 (3), 1287-1295 (1964).

[59] Zhong ZK., Guo Q.P.: Miscibility and cure kinetics of nylon – epoxy resins active blends. *Polymer* 39 (15), 3451-3458 (1998).

[60] Langmaier F., Mokrejs P., Kolomaznik K., Mladek M., Karnas R.: Cross-linking epoxide resins with hydrolysates of chrome-tanned leather waste. *J. Thermal Anal.* Calor. 88, 857-862 (2007).

[61] Draye J.P., Delaney B., van der Voorde A., Van den Buckle A., Bagsanov B., Schacht E.: In vitro release characteristics of bioactive molecules from dextran dialdehyde cross- linked gelatin films. *Biomaterials* 19, 99-107 (1998).

[62] Balakrishnan B., Jayakrishnan A.: Self-cross-linking biopolymers as injectable in situ forming biodegradable scaffolds. *Biomaterials* 26, 3941-3951 (2005).

[63] Balakrishnan B., Mohanty M., Umashankar P.R., Jayakrushnan A.: Evaluation of in situ forming hydrogels wound dressing based on oxidized alginate and gelatin. *Biomaterilas* 26, 6335-6342 (2005).

[64] Wilson R.H.: Utilization and toxicity of dialdehyde and dicarboxyl starches. *Proc. Soc. Exp. Biol. Med.* 102, 735-737 (1959).

[65] Onishi Hiraku, Nagai Tsuneiu: Characterisation and evaluation of dialdehyde starch as erodible medical polymer and drug carrier. Int. *J. Pharmaceutics* 30, 133-141 (1986).

[66] Gennadios A., Hanna A., Froning G.W., Weller C.L. Hanna M.A.: Physical Properties of egg white-diladehyde starch films. J. Agric. *Food Chem.* 46, 1297-1302 (1998).

[67] Yannas I.V., Tobolski A.V.: High temperature transformation of gelatin. Eu. *Polymer J.* 4, 257-264 (1968).

[68] [Langmaier F., Mladek M., Mokrejs P., Kolomaznik K.: Biodegradable packing materials based on waste collagen hydrolysates cured with dialdehyde starch. *J. Thermal Anal.* Calor. 93, 547-552 (2008).

[69] Nijenthuis Te.K. : Viscoelastic properties of thermoreversible hydrogels . In Burchard W., Ross/ Murphy S.B. (Eds.) Physical network polymers and gels. Elsevier, N.Y. 1990, pp 15-34.

[70] Kozlov P.V., Burdygina G.I.: The structure and properties of solid gelatin. The principles of their modification. Polymer 24, 651-666 (1999).

[71] Langmaier F., Mokrejs P., Kolomaznik K., Mladek M.: Biodegradable packing materials from hydolysates of collagen waste proteins. *Waste Management* 28, 549-556 (2008).

[72] Langmaier F., Mokrejs P., Mladek M., Kolomaznik K. Biodegradable films of collagen hydrolysate cross-linked with dialdehyde starch. (Still unpublished communication).

In: Encyclopedia of Environmental Research ISBN: 978-1-61761-927-4
Editor: Alisa N. Souter © 2011 Nova Science Publishers, Inc.

Chapter 18

BIODIVERSITY CONSERVATION AND MANAGEMENT IN THE BRAZILIAN ATLANTIC FOREST: EVERY FRAGMENT MUST BE CONSIDERED

Roseli Pellens[1], Irene Garay[2]+ and Philippe Grandcolas[1]*

[1] UMR 5202 CNRS, Département Systématique et Evolution, Muséum national d'Histoire naturelle, 45, rue Buffon, 75005 Paris, France.
[2] Laboratório de Gestão da Biodiversidade, Instituto de Biologia, UFRJ, CCS, Bl. A, Ilha do Fundão, CEP 21941-590, Rio de Janeiro, Brazil.

ABSTRACT

The Brazilian Atlantic forest is a hotspot of biodiversity, characterized by its high species richness, level of endemism and danger of extinction. The remnants of this biome stand in the most populated and developed region of the country and their conservation is in constant conflict with the strategies of development. Our study was developed in the State of Espírito Santo, in the municipalities of Linhares and Sooretama. Located at the north of Rio Doce, it was one of the last regions of the Atlantic forest to be deforested, and forest fragmentation took place mostly after 1950. Our purpose were: 1) characterize the evolution of the forest fragmentation in this region, the present situation of its remnants, and the main strategies, conflicts and potentials for conservation; 2) evaluate the conservation value of the forest fragments based on a study of the community of a group of insects – the community of Blattaria; 3) integrate these information to propose guidelines to resolve potential conflicts based on positive and non conflictual conservation recommendations. The landscape of Linhares and Sooretama is dominated by plantations, among which can be found two large reserves and fragments of diverse sizes, shapes and degrees of disturbance. Only in Sooretama there are 213 fragments, which, with the Sooretama Biological Reserve, cover 44% of the municipality area. Similarly to the reserves, the fragments were preserved by the imposition of the law and due to the interest of people of keeping the forest as a reserve of value. Fragments are used for several purposes, with some of them provoking strong disturbances that could

* E-mail: pellens@mnhn.fr.
+ E-mail: garay@biologia.ufrj.br.

lead to their massive disappearance in a near future. The analysis of the community of Blattaria showed that the number of species shared among sites is very low (between 22.9% and 39.6%), with a high number of species found only in single fragments and nowhere else; the richness per area in small fragments is not smaller than in large reserves; habitat diversity is stronger than the effect of area *per se*. These results show that even after more than 50 years of fragmentation and isolation, the situation of the biodiversity in this region is not lost: small and very disturbed fragments still shelter important biodiversity of insects, and can be of extreme interest for composing a set of remnants that maximizes the number of species to be protected. Based on it, the management and conservation of the biodiversity should first of all include the fragments, and for this reason should be conceived in the landscape or regional scale. At this scale, fragments should be envisaged in a management plan, in which restrictions and uses allowed are explicit. In such a context, the choice of priority fragments should be done taking in account the landscape diversity: it should include fragments from different hydrographic basin, soil types, vegetation *facies*, as well as the diversity of use.

INTRODUCTION

In many regions of the world the management of natural resources is at the center of never ending conflicts between the desire of keeping a part of the territory for conservation and the immediate need of using the same space for cultures, industries, cities, etc, due to cultural reasons and/or regional or national strategies of development. This situation is very common in Brazil, particularly in the region of the Brazilian Atlantic forest.

This biome is one of the richest in the world in terms of species richness and endemism: it harbors about 8.000 species of endemic plants and 567 species of endemic vertebrates. Nevertheless, the conservation of its remnants is a challenge (Gentry, 1992; Mittermeier, 1997; Myers, 1997). Originally it covered an area of about 1,400,000km^2, stretching along the Atlantic coast from 4° to 32°S, in altitudes ranging from sea level to 2,900m, covering a wide range of climatic belts and vegetation formations (Rizzini, 1992; Oliveira-Filho & Fontes, 2000). But it was by this coast that started the colonization of the country, being the first region of Brazil deforested to give place to modern civilization. Presently, the original area of the Atlantic forest has about 100 million inhabitants distributed in more than 3,000 cities, among which Rio de Janeiro and São Paulo, two of the biggest metropolis of the world. As a result, the forests are reduced to less than 8% of its original surface, distributed in scattered fragments imbedded in the most populated and developed region of the country (SOS Mata Atlântica/INPE/ISA, 1998; IBGE, 2000).

Due to its enormous surface, the land use as well as the history of deforestation/ conservation can be very different from one region to another. So must be the strategies for solving the conflicts in order to establish a long-term strategy of conservation. In this chapter we will focus at the north of Rio Doce, State of Espirito Santo, in the municipalities of Linhares and Sooretama, a part of the Atlantic forest considered of extreme biological importance for mammals, reptiles, invertebrates and vegetation (MMA/SBF, 2000). In this region the deforestation and fragmentation of the ecosystems occurred mainly after 1950, and apparently most of forest remnants still harbour an important fraction of its original diversity.

This chapter was written with a triple aim: the first was to characterize the evolution of the forest fragmentation in this region, the present situation of its forest remnants, and the main strategies, conflicts and potentials for conservation. This study was based on data from geographic censuses, on some of our observations in the field, and on published researches about the conservation of the biodiversity in this region. The second aim was to evaluate the importance of the forest fragments for the conservation of the biodiversity. This part was based on a study of the community of a group of insects – the community of Blattaria. Last but not least, our third aim was to integrate the information of these two studies to propose guidelines to resolve potential conflicts based on positive and non conflictual conservation recommendations.

A Brief Characterization Of The Landscape In The State Of Espírito Santo In The 20th Century

When compared to Rio de Janeiro, São Paulo and Minas Gerais, the three other States of the southeast region of Brazil, Espírito Santo is very peculiar in several aspects. It stayed out of the main development of Brazil, which was always more intense in the southeast region, until very recently, being the last one to be included in this process (Becker, 1969); its population density is much lower than the mean of the region; and the density of rural population is rather high (Table 1).

Table 1. Area and population of the municipalities of Linhares e Sooretama, the state of Espirito Santo, the southeast region, and Brazil in the year 2000.

	Área (km^2)	Population	Demographic density (inhabitant/km^2)	Rural population	% of rural population	Urban Population	% of urban population
Linhares	6,885.90	250,300	32.6	58,469	23.4	191,831	76.6
Sooretama	585.6	18,269	31.2	6,850	37.5	11,419	62.5
Espírito Santo	46,047.30	3, 097,232	67.2	634,183	20.5	2,463,049	79.5
Southeast	924,573.80	72, 412,411	78.2	6,863,217	9.5	65,549,194	90.5
Brazil	8,514,204.90	169,799,170	19.9	31,845,211	18.8	137,953,959	81.2

Data from IBGE – Population count 2000.

At the beginning of the 20th century most of the surface of Espírito Santo was covered by forests, and *Botocudos* Indians from the linguistic branch of Botocudos Macro-Jê, the first inhabitants of this region, were seen until the 40's (Aguirre, 1951; Zunti, 1982; Egler, 1951; Garay, 2006). We do not pretend that the forests were intact until this date, but after the 40's and particularly during the 50's the landscape changed markedly. During this period 35% of the forests disappeared, first giving place to the culture of coffee and later to more diversified land use (Egler, 1951; Becker, 1969; SOS Mata Atlântica *et al.*, 1998; Pellens, 2002). Presently, less than 9% of the original surface of the state is covered by remnants of the Atlantic forest and its associated ecosystems (e.g., mangroves, restinga), and the land use comprise urban areas in addition to large surface of rural landscape mostly used by pastures and plantations of coffee, tropical fruits, *Eucalyptus*, *Cocoa*, *Hevea* and black pepper (Figure 1) (IBGE, 1995-96; SOS Mata Atlântica *et al.*, 1998).

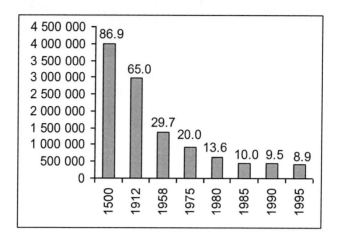

Figure 1. Area covered by vegetation (ha) in the State of Espírito Santo from 1500 to 1995. The values on top of the bars correspond to the percentage of natural forests in relation to the total area of the state. Data from SOS Mata Atlântica *et al.*, 1998.

Linhares and Sooretama: a region of beautiful forests that changed very recently

In the construction of the history of Espírito Santo, the Rio Doce, that divides the state into two portions, one austral and the other septentrional, served during a long period as a natural limit between the populated zone in the South, and the unknown zone in the North (Egler, 1951). It is in this unknown zone at the North of low Rio Doce that the municipality of Linhares and the recently created municipality of Sooretama[1] are located.

According to Egler (1951), the colonization of the North of lower Rio Doce was caused by a displacement of the population in the space and not by an expansion in terms of number of individuals, since it was not verified an increase of the population of the State in this period but a simple displacement to the North. The new population was mainly composed by Italians of third generation coming from the mountains of Espírito Santo, and by people from the Northeast escaping from the dryness. In Egler's opinion, this displacement was caused by the need of new areas of "virgin lands", because the usual habits of slash-and-burn followed by intensive culture used to lead to the exhaustion of the soil resources, forcing the search for new lands. Searching for new lands, escaping from strong dryness in Northeast, and from many other sources of poverty and difficulties existing in Brazil at that moment, the fact is that Linhares population increased more than three times between 1950 and 1970, and more 32% between 1970 and 1980 when population growth started to stabilize (Data of the Census of Linhares). In the same period, the number and the area of agricultural establishments doubled (Figure 2).

---

[1] The municipality of Sooretama was created in March 30th, 1994 by the State Law N° 4,593/1994. It was created as a municipality in the same area of the previous District named "Distrito de Córrego D'água" that belonged to the municipality of Linhares. Sooretama was installed as a municipality in 1997. For this reason, in this first part of our analysis we use the data of Census concerning the Municipality of Linhares, since there was not similar data by districts during all this period.

The major attractor of the first inhabitants was the wood of very high quality from the Atlantic forests of the table-lands, and the available lands that allowed the survival of the families with peasant farming. This first structure was followed by the production of coffee stimulated by the federal government, which failed at the end of 60's leading this population to a strong poverty. In her study about the transformation of this region, Becker (1969) showed that the colonization was marked by strong patchiness. The landscape was characterized by the small size of the coffee plantations due to the restrictions imposed by the topography and the small dimension of the properties. This situation remained until the beginning of the 70's, when "the settlements used to occur in small encroachments interrupted by large areas of forests, areas recently deforested, and decadent coffee plantations".

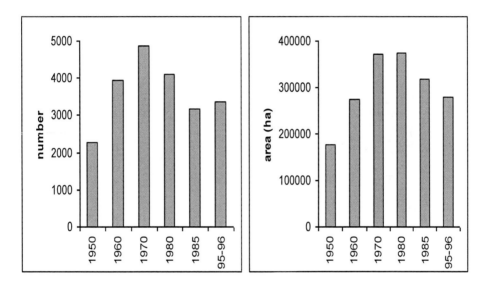

Figure 2. Number and area occupied by agricultural establishments in Linhares between 1950 and 1996. Data from IBGE, Census of Agriculture (1950, 1960, 1970, 1980, 1985, 1995-96). These data include the municipality of Sooretama that was installed as an independent municipality in 1997.

When we look at the dynamic of the area occupied by agricultural establishments, we see that if in a first moment there was an appropriation of the territory by the population, marked by a significant increase in number and size of their area in the municipality of Linhares. In a second period started in 1970 the transformations are mainly associated to shifts in the new forms of land use. Between 1970 and 1980 the area of agricultural establishments remained the same but their number was reduced, showing an expansion of the size of rural properties. Finally, between 1980 and 1995 both the number and the total area occupied by rural establishments were markedly reduced, now indicating the reduction of the rural area, or in other words, the expansion of urban areas in the municipality (Figure 2).

From the eight categories of land use considered in the Agricultural Census, only native forests, permanent cultures, natural and artificial pastures occupied a significant surface of the agricultural establishments in Linhares between 1950 and 1995-96 (Figure 3).

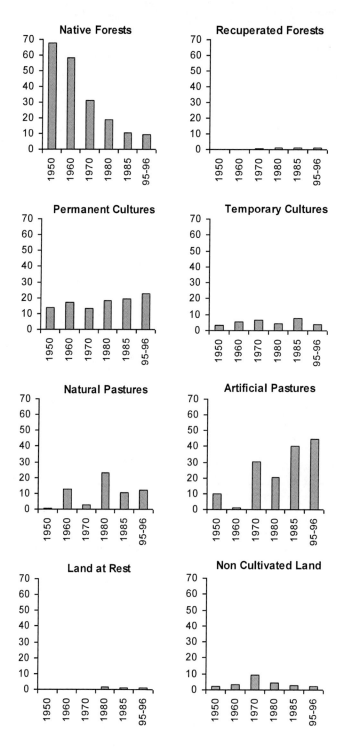

Figure 3. Area (in percentage) of agricultural establishments in Linhares – ES occupied by eight main land uses from 1950 to 1995-96. Data from IBGE Census of Agriculture (1950, 1960, 1970, 1980, 1985, 1995-96). These data include the municipality of Sooretama that was installed as an independent municipality in 1997.

As expected in a context of occupation of the territory, the main transformation in this landscape during the last 50 years corresponds to the reduction of native forests. From 1950 to 1995-96, the area of forest in the totality of agricultural establishments of Linhares was reduced in 58%, and this reduction was sharp between 1960 and 1985 (47%). At the same time there was a slow expansion of the area used to permanent cultures (23% of the total area of agricultural establishments in 1995-96), and a more intense expansion of the area for artificial pastures (45% of the total area of agricultural establishments in 1995-96) (Figure 3).

Presently, the landscape of Linhares and Sooretama is marked by the presence of an important area of continuous forest protected in the Sooretama Biological Reserve, in Natural Reserve of Companhia Vale do Rio Doce, commonly known as Linhares Reserve, and in some private properties linking them (IBDF & FBCN, 1981; Jesus, 1987; SOS Mata Atlântica/ISA, 2002; Garay & Rizzini, 2004). This green heart of about 50,000ha is surrounded by pastures and plantations of coffee, *Eucalyptus*, papaya and black pepper within which a multitude of forest fragments of diverse size, shape and distance to the reserves can be found (Figure 4).

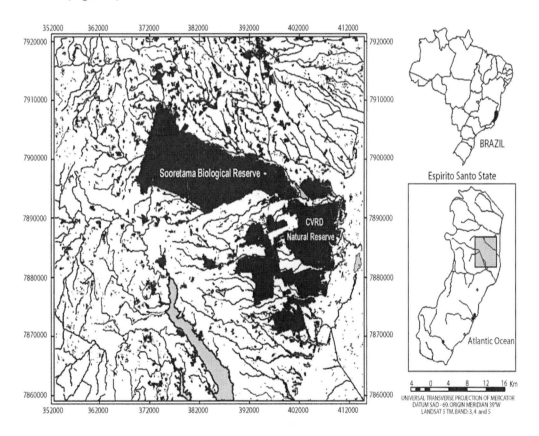

Figure 4. Localization of the remnants of Atlantic forest in the municipalities of Sooretama and Linhares, with the main water bodies. The Sooretama Biological Reserve is formed by a continuous forest, while the Linhares Reserve (CVRD Natural Reserve) is formed by multiple lots of forests not always contiguous. Around the two reserves one can see the multiple forest fragments, most of them very small.

In a characterization of these fragments in the municipality of Sooretama, Agarez *et al.*, 2004 showed that there are 213 forest fragments with area between 1 and 200ha making a total of 3,000ha of forest (Figure 5). These fragments with the Sooretama Biological Reserve cover a total of 44% of the municipality surface. Unfortunately, we do not have similar data for the municipality of Linhares.

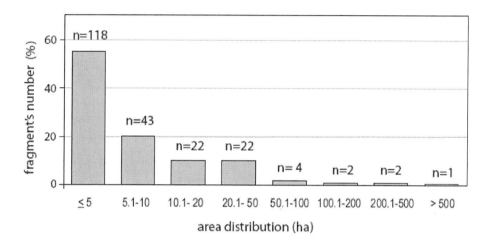

Figure 5. Number of forest fragments in different classes of sizes in the municipality of Sooretama (modified from Agarez *et al.*, 2004). The fragment with more than 500 ha corresponds to the Sooretama Biological Reserve.

The forests protected in Linhares and Sooretama constitute the biggest remnant of Atlantic forest on lowlands. The greatest part of it is represented by the table-land Atlantic forest, a semideciduous forest located on the table-lands. This geomorphological unit is characterized by a gently rolling relief with table columns with mean altitude of about 70-80m formed by deposits of sediments during the high Tertiary and beginning of the Quaternary period (Suguio *et al.*, 1987). The vegetation is exuberant, distributed in three superposed strata, with emergent trees of about 40m (Figure 6). The forest is exceptionally marked by the richness and endemism of plants, as well as for mammals, reptiles and invertebrates (Thomas *et al.*, 1998; MMA/SBF, 2000) being for this reason one priority area for conservation within Atlantic forest (MMA/SBF, 2000). Only in Linhares Reserve about 650 tree species were already identified (Linhares Reserve, 2004; Jesus & Rolim, 2005). Agarez (2002) studying tree species with DBA $\geq$ 5cm found 202 species in one hectare of forest in Sooretama Biological Reserve, and between 168 and 197 species in one hectare of forest within the fragments. In addition to that, the forests of this region are also characterized by the richness and density of woody vines as well as by the rarity or absence of epiphytes (Jesus, 1987; Gentry, 1992; Peixoto *et al.*, 1995; Peixoto & Gentry, 1990; Rizzini, 1992). The forests in the large reserves shelter a very diverse fauna, and important populations of big mammals were recently observed (Chiarello, 2000).

The Conservation Of The Forests: Its Main Historical Reasons And Present Conflicts

In the same period of large scale deforestation, there were some efforts on the opposite way, resulting in the creation of the present Sooretama Biological Reserve and Linhares Reserve. A brief context of the creation and aims of these two reserves are presented in Box 1. We will not further detail their history here because it was carefully done elsewhere (see Aguirre, 1951; IBDF & FBCN, 1981; Jesus, 1987; Pellens, 2002; Garay, 2004; Garay, 2006).

Figure 6. A picture of the understory of the table-land Atlantic forest taken at the site named Mata Alta in Linhares Reserve.

Nevertheless, it is important to remark that in spite of the efforts of some well-dedicated conservationists, the main reason of the present existence of the forest protected in these two reserves was the high quantity of wood they had. These forests represented reserves of values to be used in a near future mostly by Companhia Vale do Rio Doce, the greatest Brazilian miner that, at that time, needed wood to build railroads to transport the minerals exploited in the inlands of the state of Minas Gerais to the port in Vitória, state of Espírito Santo, to be exported. In fact, as indicated Jesus (1987) (Box 1) the aim of making reserves of wood for using in the construction of the Railroad Vitória-Minas is easily seen when reading the documents of buying the lands that became the Linhares Reserve.

Box 1

Characterization of the reserves

The Sooretama Biological Reserve is a national reserve, protected by determination of the Ministry of Environment under the responsibility of IBAMA - The Brazilian Institute of Environment and Renewable Natural Resources. The creation of this reserve started in 1940 with the creation of the Parque de Reserva, Refúgio e Criação de Animais Silvestres Sooretama (with 12,000ha) and the contiguous Reserva Florestal de Barra Seca (with 10,200 ha). The Parque de Reserva, Refúgio e Criação de Animais Silvestres Sooretama was created with the aim of preserving fauna, flora and geological formations, creation of wild animals, facilitation of studies and knowledge about the nature, and stimulation of touristic excursions. We have no information about the explicit objectives of the creation of the Reserva Florestal Barra Seca. In 1971, these two reserves became the Sooretama Biological Reserve with a surface of 22,168 ha (IBDF & FBCN, 1981). Since 1982, this reserve has the aim of integral protection of the fauna and flora. Now only scientific research on its biodiversity and educational programs are allowed.

The Linhares Reserve is private. It belongs to Companhia Vale do Rio Doce, and has its origin in a program of buying land in several parts endeavoured by this company from the beginning of 1950's, mostly in 1951. According to Jesus (1987) the aims of buying this forest is not explicit in any document. But his analysis of reports and forestry inventories always bring significant information about forestry production and the quantity of wood to be used in railroad structure, which suggests that this forest was envisaged as a reserve of wood to be used in the construction of the railroad Vitória-Minas. The decision of maintaining this forest was taken with the result of studies and the increase of information about the potential of wood production and the sustainable use of the property. Even though, at that moment none of these programs of wood extraction and of sustainable use of the forest were implemented and the present vegetation cover is practically the same existing when the land was bought. Presently the Natural Reserve of Linhares protects an area of 21,780 ha of Atlantic forest and associated ecosystems, and is used for ecological tourism and production of seedling of Atlantic forest trees to be used in programs of forest regeneration all over the country. See http://www.vale.com/hot_sites/linhares/reserva.htm.

The case of Sooretama Biological Reserve is more complex. On the one hand, the Parque de Reserva, Refúgio e Criação de Animais Silvestres Sooretama was created with the aim of preservation and associated objectives such as education and scientific knowledge. On the other hand, the Reserva Barra Seca seems to have been created with the aim of preserving a forest that contained an important reserve of wood. It came to the point that in 1968 the government of the State of Espírito Santo demanded the repeal of the donation of the lands of Reserva Barra Seca, previously donated by this State to the National Government in order to create a national reserve. This repeal aimed the used of the lands to exploit the forest (Aguirre, 1951; IBDF & FBCN, 1981). In 1971, Sooretama Biological Reserve was finally created in the form it exists nowadays. About 30 or 40 families used to live within the limits of the Parque de Reserva, Refúgio e Criação de Animais Silvestres Sooretama at the moment of its creation. With the creation of the Park, logging, hunting, and creation of domestic animals became forbidden. Apparently these prohibitions lead the families that lived within the park to search for new areas of forest still existing in the region (Aguirre, 1951). In the

Linhares Reserve there was a small settlement named community São João Batista, from which remained a bandstand that was recently destroyed and a cemetery. Their history was recently studied by Machado-Guimarães (1998) (see also Garay, 2004).

Presently the Sooretama Biological Reserve is protected by IBAMA[2], and the main problems for its conservation are the fire and the presence of hunters. The Linhares Reserve is a private property used for education, ecological tourism, and production of seedlings of forestry species used for reforestation.

The history of preservation of the small forest fragments is probably more complex than that of the two large reserves, particularly due to their high number, to the fact that they belong to different owners, and to that the land use around them can be quite different from one property to another and can change along time. Nevertheless, at least two main reasons are at the basis of their conservation.

One is the obligation by the law, first formalized in the Forest Code, the law n° 4,771 from 15.IX.1965. This law limited the right of property concerning the native vegetation existing in the national territory. It qualified the forest as wells of common interest to every inhabitant of the country subordinating its exploitation to the interest of the population. The Forest Code created the status of Permanent Preservation Area[3] and Legal Reserve[4], two kinds of situation in which the forests must be protected in rural properties or possessions. In 1993, forest logging, as well as exploitation and suppression of the vegetation in the Atlantic forest became very restricted by the Decree 750 from 10.II.1993. In addition to that, since 1998 the remnants of the Brazilian Atlantic forest are protected by the Federal Constitution that established that the Atlantic forest is a national patrimony, and its utilisation must be done in conditions that assure the preservation of the environment, including the utilization of natural resources. Presently tree cutting is mostly forbidden, and the only cases permitted need the specific authorization of the relevant environmental institutions.

In spite of the importance of the legal apparatus for avoiding the total deforestation, a close look to the distribution of the fragments in the landscape indicates that, although it obviously contributed to its maintenance, it was not sufficient. The majority of the properties do not have 20% of forest that should constitute the Legal Reserve, and many Permanent Preservation Areas (margins of rivers, lagoons and lakes, springs, hilltops, and slopes greater than 45°) are not covered by native vegetation.

The other reason for the present existence of the forest fragments in the landscape is that they have good quality wood and other natural resources, representing for their owners a reserve of values that continue to be exploited until our days. In several of them one can

---

[2] IBAMA - The Brazilian Institute of Environment and Renewable Natural Resource.

[3] The Permanent Preservation Area is the area covered or not by native vegetation, with the environmental function of preserving the hydric resources, the landscape, the geologic stability, the biodiversity, the gene flow of the fauna and flora, of protecting the soil and of assuring the well-being of human populations. It is considered of Permanent Preservation, a) the forests and any other form of vegetation located along the rivers and any body of water from its highest level in a marginal strip of width varying between 30m and 500m according to the width of the water body; b) in the edges of lagoons, lakes and natural or artificial water reservoir; c) in springs, independently of their topographic situation, in a radius of 50m of width; d) in hilltops; e) in slopes greater than 45°; and in a few other situations that do not concern this region (Forestry code – federal law n° 4,771/1965; Art. 2°).

[4] The Legal Reserve is an area localized in a rural property or possession, except the Area of Permanent Preservation, necessary to the sustainable use of natural resources, to the conservation and recuperation of ecological processes, to the conservation of the biodiversity, and to shelter and protect the native flora and fauna. In the Atlantic forest 20% of the area of the rural property or possession must be kept as Legal Reserve (Forestry code – federal law n° 4,771/1965; Art. 1°; Paragraph III).

easily see tracks, trails, traces of logging, traps, and many other indicators of an active use (Pellens *et al.*, *in prep.*) (Table 2).

Table 2. Present uses of the forest fragments in Linhares and Sooretama, based on observations of the authors of this chapter, and on reports of workers of IBAMA acting on the region.

Uses	Some Details	Indicators
Hunting	Birds, lizards, small mammals, big mammals	Traps Riffles Guards of the reserves aggressed and shot by hunters Techniques of hunting, recipes and techniques of preparation of the meat of wild animals common in the speech of the population
Wood extraction	Hardwood trees, softwood trees, bushes, young trees, etc.	Piles of trunks recently cut on the ground of the forest Firewood oven/stove Coffee dried with firewood Fences made with forestry wood Houses/stables joists made with wood Tool's handle
Herbs extraction	Medicinal Decorative	Traditional use of medicinal plants that are found in the forests – seen by people's indications when you are sick. Forestry species observed in gardens (bromeliads, orchids, ferns, palms, short trees, etc)
Protection of springs, ponds and rivers		Location of the fragment in relation to the water body Fishing Washing clothes
Protection against the wind		Long fragments usually left as a barrier against the dominant wind and the house of the property or the plantation
Protection for cattle		Cows resting at the shadow in small fragments during the sunny hours of the day
Leisure		People walking for doing exercises Dating Hunting
Scientific research		Tracks, tags on trees, ropes used to delimit stands, cameras for filming mammal's movements, holes for studying the soil, thesis sustained, scientific results published.

Forest is part of people's everyday life. It provides wood for drying the coffee, for the structure of the houses, and many other resources and services (Table 2). Although these resources and services were factors that contributed to the protection of the forest fragments, its continuation in time without a planning for the entire landscape can be dangerous to the preservation of the biodiversity. As a matter of example, we show one study of Agarez *et al.*,

(2004), in which they compared the density of the most common tree families in Sooretama Biological Reserve and in fragments submitted to different logging intensity. Their results show that the density of the families with higher wood quality, as is the case of the species of Myrtaceae, Sapotaceae, Leguminosae and Lecythidaceae, is markedly reduced in sites with higher logging intensity. Conversely, the density of the families of fast growing trees with low quality wood, such as Anacardiaceae, Moraceae and Euphorbiaceae, tends to increase in fragments with higher logging intensity (Figure 7).

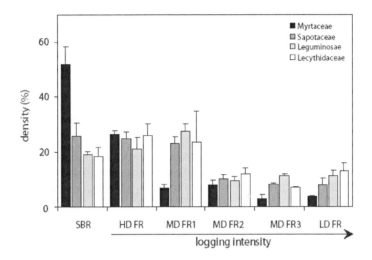

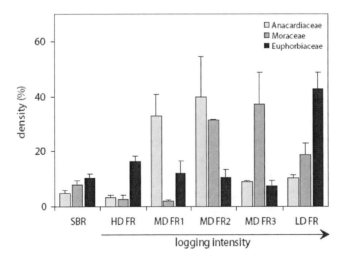

Figure 7. Density of the most abundant tree families in Sooretama Biological Reserve and in four fragments with different logging intensity (SBR: Sooretama Biological Reserve; HD FR: fragments of high density; MD FR: fragments of medium density; LD FR: fragment of low density). Modified from Agarez et al., 2004.

New perspectives for conservation: PROBIO and Ecological Corridor of Atlantic Forest

During the last twenty years, many demands of political and social character related to the management of natural resources were developed in Brazil. These demands were deeply rooted in the political priorities of the country and materialized in several governmental programs. Among them was the National Program of Biological Diversity (PRONABIO-PROBIO), created in 1994 to answer the commitments of the country to the Convention of Biological Diversity. One of its main integrative subprojects was developed in Linhares and Sooretama: the demonstrative subproject *"Conservation and Recuperation of Atlantic forest in Linhares – ES, based on the functional evaluation of the biodiversity"*. This subproject represented a study case. Its implementation constituted a search of methodologies aiming the construction of an interinstitutional and participative model of conservation that includes the transfer of technologies of forest recuperation and environmental education for conservation to the table-land Atlantic forest (Garay, 2006). The aims, methodologies, results and details of the implementation of this main subproject were carefully presented in Garay (2006). In the present chapter we simply want to characterize its existence and to show a few results related to the recuperation of the forest. However, the most significant result of this interdisciplinary research was the change of perception of the conservation value of the local biodiversity, reversing the tendency of indiscriminate deforestation of the forest remnants existing outside the reserves.

In this perspective, the central strategy consisted in relating the problems of maintenance of hydric resources necessary to the local agriculture and of controlling the soil erosion of the edges of numerous streams whose waters are essential to the agriculture as well as to the forest vegetation (an ecosystem service). It is important to remark the existence of periodic cycles of drought that not only facilitate fire in the forest remnants being also catastrophic to the local agriculture.

The sites for the implementation of the subproject were chosen based on the interdisciplinary approach shown in Figure 8. Following this interdisciplinary approach, the areas for future plantation of forests were located in the buffer zone of the reserves (nuclear areas). As far as possible they linked fragments with the aim to constitute micro-corridors, and surrounded springs and the edges of the streams, sites protected by law (Permanent Preservation areas). The socio-economic diagnostic showed the importance of small and medium properties in the productive rural system of the region, once more putting in evidence the need of restoration of the edges of water bodies (service of the forest cover). Parallel studies about the status of the biodiversity and the diffusion of part of these results were instruments of valorisation of native ecosystems. These results are presently used to evaluate the efficiency of the restoration concerning the recuperation of the biodiversity.

The project was developed in two main phases: first with local leadership and after with rural producers and other members of the community. Regarding the restored areas, a total 110ha was planted with 270,000 seedlings in a basis of 40 native species/ha; 38 rural producers were involved and about 150 people were trained; the strategy adopted lead to 90% cost reduction when compared to the costs of forest recuperation in the market.

In the first phase the project was developed in seven rural properties and comprised a total of 120,000 seedlings planted in 48ha. The recuperation was efficient, and in practically every site planted plants grew efficiently in way that in nine years the plantations had the general aspect of a young forest (Figure 9). The first phase was developed with local leaders and was characterised by significant demonstrative areas (several hectares planted in each property). The main action was the transfer of technology of maintenance of the forest plantations. The positive aspects were: higher exemplar and demonstrative value; massive participation of local workforce, training *in locus* and production of seedlings for replacement in farms. The limits were: high-cost of implementation, centralized social responsibility, limited number of social agents involved, restricted benefits to the community.

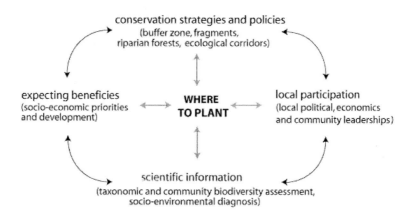

Figure 8. Interdisciplinary approach for choosing restoration areas.

The second phase was characterized by the creation of a community NGO, named Bionativa, born from a partnership between government and society (MMA[5]/UFRJ[6], Municipality, and Community). The exemplar value given by the participation of local leaders and by the creation of a local space of public access that centralized the production of low cost seedlings facilitated the diffusion of the actions of restoration to the low and medium producers (Garay, 2006). The positive results of this second phase were: dissemination of technology across the land and also in other countries (dispersion pole), shared social responsibility, broad spectrum of social agents involved, significant decrease of costs (90 %). The aims and results of conservation of the ONG Bionativa are synthesized in Box 2. In the process of creating this ONG a forest fragment was acquired with the aim of creating a park for the community. It represents the last Atlantic forest fragment in periurban zone of this region and is destined to research and environmental education.

The results obtained with the implementation of this subproject shows up the need of: 1) a change of perception of the social agents aiming to develop an active attitude concerning the conservation of the biodiversity; 2) the establishment of partnerships to develop consensual solutions that reconcile productive activities and conservation; 3) the capacity of social intervention of non governmental organization as facilitators of conservation actions, including the intervention in the reduction of costs of conservation and in exemplar demonstrative actions; and, finally, 4) the importance of developing scientific

---

[5] Ministério do Meio Ambiente = Ministry of Environment.
[6] Federal University of Rio de Janeiro.

research regarding the status of the biodiversity, with the main objective of evaluating the sustainability of experiments of forest restoration and the values of remnants as reservoirs of biodiversity and source of species in areas of recuperation.

Figure 9. Pictures of an experimental site of forest restored using 40 native species/ha and its evolution in nine years. a) The site recently planted (1998); b) three years later (2001); c) nine years later the plantation had already the aspect of a young forest.

Box 2

Results obtained and objectives relative to the transfer of technology and environmental education for conservation with the creation of the NGO Bionativa.

technological transfer for conservation purposes

Developing and assessing new technologies of sustainable use of biological resources (*research*)

Centralizing local wisdom and scientific-technical knowledge (*forest restoration activities*)

Making the new technologies of the native germoplasm use accessible to the community (*relationship between transfer and education*)

Contributing to the equilibrium between the rural development (direct and indirect benefits of biodiversity) and forest conservation (*landscape management and public policies*)

ONG education goals for biodiversity conservation

Institutionalization of the diffuse participation of the social agents and landers in conservation processes (*organization*)

Transfer of technologies of sustainable use of biodiversity in the socio-bio-cultural context (*adapting transfer*)

Public access to the restoration technologies for permanent conservation areas (degraded riparian forests) (*transfer facility*)

Field support and regional diffusion for research on forest biodiversity (*linking restoration activities and biodiversity knowledge*)

Possibility of the continuity of conservation actions/use/management (*integrating conservation purposes and future needs*)

Presently, the project Ecological Corridors of the Ministry of Environment centralizes its activities in the central corridor of Atlantic forest that goes from the South of Espírito Santo to the South of Bahia with more than 1,000km of large. It is a corridor of management in an integrated perspective of management of the territory that considers the problems related both to forest conservation and the human activities, favouring the sustainability of these activities in the interstitial areas.

Linhares and Sooretama are located in this corridor, composing the micro-corridor Linhares-Sooretama-Comboios. Our previous experience of conservation and research in this focus area enabled us to take part in this project. The program of interdisciplinary research and solidary restoration of the Atlantic forest to be developed considers the implantation of experiences of forest restoration, associating the existing fragments with strong community participation aiming the dissemination of techniques of reforestation. Presently the focus is at the potential utilization of the chemical diversity of the arboreous species of the. However, the exploitation of natural resources is totally forbidden in Sooretama Biological Reserve, and is made in Linhares Reserve with the structure of a big national company that do not represent the dominant structure of land use in the region (Box 1).

In this context, the perspective of conserving the existing fragments gains a new dimension. Being the only remnants with possibility of utilization, they are the source of genes and species that can assure the sustainability of the arboreous plantations that will emerge, protecting the forest and amplifying the ecological services that they do to the territorial management and to the economic and social development as well.

Evaluating The Conservation Value Of The Fragments With The Community Of Blattaria

The conservation and management of the biodiversity requires an integration of the human dimensions, such as the ones we described above, with the knowledge of the ecology of populations, communities and ecosystems that compose that landscape. Tropical forests are characterised by the enormous diversity of species, by a great complexity of community organization and by a multitude of ways ecosystems function. In such a context the impact of deforestation, fragmentation as well as of the different uses made in forest fragments can be very high. Deforestation can create enormous disturbances on populations, communities and ecosystems. Immediate extinctions can occur if the habitats of species are eliminated. One can imagine that it can be mostly the case of short range endemics, but at the present rates deforestation all over the world it can be also the case of species with much larger range of distribution. In this context, the survival of a great part of species becomes highly dependent on their ability of surviving in a landscape mainly formed by fragments. But the long term persistence of the populations and communities depends on factors associated to the demography of the populations, on the possibility of rescue effects among sets of fragments, on the ability to resist shifts in the environment, or on the capacity to live in an environment affected by disturbances. Above all, it depends on the long term existence of the forest fragments, on the possibility of flow of individuals among them, and on the existence of a nuclear area that can contribute to enrich the forest fragments with individuals that could assure gene flow (Saunders *et al.*, 1991; Laurance & Bierregaard Jr., 1997; Thomas, 2000).

The study of the community of Blattaria that we will show right away was developed with the aim to evaluate the value of these fragments to the conservation of the biodiversity (Turner, & Corlett, 1996; Pellens & Grandcolas, 2007). In tropical forests, insects constitute one of the richest groups in terms of number of species and diversity of roles they play in the ecosystem (Hammond, 1995). They represent an important fraction of the diversity of any tropical forest, and contribute crucially to several functions essential to the functioning of the ecosystem, such as pollination, dispersion of microorganisms, fragmentation of organic debris, etc. For that reason, insects can be good indicators of the status of the ecosystem, and the analysis of their community can give basic information about the management of an important fraction of the biodiversity in a given region.

Blattaria is a worldwide distributed insect order, but with diversity and abundance much higher in tropical forests. It comprises more than 4,000 species described (Princis, 1962-1971, and later studies), and the total number of extant cockroach species in the world is up to 20.000 (Grandcolas & Pellens, 2009). In a recent evaluation, we showed that there are 644 species of cockroaches recorded in Brazil. Nevertheless, the analysis of the number of species accumulated along time showed that the accumulation curve is still in its ascending part, which indicates that the total number of species in this wide country can be at least five or six times higher than presently known (Pellens & Grandcolas, 2008). The local richness in a forest ecosystem can vary from a few tens to more than one hundred (Fisk, 1983; Grandcolas, 1991, 1994a; Pellens, 2002), and, as a rule, a high number of new species is found in each new region prospected (Grandcolas, 1991; 1997a; Grandcolas & Pellens, pers. obs.).

Individuals of Blattaria are involved in the decomposition of organic matter. Most of the species are saprophagous *sensu lato*, feeding on dead leaves, dead wood, carcasses, and any sort of organic debris, which are probably enriched by algae, fungi and bacteria normally found in natural ecosystems. In case of wood-eating cockroaches, the dead wood that is poor in nutrients is transformed in the digestive tube sometimes with the help of commensal or symbiotic protista (see for example, Pellens *et al.*, 2002; 2007a). Cockroaches live in a very wide and complex set of microhabitats from the soil to the canopy with a multitude of ways of exploring space and resources (Grandcolas, 1993a). For example, in the Brazilian Atlantic forest, in a same dead trunk laying on the ground one can find *Parasphaeria boleiriana* living in galleries they make in the wood, *Petasodes dominicana* under the bark, *Monastria biguttata* on the bark at the underside, and *Minablatta bipustulata* in the dust under the dead trunk (Pellens *et al.*, 2002, 2007b; Pellens & Grandcolas, 2003, 2007).

Obviously, the community of Blattaria is only one of the components of the tropical forest biodiversity, and of insect diversity as well. Nevertheless, it is a group sufficiently rich for allowing comparisons of several parameters of the community composition and structure, but not exaggeratedly rich, permitting to anchor the ecological analysis on a clear taxonomy, since it is possible to identify every individual sampled. In addition, individuals of Blattaria are sampled directly (i.e., without use of traps), which permits to include observations of their habitats and some elements of their biology (e.g. long wings *vs* short wings) or behaviour in the analysis of community. We designed this study with the main aim of evaluating the conservation value of small and disturbed fragments to the conservation. The questions we asked were: 1) Do the disturbances associated to forest fragmentation lead to differences in the richness and composition Blattaria community? 2) What is the contribution of the effects of area *per se* and of habitat diversity to the richness of cockroach community?

A methodological note

Six study sites were selected to develop this research: three in the large reserves and three in small forest fragments. They were carefully chosen to be on similar type of soil and in forests with similar kind of vegetation. Sampling was done in two seasons, in March-April 2000 and November 2000. Each study site had one hectare and was selected in the less disturbed areas within the forest, where canopy had three strata, avoiding the edges and paths or trails commonly found in the interior of forest fragments.

Individuals of Blattaria from the understory were collected by direct sampling in the forest understory from 6:00 PM to midnight. The search was made in the forest understory, on the ground (litter, dead trunks, rocks), on standing trunks, aerial litter, on leaves of plants until 2m high. Each sample corresponds to one hour per person of observation and collection. The following analysis are based on 3351 individuals corresponding to 48 species, 25 genera and four families of Blattaria collected during 131 hours of study (for more details, see Table 3 and Table 4). We also made similar collection in sites within the matrix near to the reserves and to each fragment studied. Only seven of the 48 species collected were observed in the matrix, and only one of these species (*Pycnoscelus surinamensis*) is typically from the matrix, being taken to the fragments by man (Pellens & Grandcolas, 2002). These species are indicated in Table 4.

Table 3. Characterization of the study sites with the total area of the reserve or fragment, distance of the fragments to the nearest reserve, and number of hours sampled in each season. BRS: Sooretama Biological Reserve; LR – MA: Linhares Reserve, Mata Alta; LR – E: Linhares Reserve, Entrance; FPN: Pasto Novo Fragment; FBB: Bioparque Bionativa Fragment; FSP: São Pedro fragment.

			Number of sampling hours		
Site	Area	Distance to the nearest reserve	March-April 2000	November 2000	Total
BRS	21,78		14	12	26
LR –	22,16		15	12	27
LR –	22,16		-	12	12
FPN	66.7	3.47 km	-	12	12
FBB	32.7	4.0 km	15	12	27
FSP	2.4	7.0 km	15	12	27
					131

Do the disturbances associated to forest fragmentation lead to differences in the composition and structure of Blattaria community?

One of the main characteristic of forest fragments is that they are more exposed to disturbances than large reserves. Their small size expose organisms to any sort of edge effects, such as augmentation of dryness, higher exposition to the sun, to the wind, to invasive species, and to the penetration of man that by itself can also promote other disturbances (Murcia, 1995). A simple way to evaluate the effect of disturbances of fragmentation in a community is by comparing certain parameters of the community composition and structure in a standard area of their habitat in sets of sites within small fragments and large reserves. Here the questions we asked were: a) is the species *richness* of sites from fragments different of those from the reserves? b) is the species *composition* of the sites within the fragments different of that from the reserves?

A) Is the species richness of sites from fragments different of those from the reserves?

To answer this question we used the data of species richness obtained in each sampling hour in the field treated by models of simulation emulated by the software Estimate S 5 (Colwell, 1997) (Box 3).

The results obtained with this analysis indicate that there are sites with high and low richness in both reserves and fragments (Figure 10). The richest site is in FPN, in which we found 30 species with only 12 samples, and the poorest sites were the ones within the Linhares Reserve. Thus, the answer to our question is negative.

Table 4. Species of Blattaria found in the remnants of Brazilian Atlantic forest in Linhares and Sooretama. * indicates that the species was also found in the matrix. The classification follows Grandcolas (1993b; 1994b; 1996; 1997b).

Family Polyphagidae	*Cariblatta sp2*
Subfamily Latindiinae	*Cariblatta sp3*
Latindia sp	*Dendroblatta sp1*
Gen.1 sp.1	*Dendroblatta sp2*
Family Anaplectidae	*Neoblattella sp1*
Anaplecta sp	*Neoblattella sp2*
Family Blattellidae	*Neoblattella sp3*
Subfamily Blattellinae	*Neoblattella sp4*
Isoldaia sp1	*Trioblattella sp1*
Isoldaia sp2	*Trioblattella sp2*
Isoldaia sp3	*Trioblattella sp3*
*Isoldaia sp4 **	*Trioblattella sp4 **
*Isoldaia sp5 **	
Blattellidae sp1	**Family Blaberidae**
Blattellidae sp2	**Subfamily Epilamprinae**
Blattellidae sp3	*Epilampra sp1*
*Blattellidae sp4 **	*Epilampra sp2*
Blattellidae sp5	*Epilampra sp3*
Dasyblatta sp	*Epilampra sp4*
Ischnoptera sp1	
Xestoblatta sp	**Subfamily Blaberinae**
Xestoblatta sp 1	*Minablatta sp1 **
Xestoblatta sp 2	*Minablatta sp2 **
Xestoblatta sp 3	*Monastria biguttata*
Xestoblatta sp 4	*Petasodes reflexa*
Xestoblatta sp 5	
	Subfamily Zetoborinae
Subfamily Nyctiborinae	*Parasphaeria boleiriana*
Nyctibora sp1	*Zetobora sp*
Nyctibora sp2	*Zetoborine sp*
Family Pseudophyllodromiidae	**Subfamily Pycnosceliinae**
Subfamily Pseudophyllodromiinae	*Pycnoscelus surinamensis **
Cariblatta sp1	

In what concerns species richness, our data indicate that sites in forest fragments can be richer than the sites within the reserves (as is the case of FPN), have similar number of species as the poorest sites within the reserves (as is the case of FBB that is not different of LR-MA and LR-E), or be in an intermediary situation as is the case of FSP (Table 5).

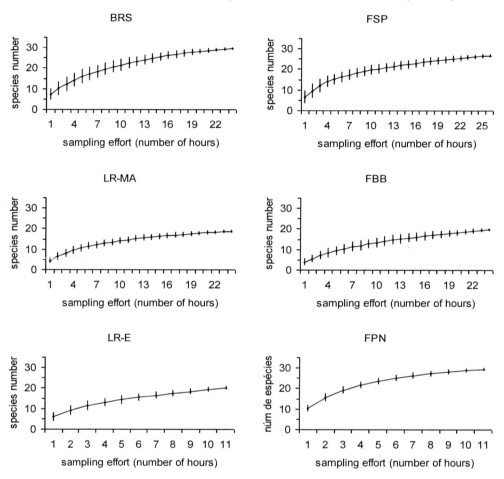

Figure 10. Cumulative curves of species number of Blattaria collected in sites within reserves (BRS, LR-MA, LR-E) and within fragments (FSP, FBB, FPN). The curves were elaborated with a method of randomization of sampling order made with the software Estimate S 5 (COLWELL, 1997). Each point of the curve is the mean and standard deviation of 100 randomizations of the sampling order without replacement.

B) Is the species composition of the sites within fragments different of those from reserves?

To answer this question we analysed some parameters of the species composition in relation to its distribution in the different study sites, and to the species shared among the sites. Besides, in the set of 48 species collected in this study, 12 of them had apterous adult females and 36 had winged adult females. Considering that it is the adult females that can disperse the brood, we also analysed their distribution in the different sites.

Regarding the species distribution we first verified if fragments or reserves have different numbers of species that are found in only one kind of them (Table 6; 8), and in second place we tested if sites within fragments have more ubiquitous species than sites within the reserves (Table 7; 9). These parameters can indicate if fragments have more species that are able to survive in disturbed sites, or more species that can be found everywhere, respectively.

Box 3

Two models of simulation emulated by Estimate S 5 (Colwell, 1997) based on the data we obtained in each sampling hour.

The first was the randomization of the sampling order. "By randomizing many times, the effect of sample order can be removed by averaging over randomizations, producing a smooth species accumulation curve and allowing a comparison of estimators for your data set that does not depend on the particular order that samples were collected or added to the analysis" (Colwell, 1997). We produced the cumulative curves with the mean and standard deviation of 100 randomizations made with this method.

The second model we used was the Michaelis Menten (MM) richness estimator named MMRuns computed by EstimateS that permits the comparison among sites. Based on the data obtained in each sample, the data the program produces represent the estimated MM asymptote based on one, two, three...QdMax samples. The MMRuns computes estimates for values for each pooling level, for each randomization run, and then averages over randomization runs (Colwell, 1997).

Table 5. Comparison of the maximal species richness estimated by the simulation method of Michaelis-Menten for each site from the data of richness per sample obtained for cockroaches in sites within the reserves and within forest fragments.
The comparison is made with Mann-Whitney U test: ns: non significant p≥0.05; *: significant p≤0.05; **: significant p≤0.01;*: significant p≤0.001 (n =20).**

Reserves	BRS > LR-MA p=0.0000 ***	BRS > LR-E p=0.0012 ***	LR-MA < LR-E p=0.0223 *
Fragments	FSP > FBB p=0.0000 ***	FBB < FPN p=0.0000 ***	FSP < FPN p=0.0005 ***
Fragments vs BRS	FSP < BRS p=0.0337 *	FBB < BRS p=0.0000 ***	FPN > BRS p=0.0166 *
Fragments vs LR–MA	FSP > LR-MA P=0.0000 ***	FBB ≈ LR-MA p=0.5075 ns	FPN > LR-MA p=0.0000 ***
Fragments vs LR-E	FSP > LR-E P= 0.0133 *	FBB ≈ LR-E p= 0.0531 ns	FPN > LR-E p= 0.0001 ***

As can be seen in table 6, the greatest part of the species studied was found in at least one site within the reserve and one in the fragments, and no species was found in every site within reserves or in every site within fragments. Five species (10.4%) were observed in up to two sites sampled in the reserves and nine (18.8%) in fragments. These figures are not significantly different of 7, the mean between 5 and 9 (χ^2 p = 0.445) (Table 6).

Table 6. Distribution of species in the sites within reserves and within forest fragments.

Study sites	Species number
The three sites within the reserves	0
Until two sites within the reserves	5
The three sites within the fragments	0
Until two sites within the fragments	9
At least one site within the reserves and one within the fragments	34

A similar trend was observed in the analysis of the number of species of the different categories of ubiquity. As a whole, a similar number of species of the different categories of ubiquity was found in every study site, independently if they were within reserves or within fragments (Table 7). These values were not significantly different (Kruskal-Wallis H=5.32 ; P>0.05. df = 5). Thus, concerning the species distribution, our analyses indicate that the sites within fragments do not have different number of particular species or of ubiquitous species than the sites within the reserves. It signifies that the fauna of the fragments has as many species that are peculiar as the sites within the large reserves, and that these species are not particularly restrict to one site, nor especially ubiquitous.

Table 7. Number of species belonging to six categories of ubiquity found in each study sites. The categories of ubiquity were defined according to the number of sites the species were found (1 to 6).

	1	2	3	4	5	6
BRS	2	6	8	4	5	5
LR-MA	1	1	4	3	5	5
LR-E	0	4	5	4	4	5
FPN	3	4	7	5	6	5
FBB	1	3	3	4	5	5
FSP	2	6	3	4	5	5

Species with apterous and winged adult females

The analysis of the distribution of the species with adult females winged or apterous showed that most of the species in both groups can be found in at least one within fragments or within the reserves (Table 8).

Winged species are found in every site, and only 3 species are restricted to until sites in the reserves while 7 species are found in until 2 sites within fragments. Conversely, 2 apterous species are restrict to two fragments and 1 to two sites within the reserves.

Table 8. Distribution of species with apterous and winged adult females in the sites within reserves and within forest fragments.

Study sites	Number of apterous species	Number of winged species
The three sites within the reserves	0	0
Until two sites within the reserves	1	3
The three sites within the fragments	0	0
Until two sites within the fragments	2	7
At least one site within the reserves and one within the fragments	9	26

Concerning the ubiquity of species with adult females apterous and winged, we verified that our sample comprised apterous species with every degree of distribution, with no difference between species in the categories of ubiquity 1,2,3 vs 4,5,6 (Mann-Whitney U test $p > 0.05$). Conversely, winged species tend to be more restrict in their distribution, and from the 36 species with winged adult females, 26 (72%) are present in until three study sites (Mann-Whitney U test $p \leq 0.05$).

Table 9. Number of species with apterous and winged adult females belonging to six categories of ubiquity. The categories of ubiquity were defined according to the number of sites the species were found (1 to 6).

	1	2	3	4	5	6
Apterous	3	2	1	2	1	3
Winged	6	10	9	4	4	3

In sum, these results show that the presence and absence of species in the studied sites do not depend on the ability of the adult females to fly for recolonizing the fragments.The number of species shared among sites was very low. It varied between 23% (LR-E and LR-MA; FBB and LR-MA), the sites with lowest number of species, and 40% (FPN and BRS) the richest sites. As a whole, there was not higher nor lower number of species shared by sites within fragments or by sites within the reserves (Table 10).

When the number of species shared was analysed in relation to the total number of species found in each site, these percentages were much higher, but varying between 43.3% (RL-E *vs* RBS) and 78.9% (RL-MA *vs* FPN). In this case the richest sites (RBS and FPN) were the ones that have higher number of species shared with other sites. Nevertheless, this figure was rarely higher than 70%, which signifies that at least 30% of the species found in any site were not shared (Table 11). This result also shows that the fauna of the sites within very small fragments are not perfect subset of the fauna of the richest sites, putting in evidence the importance of each of these sites to the total fauna of the region.

Table 10. Number of species shared by the different sites studied (with its respective percentage values) in relation to the total number of species in the 6 sites (i.e., 48 species).

	BRS	LR-MA	LR-E	FSP	FBB	FPN
BRS		14 – 30%	13 – 27%	17 – 35%	16 – 33%	19 – 40%
LR-MA			11 – 23%	12 – 25%	11 – 23%	15 – 31%
LR-E				12 – 25%	12 – 25%	16 – 33%
FSP					14 – 30%	18 – 38%
FBB						14 – 30%
FPN						

Table 11. Number of species shared by different combinations of two study sites and respective percentages in relation to the number of species in the site.

	BRS	LR-MA	LR-E	FSP	FBB	FPN
BRS		14 –74%	13 – 62%	17 – 65%	16 – 76%	19 – 63%
LR-MA	14 – 47%		11 – 53%	12 – 46%	11 – 52%	15 – 50%
LR-E	13 – 44%	11 – 58%		12 – 46%	12 – 57%	16 – 53%
FSP	17 – 57%	12 – 64%	12 – 57%		14 – 67%	18 – 60%
FBB	16 – 53%	11 – 58%	12 – 57%	14 – 54%		14 – 47%
FPN	19 – 63%	15 – 79%	16 – 76%	18 – 69%	14 – 67%	
Species number within the site	30	19	21	26	21	30

Thus, as a conclusion, our results indicate that both the species richness and the species composition of the fauna of Blattaria from small forest fragments are not different of that from large reserves. The species richness observed in the fragments is within the range observed in the reserves. The same way, the sites within fragments have similar number of species in the different categories of distribution analysed as the sites within the reserves and do not have different number of ubiquitous species. In addition, they do not share different number of species than sites within the reserves, but as these sites, they have at least 30% of the species that are not shared. As a whole, these results show that the community of Blattaria from the sites within fragments is not particularly affected by perturbations associated to forest fragmentation.

What Is The Contribution Of The Effects Of Area *Per Se* And Of Habitat Diversity To The Richness Of Cockroach Community?

The conceptual basis for this analysis is the well accepted theory in ecology that the number of species increase with the area sampled, which in fact is a theory that explains the scale effect for explaining the number of species (Lomolino, 2000). According to MacArthur & Wilson (1963; 1967), Connor & McCoy (1979 ; 2001), Haila (1983) and McGuiness (1984) there are four hypotheses to explain this pattern : (1) the random sampling hypothesis; (2) the habitat diversity hypothesis; (3) the area *per se* hypothesis that was better developed as the island biogeography theory; and (4) the disturbance hypothesis. In addition to that, a factor that is not related to the area but is important to the species-area relationship is the distance to a colonization source and the existence of multiple sites. Its contribution to the species-area relationship is due to the *rescue effect*, which reduces local extinctions (Brown & Kodrick Brown, 1977). We will not detail these theories here, for it is one of the most studied problems in ecology (for a good synthesis see Connor & McCoy, 1979, 2001). The important thing to remark here is that these mechanisms can act solely or in combination to determine the number of species found in a given site, and that separating the contribution of the different components allows understanding the contribution of the sets of fragments in the landscape to the totality of the regional richness.

In the previous section we came to demonstrate that there is no effect of disturbances associated to forest fragmentation on the Blattaria community. Thus we can consider that these effects are constant in the sites studied and go further to analyse the effect of area *per se* and habitat diversity. We want to remark that the effects of area *per se* and of random sampling are very intricate and not possible to separate with the methods we propose here. For this reason, we will treat them as one thing aiming to separate these mechanisms from the habitat diversity. Our interest in verifying the contribution of these two components in the species area relationship is to understand the contribution of the sets of fragments in the landscape to the totality of the regional richness, and based on that to propose strategies of conservation and management of the biodiversity.

To separate the effects of area *per se* and habitat diversity we developed the conceptual model that is synthesized in the Figure 11. In this model, the first step is the verification of effects related to the area *per se*. It is tested by the simple comparison of the richness in single sites with the richness observed in the sum of three sites within the reserves or three sites within fragments. If the sum of the richness in a set of sites is not higher than that observed in a single site, we show that there is no effect of area (we mean all the effects associated to area, area *per se*, random sampling, and habitat diversity) on the community. Conversely, if the sum of the samples of a set of sites gives a higher number of species than in each site two results are possible: the first is area *per se* effect; the second is the habitat diversity effect. To separate these two effects we propose to compare samples obtained in single sites with a mix of random and proportional samples from the set of sites (in the present case, from the set of three sites within fragments or three sites within reserves). This way, if with equal sample size, the number of species from the mix of samples is higher than for each site isolated, we verify the effect of habitat diversity. Conversely, if the number of species from the two sets of samples is not different we refute the hypothesis of habitat diversity.

These analyses were performed with simulations of increasing the sample size by adding the samples from different study sites with the model of simulation Michaelis Menten richness estimator (MMRuns) computed by Estimate S 5 (Colwell, 1997) (Box 3).

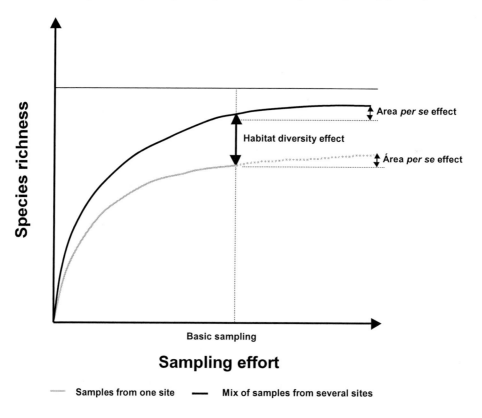

Figure 11. Conceptual model used to separate the Area *per se* effect and the habitat diversity effect.

As commonly observed in nature, the number of species increased as we increased the sampling effort: 39 species were collected in the three sites within the reserves and 43, in those from fragments (the maximum observed in a single site was 30 species in BRS and FPN). When comparing the curves concerning reserves and forest fragments, the simulation indicates differences concerning the maximum number of species, and the intensity of the sampling effort (Figure 12). In the reserves the species number increases progressively with the sample size and the 65 samples estimated for the three sites within the reserves have a higher number of species than a mix of 25 samples (Figure 12*b*) (Mann-Whitney U test, p= 0.0001). Conversely, in the set of fragments the richness estimated to the 64 samples is not significantly different than that estimated for a single site (Figure 12*a*) (Mann-Whitney U test, p=0.5074), suggesting that in the set of fragments the species richness is close to a maximum. This result suggests that the effect of area *per se* is important to explain the increase in species richness within the reserves, but not in the fragments. However, the hypothesis of habitat diversity within the reserves can not yet be refuted.

As explained in the beginning of this section, for testing the effect of habitat diversity we compared the samples obtained in each site with a mix of random and proportional samples of the three sites within fragments and three sites within reserves (Figure 13).

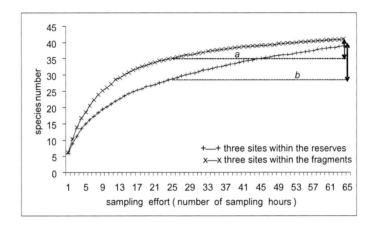

Figure 12. Verification of the effects of Area *per se* and Habitat Diversity on the species richness of Blattaria in Linhares and Sooretama, Brazil. The curves were built with the richness estimated with the simulation model Michaelis-Menten to a mix of samples from three sites of each kind. *a*) difference between the richness estimated to three sites and one site in the reserves; *b*) difference between the richness estimated to three sites and one site in the fragments.

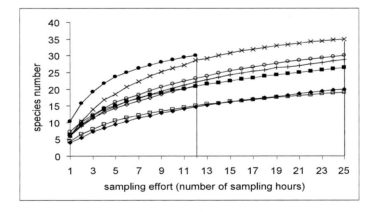

Figure 13. Verification of the effect of Habitat Diversity on the species richness of Blattaria in Linhares and Sooretama, Brazil. The richness estimated with the simulation model Michaelis-Menten to a random and proportional mix of samples obtained to the set of sites within fragments and within reserves is compared to the richness estimated to each site independently. O—O BRS; □—□ LR-MA; ◇—◇ LR-E; ●—● FPN; ■—■ FSP; ◆—◆ FBB; +—+ three sites within the reserves; x—x three sites within the fragments.

The comparison of the richness estimated from a mix of samples of the three study sites with the richness estimated to each study site shows different responses for reserves and fragments. The mix of samples from the reserves has higher species richness than that observed in LR-MA and LR-E, but not than that observed to BRS (Table 12). It indicates that the differences of species richness in the reserves are not strong enough to support the hypothesis of habitat diversity. This result plus the progressive and significant increase in species richness with the increase of sampling effort (Figure 12a) indicate that the effect of area *per se* is important to explain the species richness in the reserves. In opposition, the richness estimated with a mix of samples from the three sites within fragments is always higher than that estimated to each site within fragments. This difference indicates clearly the effect of habitat diversity on the species richness of the fragments.

Table 12. Comparison of the richness estimated with the Michaelis-Menten method to a mix of random and proportional samples from the three sites within the reserves and the three sites within the fragments with the samples obtained to each site. In this comparison we kept the number of samples as obtained in the field (12 samples in the comparison with FPN and LR-E and 25 for the other sites). P values of Mann-Whitney U test: ns: non significant p≥0.05; *: significant at p≤0.05; **: significant at p≤0.01; ***: significant at p≤0.001 (n =20).

Mix of samples from sites within reserves	Mix of samples from sites within reserves	Mix of samples from sites within reserves
≈ RBS	> RL-MA	> RL-E
P= 0.3867	P= 0.0000	P= 0.0071
ns	***	**
Mix of samples from sites within fragments	Mix of samples from sites within fragments	Mix of samples from sites within fragments
> FSP	. > FBB	> FPN
P= 0.0000	P= 0.0000	P= 0.0222
***	***	*

CONCLUSION

In this chapter we showed that the landscape at the north of the State of Espírito Santo changed markedly during the last 50-60 years, and that this transformation had strong impact on the Atlantic forest creating a new problem, a new context for the conservation of the biodiversity. The intense deforestation gave place to a rural productive landscape in which the remaining forests were confined to two large reserves and several fragments protected in private properties. Even so these remnants represent the largest area of Atlantic forest from the south of Rio de Janeiro to the south of Bahia, and more than 50% of the forests remaining in the State of Espírito Santo. Moreover, they harbour a very high species richness and endemism, attracting the attention of the Ministry of Environment in the strategy of biodiversity conservation of this biome. For this reason, Linhares and Sooretama were chosen to be the sites of implementation of one of the main subprojects of PROBIO, in the first wave of Brazilian projects to answer the commitments to the Convention of Biological Diversity:

the project *"Conservation and Recuperation of Atlantic forest in Linhares – ES, based on the functional evaluation of the biodiversity"*. As a consequence of this first endeavour, this region is presently at the centre of the project Ecological Corridor of Atlantic forest.

Our analysis of the conservation of the forest fragments showed that the most important factors were the obligations by the law and the will of putting aside a piece of forest that represented a reserve of value, due to the wood they had. However, neither the law nor the quality of wood are sufficient to assure the preservation of these fragments in the near future. The number of fragments in the landscape is below the values recommended by the law; and fragments continue to be exploited in a way that is changing the composition of their communities. The results of this exploitation can already be measured in the composition of tree families (Agarez *et al.*, 2004), and if it does not stop it could lead to a biomass collapse (Laurance *et al.*, 1997), or a total impoverishment of these remnants (Cardoso da Silva & Tabarelli, 2000) in the forthcoming decades.

Nevertheless, our evaluation of the conservation value of these fragments for conservation indicates that the situation is not lost. Even small and very disturbed fragments harbor important number of species of Blattaria. Sites within fragments are not different of sites within the large reserves concerning the species richness and composition: they do not harbor higher number of species found only in fragments or of ubiquitous species, neither a lower number of species found typically in sites within large reserves. It shows that the community of fragments is not peculiar; they do not have more species of disturbed areas or less forestry species. In addition to that, habitat diversity plays an important role in the distribution of the species in the set of fragments. It indicates that fragments permit the conservation of a higher number of species by increasing the possibility of sampling those with particular types of distribution, particular habitats, thus incorporating other aspects of the regional diversity by permitting the protection of a portion of organisms not found anywhere else. In sum, the results obtained with the analysis of the community of cockroaches bring light to the importance of fragments for increasing the total number of species protected in the region, indicating thus the high value of fragments for biodiversity conservation. This result is particularly remarkable when compared to that obtained in the *Biological Dynamics of Forest Fragments Project* in Amazon. A synthesis of their study indicates that in less than 22 years after fragmentation, the fragments contain less species per area than intact forest for several groups of vertebrates and invertebrates. In addition, a high number of species disappeared even from the largest (100ha) forest fragment, and for several groups the fragmentation increased the number of species of disturbed habitats (Laurance *et al.*, 2002).

It places the fragments of this region of the Atlantic forest in a positive context, indicating the need to look at fragments through a new perspective. Fragments do not replace large reserves; they contribute to enlarge the surface of remnants of Atlantic forest in the landscape by including other aspects of the regional environment, thus contributing to the protection of a higher number of organisms. Furthermore, they can contribute to the long-term persistence of the different species by serving as step stones, increasing the probability of dispersion and/or reducing the local competition. Nevertheless, the conservation of the biodiversity including forest fragments requires an approach considering the fragility of being numerous, small, scattered in the landscape, and susceptible to be simultaneously exploited. Recent studies indicate clearly that this kind of indiscriminate exploitation might lead to the simultaneous impoverishment of their biota, which could promote a clash of the regional biodiversity (Laurance *et al.*, 1997; Agarez *et al.*, 2004; Cardoso da Silva & Tabarelli, 2000), a "tragedy of commons" (Hardin, 1968).

The conservation of the biodiversity in this perspective requires shifting the scale: it must be envisaged at the landscape or regional scale, in which fragments shall be included in the strategy of conservation. In the present case it means that the conservation of the regional biodiversity depends on the engagement of the rural producers, owners of the fragments, in the project of conservation of the Atlantic forest. In this context, a way to avoid the "tragedy of commons" would be through the elaboration of a plan of management of the forest fragments, in which uses could be defined, and results monitored along time.

Our results of the community of cockroaches are in accordance with the study of plants in the same region (Agarez, 2002), which can be taken as an indicator that that it might be the case for several other biological groups of organisms. Based in these studies we can go further and state on the definition of priority fragments for conservation. First of all, these results indicate that the set of fragments existing in the landscape are important for increasing the protection of the biological richness of this region. As a consequence, all fragments should be protected, many of them should be recuperated in order to assure its permanence in the long run, and certain disturbances common to several fragments should be immediately stopped, as for example, closing roads or trails that permit the easy movement of people through the fragments. In a second step, the management of the set of fragments should give priority to diversity: either geologic, geomorphologic, edaphic, and ecologic; or the diversity of uses, which promotes differential survival of species and could help to protect certain populations.

In conclusion, in Linhares and Sooretama the population is becoming conscientious of the fragment's importance due to the ecosystem services provided, above all by their contribution to the protection of springs and streams, assuring the availability of the so precious water for the agriculture during the years of drought. They are also beginning to recognize the fragment's potentialities in creating a system of sustainable development using forest resources, since fragments represent the only possibility of using forest resources in the region. Our study emphasizes the potential of the set of fragments remaining in this landscape for the protection of the regional biodiversity.

ACKNOWLEDGMENTS

This research was funded by several Brazilian institutions (FAPERJ, CNPq, UFRJ) and by an international cooperation program between CNPq (UFRJ, Brazil) and CNRS (MNHN, France). We are very grateful to these institutions. The fieldwork was authorized by IBAMA. We thank Guanadir Gonçalves, chief of Sooretama Biological Reserve, Renato de Jesus, administrator of Natural Reserve of Companhia Vale do Rio Doce, in Linhares, and the owners of the forest fragments studied, for authorizing the development of the research in the sites under their responsibility. We are very indebt to Gilson Lopes Farias (*in memoriam*) Nivaldo del Piero, Renato de Jesus, Eric Guilbert, Claire Vilemant and Andréia Kindel, for their help in field work, to José Caldas and Renato de Jesus for authorizing the use of their pictures, and to Robert Barbault and Xavier Bellés, for kindly reviewing this chapter and giving suggestions that contributed to make it clearer.

REFERENCES

Agarez, F.V. (2002) Contribuição para a gestão de fragmentos florestais com vista à conservação da biodiversidade em Floresta Atlântica de Tabuleiros., p. 237. *In:* Programa de Pós-Graduação em Geografia. Federal University of Rio de Janeiro, Rio de Janeiro.

Agarez, F.V. & Garay, I. & Vicens, R.S. (2004) A floresta em pé: conservação da biodiversidade nos remanescentes de Floresta Atlântica de Tabuleiros. In: A Floresta Atlântica de Tabuleiros: diversidade funcional da cobertura arbórea, Garay, I. & Rizzini, M.C. (org.). Editora Vozes, Petrópolis, pp. 27-34.

Aguirre, A. (1947) Sooretama. Estudo sobre o Parque de Reserva, Refúgio e Criação de Animais Silvestres, "Sooretama", no Município de Linhares, Estado do Espírito Santo. *Boletim do Ministério da Agricultura*, Rio de Janeiro, 36 (4-6):1-52.

Aguirre, A. (1951) *Sooretama. Estudo sobre o Parque de Reserva, Refúgio e Criação de Animais Silvestres, "Sooretama", no Município de Linhares, Estado do Espírito Santo.* Ministério da Agricultura, Serviço de Informação Agrícola, Rio de Janeiro, 50p.

Becker, B.K. (1969) O norte do Espírito Santo: região periférica em transformação., p. 130. *In:* Instituto de Geociências. Federal University of Rio de Janeiro, Rio de Janeiro.

Cardoso da Silva, J.M. & Tabarelli, M. (2000) Tree species impoverishment and the future flora of the Atlantic forest of northeast Brazil. *Nature*, 404, 72-74.

Chiarello, A.G. (2000) Density and population size of mammals in remnants of Brazilian Atlantic Forest. *Conservation Biology*, 14, 1649-1657.

Colwell R. K. (1997) Estimate S 5. Statistical Estimation of Species Richness and Shared Species from Samples. User's Guide and application published at: http://viceroy. eeb.uconn.edu/estimates.

Egler, W.A. (1951) A zona pioneira ao norte do Rio Doce. *Revista brasileira de Geografia*, 2, 224-263.

Fisk F.W. (1983) Abundance and diversity of arboreal Blattaria in moist tropical forests of the Panama canal area and Costa Rica. *Transactions of the American Entomological Society,* 108, 479-490.

Garay, I. (2004) Uma história recente. *In:* Garay, I. & Rizzini, C.M. (Eds.) *A Floresta Atlântica de Tabuleiros. Diversidade funcional da cobertura arbórea.* Vozes, Petrópolis, pp. 1-6.

Garay, I., (2006) Construir as dimensões humanas da biodiversidade. Uma abordagem transdisciplinar para a Floresta Atlântica de Tabuleiros. In: *As dimensões humanas da biodiversidade. O desafio de novas relações sociedade-natureza no século XXI*, Garay, I., Becker, B. (Eds.). Ed. Vozes, Petrópolis, pp. 413–445.

Garay, I. & Rizzini, C.M. (2004) *A Floresta Atlântica de Tabuleiros. Diversidade funcional da cobertura arbórea.* Petrópolis, Vozes, 1255 p.

Gentry, A.H. (1992) Tropical forest biodiversity: distributional patterns and their conservational significance. *Oikos*, 63, 19-28.

Grandcolas, P. (1991) Les Blattes de la forêt tropical de Guyane Française: Structure du peuplement et étude éco-éthologique des Zetoborinae, PHD thesis, Université de Rennes, Rennes, 295p.

Grandcolas, P. (1993a) Habitats of solitary and gregarious species in the neotropical Zetoborinae (Insecta, Blattaria). *Studies in Neotropical Fauna and Environment*, 28, 179-190.

Grandcolas, P. (1993b) Monophylie et structure phylogénétique des [Blaberinae + Zetoborinae + Gyninae + Diplopterinae] (Dictyoptera : Blaberidae). *Annales de la Société entomologique de France (N.S.)*, 29, 195-222.

Grandcolas, P. (1994a) Les Blattes de la forêt tropicale de Guyane Française: structure du peuplement (Insecta, Dictyoptera, Blattaria). *Bulletin de la société Zoologique de France*, 119, 59-67.

Grandcolas, P. (1994b) Phylogenetic systematics of the subfamily Polyphaginae, with the assignment of *Cryptocercus* Scudder, 1862 to this taxon (Blattaria, Blaberoidea, Polyphagidae). *Systematic Entomology*, 19, 145-158.

Grandcolas, P. (1996) The phylogeny of cockroach families: a cladistic appraisal of morpho-anatomical data. *Canadian Journal of Zoology*, 74, 508-527.

Grandcolas, P. (1997a) Systématique phylogénétique de la sous-famille des Tryonicinae (Dictyoptera, Blattaria, Blattidae). *In:* Najt, J. & Matile, L. (Eds.) *Zoologia Neocaledonica, Volume 4*. Mémoires du Muséum national d'Histoire naturelle 171, Paris, pp. 91-124.

Grandcolas, P. (1997b) The monophyly of the subfamily Perisphaeriinae (Dictyoptera: Blattaria: Blaberidae). *Systematic Entomology*, 22, 123-130.

Grandcolas, P. & Pellens, R. (2009) Blattaria. *In:* Rafael, J.A., Rodrigues de Melo, G.A., Barros de Carvalho, C.J. & Casari, S.A. (Eds.) *Insetos do Brasil: Diversidade e Taxonomia*. INPA, Manaus, (in press).

Hardin, G. (1968) The tragedy of commons. *Science*, 162, 1243-1248.

IBDF & FBCN (1981) *Reserva Biológica de Sooretama: plano de manejo*. Brasília, DF 69p.

IBGE (2000) Censo Demográfico.

Jesus, R.M. (1987) Mata Atlântica de Linhares: aspectos florestais, p. 35-71. *In:* Seminário sobre Desenvolvimento Econômico e Impacto Ambiental em áreas do trópico úmido brasileiro: a experiência da CVRD, Rio de Janeiro.

Jesus, R.M., Rolim, S.G. 2005. Fitossociologia da floresta atlântica de tabuleiro em Linhares (ES). Boletim Técnico SIF 19, Viçosa, 149 p.

Laurance, W.F. & Bierregaard Jr., R.O. (1997) *Tropical forest remnants: ecology, management, and conservation of fragmented communities*. The University of Chicago Press, Chicago, 616+XV p..

Laurance, W.F., Laurance, S.G., Ferreira, L.V., Rankin-de Merona, J.M., Gascon, C. & Lovejoy, T.E. (1997) Biomass collapse in Amazonian Forest Fragments. *Science*, 278, 1117-1118.

Laurance, W.F., Lovejoy, T.E., Vasconcelos, H.L., Bruna, E.M., Didham, R.K., Stouffer, P.C., Gascon, C., Bierregaard, R.O., Laurance, S.G. & Sampaio, E. (2002) Ecosystem decay of Amazonian forest fragments: a 22-year investigation. *Conservation Biology*, 16, 605-618.

Linhares Reserve (2004) *Florística arbórea da RNVRD ordenada por família: 70th approximation*. Manucript Linhares Reserve, 17p.

Lomolino, M.V. (2000) Ecology's most general, yet protean pattern: the species-area relationship. *Journal of Biogeography*, 27, 17-26.

Machado-Guimarães, E.M. (1998) *Uma abordagem etnoecológica como base para as atividades de transferência e educação*. Report of the subproject from PROBIO "Conservação e recuperação da Floresta Atlântica de tabuleiros, em Linhares, ES, com base na avaliação funcional da biodiversidade. Manuscript Federal University of Rio de Janeiro, IB., Lab. de Ecologia de Solos, 9p.

Mittermeier, R. (1997) Diversidade de primatas e a floresta tropical: estudos de casos do Brasil e de Madagascar e a importância dos países de megadiversidade. *In:* Wilson, E.O. (Ed.) *Biodiversidade*. Ed. Nova Fronteira 657p.

MMA/SBF. (2000) *Avaliação de ações prioritárias para a conservação da biodiversidade da Mata Atlântica e Campos Sulinos*. Brasilia, 40 p.

Murcia, C. (1995) Edge effects in fragmented forests: implications for conservation. *Trends in Ecology and Evolution*, 10, 58-62.

Myers, N., Mittermeier, R.A., Mittermeier, C.G., Fonseca, G.A.B. & Kent, J. (2000) Biodiversity hotspots for conservation priorities. *Nature*, 403, 853-858.

Peixoto, A.L. & Gentry, A.H. (1990) A diversidade e a composição florística da Mata de Tabuleiros na reserva florestal de Linhares (Espírito Santo, Brasil). *Revista Brasileira de Botânica*, 13, 19-25.

Peixoto, A.L., Rosa, M.M.T. & Joels, L.C.M. (1995) Diagramas de perfil e de cobertura de um trecho da floresta de tabuleiro na Reserva Florestal de Linhares (Espírito Santo, Brasil). *Acta Botanica Brasilica*, 9, 177-193.

Pellens, R. (2002) Fragmentação florestal em Mata Atlântica de Tabuleiros: os efeitos da heterogeneidade da paisagem sobre a diversidade de artrópodos edáficos. Programa de Pós-Graduação em Geografia, Universidade Federal do Rio de Janeiro, Rio de Janeiro, 198p.

Pellens, R. & Grandcolas, P. (2002) Are successful colonizers necessarily invasive species? The case of the so-called invading parthenogenetic cockroach, *Pycnoscelus surinamensis*, in the Brazilian atlantic forest. *Revue d'Ecologie (Terre Vie)*, 57, 253-261.

Pellens, R., Grandcolas, P. & Silva-Neto, I.D. (2002) A new and independently evolved case of xylophagy and the presence of intestinal flagellates in cockroaches: *Parasphaeria boleiriana* (Dictyoptera, Blaberidae, Zetoborinae) from the remnants of Brazilian Atlantic Forest. *Canadian Journal of Zoology*, 80, 350-359.

Pellens, R. & Grandcolas, P. (2003) Living in Atlantic forest fragments: life habits, behaviour and colony structure of the cockroach *Monastria biguttata* (Dictyoptera, Blaberidae, Blaberinae) in Espirito Santo, Brazil. *Canadian Journal of Zoology*, 82, 1929-1937.

Pellens, R. & Grandcolas, P. (2007) The conservation refugium value of small and disturbed Brazilian Atlantic forest fragments for the endemic ovoviviparous cockroach *Monastria biguttata* (Insecta: Dictyoptera, Blaberidae, Blaberinae). *Zoological Science*, 24, 11-19.

Pellens, R., D'Haese, C., Bellés, X., Piulacs, M.D., Legendre, F., Wheeler, W. & Grandcolas, P. (2007a) The evolutionary transition from subsocial to eusocial behavior in Dictyoptera: phylogenetic evidence for modification of the "shift-in-dependent-care" hypothesis with a new subsocial cockroach. *Molecular Phylogenetics and Evolution*, 43, 616-626.

Pellens, R., Legendre, F. & Grandcolas, P. (2007b) Phylogenetic analysis of social behavior evolution in [Zetoborinae + Blaberinae + Gyninae + Diplopterinae] cockroaches: an update with the study of endemic radiations from the Atlantic forest. *Studies on Neotropical Fauna & Environment*, 42, 25-31.

Pellens, R. & Grandcolas, P. (2008) Catalogue of Blattaria (Insecta) from Brazil. *Zootaxa*, 1709. 109p.

Princis, K.. (1962-71) Blattariae. Pars 3, 4, 6, 7, 8, 11, 13, 14. *Orthopterorum Catalogus*. M. Beier. Junk's-Gravenhage, The Hague. 1224 p.

Rizzini, C.T. (1997) *Tratado de fitogeografia do Brasil. Aspectos ecologicos, sociologicos e floristicos*. Ambito Cultural Ediçoes Ltda., Rio de Janeiro, 747 p.

Saunders, D.A., Hobbs, R.J. & Margules, C.R. (1991) Biological consequences of ecosystem fragmentation: a review. *Conservation Biology*, 5, 18-32.

SOS Mata Atlântica, INPE, ISA (1998) *Atlas da evolução dos remanescentes florestais e ecossistemas associados no domínio da mata atlântica no período de 1990-1995*. São Paulo, 58p.

SOS Mata Atlântica/ISA (2002) *Atlas dos remanescentes da Mata Atlântica 2000*. http://www.sosmatatlantica.org.br/atlas2001/

Suguio, K., Martin, L. & Dominguez, J.M.L. (1982) Evolução da planície costeira do Rio Doce (ES) durante o quaternário: influência das flutuações no nível do mar., p. 93-116. *In:* Atas do Simpósio do Quaternário no Brasil. Vol. 4.

Turner, I.M. & Corlett, R.T. (1996) The conservation value of small, isolated fragments of lowland tropical rain forest. *Trends in Ecology and . Evoution.*, 11, 330-333.

Thomas, C.D. (2000) Dispersal and extinction in fragmented landscapes. *Proceedings of the Royal Society Academy of London B*, 267, 139-145.

Thomas, W.W., Carvalho, A.M.V., Amorim, A.M.A., Garrison, J. & Arbeláez, A.L. (1998) Plant endemism in two forests in southern Bahia, Brazil. *Biodiversity and Conservation*, 7, 311-322.

Zunti, M.L.G. (1982) *Panorama Histórico de Linhares*. Prefeitura Municipal de Linhares, Espírito Santo, 203p.

Reviewed by :
Prof. Robert Barbault, *Département Écologie et Gestion de la Biodiversité, Muséum national d'Histoire naturelle, 36, rue Geoffroy Saint-Hilaire 75005 Paris, France.*
Prof. Xavier Bellés, *Department of Physiology and Molecular Biodiversity, Institut de Biologia Molecular de Barcelona (CSIC), Jordi Girona 18, 0834 Barcelona, Spain.*

In: Encyclopedia of Environmental Research
Editor: Alisa N. Souter

ISBN: 978-1-61761-927-4
© 2011 Nova Science Publishers, Inc.

Chapter 19

CONSERVATION OF ARAUCARIA FOREST IN BRAZIL: THE ROLE OF GENETICS AND BIOTECHNOLOGY

Valdir Marcos Stefenon* and Leocir José Welter

Federal University of Santa Catarina, Florianópolis, SC, Brazil

INTRODUCTION

The *Araucaria* forest, also called mixed ombrophilous forest (Figure 1), is a particular ecosystem in the southern Brazilian highlands, characterized by the admixture of two distinct vegetations: the tropical afro-Brazilian and the temperate austro-Brazilian floras [1].

Figure 1. *Araucaria* Forest in Rio Grande do Sul State, in southern Brazil.

* Corresponding author: e-mail: gene_mol@yahoo.com.br

During at least 1000 years, after the post-glacial climate changes of the Quaternary [2], this ecosystem dominated the landscape of around 200,000 km^2 of southern Brazilian highlands. However, the intensive exploitation process occurred mainly between the years 1940's to 1970's reduced the Brazilian *Araucaria* forest to about 3% of their original area.

The conservation of the forest remnants, the protection of the wild animals and the natural regeneration of many plant species may be promoted by the economical valorization of *Araucaria* forest, mainly through its sustainable management [1].

Similarly, the establishment of natural reserves and agroforestry systems can also promote forest conservation. In addition, these *in situ* conservation areas may serve as source of genetic material for *ex situ* conservation, plantation establishment and for the exploration of the adaptive capacity of different genetic resources through provenance trials [3].

However, a sustainable management of the forest remnants implies comprehension about the distribution of the genetic diversity of keystone species, allowing making decisions which circumvent the reduction of the forest adaptability to environmental variations, such as the climatic changes. Furthermore, it has been shown that biotechnological tools may be very useful in promoting the preservation of the remaining genetic diversity of the forest species [4].

NATURAL DISTRIBUTION AND DECLINE OF THE *ARAUCARIA* FOREST

The natural distribution of the *Araucaria* forest is predominant in altitudes between 500 and 1800 m, from 19°15' to 31° southern latitude, mainly in Brazil (Figure 2), with small stands in Argentina, at the Brazilian board [5].

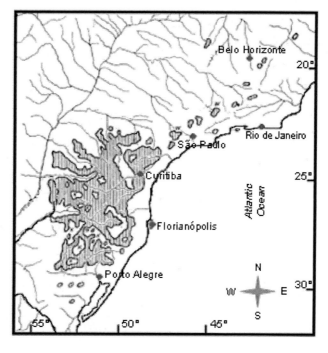

Figure 2. Area of natural occurrence of *Araucaria* Forest in South America. After [7].

Following the classification of Köppen, the climate in these regions falls into the categories *Cfa* (humid subtropical climate without dry season and with hot summers) and *Cfb* (humid subtropical climate without dry season and with mild summers) with a mean annual precipitation higher than 1250 mm [6], without dry seasons.The mean temperatures are of 20° to 21°C in summer and 10° to 11° in winter, with frequent occurrence of ice and snow. This forest is dominated by *Araucaria angustifolia* (Bert.) O. Kuntze, an ancient southern conifer species, with very important ecological and cultural significance.

At the beginning of the 20th century, this forest covered extensive areas in southern and southeastern Brazil, with around 40% of its vicinity in the Paraná State, 31% in Santa Catarina State, 25% in Rio Grande do Sul State, 3% in São Paulo State and 1% in some elevated areas in the states of Minas Gerais and Rio de Janeiro. However, the high quality of the *Araucaria*'s wood for many proposes led this species to an intensive exploitation process, which reduced the occurrence area of this ecosystem to about 3% of its original vicinity in less than a century [1]. The reduction of the raw-material as effect of the non-sustainable exploitation of the forest mainly aiming timber production (Figure 3a) resulted in the establishment of fast-growing exotic trees plantations, essentially after the 1970's. Because of the scarcity of *A. angustifolia* wood as raw material, traditional sawmill companies which did not follow the tendency of changing machineries become obsolete (Figure 3b-d).

Figure 3. Climax and decline of the *A. angustifolia* exploitation in southern Brazil. (A) Truck read to transport *Araucaria*'s timber. (B) Building of abandoned sawmills company in southern Brazil. (C and D) Abandoned machineries of the sawmill company.

At the same time, the growth of the agricultural lands also reduced the vicinity of the forest, creating "islands" of forest fragments merged in an "ocean" of crops (Figure 4).

Figure 4. Exemplars of *A. angustifolia* trees, surrounded by crop plantations in southern Brazil.

ARAUCARIA ANGUSTIFOLIA: THE KEYSTONE OF THE ARAUCARIA FOREST

Inside the forest, *A. angustifolia* generates a particular micro-environment which allows the growth and survival of many shade-tolerant plant species (Figure 5a). Many small vertebrates and invertebrates take advantage of the trees trunk and branches as housing and reproductive points (Figure 5b). Additionally, seeds feed the wild fauna, supplying the most important source of food during the winter for several mammalians and birds (see Figure 5e). Seeds are also source of food and income to small farmers, which live in the region of the forest remnants.

A. angustifolia is the unique representative of family Araucariaceae in Brazil and together with the closely related species *A. araucana* [8, 9], the unique extant representative of the family in the American continent. The genus *Araucaria* (de Jussieu) includes 19 species, with current geographic distribution restricted to the Southern hemisphere [10]. Ten botanical varieties of *A. angustifolia* are described, based on ripening time and seed colour: 1) *elegans*; 2) *sancti josephi*; 3) *angustifolia*; 4) *caiova*; 5) *indehiscens*; 6) *nigra*; 7) *striata*; 8) *semi-alba*; 9) *alba*; and 10) *catarinensis* [5, 11].

A. angustifolia is a long-lived dioecious species with adult trees displaying a characteristic umbrella-shaped crown (Figure 5a), lacking branches up to two thirds of their height. Young trees exhibit a pyramidal form typical from conifer species. The stem is cylindrical and straight reaching 20 to 50 meters of height and 1 to 2 meters of diameter (Figure 5c-d). Seeds are dispersed mainly by gravity (barochory) and the pollen is dispersed by wind (anemophily). Alternatively, seeds may be dispersed by vertebrates. However, the transported seeds are often damaged by these animals (Figure 5e) and not able to germinate [12, 13]. The pollination occurs between September and October and the complete development of the cones (Figures 5f-g) takes around 22 months, from pollination to ripening, which occurs from March to June [14].

Figure 5. Morphology and ecology of *A. angustifolia*. (A) Shade-tolerant epiphyte plant growing over *Araucaria*'s trunk. (B) Housing of a bird, constructed on a branch of an adult tree of *A. angustifolia*. (C) Characteristic umbrella-like shape of the *Araucaria*'s crown. (D) Straight shape of the *Araucaria*'s trunk, which makes the tree very interesting at economical level. (E) Seeds of *A. angustifolia*. The first one is intact, while the second was damaged by a rodent and the third by a bird. (F) Immature cones of *A. angustifolia*. (G) The same cones, showing the seeds.

ON THE GENETIC DIVERSITY OF *A. ANGUSTIFOLIA*

One of the major objectives of genetic studies is the characterization of genetic diversity within and among populations. In *A. angustifolia*, such studies started in the 1980's using morphological traits and followed the advances of the molecular markers [15]. A general agreement obtained by different studies using different techniques is the evidence of geographical races occurring among natural populations. This pattern was firstly revealed by provenance trails based on quantitative traits (diameter, height and volume), revealing statistically significant correlation between latitude of the site where seeds were sampled and the growth traits [16, 17, 18].

Equivalent conclusions were obtained from the analysis of a DNA fragment of unknown genomic origin [19]. The lack of this fragment in embryo and endosperm revealed a south-to-north gradient, suggested to be linked to adaptation to frost, which is less intense in southeastern Brazil. RAPD [20] and isoenzymes analyses [21] also revealed markedly differentiation among populations from southeastern and southern Brazil. Finally, studies based on microsatellite and AFLP markers revealed as well a clear pattern of geographic differentiation among natural populations. Microsatellite data suggested the presence of geographical groups of populations, with significant correlation between geographical distance and genetic differentiation over all populations [22].

While among population analyses suggest the occurrence of geographical ecotypes, within population analyses revealed the occurrence of family structure [23, 24, 25, 26, 27] suggesting the incidence of mating among relatives. Given that *A. angustifolia* is a dioecious species (i.e. trees are males or females, with very rare occurrence of hermaphrodite individuals), biparental inbreeding is the only source of family structure. Although dioecious completely lack self fertilization, the difference between the multilocus and single-locus outcrossing rates is used as an inference of the biparental inbreeding within a population [28]. Estimations of biparental inbreeding computed from isoenzymes data revealed values ranging from 1.8 to 6.1% of biparental mating [23, 28]. Using microsatellites [27], 11% of biparental mating was estimated for the same area investigated with isoenzymes [23] in southeastern Brazil.

The occurrence of geographical differentiation and family structure within population are strong evidences of restricted gene flow. Additional insights about gene dispersal can be obtained analyzing genetic markers with contrasting patterns of inheritance, such as chloroplast and nuclear DNA. This contrasting inheritance pattern is expected to result in different distribution of genetic diversity both within and among populations [29], but can be used to infer the relative importance of the seed and pollen dispersal. Following this principle the pollen-to-seed migration ratio was compute in seven natural populations of *A. angustifolia,* based on data obtained from isoenzymes (biparentally inherited) and cpDNA (paternally inherited) [15]. Assuming strict paternal inheritance of the chloroplast in *A. angustifolia* (i.e. just through pollen), the pollen-to-seed migrations ratio equaled 8.05, supporting a prevalence of pollen dispersal over seed dispersal. Indeed, analysis of within population spatial genetic structure revealed strong family structure in a population with elevated amount of juvenile individuals, as expected for species with clumped seed dispersal [26]. Congruent with a system dominated by short-distance gene dispersal, the estimation of the effective number of migrants between two neighboring populations obtained from microsatellite data and coalescent-based analysis was around one individual per generation [26], confirming the premise that, in comparison with other anemophylous species, pollen dispersal is relatively limited in *A. angustifolia* [30, 31]. However, absence or very low family structure revealed in some studies suggests also the occurrence of somewhat extended gene flow in some populations of *A. angustifolia* [26]. In addition, relatively low genetic differentiation at microsatellite and AFLP markers revealed among populations from the two southernmost Brazilian states suggests a reasonably effective gene flow, likely by means of barochorous/anthropogenic seed carrying and pollen transport through a stepping-stone model [22].

The pollination of a single female *A. angustifolia* tree inside a plot of tropical Atlantic forest in Santa Catarina State was already reported [32]. Such pollination likely occurred with pollen transported as far as 25 km, from the *Araucaria* forest in the neighboring highlands, suggesting occurrence of long distance dispersion under particular environmental conditions. In fact, pollen dispersion by wind is quite efficient, but greatly dependent on climatic conditions like wind direction and atmospheric temperature and humidity. Among other factors, genetic neighborhood, synchronization of flowering, spatial distribution of the plants and physical characteristics of the pollen grain greatly interfere anemophylous pollen dispersion.

THE ROLE OF GENETICS AND BIOTECHNOLOGY IN THE CONSERVATION OF *ARAUCARIA* FOREST

Conservation of endangered ecosystems requires sound knowledge about genetic diversity of keystone species and the development of pertinent biotechnological tools. Since *A. angustifolia* is the main plant species in the *Araucaria* forest, the distribution of the genetic diversity of this species is of most importance when discussing the conservation of this ecosystem. Similarly, the development of biotechnological tools aiming preservation of the existent diversity is decisive. Considering the genetic diversity characterized in *A. angustifolia* populations and the biotechnological advances obtained in the last years [for a review on this topic, see 33], four main statements may be listed, which are crucial for structuring conservation of *Araucaria* forest.

1. The genetic system of A. angustifolia seems to be very efficient to protect the species against rapid losses of its genetic diversity, even if reduction of genetic diversity by forest exploitation occurred.

Different techniques have been used to asses the levels of genetic diversity of *A. angustifolia* populations, revealing values within the range or higher than recorded in the literature for species with the same life history traits and for species of the family Araucariaceae (Table 1). These results are evidence for the capacity of *A. angustifolia* in maintaining relatively high levels of genetic diversity, even after the exploitation.

Despite the very intensive and non-sustainable exploitation of the forest, some of the studied populations reveal relatively high levels of gene diversity. Such populations must be considered when selecting areas for *in situ* protection or collecting seeds. Analogous conclusions were obtained from the analysis of genetic diversity in planted populations, which revealed genetic structure and level of diversity very similar to natural populations of the same geographical area [34]. Even a plantation established in 1992, i.e. after the massive exploitation, revealed comparatively high levels of genetic diversity at microsatellite and AFLP marker loci.

In conclusion, the genetic system of *A. angustifolia* seems to be very efficient to protect the species against rapid losses of its genetic diversity, even though we have no information about the original genetic diversity of the species prior to the forest exploitation.

Table 1. Mean gene diversity (*H*) reported for plant species according to its life history traits and for species of the family Araucariaceae. For microsatellites and isoenzymes, gene diversity estimations range from 0.0 to 1.0. For AFLPs and RAPDs, gene diversity estimations range from 0.0 to 0.5.

Trait	*H*			
	Isoenzymes[1]	SSRs[2]	AFLPs[2]	RAPDs[3]
Long-lived perennial	0.18	0.68	0.25	0.24
Regional distribution	0.15	0.65	0.21	0.22
Outcrossing	0.16	0.65	0.27	0.26
Attached seed dispersal	0.20	0.56	0.16	0.17
Seed dispersal by gravity	0.14	0.47	0.19	0.21
Mid successional	0.14	0.63	0.21	0.19
Araucaria columnaris[4,5]	-	0.53 – 0.65	-	-
Araucaria nemorosa[5]	-	0.68	-	-
Araucaria cunninghamii[6]	-	0.61	0.14	-
Araucaria araucana[7]	0.15	-	-	-
Agathis robusta[6]	0.47	-	0.19	-
Agathis borneensis[8]	0.11	-	-	-
Araucaria angustifolia[9]	0.13	0.65	0.24	0.29

[1] [35]; [2] [36]; [3] [37]; [4] [38]; [5] [39]; [6] [40]; [7] [41]; [8] [42]; [9] mean values obtained from published studies (details, see 15).

2. Conservation measures starting now may be efficient, because the genetic structures of existing populations is more largely influenced by the evolutionary past than by human disturbance.

Although the highlands in southern Brazil have never been covered by ice sheets, the cold and dry conditions during the Last Glacial Maximum until the late Holocene barred the survival of forest formations in these regions (Behling 2002). It is believed that species of the *Araucaria* forest were found just in protected valleys and/or slopes where the moisture was elevated, while the highlands were covered by grasses. Just after the early Holocene, the increase of the rainfall allowed the expansion of forest species from refugia and *A. angustifolia* migrated onto the highlands, substituting the grassland.

The current genetic structure of *A. angustifolia* populations was much likely determined by different times of species' migration from refugia onto highlands [22, 43]. Pollen records from southern and southeastern Brazil [2, 44, 45, 46, 47] suggest an earlier colonization of the southeastern highlands, about 3000 years ago (uncalibrated ^{14}C dating). The highlands of southern Brazil were colonized by *A. angustifolia* about 1500 years later than in southeastern. As result of the climatic changes occurred during the late Holocene, such as the retraction of the polar fronts [48], populations from southern and southeastern Brazil likely experienced the effect of evolutionary forces in different spectra, reflected in the current genetic structure of *A. angustifolia*.

Differences in the time of highlands colonization and the effects of evolutionary forces acting in each region may explain the notable genetic differentiation among natural populations from Campos do Jordão (southeastern Brazil) and populations from southern

Brazil reported in different studies [e.g. 21, 22]. Similarly, molecular signatures of population dynamics extracted from the analyses of bottleneck events point out the occurrence of slow recovery of effective population size in southeastern region, while a comparatively quicker expansion of effective population size took place in the southern populations [43]. Such difference in rate of population size expansion likely results from climatic conditions after highlands colonization. Comparing Brazilian southern and southeastern highlands, the latter region presents in the current climatic conditions, a comparatively dryer period of around four months between May and August [46]. Thus, populations in southeastern region likely experienced a dryer period after highland colonization, in comparison to the southern region.

Recent human interference in the forests may have caused less impact in the current genetic structure of *A. angustifolia,* which reflects its evolutionary past. Thus, conservation measures starting now may be efficient, considering the present genetic diversity and structure of the species.

3. Taking simple measures to safeguard the continued existence of the few remaining forests, there are real opportunities to safe the genetic resources of the species.

Recently, the obstacles for conservation of forest genetic resources were discussed, with focus in Europe [49]. In this study, the major barriers were identified mainly as political issues. Besides to these political difficulties, the (vague) view of the non-specialist public about forest genetic resources rises as an important factor. In addition, the collaboration between geneticists and conservationists "has been anything but common" and the examples show that overall in the world, initiatives toward genetic resources conservation "are mediocre at best" failing to preserve them [49]. The use of sound scientific arguments supported by different areas of the Science may be a very powerful instrument in persuading decision-makers towards considering forest genetic resources as a primary issue also by governments [15].

Considering that the genetic system of *A. angustifolia* seems very efficient to protect the species against rapid losses of its genetic diversity (statement 1) and that the genetic structures of existing populations is more substantially influenced by the evolutionary past (statement 2) than by recent human disturbance, there are opportunities to safe the genetic resources of the species if proper measures are took to safeguard the continued existence of the few remaining forests.

These conservation measures include mainly:

- *recuperation of degraded fragments*: due to the central importance of the *A. angustifolia* in the *Araucaria* forest, recuperation of degraded fragments and impoverished stands through planting this species implies also the recovery of the associated flora and fauna.

- *support of connectivity among fragments through reforestation*: connectivity among neighboring fragments through corridors has important effects since even limited gene flow among small populations at equilibrium has the tendency of reducing negative effects of small population size. In contrast, isolation of small forest fragments tends to difficult the genetic recovery of effective population size [43].

- *promotion of the natural regeneration*: Promotion of the natural regeneration is crucial, since *A. angustifolia* regenerates almost exclusively on forest board and gaps

(Figure 5). Natural regeneration rarely occurs inside the forest, in competition with broad leaf species.

- *encouragement of sustainable management of the forests*: Incitement of sustainable management of the forests should also be discussed when planning programs of *Araucaria* forest preservation, since it can contribute to the natural regeneration and establishment of planted forests.

Figure 5. Natural regeneration of A. angustifolia. (A) Regeneration at forest board. (C and D) Regeneration inside the forest, within a gap.

4. Biotechnological tools such as somatic embryogenesis and cell culture cryopreservation are reliable techniques towards ex situ conservation strategies

Tissue culture techniques comprise important tools towards the improvement and conservation of tree species, particularly for species with recalcitrant seeds, such as *A. angustifolia*. Investigations on organogenesis [50, 51] and on somatic embryogenesis [52, 53, 54, 55, 56, 57] of *A. angustifolia* have shown a high dependence of the tissue culture protocols on the genotype of the mother tree. In addition, the developmental stage of the explant significantly affects the induction rate of embryogenic cultures [53, 54]. The embryogenic competence of young pre-cotyledonary zygotic embryos in *A. angustifolia* is limited to seeds collected from December to February, decreasing subsequently [52].

Since species with recalcitrant seeds are not feasible to be maintained in seed banks, cryopreservation of embryogenic cultures and embryos associated to somatic embryogenesis is an appealing strategy for germplasm conservation of such species. Cryopreservation of *A. angustifolia* pro-embryos can be successfully reached through dehydration, cryoprotection, and vitrification techniques [58]. Such pro-embryos have been shown to be competent for somatic embryogenesis [59].

Ex situ conservation of *A. angustifolia* germplasm may be promoted through this approach, as proposed in Figure 6. Selected plants serve as mother trees (Figure 6A), which will be the source of precotiledonary zygotic embryos used as explant for the somatic embryogenesis. Embryogenic cultures in the earlier stages of the somatic embryogenesis (Figure 6B) may be cryopreserved (Figure 6C, route 2) or matured using controlled culture media (Figure 6D, route 3). These embryogenic cultures may also be retained in a maintenance medium and replicated (Steiner et al. 2008). The cryopreserved embryogenic cultures may be thawed and also follow the maturation (route 4 in Figure 6) or be replicated in maintenance medium. Mature cultures may originate plantlets (Figure 6E), which can be used as germplasm bank or as material for plantations establishment and forest recovery. Some technical constraints are precluding the large-scale use of this approach, since obtaining mature embryos is infrequent and their conversion into plantlets is still a trouble. Ongoing studies promise to overcome these troubles shortly, allowing the extensive application of these biotechnological tools towards conservation of this species.

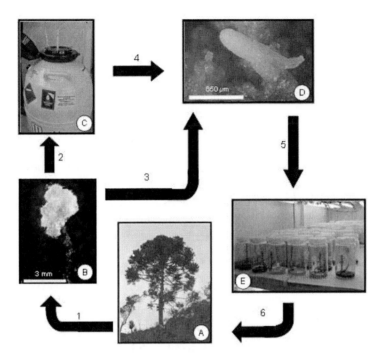

Figure 6. Integration of cryopreservation and somatic embryogenesis aiming species conservation. (A) Selected adult tree to be propagated. (B) Embryogenic cultures able to be cryopreserved or induced through somatic embryogenesis. (C) Cotyledonal stage of the somatic embryogenesis route of *A. angustifolia*. (E) Growing plantlets of *A. angustifolia* in a growth room. See text for details about the techniques and their constraints. Pictures B and D were kindly provided by Neusa Steiner (Laboratory of Developmental Plant Physiology and Plant Genetics – Federal University of Santa Catarina - Brazil).

ON FARM CONSERVATION OF THE ARAUCARIA FOREST

Although *in situ*, *ex situ* and *in vitro* strategies are constantly proposed as alternatives towards conservation of the *Araucaria* forest, the discussion about *on farm* approach is just on its infancy. *On farm* conservation has been proposed as an approach able to preserve the progression of evolution and adaptation of crops [60]. Although this concept was developed in reference to cultivated and domesticated species, *on farm* strategies of genetic resources conservation may be adopted into programs of conservation of non-domesticated tree species.

In general, *on farm* conservation is very close to agroforestry practices. In agroforestry, wild species are preserved *in situ*, coupled with production of crops or other agricultural products. Such agricultural products are exploited, while the uncultivated species are maintained non-managed. In a different way, *on farm* conservation strategies comprise the management of wild species aiming production and the selection of individuals based on traits considered of economical importance. Such management and selection is performed by the farmers, which will control the characteristic which are important and should be maintained, removing plants that presents lower production or quality. Despite the rejection of some individuals, the *on farm* conservation strategy maintains the evolutionary process and promotes the domestication of species (*in situ* domestication). As result of removing individuals with lower production or quality, the production/quality of the population tends to increase due to the higher quality of the progenies provided by the selected plants.

Applied to *A. angustifolia,* the *on farm* strategy may focus initially on seed production. A non-domesticated population of adult trees covering around 10.000 m^2 can produce more than 100 Kg of seeds per year, but with significant variation among years [14]. In a region of savanna with *A. angustifolia*, in 20 years, the commercialization of seeds (considering a mean production of 1500 Kg of seeds per tree in this period) provides an economical gain 22% higher then the commercialization of the timber (considering a tree with 25 cm DBH and 4.4 m of height) [1]. Thus, well planned *on farm* conservation programs can promote the maintenance of the forest relicts as source of income, as well as *in situ* domestication through the selection of more productive trees. Additionally, other non-timber products (fruits, honey, tee, medicinal plants, etc.) may be exploited, increasing the economical and social importance of the forest through a sustainable management. Furthermore, biotechnological tools can be used for the characterization of species genetic diversity and for propagation of selected plants, accelerating and improving this process.

CONCLUSION

Conservation of *Araucaria* forest depends mainly on the conservation of *A. angustifolia*, due to the very important ecological function of this species within the ecosystem. Despite the very intensive non-sustainable exploitation of *A. angustifolia* in the last century, it is possible to recover and to preserve this species, if pertinent measures initiate at this moment. The current knowledge about genetic structure and diversity jointly with the advances on the biotechnology of *A. angustifolia* provides powerful tools towards conservation of this very important species and ecosystem.

REFERENCES

[1]   Guerra MP, Silveira V, Reis MS, Schneider L (2002) Exploração, manejo e conservação da araucária (*Araucaria angustifolia*). In: Simões LL and Lino CF (eds) Sustenável mata atlântica: a exploração de seus recursos florestais. Editora SENAC, São Paulo

[2]   Behling H (2002) South and Southeast Brazilian grasslands during Late Quaternary times: a synthesis. Palaeogeogr Palaeoclimatol Palaeoecol 177:19-27

[3]   Stefenon VM, Gailing O, Finkeldey R (2006) Searching natural populations of *Araucaria. angustifolia*: conservation strategies for forest genetic resources in southern Brazil. In: Bohnes J and Paar E (eds) Forstliche Genressources als Produktionsfaktor.

Proceedings of the 26th Tagung der Arbeitsgemeinschaft Forstgenetik und Forst-pflanzenzüchtung. Hessen-Forst Forsteinrichtung, Information, Versuchwesen.

[4]   Finkeldey R and Hattemer HH (2007) Tropical Forest Genetics. Springer, Berlin Heidelberg

[5]   Reitz PR and Klein RM (1966) Araucariaceae: Flora ilustrada catarinense. Herbário Barbosa Rodrigues, Itajaí

[6]   Machado SA and Siqueira JDP (1980) Distribuição natural da *Araucaria angustifolia* (Bert.) O. Ktze. In: IUFRO Meeting on Forestry Problems of the Genus *Araucaria*, Curitiba, 21-28 October 1979

[7]   Hueck K (1954) Verbreitung und Standortansprüche der brasilianischen Araukarie (*Araucaria angustifolia*). Forstwissenchaftliches Centralblatt 71:272-289

[8]   Setoguchi, HH, Osawa TA, Pintaud J-C, Jaffré T and Veillon J-M (1998): Phylogenetic relationships within Araucariaceae based on *rbcL* gene sequences. American Journal of Botany 85: 1507–1516.

[9]   Stefenon VM, Gailing O, Finkeldey R (2006 Phylogenetic Relationship Within Genus *Araucaria* (Araucariaceae) Assessed by Means of AFLP Fingerprints. Silvae Genetica 55:45-52[11]

[10] Shimizu JY and Higa AR (1980) Variação genética entre procedências de *Araucaria angustifolia* (Bert.) O. Ktze. na região de Itapeva-SP, estimada até o 6.° ano de idade. In: IUFRO Meeting on Forestry Problems of the Genus *Araucaria*, Curitiba, 21-28 October 1979

[11] Sebbenn AM, Pontinha AAS, Giannotti E, Kageyama PY (2003) Genetic variation in provenance-progeny test of *Araucaria angustifolia* (Bert.) O. Ktze. in São Paulo, Brazil. Silvae Genetica 52:181-184

[12] Sebbenn AM, Pontinha AAS, Freitas SA, Freitas JA (2004) Variação genética em cinco procedências de *Araucaria angustifolia* (Bert.) O. Ktze. No sul do estado de São Paulo. Revista do Instituto Florestal 16:91-99

[13] Hampp R, Mertz A, Schaible R, Schwaigere M, Nehls U (2000) Distinction of *Araucaria angustifolia* seeds from different locations in Brazil by a specific DNA sequence. Trees 14:429-434

[14] Mazza MCM (1997) Use of RAPD markers in the study of genetic diversity of *Araucaria angustifolia* (Bert.) populations in Brazil. International Foundation for Science. Florianópolis

[15] Sousa VA, Robinson IP, Hattemer HH (2004) Variation and population structure at enzyme gene loci in *Araucaria angustifolia* (Bert.) O. Ktze. Silvae Genetica 53:12-19

[16] Stefenon VM, Gailing O, Finkeldey R (2007) Genetic structure of *Araucaria* angustifolia (Araucariaceae) in Brazil: Implications for the *in situ* conservation of genetic resources. Plant Biology 9:516-525

[17] Mantovani A, Morellato APC, Reis MS (2006) Internal genetic structure and outcrossing rate in a natural population of *Araucaria angustifolia* (Bert.) O. Kuntze. The Journal of Heredity 97:466-472

[18] Sousa MIF (2006) Análise da diversidade genética em populações de *Araucaria angustifolia* (Bertol.) Kuntze utilizando marcador AFLP. Dissertation. Universidade Federal do Rio de Janeiro

[19] Bittencourt JVM and Sebbenn AM (in press) Pollen movement within a continuous forest of wind-pollinated *Araucaria angustifolia*, inferred from paternity and TwoGener analysis. Conservat Genet. DOI 10.1007/s10592-007-9411-2

[20] Stefenon VM, Gailing O, Finkeldey R (2008) The role of gene flow in shaping genetic structures of the subtropical conifer species *Araucaria angustifolia*. Plant Biology 10:356-364

[21] Patreze CM (2008) Análise molecular da diversidade genética em una população natural de *Araucaria angustifolia* (Bertol.) Kuntze no estado de São Paulo. Dissertation. Universidade de São Paulo

[22] Sousa VA, Sebbenn AM, Hattemer HH, Ziehe M (2005) Correlated mating in populations of a dioecious Brazilian conifer, *Araucaria angustifolia* (Bert.) O. Ktze. Forest Genetics 12:107-119

[23] Petit RJ, Duminil J, Fineschi S, Hampe A, Slavini D, Vendramin GG (2005) Comparative organization of chloroplast, mitochondrial and nuclear diversity in plant populations. Molecular Ecology 14:689-701

[24] Sousa VA and Hattemer HH (2003) Pollen dispersal and gene flow by pollen in *Araucaria angustifolia*. Australian Journal of Botany 51:309-317

[25] Behling H and Negrelle RRB (2006) Vegetation and pollen rain relationship from the Tropical Atlantic Rain forest in Southern Brazil. Braz Arch Biol Technol 49:631-642

[26] Stefenon VM, Gailing O, Finkeldey R (2008c) Genetic structure of plantations and the conservation of genetic resources of Brazilian pine (*Araucaria angustifolia*). Forest Ecology and Management. 255:2718-2725

[27] [35] Hamrick JL and Godt MJW (1989) Allozyme diversity in plant species. In: Brown AHD, Clegg MT, Kahler AL, Weir BS (eds) Plant population genetics, breeding and genetic resources. Sinauer, Sunderland

[28] [36] Nybom H (2004) Comparison of different nuclear DNA markers for estimating intraspecific genetic diversity in plants. Molecular Ecology 13, 1143 – 1155

[29] Nybom H and Bartish IV (2000) Effects of life traits and sampling strategies on genetic diversity estimates obtained with RAPD markers in plants. Perspectives in Plant Ecology, Evolution and Systematics 3:93-114

[30] Kettle CJ, Hollingsworth PM, Jaffré T, Moran B, Ennos RA (2007) Identifying the early genetic consequences of habitat degradation in a highly threatened tropical conifer, *Araucaria nemorosa* Laubenfels. Molecular Ecology 16, 3581-3591

[31] Peakall R, Ebert D, Scott LJ, Meagher PF, Offord CA (2003) Comparative genetic study confirms exceptionally low genetic variation in the ancient and endangered relictual conifer, *Wollemia nobilis* (Araucariaceae). Molecular Ecology 12, 2331-2343

[32] Ruiz E, Gonzáles F, Torres-Díaz C, Fuentes G, Mardones M, Stuessy T, Samuel R, Becerra J, Silva M (2007) Genetic diversity and differentiation with and among Chilean populations of *Araucaria araucana* (Araucariaceae) based on allozyme variability. Taxon 56, 1221-1E(-1219)

[33] Kitamura K and Rahman MYBA (1992) Gentic diversity among natural populations of *Agathis borneensis* (Araucariaceae), a tropical rain forest conifer from Brunei Darussalam, Borneo, Southeast Asia. Canadian Journal of Boany 70:1945-1949

[34] Stefenon VM, Behling H, Gailing O, Finkeldey R (2008b) Evidences of delayed size recovery in *Araucaria angustifolia* populations after post-glacial colonization of highlands in Southeastern Brazil. Anais da Academia Brasileira de Ciências. 80:433-443

[35] Behling H (1995) Investigations into the Late Pleistocene and Holocene history of vegetation and climate in Santa Catarina (S Brazil). Veget Hist Archaeobot 4:127-152

[36] Behling H (1997) Late Quaternary vegetation, climate and fire history from the tropical mountain region of Morro de Itapeva, SE Brazil. Palaeogeogr Palaeoclimatol Palaeoecol 129:407-422

[37] Behling H (1997) Late Quaternary vegetation, climate and fire history of the *Araucaria* forest and campos region from Serra Campos Gerais, Paraná State (South Brazil). Rev Paleobot Palynol 97:109-121

[38] Behling H, Bauermann SG, Neves PC (2001) Holocene environmental changes in the São Francisco de Paula region, southern Brazil. J South Am Earth Scienc 14:631-639

[39] Ledru M-P, Salgado-Labouriau ML, Lorscheitter ML (1998) Vegetation dynamics in southern and central Brazil during the last 10,000 yr B.P. Review of Paleobotany and Palynology 99:131-142

[40] Geburek Th and Konrad H (2008) Why the conservation of forest genetic resources has not worked. Conservat Biol 22:267–274

[41] Iritani C, Zanette F, Cislinkski J (1992) Aspectos anatômicos da cultura *in vitro* da *Araucaria angustifolia*. I. Organização e desenvolvimento dos meristemas axilares ortotrópicos de segmentos caulinares. Acta Botânica Paranaense 21:57-76

[42] Iritani C, Zanette F, Cislinkski J (1993) Aspectos anatômicos da cultura *in vitro* da *Araucaria angustifolia*. II. O enraizamento dos brotos axilares. Acta Botânica Paranaense 22:1-13

[43] Astarita LV and Guerra MP (1998) Early somatic embryogenesis in *Araucaria angustifolia* – induction and maintenance of embryonal-suspensor mass cultures. Braz J Plant Phys 10:113-118

[44] Santos ALW, Silveira V, Steiner N, Vidor M, Guerra MP (2002) Somatic embryogenesis in Parana Pine (*Araucaria angustifolia* (Bert.) O. Kuntze). Brazilian Archives of Biology and Technology 45:97-106

[45] Silveira V, Steiner N, Santos ALW, Nodari RO, Guerra MP (2002) Biotechnology tolls in *Araucaria angustifolia* conservation and improvement : inductive factors affecting somatic embryogenesis. Crop Breeding and Applied Biotechnology 2:463-470

[46] Silveira V, Santa-Catarina C, Tun NN, Scherer GFE, Handro W, Guerra MP, Floh EIS (2006) Polyamine efects on the endogenous polyamine contents, nitric oxide release, growth and differentiation of embryogenic suspension cultures of *Araucaria angustifolia* (Bert.) O. Ktze. Plant Science 171:91-98

[47] Steiner N, Vieira FN, Maldonado S, Guerra MP (2005) Effect of carbon source on morphology and histodifferentiation of *Araucaria angustifolia* embryogenic cultures. Brazilian Archives of Biology and technology 48:895-903

[48] Steiner N, Santa-Catarina C, Silveira V, Floh EIS, Guerra MP (2007) Polyamine effects on growth and endogenous hormones levels in *Araucaria angustifolia* embryogenic cultures. Plant Cell and tissue Culture 89:55-62

[49] Demarchi G (2003) Criopreservação de culturas embriogênicas de *Araucaria angustifolia* Bert O. Kuntze. Dissertation, Universidade Federal de Santa Catarina

[50] Steiner N (2005) Parâmetros fisiológicos e bioquímicos durante e embriogênese zigótica e somática de *Araucaria angustifolia* Bert O. Kuntze. Dissertation. Universidade Federal de Santa Catarina

[51] Jarvis D and Hodgkin T (1999) Farmer decision making and genetic diversity: linking multidisciplinary research to implementation on-farm. In: Brush SB (ed) Genes in the field: on-farm conservation of crop diversity. Lewis Publishers, IPGRI, IDRC. p. 261-278.

In: Encyclopedia of Environmental Research   ISBN: 978-1-61761-927-4
Editor: Alisa N. Souter   © 2011 Nova Science Publishers, Inc.

Chapter 20

MEXICAN FISH: A FAUNA IN THREAT

Marina Y. De la Vega-Salazar
Mexico

ABSTRACT

The Mexican ichthyofauna is exceptionally rich, composed of 507 species represent 6% of the fish species known around the world, and 163 of them 32% are endemic of Mexico with endemism at a family, genera and species level. Result of complex climatological and geological history, producing high isolated basins and promoting speciation. Is common between Mexican ichthyofauna species with reduced distribution area, besides the environmental conditions become unsuitable for native fish by anthropogenic impact including; pollution, introduction of exotic species, severe fragmentation and destruction of habitat. A recent study conducted in the occident of the Mesa central in 2001 reveals that the principal causes of destruction of habitat includes principally: dryness 68%, piping 20%, and build of a spa 5%. The species in the list of danger include 185 species in a category of threat and 11 species reported as extinct. The factors that promote extinction in fish are: Channel incision of the habitat, limited physiographic distribution, biological specialization and fragmentation principally.

Mexico's evolutionary history has resulted in a mega-diverse and highly endemic fauna, including a rich and diversified freshwater fish fauna comprising 507 species, this is the 6% of fishes in the world, of which 163 species (32) are endemic of Mexico. The high diversity of the Mexican fauna is greatly influenced by the fact that the country is at interface of temperate and tropical climates [4, 23].

Diversification of the Mexican freshwater fish fauna stems from many factors: the Highly varied physical geography, great latitudinal extent, isolation of the large tropical highland (Mesa Central) that contains the important Rio Lerma fauna, adaptation by many marine groups to freshwater and the presence in the southeast of the largest river system, the Usumacinta-Grijalva basin, that lies well witching the tropics [23].

In this revision is exposed the high diversity of freshwater fish fauna, the explanation of the country's high fish endemicity are given analyzing the influence of geographic barriers, habitat heterogeneity, and the mix of temperate and tropical forms converging in the Mesa Central of Mexico (MCM), the current situation of the conservation status of the fish fauna and the principal causes of threat [2, 5, 6].

DIVERSITY OF THE MEXICAN FRESHWATER FISH FAUNA

Physiographic History of Mexico

The Mesa Central of Mexico (Figure 1) is inhabited by a diverse and largely endemic freshwater fish fauna, have the highest number of endemic taxa, 75% of endemic species are concentrated in this area (table1) followed by the Grijalva- Usumacinta basin with 9% of the endemic taxa [2, 23].

Figure 1.Mesa Central of Mexico (dark area), and principal basins in the Central zone in Mexico 1. Ameca, 2. Balsas, 3. Coahuayana, 4. Lerma-Chapala-Santiago, 5. Mezquital, 6. Pánuco, 7. Nazas, 8. Valle of Mexico.

This diversity is believed to have resulted in large part from vicariance events during the Pliocene and Pleistocene. Vicariance in this highland environment has been caused by the compartmentalization of drainage system through the isolation of endorheic (hidrogeographically isolated) basins and headwater piracy of pluvial systems [2, 34].

Whit the establishment of the land bridge between North and South America during the Paleocene, a dispersal of South America groups towards Central America occurred. With the disappearance of this land bridge during the Eocene, fauna became isolated and differentiated. A second later dispersal led to the establishment of northern groups in the north-central region of Mexico. The formation of mountains and climatic changes during the Eocene-Pliocene further isolated these and the Mesoamericans groups from the northern elements mainly in the southeastern Unites States.

Table 1. Representative families in the MCM with the highest levels of endemicity.

FAMILY	Fauna[1]	Lerma-Santiago	Balsas	Panuco	Ameca	Nazas	Mezquital	Coahuayana-Armeria	Valle of Mexico
Ameiuridae	NT	1	1	1	-	-	1	-	-
Atherinidae	NA	18	1	-	-	-	1	-	4
Catostomidae	NA	-	-	1	-	2	2	-	-
Characinidae	NT	-	1	-	-	1	-	.	-
Cichlidae	NT	-	2	4	-	-	-	-	-
Cyprinidae	NA	8	2	3	1	7	3	12	2
Goodeidae	NA	14	5	1	1	1	1	7	1
Persidae	NA	-	-	-	-	1	1	2	.
Poecilidae	NT	2	4	5	-	-	-	6	-

1. Origin of fish fauna NA = Neartic, NT= Neotropical.

The orographic formations of the Oligocene led to further vicarious speciation. It is suggested that the climatic changes during the Pleistocene may have contributed to high diversity and endemism in the region. The highlands of Central Mexico became isolated from the central plateau and the southern United States as a consequence of increased aridity and low temperatures from about the middle of the Tertiary.

The Valley of Mexico lies in a north-south trending grabes whose origins can be traces to the early mid-Tertiary. Faulting began at this time and the drainage was probably to the south into what is now the Balsas basin. It is thought that towards the close of the Miocene, volcanic activity diminished and a cycle of erosion began. Volcanic activity resumed early in the Pliocene and continued throughout that period.

Drainage patterns have been continually changed: Lakes have formed and then filled with alluvium, sometimes to great depths, or been drainages by eroding streams. Some of the many interior basins, particularly in the south, still contain remnants of once large bodies of water, while in the north the bolsons area mostly dry [2, 5, 6].

Currently the Mexican Plateau is a roughly triangular platform tending southward from the Rio Grande into a geographical cul-de-sac broken only by the westward flowing Lerma-Santiago river system and the headwater eroding tributaries of certain coastal drainages: the Rio Panuco in the east, Rios Balsas, Tuxpan and Armeria in the south and the Rio Ameca in the West. (Figure 1) [2].

ORIGIN OF THE HIGH DIVERSITY

The Mexican diversity is greatly influenced by the fact that the country is at the interface of temperate and tropical climates, the boundary between the Neartic and Neotropical biotas. Some authors draw the irregular boundary between Neartic and neotropics through the spine of the transverse group of peaks in the Trans-Mexican Neovolcanic Belt [6].

Mexico's rugged topography has resulted in a great variety of habitats and micro habitats that are subject to variable environmental conditions, consequently, there are different ecological conditions that allow the establishment of distinct animal populations isolated in small areas.

The earliest known fossil record for Primary fishes is in the Pliocene Chapala formation, and includes one living species *Notropis sallei*, and one extinct species *Micropterus relictus*. For secondary fishes the goodeid *Tapatia occidentalis* occur in the late Miocene Santa Rosa Formation. For a marine invaders the earliest known fossils occur in the Pliocene Chapala formation, species of the genus Chrostoma. Fossil salmonids are known from late Pleistocene deposition only. Species assemblages largely follow the dividing lines of the main physiographic regions of the country [23].

The history of the Mesa Central has had a profound effect on the composition of the fish fauna. First, the region was initially colonized by the ancestor of the jordani species group prior to any appearance by the primary freshwater members of either North or South American fish faunas (Characidae, Cyprinidae, and Catostomidae): The initial advantages of the early colonizers were further strengthened by the mid-Pleistocene uplift. Slowly flowing rivers now became torrents as they plunged off the plateau onto the coastal plain or eroded back into the mountains. Environmental changes directly resulting from tectonic events, including lower temperatures and siltation from eroding lacustrine deposits must have caused extinctions among those species now isolated.

Second, the continual alteration of the aquatic environment in unpredictable ways by volcanism and mountain building created ideal conditions for the geographic isolation of population and hence speciation.

The topography and the varied environments have contributed to the differentiation and radiation of these isolated populations, making Mexican biodiversity and endemicity exceptionally rich.

In this way native species evolved; no evidence of sympatric or intralacustrine speciation in any of the large lakes was found. The cases of allopatry could be historical accidents, but the fact that most of the are the same body length suggests that this similarity has been an important factor in determining their present distribution. Given the geological history of the Lerma basin, they must have had past opportunities to coexist [2, 5, 6].

SPECIES RICHNESS COMPOSITION AND ENDEMICITY

Mexican fish fauna comprising 507 species in 47 families, The freshwater ichthyofauna is interesting because of its high level of endemicity, since 163 species (32%) are endemic of Mexico, The great geographic diversity and isolation of drainages have lies to a high degree of endemisms in the Mexican fish fauna: The major centers are: The Rio Lerma Santiago with 66% of endemicity, Rio Usumacinta-Grijalva basin 36% with major endemism among poecilidae and Cichidae. Rio Panuco basin with 30%; Rio Balsas basin 32% (cyprinids and goodeids [23].

Primary freshwater are represented by eight families with 37 genera and 132 species. Witching these families the most numerous genera are: *Notropis* (Cyprinidae) with 25 species; *Catostomus* (Catostomidae) with 10 species; *Ictaluridae* with 10 species. Secondary fishes are also represented by eight families, but have more genera (43) and species (186).

The large genera in these families are: *Cyprinodon* (Cyprinodontidae) with 18 species, *Gambusia* (Poecilidae) with 19 species, and, *Cichlasoma* (Cichlidae) with 40 species. Marine derivatives comprise 15 phylogenetically diverse families (lampreys to gobies) with 22 genera and 57 species, the most prolific genus is *Chirostoma* (Atherinidae) with 19 species and Goodeidae with 16 genera and 37 species (Table 1) [2, 4, 5 6 23].

CAUSES OF EXTINCTION FOR FRESHWATER FISH

General Patterns of Speciation and Extinction

In order to protect biological diversity is important to document the factors that promote extinctions.

Because rates of speciation and extinction are correlated, rapidly speciation taxa should suffer more mass extinction than slowly speciation groups. Recently, the concern over the imminent extinction of thousands of species is the result of pollution and habitat destruction; major reviews of species extinction have revealed that most animals' extinction is caused by humans, introduction of species, and loss of habitat. The primary anthropogenic causes of species declines produce a series of ecological and genetic affects that are finally expressed, and can be evaluated, in population dynamics and extinction risk.

Species characteristics that might promote speciation, such as numerous small, geographically isolated populations, might also leave species prone to extinction. For example species with low dispersal ability should ten to be endemic, with comparatively small population sizes more subject to extinction than widespread species, but low-dispersing species should also have more subdivided populations prone to speciation.

Most traits listed above reflect population's sensitivity to decreasing suitable habitat area and increasing isolation [28, 33].

Geographic Range

Species with large geographic range sizes may be less likely to walk randomly to extinction, and thus they persist for longer; species with traits which make them less prone to local extinction, and hence able to persist for longer, may also be enable to maintain large range sizes because of this extinction resistance and species with large range sizes may have them because they have persisted for longer.

However vicariant speciation may potentially result in any degree of asymmetry in the initial range size of daughter species, and is often portrayed as generating very similar-sized ranges. In contrast peripheral isolation results, immediately post-speciation in a highly asymmetrical split. Gastton 1998).

At one extreme lie those species constrained by ecology or by history to occupy small isolated islands of habitat or very scarce sets of environmental conditions. At the other extreme lie those species which are distributed across multiple biogeographyc regions, most species have relatively small range size and a few have relative large ones [16, 17].

Small Population Size

The smaller populations are more susceptible to extinction from various causes:

-Demographic stochasticity, which arises from chance events in the survival and reproductive success of a finite number of individuals. Demographic stochasticity is most important in small populations
-Environmental stochasticity, due to temporal variation of habitat parameters and the introduction of competitors, predators, parasites and diseases.
-Natural catastrophes, such as floods, fires, droughts, which may occur at random intervals through time; environmental stochasticity is important in both large and small populations. And
-Genetic stochasticity, resulting from changes in gene frequencies due to founder effect, random fixation or inbreeding [19, 29]
-Small population size. The more gradual the reduction in population size, the grater the opportunity for purging recessive lethal mutations and avoiding a large part of the inbreeding depletion [19].

Minimum population size is required to prevent extinction, which is in turn coupled to a minimum area. Rare species would therefore *per se* have higher extinction probabilities [29].

Ecological Specialization

There is a maximum rate of directional or random environmental change that a population can tolerate by adaptive evolution without becoming extinct, depending on the amount of genetic variability it can maintain. Most extinctions of any but the smallest populations are determined by persistent changes in the local environment, and large populations are not immune to these changes.
Species persist in regions where they are able to track the environment, and they become extinct if they fall to keep up with the shifting habitat mosaic. For many island species, the loss of suitable environment conditions throughout their limited range has resulted in species extinction [1, 33].

Fragmentation

Species richness is a function of area, this phenomenon has been widely analyzed though the theory of Island Biogeography [12, 13]. Number of species is therefore also reduced by habitat fragmentation, caused by the break-up of extensive habitats into small, isolated patches, which are insufficient to maintain the original species' richness. Habitat destruction resulting in population segregation is presently the greatest threat to the preservation of genetic diversity. Species occurring over large geographic areas tend to have greater local abundance than do more restricted species, an explanation for this result is that species able to exploit a wide range of sources become both widespread and locally abundant. Species are lost from a newly formed habitat fragment because the smaller areas can support fewer individuals and because increased isolation reduce immigration rates [1, 22]

If major changes in the distribution are driven by changes in the spatial distribution of suitable habitats, conservation must focus on maintain habitat continuity, for species occupying persistent habitat, the emphasis is on habitat preservation. For species occupying short lived or dynamic habitat, the emphasis is on preservation and management of existing habitat and preservation of sufficiently large areas [33].

Fresh Water Fish Extinction Factors

Biodiversity loss from real landscape typically is due to multiple anthropogenic impacts.

Aquatic degradation through habitat loss, introduction of exotics and pollution is causing high rates of endangerment and extinction among aquatic species. A clearer understanding of the extinction pasterns produces in complex scenarios of degradation is needed before effective programs to prevent extirpation can be developed [1, 3]. The principal patterns of fish fauna extinction are:

RESTRICTED GEOGRAPHIC RANGE

The principal characteristic of extinct species is their restricted geographic range. We regard restriction of the geographic range as the principal cause of threat to native species. Species distributed in one very limited habitat or which are geographically restricted may be considered rare. Also, they may occur at extremely low densities or be very specialized to their restricted habitat; wherefore any change in habitat characteristics can lead to their extinction [1, 28 33].

EUTROPHICATION

Eutrophication is defined as the enrichment of water with inorganic salts. The increase in nitrogen and phosphorus results in an increase in primary production (mostly algae). While some water bodies are naturally eutrophic, human activities accelerate the transport of salts into water bodies, and in consequence accelerate the eutrophication of normally oligotrophic water bodies. In general eutrophication is associated with deterioration of water quality and with the simplification of the trophic web [25].

In the localities studied, eutrophication caused an increment in richness and abundance of fish including native and exotic species, while hyper-eutrophication caused a reduction in richness and in abundance of fish. At higher levels of disturbance, species diversity declines [10, 36].

FRAGMENTATION

Insularitation is an important extinction mechanism in aquatic system. Natural aquatic systems in the region are fragmented on both broad geographic and local scales, several genetic and demographic consequences of such an isolated fragments distribution are likely including: local divergence via natural selection or genetic drift; little or no gene flow among isolated deems that might otherwise moderate losses of genetic variability after population crashes; and little or no recolonization of isolated habitats after local extinction [22].

CHANNEL INCISION

Chanel incision results from an imbalance of sediment supply and transporting capacity of streams substrates are buried or eroded away, and channel morphology, hydrology and hydraulic characteristics are transformed the streams can be abruptly increased by channel straightening or lowering base level. With effects of disturbance on stream communities focuses on warm water, sand bed system, suspended sediment loads and low concentrations of residual pesticides, heavy metals and organic mater, its potential for degrading biotic integrity [30]. Flow modification is one of the most widespread human disturbances of stream environments since reservoir building directly modifies physical and chemical habitat, these modifications are important factors influencing fish community composition.

Most native species habitat requirements are running waters, high concentration of dissolved oxygen and high transparency (De la Vega-Salazar et al.,). Several rivers in México have been dammed, turning running waters into standing waters, with sediment accumulation, high turbidity, invasion of water hyacinth (*Eichornia crassipens*), depletion of dissolved oxygen and presence of pathogens, conditions that only a few species can tolerate [8, 21]

SPECIALIZATION LEVEL

Species restricted to a single water size tended to be extirpated more frequently than species found in multiple water size, ecological specialists tended to be extirpated more frequently than ecological generalists, species with very small geographic ranges, vulnerability is exacerbated by the threat of single catastrophic events [1, 21].

INTRODUCTION OF EXOTIC FISH SPECIES

It is generally recognized that introduction of exotic species of animals can reduce diversity, especially if those species are generalists and of a large size, as they may be predators of native species, produce trophic change, compete for food and introduce new parasites [12]. The introduction of exotic species in water bodies in México is common. The abundance of exotic species of fish bears an inverse relationship with the richness and abundance of native species. Richness and abundance of exotic species are higher in localities with recent environmental change, principally by damming of rivers.

Fox and Fox (1986) [15] reported that there is no invasion of natural communities without disturbance. Where there has been successful invasion of natural communities by introduced species, it is considered that there has been at least a subtle alteration or some endogenous disturbance, and the principal effect of disturbance is to make available some resources. Accordingly, rich communities are less susceptible to invasion than poor ones.

The elimination or drastic reduction of common native species has been documented as resulting from the introduction of top carnivores. The extinction may not occur immediately, because mixed competition-predation interactions occur when piscivorous species are introduced. Most piscivorous species, for instance, begin life by feeding for some period on zooplankton or small invertebrates before becoming large enough to be piscivors [36].

VIVIPARITY IN FISH

Whiting endemic fish in Mexico there are two important viviparous families; Poecilidae and Goodeidae, they are sexually dimorphic with complex courtship displays; this represents high cost to males because production of testosterone can compromise the immune system of males [14]. Additionally, conspicuous courtship displays and ornaments may result in predation of males [10]. During gestation most goodeids develop the structure termed "trophotaniae", an embryonic trophic adaptation consisting of a simple surface epithelium surrounding a highly vascularized core of loose connective tissue. Trophotaniae is the chief site of nutrient absorption in embryo goodeids and accounts for their massive increase in weight [18]. This implies that the Goodeidae are potentially exposed to pollutants ingested by the mother through the entire gestational period. The benthic feeding habits of some species, the viviparous breeding system, and other physiologic characteristics may promote pollutant accumulation in gonads and embryos, and high bioaccumulation of pollutants can affect the embryo's viability [9]

The posses that create differences between actual population size and effective population size are determined by genetic, demographic, mating system and life historic parameters of populations, differences between apparent and actual patterns of reproductive success. The breading sex ratio and distribution of reproductive success among individuals witching populations or their mating system, may have large effects on effective population size, and is the most sensitive predictor of the ability of any population to maintain genetic variability and persist through time [27].

CURRENT CONSERVATION STATUS OF THE MEXICAN FRESHAWATER FISH

Situation of Freshwater Fish Fauna

In order to protect the native freshwater ichthyofauna it is important first to analyze changes in the status of native fishes and document the extent of its decline, and then describe the principal factors that causing threat or extinction of fish.

The native component of the fish fauna has been catastrophically altered by habitat destruction, introduction of exotic fish species, environmental pollution and overexploitation principally. The native Mexican fish fauna, and most particularly the endemic freshwater community, has recently undergone severe reduction throughout several water bodies [4, 11, 20, 24, 32, 35], many fish species disappear locally or live in danger of extinction, resulting in the extinction of species and a important number of species are now enlisted in a any level of threat.

This situation is reflected in the fact that in the 1960's four species were reported as recent extinction and 36 clearly in danger. In 1979 The American Fisheries Society's reported 67 species in danger, one decade later the number increased to 123 species The last official listed from 2001 reported 185 species in danger, (Table 2), including 147 of the endemic species (90% of the endemic species) [4, 26].

The common characteristics of extinct species are: (1) be endemic, (2) occupy a relatively small area (3) Occupy just one drainage basin, (4) occur in isolated springs or fragmented habitat [7].

Table 2. Number of species enlisted in threat in several years reported.

YEAR	Number of species reported as extinct	Number of species reported in threat	Proportion of increment (%)
1960	4	36	-
1979	-	67	186
1989	-	123	341
1993	16	135	371
2001	11	185	514

Factors of Threat for Native Fish

Mexican fish fauna is naturally vulnerable because many species are distributed in very limited habitat or are geographically restricted or are very specialized to their restricted habitat [7, 8]

While the decline of the native freshwater fauna worldwide has been strongly correlated with two primary factors; the habitat degradation and introduction of exotic species [1, 33, 36], the causes of extinction are different for each species, and seldom a consequence of a single factor [7, 8].

Many rivers, lakes and springs in Central Mexico have been fragmented, polluted or dammed; consequently entire communities of native fishes now appear to be endangered. Fish are especially vulnerable by the conflict with humans over increasingly valuable water resources. More than a half of Mexico includes desert and semi desert areas that comprise a water deficient, ecologically unstable. The species area relationship predicts quite accurately the number and distribution of endemic species already in serious danger of extinction through habitat loss [7, 8, 24, 35].

An early symptom of regional water deficiency is the drying of springs. Loses of springs have accelerated during the last years. Recent study of the current situation of springs, conducted in springs of the occidental region of the Mesa Central of Mexico reveals that only

20 % of sites were considered underexploited, while the 68% were overexploited. the overall decline of the aquifers and obvious rise in regional development and grown water use, resulting in the drying of springs and rivers in the area (Table 3).

Table 3. Current situation of springs in the Occidental region of the MCM.

CONDITION OF SPRING	Proportion %
Dry	68
Piped	7.5
Used as Spa	5.0
Good conditions for fish habitat	20

For most extinct fishes, more than single factor was responsible for their decline, inadequately regulated competition among resources extractors, especially in open-access fisheries, is one of the major causes of resource overexploitation and depletion. The mean threat is habitat degradation, principally due to urbanization of Central Mexico; the MCM is the most populated zone in the country (Figure 2) and is characterized by high levels of pollution spillage and overexploitation of natural resources by economic activities which increasingly divert water for human consumption, and pollution from urban, industrial and agricultural waste water, resulting in the disappearance of important fauna.

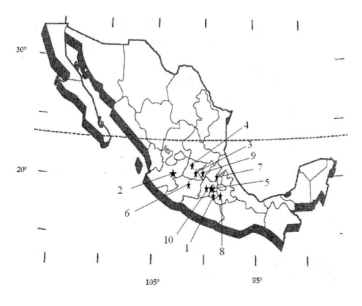

Figure 2 The most populated cities with several millions of inhabitants in the Mesa central of Mexico.
1.- Cuernavaca, 2. Guadalajara, 3. Guanajuato, 4. Leon, 5. Mexico D. F., 6.- Morelia, 7. Pachuca, 8 Puebla,   9. Queretaro, 10. Toluca.

As the human population of the Central Mexico region expands, the native fish fauna is becoming increasingly fragmented, leading to the disruption of fish assemblages and, inevitably, to the extinction of species [7, 8, 10, 11, 20, 32].

In order to know the situation in the MCM was conducted the study of several localities in the MCM from 1995 to 2005 were the physical, chemical and limnological characteristics were measured at those places (Table 4) and assessed their correlation with the presence and abundance of native and exotic fish species (Table 5).

Table 4. Europhization level, based in physicochemical characteristics in 53 localities distributed en the Mesa central of Mexico studied from 1995 to 2005.

EUTROPHIZATION LEVEL	PROPORTION (%)
O	54.7
E	24.5
HE	20.8

O = Oligotrophic, E = Eutrophic, HE = Hiper Eutrophic.

The Mesa Central of Mexico fish assemblage has been utterly transformed by a combination of exploitation, species introduction, and physical and chemical habitat change, where 21% of the rivers were dammed, almost 21% of the water bodies are hiper-eutrophic, and in 75% o the water bodies studied were introduced non native fishes. All this change cause the extinctions of expected native fish in 70% of localities studied, principally when species are introduced into isolated spring system (Table 5).

Table 5. Current characteristics of the fish fauna composition in localities studied in the Mesa Central of Mexico during the years 1995-2005.

CONDITIONS OF FAUNA PRESENT	Proportion (%)
Localities with exotic fish introduced	75.5
Localities with native fish extirpated	53
Localities without any kind of fish	17

Many species have disappear from great number of historical localities, while taxa in trouble can occur in almost any habitat, they are most frequent in habitats very limited in area, and particularly lakes, springs or headwater streams in arid areas, habitat alteration, and introduction of species are largely responsible for the decline. If diversity depends on area, drastic alteration of drainages can be expected to reduce diversity, thus most endemic species are geographically restricted and numerically rare where they occur.

Anthropogenic factors constitute the primary causes of endangerment and extinction insularization, overexploitation, species translocation and introduction and pollution. These primary factors have ramifying ecological and genetic effects that contribute to extinction risk [7, 8, 10, 20].

Consequences of the Lost of Species

Changes of the physical and biological characteristics including habitat alteration, introduction of exotic fish species and pollution of the environment are now the leading cause of the severe decline of the Mexican fish fauna. Fishes are being extirpated by complex patterns, and the only possibility for their long term survival is to control the negative effect of these factors [7, 8, 10, 20].

The existence of positive abundance-occupancy relationships, particularly among species that are in decline, would imply that double jeopardy pertains to individual species, with increases in extinction risk from reduction in abundance being accompanied by increases in extinction risk from reduction in range size and vice versa. Then a range reduction that increases the isolation of the remaining inhabited patches may have more detrimental effect on the population size where the isolation of remaining patches is not so greatly increased [16, 17].

Positive intraespecific relationship between abundance and occupancy or evidence suggestive of such relationship, have been documented for fish, many species have experienced not only a contraction of range size, but also a decline in density at those sites at which they have continued to persist, population can increase only if local densities increase.

A positive correlation between geographic range size and persistence may potentially result because widespread species tend to be locally more abundant, and locally more abundant species tend to be locally less likely to be come extinct. Likelihood of extinction is not an unbiased function of the range size, rather, risk of extinction declines with increasing range size [1].

Unfortunately, many species and freshwater ecosystems are severally in danger, the potential effects of such great losses of biotic diversity are unknown. Fishes are usually the best-known part of the aquatic fauna and therefore can serve as good indicators of the health of these environments [28]. What might happen if present-day biodiversity continues to decline and if humankind is ever able to stop its devastation?

Thus for every extant species of fish considered to be critically at risk of extinction, there are another two considered to be threatened (endangered or vulnerable), and three considered to be near-threatened or conservation-depend [26].

Alteration in biotic community structure reduce species diversity increase dominance by selected species increase dominance by exotic species, shortened food chain, increase disease permanence and reduce population stability [8].

I am especially concerned for the fate of the Mexican ichthyofauna, because environmental laws are not good enough, and the human population is rising at a staggering rate, with the intense competition for water between humans and aquatic organisms.

The overexploitation and management of natural resources usually is based on social, economic and political seditions. The most reasonable approach to conserving aquatic species may be to maintain the ecological integrity of entire watersheds and drainages, including aspects of water quality, habitat structure, hydrology and biotic interactions [1]. It is necessary to promote energetic pursuit of conservation measures, including:

1. Limnological restoration of freshwater ecosystems coupled with water quality monitoring and pollution control, may effectively promote the conservation and eventual recovery of native fish
2. Proper waste management, both in quantity and quality, are essential, both for surface and ground water supplies. Conservation of biodiversity will likewise conserve human quality of life.
3. And the control of introductions principally.

REFERENCES

[1]   Angermeier P. L. (1995) Ecological Attributes of Extinction-Prone Species: Loss of Freshwater Fishes of Virginia. *Conservation Biology* 9: 143-158.

[2]   Barbour C. D. (1973) A biogeographic history of Chirostoma (Pisces: Aterinidae): A species flock from the Mexican Plateau. *Copeia* 1973 (3): 533-556.

[3]   Cambray J. A.(2000) Threatened fishes of the world' series, an update. *Environ. Biol Fish* 50: 353-357.

[4]   Castro-Aguirre J. L. y Balart E. F. (1993) La ictiologia de México: pasado, presente y futuro. Revista de la Sociedad Mexicana de Historia Natural (XLIV): 327-343

[5]   De Buen F. (1946) Icteografia continental Mexicana. *Rev. Soc. Mex. Hist. Mat.* VII: 87-138.

[6]   De buen F. (1947) Investigaciones soble la ictiologia Mexicana. *A. Ins. Biol. Mex.* XVIII: 292-335.

[7]   De la Vega-Salazar M. Y. (2006). Estado de conservación de los peces de la familia goodeidae (Cyprinodontiformes), que habitan la Mesa Central de México. Revista de biología tropical 54 (1): 163-177.

[8]   De la Vega-Salazar M Y., Avila-Luna E y Macías-Garcia C. (2003) Ecological evaluations of local extinction: The case of two genera of endemic Mexican fish, *Zoogoneticus and Skiffia. Biodiversity and Conservation* 12:2043-2056.

[9]   De la Vega-Salazar M. Martínez Tabche L. and Macías Garcia C. (1997) Bioaccumulation of Methyl Parathion and its Toxicology in Several Species of the Freshwater Community in Ignacio Ramirez dam in Mexico. *Ecotox. Environ. Safety* 38: 53-62.

[10] De la Vega-Salazar M. Y. Macías-Garcia C. (2005) Principal Factors in the Decline of the Mexican Endemic Viviparous Fishes (GOODEINAE: GOODEIDAE). (In Uribe M. Grier H. Eds.) Viviparous fishes. New Life Publications. *México.* P. P. 505-513.

[11] Díaz-Prado E. Godines-Rodriguez M. A. López-López E. y Soto-Galera E. (1993) Ecología de los peces de la cuenca del río Lerma, México. Anales de la Escuela Nacional de Ciencias Biológicas México 39: 103-127

[12] Erlich P. and Erlich A. (1984). *Extinction.* Random House, New York, pp. 103-126.

[13] Flessa K. W. and Jablonski D. (1983) Ectintion is here to stay. *Paleobiology* 9: 315-321.

[14] Folstad, Y. and Karter A. J. (1992) Parasites, bright males, and immunocompetence handicap. *American Nataturalist* 139: 603-622.

[15] Fox, M D, Fox B J (1986) The susceptibility of natural communities to invasion. In: Groves R H, Burdon J J (eds.) Ecology of Biological invasions: An Australian perspective. *Australian Academy of Science*, Canberra. pp. 57-66.

[16] Gaston K. J. (1998) Species-range size distribution: products of speciation, extinction and transformation. *Phil. Trans R. Soc. Lon.* B 353: 219-230.

[17] Gaston K. J., Blackburn T. M., Greenwood J. J. D., Gregory R. D., Quinn R. M., Lawton J. H. (2000) Abundance occupancy relationships. *J. Appl. Ecol.* 37: 39-59.

[18] Hollenberg F. Wourns J. P. (1994) Ultrastructure and Protein Uptake of the EmbryonicTrophoyaeneae of four Species of Goodeid fish (Teleostei: Atheriniformes) *Journal of Morphology* 219: 105-129.

[19] Lande R. (1999) Extinction risk from antropogenic, ecological, and genetic factors. In Landweber L. F., Dobson A. P. (Eds) *Genetics and extinction of species.* Princeton University Press. New Jersey pp. 1-22.

[20] López-López E., Paulo Maya J. (2001) Changes in the fish assemblage in the upper Rio Ameca Mexico. *J. Freshw. Ecol.* 16: 179-187.

[21] Lydeard C. and Mayden R. L. (1995) A Diverse and Endangered Aquatic Ecosystem of the Southeast united States. *Conservation Biology* 9: 800-805.

[22] Meffe G. K., Vrijenhoek R. C. (1998) Conservation genetics in the Management of Desert Fishes. *Conser. Biol.* 2: 157-170.

[23] Miller R. R. (1986) Composition and derivation of the freshwater fish fauna of Mexico. *An. Esc. Nac. Cienc. Biol. Mex.* 30: 121-153.

[24] Miller R R Williams D. and Williams J. E. (1989) Extinctions of North American fishes during the past century. *Fisheries* 14 (6): 22-38.

[25] Moss B. (1992) *Ecology of Fresh Waters.* Man and Medium. Blackwell Scientific Publications.

[26] NOM-ECOL 059-2001. SEMARNAT. Norma official Mexicana. Secretaria del Medio Ambiente y Recursos Naturales México.

[27] Parker P. G. Wayte T. A. (1997). Mating systems, effective population size, and conservation of natural populations. In Clemmons J. R., Buchholz R. (eds.) Behavioral Aproaches to Conservation in the Wild. Cambridge University Press, pp. 242-261.

[28] Sepkoski J. J. (1998) Rates of speciation in the fossil records. *Phil. Trans. R. Soc. Lon.* B353: 315-326.

[29] Shaffer M. (1987) Minimum viable populations: coping with uncertainty. In Soulé, M. E. (Ed.) Viable Populations for Conservation. Cambridge University Press. Cambridge, New York pp. 70-86.

[30] Shields F. D. Knight S. Cooper C. M. (1994) Effects of Channel Incision on Base Flow Stream Habitats and Fishes. *Environ. Management* 18: 43-57.

[31] Soulé M. E. (1985) What is Conservation Biology? *BioScience* 35,727-734.

[32] Soto-Galera E., Paulo-Maya J.. Lopez-Lopez E., Serna-Hernandez J. A. Lyon J. (1999) Changes of the fish fauna as indicator of aquatic ecosystem condition in Rio Grande de Morelia-Lago de Cuitzeo basin, Mexico. Environ. Manage. 24: 133-140.

[33] Thomas C. D. (1994) Extinction, colonization and metapopulation: Environmental traking by rare species. *Conserv. Biol.* 8: 373-378.

[34] Webb S. A., Graves J. A., Macias García C., Magurran A. E., Ó Foighil D., y Ritchie M. G. (2004) Molecular Phylogeny of the live-bearing Goodeidae (Cyprinodontiformes). *Molecular Phylogenetic and Evolution* 30: 527-544.

[35] Williams J. E., Johnson J. E., Hendrickson D. A., Contreras Balderas S. Williams J. D., Navarro-Mendoza M. McAllister D. E., Deacon J. E. (1989) Fisheries of North America endangered, threatened, or for special concern. *Fisheries* 14: 2-20.

[36] Wootton J T (1998) Effects of disturbance on species diversity: A multitrophic perspective. *Am. Nat.* 152: 803-825.

In: Encyclopedia of Environmental Research
Editor: Alisa N. Souter

ISBN: 978-1-61761-927-4
© 2011 Nova Science Publishers, Inc.

Chapter 21

APPLICATION OF LCA TO A COMPARISON OF THE GLOBAL WARMING POTENTIAL OF INDUSTRIAL AND ARTISANAL FISHING IN THE STATE OF RIO DE JANEIRO (BRAZIL)

D. P. Souza[1], K. R. A. Nunes[1+], R. Valle[1€], A. M. Carneiro[1§] and F. M. Mendonça[2±]*

[1]UFRJ (Federal University of Rio de Janeiro), COPPE – Industrial Engineering Program
Rio de Janeiro - RJ - Brazil
[2]UFSJ (Federal University of São João del Rey)
São João del Rey – MG - Brazil

ABSTRACT

One of the main environmental impacts caused by fisheries is the emission of gases from the fuels used for the vessels. These gases contribute to global warming. The aim of this chapter is to use LCA (Life Cycle Assessment) to compare the global warming potential of artisanal and industrial fishing off the northern coast off Rio de Janeiro state (municipalities of Arraial do Cabo and Cabo Frio, respectively). For this analysis, one ton of common sea bream was adopted as the functional unit. The artisanal fishing data were obtained from interviews with local fishermen, while the industrial fishing data were collected from an industrial fishery. The study made use of the Umberto[TM] software package. The result of this comparison showed that industrial fishing has a greater impact on global warming than artisanal fishing.

Keywords: life cycle Assessment, LCA, environment, fishing, sustainability.

* djsouza@pep.ufrj.br.

+ knunes@pep.ufrj.br. (corresponding author**)**.

€ valle@pep.ufrj.br.

§ carneiro@pep.ufrj.br.

± fabriciomolica@yahoo.com.br.

1. INTRODUCTION

The growth of industrialization, population and consumption over the last century has boosted the demand for natural resources across the world. This has resulted in shortages and higher levels of pollution worldwide as gases are given off into the atmosphere and excessive amounts of industrial waste are discharged without being properly treated.

Industry has responded to this state of environmental degradation in many different ways. The trend across the globe is to lay the blame on companies for the economic, social and environmental effects caused by the entire life cycle of their products. Most of them have accordingly channeled part of their financial, human, production and knowledge resources into minimizing environmental impacts, so as to continue to prosper within a sustainable production system.

This concern has even affected the extraction of marine resources, such as fish and shellfish, which are considered to be renewable natural resources. Many stocks have already been depleted because of overfishing. Not only does fishing have a direct impact on target species, but it also has other kinds of environmental impacts, including damage to the seabed (caused by fishing equipment), the disposal of fish that are smaller than the minimum accepted size, the emission of gases from burning the fuels used by the fishing vessels, and the contamination of water by the antifouling paints used on the hulls of ships. All these impacts need to be measured and taken into account before marine resources can be used in a sustainable manner.

Within this context, LCA is an increasingly useful tool for achieving sustainability, because it can be used to identify and quantify environmental impacts throughout a product's life cycle. The objective of this study was to apply LCA to a fishing area off the northern coast of Rio de Janeiro State (Brazil) with the purpose of comparing the global warming potential of artisanal fisheries (practiced in the municipality of *Arraial do Cabo*) with industrial fisheries (practiced in the municipality of *Cabo Frio*) (FAOUN, 2009).

2. LCA AND SUSTAINABLE FISHING

2.1. The Fishing Industry

Over the past fifty years, ocean fishing has grown significantly across the world. In 2006, the total volume fished reached around 157 thousand tons, as shown in Figure 1, and it is likely to continue rising in the coming decades (FAOUN, 2009).

In Brazil, as shown in Figure 2, fish production is relatively low when compared to the industry worldwide. It declined sharply in the 1990s and only started to grow again in the 2000s, with increased seafood production in the northeast and south of the country.

Carneiro (2000) holds that even with the drop in output in the 1990s, the 700,000 tons produced in 1995 not only met domestic demand in Brazil, but was responsible for generating a positive trade balance of 160 million dollars. Thus, fisheries are a source of wealth and employment for the country.

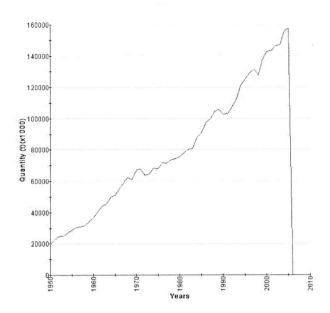

Figure 1. Global Fishing Output (FAOUN, 2009).

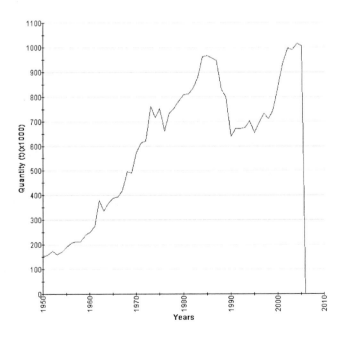

Figure 2. Fishing output in Brazil (FAOUN, 2009).

Though marine resources are renewable, they are not immune from depletion, since there is a limited number of fish that can live in the oceans. Little exploited species are becoming increasingly scarce as the result of intensive fishing and overfishing, motivated by the strength o the fishing industry (Ziegler, 2007).

In response to this worrying state of affairs, in the early 1990s, the international community addressed several issues of concern to world fisheries with a view to developing a model for sustainable fishing. At this stage, a number of issues were highlighted: *a)* a reduction of overfishing and control of intensive fishing; *b)* a reduction of accidental capture / associated fauna; *c)* a reduction of the environmental degradation of coastal areas and/or capture areas and *d)* ways of reducing the uncertainty and risks inherent to fishing.

2.2. LCA

LCA is a technique used to assess the potential environmental impacts and issues associated with the life cycle of a product. It covers all stages, from the acquisition of raw materials to production and the disposal of the product after its use. It is, in this sense, a method that covers the whole cycle "from the cradle to the grave".

LCA requires the compilation of an inventory of the inputs and outputs of a production system, an evaluation of the potential environmental impacts associated to these inputs and outputs, and an interpretation of the findings from the inventory analysis and impacts concerning the objects of study. For this reason, it has already been widely used in EU countries and the USA, as it is considered the best environmental management tool that currently exists for Industrial Ecology in the quest for sustainability (Chehebe, 1997).

Chehebe (1997) defends the use of LCA for: *a)* evaluating the whole life cycle of a product from an environmental perspective; *b)* comparing two or more products in their production chains; *c)* comparing different manufacturing processes for a given product; *d)* improving processes by reducing aspects with an environmental impact, such as cutting the amount of raw materials, natural resources or energy consumed; *e)* analyzing the feasibility of introducing alternative industrial processes and; *f)* developing the marketing of products and companies.

Around the globe, companies have to take environmental responsibility for their products throughout their life cycle. This includes not just environmental issues, but also social and economic impacts throughout these cycles to assure they are managed sustainably.

2.3. LCA in Fisheries

The new strategy of research themes in the European Union recognizes the reduction of negative environmental impacts as a key component for sustainable development in Europe and the rest of the world (Thrane, 2006). In all research areas, the strategy promotes the use of LCA to identify and quantify environmental impacts throughout products' life cycles.

Some institutions have created accreditation seals based on the LCA of fishing products. These include the Marine Stewardship Council (MSC), the Soil Association and Monterey Bay Aquarium. The last of these has created a seafood guide for consumers called Seafood Watch, which addresses typical consumer queries concerning the topic, such as where different species come from, when to buy different fish products, how fish are caught or farmed, and which fishing methods are used for different products. The guide provides the findings of research and also helps inform consumers so they can become more environmentally aware, thereby encouraging the sustainability of fishing. The results of these assessments of the fishing industry show that for most fish species, it is during the fishing stage (capture) that the greatest potential environmental impact takes place, followed by processing and consumption. Figure 3 shows the seafood life cycle.

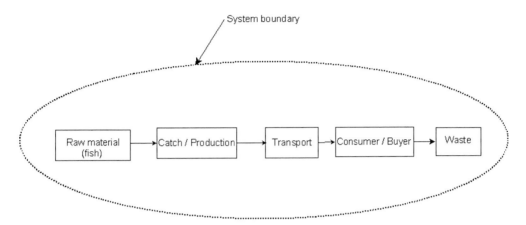

Figure 3. Fishing LCA (System boundary) (Matson, 2002).

According to Thrane (2006), the LCA of sole fishing in Denmark shows that the greatest impacts are during the capture stage, followed by the sale and consumption processes. Capture accounts for over half of the global warming potential, 90% of the ozone depletion potential, around 60% of the acidification potential and almost all the water pollution potential (ecotoxicity). These impacts are related to the intensive consumption of fossil fuels and significant emissions of biocides from the antifouling paints used on the ships' hulls.

Ziegler (2007) made an LCA study of cod and lobster fishing in Sweden. He found that in both cases, capture is the phase during which most environmental impact occurs, and that there are marked differences between the environmental impacts of different fishing methods. Hospido and Tyedmers (2005) compared the performance of tuna fishing in three oceans (the Pacific, Atlantic and Indian). The results showed that tuna fishing in the Pacific has a greater environmental impact than in the other oceans. This is because more fuel is burnt because of the greater distance traveled between the fishing site and the port of Galicia in Spain.

Ellingsen and Aanond (2006) did a life cycle assessment of farmed prawns and found that within this life cycle, the farming stage had the greatest impact. The main impacts of aquaculture are: eutrophication of the waters; destruction of benthos; disease caused by the transfer of parasites; and impacts caused by fishing, since aquaculture depends on fishing to produce the feed for the farmed species. Paradoxically, most of the regulations that exist in Europe focus on the fishing industry (production / processing). The authors therefore recommend that the scope of fishing be broadened so that regulations can be created that focus on fishing methods and energy consumption during the capture phase, where the greatest impacts in the life cycle are mostly caused.

3. Methodology

The study was conducted in the coastal areas off two neighboring municipalities in the State of Rio de Janeiro. The objective was to compare the global warming potential of industrial and artisanal fishing. Industrial fishing is carried out in *Cabo Frio* city, while artisanal fishing is done in *Arraial do Cabo* city. The data on artisanal fishing were obtained from primary sources (interviews with local fishermen) and secondary sources (results available in reports from the *Ressurgência* Project sponsored by the *Petrobras Ambiental* program[1] and conducted by Federal University of Rio de Janeiro). The data on industrial fishing were collected from an industrial fishery from *Cabo Frio*.

The data collection and analysis stages followed the methodology of LCA and made use of Umberto, a software package that serves as a technological tool capable of generating an efficient and appropriate information system for the development of LCA. This software was used to process the data gathered, taking as its parameters the distance traveled by the fishing boats and the capacity of the vessels used by each kind of fishing, and calculated the harmful gases emitted by these vessels.

In order to compare the two kinds of fishing, one ton was adopted as a standard measurement and the data was gathered on a single fish species captured by both means. Therefore, the functional unit was one ton (1t) of common sea bream *in natura*, meaning a whole fish, before it was gutted prior to sale.

The adoption of a functional unit together with the parameter as stipulated – based on the distance traveled and the kind of vessel – was decisive for the analysis. When common sea bream is caught by artisanal fishing, it is done inside the marine fishing reserve, which stretches six kilometers out towards open sea. The vessels can hold up to one ton of fish. According to the artisanal fishermen, each vessel normally catches about 400 km of common sea bream a day. Therefore, in order to capture one ton of fish using artisanal fishing, three vessels of one ton capacity are needed, traveling an average distance of 12 km, which includes the outbound and return journeys taken by the vessels, giving a total of 36 km for the three vessels. In industrial fishing, the vessels have an average capacity of between three and five tons. According to the company from *Cabo Frio*, they travel an average of 80 km, which includes the outbound and return journeys, and fishing lasts from three to five days. In this case, just one vessel is needed to capture one functional unit of study.

The process for manufacturing ice (figures 3 and 4) was not considered for the effects of the inventory calculation, since the process is similar in both kinds of fishing, as is the quantity of ice used. This means that for the purposes of comparison, the data will be cancelled out in terms of their relative global warming potential, as the same amount of emissions will be involved in each case.

In industrial fishing, the fish are wrapped in polyethylene plastic bags in 2kg-5kg batches before ice is added. The emissions calculation for the manufacture of the packaging was taken from a database that exists in the Umberto software package. These emissions were taken into account for the effects of comparing the two kinds of fishing practices.

The consumer markets for the products of these two kinds of fishing are significantly different. Most of the artisanally caught fish is destined for consumption in the city of Rio de Janeiro, while the industrially caught fish is sold on the European market. However, in this

1 Petrobras S.A. is the biggest Brazilian company.

research, for the purposes of comparison, it is supposed that the fish caught by both means is for sale in Rio de Janeiro. For this reason, in the calculations, the emissions related to the distribution and consumption stages were not included, as they would cancel each other out (figures 3 and 4). It should be noted, however, that at the consumption stage, industrial fishing generates far more solid waste because of the plastic packaging.

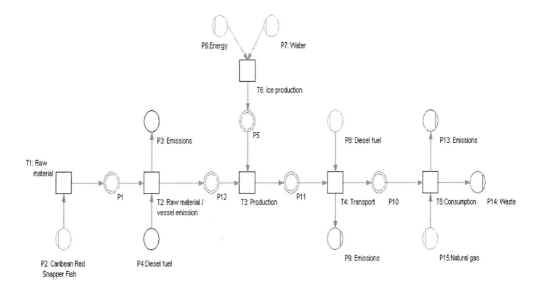

Figure 3. Artisanal Fishing Scenario.

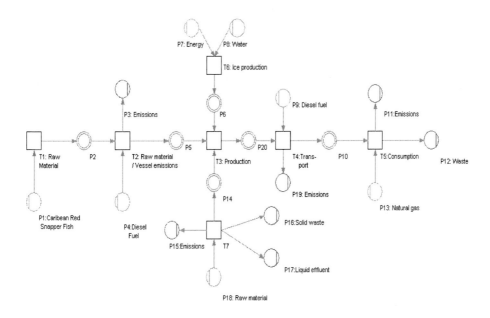

Figure 4. Industrial Fishing Scenario.

4. RESULTS AND DISCUSSION

Two material and energy flow networks, referred to as scenarios, were produced by using the work methodology in conjunction with the Umberto software package. Figure 3 shows the artisanal fishing scenario while Figure 4 shows the scenario for industrial fishing. Based on these scenarios, the gas emissions were calculated and the results expressed as a life cycle inventory (LCI), which is supplied by the software.

During the gas emission calculation and analysis, three substances with global warming potential were identified: carbon dioxide (CO_2), methane (CH_4) and nitrous oxide (N_2O). These substances are grouped in Table 1 to make it easier to compare the results.

Table 1. Quantification of the emissions with global warming potential per ton of common sea bream for both kinds of fishing.

Emissions	Artisanal Fishing Volume in kg per ton of fish (Qs)	Industrial Fishing Volume in kg per ton of fish (Qs)
CO_2	94.78	169.91
CH_4	0	0.33
N_2O	0.009	0.01

Using the graphic capability of the software, a comparison was made between the two kinds of fishing shown in the three graphics shown below. Figure 5 shows a comparison between carbon dioxide emissions for both kinds of fishing, while Figure 6 shows a comparison of methane emissions and Figure 7, a comparison of nitrous oxide emissions.

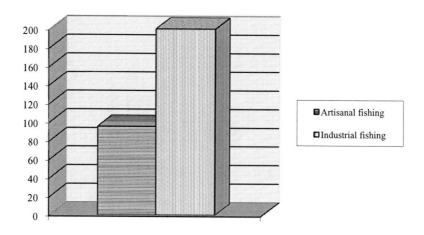

Figure 5. Comparison of carbon dioxide emissions by artisanal and industrial fishing.

The results of the study show that even though artisanal fishing makes use of three vessels to catch one ton of common sea bream, it is still responsible for far lower carbon dioxide emissions than industrial fishing.

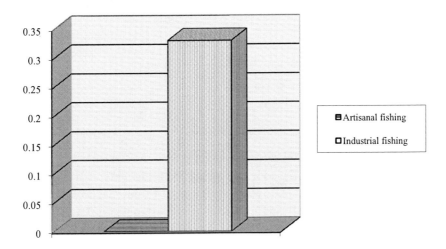

Figure 6. Comparison of methane emissions by artisanal and industrial fishing.

The methane emissions for artisanal fishing were negligible, while industrial fishing was responsible for a considerable amount emitted during the production of the polyethylene packaging used for the fish.

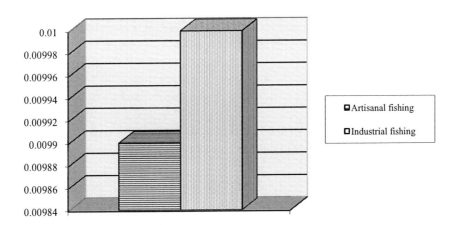

Figure 7. Comparison of nitrous oxide emissions by artisanal and industrial fishing.

Both kinds of fishing were found to emit the same amount of nitrous oxide into the atmosphere. In order to analyze the relative impacts on global warming, the global warming potential of each of the substances was calculated using the IPCC (Intergovernmental Panel on Climate Change) table of carbon dioxide equivalents (CO_{2eq}). The global warming potential of each of the gases encountered in this study is grouped in Table 2.

D. P. Souza, K. R. A. Nunes, R. Valle et al.

**Table 2. Global Warming Potential of the greenhouse gases encountered
in the study (IPCC, 2006).**

Greenhouse gases encountered	Global Warming Potential (in CO_{2eq})
CO_2	1
CH_4	25
N_2O	320

By combining the quantity of each gas emitted in kilos (Qs) per ton of common sea bream caught (shown in Table 1) with the global warming potential in terms of carbon dioxide equivalent (CO_{2eq}) supplied by the IPCC table (Table 2), we get a single CO_{2eq} indicator for the purposes of analyzing the global warming impact of both kinds of fishing. The global warming impact indicator (GWII) is obtained by formula 1,

$$GWII = \sum_{s=1}^{s=n} Qs \ x \ GWPs \qquad (1)$$

where:

GWII: Global Warming Impact Indicator;
Qs: Quantity of each substance (in kilos):
GWPs: Global Warming Potential (in CO_{2eq}) of each gas.

The results of using formula 1 to identify the GWII per ton of common sea bream caught by artisanal and industrial fishing are shown in Table 3. It can be seen that the GWII for artisanal fishing is 97.98 kg CO_{2eq}, while the same indicator for industrial fishing is 181.36 kg CO_{2eq}. This shows that industrial fishing, under the conditions studied, has an 85% higher global warming potential than artisanal fishing. These results can be visualized in Figure 8.

**Table 3. Calculation of the GWII of artisanal and industrial fishing
per ton of common sea bream caught.**

Gas (1)	Global Warming Potential (in CO_{2eq}) (2)	Artisanal Fishing		Industrial Fishing	
		Quantity (3)	Global Warming Potential (4) (4) = (2) x (3)	Quantity (5)	Global Warming Potential (6) (6) = (2) x (5)
CO_2	1	94.78	94.78	169.91	169.91
CH_4	25	0	0	0.33	8.25
N_2O	320	0.01	3.2	0.01	3.20
Σ			97.98		181.36

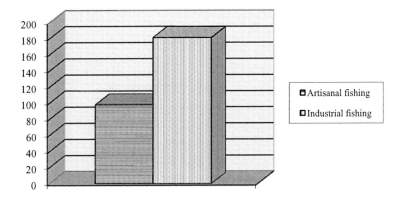

Figure 8. Comparison of the relative global warming potential of artisanal and industrial fishing.

5. FINAL CONSIDERATIONS

The objective of this study was to use LCA to study a fishing area off the northern coast of the state of Rio de Janeiro, in the municipalities of *Cabo Frio* and *Arraial do Cabo*, in order to compare the global warming potential of industrial and artisanal fishing, both of which are carried out there.

The Umberto™ software package was used to treat the data obtained by monitoring the life cycle of these activities. The results permitted comparisons to be made of the carbon dioxide, methane and nitrous oxide emissions of both kinds of fisheries. The combined results of the study showed that industrial fishing has a greater impact on global warming, and is less environmentally friendly than its artisanal counterpart.

The calculations of carbon dioxide, methane and nitrous oxide emissions from the vessels and from the manufacture of the polyethylene packaging were carried out using indicators developed in European studies (taken from the software's database) and not using specific indicators developed for Brazil. Nonetheless, these results serve as a preliminary analysis for future studies, which could focus on analyzing the impacts of the different kinds of fishing practices, analyzing the impacts of these gases on the existing stocks of species, and analyzing the impacts of waste caused by processing the fish.

As for the LCA methodology applied to fishing, it is concluded that it is a good tool for identifying and evaluating environmental aspects, as it employs a technology capable of offering a broad vision of the whole production chain, providing helpful inputs for the management of this chain.

It is hoped that this chapter will encourage the introduction of LCA to the monitoring and management of fishing in Brazil and in some developing countries with similar conditions, where the environmental impacts of fishing activities are not subject to scrutiny.

REFERENCES

ABNT (Associação Brasileira de Normas Técnicas) (2001). NBR 14040: Gestão Ambiental: Avaliação do ciclo de vida – princípios e estrutura. Rio de Janeiro: ABNT.

Carneiro, A. M. & Pimenta, E. G. (2000). O Trabalho da Pesca: Segurança, Saúde e Integração. Rio de Janeiro: PRO UNI-RIO/UNILAGOS.

Chehebe, J. R. (1997). Análise do ciclo de vida de produtos: Ferramenta gerencial da ISO 14000. Rio de Janeiro: Qualitymark.

Ellingsen, H. & Aanondsen, S, A. (2006). Environmental Impacts of Wild Caught Cod and Farmed Salmon –A Comparison with Chicken. Int. J. LCA 1 (1) 60 – 65.

FAOUN (Food and Agriculture Organization of the United Nations). Fisheries Statistics & Information. Available on: http://www.fao.org/fishery. Access on: Feb. 2, 2009.

Hospido, A.& Tyedmers, P. (2005). Life cycle environmental impacts of Spanish tuna fisheries. *Fisheries Research* 76 - 174–186.

Hunt, R. & Franklin, E. (1996). LCA How it Came About. Personal Reflections on the Origin and Development of LCA in the USA. Int. J. LCA, vol. 1 (1) 4-7. Landsberg, Germany: Ecomed.

IPCC (Intergovernmental Panel on Climate Change). (2006) Guidelines for National Greenhouse Gas Inventories, Switzerland. Available on: http://www.ipcc.ch/ipcc-reports/methodology. Access on: February 5, 2009.

Mattson, B (ed.) (2004) Environmental Assessment of Seafood Products through LCA. Copenhagen: TemaNord.

MSC (Marine Stewardship Council) (2007) Information. Available at: http://www. msc.org/. Access on July 15, 2007.

MBA (Monterey Bay Aquarium) (2007). Seafood Watch. Available at: http://www.-montereybayaquarium.org/cr/seafoodwatch.asp. Access on: July 17, 2007.

Mourad, A. L; Garcia, E. E. C.; Vilhena, A. (2002). Avaliação do ciclo de vida: Princípios e aplicações. Campinas: CETEA/CEMPRE.

Pelletier, N.; Ayer, N.; Tyedmers, P. (2006). Impact Categories for Life Cycle Assessment Research of Seafood Production Systems: Review and Prospectus. Int. J. LCA. Available at: http://dx.doi.org/10.1065/lca2006.09.275

Projeto Resurgência (2008). Relatório Parcial II. Available at: http://www.-ressurge ncia.org.br/index.php?option=com_content&task=blogcategory&id . Access on: May 3, 2008.

Soil Association. (2007). *Soil* Association Certification Limited (SA Certification). Available at: www.soilassociation.org/certification. Access on: July 5, 2007.

Thrane, M. (2004). Environmental impact from Danish fish products: Hot spots and environmental policies. Ph.D. dissertation, Department of Development and Planning, Aalborg University, Denmark.

Umberto (2008). Life Cycle Assessment Software. Germany.

Ziegler, F. & Hansson, P.A. (2003). Emissions from fuel combustion in Swedish cod fishery. J. Cl. Prod 11, 303–314.Ziegler, F; Nilsson, P.; Mattsson, B.; Walther, Y. (2003). Life Cycle Assessment of Frozen Cod Fillets Including Fishery-Specific Environmental Impacts. Int. J. LCA 8 (1) 39–47.

In: Encyclopedia of Environmental Research ISBN: 978-1-61761-927-4
Editor: Alisa N. Souter © 2011 Nova Science Publishers, Inc.

Chapter 22

THE CONSERVATION CONUNDRUM OF INTRODUCED WILDLIFE IN HAWAII: HOW TO MOVE FROM PARADISE LOST TO PARADISE REGAINED

Christopher A. Lepczyk and Lasha-Lynn H. Salbosa
Department of Natural Resources and Environmental Management,
University of Hawaii at Mānoa, Honolulu, HI, USA

"What is true of natural systems is also true of human ones: much as we might like to turn back the clock and rearrange land use as if on a tabula rasa, there are complex historical reasons why people live and work in the places they do, and it is usually best to manage with those reasons in mind rather than wishing the world were otherwise." Cronon 2000.

Prior to human settlement the only species of wildlife (i.e. terrestrial vertebrates) found in the Hawaiian Islands were birds and one mammal[1] (the Hawaiian hoary bat [*Lasirus cinereus semotus*], Beletsky 2006). Thus, due to the islands' extreme geographic isolation (ca. 3,500 km from any continental landmass), they were devoid of any mammalian herbivores, grazers, predators, amphibians, and reptiles (Beletsky 2006). The absence of these particular groups of species meant that birds, as the dominant terrestrial vertebrate, were able to exploit and adapt to the numerous tropical niches available. In fact, among the 109 endemic species of birds known to exist only in the Hawaiian Islands (Scott et al. 2001), they assumed top roles as carnivores, seed dispersers, and pollinators. When the first Polynesians arrived in the Hawaiian Islands sometime between 300 and 600 A.D., they introduced a number of wild and domestic animal species, both intentionally and accidentally, including pigs (or puaʻa; *Sus scrofa vittatus*), Pacific rats (*Rattus exulans*), and red junglefowl (or moa; *Gallus gallus*; Asquith 1995, Hawaii Audubon Society 1997). A millennium later (ca. 1778-1800) European explorers arrived and brought an additional suite of animals to the islands, including domestic

---

1 The Hawaiian monk seal [Monachus schauinslandi] is sometimes lumped into the wildlife classification, but due to its aquatic nature, we do not consider it among the terrestrial vertebrates. In addition, Tomich 1986 notes the possibility of a second bat based on fossil records, but this has not been validated to date.

livestock such as cattle, goats, sheep, and pigs, which were all large bodied mammalian grazers (i.e. ungulates). In the centuries since, people have continued to bring in new animal species for agriculture (e.g., expansion of goats and cattle), pets (e.g., cats and dogs), pest control (e.g., small Indian mongoose [*Herpestes auropunctatus*]), and hunting (e.g., mouflon sheep [*Ovis musimon*], wild turkey [*Meleagris gallopavo*]). In fact, the past two hundred years has seen an explosive increase in the number of all non-native species in Hawaii (Gagné 1988), such that today there are at least 200 (19 mammals, 53 birds, 28 amphibians and reptiles) introduced wildlife species in Hawaii (Eldredge and Evenhuis 2003).

The explosive increase in introduced wildlife, coupled with their widespread distribution across the ecosystems of Hawaii has led to a number of ecological repercussions, including the decline and/or extinction of many native plants and animals (Mountainspring and Scott 1985), habitat loss, and wholesale changes to the islands' ecosystems (Figure 1). These ecological repercussions can easily be understood in the context that native ecosystems of Hawaii were devoid of ground predators, competitors, large grazers, and many pathogens/diseases. In other words, many common types of ecological interactions present in other parts of the world (particularly continents) were either reduced or absent in Hawaii, and hence native species were not adapted to them. Consequently, birds and plants, along with the handful of insects that were able to colonize the Hawaiian Islands formed specialized relationships that were inherently susceptible to outside perturbations. Notably, though, no one knows exactly what fully intact ecosystems resembled or what all the species interactions might have been prior to human settlement, when very large bodied bird species existed and natural disturbance regimes were in place in Hawaii (Paul Banko, pers. comm.).

Figure 1. Landscape effects of ungulate presence in two different Hawaiian ecosystems. (A) Lowland dry grassland has lost much of its native shrub community due to ungulate browsing, as visible on the left side of the fence. (B) Montane wet forest illustrating encroachment of non-native grasses [larger stand of soft rush (*Juncus effuses*) shown on the left and the smaller kikuyu grass (*Pennisetum clandestinum*) shown in the foreground and right], due to soil disturbance caused by pig activity (Molokai, Hawai'i 2007; photos by L.H. Salbosa).

While the evidence continues to grow about the negative repercussions of introduced animals in Hawaii (Spatz and Mueller-Dombois 1973, Scowcroft and Giffin 1983, Scowcroft and Sakai 1983, Scowcroft and Hobdy 1987, VanderWerf 1993, Drake and Pratt 2001), there remain a number of disparate views on both the animals and their impacts to Hawaiian ecosystems. These disparate views stem from the current wide ranging attitudes, behaviors, and cultural history of the various groups of people and organizations (i.e. stakeholders) present in Hawaii, making natural resource management extremely difficult (Regan et al. 2006). For example, major stakeholders include land preservation organizations (e.g., The Nature Conservancy), large private landowners (e.g., Princess Bernice Pauahi Bishop Estate), natural resource agencies (e.g., U.S. Geological Survey, U.S. Department of Agriculture Forest Service, Hawaii Department of Land and Natural Resources), hunters, native Hawaiians, the U.S. military (all major branches), animal welfare groups (e.g., Hawaii Humane Society), conservation biologists and ecologists, ecotourists, and the general public (which itself can be divided by island or rural vs. urban; Figure 2). Because some stakeholders are able to manage their own properties without input from other stakeholders, coupled with the fact that each stakeholder may have differing goals, missions, and political agendas, there exists a great deal of disagreement and disenfranchisement about how to manage the introduced animals in Hawaii. As a result, there is little, if any, consensus on how to manage the introduced animals across the Hawaiian landscapes at present. Without any consensus or decisions, management will continue to be applied haphazardly, ultimately resulting in continued distrust among stakeholders and further degradation of the ecosystems and landscapes (*sensu* Regan et al. 2006).

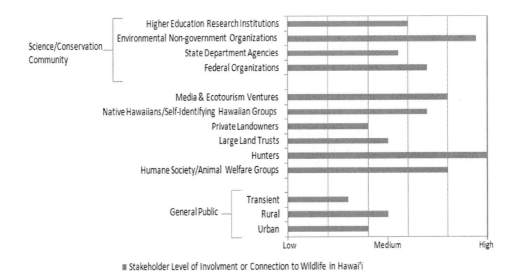

Figure 2. Stakeholders in the management of Hawaii's wildlife and their perceived level of involvement. Each stakeholder involvement level may vary due to cultural heterogeneity and value system. Moreover, stakeholder groups presented here are not necessarily mutually exclusive, but rather represent affected societal groups polarized by an emotion of shared identity and common experiences or interests.

In order to understand why such a lack of consensus exists in Hawaii, it is essential to have a basic understanding of the cultural context of the islands. In particular, it is key to understand the historical context of the Hawaiian Islands and how that has led us to our current situation. Hawaii is considered to be one of the last places on Earth that humans settled (J. Cromartie, pers. comm.), dating back to Polynesian arrival. Over a millennium later, a second influx of people began arriving in Hawaii from Europe and other western civilizations after Captain James Cook's first landing in 1778 (Bellwood 1979). European and American seafarers and explorers were met by Polynesian subjects interested in exotic goods (i.e. firearms and large livestock) and a sovereign Hawaiian Kingdom that ruled judiciously over the islands from 1810 to 1893. This influx of newcomers began as a religious and explorative conquest, subsequently fueling an agricultural economy (i.e. pineapple and sugarcane) which demanded an immigrant workforce of cheap labor to harvest various agriculture products. Over time there were cultural influxes in the immigrant workforce (e.g., from Chinese and Japanese, to Portuguese and Pilipino) as plantations sought out the least expensive labor pool. But agriculture was not strictly limited to crop production, and in fact there were extensive ranching operations throughout the Hawaiian Islands (Maly and Wilcox 2000), with 1.4 million acres of land being used for pasture in the 1900's at the height of the rangelands (Starrs 2000). Recognizing the importance of forested areas for the protection of water resources, the acting Board of Commissioners of Agriculture and Forestry limited free-ranging cattle by means of fences and sought to reforest upland areas with temperate species, including pine, spruce, and fir to support a commercial timber industry (Hosmer 1906, Blackman 1906).

The European explorers and subsequent settlers to the Hawaiian Islands did not simply transform the landscape, but had many marked impacts on the native Hawaiian population. First, the explorers and settlers inadvertently brought with them many diseases and pathogens that were novel to both the Hawaiians and the other endemic species. As a result, there was great reduction in the native Hawaiian population. Second, was the illegal overthrow of the sovereign ruler of the Hawaiian Kingdom in 1893 by a small group of men, mostly American by birth or heritage, with vested interests in sugar plantation profits and connections to U.S. military resources (most notably U.S. Minister to the Hawaiian Kingdom, John L. Stevens; Liliuokalani 1898, Castle 1913). Following the overthrow, Hawaii became a territory of the United States and with it the continental models of natural resource management were allowed to expand (Newell 1909).

The twentieth century saw continual importation of North American and European models of natural resource management (i.e. forestry and wildlife) that further altered the fragile tropical ecosystems of Hawaii. For instance, many non-native tree species were planted extensively as reforestation efforts and to stop soil erosion. Similarly, during the post World War II era there was a great deal of game species introductions to support public hunting. Thus, species such as pronghorn antelope (*Antilocapra americana*), mouflon sheep, and kalij pheasant (*Lophura leucomelana*), were introduced. However, by the second half of the twentieth century, and increasing especially in the past 30 years, the pendulum has been swinging from a more continental view of natural resource management and plantation agriculture to one of island conservation and protection. For instance, there has been protection of lands by both government agencies and non-profit organizations, a halting of the intentional introduction of new wildlife species, and the formation of invasive species eradication programs (Stone and Anderson 1988). But change does not come easy and the varied history and cultural context of Hawaii have created a perfect storm for conflicts to occur among stakeholders. So who are the stakeholders at present and what is their worldview as related to introduced wildlife species?

Today the seven main islands of Hawaii harbor a culturally diverse human population, of which nearly 30% live in the City of Honolulu (the most remote urban area in the world; U.S. Census Bureau). Its economy has switched from primarily agriculture to tourism, which brings millions of visitors to the islands each year. This economic shift towards a visitor industry, together with seven active military installations throughout Hawaii, has resulted in a large proportion of the population being transient (McDermott et al. 1980). Without the local media and agency leadership, this large transient population can become inherently disconnected from the conservation of Hawaii's natural resources. As for the host culture, it became evident in the 1970's and 1980's of a Hawaiian cultural resurgence, beginning with the formation of the Merrie Monarch Festival, the Polynesian Voyaging Society, and establishment of the State Office of Hawaiian Affairs. Similarly, native Hawaiian community groups continue to assert their hunting and gathering rights to the State of Hawaii's Department of Land and Natural Resources. In addition, a great deal of attention has been focused in the last twenty years on improving the socioeconomic well-being of many native Hawaiians, while respecting Hawaii's culturally diverse past. Meanwhile, the business community has struggled to diversify into more economically sustainable industries as the cost of living in Hawaii reaches among the highest in the nation (State of Hawaii Data Book 2007). One repercussion of this high cost of living and struggling business climate is that as financial resources are stretched to capacity the environment and natural resources suffer. Aside from the cultural and business considerations present in Hawaii, is the fact that while there are many small landowners across the state, a large portion of the total land surface area is held by a very small group of owners. These landowners include the U.S. government (in various forms from military through national parks), the Bishop Estate, and several large agricultural corporations (e.g., Dole Food Company, operating under Castle & Cooke, Inc.) and business ventures (e.g., Alexander & Baldwin, Inc.). Finally, because of the extremely unique geography, geology, and biology of Hawaii, there is a diverse scientific and conservation community comprised of research institutions, environmental non-profit organizations, state agencies, and federal organizations. All of these various stakeholder groups' influence, or are influenced by, introduced wildlife and thus must be part of the solution.

SOLUTIONS

The question we face today is what actions are needed to move forward on the problem of introduced animals in Hawaii? Some stakeholder groups assert that the topic of introduced animals in Hawaii is separate from the conservation of the islands' natural resources. However, because the former resides in and affects the later, both topics must be addressed concurrently. Furthermore, barring the exodus of people from Hawaii, any solution represents a trade-off of trying to create a balance between endemism and culture as it is present today. In other words, can some degree of endemism be retained, along with the accompanying ecosystem services they provide, while preserving a human dimension that is socially acceptable and aesthetically pleasing? This trade-off is neither simple nor easy, but nonetheless is central to moving forward. Admittedly, then, there is no perfect or correct answer to introduced animals as any direction is fraught with challenges. But from a vantage point removed from many of the stakeholders, the following are in our opinion the most apparent answers.

1. **Leadership**. Because there is no single decision on the management of introduced animals that will be socially acceptable to all stakeholders in Hawaii, it is critical to have strong and effective leaders in natural resources that are committed to long-term solutions. While leadership may seem obvious, the current state of the natural resources community suggests that there is a definitive lack of it. In fact, low morale, the transient nature of many natural resource professionals, apathy at the state of islands' landscapes, personality conflicts, and infighting among organizations is readily apparent. That said, our goal is not to point fingers or single out any stakeholder group, but rather to highlight that such problems are indicative of a lack of leadership. This leadership must occur across organizations and likely needs to be top-down. Furthermore, effective leadership requires that individuals within the organizations support and respect the decisions made at higher levels.

2. **Management Options**. Natural resources management with regard to wildlife is either non-existent or inconsistent throughout the islands. The islands have neither fully embraced a North America model of wildlife management, nor taken a complete control/eradication approach, such as occurs in New Zealand. As a result, what management of introduced animals exists in Hawaii is a haphazard quilt work dependent upon who owns the land, current laws, resources (i.e. money, people, etc.), court cases (e.g., Palila v. Hawaii Department of Land and Natural Resources, 1979[2]), and stakeholder interest. Because of this quilt work, some have even argued that management does not exist in Hawaii, only variations on control methods. Be that as it may, four management options that need to be applied in Hawaii are adaptive management, wildlife stakeholder acceptance capacity, decision theory, and evaluation of the pet trade. First, because of the harsh environmental conditions, extreme topographic variation present in the islands and uniqueness relative to other locations in the world, management must be conducted within a scientific framework proposed by adaptive management. This framework includes an *a priori* design, time lines, goals and objectives, testable hypotheses, and accepts uncertainty as part of the world (Holling 1978, Walters 1986, Lee 1993, Gunderson, 1999). Second, stakeholder acceptance capacity is needed to investigate whether commonalities exist among stakeholders for management and control options (e.g., Riley and Decker 2000, Christoffel 2007, Lischka et al. 2008). Third, if management goals and/or control options cannot be achieved through a

---

2 Landmark Hawaii federal district court case filed in 1978 on behalf of the endangered bird species Palila (*Loxioides bailleui*). Plaintiffs, including the Sierra Club, National Audubon Society, Hawaii Audubon Society, and renowned environmentalist, Alan C. Ziegler all asserted that the Hawaii Department of Land and Natural Resources were in direct violation of the Endangered Species Act by maintaining populations of feral sheep and goats within the Mauna Kea Game Management Area, which includes most of the Palila's critical habitat. Political pressure, hunter opposition, and lack of clear scientific evidence, according to court proceedings have allowed this case to continue on for over two decades, with the most recent judicial hearing occurring in 1999 (Palila v. Hawaii Department of Land and Natural Resources, 1999).

stakeholder acceptance model or facilitation, then decision theory is needed. Decision theory is valuable in general as it can be used in a variety of contexts, from creating a hierarchical decision matrix of what management decisions to make or used to create scenarios of how different management options might turn out based on competing management needs (Regan et al. 2005, Regan et al. 2006). Finally, one management decision that needs to be seriously entertained is what to do in regards to the pet trade. Certain animals (e.g., snakes) are currently barred from entering Hawaii, but there are still many species allowed in that have the potential to escape and establish themselves (e.g., numerous birds). Thus, if we are truly discussing how to deal with introduced animals, we must also consider halting entry of new pets and changing policies for existing pets (e.g., house cats).

3. **Broader Scales**. With the rise of landscape ecology and global change biology there has been a sea change in our view of the world. No longer are plots or management units the scale that we must work in. Rather we need to view the entire landscape, including landscape context, spatial relationships and temporal trends. In addition, we must also consider scales above and below the one at which we are studying or managing, as suggested by hierarchy theory (O'Neill et al. 1986). Without a broader understanding of how systems change over time and how they are spatially related to one another, successful management will not be achievable in the long run.

4. **Monitoring**. One of the greatest difficulties faced in most locations around the world is the lack of monitoring data. While Hawaii does have a number of monitoring systems in place, they are primarily related to threatened and endangered species (e.g., the Hawaiian Forest Bird Surveys) or are part of national efforts (e.g., Audubon's Christmas Bird Count). What are needed, though, are estimates of *all* wildlife species across space and over time, including introduced animals, game species, and urban wildlife. Only by understanding where the species are located and how they may be changing over time, can we begin to understand anything about the species basic population dynamics. One current approach that is quickly gaining traction in the natural resources community to deal with the need for monitoring is citizen science. Not only can citizen science greatly assist with monitoring, but it also has the added benefit of building capacity.

5. **Capacity Building**. Ultimately if we are to make any headway in managing introduced animals, we need to build capacity among the stakeholders and public. If people do not value the native systems or understand the need to control introduced species, our management goals will fall on deaf ears. Hawaii is not alone in dealing with an increasingly urban population that is removed from nature and the outdoors. But it is imperative that we not let ignorance and indifference prevail. Real world wildlife management and control costs money, requires leadership, and needs a long-term commitment to accomplish. Thus, without buy-in to management through capacity building, we will continue to tread water. To build capacity we suggest increased public outreach through Extension agents and specialists, publication of white papers by organizations, media coverage, interactions with the media and direct engagement of the public by stakeholders. Other ways to build capacity include the formation of a state Conservation

Congress where the public has a direct role in shaping management, as originated by Aldo Leopold and carried out in Wisconsin[3].

6. **Creativity**. A colleague once mentioned that the problem with Hawaii is that we do not think big and we do not dream. Regardless of whether or not one believes this point of view about Hawaii is not particularly relevant, so much as the end meaning. That is, we need to be bold, we need to think differently, we need to think large, and we need to think long term. In other words, we need *creativity*. If old models of management do not work then it is time to try something new. Simply allowing the status quo to continue is much like sitting in the restaurant at the end of the universe and watching it end (*sensu* Adams 1980).

CONCLUSIONS

Dealing with introduced animals in Hawaii is neither new nor easy. But the fact of the matter is that the longer we put it off, the worse things will become. Thus, it is both startling and yet tragic to read the words of Wayne Gagné (1988) from twenty years ago who stated that "It is difficult to rank conservation priorities, because we do not have the luxury of time to sort them step-by-step." While significant real world conservation has occurred since Gagné's article, one cannot help but feel how little distance we have moved. If we did not have the time to rank conservation priorities two decades ago, then how can we do it today? The answer is that time has always been short in Hawaii, and thus we must try new management approaches, work towards unity (or at least acceptance) among stakeholders, educate the public, and bring in new leaders.

Hawaii is truly one of a kind, both culturally and ecologically. If we are unable to change course and deal with the introduced animals in some unified and concerted effort, we will continue to bleed endemic species and ecosystems until there is little left to differentiate us from any other homogenized place in the world. However, if we can successfully deal with introduced animals, then we can serve as a role model to the rest of the world and remain a truly unique place. The choice is ours.

REFERENCES

Adams, D. 1980. The Restaurant at the End of the Universe. Harmony Books.

Asquith, A. 1995. Alien species and the extinction crisis of Hawaii's invertebrates. *Endangered Species UPDATE* 12(6).

Beletsky, L. 2006. Travellers' Wildlife Guides—Hawaii. Interlink Publishing Group, Inc. Northampton, MA.

Bellwood, P.S. 1979. Man's Conquest of the Pacific: The Prehistory of Southeast Asia and Oceania. Oxford University Press, New York.

Blackman, L.G. 1906. An offer of practical assistance to tree planters. *The Hawaiian Forester and Agriculturist* 3:83–86.

Castle Jr., W.R. 1913. Hawaii Past and Present. Dodd, Mead and Company, New York, NY.

---

3 For further details see: http://www.dnr.state.wi.us/org/nrboard/congress/ [last accessed on November 13, 2008]

Christoffel, R.A. 2007. Using human dimensions insights to improve conservation efforts for the eastern massasauga rattlesnake (*Sistrurus catenatus catenatus*) in Michigan and the timber rattlesnake (*Crotalus horridus horridus*) in Minnesota. Ph.D. Dissertation, Michigan State University, East Lansing, MI. 286 pp.

Cronon, W. 2000. Resisting monoliths and tabulae rasae. *Ecological Applications* 10:673-675.

Drake, D.R., and L.W. Pratt. 2001. Seedling mortality in Hawaiian rain forest: the role of small scale physical disturbance. *Biotropica* 33:319–323.

Eldredge, L.G., and N.L. Evenhuis. 2003. Hawaii's biodiversity: a detailed assement of the number of species in the Hawaiian Islands. Bishop Museum Occasional Papers 76, 28 pages.

Gagné, W.C. 1988. Conservation priorities in Hawaiian natural systems. *BioScience* 38:264-271.

Gunderson, L. 1999. Resilience, flexibility and adaptive management—antidotes for spurious certitude? Conservation Ecology 3(1):7. [online] URL: http://www.consecol.-org/vol3/iss1/art7.

Hawaii. Dept. of Business, Economic Development and Tourism. 2007. The State of Hawaii Data Book: a statistical abstract. Honolulu, HI.

Hosmer, R.S. 1906. Forestry reports on the lands of Makua and Keaau, Oahu 1905. *The Hawaiian Forester and Agriculturist* 3:4–8.

Holling, C. S. (ed). 1978. Adaptive Environmental Assessment and Management. Wiley, NY.

Lee, K. N. 1993. Compass and Gyroscope: Integrating Science and Politics for the Environment. Island Press, Washington, D.C.

Liliuokalani, Q. Hawaii's Story. 1898. Lothrop, Lee, and Shepard. Boston, MA.

Lischka, S.A., S.J. Riley, and B.A. Rudolph. 2008. Effects of impact perception on acceptance capacity for white-tailed deer. *Journal of Wildlife Management* 72:502-509.

Maly, K., and B.A. Wilcox. 2000. A short history of cattle and range management in Hawaii. *Rangelands* 22:21-23.

McDermott, J.F., W. Tseng, and T. Maretzki. 1980. People and Cultures of Hawaii: A Psychocultural Profile. University of Hawaii Press.

Mountainspring, S., and J.M. Scott. 1985. Interspecific competition among Hawaiian forest birds. *Ecological Monographs* 55:219-239.

Newell, F.H. 1909. Hawaii – Its Natural Resources and Opportunities for Home-making. 60th Congress, 2d Session, Senate Document No. 668. Government Printing Office, Washington, D.C.

O'Neill, R.V., D.L. DeAngelis, J.B. Waide, and T.F.H. Allen. 1986. A Hierarchical Concept of Ecosystems. Princeton University Press, NJ.

Palila v. Hawaii Department of Land and Natural Resources, 471 F. Supp. 985 (D. Hawaii 1979).

Palila v. Hawaii Department of Land and Natural Resources, 73 F. Supp.2d 1181 (D. Hawaii 1999).

Regan, H.M., Y. Ben-Haim, B. Langford, W.G. Wilson, P. Lundberg, S.J. Andelman, and M.A. Burgman. 2005. Robust decision-making under severe uncertainty for conservation management. *Ecological Applications* 15:1471-177.

Regan, H.M., M. Colyvan, and L. Markovchick-Nicholls. 2006. A formal model for consensus and negotiation in environmental management. *Journal of Environmental Management* 80:167-176.

Riley, S.J., and D.J. Decker. 2000. Wildlife stakeholder acceptance capacity for cougars in Montana. *Wildlife Society Bulletin* 28:931-939.

Scott, J.M., S. Conant, and C. Van Riper, eds. 2001. Evolution, Ecology, Conservation, and Management of Hawaiian Birds: A Vanishing Avifauna. Cooper Ornithological Society, Allen Press, Inc., Lawrence, Kansas.

Scowcroft, P.G., and J.G. Giffin. 1983. Feral herbivores suppress mamane and other browse species on Mauna Kea, Hawaii. *Journal of Range Management* 36:638-645.

Scowcroft, P.G., and H.F. Sakai. 1983. Impact of feral herbivores on mamane forests of Mauna Kea, Hawaii: bark stripping and diameter class structure. *Journal of Range Management* 36:495-498.

Scowcroft, P.G., and R. Hobdy. 1987. Recovery of goat-damaged vegetation in an insular tropical montane forest. *Biotropica* 19:208-215.

Spatz, G., and D. Mueller-Dombois. 1973. The influence of feral goats on Koa tree reproduction in Hawaii Volcanoes National Park. *Ecology* 54:870-876.

Starrs, P.F. 2000. The millennial Hawaiian paniolo. *Rangelands* 22:24-28.

Stone, C.P., and S.J. Anderson. 1988. Introduced animals in Hawaii's natural areas. *Proceedings of the Vertebrate Pest Conference* 13:134-140.

Tomich, P.Q. 1986. Mammals in Hawaii. Second Edition. Bishop Museum Press, Honolulu, HI.

US Census Bureau. 2006. 2006 Population Estimates. GCT-T1-R. Population Estimates. Hawaii—Place. U.S. Census Bureau, Washington, D.C.

VanderWerf, E.A. 1993. Scales of habitat selection by foraging 'Elepaio in undisturbed and human-altered forests in Hawaii. *Condor* 95:980-989.

Walters, C. J. 1986. Adaptive Management of Renewable Resources. Macmillan, NY.

In: Encyclopedia of Environmental Research
Editor: Alisa N. Souter

ISBN: 978-1-61761-927-4
© 2011 Nova Science Publishers, Inc.

Chapter 23

UTILIZATION OF BOVIDS IN TRADITIONAL FOLK MEDICINE AND THEIR IMPLICATIONS FOR CONSERVATION

Rômulo Romeu da Nóbrega Alves[1], Raynner Rilke Duarte Barboza[2], Wedson de Medeiros Silva Souto[3] and José da Silva Mourão[1]

[1] Departamento de Biologia, Universidade Estadual da Paraíba, Avenida das Baraúnas, Campina Grande, Brasil
[2] Pós-Graduação em Ciência e Tecnologia Ambiental, Universidade Estadual da Paraíba, Avenida das Baraúnas, Campina Grande, Brasil
[3] Programa Regional de Pós-Graduação em Desenvolvimento e Meio Ambiente (PRODEMA), Universidade Estadual da Paraíba, Avenida das Baraúnas, Campina Grande, Brasil

ABSTRACT

Animals and products derived from different organs of their bodies have constituted part of the inventory of medicinal substances used in various cultures since ancient times. Regrettably, wild populations of numerous species are overexploited around the globe, the demand created by the traditional medicine being one of the causes of the overexploitation. Mammals are among the animal species most frequently used in traditional folk medicine and many species of bovids are used as medicines in the world. The present work provides an overview of the global usage of bovids in traditional folk medicine around the world and their implications for conservation. The results demonstrate that at least 55 bovids are used in traditional folk medicine around the world. Most of species (n=49) recorded were harvested directly from the wild, and only six species of domestic animals. Of the bovids recorded, 50 are included on the IUCN Red List of Threatened Species and 54 are listed in the CITES. By highlighting the role played by animal-based remedies in the traditional medicines, we hope to increase awareness about zootherapeutic practices, particularly in the context of wildlife conservation.

Keywords: bovids, traditional medicine, wildlife conservation.

INTRODUCTION

Traditional medicinal systems play a key role in health care around the world. The World Health Organization has estimated that 60–80% of the population of non-industrialized countries rely on traditional healthcare for their basic health care needs, either on its own or in conjunction with modern medical care. Much of the world's population depends on traditional medicine, especially within developing countries and the demand for traditional medicine is increasing in many countries [1].

A tremendous variety of plants and animals are used in the preparation of traditional medicines. The pharmacopoeia of folk societies as well as of traditional (such as those of the Chinese, Ayurvedic, Unani) and western medical systems contain thousands of uses for medicines made from leaves, herbs, roots, bark, animals, mineral substances and other materials found in nature [1-4].

Since immemorial times animals and products derived from different organs of their bodies have constituted part of the inventory of medicinal substances used in many cultures, and such uses still exist in ethnic folk medicine [4]. Testimony to the medical use of animals began to appear with the invention of writing. Archives, papyruses, and other early written historical sources dealing with medicine, show that animals, their parts, and their products were used for medicine [5].

Bovids form one of the most prominent families of herbivores and constitute an important group for human beings, most of all when it comes to food, economy and religion. In traditional folk medicine, Bovids are among the mammals species most frequently used in different localities along the world. In recent years, several works reported the use of this group species for medicinal purposes [6-13] justifying the cultural lore importance of that group in different human societies, especially in traditional folk medicine.

From the historical point of view, a variety of interactions established between humans and bovids contributed, in higher or less scale, for the development of several cultures as well as the human species survival. If in a prime moment the hunt of several animals such antelopes, gazelles, buffalos and many others bovid species settled one of the prominent manners of food obtaining, equally important was the animals' domestication to the sedentarization and formation of the first urban nucleus occurred during the Holocene (Neolithic) [14-16]. Goats and sheep are among the earliest animals domesticated in southern and southwestern Asia during the early Holocene [15, 17, 18] and many evidences indicate that the first urban nucleus appeared in Asia are directly related to pastoral activities of flocks breeding (see Ponting [14]).

Unfortunately, many bovids species are found in Lists of Threatened Species, and for some of them, one of the reasons is the use and exploitation by humans. Currently, the family Bovidae includes 143 species in 50 genera represented by goats, sheep, gazelles, antelopes, goat-antelopes and cattle [19]. Because of such hunting, combined with loss of suitable habitat and other causes, most species are now considered to be endangered in the wild. According with IUCN [20], a large number of mammals are used in traditional folk medicine and many of these are of relevant to conservation. Understanding that the use of animals for medicinal purposes is part of a body of traditional knowledge which is increasingly becoming more relevant to discussions on conservation biology, public health policies, sustainable management of natural resources, biological prospection, and patents [4]. In this context, this study has focused on the global use of bovids in traditional folk medicine around the world and the implications for its conservation. We hope this chapter will serve as stimulus for further research about this use of biodiversity and its implications for bovids's conservation.

Methods

In order to examine the diversity of bovids used in traditional medicine, all available references or reports of folk remedies based on bovids sources were examined. Only taxa that could be identified to species level were included in the database. Scientific names provided in publications were updated according to the ITIS Catalogue of Life: 2008 Annual Checklist [21] and Mammal Species of the World (MSW) [19]. The conservation status of the bovids species follows IUCN [20] and CITES [22]. All the bovids sources used in traditional medicine were placed in references.

A total of 55 studies, official documents and other types of reliable publications supplied the information analyzed: Lev [5, 23], Mahawar and Jaroli [7, 8], Alves et al. [9], El-Kamali [10], Alves [11], Kakati et al. [12], Negi and Palyal [13], Adeola [24], Alakbarli [25], Almeida and Albuquerque [26], Alves and Rosa [27-29], Alves et al. [30], Apaza et al. [31], Australian Department of Environment and Heritage [32], Barboza et al. [33], Bensky et al. [34], Bryan [35], Cameron [36], Caprinae Specialist Group [37], Carpaneto and Germini [38], Chan [39], CITES [40-42], Costa-Neto [43-45], Costa-Neto and Oliveira [46], Dedeke et al. [47], Duckworth et al. [48], Enqin [49], Estes [50], Fischer and Linsenmair [51], Fitzgerald [52], Green [53], Green [54], Guojun and Luoshan [55], Homes [56, 57], Lavigne et al. [58], Lev and Amar [59], Li [60], Malik et al. [61], Marques [62], Marshall [63], Monroy-Vilchis et al. [64], Nilsson [65], Nunn [66], Pieroni et al. [67], Powell[68], Pujol [69], Rajchal [70], Ritter [71], Schaller [72], Schroering [73], Sheng [74], Silva et al. [75], Simelane and Kerley [76], Sodeinde and Soewu [77], Soewu [78], Solanki and Chutia [79], Souza [80], Stetter [81], The Chinese Materia Medica Dictionary [82], Thompson [83], Tierra [84], True [85], Truong et al. [86], Vázquez et al. [87], Walston [88], Zhang [89].

Results and Discussion

Bovids were hunted by humans from very early times and have been used for different purposes. The frequent depiction of bovids species in rock drawings attests to their importance to early hunters. Pre-historic societies used bovids and their products (primarily consumed as food), and the use of this animals has perpetuated throughout the history of humanity. In contemporary societies, bovids wild animals are used for a wide variety of finalities, such as food resources, pets, cultural activities, for medicinal and magic-religious purposes, and their body parts or sub-products are used or sold as clothing and tools. Bovids have been depicted in art and have played important roles in mythology and religion. The American bison is seen to have great spiritual importance by many Native Americans. Cattle have played an important part in many cultural and religious traditions, as they continue to do in Hinduism today. In Judaism, Islam, and Christianity, sheep and goats have symbolic roles. In Judaism, a shofar, a "trumpet" made from the horn of a ram or sometimes that of a goat or antelope, is sounded at Rosh Hashanah, the celebration of the New Year [90]. In Christianity, it is believed that the sounding of a shofar will announce the return of Christ.

The roles that bovids play in folk practices related to the healing and/or prevention of illnesses have been recorded in different social-cultural contexts worldwide. As evidenced in the present review, at least 56 species (39% of described bovids species) belonging to 29 genera and 8 subfamilies are used in traditional folk medicine. The subfamily with the largest numbers of species used were Caprinae (with 20 species), followed by Bovinae (14), and Antilopinae (9) (Table 1). These results were expected once the mentioned subfamilies are the most numerous in terms of bovid species (see Wilson and Reeder [19]).

Table 1. Felidae used in the wordwide traditional medicine.

BOVIDAE	IUCN 2008 Red List	2008 CITES Appendices
Subfamily Aepycerotinae		
Aepyceros melampus (Lichtenstein, 1812)	LC	
Subfamily Alcelaphinae		
Alcelaphus buselaphus (Pallas, 1766)	LC	
Connochaetes gnou (Zimmermann, 1780)	LC	
Subfamily Antilopinae		
Antilope cervicapra (Linnaeus, 1758)	NT	III (Nepal)
Gazella dorcas (Linnaeus, 1758)	VU	III (Algeria, Tunisia)
Gazella subgutturosa (Güldenstaedt, 1780)	VU	
Neotragus moschatus (Von Dueben, 1846)	LC	
Oreotragus oreotragus (Zimmermann, 1783)	LC	
Procapra gutturosa (Pallas, 1777)	LC	
Procapra picticaudata Hodgson, 1846	NT	
Raphicerus melanotis (Thunberg, 1811)	LC	
Saiga tatarica (Linnaeus, 1766)	CR	II

Table 1. Continued.

BOVIDAE	IUCN 2008 Red List	2008 CITES Appendices
Subfamily Bovinae		
Bison bison (Linnaeus, 1758)	NT	II
Bos frontalis Lambert, 1804		
Bos grunniens Linnaeus, 1766		
Bos javanicus d'Alton, 1823	EN	
Bos sauveli Urbain, 1937	CR	I
Bos taurus Linnaeus, 1758		
Bubalus bubalis (Linnaeus, 1758)		
Bubalus depressicornis (H. Smith, 1827)	EN	I
Pseudoryx nghetinhensis Dung et al., 1993	CR	I
Syncerus caffer (Sparrman, 1779)	LC	
Taurotragus oryx (Pallas, 1766)		
Tragelaphus euryceros (Ogilby, 1837)		
Tragelaphus scriptus (Pallas, 1766)	LC	
Tragelaphus strepsiceros (Pallas, 1766)	LC	
Subfamily Caprinae		
Ammotragus lervia (Pallas, 1777)	VU	II
Budorcas taxicolor Hodgson, 1850	VU	II
Capra falconeri (Wagner, 1839)	EN	I
Capra hircus Linnaeus, 1758		
Capra ibex Linnaeus, 1758	LC	
Capra sibirica (Pallas, 1776)	LC	
Hemitragus hylocrius (Ogilby, 1838)	EN	

Table 1. Continued.

BOVIDAE	IUCN 2008 Red List	2008 CITES Appendices
Hemitragus jemlahicus (H. Smith, 1826)	NT	
Naemorhedus baileyi Pocock, 1914	VU	I
Naemorhedus caudatus (Milne-Edwards, 1867)	VU	I
Naemorhedus crispus (Temminck, 1845)		
Naemorhedus goral (Hardwicke, 1825)	NT	I
Naemorhedus sumatraensis (Bechstein, 1799)	VU	I
Naemorhedus swinhoei (Gray, 1862)		
Ovis ammon (Linnaeus, 1758)	NT	I, II
Ovis aries (Linnaeus, 1758)		
Ovis canadensis Shaw, 1804	LC	II (Only the population of Mexico)
Ovis vignei Blyth, 1841		I, II
Pantholops hodgsonii (Abel, 1826)	EN	I
Pseudois nayaur (Hodgson, 1833)	LC	
Subfamily Cephalophinae		
Cephalophus jentinki Thomas, 1892	EN	I
Cephalophus maxwellii (H. Smith, 1827)	LC	
Cephalophus monticola (Thunberg, 1789)	LC	II
Cephalophus natalensis A. Smith, 1834	LC	
Cephalophus rufilatus Gray, 1846	LC	
Sylvicapra grimmia (Linnaeus, 1758)	LC	

Table 1. Continued.

BOVIDAE	IUCN 2008 Red List	2008 CITES Appendices
Subfamily Hippotraginae		
Hippotragus equinus (Desmarest, 1804)	LC	
Subfamily Reduncinae		
Kobus ellipsiprymnus (Ogilby, 1833)	LC	
Kobus kob (Erxleben, 1777)	LC	
Kobus leche Gray, 1850	LC	II

Legend: IUCN Red List of Threatened Species Categories – CR (Critically Endangered), DD (Data Deficient), EN (Endangered), EX (Extinct), EW (Extinct in the Wild), LR/cd (Lower risk: *Conservation dependent*), LR/lc or LC (Least Concern), LR/nt or NT (Near Threatened), VU (Vulnerable). CITES Appendices – I, II or III

Despite the fact that studies recording the use of bovids in traditional medicine are all relatively recent, an analysis of historical documents indicated that bovids have been used in traditional medicines since ancient times. Historical sources of ancient Egypt and of several civilizations of ancient Mesopotamia, mainly the Assyrian and the Babylonian, mention the medicinal uses of substances derived from bovids, for example, cattle milk, goat's skin, gazelle sinew, even sheep, and the glands of the musk deer [35, 50, 66, 68, 71, 81, 83]. Musk is used as a medicinal in Islamic countries, India and countries of the Far East [53]. Its use in medicine was recorded by Aetius, the Greek physician, in circa 520 AD (Pereira, 1857 quoted by Green [53]), and in China and India musk has been considered a superior medicine since the fifth century AD [53]. However, as early as the Han Dynasty (200 BC – 200 AD), records in Shennong Bencao Jing show the use of musk in traditional Chinese medicine (TCM) [70].

Some widespread species are used in different countries like *Naemorhedus sumatraensis* (Bechstein, 1799) in China and Lao PDR [48, 82] and probably in most of Asian Southeast countries; *Bubalus bubalis* (Linnaeus, 1758) and *Saiga tatarica* (Linnaeus, 1766) in China and India [7, 34, 84], *Ovis aries* (Linnaeus, 1758) in Brazil and Sudão [10, 27-29, 33], *Bos taurus* Linnaeus, 1758 and *Bos frontalis* Lambert, 1804 in Brazil and India [7, 8, 27-30, 44, 45].

A same species can be used in the treatment of different diseases. For instance, the Southeast musk continues to be a popular Traditional Chinese Medicine for the effective treatment of improving blood circulation and relieving ailments of the heart, nerves and breathing system. It can revive unconscious patients, stimulate circulation of vital energy and blood and it also possesses anti-inflammatory and analgesic effects. It is used in the treatment of delirium, stroke, unconsciousness, miscarriage, ejection of stillborn fetus, acute angina pectoris, acute abnormal pain, skin infection, sore throat, sprained joints, trauma, and paralysis. It occurs in three dominant forms, as oils or sprays, medicated plasters and raw musk powder [91]. The musk powder can be dissolved into water then used as a salve on parts of the body that suffer rheumatic pains. Musk powder can also be dissolved into wine or ginseng tea and consumed to improve the body's blood circulation. Preparation of musk can be used for both external and internal applications. Musk oils can be brought 'off the shelf' or blended with snake bile or various plant herbs. These are used to obtain immediate and short-term relief of rheumatic and muscular pains caused by blood stasis [70].

The domestic cattle - *Bos taurus*, perhaps the most well-known of the bovids, is an important source of remedies in traditional medicine. In Brazil, for instance, bone marrow, horns, medulla and penis of *B. taurus* are indicated for treatment of alcoholism, rib cage pain, male impotence, anaemia, dizziness, flu, pneumony, rheumatism, thrombosis, cough, asthma, sinusitis, sore throat, wounds and for removal of thorns [27-30, 44-46, 75]. In India, weakness due to fever is cure by drinking domestic cattle's urine and Ghee with black pepper are given orally to neutralize snake poison [8]. Ox gall was listed among the animals' drugs originated in Europe (see True [85]). In Viet Nam, gall, gall-stone and horn of *Bos taurus* are used in local folk medicine [92].

The effectiveness of most of the medicines from wild animals and their by-products has not been scientifically studied and proven and their potency in many cases may be questionable. As pointed out by Pieroni et al. [93], the chemical constituents and pharmacological actions of some animal products are already known to some extent and ethnopharmacological studies focused on animal remedies could be very important in order to clarify the eventual therapeutic usefulness of this class of biological remedies. However, research with therapeutic purposes into the products of the animal kingdom has been neglected until recently [4]. Another aspect that needs to be emphasized is that sanitary conditions of the zootherapeutics products generally were poor with obvious contamination risks to these products [9, 27, 29]. These observations point to the need for sanitary measurements to be taken with medicinal animal products and the importance of including considerations about zootherapy into public health programs.

Most of the species used (n = 51; 91%) are wild. In most cases remedies were prepared from dead specimens. Many of the medicinal animals are of conservation concern. Most of the recorded species (46 of 56) are included on the IUCN Red List of Threatened Species [20] (with 7 species classified as vulnerable and 6 are endangered, besides 1 critically endangered) and 21 are listed in the CITES list (Convention on International Trade in Endangered Species of Wild Fauna and Flora [22] (see Table 1).

Ingredients derived from these biological resources are not only widely used in traditional remedies, but are also increasingly valued as raw materials in the preparation of modern medicines and herbal preparations. As a result, harvest and trade of species for traditional medicines may pose a threat to their survival [94]. Traditional Chinese Medicine, for example, relies on animal and plant substances as raw ingredients for prescription medicine and manufactured patent pharmaceuticals [91].

Musk is a good example. It is considered one of the most frequently used animal products in traditional medicine practices [36]. The musk secreted by the musk gland of the males has been used in the perfumery industries for a long time for its intensity, persistence and fixative properties. In Asia, including China, it has also long been used in traditional medicine as a sedative and as a stimulant to treat a variety of ailments [53, 56, 74]. In China, Musk deer have been hunted for musk, and musk purchasing has been conducted in rural markets or via local medicine companies and the perfume industry perfume is produced based on natural musk, but production is not high at present [89]. The raw ingredients may be used directly after some preparation (grinding, washing, boiling, drying etc.) or may be made into factory processed forms such as plasters, pills or tablets and packaged in mass quantities for national or worldwide distribution.

In China, the effects of musk (*Moschus chrysogaster*) have been known in Traditional Chinese Medicine (TCM) for several thousand years, musk being used in about 300 pharmaceutical preparations [95]. China has a high domestic demand for musk [56], and this originates from both legal and illegal sources within the country. The total demand for musk is between 500 to 1,000 kg per year in China [74].

Musk (*Moschus* spp.) is currently used in as many as 400 Chinese and Korean traditional medicines to treat ailments of the circulatory, respiratory and nervous systems [96]. The demand for musk from China alone is between 500-1000 kg per annum needing extraction from pods of more than 100,000 male musk deer [96]. In Russia, 400-500 kg of raw musk was traded illegally between 1999 and 2000, needing extraction from 17,000-20,000 stags [57]. In 1987, 800 pounds of musk worth $14 million were smuggled out of China, the product of 53,000 male deer. More than 100,000 deer had been killed in the quest for these glands, since many of the dead deer were females and young which were discarded. The glands were exported to Japan [72]. An average of 700 pounds of musk are sold in world markets each year, much of it going to Hong Kong, the international center for musk; Japan is a major consumer, using musk to treat a variety of illnesses [52]. Between 1974 and 1983, Japan imported between 250 and 700 pounds of musk per year, worth an average of $4.2 million; imports increased in 1987 to 1,800 pounds, an all time high, and sold for $32,468 a pound [52].

The horns of Saiga antelope (*Saiga tatarica*), a species being slaughtered by the hundreds of thousands in Russia and Central Asia, are thought to cure many illnesses; in 1990, China imported 80 tons [72]. A trade study found that in 1994, 44 metric tons of Saiga Horn was exported illegally to China, South Korea, Japan and some European nations. One metric ton is equivalent to 5,000 horns; horn sold for as much as $30 per kilogram in East Asia [39]. In a random survey in August and September 1994, TRAFFIC International investigators found Saiga horn in 131 shops in Hong Kong, from an estimated 15,000 animals. Taiwan banned the sale of Saiga horn in 1994 [39]. Populations of this species have declined in Kazakhstan and Kalmykia and have become endangered in Mongolia. Today, the trade in Saiga horn is so uncontrolled and massive that it threatens the species' future survival [39].

Asian Red Deer (*Cervus elaphus*), known in North America as Elk, have been heavily exploited for their antlers to use in the Traditional Medicine trade, and many races are endangered [65]. A 19th century victim of the Traditional Medicine trade was Schomburgk's Deer (*Cervus schomburgki*). Discovered in eastern Thailand in 1862, no European ever saw the species in the wild [97]. They were heavily hunted for their large and many-tined antlers that supposedly possessed medicinal and magical properties [97]. In the mid-19th century, herds of Schomburgk's Deer were seen in swamps, and during floods, they were pursued by boat, marooned on small islands, and speared [97]. When swamp drainage and irrigation added to their threats, they retreated to bamboo jungles, to which they were not well adapted, until these, too, were cleared for rice fields [97].The last known Schomburgk's Deer was shot by a policeman in September 1932 [97]. The species was considered extinct and officially listed as such by the IUCN [98]. In 1991, a pair of antlers from an unknown type of deer was seen by Laurent Chazee, an agronomist with the United Nations, in a Traditional Medicine shop in a remote part of Laos [73]. Chazee photographed the antlers, which were later identified as coming from a Schomburgk's Deer; the shop owner told him that the animal had been killed the previous year [73]. Forests nearby may shelter more of these deer, and the site is considered by local people to have sacred animal spirits; hunting is prohibited there [73]. A shop in Phnom Penh, Cambodia, in February 1994 offered antlers of what were represented as Schomburgk's Deer for $10 a pair [99]. The seller was obviously unaware of the extraordinary rarity of this deer. There is still no proof that the species survives, and it has been listed as extinct in *2000 IUCN Red List of Threatened Species*.

Bovids are used in many forms of traditional medicine and uncontrolled trade may jeopardize both the long-term survival of some species of bovids and the maintenance of traditional medicine delivery. Although some information on the use of wildlife for medicinal purposes is available from published pharmacopoeias and ethno-biological studies, in most cases little is known regarding harvest and trade volumes, trade controls, market dynamics and conservation impacts [9].

Approximately 80% of the world's population relies on animal and plant based medicines for primary health care [100]. Increased demand and increased human populations are leading to increased and often unsustainable rates of exploitation of wild-sourced ingredients. The high demand for medicinal animals creates additional pressure on natural populations and most importantly on many endangered species that are noted to be in rapid decline. It is therefore imperative to record indigenous knowledge as related to the use of vertebrates and to devise strategies for sustainable utilization of these animals [47]. In this sense, as pointed by Alves et al. [30], the manner in which natural resources are used by human populations and cultural norms associated with that use are extremely relevant to the definition of possible conservation.

It is also important to consider that medicines extracted from animals and plants are significant and valuable resources since they are the unique available remedies for most of the human populations that do not have access to the industrialized drugs and medical care. The socio-cultural aspects are relevant when discussing sustainable development [4, 101]. This social perspective includes the way people become aware of natural resources, their utilization, allocation, transference, and management [102]. Thus, the inclusion of zootherapy in the multidimensionality of the sustainable development is interpreted as a fundamental component to achieve the sustainable use of faunal resources [103]. Soejarto [104] remarks that conservation permits the continuing use of the resources in ways that are non-destructive and sustainable, while from the pharmaceutical point of view, providing time to eventually demonstrate fully the medicinal value of the resources. There is an urgent need to examine the ecological, cultural, social, and public health implications associated with fauna usage, including a full inventory of the animal species used for medicinal purposes and the socio-cultural context associated with their consumption.

REFERENCES

[1]   Alves, RRN; Rosa, IML. Biodiversity, traditional medicine and public health: where do they meet? *Journal of Ethnobiology and Ethnomedicine*, 2007 3(14), 9.

[2] Good, C. Ethno-medical Systems in Africa and the LDCs: Key Issues in Medical Geography. In: Meade MS editor. *Conceptual and Methodological Issues in Medical Geography*. Chapel Hill, NC, USA: University of North Carolina; 1980.

[3] Gesler, WM. Therapeutic landscapes: medical Issues in Light of the new cultural geography. *Social Science & Medicine*, 1992 34(7), 735-746.

[4] Alves, RRN; Rosa, IL. Why study the use of animal products in traditional medicines? *Journal of Ethnobiology and Ethnomedicine*, 2005 1(5), 1-5.

[5] Lev, E. Traditional healing with animals (zootherapy): medieval to present-day Levantine practice. *Journal of Ethnopharmacology*, 2003 85, 107-118.

[6] Mahawar, MM; Jaroli, DP. Traditional knowledge on zootherapeutic uses by the Sahari tribe of Rajasthan, India. *Journal of Ethnobiology and Ethnomedicine*, 2007 3(25), 6.

[7] Mahawar, MM; Jaroli, DP. Traditional zootherapeutic studies in India: a review. *Journal of Ethnobiology and Ethnomedicine*, 2008 4(1), 17.

[8] Mahawar, MM; Jaroli, DP. Animals and their products utilized as medicines by the inhabitants surrounding the Ranthambhore National Park, India. *Journal of Ethnobiology and Ethnomedicine*, 2006 2(46), 5.

[9] Alves, RRN; Rosa, IL; Santana, GG. The Role of Animal-derived Remedies as Complementary Medicine in Brazil. *BioScience*, 2007 57(11), 949-955.

[10] El-Kamali, HH. Folk medicinal use of some animal products in Central Sudan. *Journal of Ethnopharmacology*, 2000 72, 279-282.

[11] Alves, RRN. Fauna used in popular medicine in Northeast Brazil. *Journal of Ethnobiology and Ethnomedicine*, 2009 5(1), 1-30.

[12] Kakati, LN; Ao, B; Doulo, V. Indigenous Knowledge of Zootherapeutic Use of Vertebrate Origin by the Ao Tribe of Nagaland. *Human Ecology*, 2006 19(3), 163-167.

[13] Negi, CS; Palyal, V. Traditional Uses of Animal and Animal Products in Medicine and Rituals by the Shoka Tribes of District Pithoragarh, Uttaranchal, India. *Ethno-Med*, 2007 1(1), 47-54.

[14] Ponting, C. A Green History of the World. 1st ed. London, UK: Sinclair-Stevenson Ltd, 1991.

[15] Gupta, AK. Origin of agriculture and domestication of plants and animals linked to early Holocene climate amelioration. *Current Science*, 2004 87(1), 54-59.

[16] Wyly, E. Urban Origins and Historical Trajectories of Urban Change. *Geography*, 2008(350), 1-10.

[17] Allchin, B; Allchin, R. Origins of a Civilization, The Prehistoric and Early Archaeology of South Asia. 1st ed. New Delhi: Penguin Books, 1997.

[18] MacDonald, GM. Biogeography: Introduction to Space, Time and Life. New York: John Wiley, 2003.

[19] Wilson, DE; Reeder, DM. Mammal Species of the World. Baltimore, USA: Johns Hopkins University Press, 2005.

[20] IUCN. IUCN Red List of Threatened Species [online]. 2008 [cited February 2009]. Available from URL: http://www.iucnredlist.org

[21] ITIS. ITIS Catalogue of Life: 2008 Annual Checklist [online]. 2008 [cited February 2009]. Available from URL: http://www.catalogueofife.org/search.php

[22] CITES. CITES Appendix [online]. 2008 [cited February 2009]. Available from URL: http://www.cites.org/eng/resources/species.html

[23] Lev, E. Healing with animals in the Levant from the 10th to the 18th cent. *Journal of Ethnobiology and Ethnomedicine*, 2006 2(11), 9.

[24] Adeola, MO. Importance of wild Animals and their parts in the culture, religious festivals, and traditional medicine, of Nigeria. *Environmental Conservation*, 1992 19(2), 125-134.

[25] Alakbarli, F. Medical Manuscripts of Azerbaijan. Baku, Azerbaijan: Heydar Aliyev Foundation, 2006.

[26] Almeida, CFCBR; Albuquerque, UP. Uso e conservação de plantas e animais medicinais no Estado de Pernambuco (Nordeste do Brasil): Um estudo de caso. *Interciencia*, 2002 27(6), 276-285.

[27] Alves, RRN; Rosa, IL. From cnidarians to mammals: The use of animals as remedies in fishing communities in NE Brazil. *Journal of Ethnopharmacology*, 2006 107, 259–276.

[28] Alves, RRN; Rosa, IL. Zootherapeutic practices among fishing communities in North and Northeast Brazil: A comparison. *Journal of Ethnopharmacology*, 2007 111, 82–103.

[29] Alves, RRN; Rosa, IL. Zootherapy goes to town: The use of animal-based remedies in urban areas of NE and N Brazil. *Journal of Ethnopharmacology*, 2007 113, 541-555.

[30] Alves, RRN; Lima, HN; Tavares, MC; Souto, WMS; Barboza, RRD; Vasconcellos, A. Animal-based remedies as complementary medicines in Santa Cruz do Capibaribe, Brazil. *BMC Complementary and Alternative Medicine*, 2008 8, 44.

[31] Apaza, L; Godoy, R; Wilkie, D; Byron, EH, O; Leonard, WL; Peréz, E; Reyes-García, V; Vadez, V. Markets and the use of wild animals for traditional medicine: a case study among the Tsimane' Amerindians of the Bolivian rain forest. *Journal of Ethnobiology*, 2003 23, 47-64.

[32] Department of the Environment, Water, Heritage and the Arts. Wildlife conservation and complementary medicines [online]. 2003 [cited December 2008]. Available from URL: http://www.environment.gov.au/biodiversity/trade-use/publications/traditional-medicine/index.html

[33] Barboza, RRD; Souto, WMS; Mourão, JS. The use of zootherapeutics in folk veterinary medicine in the district of Cubati, Paraíba State, Brazil. *Journal of Ethnobiology and Ethnomedicine*, 2007 3(32), 14.

[34] Bensky, D; Gamble, A; Kaptchuk, T. Chinese Herbal Medicine Materia Medica Revised Edition. 1st ed. Seattle: Eastland Press, 1993.

[35] Bryan, CP. Ancient Egyptian Medicine. The Papyrus Ebers. 1st ed. Chicago, USA: Ares, 1930.

[36] Cameron, G; Pendry, S; Allan, C; Wu(2004), J. Traditional Asian Medicine Identification Guide for Law Enforcers: Version II. 1st ed. Cambridge, UK: Her Majesty's Customs and excise, London and TRAFFIC International, 2004.

[37] Caprinae Specialist Group. Capricornis sumatraensis [online]. 1996 [cited February 2009]. Available from URL: http://www.iucnredlist.org/search/details.php/3809/all

[38] Carpaneto, GM; Germi, FP. The Mammals in the Zoological Culture of the Mbuti Pygmies in North-Eastern Zaire. *Hystrix*, 1989 1, 1-83.

[39] Chan, S; Madsimuk, AV; Zhirnov., LV. From Steppe to Store: The Trade in Saiga Antelope Horn. 1st ed. Cambridge, UK.: TRAFFIC International, 1995.

[40] CITES. List of animal species used in traditional medicine [online]. 2001 [cited February 2009]. Available from URL: http://www.cites.org/eng/com/AC/17/E17i-05Rev.doc

[41] CITES. Trade in products, possibly used as medicinals, of the species listed in document AC18 Doc. 13.1 [online]. 2002 [cited December 2008]. Available from URL: http://www.cites.org/common/com/ac/18/E18i-08.doc

[42] CITES. List of species traded for medicinal purposes [online]. 2002 [cited January 2009]. Available from URL: http://www.cites.org/eng/com/ac/18/E18-13-1.pdf

[43] Costa-Neto, EM. Faunistc Resources used as medicines by an Afro-brazilian community from Chapada Diamantina National Park, State of Bahia-Brazil. *Sitientibus*, 1996(15), 211-219.

[44] Costa-Neto, EM. Barata é um santo remédio : introdução à zooterapia popular no estado da Bahia. 1st ed. Feira de Santana, Brazil: EdUEFS, 1999.

[45] Costa-Neto, EM. Recursos animais utilizados na medicina tradicional dos índios Pankararés, que habitam no Nordeste do Estado da Bahia, Brasil. *Actualidades Biologicas*, 1999 21, 69-79.

[46] Costa-Neto, EM; Oliveira, MVM. Cockroach is Good for Asthma: Zootherapeutic Practices in Northeastern Brazil. *Human Ecology Review*, 2000 7(2), 41-51.

[47] Dedeke, GA; Soewu, DA; Lawal, OA; Ola, M. Pilot Survey of Ethnozoological Utilisation of Vertebrates in Southwestern Nigeria, Indilinga. *Afr J Indigenous Knowl Syst*, 2006 5(1), 87-96.

[48] Duckworth, JW; Salter, RE; Khounboline, K. Wildlife in Lao PDR: 1999 Status Report. 1st ed. Gland, Switzerland: IUCN, 1999.

[49] Enqin, Z. Rare Chinese Materia Medica. Shanghai, China: House of Shanghai College of Traditional Chinese Medicine, 1991.

[50] Estes, JW. The Medical Skills of Ancient Egypt. 1st ed: Science History Publications, 1989.

[51] Fischer, F; Linsenmair, KE. Demography of a West African kob (*Kobus kob kob*) population. *African Journal of Ecology*, 2002 40(2), 130-137.

[52] Fitzgerald, S. International wildlife trade: whose business is it? 1st ed. Washington, D.C.: WWF USA, 1989.

[53] Green, MJB. Aspects of the ecology of the Himalayan Musk deer. *Ph.D. thesis.* University of Cambridge, 1985.

[54] Green, MJB. Musk production from Musk deer. In: Hudson RJ, Drew KR, Baskin LM editors. *Wildlife Production System*. Cambridge, UK: Cambridge University Press; 1989.

[55] Guojun, X; Luoshan, X. The Chinese Materia Medica. Beijing: China Medicinal Science and Technology Press, 1996.

[56] Homes, V. On The Scent: Conserving Musk Deer – The Uses of Musk and Europe's Role in its Trade. 1st ed. Brussels, Belgium: TRAFFIC Europe, 1999.

[57] Homes, V. No licence to kill. The population and harvest of musk deer in the Russian Federation and Mongolia. 1st ed. Brussels, Belgium: TRAFFIC Europe, 2004.

[58] Lavigne, D; Wilson, PJ; Smith, RJ; White, BN: Pinniped penises in the marketplace: a progress report., 1999.

[59] Lev, E; Amar, Z. Ethnopharmacological survey of traditional drugs sold in the Kingdom of Jordan. *Journal of Ethnopharmacology*, 2002 82(2-3), 131-145.

[60] Li, S-Y. Habitat management and population change of Hainan Eld's deer. *Chinese Wildlife*, 2000 20(1), 2-3.

[61] Malik, S; Wilson, PJ; Smith, RJ; Lavigne, DM; White, BN. Pinniped Penises in Trade: A Molecular-Genetic Investigation. *Conservation Biology*, 1997 11(6), 1365-1374.

[62] Marques, JGW. Pescando Pescadores: Etnoecologia abrangente no baixo São Francisco Alagoano. 1 st. ed. São Paulo, Brazil: NUPAUB/USP, 1995.

[63] Marshall, NT. Searching for a Cure, Conservation of Medicinal Wildlife Resources in East and Southern Africa. 1st ed. Cambridge, UK: TRAFFIC International, 1998.

[64] Monroy-Vilchis, O; Cabrera, L; Suárez, P; Zarco-González, MM; Soto, CR; Urios, V. Uso tradicional de vertebrados silvestres en la Sierra Nanchititla, México. *INCI*, 2008 33(4), 308-313.

[65] Nilsson, G. The Endangered Species Handbook. 1st ed. Washington, D.C.: The Animal Welfare Institute, 1990.

[66] Nunn, JF. Ancient Egyptian Medicine. 1st ed: University of Oklahoma Press, 1996.

[67] Pieroni, A; Quave, C; Nebel, S; Heinrich, M. Ethnopharmacy of the ethnic Albanians (Arbereshe) of northern Basilicata, Italy. *Fitoterapia*, 2002 73, 217–241.

[68] Powell, AM. Drugs and pharmaceuticals in ancient Mesopotamia. In: I. Jacobs WJ editor. *The Healing Past*. Leiden, Netherlands: Brill; 1993, 47 - 50.

[69] Pujol, J. Natur Africa: The Herbalist Handbook. 1st ed. Durban, S. Africa: Jean Pujol Natural Healers Foundation, 1993.

[70] Rajchal, R. Population Status, Distribution, Management, Threats and Mitigation Measures of Himalayan Musk Deer in Sagarmatha National Park. 1st ed. Babarmahal, Kathmandu, Nepal: DNPWC/TRPAP, 2006.

[71] Ritter, KE. Magical-expert and physician: Notes on two complementary professions in Babylonian medicine. *Assyriological Studies*, 1965 16, 299-323.

[72] Schaller, GB. The Last Panda. 1st ed. Chicago, IL, USA: University of Chicago Press, 1993.

[73] Schroering, GB. Conservation Hotline. Swamp Deer Resurfaces. *Wildlife Conservation*, 1995 98(6).

[74] Sheng, HL. Mustela strigidorsa. In: Sung W editor. *China Red Data Book of Endangered Animals Mammalia*. Beijing, China: Science Press; 1998, 151–152.

[75] Silva, MLVd; Alves, ÂGC; Almeida, AV. A zooterapia no Recife (Pernambuco): uma articulação entre as práticas e a história. *Biotemas*, 2004 17(1), 95-116.

[76] Simelane, TS; Kerley, GIH. Conservation implications of the use of vertebrates by Xhosa traditional healers in South Africa. *South African Journal of Wildlife Research*, 1998 28(4).

[77] Sodeinde, OA; Soewu, DA. Pilot study of the traditional medicine trade in Nigeria. *Traffic Bulletin*, 1999 18(1), 35-40.

[78] Soewu, DA. Wild animals in ethnozoological practices among the Yorubas of southwestern Nigeria and the implications for biodiversity conservation. *African Journal of Agricultural Research*, 2008 3(6), 421-427.

[79] Solanki, GS; Chutia, P. Ethno Zoological and Socio-cultural Aspects of Monpas of Arunachal Pradesh. *Human Ecology*, 2004 15(4), 251-254.

[80] Souza, RF. Medicina e fauna silvestre em Minas Gerais no século XVIII. *Varia Historia*, 2008 24(39), 273-291.

[81] Stetter, C. The Secret Medicine of the Pharaohs - Ancient Egyptian Healing. 1st ed. Chigaco, USA: Edition Q, 1993.

[82] College, JNM. The Chinese Materia Medica Dictionary. 2nd ed. Shanghai, China: Jiangsu New Medical College, 2002.

[83] Thompson, RC. Assyrian medical texts from the originals in the British Museum. 1st ed. Oxford, UK: Oxford University Press, 1923.

[84] Tierra, M. Planetary Herbology, An Integration of Western Herbs Into The Traditional Chines And Ayurvedic Systems. 1st ed. Twin Lakes, Wisconsin, USA: C.A., N.D., O.M.D., Lotus Press, 1988.

[85] True, RH. Folk Materia Medica. *The Journal of American Folklore*, 1901 14(53), 105-114.

[86] Truong, NQ; Sang, NV; Tuong, NX; Son, NT. Evaluation of the wildlife trade in Na Hang District. 1st ed. Ha Noi, Viet Nam: Government of Viet Nam (FPD)/UNOPS/UNDP/Scott Wilson Asia-Pacific Ltd., 2003.

[87] Vázquez, PE; Méndez, RM; Guiascón, ÓGR; Piñera, EJN. Uso medicinal de la fauna silvestre en los Altos de Chiapas, México. *Interciencia*, 2006 31(7), 491-499.

[88] Walston, N. An overview of the use of Cambodia's wild plants and animals in traditional medicine systems. *TRAFFIC Southeast Asia, Indochina*, 2005.

[89] Zhang, B. Musk deer: Their capture, domestication and care according to Chinese experience and methods. *Unasylva*, 1983 35, 16 -24.

[90] Slifkin, RN. Exotic Shofars: Halachic Considerations. 2nd ed: Rosh Hashanah 5768, 2007.

[91] Debbie, NG; Burgess, EA. Against the Grain: Trade in musk deer products in Singapore and Malaysia. 1st ed: WWF/TRAFFIC Southeast Asia, 2005.

[92] Van, NDN; Tap, N. An overview of the use of plants and animals in traditional medicine systems in Viet Nam. 1st ed. Ha Noi, Viet Nam: TRAFFIC Southeast Asia, Greater Mekong Programme, 2008.

[93] Pieroni, A; Giusti, ME; Grazzini, A. Animal remedies in the folk medicinal practices of the Lucca and Pistoia Provinces, Central Italy. In: Fleurentin J, Pelt JM, Mazars G editors. *Des sources du savoir aux médicaments du futur/from the sources of knowledge to the medicines of the future*. Paris: IRD Editions; 2002, 371-375.

[94] Kang, S; Phipps, M. A question of attitude: South Korea's Traditional Medicine Practitioners and Wildlife Conservation. 1st ed. Hong Kong: TRAFFIC East Asia, 2003.

[95] Sheng, H; Ohtaishi, N. The status of deer in China. In: Ohtaishi N, Sheng H-I editors. *Deer of China: Biology and Management*. Amsterdam, The Netherlands: Elsevier Science Publishers; 1993, 1-11.

[96] TRAFFIC. Musk deer [online]. 2002 [cited]. Available from U RL: http://www.undp.-rg/hdr2003/indicator/indic_4_1_1.html

[97] Day, D. The Doomsday Book of Animals. A Natural History of Vanished Species. 1st ed. New York: Viking Press, 1981.

[98] WCMC (World Conservation Monitoring Centre). 1994 IUCN Red List of Threatened Animals. 1st ed. Gland, Switzerland: International Union for the Conservation of Nature (IUCN), The World Conservation Union, 1993.

[99] Martin, ES; Phipps, M. A Review of the Wild Animals Trade in Cambodia. *TRAFFIC Bulletin*, 1996 16(2), 45−60.

[100] Anon. The Wildlife (Protection) Act, (as amended upto 1991). 1st ed. Dehradun, India: Natraj Publisher, 1993.

[101] Posey, DA. Exploração da biodiversidade e do conhecimento indígena na América Latina: desafios à soberania e à velha ordem. In: CavalcantI C editor. *Meio Ambiente, desenvolvimento sustentável e políticas públicas*. São Paulo: Cortez; 1997, 345–368.

[102] Johannes, RE. Integrating traditional ecological knowledge and management with environmental impact assessment. In: Inglis JT editor. *Traditional Ecological Knowledge: Concepts and Cases*. Ottawa, Canada: International Program on Traditional Ecological Knowledge and International Development Research Centre; 1993, 33–39.

[103] Costa-Neto, EM. Implications and applications of folk zootherapy in the State of Bahia, Northeastern Brazil. *Sustainable Development*, 2004 12, 161–174.

[104] Soejarto, DD. Biodiversity prospecting and benefit-sharing: perspectives from the field. *Journal of Ethnopharmacology*, 1996 51, 1–15.

In: Encyclopedia of Environmental Research
Editor: Alisa N. Souter
ISBN: 978-1-61761-927-4
© 2011 Nova Science Publishers, Inc.

Chapter 24

THEORETICAL PERSPECTIVES ON THE EFFECTS OF HABITAT DESTRUCTION ON POPULATIONS AND COMMUNITIES

Matthew R. Falcy[*]

Department of Ecology, Evolution, and Organismal Biology,
Iowa State University Ames, IA. USA

INTRODUCTION

The geologic record indicates that species have typically persisted for 1 to 10 million years (May et al. 1995). Yet as many as 5,000 to 100,000 species are thought to go extinct each year (Wilson 1992, Eldridge 1998, Benton 2003). As human population size and standards of living increase around the world, it is anticipated that the strain on natural resources will drive an extinction rate approximately 100 to 1,000 times greater than the 'natural' background extinction rate (Pimm et al. 1995). Indeed, half of all living bird and mammal species are speculated to go extinct within 200 to 300 years (Levin and Levin 2002). This trend in biodiversity has been dubiously termed 'the sixth extinction' (Leakey and Lewin 1995, Glavin 2007), suggesting that humankind's impact on the planet is comparable to the celestial cataclysm that eradicated the dinosaurs. It has been recently estimated that 10 to 40 million years of evolution will be needed to replenish Earth's biodiversity from the current extinction crisis (Alroy 2008).

Human-induced extinctions pose an ethical problem that is probably less compelling to most people than the extent to which our welfare depends on Earth's biodiversity. Approximately one quarter of all prescription drugs taken in the United States are derived from plants. Naturally occurring ecosystem services, such as water purification, pest and climate control, and crop pollination, were valued at $33 trillion US Dollars per year in 1997 (Costanza et al. 1997). The natural world is a source of future discoveries whose worth is unknowable. For example, as-yet unidentified plants, algae, or bacteria may prove to be

---

[*] mfalcy@iastate.edu

model organisms for the industrial recovery of solar energy. Humanity's welfare is intimately linked to the intelligent use of available resources, and the urgency of implementing sustainable practices grows lock-step with the rise in global population and per-capita resource consumption.

The causes of species extinctions include overharvesting, pollution, environmental shifts (e.g. climate change, disruption of biogeochemical cycles), species introductions/invasions, habitat loss and fragmentation, and disruption of communities (i.e. coextinction). Although there is growing awareness of the severity of synergism among these factors (Ibáñez et al. 2006, Mora et al. 2007, Brook et al. 2008) and of climate change in particular (Thomas et al. 2004), habitat loss and fragmentation is historically thought to be the primary driver of species extinctions. Thus, it would be extremely useful to know how responsive populations are to habitat destruction. Does the probability of population persistence decline linearly with increasing habitat destruction or is there a critical threshold of degradation whereupon persistence declines abruptly? This chapter endeavors to derive some general conclusions about how populations and communities respond to habitat destruction. Specifically, this chapter addresses two questions fundamental to natural resource conservation: 1) What happens to the size of an animal population as its habitat is progressively fragmented? and 2) How does the relative abundance of different species of plants change under different spatio-temporal disturbance regimes? It is hoped that this kind of information can be used to help predict real-world responses to environmental change and, hence, improve management decisions regarding increasingly strained natural systems.

Any attempt to derive general conclusions is necessarily theoretical and requires some kind of model. Models are used to inductively take reasonable assumptions to novel conclusions. Such conclusions are ideally recast as hypotheses and then subjected to empirical tests. However, many theoretical modeling conclusions regarding natural systems cannot be easily tested because of logistical, financial, or ethical constraints to experimental manipulation. Indeed, modeling is often performed precisely because such constraints exist. Thousands of landscapes and hundreds of years would be needed to empirically reproduce the 'experiments' that are modeled below. Even if thousands of landscapes and hundreds of years were miraculously dedicated to a real-world study, any biologist or resource manager working with a different species (or even the same species in a different region) would be left with little more than an educated guess about how the results of such a superlative experiment apply to their particular system of interest. It is difficult to export experimental results across systems because the sensitivity of a given experiment's results to unmanipulated factors (e.g. the focal species' life history traits or the climate where the experiment is conducted) is usually unknown. Theory and models are indispensable tools that offer insight into otherwise hopelessly complex natural systems.

This chapter is divided into two parts, each corresponding to a different model designed to address a particular ecological problem related to habitat destruction. Both sections can be treated as independent investigations and are presented as such. The first section deals with the effects of habitat fragmentation on animal population size. After accounting for an associated effect due to simple habitat loss, it will be shown that the effect of fragmentation *per se* is involved in complex interactions among factors relating to landscape physiognomy and the biology of the species in question. The second section focuses on the response of a community of plants to different disturbance regimes. The results suggest that community composition is affected by an interaction between disturbance magnitude and the spatio-temporal sequence of disturbances.

THE EFFECTS OF HABITAT FRAGMENTATION ON POPULATION SIZE

Introduction

Fragmentation occurs when habitat destruction transforms a previously continuous expanse of habitat into an assemblage of disconnected patches that are immersed in a relatively inhospitable matrix environment (Wilcove et al. 1986). As fragmentation progresses, average patch size is reduced and average patch isolation increases. Concerns over habitat fragmentation lie in the possibility that these effects add significant additional risks to populations beyond the contemporaneous effect of habitat loss (Fahrig 2003).

Isolating the effect of habitat fragmentation on population size from the concurrent effect of habitat loss is notoriously difficult to achieve with field studies because of the large spatial and temporal scales involved (McGarigal and Cushman 2000). The issue has therefore been the focus of many spatially-explicit modeling efforts (Fahrig 1997; With and King 1999; Flathers and Bethers 2002; Wiegand et al. 2005; Tischendorf et al. 2005). The diversity of models has led to a variety of results regarding the magnitude of the effect of fragmentation independent of habitat loss (Fahrig 2002). Most models designed to disentangle the effect of fragmentation from habitat loss employ algorithms that enable investigators to independently vary habitat amount and degree of fragmentation. This is an important prerequisite, but many models simulate fragmentation with a single parameter (e.g. control the landscape's fractal dimension), which makes it impossible to parse-out the effects of fragmentation into its constituent elements: reduction of average patch size and increased patch isolation. Independently varying patch size and patch isolation can broaden the suite of mechanisms through which fragmentation causes population decline, and therefore provides a richer assessment of the effects of fragmentation. Does a population decline linearly with increasing inter-patch distance or is there a concave-up (or down) response? How sensitive is this relationship to the size of the habitat patches in the landscape? Related questions are: Does a population decline linearly with decreasing patch size or is there a concave-up (or down) response? How sensitive is the relationship to the average inter-patch distance in the landscape? Additional layers of complexity are added by considering the consequences of a species' propensity to leave a habitat patch and enter into the intervening matrix of non-habitat, and the mortality rate of that species in the matrix environment.

The significance of independently varying patch size and patch isolation can be appreciated when considering the consequences of individuals' propensity to leave a habitat patch and enter the relatively hostile matrix. If a habitat patch is sufficiently large to sustain a population, then a high probability of entering the relatively hostile matrix will be detrimental to the population (Tischendorf et al. 2005). However, if a patch is too small to maintain a viable population and inter-patch isolation is sufficiently small to guarantee a high probability successful dispersal to another patch (even if the destination patch is itself too small to sustain a viable population), then an increased tendency to enter the matrix will be beneficial for the population. Thus, the consequence of having a high habitat-to-matrix border crossing probability can be positive or negative, depending on patch size and patch isolation.

It is important to allow patch function to vary with patch size. If a single cell (a "cell" is a unit corresponding to a model's finest spatial resolution) represents an entire breeding habitat patch, as in Fahrig 1997 and Fahrig 1998, then fragmentation can never subdivide the patch beyond the minimum size required to carry out that function. In this case, increasing fragmentation by increasing fractal dimension does not affect a population by functionally degrading habitat; rather, it increases population risk by increasing the amount of edge in the landscape, which increases the probability that an individual leaves the habitat and dies in the relatively hostile matrix (Fahrig 2002). However, the value of real-world habitat patches can change with patch size. For example, the importance of a particular patch in a landscape to sustaining a population will drastically decline once the patch is reduced beyond the minimum size required to sustain a viable population. Failure to allow patch function to vary with patch size can result in underestimation of the effect of fragmentation by excluding mechanisms of population decline in severely fragmented landscapes.

The objective of this section is to isolate the effect of habitat fragmentation on population size from the effect of habitat loss and then further decompose the effect of fragmentation into a patch size effect, patch isolation effect, individuals' border crossing probability, and the matrix mortality rate. To this end, a computer simulation model is developed. The effect of fragmentation independent of habitat loss is estimated by making comparisons between the simulated population size established in landscapes exhibiting just habitat loss and landscapes exhibiting both fragmentation and habitat loss.

Methods

Model Demographics

The outcome of the processes described here is an individual-based, spatially-explicit model with multigenerational population dynamics. Individuals exhibit several 'realistic' behaviors such as movement, feeding, metabolism, reproduction, death, and some level of habitat and conspecific recognition. The movement and demographic components of this model were also used in a study of the effectiveness of biological corridors (Falcy and Estades 2007).

All individuals move one cell per model iteration. The turning angle of individuals, and hence displacement rate, depends on local population density. If fewer than four individuals are found within a ten-cell radius of a focal individual, then it randomly selects one of eight surrounding cells. The individual will move into that cell provided it was not already there within the previous eight steps. If the individual was recently in the selected cell, then another is randomly selected. This process is repeated until a cell that has not been visited during the previous eight steps is selected and occupied. If four or more individuals are found within a ten-cell radius of a given individual, then it randomly occupies one of the three forward-facing cells. These movement constraints result in reduced displacement rate and more efficient foraging in areas with relatively low population density.

All individuals of a given simulation have the same probability of crossing the border from habitat to matrix. However, different simulations are carried out with different habitat-to-matrix movement probabilities in order to investigate how results change for species with different ability to perceive and avoid the relatively hostile conditions of the matrix.

An energy balance approach is used to simulate birth and death. Stationary food items are randomly distributed within habitat. If an individual occupies a cell containing a food item, then it acquires 285 units of energy. The food item then disappears for 400 model iterations before regenerating. Individuals consume one unit of their energy reserve with every step. If an individual does not locate sufficient food items, it uses all of its energy and is eliminated from the program (dies). Density-dependent mortality is thus incorporated through food availability. Individuals are also randomly eliminated at a rate of 0.0005 per step, which simulates density-independent mortality factors such as aging, accidents, etc. The matrix is endowed with an additional mortality factor in order to reflect relative unsuitability. Separate simulations were run with this probability set at 0.0005, 0.001, and 0.0015 per step. Additionally, food items do not occur in the matrix.

If a female accumulates at least 900 units of energy and then comes within five cells of a male that has also acquired 900 units of energy, then five new individuals are immediately introduced (born) into the program at the cell where the female is located. The gender of each offspring is determined randomly and offspring are given an initial energy reserve of 200 units. Each parents' energy reserve is set to 300 units in order to prevent them from immediately reproducing. Individuals that surpass the reproduction threshold of 900 units of energy do not follow the movement pattern that tends to lead them out of high-density areas because it is assumed that these individuals are searching for mates, not food.

Combinations of model parameters that resulted in very large population sizes were also not considered because they required considerably more simulation time. A summary of all demographic parameters and their selected values is provided in Table 1.

Table1. Parameter values used to establish multi-generational population.

Parameter	Value
Initial density of individuals	0.005 / cell of habitat
Energy reserve of new individuals	200 units
Density of food items	0.025 / cell of habitat
Energy conferred to individual by one food item	285 units
Regeneration time of food items after consumption	400 iterations
Energy individuals expend every step	1 unit
Energy needed for reproduction	>900 units
Proximity between males and females for reproduction	<5 cells
Number of offspring per reproduction event	5 individuals
Energy of parents after reproducing	300 units
Radius around individual were local population density is measured	10 cells
Number of individuals in radius needed for"high"density movement pattern	>=3 individuals
Number of individuals in radius needed for "low" density movement pattern	<3 individuals
Probability of random mortality (all individuals)	0.005 / iteration
m, Additional probability of random mortality in matrix	0.0005, 0.0010, 0.0015 / iteration
p, Probability of crossing border from habitat to matrix	0.1, 0.5, 0.9

Population size fluctuates stochastically because individuals move in random directions and are subjected to random mortality. When the population size increases, food availability is reduced and the population size declines. When the population size becomes significantly reduced, food items regenerate more quickly than they are consumed, favoring subsequent increases in population size. Various combinations of model parameters were explored in an attempt to find a population that was relatively stable and persistent within a single square habitat patch measuring approximately 2500 cells when the border crossing probability was 0.5. Combinations of model parameters that resulted in widely fluctuating populations were not considered because of the additional simulation time and/or number of replicates required to determine the mean population size.

Model Landscapes

Three types of landscapes were used. These will be referred to as: "FRAG&LOSS", "LOSS", and "PRISTINE" (Fig 1). FRAG&LOSS landscapes are binary configurations of evenly spaced, square habitat patches immersed in matrix. Relative to "PRISTINE" landscapes, which are simply pure habitat, FRAG&LOSS landscapes manifest the effects of both fragmentation and habitat loss. LOSS landscapes have all habitat aggregated into one large block. Relative to "PRISTINE" landscapes they manifest the effect of just habitat loss. To discriminate effects of patch size and patch isolation, 18 FRAG&LOSS landscapes were modeled, each with a different combination of habitat patch size and patch isolation (Table 2).

All FRAG&LOSS landscapes have 40,000 cells of habitat (except one that had 40,401 cells of habitat. See Table 2). Thus, reducing patch size requires the addition of patches in order to ensure that 40,000 cells of habitat remain in the landscape. Landscapes are "wrapped" such that an individual leaving the landscape immediately reenters on the opposite side. This effectively converts a flat landscape into a torus. Wrapping the landscapes eliminates possibility of a "landscape edge" effects. From an individual's perspective, landscapes are infinitely large and contain an endless series of evenly spaced habitat patches.

Table 2. The size and number of habitat patches and the degree of inter-patch isolation in 18 fragmented landscapes. The size of the landscapes is given in cells. Numbers in parentheses indicate the proportion of habitat in the landscape.

Patch		Inter-patch isolation					
Size	#	2	10	18	26	34	42
25x25	64	46656 (0.86)	78400 (0.51)	118336 (0.34)	166464 (0.24)	222784 (0.18)	287296 (0.14)
40x40	25	62500 (0.64)			108900 (0.37)		
50x50	16	43264 (0.92)	57600 (0.69)	73984 (0.54)	92416 (0.43)	112896 (0.35)	135424 (0.30)
67x67	9 [a]	53361 (0.75) [a]			77841 (0.51) [a]		
100x100	4	48400 (0.83)		63504 (0.63)			

[a] 67 x 67 x 9 = 40,401cells of habitat, 1% more than the other landscapes.

Since decreasing patch size necessitates an increase in patch number (total amount of habitat in the landscape is fixed), landscape size must also increase in order for patch isolation to remain constant (Fig 1). This results in a decrease in the percentage of habitat within the landscape (Table 2). Since each of the 18 FRAG&LOSS landscapes have different percentages of habitat, the calculation of fragmentation independent of the effect of habitat loss is performed by comparing population size established in a FRAG&LOSS landscape with the population size established in a LOSS landscape containing an identical percentage of habitat. Thus, simulations are performed to estimate mean population size established on 18 different LOSS landscapes, each with 40,000 cells of habitat centered in landscapes of different size.

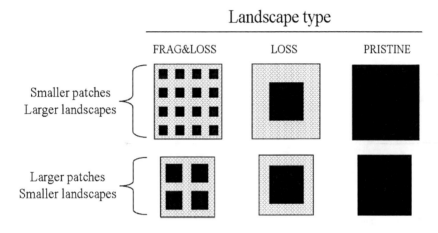

Figure 1. Columns represent three types of landscapes modeled. Eighteen FRAG&LOSS landscapes with a unique combination of patch size and isolation were modeled (two shown here). Each FRAG&LOSS and LOSS landscape contains 40,000 cells of habitat. Population size established on each landscape within a row are compared in order to estimate the effect of fragmentation independent of habitat loss. The two rows depict different patch sizes. Note that the size of the landscape must decreases as patch size increases in order for the habitat amount and patch isolation to remain constant. Since all 18 FRAG&LOSS landscapes have different sizes and percent habitat, each is compared to a corresponding LOSS landscape of identical size, percent habitat, and habitat amount. Black: habitat; stippled: matrix.

Further simulations are performed to estimate the mean population size in PRISTINE landscapes, which are composed entirely of habitat and represent the "before intervention" population size. This information is used to examine the magnitude of population decline resulting from habitat loss and habitat loss with fragmentation for landscapes with different percent cover in habitat.

Simulations

Simulations are initiated with individuals and food items randomly distributed within habitat at a per-cell density of 0.005 and 0.025, respectively. The first 4000 model iterations are used as an adjustment period, allowing the effect of initial conditions (number of individuals introduced, their initial energy reserve, the number of food items available, etc.) to dissipate. Population size is calculated as the mean number of individuals on the landscape between the 4000th and 5000th model iteration. The average lifetime of an individual is approximately 680 iterations.

Three probabilities that individuals could cross the border from habitat to matrix (0.1, 0.5, 0.9) and three probabilities of random mortality in the matrix (0.0005, 0.001, and 0.0015 per step) were crossed with the 18 FRAG & LOSS landscapes and 18 LOSS landscapes. This resulted in 324 (3 x 3 x 18 x 2) unique scenarios. Scenarios with FRAG&LOSS landscapes were replicated 10 times. Scenarios with LOSS landscapes were replicated 5 times. Since PRISTINE landscapes are pure habitat, simulations on these landscapes did not include habitat to matrix border crossing and random mortality in the matrix.

Results

The effect of fragmentation independent of the effect of habitat loss is calculated by dividing the population size established in a FRAG&LOSS landscape by the population size established in a LOSS landscape of identical size and percent habitat. This expresses the effect of fragmentation as a proportion of the population size that would have resulted if only habitat loss had occurred.

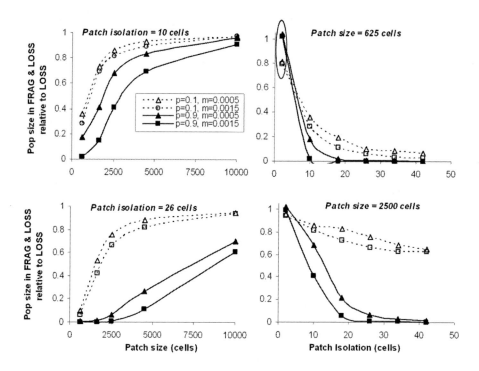

Figure 2. Population size established in FRAG&LOSS landscapes divided by the population size established in LOSS landscapes. Fragmentation independent of habitat loss increases as the y-axis approaches 0. Five different patch sizes, six degrees of patch isolation, two probabilities (p) of crossing the border from habitat to matrix, and two rates of random mortality (m) in the matrix are shown. Points in ellipse represent a greater fragmentation effect with reduced probability of entering the matrix.

Different combinations of patch size, patch isolation, border crossing probability, and mortality rate in the matrix result in population sizes ranging from 0 to100 % of population size established in LOSS landscapes (Fig 2). In general, the effect of fragmentation becomes more severe (smaller value on y-axis of Fig 2) with decreasing patch size, increasing patch isolation, increasing rate of mortality in the matrix, and increasing border crossing probability. A notable exception to these trends occurs when patch isolation is 2 cells and patch size is 625 cells. In this scenario, the effect of fragmentation increases when border crossing probability decreases (Fig 2 ellipse). Not only does the effect of border crossing reverse in this scenario, the effect of fragmentation on population size also reverses. Note that the filled points in the ellipse in Figure 2 are greater than 1, indicating that population size is greater in the FRAG&LOSS landscapes than the LOSS landscapes.

Relative to population sizes established in PRISTINE landscapes, population sizes in LOSS landscapes decline linearly as the percent of habitat in the landscape declines. However, population sizes in FRAG&LOSS landscapes decline irregularly as the percent of habitat in the landscape declines (Fig. 3).

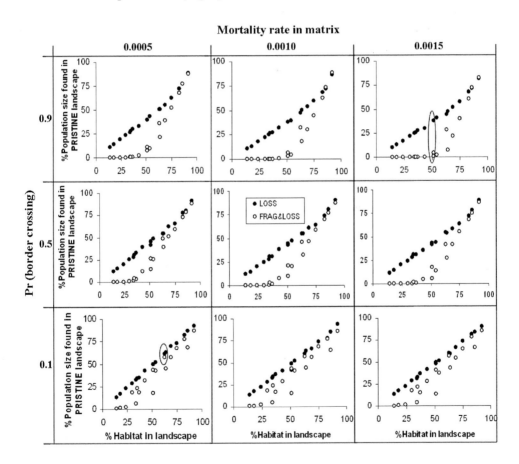

Figure 3. Percent of PRISTINE population size found in LOSS and FRAG & LOSS landscapes with different amounts of habitat in the landscape, habitat to matrix border crossing probability, and rates of mortality in the matrix. Points in ellipses show extremes in the magnitude of fragmentation effects (discussed in text).

This variation is attributable to different combinations of patch size and patch isolation. At one extreme, populations in FRAG&LOSS landscapes with habitat comprising 51% of the landscape go extinct while corresponding populations occurring in LOSS landscapes retain 40% of the PRISTINE population size (see ellipse in upper right-hand panel of Fig 3). In this scenario, patch size is 625 cells and patch isolation is 10 cells. At the other extreme (ellipse in lower left-hand panel of Fig 3), there is no difference between population sizes on LOSS and FRAG&LOSS landscapes (i.e. no effect of fragmentation) when as much as 27% of the landscape's habitat is destroyed (two tail $t = 2.16$, d.f. = 13, P = 0.059). In this scenario, patch size is 10,000 cells and patch isolation is 26 cells. In all scenarios, if habitat loss reduces population size by 80%, then the additional effect of fragmentation always results in population extinction.

An appropriate metric of fragmentation independent of habitat loss is needed for statistical tests of interactions among patch size, patch isolation, border crossing probability, and mortality rate in the matrix. A candidate metric is the difference between population size established in LOSS and FRAG&LOSS landscapes relative to the population size found in PRISTINE landscapes (i.e. $\frac{\text{LOSS}-\text{FRAG\&LOSS}}{\text{PRISTINE}}$, which is the vertical distance between open and filled points in Fig. 3). This metric expresses the effect of fragmentation independent of habitat loss. However, it is biased by an inability to capture the full effect of fragmentation when populations go extinct.

**Table 3. Type III ANOVA on factors influencing population size
in a fragmented landscape. Size = patch size, Iso = patch isolation,
Cross = border crossing probability, Mort = rate of random mortality in matrix.**

Source	DF	Type III SS	Mean Square	F Value	Pr > F
Size	4	6701509	1675377	8312	<.0001
Iso	5	10520452	2104090	10439	<.0001
Cross	2	2482689	1241344	6158	<.0001
Mort	2	100850	50425	250	<.0001
Size*Iso	8	493983	61748	306	<.0001
Size*Cross	8	1105261	138158	685	<.0001
Size*Mort	8	13724	1716	8.51	<.0001
Iso*Cross	10	764906	76491	379	<.0001
Iso*Mort	10	31535	3154	15.6	<.0001
Cross*Mort	4	7247	1812	8.99	<.0001
Size*Iso*Cross	16	443629	27727	137	<.0001
Size*Iso*Mort	16	26086	1630	8.09	<.0001
Size*Cross*Mort	16	9178	574	2.85	0.0001
Iso*Cross*Mort	20	22985	1149	5.70	<.0001
Size*Iso*Cross*Mort	32	23057	721	3.57	<.0001

Different combinations of patch size, patch isolation, border crossing probability, and mortality rate in the matrix result in population sizes ranging from 0 to 100 % of population size established in LOSS landscapes (Fig 2). In general, the effect of fragmentation becomes more severe (smaller value on y-axis of Fig 2) with decreasing patch size, increasing patch isolation, increasing rate of mortality in the matrix, and increasing border crossing probability. A notable exception to these trends occurs when patch isolation is 2 cells and patch size is 625 cells. In this scenario, the effect of fragmentation increases when border crossing probability decreases (Fig 2 ellipse). Not only does the effect of border crossing reverse in this scenario, the effect of fragmentation on population size also reverses. Note that the filled points in the ellipse in Figure 2 are greater than 1, indicating that population size is greater in the FRAG&LOSS landscapes than the LOSS landscapes.

Discussion

The strengths of this modeling approach include independently varying patch size and patch isolation, which exposes the precise spatial configuration of habitat loss with the greatest impact on the population (i.e. the least amount of habitat loss that results in the most deleterious configuration of patch size and isolation). This is opposed to modeling landscape fragmentation with a single parameter that controls the fractal dimension of habitat, which can miss particularly problematic spatial configurations of habitat. Another strength of the present modeling approach is that populations are treated as dynamic, trans-generational entities capable of responding to a minimum patch size required for population persistence. This is important since the value of a habitat patch will dramatically change as it is reduced beyond the minimum size required to sustain a population. A weakness of this modeling approach is a direct consequence of its strengths: Many parameters are required to create a spatially explicit and individual-based model with trans-generational population dynamics. Exploring the sensitivity of the results to all the parameter values would have been extremely daunting. The results presented here required the dedication of one computer for approximately one month, and simulation time increases exponentially as additional factors are explored.

Fahrig (1997, 1998) found that the effect of an index of fragmentation on population extinction became exaggerated when landscapes contained 20% habitat, suggesting that this point represents a threshold where fragmentation becomes particularly problematic. In contrast, the results presented here indicate that the effect of fragmentation can be far more severe. Extinctions occur in fragmented landscapes with up to 50% habitat, while similar landscapes that only manifest habitat loss will maintain populations at 40% of their pre-loss size. Furthermore, the results presented here indicate that the effect of fragmentation is sensitive to interactions among patch size, patch isolation, border crossing probability, and matrix mortality rate. For example, it is demonstrated that a relatively high probability of entering the matrix generally magnifies the negative effect of fragmentation, but when patches are too small to maintain a population and inter-patch isolation is minimal, then a high probability of entering the matrix reduces the effect of fragmentation. This demonstrates qualitative (not merely quantitative) dependence of fragmentation to factors describing landscape physiognomy and species biology.

Another noteworthy result is that fragmentation can actually increase population size. In a modestly fragmented landscape, the matrix can be "used" in the sense that excess individuals can go there temporarily and then return to the habitat and eventually reproduce. Returning to habitat is less likely in a LOSS landscape than a FRAG&LOSS landscape because the latter contain habitat in all directions. The "use" of matrix in this fashion is analogous to a source-sink system (Pulliam 1988) where sinks provide a net benefit to the population (Foppen 2000).

In conclusion, these results agree with assertions that the effects of fragmentation are system specific (Mönkkönen and Reunanen 1999; Wiegand et al. 2005). Population sizes in landscapes that manifest both fragmentation and habitat loss were 0 to 100% of the population sizes in landscapes representing simple habitat loss. The effect of fragmentation independent of habitat loss is highly sensitive to interactions among patch size, patch isolation, the probability of crossing the border from habitat to matrix, the rate of random mortality in the matrix. These results suggest that managing for a universal fragmentation threshold when habitat comprises 20-30% of the landscape (Boutin and Hebert 2003) is not prudent.

The minimum patch size required to sustain a viable population has not been sufficiently treated in most spatially-explicit, individual-based models of fragmentation effects. This is a consequence of excessively simplistic models that do not contain sufficient demographic complexity to model trans-generation dynamics (Fahrig 1997, Fahrig 1998) and the use of excessively coarse-scale landscapes.

THE EFFECTS OF THE SPATIO-TEMPORAL SEQUENCE OF DISTURBANCES ON COMMUNITY COMPOSITION

Introduction

Secondary succession is the process of community-level recovery from a disturbance on a given site. The identity and relative abundance of species at a site, known as the community composition, gradually changes through time as "weedy" species that readily exploit disturbed areas are replaced by "dominant" species that physiologically exclude species from previous stages. Within a given site, species biodiversity is maximized at intermediate levels of disturbance (Connell 1978, Petraitis et al. 1989). This occurs because disturbances levels (intensity and/or frequency) that are either too moderate or too severe favor the few species capable of tolerating such conditions. Intermediate levels of disturbance maximize species biodiversity by providing niches for a broader spectrum of life history strategies.

At a landscape level, species biodiversity is a function of the disturbance histories within constituent sites. Thus, landscape-level species biodiversity should depend on the average level of disturbances across the landscape. However, another potentially important factor presents itself at the landscape-level: the spatio-temporal sequence of disturbances. The spatio-temporal sequence of disturbances refers to the relative location of disturbances through time (i.e. where a given disturbance occurs relative to the previous disturbance). This can affect plant species biodiversity at the landscape-level by facilitating (or impeding) the propagation of certain species throughout the landscape.

In this section, the effect of the spatio-temporal sequence of disturbances across a landscape on plant species biodiversity is explored with a simulation model. Different disturbance magnitudes are also simulated in order to determine how much the effect of the spatio-temporal sequence of disturbances depends on disturbance magnitude. In order to make meaningful comparisons among communities affected by different disturbance regimes, nonmetric multidimensional scaling (NMDS) is performed on the resulting data.

Methods

A spatially explicit, individual-based model is used to explore how a hypothetical plant community responds to different disturbance regimes. Eight "species" of plants are distinguished by (1) the number of propagules they disperse (2) the distance their propagules can travel and (3) the probability that their propagules usurp a cell occupied by another species. Values for these characteristics (Table 4) are assigned to six of eight species in a manner that creates a continuous spectrum of life history strategies ranging from "weedy" (many propagules that disperse long distance but have a low probability of usurping occupied cells) to "dominant" (few propagules with limited dispersal range that have a high probability of usurping occupied cells). Preliminary simulations including just these six species revealed that two of them (III and IV) quickly excluded the other four from the landscape. Thus, two additional species (VII and VIII) were created that specialize in usurping cells from these otherwise super-abundant species. This resulted in a greater number of species capable of persisting in the model environment and, hence, more species depth in the resulting communities.

Table 4. Upper panel contains reproductive characteristics of simulated species. Lower panel specifies competitive interactions among species: Entries are probabilities that a propagule originating from the species given in rows usurps a cell occupied by the species given in columns. Roman numerals denote different species.

Reproduction	I	II	III	IV	V	VI	VII	VIII
# propagules dispersed	7	6	5	4	3	2	7	7
Propagule dispersal range (cells)	15	12	10	7	5	3	10	10

Competition	I	II	III	IV	V	VI	VII	VIII
I	-	0.4	0.3	0.2	0.1	0.05	0.7	0.7
II	0.6	-	0.4	0.3	0.2	0.05	0.7	0.7
III	0.7	0.6	-	0.4	0.3	0.05	0.7	0.2
IV	0.8	0.7	0.6	-	0.4	0.05	0.2	0.7
V	0.9	0.8	0.7	0.6	-	0.05	0.7	0.7
VI	0.95	0.95	0.95	0.95	0.95	-	0.95	0.95
VII	0.3	0.3	0.3	0.8	0.3	0.05	-	0.3
VIII	0.3	0.3	0.8	0.3	0.3	0.05	0.3	-

A square landscape measuring 240 by 240 cells is initialized with an equal number of species. Every individual disperses its species-specific number of propagules with each model iteration. Propagules have an equal probability of landing in any cell within the species-specific dispersal range of the parent individual. Propagules landing off the edge of the landscape are ignored. A propagule that lands in an unoccupied cell functions as a reproducing adult during the following model iteration. A propagule that lands in an occupied cell will usurp it with a probability specified by the species-to-species interaction (Table 4).

Disturbances remove all adults from affected cells. Since any propagule that lands in an unoccupied cell will become an adult, disturbances facilitate uncontested colonization. Three spatio-temporal sequences of disturbance are employed. For all three, disturbances occur every model iteration but they differ in where a disturbance occurs with respect to the location of the previous disturbance. In one spatio-temporal sequence, consecutive disturbances occur in square blocks of cells that are contiguous in space as time progresses.

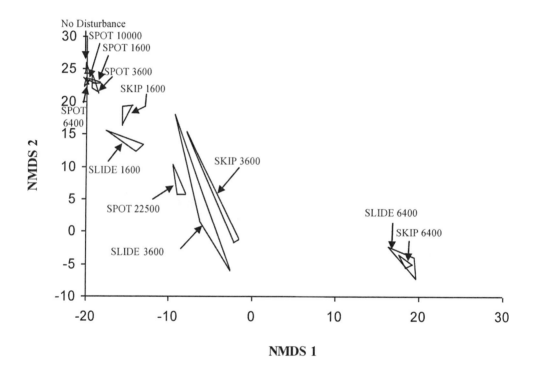

Figure 4. Nonmetric multidimensional scaling of plant communities with different disturbance regimes. "SPOT", "SKIP" and "SLIDE" refer to spatio-temporal sequence of disturbances affecting plant community (see text). Numbers refer to the magnitude of the disturbance. Vertices of triangles are replicates of indicated disturbance regime.

Thus, disturbed blocks "SLIDE" across the landscape such that a new disturbances always abuts the previous one. Another spatio-temporal disturbance sequence consists of a series of square blocks that maximizes the distance between consecutive blocks. These disturbances "SKIP" around the landscape. The final disturbance sequence does not occur in blocks; rather, randomly selected individual cells are disturbed every iteration. This results in numerous "SPOT" disturbances that are randomly distributed across the entire landscape. The magnitude of disturbances (i.e. number of cells disturbed each iteration) is also varied. SLIDE and SKIP disturbances occur in square blocks with area measuring 1600, 3600, and 6400 cells per iteration. SPOT disturbance regimes affect an equal number of randomly selected individuals. However, two additional high-magnitude SPOT disturbances were also simulated: 10,000 and 22,500 individuals per iteration.

All combinations of disturbance spatio-temporal sequence and magnitude were replicated three times. Three simulations were also conducted without disturbances in order to provide a bench mark against which the effects of the various disturbance regimes can be compared. The abundance of each species after 200 model iterations is the response variable used to make comparisons of the effect of various disturbance regimes on communities. Individuals in the landscape's outermost 50 cells were not counted in order to minimize a potential edge effect. Comparison of species abundances is performed with nonmetric multidimensional scaling (NMDS) using the 'Jaccard' distance measure.

Results

Jaccard distance is commonly used by community ecologists to quantify differences among communities. The sensitivity of the results to the underlying distance measure was explored by using both Bray-Curtis and Canberra distance measures. The results produced using these distance measures is very similar to those presented here.

A NMDS plot can be interpreted like a conventional map, except that distances between points on an NMDS plot represent dissimilarity in multidimensional space (e.g. community copsition), not 2-D space (latitude and longitude). Thus, the distance between points on Figure 4 is proportional to differences between plant communities. The plot reveals five notable relationships between disturbance magnitude and spatio-temporal sequence. 1) SPOT disturbances create communities most similar to the communities that occur when there are no disturbances.

Indeed, SPOT disturbances affecting 10,000 individuals produce communities more similar to the undisturbed communities than SKIP and SLIDE disturbances affecting 1600 individuals. Relative to undisturbed communities, SPOT disturbances affecting 22,500 individuals had less effect on community composition than SKIP and SLIDE disturbances affecting 3600 individuals. 2) For any given disturbance magnitude, SKIP and SLIDE disturbances produce communities more similar to one another than to SPOT disturbances. 3) The difference between communities affected by SKIP and SLIDE disturbances of the same intensity is much less than the difference between communities affected by the same spatio-temporal sequence of disturbances of different intensities. 4) The difference between communities affected by SKIP and SLIDE disturbances decreases as the magnitude of disturbances increases. 5) Communities affected by SKIP and SLIDE disturbances are more sensitive to the magnitude of the disturbances than communities affected by SPOT disturbances.

Discussion

Six of the eight species modeled here exhibit a spectrum of life-history strategies ranging from "weedy" to "dominant." The two remaing species are specialists at usurping cells from two other species that would otherwise over-run the landscape. The salient feature of this model is that there are numerous species with unique life history characteristics, thereby allowing the community to respond to disturbances.

Disturbances are natural processes that can enhance biodiversity. Thus, communities greatly diverging from the "no disturbance" communities presented here are not necessarily degraded communities. Degradation must be measured relative to the community composition resulting from a natural disturbance regime, which, of course, is a system-specific attribute.

The results presented here demonstrate that the spatio-temporal sequence of disturbances is an important determinant of community composition. Thus, land managers endeavoring to preserve natural communities must not focus exclusively on mimicking natural disturbance frequencies and intensities, but also the spatio-temporal sequence of disturbances in the landscape. The results also suggest that the magnitude of the spatio-temporal effect depends on an interaction between the particular spatio-temporal sequence and the magnitude of the disturbances.

General Conclusions

This chapter has explored two aspects of habitat destruction at two different levels of biological organization. The first section addresses the question: Does fragmentation impose significant risks to populations beyond the associated effect of habitat loss? The second section addresses the question: Is the spatio-temporal sequence of disturbances an important determinant of the relative abundance of plants in a landscape? The answers to both questions are conditional. The effect of fragmentation independent of habitat loss is highly sensitive to interactions among patch size, patch isolation, the probability of crossing the border from habitat to matrix, the rate of random mortality in the matrix. In one scenario modeled, fragmentation actually increases population size. In other scenarios, fragmentation has no effect on population size (population decline is exclusively attributable to habitat loss) when as much as 27% of the landscape's habitat is destroyed. Conversely, other scenarios show that fragmentation results in population extinction when habitat loss alone causes an initial decline in population size by 60%. The spatio-temporal sequence of disturbances across a landscape is shown to affect the relative abundance of plant species, but the magnitude of this effect depends on the particular spatio-temporal sequence and magnitude of disturbance. For example, the difference between communities affected by SKIP and SLIDE disturbances is much less than either of these versus SPOT disturbances. Furthermore, this pattern becomes more pronounced as disturbance magnitude increases.

It is illuminating that the answers to the questions posed above are conditional on various aspects of the systems. This implies that predicting real-world responses to habitat destruction requires detailed understanding of the systems under consideration. Unfortunate-ly, detailed information on most systems subject to human-induced habitat destruction does not exist. In lieu of additional research, it is prudent to heed Aldo Leopold's recommend-dation: "To keep every cog and wheel is the first precaution of intelligent tinkering."

REFERENCES

Alroy, J. 2008. Dynamics of origination and extinction in the marine fossil record. Proceedings of the National Academy of Sciences 105: 11536-11542.

Benton, M.J. 2003. When Life Nearly Died: The Greatest Mass Extinction of All Time. Thames and Hudson.

Boutin, S. and Hebert, D. 2002. Landscape ecology and forest management: developing an effective partnership. Ecological Applications. 12(2): 390-397.

Brook, B.W., Sodhi, N.S., and Bradshaw, C.J.A. 2008. Synergies among extinction drivers under global change. Trends in Ecology and Evolution 23(8):453-460.

Connell, J.H. 1978. Diversity in tropical rain forests and coral reefs. Science 199: 1302-1309.

Costanza, R., D'Arge, R., DeGroot, R., Farber, S., Grasso, M., Hannon, B., Limburg, K., Naeem, S., O'Neill, R.V., Paruelo, J., Raskin, R.G., Sutton, P., and Van Den Belt, M. 1997. The value of the world's ecosystem services and natural capital. Nature 387: 253-260.

Eldridge, N. 1998. Life in the balance: humanity and the biodiversity crisis. Princeton University Press. Princeton, NJ.

Fahrig, L. 1997. Relative effects of habitat loss and fragmentation on population extinction. Journal of Wildlife Management 61: 603-610.

Fahrig, L. 1998. When does fragmentation of breeding habitat affect population survival? Ecological Modelling 105: 273-292.

Fahrig, L. 2002. Effect of fragmentation on the extinction threshold: a synthesis. Ecological Applications 12(2): 346-353.

Fahrig, L. 2003. Effects of habitat fragmentation on biodiversity. Annual Review of Ecology, Evolution and Systematics 34: 487-515.

Falcy, M.R. and Estades, C.F. 2007. Effectiveness of corridors relative to enlargement of habitat patches. Conservation Biology 21(5): 1341-1346.

Flathers, C. and Bethers, M. 2002. Patchy reaction-diffusion and population abundance: the relative importance of habitat amount and arrangement. The American Naturalist 159: 40-56.

Foppen, R.P.B, Chardon, J.P., and Liefveld, W. 2000. Understanding the role of sink patches in source-sink metapopulations: reed warbler in an agricultural landscape. Conserevation Biology 14(6): 1881-1892.

Glavin, T. 2007. The sixth extinction: journeys among the lost and left behind. Thomas Dunne Books.

Ibáñez, I., Clark, J.S., Dietze, M.C., Feeley, K., Hersh, M., LaDeau, S., McBride, A., Welch, N.A., Wolosin, M.S. 2006. Predicting biodiversity change: outside the climate envelope, beyond the species-area curve. Ecology 87(8): 1896-1906.

Leakey, R.E. and Lewin, R. 1995. The sixth extinction: Patterns of life and the future of humankind. Anchor Books.

Levin, P.S. and Levin, D.A. 2002. The real biodiversity crisis. American Scientist 90(1): 6-9.

May, R.M, Lawton, J.H., and Stork, N.E. 1995. Extinction Rates. J.H. Lawton and R.M. May (Eds.) Oxford University Press. Oxford, UK. pp. 1-24.

McGarigal, K., and Cushman, S.A. 2000. Comparative evaluation of experimental approaches to the study of habitat fragmentation effects. Ecological Applications 12(2): 335-345.

Mönkkönen, M., and Reunanen, P. 1999. On critical thresholds in landscape connectivity: a management perspective. Oikos 84: 302-305.

Mora, C., Metzger, R., Rollo, A., and Myers, R.A. 2007. Experimental simulations about the effects of overexploitation and habitat fragmentation on populations facing environmental warming. Procedings of the Royal Society, Biological Sciences 274(1613): 1023-1028.

Petraitis, P.S., Latham, R.E., and Niesenbam, R.A. 1989. The maintenance of species biodiversity by disturbance. Quarterly Review of Biology 64: 393-418.

Pimm, S.E., Russell, G.J., Gittleman, J.L., and Brooks, T.M. 1995. The Future of Biodiversity. Science 269(5222): 347-350.

Pulliam, H.R. 1988. Sources, sinks, and population regulation. The American Naturalist 132(5): 652-661.

Tischendorf, L., Grez, A., Zaviezo, T., and Fahrig, L. 2005. Mechanisms affecting population density in fragmented habitat. Ecology and Society 10(1): 7.

Thomas, C.D., Cameron, A., Green, R.E., Bakkenes, M., Beaumont, L.J., Collingham, Y.C., Erasmus, B.F.N., Ferreira de Siqueira, M., Grainger, A., Hannah, L., Hughes, L., Huntley, B., van Jaarsveld, A.S., Midgley, G.F., Miles, L., Ortega-Huerta, M.A., Townsend Peterson, A., Phillips, O.L., Williams, S.E. 2004. Extinction risk from climate change. Nature 427: 145-148.

Wiegand, T., Revilla, E., and Moloney, K. 2005. Effects of habitat loss and fragmentation on population dynamics. Conservation Biology 19(1): 108-121.

Wilcove, D., McLellan, C., and Dobson, A. 1986. Habitat fragmentation in the temperate zone. Pp 236-257. in Soulé, M.E. (ed) Conservation Biology: the science of scarcity and diversity. Sinauer, Sunderland, Massachussetts.

Wilson, E.O. 1992. The diversity of life. The Belknap Press of Harvard University Press. Cambridge, Mass.

With, K.A., and King, A.W. 1999. Extinction thresholds for species in fractal landscapes. Conservation Biology 13: 314-326.

In: Encyclopedia of Environmental Research   ISBN: 978-1-61761-927-4
Editor: Alisa N. Souter   © 2011 Nova Science Publishers, Inc.

Chapter 25

FRESHWATER ECOSYSTEM CONSERVATION AND MANAGEMENT: A CONTROL THEORY APPROACH

Y. Shastri[1] and U. Diwekar[2]
[1]Energy Biosciences Institute, University of Illinois at Urbana-Champaign,
Department of Agricultural and Biological Engineering,
Urbana, IL, USA
[2]Vishwamitra Research Institute, Center for Uncertain Systems:
Tools for Optimization and Management,
Clarendon Hills, IL, USA

Abstract

Freshwater ecosystems, such as lakes, rivers, ponds, and wetlands, are a precious and critically important form of natural resource. They are important not only for humans by providing water for drinking and domestic use, but also for numerous aquatic animals that depend on water for their very existence. Moreover, freshwater ecosystems also provide valuable indirect services such as water purification and buffer against hurricanes. However, natural disasters and anthropogenic activities can unfavorably affect these ecosystems by pushing these ecosystems and the associated aquatic food webs into undesirable and potentially unstable regimes. Natural disasters such as floods and hurricanes often lead to sudden regime shifts, while changes caused by anthropogenic pollution effects are relatively slow but equally dangerous. Under such circumstances, external intervention by humans through management strategies is called for. Such strategies should aim not only to stabilize the ecosystems but also to make those inherently resilient so that destabilizing effects in future can be effectively managed. The broad theme of this work is the development of such management strategies. Since freshwater ecosystems are dynamic in nature, a time dependent management strategy is more desirable than a static one. Consequently, this work proposes to use optimal control theory for deriving time dependent management policies. The work considers an aquatic three species food chain model (Rosenzweig-MacArthur model) which has been frequently studied in theoretical ecology literature. The work identifies undesirable regimes for this model and applies optimal control theory (Pontryagin's maximum principle) to derive time dependent management strategies that achieve the desired regime change. Fisher information, a measure based on information theory, is proposed as the sustainability metric and used to formulate the objective function for the control problem. The work also compares the top-down and bottom-up control philosophies for the food chain model so that the most effective management option can be identified. Since natural ecosystems are frequently not well understood and hence associated with considerable uncertainties, the work utilizes efficient modeling techniques from finance literature for a robust analysis. An Ito process is used to model time dependent uncertainty and stochastic maximum principle is utilized to solve the optimal control problem. The results will highlight the role of systems theory in sustainable management of freshwater systems and ecosystems in general.

1. Introduction

Freshwater ecosystems are very important for the functioning of our Earth's ecosystem. The importance of water for the survival of humans as well as other species is well known. However, numerous other services provided by these freshwater ecosystems are equally important. This includes natural water purification, hurricane and flood buffer by wetlands, and commercial and sport fishing, and recreation. The level of these services is directly related to the overall health of the freshwater ecosystems, which includes not only water, but also the diverse flora and fauna that constitute these ecosystems (MEA, 2005; Science & Board, 2004). Natural controlling mechanisms within these dynamic and complex ecosystems are designed to ensure continued functioning of these ecosystem under normal circumstances. However, anthropogenic effects, such as nutrient pollution leading to eutrophication, are becoming increasingly stronger due to increased human activity. Moreover, severe natural disturbances such as hurricanes, floods and droughts, cause large disturbance in these ecosystems. In the presence of such factors, the natural regulating mechanism of these systems breaks down or is not sufficiently strong. In such circumstances, humans need to intervene to ensure sustenance of these ecosystems and ensure that they continue to provide the various services and functions.

Ecosystem management is much more complicated than the management of engineering systems. This is primarily due to the inherent complexities of the natural systems. These systems usually exhibit multiple complex regimes, and undesirable changes are often a result of nonlinear regimes shifts. This can cause intuitive solutions to fail drastically (Rosenzweig, 1971). There are a number of interwoven causal relationships in a given system, which makes the analysis of a management policies difficult through experimental studies. Moreover, these systems are often poorly understood leading to significant uncertainties, and the objectives to be achieved are quite obscure. For example, the relative importance of maintaining biodiversity and the total population at a particular trophic level is subjective, leading to lack of understanding of the appropriate targets for the given system.

The concepts of systems theory and sustainability assume importance in the wake of these issues. Systems theory enables us to model complex freshwater ecosystems using efficient modeling techniques. It allows us to focus on important aspects of the ecosystem and isolate the drivers and effects, which is often difficult in experimental studies. The simplified models capture the important ecosystem relations and functions so that the implications of different management actions can be studied. This is known as scenario analysis and this approach has been extensively used to study various ecosystems, including freshwater ones (Meadows et al., 1992; Bossel, 1998; Holling et al., 2002; Scholes et al., 2005; Stern, 2006; United Nations Environment Programme, 2007). At a higher level, techniques such as optimization and control theory can be used to systematically derive management policies to achieve appropriate objectives (Shastri et al., 2008a,b). The formulation of the right objectives is very important for the success of this approach. Sustainability has been emphasized in recent times as the most desired management objective, particularly for the natural ecosystems (Goodland & Daly, 1996; Fiksel, 2006; Ludwig et al., 1997). It is a multi-disciplinary concept that binds together the ecosystem properties of different nature as well as disparate temporal and spatial scales. The goal of the sustainability concept is to go beyond compartmentalization of scope and integrate the economic, social and environmental dimensions of a system in a holistic analysis.

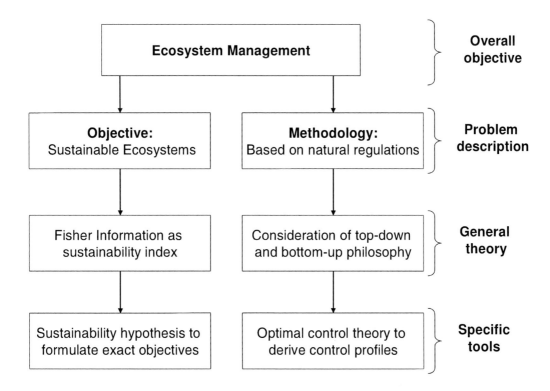

Figure 1. Integration of sustainability and ecosystem management: Theory and tools.

The work presented in this chapter builds on this concept of using a systems theory approach for the sustainable management of freshwater ecosystems. It models the freshwater ecosystem using a simplified three species food chain model. The possible undesirable (unsustainable) developments in this model system are first identified, which include not only gradual changes but also sudden regime shifts. The next goal is to employ a systematic approach to derive management policies to revert those undesirable development. Since the freshwater ecosystems are dynamic in nature, the work uses optimal control theory to derive time dependent management policies. It has been mentioned previously that natural systems such as freshwater ecosystems are often associated with significant uncertainties. Hence, it is essential to incorporate various forms of uncertainties in decision making for a robust analysis. This enhances the possibility of a successful implementation of these results on real systems. This work, therefore, includes uncertainty in the food chain model, generating a stochastic food chain model, and applies stochastic optimal control to derive the management policies. Fisher information, an information theory based measure, is used as the sustainability metric to formulate the appropriate objective functions for the optimal control problem (Cabezas & Fath, 2002). A key to effectively regulate any ecosystem is to first understand the natural regulation of the system. For this, scientists have proposed two different control philosophies: top-down control (Carpenter et al., 1985) and bottom-up control (McQueen et al., 1986). These natural regulation paths can be used to advantage when trying to externally manage the ecosystem. There has been an intense debate over the validity and relative importance of both philosophies, and the general consensus to emerge is that both regulation paths are dominant at different levels of the ecosystem represented as a food chain (Brett & Goldman, 1997). This work aims to determine the relative effectiveness of these two approach in managing the freshwater food chain model.

To summarize, this work compares top-down and bottom-up control philosophies, derived using optimal control theory, for a population model in a freshwater ecosystem (deterministic as well as stochastic), using sustainability as the objective, which is quantified by Fisher Information. Figure 1 schematically explains the relative contribution from each of the topics mentioned here.

The chapter is organized as follows. The next section reviews the theory behind the work, and section 3. gives the problem specific details. Section 4. presents and analyzes various results for a three species predator-prey model (deterministic and stochastic). The article ends with comments on the computational aspects in section 5. and important conclusions in section 6..

2. Theoretical Basics

2.1. Predator-Prey Model

The model used to represent the freshwater ecosystems is a predator-prey model, derived from the more general class of Lotka-Volterra type models. These models give a simplistic mathematical representation of the observed dynamics in natural systems. In many applications, three level food chain models are often considered to be a good enough representation of the ecosystems (Ryan Gwaltney et al., 2003). This work uses the Rosenzweig-MacArthur model, which is frequently used in theoretical ecology (Abrams & Roth, 1994; De Feo & Rinaldi, 1997; Gragnani et al., 1998). The model is given by the following set of ordinary differential equations:

$$f_1 = \frac{dx_1}{dt} = x_1\left[r\left(1 - \frac{x_1}{K}\right) - \frac{a_2 x_2}{b_2 + x_1}\right] \tag{1}$$

$$f_2 = \frac{dx_2}{dt} = x_2\left[e_2\frac{a_2 x_1}{b_2 + x_1} - \frac{a_3 x_3}{b_3 + x_2} - d_2\right] \tag{2}$$

$$f_3 = \frac{dx_3}{dt} = x_3\left[e_3\frac{a_3 x_2}{b_3 + x_2} - d_3\right] \tag{3}$$

where, x_1, x_2 and x_3 are population variables of the three species in the food chain in the ascending order of their position in the chain. These species are referred to as prey (x_1), predator (x_2) and super-predator (x_3) in the subsequent text. r and K are the growth rate and prey carrying capacity, respectively, and a_i, b_i, e_i and d_i, $i = 2, 3$, are the maximum predation rate, half saturation constant, efficiency, and death rate of the predator ($i = 2$) and super-predator ($i = 3$), respectively. $x_i(0)$ is the population of specie i at the starting time. The model parameters for the tri-trophic model are given in 1. For these parameter values, this model shows cyclic variations in the species populations (biomass), and the average population of each species remains steady.

The introductory text mentions the importance of considering uncertainty in natural systems for a robust analysis. There are many possible sources of uncertainty in the given food chain model. Most of the model parameters will not be deterministically known. In this work, the mortality rate of the predator (d_2) is considered to be uncertain. Quite often, a probability distribution is used to model uncertain variables. However, this representation is appropriate to model static uncertainties only. In the food chain model though, the predator mortality rate is expected to show time dependent variations such as seasonal variations in mortality. In such a case, the mortality rate needs to be modeled as a time dependent uncertainty, also known as a stochastic process. The finance literature is replete with examples of time dependent uncertain variables such as stock prices and interest rates. Real options theory has been developed to optimize investment decisions in the presence of these dynamic uncertainties, and presents different ways to model and forecast uncertainty using the stochastic processes. A comprehensive survey of these techniques can be found in Dixit & Pindyck (1994). A brief review of stochastic process modeling is given in Appendix A.. In this work, mean revering Ito process is used to model d_2 owing to its success in modeling various time dependent stochastic parameters, such as relative volatility (Ulas & Diwekar, 2004) and human mortality rate (Diwekar, 2005). The equation for the Ito mean reverting process to model d_2 is given as:

$$f_{ito} = \frac{dx_4}{dt} = \eta(\bar{d}_2 - x_4) + \frac{\sigma \epsilon}{\sqrt{\Delta t}} x_4 \qquad (4)$$

Here, x_4 is the Ito variable (mortality rate), $\bar{d}_2$ is the mean mortality rate, η is the speed of reversion, σ is the variance parameter, Δt is the time interval and ϵ is a normally distributed random variable with zero mean and unit standard deviation (Diwekar, 2003). The stochastic food web model is given as:

$$f_1 = \frac{dx_1}{dt} = x_1 \left[r \left(1 - \frac{x_1}{K} \right) - \frac{a_2 x_2}{b_2 + x_1} \right] \qquad (5)$$

$$f_2 = \frac{dx_2}{dt} = x_2 \left[e_2 \frac{a_2 x_1}{b_2 + x_1} - \frac{a_3 x_3}{b_3 + x_2} - x_4 \right] \qquad (6)$$

$$f_3 = \frac{dx_3}{dt} = x_3 \left[e_3 \frac{a_3 x_2}{b_3 + x_2} - d_3 \right] \qquad (7)$$

$$f_{ito} = \frac{dx_4}{dt} = \eta(\bar{d}_2 - x_4) + \frac{\sigma \epsilon}{\sqrt{\Delta t}} x_4 \qquad (8)$$

where, previously constant predator mortality rate d_2 becomes the fourth variable of the model, represented as x_4 and modeled using the Ito process. Table 1 gives the parameters for the Ito process.

2.2.　Ecosystem Management Philosophies

The earliest attempt to systematically understand the controlling effects of a natural ecosystem dates back to Hairston et al. (1960). It has since encouraged more and more research in understanding the natural regulation of ecosystems, and has led to the concept of trophic cascade hypothesis (Carpenter et al., 1985; Paine, 1980). For an aquatic ecosystem, it proposes that the predator-prey interactions are transmitted through the food webs to cause variance in the phytoplankton biomass and production at constant nutrient load (Carpenter et al., 1985), and that the responses are nonlinearly related to the strength of the interactions among the adjacent trophic levels. Another possible regulatory effect in a food web is that of the available resources on higher level species; e.g., nutrients support the phytoplankton biomass in an aquatic food web, which in turn affects the top level species that feed on it.

From the ecosystem management perspective, this has led to the formulation of two different control philosophies: top-down control and bottom-up control. Top-down control (also called as consumer control) refers to the control of the ecosystem through top level predators. Bottom-up control (also called as resource control) refers to the control of the ecosystem via available resources (e.g., nutrients). A debate has been going on over the validity of the individual control philosophies (Carpenter & Kitchell, 1988; Townsend, 1988; McQueen et al., 1989), and also, over the relative importance of each of those in a food web (Lynch & Shapiro, 1981; Vanni & Temte, 1990; Rosemond et al., 1993). The recent opinion, based on some of the published results, is that both effects are apparent in a food chain, and the relative importance of the two depends on the length of the food chain and the position of a particular species in the food chain (Brett & Goldman, 1997; McQueen et al., 1986). Thus, top-down control is more prominent in species at the top of the food chain, while the lower level species are more strongly under bottom-up control. Most of these results are based on experimental manipulations, followed by observations over a long period of time.

The existence of these two natural regulation paths in ecosystems provides two different avenues to exercise external control on these systems. Accordingly, regulation by controlling the top predator and by controlling the lowest level resources are the two options explored and compared in this work. The next section describes the optimal control theory which is used to derive the time dependent management profiles.

Table 1. Tri-trophic food chain model parameters

Prey	Predator	Super-predator	Ito process
$x_1(0) = 100$	$x_2(0) = 75$	$x_3(0) = 150$	$x_4(0) = 1.0$
$r = 1.2$	$a_2 = 2.0$	$a_3 = 0.1$	$\eta = 0.3535$
$K = 710$	$b_2 = 200$	$b_3 = 250$	$\sigma = 0.1205$
	$e_2 = 1.12$	$e_3 = 1.12$	$\bar{d}_2 = 1.0$
	$d_2 = 1.0$	$d_3 = 0.04$	

2.3. Optimal Control Theory

The application of optimization to make time independent decisions, such as those in Hof & Bevers (2002) and Mees & Strauss (1992), is easier than using a time dependent control. However, such decisions might be sub-optimal for natural systems as they ignore the inherent dynamic characteristics of these system. It may achieve the short term goals, but hamper the long term objectives. For example, limiting the nutrient or pollutant input into a lake at a constant level without giving proper credence to the environmental cycles and the aquatic life cycles in the lake can affect species diversity and their life expectancy. Such effects are not evident immediately and manifest themselves only over a longer time period. Therefore, an effective approach is to use control theory to derive time dependent management decisions. Moreover, there are multiple parameters in nature that are partially or completely under human control. In such cases, a rigorous mathematical analysis needs to replace decision making based on experience and logic. The use of advanced control strategies, therefore, might not just be an option, but rather a necessity.

Control theory aims to derive a time dependent profile of a particular system parameter, called the control variable, such that a specific objective is optimized over the considered time horizon. The development of the control theory has primarily been motivated to solve problems of engineering systems such as mechanical, electrical, chemical and others. However, environmental problems such as ecosystem management also offer an exciting avenue to implement some of the advanced control strategies. Some selected examples of this approach include: renewable resource management (Clark, 1990), food chain disaster management (Shastri & Diwekar, 2006a,b), population management through harvesting (Kolosov, 1997; Kolosov & Sharov, 1993; Sivert & Smith, 1977), lake water quality management (Ludwig et al., 2003) and forest fire management (Richards et al., 1999; Anderson, 1994). Some other applications of control theory in natural system management include Carlson et al. (1991) and Chukwu (2001).

Optimal control is at the forefront of advanced control strategies. Some advantages of optimal control over the other advanced control strategies are: it does not make any assumption about the form of the control law, it theoretically gives the best control strategy for the given objective function, and it can theoretically handle any type of system. Owing to the complexity of the natural systems and the sustainability based objectives, this work uses the theory of optimal control to derive top-down and bottom-up control profiles for the population models in an ecosystem.

Theoretical basics of optimal control are presented in standard texts such as Kirk (1970) and Lewis (1986), and the aspects important for this work are reported in Appendix B.. Briefly, for the deterministic predator-prey model, Pontryagins's maximum principle has been used (Kirk, 1970). This leads to the formulation of a set of ordinary differential equations (state and adjoint equations) and algebraic equations (optimality or stationarity condition), to be solved as a boundary value problem. For the stochastic predator-prey model, techniques based on Ito's lemma must be used. Stochastic dynamic programming is an option. However, it requires the solution of a set of partial differential equations which is computationally difficult. This work, therefore, uses the recently proposed stochastic maximum principle (Rico-Ramirez & Diwekar, 2004), which again leads to the formulation of a two-point boundary value problem, similar to the deterministic case. This method is, thus, computationally much more efficient than the stochastic dynamic programming. The next section describes the formulation of the objective function for the optimal control solution of the freshwater ecosystem model.

2.4. Sustainability and Fisher Information

The concept of sustainability came to the fore after the Brundtland commission report where sustainable development was generically defined as "the development that meets the needs of the present without compromising the ability of the future generations to meet their own needs" (Tomlinson, 1987). This has initiated research activities across various disciplines incorporating sustainability ideas (Anastas & Zimmerman, 2003; Heal, 1998; Ludwig et al., 1997; Cabeza Gutes, 1996). Goodland & Daly (1996) propose three principal dimensions of sustainability as social, environmental and ecological. Cabezas et al. (2003) refine and further discuss various perspectives of sustainability, which include, ecological (Myers et al., 2000), social (Tainter, 1988), economic (Daily et al., 2007; Ludwig, 2000; Heal, 1997), technological and systems (Fiksel, 2006). These different aspects are in continuous dynamic interaction, and consequently a truly sustainable initiative must be a multi-disciplinary effort (McMichael et al., 2003; O'Neill, 1996). Therefore, the goal of a sustainable management strategy is to promote the structure and operation of the human component of a system (society, economy, technology, etc.) in such a manner as to reinforce the persistence of the structures and operation of the natural component (i.e., the ecosystem) (Cabezas et al., 2005; Clark, 1990).

However, it is essential to have a sustainability quantifying measure from the management implementation perspective. Cabezas & Fath (2002) have proposed to use information theory in ecology to derive a measure for sustainability of a system, the hypothesis being based on the argument that information is a fundamental quantity of any system, irrespective of the discipline (Frieden, 1998). Previous applications of information theory in ecology include: using Shannon information (Shannon & Weaver, 1949) as an index of biodiversity; using entropy of information to investigate evolutionary processes (Brooks & Wiley, 35pp); using information about energy pathways to quantify interdependencies and diversity in a food web (Rutledge et al., 1976); measuring the distance of a system from thermodynamic equilibrium (based on exergy) (Mejer & Jørgensen, 1979), and developing the concept of Ascendancy (Ulanowicz, 1986). Cabezas and Fath use Fisher information as the quantity for their hypothesis.

Fisher information (FI), introduced by Ronald Fisher (Fisher, 1922), is a statistical measure of indeterminacy. One of its interpretations relevant for this work is as a measure of the state of order or organization of a system or phenomenon (Frieden, 1998). Fisher information, I, for a system with a single variable is given as (Cabezas & Fath, 2002)

$$I = \int \frac{1}{p(x)} \left(\frac{dp(x)}{dx} \right)^2 dx \tag{9}$$

where, p is the probability density function (pdf) of variable x. This definition can be extended to a system of n variables. When these n variables constitute the state variable vector of a system, it gives the Fisher information of that system. Fisher information, being a local property dependent on the derivative of the density function of the state variable vector, is sensitive to the perturbations that affect the density function, and therefore, can be used as an indicator of the organization of the system. Here, organization refers to the distribution of the system states. A system with many equally probable states is disorganized, while a system with a few preferred states is better organized. For a highly disorganized system, the lack of predictability due to the nearly uniform probability distribution of its states results in a low value of Fisher information. On the contrary, a highly organized system will have a high value of Fisher information. The central argument in the sustainability hypothesis (Cabezas & Fath, 2002) is that the stability (static or dynamic) of a system is sufficient (but not necessary) for the sustainability of the state of the system. A statically or dynamically stable system will have a high value of Fisher information due to one or a few preferred states. Thus, Fisher information, being an indicator of the system stability, is also an indicator of the sustainability of the state of the system. For the stability of an ecosystem (static or dynamic), it is important that the system is not losing or gaining species, affecting the system dimensionality and hence the value of Fisher information. The sustainability hypothesis, therefore, states that: the time-averaged Fisher information of a system in a persistent regime does not change with time. Any change in the regime will manifest itself through a corresponding change in Fisher information value (Cabezas & Fath, 2002). It should be noted that persistence of the regime does not imply system stationarity in the ecological sense. A regime can possibly have stable or unstable fluctuations, as those typically observed in natural systems. Pawlowski et al. (2005) and Cabezas et al. (2005) have successfully demonstrated that regime shifts in complex food webs are translated into corresponding changes in the Fisher information values.

Two additional corollaries to this hypothesis, stated by Cabezas & Fath (2002), are: (1) if Fisher information of a system is increasing with time, then the system is maintaining a state of organization, and (2) if Fisher information of a system is decreasing with time, then the system is losing its state of organization. These corollaries are based on the correlation between the system order and Fisher information, and they give an idea about the quality of change if the system is changing its state. An extensive review of Fisher information and the sustainability hypotheses can be found in Cabezas & Fath (2002) and Fath et al. (2003).

The sustainability hypotheses provide the theoretical basis for the work presented in this chapter. From an ecosystem management perspective, following two different objectives can be formulated based on these hypotheses:

- Maximization of time averaged Fisher information: This objective is based on the idea that higher Fisher information is a consequence of a more organized system. Hence, the objective attempts to push the system from its current state into a state that is more sustainable in the mathematical sense of Fisher information. It may demand rapid changes though.

- Minimization of Fisher information variance over time: This objective is a direct consequence of the sustainability hypothesis. Minimizing the Fisher information variance attempts to ensure the constancy of regime. The objective, thus, aims to maintain the system close to its current state. This is similar to Taguchi's approach of quality control (Diwekar, 2005; Taguchi, 1986).

3.   Optimal Control Problem Specifications

The preceding sections explained the theory behind the work presented in this chapter and the contribution of each aspect of the proposed approach. This section builds on the theoretical background and gives the optimal control problem formulations used in this work.

The objectives, as mentioned in section 2.4., are: maximization of time averaged Fisher information, and minimization of Fisher information variance over time. The definition of Fisher information given by Eq. 9 is in terms of the state variable vector x. For dynamic systems, there is a one to one correspondence between the system evolution (states) and time. Using this relationship and the chain rule of differentiation, the system pdf and Fisher information can be defined in terms of time. Time averaged Fisher information for a system with n species is thus given by:

$$I_t = \frac{1}{T_c} \int_0^{T_c} \left(\frac{a(t)^2}{v(t)^4} \right) dt \tag{10}$$

where, T_c is the cycle time of the system, and

$$v(t) = \sqrt{\sum_{i=1}^{n} \left(\frac{dx_i}{dt} \right)^2} \tag{11}$$

$$a(t) = \frac{1}{v(t)} \left[\sum_{i=1}^{n} \frac{dx_i}{dt} \frac{d^2 x_i}{dt^2} \right] \tag{12}$$

$v(t)$ and $a(t)$ are called the velocity and acceleration terms of the ecosystem, respectively. Please refer to Fath et al. (2003) for a formal derivation of Eq. 10. Based on this definition, the objectives are given as:

- Maximization of Fisher information:

$$J = \text{Max} \ \frac{1}{T} \int_0^{T} \left(\frac{a(t)^2}{v(t)^4} \right) dt \tag{13}$$

- Minimization of Fisher information variance:

$$J = \text{Min} \ \int_0^{T} (I_t - I_{constant})^2 \, dt \tag{14}$$

Here, T is the total time horizon under consideration. I_t, given by Eq. 10, is time averaged FI for one system cycle, and $I_{constant}$ is the constant around which the Fisher information variation is to be minimized.

The top-down and bottom-up control philosophies are compared by performing separate analyses using x_3 and x_1, the super-predator and the prey, as control variables, respectively. x_3 is controlled by manipulating the mortality rate d_3 of the super-predator. For freshwater ecosystems, this reflects control by managing the super-predator harvesting rate using policies such as fixed catch or fixed fishing season. x_1 is controlled by manipulating parameter

K representing the prey carrying capacity of the system. The control by manipulating K does not strictly represent bottom-up approach, which proposes control by nutrient addition. However, nutrient addition affects the prey carrying capacity of the system (Abrams, 1993). The results are, therefore, expected to give the impact of the manipulation in the lower level of the food chain on the upper level of the chain.

The starting conditions for the system (populations at the start) are known, and since the final states are free, adjoint variables at the final time are zero. The final state of the system is not constrained, and the objective does not contain a final time function. The time horizon for the control problem is considered to be large enough so that the control law is only state dependent (Kirk, 1970). The derivation of the optimal control profile requires the solution of the two point boundary value problem, which, for this highly complex differential-algebraic system of equations, is a cumbersome task. Numerical technique of steepest ascent of Hamiltonian is used to solve the boundary value problem (Kirk, 1970; Diwekar, 1996). The technique solves the problem as an optimization problem by discretising the solution horizon, using the control variable at each time instant as a decision variable, and trying to satisfy the optimality condition at each time point within a tolerance limit. The next section gives simulation results for the three species predator-prey model.

4. Results and Discussion

In this analysis, both control options and both objectives are tested on the tri-trophic food chain model (deterministic as well as stochastic). Moreover, the uncontrolled model is considered to have unstable dynamics. It simulates situations when the ecosystem needs external intervention to avoid imbalance. The analysis assesses the ability of different sustainability based objectives and control philosophies to identify and manage such situations. Two such cases are considered:

- Disturbance recovery analysis: The aim is to stabilize model dynamics after a disturbance that causes instability. Following three different cases are simulated:

    - Excessive prey and predator variation: Populations x_1 and x_2 vary excessively
    - Super-predator extinction: The super-predator population x_3 is going extinct
    - Super-predator explosion: The super-predator population x_3 is exploding

- Regime change analysis: Driving the system from an undesired to a desired regime (analysis done only for the deterministic model)

Moreover, it is important to understand the impact of uncertainty in the model on control problem results. This is achieved here through various simulation studies that are presented in section 4.3.. In the end, an interpretation of the stochastic model analysis in terms of real options theory is discussed, which presents an interesting view of analyzing the stochastic model results.

The cases are simulated by modifying the model parameters reported in Table 1. The changes are mentioned in the discussion of the respective cases. When K or d_3 is the control variable, the value reported in Table 1 is the starting guess for the steepest ascent algorithm. The control variables are constrained to avoid numerical problems. Maximum principle cannot take care of bounds on control variables. However, since the work uses steepest ascent of Hamiltonian algorithm that approximates the solution by the maximum principle, it is possible to include constraints on control variables. The constraints are $0.03 \leq d_3 \leq 0.05$ for top-down control and $500 \leq K \leq 900$ for bottom-up control. Most often, the control actions by humans are bounded by natural processes. For example, the mortality rate will be bounded by the natural reproduction rate of super-predators, and the natural input of nutrients to the lake will limit the lowest possible nutrient addition. Hence, incorporation of such bounds after considering the physical limitations will actually help in giving solutions that are optimal under the given restrictions.

The uncontrolled system is first simulated for the deterministic model, and values of the average Fisher information and Fisher information standard deviation are noted. The model is then subjected to the two different control philosophies. For the stochastic model, the uncontrolled system simulation is preceded by sampling of ϵ_t, the variable in the Ito process. The samples are stored, and the same sample set is used in all subsequent simulations of this model to allow a proper basis for the comparison. The time dependent predator mortality rate is shown in Figure 2. For FI variance minimization objective, $I_{constant}$ is given by the time averaged Fisher information for the corresponding model (deterministic or stochastic) when the model shows stable dynamic behavior, i.e., before the disturbance is introduced in the model. The following sections present results for the various cases.

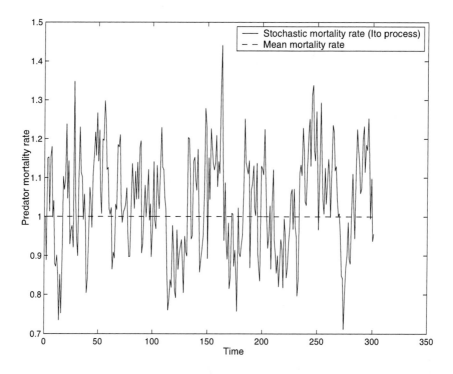

Figure 2. Time dependent random predator mortality rate for stochastic model.

Table 2. Deterministic tri-trophic food chain model: Parameter set for case 1

Prey	Predator	Super-predator
$x_1(0) = 100$	$x_2(0) = 75$	$x_3(0) = 150$
$r = 1.2$	$a_2 = 2.0$	$a_3 = 0.1$
$K = 710$	$b_2 = 235.3$	$b_3 = 250$
	$e_2 = 1.35$	$e_3 = 1.29$
	$d_2 = 1.0$	$d_3 = 0.04$

4.1. Disturbance Recovery

4.1.1. Case 1: Excessive Population Variation Analysis

Although the uncontrolled system is dynamically stable, excessive variation in prey and predator populations is undesirable. This is because when the population size is small during the cycle, the species are in a greater danger of becoming extinct due to unpredicted events such as natural disasters, external species invasion and others.

Deterministic Model

Model parameters used to simulate this case are reported in Table 2. The results show that top-down control has very little effect on the prey and predator dynamics (plots not shown). On the contrary, Figure 3 shows that bottom-up control has a significant effect on the predator-prey population dynamics. The result for FI variance minimization objective is desirable since it reduces the predator-prey variations, and achieves a dynamically stable system. Since the sustainability hypotheses argue that dynamic or static stability is a necessary condition for sustainability, the controlled system is also sustainable. Super-predator dynamics for these cases are plotted in Figure 4. The favorable result by using bottom-up control for FI variance minimization objective is accompanied by an increase in super-predator population. However, a substantial rise in the super-predator population for FI variance minimization objective using top-down control does not significantly affect the predator-prey dynamics. This observation indicates that variations in the super-predator population are the effect and not the cause of reduced predator-prey population variations. The control variable profiles for this case show that the bottom-up control variable (K) fluctuates slightly more rapidly (at a higher frequency) for the FI maximization objective than for FI variance minimization objective (plots not shown). The results for FI variance minimization objective will, therefore, be easier to implement on an actual system since the changes will be less frequent.

Stochastic Model

The model parameters used to simulate this case are reported in Table 3. Although the results are somewhat complicated to analyze, a careful analysis shows that the predator-prey variations are not affected much by top-down control (plots now shown). In comparison, bottom-up control affects the predator-prey population dynamics more severely, as shown in

Y. Shastri and U. Diwekar

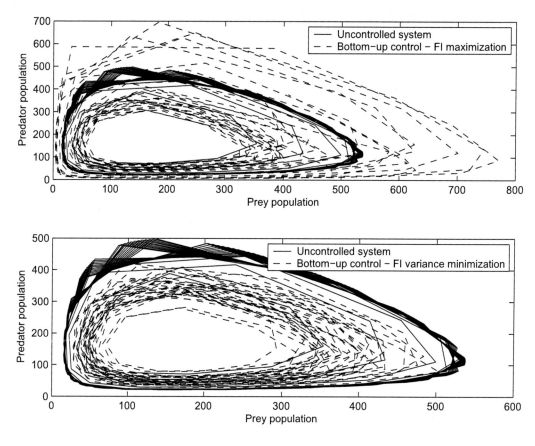

Figure 3. Case 1 for deterministic model: Predator-prey population dynamics using Bottom-up control.

Figure 5; and it is able to reduce the population variations. The result for FI variance minimization objective is particularly good, since the reduction in the variations is maximum for this case. Since the sustainability hypotheses argue that a system with few preferred states is more sustainable, the controlled system is also more sustainable than the uncontrolled one. The super-predator dynamics for these cases are plotted in Figure 6. The favorable results of variation reduction are accompanied by an increase in the super-predator population. However, a rise in the super-predator population for FI variance minimization using top-down control does not significantly affect the predator-prey dynamics. This observation indicates that the variations in super-predator population are the effect and not the cause of reduced predator-prey variations. The control variable profiles for this case again show that control variables fluctuate more rapidly (at higher frequency) for FI maximization objective than for FI variance minimization objective (plots not shown). Since FI variance minimization objective minimizes the variation between the average FI of each model cycle, control profile for this objective shows a piecewise nature.

A comparison with the deterministic model results indicates that the important conclusions are qualitatively similar. For both the models, bottom-up control achieves better reduction in the population variation of prey and predators, and control profiles obtained

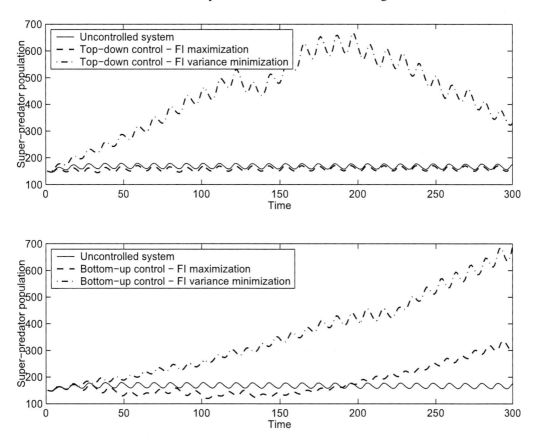

Figure 4. Case 1 for deterministic model: Super-predator population dynamics.

using FI variance minimization objectives are easier to implement.

Table 3. Stochastic tri-trophic food chain model: Parameter set for case 1

Prey	Predator	Super-predator	Ito process
$x_1(0)= 100$	$x_2(0)= 75$	$x_3(0)= 150$	$x_4(0) = 1.0$
$r = 1.2$	$a_2 = 2.0$	$a_3 = 0.1$	$\eta = 0.3535$
$K = 710$	$b_2 = 227.27$	$b_3 = 250$	$\sigma = 0.1205$
	$e_2 = 1.35$	$e_3 = 1.29$	
	$\bar{d}_2 = 1.0$	$d_3 = 0.04$	

4.1.2.   Case 2: Super-Predator Extinction

In order to simulate super-predator extinction, the value of b_2 (predator half saturation constant) is modified to 181.8181. All other parameters for the two cases are as reported in Table 1.

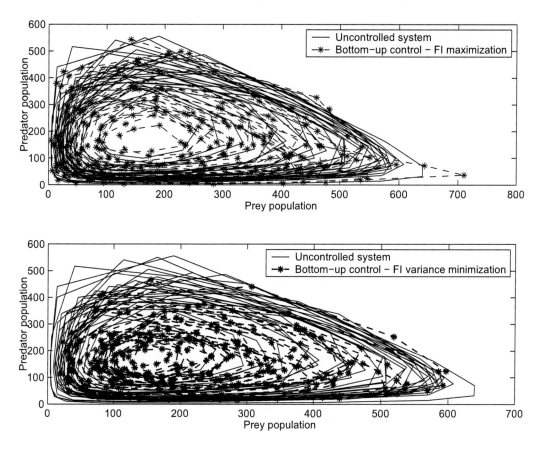

Figure 5. Case 1 for stochastic model: Predator-prey population dynamics using bottom-up control.

Deterministic Model

Super-predator dynamics for this case are shown in Figure 7. The plots show that all control options achieve the desired goal of arresting super-predator extinction. The degree of success in restricting the super-predator extinction varies for different control options. FI maximization objective exerts a stronger impact for both control options, and the super-predator achieves a much higher population for top-down control. On the other hand, FI variance minimization objective shows a slower and weaker impact. Model simulations for a longer time duration show that FI variance minimization objective is not able to elevate the super-predator population back to the initial level, but manages to avoid super-predator extinction. Since species extinction is avoided in all cases, they represent more sustainable systems as compared to the uncontrolled one. When the predator-prey dynamics for this case are analyzed, it is noticed that the effect of bottom-up control on predator-prey dynamics is stronger than top-down control. Figure 8 shows the results for bottom-up control, while the results for top-down control are not show due to their insignificant effect.

Control variable profiles for both objectives are plotted in Figure 9 (top-down control) and Figure 10 (bottom-up control). As expected, elevating the super-predator population

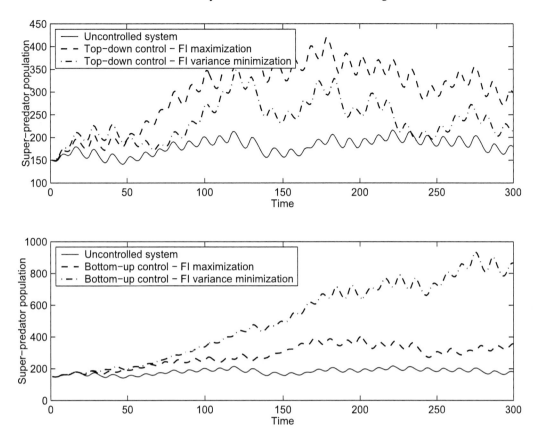

Figure 6. Case 1 for stochastic model: Super-predator population dynamics.

increases the FI value. For top-down control, this is achieved by lowering the mortality rate to the minimum possible value (0.03) for about the first half of the simulation horizon. The higher than desired increase in super-predator population is then compensated by a progressive increase in the mortality rate. Since FI variance minimization objective minimizes the variation between the average FI of each model cycle, piecewise nature of the control action is evident in Figure 9. The control actions are taken based on the dynamics of each cycle, and they keep fluctuating between the two extremes. The effect is evident in the super-predator population plot in Figure 7, where the population is maintained around the starting point for the most part. The plots for bottom-up control also show the piecewise nature to a certain extent while using FI variance minimization objective. The interpretation of these profiles is though not straightforward, since the effect on super-predator population is indirect and nonlinear. The plots also indicate that the control variable profiles for the objective of FI maximization fluctuate a little more rapidly, and hence might be difficult to implement on a physical system. Another interesting observation from these results is the relationship between the FI value and associated population dynamics for the model. For the case of FI variance minimization objective using bottom-up control, model population dynamics are clearly more sustainable than the uncontrolled case. However, average FI for this case is lower than the uncontrolled case. This illustrates that the relationship between absolute FI value and model dynamics is nonlinearity and subjective.

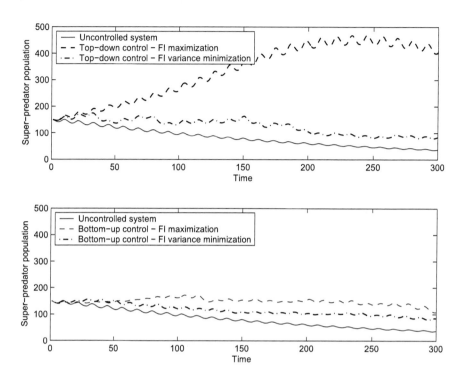

Figure 7. Case 2 for deterministic model: Super-predator population dynamics.

Stochastic Model

The super-predator population dynamics for this case are plotted in Figure 11. As for the deterministic case, the objective of super-predator elevation is achieved in all cases, and degree of success varies for different cases. Moreover, the results for all cases are qualitatively similar to those for the deterministic model. This leads to the conclusion that is similar to that for the deterministic model; i.e., FI maximization objective, particularly in combination with top-down control option, has a stronger impact than FI variance minimization. Although the predator-prey dynamics for this case are quite mixed up, careful consideration indicates that bottom-up control impacts the predator-prey dynamics more strongly than top-down control (plots not shown). Similar to the deterministic model, the super-predator response seems to be sensitive to the objective (FI maximization).

The control variable profiles for top-down and bottom-up control are shown in Figure 12 and Figure 13, respectively. Since the super-predator population can be increased by reducing its mortality rate, the mortality rate control variable is maintained around the lower bound 0.03. The bottom-up control variable varies over a greater range. Since its effect on the super-predator population is indirect, a direct interpretation is difficult. These profiles reinforce the point that FI maximization objective gives control variable profiles that are relatively difficult to implement due to more rapid fluctuations.

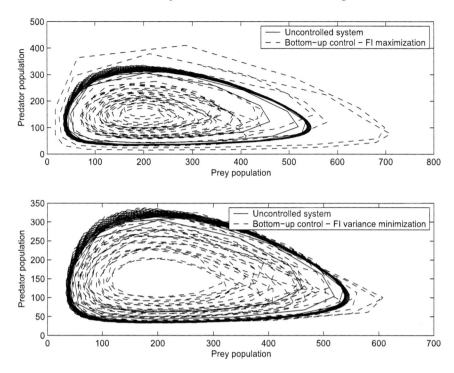

Figure 8. Case 2 for deterministic model: Predator-prey population dynamics using bottom-up control.

4.1.3. Case 3: Super-Predator Explosion

Deterministic Model

To simulate super-predator explosion, b_2 (predator half saturation constant) is modified to 204.08. All other parameters for the two cases are as reported in Table 1. Super-predator dynamics for this case are plotted in Figure 14. It is observed that FI variance minimization objective manages to restrict super-predator population explosion. However, FI maximization objective does not give the desired results. Top-down control for this objective further supports super-predator population explosion. This is because the relationship between Fisher information value and super-predator population is nonlinear, as mentioned in the discussion of case 2 results. Thus, FI variance minimization objective, which controls super-predator explosion, has a smaller average FI than the uncontrolled case. These results highlight the nonlinearity between FI value and the population dynamics, and suggest that FI maximization objective may not always lead to a desirable dynamic state of the system.

Regarding the predator-prey dynamics for this case, it is again observed that top-down control has an insignificant impact on prey and predator populations (plots not shown). On the contrary, Figure 15 showing results for bottom-up control reveals that the predator-prey dynamics are more significantly affected by bottom-up control, and this effect is more pronounced for the objective of FI maximization. The control variable profiles for both objectives are shown in Figure 16 (top-down control) and Figure 17 (bottom-up control). These plots again show the piecewise nature of the control variable when FI

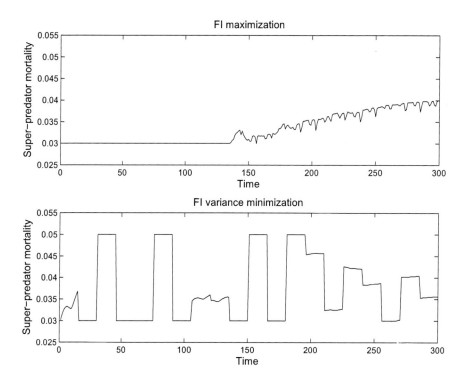

Figure 9. Case 2 for deterministic model: Top-down control variable profile.

variance minimization objective is used. In this case though, the difference in the nature of the control variable profiles for the two objectives is not as significant as in the previous cases, and hence, no specific comment about the ease of implementation can be made.

Stochastic Model

For this case, $b_2 = 208.33$ to simulate super-predator explosion. Super-predator population dynamics for this case are plotted in Figure 18. The plots indicate that all control options are able to restrict the super-predator explosion to a different extent. The objective of FI maximization with bottom-up control reduces the population excessively; however, it does not appear to take the super-predator close to extinction. For the other cases, the desired goal of controlling the super-predator extinction is achieved satisfactorily. Regarding the predator-prey dynamics, as in case B, the dynamics are mixed up, but indicate that the bottom-up control affects the predator-prey dynamics more than the top-down control (plots not shown). The control variable profiles for top-down and bottom-up control are shown in Figure 19 and Figure 20, respectively. The plots, as in the previous cases, show the piecewise nature for FI variance minimization objective. They also emphasize that objective of FI variance minimization results in control variable profiles that will be easier to implement on a physical system.

A comparison with the deterministic model shows that the results are qualitatively similar for FI variance minimization objective. However, it is different for the FI maximization objective. For the deterministic model, FI maximization objective results in a further rise in the super-predator population, while for the stochastic system, the super-predator popu-

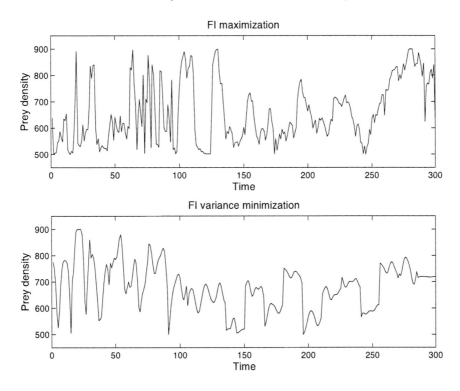

Figure 10. Case 2 for deterministic model: Bottom-up control variable profile.

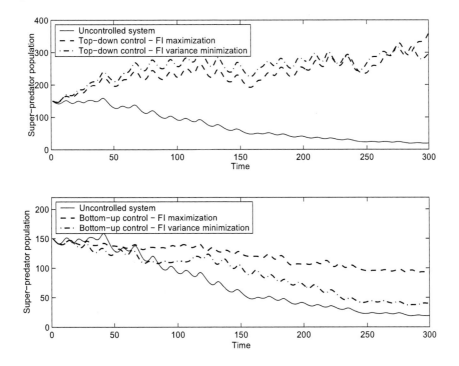

Figure 11. Case 2 for stochastic model: Super-predator population dynamics.

Y. Shastri and U. Diwekar

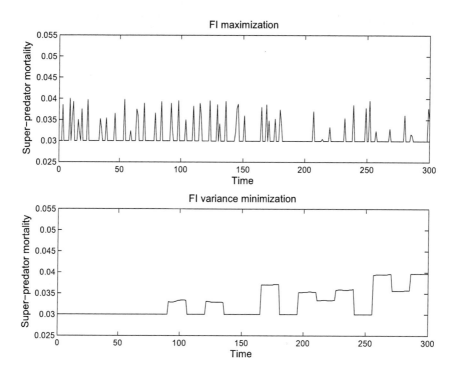

Figure 12. Case 2 for stochastic model: Top-down control variable profile.

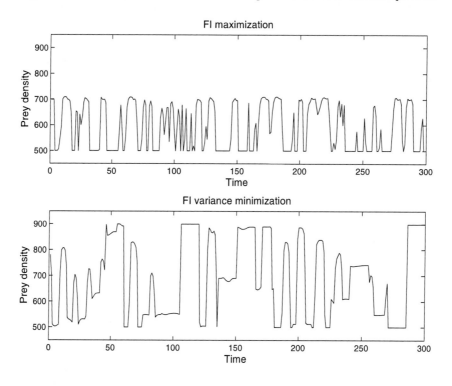

Figure 13. Case 2 for stochastic model: Bottom-up control variable profile.

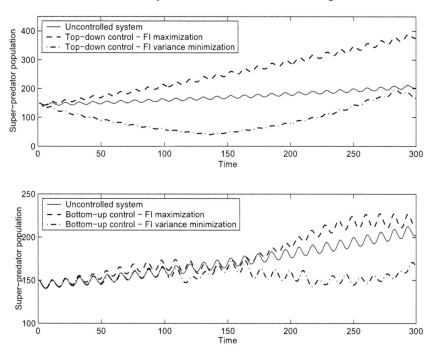

Figure 14. Case 3 for deterministic model: Super-predator population dynamics.

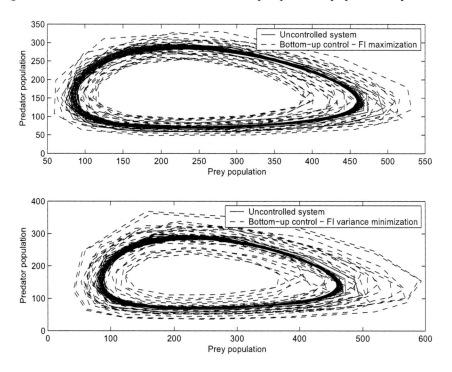

Figure 15. Case 3 for deterministic model: Predator-prey population dynamics for bottom-up control.

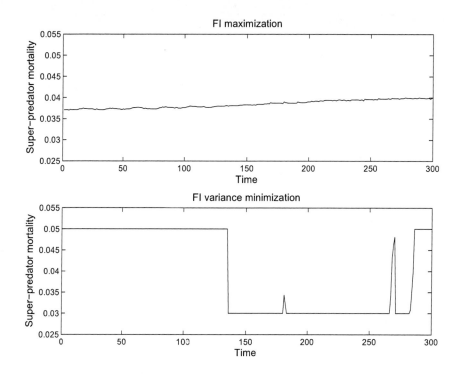

Figure 16. Case 3 for deterministic model: Top-down control variable profile.

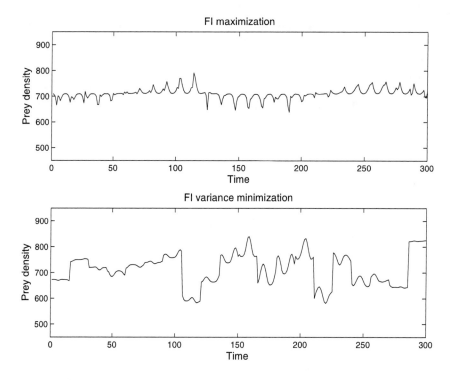

Figure 17. Case 3 for deterministic model: Bottom-up control variable profile.

lation is satisfactorily controlled. This suggests that the noise associated with uncertainty in this case reduces the destabilizing effect of FI maximization objective observed for the deterministic systems. The result shows that it might be advantageous to use the objective of FI maximization.

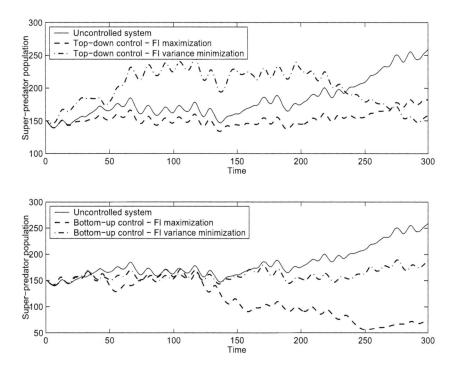

Figure 18. Case 3 for stochastic model: Super-predator population dynamics.

4.2. Regime Change Analysis

Drastic changes in the parameters of an ecosystem can cause shifts in the dynamic regimes of these systems, and the new regimes can be stable or unstable. Such changes are typically caused by natural disasters such as floods and hurricanes, or by dramatic changes in global climatic conditions (Mayer & Rietkerk, 2004). Quite often, these regime shifts are nonlinear, exhibiting phenomenon like hysteresis, meaning that the restoration of the original regime is complicated. Using time dependent manipulations of the system parameters, as discussed in this work, is a possible option to carry out this task.

The results presented in this section compare top-down and bottom-up control philosophies for the tri-trophic food chain model with the aim of shifting the population model from one dynamic regime (undesirable) to another (desirable). The same tri-trophic food chain model for a different set of parameters, reported in Table 4, is used for this analysis. The analysis is done only for the deterministic food chain model. Model parameters and different regimes for this model are discussed in Gragnani et al. (1998). The possible regimes include cyclic-low frequency regime and cyclic-high frequency regime. The parameter set reported in Table 4 results in the cyclic-low frequency regime (an undesirable regime, possibly due to a natural disaster). Population dynamics for this regime are

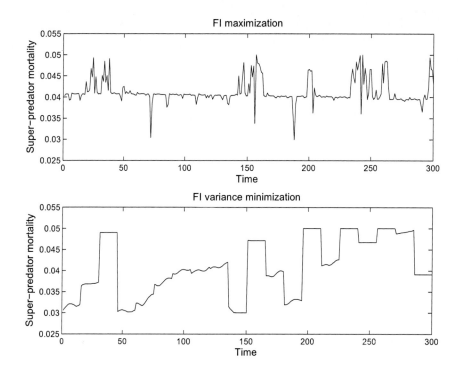

Figure 19. Case 3 for stochastic model: Top-down control variable profile.

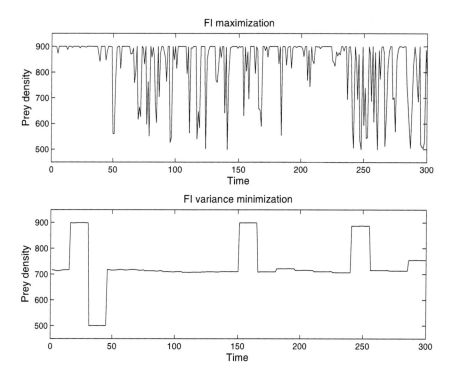

Figure 20. Case 3 for stochastic model: Bottom-up control variable profile.

Table 4. Tri-trophic food chain model: Parameter set for the regime change analysis

Prey	Predator	Super-predator
$x_1(0) = 0.75$	$x_2(0) = 0.2$	$x_3(0) = 10$
$r = 1.5$	$a_2 = 1.67$	$a_3 = 0.05$
$K = 1.0$	$b_2 = 0.333$	$b_3 = 0.5$
	$e_2 = 1.0$	$e_3 = 1.0$
	$d_2 = 0.4$	$d_3 = 0.01$

shown in Figure 21 (predator and prey populations) and Figure 22 (super-predator population). The goal is to shift the system to the desirable cyclic-high frequency regime, which is also shown in Figure 21 and Figure 22. These regimes have different average Fisher information values. The control philosophies try to achieve the regime shift by minimizing its Fisher information variance around the average Fisher information value of the desired regime (cyclic-high frequency). Thus, only FI variance minimization objective is used. Super-predator mortality rate and prey carrying capacity are again the control variables for top-down and bottom-up control, respectively. The constraints on the control variables are: $0.005 \leq d_3 \leq 0.015$ for top-down control and $0.5 \leq K \leq 1.5$ for bottom-up control.

The results for the two control options are shown in Figure 23 (predator and prey population) and Figure 24 (super-predator population). It is clearly evident that the bottom-up controlled system shifts into the desired regime. On the contrary, the top-down controlled system fails to do so in the considered time horizon. This result illustrates that bottom-up control is a better option to affect regime changes, which might be required to recover a system from natural disasters such as hurricane Katrina.

4.3.   Uncertainty Impact Analysis

Understanding the impact of uncertainty consideration and the degree of uncertainty on control problem results is important. This is done here by performing two different simulation studies for the stochastic model:

- Analyzing the dependence on Ito process

- Analyzing the impact of using stochastic control theory to derive control variable profiles

Both these cases are analyzed for the super-predator extinction case for the tri-trophic food chain model.

4.3.1.   Ito Process Dependence

Understanding the dependence of the results on Ito process parameters assumes importance because predator mortality rate representation using the Ito process is an assumption, based on its successful use to model human mortality (Diwekar, 2005). Even though an Ito process representation is valid, it is quite likely that the exact characteristics (parameters)

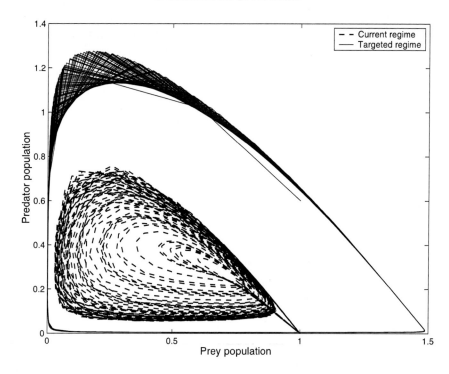

Figure 21. Regime change analysis: Predator-Prey population dynamics.

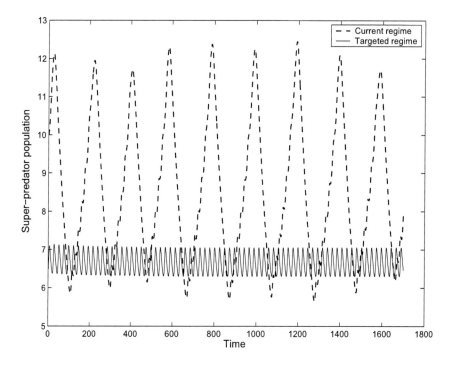

Figure 22. Regime change analysis: Super-predator population dynamics.

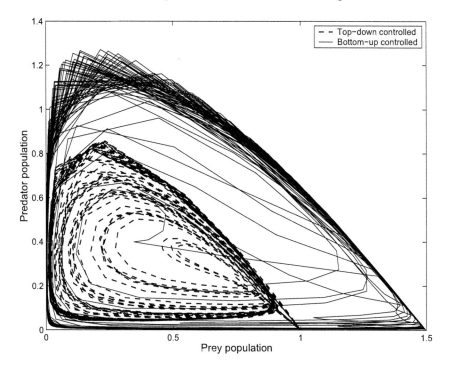

Figure 23.   Regime change analysis: Predator-Prey population dynamics for controlled system.

for predator mortality rate are different. This work analyzes the dependence on the constant variance parameter σ of the Ito process. The variance parameter for the original Ito process, $\sigma = 0.1205$, is increased to $\sigma = 0.2205$. This causes the predator mortality rate to fluctuate with higher variance, as shown in Figure 25. Case 3 (super-predator extinction) is analyzed for the new Ito process, where the aim is to avoid super-predator extinction. The super-predator dynamics for this case are shown in Figure 26. Compared to the case 2 dynamics presented before (Figure 11), the uncontrolled super-predator population falls more drastically in this case due to the increased variance of the Ito process. For the controlled cases, it is seen that bottom-up control is not able to recover the super-predator population. Interestingly, top-down control for both objectives is able to control super-predator extinction. FI maximization objective performed worse than FI variance minimization objective with the constraint of $0.03 \leq d_3 \leq 0.05$ on super-predator mortality rate (top-down control variable). This contradicts the earlier result for case 3, including those for the deterministic model, suggesting that FI variance minimization might be a better objective overall. A general conclusion is that, under severe disturbances in system dynamics, bottom-up control is not strong enough to control the system.

4.3.2.   Effect of Stochastic Optimal Control Theory

A quantitative comparison of the results for the deterministic and stochastic cases indicates that uncertainty impacts the relative extent of success or failure of a management option. It is equally important to understand the impact of uncertainty on the decisions. For

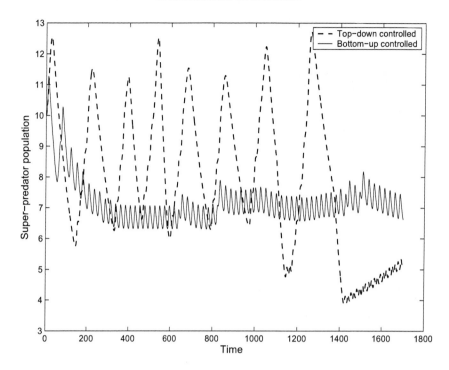

Figure 24. Regime change analysis: Super-predator population dynamics for controlled system.

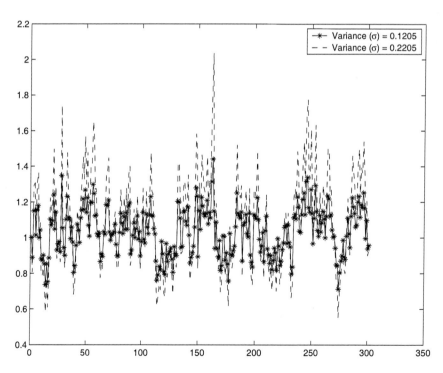

Figure 25. Comparison of Ito processes for different variance parameters.

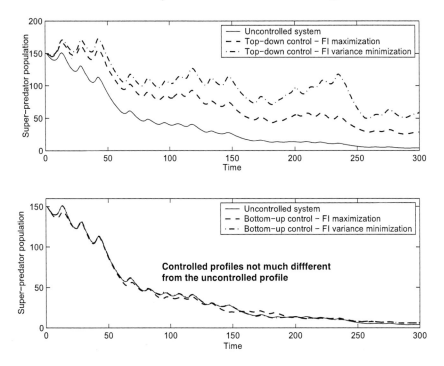

Figure 26. Super-predator population dynamics for Ito process with large variance.

the stochastic model, modeling using Ito process and using the stochastic maximum princi-ple for problem formulation complicates the analysis. Although ignoring uncertainty allows one to use simpler deterministic methods, such an approach should lead to sub-optimal re-sults since the effect of uncertainty on control variable is ignored. This is ascertained by conducting the following simulation study.

The response of the stochastic tri-trophic food chain model exhibiting super-predator extinction, when controlled by prey carrying capacity (bottom-up control) using FI maxi-mization objective, is plotted. The following two cases are considered:

- Stochastic system controlled by control variable profile generated using stochastic optimal control theory (stochastic maximum principle)

- Stochastic system controlled by control variable profile generated by using determin-istic control theory, which does not consider uncertainty to derive the control profile

The plots for the super-predator population are shown in Figure 27. They indicate that super-predator population is elevated much more using the control profile generated by stochastic optimal control theory. Although, using the control profile generated by deter-ministic optimal control theory also restricts super-predator extinction, its performance is clearly inferior to the stochastic control variable profile. One can, therefore, conclude that stochastic control gives a better result as compared to deterministic control. This trend is observed, to a greater or lesser extent, for other cases too, highlighting the importance of uncertainty incorporation in control problem solution. Figure 28 compares the determinis-tic and stochastic control variable profiles. It can be seen that the two control profiles differ, particularly during the initial half of the simulation. The profile generated by stochastic maximum principle is not only better due to the resulting dynamics, but also because its magnitude of fluctuations is less than that for the deterministic control. This emphasizes that the effect of uncertainty on decisions is significant.

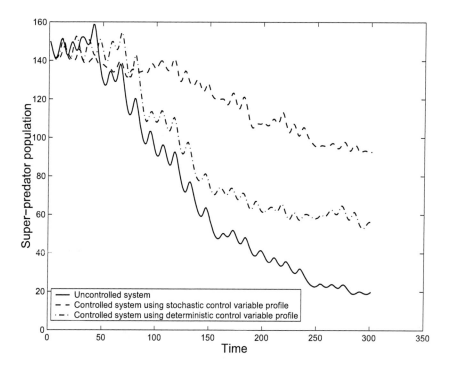

Figure 27. Comparison of deterministic and stochastic control results for stochastic model.

4.4. Uncertainty Analysis and Real Options Theory

Most investment decisions share three important characteristics in varying degrees. First, the investment is partially or completely "irreversible". In other words, the initial cost of investment is at least partially "sunk"; one cannot recover it all should one change the mind. Second, there is "uncertainty" over the future rewards from the investment. The best one can do is to assess the probabilities of the alternative outcomes that can mean greater or smaller profit (or loss) for the venture. Third, one has some leeway on the "timing" of the investment. The action can be postponed to get more information, but not with complete certainty. These three characteristics interact to determine the optimal decisions or "options" for investors. A firm with an opportunity is holding an "option" to buy an asset at some future time of its choosing. When a firm makes an irreversible expenditure, it exercises or "kills" its option to invest. This lost option value is an opportunity cost that must be included as a part of the cost of investment. Thus, not only the decisions, but also the timing of the decisions is optimized. Opportunity cost is highly sensitive to the uncertainty over the future value of the project, and options theory allows one to take decisions in the presence of forecasting of these uncertainties.

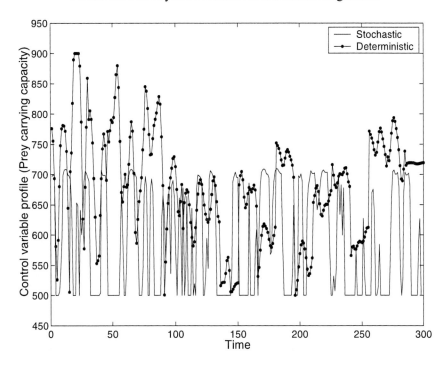

Figure 28. Comparison of deterministic and stochastic control profiles for stochastic model.

In an analogous fashion, irreversibility, uncertainty, and timing issues are also important for sustainability (Diwekar, 2003; Dixit & Pindyck, 1994). Forecasting and decision making under uncertainty in the context of sustainability is similar to financial decision making (Diwekar, 2005). The results for the stochastic food chain model presented in section 4.3. can be viewed from this perspective. When the stochastic model is controlled by a control profile derived for a deterministic model, realizations of the uncertain mortality rate are ignored. On the contrary, using the Ito process representation and stochastic maximum principle, the decisions are optimized in the presence of forecast, thereby accounting for the realizations of the uncertain parameter. The decision variable (prey carrying capacity) can be thought of as an option available to maximize the considered benefits. With stochastic control, these options were optimally utilized by changing the magnitude and timing of the decisions. The decisions will have a certain cost associated with them. For example, bottom-up control in aquatic systems is usually affected by the addition of nutrients in the water body, and will incur certain cost to the regulatory agency. Optimization considering uncertainty forecast ensures that the best use of these resources is made, consequently minimizing the cost of implementation to achieve the objectives. If optimization of the cost of these decisions is also an objective, then the proposed stochastic control method to derive the control profiles is a better alternative in the light of options theory.

The uncertainty representation used here is based on finance literature, where uncertainty realizations and their probabilities are known. However, this may not be true for natural systems, where unexpected outcomes with unknown probabilities are possible, leading to unknown values of the decisions taken at the present time. Arrow & Fisher (1974) argue that these unknown values create a quasi-option for the decisions. Conrad (1980) shows that quasi-option value is equivalent to the expected value of information, while option value is equivalent to the expected value of perfect information. In this context, incorporation of quasi-options concepts in future will broaden the scope of this work, particularly for the natural systems. Furthermore, work in finance literature is mostly restricted to linear cases. Here, however, the use of stochastic optimal control theory, based on rigorous mathematical concepts, allows one to extend the ideas to natural systems that are often nonlinear in nature.

5.   Computational Considerations

Since the system of equations being solved in this work is quite complex, computational problems need to be carefully avoided. Depending on the absolute value of Fisher information, the objective function may need to be linearly scaled to avoid numerical errors during the solution. Experience shows that this greatly reduces convergence time. Termination constant and step size for the steepest ascent method need to be carefully chosen for converging results. Most often, it is a compromise between faster convergence and running the risk of making the solution divergent. It was also observed that a good initial guess is important to have convergence.

6.   Conclusion

Sustainability of freshwater ecosystems such as lakes and rivers is important to ensure that the critical services and functions are continued to be provided. Although these ecosystems are built to be inherently resilient, excessive impact of anthropogenic effects as well as severe natural disasters make these systems vulnerable and demand external human intervention. The development of the management policies is a difficult due to the complex nature of these ecosystems. The central theme of the work presented in this chapter is to illustrate the application of systems theory based approached for sustainable management of freshwater ecosystems. The work uses the Rosenzweig-MacArthur model which is a particular form of the general Lotka-Volterra type models, to represent the freshwater ecosystem in terms of a three species predator prey model. The basic deterministic model is extended to consider uncertainty by modeling the predator mortality rate as a mean reverting Ito process. Two different control philosophies for the system, namely, top-down control and bottom up control, are compared by using super-predator mortality rate and prey carrying capacity as the policy variables respectively. Optimal control theory is used to derive time dependent profiles for these policy variables. Fisher information along with sustainability hypotheses is used to formulate different objective functions as indicators of sustainability. Section 4. reports the results for the deterministic and stochastic tri-trophic food chain models. An important conclusion to draw from those results is the manifestation of a strong correlation between Fisher information based sustainability hypotheses and ecosystem dynamics. Favorable changes in Fisher information, as suggested by the hypotheses, reflect, in general, in favorable ecosystem dynamics. The definition of favorable system dynamics is quite subjective. Here, system dynamics are considered to be favorable if species show less population variations, and/or if species do not become extinct. The classification is based on the comparison with the uncontrolled system. The results should give enough incentives for further investigations into the application of Fisher information for ecosystem management. Based on the results, following conclusions can be drawn:

- The results for the deterministic and stochastic models are qualitatively similar.

- The population dynamics and Fisher information value for a system are nonlinearly related, and hence, FI maximization is not guaranteed to improve the dynamics of the system.

- The objective of FI maximization results in an elevated super-predator population in all but one cases. For case 3 when the uncontrolled system shows super-predator explosion, this response is highly undesirable and suggests that the system might become unstable.

- The objective of FI variance minimization achieves the desired objective of super-predator population control in all cases.

- Regarding the predator-prey dynamics, bottom-up control is observed to have a greater impact on it, evident by the population dynamics considerably different from the uncontrolled case, and this impact is more significant for FI maximization objective. For top-down control, the impact on predator-prey dynamics is not significant.

- In terms of the absolute values of the objective functions, the bottom-up controlled systems obtain better values (i.e., higher time averaged FI and lower FI standard deviations) than the top-down controlled systems.

- The objective of FI maximization resulted in significantly worse values of the second objective (standard deviation), while the objective of FI variation minimization did not alter the value of the second objective (average FI) by much.

- Control profiles for FI variance minimization objective are easier to implement on a physical system than those for FI maximization objective, particulary in the presence of uncertainty.

- Bottom-up control is more effective than top-down control in changing the dynamic regime of a system.

- Incorporation of uncertainty in decision making is important for more effective management of the ecosystems.

To summarize the results, one can argue that the objective of FI variance minimization will give a stable response without much disturbance, while FI maximization objective may result in significant disturbances, and might increase the abundance of super-predator. It is also observed that FI variance minimization objective is not able to recover an uncontrolled system from significant disturbances (such as fast specie extinction). It is likely that some of these trends are system dependent. Based on the presented results though, FI variance minimization objective should be preferred in natural systems where ensuring stability might be preferred over risking significant disturbance in an attempt to improve the state. On the contrary, for systems such as fisheries, where maximization of fish harvesting is desirable, FI maximization objective should be preferred, particularly since the impacts can be kept localized, and system is under better human control than completely natural systems.

The results obtained here are based on simulations of simplified models. It is well accepted that models can never truly represent the complex natural ecosystems. One might, therefore, doubt the validity of these findings. However, experimental results come with their own set of deficiencies. The results are quite often system specific (valid only for the particular lake or river), non-reproducible, and the observations could be affected by many unknown factors not under direct human control. In such a situation, a strategic combination of theoretical and experimental approaches is needed. Theory helps import novel and proven ideas from other fields into the field of ecosystem management, while experiments help estimate the goodness of these findings. The results presented here are, therefore, to be viewed with this perspective. Application of control theory to achieve sustainable ecosystems should guide an experimentalist to try different management options. It is expected that such an approach, if replaces logic and heuristics, will simplify the task of experimental biologists.

Appendix

A.   Real Options Theory Modeling

In any stochastic process, there is a variable evolving over time t, and both, time t as well as the variable itself, can assume discrete as well as continuous nature. Wiener process, also known as Brownian motion, is a simple continuous time and continuous state stochastic process. It can be used to model a variety of continuous stochastic processes (Dixit & Pindyck, 1994; Diwekar, 2003). The Wiener process is represented as:

$$dz = \epsilon_t \sqrt{dt} \tag{A.1}$$

where, dz is the random variable, and ϵ_t is a normally distributed random variable, with zero mean and unit standard deviation. Random variable dz has the property that the expectation is zero ($E[dz] = 0$) and the variance is dt ($var[dz] = dt$). Using this definition of the Wiener process, the general form of the Ito process is given as:

$$dx = a(x, t)\, dt + b(x, t)\, dz \tag{A.2}$$

Here, $x(t)$ is the continuous time stochastic variable that is to be modeled, and dz is the Wiener increment as defined by Eq. A.1. $a(x, t)$ and $b(x, t)$ are known (nonrandom) functions. $a(x, t)$ is the drift parameter, and $b(x, t)$ is the variance parameter. The first term of the equation constitutes the deterministic part, while the second term constitutes the stochastic part. The mean of the Ito process is $E[dx] = a(x, t)\, dt$, and the variance is $var[dx] = b^2(x, t)\, dt$.

The general equation of the Ito process (Eq. A.2) is the origin of different stochastic processes, such as the simple Brownian motion with drift, the geometric Brownian motion and the mean reverting Ito process. The geometric Brownian motion is frequently used to model security prices, as well as interest rates, wages rates etc. However, this work uses the mean reverting Ito process to model the predator mortality owing to its success in modeling stochastic variables across seemingly disparate fields, including human mortality (Diwekar, 2005). The mean reverting Ito process is represented as:

$$dx = \eta \, (\bar{x} - x)dt + \sigma \, dz \qquad (A.3)$$

Here, η is the speed of reversion, σ is the constant variance parameter, and $\bar{x}$ is the normal or mean level of x to which $x(t)$ tends to revert. The expected change in x depends on the difference between x and $\bar{x}$. The characteristic of the mean reverting Ito process is that, although it models the random variable fluctuations for a short time, in the long run, the variable is drawn back to the mean value. If the variance rate is growing with x, then the Ito mean reverting process is represented as:

$$dx = \eta \, (\bar{x} - x)dt + \sigma \, x \, dz \qquad (A.4)$$

The mean reverting process has been used to model many stochastic variables, such as crude oil and copper prices (Dixit & Pindyck, 1994), relative volatility of non-ideal mixtures (Ulas & Diwekar, 2004) and also the human mortality rate (Diwekar, 2005).

B. Optimal Control Control

In general, control refers to a closed loop system, where the desired operating point is compared with an actual operating point and knowledge of difference is fed back to the system. Conventional frequency domain techniques are then used to design a controller. Optimal control problems on the other hand are defined in time domain, and their solution requires establishing an index of performance for the system and designing the course (future) of action so as to optimize the performance index (Diwekar, 1996).

B.1. Deterministic

The theory presents three possible methodologies to derive the optimal control law: dynamic programming (Hamilton-Jacobi-Bellman equation), calculus of variation (Euler-Lagrange equation) and Pontryagin's maximum principle (Kirk, 1970; Diwekar, 2003). In this work, Pontryagin's maximum principle has been used. A detailed explanation of the methodology is skipped here, and only a brief overview of the final equations is given. The interested readers are referred to Kirk (1970).

Consider a system represented by the following set of differential equations, called the state equations in the language of control theory:

$$\dot{x} = f(x, u, t) \qquad (B.1)$$

where, x is the state variable vector ($x(t) \in R^n$). The initial condition of the state variable vector is $x(t_0) = x_0$, while the final conditions is $x(T)$. u is the control variable vector ($u(t) \in R^m$). In optimal control, there is a time dependent performance index. It is represented here as:

$$J(t_0) = \int_{t_0}^{T} F(x(t), u(t), t)dt \tag{B.2}$$

where, F is the function to be optimized over the time interval $[t_0, T]$. The optimal control theory converts this integral objective into a Hamiltonian (calculated at each time step), which is defined as:

$$H(x, u, t) = F(x, u, t) + \lambda' f(x, u, t) \tag{B.3}$$

where, λ is a set of the costate or adjoint variables (R^n) (λ' represents the matrix transpose). The optimal control law is then given by the solution of the following set of equations:

State Equations

$$\dot{x} = \frac{\partial H}{\partial \lambda} = f \qquad t \geq t_0 \tag{B.4}$$

Adjoint Equations

$$-\dot{\lambda} = \frac{\partial H}{\partial x} = \frac{\partial f'}{\partial x}\lambda + \frac{\partial F}{\partial x} \qquad t \leq T \tag{B.5}$$

Optimality Conditions

$$0 = \frac{\partial H}{\partial u} = \frac{\partial F}{\partial u} + \frac{\partial f'}{\partial u}\lambda \tag{B.6}$$

This is a set of $2n$ ordinary differential equations (state and adjoint equations) and m algebraic equations (optimality or stationarity condition), to be solved as a boundary value problem. The state variables are knows at the initial time t_0 (x_0), while the adjoint variables are known at the final time T ($\lambda(T)$). The boundary values of the adjoint variables depend on the problem specification (Kirk, 1970). The control trajectory obtained using the optimality condition is optimal for the considered objective function and the starting conditions.

B.2.   Stochastic

For deterministic systems, Pontryagin's maximum principle was used to derive the optimal control equations. However, this method cannot be used in the presence of stochastic processes. For such cases, one needs to use the methods based on the Ito's Lemma. One such method, the stochastic maximum principle, has been recently proposed (Rico-Ramirez et al., 2003; Rico-Ramirez & Diwekar, 2004), and is used in this work. In this method, the stochastic dynamic programming formulation is converted into a stochastic maximum principle formulation. The main advantage of using this approach is that the solution to the partial differential equations in dynamic programming formulation is avoided. Instead, a set of ordinary differential equations needs to be solved as a boundary value problem. An exhaustive explanation of the theory is beyond the scope of this article, and interested readers are referred to Rico-Ramirez & Diwekar (2004). Given here are the final equations needed to derive the optimal control law.

Consider a system represented by the following set of differential equations.

$$dx = f(x, u, t)\, dt + g\, dz \tag{B.7}$$

where, x is the state variable vector of dimension n ($x(t) \in R^n$), and u is the control variable vector of dimension m ($u(t) \in R^m$). The starting condition for the state vector is given by $x(t_0) = x_0$, and the final condition at time T is $x(T)$. Let $1, \ldots, n_k$ be the set of deterministic states, and $n_{k+1}, \ldots, n$ be the set of uncertain states. The second part of Eq. B.7 models the uncertainty. For deterministic states, the function $g = 0$. In optimal control, there is a time dependent performance index, which, in this case, is represented as:

$$J(t_0) = \int_{t_0}^{T} F(x(t), u(t), t)\, dt \tag{B.8}$$

where, F is the function to be optimized over the time interval of $[t_0, T]$. The Hamiltonian for this stochastic case is defined as:

$$H(x, u, t) = F(x, u, t) + \lambda'\, f(x, u, t) + \frac{1}{2}\, g^2\, w \tag{B.9}$$

where, $\lambda(t)$ is the set of co-state or adjoint variables ($\lambda(t) \in R^n$) (the first derivatives of the objective function F with respect to state variables), and λ' represents the matrix transpose. $\omega(t)$ represent the second derivatives of the objective function F with respect to the state variables. This term is included due to the Ito process contribution. The optimal control law is then given by the solution of the following set of equations:

State Equation

$$\dot{x}_i = \frac{\partial H}{\partial \lambda_i} = f \qquad i = 1, \ldots, n \tag{B.10}$$

Costate Equation

$$-\dot{\lambda}_i = \frac{\partial H}{\partial x_i} = \frac{\partial f'}{\partial x_i}\lambda_i + \frac{\partial F}{\partial x_i} \qquad i = 1, \ldots, n \tag{B.11}$$

$$\frac{dw_j}{dt} = -2\, w_j\, \frac{\partial}{\partial x_j} f_j - \frac{1}{2} w_j\, \frac{\partial^2}{\partial x_j^2}(g_j^2)$$

$$- \lambda_j\, \frac{\partial^2}{\partial x_j^2} f_j \qquad j = n_{k+1}, \ldots, n \tag{B.12}$$

Stationarity Condition

$$0 = \frac{\partial H}{\partial u_p} = \frac{\partial F}{\partial u_p} + \frac{\partial f'}{\partial u_p}\lambda \qquad p = 1, \ldots, m \tag{B.13}$$

This is a set of $2n + (n - n_k)$ ordinary differential equations (state and co-state equations) and m algebraic equations (stationarity condition), and it is solved as a boundary value problem. The boundary values of the state and co-state variables depend on the problem specification, while the boundary values for w are given as $w(T) = 0$ (Rico-Ramirez & Diwekar, 2004). The control trajectory obtained is optimal for the considered objective function and starting conditions.

References

Abrams, P. (1993). Effect of increased productivity on the abundances of trophic levels. *American Naturalist*, **141**(3), 351–371.

Abrams, P. & Roth, J. (1994). The effects of enrichment of three-species food chains with non-linear functional responses. *Ecology*, **75**(4), 1118–1130.

Anastas, P. & Zimmerman, J. (2003). Design through the 12 principles of green engineering. *Environmental Science and Technology*, **37**(5), 95–101.

Anderson, D. (1994). A dynamic programming algorithm for optimization of uneven aged forest stands. *Canadian Journal of Forest Research*, **24**, 1758–1765.

Arrow, K. & Fisher, A. (1974). Environmental preservation, uncertainty and irreversibility. *The Quarterly Journal of Economics*, **88**(2), 312–319.

Millennium Ecosystem Assessment (2005). Ecosystems and human well-being: Wetlands and water. Synthesis: A report of the millennium ecosystem assessment, World Resources Institute, Washington, DC., Washington DC.

Bossel, H. (1998). *Earth at a crossroads: Paths to a sustainable future* . Cambridge University Press.

Brett, M. & Goldman, C. (1997). Consumer versus resource control in freshwater pelagic food webs. *Science*, **275**(5298), 384–386.

Brooks, D. & Wiley, E. (1986, 335pp.). *Evolution as Entropy*. University of Chicago Press, Chicago.

Cabeza Gutes, M. (1996). The concept of weak sustainability. *Ecological Economics*, **17**(3), 147–156.

Cabezas, H. & Fath, B. (2002). Towards a theory of sustainable systems. *Fluid Phase Equilibria*, **2**, 194–197.

Cabezas, H., Pawlowski, C., Mayer, A. & N.T., H. (2005). Simulated experiments with complex sustainable systems: Ecology and technology. *Resources Conservation and Recycling*, **44**, 279–291. Doi:10.1016/j.resconrec.2005.01.005.

Cabezas, H., Pawlwoski, C., Mayer, A. & Hoagland, N. (2003). Sustainability: Ecological, social, economic, technological, and systems perspectives. *Clean Technologies and Environmental Policy*, **5**, 1–14.

Carlson, D., Haurie, A. & Leizarowitz, A. (1991). *Infinite horizon optimal control: Deterministic and stochastic systems* . 2 edn. Springer-Verlag.

Carpenter, S. & Kitchell, J. (1988). Consumer control of lake productivity. *BioScience*, **38**(11), 764–769.

Carpenter, S., Kitchell, J. & Hodgson, J. (1985). Cascading trophic interactions and lake productivity. *BioScience*, **37**(10), 634–639.

Chukwu, E. (2001). *Stability and time-optimal control of hereditary systems - with applications to the economic dynamics of the U.S.*, vol. 60 of *Series on advances in mathematics for applied science*. World Scientific.

Clark, C. (1990). *Mathematical bioeconomics: The optimal management of renewable resources.* 2nd edn. John Wiley, New York, USA.

Conrad, J. (1980). Quasi-option value and the expected value of information. *The Quarterly Journal of Economics*, **94**(4), 813–820.

Daily, G., Sliderqvist, T., Aniyar, S., Arrow, K., Dasgupta, P., Ehrlich, P., Folke, C., Jansson, A., Jansson, B.-O., Kautsky, N., Levin, S., Lubchenco, J., Maler, K.-C., Simpson, D., Starrett, D., Tilman, D. & Walker, B. (2007). The value of nature and nature of value. *Science*, **289**(5478), 395–396.

De Feo, O. & Rinaldi, S. (1997). Yield and dynamics of tritrophic food chains. *American Naturalist*, **150**(3), 328–345.

Diwekar, U. (1996). *Batch Distillation: Simulation, Optimal Design and Control* . Taylor and Francis, Washington, DC.

Diwekar, U. (2003). *Introduction to Applied Optimization* . Kluwer Academic Publishers, Dordrecht.

Diwekar, U. (2005). Green process design, industrial ecology, and sustainability: A systems analysis perspective. *Resources, conservation and recycling*, **44**(3), 215–235.

Dixit, A. & Pindyck, R. (1994). *Investment under uncertainty*. Princeton University Press, Princeton, New Jersey.

Fath, B., Cabezas, H. & Pawlowski, C. (2003). Regime changes in ecological systems: An information theory approach. *Journal of Theoretical Biology*, **222**, 517–530. Doi:10.1016/S0022-5193(03)00067-5.

Fiksel, J. (2006). Sustainability and resilience: Toward a systems approach. *Sustainability: Science, Practice & Policy*, **2**(2), 14–21.

Fisher, R. (1922). On the mathematical foundations of theoretical statistics. In *Philosophical Transactions of the Royal Society of London* , A 222. pages 309–368.

Frieden, B. (1998). *Physics for Fisher Information: A Unification* . Cambridge University Press, Cambridge.

Goodland, R. & Daly, H. (1996). Environmental sustainability: Universal and non-negotiable. *Ecological Applications*, **6**(4), 1002–1017.

Gragnani, A., De Feo, O. & Rinaldi, S. (1998). Food chains in the chemostat: Relationships between mean yield and complex dynamics. *Bulletin of Mathematical Biology*, **60**(4), 703–719. Doi:0092-8240/98/040703 + 17.

Hairston, N. G., Smith, F. & Slobodkin, L. (1960). Community structure, population control, and competition. *American Naturalist*, **94**(879), 421–425.

Heal, G. (1997). Discounting and climate change; An editorial comment. *Climatic Change*, **37**(2), 335–343.

Heal, G. (1998). *Valuing the future: Economic theory and sustainability*. Columbia University Press, New York.

Hof, J. & Bevers, M. (2002). *Spatial optimization in ecological applications*. Columbia University Press, New York.

Holling, C., Gunderson, L. H. & Ludwig, D. (2002). *Panarchy: Understanding transformations in human and natural systems*, chap. In quest of a theory of adaptive change. Island Press, pages 3–22.

Kirk, D. (1970). *Optimal Control Theory: An introduction*. Prentice-Hall, Englewood Cliffs, NJ.

Kolosov, G. (1997). Size control of a population described by a stochastic logistic model. *Automation and Remote Control*, **58**(4), 678–686.

Kolosov, G. & Sharov, M. (1993). Optimal control of population sizes in a predator-prey system: Approximate design in the case of an ill-adapted predator. *Automation and Remote Control*, **54**(10), 1476–1484.

Lewis, F. (1986). *Optimal Control*. John Wiley & Sons.

Ludwig, D. (2000). Limitations of economic valuation of ecosystems. *Ecosystems*, **3**(1), 31–35.

Ludwig, D., Carpenter, S. & Brock, W. (2003). Optimal phosphorous loading for a potentially eutrophic lake. *Ecological Applications*, **13**(4), 1135–1152.

Ludwig, D., Walker, B. & Holling, C. S. (1997). Sustainability, stability and resilience. *Conservation Ecology*, **1**(1).

Lynch, M. & Shapiro, J. (1981). Predation, enrichment and phytoplankton community structure. *Limnology and Oceanography*, **26**(1), 86–102.

Mayer, A. & Rietkerk, M. (2004). The dynamic regime concept for ecosystem management and restoration. *BioScience*, **54**(11), 1013–1020.

McMichael, A., Butler, C. & Folke, C. (2003). New visions for addressing sustainability. *Science*, **302**, 1919–1920.

McQueen, D., Post, J. & Mills, E. (1986). Trophic relationships in freshwater pelagic ecosystems. *Canadian Journal of Fisheries and Aquatic Sciences* , **43**, 1571–1581.

McQueen, M., D.J. Johannes, Post, J., Stewart, T. & Lean, D. (1989). Bottom-up and top-down impacts on freshwater pelagic community structure. *Ecological Monographs*, **59**(3), 289–309.

Meadows, D. H., Meadows, L., Dennis & Randers, J. (1992). *Beyond the limits: Confronting global collapse, envisioning a sustainable future* . Chelsea Green Publishing Company, Post Mills, Vermont.

Mees, R. & Strauss, D. (1992). Allocating resources to large wildland fires: A model with stochastic production rates. *Forest Science*, **38**, 842–853.

Mejer, H. & Jørgensen, S. (1979). *State-of-the-Art in Ecological Modeling* . pp. 829-846. Pergamon Press, Oxford, NY.

Myers, N., Mittermeier, R., Mittermeier, C., Fonseca, G. & Kent, J. (2000). Biodiversity hotspots for conservation priorities. *Nature*, **403**, 853858.

O'Neill, R. (1996). Perspectives on economics and ecology. *Ecological Applications*, **6**(4), 1031–1033.

Paine, R. (1980). Food webs: Linkage interaction strength, and community infrastructure. *Journal of Animal Ecology* , **49**, 667–685.

Pawlowski, C., Fath, B., Mayera, A. & Cabezas, H. (2005). Towards a sustainability index using information theory. *Energy*, **30**, 1221–1231.

Richards, S., Possingham, H. & Tizard, J. (1999). Optimal fire management for maintaining community diversity. *Ecological Applications* , **9**(3), 880–892.

Rico-Ramirez, V. & Diwekar, U. (2004). Stochastic maximum principle for optimal control under uncertainty. *Computers & Chemical Engineering* , **28**, 2845–2845. Doi:10.1016/j.compchemeng.2004.08.001.

Rico-Ramirez, V., Diwekar, U. & Morel, B. (2003). Real option theory from finance to batch distillation. *Computers & Chemical Engineering* , **27**, 1867–1882. Doi:10.1016/S0098-1354(03)00160-1.

Rosemond, A., Mulholland, P. & Elwood, J. (1993). Top-down and bottom-up control of stream periphyton: Effects of nutrients and herbivores. *Ecology*, **74**(4), 1264–1280.

Rosenzweig, M. L. (1971). Paradox of enrichment: Destabilization of exploitation ecosystems in ecological tim. *Science*, **171**, 385–387.

Rutledge, R., Basore, B. & Mulholland, R. (1976). Ecological stability: An information theory viewpoint. *Journal of Theoretical Biology*, **57**, 355–371.

Ryan Gwaltney, C., Styczynski, M. & Stadtherr, M. (2003). Reliable computation of equilibrium states and bifurcations in food chain models. Technical report, Department of Chemical and Biomolecular engineering, University of Notre Dame, 182 Fitzpatrick Hall, Notre Dame, IN 46556, USA.

Scholes, R., Hassan, R. & Ash, N. J. (2005). Ecosystems and human well-being: Current state and trends, Volume 1. In *Millennium Ecosystem Assessment*. Island Press, Washington DC.

Science, W. & Board, T. (2004). *Valuing ecosystem services: Towards better environmental decision-making*. The National Academies Press, National Academies Press, Washington, DC 20055.

Shannon, C. & Weaver, W. (1949). *The mathematical theory of communication*. University of Illinois Press, Champaign IL.

Shastri, Y. & Diwekar, U. (2006a). Sustainable ecosystem management using optimal control theory: Part 1 (Deterministic systems). *Journal of Theoretical Biology*, **241**, 506–521. Doi:10.1016/j.jtbi.2005.12.014.

Shastri, Y. & Diwekar, U. (2006b). Sustainable ecosystem management using optimal control theory: Part 2 (Stochastic systems). *Journal of Theoretical Biology*, **241**, 522–532. Doi:10.1016/j.jtbi.2005.12.013.

Shastri, Y., Diwekar, U. & Cabezas, H. (2008a). Optimal control theory for sustainable environmental management. *Environmental Science & Technology*, **42**(14), 5322–5328.

Shastri, Y., Diwekar, U., Cabezas, H. & Williamson, J. (2008b). Is sustainability achievable? Exploring the limits of sustainability with model systems. *Environmental Science & Technology*, **42**(17), 6710–6716.

Sivert, W. & Smith, W. (1977). Optimal exploitation of a multi-species community. *Mathematical Biosciences*, **33**, 121–134.

Stern, N. (2006). *The economics of climate change: The Stern review*. Cambridge University Press, Cambridge, UK.

Taguchi, G. (1986). Introduction to quality engineering. Tech. rep., Asian Productivity Center, Tokyo, Japan.

Tainter, J. (1988). *The collapse of complex societies*. Cambridge University Press, Cambridge.

Tomlinson, C. (1987). *Our Common Future*. World Commission on Environment and Development, Oxford University Press, Oxford.

Townsend, C. (1988). Fish, fleas and phytoplankton. *New Scientist*, **118**(1617), 67–70.

Ulanowicz, R. (1986). *Growth and Development: Ecosystems Phenomenology*. Springer, NY.

Ulas, S. & Diwekar, U. (2004).   Thermodynamics uncertainties in batch processing and optimal control. *Computers & Chemical Engineering*, **28**, 2245–2258. Doi:10.1016/j.compchemeng.2004.04.001.

United Nations Environment Programme (2007). *Global Environment Outlook GEO 4: Environment for development*. United Nations Environment Programme, Progress Press Ltd., Valletta, Malta.

Vanni, M. & Temte, J. (1990). Seasonal pattern of grazing and nutrient limitation of phytoplankton in a eutrophic lake. *Limnology and Oceanography*, **35**(3), 697–709.

In: Encyclopedia of Environmental Research
Editor: Alisa N. Souter

ISBN: 978-1-61761-927-4
© 2011 Nova Science Publishers, Inc.

Chapter 26

SCIENCE AND NON-SCIENCE IN THE BIOMONITORING AND CONSERVATION OF FRESH WATERS

Guy Woodward[*,1], *Nikolai Friberg*[2] *and Alan G. Hildrew*[1]

[1]School of Biological & Chemical Sciences, Queen Mary University of London,
London, E1 4NS, U.K.
[2]NERI, Silkeborg, Denmark

ABSTRACT

We highlight the need for a strong scientific approach to biomonitoring and conservation of fresh waters and identify key shortcomings in the present approach. For instance, mismatches between the identity of the stressors that we observe and the metrics that we use to measure their impacts are becoming increasingly evident. As water quality has improved in many areas new stressors have become important and the focus purely on responses to organic enrichment and eutrophication becomes increasingly untenable. While there are notable exceptions, many biomonitoring and management schemes are inappropriate, poorly designed, or based on weak science. We argue that there is an unfortunate tendency in this field to resort to anecdotal "evidence" and unsubstantiated inference, rather than to strong experimental evidence of cause-and-effect relations based on hypothesis testing. Further, the field has often developed in isolation from recent advances in ecological theory, which has stunted our ability to progress from understanding to knowledge to prediction of ecosystem responses to stressors. Pertinent examples of this come from the rapid advances made recently on the relationship between biodiversity and ecosystem processes and in food web research, neither of which have so far informed monitoring and management practices. This disconnection has led to some notable instances of 'pseudoscience' being used at the expense of sound scientific concepts, which we identify in a set of case studies. There are also other more subtle disadvantages of an overemphasis on repetitive monitoring and data collection for the purposes simply of compliance to environmental targets, rather than for science. Vast resources are being used for such purposes with almost no improvement in our

---

[*] E-mail address: g.woodward@qmul.ac.uk

mechanistic understanding of how freshwater ecosystems operate. Even if these issues are resolved, however, legislative problems still need to be addressed. Methodological inertia inhibits the implementation of new developments in ecological biomonitoring, because of the phobia of novelty and a willingness to stick with familiar systems, even if they are no longer completely fit-for-purpose and in need of overhauling.

INTRODUCTION

The ultimate goal of any science is to progress from knowledge towards understanding and, ultimately, prediction. However, despite the enormous resources invested in biomonitoring of fresh waters over recent decades, as a scientific discipline it remains firmly rooted in the first stage of this process, with little or no predictive power. The long-standing emphasis on pattern-description, rather than developing a mechanistic understanding based on the development and testing of hypotheses, is undoubtedly one of the principal reasons for the stunted development of the field. If we are to anticipate responses of freshwaters to future environmental change, the prevailing approach needs to be overhauled and a more rigorous scientific underpinning developed. This is not to say that we necessarily need to understand all the component parts of a given ecosystem in perfect detail in order to make some educated predictions, rather that we should build on existing knowledge in a more efficient and useful manner based on sound ecological principles.

Biomonitoring is more of a technology than a scientific discipline, and was developed as a tool to assess the status of the environment. Until quite recently most monitoring of fresh waters targeted chemistry, but now biology and ecology are the focus of both national and international programmes in Europe and worldwide. There are tremendous advantages in directly assessing ecological response to perturbations, rather than relying on indirect chemical measures, and biomonitoring is therefore both a necessary and important approach. Good biomonitoring schemes, that are firmly rooted in an understanding of community and ecosystems ecology provide invaluable data, and insights that are unlikely to be offered by the molecular-based techniques (which can only detect taxa for which an appropriate "bar code" has been identified) that are being advocated in some quarters. However, it appears that there is some resistance from practitioners, both applied scientists and end-users, towards the introduction of new means of detecting and diagnosing changes in freshwater ecosystems by building on and refining classical ecological biomonitoring approaches. Rather, a large proportion of strategic funding from both national providers and (in Europe) the EU has been directed into what might be described as 'pseudo-science'. This has not, in our opinion, increased our ability to detect environmental stress and predict change. For this reason, a large proportion of ongoing biomonitoring is undertaken mainly for legal compliance. As such it is rarely cost-efficient, fit for purpose, or sufficiently sensitive to detect important, and in some cases new, environmental stressors. At the same time, resources for more fundamental research have been depleted, even though aimed at furthering understanding and prediction – the Holy Grail of environmental management. Whilst conservation biology has in many instances developed into a vibrant field that has both exchanged ideas successfully with modern ecology, this is not the case for much of biomonitoring, especially in freshwaters, some of whose major shortcomings we highlight in this commentary. We do not wish to devalue, however, the clear importance of well-constructed biomonitoring schemes, which can provide important insight into ecological problems, and we also suggest ways in which some of the current limitations might be addressed in the future.

THE QUESTION OF 'TYPOLOGY'

A persistent concept since the introduction of lake types by Naumann and Thienemann (Thienemann 1920, 1959) in the early part of last century has been that of typology. Thus, Illies and Botosaneanu (1963) divided rivers into a complex classification of distinct zones, each of which harboured well-defined biological communities. Such schemes hark back to the 'superorganism' concept of the Clements school of plant community ecology (Clements, 1936), a view in which separable communities with distinct groups of coevolved species really exist – a view hardly tenable (at least in its pure form) in modern ecology, dominated as it now is by notions of non-equilibrium processes, species turnover along gradients and community continua. Modern typologies are based on 'expert knowledge' about where, and under which environmental conditions, organisms live and then create the necessary number of distinct types to contain this information. Typologies can relate to trophic status, community composition or sets of environmental conditions but are all based on an a priori classification to which the composition of an actual community is compared. The concept of typologies has especially been used in biomonitoring, where it has formed the basis for the classification of water quality for both lakes and rivers. We believe that whole concept of compartmentalising communities into a number of distinct types is fundamentally flawed, and that the overwhelming body of evidence (and theoretical arguments) suggests there is normally a gradual adjustment of species composition along environmental gradients. Nevertheless, the use of community 'types' persists as an integral part of many monitoring systems across Europe and we fear it will significantly limit our understanding of ecosystems and hence our ability to detect degradation.

A significant advance in biomonitoring was initiated with the development of the RIVPACS methodology over 30 years ago (Wright. 2000). RIVPACS is a modelling approach, in which the species composition of the macroinvertebrate assemblage at a given site can be predicted using a number of environmental variables (themselves not subject to environmental perturbations) and then compared with the assemblage actually observed. Predictions are based on an unconstrained classification of macroinvertebrate communities (TWINSPAN) followed by a prediction procedure based on multiple discriminant analysis (MDA) from a number of minimally impacted stream sites covering all major geographically areas in the UK. There are two underlying assumptions in the RIVPACS approach that make it ecologically sound: 1) it is based on an analysis of existing communities along gradients of environmental conditions, and 2) when sampling a new site, the system tests the probability of this site belonging to one of the reference groups and its deviation from the expected list is a measure of perturbation. Hence, by using probabilities, the system acknowledges that a perfect match is unlikely, even when there is no perturbation. This is because a number of uncertainties are always at play, relating both the natural variability, stochastic factors and sampling or other errors. There has been widespread use of the RIVPACS approach worldwide, with predictive models currently used in North America, Australia, Sweden and Czech Republic (Wright at al., 2000).

Despite the clear advantages of the RIVPACS approach and the quite dated thinking behind typologies, which over the years has received considerable critique from the scientific community, our most progressive water policy to date, the Water Framework Directive (WFD) (Directive 2000/60/EC – Establishing a Framework for Community Action in the Field of Water Policy), still gives the option of using types. The Directive opens the possibility of using either a typology ("system A") approach or predictive modelling using site-specific variables ("system B"), which essentially equates to a RIVPACS-style methodology, for the characterisation of water bodies.

One can only guess, but the inclusion of both ways of water body characterisation may be the result of a late night compromise in the Commission, as typology is still at the core of monitoring programmes in Central Europe. It may also be that the modern revival of classifying communities into types, which at one time seemed to have been laid to rest, relates to the output of some multivariate statistical approaches (such as TWINSPAN, see above), which 'classify' ecological samples into groups – which to the unwary can be equated to real community types. Of course, such procedures will generate groupings even of continuously variable assemblages. Further, managers often want to simplify a bewildering array of ecological systems into a few 'types'. This can be appreciated just as long as we do not then think of such 'types' as reflecting the fundamental nature of communities. Similarly, it is evident that the understandable need by managers to compare an observed community with some relevant ecological 'reference' system. Ecologists of course realise that 'reference conditions' and 'good ecological status' do not look the same for all water bodies. For instance, a stream draining peatlands or acidic soils and geology will never approach reference conditions for a chalk stream. So, while we can appreciate the practical need to reduce what are continua of communities into a simpler series of categories, this should be an interim approach and not be confused with the existence in nature of real and distinct community 'types', as now seems to be the case.

In recent years the long-standing reliance upon community-based assessment has been questioned, and there is a growing interest in measuring "ecosystem function". This unfortunate term has now entered common scientific parlance, as the recognition of the need to include dynamical, process-based metrics has increased, rather than an over-reliance on structural measures (e.g. BMWP scores, Shannon-Wiener indices, "typologies" etc.). Ecosystem functioning is, in reality, the integral sum of multiple ecosystem processes, but in most studies only one process is measured (*e.g.*, decomposition rates [McKie et al. 2008]). Since different processes may be unrelated or even antagonistic to one another, or even non-linear in their response to environmental gradients, single measures are not necessarily good approximations of "ecosystem health" (Reiss et al., in press). Indeed, one could argue that concepts of ecosystem functioning and health again hark back to the idea of communities and ecosystems as 'superorganisms', when a more parsimonious (indeed Darwinian) view is appropriate, in which properties of systems above the level of the individual are the product of conventional natural selection.

MISMATCHES BETWEEN EMERGING STRESSORS AND INAPPROPRIATE BIOLOGICAL METRICS: THE EUTROPHICATION TIME LAG

Eutrophication (and its forerunner, organic pollution) has dominated as the principal stressor of interest in recent biomonitoring programmes, with acidification arguably in second place. The vast majority of biomonitoring schemes, therefore, are based upon detecting biological responses to nutrients and primary and secondary organic pollution (e.g. Metcalfe-Smith, 1996). Macroinvertebrates are the group of organisms most frequently used in biomonitoring of streams and rivers worldwide, with more than 50 different methods currently in use (e.g. Wright et al., 2000; Friberg et al., 2006). In essence, all existing indices are based on knowledge or assumptions about the oxygen demands of individual species (Liebmann, 1951; Sladecek, 1973), building on the original concept of the saprobic system (Kolkwitz & Marsson, 1902).

However, in recent years the importance of this stressor (and acidification) has abated (at least in western Europe, though it is increasing in many other places), as the influence of environmental legislation is manifested and water quality has improved in many areas. This means that previously secondary or tertiary stressors, such as habitat destruction, are now coming to the fore, and our biotic indices, developed with eutrophication in mind, are becoming increasingly obsolete. This is creating a time lag and a mismatch between the driver and the response we are measuring, and we need to develop new metrics for a range of different stressors, including those associated with climate change.

Another issue is that of multiple stresses on aquatic ecosystems, with an emerging number of studies showing that stressors seldom act alone (e.g. Ormerod, in press). For instance, Andersen (1994) found that an impact of BOD_5 on a macroinvertebrate-based index was reduced with increasing current velocity, and it has been demonstrated experimentally that multiple stressors can act synergistically (Folt et al., 1999, Matthaei et al., 2006). Therefore, one of the limitations of existing systems are that they were developed primarily to target one dominant stressor across an enormous gradient (typically from practically pure water to sewage) rendering them less sensitive to other types of stress. For monitoring purposes, it is essential that different stressors can be disentangled from each other, as ecological status in most cases will reflect different types of pressures.

The multi-metric approach using macroinvertebrates (e.g. Barbour and Yoder 2000), and methods using multiple indicators to assess the quality of the river (e.g. Fore et al. 1996; Weigel 2003; Hering et al. 2006), are important steps in the right direction. However, methods sensitive to multiple stressors need further development, especially with regard to detection of hydromorphological impacts. A particularly promising approach, which could be used as an extension to RIVPACS, is the development of methods based on species traits (generation times, body size, modes of feeding and respiration etc) (Woodward & Hildrew 2002). These can work across areas where there are biogeographical differences (since they are not based on the same species being present, but on traits), and potentially diagnose a variety of environmental stressors (e.g. Dolédec & Statzner, 2008). As we will see, however, uptake of new developments has been at best reluctant and extremely slow. An illustrative metaphor could be the development of analytical techniques for water chemistry and toxic compounds, where there have been significant advances in technologies that have lowered detection limits and increased processing speed: it is hard to imagine that anyone would want to stunt this development and 'freeze' analytical capabilities at the level of the 1980s, in contrast to the far more recalcitrant development of biomonitoring.

THE SQUARE PEG AND THE ROUND HOLE: WHY HABITAT DEGRADATION CANNOT BE ASSESSED

It has been apparent for decades that habitat degradation is a very important pressure on stream ecosystems. It has, with the introduction of WFD almost 10 years ago, even been dignified with its own category of impact - "hydromorphological" - which is a 'quality element' in its own right for high status sites and a 'supporting element' for all other quality classes, even though we are largely unable to detect and quantify it. One of the best known and most prominent habitat assessment tools is the River Habitat Survey (Raven et al. 1998). However, a number of studies have shown very poor relationships between RHS and macroinvertebrate indices (Vaughan et al. 2008, Friberg et al. 2009). Friberg *et al.* (2009) found only very weak relationships (maximum $r^2 = 0.21$) between various RHS scores and a number of macroinvertebrate indices in streams along a gradient of hydromorphological degradation, with other anthropogenic impacts being minimal. Moreover, the same study showed that a more detailed assessment of channel and sediment conditions improved the explanatory power of habitat conditions on macroinvertebrate indicators. Such finer-grained measurements, particularly of sediment conditions at the scale at which macroinvertebrate are sampled, are likely to increase explanatory power substantially. This contention is supported by numerous studies reporting sediments conditions as one of the primary drivers of macroinvertebrate community composition (e.g. Minshall 1984; Wood and Armitage 1997; Miyake and Nakano 2002).

Most macroinvertebrate sampling methods currently used have very crude stratification of effort by habitats and no sample replication (Friberg et al. 2006). Hydromorphological degradation, in contrast to organic pollution and pH, is likely to induce changes in the range of habitats available and, hence, in the relative abundance of species (rather than excluding major parts of the community). Novel strategies need to be developed which encompass habitat-specific sampling and appropriate replication. With regard to indices there are only very few specifically targeting hydrology and hydromorphology (Barbour et al. 1996; Extence et al. 1999; Lorenz et al. 2004), and more effort has to go into their development.

OVER-RELIANCE ON ANECDOTE AND INFERENCE AT THE EXPENSE OF MECHANISTIC UNDERSTANDING: TALES FROM THE RIVERBANK

Pattern-fitting is not hypothesis driven, nor is it in any way mechanistic. Ordination techniques, which form the core of many biomonitoring programmes, are suited to hypothesis generation, rather than hypothesis testing. In order to understand – and ultimately to predict – ecosystem responses to stressors we need to carry out experimental (or statistical) manipulations at suitable temporal and spatial scales. This can be done using traditional field and laboratory-based manipulations, in addition to in silico mathematical modelling. None of these approaches, central to fundamental ecology, is yet acknowledged in the 'science' of biomonitoring.

SPLENDID ISOLATION: THE ABSENCE OF A FIRM THEORETICAL BASIS AND THE LACK OF PREDICTIVE POWER

What determines species occurrence and abundance? This is a fundamental ecological question that is core to all of biomonitoring, and yet it is largely ignored. There is an almost axiomatic assumption that the environment provides the templet that determines which species should be present and in what numbers: if species are "missing" then something is wrong. Unfortunately, the real world is not so simple, and widely divergent suites of species can be found under identical environmental conditions, as evidenced in the alternative equilibria so familiar to shallow lake ecologists (e.g., Scheffer & Carpenter 2003). Further, any consideration of stochasticity is missing, no null models are incorporated, and biological interactions are completely ignored, even though they can exert extremely powerful influences on community structure and ecosystem functioning (Petchey et al. 2004). Much of biomonitoring is thus deterministic and phenomenological, and this undermines its potential predictive power. The use of neutral models and the introduction of Bayesian statistical approaches would be useful improvements, but this still does not address the underlying need for identifying cause-and-effect relationships via experiments.

LOSS OF OLD VIRTUES: SACRIFICING SIMPLICITY AND CREATING A PSEUDO-SCIENCE

An integral part of most assessment systems, developed initially as part of the environmental "revolution" of the 1970s, was a simplicity of approach. Sampling techniques, laboratory treatment and taxonomic resolution were all kept at the minimum that could provide sufficient information. This was based partly on cost-benefit analyses, but also to minimise the introduction of otherwise avoidable errors. The latter is a particularly key point, as biomonitoring is not a scientific discipline, but a technology used to assess the state of the environment. This requires, in practice, that samples are collected and identified by a range of surveyors with different skills, and the more complicated the system the greater the likelihood of operator error. No auditing procedures will ever quantify all sources of uncertainty and often involve disproportionate costs which, together with increased time consumption in handling each sample, will significantly limit both the spatial and temporal extent of monitoring programmes. These logical and pragmatic reasons for maximising simplicity are in stark contrast to more recent developments with elaborate sampling schemes in which species-level identification has reappeared on the agenda, with direct links to the obsolete saprobic system, and which effectively by-pass the thinking behind many of the existing assessment systems, where detail is sacrificed for a reduction in error. .

For instance, the E.U. Water Framework Directive has driven a range of large-scale pan-European research initiatives in recent years, which have accounted for the lion's share of E.U. Framework funding in fresh waters. This (largely political) desire to create a single, continent-wide approach is perhaps understandable, and in some senses desirable, but in many instances the process and end result have been less than satisfactory. This is largely because the shortcomings of the different national programmes (each cherished and defended) tend to be retained rather than ejected. A case in point can be seen by comparing the elaborate STAR-AQEM sampling protocol with a far simpler and cheaper, yet effective national programme. The STAR-AQEM method (Hering *et al.* 2003) is based on Barbour et al. (1999) and focuses on sampling major habitats according to their presence within a reach (i.e. the well-founded stratified sampling approach). Twenty Surber or kick sample-units are taken in proportion to the relative area of those habitat "types", with a minimum 5% coverage. However, these sample-units are then physically pooled, thus removing any measure of variation, and subsamples (aliquots) are then withdrawn, from each of which 700 macroinvertebrates are identified and counted. This contrasts with the replicate 1-minute kick-samples taken in the previous national scheme. There are several problems associated with the former approach. First, assigning microhabitats into categories ('types' again!) is at best extremely subjective, and is based on the assumptions that: a) such "categories" actually exist and have meaning in natural communities, b) if they do exist, they can be recognised consistently and accurately by different operators, among systems and over time, and c) the biota respond differentially to habitat types. Even if these unlikely assumptions are met, we then run into a set of statistical problems of, first, having a total lack of replication (i.e. one sample per site), which is then further compounded by introducing pseudoreplication (via subsampling of the single composite sample). This latter problem is made manifestly worse by then arbitrarily selecting 700 individuals to identify per sub-sample. This tortuous process is not only invalid scientifically – it is also extremely labour intensive, and therefore expensive, yet it offers no measurable improvement on the far cheaper kick-sampling method (Friberg et al., 2006).

CONSERVATISM, BUREAUCRATIC INERTIA AND THE FEAR OF NOVELTY

Much of biomonitoring is tightly bound up with legislative and bureaucratic inertia, and politicisation of the science has inevitably undermined its intellectual integrity. One particularly worrying trend is for pre-existing approaches to be overlain upon one another to create an unwieldy, expensive and often illogical methodology, rather than undertaking a root-and-branch overhaul of the inherent flaws from first principles. When asked about the likelihood of the uptake of newly developed methods in biomonitoring, Dr Roger Owen, Head of Ecology at the Scottish Environmental Protection Agency and chair of the EU intercalibration of WFD compliant biological methods among member states, replied:

> "While we (both nationally and in the EU) are very open on adopting new methods that can help us detecting types of stress not covered by existing methods such as a hydromorphological degradation, my view is that there will be more resistance in changing already approved, existing methods as it is a very time consuming and costly process".

This is an understandable view, given issues with time series, training of staff, re-intercalibration of methods etc. but it introduces a degree of ossification in the process of biomonitoring.

A CASE IN POINT: MYTHS AND LEGENDS OF RIVER RESTORATION

River restoration is an expensive business, and yet its scientific basis is far from solid. There are vanishingly few data of BACI-style studies that can be used to assess its success, and indeed, the evidence that the commonly used (small scale) techniques have any beneficial effect is weak (e.g. Pretty et al., 2003; Harrison et al., 2004). Many of the artificial structures introduced to enhance "flow diversity" have no measurable impact on either invertebrates or fishes, probably because inappropriate scales are involved. It is a widely-held tenet, but not one based on any hard evidence, that by introducing physical heterogeneity ("habitats-by-numbers") then biodiversity will simply follow suit. In fact, the links between biodiversity and hydrological and geomorphological diversity are still neither well-described nor well-understood (Vaughan et al. 2008), even though they are intuitively appealing.

SEEING THE WOOD FOR THE TREES: SUGGESTIONS FOR IMPROVING THE FIELD

In order for the field to advance it must become more predictive, with an emphasis on understanding mechanisms rather than simply describing patterns. This can be achieved by placing a greater emphasis on experimental and statistical manipulations and modelling. Recent advances in fundamental ecology, particularly in the fields of "biodiversity-ecosystem functioning" research and food web ecology, are potentially provide useful new perspectives for biomonitoring and conservation, especially as they both address the role of species interactions, which have been largely ignored hitherto (Woodward 2009). We also advocate the need for ecological interpretation, rather than becoming too reliant upon statistical approaches that are less objective than they may appear. It is also important to consider carefully both the responses and the drivers: are they related to one another in a meaningful way and at appropriate scales? From the practical management perspective, we argue it should be instilled as best-practice that before-and-after data are collected and analysed when carrying out restoration – this represents a small part of the total budget of such schemes, but it is the only way to assess whether they have achieved their aims. Finally, we argue that it is better to jettison ineffectual approaches that are no longer fit-for-purpose and to replace them with novel methods, than it is to retain them for the sake of familiarity. However, that is not to suggest that we should replace the taxonomic-based approach with molecular-based approaches that are being advocated in some quarters, as these are likely to provide even less ecological insight than current schemes. Rather, we should build on and apply the invaluable knowledge already gleaned from advances in community and ecosystems ecology, and to do so in a more logical and effective way.

REFERENCES

A.F.N.O.R. (1982) *Essais des euax. Détermination de l'indice biologique global normalisé (IBGN)*. Association Francaise de Normalisation NF T 90-350, France.

Agresti A. (1990) *Categorical data analysis*. Wiley and Sons. New York.

Andersen J.M. (1994) Water quality management in the River Gudenaa, a Danish lake-stream-estuary system. *Hydrobiologia*, 275/276, 499-507.

Barbour M.T. & Yoder C.O. (2000) The multimetric approach to bioassessment, as used in the United States of America. In: *Assessing the Biological Quality of Freshwaters: RIVPACS and Similar Techniques* (Eds. J.F. Wright, D.W. Sutcliffe & M.T. Furse), pp. 281-292. Freshwater Biological Association, Ambleside, UK

Biggs B.J.F (2000) Eutrophication of streams and rivers: dissolved nutrient-chlorophyll relationships for benthic algae. *Journal of the North American Benthological Society* 19, 17-31.

Biggs B.J.F, Francoeur S.N., Huryn A.D., Young R., Arbuckle C.J., Townsend C.R. (2000) Trophic cascades in streams: effects of nutrient enrichment on autotrophic and consumer benthic communities under two different fish predation regimes. *Canadian Journal of Fisheries and Aquatic Sciences*, 57, 1380-1394.

Chessman B.C. & McEvoy P.K. (1998) Towards diagnostic biotic indices for river macroinvertebrates. *Hydrobiologia*, 364, 169-182.

Clements F.E. (1936) Structure and nature of the climax. *Journal of Ecology*, 24, 252-284.

De Pauw N. & Vanhooren G. (1983) Method for biological quality assessment of watercourses in Belgium. *Hydrobiologia*, 100, 153-168.

EEA – European Environment Agency (1991) *Europe's environment. The Dobris Report*. European Environment Agency, Copenhagen, p. 676.

EEA – European Environment Agency (1994) European rivers and lakes. Assessment of the environmenal statet. European Environment Agency, *EEA Environmental Monographs* 1, Copenhagen, p. 122.

E.E.A. – European Environment Agency (1999) *Nutrients in European Ecosystems*. European Environment Agency, Copenhagen, p. 156.

European Commision (2007) *WFD intercalibration report – part 1 Rivers, Section 2 Benthic Macroinvertebrates*. Joint Research Centre, ISPRA, Italy, *in press*.

Farmer A.M (2004) *Phosphate pollution: a global overview of the problem*. Phosphorous in Environmental Technologies – Principles and Applications. IWA publishing, 174-191.

Feld C.K. & Hering D. (2007) Community structure or function: effects of environmental stress on benthic macroinvertebrates at different spatial scales. *Freshwater Biology*, 52, 1380-1399,

Folt C.L., Chen C.Y., Moore M.V. & Burnaford J. (1999) Synergism and antagonism among multiple stressors. *Limnology and Oceanography*, 44, 864-873.

Friberg N., Sandin L., Furse M.T., Larsen S.E., Clarke R.T. & Haase P. (2006) Comparison of macroinvertebrate sampling methods in Europe. *Hydrobiologia*, 566, 365-378.

Hart D.D. & Finelli C.M. (1999) Physical-biological coupling in streams: the pervasive effects of flow on benthic organisms. *Annual Review of Ecology and Systematics*, 30, 363-395.

Hynes H.B.N. (1960) *The biology of polluted waters*. Liverpool University Press, Liverpool, England.

Illies, J. and Botosaneanu, L. (1963) Problemes et méthode de la classification et de la zonation ecologique des eaux courantes considerées surtout du pointe de vue faunistique.Mitteilungen der internationale Vereingigung für theoretische und angewandte *Limnologie*, 12, 1-57.

Jeppesen E, Sondergaard M, Jensen J.P., Havens K.E., Anneville O., Carvalho L., Coveney M.F., Deneke R., Dokulil M.T., Foy B., Gerdeaux D., Hampton S.E., Hilt S., Kangur K., Kohler J., Lammens E.H.H.R., Lauridsen T.L., Manca M., Miracle M.R., Moss B., Noges P., Persson G., Phillips G., Portielje R., Schelske C.L., Straile D., Tatrai I., Willen E. & Winder M. (2005) Lake responses to reduced nutrient loading - an analysis of contemporary long-term data from 35 case studies. *Freshwater Biology*, 50, 1747-1771.

Johnston N.T., Perrin C.J., Slaney P.A. & Ward B.R.(1990) Increased salmonid growth by whole-river fertilization. *Canadian Journal of Fisheries and Aquatic Sciences*, 47, 862-872.

Kelly M, Juggins S, Guthrie R, Pritchard S., Jamieson J., Ripley B., Hirst H. & Yallop M. (2008) Assessment of ecological status in U.K. rivers using diatoms *Freshwater Biology*, 53, 403–422.

Kolkwitz R. & Marsson M. (1902) Grundsatze fur die biologische Beurteilung des Wassers nach seiner Flora und Fauna. *Mitteilungen der Prufungsansalt fur Wasserversorgung und Abwasserreining*, 1, 1-64.

Kronvang B, Jeppesen E, Conley DJ, Sondergaard M, Larsen S.E., Ovesen N.B. & Carstensen J. (2005) Nutrient pressures and ecological responses to nutrient loading reductions in Danish streams, lakes and coastal waters. *Journal of Hydrology*, 304, 274-288.

Lake P.S. (2000) Disturbance, patchiness, and diversity in streams. *Journal of the North American Benthological Society*,19, 573-592.

Liebmann H. (1951) *Handbuch der Frischwasser und Abwasserbiologie*. Munchen.

Mackay R.J. (1992) Colonization by lotic macroinvertebrates – a review of processes and patterns. *Canadian Journal of Fisheries and Aquatic Sciences,* 49, 617-628.

Matthaei C.D., Weller F., Kelly D.W. & Townsend C.R. (2006) Impacts of fine sediment addition to tussock, pasture, dairy and deer farming streams in New Zealand. *Freshwater Biology*, 51, 2154-2172.

McKie B., Woodward G., Hladyz S., Nistorescu M, Preda E., Popescu C., Giller P.S. & Malmqvist B. (2008) Ecosystem functioning in multiple stream assemblages: contrasting responses to variation in detritivore richness, evenness and density. *Journal of Animal Ecology*, 77, 495-504.

Meeuwig J.J, Kauppila P. & Pitkanen H. (2000) Predicting coastal eutrophication in the Baltic: a limnological approach. *Canadian Journal of Fisheries and Aquatic Sciences*, 57, 844-855.

Metcalfe-Smith J.L. (1996) Biological water-quality assessment of rivers: use of macroinvertebrate communities. In: *River restoration* (Eds. G. Petts & P. Calow), pp. 17-59. Blackwell Science, Oxford, UK.

Moog, O. (Ed.) 2002. *Fauna Aquatica Austriaca*. Lieferung. Wasserwirtschaftskataster, Bundesministerium fur Land- und Forstwirtschaft, Vienna.

Nilsson A. (1996) *Aquatic Insects of North Europe*. A taxonomic handbook. Volume 1+2. Apollo Books.

Perrin C.J. & Richardson J.S. (1997) N and P limitation of benthos abundance in the Nechako River, British Columbia. *Canadian Journal of Fisheries and Aquatic Sciences*, 54, 2574-2583.

Petchey O.L., Downing A.L., Mittelbach G.G., Persson L., Steiner C.F., Warren P.H. & Woodward G. (2004) Species loss and the structure and functioning of multitrophic aquatic systems. *Oikos*, 104, 467-478.

Raven P.J., Holmes, N.T.H., Dawson, F.H. & Everard, M. (1998) Quality assessment using river habitat survey data. *Aquatic Conservation: Marine and Freshwater Ecosystems*, 8, 477-499.

Reiss, J., Bridle, J.R., Montoya, J.M. & Woodward, G. (2009) Emerging horizons in biodiversity-ecosystem functioning research *Trends in Ecology & Evolution*, in press.

Robinson C.T. & Gessner M.O. (2000) Nutrient additions accelerate leaf breakdown in an alpine springbrook. *Oecologia*, 122, 258-263.

Scheffer M. & Carpenter S.R. (2003) Catastrophic regime shifts in ecosystems: linking theory to observation. *Trends in Ecology & Evolution*, 18, 648-656.

Skriver J., Friberg N. & Kirkegaard J. (2000) Biological Assessment of Running Waters in Denmark : Introduction of the Danish Stream Fauna Index (DSFI). - *Verhandlungen internationale Vereinigung für theoretische und angewandte Limnologie,* 27, 1822-1830.

Sladecek V. (1973) System of water qualification from the biological point of view. *Archiv fur Hydrobiologie Beiheft Ergebnisse der Limnologie*, 7, 1-218.

Smith A.J., Bode R.W. & Kleppel G.S. (2007) A nutrient biotic index (NBI) for use with benthic macroinvertebrate communities. *Ecological Indicators*, 7, 371-386.

Snedecor G.W. & Cochran W.G. (1989) *Statistical Methods*. 8th edition, Iowa State University Press. Ames, Iowa.

Søndergaard M., Skriver J. & Henriksen P. (2006) Aquatic Environment. Biological status. *Miljøbiblioteket* 10. National Environmental Research Institute and Hovedland. 104pp. (In Danish)

Stoltze M. & Pihl S. (1998) *Red list data book 1997 of plants and animals in Denmark*. Ministry of Environment and Energy. National Environmental Research Institute. Danish Forest and Nature Agency. 219p. (in Danish)

Thienemann, A (1920) Die Grundlagen der Biocoenotik und Monards faunistische Prinzipien. *Festschrift Zschokke* 4, 1-14.

Thienemann, A. (1959) *Erinnerungen und Tagebuchblätter eines Biologen*. Schweitzerbart, Stuttgart, pp 499.

Wiberg-Larsen P., Brodersen K.P., Birkholm S., Grøn P.N. & Skriver J. (2000) Species richness and assemblage structure of Trichoptera in Danish streams. *Freshwater Biology*, 43, 633-647.

Wiberg-Larsen P. (1984) Stoneflies and mayflies. *Danish Environmental Protection Agency,* Copenhagen, 91 p. (In Danish).

Woodward G. (2009) Biodiversity, ecosystem functioning and food webs in freshwaters: assembling the jigsaw puzzle. *Freshwater Biology* (in press).

Woodward, G., and A.G. Hildrew (2002) Food web structure in riverine landscapes. *Freshwater Biology*, 47, 777-798.

Wright J.F., Sutcliffe D.W. & Furse M.T. (eds.) (2000) *Assessing the Biological Quality of Freshwaters: RIVPACS and Similar Techniques*. Freshwater Biological Association, Ambleside, UK.

Yuan L.L. (2004) Assigning macroinvertebrate tolerance classifications using generalised additive models. *Freshwater Biology*, 49, 662-677.

Zelinka M. & Marvan P. (1961) Zur Präzisierung der biologischen Klassifikation der Reinheit fließender Gewässer. *Archiv für Hydrobiologie*, 57, 389–407.

In: Encyclopedia of Environmental Research   ISBN: 978-1-61761-927-4
Editor: Alisa N. Souter   © 2011 Nova Science Publishers, Inc.

Chapter 27

NATURAL RENEWABLE WATER RESOURCES AND ECOSYSTEMS IN SUDAN

Abdeen Mustafa Omer
17 Juniper Court, Forest Road West, Nottingham NG7 4EU, UK

ABSTRACT

The present problems that are related to water and sanitation in Sudan are many and varied, and the disparity between water supply and demand is growing with time due to the rapid population growth and aridity. The situation of the sewerage system in the cities is extremely critical, and there are no sewerage systems in the rural areas. There is an urgent need for substantial improvements and extensions to the sewerage systems treatment plants. The further development of water resources for agriculture and domestic use is one of the priorities to improve the agricultural yield of the country, and the domestic and industrial demands for water. This study discusses the overall problem and identifies possible solutions.

Keywords: Sudan, water resources development, effective water-supply management, environment

1. INTRODUCTION

Sudan is an agricultural country with fertile land, plenty of water resources, livestock, forestry resources, and agricultural residues. To ensure a better quality of life for all people, now and in the future, through the implementation of sustainable development initiatives that promote: (1) Food and water security (2) Economic efficient that helps to eliminate inequalities (3) Social equity for all, regardless of race, gender, disability or creed (4) Effective education for environmentally and socially responsible citizenship (5) Environment integrity, and environmental justice (6) Democracy, and mutual understanding between people. In fact the country can be rescued by proper organisation and utilisation of its agricultural potential. Agriculture continues to play a pivotal role in economy.

In a country with a high population density, there are extreme pressures on water and waste systems, which can stunt the country's economic growth. However, Sudan has recognised the potential to alleviate some of these problems by promoting renewable water and utilising its vast and diverse climate, landscape, and resources, and by coupling its solutions for waste disposal with its solutions for water production. Thus, Sudan may stand at the forefront of the global renewable water community, and presents an example of how non-conventional water strategies may be implemented. In Sudan, more than ten million people do not have adequate access to water supply, twenty million inhabitants are without access to sanitation, and a very little domestic sewage is being treated. The investment needed to fund the extension and improvement of these services is great. Most governments in developing countries are ready to admit that they lack the financial resources for proper water and sanitation schemes. Moreover, historically, bilateral and multilateral funding accounts for less than 10% of total investment needed. Thus, the need for private financing is imperative. Water utilities in developing countries need to work in earnest to improve the efficiency of operations. These improvements would not only lead to better services but also to enhanced net cash flows that can be re-invested to improve the quality of service. Staff productivity is another area where significant gains can be achieved. Investment and consumption subsidies have been predicated on the need to (a) help the poor, which have not an access to basic services and (b) improve the environment. Failure of subsidies to reach intended objectives is in part, from lack of transparency in their allocation. Subsidies are often indiscriminately assigned to support investment programmes that benefit more middle and high-income families that already receive acceptable service. Consumption subsidies often benefit upper-income domestic consumers much more than low-income ones. Many developing countries (Sudan is no exception) are encouraging the participation of the private-sector as a means to improve productivity in the provision of water and of wastewaters services. Private-sector involvement is also needed to increase financial flows to expand the coverage and quality of services. A key element to successful private participation is the allocation of risks. How project risks are allocated and mitigated determines the financial and operational performance and success of the project, under the basic principle that the risk should be allocated to the party, which is best able to bear it. Many successful private-sector interventions have been undertaken. Private operators are not responsible for the financing of works, nonetheless they can bring significant gains in productivity, which would allow the utility to allocate more resources to improve and extend services. Redressing productivity, subsidy and cross-subsidy issues before the private-sector is invited to participate, has proven to be less contentious. I have previously sought to encourage more private-sector involvement [1].

The government has continued to pay for the development and operation of water systems, but attempts are being sought to make the user communities pay water charges. In order to ensure the sustainability of water supplies, an adequate institutional and legal framework is needed. Funds must be generated (a) for production, (b) for environmental protection to ensure water quality, and (c) to ensure that water removal remains below the annual groundwater recharge. At present, there are private-sector providers who do not have an enabling environment to offer the services adequately. There is a need for the government to have a mechanism to assist in the regulation and harmonisation of the private-sector providers. Privatisation is part of a solution to improve services in delivery of water and of sanitation sector. At present, there is a transitional situation characterised by: (i) A resistance to payment of water charges; (ii) Insufficient suitable law and inadequate law enforcement;

(iii) Insufficient capacities to provide services; and (iv) Inadequate interaction between the private citizen, business opportunities, and government.

2. GEOGRAPHY, POPULATION AND CLIMATE

Sudan is the largest country in Africa and has a special geopolitical location bonding the Arab world to Africa south of the Sahara. Sudan is a federal republic located in the eastern Africa. It has an area of 2.5 million km^2 extending between 4° and 22°North latitudes and 22° to 38°East longitudes. Its north-south extent is about 2000 km, while its maximum east-west extent is about 1500 km. On the north-east it is bordered by the Red Sea and it shares common borders with nine countries: Eritrea and Ethiopia in the east, Kenya, Uganda and the Democratic Republic of Congo in the south, The Central African Republic, Chad and Libya in the west, and Egypt in the north. The country is a gently sloping plain with the exception of Jebel Marra, the Red Sea Hills, Nuba Mountains and Imatong Hills. Its main features are the alluvial clay deposits in the central and eastern part, the stabilised sand dunes in the western and northern part and the red ironstone soils in the south. The soils of Sudan are broadly divided into six main categories according to their locations and manner of formation: i) desert; ii) semi-desert; iii) sand; iv) alkaline catena; v) alluvial; and vi) iron stone plateau. Within these soil categories there are many local variations with respect to drainage conditions. Sudan is a relatively sparsely populated country. The total population according to the 2004 census was 34.3 x 10^6 inhabitants. The growth rate is 2.2%/y, and population density is 14 persons per square kilometres.

Sudan is known as a country of plentiful water, rich in land, with highest total and renewable supply of fresh water in the region (eastern Africa). The cultivable area is estimated at about 105 million ha (42% of the total land area), while in 2002 the cultivated land was 16.65 million ha (7% of the total land area and 16% of the cultivable area), comprising 16.23 million ha arable land and 0.42 million ha under permanent crops (Table 1). The forest resources of Sudan cover approximately 27% of the total country's area. The main forest types include: i) arid and semi-arid shrubs; ii) low rainfall savannah; iii) high rainfall savannah; iv) special areas of mountainous vegetation in Jebel Marra, the Red Sea Hills and the Imatong Mountains. Rangelands cover about 117 million ha. They spread over most ecological zones: the desert in the north, the semi-desert, the low rainfall savannah and the high rainfall woodlands in the south. Annual herbaceous plants with scattered trees and bushes dominate the northern rangelands. In the southern part, perennial herbaceous plants increase with dense stands of woody cover. The livestock population includes camels, sheep and goats, which are raised in the desert and semi-desert, and cattle, that are raised in the medium rainfall savannah and in the flood plain of the Upper Nile. Almost all livestock is raised under nomadic and semi-nomadic systems. The country has a diverse and fairly rich wildlife. Of the 13 African mammalian orders, 12 are present in Sudan. The protected wildlife areas cover around 36 million ha. There are 8 national parks, 13 game reserves and 3 sanctuaries.

Sudan is under federal rule with 26 States. Each State is governed by a Wali (Governor) with 7 to 10 State Ministers, 4 to 5 Commissioners for the different provinces and a number of localities. Each State has complete administrative and fiscal autonomy and its own State Legislative Assembly for legislative matters of the State.

Table 1. Basic statistics and population.

Physical areas			
Area of the country	2002	250 581 000	ha
Cultivated area (arable land and area under permanent crops)	2002	16 653 000	ha
• as % of the total area of the country	2002	7	%
• arable land (annual crops + temp. fallow + temp. meadows)	2002	16 233 000	ha
• area under permanent crops	2002	420 000	ha
Population			
Total population	2004	34 333 000	inhabitants
• of which rural	2004	60	%
Population density	2004	14	inhabitants/km²
Economically active population	2004	13 806 000	inhabitants
• as % of total population	2004	40	%
• female	2004	30	%
• male	2004	70	%
Population economically active in agriculture	2004	7 925 000	inhabitants
• as % of total economically active population	2004	57	%
• female	2004	38	%
• male	2004	62	%
Economy			
Gross Domestic Product (GDP)	2003	17 800	million US$/yr
• value added in agriculture (% of GDP)	2002	39.2	%
• GDP per capita	2003	518	US$/yr
Human Development Index (highest = 1)	2002	0.505	
Access to improved drinking water sources			
Total population	2002	69	%
Urban population	2002	78	%
Rural population	2002	64	%

Sudan has a tropical sub-continental climate, which is characterised by a wide range of variations extending from the desert climate in the north through a belt of summer-rain climate to an equatorial climate in the extreme south. The average annual rainfall is 416 mm, but ranges between 25 mm in the dry north and over 1600 mm in the tropical rain forests in the south. The country can be divided into three zones according to rainfall regime:

- The annual rainfall in the northern half of Sudan varies from 200 mm in the centre of the country to 25 mm northwards towards the border with Egypt. Where it rains, the rainy season is limited to 2-3 months with the rest of the year virtually dry. Rainfall usually occurs in isolated showers, which vary considerably in duration, location, and from year to year. The coefficient of variation of the annual rainfall in this northern half of the country could be as high as 100%.
- In the quarter south of the centre of the country, the annual rainfall barely exceeds 700 mm, and is concentrated in only four months, from July to October. The average annual rainfall of that region is between 300-500 mm. Rainfed agriculture in Sudan is mainly practised in this quarter. As the coefficient of variation in annual rainfall in this region is around 30% and the dry season extends for about eight months, the area cultivated and the productivity vary widely from one year to another.
- In the most southern quarter of the country, where the annual rainfall exceeds 700 mm and can go up to 1600 mm, the area is dominated by extensive wetlands some parts of which are infested by insects which are hazardous to humans and livestock.

The mean temperature ranges from 30°C to 40°C in summer and from 10°C to 25°C in winter. Potential annual evapotranspiration ranges from 3000 mm in the north to 1700 mm in the extreme south. Most of the agricultural activities are concentrated in the centre of the country, in the generally semi-arid dry savannah zone, through which the Blue Nile and the Atbara River flow. The growing season in the region is around four months. The major limiting factor is not the agricultural potential, but the short duration of the rainy season and the erratic distribution of rainfall during the growing period.

Sudan's population is 34.3 million (2004) with an annual growth rate of 2.2% (Table 1). Population density is 14 inhabitants/km^2 and 60% of the total population is rural. Most of the population lives along the Nile and its tributaries, and some live around water points scattered around the country. At the national level, 69% of the population had access to improved drinking water sources in the year 2002. In urban areas this coverage was 78%, while in rural regions it was 64% (Table 1). Displaced families have increased the total population of villages, which has placed pressure on potable water resources.

The Human Development Index ranks Sudan in 139th place among 177 countries. Poverty in the Sudan is massive, deeply entrenched and predominantly a rural phenomenon. Over two-thirds of the population and under the most favourable assumptions still around 50-70% are estimated to live on less than US$1/day. In recognition of the severity of poverty in general, and of rural poverty in particular, the government started to prepare a draft Poverty Reduction Strategy Paper (PRSP) and launched a pilot poverty-reduction programme in 2001 to improve long-neglected rural social services. The programme is financing basic education, primary health care, malaria prevention and drinking-water supply.

3. Economy, Agriculture and Food Security

The backbone of Sudan's economy is its agricultural sector. The agricultural sector determines to a great extent the economic performance of the Sudanese economy. In fact the country can be rescued by proper organisation and utilisation of its agricultural potential. Although endowed with rich natural resources, Sudan remains comparatively underdeveloped primarily as a result of protracted civil strife and poor economic management. The economy showed a limited response to reform packages during the 1980s and early 1990s. Budget deficits have been common, the average annual rate of inflation peaked at 70% for the period 1991-1995 but gradually subsided to less than 5% in 2001, then climbed to 8% in 2002. Interest rates remained negative during that period and resulted in the collapse of savings, affected the banking system adversely and eroded public confidence.

The GDP of the Sudan was US$17.8 billion (current US$) in 2003. The agricultural sector is the most dominant in the country's economy, even though its share has declined recently because of decreased agricultural production and the increased exploitation and export of mineral oil. In 2002, the sector contributed over 39% to the GDP and employed 57% of the total economically active population in 2004 (Table 1). It contributed about 90% of the Sudan's non-oil export earnings.

Sudan's agro-ecological zones support a variety of food, cash and industrial crops. Vast natural pastures and forests support large herds of livestock including cattle, sheep and goats. The main exported crops are cotton, Arabic gum, sesame, groundnuts, fruits and vegetables;

livestock is also important for exports. Within the agricultural sector, crop production accounts for 53% of agricultural output, livestock for 38% and forestry and fisheries for 9%.

Rainfed agriculture covers by far the largest area in Sudan. The area actually cultivated and total yield may, however, vary considerably from year to year depending on variability of rainfall. The rainfed farming system is characterised by a small farm size, labour-intensive cultivation techniques employing hand tools, low input level and poor yields. Crops grown in the rainfed sector include sorghum, millet, sesame, sunflower and groundnuts. According to the latest estimates, the traditional rainfed farming sector contributes all the production of millet, 11% of sorghum, 48% of groundnuts and 28% of sesame production of the country. Mechanised rainfed agriculture comprises about 10000 large farmers with farm sizes of 400-850 ha and a few large companies with holdings of 8400-84000 ha.

Sudan has the largest irrigated area in sub-Saharan Africa and the second largest in the whole of Africa, after Egypt. The irrigated sub-sector plays a very important role in the country's agricultural production. Although the irrigated area constitutes only about 11% of the total cultivated land in Sudan, it contributes more than half of the total volume of the agricultural production. Irrigated agriculture has become more and more important over the past few decades as a result of drought and rainfall variability and uncertainty. It remains a central option to boost the economy in general and increase the living standard of the majority of the population.

Sudan is generally self-sufficient in basic foods, albeit with important inter-annual and geographical variations, and with wide regional and household disparities in food security prevailing across the country. The high-risk areas are North Kordofan, North Darfur, the Red Sea, Butana and the fringes of the major irrigation schemes in addition to the Southern States. Major constraints to higher farm productivity and incomes are high marketing margins on agricultural produce and an inadequate allocation of budgetary resources and of the scarce foreign exchange earnings. As a result, the low input/low-productivity model of production continues to prevail, and small farmers' incomes remain depressed. In the wake of the food shortages experienced in the 1980s, a high priority has been given by the government to producing food crops. This has resulted in large expansions in sorghum and wheat areas and output. Much of this has been at the expense of the main cash crop, cotton, with production declining by more than 40% since the mid-1980s.

4. WATER RESOURCES AND USE

Water is one of the most fundamental of natural resources that a country must harness in its efforts for rapid economic development. The role of water in the development process cannot be over-emphasised. Sudan is rich in water (from the Nile system, rainfall and groundwater) and lands resources (Table 1). Surface water resources are estimated at 84 billion m^3. The annual rainfall varies from almost nil in the arid hot north to more than 1600 mm in the tropical zone of the south. The total quantity of groundwater is estimated to be 260 billion m^3, but only 1% of this amount is being utilised.

4.1. Water Resources

Internal renewable water resources (IRWR) include the average annual flow of rivers and the recharge of groundwater (aquifers) generated from endogenous precipitation-precipitation occurring within a country's borders. IRWR are measured in cubic kilometres per year (km^3/year). Since data were collected in different years, they may not be directly comparable.

Surface water produced internally includes the average annual flow of rivers generated from endogenous precipitation and base flow generated by aquifers. Measuring or assessing total river flow occurring in a country on a yearly basis usually computes surface water resources (Table 2).

Groundwater recharge is the total volume of water entering aquifers within a country's borders from endogenous precipitation and surface water flow. Groundwater resources are estimated by measuring rainfall in arid areas where rainfall is assumed to infiltrate into aquifers. Where data are available, groundwater resources in humid areas have been considered as equivalent to the base flow of rivers.

Overlap is the volume of water resources common to both surface and groundwater. It is subtracted when calculating IRWR to avoid double counting. Two types of exchanges create overlap: contribution of aquifers to surface flow, and recharge of aquifers by surface run-off. In humid temperate or tropical regions, the entire volume of groundwater recharge typically contributes to surface water flow. In karstic domains (regions with porous limestone rock formations), a portion of groundwater resources is assumed to contribute to surface water flow. In arid and semi-arid countries, surface water flows recharge groundwater by infiltrating through the soil during floods. This recharge is either directly measured or inferred by characteristics of the aquifers and piezometric levels.

Total internal renewable water resources is the sum of surface and groundwater resources minus overlap; in other words, IRWR = Surface water resources + Groundwater recharge − Overlap. Natural incoming flow originating outside a country's borders is not included in the total.

Per capita internal renewable water resources (IRWR) are measured in cubic meters per person per year (m^3/person/year). Per capita values were calculated by using national population data for 2001.

Natural renewable water resources, also known as Actual Renewable Water Resources, are the sum of internal renewable water resources and natural flow originating outside of the country. Natural Renewable Water Resources are computed by adding together both internal renewable water resources (IRWR) and natural flows (flow to and from other countries). Natural incoming flow is the average amount of water, which would flow into the country without human influence. In some arid and semi-arid countries, actual water resources are presented instead of natural renewable water resources.

Per capita natural renewable water resources are measured in cubic meters per person per year (m^3/person/year). Per capita values were calculated by using national population data for 2002. Water is one of the most fundamental of natural resources that a country must harness in its efforts for rapid economic development.

Table 2. Water resources and freshwater ecosystems.

Internal renewable water resources (IRWR), 1977-2001, in km^3	Sudan	Sub-Saharan Africa
Surface water produced internally	28	3812
Groundwater recharge	7	1549
Overlap (shared by groundwater and surface water)	5	1468
Total IRWR (surface water + groundwater – overlap)	30	3901
Per capita IRWR, 2001 m^3	921	5705
Natural renewable water resources includes flows from other countries		
Total 1977-2001 km^3	65	-
Per capita 2002 m^3/person	1981	-
Annual river flows:		
From other countries km^3	119	-
To other countries km^3	-	-
Water withdrawals		
Year of withdrawal data	1995	-
Total withdrawals km^3	17.8	-
Withdrawals per capita m^3	637	-
Withdrawals as a percentage of actual renewable water resources	32.1%	-
Withdrawals by sector (as a percent of total)		
Agriculture	94%	-
Industry	1%	-
Domestic	4%	-
Desalination (various years)		
Desalinated water production (million m^3)	0	-

Internally produced water resources in Sudan are rather limited. The erratic nature of the rainfall and its concentration in a short season places, has left Sudan in a vulnerable situation, especially in rainfed areas. Surface water in Sudan comprises the Nile river system (Nilotic water) and other, Non-Nilotic streams. 64% of the Nile Basin lies within Sudan, while 80% of Sudan lies in the Nile Basin. Local rainfall is the main source of the Non-Nilotic streams and of the Bahr el Ghazal basin, whereas rainfall over the Central African Plateau (Equatorial Lakes) and over the Ethiopian-Eritrean highlands is the main source of the Nile River system and other transboundary seasonal streams (Gash and Baraka). Sudan shares parts of the following basins with neighbouring countries:

- The Nile Basin, 1978506 km^2 (79% of the area of the country);
- The Northern Interior Basins, covering 313365 km^2 in the northwest part of the country (12.5%);
- The Lake Chad Basin, in the west of the country along the border with Chad and the Central African Republic, covering 101048 km^2 (4%);
- The Northeast Coast Basins, representing a strip along the Red Sea coast of the country, covering 96450 km^2 (3.8%);
- The Rift Valley Basin, in the southeast part of the country at the border with Ethiopia and Kenya, covering 16441 km^2 (0.7%).

The Nile system within Sudan comprises:

- The Blue Nile, Sobat and Atbara Rivers originating in the Ethiopian highlands;
- The White Nile system, upstream of Sobat River, originating on the Lakes Plateau;
- The Bahr el Ghazal Basin, an internal basin in southwest Sudan.

The characteristics of the Nile system tributaries are the following:

- The Blue Nile: The flow of the Blue Nile reflects the seasonality of rainfall over the Ethiopian highlands where the two flow periods are distinct. The flood period or wet season extends from July to October, with the maximum in August-September, and low flow or dry season from November to June. Therefore the annual Blue Nile hydrograph has a constant bell-shaped pattern, regardless of variation in the annual flow volumes. The average annual flow of the Blue Nile and its tributaries upstream of the confluence with the White Nile at Khartoum is about 50 km^3; the daily flow fluctuates between 10 million m^3 in April to 500 million m^3 in August (ratio of 1:50).
- The White Nile: Due to losses in the Sudd swamp area, the White Nile leaves this area with only about 16 km^3, out of 37 km^3 on entering it. The river receives about 13 km^3 from the Sobat River before joining the Blue Nile at Khartoum. The contribution of the Bahr el Ghazal basin is negligible, estimated at about 0.5 km^3. The average annual flow of the White Nile System at Malakal is about 29.5 km^3 and the daily discharge fluctuates between 50 million m^3 in April to 110 million m^3 in November (ratio 1:2). During the flood period the Blue Nile forms a natural dam that obstructs the flow of the White Nile and consequently floods the area upstream of the confluence.
- The Atbara River: This is a highly seasonal river, with an annual flow upstream of its confluence with the Nile of about 10 km^3 restricted to the flood period of July-October, the maximum occurring between August-September. The river has a steep slope and small catchments, and reflects the rainfall over the upper catchments as runoff at Sudan border within one to two days.
- The Main Nile: The reach of the Nile downstream of the confluence of the Blue Nile and the White Nile Rivers is known as the Main Nile. The Atbara River is regarded as the only and last tributary joining the Main Nile. The average annual flow of the Main Nile at the Sudan-Egypt border at Aswan is estimated at 84 km^3.

The average annual yield of the Non-Nilotic streams is estimated at about 7 km^3/yr, of which 5 km^3/yr are internally produced. The major streams are the Gash and Baraka in the east of the country, both of which are characterised by large variations in annual flow and heavy silt loads.

The major groundwater formations and basins are the Nubian Sandstone Basin and the Umm Ruwaba Basins. The Ghazal, Sudd and Sobat swamps in the south of the country represent major wetlands, from which evaporation is exceptionally high. According to an estimate from 1980, the extent of the Sudd is over 16200 km^2, but the surface area fluctuates with rainfall. Sudan's total natural renewable water resources are estimated to be 149 km^3/yr, of which 30 km^3/yr are internally produced (Table 3). In a 10th frequency dry year, the internal water resources are reduced to about 22.3 km³/yr. Of the internal water resources, 28 km³/yr are surface water and 7 km³/yr are groundwater, while the overlap between surface water and groundwater is estimated at 5 km³/yr. As a result of the Nile Waters Agreement with Egypt, total actual renewable water resources of the country amount to 64.5 km³/yr.

The high variability of river flows necessitates storage facilities. The total storage capacity of the following four main dams is estimated at 8.73 km³, reduced to about 6.90 km³ owing to sedimentation and debris:

- The Sennar Dam on the Blue Nile (design capacity 0.93 km³, present capacity 0.60 km³) is for the flood control and irrigation of the Gezira Scheme.
- The Roseires Dam on the Blue Nile (design capacity 3.0 km³, present capacity 2.2 km³; there are plans to increase the present dam height of 60 m to provide an extra capacity of 4.0 km³) is for flood control and utilises part of the country's share of the Nile waters for irrigation.
- The Jebel Aulia Dam on the White Nile (design capacity 3.5 km³, present capacity 3.5 km³) was originally designed to benefit Egypt by augmenting the supply of summer flow to the Aswan dam. After the construction of the High Aswan Dam it was no longer needed by Egypt and was officially handed over to the Sudan in 1977.
- The El Girba Dam on the Atbara River (design capacity 1.3 km³, present capacity 0.6 km³) is for flood control, irrigation of New Halfa Scheme for the benefit of the people displaced by the High Aswan Dam, and hydropower.

A small barrage was constructed on the Rahad River to divert floodwater to the Rahad Agricultural Scheme and to siphon underneath the Dinder River to augment the water supply during the dry season from the Meina Pump Station on the Blue Nile. The Jonglei Canal, between Bahr el Jebel and the White Nile, was planned to divert water from upstream of the Sudd to a point farther down the White Nile, bypassing the swamps, to make more water available for use downstream. Works on it were discontinued since 1983 after two thirds were completed. Non-conventional water sources are limited in Sudan. However, the desalination of seawater was introduced recently in Port Sudan town. Fossil groundwater resources are estimated to be 16000 km³.

4.2. Water Use

Total water withdrawal in Sudan was estimated at 37 km³ for the year 2000 (Table 3). The largest water user by far was agriculture with 36 km³ (Figure 1). The domestic sector and industry accounted for withdrawals of 0.99 km³ and 0.26 km³ respectively. Water used in Sudan derives almost exclusively from surface water resources. Groundwater is used only in very limited areas, and mainly for domestic water supply.

Water-resources assessment in Sudan is not an easy task because of uncertainty of parameters, numerous factors that affect the variables, lack of information and inaccurate measurements. However, according to seasonal water availability, Sudan can be divided into three zones: (a) areas with water availability throughout the year are the rainy regions (equatorial tropical zones); (b) areas with seasonal water availability; and (c) areas with water deficit throughout the year. The last occupy more than half the area of Sudan. Water resources in Sudan include surface water and ground water spread over large parts of the country. Both resources are utilised for agricultural production by the use of canal system from dams, pumps, embankments (flood), and wells.

Table 3. Water sources and use.

Renewable water resources			
Average precipitation		416	mm/yr
		1 042	10^9 m³/yr
Internal renewable water resources		30.0	10^9 m³/yr
Total actual renewable water resources		64.5	10^9 m³/yr
Dependency ratio		76.9	%
Total actual renewable water resources per inhabitant	2004	1 879	m³/yr
Total dam capacity	1995	8 730	10^6 m³
Water withdrawal			
Total water withdrawal	2000	37 314	10^6 m³/yr
- irrigation + livestock	2000	36 069	10^6 m³/yr
- domestic	2000	987	10^6 m³/yr
- industry	2000	258	10^6 m³/yr
• per inhabitant	2000	1 187	m³/yr
• as % of total actual renewable water resources	2000	58	%
Non-conventional sources of water			
Produced wastewater		-	10^6 m³/yr
Treated wastewater		-	10^6 m³/yr
Reused treated wastewater		-	10^6 m³/yr
Desalinated water produced	1990	0.4	10^6 m³/yr
Reused agricultural drainage water		-	10^6 m³/yr

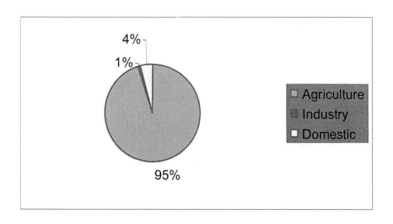

Figure 1. Surface water withdrawals by sector in Sudan (total 37.314 km³ in 2000).

Sudan's total natural renewable water resources are estimated to be 149 km³/yr, of which 30 km³/yr are internally produced (Table 2). In a 10th frequency dry year, the internal water resources are reduced to about 22.3 km³/yr. Of the internal water resources, 28 km³/yr are surface water and 7 km³/yr are groundwater, while the overlap between surface water and groundwater is estimated at 5 km³/yr. As a result of the Nile Waters Agreement with Egypt, total actual renewable water resources of the country amount to 64.5 km³/yr.

The high variability of river flows necessitates storage facilities. The total storage capacity of the following four main dams is estimated at 8.73 km^3, reduced to about 6.90 km^3 owing to sedimentation and debris:

- The Sennar Dam on the Blue Nile (design capacity 0.93 km^3, present capacity 0.60 km^3) is for the flood control and irrigation of the Gezira Scheme.
- The Roseires Dam on the Blue Nile (design capacity 3.0 km^3, present capacity 2.2 km^3; there are plans to increase the present dam height of 60 m to provide an extra capacity of 4.0 km^3) is for flood control and utilises part of the country's share of the Nile waters for irrigation.
- The Jebel Aulia Dam on the White Nile (design capacity 3.5 km^3, present capacity 3.5 km^3) was originally designed to benefit Egypt by augmenting the supply of summer flow to the Aswan dam. After the construction of the High Aswan Dam it was no longer needed by Egypt and was officially handed over to the Sudan in 1977.
- The El Girba Dam on the Atbara River (design capacity 1.3 km^3, present capacity 0.6 km^3) is for flood control, irrigation of New Halfa Scheme for the benefit of the people displaced by the High Aswan Dam, and hydropower.

A small barrage was constructed on the Rahad River to divert floodwater to the Rahad Agricultural Scheme and to siphon underneath the Dinder River to augment the water supply during the dry season from the Meina Pump Station on the Blue Nile. The Jonglei Canal, between Bahr el Jebel and the White Nile, was planned to divert water from upstream of the Sudd to a point farther down the White Nile, bypassing the swamps, to make more water available for use downstream. Works on it were discontinued since 1983 after two thirds were completed.

Non-conventional water sources are limited in Sudan. However, the desalination of seawater was introduced recently in Port Sudan town. Fossil groundwater resources are estimated to be 16000 km^3.

5. INTERNATIONAL WATER ISSUES

Surface water and groundwater resources are mostly shared with neighbouring countries. The Nile River, which is shared between 10 countries, is the primary source of Sudan's water. The four main Non-Nilotic streams are also shared with neighbouring countries. The largest groundwater aquifer, the Nubian Sandstone system, is shared with Chad, Libya and Egypt. The first Nile Waters Agreement between Egypt and Sudan was signed in 1929. It allocated to Egypt the right to use 48 km^3/yr, while it gave Sudan the right to tap only about 4 km^3/yr. The treaty does not allocate to Ethiopia any rights to use the Nile waters and also still binds Uganda, the United Republic of Tanzania and Kenya and bars them from using the Lake Victoria waters. In 1959, the Nile Waters Agreement between Egypt and Sudan assigned to Sudan 18.5 km^3/yr, measured at Aswan at the border with Egypt. The other riverside nations are still not included in this agreement.

Recently, the Nile Basin Initiative has been created and prepared a Strategic Action Programme, which consists of two sub-programmes: the Shared Vision Programme (SVP) and the Subsidiary Action Programme (SAP). The SVP is to help create an enabling environment for action on the ground through building trust and skill, while the SAP is aimed at the delivery of actual development projects involving two or more countries. Projects are selected by individual riparian countries for implementation and submitted to the Council of Ministers of the Nile Basin Initiative for approval. Sudan, Ethiopia and Egypt have also adopted a strategy of cooperation in which all projects to be launched on the river should seek the common benefit of all member states and this should be included in accompanying feasibility studies. Sudan, together with Algeria, Cameroon, the Central African Republic, Chad, Niger, and Nigeria, is located in the Lake Chad basin.

5.1. Water Management, Policies and Legislation Related to Water Use

5.1.1. Institutions

The Ministry of Agriculture and Natural Resources (MANR) supervises the Agricultural Corporations that manage the large irrigation schemes, while the Ministry of Irrigation and Water Resources (MIWR) is responsible for delivering irrigation water. The Ministry of Irrigation and Water Resources (MIWR) is the federal body in Sudan legally responsible for all water affairs. It offers technical advice and assistance to water projects within the states and the private-sector. It is in charge of the groundwater, the Non-Nilotic streams and valleys under the Groundwater and Wadis Directorate. It undertakes its task in coordination with the relevant sectors, departments and technical offices (agriculture, industry, foreign, electricity, and investment, etc.). It has the following responsibilities:

- Satisfaction of the water requirements of the various users through the country;
- Water resources planning, management and development;
- International and regional cooperation concerning the shared water sources;
- Planning, design, execution, operation and maintenance of the different irrigation schemes;
- Control of water abstraction;
- Construction of new irrigation works;
- Operation and maintenance of all large-scale irrigation structures and drinking water facilities;
- Provision of the means for hydropower generation and protection of the water-related environment.

5.1.2. Water Management

The Gezira Scheme is managed on a vertically integrated basis by the semi-autonomous Sudan Gezira Board (SGB). The MIWR is responsible for managing the Sennar Dam on the Blue Nile and the upper reaches of the irrigation system, responding to requests for water delivery from SGB's field staff. Within the scheme, the SGB serves as landlord, operates and maintains the lower reaches of the irrigation system and provides most of the inputs and services required by farmers to produce cotton, which is transported by the Board to its ginneries and sold on behalf of growers by the Sudan Cotton Company Limited. The SGB recovers the cost of advances made for inputs and services from the cotton sales before payment is made to the farmer. Tenants are wholly responsible for growing other crops in prescribed rotations with cotton (sorghum, groundnuts, forage, wheat, and vegetables), making their own arrangements for input supplies and marketing. By 2001 in the Gezira Scheme, Minor Canal Committees had been formed along the minor irrigation canals and representatives of each of these committees constitute the Irrigation Committee at the block level. In addition to the Irrigation Committee, a Financial Committee has been established that is coordinating the reimbursement of the seasonal credits and arrangements for procurement of new inputs. The Irrigation Committee with representatives of each of the minor canal committees will be responsible for the operation and maintenance of the minor irrigation system, a task presently entrusted to the SGB, with the Ministry of Irrigation and Water Resources responsible of supplying the main system.

To address some of the problems facing irrigation management and development the government has formalised a policy framework that includes:

- Transferring the operation and production of large- and medium-size irrigation schemes to the farmers and giving them full responsibility for water management on the irrigation system below the minor canals level through establishment of voluntary water users associations (WUAs).
- Fostering sustainable productivity of the large schemes through rehabilitation, combined with financial and institutional reform.
- Grouping, rehabilitating and handing over the relatively small size pump schemes in the Blue Nile and the White Nile. These schemes were originally established and run by the government. Recently, and in accordance with the economic reforms, these schemes were handed over to the private-sector represented by individual farmers, cooperatives or private companies.

5.1.3. Policies and Legislation

In 1992, the national economy was reoriented towards a free economy, a policy shift that impacted the agricultural sector profoundly. The government withdrew from the direct financing of agriculture, provision of inputs and services. The government within its policy of withdrawal from provision of goods and services handed over all the small- and medium-size irrigation schemes under its control to the farmers. The handing over policy was not successful because farmers were ill prepared and most of the schemes were in need of rehabilitation. Since 1992, the cropped areas and the productivity of many schemes have sharply declined.

Agricultural sectoral policies for the irrigated sub-sector include the following:

- To extend the market economy to all crops to allow the farmer to choose a suitable crop mix.
- To provide support to the agricultural research institutions so as to explore suitable technologies for improving crop production and productivity. In addition, to provide support to agricultural extension and services.
- To encourage the private-sector to provide agricultural inputs for the agricultural sector.
- To encourage exports through improvement of quality so as to meet international standards.

- Establishment of specialised crop committees for main crops like cotton with the objective of achieving all necessary coordination between the concerned authorities.
- Farmers participate in agricultural policy formulation.

The Sudan Comprehensive National Strategy for the Agricultural Sector (1992-2002) put food security, sustained agricultural development, efficient resource utilisation and yield enhancement on the top of the agenda. Little has been done, however, to improve the accessibility to food of the poor, the vulnerable and the marginalised strata of society.

5.1.4. Finances

Financing irrigation O&M through fees collected from the beneficiaries of the irrigation system was first introduced in Sudan with the introduction of the modern irrigation system at the El Zeidab scheme in 1909 when a private foreign company erected a pump station to irrigate local farmers' land for an agreed irrigation fee. After the success of the experiment for the first two seasons, a bad crop yield in 1911/12 meant that the farmers were unable to pay their irrigation fees. The company experienced heavy losses and decided to pull out of the scheme.

Experience from El Zeidab scheme was used in selecting the form of production relationship between the government, the Sudan Plantation Syndicate and the farmers when the Gezira scheme was commissioned. To avoid the inability of some farmers to pay irrigation fees in case of bad crop yields, a "sharing system" between the three parties was adopted. This system continued until 1981 when it was replaced by what is known as the "individual account system" in which each individual farmer is treated separately in terms of cost and profit. The objective was to create some incentive for the individual farmers to increase their productivity. The new account system failed to achieve break-even productivity. The individual account system was also applied in all the irrigation schemes run by the government at that time. Payment of irrigation fees by the farmers continued in all government schemes from 1981 to 1995. During this period irrigation fees collected were very low, averaging about 50% only. The non-recovered part of the water supply costs is borne by the government.

Starting from 1995, and as part of the liberalisation of the economy, the government withdrew from financing the cost of irrigation services, among other things. Farmers were left to pay irrigation fees to the newly established Irrigation Water Corporation (IWC), which uses these fees directly to provide water supply services to the farmers. Instead of the IWC setting up its own mechanism for collecting the fees directly from the farmers, it relies on the Agricultural Corporations (AC) managing the scheme to collect the fees from the farmers. Because these ACs were also facing considerable financial difficulties, part of the water fees collected may not reach the IWC and part of the collected fees paid to IWC is delayed for sometime as it is used for financing other urgent activities. The result of this is the inability of IWC to have the required budget that enables it to provide its services in a sustainable manner. This led to the accumulation of sediment in the irrigation canals, deterioration of the water regulation structures, machinery and pumps. By the year 2000 the IWC was dissolved and the MIWR is again responsible for the O&M of the irrigation canals up to the minor off-takes. The Ministry of Finance and National Economy provides the MIWR with the annual budgets for operation and maintenance. The Nile system travers Sudan from south to north. The Nile basin system in Sudan comprises the Blue Nile, Rahad, and Dender system. The

White Nile, Bahr El Gabal, Zaraf, Bahr El Ghazal and Sobat. In addition, the Atbara system. Water used in Sudan derives almost exclusively from surface water resources. Groundwater is used only in very limited areas, and mainly for domestic water supply. Current use for both agriculture and drinking water is about $1.2 \times 10^9 \text{ m}^3$.

6. IRRIGATION AND DRAINAGE DEVELOPMENT

Irrigation potential was estimated at about 2.78 million ha based on soil and water resources criteria. This figure does not take into account possible large-scale developments in the enormous wetlands in southern Sudan.

6.1. Evolution of Irrigation Development

Large-scale gravity irrigation started during the British colonial period (1898-1956) and the colonial agricultural policy was characterised by the promotion of cotton production in the Nile Basin. Irrigation by pumping water began at the beginning of the 20th Century, substituting traditional flood irrigation and water wheel techniques.

The Gezira Scheme is Sudan's oldest and largest gravity irrigation system, located between the Blue Nile and the White Nile. Started in 1925 and progressively expanded thereafter, it covers about 880000 ha. It receives water from the Sennar Dam on the Blue Nile and is divided into some 114000 tenancies. Farmers operate the scheme in partnership with the government and the Sudan Gezira Board, which provides administration, credit and marketing services. The scheme has played an important role in the economic development of Sudan, serving as a major source of foreign exchange earnings and of government revenue. It has also contributed to national food security and in generating a livelihood for the 2.7 million people who now live in the command area of the scheme.

In the post-colonial period, it was assumed that the only sound way to bring about development would still be through large irrigation developments. The increase in Nile water allocation through the 1959 Nile Waters Agreement with Egypt led for example to the construction of the Managil extension of the Gezira Scheme and of the New Halfa Scheme. The New Halfa Scheme is located on the upper Atbara River in the east of the country. It was partly financed by Egypt after the construction of the Aswan High Dam that created Lake Nasser, which flooded the Sudanese town of Wadi Halfa in 1964. Since then the inhabitants have been moved to the new irrigated agricultural lands where they have been growing a variety of crops.

In the 1970s, Sudan was expected to become the "bread basket" of the Arab world, and with large investments from oil-rich Gulf nations, irrigation schemes such as the Rahad Scheme, which receives its water from the Rahad River and the Blue Nile, were established. Large-scale irrigated agriculture expanded from 1.17 million ha in 1956 to more than 1.68 million ha by 1977. The 1980s were a period of rehabilitation, with efforts to improve the performance of the irrigation sub-sector. In the 1990s, some smaller schemes were licensed to the private-sector, while the four big schemes of Gezira and Managil, New Halfa and Rahad remained under government control because they were considered strategic schemes.

In 2000, the total area equipped for irrigation was 1863000 ha, comprising 1730970 ha equipped for full or partial control irrigation and 132030 ha equipped for spate irrigation. Only about 800000 ha, or 43% of the total area, are actually irrigated owing to deterioration of the irrigation and drainage infrastructures. In 1995, surface water was the water source for 96% of the total irrigated area land, and the remaining 4% were irrigated from groundwater (small tube-wells). The irrigated area where pumps are used to lift water was 346680 ha in 2000. Most irrigation schemes are large-scale and they are managed by parastatal organisations known as Agricultural Corporations, while small-scale schemes are owned and operated by individuals or cooperatives.

A number of water harvesting projects was implemented in the western part of Sudan during the 1970s, 1980s and late 1990s. The main objective was to combat the effects of drought by improving crop production and increasing domestic water use. However, few of those projects have succeeded in combining technical efficiency with low cost and acceptability to the local agro-pastoralist farmers. This is partially owing to the lack of technical know-how, but also because of the selection of inappropriate approaches with regard to the prevailing socio-economic conditions.

6.2. Role of Irrigation in Agricultural Production, the Economy and Society

The main irrigated crops are sorghum, cotton, fodder, wheat, groundnuts and vegetables. Other crops under irrigation are sugar cane, maize, sunflower, potatoes, roots and tubers and rice. Irrigated agriculture has been Sudan's largest economic investment, yet returns have been far below potential. A study by the World Bank showed that, during the period 1976-1989, yields were low and extremely variable and cultivated areas suffered a gradual decline. A study undertaken in the Rahad Scheme based on data from 1977 to 1995 shows that actual crop yields are well below potential yields. The same study also estimated the water use efficiency and found an overall efficiency of 63-68%. The distribution efficiency of the network was 93% and estimated field losses were 25-30%.

In the Gezira Scheme, a complex mix of financial, technical and institutional problems resulted in a serious fall in the productivity of the scheme and a corresponding drop in farm incomes in the late 1990s, resulting in a drop of cropping intensity from 80% in 1991/92 to 40 percent in 1998/99. About 126000 ha were taken out of production owing to siltation and water mismanagement, leading to a reduced availability of water. Because of bad water management, water supply is about 12% below crop water requirements at crucial stages in the growth cycle, while at the same time, as much as 30% of the water delivered is not used by crops. However, an initiative aimed at "Broadening farmer's choices on farm systems and water management" by Food and Agriculture Organisation (FAO) in part of the scheme, meant that productivity of sorghum, cotton and wheat could be increased to 112% for 2000/01, compared to the Gezira average of 42%.

Apart from the Gezira scheme established in 1925, most of the irrigation schemes were developed in the 1960s and 1970s. Since then, there have been no significant irrigation developments, for two reasons: any possible remaining sites would be complex and expensive to develop, and the low levels of productivity of the irrigated crops in the country make it difficult to justify further investment. As a result, in order to meet an ever-increasing demand for food and fibre, priority has been given to increasing productivity from the existing irrigation schemes. But this objective has been thwarted by the declining supply of water to farmers in these schemes. Consequently, the performance of the irrigation sector has consistently fallen short of expectations. The agricultural sector is the major source of water consumption in Sudan. Sudan is presently utilising 16.5×10^9 m^3 annually from its share in irrigated agriculture sub-sector, currently covering an area of 1.7×10^6 hectares. The potential is three-fold.

6.3. Status and Evolution of Drainage Systems

Due to excess rainfall and sometimes to misuse of irrigation water all the irrigation schemes need drainage networks to remove any excess water from the cultivated areas. In low areas, minor drains and collector drains are constructed to remove this excess water by gravity into cutout low areas or into natural drains. Sometimes, pumps are used to take water from low lands into areas outside the scheme. Also escape drains are constructed along the main canal to carry any excess water to the nearest river or natural drain. It is estimated that in 2000 about 560000 ha were drained. The development plans for the irrigation sector include the rehabilitation of the existing irrigation schemes, a shift of emphasis towards the development of small-scale irrigation schemes, and phased development and vertical expansion.

6.4. Perspectives for Agricultural Water Management

The country has an agricultural potential of 105 million ha, of which only 16.7 million ha are cultivated and only about 1.9 million ha out of an irrigation potential of 2.8 million ha are under irrigation now. Therefore, there is ample room for further developments especially in the irrigation sub-sector. However, there are three major constraints to irrigation development in Sudan:

- The ineffective process of annual maintenance of civil works (reportedly due to lack of funds), especially for the removal of silt and weeds from irrigation canals, which slows down water flow and causes continuous shrinkage of the actual irrigated area;
- The steady increase in development costs, which has been aggravated by the continuous devaluation of the local currency;
- The lack of farmer involvement in the planning and operation of the schemes and related services.

Generally, the water supply for all of the irrigation schemes is provided by dams and/or pumps and extensive networks of canals covering the whole schemes as well as drainage networks of canals (Figure 2). The overall objective of water management policies is to improve water use efficiency in agriculture, which includes efficient control of water in the irrigation networks, maintenance of the irrigation structures, provision of technical capacities capable to operate the systems, and efficient and economical maintenance of the irrigation systems. Supplementary irrigation could increase the very low or zero productivity of crops and fodder. The conjunctive use of groundwater and surface water could help to optimise the water resource productivity.

6.5. Environment and Health

The depositions of silt in irrigation canals and the subsequent built-up of aquatic weeds result in losses in production of up to 40%. Other costs of siltation include loss of hydropower potential since methods of silt removal involve measures that lower the head and interfere with generator operation. The most serious effect, however, is the loss of agricultural production. Agricultural chemicals were introduced into Sudan in 1946 in the Gezira scheme.

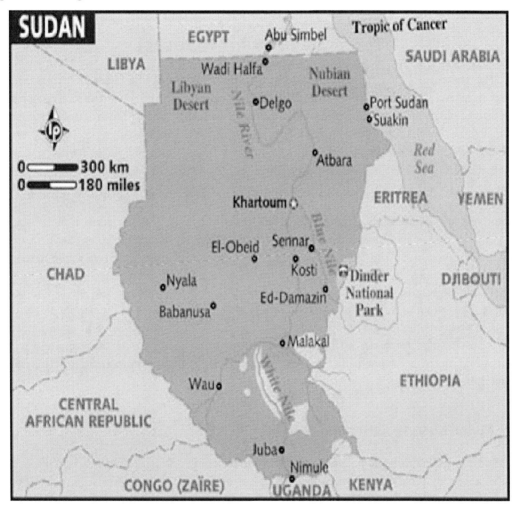

Figure 2. Map of Sudan.

Since then the application of chemicals has intensified and proliferated into other agricultural schemes and in the private vegetable and horticulture fields as well as for control of desert locusts, birds and rodents and for public health. A total of about 200 active ingredients are registered in Sudan in over 600 different formulations of pesticides. An average of about 450 tonnes of insecticides and 150 tonnes of herbicides were applied annually during the period 1993-1997. The annual consumption of fertilisers in the whole of Sudan is estimated at 80000-200000 tonnes of urea and 20000-40000 tonnes of super phosphate. The applied pesticides or the residue and degradation products can contaminate the water resources from the formulating sites, fallout from the spray, washing from contaminated clothes, empty containers, application equipment and dumping of the surplus. Serious contamination has been detected in the Gezira canals as well as in boreholes in the Qurashi area (Hassahessa Province) and the Kassala horticulture area. Fertilisers containing inorganic nitrogen as well as waste containing organic nitrogen are the two main sources of frequently reported nitrate and nitrite contamination in groundwater. Detailed studies are needed to discriminate between the two pollution sources.

7. RENEWABLE ENERGY RESOURCES

The present position for most people in Sudan for obtaining the needed energy forms (heat, light, etc.) is provided by firewood. Cooking is largely done by wood from forests or its derivative, charcoal. Cattle dung and agriculture waste being used to lesser extent. Human, animal, and diesel or gasoline engines provide mechanical power and some cooking and lighting is done by kerosene. It should be recognised that this situation is unlikely to be changed for the next one or two decades. However, because of the need to increase energy availability and also to find alternatives to the rapidly decreasing wood supplies in many rural areas. It is necessary that a vigorous programme reaching into alternative renewable energies should set up immediately. There should be much more realism in formation of such programme, e.g., it is no use providing a solar powered pump at a price competitive with a diesel for some one who cannot ever offered a diesel engine. The renewable energy technology systems (RETs) are simple, from local materials, clean energy, reliable and sustainable. Specialist on their applications carried out socio-economic and environmental studies. The output of the studies pointed out that, they are acceptable to the people and have measured remarkable impacts on the social life, economical activities and rural environment [2-3]. The demand for water in Sudan has increased tremendously over the years and will continue to increase in view of the accelerating pace of population growth, urbanisation and industrialisation.

7.1. Major Energy Consuming Sectors

Sudan is still considered between the 25 most developing African countries. Agriculture is the backbone of economic and social development in Sudan. About 80% of the population depends on agriculture, and all other sectors are largely dependent on it. Agriculture contributes to about 41% of the gross national product (GNP) and 95% of all earnings. Agriculture determines for the last 30 years the degree of performance growth of the national economy.

7.1.1. Agriculture Sector
During the last decades, agriculture contributed by about 41% to the Sudan GNP. This share remained stable till 1984/1985 when Sudan was seriously hit by drought and desertification, which led to food shortages, deforestation, and also, by socio-economic effects caused the imposed civil war. The result dropped the agriculture share to about 37%. Recent development due rehabilitation and improvement in agricultural sector in 1994 has

raised the share to 41%. This share was reflected in providing raw materials to local industries and an increased export earning besides raising percentage of employment among population.

7.1.2. Industrial Sector

The industrial sector is mainly suffering from power shortages, which is the prime mover to the large, medium and small industries. The industrial sector was consuming 5.7% of the total energy, distributed as fellows: 13.8% from petroleum products, 3.4% from biomass and 8% from electricity.

7.1.3. Domestic Use

Household is the major energy consumer. It consumed 92% of the total biomass consumption in form of firewood and charcoal. From electricity this sector consumed 60% of the total energy, and 5.5% of petroleum products.

7.1.4. Transport Sector

The transportation sector was not being efficient for the last two decades because of serious damage happened to its infrastructure. It consumed 10% of the total energy consumption and utilised 60% of the total petroleum products supplied.

7.2. Electricity Situation

Tables 4 to 9 show energy profile, consumption, and distribution among different sectors in Sudan. Sudan like most of the oil importing countries suffered a lot from sharp increase of oil prices in the last decades. The total annual energy consumed is approximately 11×10^9 tonnes of oil equivalent, with an estimated 43% lost in the conversion process [4].

Table 4. Power output of present hydropower plants (MW).

Station	Power
Rosaries	275
Sennar	15
Khashm El Girba	13
Total	303

Table 5. Annual electricity consumption in Sudan (10^6 mWh).

Sector	Energy	Percent of total (%)
Transportation	3.2	4%
Agricultural	22.4	28%
Industries	6.4	8%
Residential	48.0	60%
Total	80.0	100%

Table 6. Annual petroleum product consumption in Sudan (10^6 mWh.)

Sector	Energy	Percent of total (%)
Transportation	601	60.0%
Industries	138	13.8%
Agricultural	148	14.8%
Residential	55	5.5%
Others*	60	5.9%
Total	1002	100.0%

*Others are commercial and services.

Table 7. Percentage of the total annual electricity consumption by states.

States	Percent (%)
Khartoum, Central and East states	85.8%
Red Sea state	4.5%
Northern states	4.0%
Darfur states	3.1%
Kordofan states	2.3%
Southern states	0.3%

Table 8. Energy sources for rural area.

Source	Form
Solar energy	Solar thermal, Solar PV
Biomass energy	Woody fuels, Non woody fuels
Wind energy	Mechanical types, Electrical types
Mini & micro hydro	A mass water fall, Current flow of water
Geothermal	Hot water

Table 9. Energy required in Sudan rural area.

Rural energy	Activity
Domestic	Lighting, heating, cooking, cooling
Agricultural process	Land preparation, weaving, harvesting, sowing
Crop process and storage	Drying, grinding, refrigeration
Small and medium industries	Power machinery
Water pumping	Domestic use
Transport	Schools, clinics, communications, radio, televisions, etc.

7.3. Economic Incentives to Protect the Environment

In Some countries, a wide range of economic incentives and other measures are already helping to protect the environment. These include: (1) Taxes and user charges that reflect the costs of using the environment e.g., pollution taxes and waste disposal charges. (2) Subsidies, credits and grants that encourage environmental protection. (3) Deposit-refund systems that prevent pollution on resource misuse and promote product reuse or recycling. (4) Financial enforcement incentives, e.g., fines for non-compliance with environmental regulations. (5) Tradable permits for activities that harm the environment.

7.4. Sustainable Development in Sudan

Like most African countries, Sudan is vulnerable to climate variability and change. Drought is one of the most important challenges (Figure 3). The most vulnerable people are the farmers in the traditional rain-fed sector of western, central and eastern Sudan, where the severity of drought depends on the variability in amount, distribution and frequency of rainfall. Three case studies were conducted in Sudan as part of the project. They examined the condition of available livelihood assets (natural, physical, financial, human and social) before and after the application of specific sustainable livelihood environmental management strategies, in order to assess the capacity of communities to adapt creased resilience through access to markets and income generating opportunities.

The basic criteria for the energy choices facing the Sudan are, of course, not just environmental. Economic factors remain crucial. However, short-term economic prosperity may mean little if the health; well-being and livelihood of the populations concerned cannot be maintained. At one time a major pre-occupation of many environmentalists was the fear that supplies of key fuels would soon be exhausted. Nowadays, although the price of fuels remains a key political issue, fuel scarcity is less of an immediate concern. Nevertheless, reforestation has its attractions as interim carbon store, since it is relatively cheap, assuming land is available, and it offers other benefits, such as enhanced biodiversity.

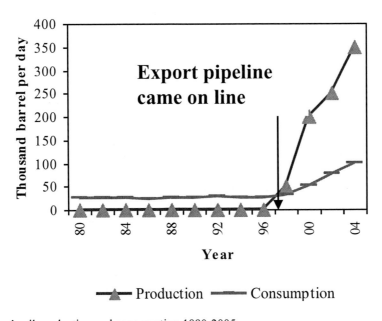

Figure 3. Sudan's oil production and consumption 1980-2005.

8. PROBLEMS AND DIFFICULTIES IN RURAL ENERGY DEVELOPMENT

The following problems are summarised:

8.1. Imbalance in Rural Energy Development

Due to the difference in economic conditions in different areas, the development of rural energy is considerably imbalanced. The main challenge to energy policymakers in the 21st century is how to develop and manage adequate, affordable and reliable energy services in a sustainable manner to fuel social and economic development. Generally, future rural energy will be oriented towards green energy, and the future development of rural energy will concentrate on biogas, small hydropower, solar energy, and wind power.

8.2. Insufficient Investment in Development of Rural Energy

Current rural energy relies mainly on charcoal, firewood and green energies, such as electricity and biogas. The bad economic situation leads to considerable difficulty in the development of rural energy, and farmers in remote areas still prefer "free firewood" for their cooking due to their low income. Hence, further development of rural energy needs significant financial support from the government at various levels. Although the work of rebuilding traditional stoves has been almost finished, most of the rebuilt fuel-saving stoves have a thermal efficiency less than 20%.

8.3. Excessive Dependence on Forests for Rural Energy

Currently, energy for rural household use comes mainly from burning of firewood. The annual consumption of forests is 1.96×10^6 m^3, and of this 0.65×10^6 m^3 is as firewood. To some extent, this pattern of energy consumption has led to environmental damage such as water and soil loss, decrease in forest cover, and air pollution. The excessive use of firewood from forests for rural energy would cause damage to sightseeing resorts, make animals lose their habitats, and lead to the extinction of some endangered plants. The future development of rural energy should be aimed at completely changing the current pattern of energy consumption, fully utilising abundant resources of hydropower, biomass, solar and wind energy, promoting economic growth through the development of rural energy and integrated utilisation of biomass.

9. ENVIRONMENT ASPECTS

Environmental pollution is a major problem facing all nations of the world. People have caused air pollution since they learned to how to use fire, but man-made air pollution (anthropogenic air pollution) has rapidly increased since industrialisation began. Many volatile organic compounds and trace metals are emitted into the atmosphere by human activities. The pollutants emitted into the atmosphere do not remain confined to the area near the source of emission or to the local environment, and can be transported over long distances, and create regional and global environmental problems.

A great challenge facing the global community today is to make the industrial economy more like the biosphere, that is, to make it a more closed system. This would save energy, reduce waste and pollution, and reduce costs. In short, it would enhance sustainability. Often, it is technically feasible to recycle waste in one of several different ways. For some wastes (i.e., agriculture) there are powerful arguments for incineration with energy recovery, rather than material recycling. Cleaner production approach and pollution control measures are needed in the recycling sector as much as in another. The industrial sector world widely is responsible for about one third of anthropogenic emissions of carbon dioxide, the most important greenhouse gas [5]. Industry is also an important emitter of several other greenhouse gases. And many of industry's products emit greenhouse gases as well, either during use or after they become waste. Opportunities exist for substantial reducing industrial emissions through more efficient production and use of energy. To make use of less energy and greenhouse gas intensive materials by fuel substitutions, the use of alternative energy technologies, process modification, and by revising materials strategies. Industry has an additional role to play through the design of products that use less energy and materials and produce lower greenhouse gas emissions.

Table 10. Annual amount of emissions from industrial processes in Sudan (10^6 tonnes).

Emissions	10^6 tonnes of CO_2 emission
Liquid	3320
Gas	N.A
Gas flaring	N.A
Cement manufacturing	84
Total	3404
Per capita CO_2 emissions	0.15

Table 11. Annual greenhouse gas emissions from different sources in Sudan (10^6 tonnes).

CO_2 emission from land use change	CH$_4$ from anthropogenic sources				Chlorofluorocarbons
	Solid waste	Oil and gas production	Agriculture	Livestock	N.A.
3800	47	N.A.	1	1100	

From the Tables (10 and 11) it is noticed that most of CO_2 emissions in Sudan were from land-use change, representing 92% of emissions. On the other hand the emissions of CO_2 from industrial represent only 8%, which is mainly from burning liquid and gas petroleum products. The per capita CO_2 emission in Sudan was estimated at 0.15×10^3 tonnes, which is considered very low compared to average of Africa which is 1.03×10^3 tonnes per capita CO_2 (world per capita is 4.21×10^3 tonnes) [5]. Gas flaring is the practice of burning off gas released in the process of petroleum extraction and processing, and the CO_2 emissions from it all-negligible. Nevertheless, due to increasing momentum in oil industry and oil products and the future increase in petroleum products consumption in Sudan. It is expected in the coming decades that the emissions of greenhouse gases from oil industry and use will certainly exceed by large figure if certain measures of mitigation are not under taken.

9.1. Environmental Policies and Industrial Competitives

The industrial development strategy in Sudan gives priority to the rehabilitation of the major industrial areas with respect to improvement of infrastructure such as roads, water supply, power supply, sewer systems and other factors. This strategy also takes into consideration the importance of incorporating the environmental dimension into economic development plans. However, the relationship between environmental policies and industrial competitiveness has not been adequately examined. For the near future, the real issue concerns the effectiveness of environmental expenditures in terms of reduction of pollution emissions per unit of output. A number of issues relevant to this central concern are presented as follows:

9.1.1. Implementing Ecologically Sustainable Industrial Development Strategies
Agenda 21 for achieving sustainable development in the 21st century calls on governments to adopt National Strategies (NS) for sustainable development that "build on and harmonise the various sectoral, social and environmental policies that are operating in the country" [6]. NS focuses almost exclusively on development issues and does not integrate industrial and environmental concerns. It does not consider industrial specific environmental objectives or time frames for achieving them. Moreover, it does not specify how specific industrial subsectors and plants will meet environmental objectives. Finally, it is formulated with minimal involvement of industrial institutions and private-sector associations. To bring together industrial development and environmental objectives it is necessary to:

- Establish environmental goals and action plans for the industrial sector.
- Develop an appropriate mix of policy instruments that support the goals of those plans.
- Design appropriate monitoring and enforcement measurements to realise those goals.

9.1.2. Applying Cleaner Production Processes and Techniques
Traditional approaches to pollution reduction have been based on the application of end of pipe technologies in order to meet discharge standards. However, the growing recognition that reduction at source is a potentially more cost effective method of abatement is resulting in replacing end of pipe technologies with cleaner production processes. Major constraints in adopting cleaner production methods relate to:

- Lack of awareness about the environmental and financial benefits of cleaner production activities.
- Lack of information about techniques and technologies.
- Inadequate financial resources to purchase imported technologies.

A coordinated effect by industry, government and international organisations can go a long way in overcoming these constraints. In this context key questions that need to be addressed are as follows:

(a) Need for local capacity building, information dissemination, training and education.
(b) Need for subsectoral demonstration projects.
(c) Need for increased cooperation with environmental market sectors in developed countries.
(d) Need for life cycle analysis and research on environmentally compatible products.

9.1.3. Implementing Environmental Management Systems

Environmental management systems (EMSs) are necessary to enable plant to achieve and demonstrate sound environmental performance by controlling the environmental impact of their activities, products and services. The basic tools to ensure compliance with national and/or international requirements and continually improve its environmental performance include:

- Environmental auditing.
- Environmental reporting, and
- Environmental impact assessments.

In addition, the adoption of EMS may require extensive training of corporate staff. A practical and effective means of doing this is through the design and support of joint capacity strengthening programmes by industry association and bilateral and multilateral agencies.

9.1.4. Managing and Conserving Water Resources

It is estimated that by year 2025, there will be a global crisis in water resources. Accelerated growth of industry will lead to increase in industrial water use. Moreover, major industrial water pollutant load is expected to increase considerably in the near future. Therefore, to better manage water resources by industry, there is a real need for integrating demand trend and use patterns. The main elements of an industrial management strategy can be identified as follows:

- Analytical services.
- Promotional services.
- Services for the development of industry and water supply infrastructure.

9.1.5. Using Market Based Instruments (MBIs) to Internalise Environmental Costs

As complements to command and control measures for resource conservation and pollution prevention in industry. MBIs represent a useful and efficient cost effective policy measures that internalise environmental costs. A plant's decision to invest in clean production depends primarily on the following factors:

(a) Relative costs of pollution control in overall production costs.
(b) Price elasticities of supply and demand for intermediary and final goods, and
(c) Competitive position of plant in a particular industrial sector.

9.1.6. Counteracting Threats from Eco-Labelling Requirements

The increasing export orientation of production makes it necessary to maintain competitive position in world markets. The emergence of a wide variety of eco-labelling requirements and lack of timely information on multitude of scheme may adversely affect certain export sectors. Needed initiatives to counteracting perceived threats could be presented as follows:

- Information dissemination.
- Life cycle analysis.
- Establishing certification centres and infrastructure support.

9.1.7. Implementing the United Nations (UN) Framework Convention on Climate Change

The UN climate change convention entered into force on 21st March 1994. The convention objective is the stabilisation of greenhouse gas concentration in the atmosphere at safe levels. For industry, responding to this convention will undoubtedly be a major challenge. Industry will be directly affected. Sudan as party to this convention is obliged to take a number of actions and cooperates effectively in order to meet this challenge. Sudan has to contribute to the common goal of reducing greenhouse gases emissions by taking precautionary measures to mitigate causes and anticipate impacts of climate change. However, there may not be adequate means to do so, and Sudan will therefore require international assistance. The main requirements are:

- Access to best energy-efficient technologies available on the world market, where such technologies are relevant to our natural resources endowments, our industrial requirements and are cost effective.
- Building an energy-efficient capital stock by accelerating the development of low energy intensity processes and equipment.
- Strengthening national capabilities for energy-efficient design and manufacturing.

Areas where technical expertise to implement the convention is necessary include:

- Preparing national communications on greenhouse gas emissions. The communications are supported to contain an assessment of the magnitudes and sources of greenhouse gases as well as identification of reduction methods.
- Supporting technology transfer for improvement in the efficiency of fuel based power generation.
- Promotion technology transfer for the use of renewable sources of energy such as biomass, wind, solar, hydro, etc.

- Developing and implementing technology transfer for energy efficiency programmes in industry, in complementarities with cleaner production/pollution prevention measures.
- Analysing the impact of climate change response measures on the economic and industrial development of the country, with the view to identifying economically viable technology options for reducing greenhouse gas emissions from the production and consumption of energy.
- Participatory approaches in planning, implementation and monitoring.
- Establishment of technical support system.
- Sensitive timing of education in hygiene and sanitation.

9.1.8. Addressing Concerns of Small and Medium Scale Industry (SMI)

Small and medium scale enterprises not only contribute to productivity growth and employment but are also important as collective sources of localised pollution loading such as organic wastes in water effluent, as well as hazardous wastes, heavy metal sludge, solvents, waste oils, acidic and alkaline wastes, photo wastes, etc. Often, these wastes are disposed of in unsafe manure and are extremely difficult to monitor. The cost of control in relation to output is too high, so even a modest increase in the costs (of environmental regulations) may threaten prevention and control may be well known and easily available, there is no guarantee that they will be adopted. Moreover, even when policy measures are in place, their enforcement and monitoring is a real problem for SMI sector on account of their large numbers and diversity. It is clear that environment problems of SMIs require special attention and special measures to address their particular problems.

9.2. Refrigeration

The public awareness of the depletion of the ozone layer has increased since 1970s. The ozone layer in the stratosphere protects life on earth against the ultraviolet radiation from the sun. Scientists and politicians argued for some years on the reasons for the reduced ozone layer, which was recognised especially over the Antarctic. The fact is that the ozone is depleted by the presence of chlorine. Further, very low temperatures as well as sunlight are required for having the process running. Vapour compression refrigeration systems using fluorocarbons, hydrocarbons or ammonia represent the established technology for household, commercial and industrial refrigeration and air-conditioning. Understanding the earth and the processes that shape it is fundamental to the successful development and sustainable management of our planet.

The evolution of the earth's crust over geological time has resulted in diverse, often beautiful, landscapes formed by earthquakes, oceans, fire and ice. These landscapes are also a source of mineral wealth and water, a base for engineering projects and a receptacle for the waste. Minerals are vital for manufacturing, construction, energy generation and agriculture. Some of the requirements can be met by increasing the use of recycled and renewable resources, but the need for new mineral resources continues. Minerals are important to maintaining the modern economy and lifestyle. Hence, everyone must make best use of these valuable assets whilst minimising the impact of their extraction on the environment. Global energy use will rise dramatically but world oil and gas production will eventually decline.

Geological hazards account for huge loss of life and damage to property. Poorer countries often suffer the most due to inappropriate land-use planning and in the future climate change may worsen these effects. Clean drinking water is a basic human need. The environmental sustainability of water resources, especially when balancing ecological and human needs with economic growth, is of major concern to governments worldwide. Increases in the world's population, the growth of cities, industrial development and ever-increasing waste and pollution can increase pressure on the environment. Understanding the effects of climate change is a key issue for society and researches focus on how to predict future climate change events and how to mitigate them. Two essential components for economic development are a soundly based knowledge of a country's natural resources, such as groundwater, minerals and energy, and an understanding of the geological environment. The latter includes the potential effects of earthquakes, volcanic eruptions, landslides and the forces of erosion and deposition on rivers and coastlines. Helping developing countries acquire this knowledge and apply it to promote economic growth, sustainable livelihoods and the protection of people. Sustainable land use involves protection of the natural environment (viable agriculture, forestry, water resources, and soil functions), the quality and character of the countryside and existing communities.

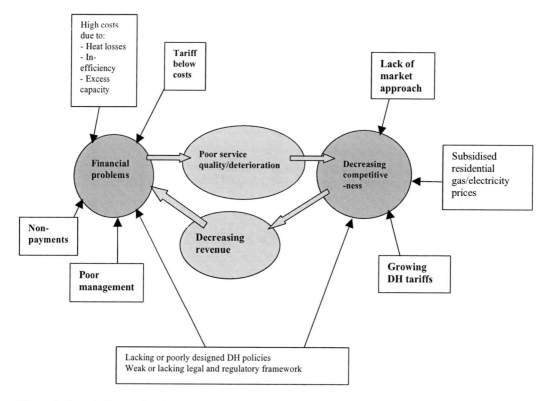

Figure 4. Key challenges for district heating systems in transition economies.

Groundwater is an essential source of drinking water (70%). Although commonly regarded as pure, it is often vulnerable to pollution or may contain natural concentrations of elements that can adversely affect health. Population growth, intensive agriculture, urbanisation and higher standards of living have resulted in increasing, often unsustainable, demands on groundwater resources. These pressures and the likely effects of climate change mean that groundwater systems must be better understood so that they can be evaluated effectively, managed sustainably and protected securely. District heating plays a very important role in transition economies. District heating systems in transition economies often face financial, technical or managerial problems largely created by an inadequate policy framework. These challenges include lack of customer focus, low efficiency, excess capacity, corruption and an uneven playing field. Figure 4 shows how these challenges are interrelated and create a vicious circle that undermines finances and competitiveness of district heating companies, and jeopardising their long-term sustainability.

9.2.1. Energy Supply

The earth is believed to be close to a state of thermal equilibrium where the energy, which is received at the surface by solar radiation, is lost again at night and the much smaller amount of energy, which is generated by the decay of unstable isotopes of Uranium, Thorium and Potassium distributed within the earth is balanced by the small continuous heat flux from the earth's interior to the pecans and atmosphere.

The use of geothermal energy involves the extraction of heat from rocks in the outer part of the earth. It is relatively unusual for the rocks to be sufficiently hot at shallow depth for this to be economically attractive. Virtually all the areas of present geothermal interest are concentrated along the margins of the major tectonic plates, which form the surface of the earth. Heat is conventionally extracted by the forced or natural circulation of water through permeable hot rock.

There are various practical difficulties and disadvantages associated with the use of geothermal power:

Transmission: geothermal power has to be used where it is found. In Iceland it has proved feasible to pipe hot water 20 km in insulated pipes but much shorter distances are preferred.

Environmental problems: these are somewhat variable and are usually not great. Perhaps the most serious is the disposal of warm high salinity water where it cannot be reinjected or purified. Dry steam plants tend to be very noisy and there is releases of small amounts of methane, hydrogen, nitrogen, amonia and hydrogen sulphide and of these the latter presents the main problem.

The geothermal fluid is often highly chemically corrosive or physically abrasive as the result of the entrained solid matter it carries. This may entail special plant design problems and unusually short operational lives for both the holes and the installations they serve.

Because the useful rate of heat extraction from a geothermal field is in nearly all cases much higher than the rate of conduction into the field from the underlying rocks, the mean temperatures of the field is likely to fall during exploitation. In some low rainfall areas there may also be a problem of fluid depletion. Ideally, as much as possible of the geothermal fluid should be reinjected into the field. However, this may involve the heavy capital costs of large condensation installations. Occasionally, the salinity of the fluid available for reinjection may be so high (as a result of concentration by boiling) that is unsuitable for reinjection into ground. Ocasionally, the impurities can be precipitated and used but this has not generally proved commercially attractive.

9.2.2. Refrigeration Application

The refrigeration industry has had to face environmental challenges. Regulation and behaviour related to the ozone layer differ from one country to another. Problems related to global warming are likely to cause similar problems in very different ways. These two kinds of problems may lead company managers to change/modernise equipment. These environmental factors must not hide other important development factors such as those related to hygiene, organoleptic quality, control methods and equipment packaging, etc. All equipment change and investment projects must take all of them into consideration.

The more developed a country, the more widely refrigeration is used. Food preservation is still the main use of refrigeration, followed by air conditioning, energy savings and transport (heat pump and liquefied gases), industrial processes and medicine. When consumers demand more and more fresh products, handling such products becomes increasingly difficult: cold stores, display cabinets, refrigerated trucks have to be much more effective. Difficulties are mostly related to:

- Very narrow ranges, in many cases, between temperatures involving microbial or chemical risk and temperatures that cause chilling injury or freezing.
- No or little overlapping between temperature ranges permitted for different products.
- No or little possibility for the consumer or the seller to evaluate the remaining life spans of the product or even to perceive any risk.
- Humidity control, vapour pressure differences have serious consequences, not only on the weight but also on the unit value of products.

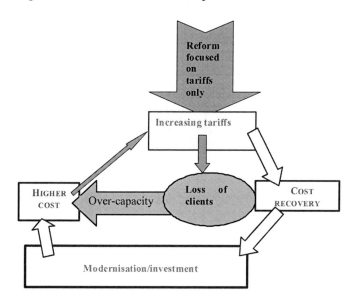

Figure 5. Unsustainable tariff growth.

The production and use of CFC refrigerants have been phased out in developed countries. HCFC refrigerants are now subject to regulation and are scheduled to be phased out in the twenty-first century. Recent developments in refrigeration technology and environment (climate) control have provided a better quality of life for mankind. As is known, refrigeration technology has played a key role in preserving the quality of food and eliminating waste (spoilage). The refrigerated transport of perishable products provides a critical link in the cold chain between the consumer and the producers, processors, shippers, distributors and retailers. The transport refrigeration industry incorporated mechanical vapour compression mechanisms into the first transport refrigeration units. Therefore, the physical, chemical and thermodynamic properties of these refrigerants had a major impact on the design on the compressors and refrigeration system components. Many refrigerants have excellent thermodynamic properties suitable for use in vapour compression refrigeration systems, but are limited by other properties such as toxicity, flammability and chemical stability within the refrigeration system. Most district heating systems in former planned economies are less efficient and oversized. In other words, their supply infrastructure is larger than necessary to meet current demand. This problem can be exacerbated when they lose customers. Reforms, particularly for tariffs, are needed to improve finances and break the vicious circle of deteriorating competitiveness. Yet policymakers need to design the reforms very carefully for them not to backfire and make matters worse, as illustrates in Figure 5.

9.2.3. Temperature Distributions

World capacity of geothermal energy is growing at a rate of 2.5% per year from a 2005 level of 28.3 GW [7]. GSHPs account for approximately 54% of this capacity almost all of it in North America and Europe [7]. The involvement of the UK is minimal with less than 0.04% of world capacity and yet is committed to substantial reduction in carbon emission beyond the 12.5% Kyoto obligation to be achieved by 2012. GSHPs offer a significant potential for carbon reduction and it is therefore expected that the market for these systems will rise sharply in Sudan in the immediate years ahead given to low capacity base at present. The heat pump recovers heat from the ventilation air, and provides water and/or space heating. There are numerous ways of harnessing low-grade heat from the ground for use as a heat pump source or air conditioning sink. For small applications (residences and small commercial buildings) horizontal ground loop heat exchangers buried typically at between 1 m and 1.8 m below the surface can be used provided that a significant availability of land surrounding the building can be exploited which tends to limit these applications to rural settings. The development of new, modern, and complete water resources information systems is one of the basic needs for the implementation of the water resources management system in Sudan. Lack of information is one of the most critical points regarding global development and implementation of new management systems.

9.2.4. Environmental Challenge

The Montreal Protocol and its amendments (London 1990, Copenhagen 1992 and Vienna 1994) have given rise to the drawing up of regulation on the phase out the substances named Ozone Depleting Substances (ODS), including CFCs such as R12, R11 and HCFCs such as R22. Alternatives to CFCs include:

- Hydro Fluoro Carbons (HFCs) such as R134a and many mixtures.
- Naturally occurring refrigerants such as NH_3, hydrocarbons, water, air CO_2. Some of them are flammable and/or toxic or should be limited to relatively small niches.
- For other technologies such as absorption, solid sorption, etc., there are also some very interesting niches. However, refrigeration as a whole is not likely to shift there very soon because of the many other challenges it has to face.

The United Nations Framework Convention on Climate Change came into effect on March 21, 1994 and the Kyoto Protocol on December 10, 1997. Global warming is mostly caused by the rising percentage of various substances in the atmosphere, particularly CO_2, CH_4 and many others including CFCs, Hydro Chloro Fluoro Carbons (HCFCs) and HFCs. Refrigeration is involved because of:

- Refrigerants themselves, if they are released into the atmosphere, this is the 'direct effect'.
- CO_2 produced by energy consumption of the system throughout its lifespan, this is the 'indirect effect'.

The Fifth Framework Programme (FP5) is designed to ensure that European research efforts are translated more effectively into practical and visible results. The programme includes four thematic and three horizontal programmes.

Thematic programmes are:

- Quality of life and management of living resources.
- Users-friendly information society.
- Competitive and sustainable growth.
- Energy, environment and sustainable development.

Horizontal programmes are:

- Confirming the international role of community research.
- Promotion of innovation and encouragement of participation.
- Improving human research potential and the socio-economic knowledge base.

Research in agro-food industry, including refrigeration, has an important role, particularly under the key action "Food, Nutrition and Health".

At present geothermal energy makes a very small, but locally important, contribution to world energy requirements. This situation will not change unless important technological advances are made. Environmentally it is probably the least objectionable form of power generation available at present with the exception of hydroelectric methods.

10. PROSPECTS FOR WATER RESOURCE PLANNING IN NILE BASIN

10.1. Common Language and Culture

A common language and similar culture simplify communication and reduce the potential for misunderstandings. In Nile basin where several languages are spoken, an international language English is used with some success by multi-jurisdictional basin management authorities.

10.2. Factors Promoting Data and Information Exchange

Data and information exchange is more probable when needs are compatible and when there is potential for mutual benefit from cooperation (Table 12). Where countries are working on developments that are beneficial to both countries as well as other riparian, there is little incentive to hide project impacts. This means that because data and information exchange are unlikely to lead to pressure from surrounding countries that might restrict developments, countries have less reason to restrict access to their data and information resources. It is important for there to be no perceived clash of interests in development plans and needs. An example of this might be in developing their part of the basin primarily for hydroelectric development, while the lower riparians are more interested in developing the irrigation potential of their portion of the basin. By constructing large storage dams in the upper part of the basin, the river Nile seasonal flow might be evened out, reducing flooding downstream while increasing irrigation water supplies and even making downstream run-of-the-river hydroelectric projects more profitable.

Table 12. Summary of the situation relating to data and information exchange in Nile basin.

River basin	Nile basin
Basin states or territories	Burundi, Democratic Republic of Congo, Egypt, Eritrea, Ethiopia, Kenya, Rwanda, Sudan, Tanzania, Uganda
Cooperative frameworks in place	Nine of the countries of basin are pursuing the development of a cooperative framework
Major languages spoken in	More than 6 official languages and numerous unofficial languages
Major water issue facing the basin	Rapid population growth, environmental degradation, under development
External funding of cooperative basin initiatives	Extensive external funding of cooperative initiative
Range of GDP *per capita* of the basin	$550-$3000
Extent of data/information exchange	Information exchange through the cooperative framework being developed is beginning to occur

10.3. Sufficient Levels of Economic Development

Sufficient levels of economic development across a basin are needed to permit joint funding of cooperative processes, particularly data collection and dissemination. Although countries with differing levels and forms of economic development may, at times, have more complementary needs than countries with similarly structured economies, the overall level of economic development is still significant. A wealthier country in a river basin may be able to assist with the funding of data collection activities in the neighbouring country with much needed data and helping to build confidence between the two countries.

10.4. Increasing Water Resources Stress

A bilateral agreement between Egypt and the Sudan in 1959 by which the two countries share the Nile flow: 55.5 billion cubic meters to Egypt, 18.5 billion to Sudan, and 10 billion were allocated to evaporation. Hopes are high for achieving a more extensive participation by the other riparian parties in what could be a multilateral treaty on the Nile encompassing the other riparian states in addition to Egypt and Sudan. The above agreement is incomplete, lack of the entry of other legitimate riparian states, lack of the water quality components, and the focus on quantity and misses important management issues. It is to be noted that regional relations, including those among the riparian parties, are not disconnected from the political, economic, and trade network of international relations. Water is not the only determinant factor in shaping the nature of bilateral, regional, or international relations. As *per capita* water resources availability decreases (Table 13), tensions between riparian nations may rise and make cooperation difficult. Stress may, therefore, reduce cooperation and data sharing rather than strife. The historical background of the basin may have a lasting effect on current negotiations. Past conflicts can have a deleterious effect on the prospects for establishing cooperative practices, such as data sharing. Where there is a history of conflict between two nations, both nations may view the present situation primarily as competitive and focus on conflicting rather than common interests. Democracies may find it easier to negotiate cooperative arrangements with other democracies. Political differences can lead to legacies of mistrust developing between countries.

Table 13. Diverse water challenge.

Country	Egypt	Sudan
Per capita annual water resources 2000 (m^3)	34	1187
Per capita annual withdrawal (m^3)	921	666
Per capita annual withdrawal for agriculture (m^3)	86	94

Source: World Resources Institute 1998 and 2000
World Resources 1998-1999
World Resources 2000-2001

Climatic and environmental changes and a rising demand have increased the competition over water resources and have made cooperation between countries that share a transboundary river an important issue in water resources management and hydro-politics. Yet in river basins around the world, international conflict and cooperation are influenced by different factors, and general conclusions about forces driving conflict and cooperation have been difficult to draw. Rivers are an essential natural resource closely linked to country's well being and economic success. But rivers ignore political boundaries, and competition over the water resources has lead to political tension between countries with Transboundary Rivers. Integrating international cooperation and conflict resolution into the water management of Transboundary Rivers has therefore become an important issue in water resources management and hydro-politics. The problem requires a good understanding of the history and patterns of conflict and cooperation among nations sharing international basins worldwide and of the different factors that have influenced their international relations. Increasing water scarcity in the downstream areas of several river basins demands improved water management and conservation in the upper reaches. Improved management is impossible without proper monitoring at various levels.

The Comprehensive National Strategy gives priority to the following strategies for achieving the set:

- Cost-effective utilisation and management of water resources.
- Introduction of low-cost appropriate technologies and encouragement of local production of equipment.
- Rehabilitation of deteriorating water sources and systems.
- An expanded programme of well-drilling and hand-pump installation, especially in priority rural areas.
- Training, capacity building and increased use of domestic technical resources, to increase cost efficiency and reduce dependence on external resources.
- Development and expansion of sanitation services.
- Increased community involvement in planning, execution and management of water supply and sanitation services encouraging cost sharing and self-help.
- Encouragement of research aimed at better water resources management, evaluation of existing schemes and identification of cost-effective alternative strategic elements.

11. WATER DISTRIBUTION

Distribution networks are old and is a potential weak link taking up contaminants, in the case of the capital Khartoum for example. Increased reliance on booster pumps and storage tanks have contributed to the depletion of the major power supply as well as offering excellent breeding sites for mosquitoes. The worst-case scenario is the city of Port Sudan suffering from chronic water shortages on the one hand and the heavy incidence of malaria on the other.

The celebrations of the city of El Obeid with the new source of water from Bara, was short lived. The old distribution system soon collapsed. Storm drains is another fiasco. It is a paradox that there is no clear link between the water supply and sanitation sectors. Even within the water sector there are poor linkages between the federal and state water corporations. The water sector itself has experienced ten major institutional changes since independence.

11.1. Water Stress in Sudan

Water stress refers to economic, social, or environmental problems caused by unmet water needs. Lack of supply is often caused by contamination, drought, or a disruption in distribution. In an extreme example, when Sudan split four years ago between the rebel-led west and government-ruled north, the conflict led to unpaid water bills, which precipitated a dangerous health threat in the region, increasing the risk of water-born diseases such as cholera. Some analysts believe the disruption of distribution was a political ploy to put pressure on the rebel-led west.

While water stress occurs throughout the world, no region has been more afflicted than sub-Saharan Africa. The crisis in Darfur stems in part from disputes over water: The conflict that led to the crisis arose from tensions between nomadic farming groups who were competing for water and grazing land-both increasingly scarce due to the expanding Sahara Desert. As Mark Giordano of the International Water Management Institute in Colombo Sri Lanka says, "Most water extracted for development in sub-Saharan Africa-drinking water, livestock watering, irrigation- is at least in some sense 'transboundary'". Because water sources are often cross-border, tension, and then conflict emerges.

Improving water and sanitation programmes is crucial to spurring growth and sustaining economic development. Because it takes time to develop these programmes, a paradox emerges: poor economies are unable to develop because of water stress, and economic instability prohibits the development of programmes to abate water stress. Developments in water storage could have prevented that drought from significantly affecting Sudan's economy. Hydropower can also spark economic development. Accordingly, some transboundary water agreements also play a clear role in fostering development, for example, by facilitating investment in hydropower and irrigation.

11.2. The Role of Agriculture in Water Stress

Agricultural development has the potential to improve African economies but requires extensive water supplies. These statics from the Water Systems Analysis Group at the Institute for the Study of Earth, Oceans, and Space at the University of New Hampshire reveal the urgent need for sustainable agricultural development:

- About 64 percent of Africans rely on water that is limited and highly variable;
- Croplands inhabit the driest regions of Africa where some 40 percent of the irrigated land is unsustainable;
- Roughly 25 percent of Africa's population suffers from water stress;
- Nearly 13 percent of the population in Africa experiences drought-related stress once each generation.

Another aspect of water-related stress is the relationship between water, soil, and agriculture. Improved access to quality water is a long-term goal that requires more than humanitarian funds.

- Because sub-Saharan Africa is subject to more extreme climate variability than other regions, it needs improved water storage capacity. Some experts say that large dam projects would create a more sustainable reserve of water resources to combat the burden of climate fluctuations, but other disagree, stating the harmful environmental impact of large dams.
- Many experts say more water treaties are needed. The transboundary water agreements have cultivated international cooperation and reduced the "probability of conflict and its intensity".
- Better donor emphasis on water development is needed. Small-scale agricultural improvements also offer a solution to water stress, including the harvest of water in

shallow wells, drip irrigation for crops, the use of pumps, and other technological innovations.

Farmers can access green water through drip irrigation systems that slowly and consistently deliver water to plant's root system, supplemental irrigation (supplementary to natural rainfall rather than the primary source of moisture during periods of drought) and rainwater harvesting (the collection of rainwater for crops, which reduces reliance on irrigation). Crops can grow poorly even during periods of rainfall, and most farms in Africa suffer from nitrogen and phosphorus depletion in soil. One way to assuage water stress in terms of food scarcity is to increase water-holding capacity with organic fertilisers that would increase availability and efficacy of green water.

11.3. Water Supply Problems in the Butana Region-Central Sudan with Special Emphasis on Jebel Qeili Area

The Butana region of central Sudan is famous for its animal wealth and extensive pastures. Yet scarcity of water resources in the area especially during the dry seasons handicaps the proper utilisation of these pastures. The area is occupied by non-water-bearing basement rocks and the only source of water is from direct run-off. Thus large numbers of small-size water reservoirs, " haffirs ", were constructed, but these are inadequate to provide enough water for the growing human and animal population. An all-year lake is here proposed to be constructed utilising the ring-structure the Jebel Qeili igneous complex, central Butana. This lake is expected to solve the present water problem and meet the future demand of central Butana at the present rate of human and animal growth.

11.4. Southern Sudan

World Vision began its work in Sudan in 1972 through a partnership with the African Committee for Rehabilitation of the Southern Sudan (ACROSS) to provide emergency relief aid to war-effected families. Efforts included the reconstruction of the Rumbek community hospital and surrounding buildings, the provision of medicine and supplies, and education in preventative health care. Other projects during this period focused on training health and social workers in general medical aid and child welfare and instruction in water development, agriculture, handcrafts, and literacy.

The 1980s brought constant turmoil to the Sudanese people as the civil war raged on and severe drought parched the country. In 1983, approximately 1,500 refugees entered Sudan daily from violence-torn neighbouring countries, straining the already limited food supply. World Vision, through the ACROSS Refugee Settlement Project, responded by distributing blankets, grain, cooking oil, medical kits, and shelter to more than 50,000 people. Supplemental feeding for children also was provided. Numerous development projects were initiated during this time that assisted communities in improved crop production, animal husbandry, health care, clean water collection, infrastructure repair, and literacy. In 1989, World Vision became a founding member of Operation Lifeline Sudan (OLS), a partnership of non governmental organisations (NGOs) and UN agencies designated to coordinate the southern relief efforts.

During the 1990s World Vision conducted operations in all major regions of southern Sudan. Project objectives included primary health care, water provision, agriculture, local grain purchase, enterprise development, and emergency relief efforts. World Vision focused on an integrated work approach that involved peace and advocacy, gender development, church support, and environment and natural resource initiatives. Some specific projects included:

- The Kapoeta Medical Supplies Project provided health and educational assistance to more than 200,000 people to help reduce incidences of disease and suffering.
- The Agriculture/Livestock Rehabilitation Project assisted the aforementioned families with food, seed, vaccinations, and agricultural consults.
- The South Sudan Relief/Church Support Project coordinated a pastors' conference for 150 pastors and religious leaders from the Western Equatoria province.
- The South Sudan Relief and Rehabilitation Project, a 10-year programme, provided 450,000 Sudanese with agriculture and economic development, food and water, health care, enterprise development opportunities, and emergency relief.

11.5. Beja People's Problems

The Beja, a semi-nomadic group of people, who live in rebel-held areas of eastern Sudan, need a huge amount of humanitarian assistance, a representative from the International Rescue Committee (IRC). Although Beja can be found throughout northeast Africa, tens of thousands are currently trapped in an area of eastern Sudan near the Eritrean border, held by Sudanese rebels since the late 1990s. Only two NGOs, both based in Eritrea, are able to access the 15,000 sq km area at the moment, one of which is the IRC. The organisation estimates the Beja population in the area to be between 45,000 and 186,000 people.

Although it did rain in the area in 2004, a shortage of water had also posed serious problems. "Fresh drinking water is incredibly hard to come by. All the settlements have just focused around dry river beds, in which people dig hand-dug wells". Locusts would eat the foliage that usually sustains the Beja's goats and camels—upon which the Beja utterly depend for survival. A few immature locust swarms have formed in northeast Sudan near the Red Sea and the border of Egypt, the UN Food and Agriculture Organisation said in March 2004. Moreover, Beja grazing areas have been severely restricted by a front line between rebel forces and Khartoum government soldiers, the second to be opened by southern rebels during Sudan's 21-year-old civil war.

Sudan is an example that projects the environmental plight of Africa, south of the Sahara – drought and desertification, floods, deforestation, loss of biodiversity, tribal and ethnic conflict and poverty are only too common. As a result, interest and commitment to environmental impact assessment practices have become mandatory by donors when executing new development projects. The ecological zones of Sudan in 1998 as:

- Deserts: cover almost 30% of the northern parts. Annual precipitation is less than 50 mm; soils are sandy. Sparse vegetation grows on seasonal 'Waddis' and the banks of the Nile.
- Semi deserts: cover above 20% south of the desert belt. Rainfall ranges from 50 to 300 mm. It is speckled with few Acacia trees and thorny bushes and zerophytes.
- Low rainfall woodland Savannah: covers about 27% of the area of Sudan with rainfall less than 900 mm, with a nine-month dry period. Annual grasses are dominant. Heavy clay soils lie on the east of the Nile and the west is sandy. Most of the 36 million feddans of rain-fed agriculture and the 4 million irrigated lands fall within this heavily populated belt.
- High rainfall woodland Savannah: 13% of the area with rainfall more than 900 mm and with broad-leafed trees in the southern parts of Sudan.
- Swamps: are probably the largest in the world but in Sudan it covers about 10% and fall in three main areas around the tributaries of the White Nile.
- Highlands: are less than 0.3% of the areas of Sudan and are scattered along the Red Sea coast, the south and the west of the country.
- The Red Sea Coast-Marine ecosystem, mangrove swamps, coral reefs and associated fauna.

Environmental problems include:

- Horizontal expansion in rain-fed and irrigated agriculture;
- The complete absence of the environmental dimensions in policies, strategies, plans and programmes of management of resources;
- Development is random and environmental evaluation does exist before or after execution of projects;
- The economy and society, in spite of the century-long attempts at 'modernisation' are still dominated by subsistence way of living;
- The economy is still affected seriously by the yearly, seasonal and geographical variability of rainfall for crop and livestock production;
- Dependence on imported seeds and agricultural chemicals has increased cost of production;
- Loss of land productivity and marketing policies decreased cash surplus;
- The civil war in the south has grave economic and social costs;
- Population distribution and rural-urban migration due to desertification and civil strife has led to deterioration of natural resources, indigenous knowledge and loss of local culture and dignity;
- Problems of poor sanitation, limited industrial pollution and food hygiene have become more complex;
- The energy crisis is aggravating desertification and affecting climate charge;
- Vast water resources are badly managed;
- Environmental education has only been recently incorporated in school curricula; and
- Laws and legislation concerning the environment are not effective and law enforcement measures are not integrated.

11.6. Western Sudan

El Fasher, Darfur region, Sudan, 24 August 2005 – Torrential rains have caused severe flooding in this city of 400,000 people and in nearby Abu Shook, a camp for people forced to flee their homes as a result of the ongoing Darfur conflict. The floods have destroyed hundreds of homes and have made El Fasher's water supply largely unsafe.

UNICEF is mounting a concerted effort to restore basic services to those affected by the flood, and to prevent the outbreak of disease. Since the flood, UNICEF has assisted with the following:

- Reinstalling pipes in Abu Shook and restoring the water supply by linking boreholes with pumps.
- Testing the water quality each day. No bacterial contamination has been found.
- Rebuilding 156 latrines and 88 bath stations.
- Renting five tankers to deliver more water.
- Repairing damaged schools and child-friendly spaces.
- Providing daily door-to-door hygiene-promotion trainings.
- Distributing jerry cans, soap, tarps, and mosquito nets.

12. CONCLUSIONS

Water is one of the most precious natural resources on earth. It is essential for human survival and development and cannot be replaced by any other resources. However, with rapid social and economic development, as well as explosive population growth, a water crisis has developed in Sudan during the 20th century. With economic development, and population growth the conflict between water supply and demand has become more and more acute in Sudan, and it has been aggravated further by the irrational utilisation of water resources. As a result, the deterioration and destruction of the eco-environment have become increasingly serious. In order to effectively protect ecosystems and improve their ecological conditions, many developments on ecological and environmental water requirements have been carried out including rivers, vegetation, lakes, wetlands and groundwater. Changes in the economy or in the population will have different impacts on the water resource availability, which in turn will impact economic output and population dynamics. Water conservation is a major challenge because of increasing competition between agricultural and non-agricultural use of water. Efficient use of water in agriculture is critical because of the large volumes of water used. Sudan is an agricultural country with fertile land, plenty of water resources, livestock, forestry resources, and agricultural residues. Energy is one of the key factors for the development of national economy in Sudan. An overview of the energy situation in Sudan is introduced with reference to the end uses and regional distribution. Energy sources are divided into two main types; conventional energy (biomass, petroleum products, and electricity); and non-conventional energy (solar, wind, hydro, etc.). Sudan enjoys a relatively high abundance of sunshine, solar radiation, and moderate wind speeds, hydro, and biomass energy resources. Application of new and renewable sources of energy available in Sudan is now a major issue in the future energy strategic planning for the alternative to the fossil conventional energy to provide part of the local energy demand. Sudan is an important case study in the context of renewable energy. It has a long history of meeting its energy needs through renewables. Sudan's renewables portfolio is broad and diverse, due in part to the country's wide range of climates and landscapes. Like many of the African leaders in renewable energy utilisation, Sudan has a well-defined commitment to continue research, development, and implementation of new technologies. Sustainable low-carbon energy scenarios for the new century emphasise the untapped potential of renewable resources. Rural areas of Sudan can benefit from this transition. The increased availability of reliable and efficient energy services stimulates new development alternatives. It is concluded that renewable environmentally friendly energy must be encouraged, promoted, implemented and demonstrated by full-scale plant especially for use in remote rural areas.

REFERENCES

[1] Abdeen M. Omer. 2004. Water resources development and management in the Republic of the Sudan. *Water and Energy International Journal* **61**(4): 27-39. New Delhi, India, October-December.

[2] Omer, A.M. 1996. Renewable energy potential and future prospect in Sudan. *Agriculture and Development in Arab World* **3** (1): 4-13.

[3] Omer, A.M. 1997. *Review of Hydropower in Sudan*. Khartoum: Sudan.

[4] National Energy Administration (NEA). 1985. *The National Energy Plan 1985-2000*. Khartoum: Sudan.

[5] World Resource Institute (WRI). 1994. World Resources: *A Guide to the Global Environment, People and the Environment*.

[6] Omer, A.M. 1998. *Renewable Energy Potential and Environmentally Appropriate Technologies in Sudan*. Khartoum: Sudan.

[7] Lund, J.W., Freeston, D.H., and Boyd, T.L. 2005. Direct application of geothermal energy: 2005 Worldwide Review. *Geothermics* **34**: 691-727.

In: Encyclopedia of Environmental Research ISBN: 978-1-61761-927-4
Editor: Alisa N. Souter © 2011 Nova Science Publishers, Inc.

Chapter 28

STRATEGY FOR SUSTAINABLE MANAGEMENT OF THE UPPER DELAWARE RIVER BASIN

Piotr Parasiewicz[1,a], Nathaniel Gillespie[2,b] Douglas Sheppard[3,c] and Todd Walter[4,d]

[1]Rushig Rivers Institute,592 Main Street, Amherst, MA 01002
[2]Trout Unlimited, 1300 N.17th Street, Suite 500, Arlington, VA 22209
[3]New York State Bureau of Habitat, Department of Environmental Conservation,
625 Broadway, Albany NY 12233-3500,
[4]Biological and Environmental Engineering, Cornell University,
211 Rilley-Rob Hall, Ithaca, NY 14853, USA

ABSTRACT

The Upper Delaware River represents a valuable ecological resource of New York and Pennsylvania. Recognized as one of the country's most scenic rivers, it is greatly appreciated for its high quality trout fisheries and the variety of outdoor recreation experiences it offers. The river system serves as a water supply for over 8 million people in New York City, via an out-of-basin transfer, and over 9 million people within the basin. The three New York City water supply reservoirs (Cannonsville, Pepacton and Neversink) are designed to withdraw more than 70% of the average annual volume from the top of the basin at three stems of the Delaware: the West Branch, the East Branch and the Neversink River. Severe reductions and fluctuations in river flows, channel alterations, and landscape modifications resulted in increased water temperature and changes in habitat suitability in the river. As a consequence of these alterations, the aquatic fauna in the upper Delaware River have been impaired.

This article summarizes (with minor modifications) a report published by Trout Unlimited in 2001 that helped galvanize efforts toward the improvement of management of the Delaware River. The complex ecological alterations caused by unnatural flow conditions, past environmental degradation, and numerous water supply demands on the basin necessitate a well-defined management scheme to optimize the use of the river as a resource and secure

---

[a] E-mail address: piotr@rushingrivers.org
[b] E-mail address: ngillespie@tu.org
[c] E-mail address: dxsheppa@gw.dec.state.ny.us
[d] E-mail address: mtw5@cornell.edu

its long-term sustainability. Rehabilitation of the upper Delaware River requires both an immediate overhaul of the flow regime below the reservoirs, monitoring and evaluation, and research. In the initial phase, acute deficits should be addressed by increasing base flows, and reducing ramping rates and peak amplitudes. In the second phase, an intensive multidisciplinary study is proposed to quantify limiting ecosystem factors, to better understand groundwater and surface water interactions, and to determine management options. This information will provide a solid basis for evaluating specific options and an integrative, long-term management plan for the upper Delaware River.

Keywords: watershed, water withdrawals, deforestation, instream flow, integrative assessment.

INTRODUCTION

With 22 million inhabitants and numerous industrial enterprises, the Delaware River Basin is an example of an intensively used resource with conflicting demands. The sparsely populated upper portion of the basin represents a valuable ecological resource for both New York and Pennsylvania. Heralded as one of the most scenic rivers in the country and highly regarded for its fisheries and outdoor experience, the river serves also as a local and regional (New York City) water supply. Today, multiple water withdrawals for domestic supplies have dramatically altered the flow of the river and, consequently, the river's character. Three New York City (NYC) reservoirs (Cannonsville, Pepacton and Neversink) are designed to withdraw over 70% of the average available annual volume at the top of the basin. The effects of flow manipulation can be observed throughout the West Branch, East Branch and Neversink watersheds and the main stem (Sheppard 1983, Hulbert 1987). Negative consequences like thermal fish kills have been observed, but many other impacts were only rarely recognized in the past. Information pertaining to the consequences of water withdrawals, such as the alteration of fauna composition, sediment transport, surface-ground water interaction, thalweg elevation and substrate composition, are commonly available for other watersheds, but sparse data exist for the Delaware watershed. Furthermore, limited knowledge of this ecosystem in its present and pristine form hampers the process of assessing human induced alterations. Hence, a deficit analysis can only be inferred from limited empirical knowledge of the system and theoretical assumptions. In the long run, a comprehensive multidisciplinary study is needed to improve our understanding of these systems and establish an explicit management plan with long-term objectives. This article is based on a report of a review published by Trout Unlimited in 2001 aimed to initiate discussion and set the stage for a new approach to the management of the Delaware River.

METHODS

This study was based on integrative, holistic analyses of the watershed that should provide directions for establishing environmental releases from NYC Reservoirs. Our observations largely build upon published literature but also include results of some recent studies conducted in the region and, where possible, simple modelling experiments. Our methodology integrates hydrological, biological and historical facts to establish a conceptual model of the original ecological character of the basin. To create a conceptual model of the Delaware River prior to colonization, we 1) reviewed available historical information and 2) estimated the expected impacts of human induced alteration and 3) "corrected" the present conditions by "removing" those impacts. Following the paradigm of maintenance of ecological integrity (Stalzer and Bloech, 2000), we can compare the conceptual model of the original basin with the present conditions to identify ecological deficits of the system, define an achievable target state accompanied by specific river management objectives and propose a plan of action.

RESULTS

Geographical, Hydrological and Historical Settings

The Upper Delaware is located in the Catskill Mountains Region about 200 kilometers northwest of New York City. The Upper Delaware system consists of 3 main fourth order rivers that flow into the Delaware's main stem: the West Branch of the Delaware, East Branch of the Delaware, and the Neversink River. The Upper Delaware is an alluvial upland river system of straightened-confined and meandering character (C and F according to Rosgen 1985) with a pluvio-nivial flow regime (i.e. high flows related to rain and snow melt in the fall and spring, and low flow in the summer (Parde 1968)). The gradient is moderate compared to headwater streams, and multiple wetlands accompany its course. The river flows over unstable glacial deposits in a U-shaped valley that was heavily forested in pre-colonial times.

The Catskill Mountain's watersheds are generally rural, topographically steep with shallow, permeable soils overlaying restrictive bedrock or fragipans. The subsurface flow in soils is generally higher than rainfall intensity and therefore the soils' infiltration capacity does not strongly influence the runoff generation, i.e., infiltration excess runoff is rare and limited to small areas in these systems (Walter et al. 2002). The runoff generated in the Catskills typically originates from localized saturated areas; this process is commonly referred to as saturation excess runoff. As with many humid temperate systems, these saturated runoff source areas expand and contract seasonally and are thus said to exhibit a variable source area hydrology (Dunne and Black, 1970; Dunne and Leopold, 1978, Hewlett and Hibbert, 1967, Hewlett and Nutter, 1970). Unlike systems with relatively deep soil, the Catskills surface hydrology is largely controlled by an interflow of shallow, perched, transient groundwater (Brown et al. 1999, Frankenberger 1996). Deep groundwater is recharged on the hillsides through cracks in the bedrock and fragipans and emerges as baseflow near or in the stream (Soren, 1963).

The Catskill region's timber resources were exploited relatively late in the history of the Northeastern United States. It survived in its maturely forested form until the beginning of the nineteenth century. This late discovery, as well as its proximity to large population centers, caused very rapid changes in the Catskill landscape. Ancient forests consisting of large white pine, hemlock and hardwood forests served as a resource for high quality wood, tannin, acid and charcoal production and were almost completely devastated over a period of 100 years. Formerly very common, animal species like the passenger pigeon became extinct within few decades. The rivers that served as transportation pathways for the log raft industry were systematically cleared from obstructions and widened. The industries, for the most part, vacated the Catskills in the late nineteenth and early twentieth centuries and the land was subsequently converted to agriculture or sold in big parcels where slow forest recovery began (Karas 1997, Kudish 2000). In the early and mid-twentieth century, multiple dams were constructed on tributaries of the Delaware River. The three biggest, located on the East and West Branches of the Delaware and on the Neversink River, created water supply reservoirs for New York City[1]. Due to the multitude of involved interests, flow in the river remains highly regulated and does not reflect ecological needs. The water supply of New York City, a salinity front in the tidal portion of the Delaware River, a flow target in New Jersey, and hydro-power generation heavily drive the complex water release scheme.

When the City of New York initially proposed to build the Delaware Subsystem which included two diverting reservoirs [Neversink on the Neversink River and Pepacton on the East Branch, Delaware River] in the Upper Delaware basin, the downstream states sued the City and ultimately obtained a U.S. Supreme Court decree in 1931 to protect their water rights. Actual construction, however, was delayed until after the Second World War; at which time, the City obtained an amendment to the original Decree in 1954 which allowed for the construction of a third diverting reservoir (Cannonsville on the West Branch, Delaware River). The amended Decree allowed for the City to divert 3 million m^3day^{-1} and established a target flow of 50 m^3s^{-1} at a gauge located in Montague (downstream of Port Jervis). While both decrees recognized the need for the baseflows to protect natural resources in the reaches immediately downstream of the reservoirs, neither provided a specific flow/release guidance. It was not until the City commenced the operation of the Pepacton Reservoir and completely shut off the releases from the Neversink Reservoir in the late 1950's that the baseflow issue was actually addressed. In the early 1960's, the City and New York State negotiated a set of interim "conservation releases" for the three Delaware Basin reservoirs (ca. 0.5 m^3s^{-1} in summer and 0.2 m^3s^{-1} in winter). These releases were to be revisited following cessation of the 1960's drought. Following a period of experimental releases during the late 1970's, both the parties to the Supreme Court Decree and the Delaware River Basin Commission chose to adopt the interim experimental releases program, namely: the Neversink Reservoir: 1.3 m^3s^{-1} and 0.7 m^3s^{-1} Pepacton Reservoir: 2.0 m^3s^{-1} and 1.4 m^3s^{-1}, Cannonsville Reservoir 1.3/9.2 m^3s^{-1} and 0.9 m^3s^{-1} for winter and summer respectively. Historical mean annual flows at outlet locations are approximately:15 m^3s^{-1} for Neversink, 21 m^3s^{-1} for Pepacton and 20m^3s^{-1} for Cannonsville. For better comparison of historic flows and releases see Figure 2.

Conceptual Model of Historical Character of the Upper Delaware

In the Upper Delaware River watershed, deforestation and agriculture is expected to have increased the rate of soil erosion and consequently a decrease in soil depth. The removed first growth (or primeval) forest in the Catskills was dominated by the sugar maple and beech, species known for growing on moderately wet sites The forest floor in the few remaining sugar maple groves at high elevations in the Catskill Mountains is covered with the herbaceous vegetation indicative of wet sites (Kudish 2000). We hypothesized that, prior to being removed, the first northern hardwood forest maintained a deep layer of top soil, which

---

[1]For map see http://www.nyc.gov/html/dep/html/wsmaps.html

effectively retained water. As mentioned above, runoff in the Catskill systems originates from areas where the soil saturates to the surface. Logic suggests that decreasing the soil depth would increase the propensity for runoff generation. To test the region's sensitivity to soil depth, we applied the Soil Moisture Routing model (SMR) (Frankenberger et al., 1999) to the Town Brook watershed (37 km^2), a sub-basin of the Cannonsville Reservoir. Mehta et al. (2002) have previously shown that SMR simulates the hydrology for this basin very well. We compared the flows modeled for the current soil conditions with those modeled for past conditions with an additional 15 cm of soil cover throughout the basin. The results of these two simulations are shown in Figure 1a and b, and confirm that deeper soils would facilitate higher summer baseflows (Figure 1a) and lower storm peak flows (Figure 1b).

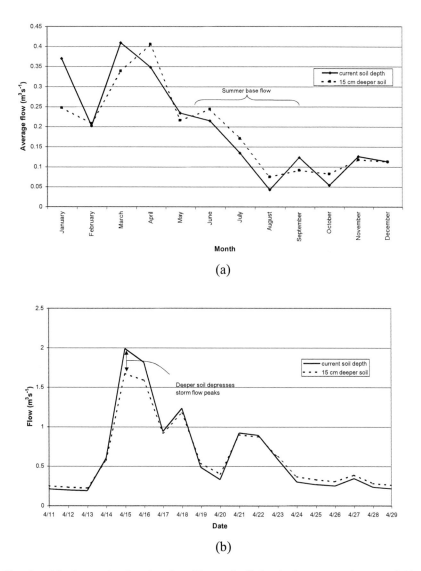

(a)

(b)

Figure 1. Simulated hydrographs showing the effects of soil depth changes on the annual (a) and storm (b) hydrology of Town Brook. Solid lines indicate current conditions and dashed lines show how the flow behaves if the soil throughout the watershed is 15 cm deeper.

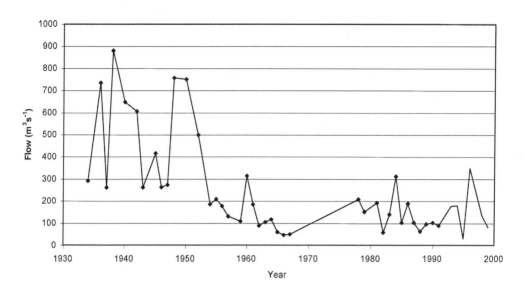

Figure 2. Historical annual peak flows at East Branch of the Delaware River, Harvard gage. The lower peaks coinside with the dam construction.

Besides storage in deeper soil cover, water flow in old growth forest is also reduced by large woody debris and massive snags, which store the water and meter it to the stream. Old growth forests have been also characterized as having much lower evapotranspiration than the young forests (Vertessy et al 2001). All these facts suggest a higher water storage capacity of the first forest, which must have had stabilizing effects on flows in the streams.

In addition to the physical removal of trees and wetlands during earlier times, most other anthropogenic modifications to the flow regime on the Delaware River act to increase the rapid transfer of water from the landscape to the stream system, exacerbating peak storm flows and reducing groundwater recharge and the associated baseflow. Examples of obvious modifications include establishing impervious areas (e.g., roofs, paved roads and barnyards), tile drainage, and drainage channels.

Subsequently, we can conclude that the historical hydrological regime of the Upper Delaware would have been characterized by higher summer baseflows and less dramatic storm peak flows than the current systems experience. The forests and wetlands, which dominated the landscape two centuries ago, have a high capacity to store rainwater and snow melt and slowly deliver water as delayed baseflow. Together with increased shading and higher infiltration of subsurface water, these circumstances would have created a flow regime that promoted cooler and more stable water temperatures. The historical evidence of coldwater fish species in the Upper Delaware confirms lower summer temperatures and/or wide availability of thermal refuges in the main stem or tributaries (Karas 1997, Van Put 1996).

In light of the above, the novelist description of pre-colonial Upper Delaware given by Rick Karas'es in his 1997 book " Brook trout" seems to be very reasonable.

"The forests and their deep duff, with topsoil intact, covered surrounding mountains like a deep sponge. Even the heaviest rain seldom raised the river's level appreciably or even clouded its waters. Flash floods were practically unknown. If they did occur, it was only in the late winter or very early spring, with the break up of river ice or when frozen ground couldn't absorb snowmelt. The river was narrower because the banks were squared and closer together, not eroded by floods. They were well defined, sharp and steep, not lensing gradually up the valley sides. The Delaware was a classic example of a freestone river, its bottom filled with glacial boulders, graded stones, and gravel. (...) Silt was unheard of. The water was cooler. Dense conifer forests interspersed with hardwoods kept tributary streams cool and helped keep water temperatures low in the main stream. Even the main river was better sheltered, because trees grew down to the banks and overhung the river. Brook trout prospered."

Ecological Deficits

The river's current condition is the result of about two hundred years of human alterations to the original system, not merely the construction of the three upper-basin reservoirs. As a result of massive deforestation in the nineteenth century, the hydrological regime has been modified in the entire basin. Runoff was destabilized, changing annual amplitudes of flow. Average spring and winter flows increased, and summer/early fall flows decreased. With the loss of retention capacity, fluctuations of flow became more frequent and intense. Unstable riverbeds formed atop of glacial till must have been altered by a common occurrence of high flows that scoured the banks. Probably the most dramatic change to the river channel, however was caused by the logging operations, which created a wider and shallower channel. In 1875, 150 miles length of the Delaware River was dredged to remove "obstacles" for log rafts, causing further reduction of river bed variability and together with flushing flows, the further over-widening of the channel (The Hancock Herald, October 29, 1876). Logically, the resulting increase in the exposure of water surface to radiation and generally lower water depth, combined with a lack of canopy cover and reduced subsurface water discharges must have led to much warmer summer flows. Increased catchment sediment yields and bank erosion must have changed substrate composition in the river and impacted spawning grounds and hyporheic refugia with silt. In response to all these factors, the faunal composition shifted from a coldwater community towards a more generalist and warm water assemblage. The New York State Department of Environmental Conservation's records, collected after the river was already heavily modified, show the presence of fallfish (Semotilus corporalis), chain pickerel (Esox americanus), golden shiner (Notemigonus crysoleucas), pumpkinseed(Lepomis gibbosus), cutlips minnow(Hybognathus regius) and sculpin(Cottidae) (New York State Department of Environmental Conservation data base). Heavy angling pressure as well as the stocking of domesticated brook trout and exotic species such as smallmouth bass (Micropterus dolomieu) or brown trout (Salmo trutta) also share the blame for loss of a regional treasure: native, long living brook trout (Karas 1997). Tanneries and acid factories also contributed to the complete destruction of the aquatic community on local levels (Kudish 2000).

In the more recent past, the construction of the Neversink, Pepacton and Cannonsville water reservoirs for New York City has continued to contribute to various deficits in the Delaware River system. At the time of this study the following deficits were expected as related to the reservoirs:

With regard to flow:

✓ Reduction of average annual volume by 50% (from nearly 400 Billion Gallons per Year (BGY) to 200 BGY). The highest flow decreases have occurred on the East

Branch and Neversink (70% and 80%, respectively). As a result, chronic low flows occurred throughout the year and water flow was severely curtailed both directly below the dams and in the lower portions of the rivers. Critical physical attributes (velocity, depth etc.) are altered, creating a habitat unsuitable for the original fish community and for nonnative cold-water fish including brown and rainbow trout, and promoting a generalist and lentic assemblage.

✓ Change of seasonal and spatial flow pattern. The magnitude of high flows is only a fraction of the pre-dam peaks (Figure 2). Average discharges over the winter and spring periods were dramatically reduced (Figure 3). Summer flows in the West Branch were unusually high, but in contrast were very low in the East Branch and Neversink. Frequently droughts occurred in the fall an winter instead of summer as a result of sudden increase in reservoir storage and the existing water release structure.

✓ Modification of daily flow pattern (increase of amplitude and ramping rate). Figure 4 compares the Walton (above Cannonsville reservoir) and Stilesvile (right below the dam) hydrographs. In the tailwater, rapid flow changes occurred more frequently and with higher intensity. Such frequent spates can result in the impoverishment of aquatic biota (Parasiewicz et al 1998). Contrary to flow increases in natural systems, these artificial flow increases are not associated with increased ground water levels that have been identified as "warning signals" for benthic fauna in alluvial rivers (Bretschko and Moog 1990). Also, the falling limb of the hydrograph fundamentally differs from a natural scenario. Such conditions can lead to the stranding of organisms and increased deposition of fine particles within the channel instead of on the adjacent floodplain areas or channel margins.

✓ Critically low flows during drought and winter seasons. During droughts, releases were reduced to almost nothing, reducing habitat and thermal refugia in both seasons. In winter, the higher volumes needed for over-wintering habitat (most frequently deep pools for adult fish and riffles for juvenile fish and fry) do not exist. These periods are very detrimental to fish.

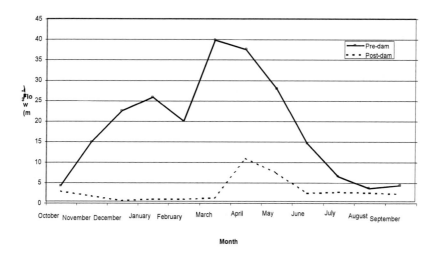

Figure 3. Median Monthly flows at the East Branch of the Delaware River, Downsville, NY under historical and present conditions.

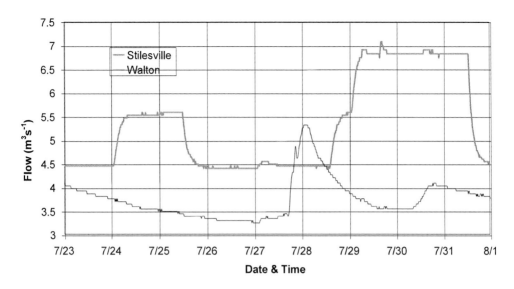

Figure 4. Comparison of daily hydrograph on the West Branch in Walton (above Cannonsville Reservoir) and Stylesville (below the dam) between 23-31 of July 2000.

With regard to the thermal regime:

✓ Higher annual amplitudes of temperature. Low flows, widened riverbeds, and reduced ground water discharge lead to higher water temperatures in the summer and lower temperatures in the winter than occurred in the pre-colonial times (see above discussion). The summer "warm ups" have already caused multiple fish kills in the Delaware River (eg. Such as documented New York State Department of Environmental Conservation in 1981 and 1983). The increased formation of ice cover and lowered water temperatures in winter reduces habitat availability in the most critical stage of the annual life cycle, and particularly impacts the survival of young fish. Anchor ice in riffle areas may damage the riverbed and impact juvenile fish and benthic fauna (Brown et. al. 1993).

✓ Extreme daily fluctuations and thermal plunges. Shallow water bodies are susceptible to highly variable water temperatures even under normal weather conditions (Figure 5). Water temperature could change by many degrees centigrade during hot summer days. Drastic changes in release of water from the Cannonsville Reservoir (e.g. 200 cfs to 1200 cfs) regularly introduced a coldwater plume that caused a thermal shock to downstream fauna.

✓ Thermal discontinuity. As described in the River Continuum Concept (Vannote et al. 1980) in a natural river ecosystem the distribution of flora and fauna varies along the river length with different organisms utilizing headwaters and lowland sections. The temperature continuum along the modified river course is interrupted by the reservoirs, disrupting the natural distribution of the flora and fauna assemblages.

✓ Longitudinal connectivity. Longitudinal connectivity is affected by a lack of fish passage facilities at the dams. The resulting habitat fragmentation impairs the migratory cycle of species and limits their access to upstream thermal refugia.

✓ Substrate recruitment is disrupted by the impoundments and could possibly lead to the downcutting of the riverbed in the tailwaters.

✓ Channel modifications that took place due to the logging and construction of roads, bank stabilization, levees and floodwalls are much more evident than potential morphological changes caused by the alteration of flow regime and substrate deficits. Nevertheless, armoring of the riverbed, together with sedimentation of the hyporheic zone, could reduce vertical connectivity and affect benthic organisms. Potential downcutting might have an influence on longitudinal and lateral connectivity of the ecosystem, affecting life cycles of dependent organisms (e.g. access to spawning grounds).

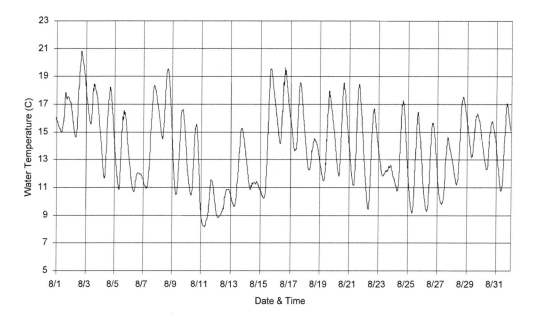

Figure 5. Water temperature on the West Branch of Delaware at Hale Eddy during August. 2000 (cold and wet summer).

These effects of reservoirs are not unique and were widely reported from other rivers (e.g. Cheslak and Carpenter 1990). Other ecological deficits exist on the Delaware that are not entirely attributable to New York City reservoirs and water withdrawals:

✓ Loss of wetlands and floodplains, increased paved surfaces, and infiltration limited by road surfaces, together with low river flows, could significantly reduce ground water contributions. Changes in water table levels could have a strong influence on the flow and temperature of the river as well as riparian areas.

✓ Exotic fish species have been introduced, but their influence on the native population is not clearly known. Brown and rainbow trout compete successfully with native brook trout. Many unlikely species (e.g. smallmouth bass, carp, and walleye) are also abundant. Additionally, many exotic plants compete with native flora.

Objectives for Sustainable Management Comment

The goal for natural resource management of the Upper Delaware River should be to optimize and sustain the existing and desired use of river resources. This can be achieved by implementing the following ecological objectives:

- Manage the upper river as a coldwater system maintaining a dynamic equilibrium and the sustainability of the associated flora and fauna.
- Maintain ecological integrity (allow only slight changes from the type specific communities in species composition and abundance).
- Focus on the recreational industry and traditional trout fisheries.
- Preserve a potable water supply for present and future generations.
- Allocate water for industry and local uses in the lower part of the basin.
- Provide appropriate flood protection for communities in the upper and lower river basin.

Proposed Target System

To fulfill management objectives the following characteristics can be envisioned as management targets:

- ➢ To allow for water withdrawals and maintain ecological integrity, the characteristics of the tailwaters of all three reservoirs should correspond to those of the natural system (but on a smaller scale). The detailed description reference natural state for present climatic conditions can be modeled incorporating new and available bio-physical data, historical information and expert knowledge of running water ecosystems. This reference state is used as a standard to develop an achievable target state the maximum realizable future condition of the Delaware River.
- ➢ The flow and temperature regime (timing, frequencies, amplitude, ramping rate) of the Delaware should correspond with the reference conditions. To mitigate the impact of increased thermal uptake (a consequence of reduced water body and increased surface area) the temperature could continue to be lowered by releases of cold bottom waters from reservoirs.
- ➢ The constant subsurface flow and ground water level should be maintained by increased infiltration and storage (by maintaing wetlands, backwaters etc.).
- ➢ Riparian vegetation should offer extensive shading and a source of woody debris, which in turn improves channel diversity.
- ➢ The substrate deficits should be controlled and minimized by adequate flow management.
- ➢ Ice damages should be of little significance and controlled by additional releases and higher ground water levels.
- ➢ The habitat diversity and connectivity should be restored.

Proposed Measures

Without conclusive historical records and available reference systems, maintenance of the target system may be vague and needs intensive monitoring and evaluation of water management, and additional study. However, immediate action is required to preserve the ecosystem. For these reasons a two-step approach is proposed. First, preliminary measures should be undertaken to solve the most urgent problems and avoid further damages. The deficits that must be addressed in the short term are:

1. Chronic low flows on the East Branch and Neversink Rivers.
2. Reduction of extreme peak flows and their impacts. While it has been speculated that the East Branch and Neversink Rivers may suffer from lack of higher flows since reservoir establishment, more data is needed concerning substrate composition, embeddedness, and channel morphology.
3. Temperature bottlenecks in summer resulting from low flow releases.
4. Winter habitat bottlenecks and anchor ice damage resulting from low winter releases.

The simplest and most assured way to solve the issues listed above is to provide a constant and adequate base flow in all three branches. It is necessary to prescribe minimum flows based on the general scientific and empirical knowledge of the ecosystem mechanisms while adding an adequate margin of safety. As we gain a better understanding of the complex ecosystem, the flow regime can be optimized.

The following factors need to be integrated to develop an interim flow regime:

- New York City's ability to withdraw the same total amount of water for water supply needs.
- Present and historical flow regime at the reservoir locations.
- Need for reduction of temperature amplitude.
- Need for riparian succession to improve the channel.
- Coldwater fisheries.

Another important issue is to avoid sudden flow and thermal fluctuations due to excessive ramping rates. For a variety of biologic and geomorphic reasons mentioned earlier in the text, flow reductions should mimic the recessional limb illustrated in Figure 4. Hence, special attention should be drawn to slow reductions of flow peaks.

These objectives can be at least partly accomplished without impacting New York City's ability to withdraw water. Analyses of the water budget over the period of record (early twentieth century until now) have shown that since the construction of the reservoirs on average 50% of the annual volume has been withdrawn from the Upper Delaware basin. Consequently, the other 50% of water flowing into reservoirs could be used for impact mitigation without affecting the current supply. The key question is how to distribute it across the year most efficiently. Until now the majority of this water has been released as spills over the reservoirs dams during times of higher flows. Relocating the same volume to times of highest ecological needs could be a simple first step without needs for negotiating water use.

To specify long term management objectives a natural target state needs to be developed to allow precise definition of the management goals, priorities and actions. Figure 6 presents a schematic of a process applied during the development of a river management plan on the River Traisen in Austria published by Eberstaller et al 2000. This procedure takes into equal account the human and ecological needs and could be used as a role model for the Upper Delaware. According to this model, investigation by a multidisciplinary team of experts is necessary to first provide an ecologically based foundation for such measures and specify objectives for resource utilization. In addition, predictive models of hydrological and thermal regimes, ground water flow, sediment transport and target-fish-community habitat should be developed for the Upper Delaware. Based on comprehensive deficit analysis and model predictions, the river management plan can be specified. It should include measures for flow management, improvements of channel diversity, connectivity restoration, or enhancement of water quality in the Cannonsville reservoir or reduction in New York City water consumption (presently 600 liters per capita[2]).

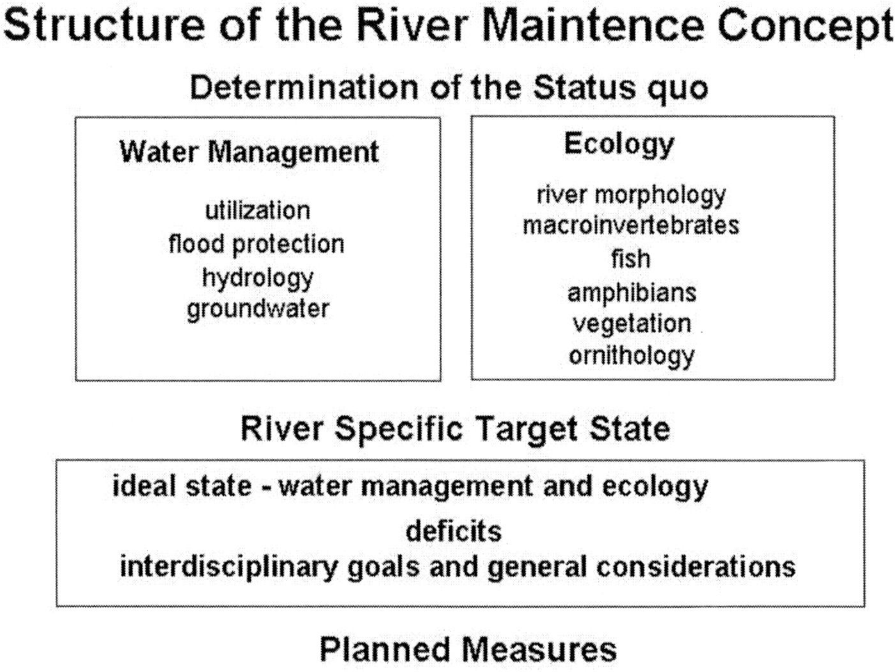

Structure of the River Maintence Concept

Determination of the Status quo

Water Management	Ecology
utilization flood protection hydrology groundwater	river morphology macroinvertebrates fish amphibians vegetation ornithology

River Specific Target State

ideal state - water management and ecology deficits interdisciplinary goals and general considerations

Planned Measures

Figure 6. Schematic of the procedure applied to develop a river management plan for Traisen River in Austria (modified form Eberstaller et al 2000).

CONCLUSION

The Upper Delaware is an impacted river system. The disturbance is a combination of its past industrial legacy, present water supply needs and reservoir operation. The above analysis is intended to document that environmental flows cannot be considered without taking into account hydro-morphological modifications and altered system characteristics. Therefore, the

---

[2](calculated from 1999 treated sewage published at http://www.ci.nyc.ny.us/html/nyw/pdf/nyw-cafr-2000-part1.pdf)

appropriate improvement measures should aim toward integrative, holistic rehabilitation rather than focusing on mitigation of individual symptoms. At this stage, interim measures need to be implemented that may be unnatural (e.g. elimination of high flows) to initialise bio-physical processes that will help restore the basin's original character. These measures have to be part of a multiphase, monitored recovery program that is implemented over a number of years. Such a program requires a well-defined, scientifically sound, long-term management concept. Managing the limiting ecosystem factors, groundwater and surface water interactions, and adaptive restoration options are key to the foundation of such a plan.

DISCUSSION

Since the publication of the summarized report, many actions have been introduced to improve the ecological status of the Upper Delaware The Delaware River Basin Commission, and the Parties to the 1954 Supreme Court Decree, have begun by creating the Subcommittee on Ecological Flows (SEF). This is a representative group of resource professionals from public agencies, academia, and the non-profit sector established to advise basin water managers on ecological flow issues. SEF worked with the U.S. Geological Survey (USGS) to complete an Upper Delaware habitat assessment and design a decision-support system, which has influenced recent water management decision-making. SEF has also provided an input to the New York City Department of Environmental Protection on limiting the ramping rate for flow releases from the water-supply reservoirs. It has also been coordinating with the U.S. Fish and Wildlife Service and Northeast Instream Habitat Program on a study focused on the instream flow needs of the dwarf wedgemussel in the upper Delaware River.

Alternative operating rules for New York City's reservoirs have been proposed by the Delaware River Basin Commission and the Conservation Coalition for the Delaware River. The Flexible Flow Management Program by DRBC is based on a reservoir-level adaptive release program that designates variable minimum releases based on reservoir levels and seasonal differences. Higher water release volumes "Adaptive Release Policy CP2" were proposed by the Conservation Coalition's to further help address additional ecological concerns. Both FFMP and CP2 have been presented to basin decision makers and discussed at public forums. Details of CP2 and the current problems with the management of New York City's reservoirs can be found at http://www.drarp.org.

However, additional aspects reaching beyond flow management become more acute in recent years. Severe flood damages in 2005 and 2006 uncovered an urgent need for better floodplain management and consideration of the impact of global warming in planning. The impact of invasive flora and fauna and need for protection of federally endangered species such as the dwarf wedgemussel Alasmidonta heterodon call for organizing all the above activities into a holistic planning process, similar to one created in Austria. It is a logical next step that would assure a focused structure and sustainability of the plan. It would help organize priorities, uncover areas with most prevalent information needs while at the same time allow the expectation of reasonable results in a relatively short time. This is particularly necessary in light of recent flooding events, which call for quick but sustainable solutions balancing the protection of human uses with the need for the protection of endangered species and maintenance of ecological integrity.

ACKNOWLEDGMENTS

Trout Unlimited sponsored the study which resulted in the described report for which the authors are grateful. The authors would like to thank the following people for their help and expertise in the course of conducting that study: Whit Fosburg, Charles Gauvin, Leon Szeptycki and Jock Conyngham. The authors would like to thank the following individuals in those agencies and for their assistance: Wayne Elliot, Phil Hulbert, Dr. Muralidhar, Norm McBride, (New York State Department of Environmental Conservation, Scott Stanton, Carol Collier (Delaware River Basin Comission), Colin Apse , George Schuler of The Nature Conservancy, Al Caucci, Jim Serio, Harry Batschelet, Larry Finley, Jerry Wolland, Joe McFadden, Don Tracy, Chuck Schwartz, Steve Sugarman, Pat Schuler, Paul Weamer, Lee Hartman, Tony Lasorte, Rick Eck, Clem and Barbara Fullerton, Jim Charron, Frank Andros, Joe Demalderis, Richard Schager, Tony Ritter, John Miller, John Morris, Matt Evans, Sam Batschelet, Chris Micalizzi, Milt Sabol, Russ Mosher, Kurt Nelson and Phil Chase. Also, we would like to thank Vishal Mehta for running the Soil Moisture Routing model (SMR).

REFERENCES

Bretschko, G, and Moog, O. (1990). Downstream effects of intermittent power generation. *Water Science and Technology*, **22**, 127-135.

Brown R.S., S.S. Stanislawski, and Mackay, W.C. (1993). Effects of Frazil Ice on Fish, *Proceedings of the Workshop on Environmental Aspects of River Ice*, T.D. Prowse (Editor), National Hydrology Research Institute, Saskatoon, Saskatchewan, 1993, *NHRI Symposium Series* No. **12**, p. 261-278.

Brown, V.A., J.J. McDonnell, D.A. Burns, C. Kendall. (1999). The role of event water, a rapid shallow flow component, and catchment size in summer stormflow. *Journal of Hydrology*. **217**, 171-190.

Cheslak. E & J. Carpenter (1990). *Compilation report on the effects of reservoir releases on downstream ecosystems*. Bureau of Reclamation. REC-ERC-90-1. p 107

Dunne, T. and Leopold, L.B. (1978). *Water in Environmental Planning*. W.H. Freeman and Co. New York. pp.818.

Dunne, T., and Black, R.D. (1970*)*. Partial area contributions to storm runoff in a small New England watershed. *Water Resour. Res.* **6**: 1296-1311.

Eberstaller, J., G. Haidvogl, M. Jungwirth, H. Zottl & G. Küblbock, (2000): The Gewässerbetreuungskonzept Traisen - How to Combine Flood Protection and Restoration in an Intensively Used River, *Eigth International Symposium on Regulated Rivers*, 17.-21.7.2000 in Toulouse.

Frankenberger, J.R., (1996). *Identification of critical runoff generating areas using a variable source area model*. PhD Thesis, Cornell University, Ithaca NY.

Frankenberger, J.R., E.S. Brooks, M.T. Walter, M.F. Walter, T.S. Steenhuis. (1999). A GIS-Based variable source area model. *Hydro. Proc. J.* **13**(6), 804-822.

Hewlett, J.D. and Hibbert, A.R. (1967). Factors affecting the response of small watersheds to precipitation in humid regions. In. *Forest Hydrology* (eds. W.E. Sopper and H.W. Lull). Pergamon Press, Oxford. pp. 275-290.

Hewlett, J.D. and Nutter, W.L. (1970). The varying source area of streamflow from upland basins. Proceedings of the Symposium on Interdisciplinary Aspects_of Watershed Management. held in Bozeman, MT. August 3-6, 1970. pp. 65-83. ASCE. New York.

Hulbert, P. J. (1987). *Impact of drought conditions on selected fishery resources in the upper Delaware River basin in 1985 – an overview.* New York State Department of Environmental Conservation.

Karas N. (1997): *Brook Trout.* Lyons & Bufford. New York.

Mehta, V.K., M.T. Walter, E.S. Brooks, T.S. Steenhuis, M.F. Walter, M. Johnson, J. Boll, D. Thongs. (2002). *Evaluation and application of SMR for watershed modeling in the Catskills Mountains of New York State.* Envir. Modeling & Assessment.

Parasiewicz, P. (2001). *Strategy for Sustainable Management of the Upper Delaware River Basin.* Trout Unlimited. NY. 21pp

Parasiewicz, P., S. Schmutz & O. Moog, (1998). The effects of managed hydropower peaking on the physical habitat, benthos and fish fauna in the Bregenzerach, a nival 6th order river in Austria, *Fisheries Management and Ecology*, **5**, 403-417.

Parde, M., (1968). *Flueves at rivieres.* A.Collin, Paris.

Rosgen, D., L, (1996). *Applied river morphology.* Wildland Hydrology.

Soren, J, (1963). *The Groundwater Resources of Delaware County, New York.* USGS/Water Res. Comm. Bull. GW-50. Albany, NY.

Sheppard J. D. (1983): *New York reservoir releases monitoring and evaluation program – Delaware River.* Summary report. New York Department of Environmental Conservation. TR. No. 83-5

Stalzer, W., and Bloech, H. (2000). The Austrian approach. *Hydrobiologia* 422/423, p. xix-xxi.

Van Put E. (1996): *The Beaverkill.* Lyons & Bufford. New York.

Vannote, R. L., G.W. Minschal, K.W. Cummins, R.J. Sedell, and Cushing, C. E. (1980). The river continuum concept. *Canadian Journal of Fisheries and Aquatic Sciences*, 37: 130-137.

Vertessy, R.A., F.G.R. Watson, and S.K. O'Sullivan. (2001). Factors determining relations between stand age and catchment water balance in mountain ash forests. *Forest Ecology and Management* 143: 13-26.

In: Encyclopedia of Environmental Research
Editor: Alisa N. Souter

ISBN: 978-1-61761-927-4
© 2011 Nova Science Publishers, Inc.

Chapter 29

OUTCOMES OF INVASIVE PLANT-NATIVE PLANT INTERACTIONS IN NORTH AMERICAN FRESHWATER WETLANDS: A FOREGONE CONCLUSION?

Catherine A. McGlynn

Hudsonia Ltd., P.O. Box 66, Red Hook, NY 12571, USA

ABSTRACT

Freshwater tidal wetlands are productive and often support high biodiversity. While there have been some quantitative studies of the effects of many invasive plants in North American freshwater wetlands, much is still assumed for a number of invasive species. The outcomes of invasive-native plant interactions, and the factors involved, are more subtle than assumed. I reviewed the statistical and anecdotal results of published studies on the impacts of invasive plants with large potential ranges in freshwater tidal wetlands of North America. A number of studies reported little or no change in native species richness and diversity, and found outcomes differed depending upon several specific factors. I recommend that more empirical research be conducted on the interactions of these species (both in field and greenhouse) with native plants in freshwater tidal wetlands because very little information is available and yet many management decisions have already been made.

INTRODUCTION

Biological invasions are responsible for billions of dollars in economic losses and an untold number of ecological impacts throughout the world (Mack et al. 2000). In the United States alien species are the second highest threat, after habitat loss and deterioration, to endangered or threatened native species (Wilcove et al. 1998). Along with changes in land-use and climate, biological invasions are drivers for environmental change on a global scale. In aquatic systems, however, the effects of biological invasions often remain unquantified.

Wetland plants (floating, submerged, and emergent) are a vital part of the aquatic ecosystem. They provide habitat and food for many taxa, both vertebrate and invertebrate; and contribute to nutrient cycling and many other ecological functions (Carpenter and Lodge 1986). Living in the diverse and dynamic habitat of freshwater tidal wetlands, wetland plants are subjected to considerable natural and human disturbance: flooding, tides, ice scouring, extremes in temperature, sediment accumulation, build up of debris; fluxes of nutrients, pollutants or sewage, and changes in hydroperiod (Dickinson and Miller 1998; Zedler and Kercher 2004).

Invasive plants find opportunities for potential establishment in the conditions created by these disturbances (Davis et al. 2000). The impacts of these invaders upon the native plant communities and the vulnerable wetlands they inhabit (Houlahan and Findlay 2004; Zedler and Kercher 2004) are not consistently studied in a quantitative manner. Of all the communities of organisms in freshwater tidal wetlands, the native plant community has the potential to be affected the most by invasive plants because both the community and the invader have similar resource requirements. Invasive plants co-opt space (Hager and McCoy 1998), and compete for light (Weihe and Neely 1998) and nutrients (Pfeifer-Meister et al. 2008, Sala et al. 2007). In addition, several studies have found that invasive plants have allelopathic effects on native plants (Nakai et al. 1999; Bais et al. 2003; Gross 2003; Hierro and Callaway 2003; Callaway et al. 2005; Morgan and Overholt 2005; Hilt 2006; Thelen et al. 2005; Rudrappa et al. 2007). Invasive-native plant interactions also present the possibilities of hybridization, parasitism, and the transfer of pathogens (Baker 1986; Simberloff 2003). Factors that affect the perceived outcome of each invasion are, in part, revealed through published studies of lab and field experiments. I review existing literature on the ecological effects of invasive wetland plants on native ecosystems and plant communities. This chapter discusses several common findings of those studies.

METHODS

To access published studies, I created a list of invasive plants that are found in freshwater wetlands in North America based on information provided by the USDA Natural Resource Conservation Service website (http://plants.usda.gov) and by searching Web of Science using the search terms "invasive plants," introduced plants," "exotic plants," in combination with "wetlands." I chose to review studies published after 1995 on wetland plant species that had the largest potential ranges in North America wetlands and were of verified non-native origin, which produced a list of 15 species (Appendix A). I chose studies by searching Web of Science using scientific names or common names of the 15 species in combination with the search terms "species diversity," species richness," and "native plants." I previewed over one hundred records for the selected species and reviewed the papers (27) that had abstracts which mentioned that part or all of the study's focus was on the effects of invasive plants on native plants. In the case of several species, I was unable to find studies of their impacts on native plant communities in North America or in greenhouse experiments.

RESULTS

The review of these publications revealed several important aspects of invasive plant-native plant interactions. Effects and factors are often difficult to distinguish from one another or may be quite subtle. However, I consistently encountered several variables that were associated with invasion impacts and outcomes in freshwater tidal wetlands. I have grouped these variables into two categories: experimental design and species-specific characteristics.

Experimental Design

Context

The context of study sites influences the ability of an invasive plant to become established or dominant. Land-use in neighboring areas can create disturbances (Waldner 2008; Gassó et al. 2009; Milbau et al. 2009), and impact nutrient availability and habitat vulnerability to invasion. Current and historical land-use of invaded areas and their surroundings can also affect the ability of invasive plants to become established (Hulme 2009; Mack et al. 2000; Waldner 2008; Wania et al. 2006). McGlynn (2006) found that human activity (proximity to an urban area) could have reduced or altered the native plant species composition of several wetland sites and affected the invasibility of wetlands surveyed. The urban wetlands studied by Ehrenfeld (2008) showed a decreased rate of plant invasions as industrialization in adjacent areas increased. Upland studies also confirm the importance of land-use in invasion outcomes (Endress et al. 2007; Aboh et al. 2008).

Seasonal Effects

Invader dominance seems to have a strong seasonal effect on native plants or its effects are modulated by seasonal effects that are related to light, nutrient availability, and possibly other abiotic factors. An invasive plant that arrives in the latter part of the growing season when many native plants are already adults might be unable to become established. Yakimowski et al. (2005) found that invasive *Lythrum salicaria* seedlings encountered limited light in *Typha*-dominated marshes (adult *Typha* spp. and plant litter) and their establishment was negatively affected. Unlike *Typha* spp., invasive *Crassula helmsii* inhibits growth of other plants at an even earlier stage—that of germination (Langdon et al. 2004). The growth rate of both native and invasive plants that are already established can also influence interactions. A greenhouse study performed by Mony et al. (2007) showed that invasive *Egeria densa* did not alter its biomass production and reproductive allocations as the seasons changed, but *Hydrilla verticillata* did and was able to outcompete *E. densa* for nutrients. Presumably *H. verticillata* outcompetes a number of native plants as well.

Temporal Scale

The length of time during which a study is conducted may determine the outcome observed. Over time the effects of an invasive plant may change. However, few of the studies I found occurred over a period of more than one or two seasons and all such studies focused on *L. salicaria*. Denoth and Meyers (2007) removed the invasive *L. salicaria* from their study plots in the first year and found a significant increase in the vegetative performance of the rare native plant *Sidalcea hendersonii* in the first year, but not in the second year. In a three year study, Farnsworth and Meyerson (1999) found that native species richness and abundance increased the first year after removal of *P. australis*, but in the remaining years *Typha* sp., and then *P. australis*, became dominant. Mal et al. (1997) found that *L. salicaria* took four years to attain dominance in the study wetland. A microcosm experiment involving 20 plant species was dominated by *L. salicaria* in its fifth year (Weiher et al. 1996). In grasslands, native plant species exhibit a pattern in which their establishment is first dependent upon disturbance and then upon nutrients available for accumulation of biomass (Thompson et al. 2001); invasive plants in wetlands may follow a similar pattern.

Spatial Scale

Like temporal scale, landscape scale could also affect the perceived outcome of an invasion (Brown and Peet 2003; Kennedy et al. 2002; Milbau et al. 2009). Small scale studies found a negative association between native and invasive diversity with some researchers surmising that high native diversity decreases invasibility (Tilman 1997; Stachowicz et al. 1999; Levine 2000; Naeem et al. 2000; Lyons and Schwartz 2001; Fridley et al. 2007). Studies performed on large scales found positive relationships between native and invasive diversity (Lonsdale 1999). Davies et al. 2005 attributed this pattern to the spatial heterogeneity in the large landscapes (including heterogeneity of abiotic factors) that allowed for spatial heterogeneity in species composition while Shea and Chesson (2002) and Levine (2000) hypothesized that high native species diversity may also create more niche opportunities for invasive plants. Whatever the mechanisms behind the relationship between native and invasive diversity may be, studies in freshwater tidal wetlands occurred at both spatial scales and exhibited the predicted patterns in species richness and diversity. Hager and Vinebrook (2004) found a positive relationship between *L. salicaria* and native species diversity. Treberg and Husband (1999) found no evidence of a decrease in native plant diversity in wetlands invaded by *L. salicaria* and Massachusetts (United States) marshes dominated by both *L. salicaria* and native *Typha* spp. had high alpha diversity (Keller 2000). McGlynn (2006; 2009) found that species composition and diversity were significantly different among plots dominated by invasive *L. salicaria* or *P. australis* or native *Typha* spp. Native plant species richness decreased with increasing cover of *L. salicaria* (Gabor et al. 1996) and *L. salicaria* dominated the seed bank in experimental areas (Welling and Becker 1990). An important factor to consider before attributing an outcome to scale or species diversity is that the disturbance which created an opportunity for an invasion may have produced changes in the native plant community (McGlynn 2006; Gurevitch and Padilla 2004).

Metrics

Measures used to quantify invasive plant effects on native plants influence perceived outcomes of an invasion. Farnsworth and Ellis (2004) found that *L. salicaria* did not necessarily impact species richness and diversity, but did affect native plant biomass. Morrison (2002) also found that native plant cover and biomass was reduced in the presence of *L. salicaria*, but plant species richness did not change significantly. However, species richness and diversity were significantly affected by *Myriophyllum spicatum* even though there was no significant relationship between biomass of *M. spicatum* and native aquatic plants (Madsen et al. 2008). In two studies (Larsen 2007; Zhonghua et al. 2007) *Nymphoides peltata* negatively affected the growth (biomass) of several other species, *Elodea canadensis*, *Ranunculus circinatus*, and *Trapa spinosa*; but no species was eliminated. It may be that, as in other habitats, invasive plants in freshwater tidal wetlands do not generally cause native plants to become extinct (resulting in a change in species richness), but they can be associated with decreased abundance of a particular native species (possibly resulting in a change in species diversity) (Brown et al. 2006).

Species-Specific Characteristics

Zonation

The different vegetation zones in freshwater tidal marshes have different flooding regimes, elevations, and soils as well as different resident native plant species. An invader to a particular zone would have to possess particular characteristics in order to become established there (McGlynn 2006; 2009). For example, the ability of *Phragmites australis*, an invader to both brackish and freshwater wetlands, to become established was reduced as marsh elevation decreased and flooding frequency increased (Silliman and Bertness 2004). In fact, many of the species in the native plant communities that were invaded may have had not only similar tolerances for flooding, but also similar requirements for soil composition (Silliman and Bertness 2004). Farnsworth and Meyerson (2003) found that some species, e.g. *Leersia oryzoides*, had very specialized microenvironment requirements, while the dominant *Typha* sp. and *P. australis* did not. Thus the effect of zonation on plant invasions is very closely tied to species-specific characteristics.

Form and Function

Species-specific traits can determine the nature of interactions between invasive plants and the native plant communities they invade (Alvarez and Cushman 2002; Kourtev et al. 2003). They can affect how some plants' characteristics change when the plants become invasive (Willis et al. 2000), how closely the characteristics of an invader match those of native plants already present (Strauss et al. 2006), and how an invader affects an invaded community (Mason and French 2008). Some characteristics that are important in determining the outcome of native-invasive plant interactions include functional group (Mahaney et al. 2006), growth form (Hager 2004), and allelopathic ability (Gross 2003). If an invasive species becomes established in an area where it is the only species with a particular growth form or from a particular functional group it is likely to produce a more easily quantifiable effect on the native plant community (Vitousek et al. 1987; D'Antonio and Vitousek 1992; Vitousek 1992; Gordon 1998; Kourtev et al. 2003; Lesica and DeLuca 2004; Mahaney et al. 2006; Windham and Meyerson 2003). The heights of *L. salicaria* and *P. australis* enable them to limit the light available to other plant's seedlings (Odum et al. 1984; Rawinski and Malecki 1984; Gaudet and Keddy 1988; Edwards et al. 1995; Minchinton and Bertness 2003), while *H. verticillata* is able to shade and reduce root growth of fellow invasive *Myriophyllum spicatum* (Wang et al. 2008)

Dominance

Houlahan and Findlay (2004) concluded from their study of wetland plant communities in Southeastern Ontario, Canada that dominant invasive plants do not impact native plant diversity and species richness any more than native dominant plants. McGlynn (2006) also found that plots invaded by *L. salicaria* had the highest species richness relative to plots with the more dominant native *Typha* spp. or invasive *P. australis*, although the differences were not statistically significant. Morrison (2002) found that in plots where *L. salicaria* was dominant it did not form monospecific stands. When *L. salicaria* was removed from plots, several native species became dominant (Morrison 2002).

CONCLUSION

Many factors influence the outcomes of invasions and these outcomes are variable. A number of these factors are part of an experiment's design and the researcher can choose which of these factors will influence the outcome. Assuming that each invasion follows a predestined course with a negative outcome is not an effective approach for understanding the ecology of individual invasions or for making informed management decisions. Such assumptions also prevent ecologists from making use of the opportunity to study basic questions about community and ecosystem ecology that invasions provide.

Invasion science and management require empirical research that is conducted in both field and greenhouse, thoroughly quantifies actual ecological effects of each invasive plant species; and incorporates different measures of effect, multiple temporal scales, changes in the physical environment (i.e. zonation or changes in elevation that are characteristic of many tidal wetlands) and knowledge of the importance of species characteristics of all plants (both native and invasive) involved in the interactions.

Even though *Alternanthera philoxeroides*, *Butomus umbellatus*, *Eichhornia crassipes*, *Hydrocharis morsus-ranae*, *Najas minor*, *Panicum repens*, *Salvinia molesta*, and *Trapa natans* have extensive potential ranges for invasion in North America, surprisingly few studies have been published about invasions involving these floating and submerged species. This may be due to the difficulty of conducting experiments in tidal open water habitats. Much information is needed about these species' interactions with native plants, native vertebrates and invertebrates, as well as their potential ecosystem effects. Once this information has been collected it should be made widely available so that decisions can be made about whether or not a particular invasive plant is an imminent threat and what can be done about it. Invasion biology and its associated management needs have evolved beyond anecdotal data.

ACKNOWLEDGMENTS

My thanks to Dr. Kerry Brown, Dr. Elizabeth Farnsworth, Dr. Eliza Woo, and Dr. Helen Bustamante Wood for their comments on this chapter.

APPENDIX A. INVASIVE SPECIES WITH LARGEST POTENTIAL WETLAND RANGES IN NORTH AMERICA AND OF VERIFIED NON-NATIVE ORIGIN

Scientific Name	Location
Alternanthera philoxeroides (Mart.) Griseb.	15 U.S. states
Butomus umbellatus L.	16 U.S. states and 8 Canadian provinces
Crassula helmsii A. Berger	At least 3 U.S states
Eichhornia crassipes (Mart.) Solms	24 U.S. states and 1 Canadian province
Egeria densa Planch.	37 U.S. states and 1 Canadian province
Hydrilla verticillata (L.f.) Royle	20 U.S. states

Hydrocharis morsus-ranae L.	1 U.S. states and 2 Canadian provinces
Lythrum salicaria L.	44 U.S. states and 10 Canadian provinces
Myriophyllum spicatum L.	43 U.S. states and 4 Canadian provinces
Najas minor All.	26 U.S. states and 1 Canadian province
Nymphoides peltata (S.G. Gmel.) Kuntze	25 U.S. states and 2 Canadian provinces
Panicum repens L.	10 U.S. states
Phragmites australis (Cav.) Trin. ex. Steud. Invasive status (haplotype) not verified in each state or province	49 U.S. states, Puerto Rico and 11 Canadian provinces
Salvinia molesta Mitchell	11 U.S. states
Trapa natans L.	9 U.S. states and 1 Canadian province

APPENDIX B. THE PAPERS REVIEWED

Citation	Invasive species
Denoth and Meyers 2007	*Lythrum salicaria*
Farnsworth and Ellis 2001	*Lythrum salicaria*
Farnsworth and Meyerson 1999	*Phragmites australis*
Farnsworth and Meyerson 2003	*Phragmites australis*
Gabor et al. 1996	*Lythrum salicaria*
Hager 2004	*Lythrum salicaria*
Hager and Vinebrook 2004	*Lythrum salicaria*
He et al. 2008	*Najas minor*
Houlahan and Findlay 2004	*Lythrum salicaria*
Keller 2000	*Lythrum salicaria* and *Phragmites australis*
Langdon et al. 2004	*Cressula helmsii*
Larson 2007	*Nymphoides peltata*
Madsen et al. 2008	*Myriophyllum spicatum*
Mahaney et al. 2006	*Lythrum salicaria*
Mal et al. 1997	*Lythrum salicaria*
McGlynn 2009	*Lythrum salicaria* and *Phragmites australis*
Mony et al. 2007	*Egeria densa*
Morrison 2002	*Lythrum salicaria*
Rawinski and Malecki	*Lythrum salicaria*
Treberg and Husband 1999	*Lythrum salicaria*
Wang et al. 2008	*H. verticillata*
Weihe and Neely 1998	*Lythrum salicaria*
Weiher et al. 1996	*Lythrum salicaria*
Welling and Becker 1990	*Lythrum salicaria*
Yakimowksi et al. 2005	*Lythrum salicaria*
Zhonghua et al. 2007	*Nymphoides peltata*

REFERENCES

Aboh, B.A., M. Hovinato, M. Oumorou, and B. Sinsin. 2008. Invasiveness of two exotic species, *Chromolaena odorata* (Asteraceae) and *Hyptis suaveolens* (Lamiaceae), in relation with land use around Betecoucou (Benin). *Belgian Journal of Botany* 141(2): 125-140.

Alvarez, M.E. and J.H. Cushman. 2002. Community level consequences of a plant invasion: effects on three habitats in coastal California. *Ecological Applications* 12(5): 1434-1444.

Bais, H.P., R.Vepachedu, S.Gilroy, R.M. Callaway, and J.M.Vivanco. 2003. Allelopathy and exotic plant invasions: From molecules and genes to species interactions. *Science* 301(5638): 1377-1380.

Baker, H.G. 1986. Patterns of plant invasions in North America. In: Mooney, H.A. and Drake, J.A., editors. Ecology of Biological Invasions of North America and Hawaii, New York: Springer-Verlag. 44-57.

Brown, R.L. and R.K. Peet. 2003. Diversity and invasibility of southern Appalachian plant communities. *Ecology* 84(1): 32-39.

Brown, K., F.N. Scatena, and J. Gurevitch, 2006. Effects of an invasive tree on community structure and diversity in a tropical forest in Puerto Rico. *Forest Ecology and Management.* 226(1-3): 145-152.

Callaway, R.M., W.M. Ridenour, T. Laboski, T. Weir, and J.M.Vivanco. 2005. Natural selection for resistance to the allelopathic effects of invasive plants. *Journal of Ecology* 93(3): 576-583.

Carpenter, S.R. and D. M. Lodge. 1986. Effects of submersed macrophytes on ecosystem processes. *Aquatic Botany* 26: 341-370.

D'Antonio, C.M. and P.M. Vitousek. 1992. Biological invasions by exotic grasses, the grass/fire cycle, and global change. *Annual Review of Ecology and Systematics.* 23: 63-87.

Davies, K., P. Chesson, S. Harrison, B.D. Inouye, B.A. Melbourne, and K. J. Rice. 2005. Spatial heterogeneity explains the scale dependence of native-exotic diversity relationship. *Ecology* 86 (6): 1602-1610.

Davis, M.A., J.P. Grime, and K. Thompson. 2000. Fluctuating resources in plant communities: a general theory of invasibility. *Journal of Ecology* 88:528-534.

Denoth, M. and J.H. Meyers. 2007. Competition between *Lythrum salicaria* and a rare species: combining evidence from experiments and long-term monitoring. *Plant Ecology* 191: 153-161.

Dickinson, M.B. and T.E.Miller. 1998. Competition among small, free-floating, aquatic plants. *American Midland Naturalist* 140 (1): 55-67.

Edwards, K. R., M.S. Adams, and J. Květ. 1995. Invasion history and ecology of *Lythrum salicaria* in North America. In: Plant Invasions: General Aspects and Special Problems, eds. P. Pyšek, K. Prach, M. Rejmánek and M. Wade. Pp 39-60. SPB Academic Publishing, Netherlands.

Ehrenfeld, J.G. 2008. Exotic invasive species in urban wetlands: environmental correlates and implications for environmental management. *Journal of Applied Ecology* 45: 1160-1169.

Endress, B.A., B.J. Naylor, G.G. Park, and S.R. Radosevich. 2007. Landscape factors influencing the abundance and dominance of the invasive plant *Potentilla recta. Rangeland Ecology and Management.* 60(3): 218-224.

Farnsworth, E.J. and D.R. Ellis, 2001. Is purple loosestrife (*Lythrum salicaria*) an invasive threat to freshwater wetlands? Conflicting evidence from several ecological metrics. *Wetlands* 21(2): 199-209.

Farnsworth, E.J. and L.A. Meyerson. 1999. Species composition and inter-annual dynamics of freshwater tidal plant community following removal of the invasive grass, *Phragmites australis*. *Biological Invasions* 1:115-127.

Farnsworth, E.J. and L.A. Meyerson. 2003. Comparative ecophysiology of four wetland plant species along a continuum of invasiveness. *Wetlands* 23(4): 750-762.

Fridley, J.D., J.J. Stachowicz, S. Naeem, D.F. Sax, E.W. Seabloom, M.D. Smith, T.J. Stohlgren, D. Tilman, and B. Von Holle. 2007. The invasion paradox: Reconciling pattern and process in species invasions. *Ecology* 88(1): 3-17.

Gabor, T.S., T. Haagsma, and H.R. Murkin. 1996. Wetland plant responses to varying degrees of purple loosestrife removal in Southeastern Ontario, Canada. *Wetlands* 16:95-98.

Gassó, N., D. Sol, J. Pino, E.D. Dana, F. Lloret, M. Sanz-Elorza, E. Sobrino, and M. Vilà. 2009. Exploring species attributes and site characteristics to assess plant invasions in Spain. *Diversity and Distributions* 15:50-58.

Gaudet, C.L. and P.A. Keddy. 1989. A comparative approach to predicting competitive ability from plant traits. *Nature* 334(6169): 242-243.

Gordon, D.R. 1998. Effects of invasive, non-indigenous plant species on ecosystem processes: Lessons from Florida. *Ecological Applications* 8(4): 975-989.

Gross, E. 2003. Allelopathy of aquatic autotrophs. *Critical Reviews in Plant Science* 22: 313-339.

Gurevitch, J. and D. Padilla. 2004. Are invasive species a major cause of extinctions? *Trends in Ecology and Evolution.* 19(9): 470-474.

Hager, H.A. 2004. Competitive effect versus competitive response of invasive and native wetland plant species. *Oecologia* 139:140-149.

Hager, H.A. and K.D. McCoy. 1998. The implications of accepting untested hypotheses: a review of the effects of purple loosestrife (*Lythrum salicaria*) in North America. *Biodiversity and Conservation* 7(8): 1069-1079.

Hager, H.A. and R.D. Vinebrook. 2004. Positive relationships between invasive purple loosestrife (*Lythrum salicaria*) and plant species diversity and abundance in Minnesota wetlands. *Canadian Journal of Botany* 82(6): 763-773.

He, F., P. Deng, X.H. Wu, S.P. Cheng, Y.N. Gao, and Z.B. Wu. 2008. Allelopathic effects on *Scenedesmus obliquus* by two submerged macrophytes *Najas minor* and *Potamogeton malaianus*. *Fresenius Environmental Bulletin* 17(1): 92-97.

Hierro, J.L. and R.M. Callaway. 2003. Allelopathy and exotic plant invasion. *Plant and Soil* 256(1): 29-39.

Hilt, S. 2006. Allelopathic inhibition of epiphytes by submerged macrophytes. *Aquatic Botany* 85(3): 252-256.

Houlahan, J.E. and C. S. Findlay. 2004. Effect of invasive plant species on temperate wetland plant diversity. *Conservation Biology* 18(4): 1132-1138.

Hulme, P.E. 2009. Relative role of life-form, land use and climate in recent dynamics of alien plant distributions in the British Isles. *Weed Research* 49: 19-28.

Keller, B.E.M. 2000. Plant diversity in *Lythrum*, *Phragmites*, and *Typha* marshes, Massachusetts, U.S.A. *Wetlands Ecology and Management* 8: 391-401.

Kennedy, T.A., S. Naeem, K.M. Howe, J.M.H. Knops, D. Tilman, and P. Reich. 2002. Biodiversity as a barrier to ecological invasion. *Nature* 417: 636-638.

Kourtev, P.S., J.G. Ehrenfeld, and M. Haggblom. 2003. Experimental analysis of the effect of exotic and native plant species on the structure and function of soil microbial communities. *Soil Biology and Biochemistry* 35(7): 895-905.

Langdon, S.J., R.H. Marrs, C.A. Hosie, H.A., K.M. Norris, and J.A. Potter. 2004. *Crassula helmsii* in UK ponds: Effects on plant biodiversity and implications for newt conservation. *Weed Technology* 18: 1349-1352.

Larsen, D. 2007. Growth of three submerged plants below different densities of *Nymphoides peltata* (S.G. Gmel.) Kuntze. *Aquatic Botany* 86: 280-284.

Lesica, P. and T.H. DeLuca, 2004. Is tamarisk allelopathic? *Plant and Soil* 267(1-2): 357-365.

Levine, J.M. 2000. Species diversity and biological invasions: Relating local process to community pattern. *Science* 288(5467): 852-854.

Lonsdale, W.M. 1999. Global patterns of plant invasions and the concept of invasibility. *Ecology* 80:1522-1536.

Lyons, K.G. and M.W. Schwartz. 2001. Rare species alters ecosystem function – invasion resistance. *Ecology Letters* 4: 358-365.

Mack, R.N., D. Simberloff, W.M. Lonsdale, H. Evans, M. Clout, and F. Bazzazz. 2000. Biotic Invasions: Causes, epidemiology, global consequences and control. *Issues in Ecology Number* 5. 20 pp.

Madsen, J.D., R.M. Stewart, K.D. Getsinger, R.L. Johnson, and R.M. Wersal. 2008. Aquatic plant communities in Waneta Lake and Lamoka Lake, New York. *Northeastern Naturalist* 15(1): 97-110.

Mahaney, W.M., K.A. Smemo, and J.B. Yavitt. 2006. Impacts of *Lythrum salicaria* on plant community and soil properties in two wetlands in central New York, USA. *Canadian Journal of Botany* 84: 477-484.

Mal, T.K., J. Lovett-Doust, and L. Lovett-Doust. 1997. Time-dependent competitive displacement of *Typha angustifolia* by *Lythrum salicaria*. *Oikos* 79:26-33.

Mason, T.J. and K. French. 2008. Impacts of a woody invader vary in different vegetation communities. *Diversity and Distributions* 14: 829-838.

McGlynn, C.A. 2006. *The effects of two invasive plants on native communities in Hudson River freshwater tidal wetlands*. Dissertation. State University of New York, Stony Brook, NY. 218 pp.

McGlynn, C.A. 2009. Native and invasive plant interactions in wetlands and the minimal role of invasiveness. *Biological Invasions*. (In press)

Minchinton, T.E. and M.D. Bertness. 2003. Disturbance-mediated competition and the spread of *Phragmites australis* in a coastal marsh. *Ecological Applications* 13(5): 1400-1416.

Milbau, A., J.C. Stout, B.J. Graae and I. Nijs. 2009. A hierarchical framework for integrating invasibility experiments incorporating different factors and spatial scales. *Biological Invasions* 11: 941-950.

Mony, C., T.J. Koschnick, W.T. Haller, and S. Muller. 2007. Competition between two invasive Hydrocharitaceae (*Hydrilla verticillata* (L.f) (Royle) and *Egeria densa* (Planch) as influenced by sediment fertility and season. *Aquatic Botany* 86: 236-242.

Morgan, E.C. and W.A. Overholt. 2005. Potential allelopathic effects of Brazilian pepper (*Schinus terebinthifolius* Raddi, Anacardiaceae) aqueous extract on germination and growth of selected Florida native plants. *Journal of Torrey Botanical Society* 132(1): 11-15.

Morrison, J.A. 2002. Wetland vegetation before and after experimental purple loosestrife removal. *Wetlands* 22(1): 159-169.

Naeem, S., J.M.H. Knops, D. Tilman, K.M. Howe, T. Kennedy and S. Gale. 2000. Plant diversity increases resistance to invasion in the absence of covarying extrinsic factors. *Oikos* 91: 97-108.

Odum, W. E., T.J. Smith III, J.K. Hoover, and C.C. McIvor. 1984. The ecology of tidal freshwater marshes of the United States East Coast: A community profile. U.*S. Fish and Wildlife Service*, FWS/OBS-83/17.

Pfeifer-Meister, L., E.M. Cole, B.A. Roy, and S.D. Bridgham. 2008. Abiotic constraints on the competitive ability of exotic and native grasses in a Pacific Northwest prairie. *Oecologia* 155(2): 357-366.

Rawinski, T.J., and R.A. Malecki. 1984. Ecological relationships among purple loosestrife, cattail and wildlife at the Montezuma National Wildlife Refuge. *New York Fish and Game Journal* 31:81-87.

Rudrappa, T., J. Bonsall, J.L. Gallagher, D.M. Seliskar, and H.P. Bais. 2007. Root-secreted allelochemical in the noxious weed *Phragmites australis* deploys a reactive oxygen species response and microtubule assembly disruption to execute rhizotoxicity. *Journal of Chemical Ecology* 33(10): 1573-1561.

Sala, A., D. Verdaguer, and M. Vila, 2007. Sensitivity of the invasive geophyte *Oxalis pes-caprae* to nutrient availability and competition. *Annals of Botany* 99(4): 637-645.

Shea, K. and P. Chesson. 2002. Community ecology theory as a framework for biological invasions. *Trends in Ecology and Evolution* 17(4): 17-176.

Silliman, B.R. and M.D. Bertness. 2004. Shoreline development drives invasion of *Phragmies australis* and the loss of plant diversity on New England salt marshes. *Conservation Biology* 18(5):1424-1434.

Simberloff, D. 2003. Confronting introduced species: a form of xenophobia? *Biological Invasions* 5:179-192.

Stachowicz, J.J., R.B. Whitlatch and R.W. Osman. 1999. Species diversity and invasion resistance in a marine ecosystem. *Science* 286: 1577-1579.

Strauss, S.Y., C.O. Webb, and N. Salamin. 2006. Exotic taxa less related to native species are more invasive. *Proceedings of the National Academy of Science* 103(15): 5841-5845.

Tilman, D. 1997. Community invasibility, recruitment limitation and grassland biodiversity. *Ecology* 78(1): 81-92.

Treberg, M.A. and B.C. Husband. 1999. Relationship between the abundance of *Lythrum salicaria* (purple loosestrife) and plant species richness along the Bar River, Canada. *Wetlands.* 19(1): 118-125.

Thelen, G.C., J.M. Vivanco, B. Newingham, W. Good, H.P. Bais, P. Landres, A. Ceasar, and R.M. Callaway. 2005. Insect herbivory stimulates allelopathic exudation by an invasive plant and the suppression of natives. *Ecology Letters* 8(2): 209-217.

Thompson, K., J.G. Hodgson, J.P. Grime, and M.J.W. Burke. 2001. Plant traits and temporal scale: evidence from 5-year invasion experiment using native species. *Journal of Ecology* 89:1054-1060.

Vitousek, P.M., L. R.Walker, L.D. Whittaker, D. Mueller-Dombois, and P.A. Matson. 1987. Biological invasion by *Myrica faya* alters ecosystem development in Hawaii. *Science* 238: 802-804.

Vitousek, P.M. 1992. Effects of alien plants on native ecosystems. In: Stone, C.P.,, Smith, C.W. , and Tunison, J.T., editors. Alien plant invasions in native ecosystems of Hawaii.

University of Hawaii Cooperative National Park Resources Studies Unit, 3190 Maile Way, Honolulu. 29-41.

Waldner, L.S. 2008. The kudzu connection: Exploring the link between land use and invasive species. *Land Use Policy* 25: 399-409.

Wang, J.W., D. Yu, W. Xiong, and Y.Q. Han. 2008. Above- and belowground competition between two submersed macrophytes. *Hydrobiologia* 607:113-122.

Wania, A., I. Kühn, and S. Klotz. 2006. Plant richness patterns in agricultural and urban landscapes in Central Germany-spatial gradients of species richness. 2006. *Landscape and Urban Planning* 75: 97-110.

Weihe, P.E. and R.K. Neely. 1998. The effects of shading on competition between purple loosestrife and broad-leaved cattail. *Aquatic Botany* 59(1-2): 127-138.

Weiher, E., I.C. Wisheu, P.A. Keddy, and D.R.J. Moore. 1996. Establishment, persistence, and management implications of experimental wetland plant communities. *Wetlands.* 16(2): 208-218.

Welling, C.H. and R.L. Becker. 1990. Seed bank dynamics of *Lythrum salicaria* L.: Implications for control of this species in North America. Aquatic Botany 38: 303-309.

Wilcove, D.S., D. Rothstein, J. Dubow, A. Phillips, and E. Losos. 1998. Quantifying threats to imperiled species in the United States. *BioScience* 48(8): 607-615.

Willis, A.J., J. Memmott, and R.I. Forrester. 2000. Is there evidence for the post-invasion evolution of increased size among invasive plant species? *Ecological Letters* 3(4): 275-283.

Windham, L. and Meyerson, L.A. 2003. Effects of common reed (*Phragmites australis*) expansions on nitrogen dynamics of tidal marshes of the Northeastern U.S. *Estuaries* 26(2B): 452-464.

Yakimowski, S.B., H.A. Hager, and C.G. Eckert. 2005. Limits and effects of invasion by the nonindigenous wetland plant *Lythrum salicaria* (purple loosestrife): a seed bank analysis. *Biological Invasions* 7:687-698.

Zedler, J.B. and S. Kercher, 2004. Causes and consequences of invasive plants in wetlands: Opportunities, Opportunists, and Outcomes. *Critical Reviews in Plant Sciences* 23(5): 431-452.

Zhonghua, W., Y. Dan, T. Manghui, W. Qiang and X. Wen. 2007. Interference between two floating-leaved aquatic plants: *Nymphoides peltata* and *Trapa bispinosa*. *Aquatic Botany* 86: 316-320.

In: Encyclopedia of Environmental Research
Editor: Alisa N. Souter

ISBN: 978-1-61761-927-4
© 2011 Nova Science Publishers, Inc.

Chapter 30

MORPHOLOGY AND BIOMASS ALLOCATION OF PERENNIAL EMERGENT PLANTS IN DIFFERENT ENVIRONMENTAL CONDITIONS- A REVIEW

Takashi Asaeda[*], P.I.A. Gomes, H. Rashid and M. Bahar

Department of Environmental Science and Technology, Saitama University,
255 Shimo-okubo, Sakura-ku, Saitama, 338-8570

ABSTRACT

Emergent macrophytes are very important contributors to sustenance of faunal communities and nutrient cycling. They also form prime feature of the landscape of almost every aquatic habitat. Material translocation dynamics between above- and below-ground components describe the life cycling and survival strategies of emergent plants. Trends of material translocation of emergent plants can be investigated by way of field observations and subsequently by the models for organ-specific growth. This paper reviews and describes the morphology and material translocation of emergent plants under a set of different environmental conditions, namely, seasonal, spatial, sediment, water depth and harvesting.

Key words: Biomass; Emergent plants; Environmental variables

INTRODUCTION

Wetland and/or aquatic macrophytes can be distinguished on the basis of morphology and physiology, such as emergent, submerged, floating, and floating leaved (Wetzel 1988, Marshall & Lee 1994, Gibbons et al. 1994, Vymazal 1995, Wetzel 2001, Cronk & Fennessy 2001, USDA 2008). Most members of the emergent group are herbaceous perennials with typically rhizomatous or cormous root systems (Moshiri 1993, Wetzel 2001). The basal

---

[*] E-mail address: asaeda@mail.saitama-u.ac.jp. Tel: +81-48-858-3563, Fax: +81-48-858-9574. (Corresponding author)

portions typically grow beneath the surface of the water, but leaves, stems (photosynthetic parts), and reproductive organs are aerial (Cronk & Fennessy 2001). Most common emergent species are found in the large families of monocotyledons that tend to dominate both freshwater and saltwater marshes: Poaceae (grasses, e.g. *Phragmites australis*), Cyperaceae (sedges, e.g., *Cartex, Cyperus*), Juncaceae (rushes), and Typhaceae (cattail). Other families with frequently encountered emergent species are Alismataceae (water plantain), Araceae (arum), Asteraceae (aster), Lamiaceae (mint, e.g., *Lycopus, Mentha*), Polygonaceae (smartweed), and Sparganiaceae (bur reed).

According to Vymazal (1995) and Wetzel (2001), emergent plants grow in water-saturated or submersed soils, from where the water table is about 0.5m below the soil surface to where the sediment is covered with approximately 1.5m of water. Figure 1 illustrates a typical profile of an aquatic ecosystem. Some of the emergent plants are known to flourish in fen, damp ground (e.g., *Lythrum salicaria, Panicum clandestinum*), and even in areas where there is very little moisture. Furthermore, some species like *Phragmites japonica* colonise relatively disturbed riparian areas, such as in channel water, and also in elevated sandy areas where moisture is scarce (Asaeda et al. 2008b). This observation is somewhat remarkable as many species of the genus *Phragmites* (such as the most common reed, *P. australis*) are known to occupy calm wetland habitats (Kadono 1994, Asaeda et al. 2008b).

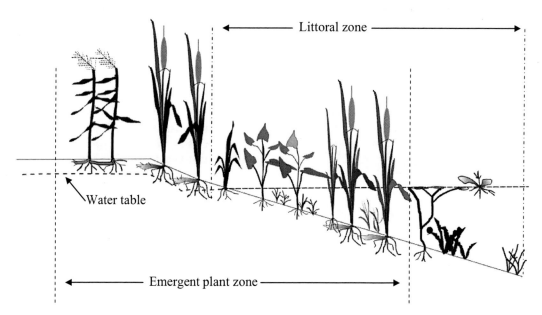

Figure 1. Typical cross sectional profile of an aquatic ecosystem based on information range of sources (see text).

Generally, emergent plants, with their high productivity, outcompete other wetland plants, first and foremost due to their ability to intercept light before it reaches the water's surface. The root and rhizome systems of emergent plants are relatively tolerant in permanently anaerobic sediments, and obtain oxygen from the aerial organs using their ventilation system (Wetzel 2001). Furthermore, the young foliage must respire anaerobically until the aerial habitat is reached, since the oxygen content of the water is extremely low compared to the air (Vymazal 1995, Wetzel 2001).

Emergent macrophytes are very important contributors to the sustenance of faunal communities (Bosschieter & Goedhart 2005) and nutrient cycling. They also form prime features of the landscape of almost every aquatic habitat (Asaeda et al. 2006a). In wet-dry tropics, when water recedes in the dry season, the emergent vegetation of wetlands is a major source of foraging for large grazers (van der Valk 2006). However, as per some literature, the overgrowth of emergent vegetation (especially *Typha* and *Phragmites*) that dominates aquatic habitats is often viewed as a potential threat to shallow ecosystems (Marks et al. 1994, Weinstein & Balletto 1999, Keller 2000, Lavoie et al. 2003, Gratton & Denno 2005, Asaeda et al. 2006b, Maheu-Giroux and de Blois 2007, USDA 2008). Species such as *P. australis* – due to its robust growth, together with morphological characteristics like longer shoots and higher density of stems compared to other species – tend to negatively affect the habitat they occupy (Asaeda et al. 2006b). Further, its elaborate and rapidly expanding rhizome system (Haslam 1969) is unfavorable to other components of an ecosystem in which they dominate (Asaeda et al. 2006b, Hung et al. 2007). Similarly, some emergent species such as *Typha* consider it a nuisance as it can invade and block drainage and supply ditches, causing various problems in rice paddies (Inoue & Tsuchiya 2006, Hai et al. 2006), and without control it will form dense stands that eventually outcompete other valuable wildlife food and cover species (Beule 1979).

Besides the threat to the ecosystem, the usage of emergent plants has been the focus of various engineering applications, ranging from bank protection from boat wash (Boyd & Hess 1970) to wastewater treatment. Emergent plant wetlands have been proposed as sites for wastewater treatment because they appear to assimilate inorganic and organic constituents of wastewater (Kang et al. 2002, Pilon-Smits 2005). Interestingly, recent studies have assessed the removal of heavy metals from sediment using various emergent species such as *Typha latifolia* and *P. australis* (Xu & Jaffe 2006). Considering the research trends in wastewater treatment, both natural and constructed wetlands with emergent species seems to be a rational option.

There are many pioneering studies on emergent plants. Haslam (1972) reviewed the sparse spatial distribution and morphological features of *Phragmites* in areas such as Polynesia, Northern Australia, tropical Asia, North-West and West Africa, with particular focus upon Britain. Vymazal (1995) provided a comprehensive set of data for standing crops, productivity, element concentrations, uptake rates, and standing stocks in aquatic and wetland plants, which also included emergent plants. The review by Barrat-Segretain (1996) targeted strategies of reproduction, dispersion, and competition by aquatic plants (including emergent plants), referring mostly to European rivers. Bart et al. (2006) reviewed the human facilitation of *P. australis* invasions in tidal marshes. Asaeda et al. (2008a) discussed the conceptual structure of several organ-specific growth models and explained their applications, with particular emphasis upon material translocation between above- and below-ground systems and the role of aquatic plants (emergent) in element cycles. Another recent review (Engloner et al. 2008) discussed the anatomy, morphology, and growth of above- and below-ground parts of *P. australis*.

Emergent plants produce organic matter in their above-ground tissues by absorbing carbon from the atmosphere (Westlake 1975, Vymazal, 1995, Wetzel 2001) and simultaneously uptake nutrients and other elements from the water or interstitial water in the sediment. They then allocate the photosynthetic products into both above- and below-ground systems to produce organs and to stock as resources (Asaeda et al. 2008a) during growth periods of individual plant cohorts (Wetzel 2001). When the above-ground organs die, they accumulate in the sediment and are then mineralized in subsequent decomposition processes. This type of element cycle functioning is highly dependent upon primary production, and translocation between above- and below-ground systems.

Compared with submerged macrophytes, which have relatively small root systems, emergent macrophytes have greater efficiency in the element cycle, not only because of their higher productivity, but also due to their distinctive resource translocation between the shoot and rhizome systems, which are particularly large (Wetzel 2001, Asaeda et al. 2006a, Asaeda et al. 2008a). Westlake (1975) stated that underground parts are often two to five times the weight of the aerial parts. Vymazal (1995) also stated that these values can approach as high as 90% (30–90%) of total biomass relative to the 10–40% for submerged macrophytes and 30–70% for floating-leaved macrophytes. This makes the discussion of the biomass allocation of emergent plants of paramount importance. These processes, namely biomass partitioning, are highly spatial and temporal (seasonal) dependent. Furthermore, they are notably species-specific and also depend upon environmental conditions such as the substrate, water depth, and management (e.g., harvesting). Trends of the material translocation of emergent plants can be investigated by way of field observations and subsequently by the models for organ-specific growth. This chapter reviews and describes the morphology and material translocation of emergent plants under a set of different environmental conditions; namely, seasonal, spatial, sediment, water depth, and harvesting.

SEASONAL VARIATIONS

The carbohydrate translocation between above- and below-ground organs is one of the crucial processes in order to understand the life cycle of emergent macrophytes (Whigham 1978, Westlake 1982, Granéli et al. 1992, Čížková et al. 1996, Asaeda & Karunaratne 2000, Asaeda et al. 2006b, 2008c). The phenological cycles of perennial emergent plants are essentially based on the carbohydrate translocation between shoots and rhizomes (Asaeda et al. 2008c). Materials stocked in rhizomes are translocated upward and consumed in the formation of new shoots during early spring (Asaeda et al. 2008c). Under stable and steady-state conditions, the biomass increase in rhizomes should be equal to the carbohydrates translocated to growing shoots in spring and to cover losses due to rhizome mortality and metabolic losses at different rates and times throughout the year (Westlake 1982).

Researchers such as Hocking (1989a, 1989b), Schierup (1978), Dykyjová and Hradecká (1976), and Fiala (1973) investigated the seasonal production dynamics of *P. australis* under various field and experimental conditions, paying particular attention to rhizome dynamics. Studies of *Phragmites* rhizomes are challenging because rhizomes live for several years (longer than that of other emergent plants such as *Typha, Zizania*, etc.) and because rates of respiration depend on rhizome age (Čížková & Bauer 1998). Čížková & Bauer (1998) observed an inverse relationship between respiration and rhizome age, with only exception been for the month May (respiration rates of 3-year old and 4-year old rhizomes surpassed those of 1-year old). Furthermore, starch and sugar accumulation (Fiala 1976, Čížková et al. 1992, 1996, Kubin et al. 1994, Kubin & Melzer 1996) depend on rhizome age (Asaeda et al. 2006b).

The reduction of rhizome biomass involves losses due to metabolism and mortality, as well as translocation to the above-ground system of *Phragmites* (Granéli et al. 1992). Rhizome biomass and rhizome standing stocks of nonstructural carbohydrates and mineral nutrients have been shown to decrease early in the growing season and increase later in the year in *Phragmites* (Dykyjova & Hradecka 1976, Schierup 1978, Linden 1980, 1986, Granéli & Solander 1988, Hocking 1989a). The reduction of rhizome biomass in spring was apparently due to the upward translocation of reserves to form foliage (Haslam 1969, Fiala 1976, Schierup 1978, Westlake 1982, Hocking 1989a). However, the processes differ between the young and old rhizomes. A greater carbohydrate translocation from old rhizomes in spring with a comparatively small storage in autumn of the same species has been reported by Fiala (1976) and Asaeda et al. (2006b). The deployment of reserve materials from rhizomes for the rapid initial shoot growth results in almost complete formation of the above-ground stand during early spring, in which the amounts of upward translocation from rhizomes are determined by the initial growth of shoots. Subsequently, rhizome total non-structural carbohydrate (TNC) and water-soluble carbohydrate (WSC) concentrations also decline in spring (Granéli et al. 1992), but only with old components (Asaeda et al. 2006b). Asaeda et al. (2006b) observed that TNC stocks increased markedly by about 700 g/m^2 (1,938 g/m^2 dry mass) from May to August, which is much larger than the rhizome biomass increment before summer reported by Westlake (1982). As suggested by Fiala (1976) and Asaeda et al. (2006b), from May to August the reserve material accumulates not only in the new rhizomes mostly formed in June, but also in older living rhizomes.

Figure 2 shows the temporal variation in dry mass and TNC reserves in living rhizomes of *P. australis* for each of the seven age-classes (modified from Asaeda et al. 2006b). The study was conducted in a swampy section of Akigase Park (35°51′10′′N, 139°35′48′′E), located on the flood plain of Arakawa River in central Japan in the year 2001. The seven age-classes had similar growth patterns, but differed when translocation occurred. Before summer, the photosynthates were evenly allocated to each age-class, while the recovery of TNC in segments was earlier with younger ones. The allocation in the fall was mainly to young rhizome segments. Older segments became severely depleted during the winter, but the young rhizome segments survived, which can be interpreted as an internal redistribution of resources. Some previous studies (e.g., Granéli et al. 1992) have reported that the downward translocation began when the shoot structure had been established, while others (Fiala 1976, Bjorndahl 1983) reported it at the end of the growing season.

Čížková et al. (1996) also studied the seasonal dynamics of TNC and carbohydrate species (starch, sucrose, glucose) in the rhizome of *P. australis*, comparing the dynamics for an oligotrophic (Branná sands pit) and a eutrophic lake (Rožmberk fishpond, western shore) in the Czech Republic. The levels of non-structural carbohydrates were mainly determined by the levels of starch and sucrose, while glucose and fructose were present at comparatively low levels. The most conspicuous differences between the sites were associated with fall and March levels of carbohydrates. In March, at the beginning of vegetative development, TNC and starch levels were lower at the hyper-eutrophic site, but not at the oligotrophic site. At the low water depth investigated, the difference in carbohydrate levels between stands did not seem to be large enough to account solely for the difference in vigor.

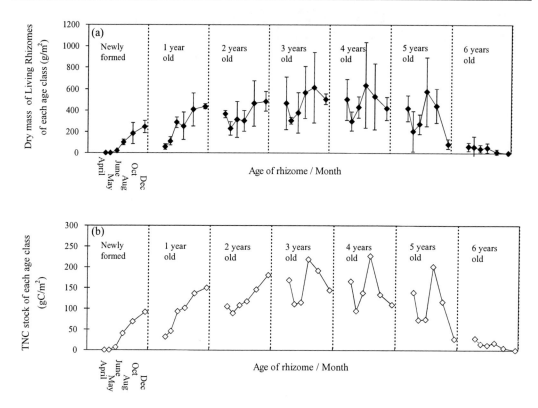

Figure 2. Temporal variation in (a) dry mass (mean ± SD) and (b) Total non structural carbohydrates (TNC) reserves of living rhizomes of each age-class in 2001. (*Note*: The biomass data for each age-class are presented separately.)

The pattern of successive upward and downward translocations is inherently common to perennial emergent plants like *Phragmites* or *Typha*, although the life span of rhizomes is different. Most rhizome segments of *P. australis* survive for five years (Čížková & Bauer 1998, Asaeda et al. 2006b), but their biomass varies seasonally based on in- and outflows of TNC and metabolic loss, then peaks when they are three to four years old (Asaeda et al. 2006b). In contrast, the horizontal rhizomes of *Typha angustifolia* only live for three years, and then their biomass reaches a maximum during the second year, supplying carbohydrates essentially from May to November, while there is almost no downward translocation in the third year. In addition, the largest component of rhizomes is the vertical one, which shares one quarter to one half of the total rhizome biomass; in the case of *P. australis* this fraction is mostly less than one quarter. Therefore, the low rhizome biomass of *T. angustifolia* compared with shoots and low seasonal variation is attributed firstly to the short life-span of rhizomes, and secondly to the large translocation rate to the vertical rhizomes, which are directly connected to shoots, and largely die associated with the mortality of shoots rather than the horizontal ones (Asaeda et al. 2008c). In general, *Typha* spp. shoots start to grow rapidly from April and the biomass accumulates at rates as high as 30 g/m^2/day (Sharma et al. 2006). Accompanying the initial growth of the above-ground parts, the rhizome biomass declined substantially. This reduction in the below-ground biomass is likely to be due to upward translocation in order to support the high growth rates of new shoots (Jervis 1969, Gustafson 1976, Smith et al. 1988).

Like *P. australis*, *Zizania latifolia* is a rhizomatous perennial plant that depends on its rhizome system for the survival and expansion of colonies (Chapin et al. 1990). However, its relatively low root:shoot biomass ratio compared with *P. australis* or *Typha* spp. (Tsuchiya et al. 1993, Asaeda & Siong, 2008) suggests that the plant has more efficient resource translocation and less dependence upon rhizome resources. *Z. latifolia* normally occurs under similar conditions to *P. australis*. Asaeda & Siong (2008) reported the intensive self-thinning of new shoots in spring, although the number of new shoots is large. In their study, they observed the intensive self-thinning of new shoots that emerged in spring, and then the dead shoot biomass increased gradually from the end of May to the end of July, although the shoot biomass peaked at the end of August before markedly declining due to the senescence of grown shoots from October to December.

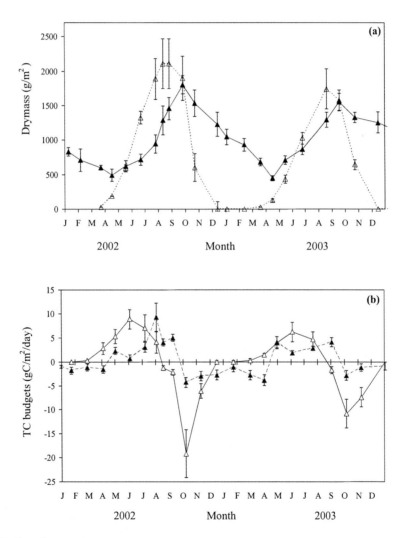

Figure 3. Continued on next page.

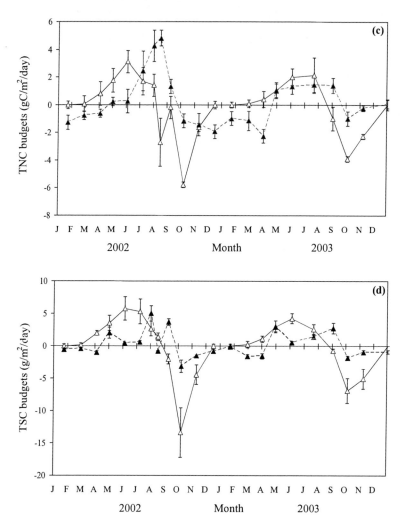

Figure 3. (a) Temporal variation of biomass for the years 2002 and 2003; (b) total carbon (TC) budgets; (c) Daily synthesis rate of total non-structural carbohydrates (TNC); (d) daily synthesis rate of total structural carbohydrate (TSC) in the above-ground (solid line with hollow symbols) and below-ground (discontinued line with filled symbols) biomass.

Figure 3(a) shows the changing pattern in the dry weight of above- and below-ground *Z. latifolia* biomass during 2002 and 2003 (modified from Asaeda & Siong, 2008). The study site was a freshwater marsh 40 km North of Tokyo ($35°59'00''$N, $139°40'53''$E). The shoot biomass increased from the end of March and continued until the end of August. The total below-ground biomass declined from the end of March to April and then substantially increased and peaked at the end of September, which was a month after the above-ground biomass peak in August. Figure 3(b)–(d) provides the component budget of total carbohydrates (TC) and the daily synthesis rate of TNC and total soluble carbohydrates (TSC) in the above- and below-ground biomass, with similar patterns in both 2002 and 2003.

SPATIAL VARIATIONS

Spatial distribution is a distribution or set of geographic observations representing the values of behavior of a particular phenomenon or characteristic across many locations on the surface of the earth. The spatial distribution of plants reflects the spatial distribution of the conditions upon which their growth depends (Goodall 1970, Gomes & Asaeda 2009). The term can also refer to the variation in characteristics over a defined area, ranging from a few square meters to several kilometers. However, in general, the area is often so large that it needs to be accounted for in latitudes, to which we will refer hereafter.

Some of the emergent species, such as *Typha* and *Phragmites*, grow in cold, temperate, and tropical climates, with their productive and morphological characteristics varying accordingly. Along the latitudinal gradient, plant species have to cope with gradually changing climatic conditions: (i) the amount of solar radiation; (ii) the contrast between the seasons (e.g., the time lag between spring in the northern and southern hemispheres); and (iii) relative day length (Clevering et al. 2001, Karunaratne et al. 2003). Further latitudinal differences can result in different photosynthetic rates (Knapp & Yavitt 1995, Karunaratne et al. 2003) due to differential exposure to light and temperatures. In high latitudes, the mortality of young shoots by late frost is also a high risk (van der Toorn 1972). Furthermore, rhizome mortality is also high (McNaughton 1975), making asexual fecundity a risk. In fact, rhizome fecundity is considered to be the main mechanism of maintaining the local population of emergent species such as *Typha* (McNaughton 1966, 1975). Further, sparse latitudinal distribution also leads to significant differences in respiration and mortality losses. Latitudinal differences in growth characteristics are specifically shown in the biomass partitioning to shoots and roots, although this is largely unknown due to difficulties in obtaining simultaneously coordinated observations (McNaughton 1975, Asaeda et al. 2005). Thus, much latitudinal-related research is based on comparative studies and/or mathematical modeling.

Considering the vast number of studies available in the literature for species like *Typha* and *Phragmites* (other than a few, many of these were carried out independent of each other), McNaughton (1975, 1966) and Asaeda et al. (2005) will be useful when discussing morphological and biomass-portioning responses against latitudinal gradient.

McNaughton (1966) studied ecotype function in *Typha*, and in 1975 reported the growth responses (r- and K-selection) of the same over a latitudinal gradient in the USA (high latitude: North Dakota; low latitude: Texas). McNaughton (1966, 1975) discussed the latitudinal difference based on growing season durations, such as short-term growing season genotypes (high latitudes) and long-term growing season genotypes (low latitudes). The results indicated that the number of rhizomes produced was considerably high in high-latitudinal populations of all *Typha* species (*T. latifolia*, *T. angustifolia* and *T. domingensis*). However, no significant difference (ANOVA $P > 0.05$) was observed in total rhizome biomass. McNaughton (1975) concluded that high-latitudinal sites are subject to r-selection, whereas low-latitudinal sites are more subject to K-selection. Based on his extensive studies (1966, 1968, 1975), McNaughton concluded that the physical and botanical components of the *Typha* ecosystem-type vary spatially synchronously in order to maintain the efficient integration of the two components.

Figure 4 illustrates (modified from Asaeda et al. 2005) the latitudinal distribution of the above- and below-ground biomass allocation of *Typha* obtained from a range of sources: Penfound (1956), McNaughton (1966), Jervis (1969), Boyd and Hess (1970), Gustafson (1976), Klopatek and Stearns (1978), Grace and Wetzel (1981), Lieffers (1983), Ulrich and Burton (1985, 1988), Dickerman and Wetzel (1985), Reddy and Portier (1987), Smith et al. (1988), Garver et al. (1988), and Vymazal (1995). Further, Figure 4 shows the modeled biomass variation for two distinct nutrient availability (K_{NP}) constants (0.7 and 1.0). When $K_{NP} = 1.0$, this suggests that nutrients are sufficient, and as it decreases, otherwise (Cary & Weerts 1984). According to Figure 4, above-ground biomass was relatively constant up to the latitude 35°. Below-ground biomass increased up to 45°, and from there onwards decreased.

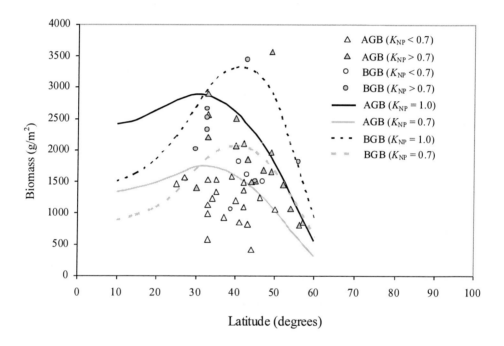

Figure 4. Maximum above-ground biomass (AGB) and below-ground biomass (BGB) for *T. latifolia* modeled (smooth lines) for $K_{NP} = 0.7$ and 1.0 at latitudes 60° to 10°, with observed data (symbols).

Asaeda et al. (2005) discussed the biomass allocation of *Typha* in detail and developed a dynamic growth model to examine the effects of latitudinal changes in temperature and radiation. Asaeda et al. (2005) and Hai et al. (2006) stated that regardless of initial rhizome biomass, both above- and below-ground biomass values conveyed a latitude-specific equilibrium produced by the balance between total production, respiration, and mortality. Further, Asaeda et al. (2005) showed that despite the high mortality, the above-ground biomass of *Typha* was high as a result of latitudes 10–30° receiving sufficient radiation. At higher latitudes, however, a sharp reduction in above-ground biomass was noticed due to low photosynthetic rates. Below-ground biomass increased with latitude up to 40° due to decreasing metabolic losses, and decreased in higher latitudes. Irrespective of latitudinal gradients, above-ground biomass was enhanced with an increasing number of cohorts. Although more cohorts resulted in a larger below-ground biomass at low latitudes, a smaller number of cohorts provide a larger below-ground biomass at high latitudes. This difference is due to low production rates of late-season cohorts in high latitudes, in spite of the higher respiration loss by a large amount of rhizomes.

Phragmites australis can be regard as one of important species for a study in the context of geographic variation, due to its wide distribution. It considers not only most widely (can observe in all continents except Antarctica) distributed emergent plant, but might also the flowering species (Tucker 1990, Asaeda et al. 2003). Clevering et al. (2001) studied the geographical variation in growth responses of *P. australis*. Plants collected from different latitudinal regions in the northern hemisphere (66° 14′N to 40° 40′N) grew in common environments situated in Denmark (56° 13′N), the Netherlands (51° 58′N), and the Czech Republic (49° 00′N). According to their results, *P. australis* originating from higher latitudes showed longer growth rates and flowered earlier than those from lower latitudes. However, the higher allocation of biomass to below-ground parts at higher latitudes observed in many past studies (Chapin & Chapin 1981, Kudo 1995, Li et al. 1998) was not observed by Clevering et al. (2001). Clevering et al. (2001) observed that with increasing latitudes, the allocation of leaves increased at the expense of stems. However, the low latitudes produced larger leaves in relation to the production of taller stems.

According to Karunaratne et al. (2003), overall crop growth rate, net aerial primary production, net above- and below-ground production, net photosynthesis, remobilization of stored rhizome reserves for spring shoot growth, and basipetal translocation to recharge rhizomes of *P. australis* were higher with increasing latitudes. Further, they observed shoot growth and panicle formation starting earlier in northern latitudes (European continent) and later in the southern continent (Australian continent) than in Japan (Asian continent).

Apart from *Typha* and *Phragmites*, Chapin et al. (1981) investigated spatial variation based on plant size, growth rate, reproduction, and survival of *Carex aquatilis* along latitudinal gradients based on simultaneously conducted observations. Further, Chapin et al. (1981) noticed the highest root:shoot ratio of *C. aquatilis* in arctic (71° 18′ N) populations (the lowest ratio was observed in a subalpine population at 40° 00′ N), and suggested low photosynthesis rates as the reason due to low temperature and nutrient absorption by cold sterile soils. In a previous study, Chapin (1974) discussed the phosphate absorption in latitudinal gradients and showed high latitudinal areas with low temperatures resulting in high root:shoot ratios, as well as relatively small root diameters. Smaller root diameters in northern habitats (high latitudes) are important in maximizing surface area in order to overcome the slow absorption of nutrients and water. Furthermore, Chapin (1974) showed that emergent marsh plants evolving in colder climates (high latitudes) differed from their warm (low latitude) adapted counterparts in having lower temperature optima for root initiation, elongation, production, and the large surface-to-volume ratios of the roots.

In general, the spatial-induced biomass partitioning of emergent plants seemed to follow the general patterns of other plants, especially with respect to large biomass allocation to below-ground parts with relatively less above-ground allocation.

EFFECTS OF SEDIMENT

Sediments are an intrinsic part of the ecosystem, and nutrients therein are in equilibrium between the input and binding capacity of the system. The position of the equilibrium controls whether input or output dominates, which may vary over the course of time, depending on the interaction between various factors, some of which are easily altered by anthropogenic causes (Golterman 1995, 2004).

Barko et al. (1991) reported that sediments are the primary source of nitrogen (N), phosphorus (P), iron, manganese, and micronutrients, with the remainder supplied by the surrounding water. However, since emergent plants grow with their roots in the sediments and most of their shoots above water, they obtain both macro- and micro-nutrients from the sediments. Sutton and Dingler (2000) found that emergent *Hygrophila* cultured with sand and fertilizer produced as much as 20 times the biomass as plants cultured with sediments collected from canals in Broward County, Florida. They concluded that the sediments lacked either the nutrient quality or quantity required for optimum growth. Their findings showed the close relationship between the emergent growth habit of *Hygrophila* and sediment nutrients. Sediment nutrient status can affect the respiration and carbohydrate dynamics of emergent species through changes in storage and the reutilization of resources (Čížková & Bauer 1998). Many emergent species like *Typha, Zazania*, and *Phragmites* function as nutrient pumps, absorbing a large amount of nutrients from the soil and accumulating them in the above-ground part as biomass (Klopatek 1970, Sharma et al. 2006).

Smart and Barko (1978) investigated the growth of four emergent salt marsh macrophytes (*Spartina alterniflora, S. foliosa, S. patens, Distichilis spicata*) on three different kinds of sediment: sand, silty clay, and clay. They found that measured nutrient levels were lowest in sand. N levels increased from 0.3 g/kg in sand to 5.2 g/kg in silty clay, and P levels increased from <0.05 g/kg in sand to 1.65 g/kg in silty clay. They grew these plants in a simulated tidal regime, and found that growth rates were an order of magnitude higher on clay than on sand, with intermediate values on silty clay. Smart and Barko (1978) concluded that the growth was limited by nutrients in sandy substrates. Some researchers have suggested N as a limiting nutrient for emergent macrophytes (Koerselman & Meuleman 1996, Asaeda et al. 2008b), while a few studies have also reported that plants growing in coastal sand dune areas were limited by N rather than P (Kachi & Hirose 1983, Koerselman 1992). However, besides the fact that a specific nutrient element is limiting for a particular group of emergent species, an entire ecosystem can be limited to a specific nutrient. Temperate wetland ecosystems are N limited (Vitousek & Howarth 1991, Bridgham et al. 1995, Bedford et al. 1999, Rejmánková 2001), while tropical and arctic ecosystems are limited to P (Feller et al. 1999, Richardson et al. 1999, Shaver et al. 1998, Vitousek 1984).

Asaeda et al. (2008b) studied the adaptation and biomass allocation of *P. japonica* on sandbars in a river channel, characterized by low N concentration in the soil. The results showed that the total biomass and the ratio of above- to below-ground biomasses were larger at the sandy sites than the stony sites. However, the fraction of root biomass (i.e., root: rhizome ratio) in the below-ground biomass at the stony sites was higher than that at the sandy sites. The higher above-ground and total plant biomass in the sandy areas was because the plants had better access to water and nutrients than plants growing in stony beds. A higher proportion of root biomass, at the expense of rhizome biomass, in the below-ground biomass of the plants growing in stony sites was due to the difficulty in accessing water and inorganic nutrients. The ratios of N to P in the plant tissues were mostly between 5 and 10, indicating that plant growth was N limited, subjected to low ratios of total N (TN) to total P (TP) in soil samples, where TN was 300–700 mg/kg and TP was 300–350 mg/kg . A wide phenotypic plasticity observed in *P. japonica* was an important factor in maximizing plant survival and fitness in frequently disturbed habitats, and this plasticity was the result of both true adjustment and ontogenetic drift.

The responses of *P. australis* to sediment fertility and nutrient availability have been studied by a number of scientists. Different vegetative parameters like height, weight, number, and diameter of reed shoots have been investigated by many researchers, and all these parameters were reported to increase with higher sediment nutrients (Gorham & Pearsall 1956, Ho 1979, Bornkamm & Raghi-Atri 1986, Young et al. 1991, Kubin et al. 1994, Saltonstall & Stevenson 2007). Ksenofontova (1988) found that basal stem diameter does not depend upon the nutrient status of the site, contrary to stem length. Ksenofontova (1989) reported an increase in thickness and length, but also the weakness of the shoots, with rising nutrient level. Bastlova et al. (2004) revealed that shoot numbers per plant, that of nodes and living leaves of shoots, correlates positively with nutrient supply. Nutrient availability affects not only plant biomass, but also the architecture of clonal plants (Piqueras et al. 1999), by variation in spacer length (Cain 1994), and by affecting frequency and angle of branching (Dong and de Kroon 1994).

Biomass, density, shoot length, and diameter were reported to be higher from constructed wetlands flooded with sewage sludge than from natural sites (Hardej & Ozimek 2002), and from reed stands subjected to a high level of nutrient treatments (Hung et al. 2007). Positive relationships were also detected between the nutrient supply (or N alone) and relative growth rate (RGR; Clevering 1999, Romero et al. 1999, Morris & Ganf 2001, Saltonstall & Stevenson 2007), stalk and rhizome growth (Votrubová & Pechacková 1996), above- and below-ground biomass (Bastlová et al. 2004, Saltonstall & Stevenson 2007), and the ratio of above-ground to below-ground biomass of *P. australis* (Allen & Pearsall 1963, Boar et al. 1989, Granéli 1990, Weisner 1996, Rickey & Anderson 2004, Saltonstall & Stevenson 2007). In contrast, the addition of nutrients to fast-growing species frequently results in a decline in the root weight ratio (RWR; Chapin III 1980, Chapin III et al. 1987).

Barko and Smart (1983) suggested that (semi-)emergent macrophytes are less susceptible than submerged species to the growth-reducing effects of added labile organic matter. Although many studies have been performed with a variety of organic matter types, in the belts of emergent macrophytes, the standing vegetation itself is the primary source of organic matter (Ostendorp 1992, van der Putten et al. 1997). The negative effects of litter on growth have been reported for a number of emergent macrophytes, such as *T. latifolia* (McNaughton 1968), *Scirpus maritimus* L. (Cleving & van der Putten 1995), *P. australis* (Cav.) Steudel (van der Putten 1993, Clevering 1998), *Z. latifolia* (Lan et al., 2005.), and *Eleocharis sphacelata* (Rajapakse et al. 2006). Growth reduction occurs in a number of emergent macrophytes when cultured on peat, or on soils otherwise enriched with organic matter (Spencer 1991, van der Putten 1993, Armstrong et al. 1996b, Armstrong et al. 1996a). Spence (1964) studied seedling growth of *P. communis* on two types of soil using three types of water (silty sand and peat; distilled water, hard water and intermediate) and found that seedlings grew best with the highest nutrient status, and seedlings on peat, without additional nutrients from the water, were stunted. Haslam (1972) also reported that germination of *P. communis* is very poor on organic soils. She further concluded that stunting of seedlings on peat was due to phosphate deficiency.

In their study of *E. sphacelata*, Rajapakse et al. (2006) found that in a deep water population, the accumulation of excessive amounts of autogenous shoot litter coupled with slower decomposition rates under anaerobic conditions led to higher nutrient enrichment in sediments and the overlying water column, causing subsequent eutrophication with signs of growth inhibition, including typical stress symptoms like stunted growth and chlorotic shoots. However, after experiencing an alternative inundation-drawdown pattern, the shallow water population showed continuation of seasonal growth. Clevering (1998) also found a similar response of *P. australis* in relation to litter and water depth in the habitat. She concluded that the morphological responses of *P. australis* to litter strongly constrain its ability to maintain itself in deep water when the substrate contains litter.

Sedimentation can affect virtually all wetland inhabitants, but particular emphasis has been placed upon questions regarding the response of emergent vegetation to sediment deposition. Sedimentation markedly affects emergent plants by inhibiting germination from soil seed banks (Neely & Wiler 1993, Wang et al. 1994), suppressing plant productivity (Van der Valk et al. 1983, Ewing 1996), decreasing plant community richness and diversity, and perhaps influencing plant succession (Jurik et al. 1994).

EFFECTS OF WATER DEPTH

Depending upon water availability and the duration of inundation, water level results in several stresses that either affect root and/or rhizome function – for example, respiration, physical support provided for shoots, reserve storage, nutrient uptake – or affect above-ground biomass and photosynthetic activity (Grace 1989, Blanch et al. 1999). Cosmopolitan species of emergent macrophytes such as *Phragmites* and *Typha* are well adapted to water logging because of the physiological tolerance of their rhizomes to anoxia (Hellings & Gallagher 1992) and their aeration capabilities (Brix et al. 1992). Despite this, underwater growth of these plants seems to be reduced (Hellings & Gallagher 1992, Weisner & Strand 1996).

Water depth has been used as a tool to control the propagation of emergent species, especially in constructed wetlands. Hummocks are shallow emergent plant beds within the wetland, positioned perpendicular to the water flow path and surrounded by water sufficiently deep to limit further emergent vegetation expansion (Thullen et al. 2005, Álvarez & Bécares 2008). The seed germination of wetland plants depends upon water depth and can only occur when shallow water prevails (Haslam 1972).

Water depth affects the biomass allocation of emergent plants. Many species, including *T. latifolia*, increase in height with increasing depth by increasing biomass allocation to leaves, whereas some species, for example *T. domingensis*, accomplish this by changing ramet size and number. Thus, for the former, the root:shoot ratio decreases with increasing depth (Grace 1989); there is not such a clear trend for the latter. Leaf height usually increases with increased water depth in emergent macrophytes (Coops & van der Velde 1996), which can obviously be adapted to maintain a large proportion of the leaves above the water surface. Thus, biomass allocation patterns reflect the influence of environmental constraints upon emergent macrophytes (Rajapakse et al. 2006).

Among *Typha* species, *T. angustifolia* is more tolerant of deep water than *T. latifolia* (McDonald 1955, Grace & Wetzel 1981, Shay & Shay 1986, Djebrouni & Huon 1988, Weisner 1991), probably because shoot elongation through the water column is facilitated by the tall, narrow leaves and large energy stores in the rhizomes (Grace & Wetzel 1982). Several previous studies have emphasized the functional responses of *E. sphacelata* in proportionate biomass allocation to above-ground organs at the cost of diminishing rhizome structure in deep water (Sorrell et al. 1997, Sorrell & Tanner 2000). Asaeda and Rajapakse (2008) reported that the above-ground biomass of *P. japonica* declined in the sandbar of a flow-regulated river. The probability of a reed stand being seriously damaged depends significantly on its elevation relative to the mean water level (Ostendrop et al. 2003). Flooding inflicts mechanical damage, causing loss of culms, then reduces convective gas flow to the basal parts of the reed plants, subsequently negatively affecting the vitality of stands in the long term (Armstrong et al. 1996c).

The morphological characteristics of *E. sphacelata* were compared between two populations, one in shallow water in an inland, harsh climate, and a deep pond at a warm site in Australia (Asaeda et al. 2006a). Shoot density was greater in shallow water, but with shorter shoot length and a lower maximum above-ground biomass density than for plant stands in deep water. However, the fraction of rhizome biomass was much larger in shallow water. These observations, which revealed distinct patterns of shoot production and demography in shallow water, are in accordance with the previous observations of Sorrell et al. (2002) and Edwards et al. (2003) of *E. sphacelata* and *E. cellulosa*. The lignin content of emergent plants increases with the increasing depth and age of the plant (Ardler 1977). Martin (2008) investigated the shoot morphology and lignifications of *T. angustifolia* at different water depths and observed that the greater the above-ground biomass allocation, the higher the lignin deposits in the lower stem segments. More lignin deposits confer greater mechanical strength upon the stem segments (Boudet 1999).

Aquatic macrophytes survive in flooded soils by virtue of their natural gas transport and the ability of their rhizomes to survive prolonged hypoxia and anoxia (Armstrong et al. 1991, Crawford 1992), but these strategies do not always provide the unlimited tolerance of highly reducing sediments. Relative to many genera of wetland plants, *Typha* species have well-developed internal gas transport pathways with low internal resistances and effective gas transport physiology (Brix et al. 1992, Bendix et al. 1994), and they tolerate deep water and reducing sediments well (Grace 1988, Callaway & King 1996). Jespersen et al. (1998) examined the growth of *T. latifolia* L. and its effects upon sediment methanogenesis in natural organic sediment and sediment enriched with acetate in the interstitial water. They found that the lower redox potential and higher oxygen demand of the acetate-enriched sediment did not significantly impede the growth of *T. latifolia*, despite some differences in the growth pattern and root morphology. Short, thick roots are favored when plants are under oxygen-depletion stress because they have low axial resistances to oxygen diffusion, whereas lateral roots are more difficult to support, having limited oxygen transport capacities due to their narrow diameter and low porosity (Armstrong et al. 1990, Sorrell & Armstrong 1994). In the littoral gradient of lakes, *Z. latifolia*, possessing a highly efficient ventilation system, occupies a deeper zone than *P. australis*, although its tolerance to reduced soil is less than *P. australis* (Yamasaki 1984). However, in spite of their inferior morphology (Szczepanska & Szczepanski 1976) to *P. australis* in terms of the reception of light, stable *Z. latifolia* colonies are also found in very shallow water, which is otherwise favorable for *P. australis*.

EFFECTS OF HARVESTING

Mechanical harvesting has been an attractive approach in the management of aquatic macrophytes for a variety of ecological and practical reasons, and has attracted much attention (Grinwald 1968, Engel 1990). The rationale behind cutting as a potential control mechanism stems from the fact that it will retard the subsequent growth and development of the stand (Marks et al. 1994, Hai et al. 2006), because previously produced reserves are removed by cutting the aerial part of the plant, thus reducing its vigor. Although nutrient uptake and growth rates are higher in young vegetation stands (Batty & Younger 2004), other factors such as nutrient loading and hydraulic retention time (HRT) may significantly affect the measured rates (Hardej & Ozimek 2002, Solano et al. 2004).

Studies of plant growth and biomass after harvesting have shown that for some species (e.g., *Schoenoplectus*), harvesting retards plant growth, whereas other species have shown invigorated growth (e.g., *Eleocharis*; Greenway & Woolley 2001, Kim & Geary 2000). Greenway and Woolley (2001) reported that cropping appeared to invigorate plant growth in *Eleocharis*, which took only two months to achieve the initial biomass standing stock (biomass prior to cropping), whereas cropping *Schoenoplectus* appeared to retard growth, taking 10 months to achieve the initial biomass standing stock.

The effects of management (mowing, grazing, or burning) on the performance of *Phragmites* and its ability to suppress other plant species is known to depend upon the type and timing of measures. Vymazal et al. (1999) reported that the harvesting of *P. australis* during periods of peak biomass can inflict considerable damage upon plants. Mowing or burning in winter generally does not negatively affect stands of *Phragmites* as nutrients are commonly translocated to below-ground biomass in winter (Breen 1990). Harvesting in winter may even contribute to their conservation (Granéli 1990, Buttler 1992, Güsewell et al. 2000). By contrast, mowing in summer or fall can reduce *Phragmites*, particularly if the measure is associated with flooding (Weisner & Granéli 1989, Hellings & Gallagher, 1992). IWA (2000) suggested that the ideal timing for cropping *Phragmites* is at the start of spring, although cropping in either spring or winter will not maximize the removal of nutrient storage (Greenway 2002). At nutrient-rich sites, mowing twice during the summer may be necessary for a substantial reduction of *Phragmites* (Gryseels 1989, Schütz & Ochse 1997). Grazing by cattle or horses can cause a drastic reduction in the abundance of *Phragmites* within only two or three years (van Deursen & Drost 1990, Rozé 1993).

The cutting of reeds in late summer as a means of controlling reed encroachment was investigated by Russell and Kraaij (2008) under three different inundation regimes, termed 'wet zone' (permanently inundated), 'moist zone' (infrequently inundated), and 'dry zone' (rarely inundated). The effects of a single annual cut were also compared to those of two successive annual cuts. Without cutting, wet zones had thinner and shorter but more abundant reeds than drier zones. Cutting in dry and moist zones resulted, after one year, in more but shorter and thinner reeds, whereas in wet zones reeds were almost eliminated. After two years, reeds in wet zones had not recovered from the first annual cut. In moist and dry zones, a second annual cut did not result in amplified detrimental effects upon reeds. Throughout the experiment, the moisture zone was the factor with the largest effect, cutting had the second largest impact, and inter-annual variation was relatively unimportant.

Mowing experiments were carried out by Güsewell (2003) from 1995 to 2001 in Swiss fen meadows in order to investigate whether the abundance of *P. australis* was reduced by mowing in early summer in addition to mowing in the fall. The results suggested that a depletion of below-ground stores caused *Phragmites* to decrease after several years of additional mowing in June. The above-ground biomass of *Phragmites* increased during this time in 49 out of 80 plots, with a mean relative difference of +35.5%. Thus, even if additional mowing in early summer only slightly reduced the performance of *Phragmites* compared to plots mown only in September, this treatment might help to prevent the species from spreading under the current conditions in Swiss fen meadows.

Asaeda et al. (2006c) studied the effect of summer harvesting of *P. australis* on growth characteristics and rhizome resource storage. They monitored the seasonal changes in rhizome biomass and TNC in seven age categories, from newly formed to six years old, for two treatment stands (June or July cut) and a control stand. The growth of the stands, as indicated by the above-ground biomass, showed a significant decline due to cutting in June, but did not show any significant difference from cutting in July compared to that of the control stand. The timing of harvesting of above-ground biomass affected the annual rhizome resource allocation. A similar trend was observed for the pattern of resource allocation, as described by the biomass variation of different rhizome-age categories for the July-cut and control stands. However, the biomass of June-harvested rhizome categories tended to be smaller than the other two stands, indicating substantially reduced resource storage as a direct result of harvesting the above-ground biomass during the previous growing season. This implies that the cutting of above-ground biomass in June is a better option for the control of *P. australis* stands than cutting later in summer.

The effect of harvesting depends upon the latitudinal gradients and geographical conditions. Harvesting is a recommended practice in wetland systems treating diluted wastewater, especially in productive areas like the Mediterranean (Álvarez & Bécares 2008). Álvarez and Bécares (2008) suggested that it could be applicable to a broader geographical area. However, Healy et al. (2007) suggested that under Irish climatic conditions, harvesting does not have any beneficial effect upon biomass production for a study based on *P. australis*. Furthermore, they stated that June cutting resulted in significantly lower values for below-ground biomass production, shoot N and P. In a study by Jinadasa et al. (2008), harvesting was carried out whenever shoot heights reached the maximum (of *Scirpus grossus* and *T. angustifolia*), which suggested the inability of *T. angustifolia* to sustain repeated harvesting (four times) as above-ground biomass production fell.

Karunaratne et al. (2004) investigated the effects of harvesting *P. australis* in a wetland in central Japan and found that biomass levels in an uncut section rose to a maximum of 1,250 g/m^2 in July, whereas June-cut and July-cut sections rose to levels of approximately 400 g/m^2 in October and November. The maximum rate of dry matter production (RDP = the mean rate of dry matter production per unit ground area per day between sampling, $g/m^2/d$) was approximately the same in the uncut and July-cut stands, attaining a rate of 18 $g/m^2/d$ and 12 $g/m^2/d$, respectively. In a laboratory study in Australia, Kim and Geary (2001) found that there was no statistical difference in TP uptake between harvested and unharvested *Schoenoplectus mucronatus* shoots. The effect of harvesting on the succeeding biomass for *Typha* spp. has been estimated using numerical simulations in some studies (Tanaka et al. 2004, Hai et al. 2006).

Biomass harvesting is often considered to remove nutrients (Reed et al. 1995, Kim & Geary 2001). Reed et al. (1995) found a direct correlation between the frequency of harvesting and P removal. It has been suggested by Richardson and Craft (1993) that wetlands can fully assimilate P at low loadings of up to $1g/m^2/yr$. Greenway and Woolley (2001) reported up to 65% and 47% P and N removal respectively by plant tissue for a low loaded surface flow system. When loading rates are high, the amount of P and N taken up by plants accounts for only a small proportion of nutrient removal (Mann & Bavor 1993, Brix 1994, Geary & Moore 1999, Tanner et al. 1999, Tanner 2000, Tanner & Kadlec 2002). This is due to plants only being able to remove a finite amount of the nutrients needed to sustain growth. As nutrient uptake by plants is often not considered to be a significant removal mechanism, biomass in a highly loaded system is often not cropped (Vymazal et al. 1998).

Kim and Geary (2000) reported that biomass removed by harvesting only accounted for less than 5% of TP removal from the system, whereas Mann (1990) reported 92% N removal for a small-scale (bucket-size) cropping experiment. Asaeda et al. (2002) developed a model of *P. australis* growth and decomposition in order to evaluate the material budget and nutrient cycles of a reed stand in Neusiedlersee, Austria. The model predicted that between 33% (six months aerated) and 48% (24 months aerated) of the annual above-ground production would decompose within one year, while the rest would remain in the anaerobic substrate because of the low decomposition rate of the species, especially with rhizomes (Asaeda & Nam 2002). It follows that nearly half of the nutrients can be removed from water even without harvesting; however, the surface sediment level is gradually elevated. Yet, research is scant about the impact of biomass harvesting on nutrient removal.

CONCLUSIONS

The significance of the environmental conditions discussed above may vary depending on the species, its habitat and season, making some findings more or less conflicting. This elucidates the need for more coordinated studies, as some of the studies for the same species have used rather different conditions (or with only a slight overlapping of the ranges used), making comparison somewhat flawed. In that case it is recommended that the effects are analyzed based on models that allow the use of different conditions as their inputs. If no such models are available it is recommended that such models are developed, as weighty approaches like field observations and/or pilot investigations may not be a rational option. The results highlighted in this chapter clearly elucidate the prominence given to some emergent species. *Phragmites* and *Typha* are undoubtedly the emergent species most frequently discussed. This is probably due to their sparse distribution worldwide. We consider this to be an area that needs further consideration in future research.

REFERENCES

Álvarez, J. A. & Bécares. E. (2008). The effect of vegetation harvest on the operation of a surface flow constructed wetland. *Water SA,* 34, 645–649.

Allen, S. E. & Pearsall, W. H. (1963). Leaf analysis and shoot production in *Phragmites. Oikos,* 14, 176–189.

Adler, E. (1977). Lignin chemistry- past, present and future. *Wood Science and Technology,* 11, 169–218.

Armstrong, J. Armstrong, W, Zenbin, Wu & Afreen-Zobayed, F. (1996a). A role for phytotoxins in the *Phragmites* die-back syndrome? *Folia Geobotanica Phytotaxonomica,* 31, 127–142.

Armstrong, J., Armstrong, W. & Van der Putten, W. H. (1996b). *Phragmites* die-back: bud and root death, blockages within the aeration and vascular systems and the possible role of phytotoxins. *New Phytologist,* 133, 399–414.

Armstrong, J., Armstrong, W., Armstrong, I. B. & Pittaway, G. R. (1996c). Senescence, and phytotoxin, insect, fungal and mechanical damage: factors reducing convective gas-flows in *Phragmites australis. Aquatic Botany,* 54, 211–226.

Armstrong, W., Armstrong, J. & Beckett, P. M. (1990). Measurement and modelling of oxygen release from roots of *Phragmites australis.* In: P. F. Cooper, & B. C. Findlater (Eds.), *Constructed Wetlands in Water Pollution Control* (pp. 41–51). Oxford, Pergamon Press.

Armstrong, W., Beckett, P. M., Justin, S. H. F. W. & Lythe, S. (1991). Convective gas-flows in wetland plant aeration. In: M. B. Jackson, D. D. Davies, & H. Lambers (Eds.), *Plant Life under Oxygen Deprivation* (pp. 283–302). The Hague, SPB Academic Publishing.

Asaeda T. & Nam L.H. (2002). Effects of rhizome age on the decomposition rate of *Phragmites australis* rhizomes. *Hydrobiologia,* 485, 205-208.

Asaeda, T. & Karunaratne, S. (2000). Dynamic modeling of the growth of P*hragmites asutralis*: model description. *Aquatic Botany,* 67, 301–318.

Asaeda, T. & Rajapakse, L. (2008). Effects of spates of different magnitudes on a *Phragmites japonica* population on a sandbar of a frequently disturbed river. *River Research and Applications,* 24, 1310–1324.

Asaeda, T. & Siong, K. (2008). Dynamics of growth, carbon and nutrient translocation in *Zizania latifolia. Ecological Engineering,* 32, 156–165.

Asaeda, T. Hai, D. N., Manatunge, J., Williams, D. & Roberts, J. (2005). Latitudinal characteristics of below- and above-ground biomass of *Typha*: a modelling approach. *Annals of Botany,* 96, 299–312.

Asaeda, T., Manatunge, J., Fujino, T. & Sovira, D. (2003). Effects of salinity and cutting on the development of *Phragmites australis. Wetlands Ecology and Management,* 11, 127–140.

Asaeda, T., Manatunge, J., Rajapakse, L. & Fujino, T. (2006a). Growth dynamics and biomass allocation of *Eleocharis sphacelata* at different water depths: observations, modeling, and applications. *Landscape and Ecological Engineering,* 2, 31–39.

Asaeda, T., Manatunge, J., Roberts, J. & Hai, D. N. (2006b). Seasonal dynamics of resource translocation between the aboveground organs and age-specific rhizome segments of *Phragmites australis. Environmental and Experimental Botany,* 57, 9–18.

Asaeda, T., Nam, L. H., Hietz, P., Tanaka, N. & Karunaratne, S. (2002). Seasonal fluctuations in live and dead biomass of *Phragmites australis* as described by a growth and decomposition model: implications of duration of aerobic conditions for litter mineralization and sedimentation. *Aquatic Botany,* 73, 223–239.

Asaeda, T., Rajapakse, L. & Fujino, T. (2008a). Applications of organ-specific growth models; modeling of resource translocation and the role of emergent aquatic plants in element cycles. *Ecological Modeling,* 215, 170–179.

Asaeda, T., Rajapakse, L., Manatunge, J. & Sahara, N. (2006c). Effect of summer harvesting of *Phragmites australis* on growth characteristics and rhizome resource storage. *Hydrobiologia,* 553, 327–335.

Asaeda, T., Sharma, P. & Rajapakse, L. (2008c). Seasonal patterns of carbohydrate translocation and synthesis of structural carbon components in *Typha angustifolia. Hydrobiologia,* 607, 87–101.

Asaeda, T., Siong, K., Kawashima, T. & Sakamoto, K. (2008b). Growth of *Phragmites japonica* on a sandbar of regulated river: morphological adaptation of the plant to low water and nutrient availability in the substrate. *River Research and Applications,* (DOI: 10.1002/rra.1191)

Barko, J. W. & Smart, R. M. (1983). Effects of organic matter additions to sediment on the growth of aquatic plants. *Journal of Ecology,* 71, 161–175.

Barko, J. W., Gunnison, D. & Carpenter, S. R. (1991). Sediment interactions with submersed macrophyte growth and community dynamics. *Aquatic Botany,* 41, 41–65.

Barrat-Segretain, M. H. (1996). Strategies of reproduction, dispersion, and competition in river plants: A Review. *Vegetatio,* 123, 13–37.

Bart, D., Burdick, D., Chambers, R. & Hartman, J. M. (2006). Human facilitation of *Phragmites australis* invasions in tidal marshes: a review and synthesis. *Wetlands Ecology and Management,* 14, 53–65.

Bastlová, D., Čížková, H., Bastl, M. & Květ, H. (2004). Growth of *Lythrum salicaria* and *Phragmites australis* plants originating from a wide geographical area: response to nutrient and water supply. *Global Ecology and Biogeography,* 13, 259–271.

Batty, L. C. & Younger, P. L. (2004). Growth of *Phragmites australis* (Cav.) Trin ex. Steudel in mine water treatment wetlands: effects of metal and nutrient uptake. *Environmental Pollution,* 132, 85–93.

Bedford, B. L., Walbridge, M. R. & Aldous, A. (1999). Pattern in nutrient availability and plant diversity of temperate North American wetlands. *Ecology,* 80, 2151–2169.

Bendix, M., Tornbjerg, T. & Brix, H. (1994). Internal gas transport in *Typha latifolia* L. and *Typha angustifolia* L. 1. Humidity-induced pressurization and convective throughflow. *Aquatic Botany,* 49, 75–89.

Beule, J. D. (1979). Control and management of Cattails in Southeastern Wisconsin Wetlands. *Technical Bulletin, Department of Natural Resources, Madison, Wisconsin,* 112, 1–41.

Blanch, S. J., Ganf, G. G. & Walker, K. F. (1999). Growth and resource allocation in response to flooding in the emergent sedge *Bolboschoenus medianus. Aquatic Botany,* 63, 145–160.

Boar, R. R., Crook, C. E. & Moss, B. (1989). Regression of *Phragmites australis* reed swamps and recent changes of water chemistry in the Norfolk Broadland, England. *Aquatic Botany,* 35, 41–55.

Bornkamm, R. & Raghi-Atri, F. (1986). On the effects of different nitrogen and phosphorus concentrations on the development of *Phragmites australis* (Cav.) Trin.ex Steudel. *Archiv für Hydrobiologie,* 105, 423–441.

Bosschieter, L. & Goedhart, P. W. (2005). Gap crossing decisions by reed warblers (*Acrocephalus scirpaceus*) in agricultural landscapes. *Landscape Ecology,* 20, 455–468.

Boudet, M. (1999). Lignins and lignifications: selected issues. *Plant Physiology and Biochemistry,* 38, 81–96.

Boyd, C. E. & Hess. L. W. (1970). Factors influencing shoot production and mineral nutrient levels in *Typha latifolia*. *Ecology, 51,* 296–300.

Breen, P. F. (1990). A Mass Balance Method for Assessing the Potential of Artificial Wetlands for Wastewater Treatment. *Water Research, 24,* 689-697.

Bridgham, S. D, Pastor, J., McClaugherty, C. A. & Richardson, C. J. (1995). Nutrient-use efficiency: a litterfall index, a model, and a test along a nutrient-availability gradient in North Carolina peatlands. *The American Naturalist, 145,* 1–21.

Brix, H. (1994). Functions of Macrophytes in Constructed Wetlands. *Water Science and Technology, 29,* 71-78.

Brix, H., Sorrell, B. K. & Orr, P. T. (1992). Internal pressurization and convective gas flow in some emergent freshwater macrophytes. *Limnology and Oceanography, 37,* 1420–1433.

Bjorndahl, G. (1983). Structure and biomass of Phragmites stands. Ph.D. Thesis. University of Goteborg, Sweden.

Buttler, A. (1992). Permanent plot research in wet meadows and cutting experiment. *Vegetatio, 103,* 113–124.

Cain, M. L. (1994). Consequences of foraging in clonal plant species. *Ecology, 75,* 933–944.

Callaway, R. M. & King, L. (1996). Temperature-driven variation in substrate oxygenation and the balance of competition and facilitation. *Ecology, 77,* 1189–1195.

Cary, P. R. & Weerts, P. G. (1984). Growth and nutrient composition of *Typha orientalis* as affected by water temperature and nitrogen and phosphorus supply. *Aquatic Botany, 61,* 239–253.

Čížková, H., & Bauer, V. (1998). Rhizome respiration of *Phragmites australis*: effects of rhizome age, temperature, and nutrient status of the habitat. *Aquatic Botany, 61,* 239–253.

Čížková, H., Lukavská, J., Přibáň, K., Kopecký, J. & Brabcová, H. (1996). Carbohydrate levels in rhizomes of *Phragmites australis* at an ougotrophic and a eutrophic site: a preliminary study. *Folia Geobotanica et Phytotaxonomica, 31,* 111–118.

Čížková-Koncalová, H., Kvet, J. & Thompson, K. (1992). Carbon starvation: a key to reed decline in eutrophic lakes. *Aquatic Botany, 43,* 105–113.

Chapin, F. S. III. (1974). Morphological and physiological mechanisms of temperature compensation in phosphate absorption along a latitudinal gradient. *Ecology, 55(6),* 1180-1198.

Chapin, F. S. III. (1980). The mineral nutrition of wild plants. *Annual Review of Ecology and Systematics, 11,* 233–260.

Chapin, F. S. III., Bloom, A. J., Field, C. B. & Waring, R. H. (1987). Plant responses to multiple environmental factors. *Bioscience, 37,* 49–57.

Chapin, F. S. III., Schultze, E. D. & Mooney, H. A. (1990). The ecology and economics of storage in plants. *Annual Review of Ecology and Systematics, 21,* 423–447.

Chapin, F. S. III. & Chapin, M .C. (1981). Ecotypic differentiation of growth processes in *Carex aquatilis* along latitudinal and local gradients. *Ecology 62,* 1000–1009.

Clevering, O. A & Van der Putten, W. H. (1995). Effects of detritus accumulation on the growth of *Scirpus maritimus* under greenhouse conditions. *Canadian Journal of Botany, 73,* 852–861.

Clevering, O. A. (1998). Effects of litter accumulation and water table on morphology and productivity of *Phragmites australis*. *Wetlands Ecology and Management, 5,* 275–287.

Clevering, O. A. (1999). Between- and within- population differences in *Phragmites australis* 1. The effects of nutrients on seedling growth. *Oecologia, 121,* 447–457.

Clevering, O. A., Brix, H. & Lukavská, J. (2001). Geographic variation in growth responses in *Phragmites australis. Aquatic Botany,* 64, 185-208.

Coops, H. & van der Velde, G. (1996). Effects of waves on helophyte stands: mechanical characteristics of stems of *Phragmites australis* and *Scirpus lacustris. Aquatic Botany,* 53, 175–185.

Crawford, R. M. M. (1992). Oxygen availability as an ecological limit to plant distribution. *Advances in Ecological Research,* 23, 93–185.

Cronk, J. K. & Fennessy, M. S. (2001). *Wetland Plants: Biology and Ecology.* Boca Raton, CRC Press /Lewis Publishers.

Dickerman, J. A. & Wetzel, R. G. (1985). Clonal growth in *Typha latifolia*: population dynamics and demography of the ramets. *Journal of Ecology,* 73, 535–552.

Djebrouni, M. & Huon, A. (1988). Structure and biomass of a *Typha* stand revealed by multidimensional analysis. *Aquatic Botany,* 30, 331–342.

Dong, M. & de Kroon, H. (1994). Plasticity in morphology and biomass allocation in *Cynodon dactylon,* a grass species forming stolons and rhizomes. *Oikos,* 70, 99–106.

Dykyjová, D. & Hradecká, D. (1976). Production ecology of *Phragmites australis* 1: relations of two ecotypes to the microclimate and nutrient conditions of habitat. *Folia Geobotanica et Phytotaxonomica,* 11, 23–61.

Edwards, A. L., Lee, D. W. & Richards, J. H. (2003). Responses to a fluctuating environment: effects of water depth on growth and biomass allocation in *Eleocharis cellulosa* Torr. (Cyperaceae). *Canadian Journal of Botany,* 81, 964–975.

Engel, S. (1990). Ecological impacts of harvesting macrophytes in Halverson Lake, Wisconsin. *Journal of Aquatic Plant Management,* 28, 41-45.

Engloner, A. I. (2008). Structure, growth dynamics and biomass of reed (Phragmites australis)- A review. *Flora, In press.*

Ewing, K. (1996). Tolerance of four wetland plant species to flooding and sediment deposition. *Environment and Experimental Botany,* 36, 131–146.

Feller, I. C., Wigham, D. F., O'Neill, J. P. & McKee, K. L. (1999). Effects of nutrient enrichment on within-stand cycling in a mangrove forest. *Ecology,* 80, 2193–2205.

Fiala, K. (1973). Growth and production of underground organs of *Typha angustifolia* L., *Typha latifolia* L. and *Phragmites communis* Trin. *Polish Archives of Hydrobiology,* 20, 59–66.

Fiala, K. (1976). Underground organs of *Phragmites australis,* their growth, biomass and net production. *Folia Geobotanica et Phytotaxonomica,* 11, 225–259.

Garver, E. G., Dubbe, D. R. & Pratt, D. C. (1988). Seasonal patterns in accumulation and partitioning of biomass and macronutrients in *Typha* spp. *Aquatic Botany,* 32, 115–127.

Geary, P. M. & Moore, J. A. (1999). Suitability of a Treatment Wetland for Dairy Wastewaters. *Water Science and Technology,* 40, 179-185.

Gibbons, M. V., Gibbons, H. L. & Sytsma, M. D. (1994). A Citizen's Manual for Developing Integrated Aquatic Vegetation Management Plans. First edition. Washington State Dept. of Ecology. Ecology, Olympia, WA.

Golterman, H. L. (1995). The labyrinth of nutrient cycles and buffers in wetlands: result based on research in the Camargue (Southern France). *Hydrobiologia* 315, 39–58.

Golterman, H. L. (2004). *The Chemistry of Phosphate and Nitrogen Compounds in Sediment.* Dordrecht/ Boston/London: Kluwer Academic Publishers.

Gomes, P. I. A. & Asaeda, T. (2009). Spatial and temporal heterogeneity of *Eragrostis curvula* in the downstream flood meadow of a regulated river. *Annales de Limnologie-International Journal of Limnology, 45,* 1-13.

Goodall, D. W. (1970). Statistical plant ecology. *Annual Review of Ecology and Systematics, 1,* 99–124.

Gorham, E. & Pearsall, W. H. (1956). Production ecology III. Shoot production in *Phragmites* in relation to habitat. *Oikos, 7,* 206–214.

Grace, J. B. & Wetzel, R. G. (1982). Niche differentiation between two rhizomatous plant species: *Typha latifolia* and *Typha angustifolia. Canadian Journal of Botany, 60,* 46–57.

Grace, J. B. (1988). The effects of nutrient additions on mixtures of *Typha latifolia* L. and *Typha domingensis* Pers. along a water-depth gradient. *Aquatic Botany, 31,* 83–92.

Grace, J. B. (1989). Effects of water depth on *Typha latifolia* and *Typha domingensis. American Journal of Botany, 76,* 762–768.

Grace, J. B. and Wetzel, R. G. (1981). Habitat partitioning and competitive displacement in cattails (Typha): experimental field studies. *The American Naturalist, 118,* 463–474.

Granéli, W. & Solander, D. (1988). Influence of aquatic macrophytes on phosphorus cycling in lakes. *Hydrobiologia, 170,* 245–266.

Granéli, W. (1990). Standing crop and mineral content of reed, *Phragmites australis* (Cav.) Trin. ex Studel, in Sweden–Management of reed stands to maximize harvestable biomass. *Folia Geobotanica Phytotaxonomica, 25,* 291–302.

Granéli, W., Weisner, S. E. B. & Systma, M. D. (1992). Rhizome dynamics and resource storage in *Phragmites australis. Wetlands Ecology and Management, 1,* 239–247.

Gratton, C. & Denno, R. F. (2005). Restoration of arthropod assemblages in a Spartina salt marsh following removal of the invasive plant Phragmites australis. *Restoration Ecology, 13,* 358–372.

Greenway, M. & Woolley, A. (2001). Changes in Plant Biomass and Nutrient Removal Over 3 Years in a Constructed Wetland in Cairns, Australia. *Water Science and Technology, 44,* 303-310.

Greenway, M. (2002). Seasonal Phragmites Biomass and Nutrient Storage in a Subtropical Subsurface Flow Wetland, Receiving Secondary Treated Effluent in Brisbane, Australia. *8th International Conference on Wetland Systems for Water Pollution Control*, Arusha, Tanzania.

Grinwald, M. E. (1968). Harvesting aquatic vegetation: *Hyacinth Control Journal, 7,* 31-32.

Gryseels, M. (1989). Nature management experiments in a derelict reed marsh. II. Effects of summer mowing. *Biological Conservation, 48,* 85–99.

Gustafson, T. D. (1976). Production, photosynthesis and the storage and utilization of reserves in a natural stand of *Typha latifolia* L. *Ph.D. Thesis,* University of Wisconsin-Madison. pp.102.

Güsewell, S. (2003). Management of *Phragmites australis* in Swiss fen meadows by mowing in early summer. *Wetlands Ecology and Management, 11,* 433–445.

Güsewell, S., Le Nédic, C. & Buttler, A. (2000). Dynamics of common reed (*Phragmites australis* Trin.) in Swiss fens with different management. *Wetlands Ecology and Management, 8,* 375–389.

Hai, D. N., Asaeda, T. & Manatunge, J. (2006). Latitudinal effect on the growth dynamics of harvested stands of *Typha*: A modeling approach. *Estuarine, Coastal and Shelf Science, 70,* 613–620.

Hardej, M. & Ozimek, T. (2002). The effect of sewage sludge flooding on growth and morphometric parameters of *Phragmites australis* (Cav.) Trin. ex Steudel. *Ecological Engineering,* 18, 343–350.

Haslam, S. M. (1969). The development and emergence of buds in *Phragmites australis* Trin. *Annals of Botany,* 33, 289–301.

Haslam, S. M. (1972). *Phragmites communis* Trin. 1972. Biological Flora of the British Isles 128. *Journal of Ecology,* 60, 585—610.

Healy, M. G., Newell, J. & Rodgers, M. (2007). Harvesting effects on biomass and nutrient retention in *Phragmites australis* in a free-water surface constructed wetland in western Ireland Biology and Environment: *Proceedings of the Royal Irish Academy,* 107B,139–145.

Hellings, S. E. & Gallagher, J. L. (1992). The effects of salinity and flooding on *Phragmites australis. Journal of Applied Ecology,* 29, 41–49.

Ho, Y. B. (1979). Shoot development and production studies of *Phragmites australis* (Cav.) Trin. ex Steudel in Scottish Lochs. *Hydrobiologia,* 64, 215–222.

Hocking, P. J. (1989a). Seasonal dynamics of production and nutrient accumulation and cycling by *Phragmites australis* (Cav) Trin. ex Steudel in a nutrient-enriched swamp in inland Australia. I. Whole plants. *Australian Journal of Marine and Freshwater Research,* 40, 421–444.

Hocking, P. J. (1989b). Seasonal dynamics of production and nutrient accumulation and cycling by *Phragmites australis* (Cav) Trin. ex Steudel in a nutrient-enriched swamp in inland Australia. II. Individual shoots. *Australian Journal of Marine and Freshwater Research,* 40, 445–464.

Hung, L. Q., Asaeda, T., Fujino, T. & Mnaya, B. J. (2007). Inhibition of *Zizania latifolia* growth by *Phragmites australis*: an experimental study. *Wetlands Ecology and Management,* 15, 105–111.

Inoue, T. M., & Tsuchiya, T. (2006) Growht strategy of an emergent macrophytes, *Typha orientalis* Presl, in comparison with *Typha latifolia* L. and *Typha angustifolia* L.. *Limnology,* 7, 171–174.

IWA (International Water Association). (2000). *Constructed Wetlands for Pollution Control: Processes, Performance, Design and Operation.* IWA Specialist Group on use of Macrophytes in Water Pollution Control. IWA Publishing.

Jervis, R. A. (1969). Primary production in the freshwater marsh ecosystems of Troy Meadows, New Jersey. *The Bulletin of the Torrey Botanical Club,* 96, 209–231.

Jespersen, D. N., Sorrell, B. K. & Brix, H. (1998). Growth and root release by *Typha Latifolia* and its effects on sediment methanogenesis. *Aquatic Botany,* 61, 165–180.

Jinadasa, K. B., Tanaka, N., Sasikala, S., Werellagama, D. R., Mowjood, M. I. & Ng, W. J. (2008). Impact of harvesting on constructed wetlands performance-a comparison between *Scirpus grossus* and *Typha angustifolia. Journal of Environmental Science and Health,* 43, 664–671.

Jurik, T.W., Wang, S.C. & van der Valk, A. G. (1994). Effects of sediment load on seedling emergence from wetland seed banks. *Wetlands,* 14, 159–165.

Kachi, N. & Hirose, T. (1983). Limiting nutrients for plant growth in coastal sand dune soils. *Journal of Ecology,* 71, 937–944.

Kadono, Y. (1994). *Aquatic Plants in Japan.* Tokyo, Bunichi-Sogo Press. (in Japanese).

Kang, S., Kang, H., Ko, D. & Lee, D. (2002). Nitrogen removal from a riverine wetland: a field survey and simulation study of *Phragmites japonica*. *Ecological Engineering*, 18, 467–475.

Karunaratne, S. Asaeda, T. & Yutani, K. (2003). Growth performance of *Phragmites australis* in Japan: influence of geographic gradient. *Environmental and Experimental Botany*, 50, 51–66.

Karunaratne, S., Asaeda, T. & Yutani, K. (2004). Shoot regrowth and age-specific rhizome storage dynamics of *Phragmites australis* subjected to summer harvesting. *Ecological Engineering*, 22, 99–111.

Keller, B. E. M. (2000) Plant diversity in *Lythrum*, *Phragmites*, and *Typha* marshes, Massachusetts, USA. *Wetlands Ecology Management*, 8, 391–401.

Kim, S. Y. & Geary, P. M. (2000). The Impact of Biomass Harvesting on Phosphorus Uptake by Wetland Plants. *7th International Conference on Wetland Systems for Water Pollution Control*. University of Florida, Florida, America.

Kim, S. Y. & Geary, P. M. (2001). The impact of biomass harvesting on phosphorus uptake by wetland plants. *Water Science and Technology*, 44, 61–67.

Klopatek, J. M. (1970). Nutrient Dynamics of freshwater riverine marshes and the role of emergent macrophytes. In: R. E. Good, D. F. Whigham, & R. L. Simpson (Eds.), *Freshwater Wetlands' Ecological Processes and Management Potential* (pp. 195–216). New York, Academic Press.

Klopatek, J. M. & Stearns, F. W. (1978). Primary productivity of emergent macrophyte in a Wisconsin freshwater marsh ecosystem. *American Midland Naturalist*, 100, 320–334.

Knapp, A. K. & Yavitt, J. B. (1995). Gas exchange characteristics of *Typha latifolia* L. from nine sites across North America. *Aquatic Botany*, 49, 203–215.

Koerselman, W & Meuleman, A. F. M. (1996). The vegetation N: P ratio: a new tool to detect the nature of nutrient limitation. *Journal of Applied Ecology*, 33, 1441–1450.

Koerselman, W. (1992). The nature of nutrient limitation in Dutch dune slacks. In R. W. G. Carter, T. G. F. Curtis, M. J. Sheehy-Skeffington, & A. A. Balkema (Eds.), Coastal Dunes: Geomorphology, Ecology and Management for Conservation (pp. 189–199). Rotterdam, AA Balkema.

Ksenofontova, T. (1988). Morphology, production and mineral contents in *Phragmites australis* in different water bodies of the Estonian SSR. *Folia Geobotanica Phytotaxonomica*, 23, 17–43.

Ksenofontova, T. (1989). General changes in the Matsalu bay reedbeds in this century and their present quality (Estonian SSR). *Aquatic Botany*, 35, 111–120.

Kubin, P. & Melzer, A. (1996). Does ammonium affect accumulation of starch in rhizomes of *Phragmites australis* (Cav.) Trin. ex Steudel. *Folia Geobotanica et Phytotaxonomica*, 31, 99–109.

Kubin, P., Melzer, A. & Čížková, H. (1994). The relationship between starch content in rhizomes of *Phragmites australis* (Cav.) Trin. ex Steud. And trophic conditions of habitat. *Proceedings of the Royal Society of Edinburgh*, 102B, 433–438.

Kudo, G. (1995). Leaf traits and shoot performance of an evergreen shrub, *Ledum palustre* ssp. *decumbens*, in accordance with latitudinal change. *Canadian Journal of Botany*, 73, 1451–1456

Lan, N. K., Asaeda, T. & Manatunge, J. (2005). The effects of litter cover on the early growth of Manshurian wild rice (*Zizania latifolia*). *Journal of Freshwater Ecology*, 20, 263–267.

Lavoie, C., Jean, M., Delisle, F. & Letourneau, G. (2003). Exotic plant species of the St. Lawrence River wetlands: a spatial and historical analysis. *Journal of Biogeography, 30,* 537–549.

Li, B., Suzuki, J.-I. & Hara, T. (1998). Latitudinal variation in plant size and relative growth rate in *Arabidopsis thaliana. Oecologia* 115, 293–301.

Lieffers, V. J. (1983). Growth of *Typha latifolia* in boreal forest habitats, as measured by double sampling. *Aquatic Botany,* 15, 335–348.

Linden, M. J. H. A. van der. (1980). Nitrogen economy of reed vegetation in the Zuidelijk Flevoland polder. I. Distribution of nitrogen among shoots and rhizomes during the growing season and loss of nitrogen due to fire management. *Acta Æcologica/ Æcol. Plant.* 1, 219–230.

Linden, M. J. H. A., van der. (1986). Phosphorus economy of reed vegetation in the Zuidelijk Flevoland polder (The Netherlands): seasonal distribution of phosphorus among shoots and rhizomes and availability of soil phosphorus. *Acta Æcologica/ Æcol. Plant.* 7, 397–405.

Maheu-Giroux, M. & de Blois, S. (2007). Landscape ecology of *Phragmites australis* invasion in networks of linear wetlands. *Landscape Ecology,* 22, 285–301.

Mann, R. A. & Bavor, H. J. (1993). Phosphorus Removal in Constructed Wetlands Using Gravel and Industrial Waste Substrata. *Water Science and Technology,* 27, 107-113.

Mann, R. A. (1990). Phosphorus removal by constructed wetlands: Substratum Adsorption. In P. Cooper, & B. C. Findlater (Eds.), *Constructed Wetlands in Water Pollution Control* (pp. 85-100). Pergamon Press.

Marks, M., Lapin, B. & Randall, J. (1994). *Phragmites australis* (*P. communis*): threats, management, and monitoring. *Natural Areas Journal,* 14, 285–294.

Marshall, T. R. & Lee, P. F. (1994). Mapping aquatic macrophytes through digital image analysis of aerial photographs: an assessment. *Journal of Aquatic Plant Management,* 32, 61–66.

Martin, K. (2008). Ecological strategies of three aquatic macrophytes in emerged, submerged and floating-leaved plant ecosystems; selected insitu-experiment scenarios. *PhD thesis,* Saitama University, Japan.

McNaughton, S. J. (1966). Ecotype function in the *Typha* community-type. *Ecological Monographs,* 36, 297–325.

McNaughton, S. J. (1975). r- and K-selection in *Typha. The American Naturalist,* 109, 251–261.

McDonald, M. E. (1955). Cause and effects of a die-off of emergent vegetation. *Journal of Wildlife Management,* 19, 24–35.

McNaughton, S. J. (1968). Autotoxic feedback in relation to germination and seedling growth in *Typha latifolia. Ecology,* 49, 367–369.

Morris, K. & Ganf, G. G. (2001). The response of an emergent sedge Bolboschoenu medianus to salinity and nutrients. *Aquatic Botany,* 70, 311–328.

Moshiri, G. A. (1993). *Constructed wetlands for water quality improvement.* Boca Raton, FL, Lewis Publishing.

Neely, R. K. & Wiler, J. A. (1993). The effect of sediment loading on germination from the seed bank of three Michigan wetlands. *The Michigan Botanist,* 32, 199–207.

Ostendorp, W. (1992). Sedimente and Sedimentbildung in Seeuferrohrichten des Bodensee-Untersees. *Limnologica,* 22, 16–33.

Ostendorp, W., Dienst, M. & Schmieder, K. (2003). Disturbance and rehabilitation of lakeside *Phramites* reeds following an extreme flood in Lake Constance (Germany). *Hydrobiologia,* 506/509, 687–695.

Penfound, W. T. (1956). Production of vascular aquatic plants. *Limnology and Oceanography,* 1, 92–105.

Pilon-Smits. (2005). Phytoremediation. *Annual Review of Plant Biology,* 56, 15–39.

Piqueras, J., Klimes, L. & Redbo-Torstensson, P. (1999). Modelling the morphological response to nutrient availability in the clonal plant *Trientalis europaea* L. *Plant Ecology* 141, 117–127.

Rajapakse, L., Asaeda, T., Williams, D., Roberts, R. & Manatunge, J. (2006). Effects of water depth and litter accumulation on morpho-ecological adaptations of *Eleocharis sphacelata. Chemistry and Ecology,* 22, 47–57.

Reed, S. C., Crites, R. W. & Middlebrooks, E. J. (1995). *Natural systems for waste management and treatment* (2nd Ed.). McGraw Hill Inc.

Reddy, K. R. & Portier, K. M. (1987). Nitrogen utilization by *Typha latifolia* L. as affected by temperature and rate of nitrogen application. *Aquatic Botany,* 27, 127–138.

Rejmánková, E. (2001). Effect of experimental phosphorus enrichment on oligotrophic tropical marshes in Belize, Central America. *Plant and Soil,* 236, 33-53.

Richardson, C. J. & Craft, C. B. (1993). Effective phosphorus retention in wetlands: fact or fiction? In G. A. Moshiri (Ed.), *Constructed Wetlands for Water Quality Improvement* (pp 271–282). Boca Raton,Lewis Publishers.

Richardson, C. J., Ferreli, G. M. & Vaithiyanathan, P. (1999). Nutrient effects on stand structure, resorption efficiency, and secondary compounds in Everglades sawgrass. *Ecology,* 80, 2182–2192.

Rickey, M. A. & Anderson, R. C. (2004). Effects of nitrogen addition on the invasive grass *Phragmites australis* and a native competitor *Spartina pectinata. Journal of applied ecology,* 41, 888–896.

Romero, J. A., Brix, H. & Comin, F. A. (1999). Interactive effects of N and P on growth, nutrient allocation and NH_4 uptake kinetics by *Phragmites australis. Aquatic Botany,* 64, 369–380.

Rozé, F. (1993). Successions végétales après pâturage extensif par des chevaux dans une roselière. *Bulletin d'Ecologie,* 24, 203–209.

Russell, I. A. & Kraaij, T. (2008). Effects of cutting *Phragmites australis* along an inundation gradient, with implications for managing reed encroachment in a South African estuarine lake system. *Wetlands Ecology and Management,* 16, 383–393.

Saltonstall, K. & Stevenson, J. C. (2007).The effect of nutrients on seedling growth of native and introduced *Phragmites australis. Aquatic Botany,* 86, 331–336.

Schierup, H. H. (1978). Biomass and primary production in a *Phragmites communis* Trin. swamp in North Jutland, Denmark. *Verh. Int. Ver. Limnol.,* 20, 94–99.

Schütz, P. & Ochse, M. (1997). Effizienzkontrolle von Pflege- und Entwicklungsplänen für Schutzgebiete in Nordrhein-Westfalen. *Natursch. Landsch'plan,* 29, 20–31.

Sharma, P. Asaeda, T., Manatunge, J. & Fujino, T. (2006). Nutrient cycling in a natural stand of *Typha angustifolia. Journal of Freshwater Ecology,* 21, 431-438.

Shaver, G. R., Johnson, L. C., Cades, D. H., Murray, G., Laundre, J. A., Rastetter, E. B., Nadelhoffer, K. J. & Giblin, A. E. (1998). Biomass and CO_2 flux in wet sedge tundras: responses to nutrients, temperature and light. *Ecological Monographs,* 68, 75–97.

Shay, J. M. & Shay, C. T. (1986). Prairie marshes in western Canada, with specific reference to the ecology of five emergent macrophytes. *Canadian Journal of Botany, 64,* 443–454.

Smart, R. M. & Barko, J. W. (1978). Influence of sediment salinity and nutrients on the physiological ecology of selected salt marsh plants. *Estuarine, Coastal and Marine Science, 7,* 487–495.

Smith, C. S., Adams, M. S. & Gustafson, T. D. (1988). The importance of belowground minerals element stores in cattails (*Typha latifolia* L.). *Aquatic Botany, 30,* 343–352.

Solano, M. L., Soriano, P. & Ciria, M. P. (2004). Constructed wetlands as a sustainable solution for wastewater treatment in small villages. *Biosystems Engineering, 87,* 109–118.

Sorrell, B. K, Brix, H. & Orr, P. T. (1997). *Eleocharis sphacelata*: internal gas transport pathways and modelling of aeration by pressurized flow and diffusion. *New Phytologist, 136,* 433–442.

Sorrell, B. K. & Armstrong, W. (1994). On the difficulties of measuring oxygen release by root systems of wetland plants. *Journal of Ecology, 82,* 177–183.

Sorrell, B. K. & Tanner, C. C. (2000). Convective gas flow and internal aeration in *Eleocharis sphacelata* in relation to water depth. *Journal of Ecology, 88,* 778–789.

Sorrell, B. K., Tanner, C. C. & Sukias, J. P. S. (2002). Effects of water depth and substrate on growth and morphology of *Eleocharis sphacelata*: implications for culm support and internal gas transport. *Aquatic Botany, 73,* 93–106.

Spence, D. H. N. (1964). The macrophytic vegetation of freshwater lochs, swamps, and associated fens. In J. H. Burnett (Ed.), *The Vegetation of Scotland* (pp. 306–425). Edinburgh, Oliver & Boyd.

Spence, D. H. N. (1982). The zonation of plants in freshwater lakes. In: A. MacEayden, & E. D. Eord (Eds.), *Advances in Ecological Research* (pp. 37–125). London, Academic Press.

Spencer, D. F. (1991). Influence of organic sediment amendments on growth and tuber production by *Potamogeton pectinatus* L. *Journal of Freshwater Ecology, 5,* 255–263.

Sutton, D. L. & Dingler, P. M. (2000). Influence of sediment nutrients on growth of emergent hygrophila. *Journal of Aquatic Plant Management, 38,* 55–61.

Szczepanska, W. & Szczepanski, A. (1976). Growth of *Phragmites communis* Trin., *Typha Latifolia* L., and *Typha angustifoila* L. in relation to the fertility of soils. *Polish Archives of Hydrobiology, 23,* 233–248.

Tanaka, N., Asaeda, T., Hasegawa, A &, Tanimoto, K. (2004). Modelling of the long-term competition between *Typha angustifolia* and *Typha latifolia* in shallow water- effects of eutrophication, latitude and initial advantage of belowground organs. *Aquatic Botany, 79,* 295-310.

Tanner, C. C. & Kadlec, R. H. (2002). Oxygen Flux Implications of Observed Nitrogen Removal Rates in Subsurface-Flow Treatment Wetlands. *In: Proc. of 8th International Conference on Wetland Systems for Water Pollution Control.* Arusha, Tanzania.

Tanner, C. C. (2000). Plants as Ecosystem Engineers in Subsurface-Flow Treatment Wetlands. *In: Proc. of 7th International Conference on Wetland Systems for Water Pollution Control.* University of Florida, Florida.

Tanner, C. C., Sukias, J. P. S. & Upsdell, M. P. (1999). Substratum Phosphorus Accumulation During Maturation of Gravel-Bed Constructed Wetlands. *Water Science and Technology, 40,* 147-154.

Thullen, J. S. Sartoris, J. J. & Nelson, S. M. (2005). Managing vegetation in surface-flow wastewater-treatment wetlands for optimal treatment performance. *Ecological Engineering, 25,* 583–593.

Tsuchiya, T., Shinozuka, A. & Ikushima, I. (1993). Population dynamics, productivity and biomass allocation of *Zizania latifolia* in an aquatic–terrestrial ecotone. *Ecological Research, 8,* 193–198.

Tucker, G.C. 1990. The genera of Arundinoideae (Gramineae) in the southeastern United States. *Journal of the Arnold Arboretum 71,* 145–177.

Ulrich, K. E. & Burton, T. M. (1985). The establishment and management of emergent vegetation in sewage-fed marshes and the effects of these marshes on water quality. *Wetlands, 4,* 205–220.

Ulrich, K. E. & Burton, T. M. (1988). An experimental comparison of the dry matter and nutrient distribution patterns of *Typha latifolia* L., *Typha angustifolia* L., *Sparganium eurycarpum* Engelm. and *Phragmites australis* (Cav.) Trin. ex Steudel. *Aquatic Botany, 32,* 129–139.

USDA (United States Department of Agriculture). (2008). Wetland Plants: their function, adaptation and relationship to water levels. *Riparian/ Wetland Project Information Series No. 21.* (http://www.plant-materials.nrcs.usda.gov/pubs/idpmcar7242.pdf).

Van Deursen, E. J. M. & Drost, H. J. (1990). Defoliation and treading by cattle of reed *Phragmites australis. Journal of Applied Ecology, 27,* 284– 297.

Van der Putten, W. H. (1993). Effects of litter on the growth of *Phragmites australis.* In: W. Ostendorp, & P. Krumpscheid-Plankert (Eds.), *Seeuferzerstorung and Seeuferrenaturierung in Mitteleuropa* (pp. 19–22). Stuttgart: Gustav Eischer Verlag, Limnologie Aktuell.

Van der Putten, W. H., Peters, B. A. M. & Van der Berg, M. S. (1997). Effects of litter on substrate conditions and growth of emergent macrophytes. *New Phytologist, 135,* 527–537.

Van der Toorn, J. (1972). Variability of *Phragmites australis* (Cav) Trin ex Steudel in relation to the environment. *Van Zee tot Land, 48,* 1–122.

Van der Valk, A. G. (2006). *The Biology of freshwater wetlands.* Oxford, UK, Oxford University Press.

Van der Valk, A. G., Swanson, S. D. & Nuss, R. F. (1983). The response of plant species to burial in three types of Alaskan wetlands. *Canadian Journal of Botany, 61,* 1150–1164.

Vitousek, P. M. & Howarth, R. W. (1991). Nitrogen limitation on land and in the sea: How can it occur? *Biogeochemistry, 13,* 87–115.

Vitousek, P. M. (1984). Litterfall, nutrient cycling, and nutrient limitation in tropical forests. *Ecology, 65,* 1476–1490.

Votrubová, O. & Pecháčková, A. (1996). Effect of nitrogen over-supply on root structure of common reed. *Folia Geobotanica Phytotaxonomica, 31,* 119–125.

Vymazal, J. (1995). *Algae and element cycling in wetlands.* Chelsea, MI, Lewis publishers.

Vymazal, J., Brix, H., Cooper, P., Haberl, R., Perfler, R. & Laber, J. (1998). Removal mechanisms and types of constructed wetlands. In, J. Vymazal, H. Brix, P. F. Cooper, M. B. Green, & R. Leiden (Eds.), *Constructed Wetlands for Wastewater Treatment in Europe* (pp. 17-66). The Netherlands, Backhuys Publishers.

Vymazal, J., Dusek, J. & Kvet, J. (1999). Nutrient uptake and storage by plants in constructed wetlands with horizontal sub-Surface flow: A Comparative Study. In J. Vymazal (Ed.),

Nutrient Cycling and Retention in Natural and Constructed Wetlands (pp. 85-100). The Netherlands, Backhuys Publishers.

Wang, S. C., Jurik, W. C. & Van der Valk, A. G. (1994). Effects of sediment load on various stages in the life and death of cattail (*Typha × glauca*). *Wetlands, 14,* 166–173.

Weigham, D. F. (1978). The relationship between aboveground and belowground biomass of freshwater tidal wetland macrophytes. *Aquatic Botany, 5,* 355–364.

Weinstein, M. P. & Balletto, J. H. (1999). Does the common reed, *Phragmites australis,* affect essential fish habitat? *Estuaries 22,* 793–802.

Weisner, S. E. B. & Granéli, W. (1989). Influence of substrate conditions on the growth of *Phragmites australis* after a reduction in oxygen transport to below-ground parts. *Aquatic Botany, 35,* 71–80.

Weisner, S. E. B. & Strand, J. A. (1996). Rhizome architecture in *Phragmites australis* in relation to water depth: implications for within-plant oxygen transport distances. *Folia Geobotanica et Phytotaxonomica, 31,* 91–97.

Weisner, S. E. B. (1991). Within-lake patterns in depth penetration of emergent vegetation. *Freshwater Biology, 26,* 133–142.

Weisner, S. E. B. (1996). Effectsofanorganicsedimenton performance of young *Phragmites australis* clones at different water depth treatments. *Hydrobiologia, 330,* 189–194.

Westlake, D. F. (1975). Primary production of freshwater macrophytes. In J. P. Cooper (Ed.), *Photosynthesis and Productivity in Different Environments* (pp. 189-206). Cambridge, UK: Cambridge Univ. Press.

Westlake, D. F. (1982). The primary productivity of water plants. In J. J. Symoens, S. S. Hooper, & P. Compere (Eds.), *Studies on Aquatic Vascular Plants* (pp. 165–180). Brussels, Royal Botanical Society of Belgium.

Wetzel, R. G. (1979). The role of the littoral zone and detritus in lake metabolism. *Archiv für Hydrobiologie, 13,* 145–161.

Wetzel, R. G. (1988). Water as an environment for plant life. In J. J. Symoens (Ed.), *Vegetation of Inland waters: Handbook of Vegetation Science* (pp. 1–30). Dordrecht, Netherlands, Dr. W. Junk Publishers (Springer).

Wetzel, R. G. (2001). *Limnology, Lake and River Ecosystems (3rd edition).* California, USA, Academic Press.

Xu, S. & Jaffe, P. R. (2006). Effects of plants on the removal of hexavalent chromium in wetland sediments. *Journal of Environmental Quality, 35,* 334–341.

Yamasaki, S. (1984). Role of plant aeration in zonation of *Zizania latifolia* and *Phragmites australis. Aquatic Botany, 18,* 287–297.

Young, S. W., Davies, D. H. & Milligan, P. J. (1991).The potential of anatomical features of the common reed *Phragmites australis* (Cav.) Trin. ex. Steudel as abiotic indicator of adjoining landuse. *Archiv für Hydrobiologie, 122,* 297–304.

In: Encyclopedia of Environmental Research
Editor: Alisa N. Souter

ISBN: 978-1-61761-927-4
© 2011 Nova Science Publishers, Inc.

Chapter 31

LIFE-HISTORY STRATEGIES: A FRESH APPROACH TO CAUSALLY LINK SPECIES AND THEIR HABITAT

Wilco C.E.P. Verberk[*]

Department of Animal Ecology / Bargerveen Foundation, Faculty of Science,
Mathematics and Computing Science, Radboud University, Nijmegen
Postbus 9010, 6500 GL Nijmegen

ABSTRACT

Community ecology searches for general rules to explain patterns in species' distribution, but to date, progress has been slow. This lack of progress has been attributed to the fact that ecological rules - and the mechanisms that underpin them - are contingent on the organisms involved, and their environment. Coupled with the vast complexity of biological systems, the result is that there are few rules that are universally true in community ecology. However, effective restoration management requires a thorough understanding of the ecological effects of degradation. This sets a challenge for ecological research to unveil the causal mechanisms underlying patterns in species' distribution. There is broad consensus that the way forward is to link pattern and process through species traits, as these provide the causal mechanisms explaining how abiotic and biotic factors set limits to species occurrences.

In a recently developed method, species traits were used to define groups of aquatic macroinvertebrate species with similar causal mechanisms underlying the species-environment relationships. By investigating interrelations between traits and interpreting their function, it was possible to define 'sets of co-adapted species traits designed by natural selection to solve particular ecological problems', which are termed life-history strategies.

In this chapter we discuss the position of life-history strategies in community ecology and its merits to conservation and nature management. By including the causal mechanisms for a species' survival under particular environmental conditions, life-history strategies can be used to explain species occurrences and generate testable predictions. As a species' identity is made subordinate to its biology, they may be used to compare water bodies found at a large geographical distance, which may comprise different regional

---

[*] E-mail address: wilco@aquaticecology.nl

species pools or span species distribution areas. By aggregating species in life-history strategies, biodiverse assemblages can be compressed into a few meaningful, easily interpretable relationships.

When the field of community ecology is envisaged as a continuum, two approaches can be considered the end points, focusing either on communities or individual species. While research on individual species is strongly rooted in causal mechanisms, the results are difficult to generalize to other species, complicating the extrapolation to whole communities. However, on the other end of the continuum, results are aggregated (*e.g.* indices of diversity, evenness and similarity) to such an extent that the causal mechanisms are obscured. The fresh approach of life-history strategies may provide the best of both worlds, aggregation information over many different species without sacrificing information on the causal mechanisms underlying a species' presence or absence.

1. INTRODUCTION

1.1. Getting More out of Your Biodiversity Data with Life-History Strategies: A Fresh Approach to Causally Link Species and Their Habitat

Collecting biodiversity data axiomatically takes a lot of effort. After all the material has been sorted and identified and all the records have been digitized there is a choice about how to analyse the data. Do we calculate indices of diversity or evenness, derive species-abundance distributions or plot species accumulation curves? Surely there must be better ways. After all, species aren't comparable units which can be added and interchanged. At least that is the message we keep telling the general public. The need for all those different critters, crawlers, flyers and swimmers is frequently both an important starting point and an end conclusion. Each of these species has a unique evolutionary history and this information can be used in analysing biodiversity data.

In this chapter I argue that explaining species occurrences is important both from a theoretical and an applied point of view in § 1.2. In § 1.3., I stress the need to incorporate species traits in our analyses. Additionally I explain the limited success of current approaches that have been employed by summarizing their major difficulties. Next, I briefly outline a fresh approach that deals with these difficulties in § 1.4. Strengths of this approach pertain to causality and aggregation. The drawback is that its development and extrapolation to other species groups requires large efforts (§ 1.5.). Nevertheless such an approach can bridge two extremes within the field of community ecology (§ 1.6.) and has much to offer the field of restoration ecology (§ 1.7.).

1.2. Species-Environment Relationships

What constitutes a favourable environment differs between species and depends on their species-specific requirements. In addition, species may use their environment on different spatial scales (Verberk *et al.* 2005). Such differences may also exist between different life stages of the same species. Likewise, environmental heterogeneity is a key element of a landscape and may occur at any spatial scale. Additionally, environmental conditions vary in time. Thus, both the requirements of a species and the environmental conditions often vary in time and space (Southwood 1977; Wiens 1989; Levin 1992). Understanding the relationship between species and their environment, which includes other species, allows one to explain why a certain species occurs where it occurs. With human activities profoundly influencing our landscapes today, this topic is far from being merely an academic one (Kerr & Currie 1995; Heywood & Watson 1995; Dudgeon *et al.* 2006; Rosenzweig *et al.* 2008).

Species are not randomly distributed in nature, but occur in regular, predictable patterns. Nevertheless, when relating variation in species composition to differences in environmental conditions, scatter is frequently observed (Figure 1; Verberk *et al.* 2006a). This variation may arise from different sources. Adjacent water bodies may be more alike with respect to species assemblage than expected from the differences in environmental conditions. This indicates that an exchange of individuals takes place and the magnitude of exchange will depend on the spatial configuration of the water bodies (Cottenie *et al.* 2003; Leibold *et al.* 2004; Verberk *et al.* 2006a). Other factors contributing to scatter around the species-environment relationship may be related to natural dynamics or sampling. For example, environmental conditions may have become recently suitable but the species have not yet colonized (Ozinga *et al.* 2005), or species may have been missed during sampling, merely appearing to be absent (Niggebrugge *et al.* 2007). Due to these sources of variation and different methods for analyzing species-environment relationships, different results can be obtained with little indications for which result best reflects reality (Nijboer 2006). Although it may be important in generating hypotheses, successfully relating species and environmental conditions statistically does not provide explanations. This requires information on the life-history of a species.

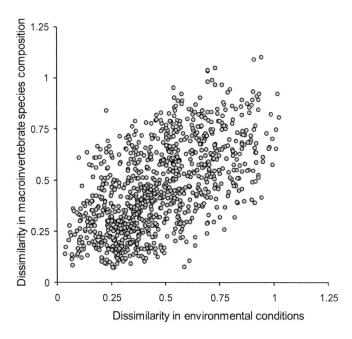

Figure 1. Pairwise comparisons of both conditions and species composition for 45 different sites (990 pairs) illustrating that relationships between the abiotic conditions and macroinvertebrate species composition can show considerable scatter (from Verberk *et al.* 2006a).

1.3. The Match between Species and Their Environment

In general, individuals of a species need to obtain resources to survive until they have successfully reproduced. This requires a species' biology to match with what is supplied by the environment (Southwood 1977; Figure 2). In the course of evolution, species have developed physiological, morphological and behavioural traits to deal with environmental conditions, consisting of abiotic conditions and other species (*e.g.* high acidity, hypoxia, competitors or predators). A species trait is here defined as any morphological, physiological or phenological feature measurable at the individual level, from cell to the level of the whole organism, without reference to the environment or any other level of organisation (Violle *et al.* 2007). Examples include a chitinous exoskeleton, diapauzing eggs, parental care, a large body size and a short development time. Thus, they are similar to what is frequently referred to as biological traits (Statzner *et al.* 1994) and opposed to ecological traits, as these latter reflect a species' habitat preference and require environmental conditions to be measured. Species traits determine a species' ability to deal with environmental conditions and changes therein. Therefore, species traits can be seen as the causal mechanisms underlying species-environment relationships and offer the potential to explain patterns in species' distribution (e.g. Keddy 1992; Statzner *et al.* 2004; McGill *et al.* 2006).

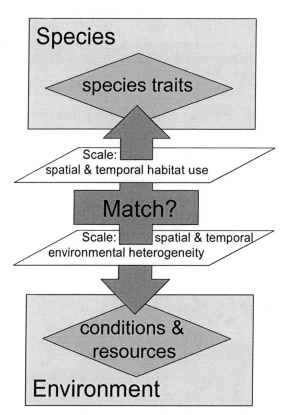

Figure 2. Schematic representation of the match between species and environment. What constitutes a favourable environment differs between species, depending on their requirements, which are rooted in species traits. Both species' requirements and environments' resources are scale dependent.

Many studies have concluded that the habitat acts as a templet for species traits, by demonstrating that patterns in the traits of species are significantly related to differences in the environmental conditions where these species occurred (Statzner *et al.* 1994; Richards *et al.* 1997; Bremner *et al.* 2006; Van Kleef *et al.* 2006). However, elucidating the causal mechanisms by which species traits provide a link between occurrence and particular environmental conditions has proved difficult. *A priori* predictions regarding cause-effect relationships for separate traits performed poorly and the magnitude of differences among traits is often very small, even if highly significant (Resh *et al.* 1994; Townsend *et al.* 1997; Statzner *et al.* 2001; but see Bonada *et al.* 2007). Several reasons for the poor performance of species traits have been forwarded. Many authors highlight the importance of considering trade-offs between traits and the functional equivalence of alternative suites of traits (e.g. Stearns 1976; Resh *et al.* 1994; Mérigoux *et al.* 2001). A specific combination of traits may be more adaptive than the constituent traits separately. For example, in a habitat with predictable but short periods of suitable conditions, the combination of rapid growth and obligate diapause provides an effective adaptation. Thus, for a given species, the adaptive value of a specific trait is context-dependent, being related to the remainder of its biology (Gould & Lewontin 1979; Verberk *et al.* 2008a). Furthermore investments in species traits are not independent, but may be interrelated through trade-offs, with investments in one trait leaving less resources available for investments in another trait. Examples include the trade-off between egg size and egg number, with animals laying either many small eggs or a few large eggs (e.g. Berrigan 1991) and the trade-off between body size and development time, giving rise to either small individuals with an early maturation, or large individuals with a late maturation (e.g. Abrams *et al.* 1996). Alternatively, investments in one trait may reduce costs or increase benefits of investments in another trait (spin-offs). For example, the lamellae of larvae of damselflies are used for both respiration and locomotion.

Given that causal mechanisms may depend on the combined effect of multiple traits, the limited success of trait-based analyses so far can be understood. These range from clustering species with similar suites of traits (Usseglio-Polatera *et al.* 2000; Ilg & Castella 2006), to analyzing the co-structure between patterns in species traits and patterns in habitat use with multivariate analytical techniques (Gayraud *et al.* 2003; Finn & Poff 2005). In these analyses, species traits are treated as independent units. In other words, the analyses assume that when a species possesses a certain trait this has no relation to any of the other traits possessed by that species. This has two important consequences. Firstly, a species may possess a trait for different reasons. For instance, a trait can be adaptive or it can result from body plan constraints. However, these different reasons are not distinguished. Consequently, an explanation for the adaptive value of a certain trait may be erroneous while the importance of another trait may be masked. This may happen when the adaptive value is in fact conferred by another trait that is strongly related to the first trait or when a the adaptive value arises from a combination of traits. For example, multivariate analyses by Van Kleef *et al.* (2006) did not reveal flight capability to be important in the recolonisation of shallow lakes. This counterintuitive result arose because many species with active flight were also carnivorous and their recolonisation was delayed due to the scarcity of prey. Secondly and related to the first, when relating a certain trait to particular environmental conditions the trait-environment relation is averaged across all species, although the relation may actually be relevant in only a subset of species. Such averaging across all species causes the magnitude of differences in trait composition between different environments to be small. Explicitly recognizing how species traits are interrelated may be more successful in unravelling the causal mechanisms underlying species-environment relationships.

1.4. Life-History Strategies: A Fresh Approach

A fresh approach to causally link aquatic macroinvertebrate species and their habitat has recently been developed (Verberk *et al.* 2008a). Functionally equivalent groups were *a priori* defined based on combinations of basic biological traits of species and their functional implications. Trade-offs between traits and the functional equivalence of alternative suites of traits are at the core of this approach rather than seeing these as obstacles which must be circumvented. To emphasize the functional interlinkage between traits, these groups are termed life-history strategies, rather than 'suites of traits'. Analogous to the definition of Stearns (1976) of life-history tactics, we define life-history strategies as a set of co-evolved traits which enable a species to deal with a range of ecological problems. Whereas Stearns mainly focussed on traits related to reproduction, here traits related to development, synchronisation and dispersal are also taken into account. Therefore the somewhat broader term life-history strategies is used here.

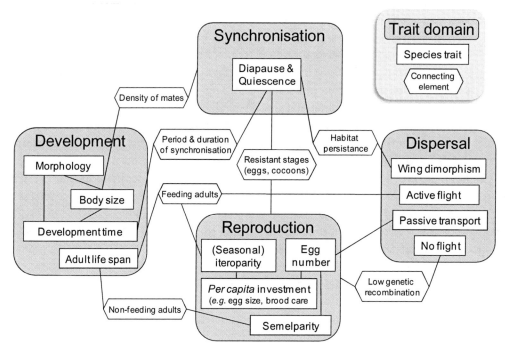

Figure 3. Schematic overview illustrating the diversity in relationships between species traits (adapted from Verberk 2008). Species traits are presented in four main domains of traits (Siepel 1994): (i) reproduction, (ii) development, (iii) dispersal and (iv) synchronisation, which are shown in grey. Species traits are interrelated, both within and between domains. Consequently, species traits of all domains together shape a life-history strategy and therefore combinations of traits (as opposed to separate traits) provide a more complete understanding of their adaptive value.

We have based life-history strategies on basic biological traits of species and their interrelationships known from life-history theory (shown in Figure 3). These interrelations can explain why high investments in one species trait constrain investments in another, or why investments in two species traits are coupled. As a result, when comparing various traits over different species, some traits may be more strongly developed or pronounced compared to other traits. As life-history studies typically deal with comparisons rather than absolutes (Begon *et al.* 1998), these differences actually refer to a relative scale (e.g. faster-slower, shorter-longer, higher-lower). Whether or not some trait is strongly pronounced becomes clear in view of the constraints and opportunities set by the remainder of a species' life-history, body plan and feeding guild (Gould & Lewontin 1979), which requires comparisons. As high investments in one trait constrain investments in another, such investments are probably relevant to overcome a particular environmental problem. By focussing on the adaptive value of combinations of traits (rather than on the identity of separate traits), different suites of traits may be equivalent, presenting a similar solution to an ecological problem. Based on these principles, thirteen theoretical life-history strategies have been defined elsewhere (Verberk *et al.* 2008a). Application of life-history strategies provided a functional classification of macroinvertebrates spanning different systematic groups (Verberk *et al.* 2008b). Different species have different species traits but may be assigned to the same life-history strategy, provided that the combination of species traits solves problems for survival in a similar way. For example, the protection of egg stages is achieved in different ways in different systematic groups, including endophytical oviposition (e.g. the dragonfly *Aeshna cyanea*), a gelatinous matrix (e.g. the caddisfly Limnephilus luridus) and skipping the vulnerable egg stage through ovoviviparity (e.g. the mayfly *Cloeon dipterum*).

The approach of life-history strategies was successfully applied to field data on species occurrences and environmental conditions. Species were assigned to the different strategies and differences in strategy composition were related to the prevailing environmental conditions in a logical fashion (Verberk *et al.* 2008b). Clear responses were observed in relation to temporal predictability and stability (dimensions and proneness to desiccation) and habitat favourability (nutrient availability, acidity and oxygen conditions). The distribution of species numbers over the various life-history strategies in a water body can therefore give direct information about how a particular environment is experienced by the species present.

1.5. Strengths and Weaknesses of Life-History Strategies

Table 1. Overview of strong and weak points pertaining to the causality (C) or aggregation (A) of life-history strategies.

Strong points	Weak points
incorporates a species' natural history (C)	detailed knowledge on species' biology required (C & A)
based on causal mechanisms (C)	no fixed guidelines for trait function (C & A)
generates testable predictions (C)	no fixed criteria for level of aggregating (A)
new species may be added relatively simply (A)	
reduces complexity (A)	

Analysing species-environment relationships from a life-history strategy approach has several strong points and a few weak points (Table 1). These strengths and weaknesses pertain to causality and aggregation. They will each be discussed separately, although in fact they are intimately related.

1.5.1. Causality

The core of the life-history strategy approach is that separate traits are combined to 'sets of co-evolved species traits'. By incorporating the causal mechanisms for a species' survival under particular environmental conditions, life-history strategies can be used to explain differences in species assemblages between locations or periods. The caveat is that the strategies have to be based on relevant species traits and species traits have to be combined in a meaningful way in order to include the main causal mechanisms that operate under the particular environmental conditions of interest. This requires detailed knowledge on (i) the traits possessed by a species, (ii) interrelations between these traits, (iii) and how the functional relevance of such trait combinations should be interpreted. Given the huge range of species traits, there are no strict guidelines for how to interpret the function of traits or combinations of traits. Consequently, such interpretations may be criticised for being subjective and lacking scientific rigour. Although the use of species traits makes sense intuitively, the lack of strict guidelines may explain why it has proven so difficult to apply species traits in analysing biodiversity data and may well explain why the rich body of literature on the natural history of many species has been largely ignored in community ecology. By rooting interpretations in fundamental trade-offs known from life-history theory and by viewing trait combinations as a part of the entire organism's biology (Gould & Lewontin 1979), it is possible to minimize subjectivity in interpreting the function of combinations of species traits (Siepel 1994; Verberk et al. 2008a). Taking the function of species trait combinations in account presents an advantage to other methods, such as multivariate analytical techniques and the use of phylogenetic independent contrasts (e.g. Resh et al. 1994; Felsenstein 1985; Westoby et al. 1995). In the former, species traits are weighted equally (i.e. independent from other traits possessed), while in the latter all species traits that can be related to phylogeny are controlled for, although they may be of adaptive value in dealing with ecological problems (explaining why they are phylogenetically conserved).

Life-history strategies allow information on the natural history of a species (pertaining to species traits) to be integrated and used to derive the most likely explanations and generate testable predictions. An example of testing predictions generated by life-history strategies is related to the widely observed pattern that widespread species are likely to occur in high densities whereas restricted species tend to be scarce (Brown 1984; Gaston et al. 1997). Although this abundance-occupancy relationship may also arise from broad statistical generalisations involving large sets of species (Hartley 1998), the underlying mechanisms pertain to differences in dispersal and reproduction capacity of species, providing a link to metapopulation dynamics and niche breadth (Gaston et al. 1997). A species' life-history strategy was used to predict its residual from the abundance-occupancy relationship (Verberk 2008). These predictions corresponded well with observed residuals derived from field data (Figure 4).

For example, species without active dispersal and a high intrinsic rate of increase showed high residual abundance, being more abundant than expected based on their occupancy, whereas species with widely scattered oviposition and moderate or strong active dispersal showed low residual abundance, being less abundant than expected based on their occupancy. This provides evidence that the causal mechanisms that are most likely to facilitate or constrain species on a local and regional scale, were incorporated in the life-history strategies (Verberk 2008). Consequently species differing in life-history strategy are expect to respond differently to either local habitat destruction or regional decline of habitat quality (Lawton 1993).

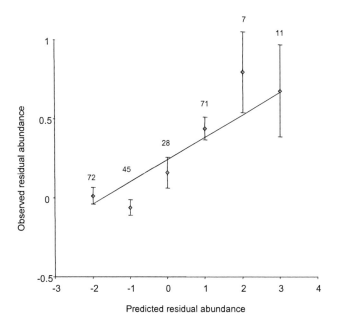

Figure 4. Relationship between the observed (averages ± standard error) and predicted residual abundance from the abundance-occupancy relationship (see Verberk 2008 for background information). Averages are calculated over all species (numbers indicate the number of species) with the same predicted residual abundance. A linear regression between predicted and observed residual abundance was highly significant (R2=0.144; Beta 0.380; P<0.0001).

By providing the most likely explanations -or by challenging previously held notions- life-history strategies may focus subsequent research and management efforts. Further research may focus on obtaining more information on species traits or on expanding the life-history strategies to include other species groups and ecosystems, such as aquatic macroinvertebrates from lotic waters and waters with fish.

1.5.2. Aggregation

Theoretically, species traits can be combined in an almost infinite number of ways. For example, 15 different traits with each three modalities (for instance being absent, weakly developed or strongly developed) already yields 14,348,907 combinations (3^{15}). In reality, species traits are not randomly distributed across species (Harvey 1996), greatly reducing the number of combinations possible (Verberk et al. 2008a). Therefore, once sets of co-evolved species traits have been delineated as separate life-history strategies, new species may be added relatively simple. Due to the strong interrelations, exhaustive data on every aspect of a species' biology does not seem necessary, rather information on a number of key traits is sufficient to confidently assign a species to a certain strategy.

The number of trait combinations is further reduced in life-history strategies by recognising that certain trait combinations may be functional equivalent: species may possess different combinations of species traits, for instance due to differences in basic morphology, but different trait combinations can solve the same environmental problem (Koehl 1996). Species identification will remain necessary for assigning species to a certain strategy. Nevertheless, life-history strategies reduce assemblages of many species belonging to many different systematic groups to a small number of strategies, representing meaningful and interpretable relationships between species and their environment.

By grouping species in life-history strategies, the signal to noise ratio increases as differences which may be regarded as noise are aggregated, while relevant differences are kept separated. An example of 'aggregating noise' is given in Verberk et al. (2008b). Here different types of bog pools were distinguished and these also differed in species composition. However, the bog pools were similar with respect to being acid and nutrient poor. Consequently, many of the more abundant species were similar in life-history strategy, having a (relatively) small body size, weak dispersal, long development time or any combination thereof, resulting from high investments in physiological and morphological adaptations (e.g. haemoglobin, gills) to combat environmental hardship. An example of 'keeping relevant differences separated' is given in a case-study on the effects of rewetting bog remnants (described in Verberk 2008). Here species assemblages were studied before and after restoration measures were taken. Although the number of species had not changed significantly, changes in strategy composition revealed a clear difference between locations where the hydrology was altered (increased or decreased groundwater influence). Moreover, these changes were consistent as similar changes were observed in different types of water bodies (bog pools and wet forests).

Determining what level of aggregation is most appropriate presents a problem when grouping species. Although it is impossible to give fixed criteria, ideally the level of aggregation should strike a bargain between maximising the signal while at the same time minimising the noise. Therefore, the appropriate level of aggregation depends on both the range in interspecific biological differences and the range in environmental conditions. Thus, when considering a single systematic group, or a narrower range in environmental conditions, species groups may be further refined to achieve a higher resolution. For example, in their study on the response of Chironomidae to natural recovery and recovery following restoration measures in moorland pools, van Kleef et al. (in preparation) distinguished six tactics for Chironomidae, rather than the four distinguished in Verberk et al. (2008a; 2008b).

1.6. Position of Life-History Strategies in Community Ecology

Community ecology searches for general rules to explain patterns in species' distribution, but to date, progress has been slow (Keddy 1992; Weiner 1995; Lawton 1999; McGill et al. 2006). This lack of progress has been explained by the fact that ecological rules - and the mechanisms that underpin them - are contingent on the organisms involved, and their environment (Lawton 1999; Simberloff 2004). Coupled with the vast complexity of biological systems, the result is that there are few rules that are universally true in community ecology. Life-history strategies are here positioned within in the field of community ecology, which is best envisaged as a continuum. The merits of the life-history strategies are compared with two approaches in community ecology, which can be considered the end points of this continuum, focussing either on individual species or aggregating information spanning multiple species (Table 2).

Table 2. Causality and aggregation of species approach, life-history strategy approach and community approach. The different approaches are visualised in figure 1 and further explained in the text.

Strong points	Species approach	Life-history strategy approach	Community approach
Aggregation	-	++	+++
Causality	+++	++	-

1.6.1. Species Approach and Community Approach

When focussing on a single or a few species, it is possible to achieve a high degree of causality in explaining a species' success. In these cases it is feasible to derive causal mechanisms from the environmental context and a species' natural history. The drawback of a species approach is that the results are difficult to generalise to other communities, consisting of other species. In contrast at the other end point, biodiverse communities are analysed and compared. Their complexity is simplified in order to enable the comparison of different communities by aggregating information spanning multiple species. Examples of aggregated community attributes include various indices (e.g. diversity, evenness and similarity indices) and plots (e.g. rank abundance plots, species accumulation curves). Such an approach treats communities as a frequency distribution of index scores and ignores a species' rich natural history. In effect, species are treated as faceless entities and communities as black boxes, which may explain the success of neutral theory in accurately describing communities in terms of indices of aggregated information (Bell 2001; Chave 2004). The focus on aggregated community attributes in the community approach obscures the causal mechanisms. This hampers scientific understanding, which ultimately comes from explanations based on causal mechanisms (Weiner 1995). For example, a correlation between the species diversity and the primary productivity (e.g. Waide *et al.* 1999), does not explain why certain species go extinct nor does it allow extrapolation to derive predictions concerning the fate of the remaining species. Consequently, its use for application to any specific situation is limited, as specific environmental conditions and the associated species, may represent exceptions to the observed pattern. The heart of the problem is that environmental conditions are used to explain the aggregated information spanning multiple species, while in fact explanations are species specific. In other words, there is a mismatch between the phenomenon to be explained and the level of causation. Although the search for general rules in community ecology to arrive at a more explanatory and predictive science is admirable, simplifying the complex reality for reasons of comparison, may be the cause, rather than the cure for the lack of progress in community ecology.

Life-history strategies may provide the best of both worlds, aggregation information over many different species without sacrificing information on the causal mechanisms underlying a species' presence or absence (Table 2). Figure 5 visualises the differences between the life-history strategies and the species approach. When focussing on individual species, it is difficult to judge the importance of the various causal mechanisms relating a species' occurrence to its environment, as there is no point of reference (i.e. other species). Causal mechanisms act through species traits, which together with the spatial and temporal variation in environmental conditions (including both abiotic and biotic factors) determine the suitability and connectivity of the environment from that species' point of view. Life-history strategies represent a species' integrated response to the environment as many traits are jointly considered. Furthermore, interspecific differences in such responses are incorporated in life-history strategies. This provides the contrasts, which are necessary to ascertain which set of causal mechanisms are most important for a given group of species in the environment under study.

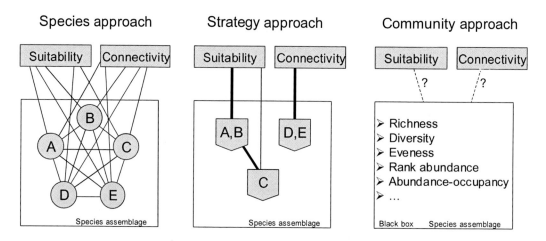

Figure 5. Visualisation of the differences between life-history strategies, the species approach and the community approach. In the species approach causal mechanisms are not differentiated in importance. With life-history strategies, species are grouped and the most important causal mechanisms are shown for a given species group. In the community approach, information spanning multiple species is aggregated, obscuring the causal mechanisms. See text for further explanation.

1.6.2. Life-History Strategies and Macroecology

Life-history strategies may be used in further research to provide insight into the causal mechanisms underpinning statistical patterns involving large sets of species. By taking many different species into account, the variation resulting from contingency is averaged out and broad statistical generalisations have been found, including for example allometric relationships (Hendriks 1999) and relationships between abundance and occupancy (Brown 1984). Life-history strategies may explain a significant part of the scatter around the observed generalisation, as was shown for the relationship between abundance and occupancy (Figure 4). Similarly, strong allometric relationships are found for reproduction rate or life span, while relationships are more variable for a species' density or territory size (Hendriks 2007). This difference between allometric relationships is perhaps not surprising given that traits such as body size, reproduction or life span are basic, measurable properties of a species, whereas a species' density or territory size are more contingent on their surroundings. Furthermore, in addition to body size, the density of a species or the size of its territory typically results from the interplay between the prevailing environmental conditions and various other species traits. When relationships between these other species traits and body mass differ between species, allometric relationships will be more variable. For example, body mass is a poor predictor for development time in species with a larval diapause. Life-history strategies may improve allometric relationships by providing insight in how body mass affects other species traits for a given life-history strategy.

1.7. Merits of Life-History Strategies in Restoration Ecology

A clear understanding of the functioning of the system of interest and changes therein due to anthropogenic impacts is vital for the effectiveness of restoration management. Such an understanding requires a framework of causal mechanisms (Bradshaw 1996; Jansson *et al.* 2005), which may explain how the system naturally functions, how it is impacted and how it can be restored (Hobbs & Norton 1996; Brouwer *et al.* 2005). Thus achieving the goals set in community ecology is what is required for effective restoration management (Palmer *et al.* 1997), blurring the line between fundamental and applied research. Therefore, the lack of progress in community ecology also hampers restoration ecology. For the use of macroinvertebrates in ecological assessment and biomonitoring, water managers have a wide variety of tools at their disposal, ranging from simple diversity indices to complex methods (see Verdonschot 2000 for a review). However, the complex methods (e.g. multivariate analysis techniques) used in ecological assessment and biomonitoring may lead to different conclusions, depending on subjective choices made during the data analysis (Nijboer 2006). These complex methods generate large amounts of information (e.g. multimetrics), making it difficult to identify key mechanistic explanations underpinning patterns in species occurrence (lack of aggregation), while aggregating information into single indices does not incorporate the causal mechanisms (lack of causality).

By providing insight how a particular environment is experienced by the species present life-history strategies may explain why current restoration practices are not or only partially successful and may provide handholds for alternative, more promising ways to achieve restoration. For example, in a case-study on the effects of rewetting bog remnants, life-history strategies were used to relate changes in species composition to the effects of restoration measures in a bog remnant. Water retention favoured species that were adapted to unpredictable conditions, while species adapted to more stable and predictable conditions were constrained. More general, the negative effects of desiccation are not remedied by large scale rewetting, because both degradation and large scale restoration result in a loss of environmental heterogeneity (Van Duinen *et al.* 2003; Verberk *et al.* 2006b). Life-history strategies have also been successfully applied to other species and other systems, including soil arthropods and ants in chalk grasslands (Siepel 1995; van Noordwijk in preparation). Life-history strategies may thus present a useful tool to evaluate and direct restoration management.

1.7.1. The Relation between Heterogeneity and Species Diversity

Environmental heterogeneity is associated with a high species diversity in many terrestrial and aquatic habitats (e.g. Huston 1994; Heino 2000; Kerr *et al.* 2001;). However, what constitutes a heterogeneous environment depends on the species in question, governing the appropriate scale level and environmental factors. This may explain why attempts to formulate general mechanisms underlying the observed relationship have been largely restricted to theoretical models. Theoretical models reduce the issue to a problem of coexistence and formulate ways in which environmental heterogeneity can prevent competitive exclusion (e.g. Tilman 1994; McPeek 1996; Chesson 2000; Amarasekare 2003). Theoretically, two species can coexist if both species can maintain positive growth when they are least abundant, which requires intraspecific competition to outweigh interspecific competition (Chesson 2000).

Although heterogeneous environments with a higher diversity in habitats can axiomatically harbour more species, they may harbour more species compared to the total number of species that would be present if the individual habitats were separated; i.e. the whole is more than the sum of its parts. Explanations for the occurrence of these additional species in heterogeneous landscapes are (i) species requiring certain conditions which only occur in a gradient between two different habitats, (ii) species depending on a combination of two or more different habitats, and (iii) species being able to better persist as a result of lower extinction (Verberk *et al.* 2006a).

The life-history strategy of a species gives information on the likely importance of each of the above explanations. The first explanation (gradient conditions) is a likely candidate for species with a long juvenile growth period (for stable gradients) and/or a high degree of synchronisation (for predictable gradients). The second explanation (multihabitat use) is a likely candidate for species with an active dispersal, having a short juvenile growth period, and/or (relatively) long lived adults. The third explanation (higher persistence) is a likely candidate for species with a high dispersal and a high reproduction rate. Especially semelparous species with high numbers of eggs are expected to achieve high persistence by avoiding competition through strong aggregation (Shorrocks *et al.* 1984).

1.7.2. Restoration of Heterogeneous and Biodiverse Ecosystems

Species are not only passengers, but can also be drivers (sensu Walker 1992). Species aggregate and store resources, depriving other nearby patches from a similar supply. This retention results in decreasing temporal variation (variability) and increasing spatial variation, creating patterns (Rietkerk *et al.* 2002), such as the hummock-hollow structure of a raised bog (Couwenberg & Joosten 2005). In turn, this may promote biodiversity on an ecological timescale, through a further partitioning of the environment in space or by allowing the build up of sufficient population size in between disturbance events. In short, biodiversity begets biodiversity, which has been reported for a range of ecosystems (Estes *et al.* 1978; Knops *et al.* 1999; Ritchie & Olff 1999; Janz *et al.* 2006).

Given the intimate links between species diversity and environmental heterogeneity, it is apparent that environmental heterogeneity plays an important role in the conservation of biodiversity. Environmental heterogeneity is the result of many individuals of many species growing, reproducing and dying (generally living their lives), thus interacting with their surroundings and in the process aggregating, redistributing resources, modifying ecosystem processes, thus shaping the landscape into a mosaic of different patches and gradual transitions in between (Baaijens 1985). Therefore, restoring heterogeneity directly is difficult. Rather, the restoration of the natural processes driving heterogeneity will often be the best option. This may require the restoring of abiotic boundary conditions, active species restoration, or both. Results from the case-study on the effects of rewetting bog remnants (described in Verberk 2008) indicate that restoring the regional hydrology offers a more promising restoration strategy in bog landscapes, leading to a gradual improvement of environmental conditions and strengthening of environmental heterogeneity. Future research on bog restoration may characterise a pristine bog landscape in terms of life-history strategies. This would enhance our understanding of how different parts of a pristine bog landscape function for aquatic invertebrates and provide additional handholds for restoration. Also in riverine landscapes strengthening environmental heterogeneity requires abiotic boundary conditions to be restored. Here protected and endangered riverine species will benefit from restoration of the spatial gradient in hydrodynamics (Ward *et al.* 1999; De Nooij *et al.* 2006). In contrast, the collapse of many coastal systems is attributed to overfishing, and consequently restoration is most likely to succeed through recovery of fish stocks (Jackson *et al.* 2001). Life-history strategies provide insight in the functioning of ecosystems and may help to unravel the importance of abiotic boundary conditions and species interactions. Successful conservation of biodiversity will strongly depend on this ability to identify and subsequently strengthen the processes underlying landscape heterogeneity.

CONCLUSION

The use of life history strategies allows species to be grouped while retaining causal information. This will help the search in community ecology for general relationships. In addition, this is also what is required for effective restoration management, where the focus is mainly on restoring biodiverse landscapes. Consequently this frequently involves large data sets with many different species. Therefore the approach of life-history strategies has much to offer for both community ecology and restoration ecology.

REFERENCES

Abrams, P.A., Leimar, O., Nylin, S. & Wiklund, C. (1996) The effect of flexible growth rates on optimal sizes and development times in a seasonal environment. *The American Naturalist*, 147, 381-395.

Amarasekare, P. (2003) Competitive coexistence in spatially structured environments: a synthesis. *Ecology Letters,* 6, 1109-1122.

Baaijens, G.J. (1985) Over grenzen. *De Levende Natuur,* 87, 102-110.

Begon, M., Harper, J.L. & Townsend, C.R. (1996) *Ecology: Individuals, Populations, and Communities. Third edition.* Oxford, Blackwell Science.

Bell, G. (2001) Ecology - Neutral macroecology. *Science,* 293, 2413-2418.

Berrigan, D. (1991) The allometry of egg size and number in insects. *Oikos,* 60, 313-321.

Bonada, N., Dolédec, S. & Statzner, B. (2007) Taxonomic and biological trait differences of stream macroinvertebrate communities between mediterranean and temperate regions: implications for future climatic scenarios. *Global Change Biology,* 13, 1658-1671.

Bradshaw, A.D. (1996) Underlying principles of restoration. *Canadian Journal of Fisheries and Aquatic Sciences,* 55 (suppl. 1), 3-9.

Brouwer, E., van Duinen, G.A., Nijssen, M.N. & Esselink, H. (2005) Development of a decision support system for LIFE-Nature and similar projects: from trial-and-error to knowledge based nature management. In: Herrier J-L, Mees J, Salman A, Seys J, van Nieuwenhuyse H & Dobbelaere I (Eds.) *Proceedings International Conference on Nature Restoration Practices in European Coastal Habitats,* VLIZ Special Publication 19, Oostende. Pp. 229-238.

Brown, J.H. (1984) On the relationship between abundance and distribution of species. *The American Naturalist,* 124, 255-279.

Chave, J. (2004) Neutral theory and community ecology. *Ecology Letters,* 7, 241-253.

Chesson, P. (2000) Mechanisms of maintenance of species diversity. *Annual Review of Ecology and Systematics,* 31, 343-358.

Cottenie, K., Michels, E., Nuytten, N. & de Meester, L. (2003) Zooplankton metacommunity structure: Regional vs. local processes in highly interconnected ponds. *Ecology,* 84, 991-1000.

Couwenberg, J. & Joosten, H. (2005) Self-organization in raised bog patterning: the origin of microtope zonation and mesotope diversity. *Ecology,* 93, 1238-1248.

de Nooij, R.J.W., Verberk, W.C.E.P., Lenders, H.J.R., Leuven, R.S.E.W. & Nienhuis, P.H. (2006) The importance of hydrodynamics for protected and endangered biodiversity of lowland rivers. *Hydrobiologia,* 565, 153-162.

Dudgeon, D., Arthington, A.H., Gessner, M.O., Kawabata, Z.I., Knowler, D.J., Leveque, C., Naiman, R.J., Prieur-Richard, A.H., Soto, D., Stiassny, M.L.J. & Sullivan, C.A. (2006) Freshwater biodiversity: importance, threats, status and conservation challenges. *Biological Reviews,* 81, 163-182.

Estes, J.A., Smith, N.S. & Palmisano, J.F. (1978) Sea otter predation and community organization in western Aleutian islands, Alaska. *Ecology,* 59, 822-833.

Felsenstein, J. (1985) Phylogenies and the comparative method. *The American Naturalist,* 125, 1-15.

Finn, D.S. & Poff, N.L. (2005) Variability and convergence in benthic communities along the longitudinal gradients of four physically similar Rocky Mountain streams. *Freshwater Biology,* 50, 243-261.

Gaston, K.J., Blackburn, T.M. & Lawton, J.H. (1997) Interspecific abundance-range-size relationships: an appraisal of mechanisms. *Journal of Animal Ecology,* 66, 579-601.

Gayraud, S., Statzner, B., Bady, P., Haybachp, A., Schöll, F., Usseglio-Polatera, P. & Bacchi, M. (2003) Invertebrate traits for the biomonitoring of large European rivers: an initial assessment of alternative metrics. *Freshwater Biology,* 48, 2045-2064.

Gould, S.J. & Lewontin, R.C. (1979) The spandrels of San Marco and the panglossian paradigm: A critique of the adaptationist programme. *Proceedings of the Royal Society of London. Series B, Biological Sciences,* 205, 581-598.

Hartley, S. (1998) A positive relationship between local abundance and regional occupancy is almost inevitable (but not all positive relationships are the same). *Journal of Animal Ecology,* 67, 992-994.

Harvey, P.H. (1996) Phylogenies for Ecologists *Journal of Animal Ecology,* 65, 255-263.

Heino, J. (2000) Lentic macroinvertebrate assemblage structure along gradients in spatial heterogeneity, habitat size and chemistry. *Hydrobiologia,* 418, 229-242.

Hendriks, A.J. (1999) Allometric scaling of rate, age and density parameters in ecological models. *Oikos, 86,* 293-310.

Hendriks, A.J. (2007) The power of size: a meta-analysis of allometric regressions reveals macro-ecological consistency. *Ecological Modelling, 205,* 196-208.

Heywood, V.H. & Watson, R.T. (Eds.) (1995) *Global biodiversity assessment.* Cambridge University Press, Cambridge.

Hobbs, R.J. & Norton, D.A. (1996) Towards a conceptual framework for restoration ecology. *Restoration Ecology, 4,* 93-110.

Huston, M.A. (1994) *Biological Diversity. The coexistence of species on changing landscapes.* Cambridge University Press, Cambridge.

Ilg, C. & Castella, E. (2006) Patterns of macroinvertebrate traits along three glacial stream continuums. *Freshwater Biology, 51,* 840-853.

Jackson, J.B.C., Kirby, M.X., Berger, W.H., Bjorndal, K.A., Botsford, L.W., Bourque, B.J., Bradbury, R.H., Cooke, R., Erlandson, J., Estes, J.A., Hughes, T.P., Kidwell, S., Lange, C.B., Lenihan, H.S., Pandolfi, J.M., Peterson, C.H., Steneck, R.S., Tegner, M.J. & Warner, R.R. (2001) Historical overfishing and the recent collapse of coastal ecosystems. *Science, 293,* 629-638.

Jansson, R., Backx, H., Boulton, A.J., Dixon, M., Dudgeon, D., Hughes, F.M.R., Nakamura, K., Stanley, E.H. & Tockner, K. (2005) Stating mechanisms and refining criteria for ecologically successful river restoration: a comment on Palmer *et al.* (2005). *Journal of Applied Ecology, 42,* 218-222.

Janz, N., Nylin, S. & Wahlberg, N. (2006) Diversity begets diversity: host expansions and the diversification of plant-feeding insects. *BMC Evolutionary Biology, 6,* 4.

Keddy, P.A. (1992) A pragmatic approach to functional ecology. *Functional Ecology, 6,* 621-626.

Kerr, J.T. & Currie, D.J. (1995) Effects of human activity on global extinction risk. *Conservation Biology, 9,* 1528-1538.

Kerr, J.T., Southwood, T.R.E. & Cihlar, J. (2001) Remotely sensed habitat diversity predicts butterfly species richness and community similarity in Canada. *Proceedings of the National Academy of Sciences USA, 98,* 11365-11370.

Knops, J.M.H., Tilman, D., Haddad, N.M., Naeem, S., Mitchell, C.E., Haarstad, J., Ritchie, M.E., Howe, K.M., Reich, P.B., Siemann, E. & Groth, J. (1999) Effects of plant species richness on invasion dynamics, disease outbreaks, insect abundances and diversity. *Ecology Letters, 2,* 286-293.

Koehl, M.A.R. (1996) When does morphology matter? *Annual Review of Ecology and Systematics, 27,* 501-542.

Lawton, J.H. (1993) Range, Population Abundance and Conservation. *Trends in Ecology and Evolution, 8,* 409-413.

Lawton, J.H. (1999) Are there general laws in ecology? *Oikos, 84,* 177-192.

Leibold, M.A., Holyoak, M., Mouquet, N., Amarasekare, P., Chase, J.M., Hoopes, M.F., Holt, R.D., Shurin, J.B., Law, R., Tilman, D., Loreau, M. & Gonzalez, A. (2004) The metacommunity concept: a framework for multi-scale community ecology. *Ecology Letters, 7,* 601-613.

Levin, S.A. (1992) The problem of pattern and scale in ecology: The Robert H. MacArthur Award lecture. *Ecology, 73,* 1943-1967.

McGill, B.J., Enquist, B.J., Weiher, E. & Westoby, M. (2006) Rebuilding community ecology from functional traits. *Trends in Ecology and Evolution,* 21, 178-185.

McPeek, M.A. (1996) Trade-offs, food web structure, and the coexistence of habitat specialists and generalists. *The American Naturalist,* 148, S124-S138.

Mérigoux, S., Dolédec, S. & Statzner, B. (2001) Species traits in relation to habitat variability and state: neotropical juvenile fish in floodplain creeks. *Freshwater Biology,* 46, 1251-1267.

Niggebrugge, K., Durance, I., Watson, A.M., Leuven, R.S.E.W. & Ormerod, S.J. (2007) Applying landscape ecology to conservation biology: Spatially explicit analysis reveals dispersal limits on threatened wetland gastropods. *Biological Conservation,* 139, 286-296.

Nijboer, R.C. (2006) *The myth of communities. Determining ecological quality of surface waters using macroinvertebrate community patterns.* PhD-thesis, Alterra Scientific Contribution nr 17, Wageningen.

Ozinga, W.A., Schaminée, J.H.J., Bekker, R.M., Bonn, S., Poschlod, P., Tackenberg, O., Bakker, J. & van Groenendael, J.M. (2005) Predictability of plant species composition from environmental conditions is constrained by dispersal limitation. *Oikos,* 108, 555-561.

Palmer, M.A., Ambrose, R.F. & Poff, N.L. (1997) Ecological theory and community restoration ecology. *Restoration Ecology,* 5, 291-300.

Resh, V.H., Hildrew, A.G., Statzner, B. & Townsend, C.R. (1994) Theoretical habitat templets, species traits, and species richness - a synthesis of long-term ecological research on the upper Rhône river in the context of currently developed ecological theory. *Freshwater Biology,* 31, 539-554.

Richards, C., Haro, R.J., Johnson, L.B. & Host, G.E. (1997) Catchment and reach-scale properties as indicators of macroinvertebrate species traits. *Freshwater Biology,* 37, 219-230.

Rietkerk, M., Boerlijst, M.C., van Langevelde, F., Hille Ris Lambers, R., van de Koppel, J., Kumar, L., Prins, H.H.T. & de Roos, A.M. (2002) Selforganization of vegetation in arid ecosystems. *The American Naturalist,* 160, 524–530.

Ritchie, M.E. & Olff, H. (1999) Spatial scaling laws yield a synthetic theory of biodiversity. *Nature,* 400, 557-560.

Rosenzweig, C., Karoly, D., Vicarelli, M., Neofotis, P., Wu, Q., Casassa, G., Menzel, A., Root, T.L., Estrella, N., Seguin, B., Tryjanowski, P., Liu, C., Rawlins, S. & Imeson, A. (2008) Attributing physical and biological impacts to anthropogenic climate change. *Nature,* 453, 353-358.

Shorrocks, B., Rosewell, J., Edwards, K. & Atkinson, W. (1984) Interspecific competition is not a major organizing force in many insect communities. *Nature,* 310, 310-312.

Siepel, H. (1994) Life-history tactics of soil microarthropods. *Biology and fertility of soils,* 18, 263-278.

Siepel, H. (1995) Applications of microarthropod life-history tactics in nature management and ecotoxicology. *Biology and Fertility of Soils,* 19, 75-83.

Simberloff, D. (2004) Community ecology: Is it time to move on? *The American Naturalist,* 163, 787-799.

Southwood, T.R.E. (1977) Habitat, the templet for ecological strategies? *Journal of Animal Ecology,* 46, 337-365.

Statzner, B., Resh, V.H. & Dolédec, S. (1994) Ecology of the upper Rhône river: a test of habitat templet theories. *Freshwater Biology*, 31, 253-554.

Statzner, B., Hildrew, A.G. & Resh, V.H. (2001) Species traits and environmental, constraints: Entomological research and the history of ecological theory. *Annual Review of Entomology*, 46, 291-316.

Statzner, B., Dolédec, S. & Hugueny, B. (2004) Biological trait composition of European stream invertebrate communities: assessing the effects of various trait filter types. *Ecography*, 27, 470-488.

Stearns, S.C. (1976) Life-history tactics: A review of the ideas. *Quarterly Review of Biology*, 51, 3-47.

Tilman, D. (1994) Competition and biodiversity in spatially structured habitats. *Ecology*, 75, 2-16.

Townsend, C.R., Dolédec, S. & Scarsbrook, M.R. (1997) Species traits in relation to temporal and spatial heterogeneity in streams: A test of habitat templet theory. *Freshwater Biology*, 37, 367-387.

Usseglio-Polatera, P., Bournaud, M., Richoux, P. & Tachet, H. (2000) Biological and ecological traits of benthic freshwater macroinvertebrates: relationships and definition of groups with similar traits. *Freshwater Biology*, 43, 175-205.

van Duinen, G.A., Brock, A.M.T., Kuper, J.T., Leuven, R.S.E.W., Peeters, T.M.J., Roelofs, J.G.M., van der Velde, G., Verberk, W.C.E.P. & Esselink, H. (2003) Do restoration measures rehabilitate fauna diversity in raised bogs? A comparative study on aquatic macroinvertebrates. *Wetlands Ecology and Management*, 11, 447-459.

van Kleef, H.H., Verberk, W.C.E.P., Leuven, R.S.E.W., Esselink, H., van der Velde, G. & van Duinen, G.A. (2006) Biological traits successfully predict the effects of restoration management on macroinvertebrates in shallow softwater lakes. *Hydrobiologia*, 565, 201-216.

van Kleef, H.H., Kimenai, F.F.P., Leuven, R.S.E.W., Verberk, W.C.E.P., van der Velde, G. & Esselink, H. (*in preparation*) Functional response of chironomids to restoration and decreased acidification over a 21 year period in formerly acidified shallow softwater lakes.

van Noordwijk, C.G.E., Boer, P., Mabelis, A.A., Verberk, W.C.E.P. & Siepel, H. (*in preparation*) Life-history tactic analysis reveals fragmentation and low soil temperature as main bottlenecks for ant communities in Dutch chalk grasslands.

Verberk, W.C.E.P., van Kleef, H.H., Dijkman, M., van Hoek, P., Spierenburg, P. & Esselink, H. (2005) Seasonal changes on two different spatial scales: response of aquatic invertebrates to water body and microhabitat. *Insect Science*, 12, 263-280.

Verberk, W.C.E.P., van Duinen, G.A., Brock, A.M.T., Leuven, R.S.E.W., Siepel, H., Verdonschot, P.F.M., van der Velde, G. & Esselink, H. (2006a) Importance of landscape heterogeneity for the conservation of aquatic macroinvertebrate diversity in bog landscapes. *Journal for Nature Conservation*, 14, 78-90.

Verberk, W.C.E.P., Kuper, J.T., van Duinen, G.A. & Esselink, H. (2006b) Changes in macroinvertebrate richness and diversity following large scale rewetting measures in a heterogeneous bog landscape. *Proceedings of the Section Experimental and Applied Entomology of the Netherlands Entomological Society (NEV)*, 17, 27-36.

Verberk, W.C.E.P. (2008) *Matching species to a changing landscape – Aquatic macroinvertebrates in a heterogeneous landscape.* Proefschrift, Radboud Universiteit, Nijmegen.

Verberk, W.C.E.P., Siepel, H. & Esselink, H. (2008a) Life history tactics in freshwater macroinvertebrates. *Freshwater Biology,* 53, 1722-1738.

Verberk, W.C.E.P., Siepel, H. & Esselink, H. (2008b) Applying life history tactics for freshwater macroinvertebrates to lentic waters. *Freshwater Biology,* 53, 1739-1753.

Verdonschot, P.F.M. (2000) Integrated ecological assessment methods as a basis for sustainable catchment management. *Hydrobiologia,* 422/423, 389-412.

Violle, C., Navas, M.-L., Vile, D., Kazakou, E., Fortunel, C., Hummel, I. & Garnier, E. (2007) Let the concept of trait be functional! *Oikos,* 116, 882-892.

Waide, R.B., Willig, M.R., Steiner, C.F., Mittelbach, G., Gough, L., Dodson, S.I., Juday, G.P. & Parmenter, R. (1999) The relationship between productivity and species richness. *Annual Review of Ecology and Systematics,* 30, 257-300.

Walker, B.H. (1992) Biodiversity and ecological redundancy. *Conservation Biology,* 6, 18-23.

Ward, J.V., Tockner, K. & Schiemer, F. (1999) Biodiversity of floodplain river ecosystems: ecotones and connectivity. *Regulated Rivers: Research & Management,* 15, 125-139.

Westoby, M., Leishman, M.R. & Lord, J.M. (1995) On misinterpreting the 'phylogenetic correction'. *Journal of Ecology,* 83, 531-534.

Weiner, J. (1995) On the practice of ecology. *Journal of Ecology,* 83, 153-158.

Wiens, J.A. (1989) Spatial scaling in ecology. *Functional Ecology,* 3, 385-397.

In: Encyclopedia of Environmental Research   ISBN: 978-1-61761-927-4
Editor: Alisa N. Souter   © 2011 Nova Science Publishers, Inc.

Chapter 32

NUTRITIONAL APPROACHES FOR THE REDUCTION OF PHOSPHORUS FROM YELLOWTAIL (*SERIOLA QUINQUERADIATA*) AQUACULTURE EFFLUENTS

Pallab Kumer Sarker[1] and Toshiro Masumoto[2]
[1]Department of Animal Sciences, Laval University, Quebec, Cananda
[2]Faculty of Agriculture, Kochi University, Nankoku, Japan

ABSTRACT

The environmental impact of phosphorus (P) waste from the aquaculture industry is increasingly a matter of concern in Japan and elsewhere around the world. Over the past two decades, increasing concerns over excessive P loading have resulted in a large number of studies aimed at better understanding issues related to P output from aquaculture production. Most of these studies were fresh water fish species, however, particular attention of P loading in marine aquaculture species and adoption of nutritional approaches for overcoming that dilemma has not been extensively studied. Yellowtail aquaculture in Japan is the highest of any farmed fish species and 57% of total Japanese fish culture production consists of yellowtail. The reduction in P from yellowtail culture will lead to a great contribution towards the goal of reducing P from the aquaculture industry in Japan. There are some approaches have been adopted to reduce P effluent from yellowtail aquaculture based on series of long term research and this recent research demonstrates that technology can be effective to meet this end. Feed formulation improvements aimed at improving digestibility and retention of nutrients by fish are key to reducing dissolved and solid waste outputs. Careful selection of more highly digestible practical ingredients, highly available inorganic P sources and judicious use of additives, such as phytase can contribute to improve digestibility of certain feed nutrients and potentially reduce solid and P waste outputs. Excretion of non-fecal soluble P is greatly increased as dietary available P is increased above the requirement level and that excess P in diets cause environmental pollution is increasingly recognized as a serious ecological problem.   Therefore, it is of importance not to feed more P than the fish require. In fish culture operations, the majority of feed is fed to larger fish. The dietary requirement of P for larger fish is less than to smaller fish. P requirement for large fish through non-fecal (balance study) approach may greatly help to formulate low P loading diet and ultimately contribute to reduce effluent P from aquaculture operation. Fish meal is the source of the most dietary P in fish commercial diets, wherein it exists as hydroxyapatite (bone P) or tricalcium phosphate which is poorly absorbed in fish intestine. Due to that low absorption of fish meal, currently feed manufacturer are interested to supplement readily available inorganic P in commercial feed along with fish meal source. However, for larger fish, P is needed only for the physiological maintenance rather than the growth. Study revealed that P supplementation is not necessary for the fish meal-based diet for large yellowtail and elimination of P supplementation to the fish meal-based diet resulted in a 34.5% reduction in non-fecal soluble P from yellowtail aquaculture effluent. The overall study suggested the key formulation of low P diet for yellowtail that will contribute to reduce the pollution in aquaculture effluents.

INTRODUCTION

Aquaculture as an industry has been generating significant economic activity in rural and coastal communities in Japan and elsewhere around the world. The environmental impact of fish culture operations is becoming a matter of close scrutiny by the public and Government. Long-term sustainability of aquaculture is, therefore, very much related to how effectively culture operations can manage or minimize their release of waste. Phosphorus (P) is an essential nutrient for fish, being a major constituent of skeletal tissues, nucleic acids DNA and RNA, energy transport compounds such as ATP, and of phospholipids in cell membranes [Lall 1991]. Phosphorus deficiency typically reduces growth rate and causes skeletal deformities, both of which impair productivity.

However, surplus P or indigestible P contained in the feed is excreted as inorganic phosphate mainly as non-fecal or fecal form [Bureau et al. 1999; Coloso et al. 2003]. The main concern is the release of solid and dissolved P and nitrogen (N) waste by fish culture operations since these wastes can impose constraints to the productivity of operations and may lead to environmental degradation. The N waste can be minimized by decreasing the dietary digestible protein (DP)/dietary energy (DE) and increasing dietary non protein energy content which enhance the nitrogen retention efficiency. However, P is treating as rate limiting nutrient in the eutrophication and one of the most troublesome component of the aquaculture effluents that is very difficult to manage easily. The excessive P into the effluents causes algal blooms, depleting dissolved oxygen in water and killing aquatic organisms.

Commercial diets for yellowtail have traditionally been formulated to contain high levels of fish meal (more than 50%) and other sources of animal by-products. These ingredients usually contain surplus P as available or complex form, and fish will excrete any excess or unavailable P as inorganic phosphorus mainly in the form of non-fecal or feces. High P waste in effluents from yellowtail culture causes the pollution of farm sites becomes more frequent. Therefore, reducing output of P wastes is considered a key element for the long-term sustainability of yellowtail aquaculture in Japan.

PROFILE OF EFFLUENT PHOSPHORUS IN AQUACULTURE

The rapid expansion and intensification of the aquaculture industry over the past two decades has brought about increasing concerns related to its environmental impact. Despite the significant reduction of P discharge from aquaculture, P retention in fish remains relatively low. P retention estimates as high as 55 % have been reported [Azevedo et al. 1998], although a summary of work done in this area [Dosdat 1992] shows that P retention rarely exceeds 30 % under practical conditions. Variation of P retention between studies can be attributed to differences in diet composition (digestibility and P content) and the size and physiological status of the fish. Non-assimilated phosphorus in the feces represents 40-50% of that supplied by the diet with 60-80% being in of particulate form [Cripps and Bergheim 2000]. Approximately 20% of the dietary P is excreted by the kidneys, but can account 90% of the unretained P excreted as soluble inorganic phosphate as non-fecal form. An important potential source of P is that found in uningested feed, which is particulate in nature, but may represent an important factor in the accounting of P loss from feeding fish.

APPROACHES OF PHOSPHORUS WASTE REDUCTION

Excess P loading from aquaculture industry arises uniquely from the feed, either as uningested feed or unassimilated and excreted fractions. This fact provides a clear opportunity in that it limits, and thereby simplifies the possible approaches to address the dilemma. To date approaches to reduce P loading from aquaculture have centred on technologies that improve production efficiency or reduce feed waste, those that remove the solid and dissolved fractions of waste P from effluent water. However, a fundamental and key aspect to any waste management plan is its reduction at the source. Lall [1991] emphasized several aspects related to nutritional approaches to reduce P out put from aquaculture industry. Our recent studies have indeed demonstrated that the nutritional strategies that reduce P at its source using a number of approaches:

Highly Digestible Practical Ingredients

The determination of nutrient availability is the first step in evaluating the potential of an ingredient for use in the diet of an aquaculture species [Allan et al. 2000]. Information on nutrition availability of feed ingredients is very useful not only to formulate diets that maximize the growth of fish by providing appropriate amounts of available nutrients but also to limit waste products from fish [Lee 2002]. In commercial yellowtail culture, most feed is used for growing larger fish (> 1 kg) farmed in sea cages. However, there is no published information presenting detailed ingredient availability for these fish. Small improvements in feed efficiency at this stage of production can greatly enhance the economic viability of yellowtail aquaculture and significantly reduce the environmental impacts associated with feed ingredients and feeding of fish. Two fish meal sources (anchovy, AF and jack mackerel, JMF), defatted soybean meal, SBM and corn gluten meal, CGM were tested. Study revealed that P digestibility of anchovy meal was better than jack mackerel meal in yellowtail and the value was 42.6 and 34.4 % respectively (Fig. 1). These digestibility values were higher than two plant protein sources (SBM and CGM). Digestibility was similar in SBM 30.0% and CGM 31.3%. Digestible P content of AF, JMF, SBM and CGM were found 0.77, 0.66, 0.09 and 0.06 % respectively.

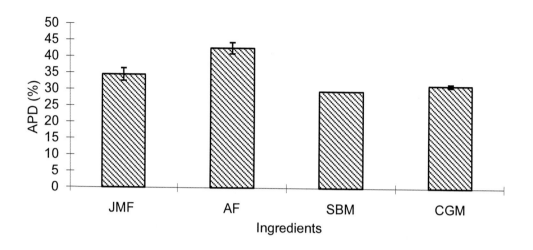

Figure 1. Apparent P digestibility of selected feed ingredient (Sarker 2007).

The reason why better digestibility found in JMF, this might be due to lower total P content in AF compared to JMF. The total P content in AF was about 1.5 times lower than JMF (Table 1). This could indicate that the lower the dietary concentration of phosphorus, the better is its availability. Similar results have been reported in rainbow trout [Vielma and Lall 1997; Burel et al. 2000] and turbot [Burel et al. 2000]. Furthermore, ash content of diet may be another factor that can modulate the availability. Ash content of JMF was higher than that of AF. The increase amount of ash in the diet is most likely due to the presence of bone and scale that may be attributable to differences in mineral composition of meals due to differences in ash content of fish itself or in processing method. It is commonly known that bone P is poorly absorbed in fish intestine, especially at high dietary inclusion levels. Sugiura [1998] also stated that if the level of fish meal or fish bone in the diet is reduced, digestibility of P increases. This indicates that the high ash containing fish meal diet is conditionally available depending on the concentration in the diet.

Table 1. Total, available and undigestible P content of selected test ingredients.

Ingredient	Total P (g/kg)	Digestible P (g/kg)[1]	undigestible P (g/kg)[2]
JMF	22	6.6	15.4
AF	16	7.7	8.3
SBM	2	0.9	1.5
CGM	2	0.6	1.4

[1] Available P = (total Pin test diet × APA)/100 (Sarker et al. 2007)
[2] Unavailable P = total P - available P

The APD value of CGM for yellowtail is similar to that observed in rain bow trout [Riche & Brown, 1996] and Nile tilapia [Köprücü et al 2005]. The APD value of SBM in our study was similar to that reported in rainbow trout [Satoh et al. 2002]. Comparatively lower APD values of CGM and SBM meal might be due to the presence of higher phytate content. P present in phytate is known to be unavailable to fishes due to the lack of endogenous or microbial phytase in their intestinal tracts, which is required to digest phytate [Lall, 1991]. In order to improve the plant P availability, some investigators have been suggested that supplementation or treatment of phytase is effective. Phytase has effectively been increased the P availability in rain bow trout fed SBM and CGM [Ramseyer et al. 1999]. Since most of the unavailable P from fish meal and plant protein source of commercial diet are becoming enormous increase the effluent P excretion in yellowtail aquaculture. Therefore, the availability of SBM and CGM in the present study indicated that we need to improve the availability of both the plant sources and then we can use SBM and CGM in diet formulation that may play an important role to reduce effluent P from yellowtail aquaculture. It has found that even within a fish meal, there was a considerable difference in P digestibility, yellowtail demonstrated higher digestibility in anchovy than that of Jackmackerel meal. Lower digestibility of SBM and CGM in the present study indicated that it may better to improve the digestibility of both the plant sources that in turn help to reduce effluent P from yellowtail aquaculture.

Use of Feed Additives

Dietary supplementation of citric acid, Na citrate, and EDTA was able to improve fish meal P digestibility in rainbow trout [Sugiura et al. 1998]. The effect was probably due to the solubilization of bone minerals in fish meal, as well as a chelating effect that reduces the antagonistic interaction between Ca and P that could precipitate Ca and P at the intestinal brush border. Vielma and Lall [1997] utilized formic acid 4 and 10 ml/kg in diet to significantly improve P digestibility from 70% to 74%, 75% for fish meal based diet. Vielma et al. [1999] found that supplementing citric acid 4, 8, 16g/kg diet to 28% of herring bone meal linearly increased body ash concentration but had no significant effect on body P concentration. They cautioned the use of acidified diets because of the possible disturbance of acid-base balance and mineral homeostatis. However, more research is also warranted in this aspect.

Phytic acid has long been known as the major storage form of phosphorus (P) in plants especially in legume, cereal and oil seeds. Phytic acid has been shown to have a strong antinutritive effect [Kerovuo 2000] which forms insoluble complexes with di and trivalent minerals, rendering these minerals unavailable to fish. Approximately two-thirds of total phosphorus in various plant ingredients is present as phytic acid [Ketola 1994]. It has been known that fish can not utilize P bound in the phytic acid complex molecule because they lack the phytase enzyme needed to hydrolyse bound phytatic acid P to available form of P.

Numerous studies have shown that dietary incorporation of commercial phytase which can hydrolyse the phytate complex bound to simpler form thereby increase the bioavailability of mineral elements [Persson et al. 1998] and holds promise for reduced effluent P and less P pollution. Use of commercial phytase has been reported to improve phytate phosphorus bioavaliability in poultry, swine and fish which should decrease the amount of phosphorus excreted [Nelson et al. 1971; Simons et al. 1990; Crowell et al. 1993; Rodehutscord et al. 1995; Schafer et al. 1995; Lanari et al. 1998; Veilma et al. 1998; Vielma et al. 2000].

Supplementation of phytase in plant based diet is one of the promising approaches for reducing N and P discharge in aquaculture operation by improving P and protein utilization. We conducted a preliminary study using Japanese flounder (*Paralichthys olivaceus*) and results indicated that phytase in the diets improved P and protein retention [Sarker et al. 2006]. Two isonitrogenous and isoenergetic diet were prepared in this study (Table 2). The two dietary groups were: diet containing 0 FTU/ kg phytase (Natuphos [TM], phytase activity 10,000 U/g, BASF corporation) (control) and diet containing 3000 FTU/kg diet (phytase).

Table 2. Ingredient and analytical composition of experimental diets.

Dietary group	Control	Phytase
Phytase (FTU/kg diet)	0	3000
Brownfish meal	400	400
Defatted soybean meal	290	290
Krill meal	100	100
Blood meal	50	50
L-Lysine	2	2
L-Methionine	4	4
Fish oil	20	20
Vitamin mixture[1]	20	20
P freemineral mixture[2]	20	20
Phytase	0	0.3
α Cellulose	44	437
αStarch	50	50
Total	1000	1000
Dry matter	642	64.3
Crude protein	544	54.2
Crude lipid	77	7.7
Crude sugar	100	10.2
Crude ash	96	9.6
Energy (kcal/kg)	3541	354.0
P content		
Total P	13	13
Soluble P	5	6

1 Contains (mg/100g): Thiamine HCl 2.2; Riboflavin 2.2; Pyrodoxin HCl 2.3; Nicotinic acid 9.6; Ca-pantothenate 7.2; Inositol 60.0; Biotin 0.14; Folic acid 2.4; Cholin chloride 300.0; Cyanocobalamin 0.04; Ascorbic acid 21.6; Vitamin A palmitate 1.1; α- Tocopherol 20; α-Cellulose 1571.22

2 Contains (mg/100g): NaCl 87; $MgSO_4$. $7H_2O$ 274; Fe-citrate 59.4; Ca-lactate 654; $AlCl3$. $6H_2O$ 0.36; $ZnSO4$. $7H_2O$ 7.14; $MnSO4$. $4H_2O$ 1.6; CuCl 0.22; KI 0..34; $COCl2$. $H2O$ 2.1; α- Cellulose 925.6 Source: [Sarker et al. 2006]

Table 3. Apparent retention of nutrients of Japanese flounder fed control and phytase supplemented diet.

Dietary group	Control	Phytase
Phytase (FTU/kg diet)	0	3000
Phosphorus (%)	28.9 ± 5.1 [a]	41.3 ± 6.8[ab]
Protein (%)	20.1 ± 0.3 [a]	28.5 ± 2.5 [b]

*Mean values with standard deviation for five fish per replication. Values within a row with same superscript letters are not significantly different, $P > 0.05$. Source: [Sarker et al. 2006]

It is reported that improved P retention was found with phytase (2000 FTU/kg diet) supplemented in soybean meal and canola meal containing diet in stripped bass [Papatryphon, 1999] and in Atlantic salmon [Sajjadi et al, 2004], respectively. In our study, phytase could enhance the protein retention. The improvement of protein retention might be explanations of the hypothesis that phytase reduces phytate-protein complexes in the gut and cause the improvement of the utilization.

Table 4. Performance of Japanese flounder fed control and P supplemented diets.*

Dietary group	Control	Phytase
Phytase (FTU/kg diet)	0	3000
Initial mean body wt. (g)	9.0 ± 0.0	9.0 ± 0.0
Final mean (g)	23.7 ± 0.7^a	28.2 ± 3.3^{ab}
Mean body wt. gain (%)	163.8 ± 8.5^a	213.25 ± 36.2^{ab}
[1]SGR (%)	2.3 ± 0.0^a	2.8 ± 0.2^b
[2]FCR	1.4 ± 0.2^b	1.1 ± 0.1^{ab}
[3]Daily feeding rate (%)	3.3 ± 0.6^a	2.7 ± 0.1^a
[4]Feed efficiency (%)	40.0 ± 5.9^a	52.9 ± 0.8^{ab}
[5]Protein efficiency ratio	0.73 ± 10.8^a	0.95 ± 1.4^{ab}
Survival rate (%)	97.5^a	100.0^a

*Values are means of duplicate tank and values in the same row containing different letter superscripts were significantly different ($P<0.05$). [1] SGR is the specific growth rate and measured as SGR= $100\times$ {In (Final wt.) − In (initial wt.)}/Duration (days). [2] FCR is the feed conversion ratio and measured as FCR= feed intake/ weight gain. [3] Daily feeding rate= $100\times$ daily feed intake/ body weight
[4] Feed efficiency (%) = $100\times$ weight gain/feed intake
[5] Protein efficiency ratio = g biomass gain/ g protein fed
Source: [Sarker et al. 2006]

The advantageous effects of phytase on growth performance were also confirmed in this experiment (Table 4). The SGR of fish fed phytase diet had significantly higher value than fish fed control diet. It has been reported that dietary phytase improved growth and overall performance with reducing P excretion in rainbow trout [Rodehutscord and Pfeffer 1995, Sugiura et al. 2001]; carp [Schafer et al. 1995]; and stripped bass [Papatryphon et al. 1999]. In this study, soybean meal based diet with phytase supplementation could significantly reduce P discharge (Fig. 2).

The reduction of P discharge by feeding phytase supplemented soybean meal diet might be due to increase of available P and ensured growth of the fish. Based on the data presented here, it is demonstrated that phytase could significantly be reduced the discharge in marine fish fed soybean meal containing diet and that ultimately eliminated P waste in aquaculture effluents. This preliminary study suggested that phytase might also play a pivotal role to reduce P wastage from yellowtail aquaculture.

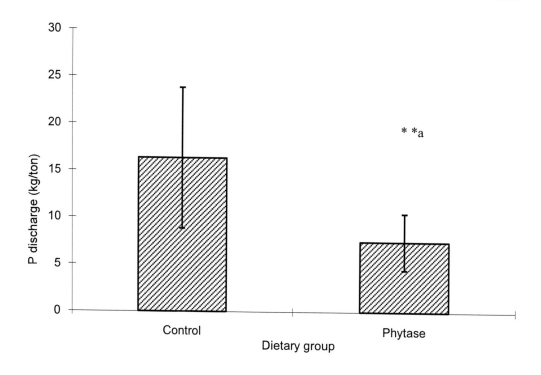

Figure 2. Discharge of P from Japanese flounder fed the experimental diets. It was significant differences ($P < 0.05$) between dietary groups. Values are means ± SD (n = 2). Source: [Sarker et al. 2006].

HIGHLY AVAILABLE INORGANIC P SOURCES

Lower supply of fish meal in aquaculture feed industries lead to increase use of plant proteins as an alternative fish meal sources. P content of plant protein sources is much less than fish meal and P in the plant ingredients is mostly in unavailable form of phytic acid. In order to ensure the satisfactory level of available P content in the fish diet with plant ingredients, supplementation of inorganic P to the feeds is necessary. Therefore, the development of effective nutritional strategies to maximize the utilization of P and minimize the P load in aquaculture effluents requires an accurate estimation of availability and available P content inorganic P sources. Five supplemental inorganic P sources (Wako pure chemical industries ltd., Japan) were chosen for determining the apparent P availability (APA) in yellowtail. They were potassium phosphate monobasic, KH_2PO_4; sodium phosphate monobasic, $NaH_2PO_4 \cdot 2H_2O$; calcium phosphate monobasic, $Ca(H_2PO_4)_2 \cdot H_2O$; calcium phosphate dibasic, $CaHPO_4 \cdot 2H_2O$; and calcium phosphate tribasic, $[Ca_3(PO_4)_2]_3 \cdot Ca(OH)_2$. There are some methods used to determine bioavailability of elements to fish, such as digestibility studies, retention studies, and primary and secondary dose response indicators. There are remarkable variations in availability values by researchers even within the same fish species. An accurate evaluation of P availability is critically difficult due to the differences in species, experimental condition as well as methodological approaches of fecal collection etc. Some investigators applied the classical balance study concept to know the effect of diet on the non-fecal soluble P excretion and they investigated that when fish intake surplus available P, excess P mostly excluded through gills and kidney as non-fecal soluble P [Bureau and Cho, 1999; Rodehutscord et al., 2000; Sugiura et al., 2000; Coloso et al., 2003]. The non-fecal soluble P excretion is also dependent mainly on the sources of P, i.e., when fish ingest surplus highly available inorganic P, the non-fecal soluble P excretion would markedly be higher compared with low or modestly available sources of P [Coloso et al., 2003; McDaniel et al., 2005]

In this study, along with the measurement of apparent availability of P, the non-fecal soluble P excretion approach was also applied to select appropriate inorganic P source with higher P availability in yellowtail fed five inorganic P sources.

Table 5. Availability (±SE) values in yellowtail fed five inorganic P source containing experimental diets.

Inorganic P source	Availability (%)
K- mono- P	96.1 ± 0.5^a
Na- mono- P	94.8 ± 1.0^a
Ca- mono- P	92.4 ± 1.0^a
Ca- di- P	59.2 ± 1.6^b
Ca- tri- Pu	48.8 ± 6.6^b

*Values (SE, n = 3 tanks per diet and triplicate samples for each tank) in column with similar superscripts are not significantly different ($P > 0.05$).

[**Source**: Sarker 2009b]

Our study revealed that the inorganic P sources significantly influenced P availability in yellowtail (Table 5). Among all dietary inorganic P sources, the availability values of K (96.1%), Na (94.8%) and Ca (92.4%) monobasic were significantly higher ($P < 0.05$) than those of di (59.2%) and tri (48.8%) calcium sources. It has been reported that for rainbow trout, apparent availability values of inorganic P were 71% for Ca di, 93% to 98% for Ca, K and Na mono P [Ogino et al. 1979; Gregus 2000]. Lovell [1978] and Lall [1991] have also been stated that Ca and Na mono P were highly available source than di and tri basic source for Atlantic salmon and channel catfish. Our results are also consistent with their availability of all monobasic sources. For carp, Ogino et al. [1979] reported that availabilities of Ca di and tri basic had 46 and 13%, respectively. However, those values were lower than the values reported in rainbow trout (71 and 64% respectively). Availabilities of P from Ca di and tri basic sources in this study were 59 and 48% respectively in yellowtail. These values in yellowtail were slightly lower than in rainbow trout and higher than in carp. The reason why lower availability of Ca di and tri basic sources in carp compare to rainbow trout and yellowtail has been thought that the absence of acidic gastric juice from stomach in carp because the availability of P is affected by solubility [Lall 2002; Hua and Bureau 2006] and these P sources are soluble in acidic condition [Ogino et al., 1979]. However, the availability of Ca di and tri basic sources were lower in yellowtail than in rainbow trout, although both fish species possess stomach. Probably, there were different acidity in the stomach of these two fish species and it might influence the solubility of inorganic P and ultimately modulate the availability. The digestive acidity of rainbow trout as fresh water fish is expected to be stronger than marine water fish like yellowtail. Kofuji et al. [2004] has been found that the stomach pH of yellowtail was about 5.5 that is suspected to be weaker acidic condition than rainbow trout.

In line with the availability, the non-fecal P excretion supported availability estimation of different P sources. Every h for 24 h after feeding of non-fecal P excretion pattern showed that the peak height at 5 h after feeding differed and depends on the availability of inorganic P sources (Fig. 3).

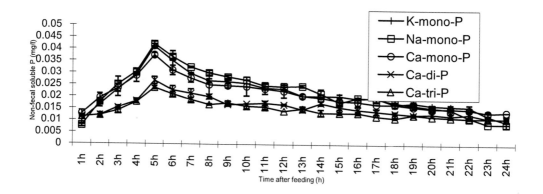

Figure 3. Daily pattern of non-fecal soluble P in effluent water yellowtail fed five inorganic P sources. Each point represents the average (± SE) of three tanks per diet. The amount of non-fecal P excretion was determined based on P concentration in water and the water flow for every h after feeding over 24 h period on feeding day 9. [Source: Sarker 2009b]

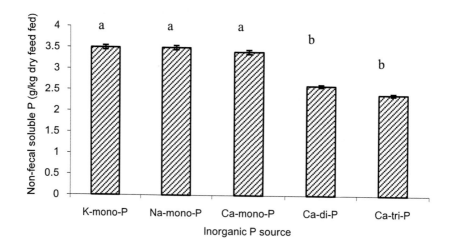

Figure 4. Non-fecal soluble P content in effluent water yellowtail fed five inorganic P sources. Each point represents the average (± SE) of three tanks per diet. Ca-di-P and Ca-tri-P were significantly lower than other monobasic sources. The amount of non-fecal P content was determined based on P concentration in water and the water flow for every h after feeding over 24 h period on feeding day 9. [Source: Sarker 2009b]

Ca di and tri basic sources showed small peak at 5 h after meal ingestion. This means that the lower availability of these two sources reduced the absorption of P by fish, appeared the weak peak in the effluent at 5 h after feeding. Available P content calculated from the availability value of different inorganic P sources showed that all mono basic sources contain higher amount of available P than Ca di and tri basic sources.

The higher contents of non-fecal (g/kg dry diet) and lower contents of fecal (g/kg dry diet) P in K, Na and Ca mono basic sources compared to Ca di and tri basic sources, indicating that K, Na and Ca mono basic sources are highly available form of inorganic P for yellowtail and that are readily absorbed by the intestine and therefore lower content of P found in the feces (Fig. 4 and 5).

In summary, this study suggests that the apparent availability of K (96.1%), Na (94.8%) and Ca (92.4%) phosphate mono basic sources are better and highly available for yellowtail than di (59.2%) and tri (48.8%) basic sources of Ca. This study also shows that the estimation of non- fecal soluble P excretion is useful index for comparing the apparent availability of P from different P sources.

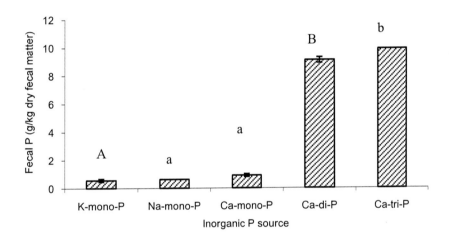

Figure 5. Fecal P content of yellowtail fed five inorganic P sources. Fecal samples were collected from tank outlets using nylon monofilament bags (200-μm mesh size plankton net) every 12 h over a 24 h period. Each point represents the average (± SE) of three tanks per diet and triplicate samples for each tank. Ca-di-P and Ca-tri-P were found to be significantly lower than other monobasic sources. [Source: Sarker 2009b]

Available P Requirement for Large Fish

Estimates of dietary nutrient requirements and availability of nutrients with feed ingredients are essential information required to formulate feeds. The precise formulation of feed allows us to minimize feed cost and nutrients loading into environment. Most nutritional requirements have been determined for juvenile (smaller size) fish but not for post-juvenile or grow-out (larger size) fish. The growth rate of larger fish is slower than that of smaller fish making it difficult to estimate dietary requirement of large fish because a longer time is needed to cause different growth responses associated with feeding different levels of dietary nutrients. Sugiura et al. [2000] applied a classical balance study concept of a minimum P requirement for large rainbow trout (*Oncorhynchus mykiss*). This concept was based on the observation that soluble P excretion occurs only when the diet contents excess available P. This concept has been supported by other investigators [Bureau and Cho 1999; Rodehutscord et al. 2000; Coloso at al. 2003]. Sugiura et al. [2000] found the minimum dietary requirement of available P for large rainbow tout was 5.54 g/kg dry diet and suggested that the non-fecal P measurement approach could be an alternative way to determine the dietary P requirement of larger fish.

We estimated the minimum P requirement for large yellowtail by measuring non-fecal P when feeding diets containing graded levels of P (Table 6).

Table 6. Ingredients and analytical composition of the experimental diets.

Ingredients (g kg^{-1})	Diet code					
	P 0.2	P 2.3	P 4.5	P 6.5	P 7.9	P 10.0
Egg white	600	600	600	600	600	600
Pollock liver oil	150	150	150	150	150	150
Dextrin	50	50	50	50	50	50
Vitamin mixture[1]	20	20	20	20	20	20
P-free mineral mixture[2]	80	80	80	80	80	80
Feeding stimulant[3]	5	5	5	5	5	5
CMC-Na	25	25	25	25	25	25
Guar gum	5	5	5	5	5	5
Cr_2O_3	5	5	5	5	5	5
$NaH_2PO_4 \cdot 2H_2O$	0	10	20	30	40	50
A-Cellulose	60	50	40	30	20	10
Total	1000	1000	1000	1000	1000	1000
Chemical analysis (g kg^{-1})						
Crude protein	527	523	520	522	526	525
Crude ash	66	65	79	67	82	68
Total phosphorus	0.3	3	5	7	9	11
Available P	0.2	2.3	4.5	6.5	7.9	10

1 Contains (mg kg -1): Thiamine HCl 22; Riboflavin 22; Pyrodoxin HCl 23; Nicotinic acid 96; Ca-pantothenate 72; Inositol 600; Biotin 1.4; Folic acid 24; Cholin chloride 3000; Cyanocobalamin 0.4; Ascorbic acid 216; Vitamin A palmitate 11; α- Tocopherol 200; α- Cellulose 15712.2
2 Contains (g kg -1): NaCl 870; MgSO4.7H2O 2740; Fe-citrate 594; Ca-lactate 6540; AlCl3.6H2O 3.6; ZnSO4.7H2O 71.4; MnSO4.4H2O 16; CuCl 2.2; KI 3.4; COCl2.H2O 21; α- Cellulose 9256
3 Contains (g kg -1): Proline 1.75; Alanine 1.12; 5'-IMP 2.12
[**Source**: Sarker 2009a]

The daily non-fecal P excretion pattern on day 9 of fish fed graded levels of P for every h over 24 h after feeding is shown in Fig. 6. There was no appreciable peaks in fish fed diets with P concentrations lower than 4.5 (g/kg) but a clear peak was noted at 5-6h after feeding in fish fed diets more than 4.5 g P /kg. The peak height was higher and duration of peak was longer for fish fed higher concentrations of dietary P. Broken line and nonlinear polynomial regression analysis ($y = 0.0557x^2 - 0.1676x + 1.4313$, $r^2 = 0.9398$) were performed with soluble P excretion (g soluble P/ kg dry diet) from yellowtail fed different levels of available P (g available P/kg dry diet) (Fig. 7). Based on the 24 h excretion amount of non-fecal soluble P in the effluent water, the minimum P requirement for yellowtail was estimated. The non-fecal soluble P excretion (g/kg dry diet) was unaffected by P intake up to 4.4 g available P/kg dry diet as estimated by broken line analysis. This means that 4.4 (g available P/kg dry diet) for 1 kg yellowtail reached maximum retention, indicating the requirement was met.

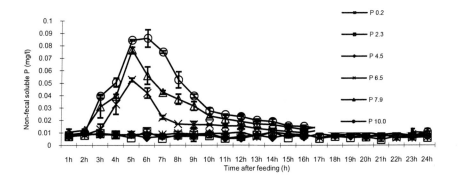

Figure 6. Daily pattern of non-fecal soluble P in effluent water yellowtail fed graded level P. Each point represents the average (± SE) of two tanks per diet. The amount of soluble P excretion was determined based on P concentration in water and the water flow for every h after feeding over 24 h period on feeding day 9. . [Source: Sarker 2009a]

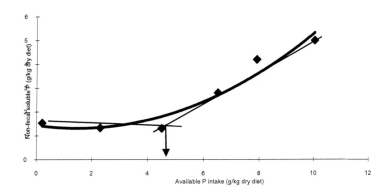

Figure 7. Broken and non linear polynomial regression line between non-fecal soluble P excretion and graded level P intake (Experiment 1). Each point represents the average (± SE) of two tanks per diet. The amount of soluble P excretion was determined based on P concentration in water and the water flow for every h after feeding over 24 h on feeding day 9. Non-fecal soluble P excretion is described by the function $y = 0.0557x^2 - 0.1676x + 1.4313$, $r^2 = 0.9398$. Least square method was adopted to make the break point of broken line analysis. The point of intersection between a linear line with an increasing slope and a horizontal line denoted the minimum dietary requirement (4.4 g/kg dry diet) of available P based on non-fecal soluble P excretion. [Source: Sarker 2009a]

The dietary P requirement for small yellowtail has been reported as 6.7 g available P/kg dry diet [Shimeno 1991; Shimeno et al. 1994]. Thus, requirement value estimated by the non-fecal method was lower than that estimated by the growth response method. It is not clear the lower value in this study was due to larger fish size used (assuming larger fish require less P) or to the use of different method (estimation by non-fecal method determine only minimum requirement). Sugiura et al. [2000] compared the P requirement values for rainbow trout between their study and other studies. They reported that there were close relationships between estimates made using conventional growth methods and the non-fecal method. So, this different P requirement values due to different fish sizes.

Yellowtail excrete or retain of dietary P before full absorption of the diet is complete. Gastric evacuation of feed for yellowtail is about 6 h when extruded pellets are fed [Murashita et al. 2007]. The non-fecal P peak appeared at 5-6 h after feeding in this study and the peak height already differed according to the dietary P levels at that time.

Other investigators have reported the comparable peak time of non-fecal P excretion. Soluble P excretion reached maximum levels at 6 h after feeding in milkfish (*Chanos chanos*) [Sumagaysay 2003] and at 3-4 h after feeding in rainbow trout [Coloso et al. 2003].

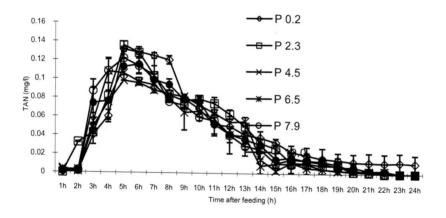

Figure 8. Daily pattern of TAN (total ammonia nitrogen) in effluent water yellowtail fed graded level P (Experiment 1). Each point represents the average (± SE) of two tanks per diet. The amount of TAN excretion was determined based on TAN concentration in water and the water flow for every h after feeding over 24 h on feeding day 9. . [Source: Sarker 2009a]

The daily non-fecal total ammonia nitrogen (TAN) excretion pattern of fish fed the experimental diets is shown in Fig. 8. All dietary groups had peaks at 5-6h after feeding and the values were similar irrespective of the dietary P contents. The major excretion route of TAN is known to be through the gills but not the kidney [Smith 1929; Wood 1958; Goldstein et al. 1964]. Accordance with peak time and tendency of the pattern of TAN and non-fecal P excretion in this study suggested that major excretion route of non-fecal soluble P also might not be the kidney. In line of our hypothesis, previously it has been reported that only 19% of dietary P excrete through the urine in gilthead sea bream [Roy & Lall 2004]. It means that the majority of dietary P does not excrete through kidney as urine and that amount might not be the representative part of the P excretion from the fish body to estimate. On the other hand, in the case of non-fecal P excretion estimation, it takes into account both the kidney (urinary) and gills excretions together. Non-fecal estimation approach plays a pivotal role in studies designed to estimate the concentration of soluble excretory products that originate from the absorption of nutrients or through metabolic transformation of nutrient reserves in the body.

Based on the non-fecal P estimation approach, 4.4 g available P /kg dry diet can meet the P requirement of large yellowtail. As the majority of feed is fed to larger fish in fish culture operation, therefore, determining dietary P requirement for large fish would lead to a substantial reduction of P loading from fish culture operation.

Fine-Tuning of Feed Formulation and Evaluate through Non Fecal Approach

Currently, fish meal consists of more than 50% of feed formulations for yellowtail feed in Japan. The P content of fish meal ranges between 1.6 to 4.2 % [Hua et al. 2005] and availability varies from 17-81 % in rainbow trout [Hua and Bureau 2006]. Thus, there is substantial variation depend on the ingredient and diet formulation. This makes difficult to decide whether or not P supplementation to fish meal-based diet is necessary. It is now the general practice to add some inorganic P to diets as a safety reason. If we have reliable techniques to determine the available P content of fish meal for yellowtail, manufactures may not need to supplement P just for safety. If fish meal has enough available P, it can be hypothesized that non-fecal P amount from fish fed fish meal based diet will be excreted at a higher levels. We determined if inorganic P supplementation in fish meal-based diets is necessary. The dietary composition of the fish meal (anchovy)-based diet with (coded as F1) and without (coded as F0) supplementation of 1% monocalcium phosphate is shown in Table 7. The purified diet (coded as PR) was the same diet of P 6.5 that was used in previous requirement study.

Table 7. Ingredient and analytical composition of experimental diets.

Ingredients (g kg⁻¹)	Diet code		
	PR	F0	F1
Wheat flour		180	170
Tapioca starch		47.6	47.6
Soybean meal		47.6	47.6
Corn gluten meal		47.6	47.6
Anchovy meal	Purified diet[1]	467	467
Soy oil		38	38
Feed oil		142.2	142.2
Ca(H$_2$PO$_4$)H$_2$O		0	10
Vitamin mix		20	20
Mineral mix		10	10
Total		1000	1000
Chemical analysis (g kg⁻¹) Crude protein	522	400	400
Crude ash	67	84	90
Total phosphorus	7	13	15

1 Purified diet (g kg-1): The combination and composition of this diet is similar to P 6.5 diet in previous requirement study. [Source: Sarker 2009a]

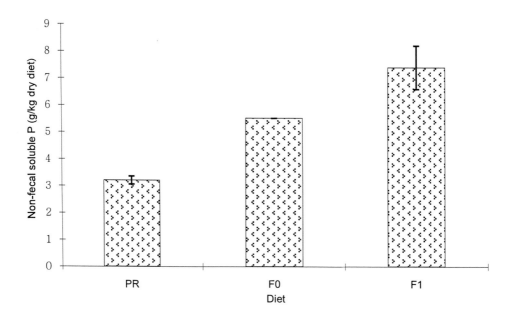

Figure 9. Comparison of non-fecal soluble P excretion (g/kg dry diet) among purified diet (PR), fishmeal based diet with (F1) and without (F0) supplementation of 1% mono calcium phosphate (Experiment 2). Each bar represents the average (± SE) of two tanks per diet. There were no significant difference between PR and F0 diet but F1 was significantly difference from PR diet. The amount of soluble P excretion was determined based on P concentration in water and the water flow for every h after feeding for 12 h, then every 4 h for 24 h period. [Source: Sarker 2009a]

There were no significantly differences between yellowtail fed required amount of P supplementing purified diet, PR (3.2 g soluble P/kg dry diet) and without P supplemented fish meal based diet, F0 (5.5 g soluble P/kg dry diet) (Fig. 9). In contrast, supplementing 0.2 % of P to the fish meal diet (F1) further increased non-fecal soluble P excretion (7.4 g soluble P/kg dry diet) that was significantly different from the PR diet group. The non-fecal P excretion of fish fed the F0 diet was higher than that of fish fed PR diet but the difference was insignificant. This means that available P content of the fish meal-based diet without P supplementation met both saturation level of P requirement (6.7 g available P/kg dry diet) and minimum requirement (4.4 g available P/kg dry diet) for large yellowtail.

Therefore, P supplementation is not necessary for fish meal-based diet used in this study for large yellowtail. Supplementation of inorganic P to the fish meal diet resulted a level an excess of the needs of the fish. Supplementation of 0.2 % additional phosphorus to the fish meal-based diet produced 34.5% excess of soluble P excretion in the effluent water. In other words, eliminate P supplementation to the fish meal-based diet resulted in a 34.5% reduction of non-fecal soluble P from yellowtail aquaculture.

CONCLUSION

Based on the above nutritional approaches the overall study suggested that anchovy meal based diet with lower dietary inclusion of soybean meal and corn gluten meal without addition of inorganic phosphorus might be the key formulation of low P loading diet for larger yellowtail that will contribute to reduce the pollution in aquaculture effluents. In the case of incremental dietary inclusion of plant protein, additives like phytase and inorganic P sources like mono basic sources of K, Na and Ca phosphate might be taken into consideration to apply. In fact successful nutritional approaches to reduce nutrient output originating from the aquaculture should revolve around basic concepts governing fundamental principles of nutrition. This includes careful formulation of diets that match the well-defined nutritional requirements of the animal. This combined with the utilization high quality and highly digestible ingredients will result in diets that promote optimal growth rate, feed efficiency and result in minimal P excretion. Nutritional strategies must, of course, be one aspect of an overall strategy to reduce P coming from aquaculture facilities, which includes coupling sound nutritional approaches with technologies that minimize feed feed waste and those that remove solids and treat the effluent to remove particulate and soluble P.

REFERENCES

Allan, G. L., Parkinson, S., Booth, M. A., Stone, D. A. J., Rowland, S.J., Frances, J. & Warner-Smith, R. (2000) Replacement of fish meal in diets for Australian silver perch, Bidyanus bidyanus: I. Digestibility of alternative ingredients. *Aquaculture*, 186, 293–310.

Azevedo, P. A., cho, C. Y., Leeson, S. & Bureau, D. P. (1998). Effects of feeding level and water temperature on growth nutrient and energy utilization and waste outputs of rainbow trout (*Oncorhynchus mykiss*). *Aquat. Living Resour.* 11, 227-238.

Bureau, D. P. & Cho, C. Y. (1999) Phosphorus utilization by rainbow trout (*Oncorhynchus mykiss*): estimation of dissolved phosphorus waste output. *Aquaculture*, 179, 127-140.

Burel, C., Boujard, T., Tulli, F. & Kaushik, S. J. (2000) Digestibility of extruded peas, extruded lupin, and rapeseed meal in rainbow trout (*Oncorhynchus mykiss*) and turbot (*Psetta maxima*). *Aquaculture*, 188, 285–298.

Coloso, R. M., King, K., Fletcher, J. W., Hendrix, M. A., Subramanyam, M., Weis, P. & Ferraris, R. P. (2003) Phosphorus utilization in rainbow trout (*Oncorhynchus mykiss*) fed practical diets and its consequences on effluent phosphorus levels. *Aquaculture,* 220, 801-820.

Cripps, S. & Berghem, A. (2000). Solids managements and removal for intensive land based aquaculture production systems. *Aquacult. Eng.* 22, 33-56.

Crowell, G. L., Stahly, T. S., Coffey, R. D., Monegue, H. J. & Randolph, J. H. (1993) Efficacy of phytase in improving the bioavailability of phosphorus in soybean meal and corn-soybean meal and corn-soybean meal diets for pigs. *J. Anim. Sci.*, 71, 1831-1840.

Dosdat, A. (1992) L'excrétion chez les poisons téléostéens II-le phosphore. *La pisciculture française.* 109, 18-29.

Goldstein, L., Foster, R. P. & Jr Faneilli, G. M. (1964) Gill blood glow and ammonia excretion in the marine teleosts, *Myoxocephalus scorpius*. *Comp. Biochem. Physiol.*, 12, 489-499.

Gregus, Z. (2000) Untersuchung zur Verdaulichkeit des phosphorus in futtermitteln mineralischer oder tierischer herkunft bei der regenbogenforelle (*Oncorhynchus mykiss*). Ph.D. thesis, University of Bonn, shaker Verlag, Aachen, Germany.

Hua, K. & Bureau, D. P. (2006) Modeling digestible phosphorus content of salmonid fish feeds. *Aquaculture*, 254, 455-465.

Hua, K., Liu, L. & Bureau, D. P. (2005) Determination of phosphorus fractions in animal protein ingredients. *J. Agric. Food Chem.*, 53, 1571-1574.

Ketola, H. G. & Richmond, M. E. (1994) Requirement of rainbow trout for dietary phosphorus and its relationship to the amount discharged in hatchery effluent. *Trans. Am. Fish. Soc.*, 104, 587-594.

Ko"pru"cu", K. & Özdemir, Y. (2005) Apparent digestibility of selected feed ingredient for Nile tilapia (*Oreochromis niloticus*). *Aquaculture*, 250, 308-316.

Kofuji, P. Y. M., Michihiro, Y., Hosokawa, H. & Masumoto, T. (2004) Comparisons of corn gluten meal and fish meal digestion in Yellowtail (*Seriola quinqueradiata*) by in vivo and in vitro approaches. *Aquac. sci.*, 52, 271-277.

Lall, S. P. (1991) Digestibility, metabolism and excretion of dietary phosphorus in fish: Nutritional Strategies and Aquaculture Wastes. In C. B. Cowey, & C.Y. Cho (Eds.), *Proceedings of the First International Symposium on Nutritional Strategies in Management of Aquaculture Wastes*. (pp. 21– 26).University of Guelph, Guelph, Ontario, Canada.

Lall, S. P. (2002) The minerals, Fish nutrition, 3rd edition. Elsevier, p. 259.

Lanari, D., D'Agaro, E. & Turri C. (1998) Use of nonlinear regression to evaluate the effects of phytase enzyme treatment of plant protein diets for rainbow trout (Oncorhynchus mykiss). *Aquaculture* 161, 345-356.

Lee, S. M. (2002) Apparent digestibility coefficients of various feed ingredients for juvenile and grower rockfish (*Sebastes schlegeli*). *Aquaculture*, 207, 79–95.

Murashita, K., H. Fukada, H. Hosokawa & Masumoto, T. (2007) Changes in cholecystokinin and peptide Y gene expression with feeding in yellowtail (*Seriola quinqueradiata*): relation to pancreatic exocrine regulation. *Comp. Biochem. Physiol.*, 146 *B*, 318-325

Nelson, T. S., Shieh, T. R., Wodzinski, R. J. & Ware, J. H. (1971) Effect of supplemental phytase on the utilization of phytate phosphorus by chicks. *J. Nutr.*, 101, 1289-1294.

Ogino, C., Takeuchi, L., Takeda, H. & Watanabe, T. (1979) Availability of dietary phosphorus in carp and rainbow trout. *Bull. Jpn. Soc. Sci. Fish.* 45, 1527-1532.

Papatryphon, E., Howell, R. A. & Soares, J. H. (1999) Growth and mineral absorption by stripped bass *Morone saxatilis* fed a plant feedstuff based diet supplemented with phytase. *J. World Aquac. Soc.*, 30, 161-173.

Persson, H., Turk, M., Nyman, M. and Sanberg, A. S. (1998) Binding of Cu^{2+}, Zn^{2+}, and Cd^{2+}, to inositol tri‾, tetra‾, penta‾, and hexa‾aphosphates. *J. Agric. Food Chem.*, 46, 3194-3200.

Ramseyer, L., Garling, Hill, G. & Link, J. (1999). Effect of dietary zinc supplementation and phytase pre-treatment of soybean meal or corn gluten meal on growth, zinc status and zinc-related metabolism in rainbow trout, *Oncorhynchus mykiss*. *Fish Physiology and Biochemistry*, 20, 251–261, 1999.

Riche, M, & Brown, P. B. (1996) Availability of phosphorus from feedstuffs fed to rainbow trout, *Oncorhynchus mykiss*. *Aquaculture*, 142, 269-282.

Rodehutscord, M., Becker, A. & Pfeffer, E. (1995) Einfluss des zusatzes einer *Aspergillus niger*- Phytase aufdie Verwerting pflantlichen Phosphorus durch die Forelle (*Oncorhynchus mykiss*). *Arch. Anim. Nutr.*, 48, 211-219.

Rodehutscord, M., Gregus, Z. & Pfeffer, E. (2000) Effect of phosphorus intake on faecal and non-faecal phosphorus excretion in rainbow trout (*Oncorhynchus mykiss*) and the consequences for comparative phosphorus availability studies. *Aquaculture*, 188, 383-398.

Roy, P. K. & Lall, S. P. (2003) Dietary phosphorus requirement of juvenile haddock (*Melanogrammus aeglefinus* L.). *Aquaculture*, 221, 451-468.

Roy, P. K. & Lall, S. P. (2004) Urinary phosphorus excretion in haddock, *Melanogrammus aeglefinus* (L.) and Atlantic salmon, *Salmo salar* (L.). *Aquaculture*, 233, 369-382.

Sajjadi, M. & Carter, C. G. (2004) Dietary phytase supplementation and the utilization of phosphorus by Atlantic salmon (Salmo salar L.) fed a canola-meal-based diet. *Aquaculture*, 240, 417-431.

Sarker, P. K. (2007) Development of low phosphorus loading diet for yellowtail (*Seriola quinqeradiata*) diet. Ph.D. thesis, *United Graduate School of Agricultural Sciences, Ehime University*, Japan. 97 pp.

Sarker, P. K., Fukada, H., Hosokawa, H. & Masumoto, T. (2006). Effects of phytase with inorganic phosphorus supplement diet on nutrient availability of Japanese flounder (*Paralichthys olivaceus*). *Aquaculture Science*, 54 *(3)*, 391-398.

Sarker P. K., Shuichi S., Fukada H., Masumoto T. (2009a). Effects of dietary phosphorus level on non-faecal phosphorus excretion from yellowtail (*Seriola quinqueradiata* Temminck and Schlegel) fed purified and practical diets. *Aquaculture Research*, 40 (2), 225 -232.

Sarker P. K., Fukada H., Masumoto T. (2009b). Phosphorus availability from inorganic phosphorus sources in yellowtail (*Seriola quinqueradiata* Temminck and Schlegel) *Aquaculture*, 289, 113-117.

Satoh, S., Takanezawa, M., Akimoto, A., Kiron, V. & Watanabe, T. (2002) Changes of phosphorus absorption from several feed ingredients in rainbow trout during growing stages and effect of extrusion soybean meal. *Fish. Sci.*, 68, 325-331.

Schäfer, A., Koppe, W. M., Meyer-Burgdorff, K. H. & Günther, K. D. (1995) Effects of microbial phytase on utilization of native phosphorus by carp in a diet based on soybean meal. *Water Sci. Technol.*, 31, 149-155.

Shimeno, S. (1991) Yellowtail, *Seriola quinqueradiata*. In: R.P. Wilson (Ed.), *Handbook of Nutrient Requirements of Finfish*. (pp. 181-191). CRC Press, Boston.

Shimeno, S., Hashimoto, A. & Ukawa, M. (1994) Effects of supplemental levels of dietary monopotassium phosphate on growth, feed conversion, and body composition of fingerling yellowtail. *Suisanzoshoku*, 49, 191-197 (in Japanese).

Simons, P. C. M., Versteegh, A. J., Jangbloed, A. W., Kemme, P., Slump, A. P., Bos, K. D., Wolters, M. G. H., Beudeker, R. F. & Verschoor, G. J. (1990) Improvement of phosphorus availability by microbial phytase in broilers and pigs. *Br. J. Nutr.*, 64, 525-540.

Smith, H. W. (1929) The excretion of ammonia and urea by the gills of fish. *J. Biol. Chem.* 81, 727-742.

Sugiura, S. H. & Hardy, R. W. (2000) Environmental friendly feeds. In R. R. Stickney (Ed.), *Encyclopedia of aquaculture* (pp. 299-310) Wiley, New York.

Sugiura, S. H., Dong, F. M. & Hardy, R. W. (2000) A new approach to estimating the minimum dietary requirement of phosphorus for large rainbow trout based on non fecal excretions of phosphorus and nitrogen. *J. Nutr.* 130, 865-872.

Sugiura, S. H., Dong, F. M., Rathbone, C. K., & Hardy, R. W. (1998) Apparent protein digestibility and mineral availabilities in various feed ingredients for salmonid feeds. *Aquaculture*, 159, 177-202.

Vielma, J. & Lall, S.P. (1997) Dietary formic acid enhances apparent digestibility of minerals in rainbow trout *Oncorhynchus mykiss* (Walbaum). *Aquac. Nutr.* 3, 265– 268.

Vielma, J., Lall, S. P., Koskela, J. F., Schoner, J. & Mattila, P. (1998) Effects of dietary phytase and cholecalciferol on phosphorus bioavailability in rainbow trout (*Oncorhynchus mykiss*). *Aquaculture*, 163, 309-323.

Vielma, J., Makinen, T., Ekholm, P. & Kosekela, J. (2000) Influence of dietary soy and phytase levels on performance and body composition of large rainbow trout (*Oncorhynchus mykiss*) and algal availability of phosphorus load. *Aquaculture*, 183, 349-362.

In: Encyclopedia of Environmental Research
Editor: Alisa N. Souter

ISBN: 978-1-61761-927-4
© 2011 Nova Science Publishers, Inc.

Chapter 33

INFLUENCE OF VITAMIN E SUPPLEMENTATION ON DERMAL WOUND HEALING IN TILAPIA, *OREOCHROMIS NILOTICUS*

Julieta Rodini Engrácia de Moraes[1,2], Marina Keiko Pieroni Iwashita[1], Rodrigo Otávio de Almeida Ozório[3], Paulo Rema[3,4] and Flávio Ruas de Moraes[1,2]

[1]Aquaculture Centre of São Paulo State University (CAUNESP), Jaboticabal, São Paulo, Brazil
[2]Department of Veterinary Pathology, São Paulo State University (UNESP), Jaboticabal, São Paulo, Brazil
[3]CIMAR/CIIMAR, Centro Interdisciplinar de Investigação Marinha e Ambiental; Universidade do Porto, Porto, Portugal
[4]University of Trás-os-Montes e Alto Douro (UTAD), Vila Real, Portugal

ABSTRACT

Aquaculture is responsible for a 20 million tonnes increase in fish production over the last decade, with an estimate of 48 million tones in 2005. The global increase in aquaculture production was made possible through the introduction of new areas, species, and practices and through increased production from existing systems. Nutrient pollution from aquaculture and intensification of fish production, in turn, may cause declines in aquaculture productivity by promoting outbreaks of disease among the fish. Antibiotics as feed additives, which played an important role for controlling diseases in the past, have been widely criticized for the negative impacts to the surround aquatic systems. The potential substitute for antibiotics are the so called nutraceuticals, multi-physiological, bioactive and pollution-free additives, acting as immuno-stimulants that improve resistance to diseases by enhancing non-specific defence mechanisms. In the present chapter we studied the effects of vitamin E supplementation in the process of induced wound healing in Nile tilapias *Oreochromis niloticus*. During a period of 60 days, fish (initial body weight=30g) were fed two experimental diets, supplemented (450 mg/kg diet) or not with vitamin E. Thereafter, all animals were anaesthetized and submitted to dermal wounds. The histomorphometric assessment was checked after 3, 7, 14, 21 and 28 days post-wounding. The cicatricial retraction and appearance of the wounds were monitored during the trial. Moreover, the histomorphometry of the mucous cells, chromatophores, inflammatory cells, fibroblasts, collagen fibers and scales were also used as indicators for the wound healing capacity. The rate of wound retraction was significantly higher in the vitamin E supplemented group. Such healing was a result of an increase of inflammatory cells, mucous cells, chromatophores and collagen fibers. The results indicate that fish fed vitamin E rich diet have enhanced dermal wound healing capacity.

INTRODUCTION

Aquaculture is a type of animal production which is a very important source of protein for human consumption. Brazil falls in the international context as one of the countries with great potential for fish production, as well as having a vast rainfall territory; the climatic conditions are conducive to implement the freshwater aquaculture. The global, and Brazilian, increase in aquaculture production was possible by the intensification of the culture practices, diversification of the cultured species, and area enlargement of the production. The diversification of the cultured species was made possible by carrying out intensive research on the biology, i.e. reproduction, nutrition and the welfare, of the "new aquaculture species".

The sanitary problems encountered by fish-breeders are from the action of several factors, including management. Changes in conditions appropriate to the concentration of dissolved oxygen, pH, temperature, salinity, high concentrations of ammonia and CO_2, and the presence of pollutants in water and livestock management that involves high population density, capture and transport are part of wide range of situations capable of inducing stress in fish (Wedemayer; 1997).

The stress is present in fish farms, according to management, such as high density storage and consequently the deterioration of water quality and greater competition among individuals to reduce the use of nutrients from the diet. Whatever the source of stress, the hit bodies become more susceptible to pathogens (Wedemayer, 1997).

Substances that added to the fish diet and benefit the response of the defense mechanism of animals are known as immunostimulants. The use of immunostimulants shows promising prospects for the sanitary management, having a coadjuvant role in chemotherapy, but also can be used for prevention of diseases. (Sakai, 1999). Immunostimulants are considered a group of synthetic or biological compounds that rapidly increase the efficiency of defense mechanisms specific and non-specific, such as combinations vitamins, trace minerals and products derived from plants or animals that become effective in the prevention of diseases (Siwicki et al., 1990; Garcia et al., 2005, 2007; Belo et al.; 2006). Balanced diet with the addition of immunostimulants improve resistance to stress and consequently to infections of different etiology by the increase in defense responses in several fish species (Reque, 2005; Val et al., 2006; Sakabe, 2007). Whatever is the source of stress, physiological or environmental, the animal become more susceptible to infections (Pickering, 1989; Fevolden et al., 1991, 1993; Fevolden and Roed, 1993).

The nonspecific stimulation of the immune system helps to increase the protection against opportunistic pathogens and prophylactic administration of immunostimulants can prevent infections by increasing the efficiency of barriers to protect fish (Reque, 2005; Sakabe, 2007).

REVISION OF LITERATURE

The Inflammation and the Dermal Repairing in Fish

The inflammatory and cicatricial processes and its mechanisms of control are well known in mammals. However, little is known of these processes in fish, which deserves special attention for being one of the most animal protein sources of high quality explored commercially. The cellular mode of action in the inflammatory reply in fish seems to be bifasic, initiating with an influx of neutrophyls followed of monocytes and macrophages (Reite et al., 2006). The first studies on the inflammation in fish had been carried by Metchinikoff (1893) and are related to the studies on phagocytosis of erythrocytes injected in the peritoneal cavity of goldfish (*Carassius auratus*).

When the animal succeeds to remove a pathogen agent, by acute or chronic inflammation, it follows a "demolition phase", where the elimination of cellular debris to the repair process can start (Bereiter-Hahn et al., 1986; Moraes et al., 2003).

The healing process in higher vertebrates is based mainly in re-vascularization, re-epithelialization and fibroplasia, which occur in late stages of chronic inflammation (Cheville, 1994; Kumar et al., 2005). The re-epithelialization in teleosts happens in two stages: the first, initial and early, is the removal of injured tissue and re-epithelialization of the wound, and second, later stage, includes the re-organization of the dermal connective tissue (Bereiter-Hahn et al., 1986; Moraes et al., 2003). In carp (*Cyprinus carpio*), the re-epithelialization is the migration of epithelial cells (Iger et al., 1988), after the phagocytosis of cellular debris (Mittal and Munshi, 1974; Phromsuthirak, 1977; Iger and Abraham, 1990). In pacu, *Piaractus mesopotamicus*, according to Moraes et al. (2003), re-epithelialization was complete in less than 24 hours, and the cover of the wound was carried out by the growth of neighboring healthy skin cells (Mittal and Munshi, 1974; Quilhac and Sire, 1998). Such rapid re-epithelialization is important to prevent osmotic imbalance and/or attack by opportunistic pathogens (Moraes et al., 2003). Parasitic infestations, such as those caused by *Gyrodactylus salaries*, can be severe and result in a significant increase in mortality of infected fish (Johnsen, 1978; Dawson et al., 1998; Nolan et al., 1999 in Wahli et al., 2003). Mechanically induced lesions on the skin are predisposing factors for the development of ulcers. It was shown that the bacterium *Vibrio viscosus* is associated with these lesions in salmonids, but it was not clear whether this pathogen is the cause of injuries or develops secondarily in the injured area (Lunder et al., 1995).

On the dermis, the repair process begins with the rupture of capillaries, with the events of the coagulation cascade, such as increased vascular permeability and migration of inflammatory cells to the central region and the vascular wound. The process continues with the proliferation of fibroblasts, synthesis of collagen and extra-cellular matrix and endothelial cells. During the first hour after injury, different types of leukocytes migrate to the inflammatory focus. The main role of neutrophils is the phagocytosis of cellular debris, foreign bodies and microorganisms, although these precursor cells are not essential for the activity of macrophages. Neutrophils arrive at the focus of the wound one hour after injury and return their number to increase again after 48 hours (Iger and Abraham, 1990).

The collagen is the predominant protein in the body (Woo et al., 2008; Nagai et al., 2002), and one of the main components of intramuscular connective tissue (Eckhoff et al.,

1998). Its fibers are arranged in the skin oriented in one only direction (Junqueira et al., 1982). The type I collagen is the main component of the extra-cellular matrix and one of its function include mechanical protection of tissues and organs of physiological regulation of the cell (Ikoma et al., 2003). Moreover, it is an important constituent of the dermis and epidermis, fish bones and scales (Lall and McCrea, 2007). The type I collagen fibers show a birefringent, yellow and red in the histochemical reactions by picro-sirius method (Junqueira et al., 1982).

Effect of Vitamins in Fish

Immunostimulants combinations are considered as vitamins, minerals traces and products derived from plants or animals that become effective in the prevention of diseases. Although the way of action of these substances is not well characterized, they can be injected or mixed in food and the results can be tested through experimental animal models suitable for this purpose, as the determination of immune parameters or by means of challenge with pathogens (Siwicki et al., 1990; Sakai, 1999; Belo et al., 2005).

An effective dosage of vitamins and minerals increase the cell viability (Hung et al., 2007). Non-specific immune responses are very important for the protection of fish against infectious agents. In particular, monocytes, macrophages derived from peripheral blood, macrophages derived from cranial kidney and liver are the key components of the non specific cellular response (Hung et al., 2007).

Moraes et al. (2003) found that fish supplemented with 100mg of vitamin C showed, macroscopically, diffuse red dots in the wound during the first 24 hours of injury. Seven days later, a network of blood vessels formation is viewed microscopically around the wound. At 21 days, the wound has a yellow shade with slight formation of scales on the outside of the injured area, indicating the effectiveness of the dietary vitamin C supplementation in accelerating the healing process.

Hibyia (1982) described that the vitamin deficiency causes cellular breakdown in the epidermis of fish, because the basal cells and not the mitotic activity are hampered by the hypovitaminosis.

Vitamin E (tocopherol) is a liposoluble antioxidant that protects lipoproteins from biological membranes against oxidation (Montero et al., 1999; Hung et al., 2007; Martins et al., 2008) and increases the absorption of vitamin A (Conn and Stumpf, 1984). Vitamin E is considered the first line of defense against lipid peroxidation, which aims to preserve the cell membrane from injury by free radicals. This important function of vitamin E on resistance to oxidation, particularly in leukocytes, is tied to the release of radicals involved in the microbicide activity of macrophages. In fish, vitamin E is also assessed as being very important for its non-specific immune response and unsaturated fatty acid. It also prevents the immuno-suppression in response to the stress of high density (Montero et al., 1999; Belo et al., 2005). According to Belo et al. (2005), diets supplemented with 450mg of vitamin E were able to reduce plasma levels of cortisol in pacu (*Mesopotamicus piaractus*), promoting the accumulation of macrophages and the formation of giant cells in glass little laminas implanted in the subcutaneous tissue. Martins et al. (2008) found that tilapia fed with supplemented diets with 500 mg / kg of vitamins C and E showed a reduction in the total number of neutrophils circulating in the blood. In animals injected with lipopolysaccharides

(LPS) from *Escherichia coli* and, despite the variation among treatments, there was no significant difference in the number of monocytes, lymphocytes and basophils in the circulating blood.

The hypovitaminosis E causes muscular dystrophy in guinea pigs and rabbits, and in chickens, there are vascular abnormalities (Conn and Stumpf, 1984). Fish fed with deficient diets in vitamin E showed loss of appetite, decreased feed conversion, and reduction in the value of the hematocrit, signs of muscular dystrophy and shortening of the operculum (Bai and Lee, 1998). Chen et al. (2004) found that vitamin E showed signs in different times of feeding. Fish fed for 10 weeks with vitamin deficient diet showed darkening of the body surface, slow response to food, cachexia associated with the loss of the epaxial muscle and bleeding. Between 10-14 weeks, the number of fish with darkening body increased gradually. Other signs of deficiency were developed after 14 weeks being fed with vitamin deficient diet, including atrophy and loss of muscle tissue, accompanied by erratic swimming. Histological sections revealed that some of the muscle fibers were severely atrophic, necrotic and inflammatory infiltrates of leukocytes in the connective tissues. This deficiency also induces fragility of erythrocytes causing its degeneration in many species such as the channel catfish, *Ictalurus punctatus* (Wilson et al., 1984; Wise et al., 1993), sunshine bass, *Morone chrysops* x *M. saxatilis*, (Hemre et al., 1994) and rainbow trout *Oncorhynchus mykiss* (Furones, 1992).

The purpose of the current study was to evaluate the nutraceutical effect of vitamin E in juvenile Nile tilapia on the kinetics of the induced process of tissue repair.

MATERIAL AND METHODS

Juveniles of Nile tilapia (n = 180), with average weight of 30g were randomly allocated into 18 tanks of 250L (10 fish/tank), with continuous flow-through system, and acclimated for 40 days (November-December 2007) prior the feed experiment. The experiment was conducted in CPPAR - Center for Research in Animal Health UNESP, Campus of Jaboticabal, SP. During this period, fish were fed twice daily to 6% body weight/day with a pelleted diet (28% crude protein) (Table 1).

The physical and chemical parameters of water were controlled on weekly basis. The water temperature was maintained at 30.20 ± 0.8 ° C, the dissolved oxygen was 7.96 ± 2.48 $mg.L^{-1}$, pH 7.96 ± 0.28, electrical conductivity 179 ± 69 $\mu S.cm^{-1}$, amount of total dissolved solids was 89 ± 34 ppm and 0.08 ± 0.01 ppm salinity.

After acclimation period, nine tanks received a vitamin-free diet (control) and the rest were offered a diet supplemented with 450mg of vitamin E. kg^{-1} diet. The vitamin (Rovimix E-50 Adsorbato, Roche®) was blended with the other ingredients, ground and pelleted (Cuesta et al., 2002) in the facilities of the CAUNESP. The diet was stored at a temperature of -4 ° C in dark plastic bag until prior to feeding. After 60 days of supplementation, it was carried out the induction of surgical wounds and intermediate sampling at 1, 3, 7, 14, 21 and 28 days after the injury. Thereafter, the animals were collected and subjected to selected biometrics, qualitative and quantitative, analysis of the healing process using the histomorphometric and histochemistry methods.

Table 1. Basal formulation and chemical composition of the experimental diets.

Ingredient	(%)
Soybean	43
Maize	22.6
Wheat bran	17
Rice bran	10
Yeast	4
L - Lysine	0.2
DL - Methionine	0.4
Dicalcium phosphate	1
Premix (without vitamin E)*	0.5
Limestone	1
Chemical composition	
Crude protein	28 %
Crude fat	4.18%
Gross energy	3900 kcal
Crude fibre	5.74%

*Premix: vitamin and mineral was supplied by Rações Fri Ribe S.A. Vitamin E was added according to the treatment (450 mg tocopherol acetate / kg of feed).

Before the skin lesion, the animals were anesthetized with an aqueous solution of benzocaine (1:10,000), until the loss of balance and decrease of opercular movements was detected. Firstly, the skin lesion was induced in the lateral- dorsal right area, just below the dorsal fin and above the lateral midline, the second model proposed by Freitas (2001), using polyethylene molds of 20 mm x 10 mm standard size, composing approximately 3% of body surface of fish. Secondly, the dermis and epidermis were removed with the aid of scalpel. After the surgery, the animals were immersed in aqueous solution based on sodium chloride $10g.L^{-1}$ for 15 minutes to disinfect and return from the anesthesia, and then placed to its original condition.

The percentage of contraction of the wound was performed with the aid of a planimeter and was calculated by the equation: RI= (AI-AF/AI) X 100, where RI - retraction index; AI - initial area; AF - area of contraction of the wound.

Then, the fish were overdosed and sacrificed by a high concentration solution of benzocaine, and small rectangle sample of skin, including the wound, and muscle layer adjacent edges were collected for histological analysis. Part of the material collected was fixed in Bouin solution and processed according to the methods for inclusion in paraffin for histological observation. Another portion of the fragments of skin samples was fixed in buffered formalin to 10%, followed by seven days of storage in phosphate-buffered saline (PBS). Subsequently, these samples were dehydrated in increasing series of alcohol solutions and immersed in resin, according to the methods described. In brief, resin blocks were cut with a thickness of two to five micrometers, and then stained with hematoxylin-eosin plus the periodic acid-schiff (PAS) reaction and Sirius red (Junqueira et al., 1982) and blue of 1% toluidine (Hopwood, 1990; Santos and Oliveira, 2007), for inclusion in histo-resin and paraffin, respectively.

The slide-cuts were examined in the light optical microscope for histomorphometry of mucous cells (MC), chromatophores (CRM), neo-vascularization (NEO), inflammatory cells

(IC), fibroblasts (FB), collagen (CG) and scales (ESC) in skin of tilapia. There were considered five microscopic fields per section of five layers representing the replicates for each of the five collection times, and for the two treatments. The histomorphometry was performed in the dermis. All results of biometrics, healing area and histomorphometry were submitted to analysis of variance (ANOVA), applying the F test ($P < 0.01$) to verify the differences among treatments. For comparison of averages, it was used the Tukey test. All statistical analyses were performed using SAS program (SAS version 9.1).

RESULTS

The biometric values for each evaluation day (intermediate sampling) were not statistically different between the control and supplemented groups ($P < 0.01$). All fish grew and gained weight gradually, regardless of the dietary supplementation.

In the macroscopic evaluation, the skin wounds of the control group maintained the original rectangular shape throughout the observation period. In the supplemented group, on the 28th day, the wounds were initially elliptical shape and then oval.

The color of wounds in the control group ranged from pinkish (day 3) to reddish (day 7) and gradually become grayish until the final time of observation. Furthermore, the wound in the supplemented group showed reddish staining (day 3) with abundant vascularization and diffuse in the central area. Then, the color became similar to the color of a healthy muscle (day 14). From day 14 to 28 post injury, the wound was gradually covered by gray skin color, typical to teleosts. In both groups, the edges of the wound were protruding and well defined in the first days and then gradually become flat, with blurred boundaries, hyper-pigmented, dark-gray color on the edges. In the supplemented group, the cicatricial retraction was evident at day 28 after lesion. The wound proved to be fully pigmented, silver in color, with scales in the region of the skin edges and the injured epidermis confused with the entire epidermis.

The retraction index in the areas of scarring was statistically difference ($p < 0.01$) between the supplemented and control group. The rate of retraction of the wound was greater in animals supplemented with vitamin E (Table 2). In the group supplemented with vitamin E, the scars were significantly lower from day 7 to 28 post injury. The percentage of retraction has also proved to be higher in fish fed vitamin E supplemented diet (24.04%) than in fish from the control group (22.95%) (Table3). In histological evaluation at day 3 post-lesion, lesions covered with stratified epithelium and re-organization of the basal layer were observed in both groups, but more evident in the supplemented group. Small amount of mucous cells was identified in both groups, but in greater numbers in the supplement (Table 4).

Table 2. Comparison of scarring area (cm^2) in Nile tilapia between control and supplemented groups, after induction of the healing process.

Variable	Treatment	Time				
		3	7	14	21	28
Area (cm^2)	Control	2.29	2.16	1.83	1.73	1.69
	Vitamin E	2.40[b]	1.95[a]	1.56[b]	1.38[a]	1.27[b]

Note: In each row, means followed by same capital letter do not differ. In each column, same lowercase letters do not differ.

Table 3. Comparison of Retraction Index (RI) of scarring in Nile tilapia between supplemented and control groups, after induction of the healing process.

Experimental groups	Percentage
Control	22.95%
Vitamin E	24.04%
Differences between groups	4.46%

At day 3 post-lesion, the amount of inflammatory cells (x = 3.29) and vascular neo-formation (x = 3.68) were maximum for fish fed vitamin E supplemented diet, when compared with the control group (Figure 1). The loose connective tissue of the dermis showed dissociation of collagen fibers, showing signs of edema in both groups, more pronounced in the control group. Over 7 days post-lesion, the supplemented group showed epidermal and dermal organization more evident than in the control group, with a greater number of mucous cells (MC) and collagen fibers parallel to the basal layer (Table 4 and 5). Hyperplasic epidermis was only observed in the control group showed that hypovitaminosis E increased the hyperplasia of these cells. Fibroblasts and collagen fibers were observed in lesions in all groups at all times.

At day 14, the amount of inflammatory cells (IC) was the highest (x = 2.79) in the control group (Table 4). The epidermal and dermal hyperplasia was shown to be evident. The collagen fibers in the control group still remained disorganized while the configuration of the fiber supplemented group showed parallel provision in relation to the basal layer (Figure 1).

At day 21, the amount of mucous cells was the highest in the group supplemented with vitamin E (x = 4.92) (Table 4), with the alignment of collagen fibers in the dermis. In addition, the amount of chromatophores reached their maximum for the control group at this stage (x = 8.76) (Figures 1 and 2).

Table 4. Comparison of counts of mucous cells (MC), chromatophores (CRM), neovascularization (NEO), inflammatory cells (IC) and scales (ESC) in skin of Nile tilapia, Oreochromis niloticus, between control and supplemented groups after induction of the healing process.

Parameter	Groups	Time				
		3	7	14	21	28
MC	Control	1.87^{Ea}	2.71^{Da}	3.32^{Ca}	4.06^{Bb}	5.22^{Aa}
	Vit. E	1.67^{Ea}	2.87^{Da}	3.55^{Ca}	4.92^{Aa}	4.13^{Bb}
CRM	Control	2.12^{Ea}	2.86^{Db}	3.76^{Cb}	8.76^{Aa}	6.17^{Bb}
	Vit. E	2.12^{Ea}	3.82^{Da}	4.38^{Ca}	6.34^{Bb}	6.79^{Aa}
NEO	Control	2.63^{Eb}	3.10^{Bb}	3.25^{Aa}	2.86^{Ca}	2.64^{Da}
	Vit. E	3.51^{Ba}	3.68^{Aa}	3.08^{Ca}	2.61^{Da}	2.27^{Ea}
IC	Control	2.09^{Db}	2.61^{Ba}	2.79^{Aa}	2.27^{Ca}	2.07^{Eb}
	Vit. E	3.29^{Aa}	2.53^{Ca}	2.69^{Ba}	1.87^{Eb}	2.42^{Da}
ESC	Control	1.03	1.08	1.08	1.07	1.00
	Vit. E	1.01	1.07	1.03	1.03	1.00

Note: In each row, means followed by same capital letter do not differ. In each column, same lowercase letters do not differ.

At day 28, the maximum amount of mucous cells (MC) was observed in the control group (x = 6.79), so as the counting of chromatophores (CRM), which was maximum for the supplemented group (x = 5.22) (Table 4, Figure 1).

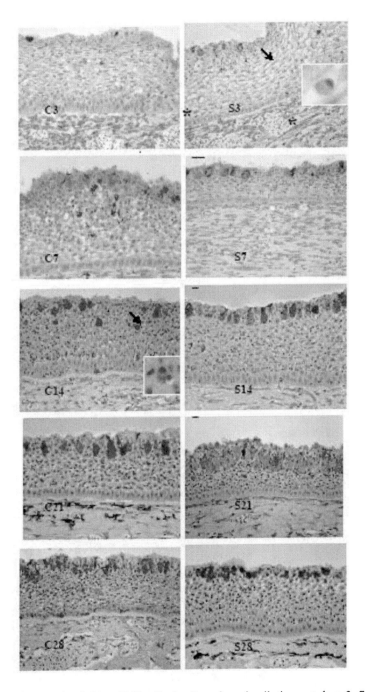

Figure 1. Photomicrograph of skin of Nile tilapia, Oreochromis niloticus, at days 3, 7, 14, 21 and 28 post-injury. Control group (C), Supplemented group (S) with 450 mg vitamin E.kg^{-1} diet. Basal layer (*), Mucous cell (Δ), Neovascularization ($\dagger$), Inflammatory cells ($\rightarrow$), Chromatophores ($\ddagger$). Blue toluidine. Scale = 100 μm.

Table 5. Percentage of collagen fibers in skin of Nile tilapia, Oreochromis niloticus, in control and supplemented groups after induction of the healing process.

Days	Collagen fibers (%)			
	Organized		Not organized	
	Control	Vitamin E	Control	Vitamin E
3	8.5	10.8	92.5	89.2
7	14.5	12.5	62.5	77.5
14	21.33	28.66	78.66	71.33
21	22.75	29.57	77.25	70.43
28	27.8	46.16	72.2	53.83

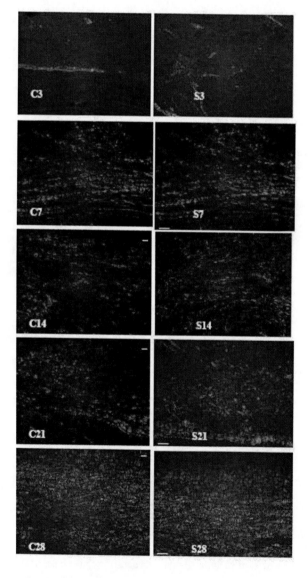

Figure 2. Photomicrograph of skin of Nile tilapia, Oreochromis niloticus at days 3, 7, 14, 21 and 28 post-injury. Control group (C), Supplemented group (S) with 450 mg vitamin E.kg^{-1} diet. Sirius red submitted to polarization. Scale = 100 µm.

In comparing the values of mucous cells between sampling period, it was observed a significant difference (P <0.01) between the day 21 and 28, with and interaction effect between treatment and time (Table 4). The overall chromatophores count was statistically different between the supplemented and control groups on days 7, 14, 21 and 28, with higher values for the supplemented group on days 3, 14 and 28.

The results of histomorphometry showed that the formation of neo-vessels (NEO) were statistically different between the treatments (P <0.01). The average count of NEO differed between treatments in days 3 and 7, where values were larger in the supplemented group (Table 4). The counting of inflammatory cells (IC) was significantly different between treatments, showing that the vitamin supplementation promoted the migration of inflammatory cells to the focus of the lesion in 3, 21 and 28 days of collection. No significant differences were observed for the counting of scales (ESC).

The current study showed that during the healing process of the skin lesion, the organization of collagen fibers was higher in the supplemented group, demonstrating that the parallelism between fibers is directly proportional to the degree of organization (Figure 2). Furthermore, the vitamin E supplementation accelerated the process of re-organization of the fibers, and promoted more quickly the process of tissue repair in tilapia (Table 5).

DISCUSSION

The analysis of physico-chemical parameters of water were within the standards levels specified species under study (Boyd, 1990), which leads us to infer that these parameters do not influence the healing response of fish.

The biometric analysis showed no effect on growth and weight gain during the 60-day supply. This result confirms those observed by Belo (2005) and Belo et al. (2006) in their studies with pacu, *Piaractus mesopotamicus*, fed for 60 days with diets supplemented with 450mg.kg^{-1} of vitamin E.

The morphometric analysis of the healing area showed that there was no variation in shape and size until the day 7 post-injury, regardless of dietary treatment. From the day 14 onwards, the supplemented group showed a difference in shape and size of the wounds, suggesting that the nutraceutical effect of vitamin E may had a positive effect on the cicatricial retraction of the lesion and therefore on the healing process.

In this study, the re-epithelialization was evident in day 3 post skin injury in both groups, corroborating with the results of induced healing in pacu (Freitas, 2001; Moraes et al., 2003) and rainbow trout, *Onchorhyncus mykiss* (Wahli et al., 2003) supplemented with with different concentrations of vitamin C. The comparison of re-epithelialization data between control and supplemented group showed no difference, allowing inferring that the amount of vitamin E supplemented did not interfere in this process. However, the early re-epithelialization immediately acts as a barrier and prevents the exposure of animals to opportunistic infections and osmotic imbalance (Moraes et al., 2003; Roubal and Bullock, 1988).

Comparing the rates of cicatricial retraction, we can observed that in the supplemented group, the retraction index was higher, which allows us to deduce that this nutraceutic influenced the speed of healing process, providing a faster development of the protective mechanical barrier, which is essential in protecting the regenerated tissue (Silva et al., 2005).

The amount of mucous cells increased gradually over the post injury stages, and in different ways between groups. There was early development of these cells in the supplemented group, when compared to the control group reached maximum at day 3. These data confirm those described for pacu (Freitas, 2001; Moraes et al., 2003) and rainbow trout (Wahli et al., 2003), supplemented with vitamin C. Several authors report that dietary supplementation with vitamin E in fish helps the inflammatory process and triggers the phenomenon of immuno-suppression in response to the stressor stimulus, because it increases significantly the quantity of defense cells (Ortuño et al., 2000; Belo et al., 2005; Martins et al., 2008).

In the current study, we observed an increase of mucous cells in fish supplemented with vitamin E. The mucus cells are rich in lysozyme and antibodies and have an important role in the defense mechanism against pathogens. It appears that vitamin E increased the effect attributed to these cells, as described by other authors (Noga, 1996).

The chromatophores were identified from day 1 post lesion, and increased progressively and rapidly in the supplemented group. Greater quantity of pigment was noted in the wound at day 14, which confirm the data found in histological analysis. Such increase is responsible for the color change of the wound, since there were intense chromatophores and melanocytes replacement in the epidermis and subsequently in the dermis. Such phenomenon was also described in *Notothenia coriiceps* (Silva et al., 2004, 2005). The presence of chromatophores was observed in Atlantic salmon, Salmo salar, only from the day 4 post injury (Anderson and Roberts, 1975). The presence of chromatophores is related to immune and inflammatory responses of chronic lesions (Angius and Roberts, 2003). The increase of this pigment may be related to the nutraceutical effect of vitamin E, increasing the migration of these cells to the focus of the lesion, to activate the defense mechanism during the inflammatory process and healing.

The results of the increase in the migration of inflammatory cells (IC) agree with those described in the study of chronic inflammation induced by the subcutaneous implantation of the glass slides in pacus (Belo, 2006), where the vitamin E increased the migration of these cells. Thus, we can infer that vitamin E supplementation accelerated the migration of these cells for defense, helping to shorten the healing process.

The adequate blood perfusion is essential for the healing of tissues (Silva et al., 2005). The vascular neo-formation was higher at days 7 and 14 for the supplemented and control groups, respectively, decreasing gradually from the latter. These data confirm those described in pacu (Freitas, 2001; Moraes et al., 2003) and Atlantic salmon (Anderson and Roberts, 1975) pre-conditionally fed with vitamin C and then induced to skin lesions. The analyzed data lead us to conclude that the nutraceutical effect of certain vitamins, such as vitamin E and C, accelerate and encourage angiogenesis at the injured area.

In the current study, the migration of fibroblasts was observed at day 3 and increased progressively in both groups, with greater emphasis on the supplemented group. It was possible to verify in both groups larger amount of collagen fibers from day 3 onwards, but disorganized. The percentage of fibers organized, parallel arranged to the surface, was higher in the supplemented group. These results confirm those described in skin of tilapia, where these fibers were identified in the dermis, parallel to the surface epithelium (Souza et al., 2002). The results of this study suggest that vitamin E had beneficial effects in the production of collagen, improving the process of tissue repair in Nile tilapia. Vitamin E has been shown to decrease the re-absorption of bone matrix and stimulated the proliferation of collagen, possibly by inhibiting the cytokines IL 1 and 6 in mammals (Lall and McCrea, 2007). These results were also described by Freitas (2001) and Moraes et al. (2003) in their studies on pacu supplemented with vitamin C. Moreover, fish fed with low levels of vitamin C showed impairment of tissue repair by interference of the synthesis of collagen (Jauncey et al., 1985).

Based on the results of the current study, we can conclude that the experimental model used was appropriate for assessing the kinetics of the healing process and suggest that dietary supplementation with 450mg vitamin E.kg^{-1} diet promote the healing process of skin lesions in tilapia Nile.

REFERENCES

Anderson, C.D. and Roberts, R.J. 1975. A comparison of the effect of temperature on wound healing in a tropical and temperate teleost. *Journal of Fish Biology*, 7: 173-182.

Angius, C. and Roberts, R.J. 2003. Melano-macrophage centres and their role in fish pathology. *Journal of Fish Diseases*, 26: 499-509.

Bai, S.C. and Lee, K. 1998. Different levels of dietary DL-alpha-tocopherol acetate affect the vitamin E status of juvenille Korean rockfish, *Sebastes schlegeli*. *Aquaculture*, 161: 405-414.

Belo, M.A.A. 2006. Recrutamento de macrófago e formação de gigantócitos em *Oreochromis niloticus*, submetidas a diferentes estímulos moduladores. PhD Dissertation, Departamento de Patologia Veterinária, FCAV-UNESP, Jaboticabal, 146pp.

Belo, M.A.A., Schalch, S.H.C., Moraes, F.R., Soares, V.E., Otoboni, A.M.M.B. and Moraes, J.R.E. 2005. Effect of dietary supplementation with vitamin E and stocking density on macrophage recruitment and giant cell formation in the teleost fish, *Piaractus mesopotamicus*. *Journal of Comparative Pathology*, 133: 146-154.

Bereiter-Hahn, J. 1986. Epidermal cell migration and wound repair. Biology of the Integument, V.2 – Vertebrates. Heidelberg: Springer. P.443-471.

Boyd, C.E. 1990. Water quality management for pond fish culture. Birmingham Publishing Co. Birmingham, Alabama.

Chen, R., Lochmann, R., Goodwin, A., Praveen, K., Dabrowski, K. and Lee, K. 2004. Effects of dietary vitamins C and E on alternative complement activity, hematology, tissue composition, vitamin concentrations and response to heat stress in juvenile golden shiner (*Notemigonus crysoleucas*). *Aquaculture*, 242: 553–569.

Cheville, N.F. 1994. *Introdução à Patologia Veterinária*. São Paulo: Editora Manole, 556p.

Conn, E.E. and Stumpf, P.K 1984. *Introdução à Bioquímica*. 4 ed. São Paulo: Edgard Blucker LTDA,. 531pp.

Cuesta, A., Ortuno, J., Rodriguez, A., Esteban, M.A.E and Meseguer, J. 2002. Changes in some innate defense parameters of seabream (*Sparus aurata* L.) induced by retinol acetate. *Fish and Shellfish Immunology*, 13: 279-291.

Dawson, L.H.J., Pike, A.W., Houlihan, D.F. and McVicar, A.H. 1998. Effects of salmon lice Lepeophtheirus salmonis on sea trout Salmo trutta at different times after seawater transfer. *Disease of Aquatic Organisms*, 33: 179– 186.

Eckhoff, K.M., Aidos, I., Hemre, G.I. and Lie, O. 1998. Collagen content in farmed Atlantic Salmon (*Salmo salar,* L.) and subsequent changes in solubility during storage on ice. *Food Chemistry*, 62: 197-200.

Fevolden, S.E., Refstie, T. and Roed, K.H. 1991. Selection for high and low cortisol stress response in Atlantic salmon *(Salmo salar)* and rainbow trout *(Oncorhynchus mykiss)*. *Aquaculture*, 95: 53-65.

Fevolden, S.E. and Roed, K.H. 1993. Cortisol and immune characteristics in rainbow trout *(Oncorhynchus mykiss)* selected for high or low tolerance to stress. *Journal of Fish Biology*, 43: 919-930.

Freitas, J.B. 2001. Cinética do processo inflamatório e reparação tecidual em pacus *Piaractus mesopotamicus* Holmberg, 1887 alimentados com ração suplementada com diferentes concentrações de vitamina C. MSc Dissertation– Faculdade de Ciências Agrárias e Veterinárias, Universidade Estadual Paulista, Jaboticabal.

Furones, M.D., Alderman, D.J., Bucke, D., Fletcher, T.C., Knox, D. and White, A. 1992. Dietary vitamin E and the response of rainbow trout, *Onchorhynchus mykiss* (Walbum), to infection with *Yersinia ruckeri*. *Journal of Fish Biology* 41: 1037–1041.

Garcia, F. 2005. Pacus *Piaractus mesopotamicus* (holmberg, 1887) alimentados com dietas suplementadas com vitaminas C e E e submetidos ao desafio com *Aeromonas hydrophila*. MSc Dissertation, Centro de Aqüicultura da Unesp, Jaboticabal, 76 pp.

Garcia, F., Pilarsky, F., Onaka, E.M., Moraes, F.R. and Martins, M.L. 2007. Hematology of *Piaractus mesopotamicus* fed diets supplemented with vitamins C and E, challenged by *Aeromonas hydrophila*. *Aquaculture*, 271: 39–46.

Hemre, G., Deng, D.F., Wilson, R.P. and Berntssen, M.H. 2004. Vitamin metabolism and early biological responses in juvenile sunshine bass *(Morone chrysops* x *M. saxatilis)* fed graded levels of vitamin A. *Aquaculture*, 235: 645-658.

Hibyia, T. 1982. *An atlas of fish histology normal and pathological features*. Tokio, Kodansha ltd., 147pp.

Hopwood, D. 1990. Fixation and fixatives. p. 21-42. *In* J. D. Bancroft e A. Stevens Theory and Practice of histological techniques. Nova Iorque, Churchil Livingstone, 3ª. Ed., 726pp.

Hung, S., Tu, C. and Wang, W. 2007. In vitro effects of singular or combined anti-oxidative vitamins and/or minerals on tilapia *(Oreochromis* hybrids) peripheral blood monocyte-derived, anterior kidney-derived, and spleen-derived macrophages. *Fish and Shellfish Immunology*, 23: 1-15.

Iger, Y. and Abraham, M. 1990. The process of skin healing in experimentally wounded carp. *J. Fish Biol.*, 36: 421-437.

Iger, Y., Abraham, M., Dotam, A., Fattal, B. and Rahamin, E. 1988. Cellular responses in the skin of carp maintained in organically fertilized water. *Journal of Fish Biology*, 33: 711-720.

Ikoma, T., Histoshi, K., Tanaka, J., Walsh, D. and Mann, S. 2003. Physical properties of type I collagen extracted from fish scales of *Pagrus major* and *Oreochromis niloticas*. *International Journal of Biological Macromolecules*, 32: 199–204.

Jauncey, K., Soliman, A. and Roberts, R.J. 1985. Ascorbic acid requirements in relation to wound healing in the cultured tilapia *Oreochromis niloticus* (Trewasas). *Aquaculture and Fisheries Management*, 16: 139-149.

Johnsen, B.O., 1978. The effect of an attack by the parasite *Gyrodactylus salaris* on the population of salmon parr in the River Lakselva, Misvaer in northern Norway. *Astarte*, 11: 7-9.

Junqueira, L.C.U., Montes, G. S. and Sanchez, E. M. 1982. The influence of tissue section thickness on the study of collagen by picrosirius-polarization method. *Histochemistry*, 74: 153-156.

Kumar, V., Abbas, A. and Fausto, N. 2005. Pathologic Basis of Disease. 7th ed., Elsevier Saunders, Pennsylvania, 1504 pp.

Lall, S.P. and Lewis-McCrea, L.M. 2007. Role of nutrients in skeletal metabolism and pathology in fish, an overview. *Aquaculture*, 267: 3–19.

Lunder, T., Evensen, O., Holstad, G. and Hastein, T. 1995. 'Winter ulcer' in the Atlantic salmon Salmo salar. Pathological and bacteriological investigations and transmission experiments. *Diseases of Aquatic Organisms*, 23: 39–41.

Martins, M.L., Miyazaky, D.M.Y., Moraes, F.R., Ghiraldeli, L., Adamante, W.B. and Mouriño, J.L.P. 2008. Ração suplementada com vitaminas C e E influencia a resposta inflamatória aguda em tilápia do Nilo. *Ciência Rural*, 38: 213-218.

Metchnikoff, E. 1893. Lectures on the comparative pathology of inflammation delivered at the Pasteur Institute in 1891. English translation by F.A. and E.H. Starling. London: Kegan, Paul, Trench, Trübner and Co.

Mittal, A.K. and Munshi, J.S.D. 1974. On the regeneration and repair of superficial wound in the skin of *Rita rita* (Ham.) (Bagridae, Pisces). *Acta Anatómica*, 88: 424-442.

Montero, D., Marrero, M., Izquiierdo, M.S., Robaina, L.,Vergara, J.M. and Tort, L. 1999. Effect of vitamin E and C dietary supplementation on some immune parameters of gilthead seabream (*Sparus aurata*) juveniles subjected to crowding stress. *Aquaculture*, 171: 269-278.

Moraes, J.R.E., Freitas, J. B., Bozzo, F.R., Moraes, F.R. and Martins, M. L. 2003. A Suplementação alimentar com vitamina C acelera a evolução do processo cicatricial em *Piaractus mesopotamicus* (HOLMBERG, 1887). *Boletim do Instituto de Pesca*, 29: 57-67.

Nagai, T., Araki, Y. and Suzuki, N. 2002. Collagen of the skin of ocellate puffer fish (*Takifugu rubripes*). *Food Chemistry*, 78: 173–177.

Noga, E.J. 1996. Fish Diseases. *Diagnosis and Treatment*. 1a ed. Raleigh: Mosby. 367pp.

Ortuño, J., Esteban, M.A. and Meseguer, J. 2000. High dietary intake of tocopherol acetate enhances the non-specific immune response of gilthead seabream (*Sparus aurata* L.). *Fish and Shellfish Immunology*, 10: 293–307.

Phromsuthirak, P. 1977. Electron microscopy of wound healing in the skin of *Gasterosteus aculeatus*. *Journal of Fish Biology*,.11: 193-206.

Pickering, A.D. and Pottinger, T.G. 1989. Stress responses and disease resistance in salmonid fish: effects of chronic elevation of plasma cortisol. *Fish Physiology and Biochesmistry*, 7: 253-258.

Quilhac, A. and Sire, J.Y. 1998. Restoration of the subepidermal tissues and scale regeneration after wounding a cichlid fish, *Hemichromis bimaculatus*. *The Journal of Experimental Zoology*, 281: 305–327.

Reite, O.B. and Eversen, O. 2006. Inflammatory cells of teleostean fish: A review focusing on mast cells/eosinophilic granule cells and rodlet cells. *Fish and Shellfish Immunology*, 20: 192-208.

Reque, V.R. 2005. Suplementação alimentar com *Saccharomyces cerevisiae* na inflamação induzida por *Aeromonas hydrophila* inativada em tilápias do Nilo *(Oreochromis niloticus)*.. MSc Dissertation, Centro de Aqüicultura da Unesp, Jaboticabal, SP.

Roubal, F.R. and Bullock, A.M. 1988. The mechanism of wound repair in the skin of juvenile Atlantic salmon, *Salmo salar* L., following hydrocortisone implantation. *Journal of Fish Biology*, 32: 545-555.

Sakabe, R. 2007. Suplementação alimentar com ácidos graxos essenciais para tilápias do Nilo: desempenho produtivo, hematológico e granuloma por corpo estranho. MSc Dissertation, Centro de Aqüicultura da Unesp. 78 F. Jaboticabal, SP.

Sakai, M. 1999. Current research status of fish immunostimulants. *Aquaculture*, 172: 63-92.

Santos, L.R. and Oliveira, C. 2007. Morfometria testicular durante ciclo reprodutivo de *Dendropsophus minutus* (Peters) (Anura, Hylidae). *Revista Brasileira de Zoologia*, 24: 67-70.

Silva, J.R.M.C., Sinhorini, I.L., Jensch-Junior, B.E., Porto-Neto, L.R., Hernandez-Blazquez, F.J., Vellutini, B.C., Pressinotti, L.N., Pinto, F.A.C., Cooper, E.L. and Borges, J.C.S. 2004. Kinetics of induced wound repair at 0°C in the Antarctic fish (Cabeçuda) *Notothenia coriiceps. Polar Biology*, 27: 458–464.

Silva, J.R.M.C., Sinhorini, I.L., Jensch-Junior, B.E., Porto-Neto, L.R., Hernandez-Blazquez, F.J., Vellutini, B.C., Pressinotti, L.N., Pinto, F.A.C., Cooper, E.L. and Borges, J.C.S. 2005. Microscopical study of experimental wound healing in *Notothenia coriiceps* (Cabeçuda) at 0°C. *Cell Tissue Research*, 321: 401–410.

Siwicki, A.K., Anderson, D.P. and Dixon, O.W. 1990. In vitro immunostimulation of rainbow trout (*Oncorhynchus mykiss*) spleen cells with levamisole. *Developmental and comparative immunology*, 14: 231-237.

Souza, M.L.R, Casaca, J. M., Ferreira, I.C., Ganeco, L. N., Nakagki, L.S., Faria, R.H.S., Macedo-Viegas, E.M. and Rielh, A. 2002. Histologia da pele e determinação da resistência do couro da tilápia do Nilo e carpa espelho. *Revista do Couro, Estância Velha*, 159: 32-40.

Val, A.L., Menezes, A.C.L., Ferreira, M.S., Silva, M.N.P., Araújo, R.M. and Almeida-Val, V.M.F. 2006. Estresse em peixes: respostas integradas para a sobrevivência e a adaptação. In: Silva-Souza, A.T. (org.). Sanidade de organismos aquáticos no Brasil. Maringá: Abrapoa, p.211-228.

Wahli, T., Verlhac, V., Girling, P., Gabaudan, J. and Aebischer, C. 2003. Influence of dietary vitamin C on the wound healing process in rainbow trout (*Oncorhynchus mykiss*). *Aquaculture,* 225: 371–386.

Wedemeyer, G.A. 1997. Effects of rearing conditions on the health and physiological quality of fish in intensive culture. In *Fish Stress and Health in Aquaculture* (G. K. Iwama, A. D. Pickering, J. P. Sumpter and C. B. Schreck, eds) pp. 35–71. Cambridge: Cambridge University Press.

Wilson, R.P., Bowserazand, P. and Poe, W.E. 1984. Dietary Vitamin E Requirement of Fingerling Channel Catfish. *Journal of Nutrition*, 114: 2053-2058.

Wise, D.J., Tomasso, J.R., Schewedler, T.E., Gatlin, D.M., Bai, S.C. and Blazer, V.S. 1993. Effecto of vitamin E on the immune response of channel catfish to Edwardissiella ictaluri. *Journal of Aquatic Animal Health*, 5: 183-188.

Woo, J.W., Yu, S.J., Cho, S.M., Lee, Y.B. and Kim, S.B. 2008. Extraction optimization and properties of collagen from yellow tuna (*Thunnus albacares)* dorsal skin. *Food Hydrocolloids*, 22: 879-887.

In: Encyclopedia of Environmental Research   ISBN: 978-1-61761-927-4
Editor: Alisa N. Souter   © 2011 Nova Science Publishers, Inc.

Chapter 34

FISHERIES ECONOMICS IMPACTS ON COOPERATIVE MANAGEMENT FOR AQUACULTURE DEVELOPMENT IN THE COASTAL ZONE OF TABASCO, MEXICO

Eunice Pérez-Sánchez[*,1], *James F. Muir*[2], *Lindsay G. Ross*[2], *José M. Piña-Gutiérrez*[1] *and Carolina Zequiera-Larios*[1]

Universidad Juárez Autonoma de Tabasco. Carretera Villahermosa-Cárdenas Km 0.5
entronque a Bosques de Saloya, Villahermosa, Tabasco, 86150 México[1]
Institute of Aquaculture, University of Stirling, FK9 4LA, Scotland, UK[2]

ABSTRACT

Coastal zones have been focal points of social conflicts and environmental problems in Tabasco, Mexico. The cumulative conflicts arising from a heterogeneous regional development had led to an increased lack of coastal resources management, increased population, reduction of stocks of valuable species due to over-fishing, and environmental deterioration. The aim of this study was to assess fishermen access to resources in the coastal zone of Tabasco in terms of income and costs of fisheries production and determine the effect on aquaculture development. Results showed that fishing cooperatives have lost competitiveness due to a lack of integration into a market-oriented production, making fishery entrepreneurship a high-risk activity and incapable of producing efficiently. The conditions fishermen experience suggest that the hierarchical approach for resources management influenced by political decisions have trapped resource managers in programs that address symptoms rather than causes of basic fisheries management problems. Therefore, economic diversification in and outside the fishery sector needs to be implemented. Aquaculture could be developed as a means to improve the role of fishing cooperatives to support production. Although state intervention is not the only solution, the devolution of regulatory functions to local communities may help to restore the crucial qualities of collective action and sustain aquaculture.

Keywords: Socio-economics, fisheries, coastal communities, resources management

---

[*] E-mail address: eunice.perez@dacbiol.ujat.mx Tel./Fax: +52 993-3544308

INTRODUCTION

Fisheries of developing regions typically face over-exploited stocks, a lack of alternative employment outside the fishery and an over-extended fleet. The increase of both human population and more efficient technology are the most serious among many causes for the decline in landings and the over-exploitation of coastal fisheries, thus reducing resources access to those dependent on the fisheries in spite of the productive potential of coastal environments. Fishermen without other options will shift from one activity to another regardless of legal conditions and management frameworks to maintain their livelihood. Thus, according to Charles and Herrera (1994) 'fishery development can be viewed as sustainable only if it enhances, or at least maintains ecosystems, communities, socio-economics, culture, and institutional structures'. In these terms, sustainable development could be defined as change that can be sustained in terms of social systems and cultures, and their relation to ecology and economics by understanding how economies grow and why some fail (Hundloe, 2000).

The social and economic organization of the artisanal fishermen is influenced by the exploitation strategy of available resources. The sharing of fish stocks, at local level, is directly linked to the way resources are accessed with spatial patterns of exploitation of the resources and with the conflicts between community objectives and national goals (Terrebone, 1995; McCay & Jentoft, 1998).

The southern Mexican states on the coast of the Gulf of Mexico produce approximately 85% of the total national oil production. Extraction is mainly carried out in the states of Tabasco and Campeche in the marine and terrestrial areas with important impacts on the coastal zone (e.g. coastal erosion, accidental spills, deforestation by drilling and exploration activities) due to the lack of government legislative controls and mismanagement of the industry (Gold-Bouchot et al., 1999). The region has also attracted attention due its important economic influence in terms of fisheries, aquaculture and agriculture, and for its ecosystem diversity. These diverse resources and their use have led to conflicts among different users (Negrete-Salas, 1984; Gold-Bouchot et al., 1995). Coastal zones have been focal points of social conflicts and environmental problems in the last 25 years in Tabasco. Most of these have also resulted from the hasty shift in regional economic development from a predominantly agrarian production to a high technology oil industry (Garcia, 1984; Rivera-Arriaga & Villalobos, 2001).

Approximately 20,954 artisanal fishermen depend on Tabasco water resources to endure their livelihoods. They represent only the 1.9% of the economically active population (people 15 to 60 years old capable of working) compared to the 17% (117,701 people) and 54% (367,376 people) of those working in the industrial and services sector, respectively (INEGI, 2002). The first fishing cooperatives in the Mecoacan lagoon were organized to provide local fishermen with institutional resources and develop the management of estuary fisheries. "Andres Garcia" was established in 1941 and was granted a fish, oyster, and shrimp fishing permit. This is the largest cooperative, with 229 members. The second cooperative "Chiltepec" was established in 1944, but due to conflicts among its members regarding financial and management issues the membership decided in 1990 to reorganize the cooperative into two new organizations "Puente de Ostion" with 54 members and "Boca de los Angeles" with 90 members. The last cooperative to be established was "Mecoacan" in 1969 and has a membership of 117.

The coastal resources of Tabasco are enclosed in 11 800 km2 of continental platform, 40 km2 of mangrove forest and sixteen coastal lagoons, though Carmen-Pajonal-Machona, Tupilco-Las Flores and Mecoacan systems are the most important because of the area covered (29,800 ha) (Velazquez, 1994). Tabasco holds an important share of Southeast Mexico's fisheries, particularly for oyster production, which contributes 30% of the state fisheries production, based on one species, the American oyster Crassostrea virginica, which is fished and cultured in 9.3% of Tabasco's coastal lagoons area by twelve Cooperatives, four in the Mecoacan lagoon and eight in the Carmen-Pajonal-Machona lagoon system. Oceanic fishing is limited to a less than 1 km from the coast shoreline with a total production of 56,888 t/yr; this represents a total value of $ 598.46 million.

Fishery production in Tabasco represents a relatively important part of the production on the Gulf of Mexico region. Fishery sector development in Tabasco has been one of the most severely affected by oil industry impacts and unmanaged coastal lagoon ecosystems. Since 1985 a constant reduction in fishery production has been recorded, particularly for oyster (Moguel, 1994). The volume of coastal fisheries is considerably larger than inland fisheries, as the most representative coastal fisheries are Bandera (*Bagre marinus*) 3,662 t/yr, Cintilla (*Trichiurus lepturus*) 3,859 t/yr, Robalo (*Centropomus undecimalis*) 2,890 t/yr, and Osyter (*Crassostrea virginica*) 20,765 t/yr, and Cichlids (5,715 t/yr) are the most representative of inland fisheries. 44% of Tabasco fisheries come from small-scale farming of Oyster *C. virginica*, which is mostly an aquaculture-based fishery with a total production of 20,765 t/yr, Tilapia *Oreochromis* spp (4,292 t/yr), shrimp (77 t/yr) and others (13 t/yr), with a total value of $ 109.71 million (SAGARPA, 2003). Despite the fact that Tabasco has a coastline of 190 km and excellent climatic conditions, the fishing grounds distributed along its coast are not large and the demersal stocks are regarded as underused.

Based on the above, the aim of this study was to develop a comparative assessment of independent and organized fishermen in the Mecoacan lagoon in terms of income and costs of fisheries production.

AREA DESCRIPTION

Mecoacan lagoon is located at the North of the state between latitude 18°15' – 18°26' and longitude 93°01' – 93°15'. It is shallow with an average depth of 1.50 meter and a total area of 5,160 ha. The system has permanent communication with the sea through the Dos Bocas natural mouth in its northern extreme, which stretches to the sea trough a narrow bar of 400 m (Fig. 1). The seaward side of the estuary is formed by a chain of small islands with the largest (3 km in length) located in the middle of the mouth. Along with its tributaries Seco, Escarbado, Cucuchapa, Cunduacan and Arrastradero rivers, the Mecoacan area is considered the second most important coastal lagoon system of Tabasco, with extensive oyster banks at its bottom and bordered by mangrove areas, which have been reduced due to timber exploitation (Dominguez-Dominguez, 1991; Valdes, 1998).

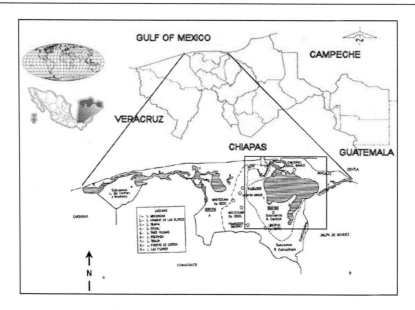

Figure 1. Location of Mecoacan lagoon, Paraiso, Tabasco, México.

To define the main physic-chemical parameters, the average water temperature is 24-30.2°C, dissolved oxygen is about 3.0-4.5 mg/l and salinity 1.3-14 ppt, precipitation averages about 1,800 mm/yr, evaporation is 1,600 mm/yr and estimated runoff is 300 mm/yr (Arredondo et al., 1993; Rodriguez, 1998; Valdes, op cit.). The major physical changes in the lagoon have been the reduction in depth by sedimentation from the rivers and the reduction of seawater inflows due to the installation of the Dos Bocas seaport and petrochemical plant breakwaters (Alvarado-Azpeitia, 1996).

The Mecoacan lagoon is Federal property, where cooperative and independent fishermen (free riders) operate indiscriminately using different fishing gear, mainly with cast and gill nets and oyster grabs. Fish and shellfish are captured year through and selected mainly by size and market value.

SAMPLING AND ANALYSIS

This study is focused on the four fishing cooperatives located in Mecoacan lagoon, which represent about 2016 people in the municipality of Paraiso. These four fishing cooperatives established in Mecoacan were selected as they are entitled by law to extract oyster species from the wild to manage culture sites and to participate directly in the management of coastal lagoon resources. Independent fishermen (free riders) were selected for the study as they represent the strongest competition to organized fishermen in terms of economic benefit of resources use.

According to the political division of the Municipality, fishermen households are located in the boundaries of the coastal lagoon area in the following towns: Jose Maria Morelos, Andres Garcia Island, Bellote, Chiltepec, Banco, Tanque, Puerto de Chiltepec, Libertad (El Chivero section), Carrizal, Puerto Ceiba (Village and Carrizal section), Nuevo Torno Largo and Miguel de la Madrid. The collection of data was based on rapid rural appraisal (RRA) approach using questionnaires (households and fishermen groups data) and interviews (cooperatives' administrative committees and government officers) to carry out a socio-economic survey (McCrossan, 1991; McArthur, 1994; Townsley, 1996). The sample size calculated was N equal to 440 households. All questionnaires were pre-tested using a sub-group of 30 individuals in order to rephrase sentences according to local expressions and information was classified according to the group from which data were collected. The final sample size was 421 organized fishermen and 21 independent fishermen. The sample size for independent fishermen was reduced by the reluctance of fishermen to participate in the survey.

The sustainability of fishery production based on economic criteria was defined by using financial indicators to separate out a range of variables that determine profitability and vulnerability to shocks (Hundloe, 2000), using a multiple regression analysis. A dummy variable (a variable that can take on only two values) controlling for the association of fishermen were used in both analysis, assuming that 0 correspond to independent fishermen (free riders) and 1 to fishermen holding a membership in a cooperative. This type of analysis is generally referred as a Single Factor ANOVA, performing an F-test for the significance of the regression using the statistical software SPSS. A Beta value was used to estimate the relationship between the variables and the predictor (dummy variable), if the value is positive, there is a positive relationship between the predictor and the outcome, whereas a negative coefficient indicates a negative relationship. This value also indicates the degree at the predictor affects the outcome comparing the value against the standard error to determine whether or not the value differs from zero and the associated t-test and P value. All the standardized Beta values are measured in standard deviation units and so are directly comparable, providing a better insight into the importance of a predictor in the regression model (Fields, 2000; Koop, 2000).

The indicators used to run the analysis included those of purchasing and/or maintaining boat and gear, operating costs such as crew wages, fees, fuel, fishing equipment and gear maintenance, gasoline and equipment rent and total amount earned per capture volume on a monthly basis. The indicators organized to test costs variables included: the type of propulsion, ownership of the fishing equipment, fishing ground, number of fishing days, storing and type of fishing gear. The variables to define the income variability were selected according to: capture volume, market price, type of processing, processing labor input, grading of produce and market channel for fish, shrimp, oyster and crab.

RESULTS

The current status of fishing cooperatives in Mecoacan is of low profitability, which has led to a run down of equipment and facilities and clear member dissatisfaction. These conditions within cooperatives have affected negatively independent fishermen's perception of associations, promoting disbelief and reluctance to join any fishing organization and increasing conflicts due to open-access conditions of fisheries resources. From an economic perspective the open-access problem of Mecoacan fisheries is a pervasive externality, where all individuals do the same, racing to catch as much as possible before someone else does. Fishing quotas were allocated for each fishing cooperative, for example, in the case of oyster as follows: 22 t for Mecoacan, 15 t for Puente de Ostion and 36 t for Andres Garcia. However, quotas are not observed by the membership of cooperatives due to the racing impose by independent fishermen.

The composition of the catches varies seasonally and brackish water species are target species, not for their market value but for their abundance in the lagoon. According to specific fisheries, one oyster fisherman catches 230 kg on average, with a range between 73-387 kg fishing five days per week. They use a thousand oysters as a unit, locally called arpilla, to sell the catch. Fishermen spend an average of 6 hours to catch 1000 shell-on oysters, some 45 kilos, from the lagoon. The largest number of fishermen involved in oyster fishing was found in Andres Garcia cooperative. Prices vary between $ 39.00 and $ 47.00, with an average of $ 44.00. Mussel species are also caught along the coastline of the lagoon, and these obtain better prices than oysters in the local market at $ 2.50 each in restaurants and ¢0.60- $1.00 in the trading market. The main species are *Mercenaria campechiensis* locally known as "almeja de laguna", *Rangia cuneata* (almeja de rio) and *Brachidontes recurvus* (almeja negra).

Regarding fish capture, 22% of the respondents were involved in marine fishing. The locals stated that marine fishing started approximately in 1996 with the arrival of marine fishermen from Veracruz State, who were looking for fish stocks in Tabasco's coastline as fisheries were declining in the southern part of Veracruz. They also stated that pollution and hydrological changes in the Mecoacan lagoon had reduced fish stocks and they have had to move out of the lagoon to gain a livelihood. Throughout the year, marine species with higher market prices are sought. Hence, less equipped fishermen seeking to maximize their income have been motivated to harvest as many coastal lagoon resources as possible in response to rising prices of marine fish. The main marine species are spanish mackerel *Scomberomorus maculatus* locally know as sierra, king mackerel *S. cavalla* (peto), belted bonito *Sarda sarda* (bonito), red snapper *Lutjanus campechanus* (huachinango), *Bagre marinus* (Bandera) and cutlassfish *Trichiurus lepturus* (cintilla).

Catch prices vary according the specie and season, fluctuating from $ 1.00 to $30.00 (recorded during Easter). The rest of the year prices fluctuated between $ 1.15-5.50 with an average of $ 3.41 per kilogram (according to prices established by brokers in Mexico City and Veracruz State middlemen). The fishing season in coastal waters lasts from February to November, subdivided into three periods according to the most important species, cutlass fish (Jan-Jul), snapper (May-Oct) and mackerel (Feb-Jun/Oct-Nov).

In the case of crab and shrimp species, fishermen capture these as alternative source of income, when weather conditions are not good enough for marine fishing; but crab captures are mainly done by elderly fishermen and those who have not been able to get involved in marine fishing. Prices for shrimp fluctuated between $ 10.00-30.00 per kilo. The fishermen stated that high prices for shrimp are only available when dealing with middlemen or during the veda (out of season fishing). Significant fluctuations for crab prices were not recorded ($4.80 - 6.40). On average, a fisherman catches about 8-20 kilos of crab and 8-10 kilos of shrimp per week, with an average income of $ 66.70 from crab and $131.60 from shrimp.

In general, household income accounts an average of $350.00 per week, including the income generated by women and children. The kind of fishing practiced was found to be a significant influence on net income. Those involved in marine fisheries obtained $ 471.00 per week on average ($ 381.00- $660.00), while coastal lagoon fishing provided an average income of $166.00 ($ 117.00-$260.00). However, catching volume fluctuates and total earning has to be shared with two or three other people in the fishing team.

The fishermen stated that from catch proceeds, the patron (usually a middlemen or other fisherman who has lend money to buy a boat and outboard engine and cover fishing cost), takes out the amount they spent in fuel and a percentage corresponding to either rent or payment for boat and outboard engine. After this discount has been made, the patron takes out his share, of 60%, leaving the rest for the fishermen.

Mecoacan lagoon fisheries were reported by fishermen as a low-income resource, as the species captured do not obtain good market prices. Most of lagoon fishermen expressed their interest in getting involved in marine fisheries. However, they also commented that the significant initial investment capital, the right contact to sell the product, and the requirement of physical strength and skills were the main constraints to access marine fisheries. In the case of marine fishermen, many of them are still involved in estuary fishing, but in a less extend. Thus, the general movement between the fisheries is moderate and highly dependent on fishermen access to financial resources.

Approximately, 25% of fishermen net income is spent on operating costs. Of these, gasoline and maintenance of equipment dominate. A regression model (Table 1) was used to determine the marginal costs of fishing per month related to the association of fishermen (fishing cooperative member or independent). The variables selected for the analysis of costs included the number of fishing days, fishing ground, ownership of the fishing equipment, type of propulsion, storing and type of fishing gear (e.g. cast net, gillnet, grab). The variables related to fishing effort and equipment ownership were not significant as most fishermen used cast nets or fishing rods and own the fishing equipment ($t=0.469$, P 0.640). The main factor found in the analysis of costs was the use of motors and equipment maintenance, both in the dry and wet season ($t=21.98$, P 0.000). In some instances this was prejudicial financially, as fishermen used fuel but did not capture enough fish to cover the costs.

Though there were no conflicts among fishermen regarding locations, fishing ground selection effect was found to be a significant variable when tested in the regression model ($t=2.048$, P O.041). Fishermen operating both in coastal lagoon and coastal waters tend to increase their costs by $116.73 per month with a marginal cost of at least $ 4.61 and at most $228.85. Labor costs tend to increase by $166.96. Thus, it is difficult to understand on the basis of stated catch value how fishermen make money, mostly because the cost of propulsion tends to increase by $ 1,065.87 per month.

The way fish was preserved was significant ($t=2.08$, P 0.038). The marginal costs of storing are between $8.40 and $ 292.36 per month, requiring fishermen to sell catch directly at whatever price is available, so being subject to opportunity cost. The wide range of storing costs is related to catch volume and marketing of product. Storing is a common practice for marine species and shrimp, and it is employed by fishermen not only to preserve the product but to be able to collect batches of at least one tonne, as requested by middlemen and make the shipment costs worthwhile. Some fishermen family groups have invested in storing facilities to collect the required load. However, most of the fishermen did not have the capital to invest, thus having to take their catch to the facilities of middlemen.

The amounts spent on fuel and other fishing costs have risen more than the general price level for fish, but without increases in catches, probably evidence that greater effort is being applied. The dummy variable used for the association of fishermen to cooperative organizations was found to be significant ($t=2.70$, P 0.007). The organized fishermen tend to increase their costs significantly more than independent fishermen by $ 411.22 with a marginal cost between $ 112.39 and $ 710.03, indicating a considerable uncertainty.

Table 1. Analysis of the monthly production costs of fishing. Dependent Variable: COSTS. Independent variables: LABOUDAY= Number of labor days, FISHGROU= Fishing ground, EQOWNER= Equipment ownership, PROPULS= Propulsion, DUMMY= Fishermen association.

Parameter	Beta	Std. Error	t	P	95% Confidence Interval	
					Lower Bound	Upper Bound
Intercept	-2,199.78	215.19	-10.22	0.000	-2,623.11	-1,776.45
LABOUDAY	166.96	16.26	10.26	0.000	134.96	198.97
FISHGROU	116.73	56.99	2.04	0.041	4.61	228.85
EQOWNER	14.74	31.45	0.46	0.640	-47.13	76.62
PROPULS	1,065.87	48.60	21.92	0.000	970.25	1,161.49
STORING	150.38	72.17	2.08	0.038	8.40	292.36
CASTNET	-27.39	63.99	-0.42	0.669	-153.29	98.49
GILLNET	-31.13	65.79	-0.47	0.636	-160.55	98.29
GRAB	-97.66	76.72	-1.27	0.204	-248.59	53.27
DIVING	-99.13	61.18	-1.62	0.106	-219.48	21.21
NAZANET	142.26	74.41	1.91	0.057	-4.11	288.65
ANGLING	40.00	105.29	0.38	0.704	-167.12	247.12
DUMMY	411.21	151.90	2.70	0.007	112.39	710.03

Tests of Between-Subjects Effects

Source	Type III Sum of Squares	df	Mean Square	F	P
Corrected Model	271677453.756	12	22639787.813	81.497	0.000
Intercept	29027746.568	1	29027746.568	104.491	0.000
LABOUDAY	29259348.066	1	29259348.066	105.325	0.000
FISHGROU	1165305.499	1	1165305.499	4.195	0.041
EQOWNER	61023.262	1	61023.262	0.220	0.640
PROPULS	133574522.744	1	133574522.744	480.829	0.000
STORING	1206017.112	1	1206017.112	4.341	0.038
CASTNET	50916.417	1	50916.417	0.183	0.669
GILLNET	62194.086	1	62194.086	0.224	0.636
GRAB	450077.861	1	450077.861	1.620	0.204
DIVING	729365.361	1	729365.361	2.626	0.106
NAZANET	1015412.075	1	1015412.075	3.655	0.057
ANGLING	40095.387	1	40095.387	0.144	0.704
DUMMY	2035825.239	1	2035825.239	7.328	0.007
Error	91951875.245	331	277800.227		
Total	671294572.063	344			
Corrected Total	363629329.001	343			

R Squared = 0.747 (Adjusted R Squared = 0.738). Exchange rate Pesos$9.30 = USD$ 1.00

The profits from fishing are extremely variable, as once marketable catches are attained fishermen market fish as quickly as possible to minimize losses due to costs, and to maximize potential benefits. The variables selected to define the income variability were based on capture volume, market price, processing labor input, grading of produce, type of processing, and market channel for fish, shrimp, oyster and crab (Table 2). Though processing is considered a potential means of diversification and improvement of produce quality, it was found insignificant and it did not offer benefits (t= -1.54, P 0.123), neither to those who process their catch in any way nor to those who uses family labor to reduce costs of hiring labor for processing (t= -0.92, P 0.356).

Table 2. Monthly income analysis of Mecoacan fishermen. Dependent Variable: GROSS INCOME. Independent variables: FISHCAPV=Fish capture volume, SHRICAPV=Shrimp capture volume, OYSCAPV=Oyster capture volume, CRABCAPV=Crab capture volume, LABPROC= Labor invest in processing, DISTCHAN= Distribution channel, DUMMY=Fishermen association.

Parameter	Beta	Std. Error	t	P	95% Confidence Interval	
					Lower Bound	Upper Bound
Intercept	2,571.68	3,279.59	0.78	0.433	-3,875.65	9,019.02
FISHCAPV	16.15	1.94	8.30	0.000	12.33	19.98
SHRICAPV	23.02	1.40	16.35	0.000	20.25	25.79
OYSCAPV	44.73	4.44	10.06	0.000	35.99	53.46
CRABCAPV	-31.59	35.78	-0.88	0.378	-101.95	38.75
FISHPRICE	122.27	20.14	6.06	0.000	82.66	161.88
SHRIPRICE	26.53	14.15	1.87	0.062	-1.30	54.36
OYSPRICE	45.00	3.78	11.87	0.000	37.55	52.45
CRABPRICE	-8.05	61.08	-0.13	0.895	-128.13	112.02
LABPROC	-939.37	1,016.98	-0.92	0.356	-2,938.66	1,059.91
GRADING	472.85	227.32	2.08	0.038	25.95	919.74
PROCESSING	-3,108.86	2,012.00	-1.54	0.123	-7,064.26	846.54
DISTCHAN	284.47	121.40	2.34	0.020	45.80	523.14
DUMMY	-198.67	985.02	-0.20	0.840	-2,135.13	1,737.78

Tests of Between-Subjects Effects

Source	Type III Sum of Squares	df	Mean Square	F	P
Corrected Model	12604912781.822	13	969608675.525	55.391	0.000
Intercept	10763453.425	1	10763453.425	0.615	0.433
FISHCAPV	1206051825.623	1	1206051825.623	68.899	0.000
SHRICAPV	4679826867.438	1	4679826867.438	267.346	0.000
OYSCAPV	1773690357.972	1	1773690357.972	101.326	0.000
CRABCAPV	13646263.895	1	13646263.895	0.780	0.378
FISHPRIC	644732099.513	1	644732099.513	36.832	0.000
SHRIPRIC	61476761.144	1	61476761.144	3.512	0.062
OYSPRIC	2469933165.044	1	2469933165.044	141.101	0.000
CRABPRIC	304310.876	1	304310.876	0.017	0.895
LABPROC	14934887.906	1	14934887.906	0.853	0.356
GRADING	75738679.956	1	75738679.956	4.327	0.038
PROCESSING	41792539.650	1	41792539.650	2.387	0.123
DISTCHAN	96107204.872	1	96107204.872	5.490	0.020
DUMMY	712113.938	1	712113.938	0.041	0.840
Error	7019396472.859	401	17504729.359		0.000
Total	40219106508.161	415			
Corrected Total	19624309254.681	414			

R Squared = 0.642 (Adjusted R Squared = 0.631). Exchange rate Pesos$9.30 = USD$ 1.00

Profits tend to decline quickly and fishermen lack the flexibility to shift to other income generating pursuits due to their heavy investment debt in fixed capital as encouraged by middlemen. The regression model showed a significant effect on income by the channel of product distribution (t=2.34, P 0.020), with an income increase at $ 284.00 per month and a variation of $ 45.00-$ 523.15. Product grading increases $ 472.85 fishermen income in average per month, but there was found a significant marginal effect on income with a variation from $ 25.96 to 919.75 (t=2.08, P 0.038).

Significant effects on income from capture volume of fish (t=8.30, P 0.000), shrimp (t=16.35, P 0.000), and oyster (t=10.06, P 0.000) were found. Though catches vary seasonally, the marginal effect found is related to the price fluctuation showed by fish, shrimp and oyster, which significantly constitute the major sources of income regardless of the affiliation status of fishermen (t= -0.20, P 0.840). The effect of prices was found to be significant for fish (t=6.06, P 0.000) and oyster (t=11.87, P 0.000). Income variations were found to be at least $ 82.67 and at most $ 161.89 for fish, and $37.56 to $ 52.45 for Oyster. However, the low effect of market prices for shrimp may be an indication of high levels of fishing effort, as fishermen have to capture a large volume to obtain significant gains.

CONCLUSION

Fishermen have to cope with rising inflation and with a marked reduction of government intervention in the fishery sector in terms of subsidies and other forms of support including credit, extension and research. Fishery produce has quality problems and fishermen sell it in small quantities. Wholesalers buy in bulk from middlemen out competing attempts by fishing cooperatives to engage in a demanding market with reduced supplies offered by their members.

Cooperative work has been almost paralyzed since the depletion of oyster production in 1989, for two reasons. Firstly, because there is insufficient support from members to take risks and develop work due to mistrust, and the members unwillingness to provide collateral to secure a bank loan. Secondly, production from members has been decreasing, as they increasingly bring their catch to middlemen.

A small group of fishermen that took advantage of past cooperative leadership, or come from neighboring states act as middlemen between fishermen and wholesalers. Middlemen have lived in the same area for many years, knew each other well and in many cases are kinsmen. They are generally wealthy and have established strong financial and social ties with wholesalers in Mexico City and Veracruz State, and with Mecoacan fishermen through credits. Therefore, they are able to capture labor by lending money to fishermen to cover fishing costs. Often they offer short-term credit for a continued operation. Once fishermen establish themselves, product is brought to middlemen and advances are paid to cover immediate expenses to continue fishing, in recognition that full payment of product will be sent when fish is sold. Profits are unpredictable as payment from middlemen is conditional. These middlemen collect the catch every day depending on the number of fishermen teams, for storage in nearly facilities providing simple refrigeration, or for direct transportation to markets. A single middleman may work with 10 or more fishing teams, 45-50 fishermen approximately.

Fishing cooperatives are financially weak in providing sufficient working capital. Problems due to lack of consensus with regard to management practices, profit distribution and resource-organizational issues have lead to ever-changing organizational dynamics, and despite the potential for increasing their income by managing their own production, cooperative administrators seemed to be complacent about the current arrangement that their membership have with middlemen. However, individual fishermen had become exasperated with this group of middlemen who they felt were exploiting them, as commented "middlemen drop the price whenever they want without authorization from anyone making our effort to increase our income, to pay our debts and accomplish our own operations worthless".

The existence of cooperatives has not resulted in a reduction of external transaction costs, due to increasing opportunism of middlemen in an inefficient market, as the economic analysis shows. The large size of cooperatives is also a main factor that has led to a lack of trust among members, eventually leading to a disbanding of groups and it seems that the scale of cooperative transactions have result in increased internal costs and reducing the positive effect that cooperative membership has on community development as shown by the effect of the dummy variable (cooperative membership) on income.

The fishery sector faces distinct problems in terms of sustainability. In the Mecoacan lagoon, fishing is structured around artisanal and small-scale operations based on pelagic species and inshore fisheries. Federal fishing licenses are issued under fishing acts that allow organized fishermen to exploit all the waters within the jurisdiction of the municipality. However, problems related to the condition of open access and high sensitivity to market prices are common to the fisheries of Mecoacan. These have increased because independent fishermen are not subject to any regulation. The undeveloped market conditions favored by local middlemen also imposes a low added value to fisheries produce, as showed by the effect of the market channel variable on fishermen income.

Regarding gender issues, fishing is a male dominated activity in Mecoacan communities. Although respondents noted that two females were significantly involved in trading fisheries products in and out of the state, the role of women in fishing is limited to processing and in local marketing of product. Although, female participation in fishing cooperatives has been partial or insignificant, their labor input for processing is significant for household income (Table 2). In the case of fisheries related activities such as aquaculture, this fall within local cultural norms in the Mecoacan lagoon, but the participation of women in these activities has also been limited to the assembling of oyster seed collectors and the processing of the harvest. This suggests that shifts in the livelihoods of Mecoacan fishing communities may occur due to significant changes in the availability of employment for males with important implications for patterns of social and economic organization in local fisheries.

Fishermen have moved their activities from coastal lagoon to marine fishing due the lesser surveillance from local fisheries authority and perceived higher income. However, they have faced fluctuating and diminishing stocks, substandard and unpredictable incomes, and increased operating costs. Due to these conditions they have had to engage in non-fishing activities (i.e. agriculture, tourism services, oil industry).

Kinship relationships between fishermen and middlemen were observed to be significant in fisheries trading and in getting a job. However, the predetermined shares from the proceeds of catch established by middlemen have been a negative factor making income from fishing less certain, as fishermen may experience prolonged poor catches that prompts a higher dependence on credit offered by middlemen. McGoodwin (1990) states that the social and cultural patterns of this way of life make fishermen continue in a fishery even when it does not provide any significant economic return. These conditions may be the result of what McCay and Jentoft (1998) describe as community failure, where social relations are embedded in the economic system instead of the economy being embedded in social relations. In these circumstances, the daily lives of producers become increasingly dominated by money transactions and bureaucratic control, and social relations become basically instrumental and utilitarian.

The analysis of incomes and costs for Mecoacan fishermen suggests that rural problems have not yet been engaged in progressive policies. It seems that previous forms of governance have been maintained to shore up power instead of laying the groundwork for viable fishery production, as it is clear that some fishermen are competitive while others are not, regardless of whether or not they are associated in cooperatives. Amid their confusion about and lack of control of the state and local economic changes fishermen are confronting and are actively struggling to survive and attempting to understand and cope with changes in the sector.

It has been asserted that the survival of cooperatives has important implications for the sustainable livelihood of fishermen (Jentoft, 1989; Bojos, 1991; Roy, 1999). Firstly, through cooperatives, fishermen could achieve some economies of scale in the purchase of fixed capital and therefore lower the costs of production. Secondly, by cutting out the middlemen, bulk fishery production is likely to increase profits and make cooperative marketing more predictable and stable. This does not mean a call for the return of the protectionist state, as it has been pointed out by Blejer and Del Castillo (1998), but rather a progressive state that creates an economic environment that encourages and supports competition as fishermen could look for greater efficiency and better product quality that would make them economically viable.

The cooperatives of Mecoacan have lost competitiveness due to a lack of integration into a market-oriented production, a high-risk fishery entrepreneurship activity and the incapability of producing efficiently. It may be assumed that an important implication of cooperative organizations is to stabilize prices and to reduce seasonality of prices paid to fishermen. In this regard, development approaches within the fishery should seek to increase the harvest added value by improving handling, as well as the development of ancillary fishery-related services in the communities. Economic diversification in and outside the fishery sector needs to be implemented to maintain community sustainable development in the face of market-orientated policies and, as a means to improve the role of fishing cooperatives in marketing. This represents a political demand for institutional reform as a key element in resource allocation.

The conditions fishermen experience suggest that the hierarchical approach for resources management influenced by political decisions have trapped resource managers in programs that address symptoms rather than causes of basic fisheries management problems. On the other hand, if these experiences can be identified as a market failure, there is justification for the role of public authorities to intervene. On this regard, Harrison (1996) points out that development action must be sensitive to the rural economy context.

Thus, information about markets, credit and new techniques are important, indicating that extension needs to be revived in order to: a) support those fishermen organizations willing to stay in the fishery sector, b) allocate resources, c) develop the concept of social and economic value of resources amongst communities, and d) organize selling within a predictable market that pays a just price. This would be assisted through the partnership of national agencies, non-governmental organizations and research institutions in terms of structures and staff and the state would have the choice to foster and support competition rather than simply and inevitably abandon its rural sector.

The occupational diversity in coastal zones provides communities with a flexible range of opportunities to respond to changing conditions. Alternative economic activities, which would not affect fishermen's principal activity of fishing and which also would provide paid work to those women who wish to diversify from traditional domestic work and child-care have been proven to develop well through the reorganization of fishing organizations into new management frameworks (Charles and Herrera, 1994; Hotta, 1994; Yap, 1996; Britz et al., 2000). It has also been acknowledged that improvement in management practices might be reached through fishing cooperatives and government jointly devising development plans, as cooperatives may provide self-regulation and coordination (McDonald, 1997; Cyrus and Pelot, 1998).

Although state intervention is not the only solution, in the case of collective bodies externalities do not rule out the possibility of private or collective owned production of the resource concerned, as all public goods have the potential for either private appropriation or governmental regulation of access (Jentoft, 1989; Pomeroy and Berkes, 1997; Britz et al., 2000). When fishermen were asked about past fisheries management based on aquaculture, they considered that resources could be equally distributed, as they experienced fishing effort coordination under government agencies in the coastal lagoon when the first oyster farms were established and natural banks were extended through restocking programs in the early 1940's. Therefore, any approach for fisheries management in the Mecoacan lagoon should consider the devolution of regulatory functions to local communities and aquaculture based fisheries scheme, which may help to restore the crucial qualities of collective action.

REFERENCES

Arredondo, J. L., De la lanza G. E., Gómez S. A., Rangel L. J. & Franyutti A. (1993*). Estudio de la relación medio ambiente-producción de ostión en el sistema lagunar de Mecoacan, Tabasco.* México: Universidad Autonoma Metropolitana UAM Unidad Iztapalapa. Laboratorio de Producción Acuícola. Informe Técnico 50 pag.

Blejer, M. I. & Del Castillo, G. (1998). "Deja Vu all over again?" The mexican crisis and the stabilization of Uruguay in the 1970s. *World Development* 26(3): 449-64.

Bojos, R. M. (1991). Community-based management of near shore fishery resources. In Council for Aquatic and Marine Research and Development (Ed.), *Management of near shore fishery resources* (130-180 pp). Philippines. Proceedings of the Seminar-Workshop on Management of Near shore Fishery Resources.

Britz, P., Sauer, W., Mather, D. & Philips, L. (2000). *Towards equity, sustainability and stability*: A sector planning approach to fishing and mariculture development in the Northern Cape Province, South Africa. http://oregonstate.edu/dept/iifet/2000/papers/ britz.pdf

Charles, A. T. & Herrera, A. (1994). *Development and diversification: Sustainability strategies for a Costa Rican fishing cooperative.* In Antona M, Catanzano J, Sutinen JG

(Eds.), Proceedings of the sixth conference of the International Institute of Fisheries Economics and Trade (1315-1324 pp). Paris, France. IIFET/ORSTOM.

Cyrus, J. P. & Pelot, R. (1998). A site management system for shellfish aquaculture. *Aquaculture Economics and Management* 2(3): 101-18.

Domínguez-Domínguez, M. (1991). *Estado actual del estrato arbóreo y algunos aspectos fisicoquímicos de los manglares de la Laguna de Mecoacan, Tabasco, México.* Villahermosa, Tabasco. Universidad Juarez Autonoma de Tabasco. 73 p.

Fields, A. (2000). *Discovering statistics: Using SPSS for Windows.* London. SAGE Publications Ltd., 496 p.

García, B. (1984). Dinámica ocupacional rural y urbana en el Sureste de México: 1970-1980. *Demografía y Economía* 18(59): 445-88.

Gold-Bouchot, G., Zapata-Perez, O., Noreña-Barroso, E., Herrera-Rodriguez, M., Ceja-Moreno, V. & Zavala-Coral, M. (1999). *Oil pollution in the southern Gulf of Mexico.* In Kumpf, H., Steindenger, K. & Sherman, K. (Eds.). *The Gulf of Mexico large marine ecosystem: Assessment, sustainability and management* (372-381 pp). Massachussetts, Blackwell Science.

Gold-Bouchot, G., Sima-Alvarez, R., Zapata-Perez, O. & Jimenez-Ricalde, J. (1995). Histopathological effects of petroleum hydrocarbons and heavy metals on the American oyster (*Crassostrea virginica*) from Tabasco, Mexico. *Marine Pollution Bulletin* 31(4-12):439-45.

Harrison, E. (1996). Options for small-scale aquaculture development. In Martinez-Espinosa, M. (Ed.). *Report of the expert consultation on small-scale aquaculture* (31-68 pp). FAO Fisheries Report 548.

Hotta, M. (1994). Rural enterprise development and economic diversification. In Indo-Pacific Fishery Commission (Ed.). *Proceedings of the Symposium on Socio-economic Issues in Coastal Fisheries Management* (95-107 pp). Bangkok, Thailand. RAPA Publication.

Hundloe, J. (2000). Economic performance indicators for fisheries. *Marine and Freshwater Resources* 51:485-91.

Inegi. (2002). XII *Censo general de población y vivienda* 2000. Tabulados Básicos. Tabasco http://www.inegi.gob.mx/difusion/espanol/portada.html.

Jentoft, S. (1989) Fisheries co-management: Delegating government responsibility to fishermen organizations. *Marine Policy (April):*137-54.

Koop, G. (2000). *Analysis of economic data. John Wiley and Sons*, Ltd. England. 226 p.

McArthur, H. J. (1994). Creating dialogue and generating information. In Pomeroy, R. S. (Ed.). *Conference Proceedings Community management and common property of coastal fisheries in Asia and the Pacific* (124-144 pp). Manila, Philippines. ICLARM.

McCrossan, L. (1991). *A handbook for interviewers: A manual of social survey practice and procedures on structured interviewing.* London. Her Majesty Stationary Office.

McCay, B. J. & Jentoft, S. (1998). Market or community failure? Critical perspectives on common property research. *Human Organization* 57:21-29.

McDonald, J. H. (1997). Privatizing the private family farmer: NAFTA and the transformation of the mexican dairy sector. *Human Organization* 56:321-331.

McGoodwin, J. R. (1990). *Crisis in the World's fisheries.* Stanford University Press. California, USA.

Moguel, J. (1994). La violencia del oro negro en Mecoacan, Tabasco. Friedrich Ebert Stiftung. México.

Negrete-Salas, M. E. (1984). Petróleo y desarrollo regional: El caso de Tabasco. *Demografía y Economía* 18(57):86-109.

Perez-Sanchez, E. & Muir, J. F. (2003). Fishermen perception on resources management and aquaculture development in the Mecoacan lagoon, Tabasco, Mexico. *Ocean and Coastal Management* 46:681-700.

Pomeroy, R. S. & Berkes, F. (1997). Two to tango: the role of government in fisheries co-management. *Marine Policy* 21:465-480.

Rivera-Arriaga, E. & Villalobos, G. (2001) The coast of Mexico: approaches for its management. *Ocean and Coastal Management* 44:726-56.

Rodríguez, C. F. (1998). Asociación Ecológica Santo Tomas, A. C. Cultivo del ostión americano *Crassostrea virginica* en cajas ostreofilas de plástico en suspensión en long-line en la laguna Mecoacan, Paraiso, Tabasco, México. Reporte Técnico. 40 p.

Roy, R. (1999). *Coming together to manage fisheries: Answers to frequently asked questions on the stakeholder approach to fisheries management.* http://www.dal.ca/~corr/7no2.htm.

Sagarpa (2003). *Anuario Estadístico de Pesca 2003.* Comisión Nacional de Acuacultura y Pesca. México. 265 p.

Semarnap (1999). *Recursos Pesqueros y Acuícolas: Tabasco.* http://www.semarnap.gob.mx/tabasco/informe/Recursos_pesqueros.htm.

Terrebone, R. P. (1995). Property rights and entrepreneurial income in commercial fisheries. *Journal of Environmental Economics and Management* 28:68-82.

Townsley, P. (1996). *Rapid rural appraisal, participatory rural appraisal and aquaculture.* Rome: FAO Fisheries Technical Paper 358.

Yap, H. T. (1996). Attempts at integrated coastal management in a developing country. *Marine Pollution Bulletin* 32(8-9):588-591.

Valdes, D. (1998) *Mecoacan lagoon, Tabasco.* http://data.ecology.su.se/mnode/mexicanlagoons/mecoacan/mecoacan.htm

Velazquez, V. G. (1994). *Los recursos hidráulicos del Estado de Tabasco: Ensayo monográfico.* Tabasco. Universidad Juarez Autonoma de Tabasco.